MW00761265

HIGH ENERGY SPIN PHYSICS

ELEVENTH INTERNATIONAL SYMPOSIUM

AIP CONFERENCE PROCEEDINGS 343

HIGH ENERGY SPIN PHYSICS

ELEVENTH INTERNATIONAL SYMPOSIUM

BLOOMINGTON, IN 1994

EDITORS: KENNETH J. HELLER
SANDRA L. SMITH
UNIVERSITY OF MINNESOTA

American Institute of Physics Woodbury, New York

L.C. Catalog Card No. 95-78431
ISBN 1-56396-374-4
DOE CONF-9409103

Printed in the United States of America.

CONTENTS

III. LEPTON ACCELERATORS AND POLARIMETERS

IV. STRONG INTERACTIONS

V. POLARIZED TARGETS AND SOURCES

VI. POLARIZED ELECTRON SOURCES

VII. ELECTROWEAK INTERACTIONS

VIII. NUCLEON SPIN STRUCTURE

APPENDICES

PREFACE

The Eleventh International Symposium on High Energy Spin Physics was held September 15–22, 1994, on the beautiful campus of Indiana University in Bloomington, Indiana in conjunction with the Eighth International Symposium on Polarization Phenomena in Nuclear Physics. The Symposia were sponsored primarily by the Indiana University Cyclotron Facility, Indiana University, the National Science Foundation, and the U.S. Department of Energy. Additional support was made available from the University of Minnesota, and the University of Michigan. The International Science Foundation enabled some of the scientists from the countries of the former Soviet Union to attend.

This conjunction of the two Symposia is a demonstration that these fields have both scientific and technological concerns in common. Spin represents a window through which both fields illuminate the interactions of quarks and leptons. Of the 355 people registered for both meetings, 269 participants registered for the Symposium on High Energy Spin Physics, and 260 registered for the Symposium on Polarization Phenomena in Nuclear Physics. The large registration overlap illustrates the evolution of a new field of physics drawing its strength from both "classical" nuclear physics and "classical" high energy physics.

Since the Symposium, there has been great progress in the acceleration and storage of polarized beams. Leptons clearly held center stage at this meeting, sharing that honor with the musical highlights of the evening programs. Only a theoretical possibility exhibited in disputed computer codes a few short years ago, SLC, HERA, and LEP have all achieved large electron polarizations. Meanwhile, beautiful experiments using CERN muons and SLAC electrons have measured the spin structure of nucleons. The interpretation of those measurements still engenders a lively debate. Techniques for accelerating polarized protons have also progressed with demonstrations of the effectiveness of various types of Siberian Snakes at Indiana and Brookhaven. It now seems possible to accelerate high energy polarized proton beams by slithering around the large number of depolarizing resonances. To effectively use the new polarized beams, experimentalists will also need the improvements in the technology of polarized targets, polarized sources, and polarimeters so aptly represented at the Symposium. New cxperiments using polarized proton beams at the world's highest energy accelerators loom on the horizon at RHIC, Fermilab, and UNK.

The overlap in interests of the two Symposia made it necessary to divide the topics, and the resulting papers are published in two volumes; this one and another emphasizing the nuclear physics interests. The high energy physics sessions were: strong interactions at high energy, electroweak interactions, polarized electron sources, polarized solid targets, polarized beam from both hadron and electron accelerators, and polarimeters for both hadrons and leptons. The nuclear sessions consisted of: hadron form factors, symmetries, few body systems including nucleon-nucleon scattering and resonance production, intermediate-energy hadron-induced reactions, intermediate-energy electromagnetic interactions, low-energy nuclear reactions, polarized hadron sources, and gaseous polarized targets.

Many people contributed to the success of the Eleventh International Symposium on High Energy Spin Physics. First of all, credit must go to the participants who made this a lively and interesting meeting. The chairs and rapporteurs of the parallel sessions were outstanding in their efforts to organize these sessions into coherent representations of their subject areas. The Organizing Committee arranged a stimulating program of plenary and parallel talks with helpful advice and assistance of the International Advisory Committee. The Co-Chair of the Organizing Committee, Don Crabbe, was ready to step in at a moment's notice to handle emergencies. This Symposium would not have been possible without the tireless efforts of Alan Krisch, the not-so-retiring chairman of the International Advisory Committee. Charles Prescott, the incoming chairman was also instrumental to the success of the Symposium.

Organizing this Symposium at long range from Minnesota required the immense efforts and bottomless understanding of the Indiana organizers. Ed Stephenson was a tower of strength; he was everywhere at once. John Cameron kept the whole show progressing under the stress of two separate organizing committees, and Steve Vigdor kept us all calm and on track. Also, we must recognize Peter Schwandt for local transportation and the companions' program, Ben Brabson for organization of the now traditional boat excursion, Bob Bent for social activities including the banquet, Tom Rinckel for computer setup, and Phil Thompson, Joyce Pace, and a number of workers that ran much-needed errands. We also wish to gratefully acknowledge Margie Rietel and the Indiana University Conference Bureau for attending to the numerous details of local arrangements. A special word of thanks is due to Sharon Herzel and Janet Meadows for providing essential secretarial help with the meeting and during the months before and after. SPIN'94 would not have been possible without the dedicated support of many people.

Ken Heller
Sandy Smith
Editors

I. GENERAL INTEREST

Summary of 11th International Symposium on High Energy Spin Physics

A. D. Krisch

Randall Laboratory of Physics, University of Michigan
Ann Arbor, Michigan 48109-1120 USA

I greatly enjoyed attending this joint meeting of the 11th International Symposium on High Energy Spin Physics and the 8th International Symposium on Polarization Phenomena in Nuclear Physics. This SPIN '94 meeting provided an excellent opportunity for nuclear and high energy spin physicists to learn something from each other in one of my favorite towns, Bloomington. I will mostly stick to my job of summarizing the High Energy Spin Symposium, but I may accidentally mention a few exciting Nuclear Spin topics.

There was an excellent historical introduction to Electron Spin Physics in the talk of Professor Ternov. In 1963, along with Professor Sokolov, he discovered the Sokolov-Ternov effect of self-polarization,[1] where electrons and positrons become polarized along the accelerator magnets' vertical field direction because of their different spin-up and spin-down synchrotron radiation rates. This self-polarization has recently become very important to our field. In the 1960's, self-polarization seemed a clever abstraction, which was only interesting to theorists. Now HERA and LEP, two of the world's largest electron facilities, both operate with polarized beams using the Sokolov-Ternov self-polarization effect. It was a pleasure and an honor to have Professor Ternov lecture at this Symposium.

Turning to the history of proton-proton spin effects, I will show Figure 1, which seems especially appropriate since this meeting includes both high energy and nuclear spin physicists. It displays the spin-spin correlation parameter for 90°_{cm} proton-proton elastic scattering from the lowest up to the highest measured energy;[2] Professor Haeberli helped me to make this compilation. When I started studying spin around 1970, most people were quite sure that there would be no spin effects at high energy. This graph certainly does not support that belief.

After these large two-spin effects were discovered at the ZGS,[3] many people said, "Perhaps there are two-spin effects when the beam and the target are both polarized, but, surely there will be no one-spin effects at high energy." The talks by Professors Devlin and Pondrom[4] on inclusive hyperon polarization and hyperon magnetic moments referred to this perturbative QCD prediction that A should go to zero at high energy and high $P_\perp$ for all hadronic reactions. Figure 2 shows the hyperon polarization plotted against transverse momentum at 12 GeV, at 400 GeV, and at 2000 GeV; this data certainly does not support the A = 0 prediction. Moreover, it seems to me that 2000 GeV is a fairly high energy.

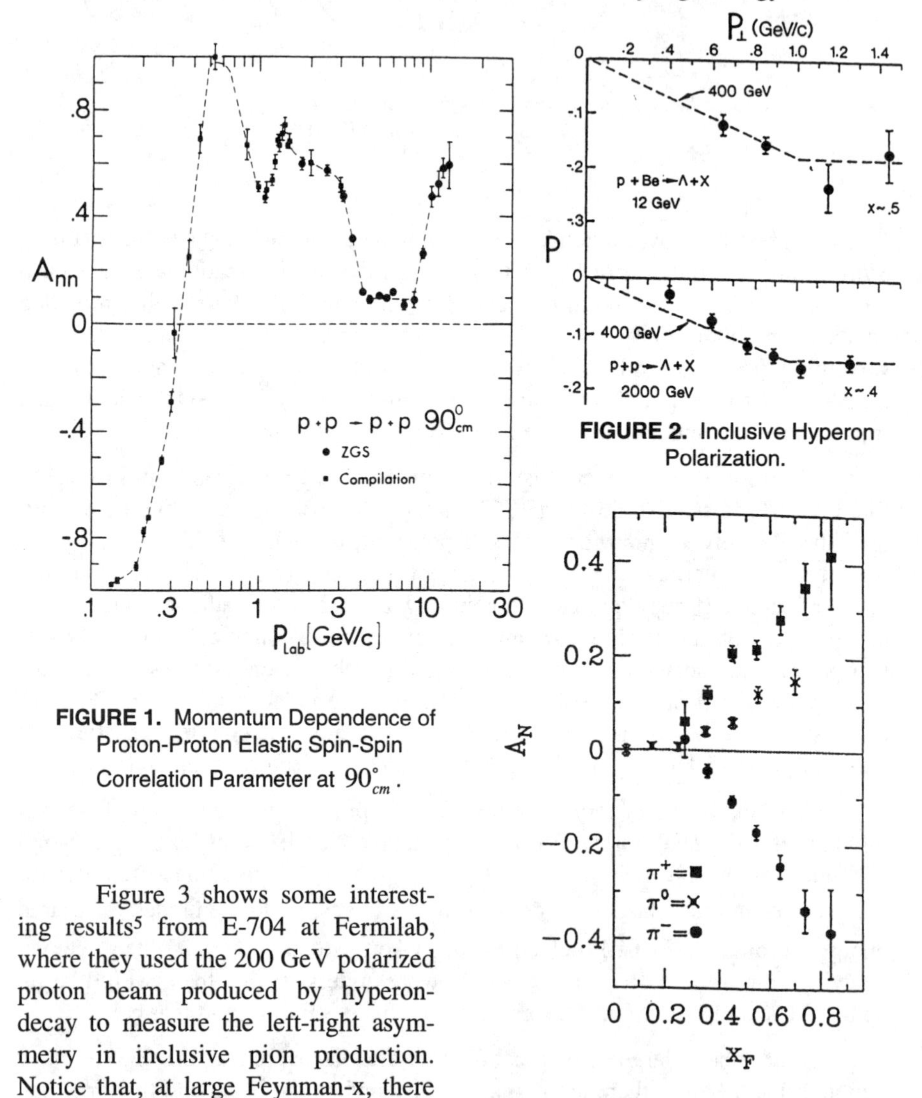

FIGURE 1. Momentum Dependence of Proton-Proton Elastic Spin-Spin Correlation Parameter at 90°_{cm}.

FIGURE 2. Inclusive Hyperon Polarization.

FIGURE 3. Fermilab E-704: Left-Right Asymmetry in Inclusive Pion Production.

Figure 3 shows some interesting results[5] from E-704 at Fermilab, where they used the 200 GeV polarized proton beam produced by hyperon-decay to measure the left-right asymmetry in inclusive pion production. Notice that, at large Feynman-x, there are certainly large left-right asymmetries which should not exist, according to PQCD.

Some people noted that, in these experiments, $P_\perp$ was only about 1 GeV/c; they said that perturbative QCD would surely force one-spin asymmetries to go to zero at higher $P_\perp$. Figure 4 shows some elastic one-spin data from the AGS and CERN. The prediction of perturbative QCD for elastic scattering in this region was again that A should be zero; the AGS data[6] certainly do not agree with this A = 0 prediction at $P_\perp^2 = 7$ $(GeV/c)^2$. Thus, the predictions of the PQCD theory of hadronic interactions do not agree with these four hadronic spin experiments. I was quite amused by the earlier quoted comment of Bjorken that perhaps theorists on Program Committees should ban spin experiments to help protect PQCD.

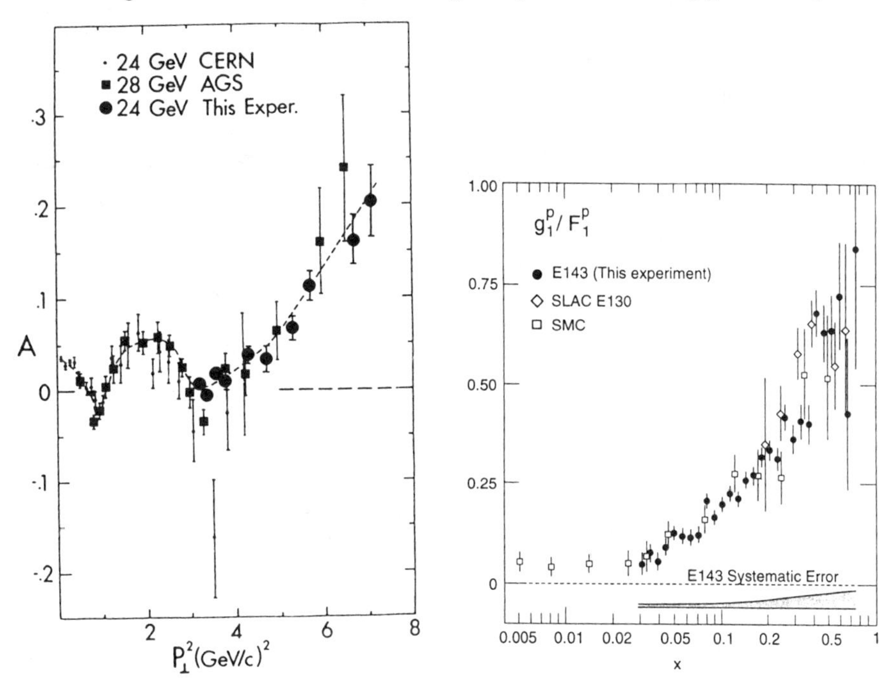

FIGURE 4. Spin Asymmetry in p-p Elastic Scattering at High $P_\perp$.

FIGURE 5. "Spin Crisis" Form Factor Ratio Plotted Against x.

Our most recent excitement has been the "Spin Crisis"; we heard many comments about this in the talks of Thomas, R. Voss, Windmolders, Day, Jackson, Soffer, and Anselmino.[7] We heard from Professor Thomas that perhaps now there is not so much of a spin crisis. I then asked him, "What changed? the data, the theory, or the definition of 'crisis'?" Apparently, there have been some small changes in the data, but Figure 5 shows that the recent SMC data using a CERN polarized muon beam on a polarized proton target agrees rather well with the SLAC polarized electron data from experiments E-130 and E-143. Turning to the theory, perhaps there was a bit of overconfidence in the validity of some sum rules and some extrapolations to very small values of x, which had not been experimentally tested; now they have been tested. But probably the biggest change was in our definition of the word "crisis"; the cancellation of the SSC helped us to better understand the word crisis. In any case, the data now indicates that probably each proton does not contain three simple quarks which carry most of the proton's spin; Professor Prescott told me that the best present estimate is that the quarks and anti-quarks together carry about $\frac{1}{3}$ of the proton's spin.

There were reports about three different workshops. Professor Anderson reported on the May 1993 Workshop on Polarized Ion Sources and Polarized Gas Targets in Madison, Wisconsin;[8] Professor Mori[9] also reviewed this subject. Many exciting results were presented, but I will only mention Figure 6 which shows the intense atomic beam sources at Heidelberg and Wisconsin. The present source technology is very impressive, especially the high-gradient permanent magnet sextupoles of 4 $T \cdot cm^{-1}$. These new polarized sources work very well; they produce intensities of $4 \cdot 10^{16}$ per second with a polarization of over 80%. The situation is now very different from 1970 when Hilton Glavish from New Zealand sold us the world's first commercial polarized proton source; it cost about $250,000 and produced 6 μA. I am very impressed by the progress in polarized ion sources.

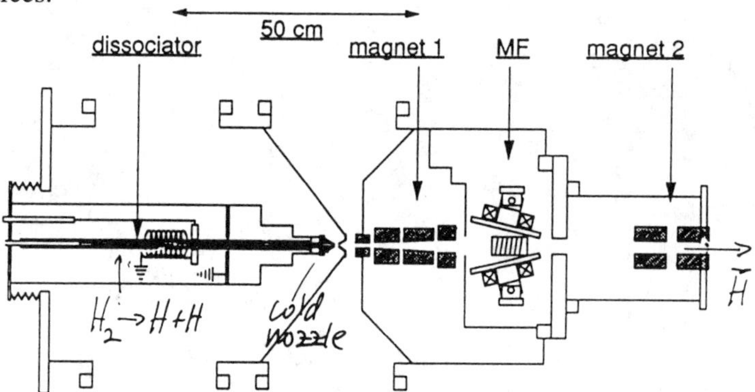

FIGURE 6. Wisconsin-Heidelberg-Marburg-Munich Source of Polarized Hydrogen Atoms.

Professor Nurushev discussed the 5th International Workshop on High Energy Spin Physics (SPIN '93).[10] These Workshops, which he organizes in the odd-numbered years at Protvino, are really small symposia which allow young physicists from the former Soviet Union to hear what is happening in spin physics. Spin physics is an area of great activity in the former Soviet Union, but the currency exchange problems make travel to foreign scientific meetings very difficult. I thank Professor Nurushev for organizing these valuable workshops.

Werner Meyer reviewed the 7th Workshop on Polarized Target Materials and Techniques,[11] which had 49 participants from many institutions. One highlight was the successful Virginia-Basel-SLAC polarized target, shown in Figure 7, which is now being used in fixed-target experiments at SLAC. Its clever arrangement of magnetic fields allows a longitudinal polarization. The target uses frozen ND_3 or NH_3; some recent results are shown in Figure 8: the deuteron polarization was over 30% and the proton polarization was over 70% in an intense beam of several 10^{11} electrons per second. It is impressive that a polarized target can work so well in such extreme conditions.

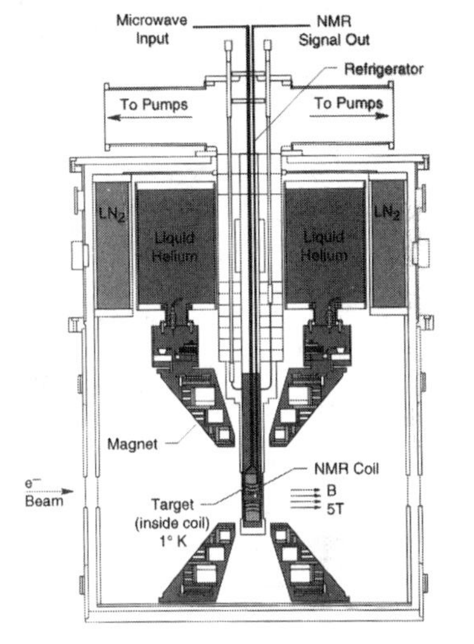

FIGURE 7. Virginia-Basel-SLAC Polarized Target.

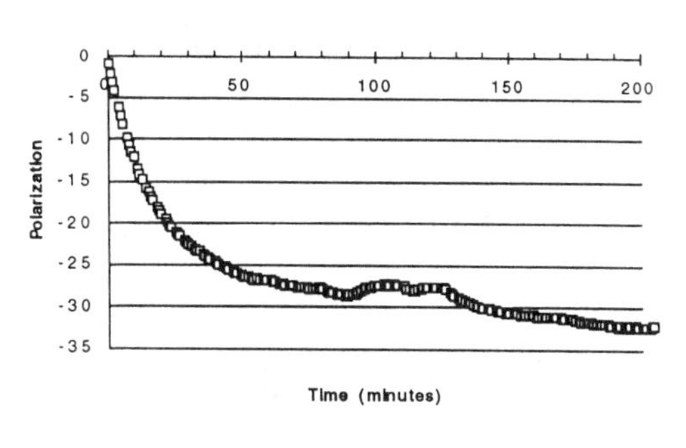

FIGURE 8. ND_3 Polarization with Beam on.

One major highlight of this High Energy Spin Physics Symposium was the polarization work at the large electron facilities: SLC, HERA, and LEP. Dr. Placidi[12] gave a very nice talk on polarization at LEP; his Figure 9 shows the transverse polarization obtained at LEP in August 1993 plotted against time. Note that the polarization reached about 55%; this clearly demonstrates the success of the

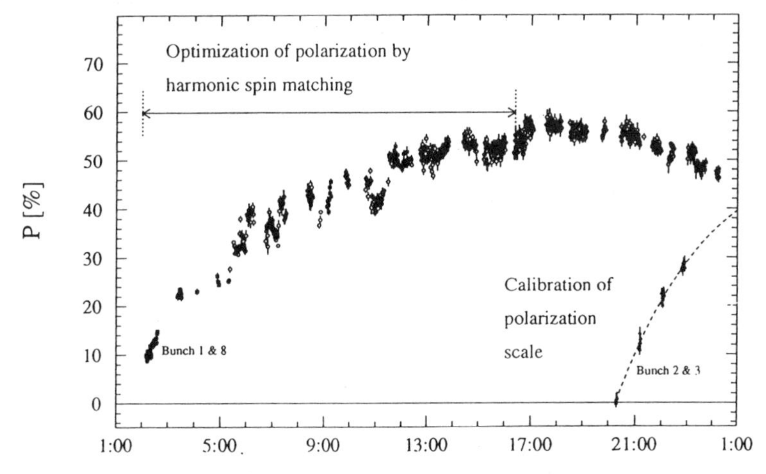

FIGURE 9. High Transverse Beam Polarization in LEP Plotted against Hours.

Sokolov-Ternov self-polarization mechanism.[1] This electron polarization was used to calibrate the LEP energy, which then provided a precise calibration of the Z mass. This was a rather significant contribution to high energy physics.

There has been a strong emphasis on polarization at HERA. In their talks, G. Voss, Barber, and Jackson[13] each discussed various aspects of polarization at HERA. The HERA ring is shown in Figure 10 along with the H-1 experiment, the Zeus experiment, the HERMES experiment, and the spin direction at each experiment.

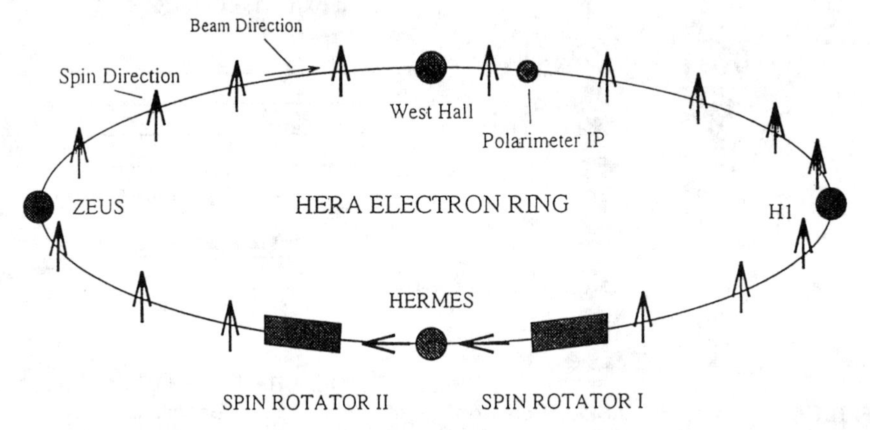

FIGURE 10. HERA Electron Ring.

I obviously cannot discuss every detail of the HERA polarization program, but I will show the HERA polarimeter in Figure 11. The incoming laser light interacts with the polarized electrons, and then the analyzer detects the scattered photons. Measuring the differences in the event rate for each spin direction gives a precise determination of the electron beam polarization. This polarimeter contains some very impressive new technology.

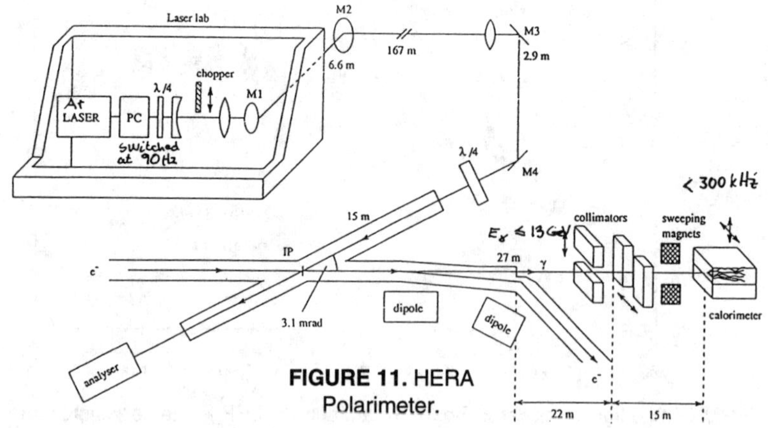

FIGURE 11. HERA Polarimeter.

Figure 12 shows the beam polarization measured at HERA plotted against time; the transverse polarization reached about 70% and the longitudinal polarization reached about 55% in about one hour. This was another success of the Sokolov-Ternov self-polarization mechanism.[1] Also plotted is something that I named the "Soergel Limit." In a comment[14] at the 1990 Bonn Spin Symposium, Professor Soergel said something like, "We decided that HERMES is approved if the HERA polarization reaches 50%. If it does not reach 50%, it is not approved." This seemed a wise thing for a Director to do; it eliminated the need for a decision and encouraged polarizers to work harder. There were about 300 witnesses to his comment, so apparently DESY decided to approve HERMES.

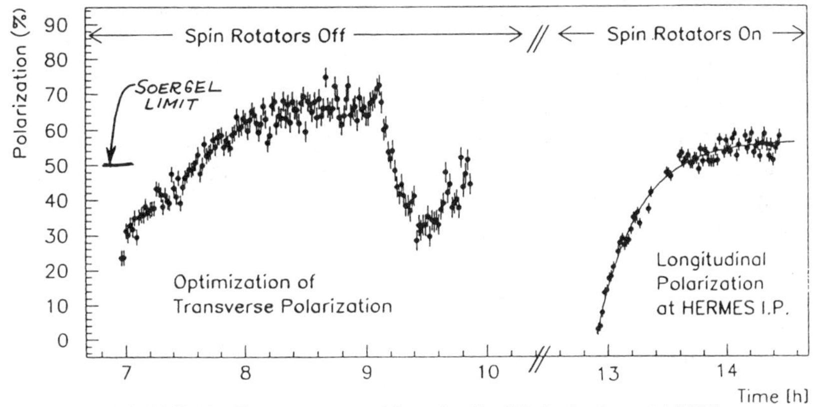

FIGURE 12. Transverse and Longitudinal Polarization at HERA.

As shown in Figure 13, the HERMES experiment uses a "storage cell" internal target which is a bit different than a Jet. This storage cell idea is very clever. The storage cell has an incoming polarized jet, but the jet does not pass through the beam. Instead it is trapped inside the storage cell, which is open only at the ends so that the proton-polarized hydrogen atoms can only escape slowly. Therefore, most of them remain in the storage cell for some time; this increases the target thickness by a factor of 10 to 100. This storage cell is another indication that hard-working, clever, and persistent physicists can solve most problems.

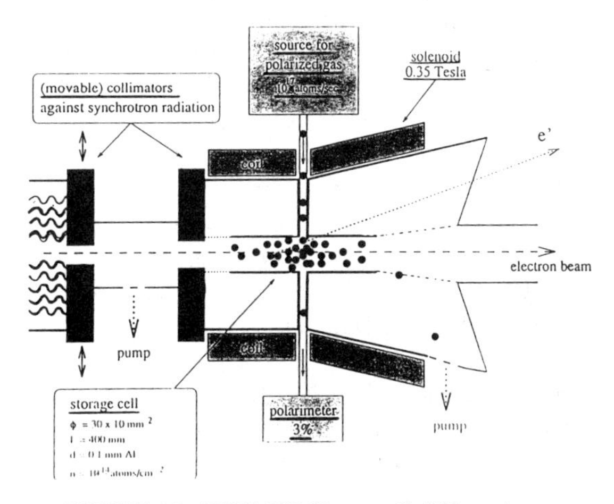

FIGURE 13. HERMES Storage Cell Target.

I will next discuss SLAC, which now has a very nice facility called SLC. We have heard talks about SLAC and SLC from Woods, Steiner, and others.[15] In recent years, SLAC, which is certainly one of the largest electron facilities in the world, has become almost exclusively a polarized electron facility. The hardware items associated with the SLAC polarized beam are shown in Figure 14. The polarized electrons are first produced in a polarized source; then one must maintain the polarization in the accumulator rings. Next the polarized electrons are accelerated in the LINAC and pass through SLC's somewhat complex non-planar arcs. Finally, one must measure the polarization near the interaction region point. So far there has been no attempt to polarize the positrons; that would be an exciting goal. Perhaps someday, we will figure out how to polarize positrons and even antiprotons.

During the past few years, the SLAC polarized source team[15] made great progress with the polarized electron source, which is shown in Figure 15. The electrons are emitted by the gallium-arsenide cathode. Then the polarized laser pumps them into the proper spin-polarization state; the laser polarizes the electrons very well. In 1992, the polarization reached 40% for fixed-target running and 22% at SLC. In 1993, it reached 85% and 63%. In 1994, the polarization reached 85% for fixed-target running and 80% at SLC. These polarizations are impressive.

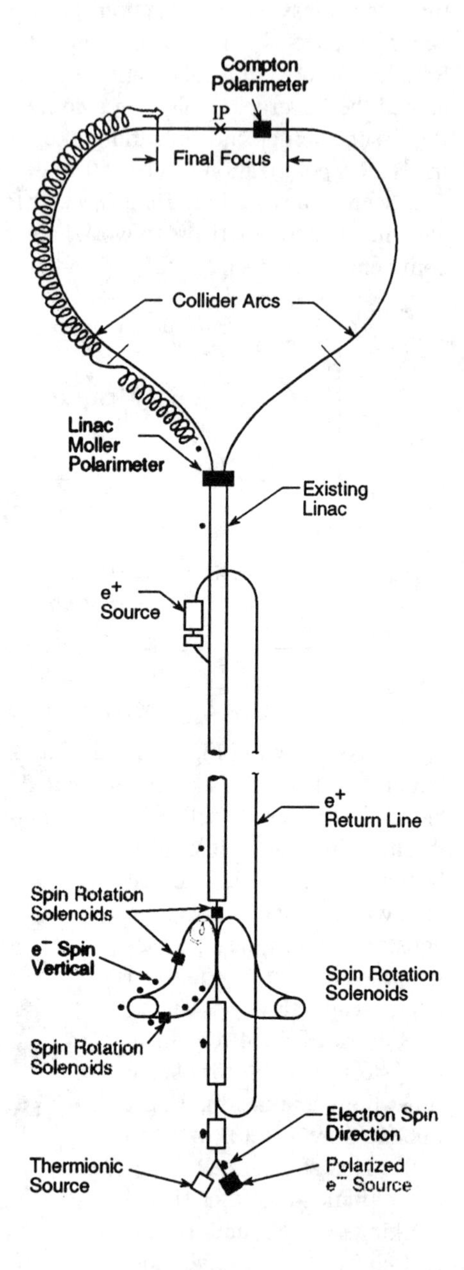

FIGURE 14. Polarization Hardware at SLAC.

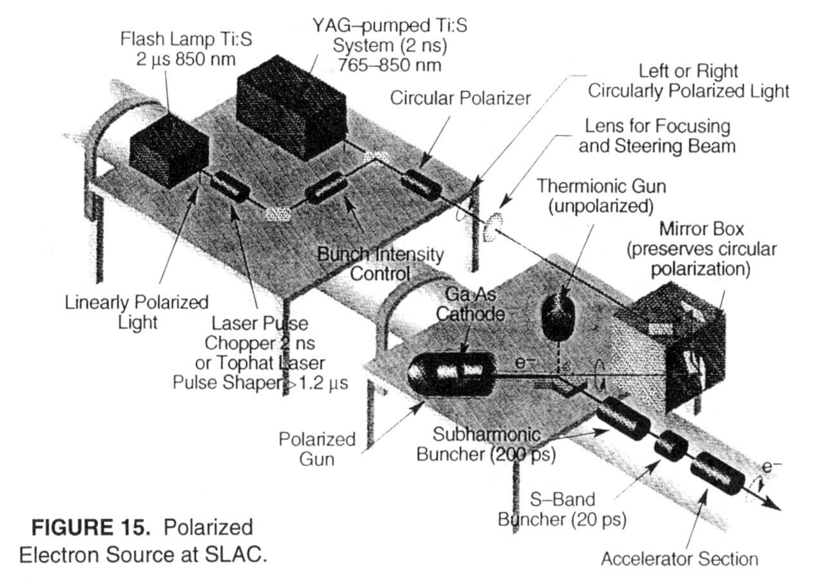

FIGURE 15. Polarized Electron Source at SLAC.

Perhaps the most significant physics result from SLAC is shown in Figure 16, where $\sin^2$ of the Weinberg angle is plotted for various experiments. The two parallel horizontal lines represent the error bars for the average of all the LEP data. The first square point on the left represents the data from SLD at SLC. All of the LEP data taken together still has a better error than the SLAC data, but only about 2.5 times better. Note that there may be some difference between the SLAC and LEP results. Using my earlier definition, I am not yet ready to call this difference a crisis, but it certainly seems interesting. Perhaps in the future, we should use the word "crisis" more sparingly and use "interesting" more often.

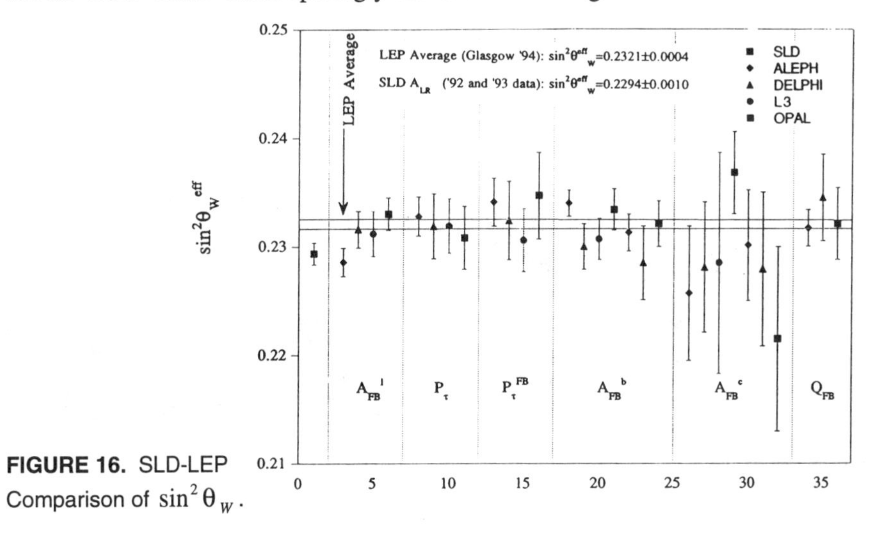

FIGURE 16. SLD-LEP Comparison of $\sin^2\theta_W$.

I will now discuss the proton facilities in slightly more detail; Professor Heller asked me to do this because Professor Ado was unable to give the requested lecture on polarized beams at Fermilab and UNK. I will begin with the development and testing of Siberian snakes. The Siberian snake is an extraordinarily clever idea which was invented by Derbenev and Kondratenko;[16] moreover, it works.

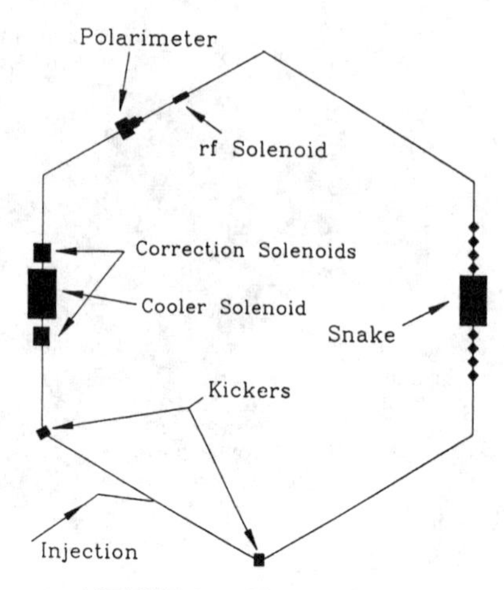

FIGURE 17. Siberian Snake at the IUCF Cooler Ring.

The Siberian snake was first tested at the IUCF Cooler Ring, which is shown in Figure 17; to some people it looks like an accelerator, but we Siberian snake people think of it as an experiment.[17] Polarized protons are injected into the Cooler Ring from the IUCF cyclotrons. The kicker magnets are sometimes used for injection, and sometimes used as rf dipoles. Notice the Siberian snake, the polarimeter, the rf solenoid, and the cooling magnets which we also use to create imperfection magnetic fields for our experiments. Thus, we are using many existing hardware items for extra jobs.

The Siberian snake itself[17] is shown in Figure 18; its heart is an aging and temperamental superconducting solenoid of two Tesla-meters. The four skew and four normal quadrupoles do nothing to the spin, but they correct the solenoid's rather strong focusing and beam twisting of the 100 or 200 MeV beam. On each turn around the ring, the snake rotates the spin by 180°. This makes any depolarizing effects, that may occur during one turn around the ring, cancel themselves during the next turn. The snake forces all the problems to cancel themselves; it is a very clever idea.

FIGURE 18. Siberian Snake at the Cooler Ring.

We have tested many different aspects of Siberian snakes. Recently we accelerated of polarized protons from 95 to 140 MeV through the $G\gamma$ imperfection depolarizing resonance at 108 MeV.[18] The measured polarization is plotted against the imperfection field integral in Figure 19; the circles show the polarization with a partial Siberian snake turned on, while the squares show the polarization with no snake. Clearly, with no snake, the polarization drops when the imperfection field is large; with the snake on, there is no depolarization. Thus the Siberian snake overcomes the depolarization during acceleration through an imperfection depolarizing resonance.

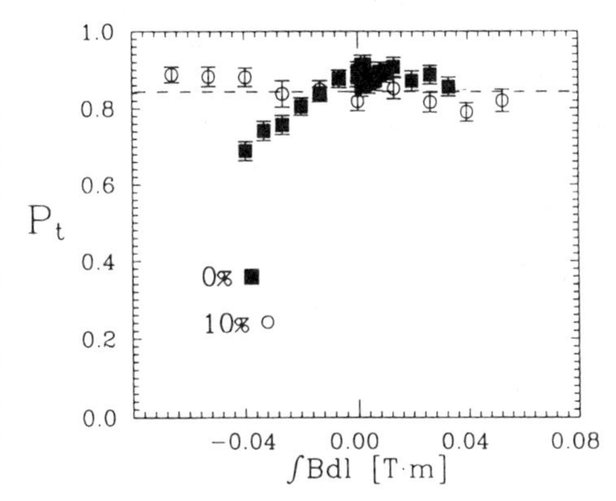

FIGURE 19. Acceleration through a Depolarizing Resonance at $G\gamma = 2$.

Since the 1992 Nagoya SPIN Symposium, we studied the adiabatic turn-on of a Siberian snake at an energy where the spin tune $G\gamma$ is a half integer; this occurs at 370 MeV at the IUCF Cooler Ring. Professor Courant had proposed that adiabatic turn-on should allow switching from one type of polarization correction to another during an acceleration cycle. The data[19] demonstrated that, within our measured precision of 2%, there was no polarization loss when the snake was turned on and off at these magic energies. At other energies, there is some depolarization during adiabatic turn-on.

Several speakers stressed the importance of flipping the spin of a stored polarized beam from up to down to discriminate against systematic errors. Dr. Phelps from Michigan has been much involved with our recent spin-flipping studies using an RF solenoid. The studies[20] showed that, with very careful tuning, there was no polarization loss within our error of $\pm$ 0.05%. This suggests that one can flip the spin perhaps 100 to 1,000 times without a significant loss of polarization. This is very important for reducing systematic errors. Professor Cameron seems to support further spin-flipping studies because there are several approved experiments at the IUCF Cooler Ring that need this capability.

Several high energy proton accelerators may use Siberian snakes because Professor Ternov's self-polarization formula[1] does not work very well for most proton accelerators. At a Coral Gables meeting, Kent Terwilliger, R.R. Wilson, and I once calculated that self-polarization would work rather well for proton storage rings at about 70 TeV. However, until Congress funds a 70 TeV SSC, we proton polarizers must either correct each resonance or use Siberian snakes.

Professor Vigdor[21] gave a very nice talk about Indiana's proposed Light Ion Spin Synchrotron (LISS) of approximately 20 GeV which would use Siberian snakes to overcome the depolarizing resonances. This proposal seems rather natural since Indiana is quite familiar with Siberian snakes. LISS is shown in Figure 20; some of its parameters are: stored polarized beams of about 20 GeV for protons and other light ions, a very high luminosity of $10^{33}cm^{-2}s^{-1}$, and long straight sections. LISS could provide a very exciting spin physics program with strong components in both High Energy Physics and Nuclear Physics. This successful SPIN '94 Meeting at IUCF should help to focus the scientific program for LISS.

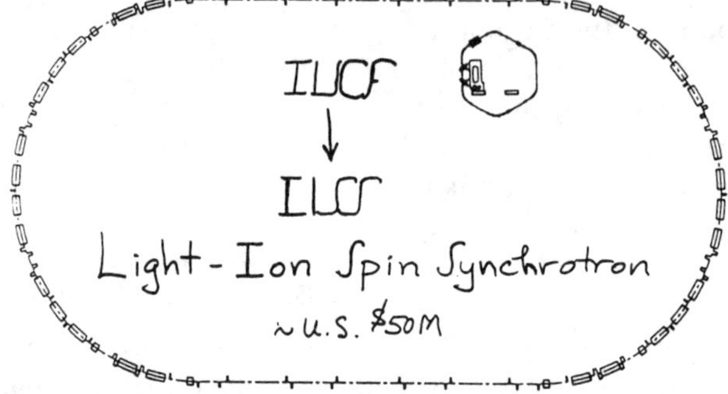

FIGURE 20. IUCF Light Ion Spin Synchrotron.

For six years, the Michigan-IUCF Siberian snake experiments had no competition anywhere in the world. This is no longer true; now there is another Siberian snake in the AGS which Huang and Roser[22] discussed. Figure 21 shows a photograph of this partial warm Siberian snake; it is a rather large power-consuming solenoid because it is quite difficult to ramp superconducting solenoids. The partial Siberian snake is rampable along with the AGS energy; it can operate as a 5% partial snake up to about 25 GeV.

FIGURE 21. AGS Partial Siberian Snake.

Some results[22] from the April 1994 AGS polarized beam run are shown in Figure 22. The upper graph shows spin-flipping with the partial Siberian snake. The lower graph plots the AGS vertical polarization against Gγ, which is the spin tune. The maximum Gγ of about 22.5 corresponds to about 11 GeV. The snake was only a partial snake, which should overcome the imperfection depolarizing resonances but not the three intrinsic depolarizing resonances; note the significant polarization loss at each intrinsic resonance. However, the partial snake did overcome the many imperfection resonances with no observable depolarization; thus, the complex system of 96 correction dipoles[23] was not needed to painfully correct them. The AGS pulsed quadrupoles were not used because their large 20 MW power supplies had no maintenance for six years; they should allow a higher polarization in the next AGS polarized beam run.

The ability to accelerate polarized protons in the AGS, without weeks of tune-up time, may revive the AGS polarized beam physics program; moreover, it could provide polarized protons for RHIC. Dr. Makdisi[24] reviewed the RHIC polarized beam program. His Figure 23 shows an overview of the polarization hardware; note the AGS's partial Siberian snake, its pulsed quadrupoles, its polarimeters, and the four full Siberian snakes and two polarimeters in RHIC, which also has eight spin rotators for the STAR and PHENIX detectors. RHIC could be the world's highest energy proton-proton collider; moreover, with Siberian snakes, both proton beams could be polarized.

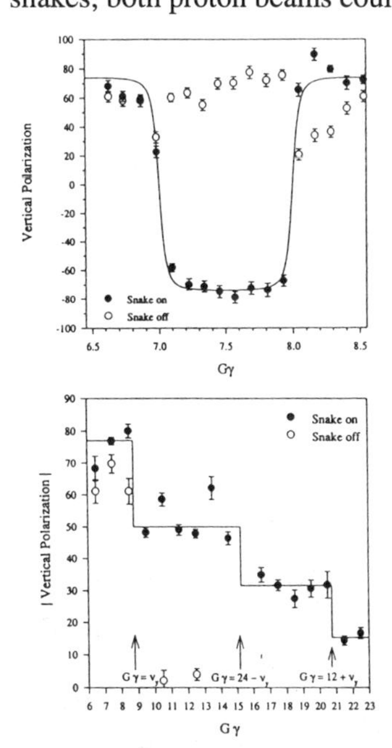

FIGURE 22. AGS Partial Snake Data.

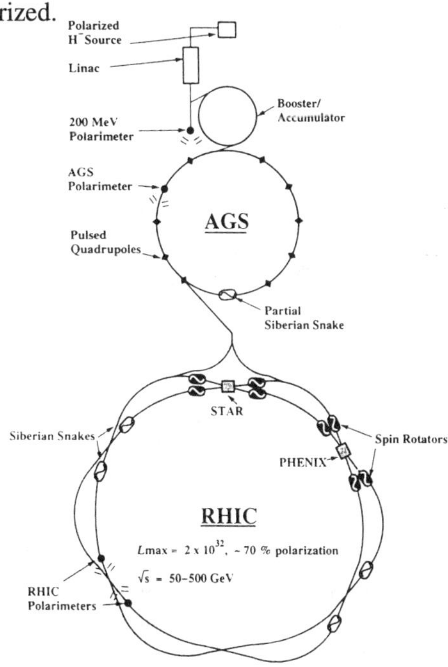

FIGURE 23. Polarized Protons at RHIC.

I hope that they accelerate polarized beams at RHIC; it could provide an excellent program of spin physics. The main goal of the first Workshop on High Energy Spin Physics, held in Ann Arbor in 1977, was to accelerate polarized protons in ISABEL. Now, ISABEL is long gone; its successor, SSC, is also gone. However, one may still get polarized protons in the old ISABEL tunnel. Spin physicists may not be fast, but we are persistent.

Fermilab has also shown some interest in polarized protons. In 1991 Fermilab first commissioned the SPIN Collaboration to do detailed studies of how to accelerate polarized protons in the various rings at Fermilab. Figure 24 shows the first page of the 1 August 1994 Polarized Tevatron Progress Report,[25] which includes detailed drawings, budgets, and schedules on how to accelerate polarized protons in both the Main Injector and the Tevatron. Figure 25 shows the new polarized hardware required in the various Fermilab rings and injectors. Two possibilities are being considered for the polarized source: an atomic beam-type polarized source (ABS) and an optically pumped polarized ion source (OPPIS). There is a competition with equal funding going to the ABS team at IUCF and the OPPIS team at TRIUMF. Both teams have made good progress; whoever develops the best source may get to build a polarized source for Fermilab.

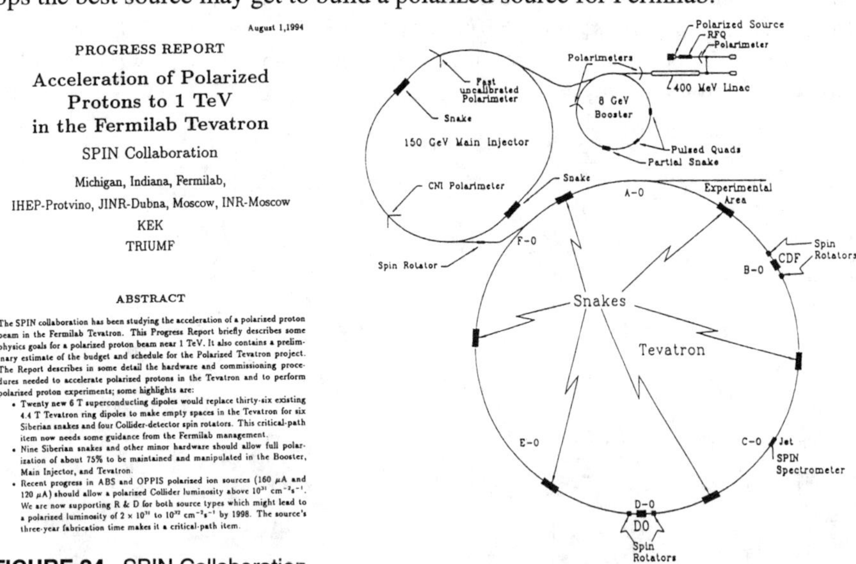

August 1,1994

PROGRESS REPORT

Acceleration of Polarized Protons to 1 TeV in the Fermilab Tevatron

SPIN Collaboration

Michigan, Indiana, Fermilab,
IHEP-Protvino, JINR-Dubna, Moscow, INR-Moscow
KEK
TRIUMF

ABSTRACT

The SPIN collaboration has been studying the acceleration of a polarized proton beam in the Fermilab Tevatron. This Progress Report briefly describes some physics goals for a polarized proton beam near 1 TeV. It also contains a preliminary estimate of the budget and schedule for the Polarized Tevatron project. The Report describes in some detail the hardware and commissioning procedures needed to accelerate polarized protons in the Tevatron and to perform polarized proton experiments; some highlights are:

- Twenty new 6 T superconducting dipoles would replace thirty-six existing 4.4 T Tevatron ring dipoles to make empty spaces in the Tevatron for six Siberian snakes and four Collider-detector spin rotators. This critical-path item now needs some guidance from the Fermilab management.
- Nine Siberian snakes and other minor hardware should allow full polarization of about 75% to be maintained and manipulated in the Booster, Main Injector, and Tevatron.
- Recent progress in ABS and OPPIS polarized ion sources (160 μA and 120 μA) should allow a polarized Collider luminosity above 10^{31} $cm^{-2}s^{-1}$. We are now supporting R & D for both source types which might lead to a polarized luminosity of 2×10^{31} to 10^{32} $cm^{-2}s^{-1}$ by 1998. The source's three-year fabrication time makes it a critical-path item.

FIGURE 24. SPIN Collaboration Report.

FIGURE 25. Polarized Tevatron

The source's polarized H^- ions would be first accelerated by an RFQ, which might be built by Professor Teplyakov's team at IHEP-Protvino. There would be no depolarization in the transfer line or LINAC. The Booster would have a polarimeter, two modest pulsed quadrupoles, and one partial Siberian snake. The Main Injector would have two polarimeters and two full helical Siberian snakes. Professor Ado and Dr. Ludmirsky at IHEP-Protvino built a 10% scale model of a Main Injector helical dipole snake; this working model was successfully tested at both Protvino and Michigan.

There might be two fixed-target experimental areas for spin physics. In the 120 GeV extracted beam area, there could be an experimental program using the Michigan solid polarized target which has a polarization of over 90% in a $10^{11}s^{-1}$ beam intensity. The C0 internal target area may have a Mark-III polarized jet, similar to the Mark-II jet for UNK (see Figure 30).

The most complex part of the Polarized Tevatron project came from the discovery that there were no places to put the Tevatron Siberian snakes with the proper symmetry defined by Professors Derbenev and Courant; I will not discuss this symmetry in detail. In any case, we decided to make spaces for the six Tevatron snakes by removing at each snake point four existing 6-m-long 4.4 Tesla dipoles. We could then install, in the resulting 24-m-long space, two 8.8-m-long 6 Tesla dipoles; this would easily create a space for one 5-m-long Siberian snake. These 6 Tesla dipoles might be slightly modified HERA dipoles, UNK-2 dipoles, or new Fermilab-designed dipoles. We are now interacting with many high-field dipole people who are usually outside the spin business.

I want to stress that this project is not yet approved. Fermilab has been funding these studies and R&D, but they certainly have not approved the required \$25 to \$30 million; we expect a decision by mid-1995. We are developing a schedule so that the Polarized Tevatron installation, which requires removing 36 existing Tevatron ring magnets, could occur during the Main Injector installation now scheduled for Fall 1997.

While preparing an EOI, Dr. Stanfield asked us to interact with the CDF and D0 people about this Polarized Tevatron project. We had several meetings, which at first were slightly painful because the collider-detector people and the polarized beam people have not talked much to each other for the last few decades. However, a good idea came out of this "painful" interaction. During a Polarized Tevatron meeting, Professor Weerts from Michigan State University showed Figure 26, where the inclusive cross-section for jet production is plotted against transverse energy. I had seen this graph

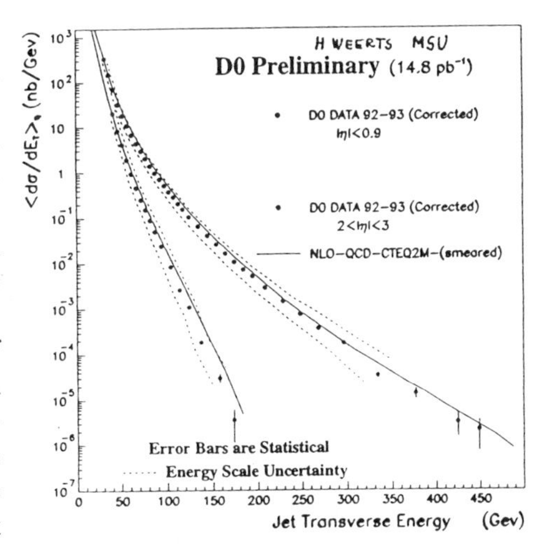

FIGURE 26. Inclusive Jet Cross Section.

at many seminars, but I never before thought of it as being related to polarization. I suddenly saw that at a huge transverse energy (or $P_\perp$) of 100, 200, or 300 GeV, there is still a very high event rate. I remembered that perturbative QCD predicted that A must eventually be zero, but essentially ignored all existing spin experiments because they were not at "a high enough energy or high enough $P_\perp$". However, Figure 26 shows high-event-rate data at $\sqrt{s}$ of almost 2 TeV and $P_\perp$ of 100 or 200 GeV/c. These inclusive Jets could easily test with great precision the perturbative QCD prediction that A must be zero. Weerts said that high precision measurements at $P_\perp$ = 100 GeV/c can be made in a few hours. If A is not zero at $\sqrt{s}$ = 2 TeV and $P_\perp$ = 100 GeV/c, then it may be difficult to say where PQCD is useful.

The last proton accelerator that I will discuss is UNK, which is now being built at IHEP in Protvino, Russia; IHEP is sometimes called Serpukhov, which is a city ten miles away. IHEP's major activity now is to get the UNK-1 accelerator built and to get the NEPTUN and NEPTUN-A polarization experiments running as soon as possible. This huge new facility is shown in Figure 27. There is also an informal proposal to develop a 70 GeV polarized proton beam at U-70 by Professor Ado and his student, Anferov, who recently proposed making spaces for the snakes in the existing U-70 ring by installing some higher field conventional magnets.

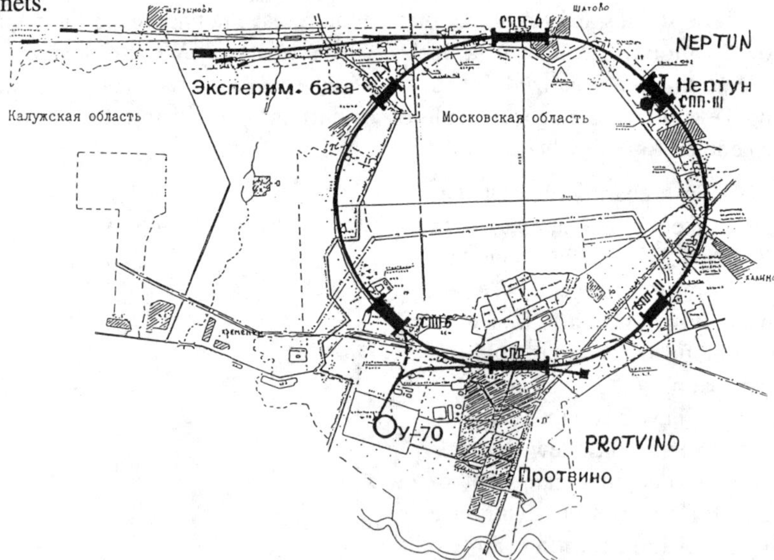

FIGURE 27. UNK and U-70 in Protvino, Russia.

In March 1994, IHEP successfully extracted a beam from the 70 GeV accelerator U-70 and transported it through the 2.7 km transfer line to the UNK tunnel.[26] The extraction and transfer efficiency was above 90%. When all of its magnets are installed, the UNK tunnel will look like this transfer line, which uses UNK-1's standard 6-m-long warm dipoles, its quadrupoles, and its standard vacuum system. Figure 28 shows the transfer line. They have not yet started installing the UNK-1 magnets in the main 21-km UNK tunnel. Although about 1500 of the 2200 dipoles and all of the 500 quadrupoles are already on site, they are waiting for the tunnel air conditioning before starting to install magnets.

FIGURE 28. Transfer Line from the U-70 Accelerator into the UNK Tunnel.

Building UNK in these complex times in Russia has been a great challenge. However, UNK still progresses at a fairly good rate, and there is good progress in some areas. Apparently it was decided to concentrate all efforts during the next few years on finishing and then operating the 400 to 600 GeV UNK-1 accelerator and the polarized internal jet target experiments NEPTUN and NEPTUN-A. NEPTUN is led by Professor Solovianov, and I lead NEPTUN-A. The large SS-3 underground hall, shown in Figure 29, is essentially finished; it

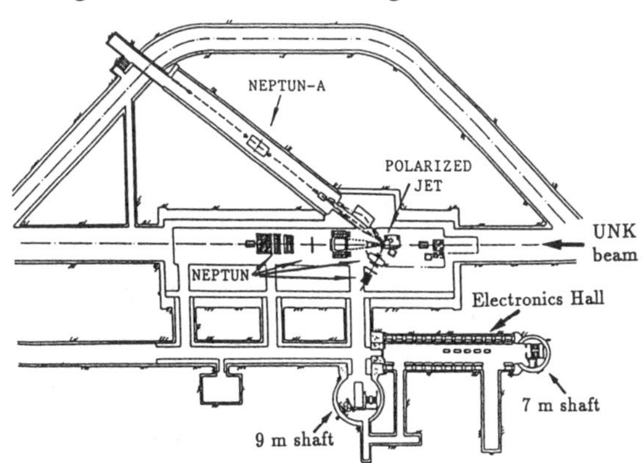

FIGURE 29. NEPTUN and NEPTUN-A Internal Polarized Jet Experiments.

was dedicated last fall. The Mark-II polarized jet target and the NEPTUN and NEPTUN-A spectrometers are shown along with the underground electronics hall. There may also be a second upstream jet target built by Professor Pilipenko at Dubna with several smaller spectrometers. Thus, for several years, the 21 km circumference UNK complex may be completely devoted to polarization experiments until the 3 TeV superconducting UNK-2 ring operates.

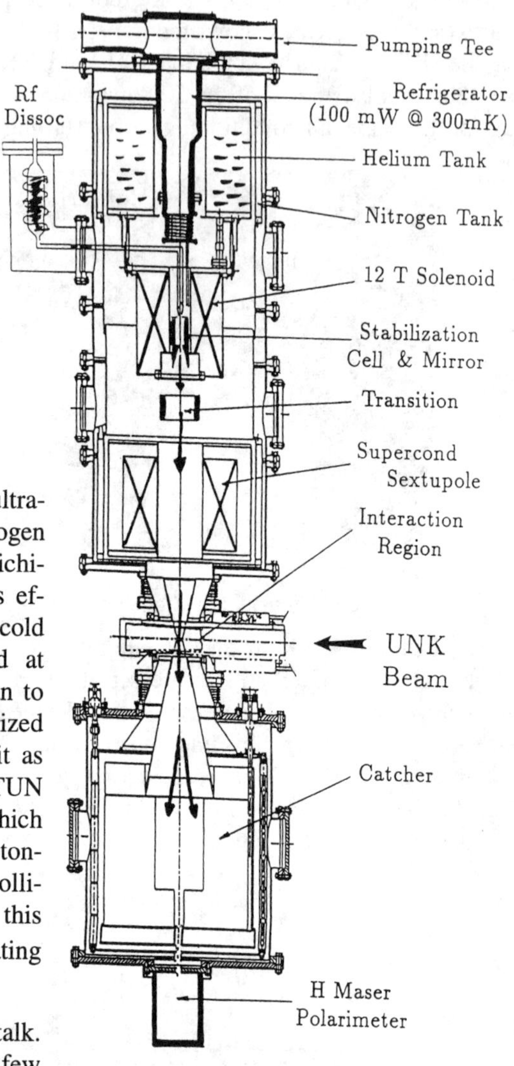

FIGURE 30. Michigan Mark-II Ultra-Cold Spin-Polarized H Jet.

Figure 30 shows the ultra-cold spin-polarized atomic hydrogen jet[27] which is being built at Michigan; Dr. Luppov is heading this effort. This jet uses the ultra-cold technique which was discussed at several of the sessions. We plan to take this Mark-II spin-polarized proton jet to Protvino and use it as the internal target for the NEPTUN and NEPTUN-A experiments, which will measure A in 400 GeV proton-proton elastic and inelastic collisions. Our goal luminosity with this jet and UNK's $10^{19}\,s^{-1}$ circulating protons is $10^{32}\,s^{-1}cm^{-2}$.

That ends my scientific talk. I would now like to make a few comments on behalf of the International High Energy Spin Physics Committee. I would first like to thank the three secretaries of these two Symposia: Sharon Herzel, Janet Meadows, and Sandy Smith. They were enormously helpful and tolerant of the strange behavior of about 340 physicists; we physicists sometimes do good research but can be a problem to deal with personally.

Figure 31 demonstrates an interesting situation where five different Committees have been organizing the two Symposia that are running together here in the Indiana Memorial Union. I have never before been to a meeting where there were five Committees. Two nights ago, I pointed out to Dr. Happer, who formerly directed research for DoE, that apparently DoE and NSF are now supporting advanced R&D in Political Science. It was certainly interesting to watch the five Committees interact with each other.

11th International Symposium
on High Energy Spin Physics

International Committee

A. D. Krisch (chair)	Michigan
C. Y. Prescott (chair-elect)	SLAC
D. P. Barber	DESY
O. Chamberlain	Berkeley
E. D. Courant	Brookhaven
G. R. Court	Liverpool
A. V. Efremov	Dubna
G. Fidecaro	CERN
W. Haeberli	Wisconsin
K. J. Heller	Minnesota
V. W. Hughes	Yale
D. Kleppner	MIT
A. Masaike	Kyoto
P. W. Schmor	TRIUMF
A. N. Skrinsky	Novosibirsk
V. Soergel	Heidelberg
J. Soffer	Marseille
L. D. Soloviev	Protvino

Organizing Committee

K. J. Heller (chair)	Minnesota
S. J. Broksky	SLAC
A. W. Chao	SSC
D. G. Crabb	Virginia
A. R. Dzierba	Indiana
R. L. Jaffe	MIT
S. Y. Lee	IUCF
D. B. Lichtenberg	Indiana
R. A. Phelps	Michigan
R. Prepost	Wisconsin
R. A. Rameika	Fermilab
J. B. Roberts	Rice
T. Roser	Brookhaven
H. M. Steiner	Berkeley
D. G. Underwood	Argonne

NUCLEAR REACTIONS AND STRUCTURE

PHYSICS BEYOND THE STANDARD MODEL

RECENT DEVELOPMENTS IN

POLARIZED SOURCES

ACCELERATION OF POLARIZED BEAMS

POLARIZED TARGETS

POLARIMETERS

SPIN '94 BLOOMINGTON

8th International Symposium
On Polarization Phenomena
in Nuclear Physics

International Advisory Committee

J. M. Cameron (chair)	IUCF
J. Arvieux	Saturne
A. S. Belov	INR Moscow
L. S. Cordman	CEBAF
C. R. Gould	North Carolina State
R. J. Holt	Argonne
R. L. Jaffe	MIT
M. Kondo	RCNP
C. Lechanoine-Leluc	Geneva
V. M. Lobashov	Moscow
J. B. McClelland	LANL
R. D. McKeown	Caltech
W. T. H. von Oers	Manitoba
H. Ohnuma	Tokyo Inst. Tech.
J. M. Richard	CERN
H. Sakai	Tokyo
F. D. Santos	Lisboa
M. Simonius	Zurich
J. Speth	KFA Julich
E. Steffens	Heidelberg
K. Yogi	Tsukuba

Organizing Committee

S. E. Vigdor (chair)	IUCF
T. B. Clegg	North Carolina
W. W. Jacobs	IUCF
L. D. Knutson	Wisconsin
J. T. Londergan	Indiana
J. W. Wissink	IUCF

Local Arrangements Committee

J. M. Cameron (chair)
E. J. Stephenson (co-chair)
A. D. Bacher
R. D. Bent
B. B. Brabson
P. Schwandt
J. Sowinski

FIGURE 31. SPIN '94 Committees.

I would especially like to praise Ed Stephenson, the Co-Chair of the Local Arrangements Committee. To appreciate his contributions, let me briefly describe a typical meeting of the Chairs of these five Committees; the meeting would usually take place in a small conference room in the IUCF building. Typically, Professor Cameron and I would be rolling around on the floor, shouting at each other politically incorrect insults about each other's energy preference. Professors Vigdor and Heller would be sitting there looking mildly embarrassed, but nevertheless cheering us on a bit. Dr. Stephenson would ignore the insults and the noise and continue working on a list of candidates for the SOROS Foundation; he just kept doing his job. We should all thank Ed Stephenson; perhaps we should make him the Permanent Chair of the Local Arrangements Committee.

Next, I would like to announce some results of yesterday's Meeting of the International Committee for High Energy Spin Physics; the Committee Members are listed in Figure 31. The first announcement is that the 12th International High Energy Spin Physics Symposium will be held during 9-14 September 1996 in Amsterdam. Professor C.W. de Jager will be the Chair of the Organizing Committee; I am looking forward to a very exciting Symposium.

Next we approved the 5th Workshop on Polarized Ion Sources and Gas Targets. It will occur during 6-9 June 1995 in Cologne. Professor H.P.g. Schieck will chair the Workshop; anyone interested in participating should contact him.

We also approved the 6th Protvino Spin Workshop; it will take place during 18-23 September 1995 in Protvino. Professor S.B. Nurushev will again chair this Workshop; anyone interested in attending should contact him.

The 8th Workshop on Solid Polarized Targets was also approved. The exact date is not yet fixed, but it will be at TRIUMF in either May or June 1996; the Chair will be Professor P. Delheij.

We also had some discussions about the 13th and 14th High Energy Spin Physics Symposia, which will occur in the years 1998 and 2000. We have a tentative proposal from Prague for the year 2000; we certainly took no action on what could happen in the year 2000, but it might be nice sometime to have the Symposium in Prague.

We talked informally about many places for the 1998 Symposium. It has been a long time since the Symposium was at one of the US national labs; it was at Brookhaven in 1982 and at Argonne in 1974, 1976, and 1978. Now SLAC, Fermilab, and Brookhaven may each have some interest. Argonne might also be interested; they no longer have an accelerator, but we should sometime return to Argonne where this Symposium series started. There was an informal proposal that the Symposium might be in Hawaii, jointly sponsored by America and Japan. This seems a bit complicated, as we again might need five Committees, but who knows? The 1998 Symposium could also be at some North American university that is heavily involved in High Energy Spin Physics. We have had very good experiences here and at Minnesota; I would note that Wisconsin is a very nice place. Another place we discussed was Lake Louise, Canada, where the 1986 Intersections Meeting occurred. One reason for thinking of Lake Louise is that we were all so pleased with the logistics here at Indiana; having the meeting rooms and many hotel rooms together in one building in a small town seems very good. Perhaps we should try to set this as a goal for the future. Lake Louise would be a change from the past in some ways, but we can think about new things. If anyone has any input on this, please send it to the Committee.

My next comment is about the International Committee Members. Four very distinguished Committee Members are retiring at the end of this year: Owen Chamberlain, Vernon Hughes, Daniel Kleppner, and Akiro Masaike. I would like to thank these four distinguished gentlemen for their long service. I am pleased to announce that there are four distinguished new Committee Members: John Cameron, Hiroyasu Ejiri, William Happer, and Yoshi Mori. These changes represent an effort to make the Committee even more international by increasing the Japanese representation. We also added two intermediate energy physicists: Professor Cameron and Professor Ejiri; we hope that this will encourage more intermediate energy and nuclear physicists to participate in the Amsterdam High Energy Spin Physics Symposium in 1996. By inviting more distinguished members of the Nuclear Polarization community to join our Committee, we hope to give a positive message that we are open to change. We look forward to many intermediate energy and nuclear physicists attending the Amsterdam meeting.

The trial partial merger of the two Spin Symposia was very exciting and exhausting but also very successful; we had a marvelous scientific meeting. I hope that John Cameron and I are still good friends, but the great success of the Spin '94 Meeting is what is really important.

Finally, this is my last Symposium as Chair; it has been an exciting, rewarding, and exhausting 20-year term. Our new Chair will be Charles Prescott of SLAC. I wish good luck to Charlie and all the rest of you.

ACKNOWLEDGMENTS

This work was partially supported by a Research Grant from the U.S. Department of Energy. I would like to thank Drs. A.M.T. Lin and L.G. Ratner and Ms. D.C. Barron for carefully proofreading this manuscript.

REFERENCES

1. I.M. Ternov, these Proceedings; A.A. Sokolov and I.M. Ternov, Sov. Phys. Dokl. **8**, 1203 (1964).
2. A.D. Krisch, Z. Phys. **C46**, S133 (1990).
3. E.A. Crosbie et al., Phys. Rev. **D23**, 600 (1981).
4. T.J. Devlin and L.G. Pondrom, these Proceedings.
5. A. Yokosawa, Proc. of 10th Int. Symp. on High Energy Spin Physics, eds. T. Hasegawa et al., p. 93 (Univ. Acc. Press, Inc. Tokyo 1993).
6. D.G. Crabb et al., Phys. Rev. Letters **65**, 3241 (1990).
7. A. Thomas, R. Voss, R. Windmolders, D. Day, H. Jackson, J. Soffer and M. Anselmino, these Proceedings.
8. Workshop on Polarized Ion Sources and Polarized Gas Targets, L.W. Anderson and W. Haeberli eds., AIP Conf. Proc. **293** (AIP, New York, 1993).
9. Y. Mori, these Proceedings.
10. Spin '93 Proceedings, S.B. Nurushev et al. eds., (IHEP, Protvino 1994).
11. Bad Honneff Workshop Proceedings, hopefully to be published; W. Meyer, these Proceedings.
12. M. Placidi, these Proceedings.
13. G. Voss, D. Barber, and H. Jackson, these Proceedings.

14. V. Soergel, Unpublished comment at plenary session of 9th Int. Symp. on High Energy Spin Physics, Bonn 1990.
15. M. Woods and H.B. Steiner, these Proceedings.
16. Ya.S. Derbenev and A.M. Kondratenko, Part. Accel **8**, 115 (1978).
17. A.D. Krisch et al., Phys. Rev. Letters **63**, 1137 (1989).
18. B.B. Blinov et al., Phys. Rev. Letters **73**, 1621 (1994).
19. R.A. Phelps et al., Phys. Rev. Letters **72**, 1479 (1993).
20. D.D. Caussyn et al., Phys Rev. Letters **73**, 2857 (1994).
21. S.E. Vigdor, these Proceedings.
22. H. Huang and T. Roser, these Proceedings; H. Huang et al., Phys. Rev. Letters **73**, 2982 (1994).
23. F.Z. Khiari et al., Phys. Rev. **D39**, 45 (1989).
24. Y. Makdisi, these Proceedings.
25. Polarized Tevatron Progress Report, unpublished University of Michigan Report UM HE 94-15 (August 1, 1994).
26. CERN Courier, June 1994, pp. 18-20.
27. V.G. Luppov et al., Phys. Rev. Letters **71**, 2405 (1993).

SUMMARY OF THE 5-TH INTERNATIONAL WORKSHOP ON HIGH ENERGY SPIN PHYSICS, SPIN-93,

SEPTEMBER 20-24, 1993, PROTVINO, RUSSIA

S.B.Nurushev

Institute for High Energy Physics, 142284 Protvino,Russia

A brief review is presented of Spin Workshop held at Protvino on September 20-24, 1993 with emphasis on new experimental results, theoretical approaches and innovative techniques. This Workshop sponsored by the Russian Ministery of Atomic Energy , Institute for High Energy Physics and the International Spin Committee presents a good source of informations for the scientists from the CIS. It gives also an opportunity to evaluate the spin physics development in the time between two subsequent International Spin Symposia.

INTRODUCTION

On a regular base, each two years , just in the middle of two subsequent International Symposia on High Energy Spin Physics the International Workshop under the same title, gathers mostly the CIS scientists at IHEP, Protvino. Started in September 14-17, 1983, in very tough political situation it aims to keep together all the CIS physicists interested in spin physics, to furnish this community with the news on recent developments in science and to give incentives for tight International Collaborations. This series of Workshops becomes now popular and traditional and they are sponsored by the Ministery of RF on Atomic Energy, the International Committee on Spin Symposia and IHEP. Five Workshops have been held and according to the requirements of the International Committee I am going to give an account to you on the V-th International Workshop on high energy spin physics held on September 20-24, 1993, in Protvino.

THE V-TH INTERNATIONAL WORKSHOP ON HIGH ENERGY SPIN PHYSICS

The V-th International Workshop on High Energy Spin physics was held on September 20-24, 1993, in Protvino, Russia. 120 participants(100 from CIS), 30 reports on plenary sessions and more than 30 reports on parallel sessions made this Workshop very alive. Further I intend to present to you the highlights of this Workshop.

THE EXPERIEMNTAL DATA.

The High Energy.

We were excited by the new results on the deuteron and neutron spin structure measurements presented by the SMC (R. Windmolders and G. I. Smirnov) and E-142 Collaboration(Y. Terrien). Both Collaborations have demonstrated impressive achievements in spin technology. Their results are in agreement with each other taking into account their error bars. For the averaged Q^2=2 $(GeV/c)^2$ the E142 experiment yields $\Gamma_1^n(x) = -0.021 \pm 0.011$ with all errors combined, while the SMC obtains at Q^2=4.6 $(GeV/c)^2$ $\Gamma_1^n = -0.08 \pm 0.04(stat) \pm 0.04(syst)$. The Ellis-Jaffe sum rule predicts $\Gamma_1^n = -0.021 \pm 0.018$ at $\overline{Q}^2 = 2(GeV/c)^2$. Therefore in frame of error bars one can see a rather good consistency between both experiments and the Ellis-Jaffe sum rule. Also much better agreement was demonstrated with the Bjorken sum rule by the both experimental data. Therefore we are wondering why these collaborations somehow came to different conclusions about the spin content of proton, about the sum rules and kept the spin crisis as an open problem. Since both experiments are continuing the data taking and also plan to make the measurements of the proton spin structure functions we believe that soon all such discrepancies will be removed.

The left-right asymmetry measurements at Z-pole became the next topic(G.Shapiro). The polarized SLC electron beam parameters were drastically improved(22% polarization in 1992 and more than 80% in 1994) and the SLD was able to gather 50,000 events and reached the following precision in the weak angle measurement : $\sin^2\theta_W = 0.2290 \pm 0.0010$. This collaboration wants to increase the statistics by one order of magnitude in a couple years, not only to achieve the highest precision in the weak angle, but to also look for some physics beyond the Standard Model.

New data from the WA89 CERN experiment opens a new avenue for spin physics by making use of the hyperon beam. The polarization of Λ produced by Σ^- beam appears to be smaller than the Λ polarization produced by the proton beam of the same energy. Authors believe that such results are qualitatively consistent with the Lund model expectation. By using the strange quarks (unpolarized at the moment, but may be polarized later) one can hope to clarify the origin of large hyperon polarization and separate contributions to polarization from the mechanisms of production and fragmentation.

The new results from the Tevatron reveal some new features in hyperon polarization as can be succinctly summarized by the following statements:

- the antihyperons $\overline{\Sigma^-}$ and $\overline{\Xi^+}$ are polarized at the same level as were Σ^- and Σ^+. These results contrast to the zero polarization in the case of $\overline{\Lambda}$

and present a challenge to the spin models;

- polarization of Σ^+ decreases with p_T for p_T >1 GeV/c;

- clear energy dependence of polarization has been observed for Σ^+ and Ξ^- : $P(\Sigma^+$,400 GeV)> $P(\Sigma^+$, 800 GeV), while $P(\Xi^-$, 400 GeV)< $P(\Xi^-$, 800 GeV);

- the Ξ^- polarization differs in x_F dependence from the Λ polarization, flattening off earlier in x_F;

- it was proven that there is a spin transfer mechanism: one needs to prepare the polarized strange quark in the initial state in order to produce the polarized Ω^- in the final state. But no quantative data on the spin transfer tensor were obtained yet.

The E-704 experiment continues to furnish a new data. The results on the measurements of the difference of the total pp- and $\bar{p}p$- cross-sections in the longitudinal polarization states showed that $\Delta\sigma_L(pp) = -[74 \pm 28(stat) \pm 36(syst)]\mu b$ and $\Delta\sigma_L(\bar{p}p) = -[61 \pm 86(stat) \pm 82(syst)]\mu b$ (A.de Lesquen). A series of E-704 results(the inclusive pion asymmetries at large x_F, asymmetry of direct photons, Λ- polarization) gives incentive to develope spin theory.

The Intermediate Energy.

The intermediate energy spin physics is a field of high scientific activity. "The full set of experiment" is a goal of many programs aiming to make model independent reconstruction of production amplitudes from data. The hadron spectroscopy studies can then be done on the level of individual amplitudes rather than spin-averaged cross-sections. Such an approach may lead to the discovery of new effects. As an example, consider the reaction $\pi + N_\uparrow \rightarrow \pi^+ + \pi^- + N$, where at 17.2 GeV/c the two important features were found out(M.Svec):1) the new scalar state $I = 0$, $J^{PC} = 0^{++}$ (750 MeV) and 2) the spin dependent mass splitting in the ρ^0 mass region. Further investigations are needed but there is no doubt that partial wave analysis by using "the full set of data" is very effective.

Neutron-proton total and differential cross-sections were measured at SATURNE II with polarized beam or/and target(presented by A. de Lesquen). First direct reconstructions of the np scattering matrix as well as of the isotopic singlet amplitudes at 0.84 and 0.94 GeV have been done showing some resonance-like structures.

The phase shift analysis of proton data in the energy region 2- 12 GeV was concluded by a statement that the spin- orbit interaction shrinks at high energies(M.Matsuda). It was proposed to do spin-correlation experiments to check such a conclusion.

Dubna and Saclay physicists are devoted to the measurements of the polarization transfer and tensor analyzing power in deuteron backward elastic scattering $dp \rightarrow pd$ to the highest possible energy(I.Sitnik). The results have been compared with previously published data for the 0° inclusive breakup of 2.1 GeV deuteron on hydrogen. Assuming the applicability of the impulse approximation and the use of S- and D- wave functions for deuteron there is a direct relation between T_{2O} and κ_0. This relation is evidently violated for both reactions indicating the possible involvement of new mechanism of interactions.

The Gatchina physicists using the high resolution magnetic spectrometer were able to measure the effective polarization P_{eff} for S- and P- shells separately in reaction ^{6}Li(p,2p)^{5}He (A.Prokofiev). According to the authors their results are inconsistent with the impulse approximation .

THE SPIN TECHNOLOGY.

This Workshop demonstrated the indisputable progress in spin technology. The reliable electron source of high intensity and polarization has been built at SLAC by using YAG-pumped Ti:Sapphire laser operating at 865 nm. A series of optical elements to control the intensity, pulse length, circular polarization, and steering was invented. Several variations on the type of GaAs material have been used in recent accelerator runs resulting in increase of electron polarization from 22% in 1992 to in excess of 80% in late 1994.

According to A.Zelenski et al., now there is a possibility to build the pulsed Optically Pumped Polarized Ion Source(OPPIS) producing up to 1.2 mA H^- ion beam with polarization in excess of 80% within 1.5 $\pi \times mm \times mrad$ normalized emittance. According to the authors the current is ten times higher than the best current available from atomic beam sources. The pulsed OPPIS is quite inexpensive in comparison with atomic beam source, and is ideally suited for high energy accelerator applications. Therefore we are approaching our dream: the intensity of polarized beam must not yield to the unpolarized one and this will be the case soon.

The technique of siberian snake has been thoroughly tested in the Indiana University Cooling Facility(IUCF) and its ability to eliminate many depolarizing resonances including the overlapping ones was demonstrated. Recently two methods of spin flipping a stored vertically polarized proton beam at the IUCF Cooler Ring have been tested (R.Phelps). One uses an RF solenoid to induce an artificial depolarizing resonance in a storage ring at the resonance frequency. If the rotation rate of the stable spin direction caused by the RF field is much slower than the spin precession frequency, the spin will follow the stable spin direction and be reversed. A more efficient spin flip method using the "free spin precession" technique was tested also. Here the RF solenoid is turned on exactly at the resonance frequency for a fixed amount of time and then turned off before the polarization is measured. The RF on time is adjusted to give complete spin reversal. At 10.3 msec RF on time full polarization flip was achieved. A study showed that 0.2% polarization loss/flip occured, so one expects more than 300 reversals before half the original polarization is lost. This method proves to be more efficient for spin flipping than the ramped frequency method. The same group begun a study of the snake adiabatic turn on and they found no polarization loss in ramping a siberian snake in 1 second from 0 to 25%. This result is encouraging since one may hope to pass the spin resonances at low energy by using such technique. Therefore the way becomes open for the acceleration of polarized proton beams up to the highest energies of current and upcoming big accelerators.The helical spin rotating magnets attract the deep attentions of physicists.The scheme based on helical undulators is more preferable as a spin rotator at medium energy range (from say 10 to 300 GeV) since it keeps a small orbit distortion ($y_{max} = 0.2$ at $\gamma = 200$ for 45° snake axis angle). This configuration was proposed for RHIC operating with polarized proton beams. (E.Perevedentsev et al.). There are suggestions to use another type of helical magnets for spin rotation in the Tevatron and small prototype was produced at IHEP for test (Spin Collaboration).

The idea to avoid the depolarizing resonances by the spin precession frequency jump rather than the betatron tune jump seems very promising(N.Golubeva et al.). Then the spin tune becomes dependent not only on energy but also on the introduced magnetic fields. There appears a possibility to organize the fast crossing not only of the intrinsic spin resonances but also the integer imperfection spin resonances. With an appropriate choice of the insertion magnetic element parameters, the orbit distortion may be localized and the betatron tunes may be set unchanged in the whole accelerator energy range.

The essential progress was demonstrated by the University of Michigan team in building the ultra-cold spin-polarized hydrogen atom source. They have shown the possibility of forming a high intensity beam of ultra-cold spin-polarized hydrogen atoms by using a quasi-parabolic polished copper mirror coated with a superfluid ^{4}He film (W.Kaufman).

The LHE JINR team built its first cryogenic polarized deuteron source POLARIS with 200 μA at the early 80's and since this source is efficiently used in 10 GeV Synchrophasotron. The source can stably run for a month without a sublimation of the condensed gas. The module variant of the set-up allows to accomplish a replacement of the RF cell for vector and tensor polarization mode, a HF acid washing of the dissociator without warming of the cryostat. The intensity of the extracted polarized deuteron beam is of order 10^9 pol.deutr./ pulse, at polarizations(measured): $P_z^+ = 0.54 \pm 0.01$, $P_z^- = -0.57 \pm 0.01$, $P_{zz}^+ = 0.76 \pm 0.02$, $P_{zz}^- = -0.79 \pm 0.02$. This team plans to increase the intensity of polarized deuteron beam up to 2mA and develope in similar way the pulsed cryogenic polarized atomic hydrogen jet source, which may be used in the new accelerator Nuclotron. In looking for possible increase of intensity they plan also to develope the ultra-cold polarized source, as the

University of Michigan team does.

The searches are made for optimal target materials for the solid polarized targets including their resistance to the high radiation (E. Bunyatova, M.Liburg).

The application of bent crystals in high energy spin physics has been discussed by V.Maisheev et al. The possible avenues are:1) extraction of polarized beam from accelerator/collider by transforming beam polarization from longitudinal to transverse one,2) the measurements of the emittance of the circulating polarized beams and 3) the polarimeters.

The possibility of producing the longitudinally polarized proton beam at IHEP has been discussed by V.Zapolsky. It was shown that the existing beam line transporting the transversally polarized proton (antiproton)beam from $\Lambda(\overline{\Lambda})$ by using the coherent precession of proton spin in the transverse magnetic field. For momentum of 40.7 GeV/c, the momentum acceptance $\frac{\Delta P}{P} \approx 2\%$, the intensity is $I_p \approx 10^7$ pol.prot/ pulse at the primary proton intensity of 10^{12} p/pulse. Polarization is $\approx 30\%$.

Two aspects of the inclusive break-up of polarized deuteron on proton have been discussed by L.S.Azhgirey et al. First, a tensor analyzing power with proton emission at non-zero angles is calculated. Such an approach proves a higher sensitivity to the small distances inside deuteron due to a large transverse momenta. Second, it is suggested to compare the nucleon polarization in the relativistic deuteron with the nucleon polarization in a deuteron at rest. Since the Lorentz boost operator depends also on an interaction between nucleons, one hopes to extract from such a study some informations about dynamics at small distance interactions.

The polarimeters become the subject of important interests. The five reports on this problem were presented. One report(G. Shapiro) presented the currently used at SLC compton polarimeter providing a fast and accurate measurement of the longitudinal polarization of electron beams available at linear colliders. Statistical accuracy of better than 1% in the absolute beam polarization is achieved routinely in runs of three minutes duration, with the laser firing at 11 Hz. This polarimeter is close to the "ideal" one. Four other reports promote different types of polarimeters for realization in electron and hadron accelerators/colliders. A method for measuring the photon beam linear polarization by means of measuring the recoil electron asymmetry in e^-e^+ pair photoproduction on electrons was proposed and tested at Kharkov electron accelerator LAE-2000 (V.Boldyshev et al.). Despite a lack of statistics, experimental data demonstrate that the recoil electron asymmetry in the process of triplet photoproduction may be used for the measurements of photon beam linear polarization. Exclusive pion production in the reactions $p_{\uparrow}p \rightarrow pN\pi$ and $e_{\uparrow}p \rightarrow eN\pi$ have been proposed as polarimeters. The analyzing powers appear in these reactions thanks to interference of the intermediate resonating and no resonating channels(A.Arbuzov et al.). According to authors the analyzing powers are expected on order of 10%. The pp- elastic scattering is considered for the the case of polarized colliders. It is shown that for the CNI region the use of two transversally polarized proton beam lead to an increase of efficiency of polarimeter by coefficient two(A.Potylitsin et al.). The elastic scattering of the polarized proton on the polarized electron as an efficient polarimeter was advocated by a group of physicists looking for the ideal polarimeter(M.Strikhanov et al.). No doubt that fresh ideas, inventions of new technology will stimulate polarization physics at the new generation of accelerators/colliders.

THE SPIN THEORY.

Theoreticians are working hard on the intepretations of SLAC, EMC, SMC, E-142 results on the spin structure functions. Value of power corrections to the integrals from polarization structure functions of proton and neutron was determined based on a model which accounts for the higher twist terms, has correct asymptotics at large Q^2 and satisfies the Gerasimov-Drell-Hearn sume rule at $Q^2 = 0$. The contributions of resonances up to W=1.8 GeV at $Q^2 = 0$ is taken into account on the basis of experimental data. It is shown that when taking into account these higher twist terms, the experimental data agree with the Bjorken sum rule and the part of the proton spin projection carried by quarks, does not disagree to the estimate of 50%(V.Burkert and B.Ioffe).Taking into account the power corrections, 8% for proton at average $\bar{Q}^2 = 10.7GeV^2$ and 11% for neutron at average $\bar{Q}^2 = 2GeV^2$ it was possible to reconcile all experimental data with the sum rules and the expected spin content of nucleons (Ioffe). It was shown also that the effect of the relativistic Fermi motion, described by the ratio of the spin structure functions $R_g^{D/N} = g_1^D/g_1^N$ reaches $\sim$ 9% in the range $x = 10^{-3} \rightarrow 0.7$ and $Q^2 = 1 \rightarrow 80(GeV/c)^2$. The dependence of the Bjorken integral on x and Q^2 was studied and shown that the nuclear effect contributes by $11 \rightarrow 16\%$ and should be taken into account in checking the Bjorken sum rule(M. Tokarev).

It was stated that the G_2 strong Q^2 dependence implied by the Burkhard-Cottingham(BC) sum rule naturally explains the sign change of the generalized Gerasimov-Drell-Hearn sum rule. The status of the BC sum rule and its implications for other spin processes were discussed (J.Soffer and O.Teryaev).

N.Kochelev emphasized that specific chirality and flavor properties of the instanton-induced interaction explain the observed violation of Ellis-Jaffe and Gottfried sum rules. These violations originate from the anomalous growth of the instanton-induced interaction between quarks at high energy. The new sum rule which connects the values of violations of Ellis-Jaffe and Gottfried sum rules has been obtained.

A.Efremov in his talk "Hunting for handedness" reviewed the status for search of longitudinal jet handedness in e^+e^-- annihilation into two jets in the Z^0-resonance region . He listed the factors diluting the effect of handedness up to the level of 0.3-0.4% making it unobservable and proposed some approaches to overcome this problem. The possible way of measuring the gluon and sea quarks polarizations by studying the Z^0 asymmetries in collision of polarized protons has been discussed by Yu.I.Arestov et al. In energy region of RHIC the asymmetries are expected to be of order 10% and due to mostly to annihilation in the $q\bar{q}$ subprocesses.

A.P.Contogouris gave a talk on "One-loop corrections to lepton pair production by transversaly polarized hadrons". A complete calculation of QCD higher order corrections at one loop for Drell-Yan lepton pair production by transversely polarized hadrons was made. Productions via both γ and Z were taken into account. K-factors well exceeding unity are found and, in particularly near the Z peak, significant rates may be anticipated.

The large value of the single-spin asymmetry, its flavor dependence, sign, energy dependence, transversity are very "hot" items on the theoreticians agenda. The different approaches have been presented on this Workshop. It was emphasized that ratio of spin flip to spin non flip amplitudes is independent of energy (Goloskokov, Selyugin). There are models not only pretending to describe the single spin asymmetry, measured in E-704 experiment, but making quantative predictions for future experiments (Artru,Meng Ta chung, Ramsey, etc.).

There was one unusual approach to the interpration of the large spin-spin correlation parameter A_{NN} in the frame of the non-associative field theory (Begeluri et al.). This might be a promising direction in spin theory.

M.Chavleishvili discussed spin asymmetries at high energy binary reactions on the bases of the general dynamic amplitude formalism and "kinematic hierarchy". The formalism provides an analysis where spin -kinematics is fully taken into account and clearly separated from dynamics. At high energies and fixed angles kinematic factors in observables can be considered as small parameters and this results in a hierarchy of amplitude contributions, connections between asymmetry parameters, and even numerical values for some of them.

We don't yet understand the importance of neutrino interactions for spin physics. But theoreticians are pushing ahead with this problem enthusiastically. The effect of possible transition of left-handed polarized neutrinos into right handed ones was estimated in external electromagnetic and gravitational fields. According to this estimate for sun's neutrino transition effect $\nu_L \rightarrow \nu_R$ will be notable for mass of neutrino $m_\nu \leq 10^{-4} GeV$.

THE FUTURE SPIN EXPERIMENTS.

The electron accelerators at KHPTI, Novosibirsk and Bonn are actively producing results on electro-and photoproductions of hadrons. The strong interaction spin physics is firmly supported by Synchrophasotron (later Nuclotron) at Dubna, Saturne II at Saclay, TRIUMF, LAMPF, SIN, also at SPINP, ITEP, IHEP, AGS, KEK. The exciting Spin Program was approved recently at RHIC (RSC). As G.Bunce noted in his talk, the spin potential of colliding beams of protons, polarized either longitudinally or transversely , at RHIC, is remarkable . A luminosity of $L = 2 \times 10^{32} cm^{-2} sec^{-1}$ with 70% polarized beams will be available with up to 250 GeV energy in each beam. The spin physics program includes measurement of gluon and sea quark polarization in the longitudinally polarized proton, measurement and then application of parity violation in W and Z production, measurement of hard scattering parton-parton asymmetries, and quark polarization or transversity in transversely polarized protons. Single spin asymmetries allow sensitive searches for parity violation(longitudinal polarization) , and correlations between quark spin and gluons(transverse). Probes include direct photons(to p_t=20 GeV/c), jets(to p_t >50 GeV/c), Drell-Yan pairs (to m_{ll}=9GeV), $W^{\pm}$, Z.

Dubna physicists proposed to extend the RHIC polarization program by adding the study of color transparency, quark-gluon correlator and transversity(Yu.Panebratsev).

A.Krisch described the status of "Spin experiments at UNK and Fermilab". In the first case it is foreseen to use the high density ultra-cold polarized jet installed in the unpolarized internal UNK proton beam. A recoil arm high resolution spectrometer equipped with Cherenkov counters will be used to identify the violent proton-proton scattering. In the Tevatron the polarized protons will be accelerated and then strike the polarized and unpolarized jet targets. The collider detectors may profit also from a polarized high intensity proton beam.

Polarization experiment(POLEX), making use of the secondary polarized beams from the IHEP 600 GeV accelerator, was described by S.Nurushev. The major part of program is devoted to the study of spin structure of nucleons. The polarization measurements in elastic and charge-exchange reactions with use of polarized targets are also foreseen.

A.Ufimtsev suggested to extend the R7 experiment at RHIC by adding to the program the total cross-section measurement in pure initial spin states, measurements of slope and ρ parameters for different orientations of initial spins.

IHEP physicists presented the program of asymmetry measurements of charged hadrons with a 40 GeV/c IHEP polarized proton beam(A.Dyshkant).

A comprehensive program for detailed study of the total cross section differences $\Delta\sigma_L$ and $\Delta\sigma_T$ with polarized beams and polarized target at the JINR accelerators has been presented by the Dubna-Saclay Collaboration(L.Strunov).

M.Tanaka(Kobe,Japan) stated that the recent progress in the construction of the polarized ^{3}He ion source makes possible to develope a program with polarized heavy ions of 100 MeV/u at RCNP.

More than 120 attendees, 30 reports on Plenary Sessions and around 30 reports on Parallel Sessions illustrate the scale of this Workshop.

This V-th International Workshop was sponsored by the Russian Ministry of Atomic Energy and Industry and by the International Committee on Spin Symposia. It was actively supported by the IHEP managers and staff members. We would like to thank all of them for invaluable help.

With my pleasure I announce that the 6-th International Workshop on High Energy Spin Physics will be held at Protvino,on September 18-23, 1995. All of you are welcome to participate in this Workshop.

Thank you .

The spin of relativistic electrons and nuclei in external magnetic field

Igor M. Ternov

Department of Theoretical Physics, Faculty of Physics, Moscow State University, Moscow 119855

1. THE PROBLEM OF SPIN DESCRIPTION IN A RELATIVISTIC MOTION OF FREE PARTICLES

The original development of the theory of spin was based on the hypothesis of Uhlenbeck and Goudsmit (1925) in accordance with which is assumed that the electron possesses spin mechanical and spin magnetic momenta that are not related to the displacement of the particle in space. In Pauli theory the orbital and spin angular momenta are integrals of motion:

$$\frac{d\vec{L}}{dt}=0, \quad \frac{d\vec{S}}{dt}=0. \tag{1}$$

The situation in relativistic case is different, because in Dirac theory only the total angular momentum is integral of motion

$$\frac{d(\vec{L}+\vec{S})}{dt}=\frac{d\vec{J}}{dt}=0. \tag{2}$$

The methods of spin description must be revised because it is impossible to separate the spin from orbital motion of particles. This is the result of special kind of Dirac electron motion: trembling motion ("zitterbewegung") in consequence of charge conjugation states interference.

For the spin description of relativistic particles we can introduce the polarization operator $\hat{\vec{S}}=\frac{\hbar}{2}\hat{\vec{O}}$. This operator possesses the properties:

1. It commutes with the Hamilton operator $\hat{\mathsf{H}}$

$$\hat{\mathsf{H}}\hat{O}-\hat{O}\hat{\mathsf{H}}=0.$$

2. In nonrelativistic limit $\hat{\vec{O}}$ corresponds to Pauli matrix $\vec{\sigma}$

$$\vec{S} \to \vec{S}^{p} = \frac{\hbar}{2}\vec{\sigma}.$$

3. It satisfies the requirement

$$\hat{O}_i\hat{O}_j - \hat{O}_j\hat{O}_i = 2i\varepsilon_{ijk}\hat{O}_k \qquad (3).$$

4. Operator $\hat{\vec{O}}$ must be unit operator: its projection on to any direction $\vec{n}$ $(|\vec{n}| = 1)$ in the space satisfies the requirement

$$(\hat{\vec{O}}\vec{n})(\hat{\vec{O}}\vec{n}) = 1.$$

The best polarization operator is three dimensional spin vector $\hat{\vec{S}} = \frac{\hbar}{2}\hat{\vec{O}}$

$$\hat{\vec{O}} = \rho_3\vec{\Sigma} + \rho_1 c\hat{\vec{p}}/E - \rho_3 \frac{c^2\hat{\vec{p}}(\vec{\Sigma}\cdot\hat{\vec{p}})}{E + mc^2} \qquad (4)$$

$(\rho_1, \rho_3, \vec{\Sigma}$ are matrix of Dirac).

In the direction of motion of the particles spin vector $\hat{\vec{O}}$ is equal to $\vec{\Sigma}$ (longitudinal polarization) and $\rho_3\vec{\Sigma}$ in the direction perpendicular to the motion (transversal polarization).

Since thus $\hat{\vec{O}}$ is unitary transformation of the ordinary spin operator (Foldy-Wouthaisen transformation) the eigenvalues $(\hat{\vec{O}}\cdot\vec{n})$ are equal to the eigenvalues of the spin operator in rest frame. The polarization remains unchanged in all spaces.

2. TWO GENERAL METHODS OF THE SPIN EFFECTS CALCULATIONS

Dirac-Schrödinger representation

For the calculation of quantum effects "Method of exact solution" is very important. In Hamilton operator [1]

$$\hat{\mathbf{H}} = c\vec{\alpha}(\hat{\vec{p}} - \frac{e}{c}\vec{A}^{ext} - \frac{e}{c}\vec{A}^{q}) + \rho_3 mc^2 \tag{5}$$

external field $\vec{A}^{ext}$ must be included in the exact solution of equation

$$\hat{\mathbf{H}}|\Psi\rangle = E|\Psi\rangle.$$

The quantum part of electromagnetic field $\vec{A}^{q}$ corresponding to quantized field of radiation can be calculate on the ground of perturbation theory (Furry representation).

To introduce in the theory of "forth" quantum number (spin number) we can choose the $\hat{\vec{O}}$ operator (q) with the generalized momentum $\vec{\mathbf{P}} = \vec{p} - \frac{e}{c}\vec{A}$.

The projection of $\hat{\vec{O}}$ onto the direction of the magnetic field is conserved:

$$\hat{O}_3\hat{\mathbf{H}} = \hat{\mathbf{H}}\hat{O}_3.$$

Choosing now $\hat{O}_3$ as the polarization operator (transverse polarization) and requiring the wave function to be an eigenfunction of this operator

$$\hat{\mathbf{H}}|\Psi\rangle = E|\Psi\rangle, \quad \hat{O}_3|\Psi\rangle = \zeta|\Psi\rangle,$$

where $\zeta = \pm 1$ determines the orientation of the electron spin along magnetic field and in the direction opposite to it, respectively, we obtain the possibility of introducing forth number (spin number) for the characteristic quantum state $|\Psi\rangle$.

All the solutions of Dirac equation can be separate according to the spin state of electron.

Heisenberg representation

In the ground of this representation is the generalization of Bargman-Michel-Telegdi (BMT) equation taking account of special character of Dirac electron motion - "zitterbewegung" [2]

$$\frac{d\hat{\vec{O}}}{dt} = \frac{ec}{E}(1 + \frac{\alpha}{2\pi}\frac{E}{mc^2})[\hat{\vec{O}}\mathbf{H}] - \frac{\alpha}{2\pi}\frac{e}{mc}\frac{c^2[\hat{\vec{O}}\hat{\vec{\mathbf{P}}}][\hat{\vec{\mathbf{P}}}\mathbf{H}]}{E(E + mc^2)} + \frac{i\hbar}{2E}\frac{d^2\hat{\vec{O}}}{dt^2}. \tag{6}$$

The last term in this equation corresponds to "zitterbewegung". If we introduce the average $\vec{\zeta} = <\hat{\hat{O}}> = <\Psi|\hat{\hat{O}}|\Psi>$ and eliminate the term $\hbar \to 0$ we can obtain BMT equation in classical approximation, because it is independence on Planck constant $\hbar$.

It is interesting that this equation (BMT) was obtained by Ya. Frenkel on the ground of classical model "rotating point top".

3. THE QUANTUM PROCESSES IN A STRONG MAGNETIC FIELD

The quantum processes in magnetic field as a rule depends on invariant dynamical parameter [3]

$$\chi = \frac{1}{H_0 mc}[-(F_{\mu\nu}\mathbf{P}^{\nu})^2]^{1/2} = \frac{H}{H_0}\frac{p_\perp}{mc} \tag{7}$$

(in a magnetic field). $F_{\mu\nu}$ - is tensor of electromagnetic field, $\mathbf{P}^{\nu}$- four momentum vector, $H_0 = m^2c^3/e\hbar = 4.41 \cdot 10^{13}\,Gs$ - Schwinger field. For the synchrotron radiation χ is quantum parameter

$$\chi = \frac{\mathbf{E}_q}{E} = \frac{\hbar\omega}{E} \cong \frac{\hbar\omega_0}{E}\left(\frac{E}{mc^2}\right)^3 = \frac{\hbar}{mcR}\left(\frac{E}{mc^2}\right)^2$$

and in practically for all the real magnetic fields χ is too small: $\chi << 1$. The quantum effects appear in this conditions, as a quantum corrections to classical theory.

But the Dirac theory is true also for the strong and superstrong fields $H \geq H_0$. For the motion of electron in homogeneous magnetic field the energy spectrum depends on general number $n = 1, 2, 3...$

$$E / mc^2 = \sqrt{1 + 2nH / H_0} \tag{8}$$

(momentum along magnetic field $\mathbf{P}_{||} = 0$). In this case parameter $\chi = (H/H_0)^{3/2}\sqrt{2n}$ and it satisfies to the condition $\chi >> 1$ including the first exiting state $n \sim 1$. This case is interesting because it corresponds to ultraquantum physics.

The recent years in the development of physics have been characterized by discoveries of sources of superstrong fields. In the first place there is superstrong magnetic field $(10^{12} - 10^{13} Gs)$ nearly the pulsars - rotating neutron stars.

Such values of the magnetic field are not exotic and arise even under laboratory conditions: in a non central collision of heavy ions (Darmstadt, UNILAC, GFR). The quantum effects in this magnetic field are unusual and some times exotic. The calculation of this effects can be made only on the basis of "exact solution".

4. THE SPIN EFFECTS IN A WEAK MAGNETIC FIELD. RADIATIVE POLARIZATION OF ELECTRONS AND POSITRONS IN STORAGE RINGS

This effect was predicted by present author (1961) /doctoral dissertation [4] and was rigorously established theoretically in collaboration with A.A. Sokolov (1963) [5]. That time it was unusual and unexpected phenomenon. On the ground of "exact solution" method of calculations it is possible to obtain the exact formula for the probability of transitions with spin flip:

$$w^{\uparrow\downarrow} = \frac{\sqrt{3}}{4\pi}\frac{e^2}{\hbar R}\xi^2 \int_0^\infty \frac{y^2 dy}{(1+\xi y)^3}[K_{2/3}(y) + \zeta K_{1/3}(y)],$$

$$\xi = \frac{3}{2}\chi = \frac{3}{2}\frac{H}{H_0}\frac{E}{mc^2} = \frac{3}{2}\left(\frac{H}{H_0}\right)^{3/2}\sqrt{2n}. \tag{9}$$

Integration of this expression of over the spectrum leads us to the final formula for the probability of transitions in 1 sec (in the case $\chi << 1$):

$$w^{\uparrow\downarrow} = \frac{1}{2\tau}(1 + \zeta\frac{8\sqrt{3}}{15}) \tag{10}$$

where the time of polarization is

$$\tau = \frac{8\sqrt{3}}{15}\frac{\hbar^2}{mce^2}\left(\frac{mc^2}{E}\right)^2\left(\frac{H_0}{H}\right)^3. \tag{11}$$

The most important: the probability depends on the initial orientation of electron spin ζ - this is the ground of radiative polarization effect.

Now we consider the statistical equations that characterize the change in the spin orientation of the particles in an electron beam. The kinetic equation for polarization process:

$$\begin{aligned} \frac{d}{dt}n^{\downarrow} &= n^{\uparrow}w_{\zeta=1} - n^{\downarrow}w_{\zeta=-1} \\ \frac{d}{dt}n^{\uparrow} &= n^{\downarrow}w_{\zeta=-1} - n^{\uparrow}w_{\zeta=1} \end{aligned} \tag{12}$$

where $n^{\uparrow} + n^{\downarrow} = n$, $n^{\uparrow}$ and $n^{\downarrow}$ - the number of electrons with spin oriented along the magnetic field and in the direction opposite to it. If in initial time $t = 0$ the beam was unpolarized $n_0^{\uparrow} = n_0^{\downarrow} = n/2$, we find

$$P(t) = \frac{n^{\downarrow} - n^{\uparrow}}{n^{\downarrow} + n^{\uparrow}} = \frac{8\sqrt{3}}{15}(1 - \exp\{-t/\tau\}) \tag{13}$$

and the limit degree of polarization is

$$P(\infty) = P_{\infty} = \frac{n^{\downarrow} - n^{\uparrow}}{n^{\downarrow} + n^{\uparrow}} = \frac{8\sqrt{3}}{15} = 0.924 \tag{14}$$

(Sokolov, Ternov , 1963, [5]).

It is well-known result for the model of homogeneous magnetic field. Later it was repeated again by other authors made by different methods: V. Baier and V. Katkov (1966) [6], E. Stork (1968) [7], J. Schwinger and Tsai (1974) [8].

It is interesting to remark that polarization effect depends on the angle of photons emission ϑ [9]

$$\frac{dw^{\uparrow\downarrow}}{d\Omega}=\frac{2\sqrt{3}}{9}\frac{e^2}{\hbar R}\frac{\xi^2}{\pi^2\varepsilon_0}\frac{1}{(1+\psi^2)^4}[1+\frac{35\sqrt{3}}{192}\frac{\pi}{\sqrt{1+\psi^2}}\zeta] \qquad (15)$$

where

$$\varepsilon_0=1-\beta^2<<1, \quad \psi=\cos\vartheta/\sqrt{\varepsilon_0}.$$

The term containing the initial spin orientation attains its maximal value, when $\psi=0$ - in the plane of the orbit ($\vartheta=\pi/2$). And then

$$P(\infty)\big|_{\vartheta=\pi/2}=\frac{35\sqrt{3}\pi}{192}=0.99.$$

The influence of the quantum fluctuations on the process of radiative polarization (recoil effect) in the plane of orbit is minimal.

Analysis of the polarization time τ shows that under the conditions of ordinary values of the magnetic field for accelerators $H\sim 10^4\,Gs$ radiative polarization can be observed only in the case of prolonged circulation of the particles in the magnetic field. Then the radiative energy losses must be compensated and the electron energy must be on the average constant.

Such a situation is realized in storage rings - the storage rings are the exclusive laboratory for the observation of quantum effects.

5. THE VISUAL OBSERVATION OF SPIN

According to quantum theory of synchrotron radiation [10] the power of radiation depends on initial spin orientation:

$$W=\frac{3\sqrt{3}}{4\pi}\frac{e^2c}{R^2\varepsilon_0^2}\int_0^\infty ydy\{(1+\xi y)\int_y^\infty K_{5/3}(x)dx-\zeta\xi yK_{1/3}(y)\}. \qquad (16)$$

The last term in this equation corresponds to mixed radiation of charge and magnetic moment [11]. It is clear that spin makes the contribution in total synchrotron radiation and the visual observation of spin depend on synchrotron radiation is very interesting: we can obtain the information about spin of particle.

This method of visual observation was proposed at the Institute of Nuclear Physics at Novosibirsk (1977) [12] (see also [13]) and was realized in 1983 using the VEPP-4 storage ring [14].

From (16) we can see that the short wavelength part of synchrotron radiation spectrum is most favorable for observation, the additional power due to the spin orientation has the form:

$$\Delta = \frac{2\xi P_\infty y K_{1/3}(y)}{\int\limits_y^\infty K_{5/3}(x)dx + K_{2/3}(y)}, \quad \xi = \frac{3}{2}\chi \quad \Delta\big|_{y\to\infty} \cong P_\infty y\xi \qquad (17)$$

Here P_∞ is the degree of beam polarization.

For fixed observation wavelength, Δ is proportional to the parameter ξ: $\xi = \frac{3}{2}(E/mc^2)(H/H_0)$- i.e. to the magnetic field and energy of the particle. The quantity Δ is the jump in the energy when the depolarizer is switched on.

We note that the measurement of radiative polarization by method which uses observation of the spin depose on SR introduces a new concept in the problem of measuring spin of a free electron (quantum theory of measurements). Here, the source of information about spin orientation is the electron itself, which is free and not bound in atom. This experiment is the first visual observation of spin.

6. THE SPIN EFFECTS IN A SUPERSTRONG MAGNETIC FIELD

Radiative effect of polarization (Electromagnetic interactions)

From the exact formula (9) for the probability of spin-flip transitions in the case of strong magnetic field $\chi >> 1\,(\xi >> 1)$ and excite states $n >> 1$ we can obtain

$$w^{\uparrow\downarrow} = \frac{1}{2\tau^*}\left\{1+\zeta\frac{5}{2}\frac{\Gamma(1/3)}{\Gamma(2/3)}\frac{1}{(3\chi)^{1/3}}\right\} \qquad (18)$$

and

$$P_{\max} = \frac{5}{2}\frac{\Gamma(1/3)}{\Gamma(2/3)}\frac{1}{(3\chi)^{1/3}}. \tag{19}$$

In the case of strong magnetic field the present is very pessimistic : the radiative effect of polarization is very small. Process of polarization disappears.

Polarization of protons and nuclei in the process of β-decay in a strong magnetic field (weak interactions)

The another situation appears in the weak interactions, for example β-decay of neutron. We consider simplest variant β-decay in local fermion interaction. The description corresponds to transition

$$n \to p + e^{-} + \tilde{\nu} \quad \varepsilon_0 = 2.58 \tag{20}$$

and for transition between mirror nuclei (for example)

$$^{3}_{1}H \to {}^{3}_{2}He + e^{-} + \tilde{\nu} \quad \varepsilon_0 = 1.036 \tag{21}$$

$\varepsilon_0 = [M(z,N) - M(z+1,N-1)]/m$ - energy release.

We shall consider the transition $n \to p$ but assume that energy release is arbitrary.

In usual variant of the universal V-A weak interaction we can obtain for the neutron decay amplitude

$$M = \frac{G}{\sqrt{2}}\left\{\overline{\Psi}_p \gamma_\mu (1 + \alpha_0 \gamma_5)\Psi_n \overline{\Psi}_e \gamma^\mu (1 + \gamma^5)\Psi_{\tilde{\nu}}\right\}, \tag{22}$$

$\Psi_p, \Psi_n, \Psi_e, \Psi_{\tilde{\nu}}$ are exact solutions of Dirac equation, $\alpha_0 = G_A/G_V$ - the ratio of axial and vector coupling constant,

$$G = e^2/(8M_W^2 \sin\theta_W) = 1.4 \cdot 10^{-15}\, erg \cdot cm^3 \tag{23}$$

is Fermi constant, θ_W Weinberg angle.

On the ground of method of "exact solution" - Furry representation we obtain [15]:

1) In a strong field $H \geq H_0$ $[N]=0$: in the sum remains only single term $n=0$ - electron can be only in the ground state $Q=\frac{1}{2}f(0)$ if $N=\frac{H}{2H_0}(\varepsilon_0^2-1)<1$. Then

$$\frac{W}{W_0}=\frac{H}{H_0}\frac{f(0)}{1+\alpha_0^2}\left[1-\zeta_n\frac{2\alpha_0(\alpha_0-1)}{1+3\alpha_0^2}\right]. \tag{24}$$

2) In weak magnetic field $H < H_0$

$$\frac{W}{W_0}=\Phi_0-\zeta_n\frac{2\alpha_0(\alpha_0-1)}{1+3\alpha_0^2}\frac{H}{H_0}\left[\frac{\delta^3}{3}+\delta-\varepsilon_0\ln(\varepsilon_0+\delta)\right], \tag{25}$$

where $\delta=\sqrt{\varepsilon_0^2-1}$.
The correction is linear in the field. It opens two possibility of experimental verification.

The polarization of nucleus in β-decay in homogeneous magnetic field

If the introduced $n^{\uparrow}$ and $n^{\downarrow}$ are numbers of neutron and $p^{\downarrow}$, $p^{\uparrow}$ protons with the polarization along magnetic field $(\uparrow)$ and opposite $(\downarrow)$ we can obtain kinetic equations:

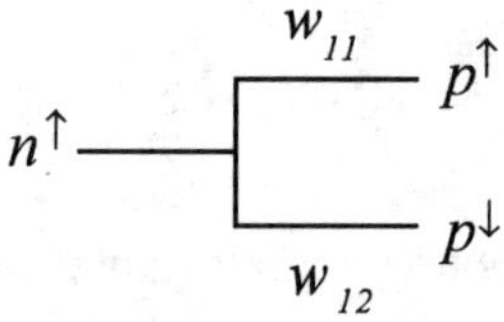

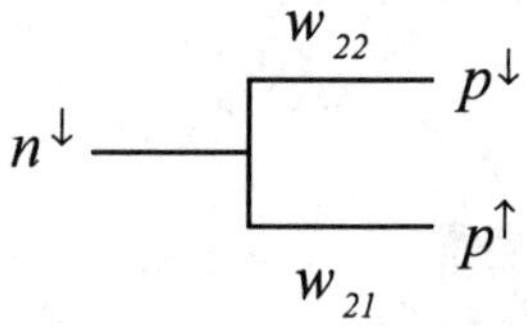

$n^{\uparrow}$ → $p^{\uparrow}$ (w_{11}), $p^{\downarrow}$ (w_{12}) $n^{\downarrow}$ → $p^{\downarrow}$ (w_{22}), $p^{\uparrow}$ (w_{21})

$$\frac{dp^{\uparrow}}{dt}=n^{\uparrow}w_{11}+n^{\downarrow}w_{21}, \qquad \frac{dp^{\downarrow}}{dt}=n^{\downarrow}w_{12}+n^{\uparrow}w_{22}.$$

For the time $t > \tau \sim 1/w_0$ (w_0 is probability of free β-decay) - the degree of polarization is

$$P = \frac{p^{\uparrow} - p^{\downarrow}}{p^{\uparrow} + p^{\downarrow}} = \frac{2\alpha_0^2}{1+3\alpha_0^2}\left[\frac{1+\Lambda}{1+\Lambda\frac{2\alpha_0(\alpha_0-1)}{1+3\alpha_0^2}} - \frac{1-\Lambda}{1-\Lambda\frac{2\alpha_0(\alpha_0-1)}{1+3\alpha_0^2}}\right] \tag{26}$$

where $\Lambda = f(0)/2Q$. This is exact formula. In the two limit cases we can obtain:

1°. The weak field $H << H_0, \quad \Lambda << 1$

$$\sum_{n=1}^{[N]} f(n) \cong [N] = [\frac{H_0}{2H}(\varepsilon_0^2 - 1),$$
$$P = \left[\frac{2\alpha_0(\alpha_0-1)}{1+3\alpha_0^2}\right]^2 \frac{H}{H_0(\varepsilon_0^2-1)} \tag{27}$$

It follows from expression (27) that the difference between the numbers of nuclei formed as a result of decay with spin directed along and opposite to a magnetic field of strength $H = 10^5\,Gs$ is $10^{-6} - 10^{-8}$, varying as a function of the energy release in the decay. Thus, in magnetic fields with an intensity that can be reached in modern laboratories the effect of the orientation of the nuclei as result of β-decay is very small, but in order of magnitude it is comparable with the values that characterize the phenomenon of nuclear paramagnetism, the observation of which can be effectively realized at the present time by means of the well-developed methods of nuclear magnetic resonance.

2°. The superstrong field $H \geq H_0 \quad \Lambda \to 1$

In this case $\sum_{n=1}^{[N]} f(n) \to 0$ because $[N] = 0$, and

$$P = \frac{4\alpha_0^2}{(1-\alpha_0)^2 + 4\alpha_0^2} \to 1 \tag{28}$$

For the neutron: if $\alpha_0 = 1.25$, $H = 2.7H_0$, $P = 99.5\%$. For realistic values $\alpha_0 \sim 1$ the degree of polarization is very close to 100%.

The obtained expressions are also valid for positron decay. In this case the nuclei in the final state are oriented in the opposite direction. We note also that with increase of the field in this region the decay time increases linearly.

For the comparison we have:

Synchrotron radiation (electromagnetic interactions)

$$P = \frac{n_e^{\downarrow} - n_e^{\uparrow}}{n_e^{\downarrow} + n_e^{\uparrow}} = \begin{cases} \dfrac{8\sqrt{3}}{15} & \chi << 1 \\ \dfrac{5\Gamma(1/3)}{2\Gamma(2/3)}(3\chi)^{-1/3} & \chi >> 1 \end{cases}$$

β-decay (weak interactions)

$$P = \frac{p^{\uparrow} - p^{\downarrow}}{p^{\uparrow} + p^{\downarrow}} = \begin{cases} \left[\dfrac{2\alpha_0(1+\alpha_0)}{1+3\alpha_0^2}\right]^2 \dfrac{H}{H_0(\varepsilon_0^2 - 1)} & \chi << 1 \\ \dfrac{4\alpha_0^2}{(1-\alpha_0)^2 + 4\alpha_0^2} \to 1 & \chi >> 1 \end{cases}$$

REFERENCES

1. Sokolov, A. A., Ternov I. M., *Radiation from Relativistic Electrons*, New York: AIP, 1986.
2. Ternov, I. M., *Zh. Eksp. Teor. Fiz.* **98**, 1169-1172 (1990) /in Russian/.
3. Ternov, I. M., Dorofeev O. F., *Fizika Elementarnikh chastits, Atomnogo Yadra*, **25**, 5-93 (1994)
/in Russian/.

4. Ternov, I. M., *Investigations in the quantum theory of radiating electrons,* Doctoral Dissertation:
Moscow, 1961, /in Russian/.
5. Sokolov, A. A., Ternov, I. M., *Dokl. Akad. Nauk SSSR*, **153,** 1052 (1963) /in Russian/;
Proceedings of the International Conference High Energy Accelerators, Moscow, Gosatomizdat,
1964, p.921 /in Russian/.
6. Baier, V. N., Katkov, V. M., *Sov. J. Nucl. Phys.* **3**, 57 (1966).
7. Stork, E., *Phys Lett.* **27A**, 651 (1968).
8. Schwinger, J., Tsai, W. *Phys. Rev.* **D9**, 1843 (1974).
9. Ternov, I. M., *Sov. J. Part. Nucl.*, **17**, 837-1079 (1986).
10. Ternov, I. M., Bagrov, V. G., Rzaev, R. A., *Sov. J. Phys. JETP*, **19**, 255 (1964).
11. Ternov, I. M., Bordovitsin, V. A., *Vestnik Moskovskogo Universiteta, Fiz. Astron.* **28** 21-24 (1987).
12. Korchuganov, V. N., Kulipanov, G. N. at al., Preprint 77-83 Inst. of Nuclear Phys. Siberian Branch USSR Ac. of Sc. Novosibirsk, 1977 /in Russian/.
13. Ternov, I. M., Khalilov, V. R., *Sov. J. Nucl. Phys.* **35**, 1441-1443 (1982).
14. Belomestnykh, S.A., Bondar, A. E. et al., Preprint 83-86 (loc. it.) 1977 /in Russian/.
15. Ternov, I. M., Rodionov, V. N., Dorofeev, O. F., *Sov. J. Part. Nucl.* **20** 22-39 (1989).

II. HADRON ACCELERATORS AND POLARIMETERS

Hadron Beams and Accelerators*

Thomas Roser
Brookhaven National Laboratory, Upton, N.Y. 11973-5000

1 Introduction

There were four sessions on Hadron Beams and Accelerators with 7 talks on Siberian Snakes and spin rotators, 3 talks on polarization build-up of unpolarized beams in storage rings and 5, 9, and 3 talks on low, medium, and high energy polarimeters, respectively. In this paper I will briefly describe a few highlights from these sessions, giving emphasis to topics which I think will play an important role in the future.

2 Lamb Shift Polarimeter

To measure the nuclear polarization of the polarized proton or H^- beams it used to be necessary to accelerate the beam to a high enough energy so that a nuclear reaction could be used to analyze the beam polarization. After an initial test by S.K. Lemieux et al.[1] a new and elegant design for a low energy polarimeter was developed. It is based on driving a three-level interaction between the two Zeeman-split metastable $2S_{1/2}$ states and the unstable $2P_{1/2}$ state in a strong magnetic field using a combination of a rf and dc electric field and a rf magnetic field. The energy level diagram for a deuterium beam is shown in Fig.1. The resonance magnetic field depends on the nuclear spin state which allows one to use such a device as a spin filter to filter out particular nuclear spin states from a polarized hydrogen or deuterium beam. For a complete polarimeter the spin filter needs to be complemented with a neutralizer for the incoming beam to generate the metastable state and a detector of the metastable state of the outgoing beam. The detector observes the light emitted when the atoms remaining in the metastable state are quenched in a high electric field. Polarimeters based on this scheme are now under construction at TUNL and IUCF.

*Work was performed under the auspices of the U.S. Department of Energy

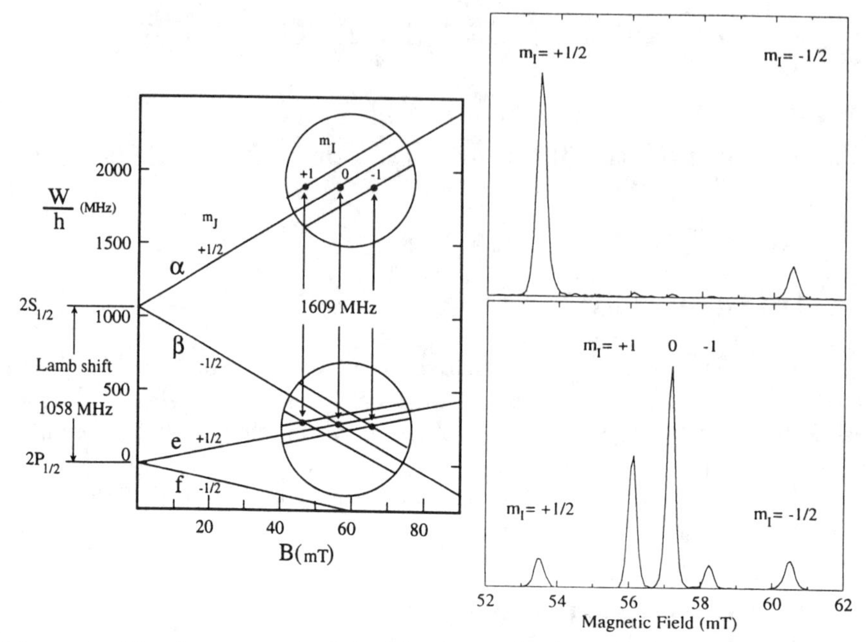

Figure 1: The figure on the left side shows the Breit-Rabi diagram for deuterium atoms in $2S_{1/2}$ and $2P_{1/2}$ states. The figure on the right side shows results from Lemieux at al.[1] for hydrogen (top) and deuterium (bottom). The measurements were obtained in about 10 s.

3 Polarization Build-up in Storage Rings

Several people have proposed schemes to build-up the polarization of a stored beam using the fact that a small spin effect can lead to a sizable polarization if it accumulates over many million revolutions. In fact, high levels of electron polarization are being achieved routinely using the very small spin flip probability during the emission of synchrotron radiation. The motivation for such schemes is two-fold: by polarizing the beam after acceleration the depolarization from passing through the many spin resonances can be avoided and, maybe more importantly, such a scheme could also be used to produce a polarized anti-proton beam. Now, for the first time, a positive result using a stored proton beam was reported by F. Rathman of Marburg. A polarized internal hydrogen gas target was inserted in the Test Storage Ring in Heidelberg and the polarization of the circulating proton beam started to build-up reaching a maximum value of about 2% after about 90 minutes. This result is shown in Fig. 2. The beam polarization was measured before and after the build-up process using proton-alpha scattering as analyzing reaction. H.O.

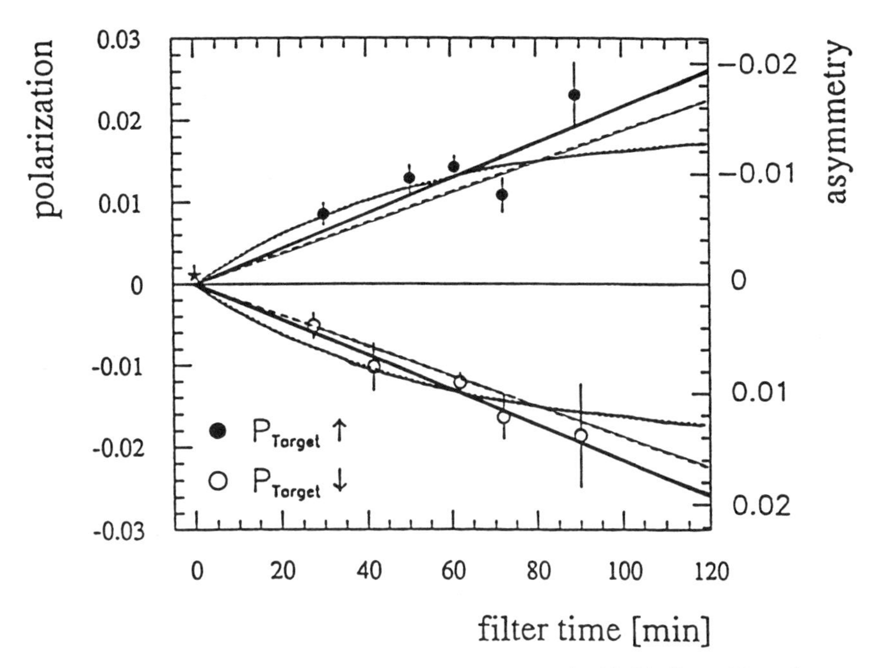

Figure 2: Polarization build-up measurements at the Heidelberg Test Storage Ring.

Meyer presented a detailed analysis of this result that included three different processes that contribute to the polarization build-up. The first process is spin dependent beam loss due to proton-proton scattering out of the storage ring acceptance. The remaining two effects are due to polarization transfer from either the proton or the electron to the beam proton. The polarization transfer from the electron has the opposite sign of the other two effects. The combination of all three effects is in very good agreement with the measured result.

There is no data available on the nuclear spin dependent effects between anti-proton and protons. However, the polarization transfer from polarized electrons to anti-protons can be calculated[2]. For the first time, we seem to have a realistic scheme to produce polarized anti-proton beams.

4 Siberian Snakes and Spin Flippers

The acceleration of polarized proton beams to energies higher than a few GeV's has been complicated by the existence of many depolarizing spin-resonances. The two main types of spin-resonances are the imperfection resonances driven

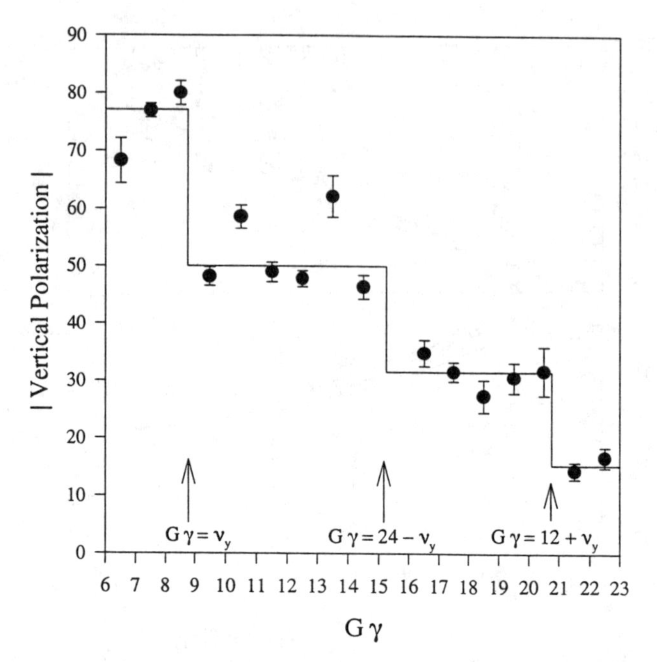

Figure 3: The absolute value of the vertical polarization as a function of $G\gamma$. Note that partial depolarization is due to intrinsic spin resonances at $G\gamma = \nu_y, 24 - \nu_y$ and $12 + \nu_y$.

by field imperfections and magnet misalignments and the intrinsic resonances driven by the focusing fields. It was the correction of the numerous imperfection resonances that have made the set-up for the acceleration of polarized beams very tedious. It has long been the hope that Siberian Snakes, which are local spin rotators inserted into a ring accelerator, would eliminate these resonances and allow acceleration of polarized beams with the same ease and efficiency that is now routine for unpolarized beams. First tests at IUCF with a full Siberian Snake showed that the spin dynamics with a Snake can be understood in detail. We now have the results of the first test of a partial Siberian Snake at the AGS accelerating the polarized proton beam to an energy of about 11 GeV. The partial Snake eliminated the effect of 18 imperfection resonances. The only depolarization observed was due to three intrinsic resonances, which in the future can be overcome with the traditional tune jump method. The results of these measurements are shown in Fig. 3 A separate study showed that a minimum of 6 degree spin rotation is required for the AGS, well below the capability of the installed 9 degree partial Snake.

The history of partial Siberian Snakes goes back to 1976 as Yuri Shatunov from Novosibirsk told us. During that year a solenoid in the VEPP-2M storage ring was used to act as partial Snake to avoid depolarization from the $G\gamma = 1$ imperfection resonance. A partial Snake has also been tested at IUCF by accelerating through the $G\gamma = 2$ resonance. In this case it was shown that the polarization becomes insensitive to additional longitudinal fields in the IUCF Cooler ring.

Of particular interest for experiments using stored polarized beams either with internal targets or in colliders are spin flippers that allow for the spin reversal of the polarization of the stored beam. Spin flippers work by introducing an artificial spin-resonance using an external rf field. Depolarization using an artificial spin-resonance has long been used to determine the spin precession frequency of stored electron beams from which one can then extract the electron beam energy. By carefully ramping the frequency through the spin-resonance condition complete spin flip of stored proton beams can be achieved. R.A. Phelps from Michigan reported a series of experiments at IUCF to measure the spin flip efficiency. Under optimal conditions a spin flip efficiency of at least 99.6% was achieved.

5 RHIC Spin and Tevatron Spin Projects

With the successful tests of Siberian Snakes the stage is set for the acceleration of polarized proton beams to much higher energies to be used in collider experiments to explore spin effects at the highest energies attainable. Two schemes have been presented here by E. Courant from Brookhaven. In the first project polarized protons from the Brookhaven AGS will be injected into the two RHIC rings to allow for up to $\sqrt{s} = 500\,GeV$ collisions with both beams polarized. Fig. 4 shows the lay-out of the Brookhaven accelerator complex highlighting the components required for polarized beam acceleration.

Of particular interest is the design of the Siberian Snakes (two for each ring) and the spin rotators (four for each collider experiment). Proposed by V. Ptitsin and Yu. Shatunov from BINP, it is based on helical dipole magnet modules each having a full 360 degree helical twist. Using helical magnets minimizes orbit excursions which is most important at injection energy.

The Tevatron project consists of accelerating polarized protons in the Tevatron replacing some of the Tevatron dipole magnets with higher filed magnets to gain space to install the six required Siberian Snakes. The project focuses on single spin collider experiments since at this time it is not feasible to polarize anti-protons, although the results from Heidelberg mentioned above could open up the possibility of polarized anti-protons in the future.

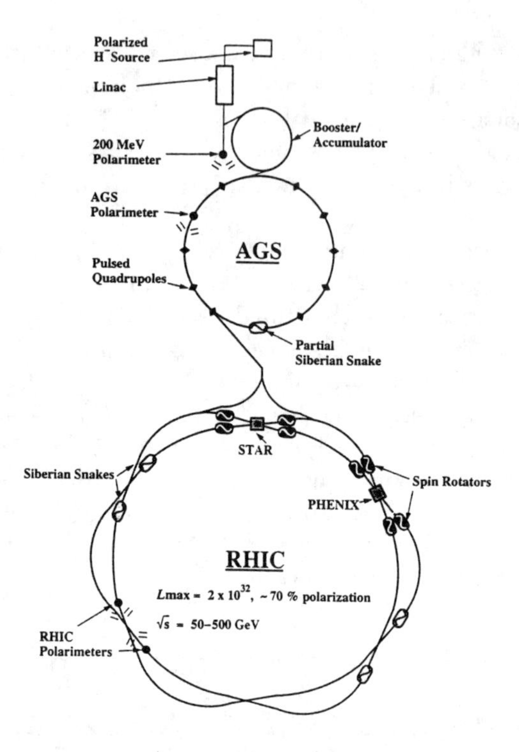

Figure 4: The Brookhaven accelerator complex with the components required for polarized beam acceleration highlighted.

6 Conclusions

In the four sessions on hadron beams and accelerators many new tools and techniques to work with polarized beams have been presented. For the immediate future the renewed availability of polarized beams in the multi GeV region at the AGS using the partial Siberian Snake hopefully stimulates an experimental spin physics program. In the more distant future we can look forward to the first polarized hadron colliders becoming reality.

References

[1] S.K. Lemieux et al. Nucl. Instr. Meth. **A333** (1993) 434

[2] C.J. Horowitz and H.O. Meyer, Phys. Rev. Lett. **72** (1994) 3981

LISS: Planning for Spin Physics with Multi-GeV Nucleon Beams at IUCF

S. E. Vigdor

Department of Physics, Indiana University and Indiana University Cyclotron Facility, Bloomington, Indiana 47405, USA

Abstract. The technology developed in recent years to facilitate experiments with stored, cooled polarized beams bombarding internal targets (including polarized gaseous targets) has natural and novel applications at multi-GeV energies. At IUCF we are preparing a proposal for a Light-Ion Spin Synchrotron (LISS) that would adapt this technology to the exploration of nucleon spin physics in the non-perturbative QCD regime from 1–20 GeV. I will describe the research goals of such a facility, with emphasis on a few contemplated experiments, chosen to illustrate both the range of physics issues to be addressed and the considerable advantages offered by storage ring techniques.

1. INTRODUCTION

Indiana University is very pleased to be hosting this first attempt at a merger of the two long-standing conference series on High-Energy Spin Physics and Polarization Phenomena in Nuclear Physics. And I am happy to have this opportunity to describe for you the future we envision for our laboratory here, because it falls squarely in the considerable overlap region between the communities, physics research programs and technological issues that have been addressed separately by these symposia in the past.

In contemplating the future of the Indiana University Cyclotron Facility (IUCF), we have decided we would like to contribute to forefront research on the nature of hadronic interactions in the "non-perturbative QCD" regime. Thus, we would like to devise experiments utilizing hadron beams to test models for handling the transition from the effective meson-exchange theories that work well at low energies to the perturbative QCD regime, where hadron-hadron interactions can be described via individual hard collisions of the constituent partons. A central requirement of models for this transition regime will be that they provide a unified framework for treating the non-trivial interplay between hadron substructure and hadronic interactions. An aspect of this interplay which interests us in particular, and which has been only lightly explored to date, is the effect of the non-perturbative *spin* structure of hadrons on the *spin*-dependence of their interactions.

Along with considerations of our physics interests, we have been seeking new ways to exploit innovative technology that we at Indiana have played a significant role in developing. Several years of experience in doing experiments in the IUCF Cooler ring have helped to crystallize our views of the types of questions and the energy regions to which storage ring and internal target techniques may be most naturally and cost-effectively applied. We are particularly impressed with the potential of new technology for handling polarized beams in storage rings, which has been tested at our laboratory: e.g., Siberian Snakes[1,2] and rf-induced spin flip.[3]

These physics and these technical issues have all led us to contemplate the transition depicted in Fig. 1 from IUCF to LISS: a Light-Ion Spin Synchrotron. The figure shows the presently imagined layout of LISS, drawn to scale with an inset of the IUCF Cooler ring. Although the two rings have quite different geometries, there would be many similarities in the philosophy of performing experiments in the two. It is these similarities that we emphasize in the figure by the subtle transition in lettering that takes us from the old to the new acronym.

Our major focus in designing LISS is on accelerating high-quality polarized beams of protons and deuterons, with flexible spin orientation, to any energy within the range 0.4 – ~20 GeV. Breakup of the polarized deuterons would be exploited to yield secondary $\vec{n}$ beams of interesting intensity and polarization. Unpolarized Z≥2 ions would also be available over corresponding energy ranges. The beams would be injected from the present IUCF Cooler. The initial emphasis, for reasons to be developed further below, would be on experiments utilizing stored beams bombarding internal targets. With this

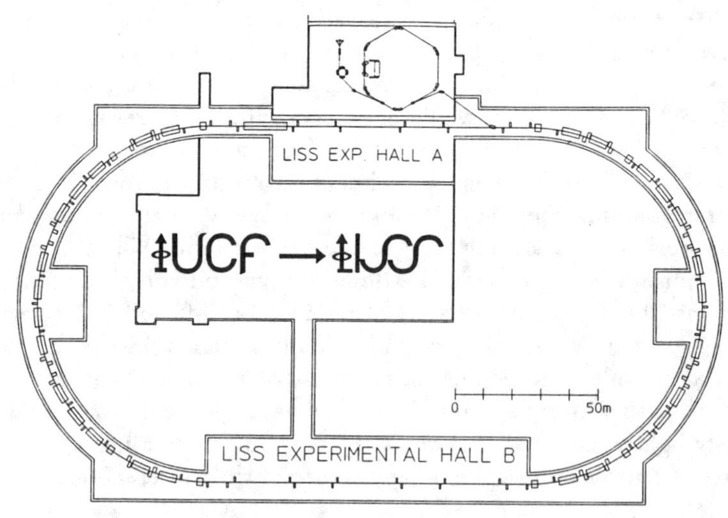

Figure 1. The presently envisioned "racetrack" configuration of the LISS ring, with the IUCF Cooler ring inset to set the scale.

approach we should be able to provide luminosities in excess of 10^{33} $cm^{-2}s^{-1}$ for polarized proton beams on an unpolarized hydrogen jet target; the accessible luminosities with polarized internal targets (e.g., of the "spin exchange" type[4]) would be an order of magnitude lower. Electron and/or stochastic cooling would be incorporated (with the balance between the two depending on technological advances made in the intervening years) to improve the stored beam emittance and resolution.

LISS would *not*, unfortunately, provide a replacement for KAON. The beam energies and luminosities are not appropriate to produce very high fluxes of secondary mesons. The good news is that elimination of most secondary beams keeps the construction cost modest: our current estimate, including an initial complement of detection systems, is ~U.S.$50 million. Our goal, then, vis-a-vis a kaon factory, is to reach an interesting fraction of the physics at a small fraction of the cost.

2. WHY A STORAGE RING?

Hadron spin physics investigations in the multi-GeV energy range have been seriously impeded in the past by technical difficulties in accelerating polarized beams through the depolarizing resonances that occur unavoidably in circular accelerators. It is now widely appreciated that the depolarization effects can be overcome by inserting Siberian Snakes[1,2] – magnetic systems that precess the beam spin by 180° without causing any net beam deflection – into ring accelerators. What is less widely appreciated is that if the beams are stored in the ring after acceleration, the same Snakes can be used to provide great flexibility in orienting the beam polarization axis. In addition, a Snake causes the "spin tune" of a ring to be half-integral, implying that effects produced by *undesired* beam polarization components can be very effectively cancelled on successive turns around the ring. These features make the environment provided by stored, polarized beams and internal targets a very attractive one for performing spin experiments, including high-precision tests of symmetry principles.[5] Such experiments also demand a mechanism for rapid reversal of the beam polarization, on a time scale which is short compared to that for refilling a storage ring. Great progress has been made on such a mechanism in recent experiments at IUCF testing the NMR method known as "adiabatic fast passage" (ramping the frequency of an rf perturbing field through the value corresponding to resonance with the beam spin precession at the location of the rf field).[3]

In addition, cooled, stored beams become, in many ways, more naturally applicable to hadron physics experiments at multi-GeV than at lower energies. An important transition occurs in the loss mechanisms for light ions stored in a ring housing an internal target. As indicated in Fig. 2, for proton energies below a few hundred MeV, these losses are dominated by single-step Rutherford scattering outside the ring acceptance. Only ~1% of the stored protons in this energy regime survive the Coulomb losses to initiate a nuclear reaction in the target. As the energy is raised above 1 GeV, the nuclear loss cross sections become larger than the Rutherford scattering losses; at 5–20 GeV, nearly *all* of the stored protons would eventually induce a nuclear reaction in an internal hydrogen target, if we waited long enough! This impressive efficiency of beam usage yields a new method for measuring the spin-dependence of total nuclear cross sections (e.g., $\Delta\sigma_L$ or $\Delta\sigma_T$, measured with polarized beam and polarized target): one would search for a spin-dependence of the stored beam lifetime as a signature. Furthermore, the much lower loss cross sections at multi-GeV energies imply that thicker internal targets, hence higher luminosities, can be used. Of course, higher luminosities are also needed, as cross sections generally decrease with increasing momentum transfer.

The advantages offered by beam cooling would also facilitate many interesting experiments in the multi-GeV energy regime. There is ample evidence already from results obtained at the IUCF Cooler[6] that the small beam energy spread allows unprecedented precision in investigations near particle production thresholds. LISS would open the way to a host of new thresholds associated with production of strangeness, charm, and various heavy mesons. The small emittance of cooled beams allows experiments probing the domain of very small momentum transfer, where high-energy hadronic interactions appear to be dominated by multiple gluon exchanges. The availability of high luminosities with pure internal polarized targets will take on added significance at multi-GeV energies. For example, they will allow spin correlation measurements for cleanly isolated, exclusive reaction channels, in an energy regime where detector resolutions would nor-

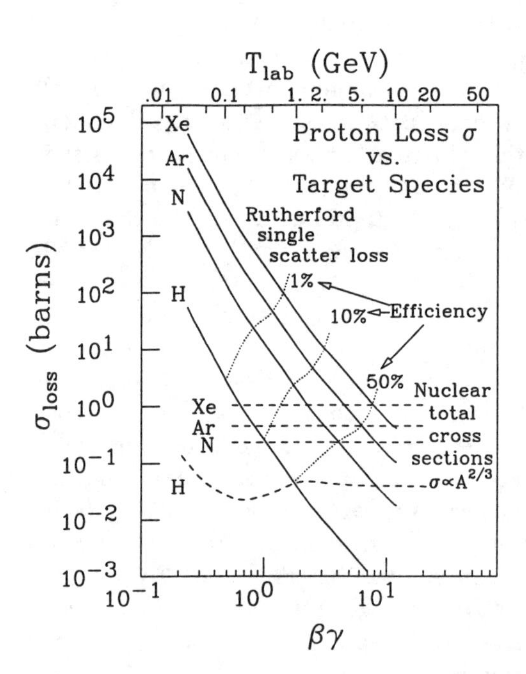

Figure 2. Rutherford scattering *vs.* nuclear reaction loss cross sections for stored proton beams interacting with internal targets of various Z-values, plotted as a function of beam energy.

mally be insufficient to distinguish quasifree reactions on contaminant nuclei from the corresponding free $\vec{p}$-$\vec{p}$ reactions. It is furthermore an energy regime where important differences between free and quasifree reaction mechanisms (e.g., color transparency) have been predicted.

A number of the advantages listed above for a storage ring design will be illustrated further via specific examples of experiments in Sec. 4. A potential drawback to a ring is the limited space and flexibility for mounting experimental equipment. We are trying to address this issue in our lattice design for LISS, with our current ideas illustrated in Fig. 1. The "racetrack" configuration shown there is characterized by two ~100 m long straight sections, within each of which the beam would have zero dispersion (facilitating use of long polarized target storage cells) and an integral vertical betatron tune (reducing significantly the number of intrinsic depolarizing resonances across which one must accelerate). Within each of these sections are contained 3 subsections with low β-functions and ~20 m insertion length, very suitable for installation of sophisticated detection equipment or Siberian Snakes, etc. In addition, each side contains 2 ~14 m sections with higher β_x. The two acceleration arcs contain, in addition, a total of 16 5–7 m long straight sections with a variety of dispersion and β values, in which one could insert beam manipulation and monitoring (e.g., intensity monitoring) equipment or special-purpose experiments. The lattice will be designed to give an imaginary transition energy, to avoid problems with acceleration. The version shown in Fig. 1 has a circumference of 608 m, seven times that of our Cooler ring, and fits around (or through!) the present IUCF building and within existing land boundaries. One of the clear advantages of the racetrack design is that it provides an easy option for future expansion of the straight sections, if demanded by high-priority experiments.

3. THE EMERGING LISS PHYSICS LANDSCAPE

The research program at LISS would emphasize hadron spin physics investigations in the 1–20 GeV energy range. These would provide sensitivity to the spin-dependence of the short-range (~0.1–1.0 fm) nucleon-nucleon interaction. At such short ranges we should expect the interactions between nucleons to be significantly influenced by the non-perturbative spin substructure of the nucleon itself. In this context, one should keep in mind a theorem put forth some years ago by Brodsky and Lepage.[7] They argued that the chiral symmetry of QCD implies that single-spin asymmetries (i.e., analyzing powers and reaction product polarizations) should vanish for hard collisions in the asymptotic limit of $\sqrt{s} \to \infty$. This theorem has often been used in the past to discourage high-energy spin physics experiments, without noting explicitly a necessary condition for the theorem's validity, namely, that the helicity of a hadron be strictly equal to the sum of its constituent parton helicities. In the wake of the "proton spin crisis," this is clearly no longer viewed as a benign assumption! Indeed, from the contemporary theoretical viewpoint,[8] a significant spin-dependence may arise naturally, even for relatively hard hadron collisions, from the interplay of the exchanges between hadrons with the internal structure of each (e.g., from the mixing of orbital and spin angular momentum within the hadrons). In addition, in the proposed LISS energy regime, one is far from the asymptotic limit, so that high-twist and "soft" collision processes can contribute significant spin-dependence to hadron interactions.

The interplay between interactions and hadron substructure implies that it will be challenging to separate mechanism from structure ambiguities in the interpretation of experiments. As I will show by examples, the constraints imposed on polarization observables by symmetry principles can be exploited very effectively to aid this separation in selected experiments.

Within my time constraints here, I can only provide you with a brief enumeration below of some of the experimental programs we are currently developing for LISS, expanding in detail on a few of them in the following section. A broader overview of the LISS program is provided in Ref. 9, and more details on some experiments in Refs. 5,10,11.

Symmetry tests via polarization measurements would play a prominent role in the LISS program. The technical advantages of spin handling in a storage ring have very interesting applications[5] to searches for parity and time-reversal violation in $\vec{p}$–p and $\vec{p}$–$\vec{d}$ scattering, respectively. The availability of polarized neutron beams would allow interesting tests of charge symmetry in a region where one might gain sensitivity, e.g., to SU(2)-violating differences between d-quark distributions in the proton and u-quark distributions in the neutron. The LISS energy regime is one where breaking of such fundamental symmetries has been only lightly explored in the past, yet where there are already indications of some anomalies (see next section).

Considerable attention would be devoted to investigations of baryon-baryon interactions. The experimental facilities would allow measurements of $\vec{p}$–$\vec{p}$ and $\vec{n}$–$\vec{p}$ elastic scattering over a wide range of momentum transfers, from $|t| \lesssim 0.01\ (\mathrm{GeV/c})^2$ – where one expects multiple-gluon (or, in alternative language, pomeron) exchange to dominate the strong interaction – to relatively hard collisions at $p_\perp^2 \sim 5\ (\mathrm{GeV/c})^2$. Outside the nucleon–nucleon sector, the study of final-state interactions with polarized beams can provide unique information on the spin-dependence of $\Delta\Delta$ (e.g., excited via the ^{2}H($\vec{d}$,$\vec{d}'$)$\Delta\Delta$ reaction) and ΛN interactions (see Sec. 4.2 for more details).

The study of nucleon knockout in A($\vec{p}$,pN) reactions can probe modifications to the NN interaction that occur inside nuclei. Such effects might arise at high momentum transfer, for example, from "color transparency".[12–13] One can also imagine distortion-induced polarization asymmetries in reactions on nuclei, of a sort reminiscent of the Maris effect[14] seen in low-energy nuclear physics, if there is an energy-dependent relaxation length for the point-like color singlets believed[12-13] to be selected in the hard scattering of quarks in a nucleus.

Polarization measurements may prove to be critical in enhancing signal-to- background ratios in investigations of nucleon resonance or more exotic hadron structure. The use of the spin- and isospin-selectivity available with polarized light-ion beams can provide a filter for N* resonances of specific quantum numbers. Efficient searches for $M \gtrsim 3$ GeV dibaryon resonances that couple to the pp system could be made at LISS by determining the spin-dependent total cross sections $\Delta\sigma_L$, $\Delta\sigma_T$ as a function of bombarding energy, via measurements of the stored beam lifetime for different spin orientations of polarized beams and polarized internal targets.

Even without secondary kaon or pion beams, LISS research can have a substantial impact on hypernuclear physics. The mechanism for large polarization effects in hyperon production would be studied in *exclusive* reaction channels for a variety of hyperons (Λ, Σ and selected Y* resonances), and for 1-, 2-, and even 3-spin observables. Symmetry constraints on spin observables can be used to produce certain light hypernuclei (e.g., ${}^{5}_{\Lambda}$He) with *a priori* known polarization via ($\vec{p}$,K^+) reactions. One could then use these systems to measure decay asymmetries for *bound* Λ's, including non-mesonic as well as pionic decay.

Many interesting hadron structure questions can be approached via experiments on heavy meson production with LISS beams. These include: searches for contributions from the possibly polarized strange quark sea to polarization transfer in the OZI-suppressed reaction $\vec{p}p \rightarrow pp\vec{\phi}$; measurements sensitive to isospin-mixing of the a_0 and f_0 mesons; searches for mesons with enhanced gluonic content in production processes at very low momentum transfer, where the NN interaction should be dominated by pomeron exchanges.

In addition to the above "core" program, we expect to develop research efforts as well in macroscopic aspects of light (unpolarized) nucleus–nucleus collisions and in accelerator physics.

4. SPECIFIC EXAMPLES OF LISS EXPERIMENTS

An alternative cut across the research program outlined above would reveal three broad classes of LISS experiments: (1) polarization tests of symmetry principles; (2) exploitation of symmetry constraints for "spin filtering" of states with specific quantum numbers; (3) probes of the mechanism for producing spin effects in non-perturbative QCD. I would like now to discuss in a bit more detail one example of each of these types, chosen to illustrate the range of physics questions one can address and the variety of technical advantages offered by the storage ring/internal target environment. The examples I choose can all be performed with relatively simple detection systems, all of which would be included within the initial complement of detectors subsumed in the $50M estimated price tag for the facility.

4.1 Parity Violation in p–p Scattering

One can test parity conservation in the $\vec{p}$–p system with a longitudinally polarized beam by measuring the dependence of the total cross section on the beam helicity (A_L^{tot}). Because the expected violations are tiny, such experiments are technically extremely challenging. The results are of considerable interest because they probe the interference of strong- and weak-interaction amplitudes in hadronic collisions. The results of the best experiments to date are summarized in Fig. 3. Exquisite low-energy experiments[15,16] have revealed violations $\sim 10^{-7}$, as expected from the ratio of weak to strong couplings. These results are interpreted in terms of effective weak meson-nucleon coupling constants,[17] and additional measurements are ongoing[18] to try to improve the constraints on all the relevant couplings in this scheme.

The most surprising result obtained to date is that at the highest energy in Fig. 3. At the now defunct ZGS accelerator, Lockyer *et al.*[19] measured $A_L^{tot} = (2.65 \pm 0.60 \pm 0.36) \times 10^{-6}$ for 6 GeV/c $\vec{p}$ bombarding an H_2O target. This value is an ***order of magnitude larger*** than those at lower energies. It is expected that the oxygen in the target ***dilutes*** the effect,[19] so that the

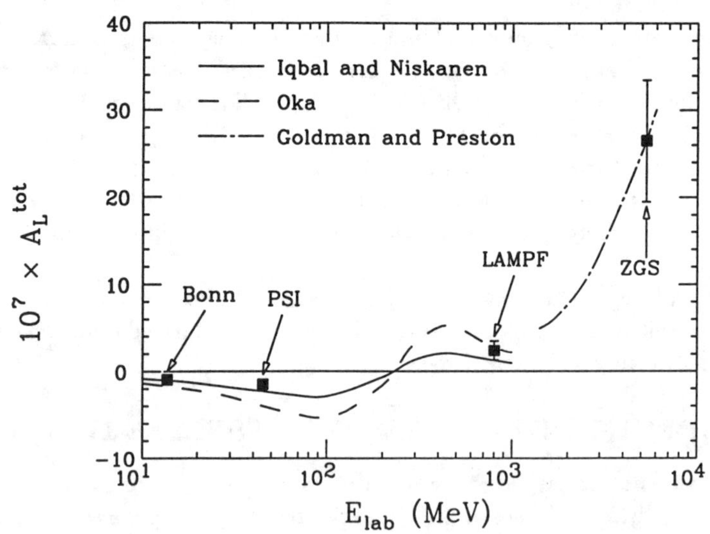

Figure 3. Compilation (from Ref. 21) of the best existing measurements of the parity-violating helicity-dependence of the total $\vec{p}$–p cross section, plotted *vs.* proton bombarding energy. The solid and dashed curves represent theoretical models[22,23] based on weak meson exchange, while the dot-dashed curve[20] is described in the text.

violation in $\vec{p}$–p scattering would be even larger! No instrumental flaw in the experiment, capable of producing such a large false signal, has been uncovered. Nor, however, has it been reproduced, since suitable facilities have been unavailable since the demise of the ZGS.

It has not been easy theoretically to explain what might lead to an order of magnitude increase in the $\vec{p}$–p parity violation between 800 MeV and 5 GeV. The only viable (though controversial) model I know of is that due to Goldman and Preston,[20] represented by the dot-dashed curve in Fig. 3. They have argued that the short range of the weak interaction introduces inherent sensitivity in parity violation to short-range quark–quark correlations within a nucleon. In particular, they find an important role for diagrams in which a weak (4-point Fermi) interaction between the members of a vector diquark in the polarized proton accompanies strong (gluon-exchange) interactions between that diquark and a quark from the target proton. Within their model, such diagrams lead naturally to a very strong energy-dependence of A_L^{tot}. They are unable to carry out a sufficiently complete calculation to predict the absolute magnitude of the parity violation, so their curve in Fig. 3 has been simply *normalized* to pass through the ZGS experimental result. If both that experiment and the Goldman–Preston model were correct, one would clearly expect quite large parity-violating signals at still higher energies ($\sim 10^{-5}$ within the LISS range, $\sim 10^{-4}$ at energies above 100 GeV), and by measuring them one would probe diquark correlations in the nucleon.

Clearly, at LISS, one would want at least to try to reproduce the ZGS result, and to make measurements as a function of bombarding energy if the large value of A_L^{tot} were confirmed. The storage ring/internal target environment allows for innovative methods of making such a measurement, quite different from the methods used for all previous parity experiments. The very substantial potential technical advantages of this new approach[5] can be discussed in the context of Fig. 4, which illustrates a sample layout for such an experiment. The polygonal ring layout in Fig. 4 corresponds to an earlier design contemplated for LISS, and has a number of pedagogical advantages for this discussion; however, the qualitative points I will make apply equally well to our now favored racetrack layout of Fig. 1.

The Siberian Snake indicated in Fig. 4 is meant to be a *full* Snake of the "first kind," which precesses the beam spin by 180° about a longitudinal axis. When it is activated, it causes the stable polarization orientation in each straight section of the ring to be horizontal, rather than vertical. The stable orientations are indicated in Fig. 4 by arrows beside each section for a bombarding energy of 6.13 GeV, corresponding to $G\gamma$=13.5 (where G is the anomalous magnetic moment of the proton and γ is the standard relativistic factor). In particular, the stable spin direction *for any bombarding energy* is *longitudinal* in the straight section directly

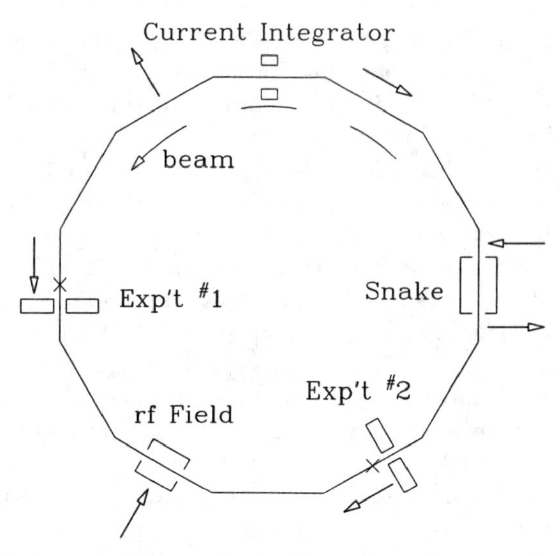

Figure 4. Schematic layout of possible p–p parity experiments in a storage ring with internal targets. The target for a transmission experiment would be mounted at the location labelled "Exp't #1", while both this location and that labelled "Exp't #2" could be used for scattering experiments. See text for details.

opposite the Snake, making this a prime location in which to perform a parity experiment. The normally dominant systematic errors in parity experiments are associated with small *transverse* components of the beam polarization (especially with their correlation with transverse position within the beam). The presence of the Snake cancels these errors very effectively, because it sets the "spin tune" of the ring to a half-integer, independent of bombarding energy, meaning that polarization components *perpendicular* to the stable orientation *reverse sign on every successive pass* through the Snake.

At its simplest level, a transmission measurement of A_L^{tot} in a storage ring can be viewed as a search for systematic helicity-dependence of the stored beam lifetime when a (windowless) hydrogen target is placed in the straight section opposite the Snake. The use of a pure hydrogen target removes complications in interpretation caused by interactions with contaminant nuclei, such as the oxygen in the ZGS water target.[19] Furthermore, since the ring can be maintained at ultra-high vacuum, one can minimize systematic errors associated with beam interactions with matter[19] in other ring sections, where the spin is not longitudinally oriented. The stored beam spin orientation can be flipped rapidly via the "adiabatic fast passage" technique mentioned in Sec. 2 (utilizing the "rf field" in Fig. 4), and independently (although on a longer time scale) by reversal at the ion source in between successive beam injection cycles. The ring itself serves as a superior spectrometer for such a transmission experiment, ensuring that only protons within a very narrow momentum range of the beam are included among the transmitted yield. This removes concerns[19] about apparent parity violations arising from contamination of the transmitted sample by protons from the weak decay of polarized hyperons produced in the $\vec{p}$–p collisions.

The parity-violating signal in such an experiment would be a change, correlated with beam helicity at the target location, in the logarithmic time derivative of the stored beam current, as monitored by some non-intrusive device. The resolution of this current monitor is one of the factors that determines the attainable precision. Most existing storage rings monitor (DC) stored current with current transformers,[24] for which Barkhausen noise appears to impose a hard resolution limit of a few hundred nA for an integration time of 1 second. At IUCF we are developing a new type of "rf current integrator" (RFI) based on capacitive beam pickup, which can improve the above resolution by 1–2 orders of magnitude for a **bunched**, stored beam.[25] To take full advantage of this excellent precision, one needs sufficiently high luminosity and long spin cycle periods that statistical fluctuations in the number of particles removed from the beam during each cycle do not impose a much more stringent limit on precision. This, in turn, demands that the RFI performance be free from significant drifts over time periods ~1 minute. With the expected RFI performance and an average luminosity of 10^{33} cm^{-2}s^{-1}, one could measure A_L^{tot} to a statistical precision of 1–2 $\times 10^{-6}$, sufficient to confirm the ZGS result,[19] in a single day of counting.

Finally, it is worth noting that the ring also offers some unique advantages for *scattering* (as opposed to transmission) tests of parity. There are "magic" beam energies at which the value of $G\gamma$ is appropriate, with a full Snake activated, to give stable longitudinal polarization **of opposite sign** in two different straight sections, such as those labeled "Exp't #1" and "Exp't #2" in Fig. 4. One can then imagine performing two *simultaneous* measurements, with nominally identical targets and detectors but with *opposite* beam helicities! Many potential systematic errors would be cancelled by looking for a spin-flip-correlated change in the *ratio* of scattered yields in the two setups. In principle, one should be able to obtain comparable statistical precision to that in a transmission experiment, but with quite different sensitivity to systematic errors.

Clearly, a program of several years would be required to explore systematic errors in these novel parity experiments, but the potential for unique and decisive measurements is exciting!

4.2 Determination of ΛN Spin-Dependent Scattering Lengths

Despite possible small and interesting violations, parity conservation and other symmetries of the strong interaction impose very useful constraints on the values of certain spin observables for exclusive reactions of simple spin structure. These constraints can often be exploited to isolate, or at least to enhance relative to background, the production of final-state particles of specific quantum numbers of interest. As an example, consider an experiment to determine the spin-dependence of the ΛN interaction at zero momentum.

This spin-dependence is of interest as part of the broader study of how baryon-baryon interactions are affected by substituting a heavy (s or c) for a light (u or d) valence quark in the baryons. It is important in probing the transition from quark–gluon to mesonic degrees of freedom to know whether meson-exchange models provide as complete a description of hyperon-nucleon interactions at low energy as they do of the NN force, despite the much greater breaking of chiral symmetry in the strange sector. ΛN scattering is an especially interesting testing ground because single pion exchange contributions are isospin-forbidden. Hypernuclear spectroscopy provides information on the *effective* ΛN interaction inside nuclear matter, but this needs to be complemented by studies of the *free* interaction.

To this date, ***absolutely no polarization measurements exist*** for hyperon-nucleon scattering, because their short decay lengths make it very hard to produce useable beams of low-energy polarized hyperons. The various meson-exchange potential models that have been developed for the YN systems give dramatically different predictions for free-scattering polarization observables,[26,27] despite common constraints from the (sparse) existing cross section data and from quark-level symmetry relations among meson-NN and meson-YN coupling constants. An illustration of these differences is given in Table 1 for the simplest spin-dependent feature of the interaction: the difference between $^1S_0$ and $^3S_1$ ΛN scattering lengths.

At LISS we could determine the spin-dependence of the YN scattering lengths via the study of final-state interactions (FSI) in nucleon-nucleon hyperon production, e.g., $\vec{p}+\vec{p}\rightarrow K^+ + (\Lambda p)$. Imagine an inclusive measurement in which we detect only the K^+ at a single reaction angle, and we concentrate on the region in the doubly differential cross section spectrum, $d^2\sigma/d\Omega_K dM_{\Lambda p}$, very near the minimum possible mass ($M_{\Lambda p}$) of the Λp system. Here, the unobserved Λp system must be in an S-state, either $^3S_1$ or $^1S_0$. For the latter state, the production reaction takes on a very simple spin

TABLE 1. Spin-dependent scattering lengths for various ΛN potential models

ΛN Potential	Reference	$a_{singlet}$ (fm)	$a_{triplet}$ (fm)
Bonn A	[26]	-1.60	-1.60
Bonn B	[26]	-0.57	-1.94
Nijmegen Λp[a]	[27]	-2.73	-1.48
Nijmegen Λn[a]	[27]	-2.86	-1.24

[a] The Λp and Λn predictions differ because charge symmetry violation via $\Lambda^0 - \Sigma^0$ mixing is included in the Nijmegen model.

structure: $\frac{1}{2}^+ + \frac{1}{2}^+ \rightarrow 0^- + 0^+$. For any such reaction, in which a unique pseudovector direction (here, $\vec{n} = \vec{p}_{inc} \times \vec{p}_K$) is defined by the beam-detector plane, parity and angular momentum conservation constrain the value of the spin correlation parameter A_{nn} to be $= +1$ at all angles![10]

The constraint can be exploited by measuring the "spin correlation" spectrum $[d^2\sigma/d\Omega_K dM_{\Lambda p}] \cdot (1 - A_{nn})$ with a polarized beam and polarized target. *Only* the $^3S_1$ Λp FSI can contribute to this spectrum near threshold! From the shape of this spectrum one should then be ablē to deduce $a_{^3S}$. That the spectrum shape is indeed sensitive to the FSI can be seen clearly from the existing *unpolarized* data and analysis shown in Fig. 5. In fact, that figure demonstrates that the opening of the ΣN channel is marked by an even more pronounced FSI effect, so that the experiment we suggest could equally well determine $a_{^3S}$ for ΣN. It is more difficult to determine $a_{^1S}$ directly, since the complementary spectrum $(d^2\sigma/d\Omega_K dM_{\Lambda p}) \cdot A_{nn}$ will contain, in general, both $^1S_0$ and $^3S_1$ contributions, in a ratio that is not known *a priori*. But if $a_{^3S}$ is determined, then $a_{^1S}$ can be deduced from fits to the existing unpolarized Λp cross sections at low energy.[26]

A great advantage of performing the suggested experiment inside a storage ring is from the use of a *pure* windowless, gaseous internal polarized proton target. The target ensures that the observed K^+ spectrum will be free from $A(\vec{p},K^+)X$ reaction background induced on nuclei whose K^+ production thresholds lie much lower than that for pp. At bombarding energies of a few GeV, the forward-going K^+ will have much smaller magnetic rigidity than the beam protons, and can be readily separated from them by a chicane in one of the ring straight sections. A magnetic spectrometer of modest resolution ($\Delta p/p \sim 10^{-3}$) and broad range following the chicane would be suitable to measure the proposed spectra.

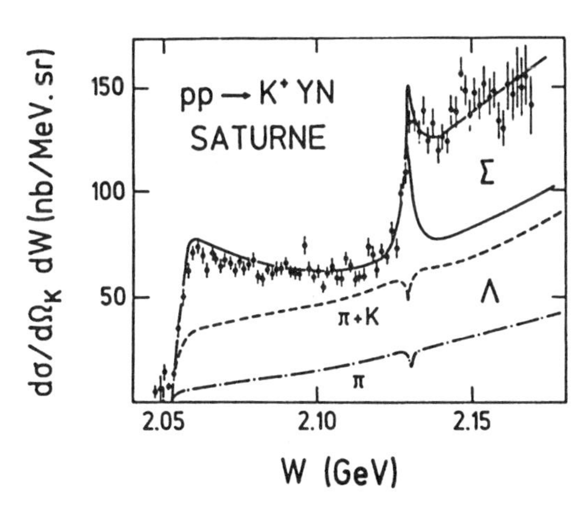

Figure 5. Doubly differential cross sections measured[28] for $pp \rightarrow K^+X$ at a bombarding energy of 2.3 GeV and a lab angle of 10°. The calculated curves[29] include π- and K-exchange and the important effects of YN final-state interactions near the Λ and Σ thresholds.

The availability of polarized *neutron* beams of several GeV at LISS would offer the opportunity to use the above technique to measure also the Λn $^3S_1$

scattering length, in a separate $\vec{n}+\vec{p}\to K^+ + (\Lambda n)$ experiment with an external cryogenic polarized target. Although in this case one would have to give up some technical advantages of stored beams and internal targets, a test of charge symmetry in the strange sector is quite interesting to pursue. As seen in Table 1, a large (~20%) *violation* of charge symmetry in the Λn *vs.* Λp scattering lengths is predicted[27] to arise from $\Lambda^0 - \Sigma^0$ mixing.

4.3 Measurement of the Strong-Interaction Spin-Flip Amplitude in Small-Angle p–p Scattering

A substantial fraction of the LISS program would be devoted to probing the mechanism for producing polarization effects in non-perturbative QCD interactions. While systematic studies spanning a variety of reaction channels, momentum transfers, and polarization observables will be valuable for this purpose, it is important to note that well-designed individual experiments can also provide definitive answers to well-posed questions The example I offer is a measurement of the absolute analyzing power for pp elastic scattering in the Coulomb-nuclear interference (CNI) region. Here, one can directly test the standard ***assumption*** that high-energy strong interactions at very low momentum transfer are dominated by multiple gluon exchanges (often characterized as pomeron exchange) that introduce no spin-flip.

Proton–proton scattering at momentum transfers 10^{-3} (GeV/c)^2 $\lesssim |t| \lesssim 10^{-2}$ (GeV/c)^2 is dominated by the interference of strong and electromagnetic amplitudes. Even in the absence of strong-interaction spin-flip, a non-zero analyzing power is expected[30] to arise from the combination of the imaginary part of the helicity-conserving strong amplitude [Im $f_{++}^{strong}(t \to 0)$] and the real part of the electromagnetic spin-flip amplitude [Re $f_{+-}^{\gamma}(t)$]. The former amplitude is directly related by the optical theorem to the total cross section, which is known to be constant ($\sigma_{pp}^{tot} \simeq 40$ mb) over the entire energy range from $\sqrt{s}$ of a few to ~40 GeV. The latter amplitude corresponds to Schwinger (magnetic) scattering, and is precisely calculable. The net result is a firm prediction[30] of an ***energy-independent*** analyzing power over the t and $\sqrt{s}$ ranges mentioned, peaking at $A_y = +0.046$ at $t = -3 \times 10^{-3}$ (GeV/c)^2. It has been proposed[31] to use this A_y to provide an absolute polarization calibration standard for high-energy proton beams.

The above prediction is based on an assumption – that $f_{+-}^{strong} \to 0$ at small t – that is neither very well tested experimentally, nor on firm grounds theoretically. It can be made plausible from a perturbative QCD viewpoint, by treating the interaction at low t as the result of multiple gluon (g) exchanges, in which each qqg vertex conserves the quark's helicity, by virtue of the chiral symmetry of QCD. However, as I have noted earlier (see Sec. **3**), this scenario guarantees ***hadron*** helicity conservation **only if** the hadron's helicity can be strictly expressed as the sum of its constituent parton helicities. In specific models of the proton's non-perturbative spin substructure, this equality can be violated substantially. Sufficiently precise absolute measurements of A_y for $\vec{p}$–p scattering in the CNI region, made as a function of $\sqrt{s}$, can test the viability of such models.

The only existing relevant measurements of A_y have been made recently at Fermilab, using a *tertiary* polarized proton beam obtained from $\vec{\Lambda}$ decay.[32] The results are shown in Fig. 6, along with the above prediction based on $f_{+-}^{strong} = 0$. Also shown are calculations of A_y carried out[33] for a generalized empirical form of the strong amplitude, including spin-flip:

$$f_{strong}^{elastic}(q^2) = f_0[1 + i(q/2m)M_{sf}\vec{\sigma}\cdot\hat{n}]. \tag{1}$$

The standard calculation corresponds to M_{sf}=0; as seen in Fig. 6, non-zero values of M_{sf} lead essentially to a renormalization of A_y in the CNI region. The measured values are consistent with M_{sf}=0, but within large uncertainties that do not rule out values as large as $|M_{sf}| \simeq 0.4$. For comparison, Kopeliovich and Zakharov show[33] that theoretical values as large as $|M_{sf}| \simeq 0.3$ can be obtained easily within models based on two-gluon exchange between protons containing dynamically enhanced compact diquarks.

It is clear from Fig. 6 that the existing data are of insufficient quality to test non-perturbative structure models, or to calibrate A_y in CNI scattering as an absolute polarization standard to better than $\sim \pm 25\%$. One would like to see measurements of $A_y(t)$ made with an absolute precision $\sim \pm 0.003$, at several c.m. energies. While the cross sections in the CNI regime are large, the ***kinematics*** of the scattering pose severe challenges for conventional experiments; in contrast, the situation is well suited to experiments with a cooled polarized beam on an internal target. At a bombarding energy of 10 GeV, the recoiling protons have laboratory energies of only $\sim 1-10$ MeV. However, as indicated in Fig. 7, these recoils can easily penetrate a gaseous internal target and be detected with good angle and energy resolution (to provide the t measurement)

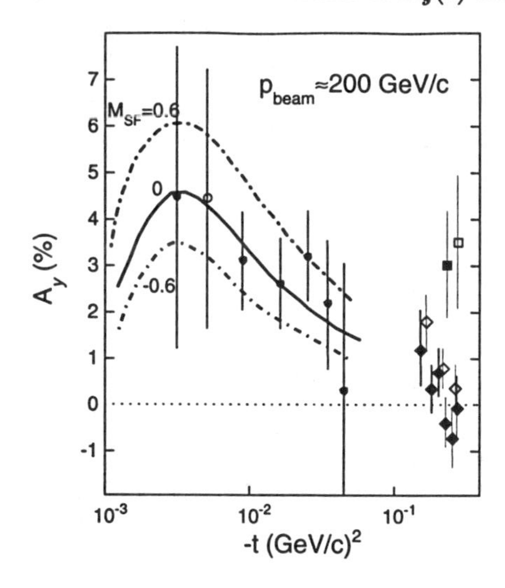

Figure 6. Measurements[32] and calculations[33] of the p–p scattering analyzing power in the Coulomb–nuclear interference region. The broken curves include a phenomenological strong-interaction spin-flip amplitude via Eq. (1).

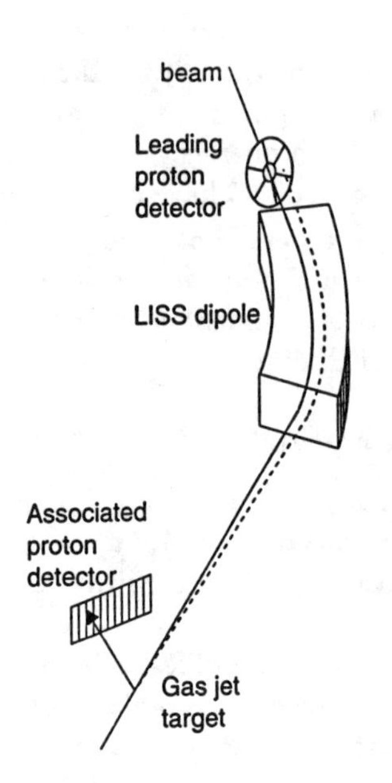

Figure 7. Schematic illustration of a LISS experiment in the Coulomb–nuclear interference region, indicating detection of the recoil proton in a silicon strip detector and of the forward proton after the next ring dipole.

in silicon microstrip detectors inside the ring vacuum. The forward protons emerge at lab angles $\lesssim 1°$, and with momenta only slightly different than that of the beam. Nonetheless, they fall outside a cooled beam's phase space, and could be detected, in coincidence with the recoils, in an annular detector at the end of a suitably long ring straight section, or after the next bending magnet encountered in the ring. The use of a storage ring + synchrotron also permits a "bootstrapping" technique for absolute calibration of the analyzing power, via p–p scattering measurements of A_y (not at small angles) with an *internal polarized target* at two different energies within each machine cycle: the injection energy of 370 MeV, where the absolute A_y is already known,[34] and the high energy of interest. The stability of the target polarization during acceleration could be monitored at the source for the target.

5. CONCLUSIONS

For many years in low- and intermediate-energy nuclear physics, polarization measurements were pursued at the vast majority of hadron beam accelerators, while they played a relatively minor role in studies of electromagnetic interactions. Now, among multi-GeV accelerator facilities, the situation is almost completely reversed! *Every* intermediate- and high-energy electron machine justifiably views polarization measurements as a very high priority aspect of their program, while few facilities exist for multi-GeV *hadron* spin physics investigations. We do not believe that the current situation is justified by the physics issues: spin is a critical degree of freedom in strong interactions, as it is in electroweak interactions. In particular, I have attempted to convince you by some examples that one must study the influence of the non-perturbative spin substructure of hadrons on the spin-dependence of their interactions, in order to make progress in treating the transition from quark–gluon to baryon–meson degrees of freedom.

Nor can the present paucity of polarized hadron facilities be attributed, any longer, to technological difficulties: there has been very rapid recent progress on the acceleration and handling of polarized high-energy nucleon beams. Nonetheless, the existing $\vec{p}$ facilities above 2.5 GeV (taking into account the imminent closure of Saturne) are simply inadequate for the research program I have outlined. The combination of stored, cooled polarized beams with internal (polarized or unpolarized) targets can provide a novel, very well-suited, and cost-effective environment for pursuing nucleon spin physics up to beam energies of 20 GeV.

To address the present imbalance in polarization facilities at electron *vs.* hadron accelerators, we are preparing to propose LISS to the U.S. National Science Foundation (NSF) in 1996, at a project cost (including an initial complement of detectors) of ~$50M. We have already achieved, or expect to achieve soon, the following milestones of this preparation: a LISS Brief, outlining the research program and technical advantages, was circulated in February, 1994 (copies are available from IUCF upon request); a proposal for a new small synchrotron to inject beams into the present IUCF Cooler, needed to achieve the eventual luminosities planned for LISS, was approved for funding by the NSF in September, 1994; a White Paper providing more details on proposed experiments, equipment and machine design will be circulated by the end of 1994. We welcome your input to our plans, and hope that by the end of this millenium some of you will be submitting experiment proposals to (or serving on!) the first LISS Program Advisory Committee.

REFERENCES

[1] Ya.S. Derbenev and A.M. Kondratenko, Part. Accel. **8**, 115 (1978).

[2] A.D. Krisch *et al.*, Phys. Rev. Lett. **63**, 1137 (1989); J.E. Goodwin *et al.*, Phys. Rev. Lett. **64**, 2779 (1990).

[3] R.A. Phelps *et al.*, in Proc. SPIN'94 Conference, Bloomington, Sept. 1994 (to be published by AIP); B. von Przewoski, private communication (1994).

[4] M. Poelker *et al.*, AIP Conf. Proc. **293**, *Polarized Ion Sources and Polarized Gas Targets*, ed. L.W. Anderson and W. Haeberli (AIP, New York, 1994), p. 125; T. Walker and L.W. Anderson, *ibid.*, p. 138.

[5] S.E. Vigdor, in *Proc. Workshop on Future Directions in Particle and Nuclear Physics at Multi-GeV Hadron Beam Facilities*, ed. D.F. Geesaman (Brookhaven National Laboratory report #BNL-52389, 1993), p. 171.

[6] H.O. Meyer *et al.*, Nucl. Phys. **A539**, 633 (1992); H. Rohdjess *et al.*, Phys. Rev. Lett. **70**, 2864 (1993).

[7] S.J. Brodsky and G.P. Lepage, Phys. Rev. D**22**, 2157 (1980) and Phys. Rev. D**24**, 2848 (1981).

[8] I.G. Aznauryan *et al.*, Phys. Lett. **126B**, 271 (1983); J.P. Ralston, in *Perspectives in the Structure of Hadronic Systems*, ed. M. Harakeh *et*

al.(Plenum, New York, in press); M. Anselmino and S. Forte, Phys. Rev. Lett. **71**, 223 (1993).
[9] S.E. Vigdor, *The LISS Brief*, IUCF Internal Report (1994).
[10] S.E. Vigdor, in *Flavour and Spin in Hadronic and Electromagnetic Interactions*, ed. F. Balestra *et al.*(Italian Physical Society, Bologna, 1993), p. 317.
[11] W.W. Jacobs, in *Proc. Workshop on Future Directions in Particle and Nuclear Physics at Multi-GeV Hadron Beam Facilities*, ed. D.F. Geesaman (Brookhaven National Laboratory report #BNL-52389, 1993), p. 241.
[12] J.P. Ralston and B. Pire, Phys. Rev. Lett. **61**, 1823 (1988) and U. Kansas preprint (1992); L. Frankfurt and M. Strikman, Prog. Part. Nucl. Phys. **27**, 135 (1991).
[13] B.G. Zakharov and B.Z. Kopeliovich, Yad. Fiz. **46**, 1535 (1987) [Sov. J. Nucl. Phys. **46**, 911 (1987)].
[14] G. Jacob *et al.*, Nucl. Phys. **A257**, 517 (1976).
[15] S. Kistryn *et al.*, Phys. Rev. Lett. **58**, 1616 (1987).
[16] P.D. Eversheim *et al.*, Phys. Lett. **256B**, 11 (1991).
[17] B. Desplanques, J.F. Donoghue, and B.R. Holstein, Ann. Phys. (N.Y.) **124**, 449 (1980); E.G. Adelberger and W.C. Haxton, Ann. Rev. Nucl. Part. Sci. **35**, 501 (1985).
[18] S.A. Page, in *Proc. 7th Intl. Conf. on Polarization Phenomena in Nuclear Physics*, Paris, 1990, ed. A. Boudard and Y. Terrien, Colloque de Physique **51**, C6-253, 1990.
[19] N. Lockyer *et al.*, Phys. Rev. D**30**, 860 (1984).
[20] T. Goldman and D. Preston, Phys. Lett. **168B**, 415 (1986).
[21] W.T.H. van Oers, in *Proc. Workshop on Future Directions in Particle and Nuclear Physics at Multi-GeV Hadron Beam Facilities*, ed. D.F. Geesaman (Brookhaven National Laboratory report #BNL-52389, 1993), p. 161.
[22] T. Oka, Prog. Theor. Phys. **66**, 977 (1981).
[23] M.J. Iqbal and J.A. Niskanen, Phys. Rev. C**42**, 1872 (1990).
[24] K. Unser, IEEE Trans. Nucl. **NS-28**, 2344 (1981), and in *Proc. 1989 Particle Accelerator Conference*, Chicago, March 1989.
[25] M.S. Ball *et al.*, *Design and Preliminary Tests of a Bunched Beam Intensity Monitor for the IUCF Cooler*, IUCF Internal Report, 1994.
[26] B. Holzenkamp, K. Holinde, and J. Speth, Nucl. Phys. **A500**, 485 (1989).
[27] P.M.M. Maessen, Th.A. Rijken and J.J. de Swart, Phys. Rev. C**40**, 2226 (1989).
[28] R. Frascaria *et al.*, Nuov. Cim. **102A**, 561 (1989).
[29] J.M. Laget, Phys. Lett. **259B**, 24 (1991).
[30] B.Z. Kopeliovich and L.I. Lapidus, Yad. Fiz. **19**, 218 (1974).
[31] *Report on Acceleration of Polarized Protons to 120 and 150 GeV in the Fermilab Main Injector*, SPIN Collaboration (FNAL Internal Report, March, 1992, unpublished); *Proposal on Spin Physics Using the RHIC Polarized Collider*, RHIC Spin Collaboration (August, 1992, unpublished).
[32] N. Akchurin *et al.*, Phys. Rev. D**48**, 3026 (1993).
[33] B.Z. Kopeliovich and B.G. Zakharov, Phys. Lett. **226B**, 156 (1989).
[34] L.G. Greeniaus *et al.*, Nucl. Phys. **A322**, 308 (1979).

SPIN AT RHIC

Yousef I. Makdisi*
Brookhaven National Laboratory, Upton, New York 11973

1 Introduction

The relativistic Heavy Ion Collider (RHIC) at BNL is in its fourth year of construction. The target date for completion is March 1999. In this report, I will describe the accelerator complex and its status with special emphasis on its capability as a polarized proton collider, the proposed physics program, the detectors, and the expected sensitivities to physics signatures.

The primary mission of RHIC is the study of nuclear matter under conditions of extreme temperatures and densities. The polarized proton capability, with a luminosity of $2x10^{32}$ cm^{-2} sec^{-1} and 70% beam polarization, provides the exciting potential to study the spin structure of nucleons with sensitivities of 5% specially in areas that are accessible only through hadronic collisions such as gluon and sea quark polarization. Transverse spin effects, at high p_T where perturbative QCD calculations are applicable, will also be addressed. These are not accessible to fixed target experiments.

The corresponding physics program will focus on measurement of longitudinal and transverse spin interactions with the STAR and PHENIX detectors under construction, utilizing such probes as direct photons, high p_T jets, $W^{+/-}$ production, and Drell Yan. Another set of heavy ion experiments; PHOBOS, BRAHMS, and a proton-proton elastic and total cross section experiment PP2PP will see transversely polarized beams only.

2 The Collider Complex

The RHIC accelerator [1] is designed to collide a wide range of ion species from protons to gold. The available center of mass energies are variable from (60-200 GeV/nucleon) for gold-gold and (60-500 GeV) for proton-proton collisions respectively. The two independent ring configuration allows collisions between different ion species, Figure 1.

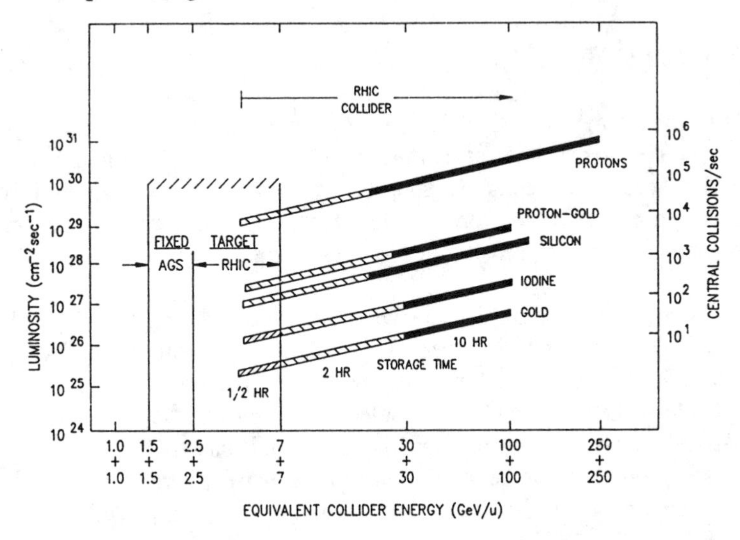

Figure1. RHIC design luminosities and frequency of collisions

The injection sequence to RHIC starts with either the Tandem, for heavy ions, or the 200 MeV Linac, for polarized protons, followed by the Accumulator/Booster then the AGS that feeds gold beams at 10.5 GeV and protons at 25 GeV respectively, Figure 2. Fill time is one minute and the expected beam lifetime is 10 hours, the turn around between stores is about 30 minutes. The construction project is proceeding smoothly. The major superconducting magnet technology transfer from BNL to industry is a success. Industry represents an important partner in this venture. The Northrop-Grumman Corporation is under contract to supply all the arc superconducting dipoles and the quadrupole cold masses which will be produced at peak rates of one and two per day respectively. Everson Electric is providing the sextupole cold masses. These and the quadrupoles as well as a corrector set will be packaged in one cryostat at BNL. As of December 1995, 41 dipoles were delivered and 26 of which were installed in the ring. An AGS extraction and transfer line test are scheduled for late 1995 and a beam test of the first sextant is scheduled for late 1996. This will exercise the full beam transport, injection, cryogenic, and control systems.

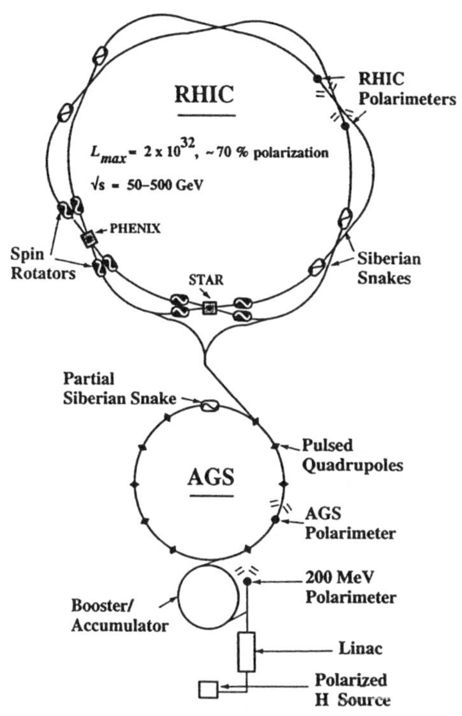

Figure 2. The Accelerator complex with the components for the polarized collider

The RHIC Spin Collaboration submitted a letter of intent [2] detailing the physics potential of a polarized proton collider at RHIC with luminosity near $2x10^{32}$ cm^{-2} sec^{-1} and 70% beam polarization. This was followed with a proposal [3] to equip the AGS with a partial snake (a solenoid spin rotator) to overcome imperfection resonances during the acceleration process. The initial test of this snake, last April, was quite a success. The AGS polarized beam was accelerated to 11 GeV overcaming 18 imperfection resonances [4] in the process. Figure 3. It was also determined that a 6 degree spin rotation was the required minimum for the AGS, well below the snake design capability of 9 degrees. A second run in December 1994 used the partial snake along with the fast tune jump quadrupoles. Intrinsic resonances were overcome in the presence of a 5% partial snake. The beam was accelerated to 25 GeV. Some polarization loss occured at two intrinsic resonances $12+\upsilon_y$ and $36+\upsilon_y$. Analysis is underway to determine the final beam polarization.

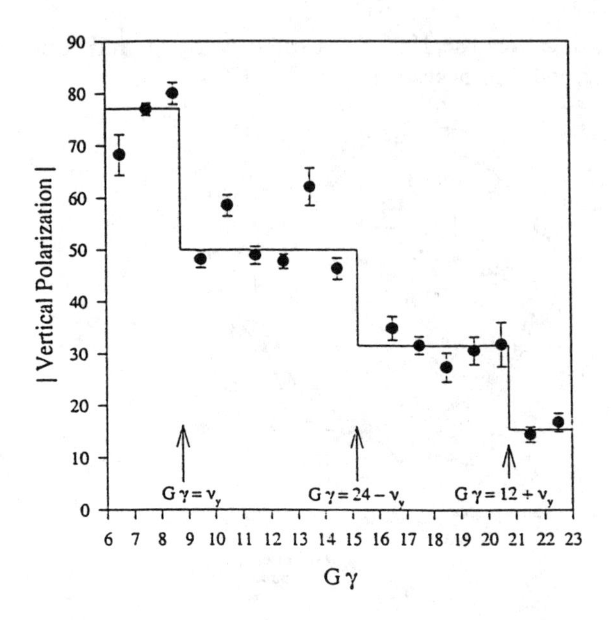

Figure 3. The vertical polarization as a function of $G\gamma$ using the partial snake in the AGS. Depolarization is due to the uncorrected intrinsic resonances.

These successful tests and implementations of Siberian snakes at IUCF, HERA, and the AGS make the prospect of polarized beam colliders more feasible. The conceptual design for the acceleration of polarized proton beams at RHIC [5] passed an external review in June 1993. The RSC proposal [6] to accelerate and carry out a spin physics program at RHIC was approved by the BNL Program Advisory Committee in November 1993.

The RHIC design calls for placing two 180 degrees Siberian snakes in each ring at opposite points in the lattice. This system will deal with intrinsic and imperfection depolarizing resonances up to 250 GeV. Their orthogonal axes of rotation are in the horizontal plane pointing at 45 degrees to the inside and outside of the ring respectively. The required magnetic strength is approximately 20 T-m. Inaddition, two intersection regions, where the STAR and PHENIX detectors are located, will be equipped with split spin rotators in order to achieve either longitudinal or transverse spins at the beam crossing points. Several snakes and spin rotator designs have been proposed (see contributions by Courant and Shatunov in these proceedings). The most promising are those that utilize helical dipole magnets with

360 degrees twist. This reduces the orbital beam excursions at injection energies, Figure 4. The RHIC magnet group has completed a design of a helical dipole magnet, Figure 5, and a hand wound prototype in under construction. Funding is provided from a BNL director's grant of $600K for three years .

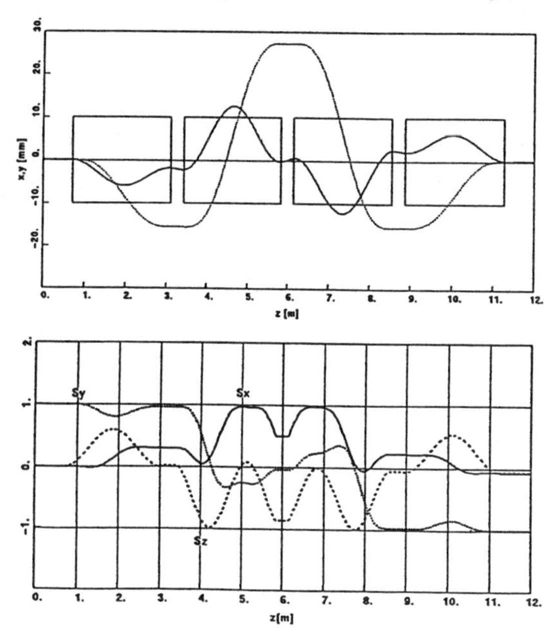

Figure 4. Orbit and spin tracking at RHIC injection energy through a Siberian Snake with four helical magnets.

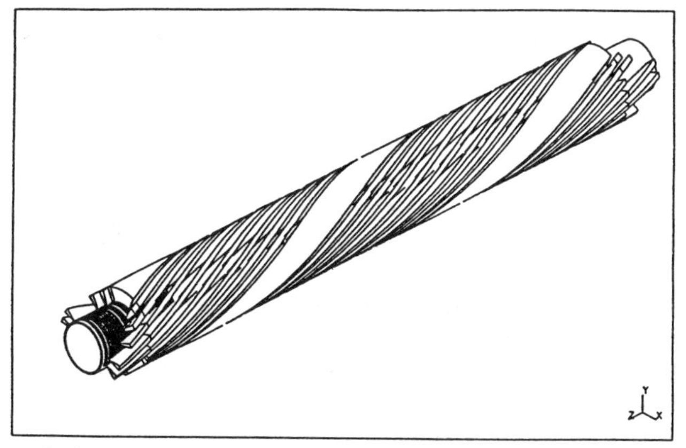

Figure 5. A conceptual design of a 360 degree helical snake magnet.

AGS bunches, 57 with $2x10^{11}$ protons each, are injected into RHIC with different beam polarization. Thus each crossing will measure interactions with specific spin-spin combinations. In order to further reduce the systematic errors, the beam polarization will be adiabatically reversed by spin flippers driven at a frequency of 40 khz that will artificially introduce a spin flipping resonance. Spin tracking programs have been written and studies of beam dynamics have commenced.

Beam polarization will be monitored during acceleration and storage in each ring. The polarimeters will use a 5μm carbon fiber target and hadron calorimeters to measure the asymmetry in inclusive negative pion production at $x_f = .5$ and $p_T =$ 0.8 GeV/c. At these parameters, the measured asymmetry at 125 GeV incident beam momentum [7] is 18% and the invariant cross section is about 100 $\mu b/GeV^2$ [8]. These values need to be measured at RHIC injection energy. A 5% polarization measurement will require a few seconds at the worst case at injection. Another absolute measurement would utilize the Coulomb Nuclear Interference method in pp elastic scattering.

3 The physics program

The polarized RHIC collider offers the combination of variable center of mass energy, high luminosity and polarization in both beams. The primary goal is to probe the nucleon spin structure using longitudinal and transverse spin configurations. The approved physics program assumes a running period of 8-10 weeks per year and integrated luminousities of $8x10^{38}$ and $3.2x10^{38}$ cm^{-2} at $\sqrt{s}$ of 500 and 200 GeV respectively with the STAR and PHENIX detectors.

The deep inelastic lepton scattering experiments (YALE-SLAC, NMC, and SMC) have pioneered the study of the spin structure of the valence quarks inside protons and neutrons. While the data show that the valence quarks are highly polarized at large x, a good percentage of the nucleon spin is carried by partons at low x (see contributions by J. Soffer in these proceedings) specially gluons and sea quarks. These are not directly or easily accessible through such experiments. The RHIC program with emphasis on hadronic interactions is complementary to the deep inelastic experiments. The questions to be addressed are: to what degree do gluons and sea quarks contribute to the proton spin? The predicted large parity violation in W and Z production will be directly measured. Are there are any surprises? Are the transverse and longitudinal structure functions different? Do the large transverse single spin asymmetries that are observed at modest p_T persist at large transverse momenta where PQCD calculations are applicable?

Two large RHIC detectors will be utilized in these measurements. STAR [9] is a large acceptance solenoid based TPC tracking detector, $|\eta| = 2$ and $\varphi = 2\pi$, will be augmented with an electromagnetic calorimeter and a shower max detector for the spin physics program. This will provide a capability for photon, lepton, as well as jet physics and $W^{+/-}$ and Z studies. PHENIX [10] is a high data rate lepton photon tracking detector with a highly segmented electromagnetic calorimeter, $|\eta| = .35$ and $\varphi = 2\pi$, and one muon endcap, $1.1 < y < 2.5$, and $\varphi = 2\pi$,. A second muon endcap will be added for the spin physics program. The focus is on photon, dilepton, $W^{+/-}$, and J/ψ production. Jet physics will be at the leading particle level.

The polarized gluon structure function can be studied using the following reactions: the measurement of the double spin asymmetry A_{LL} in direct photon production ($qg \Rightarrow \gamma q$) and inclusive single jet production ($gg \Rightarrow gg$). The direct photon measurement will yield a statistical error of $\delta A_{LL} \approx .01$ at p_T of 20 GeV/c, Figure 6, for both STAR and PHENIX. This translates to a 5% sensitivity in $\Delta G/G$ the polarized gluon structure function. Inclusive jet production will yield a more significant result at higher p_T. The STAR detector has the added advantage of being able to measure jets produced opposite the direct photons and thus, determine the x dependence of the structure function. Heavy quark production that is produced in gg fusion is another mechanism to study the gluon structure function. PHENIX will measure the process $g + g \Rightarrow J/\psi + \gamma$.

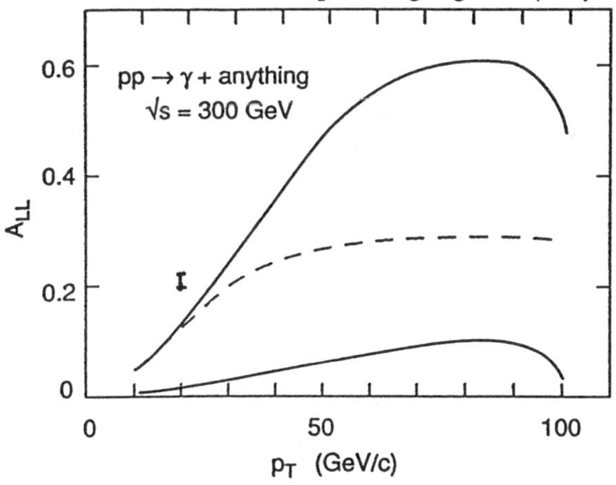

Figure 6. Asymmetry [13] in direct γ production as a function of p_T. The upper curves are for large ΔG and 45°(solid) and 90°(dashed). The lowest curve is for standard ΔG and 90°. A typical error is also shown.

The polarized sea quark distribution is accessible via the A_{LL} measurement in Drell Yan (dilepton production) which is sensitive to the product $[\Delta q/q \times \Delta\bar{q}/\bar{q}]$. This is a difficult measurement requiring higher intensities and longer running periods due to low production rates and detector reaches of the desirable kinematic coverage of low x regions for the antiquark and relatively higher x for the participating quark. Both STAR and PHENIX will detect Drell Yan pairs in complementary rapidity coverage.

Bourrely and Soffer [11] suggested a formalism for calculating the single spin parity violating asymmetry A_L in W^+ and W^- production to get another handle on the sea quark polarization. Their approximation shows that $A_L(W^+) \approx (\Delta\bar{d})/\bar{d}$ and $A_L(W^-) \approx (\Delta\bar{u})/\bar{u}$. These asymmetries are shown in Figure 7. STAR and PHENIX (with its second muon endcap) will collect 62K and 24K W^+ events , and 14K and 9K W^- events respectively.

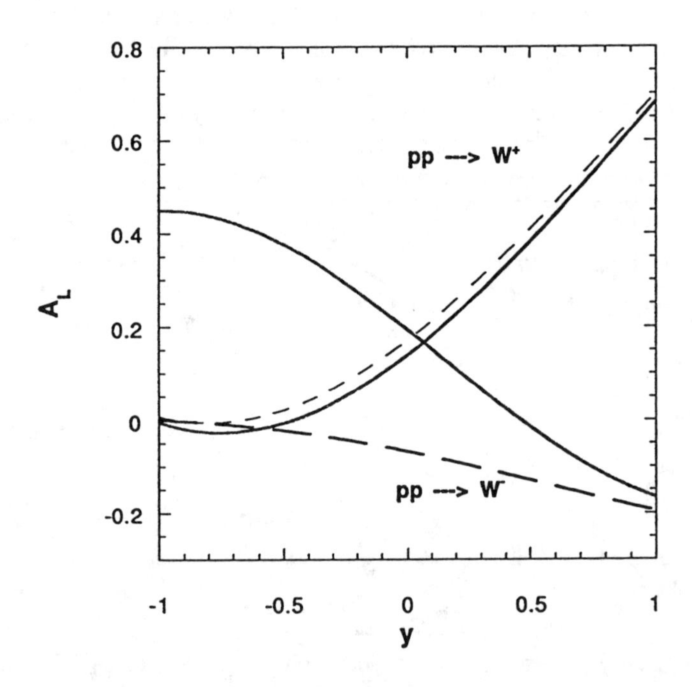

Figure 7. Parity Violating asymmetries in $W^{+/-}$ production at $\sqrt{S}$ =500 GeV. The solid lines assume sea quark polarization [11] and the dashed lines no polarization.

Transversity distributions (h_1) will also be measured. The predicted asymmetries are smaller and thus longer running is needed. The transverse spin structure functions, which are present only for the valence partons, represent new measurements. It would be interesting to compare these to the corresponding longitudinal ones. Drell Yan as well as Z production would the appropriate probes.

Transverse single spin asymmetries have been the domain of fixed target experiments. Large asymmetries have been observed in inclusive pion, and lambda production up to p_T of 4 GeV/c. Perturbative QCD calculations expect these effects to be quite small and to vanish at high momentum transfer. RHIC would provide a natural extension of these data to higher p_T. STAR and PHENIX can extend the π^o measurements up to 20 GeV/c but these detectors lack any charged particle identification for high momenta. A third detector at RHIC, BRAHMS [12] is specifically built to study single particle production spectra. It will achieve particle separation up to 25 GeV/c. The case for kaons is particularly interesting since they carry strange valence quarks.

4 Costs and Funding

The cost for the construction and installation of the accelerator hardware is estimated at $10M. The Institute of Physical and Chemical Research of Japan (RIKEN) has expressed interest in both the accelerator and the physics research aspects of this effort. They submitted a proposal for $20M to the Japan Science and Technology Agency to fund the accelerator hardware and a new muon arm in the PHENIX detector in order to enhance its coverage for spin physics. The proposal has received initial approval for a first phase of funding in the amount of $5M over the next three years starting in April 1995. It is hoped that approval for the two remaining phases [$5M and $10M] will follow in subsequent years. The construction schedule will aim at installing the accelerator hardware in RHIC in a timely fashion consistent with RHIC turn on in 1999 in order to minimize the impact to the heavy ion program.

5 Conclusion

The polarized RHIC capability provides a unique "low cost" opportunity to access this important aspect of spin physics. This will extend to large p_T a program that is uniquely accessible via hadronic interaction. This is a complementary program to the Deep Inelastic experiments at facilities such as SLAC, CERN and HERA, and judging by the theoretical interest, forms a good coupling between theory and experiment. The RIKEN initiative has put the effort on a more realistic basis. The stage is set to meet the technical challenges associated with acceleration and preservation of polarized beams at energies much higher than has been attained in the past.

6 References

* This work is Supported by the U.S. Department of Energy contract number DE-AC02-76CH00016.

[1] Conceptual Design of the Relativistic Heavy Ion Collider, BNL 52195, May 1989

[2] RHIC Spin Collaboration Letter of Intent, 5 April, 1991 submitted to the BNL Program Advisory Committee, October 1992

[3] AGS experiment E880, The Effect of a Partial Snake on Polarization at the AGS, May 1991

[4] H. Huang et al. Phys. Rev. Lett. **73**, 2982 (1994)

[5] K. Brown et al. Conceptual Design for the Acceleration of Polarized Protons in RHIC. May 1993

[6] Proposal on Spin Physics Using the RHIC Polarized Collider, August 1992. Update, September 1993

[7] D.L. Adams et al. Phys. Lett. **B264**, 462 (1991)

[8] L.G. Ratner et al. Proc. of the Rochester Meeting APS/DPF, Rochester, p 99 (1971)

[9] STAR Conceptual Design Report, LBL 5380, September 1993

[10] PHENIX Conceptual Design Report, January 1993, updated November 1993

[11] C. Bourrely and J. Soffer, Phys. Lett. **B314**, 132 (1993)

[12] Interim Design Report for the BRAHMS Experiment at RHIC, February 1994

[13] Bourrely, Guillet, and Soffer, Nucl. Phys. **B361**, 72 (1991);
Bourrely, Soffer, Renard, and Taxil, Phys. Rep. **177**, 319 (1989);
E.L. Berger and J. Qui, Phys. Rev. **D40**, 778 (1989)

First Test of a Partial Siberian Snake for Acceleration of Polarized Protons

D.D. Caussyn[*], R. Baiod[‡], B.B. Blinov[*,(a)], C.M. Chu[*], E.D. Courant[*,(b)]
D.A. Crandell[*], Ya.S. Derbenev[†,(c)], T.J.P. Ellison[†], W.A. Kaufman[*],
A.D. Krisch[*], S.Y. Lee[†], M.G. Minty[††], T.S. Nurushev[*], C. Ohmori[§],
R.A. Phelps[*], D.B. Raczkowski[*], L.G. Ratner[*,(b)], P. Schwandt[†],
E.J. Stephenson[†], F. Sperisen[†], B. von Przewoski[†], U. Wienands[||,(d)],
V.K. Wong[*,(e)]

[*] *Randall Laboratory of Physics, University of Michigan, Ann Arbor, Michigan 48109-1120*
[†] *Indiana University Cyclotron Facility, Bloomington, Indiana 47408-0768*
[‡] *Fermilab, Batavia, Illinois 60510*
[§] *Institute for Nuclear Study, University of Tokyo, Tanashi, Tokyo 188, Japan*
[††] *Stanford Linear Accelerator Center, Stanford, California 94309*
[||] *SSC Lab, Dallas, Texas*

Abstract. We recently studied the first acceleration of a spin-polarized proton beam through a depolarizing resonance using a partial Siberian snake. We accelerated polarized protons from 95 to 140 MeV with a constant 10% partial Siberian snake obtained using rampable solenoids. The 10% partial snake suppressed all observable depolarization during acceleration due to the $G\gamma = 2$ imperfection depolarizing resonance which occurred near 108 MeV. However, 20% and 30% partial Siberian snakes apparently moved an intrinsic depolarizing resonance, normally near 177 MeV, into our energy range; this caused some interesting, although not-yet-fully understood, depolarization.

INTRODUCTION

Many depolarizing resonances are encountered in accelerating a polarized proton beam to high energies in circular accelerators. Polarized proton beams have been accelerated by correcting each individual resonance at the ZGS (1), Saturne (2), KEK (3), and until recently at the AGS (4). However, correcting individual resonances becomes impractical for energies above 20 GeV.

A full Siberian snake has been proposed as way of overcoming all depolarizing resonances (5), a proposition which has been supported by experimental tests at the IUCF Cooler Ring (6–9). By rotating the proton's spin by 180° every turn, a full Siberian snake causes the cancellation of depolarizing precessions due to magnetic field errors on alternate rotations through the accelerator. Attractive designs for full Siberian snakes exist for low and high energy accelerators, but no simple solutions exist for medium energy accelerators, like the Fermilab 8 GeV Booster (10) or the Brookhaven AGS (11). In such accelerators partial Siberian snakes might be used to overcome many of the depolarizing resonances.

Without a Siberian snake, each proton's spin precesses around the ring's vertical magnetic field, which is the stable spin direction in this case. However, horizontal magnetic field errors may depolarize the beam when the spin precession tune ν_s, which is the number of times the spin precesses around the stable spin direction per turn in the accelerator, satisfies the following condition

$$\nu_s = n + m\nu_y, \tag{1}$$

where the vertical betatron tune ν_y is the number of vertical betatron oscillations per turn in the accelerator, and n and m are integers. Imperfection depolarizing resonances occur when $m = 0$, while first-order intrinsic resonances occur when $m = \pm 1$. Without a Siberian snake the spin tune is given by $\nu_s = G\gamma$ where γ is the Lorentz energy factor and $G = 1.792847$ is the proton's anomalous magnetic moment.

A recent experiment (8) confirmed that, in a ring containing a partial Siberian snake of strength s, the spin tune is found from

$$\cos(\pi\nu_s) = \cos(\pi G\gamma)\cos\left(\frac{\pi s}{2}\right), \tag{2}$$

where $s = 1$ corresponds to a full snake, for which ν_s is a half-integer and independent of the energy. On the other hand, for a weak partial Siberian snake, the spin tune is changed from $G\gamma$ but only strongly in the vicinity of the imperfection resonances. Thus, partial Siberian snakes may be useful for overcoming the many weak imperfection resonances in medium energy accelerators, while the comparatively fewer intrinsic depolarizing resonances might be overcome using pulsed quadrupoles to jump the vertical betatron tune as the resonances are approached.

We report here the results of the first acceleration of a polarized proton beam through a depolarizing resonance using a partial Siberian snake, much of which have been recently published elsewhere (12). Not long after our work, a partial Siberian snake was independently tested at the AGS (13). It should also be noted that partial Siberian snakes have been previously used to overcome an imperfection resonance for accelerated polarized positrons (14).

EXPERIMENTAL METHODS AND RESULTS

The snake strength s of a single solenoid magnet with NI ampere-turns is given by

$$s = \frac{\mu_0(1+G)}{10.479\ p} NI, \tag{3}$$

where $\mu_0 = 4\pi \times 10^{-7}$ T m A^{-1} and p is the proton's momentum in GeV/c. Thus to maintain a constant snake strength as the beam is accelerated, the solenoid current must be ramped. For our studies, we installed two rampable warm solenoid magnets symmetrically around our previously installed superconducting solenoid in the IUCF Cooler Ring, separated only by drift spaces. The superconducting solenoid, the polarimeter and the Cooler Ring's operation with polarized protons were discussed earlier (6–9,15–17). The two warm solendoids' currents were varied together so that, in combination with the superconducting solenoid operating at constant current, a constant snake

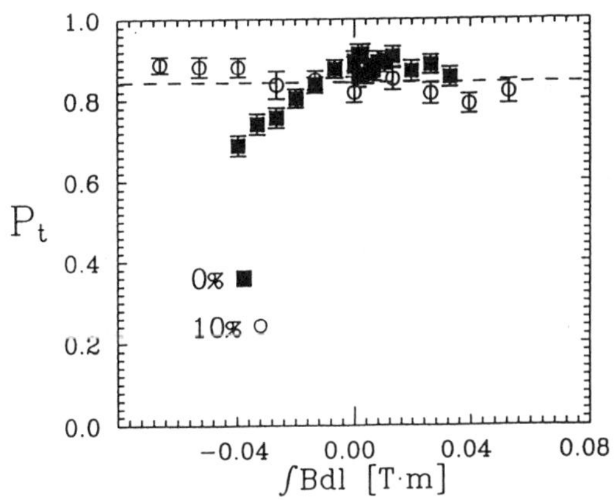

FIGURE 1. The measured transverse polarization P_t after acceleration to 140 MeV plotted against the imperfection $\int B d\ell$ with no snake and with a constant 10% partial Siberian snake. The dashed line is the best constant polarization fit to the snake-on data.

strength was maintained during acceleration (12). This combination of magnets allowed constant partial Siberian snake strengths of 10%, 20%, and 30% while accelerating from 95 MeV to 140 MeV. This energy range includes the $G\gamma = 2$ imperfection depolarizing resonance (6) near 108 MeV, but is well below 177 MeV where the $7 - \nu_y$ intrinsic depolarizing resonance normally occurs (7).

We first studied the depolarization during the acceleration with no Siberian snake (0%), and then with a 10% partial Siberian snake. In each study the beam polarization transverse to the beam motion, P_t, was measured after accelerating to 140 MeV for different values of the imperfection $\int B d\ell$, which is produced by the correction solenoid magnets in the ring's cooling section. The results of these measurements are shown in Fig. 1. It should be noted that the spin direction was actually flipped both with and without the partial Siberian snake, and there is a normalization uncertainity of about 15% for the values of P_t, which does not affect the shape of the curves.

With a 0% snake, there is a significant decrease in P_t for nonzero $\int B d\ell$ due to the $G\gamma = 2$ imperfection depolarizing resonance. Since the beam accelerated through the resonance with $\langle d\gamma/dt \rangle = 0.061\ s^{-1}$, this P_t curve is flatter than the 104 MeV fixed-energy data of Ref. (6); nevertheless, with no snake, the $G\gamma = 2$ resonance clearly depolarized the accelerated beam, or rotated its spin into the unmeasurable longitudinal direction. However, with a 10% partial Siberian snake, the beam polarization measured after acceleration to 140 MeV was almost independent of the imperfection $\int B d\ell$ within our precision of about 2%.

We also studied the effect of stronger partial Siberian snakes. In Fig. 2, the transverse beam polarization, measured after acceleration from 95 to 140 MeV, is plotted against the imperfection $\int B d\ell$ for constant partial Siberian snake strengths of 0%, 10%, 20%, and 30%. Note the behavior in the 20% and 30% snake data; the polarization was unexpectedly lost for higher $\int B d\ell$. A possible explanation for this depolarization may be deduced from Fig. 3, in which the calculated values of the spin tune are plotted versus $G\gamma$ for various snake strengths. The spin tunes at which the imperfection and intrinsic resonances occur are marked with horizontal dashed lines, and the values of $G\gamma$ for the

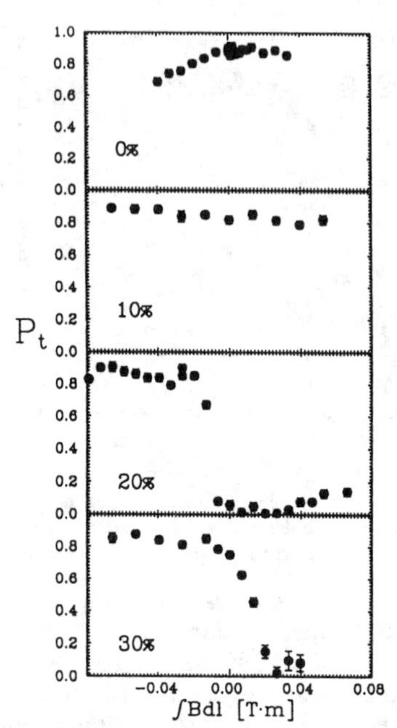

FIGURE 2. The measured transverse polarization at 140 MeV plotted against the imperfection $\int Bdl$ for a partial Siberian snake with strengths 0%, 10%, 20% and 30%.

initial (95 MeV) and final (140 MeV) energies during the ramp are marked by vertical dashed lines. As long as the spin tune (solid lines) do not pass through a resonance within the range of $G\gamma$ during the ramp, taking into account that the resonance has some width, depolarization should not be observed. Without a Siberian snake, $\nu_s = G\gamma$, and the intrinsic resonance would not be encountered until the proton kinetic energy reaches 177 MeV. As can be seen in Fig. 3, Siberian snake strengths between 20% and 30% result in spin tunes relatively near to the intrinsic depolarizing resonance within our energy range. Although this suggests how the beam may have been depolarized, we do not yet have a detailed explanation for actual depolarization we observed.

To summarize, the 10% snake data support the conjecture that a weak partial Siberian snake can maintain full beam polarization during acceleration through a weak imperfection depolarizing resonance. Therefore, partial Siberian snakes might be useful for accelerating a polarized proton beam at medium energy accelerators such as the Fermilab 8 GeV Booster (10) or the Brookhaven AGS (11). However, the effect of partial Siberian snakes near intrinsic depolarizing resonances requires further study. Experimental studies are underway which show promise for explaining the depolarization observed with partial snakes with strengths greater than 10%.

We would like to thank J. M. Cameron and the entire Indiana University Cyclotron Facility staff for the most successful operation of the Cooler Ring. We are grateful to V. A. Anferov, A. W. Chao, S. V. Gladysheva, R. S. Herdman, S. Hiramatsu, F. Z. Khiari, A. V. Koulsha, W. F. Lehrer, H.-O. Meyer, J. B. Muldavin, R. E. Pollock, T. Roser, H. Sato, D. S. Shoumkin, S. E. Sund,

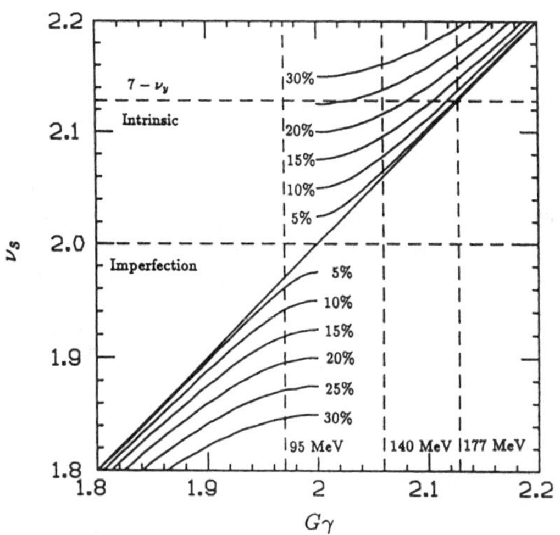

FIGURE 3. Spin tune ν_s versus $G\gamma$ (solid lines) for various partial Siberian snake strengths, as calculated from Eq. 2.

T. Toyama, and S. J. Wheeler for their help with parts of this experiment. This research was supported by grants from the U.S. Department of Energy and the U.S. National Science Foundation.

REFERENCES

[a] Also at Moscow State University, Moscow, Russia.
[b] Also at Brookhaven National Laboratory, Brookhaven, NY 11973.
[c] Also at Department of Nuclear Engineering, University of Michigan.
[d] Now at Stanford Linear Accelerator Center, Stanford, CA 94309.
[e] Also at Office of the Provost, University of Michigan at Flint, Flint, MI 48501.

1. T. Khoe *et al.*, Part. Accel. **6**, 213 (1975).
2. J. L. Laclare *et al.*, J. Phys. (Paris), Colloq. **46**, C2-499 (1985).
3. H. Sato *et al.*, Nucl. Instrum. Methods Phys. Res. Sect. A **272**, 617 (1988).
4. F. Z. Khiari *et al.*, Phys. Rev. **D39**, 45 (1989).
5. Ya.S. Derbenev and A. M. Kondratenko, Part. Accel. **8**, 115 (1978).
6. A. D. Krisch *et al.*, Phys. Rev. Lett. **63**, 1137 (1989).
7. J. E. Goodwin *et al.*, Phys. Rev. Lett. **64**, 2779 (1990).
8. V. A. Anferov *et al.*, Phys. Rev. **A46**, R7383 (1992).
9. R. Baiod *et al.*, Phys. Rev. Lett. **70**, 2557 (1993).
10. Acceleration of Polarized Protons to 120 and 150 GeV in the Fermilab Main Injector, SPIN Collaboration, Univ. of Michigan Report (March 1992), (unpublished).
11. RHIC Spin Proposal M. Beddo et al.; Brookhaven National Lab proposal (1992), (unpublished).
12. B.B. Blinov *et al.*, Phys. Rev. Lett. **73**, 1621 (1994).
13. H. Huang *et al.*, these proceedings.
14. Ya.S. Derbenev *et al.*, *Proc. of the X International Conference on High Energy Accelerators*, Protvino, 1977.
15. R.A. Phelps *et al.*, Phys. Rev. Lett. **72**, 1479 (1994).
16. M.G. Minty *et al.*, Phys. Rev. **D44**, R1361 (1991).
17. J.E. Goodwin, Ph.D. Thesis, Indiana University (1990).

Partial Siberian Snake Experiment at the AGS

H. Huang[1,3], L. Ahrens[2], J.G. Alessi[2], M. Beddo[3], K.A. Brown[2], G. Bunce[2], D.D. Caussyn[1] *, D. Grosnick[3], A.E. Kponou[2], S.Y. Lee[1], D. Li[1], D. Lopiano[3], A.U. Luccio[2], Y.I. Makdisi[2], L. Ratner[2†], K. Reece[2], T. Roser[2], H. Spinka[3], A.G. Ufimtsev[2‡] D.G. Underwood[3], W. van Asselt[2], N.W. Williams[2], A. Yokosawa[3]

[1]*Department of Physics, Indiana University, Bloomington, IN 47405*
[2]*Brookhaven National Laboratory, Upton, NY 11973*
[3]*Argonne National Laboratory, 9700 Cass Ave. Argonne, IL 60439*

Abstract A 9° solenoidal spin rotator or 5% partial Siberian snake was used to successfully accelerate polarized protons for the first time to 10.8 GeV kinetic energy in the Brookhaven AGS. It was found that a 5% partial snake can effectively overcome 18 imperfection resonances in this energy range. We also observed an interference between the spin flip induced by an intrinsic resonance and linear coupling due to the solenoid field of the partial snake.

The acceleration of polarized beams in circular accelerators is complicated by the presence of numerous depolarizing resonances. The depolarization is caused by the small spin-perturbing magnetic fields which, at the resonance condition, act coherently to move the spin away from the stable vertical direction. The resonance conditions are usually expressed in terms of the spin tune ν_s , which is defined as the number of spin precessions per revolution. For an ideal planar accelerator, where orbiting particles experience only the vertical guide field, the spin tune is equal to $G\gamma$[1], where $G = 1.7928$ is the anomalous magnetic moment of the proton and γ is the relativistic Lorentz factor. The resonances occur when $\nu_s = G\gamma$ =integer(imperfection resonances), and $\nu_s = G\gamma = kP \pm \nu_y$(intrinsic resonances), where k is an integer, ν_y is the vertical betatron tune and P is the superperiodicity. For the AGS, $P = 12$ and $\nu_y \approx 8.8$. When a polarized beam is accelerated through an isolated resonance, the final polarization P_f can be related to the initial polarization P_i by[2]

$$\frac{P_f}{P_i} = 2e^{-\frac{\pi|\epsilon|^2}{2\alpha}} - 1, \quad (1)$$

where ϵ is the resonance strength, and α is the change of the spin tune per radian of the orbit angle. Traditionally, the intrinsic resonances are overcome

*now at: Department of Physics, University of Michigan, Ann Arbor, MI 48109.
†also at: Department of Physics, University of Michigan, Ann Arbor, MI 48109.
‡permanent address: Institute for High Energy Physics, Protvino, Russia.

by using a betatron tune jump, which effectively makes α large, and the imperfection resonances are overcome with the tedious harmonic corrections of the vertical orbit to reduce the resonance strength ϵ[3,4].

By introducing a 'Siberian Snake'[5], which is a 180° spin rotator about a horizontal axis, the stable spin direction remains unperturbed at all times. Therefore the beam polarization is preserved during acceleration. For lower-energy synchrotrons, such as the Fermilab booster and the Brookhaven AGS with weaker depolarizing resonances, a partial snake[6], which rotates the spin by less than 180°, is sufficient to keep the stable spin direction unperturbed at the imperfection resonances. This paper reports the first test of the partial snake concept crossing 18 imperfection resonances.

There have been several snake experiments performed at the Indiana University Cyclotron Facility (IUCF) Cooler Ring[7]. In these experiments, the acceleration rate was so low that the imperfection resonance always induced complete spin flip and no loss of polarization could be observed without a snake. Thus, a partial snake experiment at an accelerator such as the Brookhaven AGS, where numerous imperfection resonances lead to complete depolarization, was necessary to confirm the applicability of the partial snake.

The solenoidal partial snake is installed in a 3-meter straight section in the AGS. It can be ramped to 4.7 Tm in about 0.6 s at a 33% duty cycle and is capable of achieving a 5% snake strength up to $G\gamma = 48$, or 25 GeV kinetic energy. The spin rotation ϕ in the solenoid is given by

$$\phi = \frac{e(1+G)\mu_0 NI}{cp}, \tag{2}$$

where p is the momentum of the proton, μ_0 is the permeability of vacuum, and NI the ampere-turns. The effective snake strength s of the partial snake is $s = \phi/\pi$.

The polarized beam from the atomic source[8] was accelerated through a radio frequency quadrupole (RFQ) and the 200 MeV LINAC. The beam polarization at 200 MeV was measured with elastic scattering from a carbon target[4] and the beam polarization was quite stable at $80 \pm 5\%$. The beam was then injected and accelerated in the AGS Booster up to 1.56 GeV or $G\gamma = 4.77$. The vertical betatron tune of the AGS booster was chosen to be 4.88 to avoid crossing the intrinsic resonance $G\gamma = \nu_y$ in the booster. The imperfection resonance at $G\gamma = 4$ was corrected by harmonic orbit correctors. The beam intensity in the AGS was typically 5×10^9 polarized protons per pulse. The polarized beam in the AGS was accelerated up to $G\gamma = 22.5$ or 10.8 GeV kinetic energy passing through eighteen imperfection resonances and three intrinsic resonances at $G\gamma = \nu_y, 24 - \nu_y$, and $12 + \nu_y$.

The beam polarization in the AGS was measured by scattering the polarized protons from a nylon string. At $G\gamma = 10.5$ the left-right asymmetry was measured with both the nylon target and a carbon target. This allowed the determination of the quasi-elastic background to the elastic scattering from the

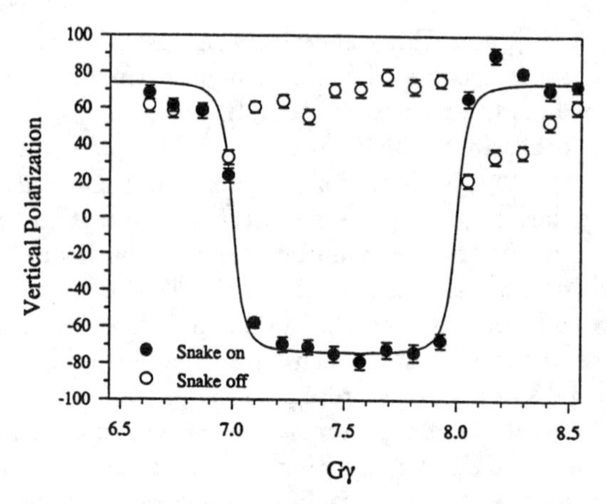

Figure 1: The measured vertical polarization as a function of the spin tune $G\gamma$ for a 10% snake is shown with and without a snake.

free hydrogen in the nylon. Using the analyzing power measurements from ref.[9], the effective analyzing power, and therefore the absolute beam polarization, was determined with a systematic error of about 10%. For other beam energies, the analyzing power was scaled according to the measured values for elastic scattering in ref. [9].

In the presence of a snake with strength s and for small imperfection resonance strength, the spin tune ν_s obeys the relation

$$\cos\left(\pi\nu_s\right) = \cos(\pi G\gamma)\cos\left(\frac{\pi s}{2}\right). \tag{3}$$

When s is small, the spin tune is nearly equal to $G\gamma$ except when $G\gamma$ equals an integer n, where the spin tune is shifted away from the integer value by $\pm\frac{s}{2}$. Thus the partial snake can overcome all imperfection resonances, provided that the resonance strengths are much smaller than the gap created by the partial snake. When passing through an imperfection resonance condition with a snake on, the stable spin direction moves slowly from up to down. For a 5% partial snake and the AGS acceleration rate $\alpha \approx 1.1 \times 10^{-5}$, the change in the stable spin direction is always adiabatic.

Fig. 1 shows the measured polarization as a function of $G\gamma$ for a 10% snake. The polarization was observed to follow the predicted spin flip in passing through each imperfection resonance without loss of polarization.

Fig. 2 shows the absolute value of the vertical polarization with a 5% snake measured at every half-integer value of $G\gamma$ between 6.5 and 22.5. As shown in Fig. 1 the sign of the polarization is actually changing at every integer value of $G\gamma$. Without the partial snake, the beam was completely depolarized after passing through the resonances at $G\gamma = 8$, 9, 10, and ν_y. Fig. 2 also shows that the depolarization resulted only from the three intrinsic resonances, located at $G\gamma = \nu_y, 24 - \nu_y$, and $12 + \nu_y$. The sizable linear betatron coupling introduced

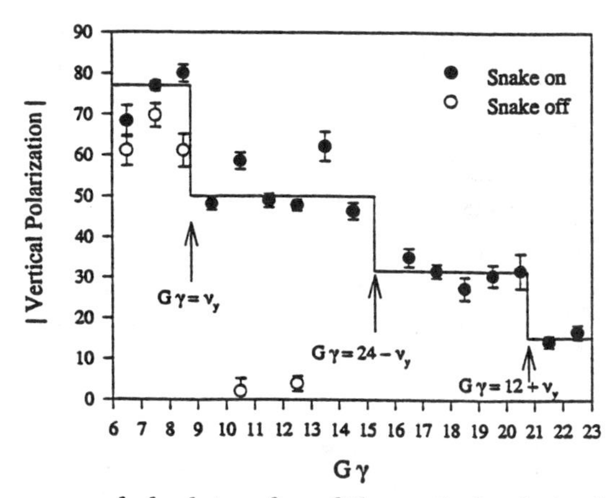

Figure 2: The measured absolute value of the vertical polarization at $G\gamma = n + \frac{1}{2}$ up to $G\gamma = 22.5$. Partial depolarization is due to intrinsic resonances at $G\gamma = \nu_y$, $24 - \nu_y$, and $12 + \nu_y$.

by the solenoid caused significant additional depolarization when $\nu_y \approx \nu_x$. Fig. 3 shows the beam polarization measured at $G\gamma = 10.5$, after passing through the $G\gamma = \nu_y$ resonance, as a function of the vertical betatron tune ν_y for $\nu_x = 8.67, 8.75$, and 8.80, respectively. The adiabatic 65% spin flip for the $G\gamma = \nu_y$ intrinsic resonance was diminished by the linear betatron coupling. We also observed less spin flip as ν_y approaches 8.5. We know of no depolarization mechanism that could account for this phenomenon. Further studies to understand in detail the depolarization mechanism are needed.

In conclusion, we have demonstrated for the first time that a 5% partial snake can effectively overcome 18 imperfection resonances without noticeable depolarization. The polarized beam in the AGS was accelerated to 10.8 GeV kinetic energy without using harmonic orbit correctors. This fact gives great optimism for the possibility of accelerating polarized proton beams to very high energies to study spin effects in polarized proton colliders.

ACKNOWLEDGMENTS

We would like to thank A. Otter and W.J. Leonhardt for their help in snake design and construction, G.J. Mahler, P.R. Cameron, L. Balka, D. Carbaugh, T. Kasprzyk, C. Keyser, T. Kicmal, J. Nasiatka for their help in the polarimeter design and construction. We are grateful to S. Hiramatsu, S. Hsieh, S. Peggs, H. Sato, L. Teng, S. Tepikian, T. Toyama, U. Wienands, K. Yokoya for their help with parts of this experiment. We are especially indebted to T.B. Kirk for his efforts to obtain funding for the snake magnet and for his continued support. The research was supported by grants from the U.S. Department of Energy, Division of High Energy Physics grant number DE-FG02-92ER40747 and the contract number W-31-109-ENG-38.

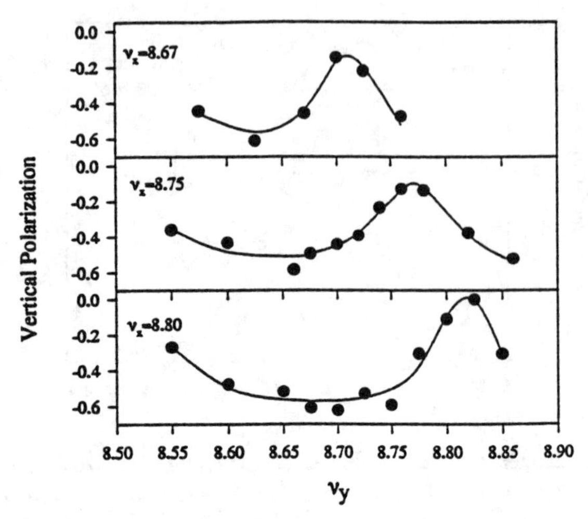

Figure 3: Polarization measured at $G\gamma = 10.5$ as a function of the vertical betatron tune for 5% snake with nominal horizontal tune $\nu_x = 8.67, 8.75$, and 8.8, respectively. Depolarization was observed at $\nu_y \approx \nu_x$ in all cases. The lines are drawn to guide the eye.

REFERENCES

1. Thomas, L.H., Phil. Mag. **3**, 1 (1927); Bargmann, V., Michel, L., Telegdi, V.L., Phys. Rev. Lett. **2**, 435 (1959).

2. Froissart,M., and Stora, R., Nucl. Instrum. Methods, **7**, 297(1960).

3. Khoe, T., et al., Part. Accel. **6**, 213 (1975); Laclare, J.L., et al., J. Phys. (Paris), Colloq. **46**, C2-499 (1985); Sato, H., et al., Nucl. Instrum. Methods, Phys. Res. Sec **A272**, 617 (1988).

4. Khiari, F.Z., et al., Phys. Rev. **D39**, 45 (1989); see also Lee, S.Y., in *Proceedings of the 1992 RCNP Kikuchi school*, Osaka, November, 1992, edited by Fujiwara, M. and Kondo, M.(RCNP-P-128,1992), pp. 123-192.

5. Derbenev, Ya.S., et al., Part. Accel. **8**, 115 (1978).

6. Roser, T., AIP Conference Proceedings No. 187, ed. Heller, K.J., p. 1442(AIP, New York, 1989).

7. Krisch, A.D., et al., Phys. Rev. Lett. **63**, 1137 (1989); Goodwin, J.E., et al., Phys. Rev. Lett. **64**, 2779 (1990); Minty, M.G., et al., Phys. Rev. D44, R1361 (1991); Anferov, V.A., et al., Phys. Rev. **A46**, R7383 (1992); Baiod, R., et al., Phys. Rev. Lett. **70**, 2557 (1993); Phelps, R.A., et al., Phys. Rev. Lett. **72**, 1479 (1994).

8. Alessi, J.G., et al., AIP Conference Proceedings No. 187, ed. Heller, K.J., p. 1221(AIP, New York, 1989).

9. Spinka, H., et al., Nucl. Instrum. Methods, **211**, 239 (1983).

Prospect for polarized beam acceleration at the KEK PS

T. Toyama and H. Sato

National Laboratory for High Energy Physics,
Oho 1-1, Tsukuba-shi, Ibaraki-ken, 305, Japan

Abstract. Polarized deuteron-beam acceleration is scheduled at the KEK PS in 1995. No depolarizing resonances are expected if the vertical betatron tune is chosen to be at 6.25. Polarimeters are now being prepared. The acceleration of polarized protons has been reconsidered in the context of a partial snake. Two possibilities, solenoid and helical snakes, have been examined.

1. INTRODUCTION

Undergoing a successful commissioning in ordinary deuteron-beam acceleration in 1991, polarized deuteron acceleration has become realistic (1). The advantages of polarized deuteron acceleration are, as is well known, a much smaller number and weaker strength of resonances compared with those of polarized proton acceleration. The proposal of an acceleration test was approved by the PS-PAC at the end of fiscal year 1993, and an experiment (T-323) is scheduled for 1995.

On the other hand, polarized proton beam acceleration was studied (2) and utilized for physics experiments at the KEK PS in the 1980's. In those studies, an incomplete resonance correction made it difficult to go up to higher energies of more than ~ 4 GeV without experiencing polarization loss. However, by utilizing a promising method, "Siberian Snake" (3), we have recently started to reconsider polarized proton beam acceleration at the KEK PS.

We describe here the basic acceleration scheme of both beams.

2. POLARIZED DEUTERON BEAM

Because the deuteron is a spin-1 particle, the spin orientation is expressed by the vector polarization (Pz) and the tensor polarization (Pzz), assuming that the bending field of the ring lies in the vertical axis and that symmetry exists about this axis. The polarization ratio before and after crossing the resonance is (4)

$$Pz(f)/Pz(i) = 2\, e^{-\pi|\varepsilon|^2/(2\alpha)} - 1, \qquad (1)$$

$$Pzz(f)/Pzz(i) = \{\, 3\, (\, 2\, e^{-\pi|\varepsilon|^2/(2\alpha)} - 1)^2 - 1\, \}/2, \qquad (2)$$

for each particle under the assumption of linear crossing of an isolated resonance. Here, (i) and (f) suggest the initial and final states, respectively, ε is the resonance strength and α is the resonance crossing speed.

In the KEK PS, a dual-optically-pumped polarized ion source (OPPIS) for the negative deuterium has been successfully developed (1, 5), which enables us to produce pure vector and tenser polarizations as a theoretical limit. The first step of an acceleration test will be performed using a vector-polarized deuteron beam. The beam intensity of the OPPIS is expected to be greater than 10 μA, and having a vector polarization of 70%.

Deuterons are accelerated from 19 MeV to 294 MeV in the Booster Ring (BR) and then up to 11.2 GeV in the Main Ring (MR) (1). There are no depolarizing resonances in the BR. In the MR, no intrinsic resonance exists in the case of ν_z=6.25 and one intrinsic resonance ($\gamma G=-8+\nu_z$) exists at 8 GeV in the case of ν_z=7.25, where ν_z is the vertical betatron tune, γ is the Lorentz factor and G is the gyromagnetic anomaly. The vertical tune for normal beam operation has been changed from ν_z=6.25 to 7.25 because of a ripple problem of the magnet power supply since 1988. At present, this problem has been eliminated (6). However, the operating point remains at ν_z=7.25. Therefore, the operation point for polarized deuteron should be changed to ν_z=6.25 in order to escape the intrinsic resonance.

The imperfection resonance $\gamma G=-1$ is located at 11.28 GeV, slightly above the top energy. The calculated resonance strength is $\varepsilon \sim 1.5 \pm 1.6 \times 10^{-5}$, using a closed-orbit distortion ($<z_{COD}>$) of 1mm rms. The harmful energy region ($\pm 7\ \varepsilon$) expected to be within ±0.01 GeV around the resonance energy. Consequently, no depolarization will occur if ν_z=6.25. On the other hand, if ν_z=7.25, depolarizations of about 8% and about 23% are expected for the vector and tensor polarization, respectively.

The polarimeters will be installed at 10 MeV, 294 MeV and the beam-extraction line of the MR. The 10 MeV polarimeter will be the modified version of the proton polarimeter between the two LINAC's, using d-C elastic scattering. The 294 MeV polarimeter will also be modified from the "internal polarimeter" installed in the short straight section of the MR. The analyzing power of d-p elastic is expected to be 40% at θ_{CM}=90 degrees (7) and only a few minuets will be needed for a polarization measurement with 1% statistical error. The final polarization will be measured by an "external polarimeter" which will be build at the East Counter Hall (EP1-B line). The reaction of d-p elastic scattering will be utilized, where the target will be a polyethylene string and carbon for a background subtraction. The existing analyzing powers are only up to 3 GeV/c. The analyzing powers at 3 GeV/c are 0.3 and 0.6 for the vector and tensor, respectively, at t=0.14 $(GeV/c)^2$ (8). To measure the beam polarization from 3 GeV/c to the top energy, the beam will be decelerated down to 3 GeV/c or less after acceleration to the top energy. The polarization will be measured before and after this cycle. The depolarization will be detected as the difference of those polarizations.

3. POLARIZED PROTON BEAM

There are two depolarizing resonances in the BR: $\gamma G=2$ and ν_z. These will be passed by spin-flip as in past operation.

In the MR, there are 10 intrinsic and 22 imperfection resonances. The calculated resonance strengths are summarized in Fig. 1. The calculation was performed using 40 sets of random field errors, producing $\langle z_{COD} \rangle=1.0$ mm for the imperfection resonances (bars). The lengths of the bars give the variation of the calculated results, from the minimum to the maximum. The resonance strengths of the intrinsic resonances were calculated assuming an emittance of 7 π μmrad (circles). The measured strengths are plotted with rectangles. The strongest imperfection resonance is $\varepsilon = 2.0\times10^{-2}$ at $\gamma G=18$ (T= 8.48 GeV) in the case of $\nu_z=6.25$. By 5-th, 6-th and 7-th harmonics corrections, the strengths will be reduced. Then, all of the strengths of the imperfection resonances will become less than 1.3×10^{-2}. On the other hand, intrinsic resonances seem not to be sufficiently strong to completely flip the spin according to past experiments. It seems to be better to change the acceleration rate much more slowly. We assume $\alpha = 2.9\times10^{-6}$ instead of $\alpha = 5.8\times10^{-6}$.

FIGURE 1. Depolarizing resonances in the MR.

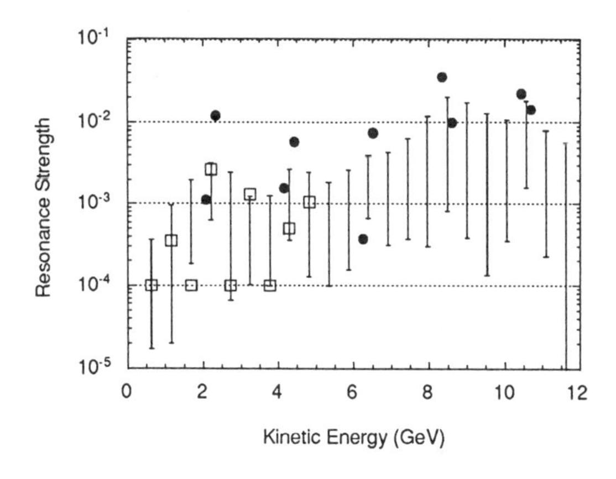

There may be two possibilities for realizing such a snake: solenoid and helical dipole. The required BL's are tabulated in Table 1. To use a helical dipole snake, other bending magnets are necessary for orbit-matching. We studied each of the possibilities, as follows.

To obtain a final polarization of 99.999% of the initial polarization after passing 22 imperfection resonances, the resonance strength of the partial snake ($\delta/(2\pi)$) should always be larger than that of all of the imperfection resonances by 5.3×10^{-3}. To preserve the polarization, a partial snake with 7 degrees is necessary. Employing a solenoid magnet of 1.0 m length, this can be installed in one of the short straight sections of the MR (1.8 m). The focusing function of this solenoid and the coupling between the horizontal and vertical motions are negligible.

TABLE 1. Integrated field of the Partial Snake

angle	%	ε ($\times 10^{-2}$)	BL_{sol} (Tm)	BL_{helix} (Tm)
5°	2.78	1.39	1.34	1.83
6°	3.33	1.67	1.61	2.00
7°	3.89	1.94	1.88	2.17

By using a helical snake as the transverse field snake, the snake axis angle can be changed (9). The strength of the imperfection resonances will be less than 0.8 x 10^{-2}, except for $\varepsilon = 1.3 \times 10^{-2}$ at γG=20. The required angle of the snake is less than that of the solenoid snake, i.e. 5 degrees for the resonances except for γG=20. For γG=20, the phase difference between the snake resonance strength and the imperfection resonance strength can be controlled by changing the snake axis, and we can escape from depolarization with the snake-axis variation of ± 20 degrees at most, as shown in Fig. 2. Using a 1.4-T helical dipole and four 1.4-T bendings, a 5.4 m straight section is necessary (Fig. 3). This can be installed in one of the long straight section.

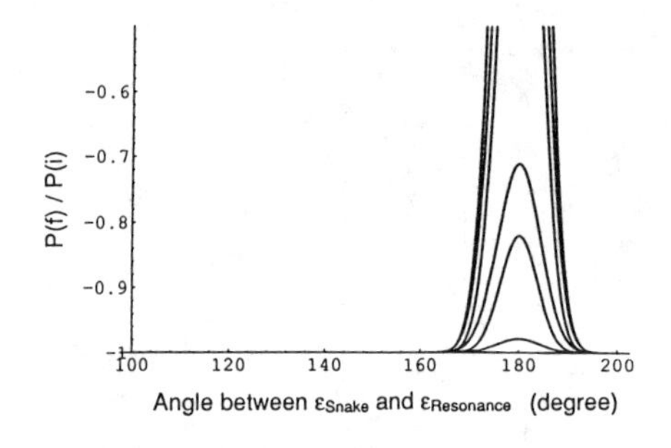

FIGURE 2. Depolarization due to a strong imperfection (snake axis variable, ±20°).

FIGURE 3. Helical partial snake orbit. Spin rotation, 5°; kinetic energy, 12GeV.

The orbit displacement and the focusing due to this helical dipole is large near injection energy. The local focusing due to helical dipole is sufficiently strong to make non-structure intrinsic resonances dangerous due to the destruction of ring symmetry. By ramping the current from 40% to 100%, the maximum orbit displacements and non-structure intrinsic resonances can be reduced to sufficiently small values. This scheme seems to be complicated, especially in manufacturing the helical dipole and replacing two of the four rf cavities at the long straight sections by the snake. However, this makes it possible to test the scheme of a variable snake axis with a helical dipole, which is indispensable for a collider ring with a crossing collision angle, such as the PS Collider.

4. SUMMARY

Vector-polarized deuterons will be accelerated in 1995. No depolarization is expected if we choose the vertical betatron tune to be 6.25. The expected beam intensity and vector polarization are 3-4x10^{10} and ~70%, respectively.

The scheme of polarized proton acceleration has been reconsidered using a partial snake. At the KEK-PS energies, a solenoid is superior due to its compactness. A helical dipole, however, can provide a variable snake-axis angle. This feature will be essential in the future collider ring. By incorporating the depolarizing effect of synchrotron oscillation, it is possible that the necessary field will increase. Another problem is to cross the intrinsic resonances. More studies should be proceeding.

ACKNOWLEDGMENT

The experiment concerning polarized deuteron acceleration is being carried out in collaboration with INS (Tokyo University), KEK, Miyazaki University, Nagoya University, Tokyo Institute of Technology, Tsukuba University and Yamagata University.

The authors would like to thank Drs. T. Iwata, M. Kinsho, Y. Mori and C. Omori for helpful discussions and suggestions.

REFERENCES

1. Mori, Y., "Polarized Deuterons in the KEK 12-GeV Proton Synchrotron," in Proc. of 10th Intern. Symp. on High Energy Spin Physics, 1992, pp. 383–386.
2. Sato, H. et al., Nucl. Instrum. Methods **A272**, 617– 625 (1988).
3. Derbenev, Ya. S. and Kondratenko, A. M., "On the Possibilities to Obtain High Energy Polarized Particles in Accelerators and Storage Rings," in Proc. of High Energy Physics with Polarized Beams and Targets, 1978, pp. 292–306.
4. Froissart, M. and Stora, R., Nucl. Instrum. Methods **7**,.297–305 (1960).
5. Kinsho, M., "A dual-optically-pumped polarized negative deuterium ion source," presented at this conference.
6. Sato, H. et. al., "Upgrade of the Main Ring Magnet Power Supply for the KEK 12GeV Proton Synchrotron," in Proc. of IEEE Particle Accelerator Conference, 1991, pp.908–910.
7. Sakamoto, N. el al ., "Measurement of the vector and tensor analyzing powers for the d+p elastic scattering at 270 MeV," presented at this conference.
8. Haji-Saied, M. et al., Physical Review **C36**, 2010–2017 (1987).
9. Toyama, T., "Compact Snake and Rotator Scheme," in Proc. 10th Intern. Symp. on High Energy Spin Physics, 1992, pp. 433–436.

Polarizing a stored, cooled proton beam by spin-dependent interaction with a polarized hydrogen gas target*

F. Rathmann[1], C. Montag[2], and D. Fick

Philipps-Universität, Fachbereich Physik, 35032 Marburg, Germany

J. Tonhäuser, W. Brückner, H.-G. Gaul, M. Grieser, B. Povh, M. Rall, E. Steffens, F. Stock[3], and Kirsten Zapfe

Max-Planck-Institut für Kernphysik, 69029 Heidelberg, Germany

B. Braun, and G. Graw

Sektion Physik der Universität München, 85748 Garching, Germany

W. Haeberli

Department of Physics, University of Wisconsin-Madison, Wisconsin 53706, USA

* Supported by the Bundesministerium für Forschung und Technologie under various contracts.

1 Department of Physics, University of Wisconsin-Madison, Wisconsin 53706, USA.

2 Now at Deutsches Elektronen-Synchrotron DESY, 22603 Hamburg, Germany.

3 Now at Physikalisches Institut der Universität Erlangen-Nürnberg, 91058 Erlangen, Germany.

We report results from an experiment, which shows, that a stored proton beam can be polarized by spin-dependent attenuation through a high-density polarized hydrogen gas target [1]. Since the total hadronic cross section differs for parallel $(\uparrow\uparrow)$ and antiparallel $(\uparrow\downarrow)$ spins of beam and target protons one of the initially equally populated beam spin states in the unpolarized stored proton beam is depleted more than the other. The method, also referred to as Spin-Filtering, was first proposed by Csonka [2] and could be of interest for the production of polarized antiprotons.

The experiment was carried out at the Heidelberg low energy Test Storage Ring (TSR), shown schematically in Fig. 1. In order to maintain a long beam lifetime, the polarized hydrogen gas target was located in the low-β-section of the TSR. Using multi-turn injection and cooler stacking, protons of 23 MeV with intensities up to 1mA were accumulated in the TSR and then left circulating through the polarized gas target for times from 30 up to 90 min. The remaining beam intensities were 50 - 100 μA. During the filtering process, the target polarization was constant and perpendicular to the ring plane.

The polarized hydrogen storage cell gas target [3] consists of a thin-walled 25 cm long, cylindrical aluminum tube of 11mm diameter, through which the beam passes. Polarized hydrogen atoms, produced

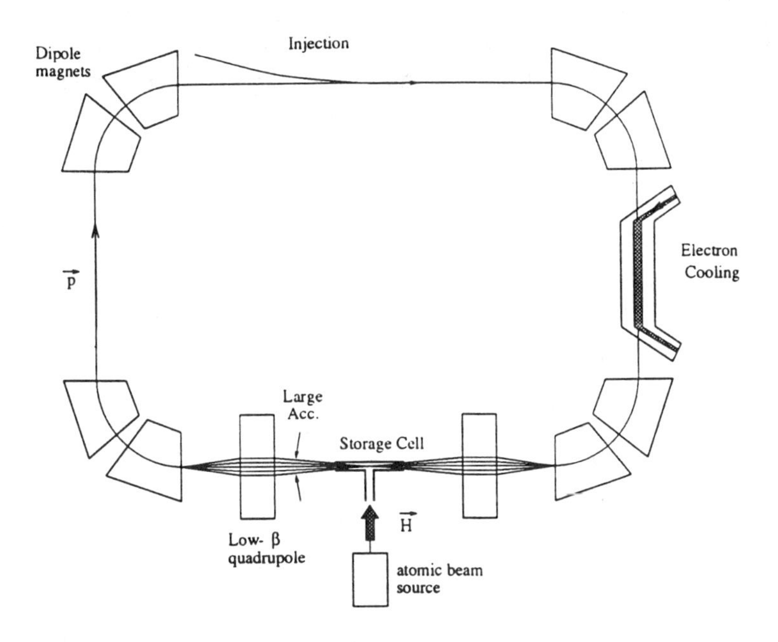

Fig. 1. Schematic diagram of the Heidelberg Test Storage Ring (TSR) with internal gas target. Polarized gas from an atomic beam source is feed into a thin-walled T-shaped storage cell, which is placed on the axis of the ring. The acceptance of the storage ring is increased by low-β quadrupoles, which also reduce the beam diameter and allow to use narrower storage cells of higher target thickness.

by the Heidelberg atomic beam source [4], are injected through a feed tube into the storage cell. The resulting target thickness is increased by cooling the storage cell from room temperature to $T \approx 115$K. The target polarization at that temperature is determined to be $P_t = 0.80 \pm 0.02$, while the target thickness is $n = (5.3 \pm 0.3) \cdot 10^{13}$ H/cm^2.

For low energy pp scattering the total hadronic cross section $\sigma_{tot}(\uparrow\uparrow)$ is smaller than $\sigma_{tot}(\uparrow\downarrow)$, therefore the beam polarization P_b after the filtering process is expected to be parallel to the target polarization P_t during filtering. The beam polarization was analyzed from the $\uparrow\uparrow/\uparrow\downarrow$ count rate asymmetry by reversing the target polarization with a frequency of ≈ 1Hz, making use of the large spin correlation coefficient $A_{xx} = -0.93$ [5]. The beam lifetime $\tau_b \approx 30$min is mainly determined by the Coulomb loss cross section $\Delta\sigma_c$. If one assumes that only those particles are lost, which undergo scattering at angles larger than the acceptance angle of the TSR ($\psi_{acc} = 4.4 \pm 0.5$mrad), a Coulomb loss cross section $\Delta\sigma_c = 6.1$barn is calculated from the measured acceptance angle.

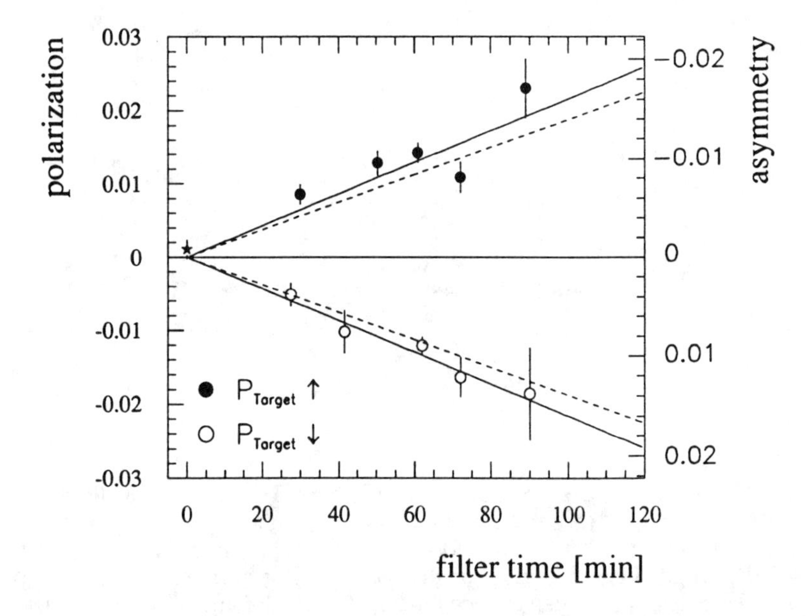

Fig. 2. Beam Polarization (left-hand scale) and asymmetry (right-hand scale) measured after filtering the beam for different times t. The full points refer to fil tering with target polarization up $(\uparrow)$, the open ones to target polarization down $(\downarrow)$. The full lines show a fit to the data assuming an infinite polarization lifetime, resulting in $\tau_1 = (-4637 \pm 228)$min ($\chi^2/$n.d.f. = 1.39). The dashed lines show the result of a recent calculation of the polarizing cross section [6,7].

Figure 1 shows the measured beam polarization for filtering with target spin up $(\uparrow)$ and down $(\downarrow)$ as a function of the filtering time t. As expected, the beam polarization changes sign if the target polarization is reversed. In the absence of depolarizing mechanisms (polarization lifetime $\tau_p = \infty$) the beam polarization after a certain time t is given by $P_b(t) = -tanh(t/\tau_1)$, where τ_1 is the polarization build-up time. The full lines in fig. 1 show a fit to the data assuming $\tau_1 = (-4637 \pm 228)\text{min}$ $(dP_b/dt = 0.0129 \pm 0.0006\,\text{h}^{-1})$. The theoretically expected polarization build-up time, calculated from the spin dependent total hadronic cross section $\sigma_1 = 122\,\text{mbarn}$ [5] is 1.7 times larger than the observed value.

Calculating the polarizing cross section from the measured polarization build-up rate for infinite polarization lifetime gives $\sigma_1 = 72.5 \pm 5.8\,\text{mbarn}$. In a recent calculation, the expected polarizing cross section was deduced (dashed lines in fig. 2) and was found in good agreement with the measured cross section [6,7]. This calculation takes into account three distinct spin-dependent mechanisms that contribute to the polarizing cross section. Particles that undergo scattering at angles larger than the acceptance angle of the ring are lost. It is incorrect to include only the spin dependent total hadronic cross section into this process, since there is a significant contribution from coulomb-nuclear interference terms. A second effect, polarization transfer in p+p scattering also contributes, since protons can be scattered into the acceptance and thus remain circulating in the ring. A third contribution comes from polarization transfer of polarized electrons that are contained in the polarized hydrogen atoms of the gas target onto the circulating protons. The angles in e+p scattering are always much smaller than the machine acceptance, therefore particles which undergo such a collision are not lost. This effect contributes significantly and with the opposite sign than the two other contributions.

A smaller β–function together with a larger acceptance of the TSR could probably lead to a significantly higher beam lifetime τ_b, which then would allow for much longer filtering times, leading to a beam polarization of $P_b = 0.07 - 0.08$.

The described experiment has shown, that at a luminosity of $3.3 \cdot 10^{28}\,\text{cm}^2\text{s}^{-1}$ and a beam polarization of a few percent, even small effects can be measured to a high accuracy. With a beam polarization of a few percent, the FILTEX experiment at LEAR, aiming for the production of polarized antiprotons and the determination of $\bar{p}p$ spin correlations, is feasible.

References:

[1] F. Rathmann et al.; Phys. Rev. Lett. **71** 1379 (1993).
[2] P.L. Csonka; Nucl. Instrum. Methods **63**, 247 (1968).
[3] K. Zapfe; *A High Density Polarized Hydrogen Gas Target For Storage Rings;* contribution to this conference.
[4] F. Stock; *The HERMES-FILTEX Target Source for Polarized Hydrogen and Deuterium*; contribution to this conference.
[5] R.A. Arndt, J.S. Hyslop, and I.D. Roper; Phys. Rev. **D35**, 128 (1987); program SAID (scattering analysis interactive dial up).
[6] H.O. Meyer; Phys. Rev. **E50, 1485** (1994); and *On the Polarization of a Stored Beam,* contribution to this conference.
[7] C.J. Horowitz and H.O. Meyer; Phys. Rev. Lett **72**, 3981 (1994); and contribution to this conference.

On the Polarization of a Stored Beam

H.O. Meyer

Department of Physics, Indiana University, Bloomington, IN, USA

Abstract. Nuclear physics experiments using stored, polarized proton beams are now routine. The stored beam polarization has been found to be quite stable, however, a number of mechanisms have been identified that can lead to a change of the polarization of an orbiting beam. In this contribution I make an attempt to summarize these mechanisms as they are known to date.

INTRODUCTION

Beam from a polarized ion source is accumulated in a storage ring by the usual injection methods. If polarized H^- ions are available, stripping injection can be used. At IUCF, polarized protons (H^+) are injected and accumulated using phase space manipulation in conjunction with electron cooling. Typical fill rates are about 50 μA/min. The beam polarization is typically 70%. When the intensity has reached several hundred μA the stored beam is accelerated to the desired energy, then the experiment starts to acquire data.

Here, we describe how the polarization of a stored beam can be manipulated. This includes change of the direction of the polarization vector, reversal of the sign, and change of the magnitude (both up and down).

Polarization in a ring is described by an equation of motion for the magnetic moment of an orbiting particle. Formal simplicity results when two-component spinors Ψ are used to express the polarization of a (spin-½) beam. The polarization vector is then given by $\vec{P} = <\Psi^+|\vec{\sigma}|\Psi>$, where $\vec{\sigma}$ are the Pauli matrices. An excellent introduction into this formalism can be found in [1]. Spin motion for one complete orbit, starting from a certain point, corresponds to a rotation around an axis $\hat{n}$. In general, $\hat{n}$ depends on the starting point. The polarization component along this direction is unchanged, while the transverse component precesses rapidly and averages to zero over time. If a storage ring would feature only vertical fields (needed to bend the beam), the stable spin direction $\hat{n}$ would be vertical everywhere. The precession angle in one revolution around this direction, the "spin tune ν_s", amounts to $\nu_s = G\gamma$, where γ is the energy-mass ratio of the stored particle and G=1.793 (for protons).

EFFECTS OF NON-VERTICAL MAGNETIC FIELDS

Depolarization Resonances

At certain beam energies the spin motion can be synchronous with non-vertical perturbing fields encountered by the orbiting particle, resulting in a depolarizing resonance. Such fields can arise from non-ideal optics, causing "imperfection resonances" at $\nu_s = m$ (m is any integer), or from the transverse field of the focussing quadrupoles, causing "intrinsic resonances" at $\nu_s = m' + m\nu_z$, where ν_z is the vertical betatron tune. For a comprehensive treatment of depolarizing resonances, see ref.[2]. In an experiment with polarized beam these resonance conditions have to be avoided. In the case of intrinsic resonances this implies a careful control of the machine tune. On the other hand, it is conceivable to make use of a resonance to depolarize the beam for a study of instrumental asymmetries.

Longitudinal polarization

At energies away from an imperfection resonance, the presence of static, non-vertical fields does not affect the magnitude of the beam polarization. However, it does change the stable spin direction $\hat{n}$ away from the vertical direction. This makes experiments possible with beams polarized in directions other than vertical. This effect has been demonstrated in the Indiana Cooler in a study of the effect of a longitudinal solenoid field on the polarization of stored protons [3].

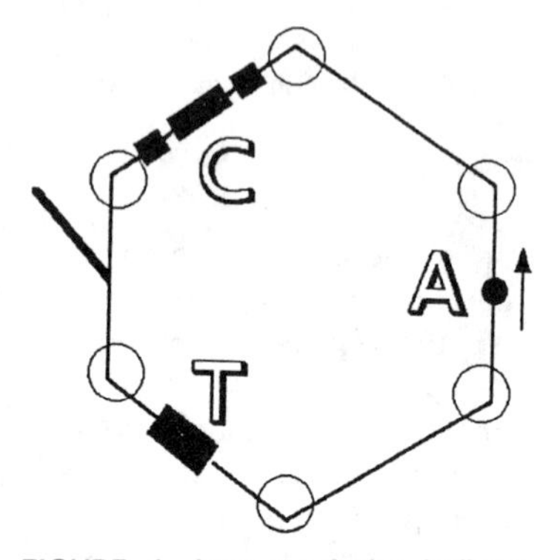

FIGURE 1. Layout of the Indiana Cooler. Shown are the target position (A), the three cooling solenoids (C), and the additional solenoid in the T-section.

In nuclear physics experiments with polarized beam on a polarized target, certain observables require longitudinally polarized beam, even if it is possible to freely choose the target polarization direction. Such experiments are planned for the near future in the A-section of the Indiana Cooler (see Fig.1). For the six-sided Cooler it has been predicted that longitudinal polarization in the target section (A) at any energy can be obtained with two solenoids mounted in symmetric, but non-adjacent straight sections (C, T). In this scheme, one can make use of the three solenoids already present in the C-section (one of these solenoids is needed to confine the cooling electron beam, the others are normally used in a compensating mode, but can be operated with the same sign as the cooling solenoid, resulting in a longitudinal field integral of up to 1 Tm). One additional solenoid with up to 3.5 Tm in the T-section is then sufficient to allow longitudinal polarization in the A-section for proton beams of up to 500 MeV. Because of the limit on the field in C-section, the polarization is not exactly longitudinal for all energies, but never deviates by more than 20% of the maximum value.

Polarization Reversal

Whenever one of the parameters that control the location of a depolarizing resonance is varied adiabatically across the resonance condition, polarization reversal occurs. Such a resonance can be induced by means of an RF solenoid that generates an alternating, longitudinal field that is synchronous to the spin tune. Then, to flip the beam polarization, the RF frequency has to be varied across the resonance frequency at the appropriate rate [4]. An RF solenoid for this purpose has been constructed for the Indiana Cooler [5], and successfully used to demonstrate polarization reversal of a 139 MeV proton beam [6].

One difficulty in the practical use of such a device is the fact that the resonance is accompanied by sidebands caused by the synchrotron motion of the beam particles. The separation between resonance and sidebands is typically 1-2 kHz, or a thousandth of the resonance frequency. To correctly position the frequency sweep requires a careful and time-consuming mapping out of the resonance and sidebands. Recently, it has been shown that the effect of the sidebands can be avoided by debunching the beam just prior to the frequency ramp. A wide frequency sweep (20 kHz) which is guaranteed to contain the resonance frequency can then be used. Repeatedly sweeping back and forth many times in a row allows to study the completeness of one flip (see Fig.2). This development work was carried out in the context of a measurement of spin correlation coefficients in pp elastic scattering at 200 MeV [7]; this experiment now makes routine use of this technique to reduce systematic uncertainties. To be able to reverse the stored polarization also helps the average luminosity since the sign of the injected polarization can always be the same and the beam still stored at the end of the cycle does not have to be discarded.

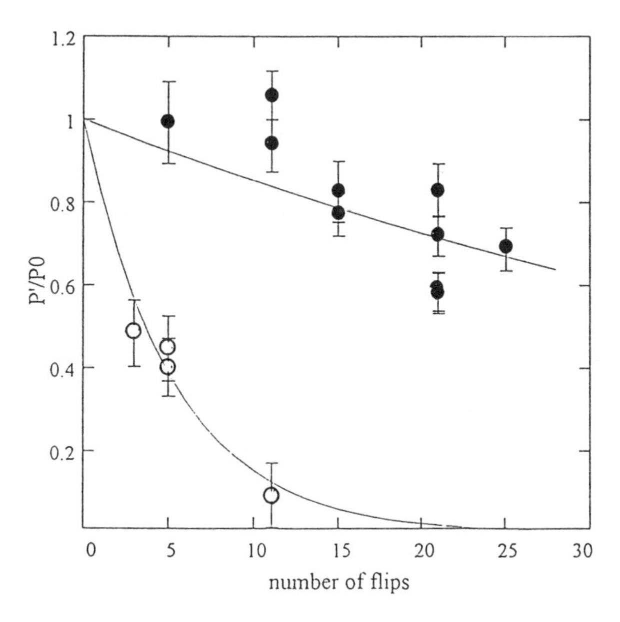

FIGURE 2. Polarization flipping by crossing an induced depolarizing resonance: fraction of polarization of a 200 MeV beam remaining after multiple flips [7]. The solid symbols (upper curve) were obtained when the beam was debunched prior to flipping by turning off the ring RF. The efficiency is $(98.4 \pm 0.4)\%$ per flip. The lower trace (open symbols) illustrates the same measurement with a bunched beam (resonance with sidebands), corresponding to an efficiency of $(83 \pm 3)\%$.

EFFECT OF A POLARIZED INTERNAL HYDROGEN TARGET

Definition of the Polarizing Cross Section

With a 23 MeV proton beam it has recently been demonstrated at the Test Storage Ring (TSR) in Heidelberg the beam becomes polarized when it circulates through a polarized internal target [8]. A detailed study of the mechanisms that gives rise to such a polarization has recently been published [9].

Consider a proton beam orbiting with frequency f_R in a ring, interacting with a polarized, internal, atomic hydrogen target of polarization P_T and thickness d. We denote the time-dependent beam polarization by $P_B(t)$. The so-called polarizing cross section $\hat{\sigma}$ is then defined by

$$\frac{dP_B}{dt} = (1-P_B^2)\ f_R\ d\ P_T\ \hat{\sigma}\ . \tag{1}$$

The sign of $\hat{\sigma}$ determines whether the change of beam polarization is in the direction of the target polarization, or opposite. We have found [9] that there are three distinct mechanisms that contribute significantly to $\hat{\sigma}$, namely spin-dependent beam removal ($\hat{\sigma}_R$), and polarization transfer from polarized target protons, or electrons, to the stored protons ($\hat{\sigma}_S$, $\hat{\sigma}_E$, respectively). Here, for simplicity, we assume a ring with only vertical fields and discuss only vertical polarization.

Spin-dependent Beam Removal ($\hat{\sigma}_R$)

A beam proton that scatters from a target proton is removed from the stored beam if the scattering angle is larger than the ring acceptance angle Θ_{acc}. Because of the spin-dependence in p+p scattering, the removal probabilities of beam protons in the ↑ and ↓ spin states differ. The transverse polarizing cross section follows from integrating $\frac{1}{2}(A_{nn}+A_{mm})(d\sigma/d\Omega)$ over solid angle for polar angles from Θ_{acc} to $\pi/2$. The A_{ii} are the usual spin correlation coefficients. At 23 MeV and an acceptance angle of $\Theta_{acc}=4.4$ mrad, for example, one finds $\hat{\sigma}_{R\perp}=83$ mb. This number is different from the (hadronic) spin-dependent, transverse total cross section of $\frac{1}{2}\Delta\sigma_T=122$ mb at this energy, because the integration does not start at $\Theta=0$, and because there is a significant contribution from Coulomb-nuclear interference terms.

Polarization Transfer in p+p Scattering ($\hat{\sigma}_S$)

Beam protons that scatter from the target by angles less than Θ_{acc} remain in the ring, carrying out betatron oscillations around the equilibrium orbit. Eventually, the amplitude of these oscillations is damped by electron cooling. During the scattering event, the target polarization may be transferred to the projectile. The resulting change $<\Delta P>$ of the polarization of the scattered particle approaches zero for small angles, $\theta \to 0$. The cross section, however, grows more rapidly, in fact, one finds that the product $<\Delta P>(d\sigma/d\Omega)$ diverges with $1/\theta^2$, as $\theta \to 0$. As a consequence, small-angle scattering does affect the polarization of a stored beam, even though the limiting scattering angle Θ_{acc} is very small (typically 1 to 10 mrad). To obtain the polarizing cross section, we note that for small angles, the spin transfer coefficients K_{ii} are nearly the same as the spin correlation coefficients A_{ii}. We then evaluate $\hat{\sigma}_S$ like $\hat{\sigma}_R$, but with opposite sign of the integrand. The integration in this case ranges from θ_{min} to Θ_{acc}. The lower bound θ_{min} on the scattering angle is a result of the atomic screening of the Coulomb potential (at 23 MeV, θ_{min} is about 0.2 mrad).

Polarization Transfer in e+p Scattering ($\hat{\sigma}_E$)

A recent calculation [10] shows that a polarized electron target can significantly affect the polarization of a stored proton beam because of a relatively large contribution from the interference between the hyperfine and the Coulomb amplitude. Note that the scattering angle of protons from electrons is much smaller than the typical machine acceptance, thus $\hat{\sigma}_E$ does not depend on Θ_{acc}. The contribution to the (transverse) polarizing cross section is

$$\hat{\sigma}_E = -\left[\frac{2\pi\alpha^2(1+G)\,m_e}{p^2 m_p}\right]\cdot C_0^2\cdot\left(\frac{v}{2\alpha}\right)\cdot\sin\left(\frac{2\alpha}{v}\ln(2pa_0)\right) . \qquad (2)$$

Here, p is the center-of-mass momentum, α is the fine structure constant, G=1.793 is the anomalous moment of the proton, and v is the lab velocity of the projectile. The second half of eq.2 arises from Coulomb distortions (for more detail, see [10]). The minus sign indicates that the induced proton polarization is *opposite* to the direction of the electron spin.

Comparison with Experiment

So far the only experimental evidence of polarizing a stored beam by target interaction is from an experiment carried out with a 23 MeV proton beam on a vertically polarized hydrogen target at the TSR in Heidelberg [8]. From the known orbit frequency and target thickness the measured polarization buildup can be expressed as a polarizing cross section of $\hat{\sigma}_{exp}$=(63±3)mb (see eq.1). The measured acceptance angle of the ring was Θ_{acc}=(4.4±0.5)mrad.

With the parameters of the TSR experiment one finds $\hat{\sigma}_R$=83mb for the predicted contribution from spin-dependent beam removal, and $\hat{\sigma}_S$=52mb for the effect of polarization transfer in pp scattering. The effect of beam removal, $\hat{\sigma}_R$, as a function of Θ_{acc} is shown as a dashed line, and the combined, hadronic effect, $\hat{\sigma}_R+\hat{\sigma}_S$, as a dash-dot line in Fig.3. In the TSR experiment, the target was prepared in a pure, atomic spin state with protons and electrons polarized in the same direction. Thus, the effect of the target electrons is opposite to that of the protons, and one finds $\hat{\sigma}_E$= -70mb, independent of Θ_{acc}. The polarizing cross section from all three sources, $\hat{\sigma}_R+\hat{\sigma}_S+\hat{\sigma}_E$ is shown as a solid line in Fig.3, together with the experimental result. As can be seen, the expected polarizing cross section is in excellent agreement with the experiment.

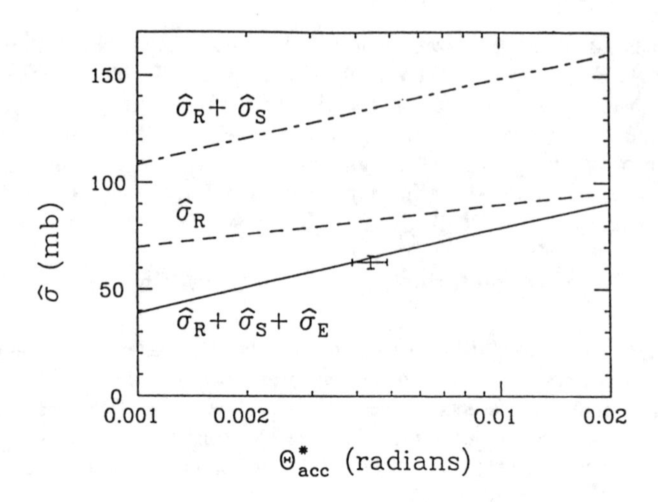

FIGURE 3. Polarizing cross section for a polarized hydrogen target and a 23 MeV proton beam as a function of the laboratory acceptance angle Θ_{acc}. The three lines represent a calculation and are dicussed in the text. The datum shown is the result of the TSR experiment [8].

DEPOLARIZATION BY TARGET SCATTERING

If care is taken to avoid the critical energies and machine tunes, the beam polarization is remarkably stable, however, from the combined data from several IUCF experiments that were carried out with polarized protons, there now seems to be some indication of a small, but finite depolarization rate, (-1/P)(dP/dt), in the presence of an internal target.

In the following we offer a possible explanation of this effect. Consider a proton on the equilibrium orbit with its magnetic moment $\vec{\mu}$ along the z direction (vertical). Since its emittance ϵ is zero, its stable spin direction $\hat{n}$ is also in the z direction. Assume that this proton scatters from a target nucleus by an angle θ_z in the vertical plane. It then acquires an emittance of $\epsilon = \theta_z^2 \cdot \beta_z$, where β_z is the vertical betatron function at the point of scattering. This suddenly changes $\hat{n}$ away from vertical by an angle α, which depends on the strength of the closest intrinsic resonance and the particle emittance ϵ, and thus, on the scattering angle θ. The magnetic moment which is still along z now precesses around $\hat{n}$, and only the projection of $\vec{\mu}$ onto $\hat{n}$ is preserved. The remaining vertical component of the moment is thus $\mu \cos^2(\alpha)$. This single scattering event consequently leads to a loss of vertical polarization of

$\Delta P/P=\alpha^2$. The polarization loss due to many scatterings is cumulative. In order to calculate the depolarization per unit time, the Rutherford scattering rate times θ_z^2 has to be integrated over angle up to the maximum angle at which the particle is still contained in the ring. The same scattering process is responsible for the beam life time. Hence, the depolarization rate can be related to the beam lifetime τ. The result is

$$\frac{1}{P}\frac{dP}{dt} = -\frac{E^2}{\delta^2}\frac{A}{\epsilon_{ref}} C\,\tau^{-1} . \qquad (3)$$

where E is the strength of the closest intrinsic resonance, evaluated at an emittance ϵ_{ref}, δ is given by $\delta=|G\gamma - \nu_z \pm m|$ with m an integer which is chosen to make δ as small as possible, A is the machine acceptance, $C=\ln(\theta_{max})-\ln(\theta_{min})$ is the so-called Coulomb logarithm (C~6), and τ is the beam lifetime. Within this model, the polarization lifetime $\tau_P=-P/(dP/dt)$ is proportional to the beam lifetime τ. For the Cooler intrinsic resonance near 200 MeV ($G\gamma=7-\nu_z$), the resonance strength is $E=7\ 10^{-3}$ for an emittance of $\epsilon_{ref}=25\pi 10^{-6}$ m. The fractional vertical tune was measured as $\nu_z=0.12\pm0.005$, or $\delta=0.055\pm0.005$, and the machine acceptance was estimated as $A=8\pi 10^{-6}$ m.

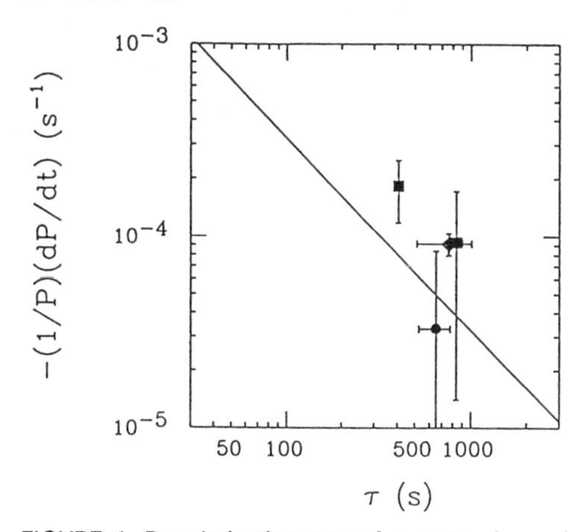

FIGURE 4. Depolarization rate of a proton beam in the presence of a target as a function of the beam lifetime τ. The data and the solid line are described in the text.

The calculation does *not* assume that all scattering events take place in the target. In fact, calculating the beam lifetime from the known target thickness gives values that are much larger than was actually observed. Thus, there are important contributions from residual gas scattering. Then some scattering events take place in regions where the dispersion is large. In this case, also the particle emittance can change due to energy loss when scattering from electrons. This effect is not included in the calculation.

The depolarization calculated with these parameters is shown as a solid line as a function of the beam lifetime in Fig.4. Also shown are depolarization rates -P/(dP/dt) observed as a byproduct during several experiments requiring polarized protons. The cycle length over which the polarization could be observed ranged between 400 s and 700 s. The data include measurements with an H_2 gas of varying thickness (squares) and He gas (circle) in a buffer cell target at 200 MeV [7]. For completeness, a measurement at 300 MeV with a H_2 gas jet target (diamond) is also shown. A dedicated experiment to study this depolarization effect is under way at IUCF.

CONCLUSIONS

The dynamics of the polarization in a storage ring provides nuclear physics experiments with important tools. These include longitudinal polarization at the target location, and the ability to quickly reverse the sign of the beam polarization or set its magnitude to zero. Some of these features are already in routine use, others need to be developed further to be useful in nuclear physics measurements.

The presence of internal targets also affects the polarization of the stored beam. In this contribution we have mentioned four distinct mechanisms by which this may happen; as we gain more experience it is likely that one finds more such mechanisms.

Target-induced polarization changes are very small. Nevertheless, they may become measurable if they are cumulative over the time interval between refills (during which a stored proton typically completes 10^9 revolutions). The continuing study of such effects is clearly of importance to future precision experiments (such as measurements of symmetry violations) with storage rings. Such experiments have been proposed, and it has been pointed out that ring experiments are less susceptible to many of the systematic errors of corresponding conventional experiments. However, polarization effects like those that are discussed here and that only occur in storage rings represent a class of new possible systematic uncertainties to precision experiments in rings. On the other hand, it is conceivable that one will find physically interesting observables that are too small to measure conventionally but become accessible because they are of the kind that accumulates in successive orbits.

REFERENCES

[1] B.W. Montague, Phys. Rep. **113**, 1 (1984).

[2] S.Y. Lee, Phys. Rev. **E47**, 3631 (1993).

[3] A.D. Krisch et al., Phys. Rev. Lett. **63**, 1137 (1989).

[4] M. Froissart and R. Stora, Nucl. Instr. Meth. **7**, 297 (1960).

[5] B.S. van Guilder, Ph.D. thesis, University of Michigan, 1993.

[6] R. Phelps, contribution to this conference, and D.D. Caussyn et al., University of Michigan preprint HE 94-17, 1994.

[7] Measurement of $\vec{p}\vec{p}$ Spin Correlation Parameters at the Indiana Cooler, experiment CE35, spokesman W. Haeberli, Univ. of Wisconsin.

[8] F. Rathmann et al., Phys. Rev. Lett. **71**, 1379 (1993).

[9] H.O. Meyer, Phys. Rev. **E50**, 1485 (1994).

[10] C.J. Horowitz and H.O. Meyer, contribution to this conference, and Phys. Rev. Lett. **72**, 3981 (1994).

A Review of High Energy Polarimetry, With a view toward RHIC

David G. Underwood
High Energy Physics
Argonne National Laboratory
Argonne Il 60439

ABSTRACT

We present four physics processes that have measured asymmetries at high energy. Some problems and some advantages of utilizing each in a polarimeter are discussed. For each we give references to examples of experimental setup and data.

INTRODUCTION

The energy dependence of the analyzing power for the processes we discuss here is shown in a general way in reference 1. The analyzing power from interference of strong amplitudes in elastic or quasi-elastic scattering falls as roughly the inverse of the beam momentum. The others, Inverse-Primakoff, Coulomb-Nuclear, and large-x pion inclusive appear to be energy independent in the relevant ranges.

The possibilities can be evaluated with the following parameters in mind:

+ Analyzing power vs Energy,
+ Usable Luminosity,
+ Quality factor, $A^2 N$,
+ The continuum from Online Scaler readout to Difficult analysis,
+ Absolute vs Relative Analyzing power,
+ Internal to an Accelerator vs in External Beamline.

These parameters are discussed in terms of the advantages and difficulties for each polarimeter.

P P Elastic and P N elastic

There are a number of problems which make elastic scattering a more difficult process to utilize at increasingly high energy. The analyzing power typically peaks near a t, momentum transfer squared, of 0.15 $(GeV/c)^2$, and falls off at higher t. The analyzing power falls roughly as 1/ p lab, so that at 100 GeV/c, the analyzing power is down to about 1 %. The next problem is that the increasing number of pions produced at high energy is a serious background to both the forward and the recoil detection. The angular separation between the forward scatter and the beam becomes small so that, for example, at 300 GeV in Fermilab experiment 61, a septum was required in a Cherenkov counter at 100 meters from the target.[2] One way around these difficulties which was used in an AGS external beamline up to 22 GeV was to utilize magnetic spectrometers for the forward and recioil arms, and to accept somewhat larger angle forward scatters and higher t at the highest energies.[3]

Elastic scattering has been used effectively for internal polarimeters in accelerators even in the face of difficulties like limited space and a quasi-elastic background from CH targets. One example is at KEK for energies up to 12 GeV.[4]

Another recent example is the polarimeter in the AGS for the partial Siberian snake tests.[5] In this case, the forward scatter was inaccessible, both because the small forward angle put it inside the beam pipe at some energies, and because of high background rates. The recoil arms were designed to use an angle-energy correlation to separate elastics from quasi-elastics, with the protons stopping in a plastic scintillator for the energy measurement. A veto counter was used to eliminate penetrating pions. A previous version of the polarimeter utilized range information in a limited way. This version has a very great improvement , almost an order of magnitude, in eliminating the quasi elastics from carbon in the fishline target, but it is not perfect. A carbon fiber was added to the target flipping mechanism so that measurements were made with C and CH2 targets, and a carbon subtraction could be made from the data.

There were some interesting technical aspects to this polarimeter. The energy-angle correlation in the recoil proton measurement was done with memory lookup units which could handle several tens of millions of events per second peak rate. The new target flipping mechanism could place the target in the beam in 50 milliseconds while the fishline was spooling at a meter per second to reduce heat and radiation damage. A baler for spooling the fishline contributed to the overall reliability during the two week run.

COULOMB-NUCLEAR INTERFERENCE

The Coulomb-nuclear interference polarimeter utilizes the interference between the spin flip part of the Coulomb elastic amplitude, and the large imaginary part of the strong amplitude. From the known amplitudes in the p p case, one can calculate the analyzing power and the t value in elastic scattering where the maximum occurs. The maximum is about 4%, almost independent of energy, at -t of 0.005 $(GeV/c)^2$. Calculations, and to some extent data, indicate similar parameters for p-Carbon elastic.[6]

One needs good kinematic constraints to separate the elastic scattering from diffraction dissociation, which has a similar low energy recoil and fast forward particles. A precise t measurement is needed, for example +- 0.001 at t of -0.003 $(GeV/c)^2$. In the fixed target experiments done to date, primarily Fermilab experiment E704,[7] the main experimental problem was to isolate and measure the recoil. For this reason, a momentum and angle measurement on the forward scattered particle was also done to help in the kinematic constraint. The forward particle is at an angle of about 1.5×10^{-4} radian at 200 GeV/c, so that it was within the beam phase space for the measurements done. A particular problem with a solid sensitive target is that the beam particles deposit energy in the target with magnitude comparable to that deposited by the recoil. For some of the Fermilab data, a Stilbene target was used, which has a different scintillation pulse time dependence for fast minimum ionizing particles, such as the beam, than for slow stopping protons, such as the recoil.

There is some interest in using this process inside an accelerator with a gas jet target. for example at RHIC. The kinematics of the forward particle are then the same as in the unpolarized experiments to measure the real part of the elastic amplitude. These experiments use Roman pot detectors in the accelerator lattice. The recoil particle detecton would be free of carbon background and energy deposition by the beam. However, precise measurements of the 30 to 100 MeV/c momentum recoil may be difficult in an environment with many pions. Many pions are produced each 110 ns when a beam bunch crosses the target. There are also beam-gas backgrounds from the accelerator.

It should also be possible to use the Coulomb -nuclear method with elastic scattering of the two proton beams in a collider, with Roman pots on either side of the interaction region. One has to consider the reduced luminosity possible at the high beta parameter needed to detect the scatters, in light of the number of events needed at the low analyzing power. High beta means the beam is almost parallel trajectories and not tightly focused.

INVERSE PRIMAKOFF

Primakoff originally proposed using the connection between S-channel and T-channel scattering as a way to measure the electromagnetic coupling, radiative width, of short lived particles. This is of interest to us because we can relate a body of low energy data on polarized photoproduction of baryon resonances to the high energy process of producing these resonances in the Coulomb field of a nucleus. For example, the asymmetry in the Coulomb dissociation of high energy protons into a proton plus pi-zero with invariant mass near 1350 MeV/c ^{2} is directly related to the asymmetry in photoproduction on a polarized proton target with photons of 400 to 500 MeV. This has been done at 200 GeV in Fermilab Experiment E704, and it works, albeit with some experimental difficulties.[7, 8]

The primary experimental difficulty in this type of polarimeter is the separation of Coulomb dissociation from diffraction dissociation. The parameter which is used for this separation is the slope of the t distribution. For Coulomb events the distribution has a slope determined primarily by experimental resolution, which has to be about e - 2000 t in order to do the separation. The slope for strong diffraction is approximately $e^{-10 A^{2/3} t}$, which is e-400 t for diffraction off a lead nucleus. The direction of the proton-pi-zero effective mass must be determined to about +- 3 x 10^{-5} radian at 200 GeV/c. This requires a rather good spectrometer for the proton and a photon detector with good position and energy resolution for the pi-zero.

In principle, an analyzing power of 90% is possible independent of energy. In practice the analyzing power is dominated by the fraction of background, which can be determined well, but not eliminated. Thus, in E704, the measured analyzing power was about 30%, and substantial analysis effort was required, at the level of a PhD thesis. This may be practical for a one-time absolute polarization measurement, but not for a beam tuning polarimeter where information is required quickly and repeatedly.

PION INCLUSIVE

Large asymmetries in the production of very forward pions from polarized proton beams have been observed over a large range of energy. There are data at 6 and 12 GeV from the ZGS, and at 200 GeV in E704.[9] There are also pion production asymmetries with polarized targets at Cern at 30 GeV and at Serpukhov at 70 GeV. The 200 GeV data are well characterized in terms of xf and pt, and will be useful at similar energies. One can speculate that the effects are largely energy independent, however there is insufficient data to connect all the energies in a useful parameterization to show this. As shown in reference 9, the pi + and pi - asymmetries are large, almost linear in xf, and approximately mirror images of each other. The pi-zero asymmetry is smaller.

The RHIC Spin Collaboration is considering this process for an internal polarimeter in the RHIC machine. Measurements are desired on a repetative basis several times per hour. Pi - production appears to have several advantages. Pi+ would require the separation of protons at forward angle and high momentum whereas the anti proton flux in this region is negligible. The Kaon flux could either be considered a dilution factor or could possibly be eliminated with a Cherenkov. Pi-zeros would require something like a lead glass array very close to the beam pipe and considerable event reconstruction effort.

The optimum pion momentum has been determined by finding the derivative of a parameterization for $A^2 N$, the quality factor which minimizes error bars. The cross secton was parameterized as a function of xf above pt of 0.7 Gev/c and the asymmetry was also so parameterized. The result is that xf of about 1/2 is optimum. The apparatus must select pions of 1/2 the proton momentum and pt of greater than 0.7 GeV/c. This may be done with magnetic spectrometers using either limited aperture or very fast electronics to correlate hodoscopes, so that primarily events of interest are scaled.

References

1) Summary of Polarimeter Session, D. G. Underwood, High Energy Spin Physics, AIP Conf. Proc. 187, Particles and Fields 37, 1352 (1988)

2) R.V. Kline et al. Phys. Rev. D 22 553 (1980)

3) F. Khiari et al. Univ of Michigan Report UM-HE-87-36 (1988)

4) Internal Polarimeters for the Polarized Proton Beam at the KEK PS, H. Sato et al. High Energy Spin Physics, AIP Conf. Proc. 187, Particles and Fields 37, 1355 (1988)

5) Preservation of Proton Polarization by a Partial Snake, H. Huang et al. to be published Phys Rev Lett Nov, 1994

6) A. K. Akchurin et al. Phys Rev D 48, 3026 (1993)

7) D. Grosnick et al. NIM A 290, 269 (1990)

8) D. C. Carey et al. Phys Rev Lett 64, 358 (1990)

9) D. L. Adams et al. Phys Lett B 264, 14 (1991)

Spin flipping a stored vertically polarized proton beam with an RF solenoid

R.A. Phelps, B.B. Blinov[d], C.M. Chu, E.D. Courant[e], D.A. Crandell, W.A. Kaufman, A.D. Krisch, T.S. Nurushev, L.G. Ratner[e] and V.K. Wong[f]

Randall Laboratory of Physics
University of Michigan
Ann Arbor, Michigan 48109-1120

D.D. Caussyn[a], Ya.S. Derbenev[b], T.J.P. Ellison[c], S.Y. Lee, T. Rinckel, P. Schwandt, F. Sperisen, E.J. Stephenson and B. von Przewoski

Indiana University Cyclotron Facility
Bloomington, Indiana 47408-0768

C. Ohmori

Institute for Nuclear Study
University of Tokyo
Tanashi, Tokyo 188, Japan

Abstract. A recent experiment in the IUCF cooler ring studied the spin flip of a stored vertically polarized 139 MeV proton beam. This spin flip was accomplished by using an RF solenoid to induce an artificial depolarizing resonance in the ring, and then varying the solenoid's frequency through this resonance value to induce spin flip. We found a polarization loss after multiple spin flips of about $0.00 \pm 0.05\%$ per flip and also losses for very long flip times. This device will be useful for reducing systematic errors in polarized beam-internal target scattering asymmetry experiments by enabling experimenters to perform frequent beam polarization reversals in the course of the experiment.

Interest is growing in polarized beam experiments in storage rings such as IUCF (1), RHIC (2), and the Tevatron (3). Any scattering asymmetry experiment which uses a polarized beam must perform frequent polarization reversals of the beam to reduce systematic errors in the measured asymmetry due to efficiency and acceptance mismatch of the detectors. In a storage ring one confronts the problem of how to reverse the polarization of a beam stored for long periods of time with a minimum of polarization loss from each reversal. In a circular accelerator ring with no Siberian Snakes, each proton's spin precesses around the vertical field of the accelerator's dipole magnets; however, any horizontal magnetic fields can depolarize the beam. This depolarization occurs when the spin precession frequency, f_s, satisfies the resonance condition

$$f_s \equiv f_c \nu_s = f_c(n + m\nu_y), \tag{1}$$

where n and m arc integers; f_c is the protons' circulation frequency; the vertical betatron tune, ν_y, is the number of vertical betatron oscillations during each turn around the ring; and the spin tune, ν_s, is the number of spin precessions during each turn around the ring. The imperfection resonances occur when m = 0, while the first-order intrinsic resonances occur when m = ±1.

With no Siberian Snake, the spin tune is proportional to the proton's energy

$$\nu_s = G\gamma, \tag{2}$$

where γ is the Lorentz energy factor and $G = 1.792847$ is the proton's anomalous magnetic moment. Combining equations 1 and 2, one arrives at the resonance condition

$$G\gamma = n + m\nu_y, \tag{3}$$

Spin flip can be achieved by slowly varying one side of equation 3 so that the resonance condition is passed through at a slow enough rate with a strong enough non-vertical magnetic field. This has been accomplished in accelerators by passing through a strong imperfection depolarizing resonance (4–9). The resonance condition given in equation 3 is then passed through as $G\gamma$ is slowly increased. In a storage ring where the energy, and hence $G\gamma$, is constant, an alternative is to vary the rate at which the protons experience non-vertical fields by introducing an RF magnet with a non-vertical field, the resonance condition is then

$$G\gamma = n \pm f_{rf}/f_c, \tag{4}$$

where f_{rf} is the applied frequency, which is slowly varied through this resonance condition to flip the spin.

We installed an RF solenoid at the IUCF cooler ring to test this spin flipping technique. The RF solenoid, polarimeter and the Cooler Ring's operation with polarized protons were discussed earlier (10–14). Before spin flipping could be accomplished, we first found the depolarizing resonance frequency of the 139 MeV vertically polarized proton beam. This procedure is outlined in a previous paper (12). The resonance frequency was $f_{rf} = 1,800230 \pm 10$Hz with a half width half max resonance width δ of 227 ± 9Hz. The RF solenoid field amplitude was 0.0014 T·m and the beam kinetic energy was fixed at 139 MeV for all results reported here.

According to theory (15), if the applied frequency starts out very far to one side of the resonance frequency, and is changed linearly in time to go through the resonance very far to the other side, the beam polarization should be reversed with an efficiency which increases as the RF field strength is increased, or the rate of change of frequency is decreased. We tested this by injecting vertically polarized protons into the cooler ring, then turned on the RF magnetic field 1.75 kHz below the resonance frequency, and linearly ramped the frequency to 1.75 kHz above the resonance frequency. In each run, only the RF ramp time was changed. The ramp time was varied from 1 msec to 1 sec. Polarization was measured after the end of the frequency ramp. A plot of the vertical polarization after spin flip vs. ramp time is given in Fig. 1. It is seen that complete spin flip occurs for ramp times of 20 msec or greater. It should be noted that the injected beam polarization was about 75%. This data shows that spin flip of a stored vertically polarized beam is possible using this RF ramp technique.

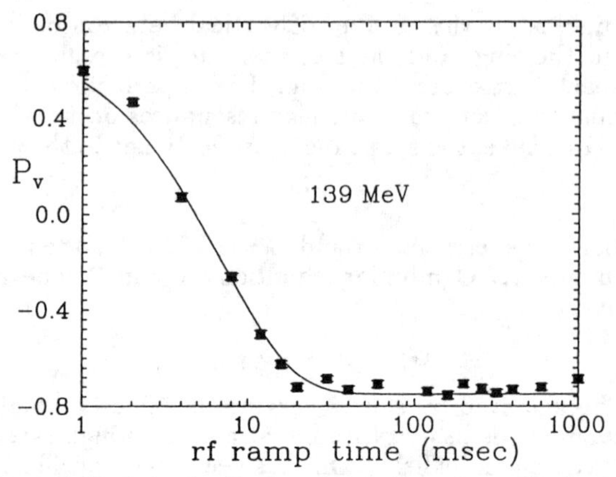

FIGURE 1. The vertical polarization P_v plotted against the time taken for the RF ramp through a 3.5 kHz range around the resonance for a single ramp.

The curve in Fig. 1 is a χ^2 fit of the Froissart-Stora formula (15), modified for the case when the energy is fixed and the resonance frequency is ramped. This formula for the vertical polarization P_v after the ramp in terms of the initial vertical polarization P_o, the resonance half width half max. δ, the frequency ramp interval Δf of 3.5 kHz and variable ramp time Δt is

$$P_v = P_o\Big\{2\exp[-(\pi\delta)^2\Delta t/\Delta f] - 1\Big\}. \tag{5}$$

Note from equation 5 that for a slow enough ramp rate $\Delta f/\Delta t$ or large enough resonance width, $P_v/P_o \to -1$, so the spin will be completely flipped. The fit parameters were $P_o = 0.741 \pm 0.011$ and $\delta = 225 \pm 4$Hz. Note that this width δ is in good agreement with the 227 Hz width obtained earlier when the depolarizing resonance frequency was found.

One expects a small loss of polarization each time the spin is flipped. In a storage ring where the spin will be flipped many times, these losses would compound from flip to flip. That is, if p is the fraction of polarization left after one flip, after n flips one would expect a polarization proportional to p^n. Note that flipping the spin many times before polarization measurement will allow for increased precision on the determination of p. We measured p for a specific ramp rate and strength. We ramped the RF many times up and down through the resonance at a fixed ramp time of 160 msec and the same 3.5 kHz frequency ramp as in the single flip case before measuring the beam polarization. Between each ramp there was a 40 msec period where the RF was fixed off resonance. A plot of the vertical polarization vs number of ramps is shown in Fig. 2, where the maximum number of flips was 200. The curve is a 2 parameter χ^2 fit of the form $P_v = P_o p^n$, where n is the number of flips, $P_o = 0.74 \pm 0.03$ is the injected polarization and the fractional surviving polarization per ramp $p = 0.996 \pm 0.001$. We therefore lost about 0.4% polarization per flip with this ramp time, resonance strength and frequency range.

Finally, we measured the polarization loss per flip as a function of the RF

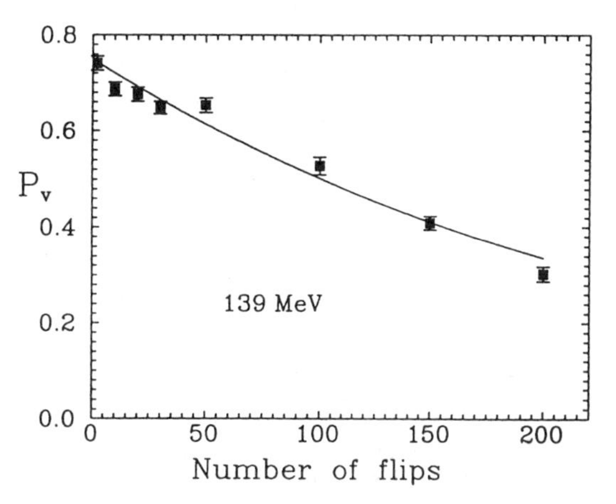

FIGURE 2. The vertical polarization P_v plotted against the number of times the RF ramped through a 3.5 kHz range around the resonance. The curve is a fit of the form $P_v = P_o p^n$ where n is the number of ramps.

ramp time by fixing the number of flips to be 50 and varying the ramp time. We again measured the polarization after the 50 flips were completed, and with the same 3.5 kHz ramp range, varied the ramp time up to 1000 msec per ramp. A plot of the vertical polarization vs the ramp time for one of the 50 ramps is shown in Fig. 3. The polarization increased as the ramp time increased for times up to 60 msec, and then started a slow decline with increasing ramp time. Analysis gives a value for p, the fraction of polarization surviving each flip, of 1.0000 ± 0.0005 for the 60 msec ramp rate. Hence to within our experimental uncertainty, no polarization was lost over 50 flips for this ramp rate. The loss for longer ramp times might be due to the synchrotron motion of the beam. The 139 MeV beam energy had a 4.1 kHz modulation. When one ramps the RF very slowly, the resonance is actually passed through many times since $G\gamma$, hence the RF resonance frequency, also has some modulation. This phenomenon has been studies for the case of passage through a depolarizing resonance by acceleration (16). The data indicates that there is an optimum ramp rate for spin flip, which is long enough to achieve full spin flip, but not so long that other depolarizing effects can occur.

This experimental data shows that spin flipping a vertically polarized beam can be accomplished with very little loss in polarization, making it a useful technique in machines where polarized beam is stored for long periods of time. It also shows that in the case of spin flipping a vertically polarized beam when no snakes are present, there is an optimum rate at which to vary the applied RF for the most efficient spin flip. This spin flip procedure will be essential to reduce the systematic errors in stored polarized proton beam scattering asymmetry experiments.

This research was supported by grants from the U.S. Department of Energy and the U.S. National Science Foundation.

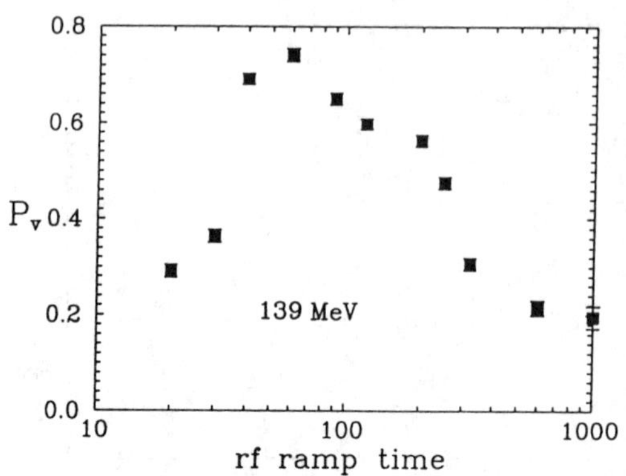

FIGURE 3. The vertical polarization P_v plotted against the RF ramp time through a 3.5 kHz range around the RF resonance for one of the 50 ramps performed.

REFERENCES

[a] Now at: Department of Physics, University of Michigan.
[b] Also at: Department of Nuclear Engineering, University of Michigan.
[c] Now at: Energy Conversion Devices Inc., Troy, MI.
[d] Also at: Moscow State University, Moscow, Russia.
[e] Also at: Brookhaven National Laboratory.
[f] Also at: Office of the Provost, University of Michigan at Flint.

1. Haeberli, W. "Polarized Hydrogen targets for Storage Rings", in New *Nuclear Physics with Advanced Techniques*, eds F.A. Beck, *et al.*, (World Scientific, Singapore 1992).
2. RHIC Spin Proposal, Beddo, M. *et al.*, Brookhaven National Laboratory proposal, 1992 (unpublished).
3. Acceleration of Polarized Protons to 120 and 150 GeV in the Fermilab Main Injector, University of Michigan Report, March 1992 (unpublished); Progress Report: Acceleration of Polarized Protons to 1 TeV in the Fermilab Tevatron, University of Michigan Report, August 1994 (unpublished)
4. Khoe, T. *et al.*, Part. Accel. **6**, 213 (1975).
5. Cho, Y. *et al.*, Proc. 1976 Workshop on High Energy Physics with Polarized Beams and Targets, AIP Conf. Proc. **35**, 396, ed. M.L. Marshak, (AIP, New York 1976)
6. Laclare, J.L. *et al.*, J. Phys. (Paris) Colloq. **46**, C2-499 (1985).
7. Grorud, E. *et al.*, Proceed. of the 1982 Symposium on High Energy Spin Physics, AIP Conf. Prod. **95**, 407, ed. G. M. Bunce, (AIP, New York 1983).
8. Sato, H. *et al.*, Nucl. Instrum. Methods Phys. Res. **A272**, 617 (1988).
9. Hiramatsu, S. *et al.*,Proceed. of the 8th Int. Symposium of High Energy Spin Physics , AIP Conf. Proc. **187**, 1077, eds. K. J. Heller,(AIP, New York 1988).
10. Krisch, A.D. *et al.*, Phys. Rev. Lett. **63**, 1137 (1989).
11. Goodwin, J.E. *et al.*, Phys. Rev. Lett. **64**, 2779 (1990).
12. Anferov, V.A. *et al.*, Phys. Rev. **A46**, R7383 (1992).
13. Baiod, R. *et al.*, Phys. Rev. Lett. **70**, 2557 (1993).
14. Minty, M.G. *et al.*, Phys. Rev. **D44**, R1361 (1991).
15. Froissart, M. and Stora, R. Nucl. Instrum. Methods **7**, 297 (1960).
16. Hiramatsu, S. *et al.*,Proceed. of the 8th Int. Symposium of High Energy Spin Physics , AIP Conf. Proc. **187**, 1436, eds. K. J. Heller, (AIP, New York 1988).

Technique for Rotating the Spin Direction at RIKEN

H. Okamura[a], N. Sakamoto[a], T. Uesaka[a], H. Sakai[a], K. Hatanaka[b], A. Goto[c], M. Kase[c], N. Inabe[c], and Y. Yano[c]

[a] *Department of Physics, University of Tokyo, Bunkyo-ku, Tokyo 113, Japan*
[b] *Research Center for Nuclear Physics, Ibaraki, Osaka 567, Japan*
[c] *The Institute of Physical and Chemical Research (RIKEN), Wako, Saitama 351-01, Japan*

Abstract. A 270-MeV polarized deuteron beam is provided with its polarization axis controlled to an arbitrary angle at the RIKEN accelerator research facility. The spin rotation method which has been established incorporated with the accelerator technique is described as well as the deuteron polarimeter at intermediate energies.

INTRODUCTION

The RIKEN Ring cyclotron, though it was originally designed as a heavy-ion accelerator, can accelerate protons and deuterons up to 210 and 270 MeV, respectively. Expecting that use of a polarized deuteron beam would provide unique opportunities for the study of the spin-dependent physics at intermediate energies, the construction of the polarized ion source started in the end of 1990.[1] The original design of the ion source was close to that of HIPIOS at IUCF[2] which was based on the source at TUNL,[3] but many modifications have been made during the development. The present level of performance is 140 μA at the ion source exit with 80% polarization of the ideal value. The 270-MeV polarized deuteron beam, though its intensity is limited to 1 μA by the radiation protection, has served for research programs since 1993.

To make the most of its unique feature at intermediate energies, a technique has been developed to freely rotate the spin direction. The importance of the control of the spin-quantization axis is obvious. For the deuteron-induced reactions, the analyzing powers A_y and A_{yy} are frequently measured but they are sensitive mainly to the spin-orbit potential. The A_{xx} and A_{xz}, on the other hand, contain physically more meaningful information such as the tensor interaction and the D-state component of the deuteron wave function. These quantities have to be measured by rotating the spin-quantization axis to an appropriate angle.

Usually, the beam is injected with its polarization axis parallel to the cyclotron magnetic field to avoid Larmour precession. A common method to rotate the *proton* spin is to employ the superconducting solenoids. Since the spin of 500-MeV proton, for example, is rotated by 90° with respect to the beam direction in the dipole magnet with the bending angle of 33°, the combination of two 2-T·m solenoids at both sides of the bending magnet allows one to rotate the spin direction to an arbitrary angle. For the 270-MeV deuteron, however, the method is not practical because a 550°-bending magnet is required as well as 6.5-T·m solenoids.

An alternative method is to use the Wien filter which is commonly employed at the Van de Graaff accelerators. The Wien filter, however, is operated only at very low energies, preferably downstream of the ion source. Since the beam is injected and accelerated with its polarization axis inclined to the cyclotron magnetic field, it is crucial to maintain the single-turn extraction so as not to reduce the polarization amplitude. It is also important to efficiently measure the polarization of the *accelerated* beam in order to determine the final direction of the spin.

MONITORING SYSTEM OF THE SINGLE-TURN EXTRACTION

At the RIKEN facility, the single-turn extraction is available both for the AVF (injector) and Ring cyclotrons, yet the drift of the main magnetic field is sometimes troublesome. A monitoring system of the single-turn extraction has been developed, which consists of a single-bunch selector and a non-destructive beam monitor. The single-bunch selector is a simple electric beam deflector which is installed downstream of the ion source and is very compact owing to the low beam energy (7.5 keV).[4] The beam monitor is installed at the exit of each cyclotron and utilizes the secondary electrons which are produced by a thin wire scanning across the beam and are detected by a micro-channel plate.[5] During the experiment, the beam with duration of 1 μs is swept out in every 100 μs by the single-bunch selector. The turn-mixing ratio is obtained in *real time* from the yields of adjacent beam bunchs at the rising edge of the signal from the single-bunch selector.

DEUTERON POLARIMETER AT INTERMEDIATE ENERGIES

Although the deuteron polarimeter at intermediate energies is not well established, the most probable candidate for the polarization analyzer is the elastic scattering on the hydrogen target. A clean trigger is obtained by the coincidence detection of the scattered deuteron and the recoiled proton with simple plastic scintillation counters. A polyethylene target is conveniently used.

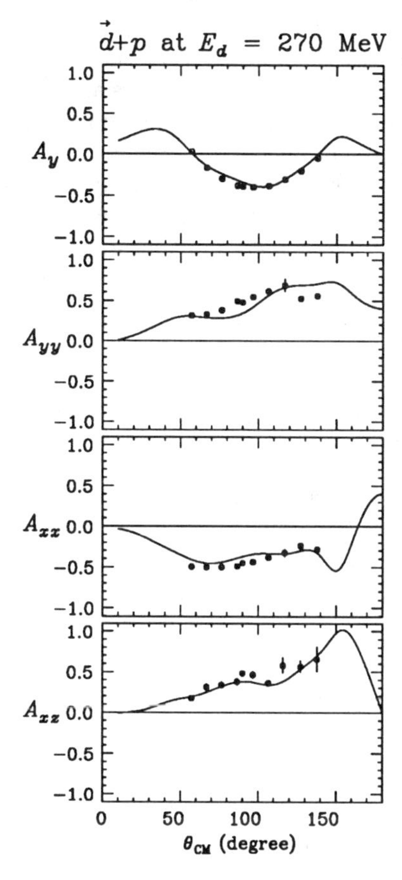

FIGURE 1. Angular distributions of the complete set of the analyzing powers, A_y, A_{yy}, A_{xx}, and A_{xz}, for the $\vec{d}+p$ elastic scattering at 270 MeV. Preliminary result of Fadeev calculation is also shown by solid curves.[7]

The analyzing powers, A_y, A_{yy}, and A_{xx}, of the $\vec{d}+p$ scattering at 270 MeV have been calibrated against the beam polarization at the exit of the AVF cyclotron which was measured by using the $^{12}C(\vec{d},p)^{13}C_{gnd}$ reaction. It is assumed that no depolarization occurs during the acceleration by the Ring cyclotron when the spin-quantization axis is parallel to the cyclotron magnetic field. For the completeness, the A_{xz} has been also measured by using the Wien filter. The result is shown in Fig. 1.[6] All the analyzing powers have flat distributions with fairly large amplitudes, indicating the reaction is useful as a polarization analyzer. It should be noted that the spherical tensor analyzing power T_{20} crosses zero at θ_{cm}=90°. There the sum of count-rates at left, right, up, and down counters becomes proportional to the beam intensity irrespective of the beam polarization and consequently the polarimeter also serves as a convenient beam-intensity monitor.

PERFORMANCE

The Wien filter is tuned by monitoring the polarization at the exit of the Ring cyclotron. Fig. 2 shows the dependence of p_x on the Wien filter angle, where the beam is purely vector-polarized and the spin direction is rotated by 90° so that it lies in the horizontal plane. The statistical errors are well within the size of the data points. The data is neatly fitted by a sine function and its amplitude shows the polarization degree, 77%, is not reduced. Once the tuning is completed, the amplitude of polarization, as well as the beam intensity, is not reduced irrespective of the spin direction. Several physics experiments utilizing this feature have already been performed and are presented in this conference.[6,8–10]

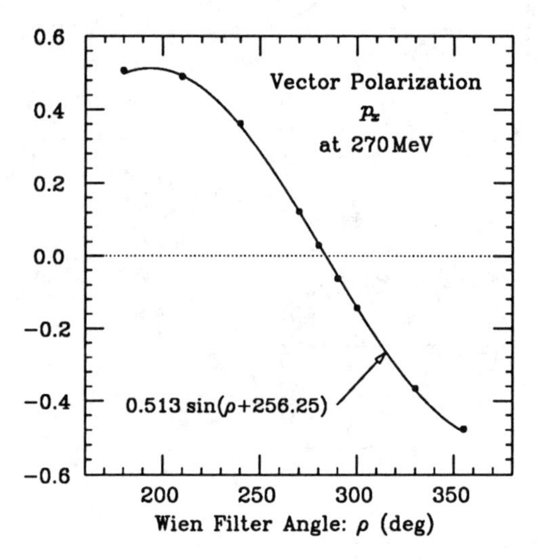

FIGURE 2. Dependence of the horizontal vector polarization p_x on the Wien filter angle. See text for details.

REFERENCES

1. Okamura, H. *et al.*, AIP Conf. Proc. **293**, 84 (1994); contribution to this conference.
2. Derenchuk, V. P., *et al.*, AIP Conf. Proc. **293**, 72 (1994); contribution to this conference.
3. Clegg, T. B., AIP Conf. Proc. **187**, 1227 (1989).
4. Inabe, N. *et al.*, RIKEN Accel. Prog. Rep. **27**, 133 (1993).
5. Kawama, T. *et al.*, RIKEN Accel. Prog. Rep. **27**, 135 (1993).
6. Sakamoto, N. *et al.*, contribution to this conference.
7. Koike, Y., private communication.
8. Okamura, H. *et al.*, Nucl. Phys. A577 (1994) 89c; contribution to this conference.
9. Uesaka, T. *et al.*, RIKEN Accel. Prog. Rep. **27**, 40 (1993); contribution to this conference.
10. Ishida, S. *et al.*, contribution to this conference.

EXTRACTION AND TRANSFORMATION OF PROTON BEAM AT *RHIC* USING BENT CRYSTALS

S.B.Nurushev

IHEP, Protvino, Russia

O.E.Krivosheev, Yu.L.Pivovarov and A.P.Potylitsin

Nuclear Physics Institute, Tomsk Polytechnic University, Russia

The effect of transformation of polarization of relativistic protons using extraction and deflection by bent crystal is investigated by means of computer simulation. The 250 GeV/c proton beam from $RHIC$ can be deflected to an angle $\Theta_p = 3$ mrad with simultaneous transformation of initial longitudinal polarization to transverse polarization.

The channeling of relativistic nuclei in the crystals let avoid the central nucleus-nucleus collisions and let study various electromagnetic interactions of relativistic nuclei with target nuclei [1]. In particular, the electromagnetic interaction of high-energy nuclear (proton) beam with bent crystals leads to the possibility of deflection of the beam to rather large angles with respect to initial beam. The various applications of bent crystals to the high energy physics, including extraction of the beams of relativistic protons and nuclei, are considered in [2]. In particular, it was demonstrated the possibility to apply the bent crystals for multiturn extraction of 500 GeV protons from SIS [3] beam halo, using High-Frequency Noise for effective deflection of beam particles to the crystal front surface.

In connection with future experiments at $RHIC$, we suggest to consider the possibilities to use the bent crystals at this accelerator. The principal questions for analysis are:

a) Is it possible to direct to the crystal about 10% of the beam, or during the short time or during beam life time (about 10 hours), under condition that the parameters of the main beam will not be distorted ?

b) What is the deflection efficiency of these 10% , in dependence on $RHIC$ parameters and parameters of the proposed crystals ?

c) What are the spin rotation angles for deflected protons ?

When one knows the parameters of the beam interacting with the crystal, the problem of relativistic particles motion in a bent crystal and spin rotation during deflection , can be solved using computer simulation method [4,5].

As is known [6], when the relativistic particles is deflected by an external electrical field to an angle Θ_p, the spin of the particle rotates to the angle Θ_S, according the relation:

$$\Delta\Theta_S = [\frac{(g-2)}{2}\frac{\gamma^2-1}{\gamma} + \frac{\gamma-1}{\gamma}]\Delta\Theta_p \quad , \tag{1}$$

where g - is the gyromagnetic relation (for our case $g = 2 \times 2.793$), $\gamma = E/Mc^2$ - is the relativistic factor, E - is the energy and Mc^2 - is the proton mass. At $RHIC$ energies $\gamma \simeq 250$, therefore the spin rotation angle is large unless the small Θ_p. The typical deflection angles for the protons with these energies could be of order of $1-10$ $mrad$, and corresponding spin rotation angles Θ_S could be of order of $\Theta_S = \pi/2$. That means, that extracted by the bent crystal proton beam having initial longitudinal polarization, will have after deflection the transverse polarization.

The deflection angle of the protons is expressed through the bend radius R and crystal length L by the simple relation:

$$\Theta_p = L/R \quad , \tag{2}$$

if the angle of incidence relative to the crystallographic planes is less than the Lindhard's critical angle and if the crystal length is less than the dechanneling length and less than the free path between successive central nuclear collisions.

For the case $\Theta_S \to \pi/2$ we performed the computer simulations of distribution function $N(\Theta_p, \Theta_S)$ of relativistic protons with initial momentum 250 GeV/c and different initial angular and momentum spread. It was suggested that the beam is deflected by (110) bent planes of Si crystal, the bend radii are $R = 8$ m and $R = 7$ m, the crystal lengths are $L = 2$ cm and $L = 2.3$ cm. The calculated spin rotation angles at this energy are about 1.2 rad and 1.57 rad, whereas the beam deflection angles are about 2.5 $mrad$ and little bit greater than 3 $mrad$.

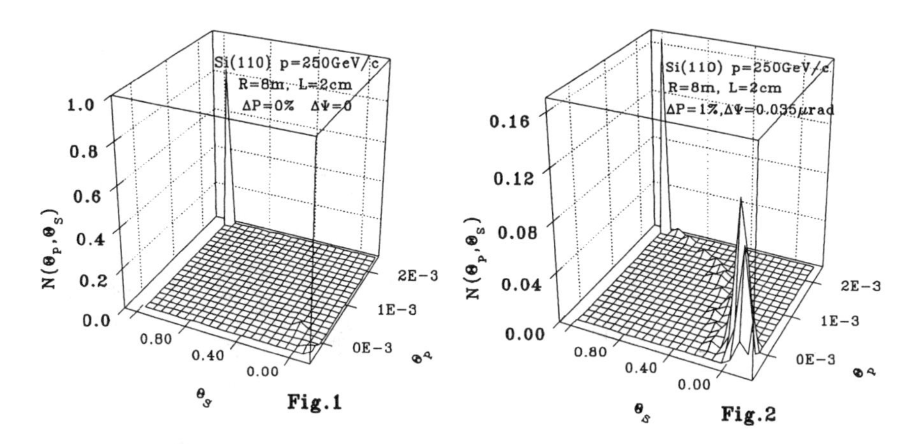

Fig.1 shows the normalized to unity distribution function of protons $N(\Theta_p, \Theta_S)$ for the case when the spread over initial angle of incidence and initial momentum vanishes, $R = 8\ m$, $L = 2\ cm$. Practically all protons experience in this case the front capture: the channeled fraction is about $\sim 80\%$ of the initial beam and has the spin rotation angle about $\sim 1.2\ rad$.

Fig.2 shows the similar distribution function for the case when the initial angular spread is of order of $3.5 \times 10^{-5}\ rad$ (the critical channeling angle is $\psi_L \sim 1 \times 10^{-5}$), and initial momentum spread is of order of $\Delta p/p = 0.01$, $R = 8\ m$, $L = 2\ cm$. It is necessary to notice, that while the channeling fraction decreased in Fig.2 more than 10 times as compared with Fig.1, the width of distributions over deflection angles and over spin rotation angles remained practically the same as in Fig.1.

Fig.3 shows the two-dimensional distribution function $N(\Theta_p, \Theta_S)$ for the case when the angular spread of initial proton beam is about $1.4 \times 10^{-4}\ rad$, and the momentum spread is about $\Delta p/p = 0.03$, $R = 8\ m$, $L = 2\ cm$. As it follows from Fig.3, the dechanneled fraction forms the broad distribution after the crystal, whereas the deflected fraction still has very narrow distribution both over deflection angles and spin rotation angles, unless the total amount of deflected particles is of order of 1% as compared with the initial beam. The protons, deflected inside a crystal (volume capture or volume deflection) formed the number of peaks in accordance with approximate linear dependence between proton deflection angle and it's spin rotation angle.

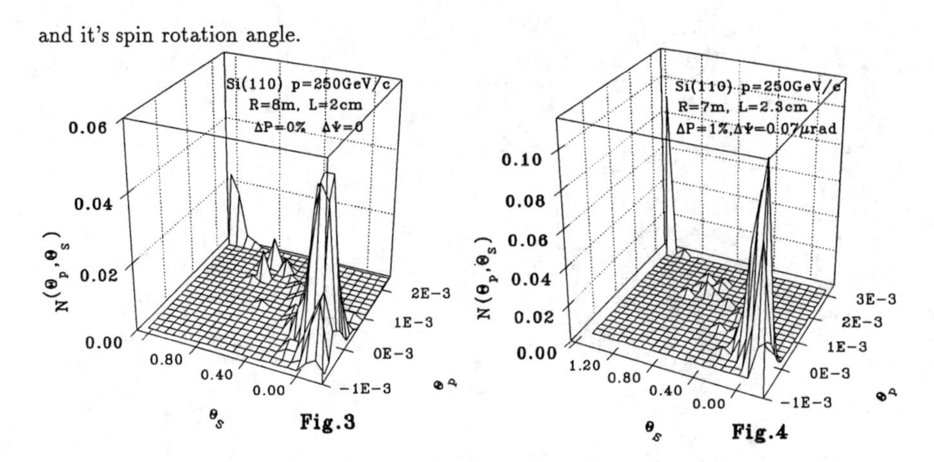

Fig.3 **Fig.4**

Fig.4 shows the two-dimensional distribution function $N(\Theta_p, \Theta_S)$ for the case when the bend radius is $R = 7\ m$, $L = 2.3\ cm$, the angular spread of initial proton beam is $\Delta\Psi = 7 \times 10^{-5}\ rad$ and the momentum spread is about $\Delta p/p = 0.01$. As it follows from Fig.4, when the bend radius is decreased down to $R = 7\ m$ and the crystal length is increased up to $L = 2.3\ cm$ as compared with Fig.3, we got the transverse polarized ($\Theta_S = \pi/2$) deflected to the angle $\Theta_p \simeq 3\ mrad$ proton beam with intensity about few percents of primary ones.

In a real experimental situation one should expect that having the deflection efficiency of order of $\epsilon \simeq 10^{-4}$ and the beam halo intensity of order of $10^{10}\ p/s$, one can prepare the deflected beam of transverse polarized protons with intensity up to $10^6\ p/s$, which is sufficient e.g. for the "fix-target" experiments and continuous monitoring of the polarization degree of the collider beam. This extraction method does not perturb the main $RHIC$ beam and simultaneously prepares the beam of extracted protons which has small angular divergence both in momentum and spin spaces. The data obtained using computer simulations show the possibility to use the bent crystals at $RHIC$ to prepare the "secondary" beam of transverse polarized protons with an intensity up to $10^6\ p/s$ and angular spread of order of $10^{-4}\ rad.$

Since the deflected beam is governed by electrical field of a crystal plane and influences only multiple scattering on electrons inside a crystal channel, no depolarization is expected. As is known [2], the radiative resistance of the usually used bent crystal allows doses up to

10^{18} p/cm^2 without the distortion of channeling properties of a crystal.

This method of proton extraction from $RHIC$ beam halo can be used:

a) For polarimetry of the basic longitudinal polarized $RHIC$ beam, without it's perturbation (using the asymmetry of elastic scattering of deflected transverse polarized proton beam).

b) For independent "fix-target" experiments using transverse polarized protons.

c) The internal crystal with goniometer might be useful for quick measurements of polarized (unpolarized) beam phase space distribution.

In conclusion, we point out that this method can be applied to extraction of heavy relativistic ions. As is known from Dubna experiments [7] and [2], P.45, the dechanneling length was the same as for 9 GeV/c protons and 72 GeV/c ^{16}O nuclei and the critical channeling angle for heavy ions is only by a factor $(Z/A)^{1/2}$ less than those for protons. The calculation of extraction of relativistic ^{197}Au nuclei using bent crystal is in progress and will be published in a separate paper.

---

[1] Pivovarov Yu.L et al., Nucl.Phys., 1990, V.A509, P.800.

[2] Proceedings of All-Union Meeting "Problems of Applications of Channeling Effects in High Energy Physics", IHEP, Protvino, 1991

[3] Taratin A.M. et al. Report SSCL 28/91-15. - Texas, 1991

[4] Krivosheev O.E., Proceedings of All-Union Meeting "Problems of Applications of Channeling Effects in High Energy Physics", IHEP, Protvino, 1991, P.41

[5] Krivosheev O.E., Izv. Akad. Nauk, Ser.Fiz., 1994, V.58, No 1, P.106

[6] Lyuboshitz V.L., Yad.Fiz., 1980, V.31, P.986.

[7] Bel'zer L.I., Bodyagin V.A., Vardanian I.N. et al, Pis'ma Zh.Exp.Teor.Fiz., 1987, V.46, P.303.

Electromagnetic Field Requirements for a Lamb-Shift Spin-Filter Polarimeter

C.D. Roper[a], T.B. Clegg[b] and A.J. Mendez[b]

*Triangle Universities Nuclear Laboratory, Durham, NC 27708-0308**
and Duke University, Durham, NC 27708-0308
and University of North Carolina at Chapel Hill, Chapel Hill, NC 27599-3255

Abstract. A polarimeter based on the spin-filter cavity is being developed to measure the absolute polarization of low energy deuterons (1 keV) and protons (.5 keV) in real time. Successful polarimetry requires an axial magnetic field uniform to ± .05 mT over the central region of the spin-filter cavity. Computer calculations of the magnetic fields were used to design a system of coils and magnetic flux return to meet the uniformity requirement with minimal disturbance of the present polarized ion source. Off-line measurements of the axial magnetic fields produced by the coil/flux return system agreed within the experimental error with the computer calculations. The uniformity attained was ± .05 mT over a 13 cm region.

INTRODUCTION

The Atomic Beam Polarized Ion Source (ABPIS) (1) produces polarized proton and deuteron beams for a variety of nuclear physics experiments at Triangle Universities Nuclear Laboratory (TUNL). Most of the experimenters rely on nuclear reaction polarimeters to tune the radio-frequency transitions of the ABPIS and monitor beam polarizations during experiments. Polarimeters based on atomic physics are much more efficient, particularly at low energies, and provide a measurement of the absolute beam polarization. A source polarimeter is being developed which is based on the atomic "three-level interaction" phenomenon of the n=2 level in hydrogen and deuterium atoms. This polarimeter, referred to as the spin-filter polarimeter (SFP), is based on the nuclear spin-filter, a specially designed rf cavity used in the Lamb-shift polarized ion source developed at Los Alamos National Laboratory (LANL) (2) and used for many years at TUNL (3). Accurate polarimetry using this method requires that the polarimeter be integrated into the existing ABPIS immediately after the charge-exchange region. Computer modeling and calculations of the electromagnetic fields were used to design a system that would meet the stringent field requirements of the spin-filter polarimeter, without disturbing the ABPIS operation.

The spin-filter is a rf cavity cut into four sectors. Two opposing sectors are biased to produce a static electric field transverse to the beam axis. The cavity resonates at 1609 MHz in the TM_{010} mode producing a longitudinal rf electric field. The $2P_{1/2}$ state of hydrogen (deuterium) has a short lifetime (τ =1.6 ns) since the transition to the $1S_{1/2}$ ground state is an allowed electric dipole transition. However, the $2S_{1/2}$ state is metastable (τ = .14 s) because the transition to the ground state is forbidden. Metastable atoms can be forced to decay to the ground state by applying an electric field to couple the $2S_{1/2}$ state with the $2P_{1/2}$ state. This effect, known as Stark-effect quenching, depends on the strength of the applied field. In the presence of a uniform axial magnetic field, the Zeeman effect causes the $2S_{1/2}$ state to split into an α–state ($m_J = +1/2$) and a β–state ($m_J = -1/2$), and the $2P_{1/2}$ state to split into an e-state ($m_J = +1/2$) and a f-state ($m_J = -1/2$). As the magnetic field approaches 57.5 mT, the energy difference of the β and e levels becomes zero, and the energy difference between the α and e levels matches the applied rf frequency. The static electric field couples the β–state with the e-state, and the rf electric field couples the α–state with the e-state. Static and rf electric fields applied simultaneously induce a resonant α-e-β three-level interaction between levels with the same m_I. This interaction preserves the α–state population for the selected m_I, while the atoms in non-resonant m_I states are quenched to the $1S_{1/2}$ ground state. Therefore, metastable deuterium atoms with m_I = +1,0,-1 can be selected by tuning the magnitude of the magnetic field to 56.5, 57.5, 58.5 mT respectively.

The SFP measures hyperfine populations of the states of $2S_{1/2}$ metastable atoms produced during cesium charge-exchange of low energy deuterons (1 keV) and protons (.5 keV). The ratio of the hyperfine state populations of the atoms reveals the nuclear spin polarization of the negative ions exiting the charge-exchange region. The selected metastable atoms exiting the spin-filter cavity drift downstream into a region of strong static electric field where they are quenched to the ground state by the Stark effect. The photons from these transitions are detected by a phototube. The population of each hyperfine state is revealed by the peaks in the photon intensity as a the magnetic field strength is varied. In order to resolve the peaks corresponding to each m_I, the magnetic field must be uniform to $\pm$ .05 mT. Preliminary tests using the spin-filter cavity and coil system from the decommissioned Lamb-shift polarized ion source have shown that the nuclear polarization of ion beams from the ABPIS can be measured in real time ($\leq$ 1 s) (4).

SPIN-FILTER POLARIMETER DESIGN

The ABPIS consists primarily of a molecular dissociator, a sequence of rf transition units and sextupole magnets, an atomic ionizer, and a cesium oven for charge-exchange of positive ions. A large axial magnetic field (~150 mT) is required in the cesium charge-exchange region of the ABPIS to maintain the nuclear spin polarization during charge-exchange. The SFP should be installed within this magnetic field region to reduce changes in the hyperfine state populations caused by the passage of the metastable atoms through large magnetic field gradients. Locating the spin-filter in this region also insures maximum metastable beam flux.

The original TUNL spin-filter cavity used a 53 cm long assembly of solenoidal coils designed to provide a 20 cm long region where B = 57.5 ± .02 mT. Although providing more than adequate field uniformity, this design is impractical for the ABPIS due to its length. The system of coils and flux return added to the ABPIS must be as short as possible to minimize loss of beam intensity, to maintain access to the ionizer and cesium oven for maintenance, and to avoid major modifications of the ABPIS required for length additions greater than ~30 cm. The coil/flux return system must also meet the uniformity requirement, and the presence of stray magnetic fields from the charge-exchange region makes it more difficult to fulfill this requirement.

The POISSON/SUPERFISH group of codes (5) was used to model various coil arrangements and magnetic flux return configurations. The POISSON calculations of the axial magnetic field for the final coil/flux return design (Fig. 1) show the magnetic field to be uniform to ± .05 mT for a 13 cm long region. This design consisted of ten water-cooled coils grouped into two end banks of three coils and one middle bank of four coils. The coil banks were enclosed in an iron box consisting of two iron end plates connected by iron bars for magnetic flux return. The 2.5 cm thick iron plate separating the charge-exchange region from the spin-filter cavity region was necessary in order to reduce the stray fields in the spin-filter region. POISSON calculations showed that the magnetic field uniformity was critically dependent on the location of the center bank of coils. Therefore, the center bank was supported on threaded rods which enabled the center coil bank position to be varied in order to compensate for changes in the stray fields from the charge-exchange region. The coils and flux return added 30.5 cm to the length of the source.

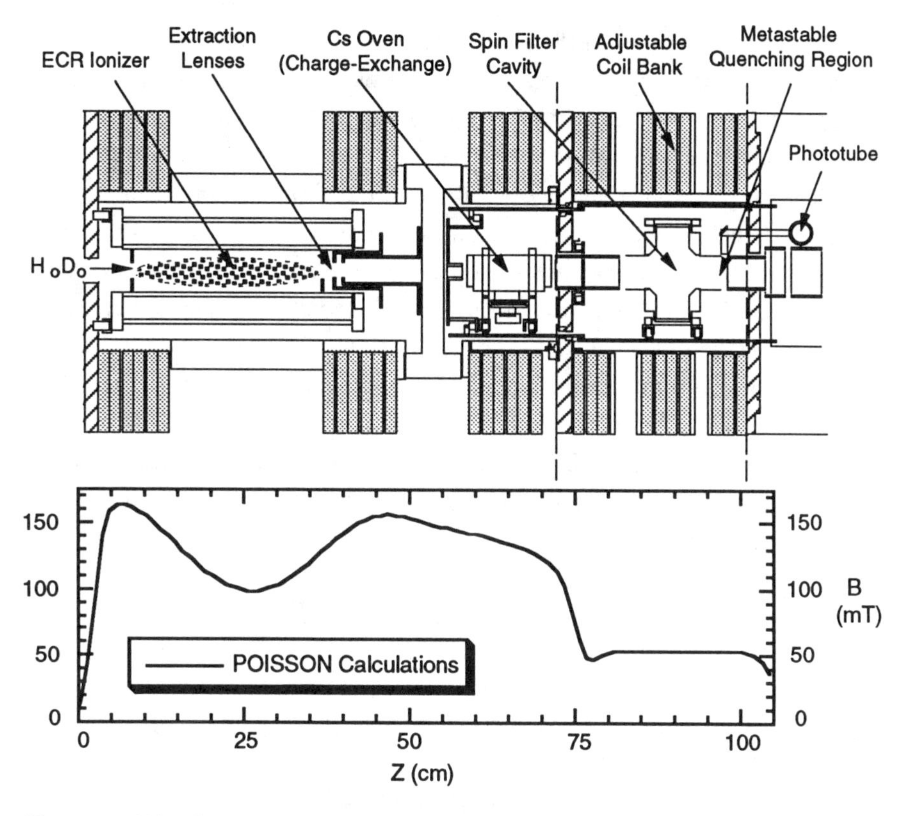

Figure 1. Side view of the Spin-Filter Polarimeter to be installed on the polarized ion source. The system of coils and flux return added to the source is indicated by the dashed lines. The POISSON calculations show the axial magnetic field produced at the axis of this system. The uniformity is ± .05 mT for 13 cm long region.

The coils and magnetic flux return system to be added to the ABPIS was assembled off-line in order to measure the axial magnetic field along the beam axis. The measurements were made using a Hall probe stepped along the axis of the coil/flux return system. This coil/flux return setup was modeled using the POISSON codes so that comparisons could be made between the measurements and POISSON calculations. The experimental measurements of the magnitude of the magnetic field agreed with the POISSON calculations to within the total Hall probe accuracy of ± .8 mT. The measured shape of the magnetic field distribution along the axis also agreed with the POISSON calculations to within the measured error of ± .05 mT. The magnetic field uniformity attained was ± .05 mT for 13cm along the axis.

The existing ABPIS electrostatic lens system was modified to accommodate the spin-filter polarimeter. An electrostatic lens located between the cesium oven and the spin-filter cavity shields the metastable atoms from ground potential to prevent quenching as they drift toward the spin-filter cavity. This lens and the spin-filter cavity will be at the cesium oven potential during polarimetry measurements; otherwise, the lens voltage and the spin-filter voltage will be tuned for maximum ion beam intensity. The acceleration gap between the spin-filter cavity and a downstream lens will be used to quench the metastable atoms exiting the spin-filter. The electrostatic fields for the modified lens system, as well as the present system, were calculated using POISSON. These fields were used to make beam emittance calculations to determine the effects of the SFP modifications on the ABPIS performance. Preliminary calculations suggest that there should be not be significant changes in the average ion beam intensities obtainable from the source (6).

ACKNOWLEDGMENTS

This work supported in part by the U.S. Department of Energy, Office of High Energy and Nuclear Physics, under Grant No. DEFG05-91-ER40619

REFERENCES

1. Clegg, T.B., Karwowski, H.J., Lemieux, S.K. , Sayer, R.W., Crosson, E.R., Hooke, W.M., Howell, C.R., Lewis, H.W., Lovette, A. W., Pfutzner, H.J., Sweeton, Wilburn, W.S., Nucl. Instr. Meth. in press.
2. McKibben, J.L., Lawrence, G.P. and Ohlsen,G.G., Phys. Rev. Lett. **20** (1968) 1180.
3. Clegg, T.B., Bissinger , G.A. and Trainor, T.A., Nucl. Instr. Meth. **120** (1974) 445.
4. Lemieux, S.K., Clegg, T.B., Karwowski, H.J., Thompson, W.J. and Crosson, E.R., Nucl. Instr. Meth. **A333** (1993) 434.
5. Obtained from the Los Alamos Accelerator Code Group, AT-6, Mail Stop H829, AT-Division, LANL, Los Alamos, NM 87545, USA.
6. Mendez, A.J., Clegg, T.B. and Roper, C.D., Spin94 Conference Proceedings, to be published.

Modeling the Hyperfine State Selectivity of a Short Lamb-Shift Spin-Filter Polarimeter

A.J. Mendez[a], C.D. Roper[b] and T.B. Clegg[a]

*Triangle Universities Nuclear Laboratory, Durham, NC 27708-0308**
[a]and Department of Physics and Astronomy,
University of North Carolina at Chapel Hill, Chapel Hill, NC 27599-3255
[b]and Department of Physics,
Duke University, Durham, NC 27708-0308

Abstract. An rf cavity, previously used as a spin filter in a Lamb-shift polarized ion source, is being adapted for use as a polarimeter in an atomic beam polarized hydrogen and deuterium ion source. Paramount among the design criteria is maintaining the current source performance while providing on-line beam polarization monitoring. This requires minimizing both the polarimeter system length and the coupling with the magnetic fields of the other ion source systems. Detailed computer calculations have modeled the four-level interaction involving the $2S_{1/2}$ - $2P_{1/2}$ states of the atomic beam. These indicate that a significantly shorter spin-filter cavity and uniform axial magnetic field than used in the Lamb-shift source do not compromise the spin-state selectivity. The calculations also predict the axial magnetic field uniformity needed as well as the gains achieved from proper shaping of the cavity rf and dc fields.

INTRODUCTION

The Atomic Beam Polarized Ion Source (ABPIS) at the Triangle Universities Nuclear Laboratory (TUNL) is used approximately 80% of the calendar days by a wide variety of experimental groups. Because of the difficulties in making accurate, efficient absolute beam polarization measurements, there is considerable interest in developing a source polarimeter to provide on-line monitoring of the beam polarization. We are currently developing a Lamb-shift spin-filter polarimeter (SFP) for installation on the ABPIS at TUNL. The Lamb-shift technique employs a resonant interaction among the $2S_{1/2}$-$2P_{1/2}$ states of the atomic hydrogen and deuterium induced in a uniform axial magnetic field by longitudinal rf and transverse dc electric fields. It has been shown to provide an efficient, accurate measure of the m_I -state population distribution (1). The spin-filter cavity used for the off-line evaluation in ref. (1) was that of the old TUNL Lamb-shift polarized ion source (2), based on the original Los Alamos design (3).

The SFP cavity design criteria differ from those used in the Lamb-shift source. The SFP system must be physically as short as possible to fit into the available space, it must detract minimally from output polarized ion beam intensities, and it must be minimally influenced by stray magnetic fields from other ABPIS components.

Because the polarized source is essentially in constant use, it was decided to model the SFP systems as realistically as possible during the design phase to minimize the source downtime during later installation. The performance of the cavity itself was modeled using the full, time-dependent quantum mechanical treatment of the atomic four-level problem, much as was done in the original Los Alamos design (4). These calculations used realistic values of the cavity electric and magnetic fields determined using the POISSON/SUPERFISH family of codes (5) from Los Alamos. Two types of calculation were performed. The first calculated the time-evolution of the individual $2S_{1/2}$-$2P_{1/2}$ state amplitudes, and the second calculated the transmitted intensity as a function of the axial dc magnetic field. The calculations were used to test the sensitivity of the hyperfine state selectivity to the SFP design parameters: spin-filter cavity length and uniformity of the axial dc magnetic field.

QUANTUM MECHANICS OF THE 4-LEVEL PROBLEM

The Schrödinger equation for the problem is

$$(H_0 + H')\psi = i\hbar\frac{\partial\psi}{\partial t} \tag{1}$$

where H' is the dipole operator. In the interaction picture the wavefunction is given by

$$\psi = \sum a_n(t)\, u_n\, e^{-iE_n t/\hbar} \tag{2}$$

where the u_n are the hydrogen wavefunctions satisfying

$$H_0 u_n = E_n u_n. \tag{3}$$

It is straightforward to show that the amplitudes $a_n(t)$ satisfy the following set of coupled, first-order differential equations:

$$i\hbar\dot{a}_k = \sum H'_{kn}\, a_n\, e^{i\omega_{kn}t} \tag{4}$$

where

$$\begin{aligned} \omega_{kn} &= (E_k - E_n)/\hbar, \text{ and} \\ H'_{kn} &= \int u_k^* H' u_n d\tau. \end{aligned} \tag{5}$$

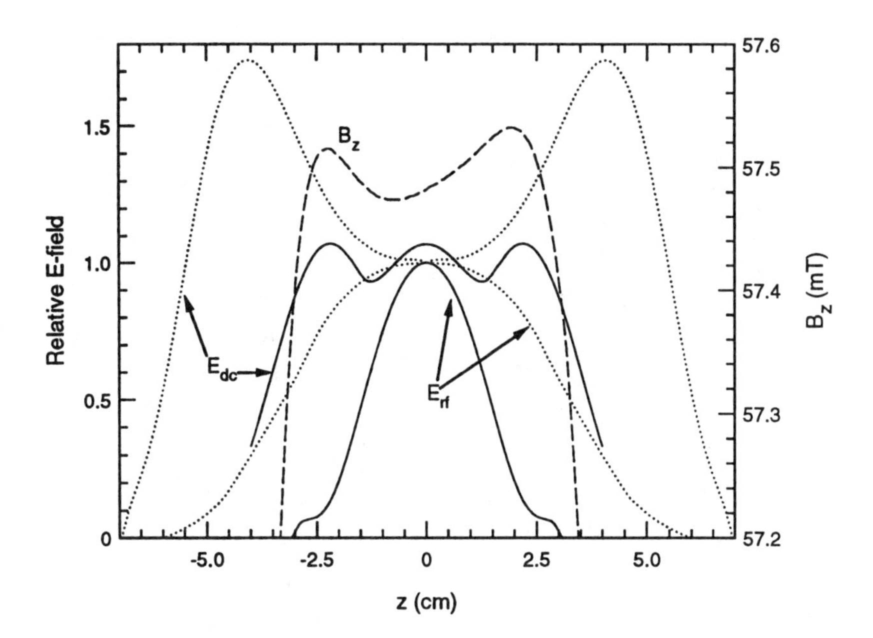

FIGURE 1. Cavity field distributions. The dashed curve (B_z) and the solid curves are for the short, TUNL cavity design. The dotted curves are for the original LANL cavity design.

The electric and magnetic dipole matrix elements H'_{kn} can be easily calculated and are given in ref. (4). Equations (4) can then be written in a form suitable for solution by standard numerical integration techniques.

RESULTS

A FORTRAN program was written to solve eqs. (4) for the four-level amplitudes $a_n(t)$ both as a function of transit time through the cavity and as a function of axial dc magnetic field. Figure 1 shows the cavity field distributions for the original Los Alamos cavity design and for the shorter TUNL cavity design. The principal effect of reducing the cavity length was to increase the rates of change of the rf and dc electric field strengths, which determine the adiabaticity of the passage of atoms through the cavity. In the left plot of fig. 2, the time evolution of the α-state ($2S_{1/2}$, m_J = +1/2) probability is shown for various configurations of the cavity fields. The effect of adiabatic variation of the rf is

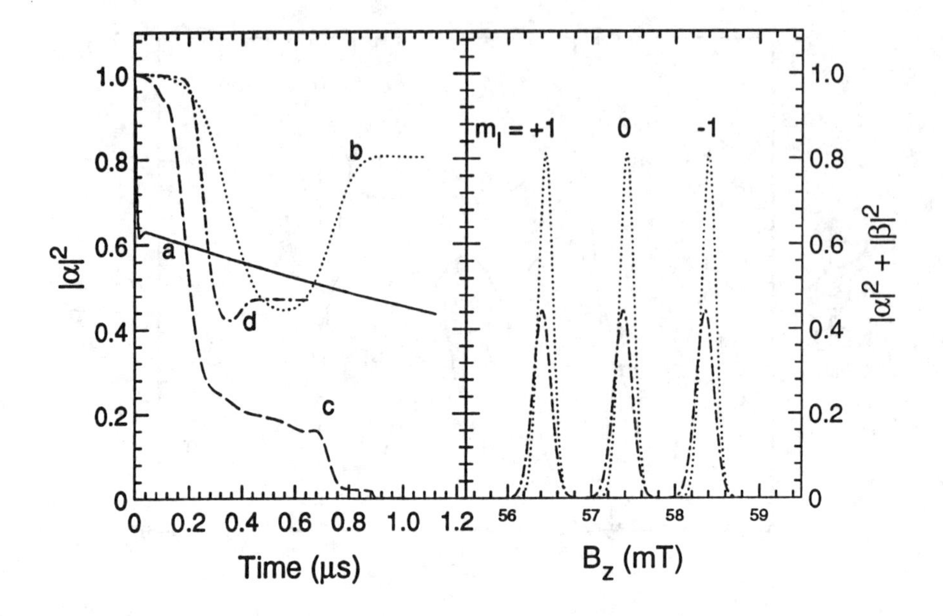

FIGURE 2. Left plot shows time evolution of α-state probability for various configurations of electric and magnetic fields for $m_I = 0$ deuterons. Calculation (a) has uniform rf and dc electric fields, and a uniform B_z = 57.4 mT. In (b) realistic rf and dc electric fields for the original LANL designed cavity (see dotted curves in fig. 1) are used with uniform B_z = 57.4 mT. In (c) the uniform B_z is replaced with the actual distribution produced by the shorter coil system designed for the TUNL polarimeter (dashed curve in fig. 1). Note that the transmitted intensity for this case is essentially zero. In (d) the LANL cavity electric fields are replaced with the short TUNL cavity field distributions (solid curves in fig. 1). Right plot shows calculated spectra for cases (a) solid curve, and (d) dot-dashed curve.

seen clearly in the gradual rise and fall in α-state probability (curve b) as the rf slowly increases and then decreases. The transmitted intensity falls rapidly if the rf electric field extends beyond the region of nearly constant magnetic field. This effect is seen clearly in curve (c). To constrain the rf electric field to the shorter axial region of uniform *B*, a new spin-filter cavity will be constructed which has a 5.1 cm long central region, as compared with the 15.24 cm long central section of the LANL cavity. The diameter of the endpipes will be reduced from 7.62 cm to 5.08 cm. The calculated electric field distributions for this geometry are the solid curves in fig. 1. With the dc *B*-field uniform to 0.025 mT over 13 cm, the calculated state selection for the new short cavity is quite adequate, as seen in curve (d) and the corresponding dot-dashed curve in the right plot of fig. 2.

* Supported in part by the U.S. Dep't of Energy, Office of High Energy and Nuclear Physics.

1. Lemieux, S.K., Clegg, T.B., Karwowski, H.J., Thompson, W.J., and Crosson, E.R., Nucl. Instr. Meth. **A333**, 434 (1993).
2. Clegg, T.B., Bissinger, G.A., and Trainor, T.A., Nucl. Instr. Meth. **120**, 445 (1974).
3. McKibben, J.L., Lawrence, G.P., and Ohlsen, G.G., Phys. Rev. Lett. **20**, 1180 (1968).
4. Ohlsen, G.G., and McKibben, J.L, Los Alamos technical report no. LA-3725 (1967).
5. Obtained from the Los Alamos Accelerator Code Group, AT-6, Mail Stop H829, AT-Division, LANL, Los Alamos, NM 87545, USA.

The IUCF/TUNL Spin-Filter Polarimeter

V.P. Derenchuk[a], A.J. Mendez[b] and T.B. Clegg[b]

[a]*Indiana University Cyclotron Facility, Bloomington, IN 47408*[*]
[b]*Triangle Universities Nuclear Laboratory, Durham, NC 27708-0308*[#]

Abstract. A nuclear spin-filter polarimeter is being installed in the high voltage terminal of HIPIOS at the Indiana University Cyclotron Facility. It will be used to measure the polarization of protons and deuterons at a source extraction energy of 10 keV. The polarimeter installation is described briefly, and the results of calculations modeling its anticipated performance at IUCF are presented.

INTRODUCTION

The high intensity polarized ion source (HIPIOS) (1) at Indiana University has been delivering beams to users since early 1994. Early experiences using this ion source are described in another contribution to this meeting (2). Until now, IUCF has used only nuclear scattering polarimeters to measure the polarization of beams from this source. These require beam acceleration in the cyclotrons to an appropriate energy. Both the polarimeters' relatively low efficiency and cyclotron scheduling restraints severely limit the efficiency of developing optimal conditions for producing the desired beam polarization. Once such conditions are found, similar constraints dictate how rapidly any desired beam polarizations can be obtained for the experiment at hand.

To overcome these difficulties, an efficient, accurate polarimeter is needed to measure the polarization of the outgoing ion beam at the ion source extraction energy of 10 keV. This device must measure proton vector polarization and both vector and tensor polarization of deuterons. Recently, a spin-filter polarimeter has been developed for this purpose at the Triangle Universities Nuclear Laboratory (TUNL) (3).

[*] Work supported in part by the National Science Foundation Grant PHY-891440.

[#] Supported in part by the U.S. Dep't of Energy, Office of High Energy and Nuclear Physics.

The primary component of the spin-filter polarimeter is an rf cavity designed at Los Alamos (4) and used for many years in Lamb-shift polarized ion sources (5). The cavity uses Stark mixing of the n = 2 levels of atomic hydrogen/deuterium with longitudinal rf and transverse dc electric fields and resonant Zeeman tuning with a longitudinal dc magnetic field to selectively pass atoms in a single nuclear hyperfine state. The polarimeter signal is obtained by subsequently quenching the transmitted atoms to the ground state and observing the emitted Lyman-α light with a photomultiplier tube. The intensity of the light is proportional to the population of the selected spin state. Thus, if the light intensity is measured as the magnetic field of the polarimeter is scanned to select sequentially each hyperfine state, the beam polarization can be extracted immediately from the relative intensities of the resulting peaks. Since the cross section magnitudes underlying the spin-filter polarimeter are atomic rather than nuclear, the intrinsic polarimeter efficiency is very high. In principle, a ~1 s measurement accurate to ~1% is possible.

Since this polarimeter measures the polarization of metastable atoms, the incident polarized ion beam must be neutralized to the $2S_{1/2}$ atomic state prior to

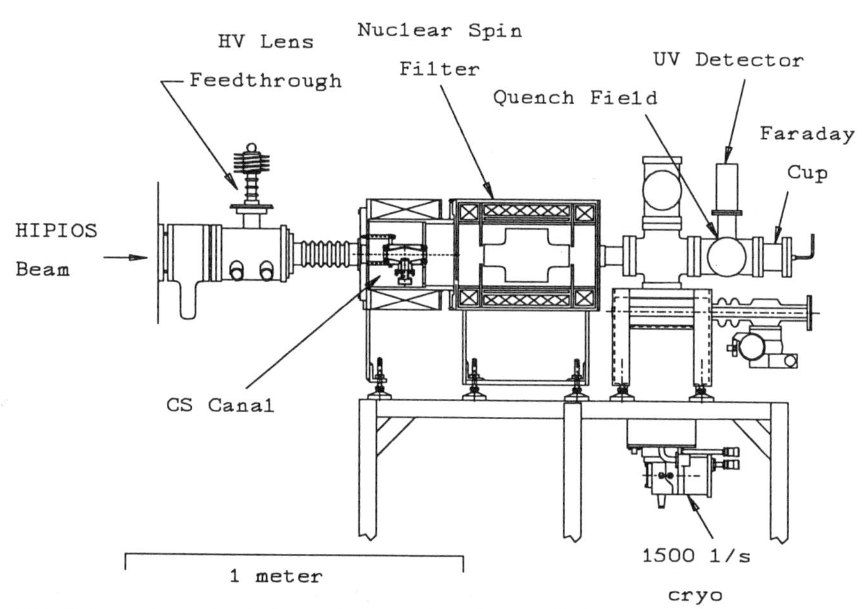

Figure 1. Schematic of the cesium canal and solenoid followed by the TUNL nuclear spin filter. The ion beam passes through the cesium canal where it is neutralized, producing metastable hydrogen or deuterium atoms.

entering the spin filter. This is achieved by charge transfer in an alkali vapor immersed in a strong axial magnetic field which decouples the electron and nuclear magnetic moments and thus minimizes depolarization by the hyperfine interaction. The cross section for producing the metastable state is strongly energy dependent, with a maximum at about 0.5 keV/u when cesium vapor is used, falling by roughly an order of magnitude at 10 keV/u.

THE IUCF/TUNL SPIN-FILTER POLARIMETER

The spin-filter polarimeter rf cavity at IUCF is identical in dimensions to that originally used at Los Alamos (4). The cavity/coil system was previously used at TUNL by Lemieux et al. (6) to test the efficacy of the spin-filter polarimeter technique. Installed in the HIPIOS high-voltage terminal at IUCF, as shown in fig. 1, it is immediately preceded by a cesium charge-exchange region which will produce the required metastable atoms from the incident 5 to 15 keV polarized H^+ or D^+ ion beam. The magnetic field distribution over the cesium charge-exchange and spin-filter cavity regions has been measured and is shown in fig. 2.

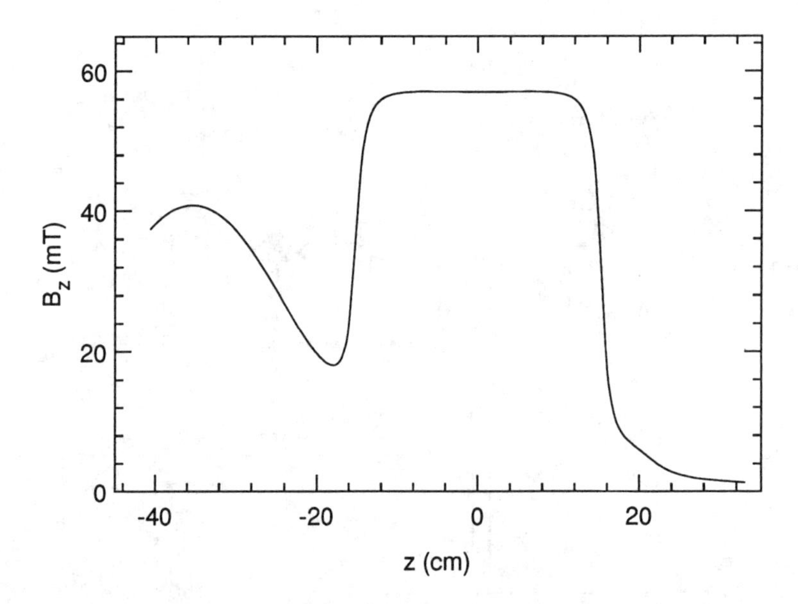

Figure 2. Axial magnetic field along the central axis of the polarimeter. The data begin at the entrance of the cesium canal and extend to the exit of the nuclear spin filter. The 25.4 cm long spin-filter cavity is centered at z = 0 in the middle of the flat *B*-field region. The field in this region is uniform to ±0.05 mT over 17.6 cm.

CALCULATIONS OF THE SPIN FILTER PERFORMANCE

Using the code described in ref. (7), the expected spin-filter polarimeter performance was calculated. This code solves the time-dependent Schrödinger equation for the behavior of the $2S_{1/2}$ - $2P_{1/2}$ amplitudes in the presence of the cavity rf and dc electric and magnetic fields. The present calculations investigate the transmission efficiency and hyperfine state selectivity of the spin filter as a function of metastable hydrogen and deuterium atomic beam energy in the range 0.1 - 20 keV/u. For simplicity, a uniform axial dc magnetic field was used, with realistic rf and dc electric fields as described in ref. (7).

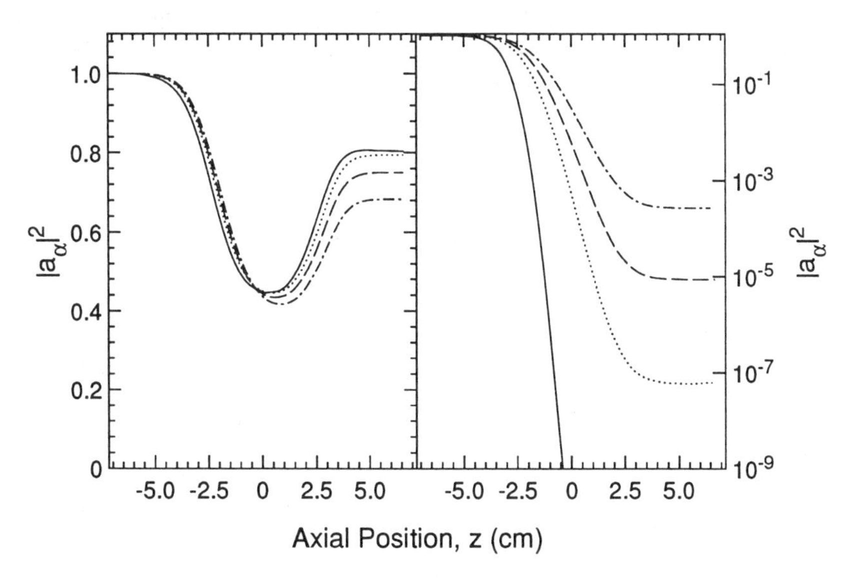

Figure 3. Time evolution of the α-state ($2S_{1/2}$, $m_J = +1/2$) probability for various energies in an axial field of 57.41 mT. The left plot shows the on-resonance $m_I = 0$ state, and the right plot shows the off-resonance $m_I = 1$ state. Note the log scale in the right plot. The four curves are: solid -- 1 keV, dotted -- 10 keV, dashed -- 20 keV, dot-dashed -- 40 keV.

Figure 3 shows, in the left plot, the time evolution of the probability for selected $m_I = 0$ deuterons at the resonant field $B_z = 57.41$ mT for several beam energies. As can be seen, the transmission efficiency for the selected state varies little with energy. However, it is relevant to consider the transmission of unselected m_I states. In the right plot of fig. 3, the probability for $m_I = 1$ deuterons at $B_z = 57.41$ mT is shown. Above 5 keV, the undesired state transmission efficiency increases rapidly with energy, so that the ratio of unselected to selected state probability varies from $< 10^{-9}$ to 4×10^{-4} over the energy range investigated. Results are similar for protons.

Adequate state selection and efficiency can thus be obtained for protons up to about 20 keV and for deuterons up to about 40 keV; however, producing a metastable beam of these higher energies may prove difficult. In the TUNL polarized atomic beam source, the metastable beam is produced by charge transfer in cesium vapor at 0.5 keV/u. The cross section is a maximum at this energy, about 3.5×10^{-15} cm^2 (8), and drops rapidly as the energy increases, by about an order of magnitude at 10 keV/u. At higher energies, charge transfer in Mg vapor is possible, but at 20 keV/u, this cross section is small, about 1.0×10^{-16} cm^2 (7). It is expected that the high ion beam intensity and the proximity of the cavity to the cesium canal will ensure an adequately strong polarimeter signal for a 10 keV beam, especially in the proton case, in spite of the low metastable atom production cross section at this energy.

REFERENCES

1. Derenchuk, V.P., Brown, R., Wedekind, M., "Polarized Ion Source Development at IUCF," in *Proceedings of the Conference on Polarized Ion Sources and Polarized Gas Targets*, 1993, pp. 72–75.
2. Derenchuk, V., Belov, A., Brown, R., Collins, J., Sowinski, J., Stephenson, E., Wedekind, M., "IUCF High Intensity Polarized Ion Source Operation," in *Proceedings of the 8th International Symposium on Polarization Phenomena in Nuclear Physics,* 1994, to be published.
3. Crosson, E.R., Roper, C.D., Clegg, T.B., "On-line, Efficient, Low-Energy Proton and Deuteron Polarimetry," in *Proceedings of the Conference on Polarized Ion Sources and Polarized Gas Targets*, 1993, pp. 115–118.
4. McKibben, J.L., Lawrence, G.P., and Ohlsen, G.G., Phys. Rev. Lett. **20**, 1180 (1968).
5. Clegg, T.B., in *Proceedings of the Conference on Polarized Proton Ion Sources*, 1981, p. 21–.
6. Lemieux, S.K., Clegg, T.B., Karwowski, H.J., Thompson, W.J., and Crosson, E.R., Nucl. Instr. Meth. **A333**, 434–442 (1993).
7. Mendez, A.J., Roper, C.D.and Clegg, T.B., "Modeling the hyperfine state selectivity of a short Lamb-shift spin-filter polarimeter," in *Proceedings of the 8th International Symposium on Polarization Phenomena in Nuclear Physics*, 1994, to be published.
8. Morgan, T.J., Olson, R.E., Schlacter, A.S. and Gallagher, J.W., J. Phys. Chem. Ref. Data **14**, 971–1040 (1985).

HYSICS WITH A FOCAL PLANE PROTON POLARIMETER FOR HALL A AT CEBAF

R. Gilman,[b,d] F. T. Baker,[e] L. Bimbot,[d] E. Brash,[a] C. Glashausser,[d]
M. Jones,[a] G. Kumbartzki,[d] S. Nanda,[b] C. Perdrisat,[a]
V. Punjabi,[c] R. Ransome,[d] and P. Rutt[d,e]

[a] *College of William & Mary, Williamsburg, VA 23187, USA*
[b] *Continuous Electron Beam Accelerator Facility, Newport News, VA 23606, USA*
[c] *Norfolk State University, Norfolk, VA 23504, USA*
[d] *Rutgers University, Piscataway, NJ 08855, USA*
[e] *University of Georgia, Athens, GA 30602, USA*

INTRODUCTION

A focal plane polarimeter intended for the CEBAF Hall A high resolution hadron spectrometer is under construction at Rutgers University and the College of William & Mary. Experiments with focal plane polarimeters are only now beginnning at electron accelerators; they play a prominent role in the list of approved experiments for Hall A. Construction of the polarimeter is in progress; it is expected to be brought to CEBAF in spring 1995. Several coincidence (e,e'$\vec{p}$) and singles (γ,$\vec{p}$) measurements by the Hall A Collaboration are expected to start in 1996. In this paper we describe the polarimeter and the physics program planned for it.

THE POLARIMETER

Intermediate energy proton polarimeters are generally based on a nuclear scattering of the proton from a carbon analyzer. The spin orbit force introduces an azimuthal asymmetry, or ϕ dependence, into the scattering differential cross section. The scattering measurement is inclusive, with no determination made of the energy of the outgoing proton. The angle-averaged analyzing power of carbon in this situation ranges from about 0.5 near 200 MeV kinetic energy to about 0.1 for energies near 2 GeV. Polarimeters generally use wire chambers before and after the analyzer to determine the scattering angles.

The CEBAF Hall A polarimeter follows the usual techniques described above. It is unusual, however, in several respects, which are discussed below. The polarimeter is physically large, makes use of straw tube drift chambers and multiplexing electronics, and will be located at an electron accelerator.

The polarimeter system is shown in Figure 1. Four straw–tube chambers are used to track the proton trajectory into and out of a graphite block. The geometrical sizes of the chambers, up to 1.4 m × 2.8 m, have been set to accept 20° scatters from almost the entire analyzer block, including the effects of divergence of the beam envelope in the detector stack. To make the response of the detector identical in two orthogonal directions, the straws are oriented along UV axes, at ±45° to the edges of the chamber. The straw diameter is slightly greater than 1 cm; each detector plane contains at least 176 straws, and is constructed so that the geometrical coverage is about 97%. These large sizes and the large numbers of wires in the chambers (~ 5500) influenced the choice of straw tubes over conventional wire chambers. Too, since the chambers are used in conjunction with a thick - up to 55 cm - carbon analyzer, multiple scattering in the chambers is not an important consideration. Position resolution of the chambers is expected to be about 150 - 200 μ. Monte Carlo simulations of the system indicate that tracking efficiency should be close to 100%, and false asymmetries less than 0.01 should be achievable at lower energies.

The design requirements for rate capability of the polarimeter arise from typical experiments in the physics program that measure the coincidence (e,e'p) reaction. Since coincidence rates can be small fractions of the singles rate in each detector arm, we require that the polarimeter have a 1 MHz singles rate capability. This leads to modest rates of 5 - 6 kHz in individual straws. However, individual wire readout was deemed to be prohibitively expensive.

The electronics design chosen uses readout boards developed at Rutgers University that amplify and discriminate the chamber signals, and multiplex the resulting logical signals. Given the needed rate capability, the boards were designed to multiplex 8 signals into an individual TDC channel. This is accomplished by using fixed width discriminator one–shots that vary across each set of eight channels from 35 to 70 ns, and fast logical ORs. Time digitization will use the multihit LeCroy 1877 TDC to measure both the leading edge of the pulse for drift time information, and the trailing edge for demultiplexing information.

The use of the polarimeter at CEBAF leads to calibration difficulties, as one lacks both a proton beam of known polarization and previously measured nuclear reactions. The calibration procedure at an electron accelerator uses elastic scattering of polarized electrons from an unpolarized hydrogen target. Insofar as the proton electric and magnetic form factors are known, this leads to protons of known polarization that can then be used to calibrate the polarimeter. This technique is discussed further below in the description of the G_{E_p} experiment. It is expected that the polarimeter will be calibrated and used for proton kinetic energies up to about 3 GeV.

PHYSICS EXPERIMENTS

Two ($\vec{e}$,e'$\vec{p}$) polarimetry experiments require a proton target. The proton electromagnetic form factors will be studied[1] in elastic scattering of polarized electrons. Spin transfer yields a longitudinal polarization that depends on the squared magnetic form factor, G_M^2, and a transverse, in plane, polarization that depends on the product of electric and magnetic form factors, $G_E G_M$. The ratio of these spin components allows one to measure precisely the poorly known G_E relative to the well–known G_M without knowing the beam polarization or the carbon analyzing power. The experiment also serves to calibrate the focal plane polarimeter, provided the beam polarization is known, using the transverse spin component. In kinematics in which the Δ resonance is excited, response functions will be separated and subjected to a multipole analysis. The varying response functions allow accurate determination of the dominant $|M_{1+}|^2$ amplitude, the interference term

$\text{Re}(S^*_{1+}M_{1+})$ involving the quadrupole amplitude, and the influence of other resonant and nonresonant backgrounds.

Studies[3,4,5] of (e,e'$\vec{\text{p}}$) are also planned on nuclei ranging from ^{2}H to ^{16}O. With unpolarized beam, the normal component of the polarization results solely from final state interactions. This should yield more precise information than simple cross section measurements about the reaction mechanism and nuclear wave functions, and allow improved tests of phenomena such as color transparency. The spin transfer with polarized beam depends, in the plane wave approximation, on known kinematic factors, the nuclear wave function, and the electromagnetic form factors of the proton. Spin transfer is generally insensitive to final state interactions. With the nuclear wave function and final state interactions constrained by cross section and induced polarization measurements, it is expected that the spin transfer data can provide the best restrictions on the modification of the proton electromagnetic form factors in the nucleus.

Studies[6,7] of the (γ,$\vec{\text{p}}$) reaction are planned on hydrogen and deuterium. These two experiments address the question of whether the scaling of the cross sections in these reactions is a real indication that quark/gluon degrees of freedom are dominant. The energy dependence of the cross section has been reproduced in both quark/gluon and meson/baryon models. The quark/gluon interpretation requires that the recoil proton polarization be zero, due to helicity conservation. The strong–interaction meson/baryon model essentially requires significant nonzero polarization due to final state interactions and the interference of baryon resonances with nonresonant contributions to the cross section.

1. C. F. Perdrisat, V. Punjabi, Mark Jones *et al.*, Electric Form Factor of the Proton by Recoil Polarization, CEBAF Proposal PR-93-027.
2. R. Lourie, S. Frullani *et al.*, High Precision Separation of Polarized Structure Functions in Electroproduction of the Δ and Roper Resonances, CEBAF Proposal PR-91-011.
3. J. M. Finn, P. Ulmer *et al.*, Polarization Transfer Measurements in the D($\vec{\text{e}}$,e'$\vec{\text{p}}$) Reaction, CEBAF Proposal PR-89-028.
4. C. Glashausser, C. C. Chang, S. Nanda, J. W. van Orden *et al.*, Measurement of Recoil Polarization in the ^{16}O($\vec{\text{e}}$,e'$\vec{\text{p}}$) Reaction with 4 GeV Electrons, CEBAF Proposal PR-89-033.
5. A. Saha *et al.*, Study of Nuclear Medium Effects by Recoil Polarization up to High Momentum Transfers, CEBAF Proposal PR-91-006.
6. R. Gilman, R. J. Holt, Z.–E. Meziani *et al.*, Measurement of Proton Polarization in the d(γ,$\vec{\text{p}}$)n Reaction, CEBAF Proposal PR-89-019.
7. R. Gilman, R. J. Holt *et al.*, Measurement of Photoproton Polarization in the H$(\gamma,\vec{\text{p}})\pi^0$ Reaction, CEBAF Proposal PR-94-012.

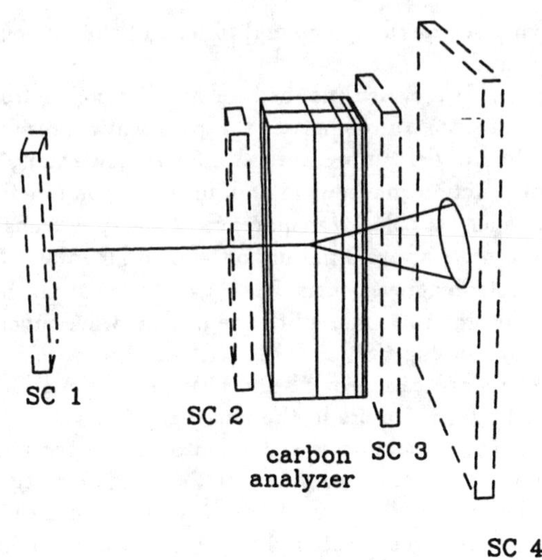

Figure 1. The CEBAF Hall A focal plane polarimeter.

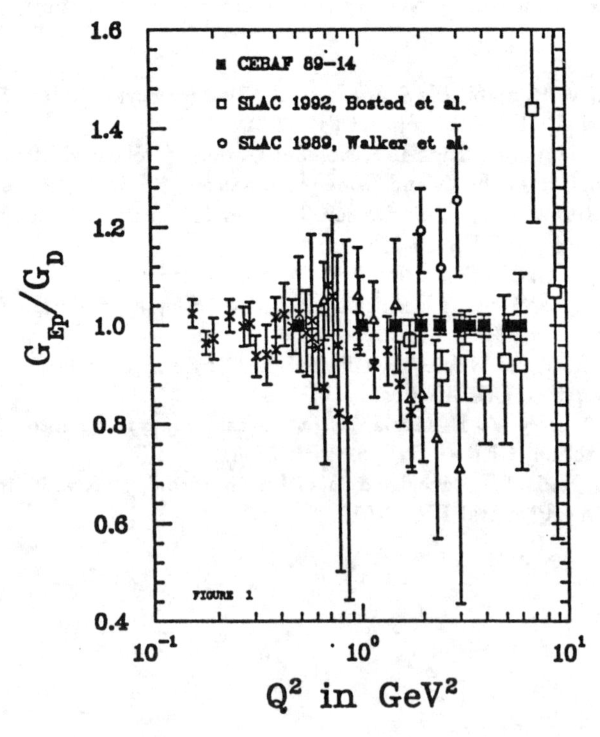

Figure 2. Expected results for the G_{E_p} experiment.

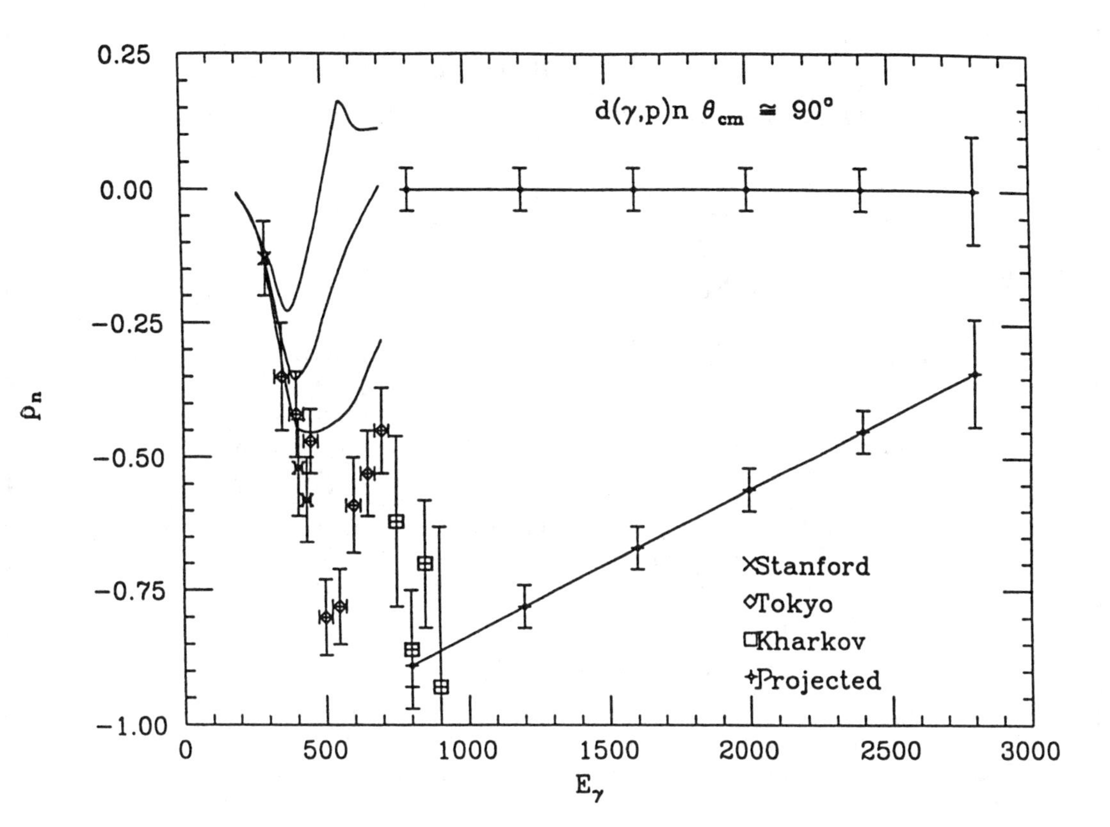

Figure 3. Expected results for the $\gamma d \rightarrow \vec{p}n$ experiment in two scenarios.

A Calibration of the IUCF K600 Focal Plane Polarimeter

S.M. Bowyer, S.W. Wissink, A.D. Bacher, T.W. Bowyer, S. Chang, W. Franklin, J. Liu, J. Sowinski, E.J. Stephenson, and S.P. Wells

Indiana University Cyclotron Facility, Bloomington, Indiana 47408

Abstract We have carried out a precise determination of the effective analyzing power of the K600 focal plane polarimeter at IUCF for incident proton energies extending from 200 MeV down to 120 MeV. Two different calibration techniques, and several analyzer thicknesses, were used over this range. As the energy decreases, the effective analyzing power falls almost linearly. The polarimeter efficiency, however, rises rapidly, thus maintaining a reasonable figure-of-merit.

INTRODUCTION

Over the past few years, several precise $(\vec{p},\vec{p}\,')$ spin-transfer measurements have been carried out at IUCF [1,2], using a polarimeter [3] mounted at the exit of the high-resolution K600 spectrometer. This 'focal plane polarimeter' (FPP) has a large (~8%) momentum bite, high efficiency and effective analyzing power (η_{FPP} and A_{FPP}, see below), provides information on both transverse polarization components, and is capable of handling high count rates via intelligent front-end processing fed into a fast second-level trigger [3]. For most of these studies, it was necessary to know the performance characteristics of the FPP over only a limited range of scattered beam energies, typically 180–200 MeV. In a recent precise measurement of the spin-transfer coefficient $D_{NN'}$ for pp elastic scattering at 200 MeV for $\theta_{lab} = 5°–38°$ [4], however, the energy of the protons incident on the FPP extended down to ~120 MeV. A new, and more accurate, calibration of the FPP down to these lower energies was therefore needed.

The FPP itself consists of a thick carbon block, followed by two sets of paired x-y multiwire proportional chambers and two planes of plastic scintillator. A high-density (1.78 g/cm^3) graphite target (thickness = 0.5″, 1.0″, or 2.0″) serves as the polarization analyzer; between 120 and 200 MeV, the p-^{12}C analyzing powers at forward angles are large in magnitude for both elastic and low-excitation inelastic scattering processes [5]. Wire chamber hit patterns are read out via the LeCroy PCOS III system, which also provides information used in the second-level trigger decision. The scintillator planes consist of sheets of 0.62 cm and 7.62 cm thick NE102 stock, and provide a measure of the proton energy after scattering in the analyzer.

EXPERIMENTAL TECHNIQUES

The methods used to calibrate the FPP have been described previously [3]. In brief, we elastically scatter vertically-polarized protons from a spin-0 target (in which case the induced polarization P and the reaction analyzing power A_y are equal, and $D_{NN'} = 1$), then take advantage of the simplifications that result in the spin-transfer equations. In particular, the normal (to the scattering plane) polarization components of the beam (p_N) and scattered ($p_{N'}$) protons are related by

$$p_{N'} = \frac{p_N + A_y}{1 + p_N A_y}, \tag{1}$$

where A_y is again the analyzing power of the primary reaction. If $p_{N'}$ is known, the effective analyzing power of the FPP can then be deduced from

$$A_{FPP} \equiv \frac{\varepsilon_{FPP}}{p_{N'}} , \tag{2}$$

where ε_{FPP} is the left/right scattering yield asymmetry observed in the FPP.

To determine $p_{N'}$, we make use of two special cases for Eq. 1. Note that if a scattering angle is chosen such that $A_y = 0$, then $p_{N'} = p_N$, *i.e.*, the polarization of the protons incident on the FPP can be determined by measuring the polarization of the incident proton beam. Using well-calibrated beamline polarimeters [6], this latter quantity can be measured to a fractional uncertainty of better than 1% near 200 MeV, and A_{FPP} can thus be determined to a comparable level of precision. Moreover, because $p_{N'}$ "flips" when p_N is reversed at the polarized ion source, determination of A_{FPP} via this method is seen to be fairly insensitive to any false (geometric) asymmetries in the polarimeter. The measured asymmetries can, in fact, be used to estimate the size of these geometric factors [3].

Between 120 and 200 MeV, there are also scattering angles at which $|A_y|$ approaches 1 [5]. We generate a second set of calibration points by noting that when $A_y \to 1$, modest values of p_N lead to values of $p_{N'}$ that are extremely close to 1; *e.g.*, if $A_y = 0.95 \pm 0.02$, and $p_N = 0.75 \pm 0.02$, then $p_{N'} = 0.993 \pm 0.003$. Thus $p_{N'}$, and hence A_{FPP}, can be determined much more accurately than either A_y or p_N may be known. The primary disadvantage of this method is that one is now quite sensitive to false asymmetries in the FPP geometry. The complementary nature of these two calibration schemes is obvious.

TABLE I. Summary of FPP calibration running conditions

Running Period	T_p (MeV)	θ_{lab} (deg)	A_y	t_C (in)
I	198	16.1	~1	2
I&II	198	24.6	0	2
I	177	26.5	0	2
II	174	18.8	~1	1,2
II	174	27.1	0	1,2
I	148	21.5	~1	1,2
I	148	32.3	0	1
II	127	25.6	~.7	$\frac{1}{2}$,1
II	127	46.7	~.9	$\frac{1}{2}$,1
II	127	58.6	0	1
I	118	25.6	~.7	$\frac{1}{2}$,1
I	118	46.7	~.9	1
I	118	58.5	0	1

RESULTS

The calibration data for this work were taken during two running periods in February and August 1993. For each run, three or four incident beam energies T_p were used; at each energy, several different values of the scattering angle θ_{lab} (or, equivalently, A_y values) were studied. In some cases, the thickness t_C of the carbon analyzer was also changed. Because the choice of analyzer thickness is always a compromise between maximizing the efficiency of the polarimeter, while minimizing energy loss and multiple-scattering within the analyzer, the optimal thickness is energy-dependent. The values of these parameters for the present work are summarized in Table I.

The calibration data were analyzed with the specific intent of minimizing sensitivity to several possible sources of systematic error. Software cuts on the FPP horizontal scattering angle θ_x were set from 6° to 23°, with the lower limit chosen to exclude the multiple scattering peak at all energies. Gates on the FPP scintillator spectra (primarily a ΔE-E correlation, with each signal corrected for energy losses and kinematic effects appropriate for p-^{12}C elastic scattering) were drawn using an algorithm that can be applied consistently in

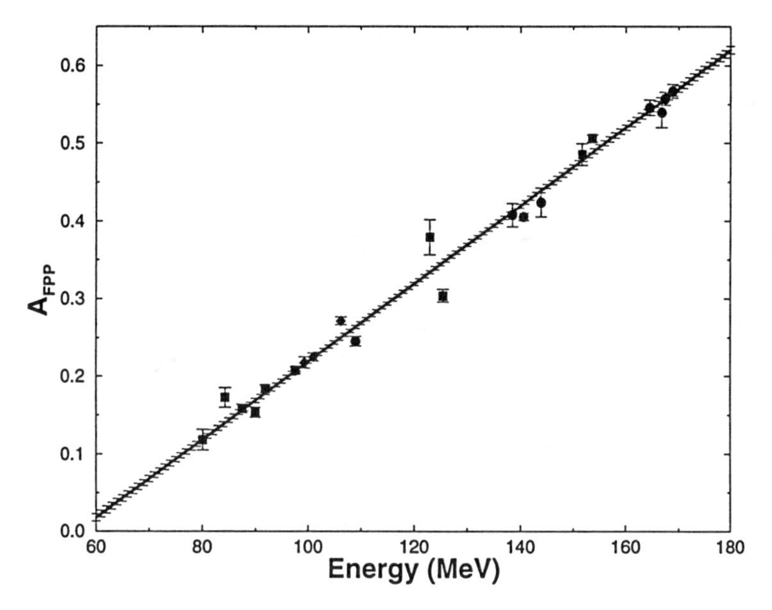

FIGURE 1. A_{FPP} as a function of average proton energy within the FPP graphite analyzer. The various plotting symbols correspond to different values of the analyzer thickness: • = 2″, ▪ = 1″, and ◆ = 1/2″.

analysis of spin observable data. Small corrections were applied to the final results to account (in first order) for deviations of A_y from its limiting value of 0 or 1, and for small differences in the 'spin-up' and 'spin-down' beam polarization magnitudes. The sensitivity of the deduced value of A_{FPP} to each of these software cuts and corrections was also investigated.

The final values determined for A_{FPP} in this work are presented in Fig. 1. The energy assigned to each point is the average of the proton energy before and after passing through the carbon analyzer. Plotted in this manner, all calibration points should lie on (approximately) the same curve, independent of the analyzer thickness used. This "universality" is borne out reasonably well by the data, although small deviations can be seen near 105 MeV between the 2″ and 1/2″ values. In practice, A_{FPP} is parameterized independently for each analyzer thickness.

The curve shown in Fig. 1 is the result of a linear least-squares fit, with the fit uncertainty indicated by hatching. The reduced chi-squared for the fit is large ($\chi^2_\nu = 3.22$), due primarily to the two points obtained near 120 MeV with the 1″-thick analyzer. In this case, the elastically scattered protons are very close to ranging out in the thicker scintillator, and the energy deposition varies rapidly with both proton energy and scattering angle. The value of A_{FPP} deduced is therefore extremely sensitive to the choice of scintillator software cuts. For actual spin transfer measurements, this condition is

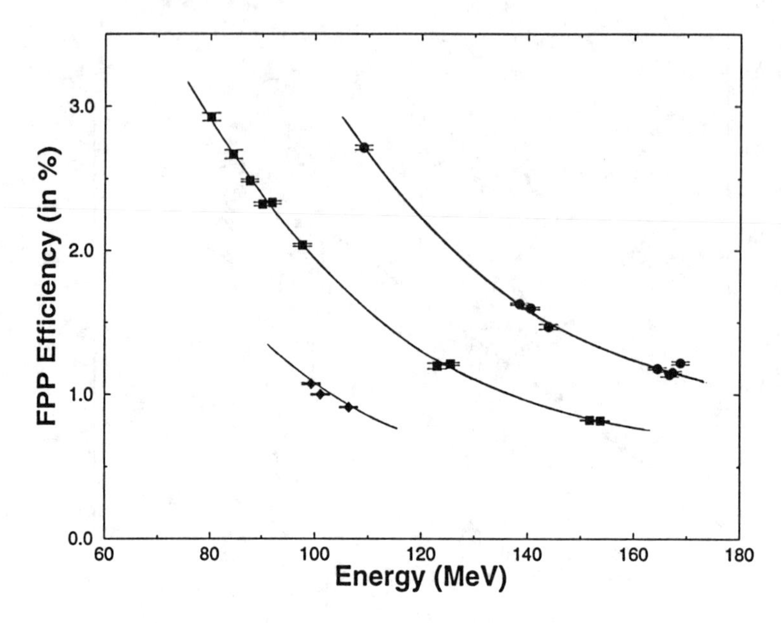

FIGURE 2. FPP efficiency as a function of average proton energy within the analyzer. The different symbols are defined in the caption for Fig. 1.

avoided through judicious choice of analyzer thickness.

The polarimeter efficiencies (defined as the ratio of the number of protons that satisfy all FPP software cuts to the number that pass through the focal plane) for the three values of t_C are shown in Fig. 2. The curves are drawn simply to guide the eye. It is seen that efficiencies of 1-2% are maintained at all energies, even as the analyzer thickness is decreased at lower energies.

With this set of calibration data, precise studies of spin transfer observables may now be carried out over a wide range of beam energies and scattering angles.

This work was supported by the U.S. National Science Foundation.

REFERENCES

[1] S.W. Wissink, in *Spin and Isospin in Nuclear Interactions*, edited by S.W. Wissink, C.D. Goodman, and G.E. Walker (Plenum, New York, 1991), p. 253.

[2] J. Liu *et al.*, and S.P. Wells *et al.*, contributions to this Symposium.

[3] A.K. Opper, Ph.D. dissertation, Indiana University, 1991; see also Ref. [1].

[4] S.M. Bowyer *et al.*, contribution to this Symposium.

[5] M.W. McNaughton *et al.*, Nucl. Instrum. Methods Phys. Res. **A241**, 435 (1985); H.O. Meyer *et al.*, Phys. Rev. C **27**, 459 (1983).

[6] S.P. Wells *et al.*, Nucl. Instrum. Methods Phys. Res. **A325**, 205 (1993).

FOCAL PLANE POLARIMETER FOR THE GRAND RAIDEN AT RCNP

M. Yosoi,[a] H. Akimune,[b] I. Daito,[b], M. Fujiwara,[b] S. Hirata,[b] T. Inomata,[c] O. Kamigaito,[a1] M. Kawabata,[b] T. Noro,[b] Y. Sakemi,[a2] T. Takahashi,[a] A. Tamii,[a] S. Toyama,[a] A. Yamagoshi,[a] M. Yoshimura,[a] and H. Sakaguchi [a]

[a]*Department of Physics, Kyoto University, Kyoto 606, Japan*
[b]*Research Center for Nuclear Physics, Osaka University, Ibaraki, Osaka 567, Japan*
[c]*Department of Physics, Osaka University, Toyonaka, Osaka 560, Japan*

Abstract. A large focal-plane polarimeter which uses inclusive scattering from carbon has been installed for the high resolution spectrometer 'Grand Raiden' at RCNP. A fast second-level trigger system can reject events with small scattering angles. A special dipole magnet is added to the final stage of the spectrometer in order to rotate the spin direction. High resolution measurements of all spin observables for proton scattering are now able to be made with high efficiency over a broad energy range. The present system has been checked by the measurement of spin rotation parameters in proton elastic scattering at 300 MeV.

INTRODUCTION

Spin transfer observables are required by many nuclear studies of considerable interest. Some examples include studies of spin transverse and longitudinal responses and relativistic effects in nuclear reactions. During the past decade, polarization transfer coefficients D_{ij}'s in proton-nucleus scattering have been measured using polarimeters at LAMPF(1), TRIUMF(2), and IUCF(3) at intermediate energies. But they are still not enough to obtain definite conclusions. In particular, more systematic and exclusive measurements of spin transfer observables with high resolution are necessary for investigating the various spin aspects in nuclear reactions and structures.

We have newly built a large focal-plane polarimeter (FPP) for the high resolution spectrograph 'Grand Raiden'(4) at the Research Center for Nuclear Physics (RCNP), Osaka University. The high resolution spin physics is one of the major subjects of the experiments using the ring cyclotron with K=400 MeV at RCNP. In this paper, we report on the present status of the polarimeter system and some preliminary results obtained from the performance test.

[1]Present address : The Institute of Physical and Chemical Research, Wako, Saitama 351, Japan
[2]Present address : Department of Physics, Tokyo Institute of Technology, Meguro, Tokyo 152, Japan

BEAM POLARIMETRY AND THE 'GRAND RAIDEN'

In the polarization transfer measurements, it is important to control the polarization direction both of the beam and of the scattered particles. For the spin-direction control of the primary beam, two superconducting solenoids are installed in the transport line between the injection AVF cyclotron and the ring cyclotron. Each solenoid is located upstream or downstream from the 45° deflecting magnet. The spin axis of protons is oriented in two different in-plane directions before the injection of the ring cyclotron. The beam polarimetry system consists of three polarimeters at different locations. A low energy polarimeter is placed in the injection line and two high energy polarimeters are installed in the beam line before the spectrometer. The p-C elastic and p-p scattering are employed for the low-energy polarimeter and for the high energy polarimeters, respectively. The single turn extraction of the proton beam from the ring cyclotron, which is essential to make the in-plane polarization transfer measurement successful, has been achieved.

Momenta of scattered particles from the primary target are analyzed by the spectrometer Grand Raiden. This spectrometer has a very high momentum resolution ($p/\Delta p = 37,000$) and until now we have achieved the 25 keV(FWHM) resolution at 300 MeV with the dispersion matching technique. The layout of the Grand Raiden is shown in Fig. 1.

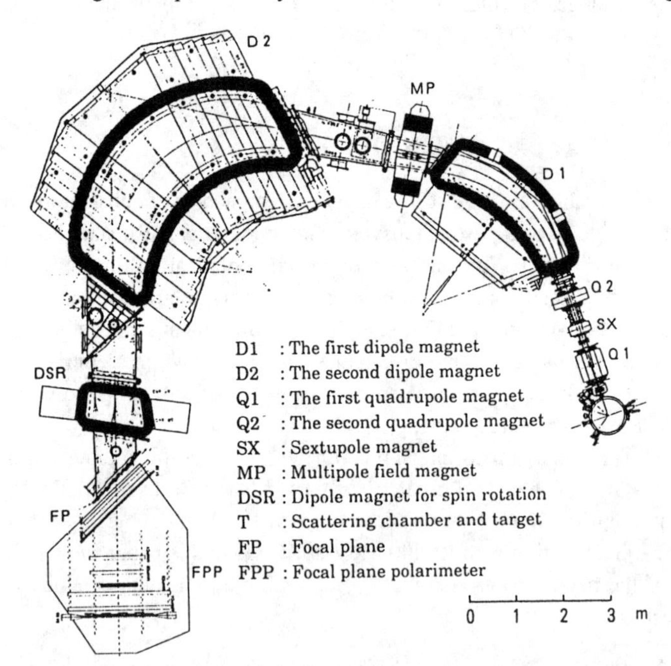

FIGURE 1. Schematic layout of the spectrometer 'Grand Raiden'.

The Grand Raiden has a special dipole magnet for spin rotation called DSR between the main QSQDMD magnet system and its focal plane. Usually, we can measure only transverse components of the polarization at the focal plane. The DSR magnet can make other two focal planes (FP+ for the DSR with positive polarity and FP− with negative polarity) in addition to the standard one (FP0) as shown in Fig. 2. This makes it possible to measure all in-plane proton polarization observables with the same accuracy in the energy region between 100 MeV and 400 MeV, continuously. For instance, at 400 MeV, when the momentum directions changed by 35° (+18° at FP+ to −17° at FP−), the proton spin axis is rotated about 90° relative to the momentum axis because of its large magnetic moment. Then, we can measure both the sideway spin component and the longitudinal spin component in the primary scattering.

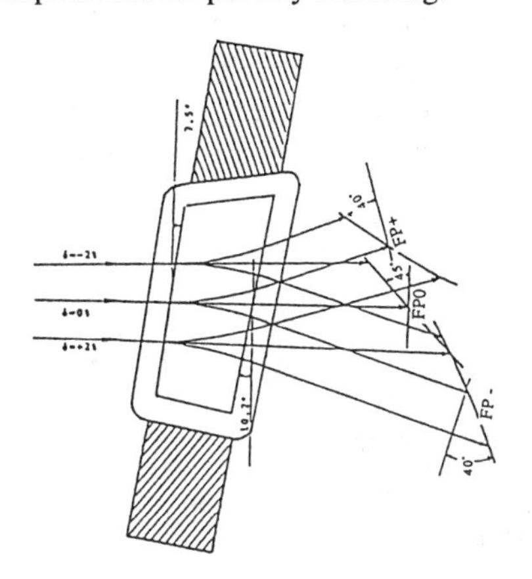

FIGURE 2. Path of the beam trajectories around the DSR magnet. There are three focal planes, corresponding to the case in which the DSR is excited positively or negatively, or is not excited.

FOCAL PLANE POLARIMETER

The last and the most important equipment for the spin transfer measurement is the focal plane polarimeter (FPP). We have designed the FPP for the Grand Raiden under the following characteristics:

- ·Polarization transfer measurements can be performed simultaneously *for the full momentum range* allowed by the Grand Raiden (±2.5% of the central momentum).
- ·The data acquisition rate can be increased *by means of the second-level trigger* for the on-line rejection of small angle scattering and *of an improved data transfer system.*
- ·Depolarization measurements of inelastic scattering are possible *even at zero degree.*

As shown in Fig. 3 schematically, the FPP consists of a thick carbon analyzer target, two pairs of multi-wire proportional chambers (MWPC's) and two plastic-scintillator hodoscopes (X and Y), and it locates downstream from the standard focal plane counters (two vertical drift chambers (VDC's) and a trigger scintillator)(5). These two VDC's determine positions and incidence angles of particles. The large MWPC's (MWPC3 and MWPC4) and the hodoscopes measure re-scattered particles from a thick carbon target.

The effective area of the largest MWPC4 is 1400mm×600mm and the total anode wires of three planes (XUV) are about two thousands. The $25\mu m^{\phi}$ gold-plated tungsten wires were wound with the tension of 90 g, and were placed on each anode plane with 2mm wire spacing. The U(V) plane has wires tilted by +45°(-45°) to the X plane. Aluminized mylar film was employed as the cathode plane of the MWPC4. Since this film was found to be weak against the breakdown, we used the carbon-Alamid film for the MWPC3.

The readout of the wire chamber information occurs via the LeCroy 4290 system for VDC's and via the LeCroy PCOSIII system for MWPC's, respectively. Digitized data are transferred to a VAX computer through the CAMAC crate using the Kinetic 3922-2922 system. The second level trigger system is constructed by fully utilizing the standard CAMAC ECLine from the LeCroy modules such as the Memory Lookup Unit (LRS2372), the Data Stack (LRS2375), and the Arithmetic Logical Unit (LRS2378). The total time to be required for the second-level trigger is only a few micro second. The MWPC1 and MWPC2 have not yet been installed, which will be used to make the second-level trigger more effective.

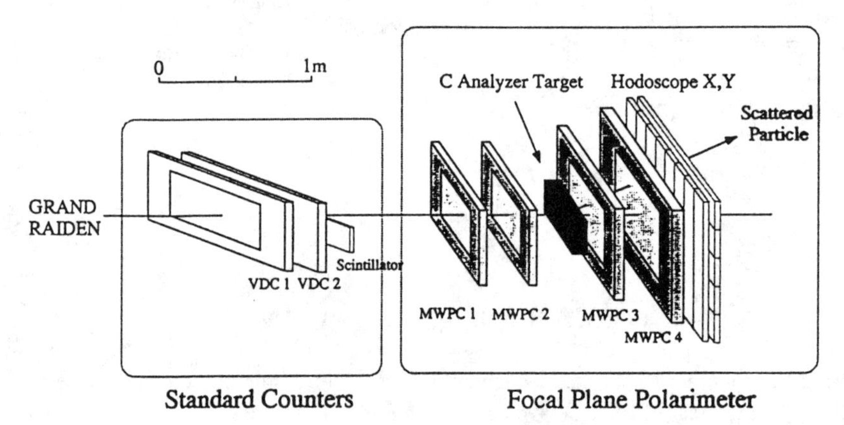

FIGURE 3. Schematic layout of the final setup of the focal plane polarimeter.

We are developing a new data transfer system to increase the data acquisition rate. The most important component is the reflective memory VME module with optical data transfer. The transfer rate of more than 2 MByte/sec has been achieved between two VME modules connected by optical cables with about 200 m lengths.

A special interest is taken in the measurement of inelastic scattering at 0° where the momentum transfer is minimized. We have already succeeded to measure cross sections of proton inelastic scattering at 0°(6) in the following manner. The primary beam was lead to a beam dump through the most high-momentum side of the focal plane without hitting any counters. In order to measure the D_{nn} at 0° for as low excitation as possible, all wire chambers (VDC's and MWPC's) have holes in their frames near the effective areas, through which a vacuum pipe for the primary beam can be placed.

EXPERIMENTAL TEST

The FPP system was checked by measuring the spin rotation parameters in proton elastic scattering from ^{58}Ni at 300 MeV at forward angles. The center scintillator of the y-hodoscope was not used in order to reduce small angle scattering events instead of using the second level trigger. The thickness of the carbon block as the second scatterer was chosen 6 cm, considering the existing A_y data at about 300 MeV (7). In elastic scattering from the spinless target, three observables (differential cross section (σ), analyzing power (A_y), and spin rotation parameter (Q)) make a complete set. A Q-parameter is deduced from each one of in-plane spin transfer observables. When we measure two of in-plane spin transfer observables using the DSR+ mode and DSR− mode, data have a redundancy and we can obtain the effective analyzing power of the FPP by itself. The experimental results are consistent with the expected values from the existing data (7) within the statistical uncertainties as listed in Table 1. The data and analysis of obtained Q-parameters are reported in the present proceedings by Tamii et al.(8).

TABLE 1. Effective analyzing powers of the FPP

DATA	Thickness of the Carbon Block	Proton Energy at the Center of the C Block	Effective Analyzing Power
E. Aprile-Giboni et al.	7cm	300MeV	0.398
E. Aprile-Giboni et al.	5cm	275MeV	0.401
D. Besset et al.	5cm	299MeV	0.395
Energy Depedent Fit by E. Aprile-Giboni et al.	3cm～7cm	284MeV	0.399
DSR±[*1)]	6cm	284MeV	0.403±0.011 (4°) 0.415±0.011 (6°) 0.394±0.011 (8°) 0.397±0.015 (10°)

[*1)] These are estimated from this experiment using both data of the DSR(+) and the DSR(-) .

As one of the future experiments, an experiment for the polarization transfer coefficients of the (p,2p) reaction has been approved, which uses also the Large Acceptance Spectrometer (LAS) together with the Grand Raiden and the FPP, and will start after finishing the check of the whole operation of the FPP and the calibration.

REFERENCES

1. Ransom, R. D., *et al.*, Nucl. Instr. and Meth. **201**, 309(1982).
2. Höusser, O., *et al.*, Nucl. Instr. and Meth. **A254**, 67(1987).
3. Wissink, S. W., *et al.*, IUCF Sci. and Tech. Report 1988, p204.
4. Fujiwara, M., *et al.*, Proc. Int. Conf. on Heavy Ion Research with Magnetic Spectrograph, edited by N. Anantaraman and B. Sherill, NSCL Technical Report No. MSUCL-685, 1989, p283.
5. Noro, T., *et al.*, RCNP Annual Report 1990, p217.
6. Sakemi, Y., *et al.*, Nucl. Phys. **A577**, 33c(1994).
7. Besset, D., *et al.*, Nucl. Instr. and Meth. **166**, 379(1979);
   April-giboni, E., *et al.*, Nucl. Instr. and Meth. **215**, 147(1983).
8. Tamii, A., *et al.*, the present proceedings.

Facility for the (p, n) Polarization Transfer Measurement at RCNP

H. Sakai[a], K. Hatanaka[b], and H. Okamura[a]

[a] *Department of Physics, University of Tokyo, Bunkyo, Tokyo 113, Japan*
[b] *RCNP, Osaka University, Mihogaoka, Ibaraki 567, Japan*

Abstract. We have designed and constructed a facility to measure the complete polarization transfer coefficient D_{ij} for the (p, n) reaction at RCNP, Osaka University.

Introduction

Polarization transfer coefficients D_{ij} for the charge-exchange (p, n) reaction

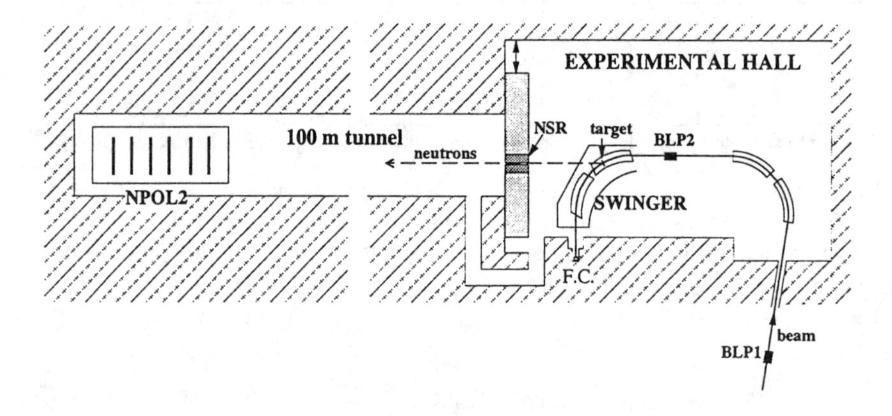

Figure 1: Schematic layout of the facility for the (p, n) polarization transfer measurement at RCNP (not in scale).

at intermediate energy provides a unique opportunity to examine spin-isospin excitations in nuclei. We have built a facility to measure D_{ij} for the (p, n) reaction over the $E_p = 200 - 400$ MeV region at the Research Center for Nuclear Physics, Osaka University.

Figure 1 shows a schematic drawing of the facility. Major equipments for the complete polarization transfer measurement are the two beam line polarimeters (BLP1 and BLP2), the neutron spin rotation magnet (NSR) and the neutron polarimeter (NPOL2) in addition to the beam swinger system and the 100 m neutron flight path for the time-of-flight (TOF) experiment.

Beams for the $(\vec{p}, \vec{n})$ Measurement

The beam line is designed to achromatically transport the beam with a resulting time resolution of typically 300-400 ps on target with the typical size of 2 mm × 2 mm. Beam pulse selection of the beam of $1/2 - 1/9$ enables a reduction of the wrap around of slow neutrons from succeeding beam pulses. The beam pulsing device, not shown in fig.1, is located in the injection beam line between the injector cyclotron and the main ring cyclotron. Two superconducting solenoids are also prepared in the injection beam line to rotate the beam polarization from the normal $(\hat{N})$ direction into sideway $(\hat{S})$ direction. These two solenoides are installed before and behind the 45° bending magnet so that they are capable of delivering two different directed proton spins in the $\hat{S} - \hat{L}$ plane to the main ring cyclotron since the spin precession in the 45° magnet is close to 86° for 63 MeV protons which corresponds to the injection energy to get a final energy of 400 MeV. Here, $\hat{L}$ indicates the longitudinal direction. To avoid a loss of the beam polarization during an acceleration, the main ring cyclotron has to be tuned so as to realize the single turn extraction.

Beam Line Polarimeters

The beam polarization $(\hat{N}, \hat{S}, \hat{L})$ can be monitored by two beam line polarimeters (BLP1,BLP2) located between the Ring cyclotron and the beam swinger target. Both BLP1 and BLP2 have four pairs of detectors (left, right, up and down) for measuring both normal $(\hat{N})$ and sideways $(\hat{S})$ components of beam polarization. A 98° bending magnet exists between these two polarimeters. The spin precession angle due to this bending is 70.6° for the 400 MeV protons. Thus the longitudinal component $(\hat{L})$ at BLP2 can be measured almost as the sideway component (≃70.6°) at the BLP1 position.

The polarimetry is based on the analyzing power of the ^{1}H$(\vec{p}, pp)$ reaction. The elastically scattered and recoiled protons are detected in coincidence with a pair of plastic scintillators. Both polarimeters use self-supporting thin CH_2 targets with a thickness of ≃250 μg/cm^2.

Beam Swinger

The beam swinger consists of two 45° bending C-shaped magnets with the orbit radius of 2 m. The gap and width of the pole are 10 cm and 40 cm, respectively. Both magnets are excited simultaneously by using one set of coil (1.53×10^5 AT) and the maximum field strength is measured to be 1.603 T at 520 A. The scattering angle ($\theta_n = -0.5° \sim 85°$) can be varied by repo-

Figure 2: Photograph of the beam swinger magnet and the neutron spin rotator.

sitioning a target along the beam trajectory inside the pole gap for a fixed neutron detectors in the 100 m tunnel.

Figure 2 is the photograph of the beam swinger magnet and the neutron spin rotation magnet.

Neutron Spin Rotation Magnet

In order to measure the $\hat{L}$ neutron polarization, an H-type, spin rotation magnet with a pole gap of 12 cm and width of 35 cm is embedded in the 1.5 m concrete shielding wall at the entrance of the TOF tunnel (see Fig.1 and 2). The maximum integrated magnetic field strength (BL) along the neutron path has been measured to be 1.96 Tm which is capable of rotating 400 MeV $\hat{L}$ neutron spins into the $\hat{N}$ direction. This magnet is also used for a measurement of the induced polarization P. In this case the $\hat{N}$ neutron spin is rotated into the $\hat{L}$ direction. To reduce the neutron background in the tunnel the aperture of the pole gap is collimated to a size of 12 cm × 12 cm with aluminum plates. The position of the concrete wall may be tracked so as to maintain its linear relationship with the movable target and fixed detector positions.

Neutron Polarimeter NPOL2

The current version of the neutron polarimeter (NPOL2) consists of 6 planes (4

Figure 3: Photograph of NPOL2.

planes of liquid scintillators BC519 and 2 planes of plastic scintillators BC408) of two-dimensional position-sensitive neutron counters [1] with a size of 1 m$\times$ 1 m $\times$ 10 cm. A picture of NPOL2 is shown in Fig.3. Since the counter has no preference in direction in the detector plane, both $\hat{N}$ and $\hat{S}$ components of the neutron polarization can be determined in similar accuracy. Figure 4 shows a typical energy spectrum for the ^{7}Li$(p,n)^7$Be reaction at 295 MeV and 0° measured with a singles mode of NPOL2 at a 100 flight path. The prominent peak at 295 MeV consists of two transitions to the ground and first excited (0.43 MeV) states each carrying similar amount of the Gamow-Teller strengths. These transitions were not resolved under the present energy resolution of $\simeq$ 1.5 MeV. Except this peak, the spectrum is made up of very smooth continuum which extends down to the neutron energy of 100 MeV. The continuum cross section is about 0.1 mb/sr·MeV at 100–200 MeV region while the continuum background yield in the same energy region is estimated to be negligibly small, less than 0.01 mb/sr·MeV, under the present experimental conditions.

A previous version (NPOL1) consisted of 4 planes of neutron counters is described elsewhere in detail [2]. We will give only outline here.

The neutron polarimeter NPOL1 has been calibrated at 300 MeV by using the ^{2}H$(\vec{p},\vec{n})pp$ reaction whose polarization transfer is known to be $D_{NN}(0°)$ $= -0.295 \pm 0.004$ [4]. The polarimetry is made utilizing the analyzing power

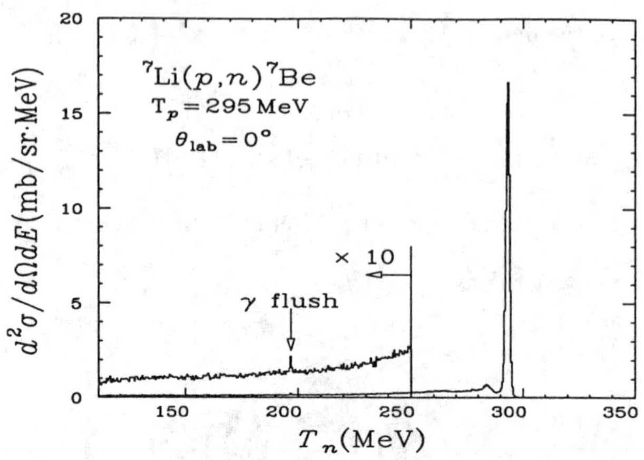

Figure 4: Neutron energy spectrum for the ^{7}Li$(p,n)^7$Be reaction at 295 MeV and 0°. See text for detail.

of n-p scattering by detecting either scattered neutrons or recoil protons. The angle averaged effective analyzing power ($< A_y^{eff} >$) values, thus obtained, are $0.292 \pm 0.012 \pm 0.009$ for the ^{1}H$(\vec{n},n)p$ channel and $0.115 \pm 0.003 \pm 0.003$ for the ^{1}H$(\vec{n},p)n$ channel where the errors are statistical and systematical, respectively. The double scattering efficiencies (DSE) are 0.0010 and 0.0194 for each channel. Based on this measurement, NPOL2 is expected to have DSE more than two times larger as compared with NPOL1.

During the past year this facility has been used to measure the $(\vec{p},n)$ quasi-elastic scattering on various targets ranging from ^{2}H to ^{208}Pb [3] and $D_{NN}(0°)$ values on p-shell targets [2]. The complete D_{ij} measurements are expected to begin after this conference.

ACKNOWLEDGEMENTS

The authors would like to thank H. Otsu, T. Wakasa, S. Ishida, N. Sakamoto, T. Uesaka, Y. Satou, S. Fujita, A. Okihana, M.B. Greenfield, S. Kamigaito, and N. Koori for their help in various stages of construction and R/D.

REFERENCES

1. H. Sakai, Nucl. Instr. and Meth. **A320** , 479(1992).
2. T. Wakasa, contribution to this conference and Master thesis, University of Tokyo, 1994.
3. H. Otsu, contribution to this conference and Master thesis, University of Tokyo, 1993.
4. M.W. McNaughton et al., Phys. Rev. **C34**, 2564(1992).

A Proton Polarimeter Using Liquid Helium Target

Y. Mukouhara, S. Nakagawa, H. Kishita, T. Katabuchi, K. Hirota, M. Masaki, Y. Aoki and Y. Tagishi

Institute of Physics and Tandem Accelerator Center, University of Tsukuba, Ibaraki 305, Japan

Abstract. The design, the calibration and the performance of a high efficiency proton polarimeter for double scattering experiments are described. The polarimeter target is liquid helium held in a conical cell. Protons scattered by helium are detected by two phototubes with CsI(Tl) scintillator, having a diameter of 50 mm and a thickness of 2 mm, and placed at a scattering angle of 60° in the left and right at a distance of 130 mm from the ^{4}He target. The effective analyzing power is about -0.5 and the detection efficiency is about 2×10^{-4} between 12 and 22 MeV. Once liquid helium is filled in the 6 ℓ reservoir tank, the liquid helium target works stably at least for three days.

INTRODUCTION

We have developed a proton polarimeter to measure the polarizations of protons emitted in reactions induced by polarized and unpolarized beams. This polarimeter is set at the focal position of the QDQ-magnetic spectrometer at the Tandem Accelerator Center, University of Tsukuba (UTTAC). Figure 1 shows the layout of the double scattering experiment. The polarized proton and deuteron beams are produced by a Lamb-shift-type ion source (1) and accelerated up to 24 MeV.

Polarimeters are usually based on the scattering from ^{12}C and ^{4}He at these energy regions. In double scattering experiments, the intensity of the scattered particles is extremely weak. Thus it becomes necessary to have a polarimeter of high efficiency. Helium is the most popular analyzer-target

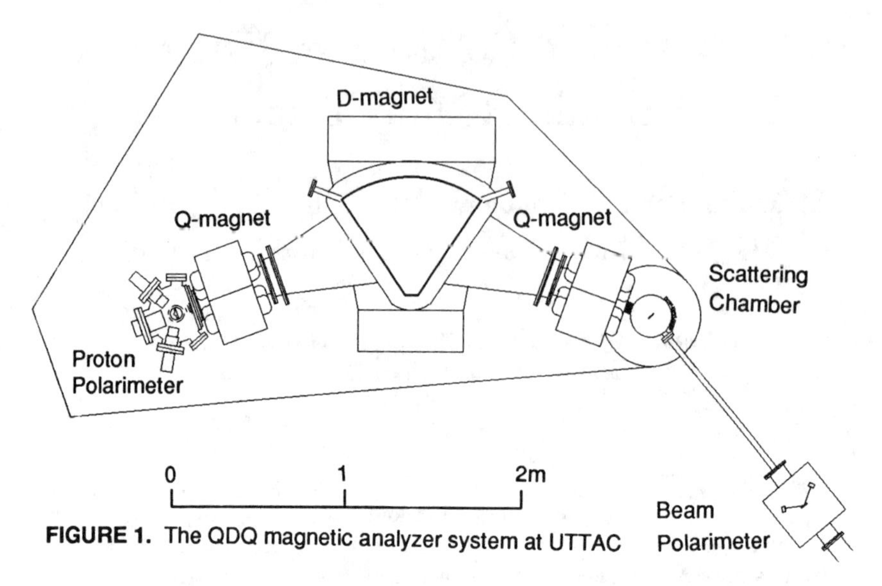

FIGURE 1. The QDQ magnetic analyzer system at UTTAC

for protons below 30 MeV because of its large cross section and slow variation of analyzing power with energy and angle. Then, we have designed a high efficiency proton polarimeter using a liquid helium target with reference to that at Kyushu University (2). We report here the design and performance of this polarimeter.

DESIGN

Figure 2 is a schematic cross section of the liquid helium cryostat. The characteristics of the present cryostat are the following: (a) The structure is simple. The liquid nitrogen and helium tanks are supported by the stainless steel pipes from the upper flanges. This results the heat loss keeps as small as possible. (b) To reduce heat input to the helium, the helium tank and target cell are completely surrounded by walls cooled to the liquid nitrogen temperature. Moreover the radiation shields of the aluminum-coated mylar foils are inserted between the tank and walls. The heat input is estimated to reduce to approximately 6.5 mW with these mylar shields compared to 13 mW without them. (c) To avoid bubbles in the target, liquid helium in the target cell can be kept in the superfluid state by decreasing the vapor pressure of the small reservoir tank (200 cc) below 37 mm Hg using a pump. At present, the whole heat input comes from the supporting pipes, which is estimated to be about 70 mW, so once liquid helium is filled in the 6 ℓ reservoir tank, the liquid helium target works stably for three days.

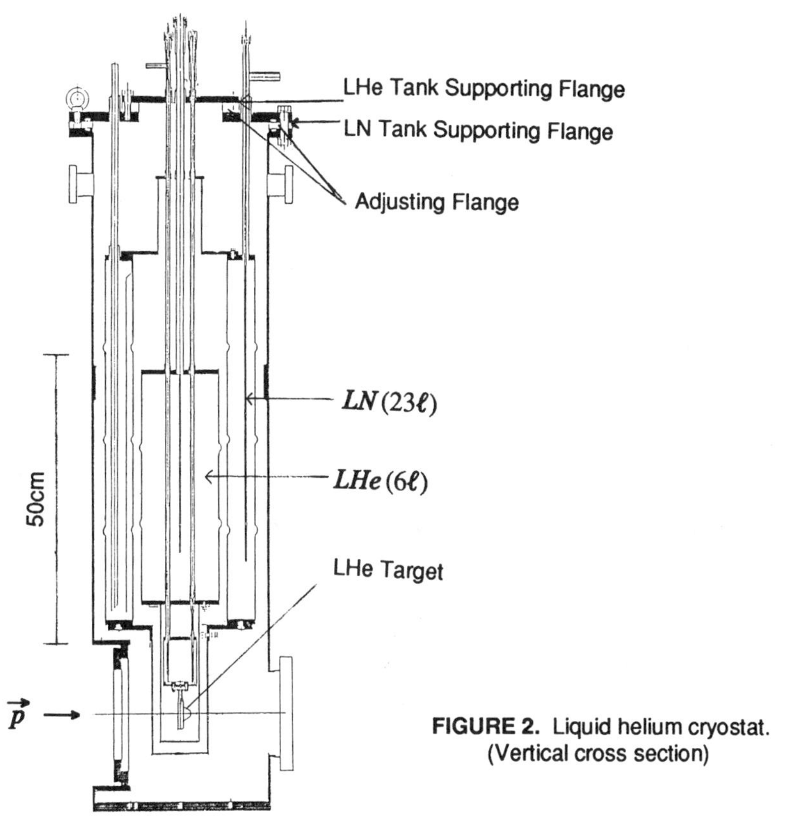

FIGURE 2. Liquid helium cryostat. (Vertical cross section)

The target and detectors are shown in Fig. 3. The target thickness is about 1 cm (146 mg/cm^2), equivalent to gaseous helium target at about 20 atm$\times$40 cm. A 40 μm-thick mylar foil is used for the entrance and exit windows of the target. The exit window is made a conical shape to minimize the energy spread of the scattering protons (2). As shown in Fig. 3, the protons focused by the QDQ analyzer, pass through a silicon transmission detector ΔE (700 μm) and through the liquid helium target in the polarimeter. The beam distribution is monitored by using a plastic scintillator of the bundled-fiber with 50 mm in diameter and 5 mm thickness, which is attached to a 3-inch position-sensitive phototube. The false asymmetry can be reduced by using a large size of detector in case of keeping a certain solid angle. The protons scattered by helium are detected by two phototubes with CsI(Tl) crystals, having a diameter of 50 mm and a thickness of 2 mm, placed at a distance of 130 mm from the center of the ^{4}He target and at a scattering angle of 60° in the left and right, which corresponds the maximum figure of merit around 20 MeV.

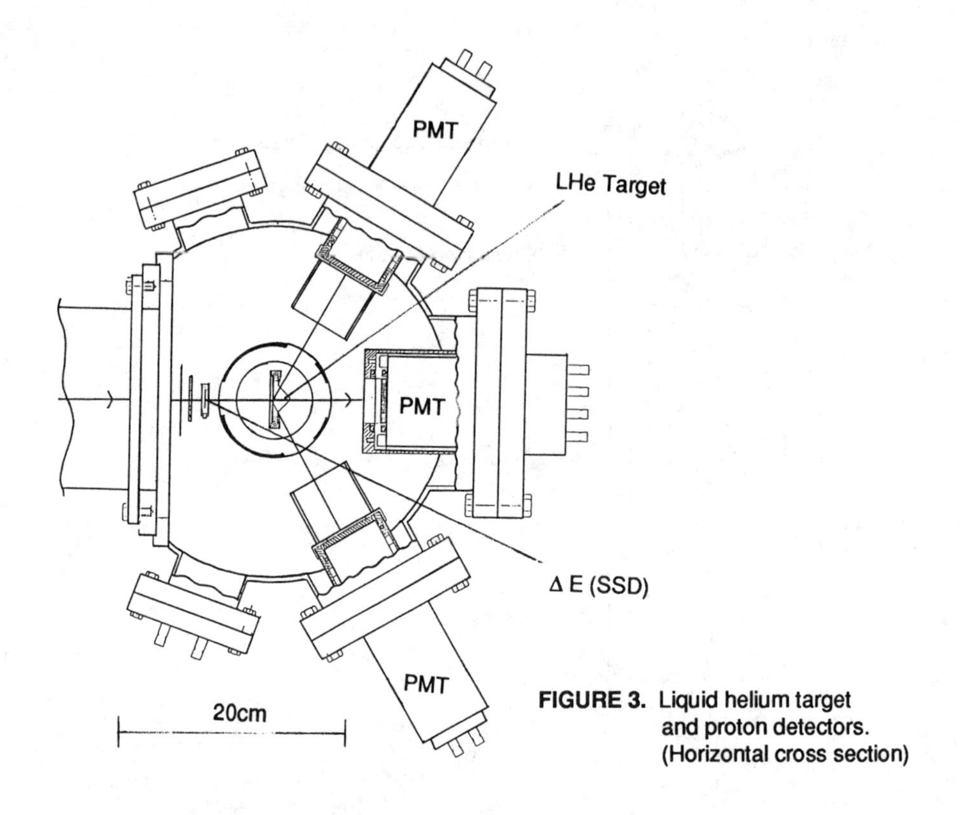

FIGURE 3. Liquid helium target and proton detectors. (Horizontal cross section)

CALIBRATION

A typical energy spectrum of the 16.5 MeV protons scattered from liquid helium target is shown in Fig. 4. The background at low energy can be rejected by making coincidence with the ΔE counter. We measured the effective analyzing powers and efficiencies for proton energies between 12 and 22 MeV using the polarized protons from the Au($\vec{p}$, $\vec{p}$) elastic scattering at 30°. The polarization of the primary beam was measured by a quench-ratio method. The accuracy of this method is estimated to be correct within 2% at UTTAC. The experimental results are shown in Fig. 5. The incident energy in the figure shows the proton energy at the entrance of the helium target. The solid curve indicates the result of the least-squares fit. In case of the incident energy is below 14 MeV, some of the scattered protons are stopped in the helium target, so the detection efficiency decreases. The efficiency can recover by using a thinner ΔE counter for protons below 14 MeV. The effective analyzing power is about −0.5 and varies smoothly with the incident energy. The efficiency is about 2×10^{-4}.

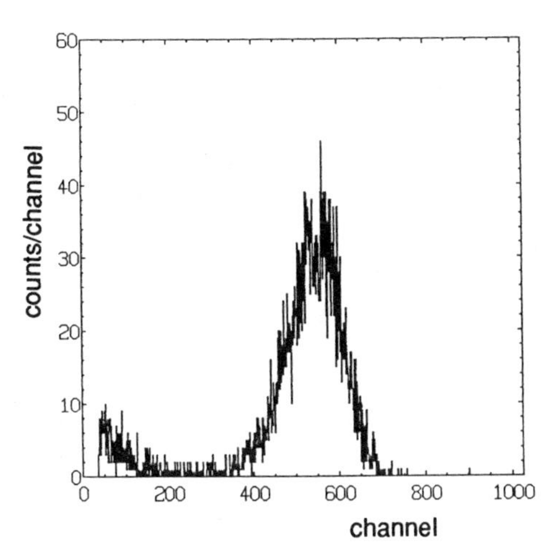

FIGURE 4. Energy spectrum of protons scattered by the liquid helium target.

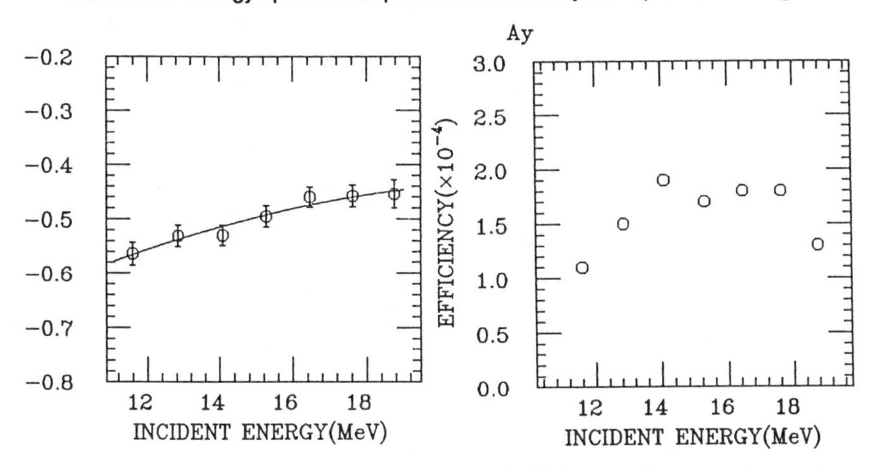

FIGURE 5. Effective analyzing power and efficiency of the secondary proton polarimeter. Solid curve shows the least-squares fit. The incident energy shows the proton energy at the entrance of the helium target.

ACKNOWLEDGEMENTS

The authors would like to thank K. Sagara and H. Ikeda for helpful discussions and comments.

REFERENCES

1. Tagishi Y. and Sanada J., Nucl. Instr. and Meth. **164**, 411 (1979).
2. Sagara K. et al., Nucl. Instr. and Meth. **A270** 450 (1988).

A Breit–Rabi Polarimeter for internal H/D–Targets and the Observation of Relaxation Processes

B. Braun[a], E. M. Gabriel[b], H. G. Gaul[b*], G. Graw[a], H. Kolster[a], A. Metz[a], K.Reinmüller[a], P. Schiemenz[a], I. Simiantonakis[b], E. Steffens[b]

[a]*Sektion Physik, Universität München, 85748 Garching, Germany*
[b]*MPI für Kernphysik, 69029 Heidelberg, Germany*
* *now at SAP, 69190 Walldorf, Germany*

Abstract. This contribution describes design and operation of a polarimeter, as it will be used in the HERMES experiment for the continuous control of the polarization of the hydrogen or deuterium target. We also report on some results about relaxation processes like spin exchange and wall collisions.

The HERMES target, as it will be installed in the HERA ring, is a thin walled storage cell, fed by a high intensity atomic beam source (ABS) of polarized hydrogen and deuterium atoms [1]. The effective target thickness, seen by the positron beam of the HERA ring, is near $1 \cdot 10^{14}$ atoms $\cdot$ cm^{-2} [2]. For polarization control a small fraction of the stored gas will be allowed to exit the cell through an additional tube forming a sample beam whose polarization is analyzed in a Breit–Rabi polarimeter (BRP) [3]. It determines the occupation probabilties n_i of the ground state hyperfine levels which are numbered according to their energy, starting with $i = 1$ for the highest energy state.

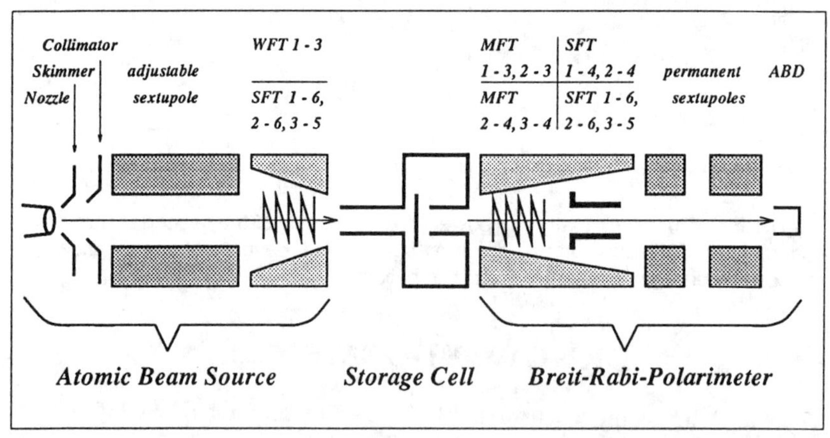

Figure 1: Experimental set–up for testing the Breit–Rabi polarimeter and for measurements concerning relaxation processes inside storage cells. The upper transitions are applied for hydrogen, the lower ones for deuterium. WFT, MFT and SFT are indicating weak, medium and strong field transitions respectively.

Fig.1 shows the test set up for the studies reported here: A colinear arrangement of a test ABS, a storage cell especially designed for highly sensitive relaxation measurements and the Breit–Rabi polarimeter.

In the BRP two high flux permanent sextupole magnets [4] separate the atoms according to their electron spin states and focuss the $m_j = +1/2$ states into the atomic beam detector (ABD). Before entering these magnets the beam passes adiabatic high frequency transition (HFT) units which are used to interchange the population between two hyperfine states of the beam with different m_j quantum numbers. This leads to changes in the atomic beam signal (detected after the sextupole magnets) which are proportional to the difference of the population of the states involved. The occupation probabilities inside the beam are obtained from the signal ratios for HFT on/off. In addition to the normalization condition $\Sigma_i n_i = 1$ three signal ratios using different transitions are needed in the case of hydrogen and five in the case of deuterium.

The BRP is mounted in a system of three differentially pumped chambers, using turbo, cryogenic, ion getter and titanium sublimation pumping. The pressure in the bakeable detector chamber (200° C) is below $3 \cdot 10^{-9}$ mbar with gas load from the probe beam and below $1 \cdot 10^{-10}$ mbar without gas load. In the detector chamber the probe beam passes a chopper with an opening period of $20\,\text{sec}^{-1}$ to allow for the subtraction of background arising from the residual gas and a subsequent electron collision ionizer. The H^+/D^+ ions selected by a quadrupole mass spectrometer are accelerated and focussed into the entrance window of a secundary electron multiplier, allowing thus time resolved single ion counting (Fig.2).

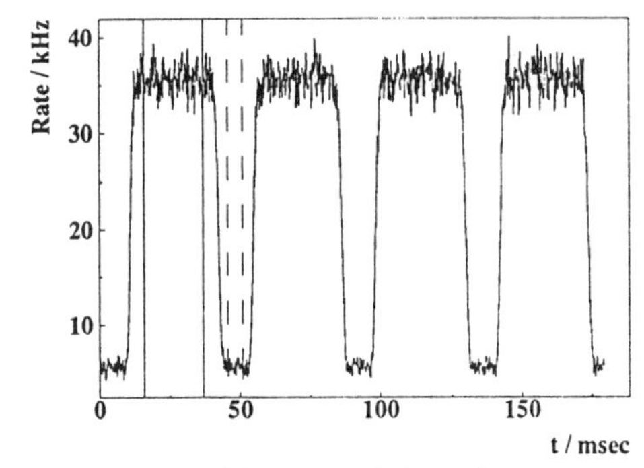

Figure 2:
Signal and background count rates of $35 \cdot 10^3\,\text{sec}^{-1}$ and $5 \cdot 10^3\,\text{sec}^{-1}$ resp. of a chopped atomic hydrogen beam in a typical cycle for an incoming flux of $5 \cdot 10^{10}$ atoms$\cdot$sec^{-1}. The effective time to measure the beam signal with the statistical accuracy shown in this figure is 50 sec.

For the calibration of the polarimeter we removed the storage cell, installed beam stoppers on the axis of the sextupoles and measured the drop of the atomic beam signal in the ABD when switching on a HFT. The decrease of the signal is proportional to the interchange efficiency of the transition. The efficiencies of the two interchanges going on in parallel in a medium field transition can be measured separately using an additional HFT in front of the medium field transition, e.g. the WFT 1–3 in case of hydrogen.

With one exception we obtained [5] efficiencies better than 99.5% with an accuracy of better than 0.5% for all hydrogen transitions used. For the deuterium transitions we obtained efficiencies near 99% with an accuracy near 2%. Due to these high efficiencies all the sextupole corrections concerning the efficiency measurement itself could be done experimentally. The sextupole corrections for a polarization measurment on a beam emerging out of a storage cell are determined by means of Monte–Carlo simulations.

In the HERMES experiment, the axes of ABS and the BRP are tilted with respect to each other by an angle of 60°. For calibration a beam consisting exclusively of $m_j = +1/2$ states is provided then inserting an additional removable sextupole magnet as a filter in front of the BRP.

The behaviour of the polarization of hydrogen and deuterium atoms in storage cells has been studied, using a test cell arrangement, designed for a mean number of wall collisions of about $N = 3500$ (or $N = 1500$), for variation of temperature ($T > 90\,\mathrm{K}$) and magnetic field ($0.01\,\mathrm{T} < B < 0.75\,\mathrm{T}$) and for a comparison of different surface materials: Drifilm, Teflon and uncoated aluminum. For the HERMES cell we have a much lower number of wall interactions $N = 300$, but due to the higher density in the target cell about the same number of atom-atom collisions in the cell.

In the ABS a weak field transition (1–3) for hydrogen and three different strong field transitions (1–6, 2–6 and 3–5) for deuterium have been used to prepare the beam before injection into the storage cell.

For hydrogen, Fig.3a shows the substate populations n_i as function of the magnetic field for a case, where predominantly substates n_2 and n_3 have been injected into the cell, i.e. WFT 1–3 switched on. The magnetic field dependence is much stronger for the mixed states $|\,2\,\rangle$ and $|\,4\,\rangle$ than for the pure

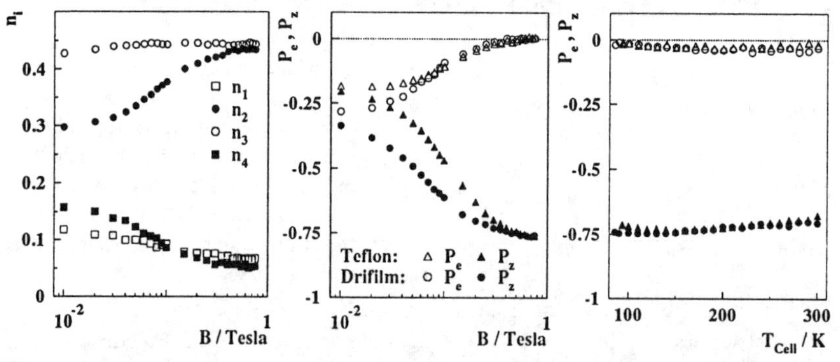

Figure 3: Relaxation of hydrogen in a cell with $N = 3500$ wall bounces. a.) Substate populations as function of the magnetic field for a Teflon coated cell at $T = 100\,\mathrm{K}$. b.) Electron and nuclear polarizations as function of the magnetic field for a Teflon or Drifilm coated cell at $T = 100\,\mathrm{K}$. c.) Electron and nuclear polarizations as function of the cell temperature for a Teflon or Drifilm coated cell at $B = 0.35\,\mathrm{T}$.

states $|1\rangle$ and $|3\rangle$. The respective electron and nuclear polarizations P_e and P_z are plotted in Fig.3b, comparing also Drifilm and Teflon coatings. Fig.3c shows at $B = 0.35\,\mathrm{T}$, the HERMES working point, the electron and nuclear polarizations as function of the cell temperature, again comparing Drifilm and Teflon coatings. Using a cell with $N = 1500$ we measured nearly identical nuclear polarization at $B = 0.35\,\mathrm{T}$ indicating a neglectable contribution of wall collisions to losses in P_z at this field.

For deuterium Fig.4 shows an example for the relaxation of the substate population of deuterium as function of the magnetic field in a Teflon coated storage cell. Substates $|1\rangle$, $|3\rangle$ and $|6\rangle$ have been injected. At high magnetic field we observe a significant increase of the population of state $|1\rangle$ over the initial value of 1/3. This surprising behaviour is understood as observation of spin exchange collisions (SEC) [6].

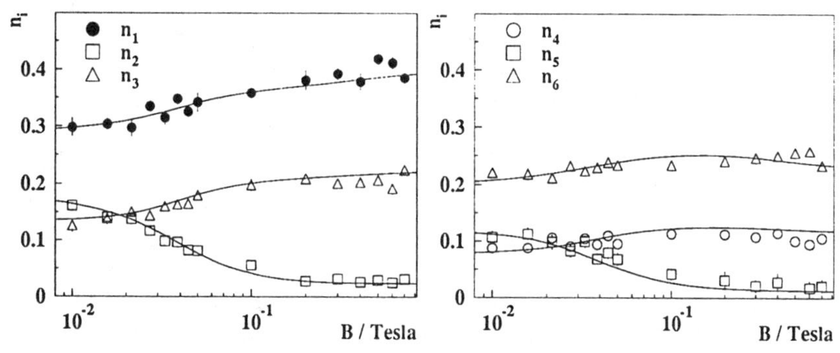

Figure 4: Substate population of deuterium as function of the magnetic field for a Teflon coated storage cell at a cell temperature of $T = 100K$ and a mean wall collision number of $N = 3500$. Left: substates $|1\rangle$ to $|3\rangle$. Right: substates $|4\rangle$ to $|6\rangle$. The lines give a calculation considering spin exchange collisions and wall bounces.

According to Purcell and Field [6] the collision of thermal hydrogen or deuterium atoms leads to an electron spin exchange with probability 1/2, independent on the external magnetic field. At high fields the nuclei are decoupled from the electrons and act as nearly pure spectators. Therefore the predominant spin exchange reaction in the presently regarded case is $|3\rangle + |6\rangle \longrightarrow |1\rangle + |4\rangle$ causing the observed increase of the already populated state $|1\rangle$. At a lower magnetic field the hyperfine mixing increases the number of states participating in SECs and thus causes a faster approach to the m_F dependent spin temperature equilibrium [7].

As shown in Fig.5 (left) wall collisions [8, 9] lead predominantly to a loss of electron polarization while SECs affect mainly the tensor polarization P_{zz} of the nuclei Fig.5 (right). We could not detect any significant effect on the vector polarization for deuterium at magnetic fields above 0.05 T which is due to the low critical field of 0.012 T.

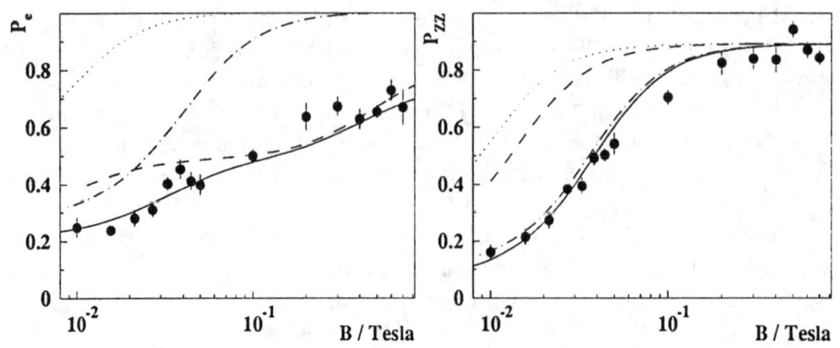

Figure 5: Relaxation of electron (left) and tensor polarization (right) of deuterium as a function of the magnetic field in a Teflon coated storage cell at a cell temperature of $T = 100\,\mathrm{K}$. The mean wall collision number is about 3500. Left: Injection of substates $|1\rangle$, $|2\rangle$ and $|3\rangle$. Right: Injection of substates $|1\rangle$, $|3\rangle$ and $|6\rangle$. The dotted lines give the injected polarizations, the dashed show the effect of wall collisions and the dashed–dotted the effect of SECs. The solid lines enclude both mechanisms.

We needed less than 30 sec for hydrogen and about 200 sec for deuterium to reach a statistical error of 3 % which is the HERMES request for the BRP. The systematic errors which are due to calibration errors are $\Delta P_z = 0.013$ for hydrogen and $\Delta P_z = 0.032$ and $\Delta P_{zz} = 0.054$ for deuterium. The systematic errors for deuterium will be improved in future by increasing the efficiencies of the HFTs to a level already obtained for the hydrogen transitions.

Work is supported by the BMFT under grant 06 ML171/3 and 056 MU22 I1.

References

[1] F. Stock et al., *The HERMES-FILTEX target source for polarized hydrogen and deuterium*, these proceedings.

[2] K. Zapfe et al., *High density polarized hydrogen gas target for storage rings*, these proceedings.

[3] H.G. Gaul and E. Steffens, *Nucl. Instr. Meth.* **A 316** (1992) 297

[4] P. Schiemenz et al., *Nucl. Instr. Meth.* **A 305** (1991) 15

[5] B. Braun, *Phd. thesis* in prep. and to be published in *Nucl. Instr. Meth.*.

[6] E.M. Purcell and G.B. Field, *Astrophys. J.* **124** (1956) 542

[7] T. Walker and L.W. Anderson, *Nucl. Instr. Meth.* **A 334** (1993) 313

[8] M.A. Bouchiat and J. Brossel, *Phys. Rev.* **147** (1966) 41

[9] D.R. Swenson and L.W. Anderson, *Nucl. Instr. Meth.* **B 29** (1988) 627

The use of the $^3He(\vec{d},p)^4He$ reaction for polarimetry at low energies.

W. Geist, Z. Ayer, A.C. Hird*, K.A. Fletcher#, H.J. Karwowski and E.J. Ludwig

Department of Physics and Astronomy, University of North Carolina, Chapel Hill, NC 27599-3255, USA, and
Triangle Universities Nuclear Laboratory, Durham, NC 27706, USA
**Present address: US AFB. Dover, Delaware, USA*
#Present address: Department of Physics and Astronomy, State University of New York at Geneseo, Geneseo, NY 14454, USA

Abstract. Angular distributions of tensor analyzing powers A_{yy} and A_{zz} have been measured in $^3He(d,p)^4He$ reaction in the 130 to 322 keV energy range. The reaction proves to be very efficient and easy to use to determine polarization of deuteron beams.

Nuclear reaction measurements in the energy range from 0 to 500 keV are of importance for studies of properties of few-body systems, mechanism of nuclear fusion and for problems relevant to nuclear astrophysics. Measurements of polarization observables provide new insights into the reaction mechanisms and nuclear structure questions. High precision measurements require efficient and accurate polarimeters. However, there are a limited number of reactions which can serve as polarization analyzers.

For deuteron polarimetry, the $^3H(\vec{d},n)^4He$ (1) and $^2H(\vec{d},p)^3H$ (2) reactions have been used as analyzers in the energy region below 1 MeV, but neutrons from $^3H(\vec{d},n)^4He$ are difficult to detect and the cross section of the $^2H(\vec{d},p)^3H$ reaction is low and highly anisotropic. The $^3He(\vec{d},p)^4He$ reaction offers an attractive alternative to these reactions in that the exiting protons have energy greater than 18 MeV and the presence of $J^\pi = 3/2^+$ resonance in ^5Li, reached when $E_d = 430$ keV, enhances the yield significantly at energies in the vicinity of the resonance. Moreover, S-wave capture at or near the resonance energy provides an A_{yy} value of +0.5 for the reaction independent of energy and angle (3) and an A_{zz} value which is close to -1 at forward angles (4), while the vector analyzing power is equal to zero. The same reaction is commonly used for polarimetry at tandem energies, where $A_{zz}(0^o)$ reaches values well below -1.0.

The efficiency of polarization analyzers depends on σA^2 where σ and A are the cross section and analyzing power, respectively. A large total cross section, close to 900 mb at the resonance, makes the $^3He(\vec{d},p)^4He$ reaction in the energy range between 100 and 600 keV a considerably more efficient analyzing reaction than $^2H(\vec{d},p)^3H$. The comparison of those two reactions is shown in Fig. 1.

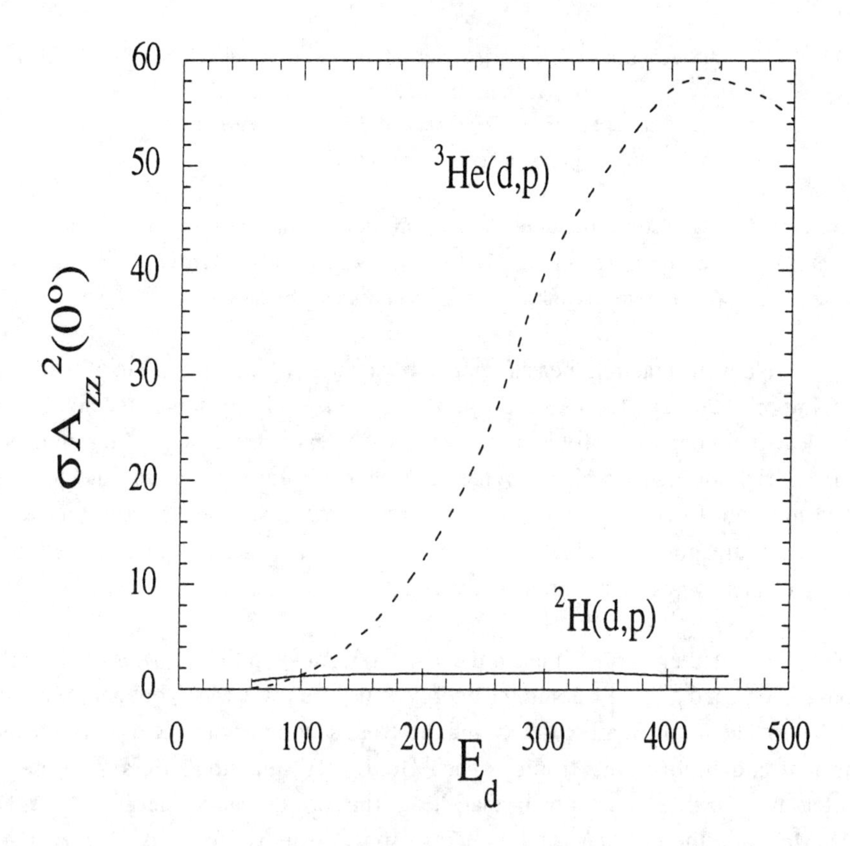

Figure 1. Comparison of the figure of merit for $^3He(\vec{d},p)^4He$ and $^2H(\vec{d},p)^3H$ reactions..

There exist few analyzing-power data for the $^3He(\vec{d},p)^4He$ reaction in the energy range of interest. Only measurements of $T_{20}(0^o)$ at 240 keV (4) and 340 keV(5) and $T_{20}(\theta)$ at 340 keV(6) exist below the resonance energy. We have therefore measured analyzing powers of the $^3He(\vec{d},p)^4He$ reaction not only to determine its usefulness as a high efficiency analyzer, but also to provide a better understanding of the d + ^3He reaction mechanism at energies below the resonance.

In the present work, measurements of $A_{yy}(\theta)$ and $A_{zz}(\theta)$ for the $^3He(\vec{d},p)^4He$ reaction were made at forward angles at deuteron energies from 130 keV to 322 keV in approximately 50 keV steps. Deuterons in the TUNL atomic beam polarized ion source were accelerated to 72 keV and then to their final energy using the TUNL minitandem accelerator (7). The beam polarization was better than 65%. The beam was incident on a target made by ion-implanting 17 keV ^3He ions into a tantalum foil. Gas cells are not suitable targets at low energies because of large energy loss and straggling of the incident deuterons while penetrating the cell walls. It is also important that implants can be deposited in the surface layer of the foil (in this case within 72 nm) which allows precise determination of the energy of the interaction. The typical spectrum is shown in Fig. 2.

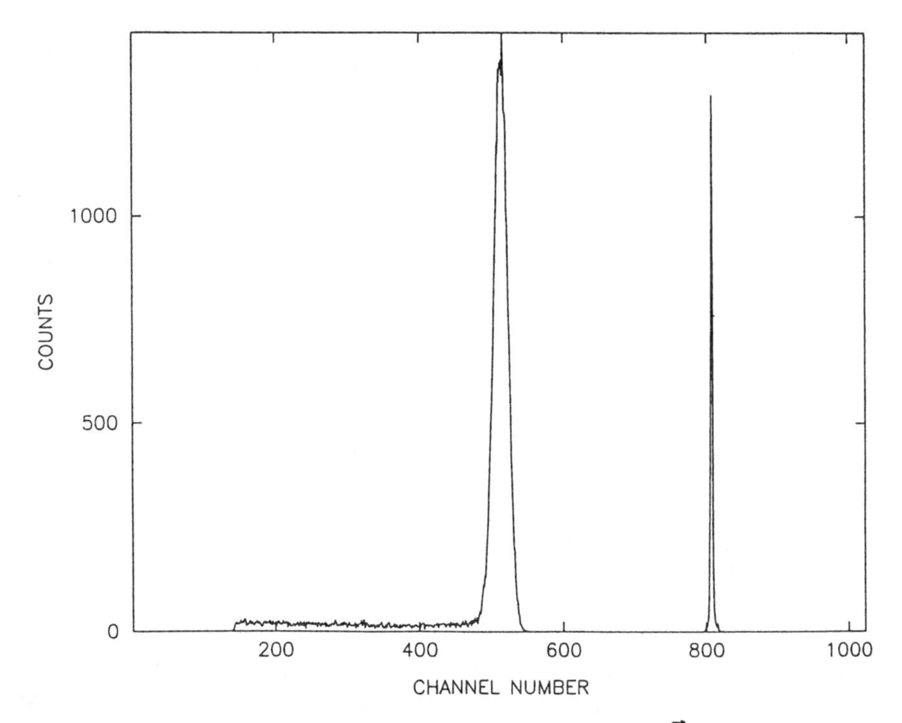

Figure 2. An example of the spectrum of protons from $^3He(\vec{d},p)^4He$ reaction at E_d = 322 keV. The narrow peak at channel 800 is the pulser.

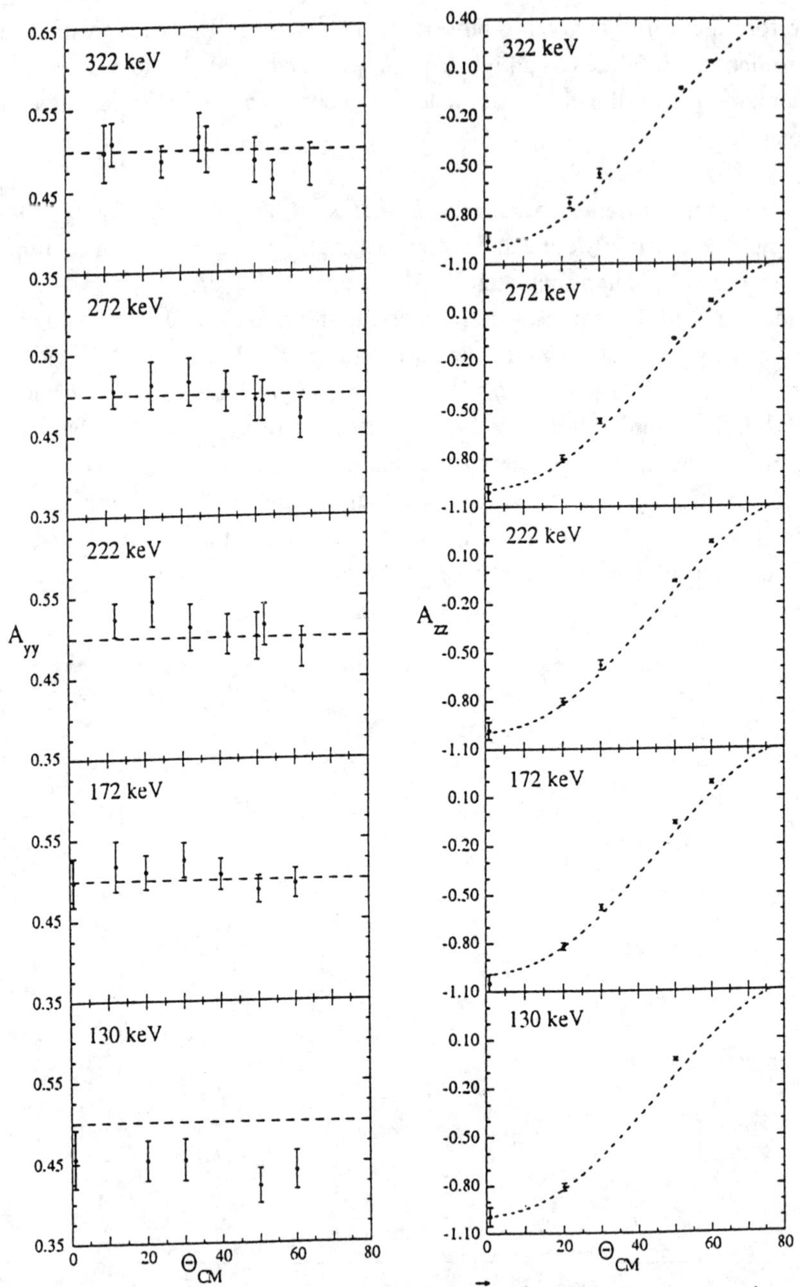

Figure 3. Tensor analyzing powers for the $^3He(\vec{d},p)^4He$ reaction at energies below the $J^\pi=3/2^+$ resonance. The dashed line represents the expected values for s-wave resonance, as described in the text.

The polarization of the incident deuteron beam was determined by accelerating the beam to 5 MeV in the tandem accelerator and sending it to a well-calibrated polarimeter. Using the normalization obtained in high-energy experiment we established that A_{yy} at E_d=322 keV is in fact equal at this energy to 0.50± 0.01 and in subsequent runs at lower energies we used the reaction at this energy to monitor the beam polarization.

The $A_{yy}(\theta)$ are consistent with the expected value of +0.5, independent of energy and angle, for energies near the resonance as shown in Fig. 3a. The average over all whole angular range is equal to 0.501±0.004. At the lowest energy of 130 keV measured A_{yy} is considerably lower and on average is equal to 0.44±0.01.

The angular distribution of A_{zz} at 322 keV follow the expected angular function:

$$A_{zz}(\Theta_{cm})=-0.5[3\cos^2(\Theta_{cm})-1]$$

as shown in Fig. 3b, although the data points at larger angles are slightly above the theoretical curve. The deviations from expected values are likely due to the deuteron D-state effects.

The close agreement with expected values over the energy range studied provides ample indication that the $^3He(\vec{d},p)^4He$ reaction can be used as an efficient polarization analyzer for deuterons with energies between 100 and 400 keV.

This work is supported by the U.S. Department of Energy Grant # DE-FG05-88ER40442

REFERENCES

1. E.R. Collins and H.F. Glavish, Nucl. Inst. and Meth. **30,** 245 (1964).
2. K.A. Fletcher *et al.*, Nucl Inst and Meth. **329**, 197 (1993).
3. H. E. Conzett, Few Body Problems in Physics, Vol II, 539 (1984).
4. L.J. Dries *et al.*, Physics Rev. **C21**, 475 (1980).
5. P.A. Schmelzbach *et al* ., Nucl. Phys. **A264**, 45 (1976).
6. R. Garrett and W.W. Lindstrom, Nucl. Phys. **A224**, 186 (1974).
7. T.C. Black *et al.*, Nucl. Inst. and Meth. **A333**, 239 (1993).

Construction of the Deuteron POLarimeter DPOL at RIKEN

S. Ishida, S. Fujita, Y. Hara*, K. Hatanaka**, T. Ichihara†, K. Katoh*, T. Niizeki*, H. Okamura, H. Otsu, H. Sakai, N. Sakamoto, Y. Satou, T. Uesaka, T. Wakasa, and T. Yamashita*

Department of Physics, University of Tokyo, Tokyo 113, Japan
**Department of Physics, Tokyo Institute of Technology, Tokyo 152, Japan*
***Research Center for Nuclear Physics, Osaka University, Osaka 567, Japan*
† The Institute of Physical and Chemical Research (RIKEN), Saitama 351-01, Japan

Abstract. The new intermediate energy Deuteron POLarimeter DPOL at RIKEN is described. DPOL is designed to measure all the vector and tensor polarization components simultaneously by utilizing the three reactions : ^{12}C(d,d_0), ^{1}H(d,^{2}He) and d+p. Preliminary results of the first test experiment with a polarized deuteron beam at 270 MeV are also presented.

INTRODUCTION

Many studies of spin excitations in nuclei have recently been performed in order to get information on the spin-dependent part of the effective nucleon-nucleon interaction. In the (p,p') reaction at intermediate energies ΔT=1 transitions are dominant because the vector-isoscalar (V_σ) component of the effective interaction is much weaker than the vector-isovector $V_{\sigma\tau}$ component. Therefore we have little information on the ΔT=0 spin excitations up to now from inelastic proton scattering or other probes. This is due both to the lack of useful probes as well as the weakness of V_σ in proton scattering.

Inelastic deuteron scattering should be one of the most efficient probes of ΔT=0 spin excitations, because it excites exclusively ΔT=0 states. Since the deuteron has a spin 1, we can define probabilities of non-spin-flip (S_0), spin-flip (S_1) and double-spin-flip (S_2), where Δm of $S_{\Delta m}$ is the change in S_y, the spin component of the deuteron perpendicular to the reaction plane (the y component in the Madison convention). S_1 will be a signature of spin excitations as S_{nn} is in the (p,p') reaction (1,2). S_1 is written, in terms of polarization observables as

$$S_1 = \frac{1}{9}(4 - P^{y'y'} - A_{yy} - 2K_{yy}^{y'y'}) \ , \tag{1}$$

where yy denotes the tensor polarization and $P^{y'y'}$, A_{yy} and $K_{yy}^{y'y'}$ are the tensor polarizing power, the tensor analyzing power and the tensor-tensor polarization transfer, respectively. Thus measurements of the tensor-polarization observables are needed to extract S_1 via the $(\vec{d},\vec{d}')$ reaction. This makes the measurement of S_1 in the (d,d') reaction much more difficult than that in the (p,p') reaction where the S_{nn} value can be derived from the vector-polarization transfer $K_y^{y'}$ alone.

DESIGN OF THE POLARIMETER

We are constructing the Deuteron POLarimeter DPOL at the RIKEN Accelerator Research Facility. It is designed to measure all the vector (it_{11}) and tensor (t_{20}, t_{21} and t_{22}) polarization components by utilizing three kinds of reactions, shown in table 1, as polarimetries. These reactions are measured semi-inclusively.

DPOL is located at the second focal plane (FP-2) of the spectrograph SMART (3) and the geometry is shown in fig. 1. Note that the reaction plane of SMART is vertical because a scattering angle is changed by using a beam swinger system.

A multiwire drift chamber (MWDC) and trigger plastic scintillators (0.5×18 ×80 cm^3 and 1.0×18×80 cm^3) are used for detecting scattered deuterons. These plastic scintillators, contain C and H, are also used as active scatterers for double scattering. The polarimeter-system consists of three hodoscopes, HOD1 (28 times 6.5×6.5×220 cm^3), HOD2 (12 times 6.5×6.5×220 cm^3) and HOD3 (2 times 1.0×13×100 cm^3), and two calorimeters, CM1 (6

TABLE 1: Reactions which are utilized as polarimetries. Corresponding components of analyzing power and triggers are also shown. Subscripts u and d in Hi=HODi denote upper and lower part of the hodoscopes, respectively.

	Reactions	Analyzing powers	Triggers
1 :	$d + {}^{12}\mathrm{C} \rightarrow d + {}^{12}\mathrm{C}$	iT_{11}	H1×CM1+H2×CM2
2 :	$d + p \rightarrow {}^2\mathrm{He}\,(2p) + n$	T_{20} and T_{22}	H1×H1*
3 :	$d + p \rightarrow d + p$	all components	$(\mathrm{H1+H2+H3})_u \times (\mathrm{H1+H2+H3})_d$

* coincidence events with two adjacent counters are excluded.

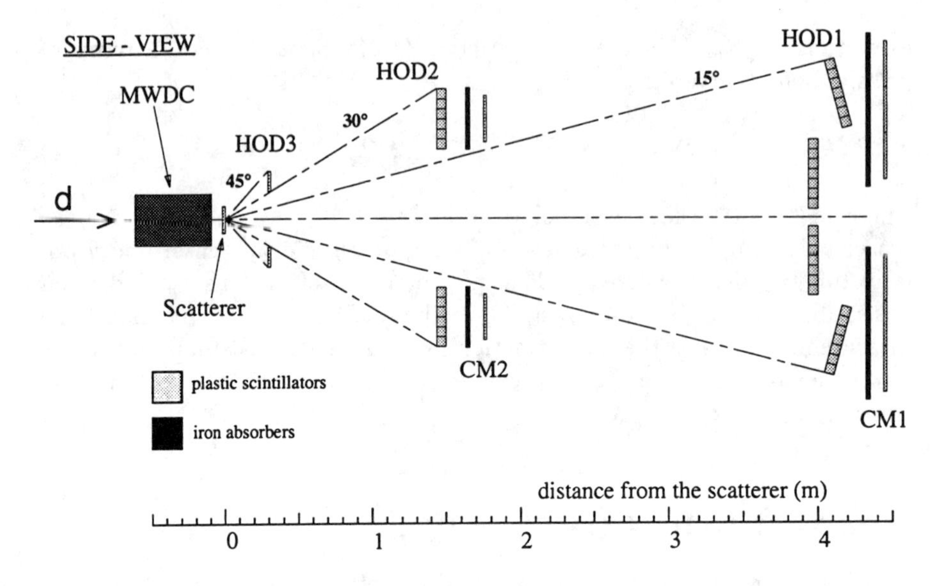

FIGURE 1: A cross sectional view of the Deuteron POLarimeter DPOL. Details are described in the text.

times $1.0\times29\times220$ cm^3 and 1.8 cm thick iron absorbers) and CM2 (2 times $1.0\times29\times220$ cm^3 and 1.8 cm thick iron absorbers). For all hodoscopes and calorimeters, plastic scintillators are used and they are viewed at both ends by photomultiplier tubes.

Three kinds of triggers for double scattering events corresponding to the three different reactions are shown in table 1. For the ^{12}C(d,d_0) event, calorimetry by using iron absorbers and scintillators is employed to identify deuterons. For the ^{1}H(d,^{2}He) and ^{1}H(d,d) event, two charged particles are detected in coincidence.

EXPERIMENT AND RESULT

The first experiment for testing the polarimeter-system has been recently performed with a 270 MeV polarized deuteron beam extracted from the RIKEN ring cyclotron.

Three combinations of the beam polarizations were used in this experiment and their ideal values of vector and tensor polarizations are $(p_Z,p_{ZZ}) = (0,-2)$, $(-1,+1)$ and $(+1,+1)$, respectively. The polarization axis was controlled with the Wien filter downstream of the ion source so that it lay on the normal or sideway direction of the scattering plane at FP-2. The beam polarization was monitored by using the $d+p$ scattering at 270 MeV and typically 60-65 %

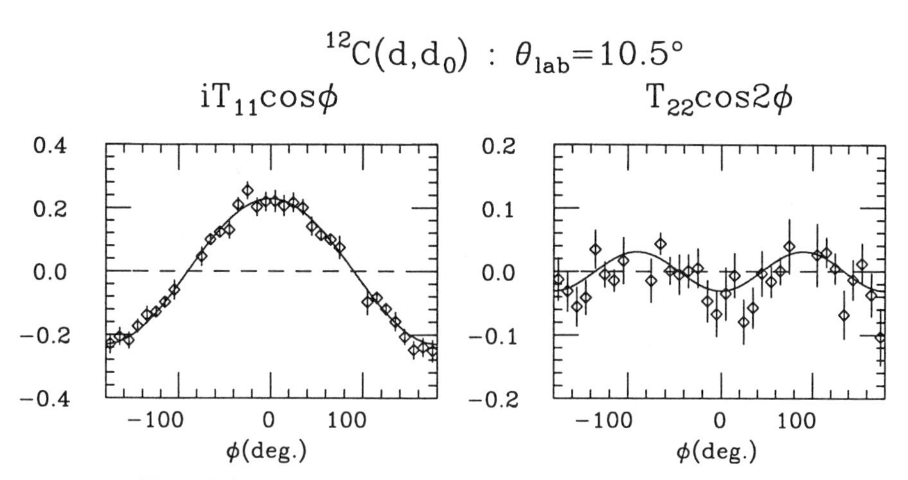

FIGURE 2: Typical fitting results for iT_{11} and T_{22} at θ_{lab}=10.5°. Only statistical errors are indicated in the figures.

polarization of the ideal value was obtained throughout the experiment.

The cross section with a polarized beam is written in terms of spherical analyzing powers as :

$$\sigma(\theta,\phi) = \sigma_0(\theta)\{1 + 2it_{11}iT_{11}(\theta)\cos\phi + t_{20}T_{20}(\theta) + 2t_{22}T_{22}(\theta)\cos 2\phi\} \quad , \quad (2)$$

where a T_{21}-term vanishes because the polarization axis is perpendicular to the direction of the incident beam. $iT_{11}\cos\phi$, T_{20} and $T_{22}\cos 2\phi$ values were obtained separately by using the $\sigma_0(\theta)$ value and the $\sigma(\theta,\phi)$ values for different combinations of the beam polarization. Finally, the effective analyzing powers were extracted by using a least squares fitting method. Fig. 2 shows typical fitting results of $iT_{11}\cos\phi$ and $T_{22}\cos 2\phi$ for the ^{12}C(d,d_0) events at θ_{lab}=10.5°.

Fig. 3 shows the preliminary results of the effective analyzing powers for the ^{12}C(d,d_0) events. Large iT_{11} values are obtained, consistent with the result of POMME (4).

SUMMARY

We are constructing the focal plane Deuteron POLarimeter DPOL at RIKEN in order to study the isoscalar spin excitations. The first test experiment has been performed to calibrate the effective iT_{11}, T_{20} and T_{22} values. The preliminary result for the ^{12}C(d,d_0) data shows large iT_{11} values. Further analysis and improvement of the polarimeter is now in progress.

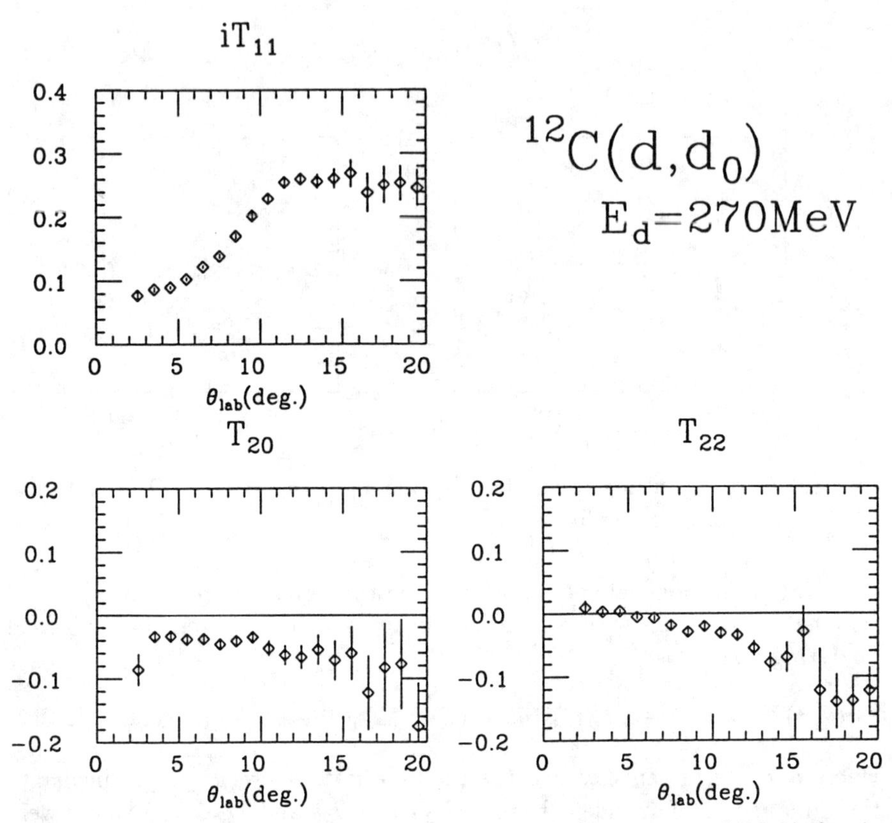

FIGURE 3: Preliminary results of the effective analyzing powers of DPOL for the $^{12}C(d,d_0)$ events.

ACKNOWLEDGEMENT

This work is supported financially in part by the Grant-in-Aid for Scientific Research No.04402004 of Ministry of Education, Science and Culture of Japan. The test experiment was performed at RIKEN under program number R138n.

REFERENCES

1. Ishida, S. *et al.*, *Phys. Lett.* **B314**, 279-283, 1993.
2. Morlet, M. *et al.*, *Phys. Lett.* **B247**, 228-232, 1990.
3. Ichihara, T. *et al.*, *Nucl. Phys.* **A569**, 287c-296c, 1994.
4. Bonin, B. *et al.*, *Nucl. Instr. Meth.* **A288**, 389-398, 1990.

A deuteron vector and tensor polarimeter up to 2 GeV

E. Tomasi-Gustafsson[a,c], J. Yonnet[a] and V. Ladygine[b]

[a] *CEA-DSM CNRS-IN2P3 Laboratoire National Saturne F-91191 Gif/Yvette cedex*
[b] *JINR-LHE Dubna-Moscow Region*
[c] *CEA-DSM, DAPNIA-SphN, F-91191 Gif/Yvette cedex*

Abstract. We present the results of simulations concerning a vector and tensor deuteron polarimeter, working in the range 1.2-2 GeV, based on d,p elastic scattering. It could be built as a modification of the existing polarimeter POMME working at Saturne. The first results of a MonteCarlo simulation show a large figure of merit. The first experiment in which it is supposed to work is the study of baryon resonances in the framework of the SPES4π project at Saturne.

Many recent programs proposed at new or existing intermediate energies accelerators like CEBAF or SATURNE require the measurement of the protons or deuterons polarizations at energies up to a few GeV. In particular at Saturne it is planned to study the spin content of the Roper resonance, by measuring the spin-flip probability in the ${}^1H(\vec{d},\vec{d}')^1H$ inelastic scattering. It requires the measurement of the polarization of the scattered deuteron, which has an energy of 1.6 to 2 GeV.

We present here the project for an extended vector and tensor polarimeter for deuterons at energies up to 2 GeV which would cover a large part of the focal plane of a spectrometer. It is based on the elastic $\vec{d},p$ reaction which has very large analyzing powers and large cross sections. This polarimeter can be built as a modification of the existing polarimeter POMME, in operation at Saturne [1].

Since 1989, POMME has been used in several experiments, as a proton polarimeter up to an energy of 2.3 GeV and as a vector deuteron polarimeter up to 700 MeV, mainly concerning nuclear structure (identification of isoscalar spin transitions in nuclei by measuring the spin-flip probability) and deuteron structure (polarization observables in $\vec{d},\vec{p}$ backward elastic scattering). The working principle of Pomme is the measurement of the azimuthal asymmetry in the inclusive $\vec{p},C$ or $\vec{d},C$ reactions.

Three ionization proportional chambers constitute the focal plane detection of a spectrometer and the detection of the incoming trajectories on the polarimeter target. The front chambers dimension is $50 * 50$ cm^2 and their resolution is better than 2 mm. The scattered particles are detected by three rear chambers which are $100 * 100\ cm^2$ wide. Their resolution is better than 3 mm. Two scintillators at the focal plane and a wall of scintillators behind the rear chambers constitute the trigger (Fig. 1).

As the energy increases the analyzing powers in inclusive reactions on Carbon become smaller. We plan to replace the Carbon target with a liquid hydrogen target and to add a detection in order to select d, p elastic scattering.

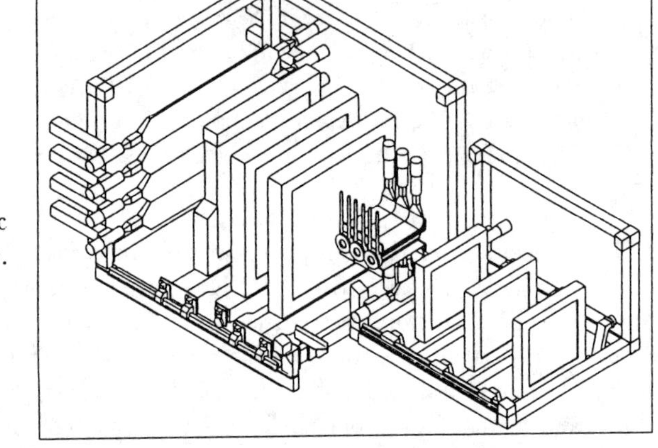

FIGURE 1. Schematic view of the polarimeter.

The new target consists in three cylindrical cells of 7 cm diameter and 50 cm length filled with liquid hydrogen. with the axis perpendicular to the beam. On the top and on the bottom the recoil protons are detected by four layers of several drift straw detectors. each of diameter 11 mm, with a geometry following the target geometry and one plane of scintillators which give the fast signal. The beam particles, incident on the polarimeter, will be counted one by one by a set of scintillators placed at the focal plane.

A MonteCarlo has been done to optimize the geometry, to calculate the efficiency and the figure of merit at 1200 MeV, 1600 MeV and 2000 MeV [2]. The main ingredients of the simulations are cross sections and analyzing powers for the elastic d,p reactions. These data are available as they have been measured in several experiments [3]. An energy dependent polynomial fit as a function of the angle has been used as input for the simulations. Multiple scattering, probability of a second reaction in the polarimeter target, as well as energy losses and detection resolution have been taken into account.

The deuterons are mainly scattered within a cone of 12 degrees and the corresponding protons between 60 and 90 degrees.

In Fig. 2 we show the angular and vertex coordinate spectra for the deuteron, from the informations given by the ionization chambers. The system of coordinates is chosen with the z-axis along the beam direction, the y-axis as the vertical direction and the x-axis such as to form a right handed coordinate system.

The scattering plane is known with good precision from the deuteron trajectories. The difference between the calculated and reconstructed polar and azimuthal angle is shown in (Fig. 2 a,b).

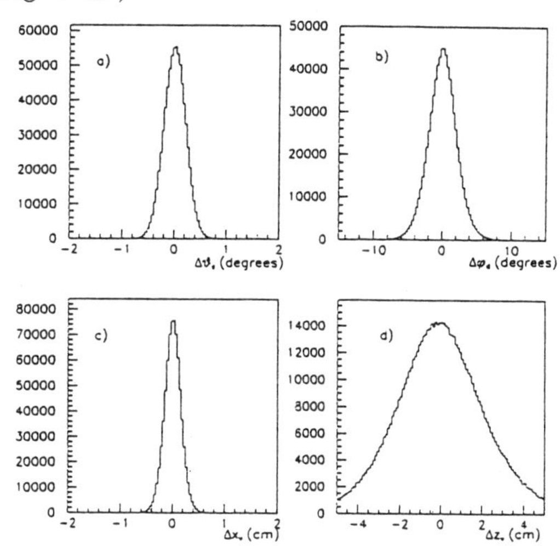

FIGURE 2. Deuteron spectra: difference between the reconstructed and the calculated polar angle (a) and azimuthal angle (b), same for the vertex coordinate in the transversal direction (c), in the longitudinal direction (d).

The precision of the x and y reconstruction of the vertex is of the order of 1 mm (Fig. 2c), but due to the small diffusion angle, the z coordinate of the vertex point is very inaccurate ($FWHM \simeq 4\ cm$) (Fig. 2d). The calculation of the trajectory for the proton, in particular its polar angle, can not be done from the vertex informations. In order to measure independently the direction of the proton the z-coordinate on two different planes has to be known. As a good efficiency is obtained with two planes of straws, due to the cylindrical geometry of the detectors, four layers of such detectors are necessary. After requiring the condition of coplanarity, the polar angle of the proton can be reconstructed and compared to the angle expected from kinematics. The width of this distribution ($\sigma \simeq 1.5^o$) is good enough in order to discriminate elastic scattering from events of other reactions (Fig. 3).

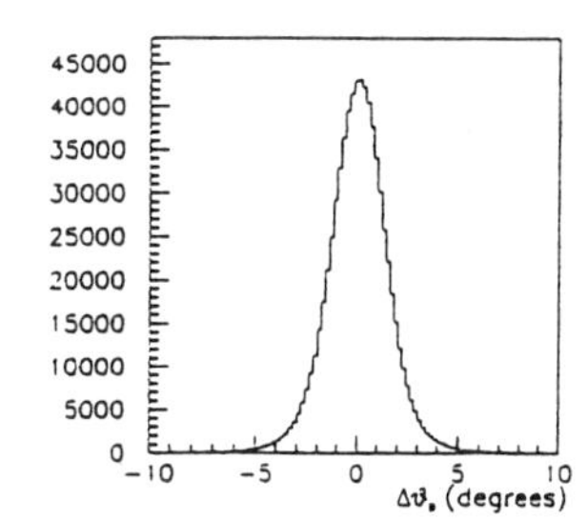

FIGURE 3. Histogram of the difference between the calculated and reconstructed polar angle for the proton.

In Fig. 4 we report the analyzing powers recalculated with a fit procedure with a polynome in $cos\phi$ and $cos2\phi$. The line is the energy dependent fit used as input in the simulations.

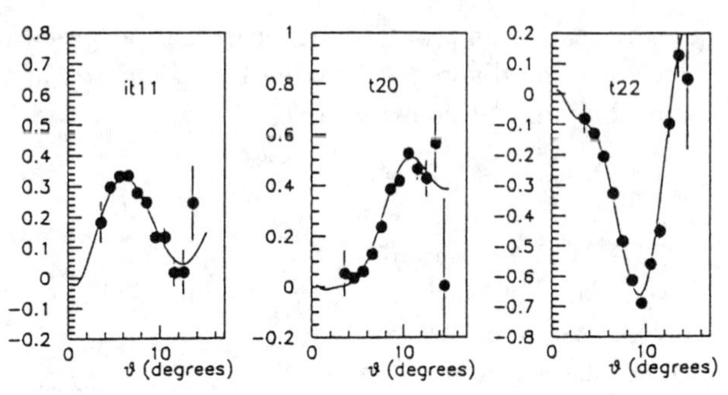

FIGURE 4. Effective analyzing powers recalculated from the azimuthal distributions obtained by the simulation and compared to the curves which are the fits of the input analyzing powers.

Most of the protons emitted beyond 80 degrees have small energies and are stopped in the target. The requirement of the proton detection makes the efficiency drop from $1^o/_o$ to $3^o/_{oo}$ but the figure of merit does not decrease too much. The protons absorption is maximum near the horizontal plane as we deal with an extended target. The lost events correspond to the small deuteron scattering angles for which the analyzing powers are low. The loss in the efficiency is then partially compensated by an increase of the average effective analyzing powers.

For each polarization state of the beam, the angular spectra are reconstructed in bins of 1^0 for the polar angle and 10^0 for the azimuthal angle. The angular spectra are corrected by a function which takes into account the unpolarized cross section and the experimental cuts. Such a correction will be experimentally determined during the calibration as well as the possible systematic errors due to the beam distribution on the target.

The polarization from the individual beam state spectra, is then calculated with a fit procedure in order to check the systematic errors due to the geometry. The individual determination is precise at 2% level (Fig. 5).

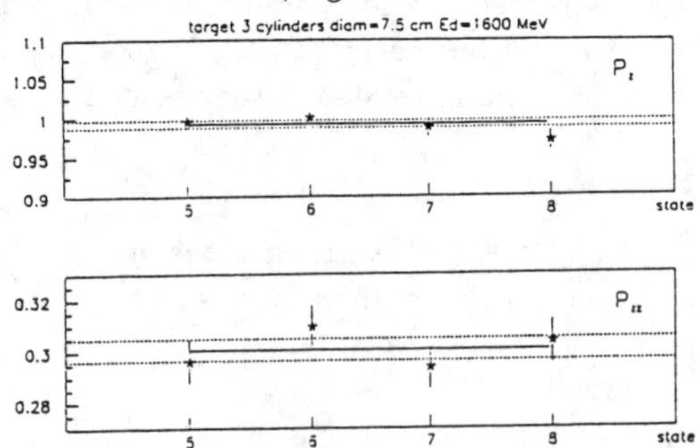

FIGURE 5. Results of the simulation for the polarization measurement. They have to be compared to the input values: P_z = 0.3 (vector polarization) and P_{zz} = 1 (tensor polarization).

The performances of a polarimeter are often expressed in terms of the figure of merit, $\mathcal{F}$. It is a function of the efficiency ε and of the analyzing power, A, and it is defined as $\mathcal{F}^2 = \int \varepsilon(\theta)A^2(\theta)d\theta$ where θ is the polar angle and the integration is over the angular domain where the polarimeter is efficient. The figure of merit allows to evaluate the counting rate, N_{inc}, necessary to obtain the aimed precision on the polarization. The statistical error on the polarization, ΔP, indeed, can be expressed as $\Delta P = \frac{\sqrt{2}}{\sqrt{N_{inc}\mathcal{F}}}$. The figure of merit of POMME as a function of energy, from the calibration data with a Carbon target up to 700 MeV (shaded area), and from the simulations with an hydrogen target above 1 GeV is shown in Fig. 6. It has a maximum around 600 MeV. The efficiency varies from 10% at the lowest energy to $3^0/_{00}$ at the higher energy.

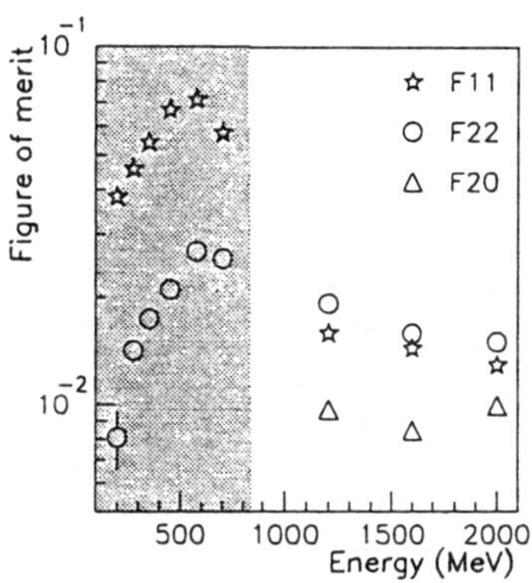

FIGURE 6. Figure of merit in the available energy range.

We have presented the project of a polarimeter which will allow the measurement of deuteron vector and tensor polarization at high energies. It can be adapted as a proton polarimeter, but the study at the moment is restricted to its use as a deuteron polarimeter. It is planned to be calibrated at Saturne and available at the end of 1995.

References

[1] B. Bonin et al. Nucl. Instr. and Meth. A288 (1991) 389

[2] E. Tomasi-Gustafsson, J. Yonnet, and V. Ladygine LNS/ph/94-07

[3] J.S.Vincent e.a. PRL 24 (1970),236
L.M.C.Dutton e.a. PRL 21 (1968),1416
F.Irom e.a. PR C28 (1983),2380
J.Banaigs e.a. PL 45B (1973),535
N.Dalkhazav e.a. Yad.Fiz.8,N2,1968,342
G.W.Bennett e.a. PRL 19 (1967),387
M.Haji-Saied e.a. PR C36 (1987),201
V.Ghazikhanian e.a. PR C43 (1991),1532

PROTON POLARIZATION DETERMINATION BY ELASTIC p-e SCATTERING

S.B.Nurushev
IHEP, Protvino, Russia

A.P.Potylitsin, G.M.Radutsky, A.N.Tabachenko
NPI, Tomsk, Russia

M.N.Strikhanov
MEPhI, Moscow, Russia

The analyzing power of the multi-GeV protons scattering on the electrons is estimated. The possibility to use the polarimeters based on ep-scattering for the measuriment of the polarization of the protons is discussed.

INTRODUCTION

In [1] the polarization measurements were performed for 180 GeV proton with polarimeters based on Primakoff effects (asymmetry ratio reaches $A \approx 0.5$) and on Coulomb-nuclear interference process (CNI) ($A < 0.05$). As proton energy increases the possibility of these absolute polarimeters seems to have some difficulties as analyzing power for strong interaction processes can not be calculated using the existing theoretical models without normalization uncertainties.For the multi-GeV-energy protons it is more convenient to utilize the p-e polarimeters based on electromagnetic interactions with well known analyzing power.In particularly, the analyzing power for the scattering of the 250 GeV protons on the rest electrons or the moving electrons with the small energy may be found quite simply as the momentum transfer is sufficiently the small value of the order $0.5(GeV/c)^2$, therefore one may to use the one photon approximation [2].

Besides the polarimeters based on p-e scattering posses also the following advantages:a) simple two-particles identification scheme,b) relatively large recoil electron exit angles (for 250 GeV proton scattering on the rest electron the maximum proton scattering angle $\theta_p =$ 0.56 mrad whereas the electron recoil angle $\theta_e = 4$ mrad).

ANALYZING POWER FOR p-e SCATTERING

The invariant amplitude of the p-e elastic scattering with $P_1(\vec{p}_1, E_1) + K_1(\vec{k}_1, e_1) = P_2(\vec{p}_2, E_2) + K_2(\vec{k}_2, e_2)$ can be written in the one photon approximation in following form:

$$M_{fi} = ie^2 t^{-1} \bar{u}(K_2)\gamma_\alpha u(K_1)\bar{u}(P_2)[\gamma_\alpha(F_1(t) + F_2(t)) + in_\alpha F_2(t)/2M]u(P_1)$$

here t is the square of the momentum transfer $t = -(K_1 - K_2)^2$, F_1 andF_2 are the proton form factors, $n_\alpha = (P_1 + P_2)_\alpha$. Whith help this amplitude one may to receive the differential cross section for the scattering of the polarized protons from polarized electrons.In a arbitrary reference frame the expresson for this differential cross section is given in referns [3].From the differential cross section for the scattering of the polarized protons from polarized electrons one may to receive the helicity asymmetry

$$A = \frac{d\sigma(\uparrow\uparrow) - d\sigma(\uparrow\downarrow)}{d\sigma(\uparrow\uparrow) + d\sigma(\uparrow\downarrow)}$$

here the arrows refer to the proton and electron spin states, respectively. In an arbitrary reference frame it may be written as

$$A = \frac{1}{X}\{2mM(F_1(t) + F_2(t))t[(\xi \cdot \zeta + \frac{\xi \cdot K \zeta \cdot K}{t})F_1(t)$$

$$+ \frac{t}{4M^2}(\xi \cdot \zeta - 2\frac{\xi \cdot P_1 \zeta \cdot K}{t})F_2(t)]\}$$

Where

$$X = \{[2a(2a - t) + tM^2](F_1^2(t) - \frac{t}{4M^2}F_2^2(t))$$

$$+ \frac{t}{2}(t + 2m^2)(F_1(t) + F_2(t))^2\}$$

connects with the Rosenbluth cross-section σ_0

$$\sigma_0 = \frac{\alpha}{t^2} \cdot \frac{1}{(a^2 - m^2M^2)^{1/2}} \cdot \frac{|\vec{k}_2|^2}{|\vec{k}_2|(e_1 + E_1) - e_2|\vec{k}_1 + \vec{p}_1| \cos\theta_2} \cdot X$$

Here a is the invariant $K_1 \cdot P_1$, the recoil angle θ_2 is the angle between $\vec{k}_1 + \vec{p}_1$ and $\vec{k}_2$, $K = K_1 - K_2$, M and m are the proton and electron masses respectively, $\alpha = 1/137$ is the fine structure constant. The unit 4-vectors ξ and ζ are electron and proton polarization vectors, which components are expressed in terms of the particle's spin in its rest frame,

$$\xi = (\hat{\vec{s}}_1 + \frac{\hat{\vec{s}}_1 \cdot \vec{k}_1}{m(m + E_1)}\vec{k}_1, \frac{\hat{\vec{s}}_1 \cdot \vec{k}_1}{m})$$

for the electron and analogous one for the proton.Here $\hat{\vec{s}}_1$ is the unit vector in the direction of the spin of the fermion in its own rest system.

If we take the dipole representation for the electric G_E and magnetic G_M form factors as it is usually done for such problems and assume that $G_E \cdot G_M > 0.$, we can to calculate any spin effects.

We calculated the spin correlation parameters A_{LL} for the longitudinally polarized proton and electron,A_{NN} and A_{SS} for the transversal polarization of the proton and electron to the direction of the spin of the proton and electron in the y-axis direction and x-axis direction respectively and the elastic scattering cross section for the angular region $0 < \theta_e < 7\ mrad$ and the several values of the energies of the proton and electron. Here θ_e is angle between the momenta of the scattering electron end the initial proton and the initial proton moves in the direction of the z-axis,the y-axis is perpendicular to the scattering plane, the x-axis lies in the scattering plane .

The results of the calculations of the analyzing powers A_{LL}, A_{NN} and cross-sections for the case when the electron is at rest before the collision are shown on the fig.1-2.

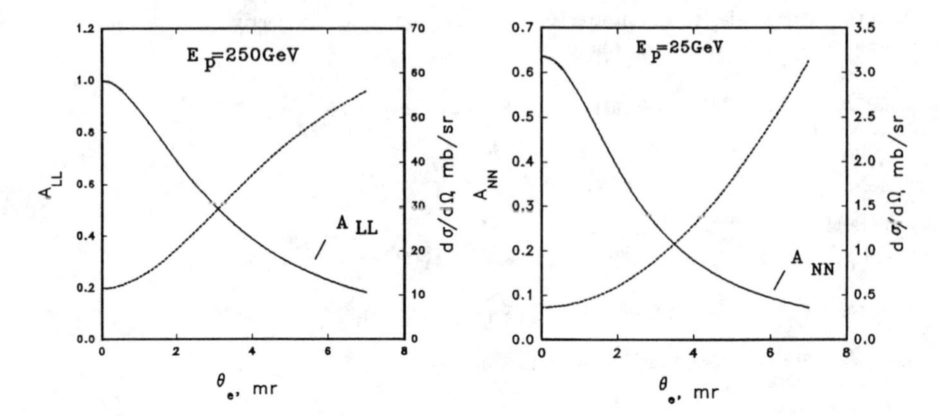

FIG. 1. Cross-sections (dashed curves) and analyzing powers A_{LL} (A_{NN}) for the elastic scattering of longitudinally(transversally) polarized proton on the longitudinally(transversally) polarized electrons for the rest electrons.

The angular region $\theta_e < 4$ mrad is more suitable for polarization measuring despite reduce of the cross-section with decrease of the recoil electron angle θ_e, because the effective analyzing power $A_{LL}^2 d\sigma/d\Omega$ has a considerable value in this region with large values of asymmetry ($A_{LL} > 0.4$, see Fig.2).

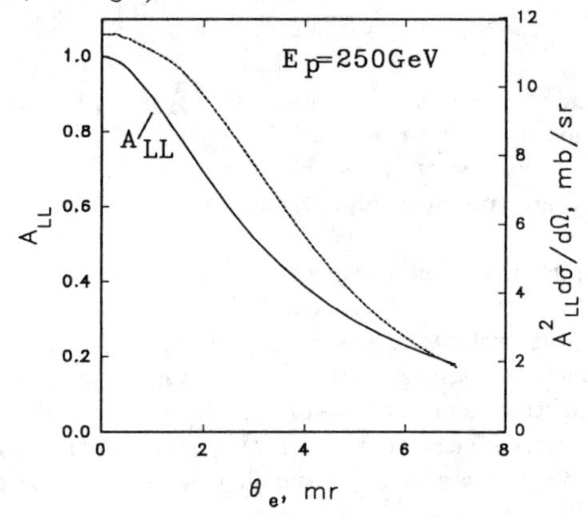

FIG. 2. The effective analyzing power $A_{LL}^2 d\sigma/d\Omega$ (dashed curve) and the analyzing power A_{LL} for the elastic scattering of the longitudinally polarized protons on the longitudinally polarized electrons for the rest electrons.

The analyzing powers and cross-section for the polarized protons and electrons for the case when the electron moves opposite the move of the proton (collider case)are shown on the Fig.3.

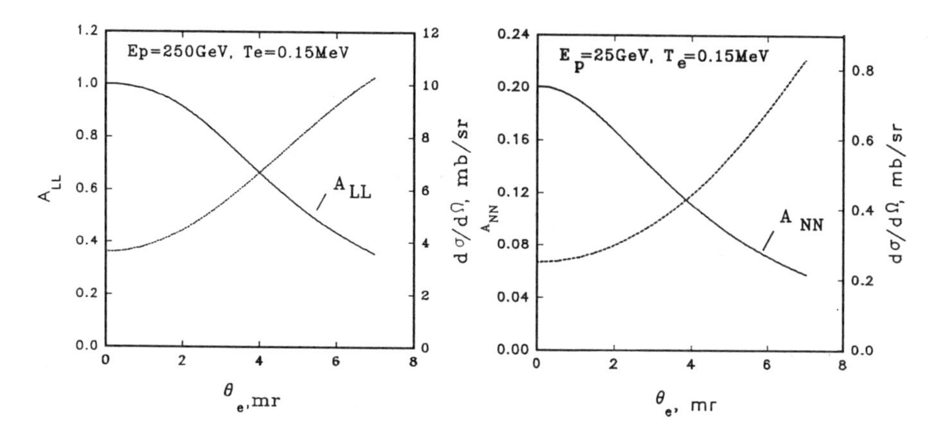

FIG. 3. Analyzing powers $A_{LL}(A_{NN})$ and cross-sections (dashed curves) for the elastic scattering of the longitudinally (transversally) polarized protons on the longitudinally (transversally) polarized electrons with $T_e = 0.15$ MeV(collider case).

ELECTRON POLARIZED TARGET

The electron polarized target (supermendur in a weak magnetic field with H = 90 Gauss) was used to measure the longitudinal polarization of 10 GeV electrons due to elastic ee scattering (Moller polarimeter) [4]. The similar method may be suggested to measure proton polarization. The mean degree of the polarization for magnetized ferromagnetic target may be estimated with help the formula:

$$\bar{P}_e \approx \frac{\mu}{\mu_0 Z}$$

where μ denotes atomic magnetic moment, Z - the atomic number, μ_0 - the Bohr magneton (for Fe one gets $P_e \approx 0.084$). The sign of electron polarization can be easily changed by reversing the external magnetic field.

Another result may be obtained using the oriented crystal. When the energetic proton is incident at small angle $\theta < \theta_c(\theta_c$ is critical angle [5]) to crystal planes, the proton undergoes steering by the crystal planes (channeling) [5]. The channeling proton moves far from nuclei in space regions with low electron density ρ_e but enhanced polarized d -electron density ρ_B Thus, the effective electron polarization degree $P_e \sim \rho_B/\rho_e$ increases.

Fig.4 shows the electron polarization degree dependence on proton incident angle θ in Fe crystal oriented with (110) plane along the beam. The dashed line corresponds to random incident. Thus for narrow beam $\Delta\theta < \theta_c$ ($\Delta\theta$ is the beam divergence) one gets $\approx 60\%$ enhancement of the polarization of the electron target polarization.

The polarimeter analyzing power may be evaluated as follows:

$$\bar{R}_{p-e} = A_{LL}\bar{P}_e$$

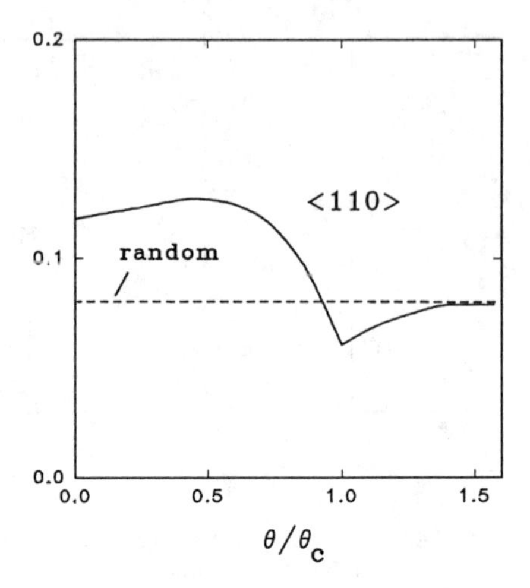

FIG. 4. The mean polarization degree of electron for aligned crystal of Fe (solid curve) and for amorphous Fe (dashed curve).

where A_{LL} is the asymmetry ratio.

Let us estimate of the polarization measurements accuracy for following parameters: proton energy E = 250 GeV, electron scattering angle $\theta_e = 3$ mrad, scattering cross section $d\sigma/d\Omega = 30$ mBarn/sterad, detector angular aperture $\Delta\Omega = 10^{-2}$ sterad, extracted beam intensity $j_{exi} = 10^9$ protons/sec, target electron density $n = 10^{23} electrons/cm^2$. The rate of the detected electrons is approximately $N = 3 \cdot 10^4$ events per sec. For 1 minute measurement with effective analyzing power $A_{LL} = 0.4$ and 12% polarized electrons one gets $\Delta P/P = 1.5\%$ accuracy.

For polarization measurements in the storage ring we suggest using polarized hydrogen jet. To estimate the accuracy of polarization measurements let us consider a 100 % polarized hydrogen jet with density $n \approx 2 \cdot 10^{11} atoms/cm^2$ interacting with collider beam. For electron recoil angle $\theta_e = 3$ mrad with detector aperture $10^{-4} sterad$ and for 1 hour measurement with proton current in ring j = 0.1 A one gets statistics $N = 3 \cdot 10^3$. Thus, the polarization measurement accuracy is defined by the relation:

$$\Delta P/P = 1/(NR_{pe}^{-2})^{1/2} = 5\%$$

At last it is worth to mention that there is another possibility to create a polarized electron target, which may be good enough to measure polarization of high-energy protons. An intensive e^- - beam with the energy $E_e > 100$ KeV and polarization $P_e > 50\%$ can be produced by acceleration of photoelectrons created by circularly polarized laser beam [6].

For case when polarized electrons move opposite the proton beam the rate of scattered electrons (or protons) may be estimated from following expression (similar to luminosity expression):

$$\dot{N}_{ep} = \frac{N_e N_p f_B}{\pi(\sigma_e^2 + \sigma_p^2)} \frac{d\sigma}{d\Omega} \Delta\Omega,$$

where f_B is the proton bunch collision frequency, N_e is the electron gun flux, N_p is the number of protons in a bunch, $\sigma_e(\sigma_p)$ is the mean radius of the electron (proton) beam.

As a rule, electron gun has a frequency $f_e = 10^3$ Hz and duration of the pulse $\Delta t_e = 10^{-6}$

s. so yield during time t may be obtained from relation:

$$N = \dot{N}_{ep} f_e \Delta t_e t.$$

If we consider the electron gun with following parameters: energy E_e =150 KeV, $N_e = 10^{11} e^-/pulse$, $\sigma_e = 0.1cm$ and detector of scattered electrons with solid angle $\Delta\Omega = 10^{-2} sterad$ placed at angle $\theta_e = 4$ mrad we may obtain the yield N $\sim$ 10 events per second (see Fig.3, for example).For this rate the accuracy of measurement of longitudinal proton polarization $\Delta P/P = 5\%$ may be achieved during 4 minutes. From Fig 1 and 3 one may see that there is a possibility to measure the transversely proton polarization too. In conclusion we would like to remark that proton polarimeter with using of polarized electron beam is more preferred because the ep scattering is the single one that influences on the proton beam life time very small. In this case the polarization measurements and monitoring may be carried out in "parasitic mode".

---

[1] D.R. Grosnik, D.A. Hill, M.R. Laghai et al., ANL Preprint ANL-HEP-PR-89-89, 1989.

[2] B.Barish et al., *Proc. of the Heidelberg Intern. Conf. on Elementary Particles at High Energies*, Amsterdam,1968.

[3] P. Scofield, Phys. Rev. **113**, 1599(1959).

[4] P.S. Cooper et al., Phys. Rev. Lett. **34**, 1589(1975)

[5] D.S. Gemmel, Rev. Mod. Phys. **46**, 129(1974)

[6] C.Y. Prescott, SLAC Preprint SLAC-PUB-6242,July 1993.

Effective Analyzing Powers of NPOL at 290 and 386 MeV

T. Wakasa, S. Fujita, M. B. Greenfield*, K. Hatanaka**, S. Ishida, N. Koori†, H. Okamura, A. Okihana‡, H. Otsu, H. Sakai, N. Sakamoto, Y. Satou, and T. Uesaka

Department of Physics, University of Tokyo, Bunkyo, Tokyo 113, Japan
**International Christian University, Mitaka, Tokyo 181, Japan*
***Research Center for Nuclear Physics, Osaka University, Ibaraki, Osaka 567, Japan*
†Faculty of Integrated Arts and Science, The University of Tokushima, Tokushima 770, Japan
‡Kyoto University of Education, Fushimi, Kyoto 612, Japan

Abstract. The effective analyzing powers ($A_{y;\mathrm{eff}}$) of a newly developed neutron polarimeter (NPOL) are reported. Polarized neutrons from the zero-degree $^2\mathrm{H}(\vec{p},\vec{n})pp$ reaction at 295 and 392 MeV were used to derive the $A_{y;\mathrm{eff}}$ at 290 and 386 MeV. The figure-of-merit (FOM) were derived from the effective analyzing powers and the double scattering efficiencies. FOM of NPOL are 3.1×10^{-4} and 3.3×10^{-4} at 290 and 386 MeV, respectively.

INTRODUCTION

The spin longitudinal (R_L) and spin-transverse (R_T) responses of the nucleus are closely related to the π- and ρ-meson correlations, respectively. At higher excitation energies ($\omega > 30$ MeV) and large momentum transfers ($q > 1$ fm^{-1}), the spin-longitudinal interaction is expected to become attractive, while the spin-transverse interaction remains still repulsive due to the heavier mass of the ρ-meson compared with that of the pion. The different dependence on the momentum transfer of the interactions may bring in the different nuclear spin-isospin responses. The spin-longitudinal response has been predicted to be strongly enhanced relative to the spin-transverse response. Such responses for the (p,p') reaction have been measured at $q = 1.75$ fm^{-1} and no enhancement has been observed. The (p,p') reaction, however, excites not only the isovector states but also the isoscaler states and the interpretation is complicated by uncertainties of the existence of the isoscaler interaction. In contrast to the (p,p') reaction, the (p,n) reaction proceeds only the isovector interaction and the enhancement in R_L/R_T might be observed.

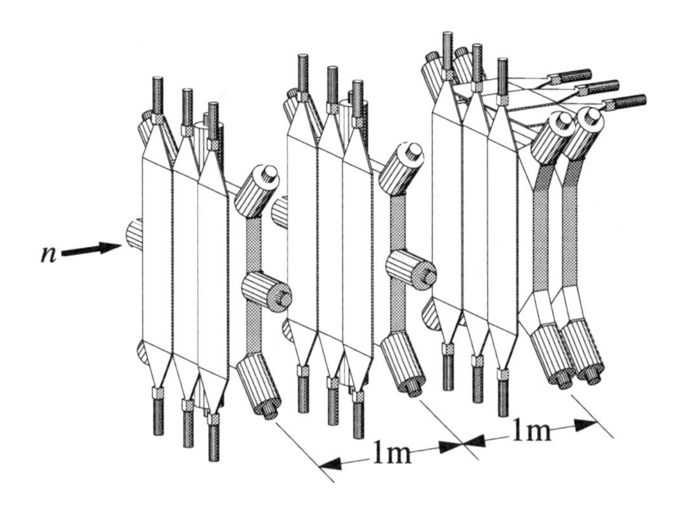

FIGURE 1: Schematic layout of NPOL. See text for details.

A complete set of the polarization transfer coefficients D_{ij} is required to extract the response functions. To get D_{ij} experimentally, the polarization of the outgoing neutrons must be measured by the double scattering technique and for this purpose, we have constructed a neutron polarimeter (NPOL) (1). A schematic layout of NPOL is presented in Fig. 1. The polarimeter consists of four planes of two-dimensional position sensitive scintillation detectors. The first two detectors are made of the liquid scintillator BC519 and the last two detectors are made of the plastic scintillator BC408. The two-dimensional detected position (x, y) is reconstructed by using the timing information derived from four photo-multiplier tubes (PMT) attached at each corner. The position resolutions are about 6 to 10 cm (4 to 8 cm) for liquid scintillators (plastic scintillators) depending on positions. Thin plastic scintillation detector placed in front of each detector serves to distinguish charged particles from neutrons. The neutron polarization can be obtained utilizing both the ${}^{1}\text{H}(\vec{n}, n){}^{1}\text{H}$ and ${}^{1}\text{H}(\vec{n}, p)n$ reactions.

The effective analysing powers of NPOL were calibrated by using polarized neutrons from the ${}^{2}\text{H}(\vec{p}, \vec{n})pp$ reaction at zero degree. The polarization transfer coefficients $D_{LL}(0°)$ of this reaction between 305 and 788 MeV were measured by McNaugton *et al.* (2). At zero degree, polarization transfer coefficients of the Gamow-Teller transition satisfy the following relation,

$$D_{LL}(0°) + 2D_{\rm NN}(0°) + 1 = 0. \qquad (1)$$

From the relation we deduce the $D_{NN}(0°)$ values to be -0.299 ± 0.004 and -0.249 ± 0.005 at 295 and 392 MeV, respectively.

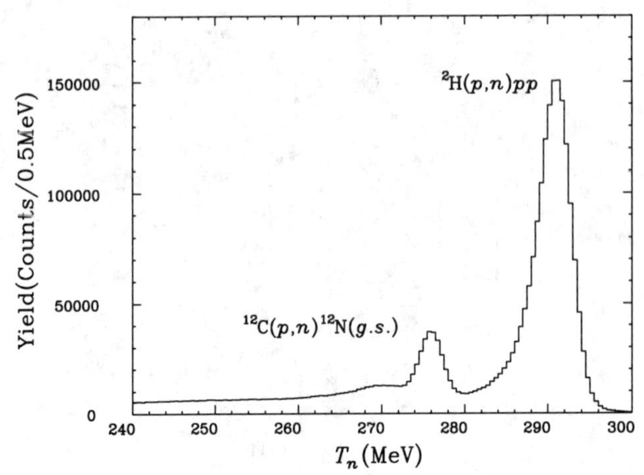

FIGURE 2: Cross section spectra for the $CD_2(p,n)$ reaction at 295 MeV and 0 degree.

EXPERIMENT

The calibration was performed at the neutron time-of-flight (TOF) facility (3) at the Research Center for Nuclear Physics (RCNP), Osaka University. Polarized proton beams of 295 and 392 MeV provided by the RCNP ring cyclotron were used to bombard deuterated polyethylene targets with thicknesses of 662 and 924 mg/cm^2 for 295 and 392 MeV, respectively. NPOL was positioned at a distance of 58 m.

ANALYSIS

The spectrum for the $CD_2(p,n)$ reaction at 295 MeV is presented in Fig. 2. The typical energy resolution of 3.5 MeV (FWHM) was obtained. The spectrum shows both the pronounced $2p$ peak from the $1^+ \rightarrow 0^+$ $^2H(p,n)$ reaction and the ^{12}N peak from the $^{12}C(p,n)$ reaction. The data with the energy losses less than 11 and 12 MeV were used at 295 and 392 MeV, respectively to derive the effective analyzing powers. The resulting average neutron energies are 295 and 386 MeV.

After selecting the $^2H(p,n)pp$ reaction, we first reconstructed the detected positions in a scatterer and a catcher. The position information allows us to deduce the neutron scattering angle or the proton recoil angle. The timing

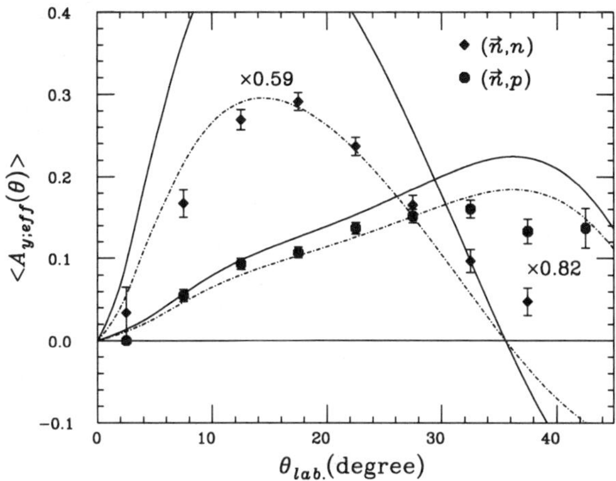

FIGURE 3: Angular distributions of the effective analyzing powers of NPOL at 290 MeV. The filled circles and the filled diamonds are for the $(\vec{n}, n)$ and $(\vec{n}, p)$ channels, respectively. The curves are described in the text.

information between a scatterer and a catcher gives the velocity (β_{exp}) of the doubly scattered neutron or the recoiled proton. The velocity also can be calculated by using the double scattering angle θ with an assumption of the $n + p$ reaction as,

$$T_N = \frac{2T_0 m_N \cos^2\theta}{T_0 \sin^2\theta + 2m_N}, \tag{2}$$

$$\beta_N = \frac{\sqrt{T_N^2 + 2m_N T_N}}{m_N + T_N}, \tag{3}$$

where m_N and T_0 are the nucleon mass and the incident kinetic energy of the neutron, respectively. If the event is due to the $n + p$ scattering, the ratio of $R_v = \beta_{\text{exp}}/\beta_N$ should become unity. The kinematical selection of the $n + p$ events from the n + C or γ-ray events was performed by using the R_v values to maximize the figure-of-merit (FOM) of NPOL. The detailed procedure (kinematical selection and the other selections) is described in Ref. 4.

RESULTS AND SUMMARY

Fig. 3 shows the measured angular distributions of $A_{y;\text{eff}}$ at 290 MeV for each reaction channel. Solid curves are the results from the phase shift analysis for the *NN* scattering (4). They are normalized to the measured values and the results are shown in Fig. 3 by the broken curves. The reduction from the free

TABLE 1: $A_{y;\mathrm{eff}}$, $\epsilon_{D.S.}$ and FOM of NPOL at E_n =290 and 386 MeV.

	E_n = 295 MeV		E_n = 386 MeV	
	$(\vec{n}, n)$	$(\vec{n}, p)$	$(\vec{n}, n)$	$(\vec{n}, p)$
$A_{y;\mathrm{eff}}$	0.292 ± 0.012	0.115 ± 0.003	0.223 ± 0.019	0.127 ± 0.005
$\epsilon_{D.S.}$	0.0010	0.0166	0.0012	0.0167
FOM	0.84×10^{-4}	2.16×10^{-4}	0.60×10^{-4}	2.69×10^{-4}

NN values is mainly due to the contribution from the quasi-free scattering on C. The effective analyzing powers for the $(\vec{n}, n)$ and $(\vec{n}, p)$ channels at 290 and 386 MeV are listed in Table 1 where uncertainties are statistical only. The systematic uncertainties are estimated to be about 0.009 and 0.003 for the $(\vec{n}, n)$ and $(\vec{n}, p)$ channels, respectively, which are mainly due to the systematic uncertainties for the polarization of incident protons and the uncertainties of the $D_{NN}(0°)$ values of the reaction. The double scattering efficiencies ($\epsilon_{D.S.}$) were also determined and the results are tabulated in Table 1. A figure of merit of a neutron polarimeter can be defined by FOM=$\epsilon_{D.S.}A^2_{y;\mathrm{eff}}$. The total FOM of NPOL are 3.0×10^{-4} and 3.3×10^{-4} at 290 and 386 MeV, respectively.

ACKNOWLEDGEMENT

This experiment was performed at RCNP under program numbers E18 and E30. This work was supported financially in part by the Grant-in-Aid for Scientific Research No.6342007 and for Special Project Research on Meson Science of Ministry of Education, Science and Culture of Japan.

REFERENCES

1. Sakai, H. *et al.*, *Nucl. Instr. Meth.* **A320**, 479-499, 1992.
2. Mcnaughton, M.W. *et al.*, *Phys. Rev.* **C45**, 2564-2569, 1992.
3. Sakai, H. and Noro, T., *RCNP Annual Report* **1987**, 171-172, 1987.
4. Wakasa, T., Master thesis, University of Tokyo, 1994, unpublished.
5. Arndt, R.A. and Roper, L.D., Scattering Analysis Interactive Dial-In (SAID) program, phase shift solution SM89, Virginia Polytechnic Institute and State University (unpublished).

THE KENT STATE "2π" NEUTRON POLARIMETER

J. W. Watson,[a] Q.-Q. Du,[a] B. D. Anderson,[a] A. R. Baldwin,[a] C. C. Foster,[b] L. A. C. Garcia,[a] X.-D. Hu,[a] R. Kurmanov,[c] D. L. Lamm,[a] R. Madey,[a] P. J. Pella,[d] E. J. Stephenson,[b] Y. Wang,[a,b] B. Wetmore[a] and W.-M. Zhang[a]

(a)Department of Physics, Kent State University, Kent, OH 44242 USA
(b)Indiana University Cyclotron Facility, Bloomington, IN 47405 USA
(c)Omsk Railway Engineering Institute, 644046 Omsk, Russia
(d)Department of Physics, Gettysburg College, Gettysburg, PA 17325 USA

Abstract. We designed, tested and calibrated a medium-energy neutron polarimeter of a new design, which we call the "2π" polarimeter because of its symmetric coverage of all 2π of azimuth for double-scattered neutrons. During calibration tests at the IUCF we observed an over all neutron time-of-flight resolution of 360 ps. The measured analyzing power is typically 39% for neutrons of both 130 and 165 MeV for optimum software cuts. The efficiency is typically 0.3%.

We report here the performance of a medium-energy neutron polarimeter of a new design. Over the past decade, the Kent State group has undertaken a series of studies[1-4] at IUCF of spin-observables for the (p,n) reaction with the neutron polarimeter described in Ref. 5. The observables we measured were the analyzing power $A_y(\theta)$, the induced polarization $P(\theta)$, and the transverse polarization transfer coefficient $D_{NN'}(\theta)$, which are the only spin observables that can be measured with a beam polarized normal to the reaction plane. IUCF now has the capability of delivering proton beams of all three polarization states, and we developed our new polarimeter to utilize this capability. As with the design of our earlier polarimeter[5] this new polarimeter utilizes the analyzing power of n-p elastic scattering from the H nuclei of organic scintillators. Figure 1 shows the laboratory differential cross section $\sigma(\theta)$, the analyzing power $A_y(\theta)$, and the product $(A_y)^2 \times \sigma(\theta)$ (which is the standard "figure of merit" for polarization analyzing reactions) for n-p elastic scattering at 130 MeV. The desired features that motivated the design of our new polarimeter were: (1) that it should be capable of measuring both normal and sideways components of polarization, simultaneously, for obtaining full sets of polarization transfer observables from

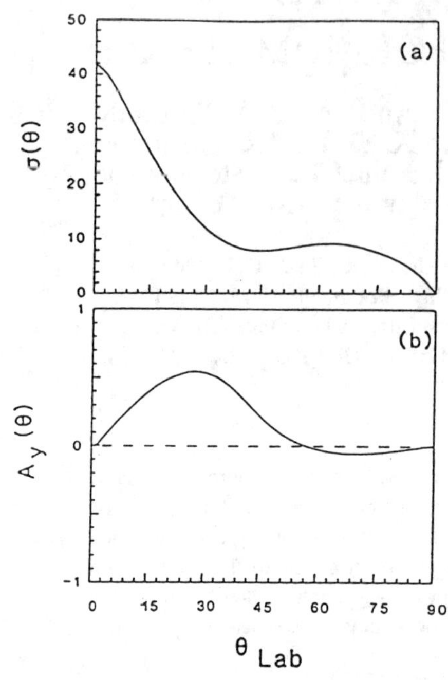

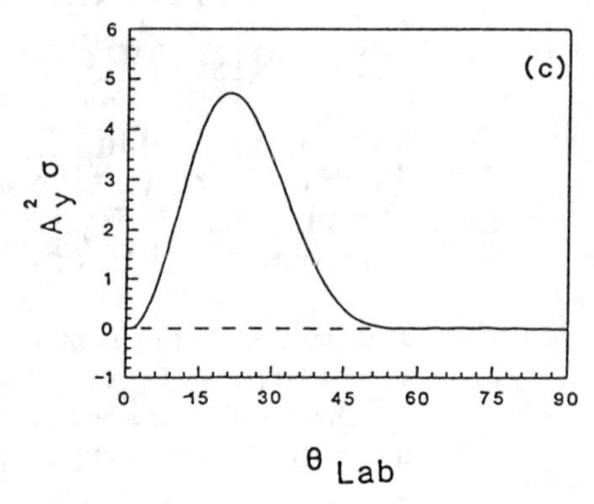

Figure 1. (a) The differential cross section $\sigma(\theta)$, (b) the analyzing power $A_y(\theta)$ and (c) the "figure-of-merit" $(A_y^2\sigma)$ versus the laboratory scattering angle for neutron-proton elastic scattering at 130 MeV.

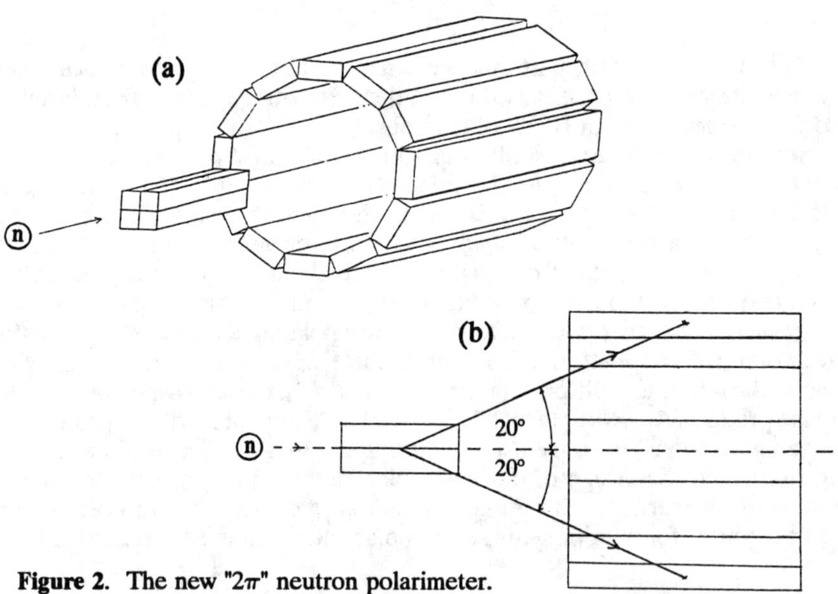

Figure 2. The new "2π" neutron polarimeter. (a) Isometric View. (b) From above.

(p,n) reactions; (2) that the scatterers present a compact face to the neutron production target, so that neutron spin-rotation magnets of modest aperture could be placed between the target and the polarimeter; and (3) that the photomultiplier tubes should be gain stabilized to reduce instrumental asymmetries.

In Fig. 2 we show the design chosen for this device. The scatterers are four 0.102 m x 0.102 m x 0.508 m BC-404 plastic scintillators. Scattered neutrons are detected with an azimuthally symmetric array of twelve 0.102 m x 0.254 m x 1.016 m BC-400 plastic scintillators at a central scattering angle of 20.0°, which is near the angle of maximum value for $A_y^2\sigma$ for n-p scattering in the energy range of 100-200 MeV. Because of the full azimuthal coverage, we call this the "2π" Polarimeter. All 16 detectors are mean timed, with a fast 50.8 mm phototube (XP-2020) on each end of the scatterers, and a fast 227 mm phototube (XP-2041 or R-1250) on each end of the 12 back detectors. Gains for all 32 phototubes are stabilized with stand-alone microprocessor-based pulser systems utilizing high-stability blue LEDs.

We measured the performance of the scatterers with cosmic rays. Using techniques similar to those described in Ref. 6, we obtained 122 ± 6 ps (fwhm) for the intrinsic time resolution, and 17 ± 2 mm (fwhm) for the position resolution. We used the $^{14}C(p,n)^{14}N$ reaction at 135 and 170 MeV at IUCF to calibrate

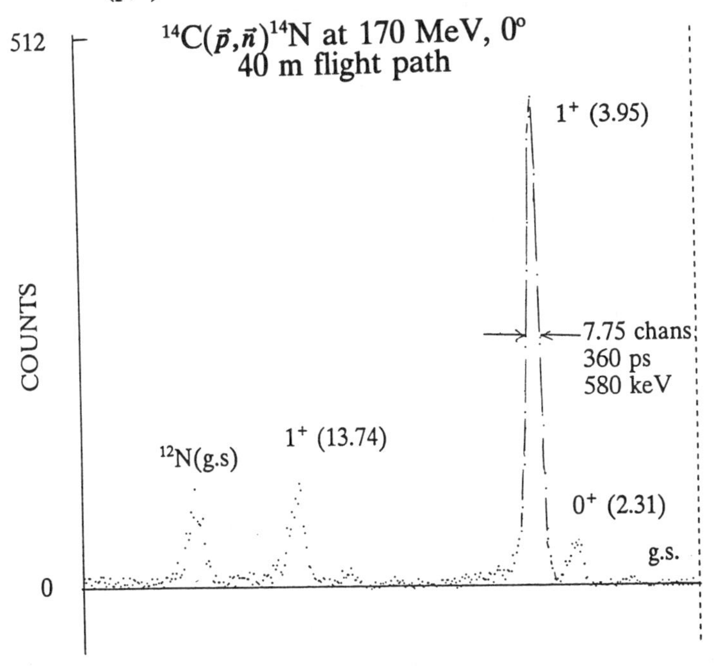

Figure 3. Flight-path corrected neutron time-of-flight spectrum measured with the "2π" polarimeter for the $^{14}C(\vec{p},\vec{n})^{14}N$ reaction at 170 MeV and 0° for a flight path of 40 m.

the polarimeter. The flight path was 40 m. Figure 3 shows the 170 MeV 0° neutron time-of-flight spectrum (for double scattering events) after event-by-event flight-path correction using position information from the scatterers. The overall time resolution is 360 ps (580 keV). This 360 ps overall time resolution is due primarily to the 300 ps beam burst width of the cyclotron. As indicated above, the intrinsic time resolution of the scatterers is 122 ps; also, the 17 mm position resolution (Δx) contributes only ($\Delta x/v$) = 109 ps to the overall time resolution.

To optimize the performance of the polarimeter, it is important to eliminate (to the extent possible) events due to reactions on the carbon in the scatterers. These are typically $^{12}C(n,np)$ quasi-elastic scattering events; at energies below 200 MeV, these events have no useful analyzing power[5,8]. Quasi-elastic events from carbon will generally produce neutrons of lower energies than n-p elastic scattering from hydrogen, and are eliminated using "velocity ratio" techniques[5,8]. For n-p elastic scattering the velocity, v_{scat}, of the scattered neutron, is related (non-relativistically) to the velocity, v_{inc}, of the incident neutron: $v_{scat} = v_{inc} \times \cos^2\theta$. For each double-scattering event we reconstruct (r, θ, ϕ) from the identity of the active scatterer and back detector, and the measured positions of interaction in each. From r, θ, and the measured times of flight (target to scatterer, scatterer to back) we then construct the velocity ratio = $v_{scat}/(v_{inc} \times \cos^2\theta)$. The velocity ratio should be 1.0 for n-p events and < 1.0 for carbon quasi-elastic events.

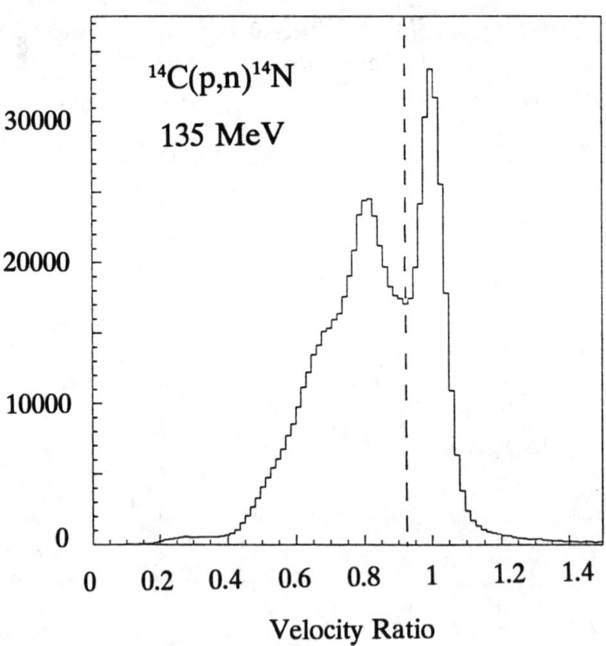

Figure 4. The velocity ratio spectrum for doubly-scattered neutrons from the $^{14}C(p,n)^{14}N$ reaction at 0° and 135 MeV.

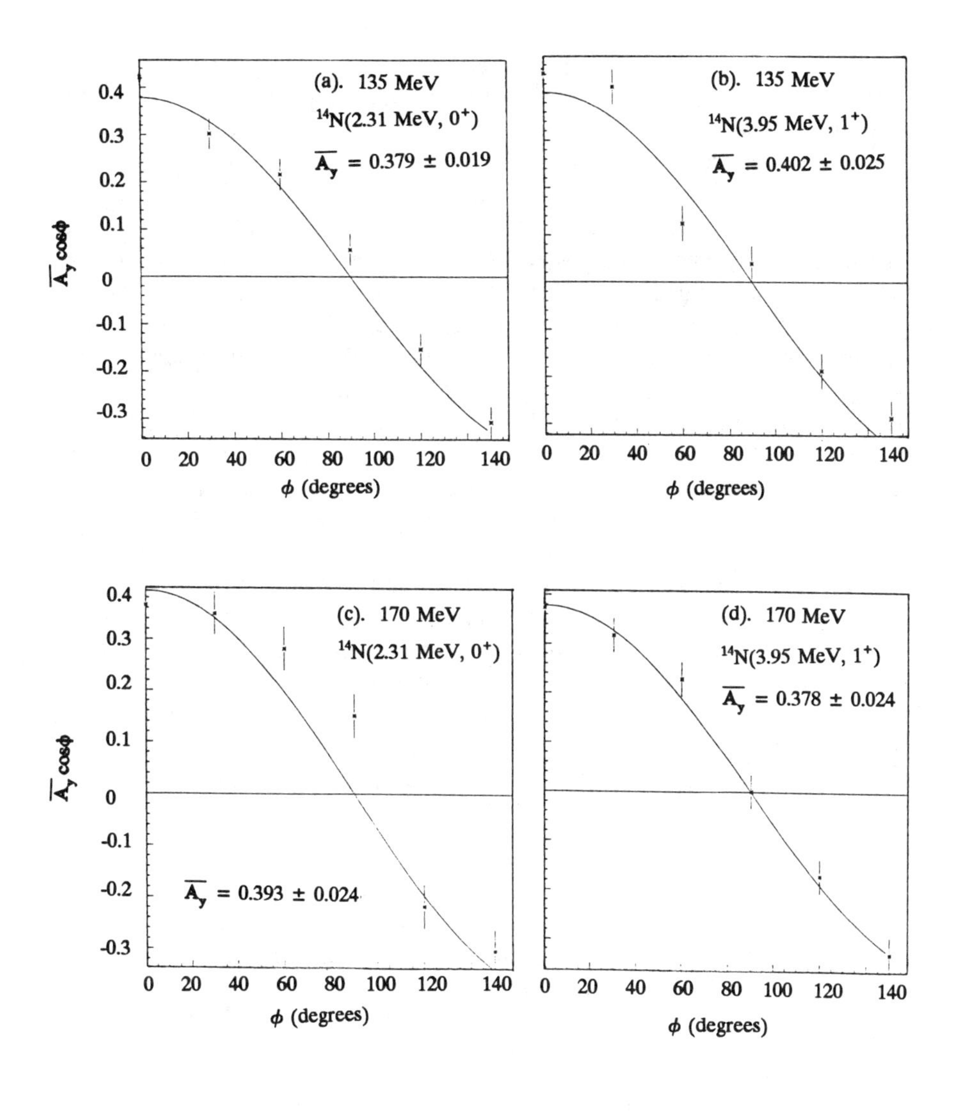

Figure 5. Data for $\overline{A_y}\cos\phi = \xi(\phi)/p_N D_{NN'}$ from the $^{14}C(\vec{p},\vec{n})^{14}N$ reaction: (a) and (b) for 135 MeV; (c) and (d) for 170 MeV.

Figure 4 presents the velocity ratio spectrum from our calibration data for the $^{14}C(\vec{p},\vec{n})^{14}N$ reaction at 135 MeV. The n-p elastic peak at 1.0 is very clear, and a cut in the vicinity of 0.9 will eliminate most of the carbon quasi-elastic events which peak around 0.8.

We used the data from the $^{14}C(\vec{p},\vec{n})^{14}N$ reaction at 135 and 170 MeV also to calibrate the analyzing power of the polarimeter. For these measurements, the proton beam was in a pure normal state, with mean polarization $p_N = 0.71$. The beam polarization was reversed every 30 s. The double scattering yield should vary as $I(\phi) = I_0(1 + p_{N'}\overline{A}_y \cos\phi)$, where the neutron polarization $p_{N'} = D_{NN'} p_N$ and $\overline{A}_y$ is the analyzing power of the polarimeter. After applying "velocity ratio" cuts, we combined the spin up and spin down yields for pairs of back detectors differing in azimuth by $\Delta\phi = \pi$ to obtain the asymmetry ξ as a function of ϕ. Since $D_{NN'}$ is known ($D_{NN'} \equiv 1$ for the 2.31 MeV state; $D_{NN'} = -0.29 \pm 0.02$ for the 3.95 MeV state[7]) fitting the asymmetry yields $\overline{A}_y$. Figure 5 shows the fits to the asymmetry data for both the 0^+ (2.31 MeV) and 1^+ (3.95 MeV) states at 135 and 170 MeV. For the 135 MeV data the weighted mean of the two fits yields $\overline{A}_y = 0.39 \pm 0.02$; for the 170 MeV data the weighted mean is 0.385 ± 0.02. From known cross sections for the 3.95 MeV state, and the total double scattering yields (after cuts) we obtained the polarimeter efficiency $\epsilon \approx 0.3\%$ for both energies. For a velocity ratio cut at 0.92, indicated by the dashed line in Figure 4, we obtained the maximum polarimeter figure-of-merit $(\overline{A}_y)^2\epsilon$ for both energies. The data in Figure 5 are for this cut.

REFERENCES

1. J. W. Watson, P. J. Pella, B. D. Anderson, A. R. Baldwin, T. Chittrakarn, B. S. Flanders, R. Madey, C. C. Foster, and I. J. van Heerden, Phys. Lett. B 181, 47 (1986).
2. J. W. Watson, B. D. Anderson, and R. Madey, Can. J. Phys., 65, 566 (1987).
3. W.-M. Zhang, B. D. Anderson, A. R. Baldwin, T. Eden, D. M. Manley, R. Madey, E. Steinfelds, J. W. Watson, P. J. Pella, and C. C. Foster, Phys. Rev. C45, 2819 (1992).
4. J. W. Watson, B. D. Anderson, A. R. Baldwin, C. C. Foster, D. L. Lamm, R. Madey, P. J. Pella, Y. Wang and W.-M. Zhang, Nucl. Phys. A577, 79c (1994).
5. J. W. Watson, Marco R. Plumley, P. J. Pella, B. D. Anderson, A. R. Baldwin, and R. Madey, Nucl. Instrum. and Meth., A272, 750 (1988).
6. R. Madey, J. W. Watson, M. Ahmad, B. D. Anderson, A. R. Baldwin, A. L. Casson, W. Casson, R. A. Cecil, A. Fazely, J. N. Knudson, C. Lebo, W. Pairsuwan, J. C. Varga and T. R. Witten, Nucl. Instrum. and Meth., A214, 401 (1983).
7. T. N. Taddeucci, T. A. Carey, C Gaarde, J. Larsen, C. D. Goodman, D. J. Horen, T. Masterson, J. Rapaport, T. P. Welch, and E. Sugarbaker, Phys. Rev. Lett. 52, 1960 (1984).
8. T. N. Taddeucci, C. D. Goodman, R. C. Byrd, T. A. Carey, D. J. Horen, J. Rapaport, and E. Sugarbaker, Nucl. Instrum. and Meth., A241, 448 (1985).

III. LEPTON ACCELERATORS AND POLARIMETERS

Polarized Beams and Polarimeters at Lepton Accelerators

D.P. Barber

Deutsches Elektronen Synchrotron, DESY,
22603 Hamburg, Germany

Abstract

The parallel sessions with the above title are summarized.

INTRODUCTION

This article summarizes the twenty-two talks given in four parallel sessions under the above title at this symposium. The topics covered included polarimetry, polarized high-energy electron beams, polarized muon beams, spin flipping and mathematical techniques.

POLARIMETRY

When polarized electron or muon beams are used for testing the Standard Model or investigating the spin structure of nucleons, it is essential that the beam polarizations be known at the 1% level (1,2). The usual way to measure the polarization of an electron beam is to have it collide with a circularly polarized visible laser beam and measure the distribution of the (Compton) backscattered high-energy photons. This is one of several methods proposed by Baier and Khoze (3). If the electron beam is vertically polarized the polarization is proportional to the shift in the vertical plane of the centroid of the scattered photon distribution when the helicity of the laser beam is reversed. If the beam is longitudinally polarized, one measures the asymmetry in the backscattering rate instead, while tagging the scattered electron energies. The two methods are susceptible to very different systematic errors. For example vertical polarization involves measuring centroid positions to a few microns and longitudinal polarization requires that the electron-photon luminosity is independent of the laser beam helicity. The required specifications are difficult to achieve but the techniques are now sufficiently mature that variations on the theme are being tried out or proposed, mostly with the aim of improving the statistical precision achievable in a given time interval.

C. Cavata (4) explained how at CEBAF, the longitudinal polarization of the 4 GeV electrons will be measured by Compton scattering them from a Nd-Yag photon field trapped in an optical cavity formed from two mirrors. This is necessary since the analysing power is small at 4 GeV and the electron currents are only about 100 microamps. The cavity amplifies the scattering rate by a factor of ten thousand.

In his plenary talk M. Placidi (5) recalled that at LEP the polarization of both the electrons and the positrons is measured by using one laser and a mirror to reflect the "once used" laser light back though the positron beam.

At Novosibirsk G. Kezerashvili (4) and the ROKK-1M collaboration working on the VEPP-4M electron-positron ring uses three high power pulsed Nd-Yag lasers to produce frequency doubled, tripled and quadrupled beams. These are used for polarization measurements and the well collimated, intense backscattered photon beams, with energies up to 1.5 GeV can also be used for photonuclear physics experiments and for investigating nonlinear QED.

As reported by F. Zetsche (4), at HERA the backscattered photon rate is about 1.5 kHz per milliamp. The tungsten-scintillator sandwich detector can record single photons. So with electron beam currents of tens of milliamps, the readout electronics must be able to process photons at up to 100 kHz. The photon energies are recorded using 1 MHz 12 bit ADCs and a microprocessor accumulates a set of diagnostic and data histograms online so that they are immediately available for display. The control software now enables the polarimeter to run unattended. The statistical error is one to two percent per minute depending on the electron current.

At the SLC (2) the longitudinal electron polarization is measured behind the interaction point using a Compton polarimeter but also using Möller polarimeters at the beginning of the arc. With the strained GaAs source over 80% electron polarization is regularly delivered to the electron–positron collision point. However, initially the Möller measurements were about 15% below the Compton measurements. In Möller polarimetry one measures asymmetries when polarized electrons are scattered in magnetized foils. As explained by H. Band (SLAC)(4) and A. Afanasev (CEBAF) (4) this discrepancy is now understood and is due to the neglect of the atomic binding energies although these are only in the tens of KeV range. Once an appropriate correction is made the Compton and Möller results are compatible. The correction is always positive and depends on the design of the polarimeter.

A completely different technique, to be applied to stored beams was suggested by Ya. Derbenev (4,6). A polarized particle bunch has a magnetic moment and this can be used to set up an electromagnetic field in a suitably tuned superconducting rf cavity. This is reminiscent of a speculation put forward in 1983 (7) that it might be possible to measure the magnetic flux from the magnetic moment of bunches using superconducting loops.

POLARIZED STORED ELECTRON BEAMS

It is well known that electrons in storage rings can become spin polarized as the result of the emission of spin flip synchrotron radiation (8,9). This, the Sokolov-Ternov effect, results in a maximum of 92.4% polarization in uniform fields. Electron polarization is obtained regularly in low-energy rings. But depolarizing effects associated with orbit excitation resulting from the stochastic nature of photon emission, increase strongly with the beam energy and the degree of misalignment of the ring. The asymptotic polarization essentially results from a balance between the Sokolov-Ternov effect and the depolarization.

Nevertheless as the result of careful adjustments of the orbit and other parameters, over 65% vertical polarization has been obtained at HERA (27 GeV) and 57% at LEP (46 GeV) which is the highest energy electron ring ever built.

For LEP, R. Assmann (4) reported on deterministic closed orbit corrections and B. Dehning (4) on beam energy calibrations. The depolarization due to closed orbit distortion appears mainly via the resulting tilt from the vertical of the equilibrium spin axis. If this is more than about 20 mrad, spin diffusion due to horizontal orbit motion becomes serious. The orbit can be corrected (flattened) with correction coils but this does not automatically mean that the spin tilt is corrected. However, the tilt is dominated by harmonics close to the spin tune in a Fourier analysis of the orbit distortion (10). So the best way to correct the tilt is to use the correction coils to generate "anti-harmonics". This can be done empirically by maximizing the measured polarization but at LEP the characteristic reaction time of the polarization to changes in the conditions is measured in hours. Luckily, at LEP the beam position monitor system is so precise that the Fourier harmonics in the orbit can actually be measured. This allows the orbit to be corrected deterministically with a corresponding large saving of time. Another source of spin axis tilt is the fields of the solenoids in the detectors. At LEP there are no local compensating solenoids but instead the tilts are compensated by generating "anti-tilts" in orbit bumps placed on either side of each solenoid. It is the systematic application of these measures that has led to the 57% polarization even at this highest of energies. At LEP the quadrupoles are correctly positioned at the 0.1 mm level and the monitor positions are known to comparable precision. The ring has a circumference of 27 km. So we are witnessing a major achievement in accelerator construction.

The vertical polarization at LEP provides a route to the very precise beam energy calibrations needed for a full evaluation of the data at the Z. If a small radial rf field running in resonance with the natural spin precession frequency is applied to a vertically polarized beam, the beam is quickly depolarized and a measurement of the required rf frequency enables the beam energy to be estimated with great precision. Thus the width and mass of the Z are now known to 3.2 MeV/c^2 and 2.5 MeV/c^2 respectively. These depolarization methods also allow the change in ring length due to tidal forces to be measured.

It is also noteworthy that these resonant depolarization runs sometimes lead to spin flip. There *is* a decrease in the absolute size of the polarization but the fact that some polarization remains is surprising given that a recent calculation by D. P. Barber et al.(4) suggests that the projections of the spin vectors on the horizontal plane should effectively become decoherent within several damping times. This is clearly a topic which would benefit from further analytical and Monte-Carlo investigation as spin flipping with an rf field would provide an attractive way to reverse the spin direction (11) when spin rotators provide longitudinal polarization at the interaction points.

As pointed out by Yu. Shatunov (4), the calculation and realisation of spin flip is particulary tricky at half integer spin tune, where the assumption of adiabaticity breaks down. In this case it is necessary to arrange that the spins effectively see a circularly polarized rf field and he made suggestions as to how this could be achieved.

On the related topic of spin flip during polarized proton acceleration through intrinsic (i.e. spin-synchrobetatron) resonances, Ya. Derbenev (4) suggested how to use an rf field to prevent loss of polarization by forcing complete spin flip.

The year 1994 brought a major advance in the manipulation of polarization in storage rings: as reported by M. Düren (4,11), spin rotators were installed in the HERA electron ring and for the first time in the history of electron storage rings it was demonstrated that stable longitudinal polarization could be provided at an interaction point. Furthermore, the longitudinal polarization was achieved on the first occasion that the rotators were activated and reached about 55% compared to about 65% immediately before the turn on. The HERA rotator design adopted at HERA is the "MiniRotator" scheme of the late Klaus Steffen of DESY and of Jean Buon of Saclay (12). These rotators are installed at the ends of the arcs on either side of the East straight section and each rotator consists of strings of interleaved horizontal and vertical bending magnets which in horizontal bending angle replace two arc dipoles. The vertical bends form a closed bump with a total vertical excursion of only twenty centimetres. Thus the distortion of the standard machine geometry is minimal. That this is possible is due to the fact that the spin precession rate in a transverse field at the chosen energy of 27.52GeV is about sixty-three times larger than the orbit deflection. Thus small commuting orbit deflections can be used to generate large noncommuting spin precessions which turn the equilibrium polarization vector from vertical in the arcs to longitudinal at the interaction point and back again on entering the arc. All rotator magnets are mounted on remotely controlled jacks so that the spin helicity at the interaction point can be reversed by reversing the fields and geometry of the vertical bends.

The achievement of longitudinal polarization can be better appreciated when it is recalled that it had not been self evident that the polarization could be maintained in the presence of rotators (8). It was pointed out above that tilts of the spin axis from the vertical of just a few tens of milliradians can strongly reduce the polarization. Thus a "designed in" tilt of ninety degrees could be fatal. However, a detailed look at the linear theory of the relation of spin diffusion to synchrobetatron shows that the depolarization taking place in the region between the rotators can in principle be strongly suppressed by a suitable design of the optics. This technique is called strong spin matching (11) and such a spin matched optic was used for the rotator tests. However, strong spin matching does not take direct account of higher order effects such as the so called synchrotron sideband resonances. But the successful generation of longitudinal polarization at HERA shows that strong spin matching is sufficient and opens up a new vista for high-energy electron spin physics.

The potential for further improvement of the polarization at HERA was discussed by M. Böge (4). So far at HERA, the standard way to correct the tilt of the spin axis is to use empirical harmonic closed orbit correction. At HERA it is difficult to apply deterministic harmonic closed orbit correction since the positions of the beam position monitors with respect to the quadrupoles are not well enough known. However, by systematically perturbing the quadrupole strengths around the ring and thereby shifting the closed orbit by small amounts it is possible to measure the relevant monitor positions. Simulations for HERA with the SITROS program (13) show that once these are known, a conventional closed orbit correction already brings a large improvement in the polarization and that when harmonic orbit correction is applied in addition, polarizations of well over 80% should be possible. SITROS simulations with HERA orbits optimized so far by the existing methods now show (13) good qualitative agreement with the measurements of both vertical and longitudinal polarization at HERA.

As explained by H. Grote (4), there is also a spin matched design for rotators at LEP and a proposal has been submitted for a test installation. These rotators are of the vertical S-bend Richter-Schwitters type. In contrast to the MiniRotators, they contain quadrupoles but the spin matching can still be achieved and high polarizations as predicted by the SODOM program (14) are even expected in the presence of the synchrotron sideband resonances and distortions.

The existence of depolarization was first predicted by Baier and Orlov in 1966 (15) in a picture based on spin diffusion caused by orbit oscillations. At high energy these effects can be very strong and those of us who have been involved know how important it is to have reliable methods of calculating the depolarization that can be expected in a storage ring with a particular set of distortions, with and without rotators. One way to proceed is to model the spin diffusion process using the numerical Monte-Carlo tracking program SITROS . First order resonance effects can be calculated analytically by the SLIM program and its derivatives (16). However, it is also very desirable to have methods available which calculate the polarization analytically to high enough order that all the strong depolarizing resonances can be included. All of the existing analytical methods are based on a formalism introduced by Derbenev and Kondratenko in 1972 (17) and reinterpreted by Mane in 1987 (18). In this approach the consideration of high order depolarizing effects is greately simplified by the recognition that at equilibrium the direction of the polarization vector depends on the position in phase space and azimuth. The *value* of the polarization is independent of these. Once the polarization axis is known as a function of phase space position and azimuth, the equilibrium polarization can be calculated with a simple formula which implicitly contains the Sokolov-Ternov effect in the limit of zero depolarization. However, with conceptual simplification comes numerical and analytical complication: this position dependent polarization axis is difficult to calculate and faster and more reliable algorithms are in demand. The SODOM algorithm has already been mentioned. This can calculate the polarization in the limit of linear orbit motion up to arbitrary resonance order given a powerful enough computer. But it would be useful to have an analytical estimate of the effect of nonlinear orbit motion and the first analytic algorithm capable of doing that, the SPINLIE algorithm of Eidelman and Yakimenko, was presented by Yu. Eidelman (4). SPINLIE uses Lie algebraic methods to construct one turn spin-orbit maps and can calculate up to second order in spin and orbit motion. Calculations for HERA are in progress.

One turn maps can also be generated in Taylor series form by the method called Differential Algebra pioneered by M. Berz and his group. As explained by Berz (4), they have now generalized their COSY INFINITY program, which was originally designed for orbital motion, to handle spin motion too. The new version of COSY can calculate the position dependent polarization axis up to high order in both spin and orbit and this extention promises to have a wide field of application both in polarized electron and in proton studies.

The simple picture of polarization build-up balancing depolarization can sometimes be too naive. This becomes clear on inclusion of all the terms in the Derbenev-Kondratenko formula (19,20) and serves both as a warning against oversimplifying the picture and as an encouragement to use rigorous methods. For example, in his talk on spin manipulation at the MIT-Bates ring Yu. Shatunov (4) pointed out that in the presence of snakes the equilibrium polarization generated by synchrotron radiation would, if it were to be used, be a strong function of energy, even changing sign!

It has long been common practice and useful for rough calculations, to view synchrotron radiation as an inverse Compton scattering process: in the electron rest frame the static magnetic field is equivalent to a gas of virtual photons which scatter off the electron and appear as synchrotron radiation in the laboratory. In his talk, R. Lieu (4) explained how he and W.I. Axford have been able to reproduce the usual expressions for the synchrotron radiation spectrum including quantum corrections by a careful coherent addition of the scattered photon field. So the connection between synchrotron radiation and inverse Compton scattering is now formally established. But that is not all: Lieu has recently also shown that the standard spin flip rates of the Sokolov-Ternov effect can be obtained by further careful use of the Klein-Nishina formula. So the value of 92.4% for the maximum value of the polarization can be understood directly from Compton scattering theory in the electron rest frame without having to appeal, for example, to the use of a mixed rest frame and laboratory frame description (21). It is to be hoped that this will provide further insights into the spin flip of accelerated electrons.

In the very strong fields of the kind found in crystals, the forces acting on the spins are no longer simply of the Stern-Gerlach kind and as explained by Y. Bashmakov (4) they can be used to spacially separate electrons in different spin states.

MUON POLARIZATION

If electrons were to have no gyromagnetic anomaly, $(g-2)/2$, then according to the BMT equation (22) spin depolarization (see above) would be much weaker and dipole spin rotators would not work. As explained by G. Bunce (4) the $(g-2)/2$ for muons has a much subtler significance: it holds many clues to the electroweak contribution to the effective muon structure. Currently the difference between the measured and calculated value is $(4 \pm 9)10^{-9}$. When the new muon storage ring is running at Brookhaven the error on the measurement will be improved by a factor of twenty.

F. Feinstein (4) described muon polarimetry at the SMC beam at CERN based on measuring the decay positron spectrum and $\mu - e$ scattering.

CONCLUSION

When we look back to the dreams and plans of the early 1980's for spin polarization in high energy colliders we realise that the persistent efforts of the intervening years have paid off handsomely. These parallel sessions have demonstrated that the technology of spin polarization in lepton colliders is now almost a standard part of the accelerator engineer's toolkit. It is now up to the experimenters and theorists to exploit the new possibilities to the full.

ACKNOWLEDGMENTS

The author wishes to thank Dr. M. Woods, Dr. M. Placidi, Prof. I.M. Ternov and the speakers at these parallel sessions for helpful discussions.

REFERENCES

1. Steiner, H., Plenary talk, these Proceedings.
2. Woods, M., Plenary talk, these Proceedings.
3. Baier, V.N. and Khoze, V.A., Sov. J. Nucl. Phys., **9**, 238 (1969).
4. Parallel Session talk in these Proceedings.
5. Placidi, M., Plenary talk, these Proceedings.
6. Derbenev, Ya., Nucl. Inst. Meth., **A336**, 12 (1993).
7. Barber, D.P., Cabrera, B. and Montague, B.W., private communications, 1983.
8. Barber,D.P.,*Proc.9th Int.Symp.High Energy Spin Physics*, Bonn,Germany,1990.
9. Barber,D.P.,*Proc.10th Int.Symp.High Energy Spin Physics*, Nagoya,Japan,1992.
10. Barber, D.P., et al., Nucl. Inst. Meth., **A338**, 166 (1994).
11. Barber, D.P., et al., DESY Report 94-171 (1994). To appear in Phys.Letts. B.
12. Buon, J. and Steffen, K., Nucl. Inst. Meth., **A245**, 248 (1986).
13. Böge, M., DESY Report 94-87 (1994).
14. Yokoya, K., KEK Report 92-6 (1992).
15. Baier, V.N. and Orlov, Yu., Sov. Phys.-Doklady, **10**, 1145 (1966).
16. Chao, A.W., Nucl. Inst. Meth. **180**, 29 (1981).
17. Derbenev, Ya. and Kondratenko, A., Sov. Phys. JETP., **37**, 968 (1973).
18. Mane, S.R., Phys. Rev., **A36**, 105 (1987).
19. Barber,D.P.,*Proc.8th Int.Symp.High Energy Spin Physics*, Minniapolis,1988.
20. Barber, D.P. and Mane, S.R., Phys. Rev. **A37**, 456 (1988).
21. Baier, V.N. and Katkov, V.M., Sov. Phys. JETP., **25**, 944 (1967).
22. Bargmann, V., Michel, M. and Telegdi, V.L., Phys. Rev. Lett., **2**, 435 (1959).

LEPTON BEAM POLARIZATION AT LEP

R. Assmann*, A. Blondel*, B. Dehning, A. Drees, P. Grosse-Wiesmann, H. Grote, R. Jacobsen°, J.-P. Koutchouk, J. Miles, M. Placidi, R. Schmidt, F. Tecker†, J. Wenninger

CERN, CH-1211 Geneva 23, Switzerland
* *SLAC, Stanford Univ., P.O.Box 4349, CA 94309, USA*
* *Ecole Polytechnique, Paris, France*
° *Dept. of Physics, U.C. Berkeley, CA 94720, USA*
† *Physikalisches Institut III A, RWTH Aachen, Germany*

Abstract. Results from studies on transverse polarization in LEP over the past two years are presented. A single beam transverse polarization level of 57% at 45 GeV was reached adopting strategies to compensate depolarizing effects originating in the four experimental solenoids and from orbit perturbations. Beam Energy Calibration was performed by Resonant Depolarization during the 1993 LEP Run for Physics at three different energies centered around the Z peak. The uncertainty on the beam energy was reduced to about 1 MeV, thus improving the accuracy on the Z–resonance mass and width with respect to previous results. Successful results obtained at the end of the 1994 LEP Run on polarization with colliding beams are reported and future plans outlined.

1 INTRODUCTION

The experiments conceived and performed in the last two years aimed at producing beam polarization in conditions similar to luminosity operation and at developing orbit correction methods capable of increasing the polarization level at the Z energies and beyond [1].
The accuracy and reliability of the beam position monitors was greatly improved for this purpose. A vertical realignment of the magnetic structure of LEP was carried out and proved very beneficial.
Polarization during luminosity operation required compensating the spin precession in the Experimental Solenoids and studies on the effects of the machine tunes [2],[3].
The spin resonance compensation method known as Harmonic Spin Matching [4] was implemented and improved [5] providing high single beam polarization level.
Polarization with colliding beams was obtained in two dedicated machine development sessions with a proper choice of the beam energy and by controlling the number of collisions per revolution of the circulating bunches.

2 SINGLE BEAM POLARIZATION STUDIES

2.1 Solenoid compensation

Precise Energy Calibration during physics runs requires polarization in presence of the very strong longitudinal magnetic fields produced by the experimental solenoids. The spin rotation around the longitudinal axis (up to 66 mrad for the 10 Tm strength of the ALEPH solenoid) can be compensated by a proper configuration of vertical closed orbit bumps at both sides of each solenoid. The spin axis of the incoming beam is on purpose counter–rotated out of the vertical direction by half of the rotation angle produced by the magnet by the closed bumps at the entrance of the solenoid. An identic bump configuration at the other side of the solenoid makes a *closed-spin-bump* antisymmetric w.r.t. the IP and compensates for the spin rotation in the second half of the magnet.

The initial scheme foreseen for the 1991 LEP optics (60 degree lattice) [2] was adapted to the recent LEP optics developments [3] and provided substantial improvement in the attained polarization level.

2.2 CONSTRAINTS FROM THE TUNES

Strong depolarizing processes occur when the spin precession is in resonance with synchrotron and betatron oscillations.

The synchrotron tune Q_s was adjusted with the criterion of making the satellites from the integer spin tunes *above* and *below* the fractional one to coincide, which defined a *polarization* $Q_s = 0.0625$.

The integer part of the betatron tunes produces systematic integer resonances [6]:

$$(\mathrm{Int}[\nu_s])_{syst} \equiv N_{syst} = 4 \cdot k \pm \mathrm{Int}[Q_x, Q_y] \tag{1}$$

which, at the present vertical tune (Q_y =76.xx) generate "unwanted lines" dangerous at PEAK CM energy ($\nu_s \sim 103.5$) but in principle less important at PEAK±2 CM energies (LEP scan campaign in 1993).

The fractional part of the betatron tunes intervenes in the $Q_x + Q_y$ resonance driven by off-axis orbits in the sextupoles. For the values of the betatron tunes used in physics the above resonance is very close to the fractional part of the spin tune at which most of the calibrations take place.

To avoid these unwanted effects the machine tunes for polarization studies were :

$$(q_x \mid q_y \mid Q_s)_{P_\perp} = 0.10 \mid 0.16 \div 0.20 \mid 0.0625 \tag{2}$$

2.3 SPIN DIFFUSION

In a perfectly flat magnetic structure the equilibrium spin vector $\mathbf{n_o}$ is aligned with the vertical magnetic field $\mathbf{B}$ of the dipoles where the radiative Sokolov–Ternov polarizing mechanism originates. Polarization builds up to a maximum value of $8/5\sqrt{3} \sim 92.4\%$

Misalignments in a *real* machine cause the beam to experience **non–vertical** magnetic fields which bend the spin away from the unperturbed direction and generate spurious vertical dispersion. The equilibrium spin vector becomes orbit–dependent and **spin diffusion** occurs due to random fluctuations of the precession axis caused by quantized energy changes from emission of synchrotron radiation, resulting in a reduced asymptotic polarization level. Spin diffusion is governed by the *spin–orbit coupling function* $\mathbf{\Gamma}(s) = \gamma \frac{\partial \mathbf{n}(s)}{\partial \gamma}$ [7] and the reduced asymptotic polarization level can be written as :

$$P_\perp^\infty = \frac{8/5\sqrt{3}}{1 + \left(\frac{\tau_p}{\tau_d}\right)_{orbit}} \tag{3}$$

where both the polarization and the depolarization rates

$$(\tau_p)^{-1} \propto \oint \frac{ds}{|\rho|^3} \quad , \quad (\tau_d)^{-1} \propto \oint \frac{|\mathbf{\Gamma}|^2}{|\rho|^3} ds \tag{4}$$

are driven by the synchrotron radiation term $\oint \frac{ds}{|\rho|^3}$.

2.3.1 Harmonic Spin Matching

Spin motion can be highly perturbed when the spin precession is in phase with energy (synchrotron) and betatron oscillations, in particular for harmonics of the perturbed orbit close to the actual spin tune. Suggested [4] in 1985, Harmonic Spin Matching (*HSM*) is a method to minimize the spurious tilt of the spin precession axis by compensating the harmonics of the Fourier expansion of the real orbit close to a specific value of the spin tune. The method relies on a lengthy procedure for the correction of the orbits, made difficult from the tiny entities of the corrections required, only monitored by the observation of the effects on the polarization itself.
The *HSM* implementation described in [5] was intended to reduce the drawback caused by the very long polarization time in LEP and consisted in deriving the amplitude of the correcting bumps for the harmonics of interest directly from the beam position information (**Deterministic** *HSM*).
Its application proved to be very beneficial allowing to improve the polarization level obtained with the solenoid compensation ($\sim 15\%$) to more than 40% [1].
A further improvement of the *HSM* method was tried by varying the correcting harmonic amplitudes in steps while recording the polarization level to search for an optimum (**Empirical** *HSM*). The experiment, performed at a spin tune $\nu_s = 101.5$ corresponding to a PEAK-2 CM energy, produced a record level of $(57 \pm 3)\%$ as shown in fig.1.

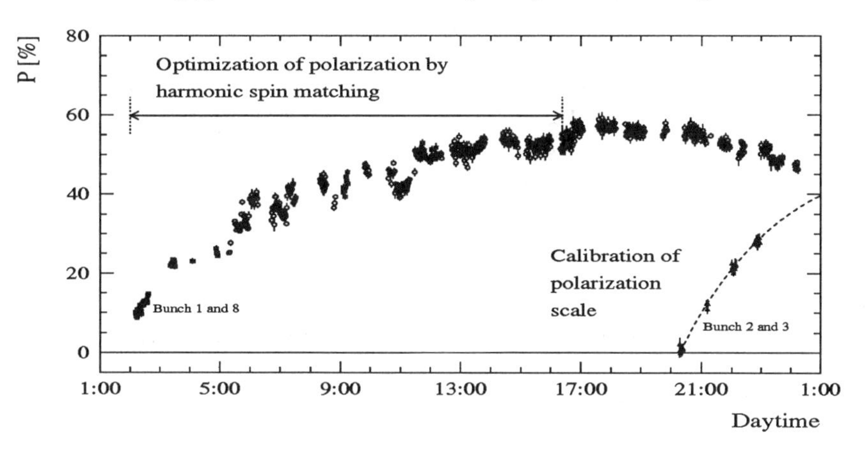

Figure 1: Maximum measured polarization level of $(57 \pm 3)\%$ at PEAK-2 CM energy with the adoption of Deterministic and Empirical Harmonic Spin Matching. Bunches 2 and 3 were on purpose depolarized to calibrate the polarimeter scale by measuring the Sokolov–Ternov radiative polarization rise time.

3 PRECISE BEAM ENERGY CALIBRATION

Operational beam Energy Calibration by resonant depolarization [8] was successfully implemented to improve the measurements of the Z–resonance mass and width [9] during the energy scan in 1993. A typical example of a spin–tune scan associated to a controlled resonant depolarization is shown in fig.2.

The limitations and the systematic errors which can affect the Energy Calibration have been thoroughly accounted for in [9] and summarized in table 1. To combine the 1993 Energy Calibration results the measurements were corrected to a reference set of parameters where the effects from terrestrial tides [10] and from temperatures changes in the LEP dipoles were accounted for.
Analysis of the orbits measured during all LEP physics fills [11] allowed the energies measured over 24 calibrations in the 1993 Physics Run to be also corrected for non–periodical ring deformations from, for example, hydrostatic pressure on the tunnel walls due to rain and changes in the water level of the Geneva lake [12].
The time evolution of the electron beam energy during the 1993 scan is shown in fig.3.

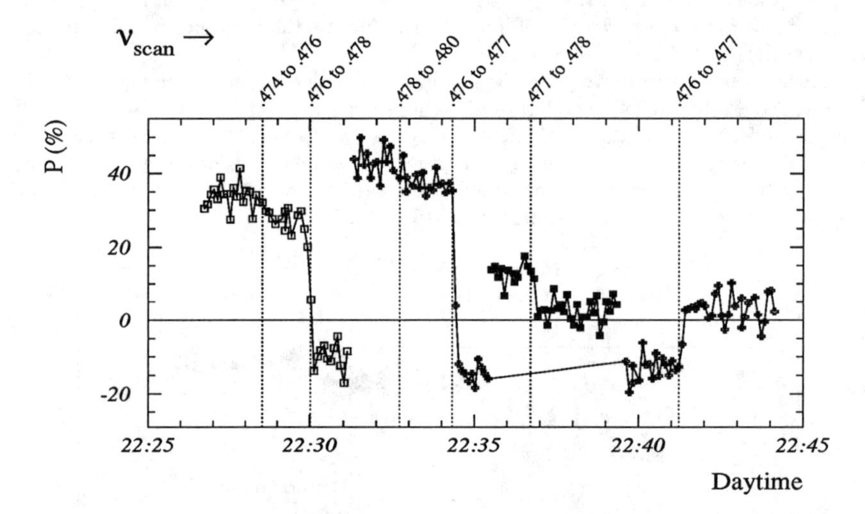

Figure 2: Example of Energy Calibration with Resonant Depolarization. The upper scale shows the non–integer part of the spin tune corresponding to the sharp drop in the polarization level at the depolarizing resonance. A partial spin–flip was checked to be real by flipping it back at the same spin tune (0.476–0.477).

3.1 Comparing Electron and Positron Beam Energies

Possible differences between electron and positron beam energies E_b were investigated by almost simultaneously calibrating the two beams with a modified setup of the polarimeter to illuminate both beams with the same laser beam back-reflected by a retro-focusing concave mirror [5]. Results shown in fig.4 are consistent with an energy difference in agreement with the upper limit of $\Delta E_b \leq 0.2$ MeV from theoretical calculations [13].

The absolute calibration of the beam energies of the off–peak points was determined with a precision of $2 \cdot 10^{-5}$ resulting in a systematic error of ~ 1.4 MeV on the Z–mass and of ~ 1.5 MeV on the Z–width [11].

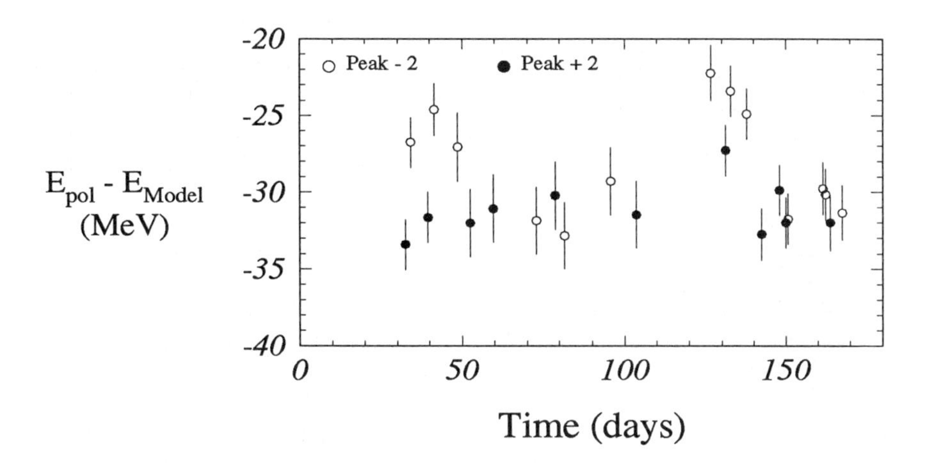

Figure 3: Evolution of the electron beam energy during the 1993 scan after correction of periodic effects (tides), dipole temperature changes and radial orbit movements.

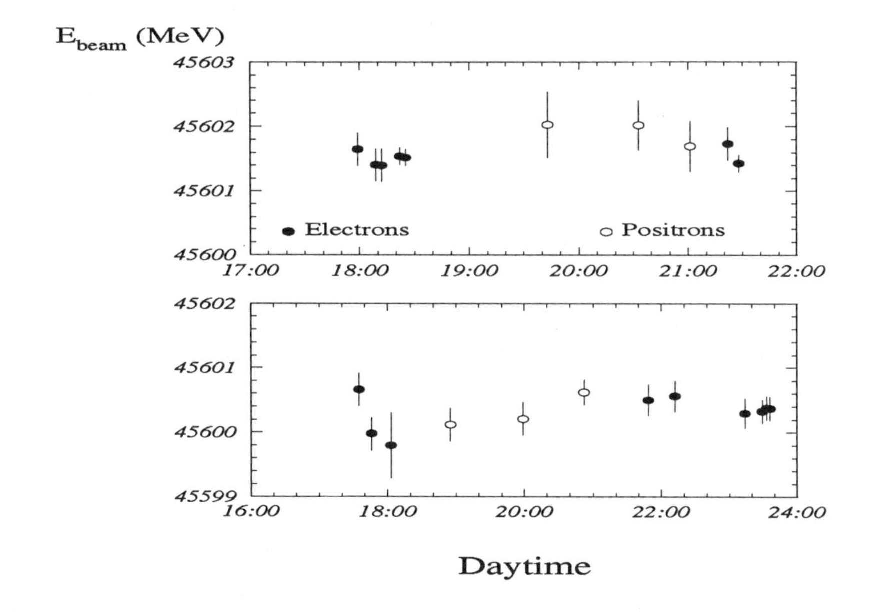

Figure 4: Electron and Positron beam energies measured almost simultaneously in two experiments performed on July 15th (top) and August 1st 1994. The energies are corrected for tides, magnet temperatures, reference magnet field and slow ring distortions. The resulting energy difference is $\Delta E_b \leq 0.2$ MeV.

SOURCE	ΔE Theoretical estimates	ΔE Experimental upper bound
Electron mass		13 keV
Revolution frequency		<1 keV
Depolarizer frequency		100 keV
Width of excited resonance		100 keV
Interference of resonances		< 100 keV
Quadratic nonlinearities	< 5 keV	< 500 keV
ν_s-shifts from long. fields	< 5 keV	< 500 keV
ν_s-shifts from rad. fields	< 100 keV	< 800 keV
Total systematic error	**100 keV**	
Total upper bound		**< 1.1 MeV**

Table 1: Systematic errors in the measurement of the beam energy by resonant depolarization assuming a well corrected vertical orbit. Quoted errors, evaluated at $E = 45.6$ GeV, are understood to be Gaussian and refer to the energy of a single beam. The contributions in the third column are experimental upper bounds used to compute the total upper bound on the systematic error.

4 POLARIZATION IN COLLISION

Two experiments were performed to assess the feasibility of achieving transverse polarization with colliding beams. A high transverse polarization level (40%) was maintained while colliding one e^+ against one e^- bunch in the OPAL detector with high luminosity (beam-beam tune shift of ~ 0.04) at a PEAK-2 CM energy.
Writing the asymptotic value for the observed polarization level as

$$P_{\perp}^{\infty} = \frac{8/5\sqrt{3}}{1 + \left(\frac{\tau_p}{\tau_d}\right)_{orbit} + \left(\frac{\tau_p}{\tau_d}\right)_{BB}} \quad (5)$$

one can phenomenologically distinguish between the depolarizing effects inherent to the quality of the orbit in the arcs and those produced by the beam–beam interaction.
The contribution from vertical orbit deviations y_{rms} in the lattice quadrupoles scales as $(\tau_p/\tau_d)_{orbit} \propto | \mathbf{\Gamma} |^2 \propto y_{rms}^2$ (4) and the chance to produce high transverse polarization levels relies on the *quality of the vertical orbit in the arcs.*
The contribution from beam–beam interaction can be reduced with a proper choice of machine tunes (betatron and synchrotron) once the beam energy has been correctly defined to avoid single beam depolarizing resonances.
The experiments were performed at a beam energy corresponding to a PEAK-2 CM energy ($\nu_s \cong 101.5$) in order to avoid systematic spin resonances of the type (1).
A *3e⁻ ON 1e⁺* bunch scheme [14] allowed colliding at *one IP at a time* and to measure the polarization level for colliding and non–colliding bunches as shown in table 2. Only the bunches $P1, E1, E2, E3$ were injected to produce collisions between $P1$ and $E2$ in $IP6$ (and later in $IP2$). The non–colliding e^- bunches $E1$ and $E3$ were used for Energy Calibration and to provide a reference value for "single beam" polarization.
The results are reported in the following.

IP	ENCOUNTERS	MODE
1 & 5	**P1·E1** + P2·E2 + P3·E3 + P4·E4	V-separation
2 & 6	**P1·E2** + P2·E3 + P3·E4 + P4·E1	**collisions**
3 & 7	**P1·E3** + P2·E4 + P3·E1 + P4·E2	V-separation
4 & 8	P1·E4 + P2·E1 + P3·E2 + P4·E3	(collisions)

Table 2: Collisions with 4 equidistant bunches per beam. In **thick**, the bunches colliding with the adopted scheme.

4.1 THE FIRST EXPERIMENT

$4 \times 220\,\mu$A were injected and accelerated at the betatron/synchrotron tunes (2) and the beam energy measured to localize the working point in the spin–tune space to avoid depolarizing resonances.
The quality of the vertical orbit was not very satisfactory ($y_{rms} = 0.45$ mm) and the application of Deterministic HSM did not help in reaching a polarization level higher than $\sim 25\%$ for the non–colliding bunch $E3$.
The polarization level for the $E2$ bunch, when colliding in $IP6$ with the $P1$ bunch, remained limited to $\sim 20\%$, but *didn't decrease* when the two bunches were made to collide also in $IP2$ (see fig.5).
The luminosity measured in one of the LEP detectors (OPAL) was consistent with a relatively high beam–beam strength parameter:

$$L_{OPAL} = (1.3 \rightarrow 1.0) \times 10^{30}\mathrm{cm}^{-2}\mathrm{s}^{-1} \quad , \quad \xi_y \sim 0.037.$$

4.2 THE SECOND EXPERIMENT

$4 \times 190\,\mu$A were accelerated in LEP with the same machine conditions as in the previous experiment, with the exception that this time the four solenoids were switched OFF. A reasonable amount of time was spent trying to improve the polarization level with separated beams adopting a kind of *educated HSM* [15] in which the *true bump amplitude* needed to compensate the four harmonics was experimentally determined by observing the separated effect of each of them. As a result the polarization level started rising from a saturation value of $\sim$25% to more than 30% (see fig.6) and **kept rising** when $E2$ was brought in collision with $P1$ in OPAL. The empirical HSM procedure was pursued with colliding beams and a 40% polarization level was attained for *both the colliding and the non–colliding e^- bunches.*
High luminosity and beam–beam strength parameter were again measured in OPAL:

$$L_{OPAL} = (1.5 \rightarrow 1.0) \times 10^{30}\mathrm{cm}^{-2}\mathrm{s}^{-1} \quad , \quad \xi_y \sim 0.040.$$

When a second interaction was added in $IP2$ the polarization level showed a tendency to decrease, but the beams were lost due to a machine problem before we succeeded in compensating the effect.

4.3 SUMMARY OF RESULTS

The results obtained in the two experiments described above are compared to the high polarization one obtained at the same beam energy in 1993 [5] in table 3.

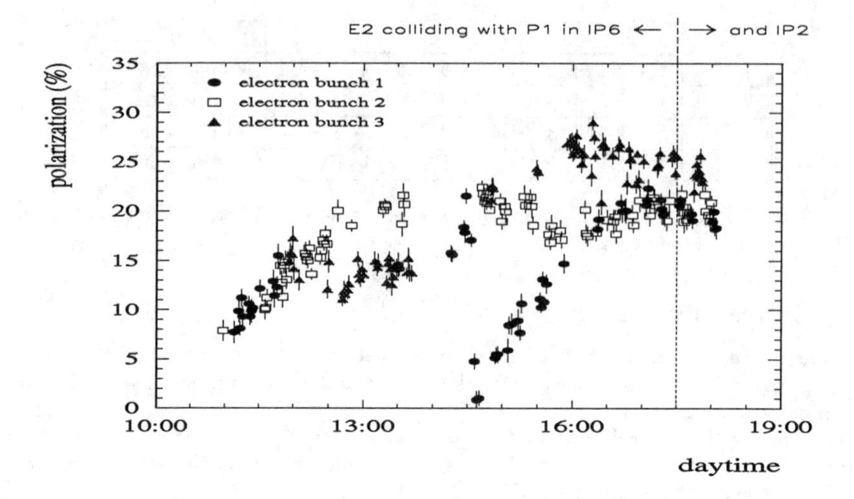

Figure 5: Polarization with colliding beams. With the adoption of a purely Deterministic *HSM* scheme the maximum polarization level attained by the non–colliding electron bunch $E3$ was about 25% while the colliding bunch $E2$ reached a maximum of ~20%. The electron bunch $E1$ was depolarized on purpose to calibrate the polarimeter scale through the Sokolov-Ternov polarization rise time.

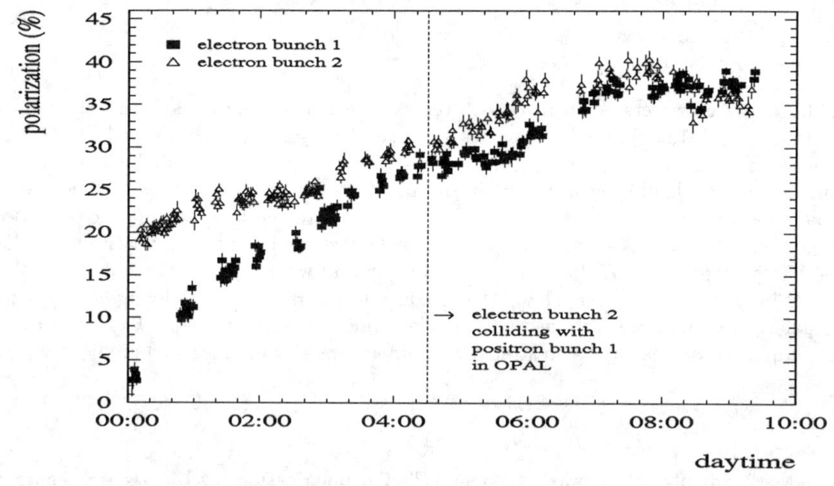

Figure 6: Polarization for colliding and non–colliding electron bunches was raised up to a 40% level during the last experiment in the 1994 LEP Run with the adoption of Empirical and Deterministic *HSM* which allowed for polarization to keep rising after the $E2$ electron bunch was made to collide with the positron bunch $P1$ in the OPAL experiment.

In particular it is shown (fifth column) that the beam-beam depolarization rate measured during the second experiment was at least a factor of three smaller than the polarization one.

Date	MODE	P_∞	$(\tau_p/\tau_d)_{orbit}$	$(\tau_p/\tau_d)_{BB}$	y_{rms}/mm	ξ_y
23.08.93	1 beam	57%	0.62	–	0.33	–
16.11.94	2 beams, Sep.	26%	2.55		0.45	
	Colliding	22%		0.65		0.037
05.12.94	2 beams, Sep.	40%	2.7		0.36	
	Colliding	38%		≤ 0.3		0.040

Table 3: Comparison between the experiments 1994 and the High Polarization experiment in 1993. All experiments performed at PEAK-2 CM energy and with *HSM*.

5 FUTURE PLANS

5.1 ENERGY CALIBRATION IN COLLISION

Differently from the Energy Calibration campaign for the 1993 Scan [11] the proposed 1995 schedule foresees to devote between 20 to 40 pb^{-1} to perform Energy Calibration at both PEAK$\pm$2 energies [16],[17]. With en expected 1995 performance of about 1 pb^{-1} per fill, about **40 fills** in the 11 weeks between July and September should be calibrated.
A strong motivation exists then to *perform the calibration during the LEP operation for physics.* To meet this target some problems will be addressed [14] namely :

1. the **polarization β-tron tunes** (2) differ from those used in "production";
2. the present *HSM* scheme affects the Luminosity and proved to be marginal with the optics used at the end of the 1994 run.

5.1.1 Residual Longitudinal Polarization at IP's

The possibility that a small amount of residual longitudinal polarization at the Interaction Points could survive due to machine optics imperfections is of some concern for the LEP experimental collaborations since it represents a bias in the measurements.
The question was addressed in [18] where, in the hypothesis that the experimental solenoids are spin–compensated, the main source for this phenomenon is traced back to beam orbit offsets in the low-β doublets close to the IP's, which can behave like small spin rotators. Quantitatively, a 1 mm orbit deviation in the strong focusing doublets can produce $\sim$ 40 mrad local rotation of the $\mathbf{n}_o$ vector which corresponds to a $\sim$ 0.4% longitudinal polarization level at the experimental IP's for a 10% transverse polarization.
This value, well within the limits acceptable by the experiments, can be drastically reduced centering the beams in the low-β doublets by means of the *K–modulation* technique [19]. Under these conditions, the effect of residual longitudinal polarization at the Interaction Points leads to negligible systematic uncertainties on precision measurements [20].

6 CONCLUSIONS

Considerable achievements have been obtained with transverse polarization at LEP in the last two years, both with single–beam operation and with colliding beams, thanks to the

successful implementation of techniques intended to compensate depolarizing effects from the experimental solenoids and to reduce the contribution to spin diffusion from machine imperfections and large closed orbit deviations.
The Energy Calibration campaign scheduled for the 1995 LEP Run is expected to absorb a considerable amount of beam time during the operation for physics.
The encouraging results obtained with polarization in collision at a PEAK-2 beam energy in 1994 are being considered to be adapted to a more efficient application of the calibration procedures in the 1995 scan to reduce the Z–production time to invest into Energy Calibration.

References

[1] Assmann, R. *et al.*, *Polarization Studies at LEP in 1993*, CERN SL/94-08, March 1994.

[2] Blondel, A., *Compensation of Integer Spin Resonances Created by Experimental Solenoids*, LEP Note 629 (1990).

[3] Grote, H., *A New Solenoid Compensation for Polarization compatible with Pretzel*, SL/Note 94–128 (AP), July 1994.

[4] Rossmanith, R. and Schmidt, R., *Compensation of Depolarizing Effects in Electron Positron Storage Rings*, NIM A 236 (1985) 231.

[5] Assmann, R., *Transversale Spin-Polarisation und ihre Anwendung für Präzisionsmessungen bei LEP*, PhD Thesis, 1994.

[6] Koutchouk, J.-P., 8th. H. E. Spin Physics Symposium, AIP Conf. Proceedings No.187, Minneapolis, MN, 1988.

[7] Ya. S. Derbenev and A. M. Kondratenko, Sov. Phys. JETP 37 (1973), 968.

[8] Arnaudon, L., *et al.*, *Measurement of LEP beam energy by Resonant Spin Depolarization*, Phys. Lett. B 284 (1992) 431–439.

[9] The LEP Polarization Team, *Accurate Determination of the LEP Beam Energy by Resonant Depolarization*, CERN SL/94–71 (BI) August 1994.

[10] The LEP Polarization Team, *Effects of Terrestrial Tides in the LEP Beam Energy*, (Submitted to NIM A).

[11] The LEP ENERGY Working Group, *The Energy Calibration of LEP in the 1993 Scan*, CERN SL/95-02 and CERN PPE/95-10, Feb. 1995.
(Submitted to Zeitschrift für Physik C).

[12] Wenninger,J., *Radial Deformations of the LEP ring*, CERN SL/Note 95-21 (OP), Feb. 1995.

[13] Drees, A., *Energy Difference of Electron and Positron Beam*, CERN SL/Note 94-100 (BI), March 1995.

[14] Placidi, M., *Transverse Polarization in Collision at LEP*, Vth Chamonix Workshop on LEP Performance, Chamonix, 13-18 January 1995.

[15] Blondel, A., Report in publication.

[16] Wenninger, J,. *What will cost in time and efficiency in order to calibrate the LEP energy every fill?*, Vth Chamonix Workshop on LEP Performance, Chamonix, 13-18 January 1995.

[17] Camporesi, T., *Overall requirements for LEP in 1995 (and 1996)*, Vth Chamonix Workshop on LEP Performance, Chamonix, 13-18 January 1995.

[18] Assmann, R. and Koutchouk, J.-P., *Residual Longitudinal Polarization at LEP*, CERN SL Note 94-37, April 1994.

[19] Dehning, B., *Do we need K–modulation in the arcs?*, Vth Chamonix Workshop on LEP Performance, Chamonix, 13-18 January 1995.

[20] Blondel, A., *Systematic Uncertainties on LEP Electroweak observables due to Unknown Residual Longitudinal Beam Polarization*, ALEPH Note 95–018, Feb. 1995.

POLARIZATION AT SLAC*

M. Woods

Stanford Linear Accelerator Center
Stanford University, Stanford, CA 94309

Abstract. A highly polarized electron beam is a key feature for the current physics program at SLAC. An electron beam polarization of 80% can now be routinely achieved for typically 5000 hours of machine operation per year. Two main physics programs utilize the polarized beam. Fixed target experiments in End Station A study the collision of polarized electrons with polarized nuclear targets to elucidate the spin structure of the nucleon and to provide an important test of QCD. Using the SLAC Linear Collider, collisions of polarized electrons with unpolarized positrons allow precise measurements of parity violation in the Z fermion couplings and provide a very precise measurement of the weak mixing angle. This paper discusses polarized beam operation at SLAC, and gives an overview of the polarized physics program.

POLARIZED PHYSICS PROGRAM

1. Fixed Target Experiments in End Station A (ESA)

The nucleon has spin 1/2, and it has a rather complex internal structure which contains valence quarks, sea quarks and gluons. Deep inelastic scattering experiments tell us that about 1/2 of a nucleon's momentum is carried by the quarks and about 1/2 is carried by the gluons.[1] But how much of the nucleon's spin is due to the quarks? How much is due to the gluons? And how much is due to orbital angular momentum? Measurements of the longitudinal proton spin structure function (g_1^p) by the SLAC experiments E80[2] and E130,[3] and by the CERN EMC experiment,[4] indicated that the quarks contribute only $(12 \pm 17)\%$ of the nucleon spin.[4] This small contribution, consistent with zero, came to be known as the 'Spin Crisis' and led to the proposal of new experiments at SLAC, CERN and DESY.

The Ellis-Jaffe sum rules[5] give predictions for $\int g_1^p dx$ and $\int g_1^n dx$, and are based on SU(3) symmetry and an assumption that the strange sea is unpolarized. Using the quark parton model, $\int g_1^p dx$ and $\int g_1^n dx$ can be used to determine the total quark contribution to the nucleon spin.[6] Bjorken has also developed a sum rule for $\int (g_1^p - g_1^n) dx$, which follows from current algebra.[7] A violation of the Ellis-Jaffe sum rules would imply that the existing model of nucleon structure is too simple. A violation of the Bjorken sum rule would be more serious, and would pose a significant challenge to current QCD theory.

Recently, SLAC experiment E142 has made the first measurement of g_1^n,[8] and E143 has measured g_1 for the proton and deuteron.[9] The CERN SMC experiment has also recently

* Work supported in part by the Department of Energy, contract DE-AC03-76SF00515

measured g_1 for the deuteron[10] and proton.[11] As summarized in other contributions to this conference,[12] the new experimental data from SLAC and CERN indicate that the Bjorken sum rule is satisfied (less than 1σ discrepancy with the data). However, the Ellis-Jaffe sum rule for the proton appears to be violated (greater than 3σ discrepancy with the data), and the quarks appear to account for only 1/3 of the proton's spin (0.31 ± 0.07).

The g_1 measurements at SLAC will continue with further measurements on the neutron by E154, and on the proton and deuteron by E155 (see Table I for relevant beam parameters and dates for running).

2. The SLD Experiment at the SLAC Linear Collider (SLC)

The Minimal Standard Model (MSM) of electroweak interactions gives the gauge structure of the theory as $SU(2)_L$ X $U(1)$. This gauge structure results in four physical gauge bosons (W^+, W^-, Z^0, γ) that mediate the interactions. Figure 1 shows the Feynman diagrams for the different gauge boson - fermion vertices and gives the couplings of these vertices separately for left- and right-handed fermions. It is clear from this that electroweak interactions treat left-handed and right-handed fermions differently.

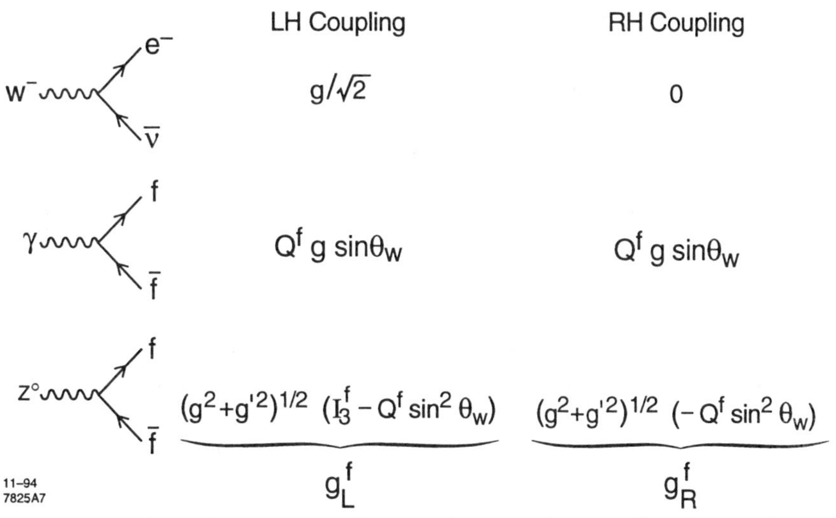

Figure 1: Left- and Right- couplings at the gauge boson - fermion vertices. g is the $SU(2)_L$ coupling constant; g' is the $U(1)$ coupling constant; Q^f is the fermion charge; I_3^f is the fermion isospin component; and θ_w is the weak mixing angle ($\tan\theta_w \equiv g'/g$).

The charged W couples *only* to left-handed fermions; the photon couples *equally* to both; and the Z couples to both, but *asymmetrically*. For the Z-fermion coupling, one can define the following asymmetries,

$$A_f \equiv \frac{(g_L^f)^2 - (g_R^f)^2}{(g_L^f)^2 + (g_R^f)^2}$$

These asymmetries are large. For example, A_e is expected to be about 15%.

The SLC is currently operating with e^+e^- collisions at the Z^0 resonance. The availability of a highly polarized electron beam gives the capability for direct measurements of the asymmetries A_f. These are now being measured by the SLD detector.[13,14]

In particular, the SLD is making a very precise measurement of the left-right asymmetry,

$$A_{LR} \equiv \frac{\sigma(e_L^- e^+ \to Z^0) - \sigma(e_R^- e^+ \to Z^0)}{\sigma(e_L^- e^+ \to Z^0) + \sigma(e_R^- e^+ \to Z^0)} = A_e$$

This measurement[14] now gives the world's best single determination of the weak mixing angle, $\sin^2 \theta_W^{\text{eff}}$,[15] and provides one of the best tests of the MSM.[16] The current SLD run is expected to reduce its error on $\sin^2 \theta_W^{\text{eff}}$ by a factor of 2.

POLARIZED BEAM OPERATION

1. Polarized Source

Polarized electrons are produced by photoemission from a GaAs photocathode as shown in Figure 2. Different laser light sources are used for the ESA and SLC physics programs due to the different pulse structures required (see Tables I and II). For ESA, a flashlamp-pumped Ti:sapphire laser[17] is used to produce a 2μs pulse. For SLC operation, two Nd:YAG-pumped Ti:sapphire lasers[18] produce two 2ns pulses separated by about 60ns. One of these pulses is used to make electrons for collisions, and the other one is used to make electrons for positron production.

The laser beams are circularly polarized by a linear polarizer followed by a Pockels Cell operating at its quarter-wave voltage. A positive HV pulse on the Pockels Cell produces one helicity, while a negative HV pulse produces the opposite helicity. The sign of the HV pulse is set by a pseudo-random number generator, which updates at 120 Hz (the SLAC machine pulse rate). This very effectively minimizes false experimental asymmetries.

The photoexcitation of electrons in the GaAs cathode from its valence band to the conduction band is illustrated in Figure 3. Consider first the situation for unstrained GaAs in Figure 3a. Photons with positive helicity and with energies greater than the band gap energy of 1.43eV, but less than 1.77eV, can excite the two indicated solid transitions from the j=3/2 valence band to the j=1/2 conduction band. Clebsch-Gordon coefficients give the relative probability for these two transitions to be 3:1. Thus, positive helicity light will produce negative helicity electrons with a net theoretical polarization of 50% ($P = \frac{3-1}{3+1}$). The extracted electrons from the GaAs cathode, however, will have the same helicity as the incident photons since they have opposite direction to the incident photons.

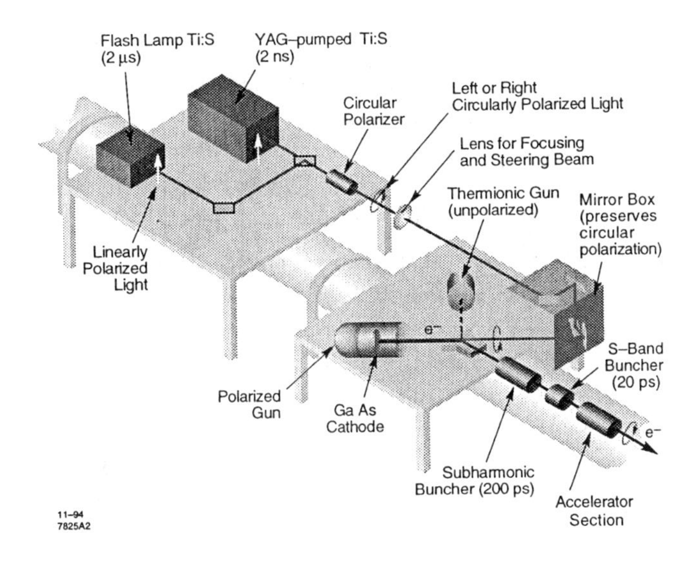

Figure 2: The Polarized Electron Source at SLAC

If one splits the j=3/2 valence band degeneracy as shown in Figure 3b, one can theoretically produce an electron beam with polarization close to 100%. In practice this can be accomplished by growing a thin layer of GaAs on GaAsP. The lattice mismatch between the two results in a strained GaAs lattice, which indeed breaks the degeneracy.[19] Such cathodes are now commercially available and have demonstrated polarizations in excess of 80%. The quantum efficiency (emitted electrons per incident photon), QE, of such cathodes is typically 0.2%.

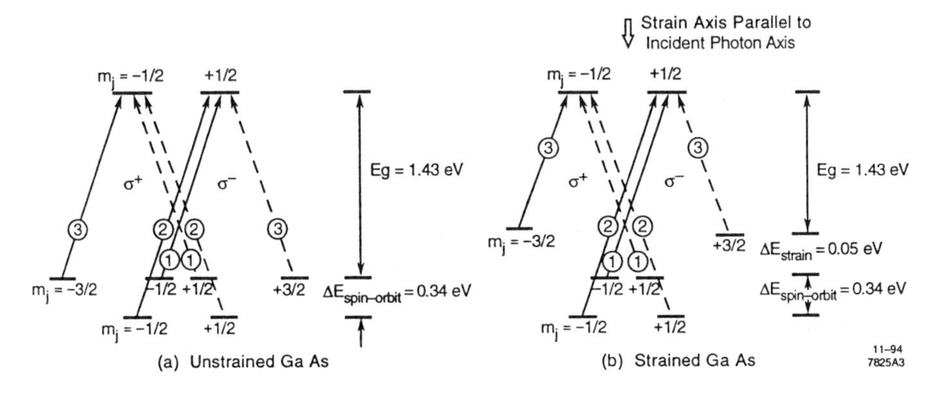

Figure 3: GaAs energy levels and allowed transitions from the valence band to the conduction band. Solid (dashed) lines indicate transitions due to positive (negative) helicity photons. Circled numbers indicate relative probabilities of transitions.

2. Beam and Spin Transport

End Station A Operation. Figure 4 illustrates the beam and spin transport from the polarized source to the End Station A experiments. The electron beam energy at the source is 60 keV. The beam is injected at this energy into the 3-km SLAC Linac, where it is bunched and accelerated to energies as high as 29 GeV. (The ESA beamline is currently being upgraded to transport beams with energies up to 50 GeV for experiments E154 and E155.)

The electron spin is longitudinal at the source and remains longitudinal upon injection into the Linac. At the end of the Linac the beam is deflected with horizontal bend magnets by an angle, θ_b, of 428 mrad into the ESA beamline. This deflection causes the spin to precess with respect to the momentum vector by $\Delta\theta = (\frac{g-2}{2})\gamma\theta_b$. When $\Delta\theta = n\pi$ the spin is longitudinal in ESA; this is achieved at beam energies of $E_b = n \cdot 3.24$ GeV. The polarized beam electrons are scattered by the polarized target and are detected in two spectrometers as shown in Figure 4. The polarized electron beam and polarized nuclear target can be set up to have their relative spins longitudinally aligned either parallel or anti-parallel. From the measured cross-section asymmetry for these two cases, the nucleon spin structure functions (g_1) can be determined.

Table I: Beam Parameters for ESA Operation

Parameter	E142	E143	E154	E155
N^-	$2 \cdot 10^{11}$	$4 \cdot 10^9$	$2 \cdot 10^{11}$	$4 \cdot 10^9$
f_{rep}	120 Hz	120 Hz	120 Hz	120 Hz
Pulse Length	1us	2us	100ns	100ns
Beam Energy	22.7 GeV	29.2 GeV	48.6 GeV	48.6 GeV
Polarization	40%	84%	80%	85%
Run time	2 months	3 months	2 months	3 months
Year	1992	1993	1995	1996

SLAC Linear Collider Operation. The beam and spin transport are each considerably more complex for SLC operation. This is shown in Figure 5. Two electron bunches are produced from the photocathode gun, which operates at 120 kV.[20] The higher voltage (than the 60 kV required for ESA operation) is needed to increase the *space charge limit* current capability of the gun above the 6 amps of peak current required for SLC operation. During early operation of the polarized gun for SLC, an unexpected cathode *charge limit* was observed below the space charge limit.[21] The cathode charge limit was observed to be proportional to the cathode quantum efficiency and posed a worry for achieving the needed high currents at the low QEs of the strained lattice cathodes. Similar behaviour of the charge limit is indeed observed with the strained lattice cathodes, but the QE scaling factor is different and adequate QEs can be achieved for the SLC current requirements.[22]

The two electron bunches produced from the photocathode gun are injected into the SLAC Linac where they are bunched and accelerated to 1.19 GeV. They are then kicked by a pulsed magnet into the Linac-to-Ring (LTR) transfer line to be transported to the electron damping ring (DR). The DR stores the beam for 8ms to reduce the beam emittance. The Ring-to-Linac (RTL) transfer line transports the two bunches from the DR and a pulsed magnet kicks them back into the Linac. These two bunches are preceded down the Linac by a positron bunch which has been extracted from the positron DR. Three bunches are then accelerated down the Linac. The trailing electron bunch is accelerated only to 30 GeV, and is then sent to the positron production target. Positrons in the energy range 2-20 MeV are collected, accelerated to 200 MeV, and transported to near the start of the Linac for transport to the positron DR, where they are damped for 16 ms. At the end of the Linac, the electron and positron energies are each 46.6 GeV. A magnet deflects the electron (positron) bunch into the north (south) collider arc for transport to the Interaction Point (IP). In the arcs, the beams lose about 1 GeV in energy from synchrotron radiation so that the resulting center-of-mass collision energy is 91.2 GeV, which is chosen to match the Z^0 mass. The beam energies are measured with energy spectrometers[23] to a precision of 20 MeV.

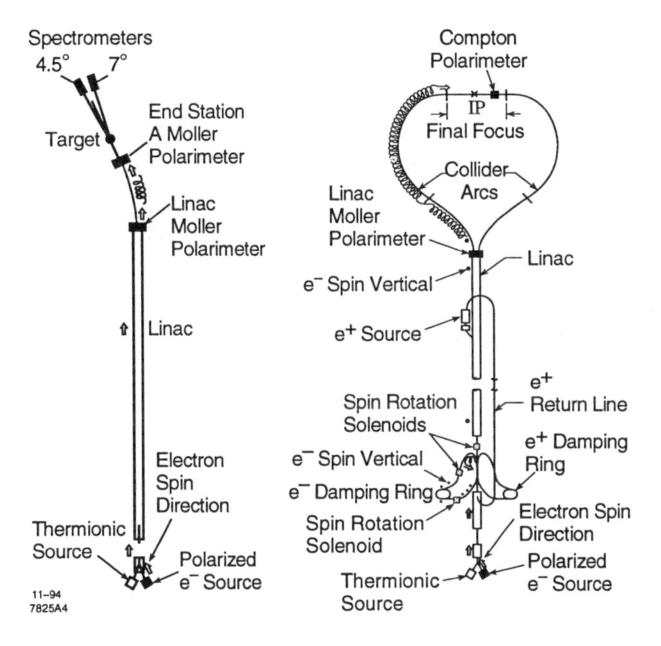

Figure 4: Polarized ESA Operation **Figure 5: Polarized SLC Operation**

The electron spin orientation is longitudinal at the source and remains longitudinal until the LTR transfer line to the electron DR. In the LTR, the electron spin precesses by 450° to become transverse at the entrance to the LTR spin rotator solenoid. This solenoid rotates the electron spin to be vertical in the DR to preserve the polarization. The spin orientation is vertical upon extraction from the DR; it remains vertical during injection into the linac and during acceleration to 46.6 GeV down the linac. The spin transmission of this system is 0.99, with the small loss resulting from the beam energy in the DR being 1.19 GeV, slightly lower than the design energy of 1.21 GeV; this causes the spin precession in the LTR to be 442° rather than 450°, and the spin transmission is the sine of this angle.

The SLC Arc transports the electron beam from the linac to the IP and is comprised of 23 achromats, each of which consists of 20 combined function magnets. At 46.6 GeV, the spin precession in each achromat is 1085°, while the betatron phase advance is 1080°. The SLC Arc is therefore operating near a spin tune resonance. A result of this is that vertical betatron oscillations in the arc's achromats (combined function magnets which bend the beam in the horizontal plane), can cause the beam polarization to rotate away from vertical; this rotation is a cumulative effect in successive achromats. (The rotation of the vertical spin component in a given achromat is simply due to the fact that rotations in x and y do not commute, while the cumulative effect is due to the spin resonance.) The resulting spin component in the plane of the arc then precesses significantly.

The arc's spin tune resonance, together with misalignments and complicated rolls in the arc,[24] result in an inability to predict the spin orientation at the IP for a given spin orientation at the end of the Linac. However, we have two good experimental techniques for orienting the spin longitudinally at the IP. First, using the RTL and Linac spin rotator solenoids one can orient the electron spin to be along the x, y, or z axis at the end of the Linac. The z-component of the arc's spin transport matrix can then be measured with the Compton polarimeter, which measures the longitudinal electron polarization as described below. The Compton measures

$$P_z^C = R_{zx} \cdot P_x^L + R_{zy} \cdot P_y^L + R_{zz} \cdot P_z^L \tag{1}$$

The experimental procedure is referred to as a 3-state measurement, and is accomplished by measuring P_z^C for each of x, y, or z spin orientations at the end of the linac. Using equation (1), the arc spin rotation matrix elements R_{zx}, R_{zy}, R_{zz} are then determined. This is sufficient to determine the full rotation matrix, which is described by three Euler angles. The matrix R can then be inverted to determine the required spin orientation at the end of the Linac for the desired longitudinal orientation at the IP. This Linac spin orientation is achieved with appropriate settings of the RTL and LINAC spin rotators.

Table II: Beam Parameters for SLC Operation

Parameter	1993	1994 (expected)
N^+	$3.0 \cdot 10^{10}$	$3.5 \cdot 10^{10}$
N^-	$3.0 \cdot 10^{10}$	$3.5 \cdot 10^{10}$
f_{rep}	120 Hz	120 Hz
σ_x	$0.8 \mu m$	$0.5 \mu m$
σ_y	$2.6 \mu m$	$2.4 \mu m$
Luminosity	$5 \cdot 10^{29}$ cm^{-2}s^{-1}	$1 \cdot 10^{30}$ cm^{-2}s^{-1}
Z/hr (peak)	50	100
Collision Energy	91.26 GeV	91.26 GeV
Polarization	63%	80%
Uptime	70%	70%
Run time	6 months	7 months
Integrated Zs	50K	100K

A second method to orient the spin longitudinally at the Compton takes advantage of the arc's spin tune resonance. A pair of vertical betatron oscillations ('spin bumps'), each spanning 7 achromats in the last third of the arc, are introduced to rotate the spin.[25]. The amplitudes of these spin bumps are empirically adjusted to achieve longitudinal polarization at the IP. Experiments have verified that the two spin orientation techniques provi le consistent results to a precision of about 1% in the longitudinal IP polarization. Thus, the two spin bumps can effectively replace the two spin rotators. This turns out to be very important for SLC operation, where high luminosity has been achieved by producing and colliding flat beams (see Table II). Flat beams are naturally produced in the damping rings if the x and y betatron tunes are different. Preserving the flat beams during acceleration and transport to the IP requires minimizing any x-y coupling in the accelerator. The coupling introduced by the RTL and Linac spin rotator solenoids proves to be unacceptable. So the *problem* of the arc spin tune resonance for modeling the arc spin transport has become a *feature* that allows both spin orientation control and flat-beam running for high luminosity.

3. Polarimetry

Three different polarimeter techniques are used at SLAC based on Mott scattering, Moller scattering, and Compton scattering. The ESA experiments rely on Moller polarimeters, while the SLD experiment relies on a Compton polarimeter. Mott polarimeters are used in test labs for polarized gun development work and for photocathode R&D. A considerable amount of work has been done and is continuing to be done at SLAC to compare results from the different polarimeters, and this will be discussed below. Comparing the results requires a degree of caution, since the polarization of a cathode has been observed to have significant dependencies on the cathode QE and the thickness of the strained layer. Also, different cathodes with similar properties may exhibit different polarization characteristics, especially if they are from different wafers. Both the E143 and SLD experiments have utilized the strained lattice cathodes. Because E143 does not require much peak current, it can run with much lower cathode QE than the SLD experiment. This provides as much as 5-10% higher relative polarization for E143. For E143 and for the 1994 run of the SLD experiment, strained-lattice cathodes of thickness 0.1μm are used. The 1993 SLD run, however, used an 0.3μm thick strained lattice cathode. The strain in the thicker cathode partially relaxes, and results in a polarization of about 65% compared to about 80% for the thinner cathode (as determined by the Compton polarimeter during SLD running). With these considerations in mind, let us consider the different SLAC polarimeters.

Mott Polarimeters.

Mott polarimeters utilize the spin-dependent cross-section asymmetry in the elastic scattering of polarized electrons from an unpolarized high Z nucleus.[26] There are three Mott polarimeters at SLAC. All of these are in test labs. The first one is PEGGY, which is being used for R&D on high polarization cathodes. It was used in the parity violation experiments at SLAC in the late 1970's, and was calibrated against a Moller polarimeter at that time. A newer polarimeter is SLAC's Cathode Test System (CTS) Mott,[27] which was built and calibrated at UC Irvine.[28] Following the 1993 SLD run, the cathode for that run was measured by the CTS polarimeter to give $P_e = (64 \pm 2)\%$. This would give $P_e \sim 62\%$ at the SLD IP, given spin transport losses in the DR and the arc. A third Mott polarimeter is now installed in SLAC's polarized gun test system (GTS), which is a mockup of the polarized source. This Mott is currently being commisioned. The PEGGY Mott and the GTS Mott use an energy filter to select elastically scattered electrons, while the CTS is a more precise retarding-field Mott. SLAC's polarized source group (in collaboration with other labs at Rice University, UC Irvine, University of Nebraska and Nagoya University) is currently undertaking a program to cross-calibrate these Mott polarimeters using a standard cathode. This standard cathode is chosen to be an 0.1μm thick active layer of unstrained GaAs. Using the standard cathode, the calibration of SLAC's CTS Mott has been re-checked against the UC Irvine Mott and found to be consistent within the quoted 2% uncertainty.

Moller Polarimeters. SLAC's Moller polarimeters[29] measure the elastic scattering cross-section asymmetry in the collision of polarized beam electrons with polarized electrons in a magnetized permendur foil (49% iron, 49% cobalt, 2% vanadium). The Linac Moller polarimeter is a single-arm device, detecting only the scattered beam electrons. This polarimeter is used as a diagnostic for both SLC and ESA operation. A schematic of it is shown in Figure 6. The ESA Moller polarimeter consists of both a single-arm Moller and a double-arm Moller, which detects both the scattered beam electrons and the scattered target electrons. Unlike the Compton polarimeter described below, Moller polarimeter operation is not compatible with normal datataking and special runs are needed. For the ESA experiments, polarimeter runs typically take 30-60 minutes to achieve 1% statistical precision, and are scheduled once per day.

The Moller polarimeters measure the cross-section asymmetry for beam and target spins aligned vs anti-aligned, and the asymmetry is proportional to the product of the beam and target polarizations. The target polarizations are typically about 8% (iron has 2 electrons out of 26 polarized) resulting in a small measured asymmetry. The uncertainty in the target polarization (about 3% relative) represents the largest systematic error in the measurement.

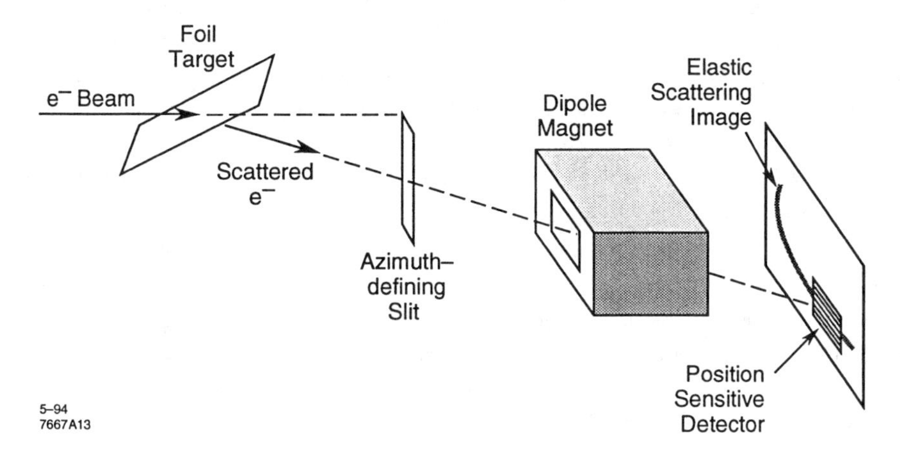

Figure 6: Moller Polarimeter

It was discovered during the 1993 SLD run that there was a large discrepancy between the beam polarization in the Linac measured by the Linac Moller and that inferred from the SLD Compton after small corrections for spin diffusion in the arc. The Compton measurements implied that $P^{Linac} = (65.7 \pm 0.9)\%$, while the initial Moller results gave $P^{Linac} \sim 75 - 80\%$. This discrepancy actually delayed announcing SLD's precise A_{LR} measurement by about 6 months until the discrepancy could be resolved.

Resolution of this discrepancy came from a correction to the Moller polarization analysis. This correction required proper accounting for the effect of atomic momenta of the electrons in the Moller target (the Levchuk Effect).[30] The atomic momentum of the inner K shell electrons in an iron nucleus is about 100 keV/c. This is small compared to the 46.6 GeV beam energy, but not so small compared to the electron mass. Proper treatment of the relativistic scattering kinematics for the Linac Moller geometry revealed that this small atomic momentum significantly broadened the elastic Moller peak in the high resolution silicon strip detector.

The polarized electrons in the target are outer-shell electrons, which have a small binding energy and hence do not cause a significant broadening of the Moller peak. The unpolarized inner-shell electrons, however, do cause a significant broadening. Thus the observed Moller lineshape in the detector is broader than what would be observed if the target electrons were free, and the observed Moller asymmetry in the center of the Moller peak is greater. A proper analysis of the Linac Moller data then gave $P^L = (66 \pm 3)\%$, in good agreement with the Compton result.[31] And the χ^2 of the lineshape agreement between the montecarlo and data improved by about a factor of 10.

The large 15% (relative) correction in the Linac Moller analysis due to the Levchuk

Effect does *not* imply a large universal correction for all Moller polarimeters. The correction depends on geometrical details of the polarimeter and also on electron beam characteristics. The correction is particularly large for the Linac Moller due to its fine resolution and also the low emittance of the SLC beam. Double arm polarimeters, for example, tend to have large acceptance and poor resolution and this effect can be negligible. Such is the case for the ESA double arm Moller polarimeter. For E143, it measured an average beam polarization of $(84 \pm 3)\%$. This result is consistent with the preliminary SLD Compton result of about 80% (and is somewhat higher due to the lower cathode QE that E143 can accommodate).

Compton Polarimeters. The longitudinal electron beam polarization ($\mathcal{P}_e$) at the SLC IP is measured by the Compton polarimeter[32] shown in Figure 7. This polarimeter detects Compton-scattered electrons from the collision of the longitudinally polarized electron beam with a circularly polarized photon beam. The photon beam is produced from a pulsed Nd:YAG laser operating at 532 nm. After the Compton Interaction Point (CIP), the electrons pass through a dipole spectrometer; a nine-channel Cherenkov detector then measures electrons in the range 17 to 30 GeV.

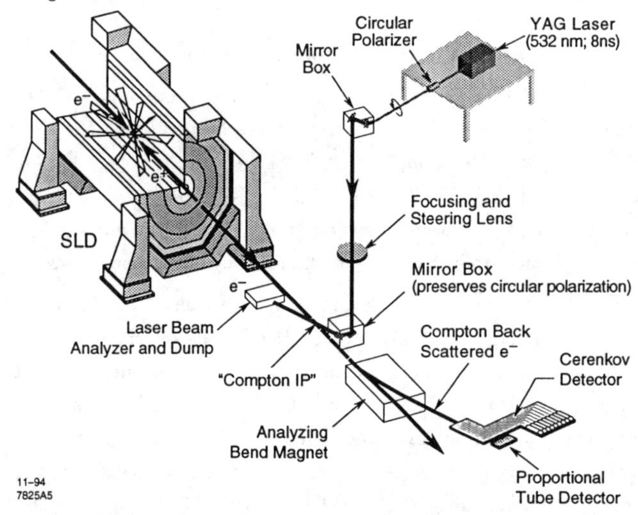

Figure 7: SLD Compton Polarimeter

The counting rates in each Cherenkov channel are measured for parallel and anti-parallel combinations of the photon and electron beam helicities. The asymmetry formed from these rates is given by

$$A(E) = \frac{R(\rightarrow\rightarrow) - R(\rightarrow\leftarrow)}{R(\rightarrow\rightarrow) + R(\rightarrow\leftarrow)} = \mathcal{P}_e \mathcal{P}_\gamma A_C(E)$$

where $\mathcal{P}_\gamma$ is the circular polarization of the laser beam at the CIP, and $A_C(E)$ is the Compton asymmetry function. The laser is polarized with a linear polarizer and a Pockels Cell, similarly to how it is done for the polarized electron source. Measurements of $\mathcal{P}_\gamma$ are made before and after the CIP. By monitoring and correcting for small phase shifts in the laser transport line, we are able to achieve $\mathcal{P}_\gamma = (99 \pm 1)\%$. $A_C(E)$ and the unpolarized Compton cross-section are shown in Figure 8. The Compton spectrum is characterized by a kinematic edge at 17.4 GeV (180° backscatter in the center of mass frame), and the zero-asymmetry point at 25.2 GeV (90° backscatter in the center of mass frame). $A_C(E)$ is modified from the theoretical asymmetry function[33] by detector resolution effects. This effect is about 1% for the Cherenkov channel at the Compton edge. Detector position scans are used to locate precisely the Compton edge. The position of the zero-asymmetry point is then used to fit for the spectrometer dipole bend strength. Once the detector energy scale is calibrated, each Cherenkov channel provides an independent measurement of $\mathcal{P}_e$. The Compton edge is in channel 7, and we use this channel to determine precisely $\mathcal{P}_e$. The asymmetry spectrum observed in channels 1-6 is used as a cross-check; deviations of the measured asymmetry spectrum from the modeled one are reflected in the inter-channel consistency systematic error (see Table III). Figure 9 shows the good agreement achieved between the measured and simulated Compton asymmetry spectrum for the 1993 SLD run. As an example of raw online data from the polarimeter, Figure 10 shows the signal height in ADC counts from channel 7 for 100 consecutive triggers. This data was taken during commissioning for the 1994 run. The polarimeter data acquisition was running at 30 Hz; the electron beam was present at 10 Hz; and compton collisions occured at 5 Hz.

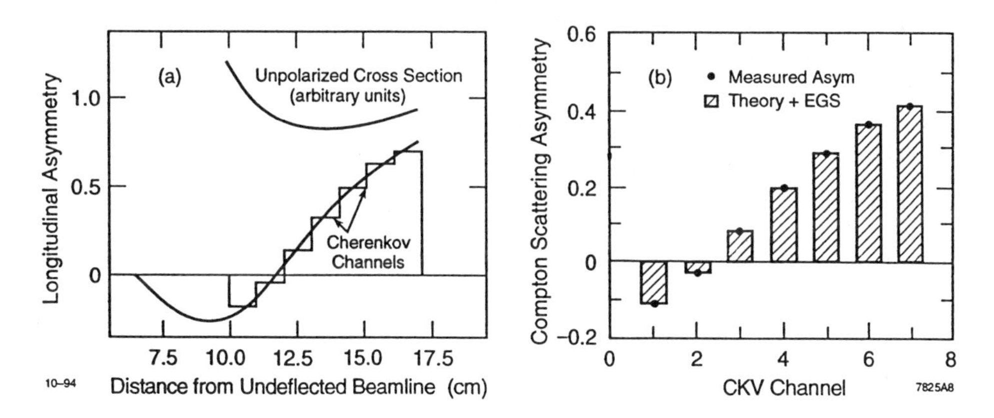

Figure 8: Compton cross-section and asymmetry

Figure 9: Measured and simulated Compton asymmetry

Clearly evident are the pedestal (no electrons), the background (electrons, but no laser), and the J=1/2 and J=3/2 compton signals. This data corresponds to $P_e \sim 80\%$.

Polarimeter data are acquired continually during the operation of the SLC. The absolute statistical precision attained in a 3 minute interval is typically $\delta\mathcal{P}_e < 1.0\%$. The systematic uncertainties that affect the polarization measurement are summarized in Table III for the 1993 run, where the IP beam polarization averaged over the run was found to be $< P_z^{IP} >= (61.9 \pm 0.8)\%$. For the 1994 run, we expect the polarization to be close to 80%, and with a somewhat smaller uncertainty than in 1993.

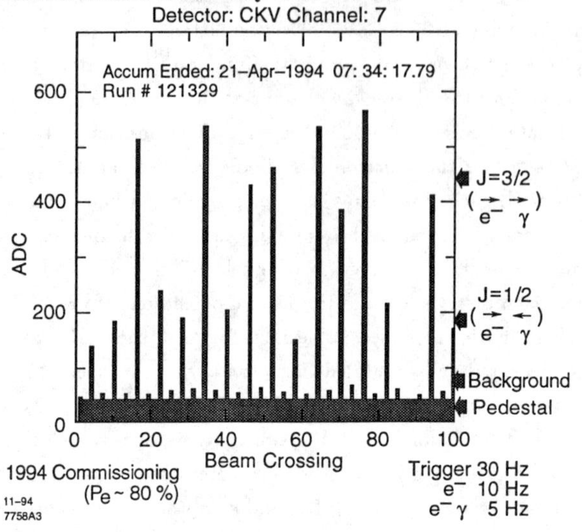

Figure 10: Online data from SLD Compton Cherenkov Detector

A second Compton polarimeter at SLAC is being commissioned in the Final Focus Test Beam (FFTB) at the end of the Linac. This is being done utilizing the laser and detector systems for experiment E144.[34] This polarimeter analyzes both the Compton-scattered gammas and electrons using silicon-tungsten calorimeters. A first test run was completed in May 1994 and a second run should occur before spring 1995.

Table III: Systematic Uncertainties for the SLD Compton Polarimeter

Systematic Uncertainty	$\delta\mathcal{P}_e/\mathcal{P}_e$ (%)
Laser Polarization	1.0
Detector Calibration	0.4
Detector Linearity	0.6
Interchannel Consistency	0.5
Electronic Noise	0.2
Total Polarimeter Uncertainty	1.3

SUMMARY

Two years ago at the SPIN92 conference in Nagoya, SLD had just completed a physics run with $P_e = 22\%$ and E142 was about to start its physics run with $P_e = 40\%$. As of SPIN94 in Bloomington, SLD is in the midst of a physics run with $P_e = 80\%$ and E143 has completed its physics run with $P_e = 84\%$. This high polarization is a tremendous accomplishment and has greatly enhanced SLAC's physics program. SLAC is now able to deliver a wide range of beam parameters with $P_e = 80\%$, including up to 6 amps of peak current.

Polarized source operation at SLAC has become routine, achieving uptimes of > 95% for roughly 5000 operating hours per year. This is encouraging the design of a polarized source for the next generation of linear colliders.[35] Yet the work for our polarized source colleagues is not yet finished. The beam polarization is still only 80-85%; they should now strive for 95-100%!

In End Station A, E154 and E155 are scheduled to run in 1995 and 1996 respectively. These experiments will further elucidate the spin structure of the nucleon. With the SLAC Linear Collider, the SLD experiment is scheduled to run through 1998. This will produce more precise measurements of parity violation in the Z-fermion couplings. SLD will continue to provide (and improve on) the world's best measurement of the weak mixing angle; and together with other precision electroweak experiments, it will continue the experimental assault on the MSM.

REFERENCES

1. See for example, F. E. Close, in *An Introduction to Quarks and Partons* (Academic Press Inc., 1979), p. 233.
2. M.J. Alguard et. al., *Phys. Rev. Lett.* **37**, 1261 (1976).
3. G. Baum et. al., *Phys. Rev. Lett.* **51**, 1135 (1983).
4. J. Ashman et. al., *Nucl. Phys.* **B328**, 1 (1989).
5. J. Ellis and R. Jaffe, *Phys. Rev.* **D9**, 1444 (1974).
6. See, for example, R.L. Jaffe and A. Manohar, *Nucl. Phys.* **B337**, 509 (1990).
7. J.D. Bjorken, *Phys. Rev.* **148**, 1467 (1966); *Phys. Rev.* **D1**, 1376 (1970).
8. P.L. Anthony et. al., *Phys. Rev. Lett.* **71**, 959 (1993).
9. The E143 proton results can be found in K. Abe et. al, SLAC-PUB-6508, August 1994; submitted to *Phys. Rev. Lett.*. See also presentations to this conference by D. Day, A. Feltham, H. Borel.
10. B. Adeva et. al., *Phys. Lett.* **B302**, 533 (1993).
11. D. Adams et. al., *Phys. Lett.* **B329**, 399 (1994).
12. For a recent review of the experimental data, see J. Ellis and M. Karliner, CERN-TH-7324, July 1994.
13. Presentations to this conference were given by M. Woods and K. Pitts. See also K. Abe et. al., SLAC-PUB-6607, submitted to *Phys. Rev. Lett.*; K. Abe et. al., SLAC-PUB-6644, submitted to *Phys. Rev. Lett.*; K. Abe et. al., SLAC-PUB-6605, submitted to *Phys. Rev. Lett.*.

14. K. Abe et. al., *Phys. Rev. Lett.* **73**, 25 (1994).

15. The weak mixing angle describes the mixing of the neutral $SU(2)_L$ and U(1) gauge bosons to form the physical γ and Z^0. Due to radiative corrections, there are different conventions for defining this angle. We follow the convention for $\sin^2\theta_W^{\rm eff}$ as used by the LEP collaborations, which is given in *Phys. Lett.* **B276**, 247 (1992).

16. For recent reviews of tests of the MSM, see M. Swartz, SLAC-PUB-6384, November 1993; and A. Blondel, CERN-PPE/94-133, August 1994.

17. K. Witte, SLAC-PUB-6443, March 1994

18. J. Frisch, R. Alley, M. Browne, M. Woods, SLAC-PUB-6165, April 1993.

19. T. Maruyama et. al., *Phys. Rev.* **B46** 4261 (1992). This technique was first demonstrated with a sample of InGaAs grown on GaAs by a SLAC-Wisconsin-Berkeley group in 1991; see T. Maruyama et. al., *Phys. Rev.* **B46**, 4261 (1992). Some of the current R&D on high polarization photocathodes was presented at this conference by T. Maruyama (T. Maruyama, SLAC-PUB-6712, November 1994).

20. D. Schultz et. al., SLAC-PUB-6606, August 1994.

21. M. Woods et. al., *J. Appl. Phys.* **73**, 12 (1993).

22. H. Tang et. al., SLAC-PUB-6515, June 1994.

23. J. Kent et al., SLAC-PUB-4922, March 1989.

24. The SLC arc does not lie in a plane; it follows the contours of the local terrain.

25. T. Limberg, P. Emma, and R. Rossmanith, SLAC-PUB-6210, May 1993.

26. For a recent review of Mott polarimetry, see T.J. Gay and F.B. Dunning, *Rev. Sci. Instrum.* **63**, 1635 (1992).

27. G.A. Mulhollan, SLAC-432 (Rev), 211 (1994).

28. The calibration of the UC Irvine Mott polarimeter is described in H. Hopster, D.L. Abraham, *Rev. Sci. Instrum.* **59**, 49 (1988).

29. H. Band gave a presentation to this conference on SLAC's Moller polarimeters (H. Band, WISC-EX-94-341, November 1994).

30. L.G. Levchuk, *Nucl. Inst. Meth.* **A345**, 496 (1994).

31. M. Swartz et. al., SLAC-PUB-6467, in preparation.

32. M.J. Fero *et al.*, SLAC-PUB-6423, in preparation.

33. See S.B. Gunst and L.A. Page, *Phys. Rev.* **92**, 970 (1953).

34. J. Heinreich et. al. SLAC-PROPOSAL-E-144, October 1991.

35. H. Tang gave a presentation to this conference on a proposed polarized source for the NLC (H. Tang, SLAC-PUB-6585, September 1994).

MØLLER POLARIMETRY AT SLAC*

H. R. Band

University of Wisconsin
Madison, WI 53706

Abstract. Four Møller polarimeters have been constructed and run at SLAC in the past three years and a fifth is planned for operation in 1995. Typical parameters and operational details are discussed. The analyzing power for all the polarimeters have been modeled, including effects due to the atomic motion of the target electrons. These corrections are sensitive to the geometry of the polarimeter design and vary in magnitude from $\leq 1\%$ for the E-143 double arm detector to $\approx 15\%$ for the Linac polarimeter.

1. SLAC MØLLER POLARIMETERS

SLAC has a long history of Møller polarimeters in SLAC's End Station A (ESA) fixed target program. The first polarimeter was constructed for E-80 in 1976 with Møller polarimeters also used in E-122 (1978) and E-130 (1982). Recently, Møller polarimeters were constructed for the E-142 (1992) and E-143(1993-94) experiments on nucleon spin structure. These polarimeters were located in ESA in front of the main experimental target. Dedicated Møller runs took typically 1 hour and were made every one to two days. The E-143 polarimeter is noteworthy in that it included both single and double arm detectors and will be described later. A similar polarimeter is planned for the E-154 and E-155 experiments which will run in 1995 and 1996.

The Linac Møller polarimeter located at the end of the SLAC Linac serves as a diagnostic polarimeter for both SLC and ESA beams. Initially commissioned in the spring of 1992, the polarimeter has operated 5 to 10 times per year under a wide range of beam conditions. A silicon microstrip detector and submillimeter beam sizes give the Linac polarimeter excellent angular resolution. The narrow width of the Møller scattering peak in the detector makes it very sensitive to broadening of the expected line shape due to motion of the target electrons. This effect is discussed later in more detail. A second Møller polarimeter (XLM) was constructed in the Extraction Line leading from the SLC interaction point. Measurements were made in August 1993 and April 1994. Parameters of the polarimeters can be found in Table 1. More information on the Linac polarimeter can be found elsewhere[1] in these proceedings.

---

* Work supported in part by the Department of Energy, contract DE-AC02-76ER00881

Table I: Beam and Møller Polarimeter Parameters

Parameter	Linac	XLM	E-142	E-143
Beam:				
e^-/pulse	$1-3\cdot10^{10}$	$1-3\cdot10^{10}$	$1-2\cdot10^{11}$	$2-4\cdot10^{9}$
Pulse Length	3ps-2μs	3ps	1μs	2μs
Energy	22.7-46.6 GeV	45.6 GeV	22.7 GeV	29.1 GeV
Target thickness	$50, 154\mu m$	$50, 154\mu m$	$20, 30, 50\mu m$	$20, 30, 40, 154\mu m$
Mask width	4.1 mm	10.2 mm	10.2 mm	0.2 radian
$\int Bdl$	47-62.0 kG-m	15.8 kG-m	14.5 kG-m	21.0 kG-m
Target to :				
Mask	3.80 m	6.96 m	7.11 m	8.80 m
Magnet center	8.2 m	8.3 m	8.7 m	11.0 m
Detector	13.2 m	19.1 m	22.4 m	27.0 m
Detector :				
θ acceptance	6.4-7.8 mrad	5.1-9.5 mrad	5.0-10.5 mrad	4.1-10.3 mrad
θ_{cm}	$92-112^o$	113°	97°	$70-110^o$
p	11-14.5 GeV/c	13.9 GeV/c	10.0 GeV/c	14.6 GeV/c
$\Delta p/p$	6.2%	10.7%	2.9%	5.8% (4)
Channel width	0.6 mm (64)	2.1 mm (28)	4.0 mm (35)	8.6 mm (12)
Peak width σ	0.13 mrad	0.23 mrad	0.26 mrad	0.23 mrad

In the e^-e^- center of mass the polarized cross section[2] is

$$\frac{d\sigma}{d\Omega} = \frac{\alpha^2}{s}\frac{(3+\cos^2\theta)^2}{\sin^4\theta}[1 - P_z^B P_z^T A_z(\theta)], \quad \text{and} \quad A_z(\theta) = \frac{(7+\cos^2\theta)\sin^2\theta}{(3+\cos^2\theta)^2}.$$

Here P_z^B (P_z^T) is the beam (target) longitudinal polarization. When $\theta = 90^o$ $A_z(\theta) = 7/9$ and both electrons in the lab frame have momenta ($E_{beam}/2$) and equal but opposite scattering angles. The measured asymmetry is

$$A_{meas.} = \frac{\sigma^{\uparrow\downarrow} - \sigma^{\uparrow\uparrow}}{\sigma^{\uparrow\downarrow} + \sigma^{\uparrow\uparrow}} = P_z^B P_z^T A_z(\theta),$$

where $\sigma^{\uparrow\uparrow}$ ($\sigma^{\uparrow\downarrow}$) is the cross section for spins aligned (anti-aligned).

The single arm SLAC Møller polarimeters are all similar. A polarized foil target (49% Fe, 49% Co, 2% Va) is mounted at a 20° angle to the beam and placed inside a 100 Gauss magnetizing field. Only about 8% of the atomic electrons are polarized. A collimator downstream of the the target selects scattering angles transverse to the bend plane of the following dipole magnet. The Møller electrons are then detected by a position sensitive strip or tube detector whose segmentation is in the scattering plane and which is behind several radiation lengths of lead. The lead convertor absorbs soft photon backgrounds and amplifies the Møller signal. The signal is integrated over the beam pulse and recorded with the sign of the beam polarization which is randomly reversed between pulses to reduce systematic errors. The number of Møllers detected per pulse varies with current and target thickness, but is typically 10-100 per pulse. Since the momenta and scattering angle of the Møller scatters are correlated, the scatters fall in a tilted stripe at the detector and are seen as an elastic scattering peak with a small radiative tail above an unpolarized background. After a data run, the average signal is calculated separately for the two beam orientations. An unpolarized signal and an asymmetry is computed channel by channel. The background is fit by using the wings of the distribution and subtracted from the region of the Møller peak.

Until recently, it was commonly assumed that the Møller asymmetry was constant across the measured peak. However, it has been pointed out by Levchuk[3] that the electron orbital motion of the target foil electrons could have a significant effect on the Møller lineshape. The atomic electrons have a momentum distribution that depends on which atomic shell the struck electron is in. Electrons in the outer shells have small momenta but those from the inner shells have momentum of $\approx$ 100 KeV/c. Although small compared to a

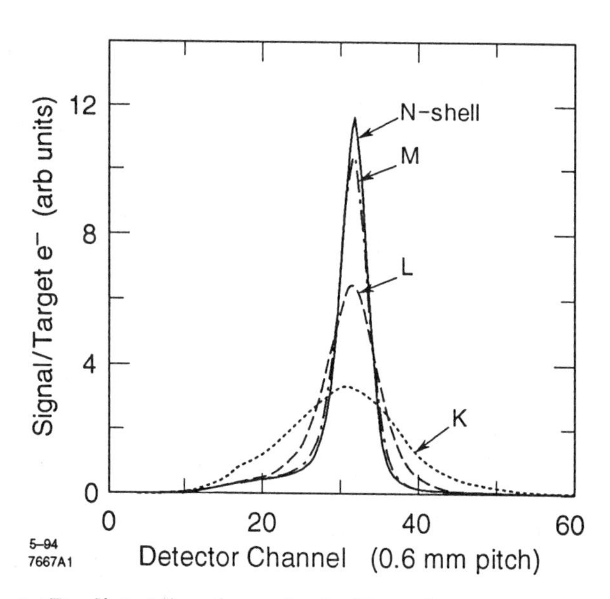

Figure 1: Predicted line shapes in the Linac detector for Møller scattering from electrons in the K, L, M, N atomic shells (from reference [4]).

beam energy of 50 GeV, these momenta are not small compared to the electron rest mass and can alter the scattering angle by up to 10 %. As shown in Figure 1, the effect causes different line shapes for scatters from different shells. Since the polarized target electrons are only in the 3d (M) shell, the fraction of signal from the polarized target electrons and thus the expected Møller asymmetry

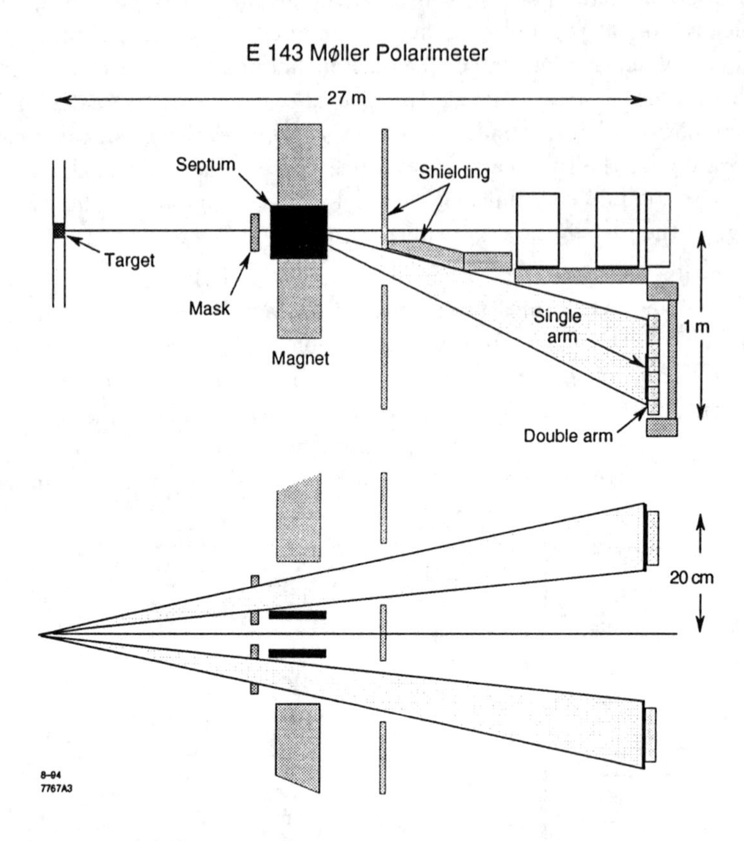

Figure 2: E-143 Møller polarimeter. The bend and scattering planes are shown. The placement of the single arm (silicon) and double arm (lead glass) detectors are shown relative to the Møller scattered electrons.

varies over the Møller scattering peak. Inclusions of these effects in a Monte Carlo simulation[2] revised the Linac Møller analyzing power upwards by 15% and were crucial in attaining good agreement between the Linac Møller and SLD Compton Polarimeter[1,2] measurements.

2. E-143 POLARIMETER

The E-143 Polarimeter, shown in Figure 2, differs from other recent SLAC polarimeters in that a double arm coincidence detector operated in parallel with a single arm detector. The single arm detector consists of 8 silicon pad detectors 12 channels wide operating in parallel. Each detector accepts a specific momentum (or θ_{cm}) range. The pad widths are large and the Møller peak is mostly contained in two channels. For this reason corrections to the line shape for target momenta are small and raise the expccted Møller asymmetries by 3%.

Behind the silicon detectors is a Pb-glass block array with seven 4 by 4 inch blocks above and below the bend plane. Both electrons from a Møller scatters are seen by looking for time coincidences between appropriate pairs of top and bottom blocks. The time resolution is $\approx$ 1 nsec, yielding sharp coincidence peaks with negligible background. The large size of the lead glass blocks contains all of the Møller lineshape. Consequently, corrections for target momenta are $\leq$ 1%. The observed raw asymmetry of the coincidence pairs only needed small corrections for deadtime and acceptance. The results of the single and double arm measurements were in good agreement.

The polarization of the 6 foil targets (3 cm wide and 30 cm long) has been measured to be 0.0803 for the 20μm foils and 0.0814 for the 30, 40, and 154μm foils with a systematic error of 1.7%. Beam polarization measurements were made daily throughout E-143 and were found to vary with the quantum efficiency(QE) of the SLAC polarized gun. The measured values of 0.83-.86 were in good agreement with Linac polarimeter measurements. The spread of the daily measurements about the fitted P^B versus QE function was somewhat larger than could be explained by statistics, either due to systematic errors in the Møller measurements or to nonreproducibility in the P versus QE behavior of the beam. Including all of these effects the overall systematic error on the E-143 beam polarization measurements was calculated to be 0.02.

REFERENCES

1. M. Woods, these proceedings and SLAC-PUB-6694(1994).
2. C. Møller, Ann. Phys. **14** (1932)532.
3. L.G. Levchuk, *Nucl. Inst. Methods* **A345**, 496 (1994).
4. M. Swartz et. al., SLAC-PUB-6467, in preparation; to be submitted to *Nucl. Inst. Methods.*

A COMPTON POLARIMETER FOR CEBAF HALL A

C. Cavata[a,b]

[a]DAPNIA/SPhN Saclay F91191 France
[b] for the CEBAF Hall A Collaboration

The Physic program at CEBAF Hall A includes several experiments using $4\,GeV$ polarized electron beam [6]: Parity Violation in electron elastic scattering from proton and ^{4}He , electric form factor of the proton by recoil polarization , neutron spin structure function at low Q^2 , ... Some of these experiments will need beam polarization measurement and monitoring with an accuracy close to $\frac{\Delta P_e}{P_e} \leq 4\%$, for beam currents ranging from $100\,nA$ to $100\,\mu A$. We present a project of a Compton Polarimeter that will meet these requirements.

With a Compton Polarimeter, the longitudinal polarization , P_e , of an electron beam is extracted from the measurement of the experimental asymmetry, A_{exp} , in the scattering of circularly polarized photons on the electron Beam : $A_{exp} = \frac{N^{\rightarrow}_{\Rightarrow} - N^{\leftarrow}_{\Rightarrow}}{N^{\rightarrow}_{\Rightarrow} + N^{\leftarrow}_{\Rightarrow}} = P_e P_\gamma A_l$. $N^{\rightarrow}_{\Rightarrow}$ (resp $N^{\leftarrow}_{\Rightarrow}$) is the number of compton scattering events when electrons are polarized parallel (resp. antiparallel) to the LASER beam polarization. These counting rates can be expressed as :

$$\frac{dN^{\rightarrow}_{\Rightarrow}}{d\rho} = \frac{d\sigma}{d\rho}\epsilon(\rho)\left(1 + P_e P_\gamma A_l(\rho)\right) \;\; ; \;\; \frac{dN^{\leftarrow}_{\Rightarrow}}{d\rho} = \frac{d\sigma}{d\rho}\epsilon(\rho)\left(1 - P_e P_\gamma A_l(\rho)\right). \quad (1)$$

In these equations, ρ is the scattered photon energy (normalized to the maximum scattered photon energy), $\frac{d\sigma}{d\rho}$ is the unpolarized Compton scattering cross section (See Eq. 2), $\epsilon(\rho)$ is the acceptance of the polarimeter, P_γ is the photon beam polarization and A_l is the longitudinal theoretical cross section asymmetry given by Eq. (3). A_{exp} and P_γ are measured quantities, and A_l is calculated in the framework of the standard model [3], so that the only unknown quantity is the electron beam longitudinal polarization P_e. This method is a well established technique [7] and is currently used at several high energy machines [5].

For an incident electron with energy E and momentum $\vec{p} = (0,0,p)$ along (z) axis , an incident photon with energy k, crossing angle α and momentum $\vec{k} = (0, -k\sin\alpha, -k\cos\alpha)$, and a scattered photon with energy k', scattering angle θ_γ , the scattered photon energy is given by $k' = k\frac{E + p\cos\alpha}{E + k - p\cos\theta_\gamma + k\cos(\alpha - \theta_\gamma)}$. This gives at $\alpha = 0$ crossing angle, using $\gamma = E/m$, $\frac{k'}{k} \simeq \frac{4a\gamma^2}{1 + a\theta_\gamma^2\gamma^2}$, with $a = \frac{1}{1 + \frac{4k\gamma}{m}}$. The maximum scattered photon energy is $k'_{max} = 4ak\gamma^2$. The photon scattering angle at which $k' = k'_{max}/2$ is $\theta_{\gamma 1/2} = \frac{1}{\gamma\sqrt{a}}$. These kinematic parameters are listed in table 1. Backscattered photons have a very small opening angle ($\simeq 150\mu rad$). In the CEBAF Hall A beam tunnel, only 10 meters are available for the Compton Polarimeter. To allow scattered photon detection one then needs to separate the scattered photons from the incident electrons. This will be done by a magnetic chicane that will deflect the electrons and let room for the photon detector. The proposed setup (See Fig. 1) consists of 4 magnets with magnetic field $B = 1\,T$, $1\,m$ length , and $1\,m$ between two successive dipoles. This gives at $4\,GeV$ a transverse deflection of $\simeq 15cm$. For $\alpha = 0$ crossing angle, the differential unpolarized

$E(GeV)$	1	4	6	8
a	0.96	0.87	0.82	0.77
$k'_{max}(GeV)$	0.035	0.500	1.058	1.776
$\theta_{\gamma 1/2}(\mu rad)$	521	136	94	73
$\sigma(barn)$	0.64	0.58	0.55	0.52
$A_l^{max}(\%)$	3.7	15.6	24.0	32.6
$< A_l > (\%)$	0.9	3.4	4.9	6.2
$\mathcal{L}(barn^{-1}s^{-1})$	300	300	300	300
$N_t(10^6)$	513	37	18	11
$t(hours)$	730	57	29	19

$E(GeV)$	1	4	6	8
a	0.98	0.93	0.90	0.87
$k'_{max}(GeV)$	0.018	0.266	0.580	0.999
$\theta_{\gamma 1/2}(\mu rad)$	515	132	89	68
$\sigma(barn)$	0.67	0.62	0.60	0.58
$A_l^{max}(\%)$	1.8	7.4	11.2	15.2
$< A_l > (\%)$	0.5	1.7	2.5	3.3
$\mathcal{L}(barn^{-1}s^{-1})$	440	440	440	440
$N_t(10^6)$	1900	135	62	36
$t(hours)$	1890	140	66	40

Table 1: *Kinematic Parameters , Cross section, maximum and mean longitudinal Asymmetry , Luminosity, number of events and time to get* 1% *statistical error on* $P_e = 50\%$. *For different electron beam energies* E *at* $100\,\mu A$ *, for a* $0.5\,W$ *Laser , at* $\alpha = 20mrad$ *crossing angle, with* $k = 2.4\,eV$ *for* 0.6*cm interacting length and a waist beam size* $d_0 = 116\mu$ *(left, Green Argon) and* $k = 1.16\,eV$ *and* 0.8*cm interacting length and* $d_0 = 164\mu$ *(right, IR NdYAg)*

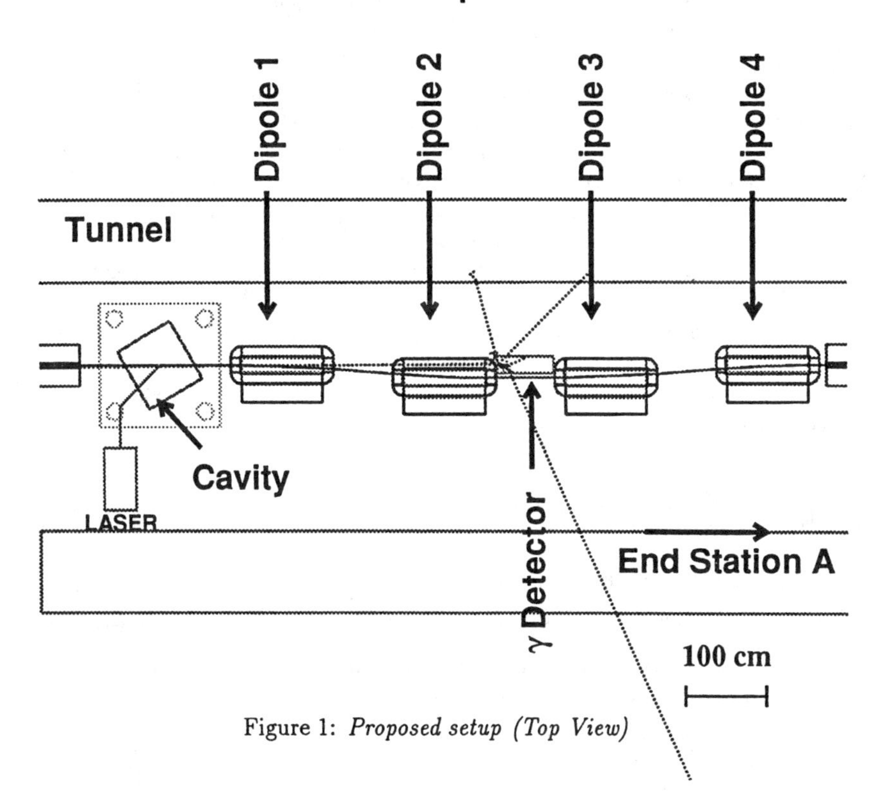

Figure 1: *Proposed setup (Top View)*

cross section is [4]

$$\frac{d\sigma}{d\rho} = 2\pi r_0^2 a \left[\frac{\rho^2(1-a)^2}{1-\rho(1-a)} + 1 + \left(\frac{1-\rho(1+a)}{1-\rho(1-a)} \right)^2 \right], \tag{2}$$

whereas the longitudinal asymmetry is given by

$$A_l = \frac{\sigma_{\Rightarrow}^{\rightarrow} - \sigma_{\Rightarrow}^{\leftarrow}}{\sigma_{\Rightarrow}^{\rightarrow} + \sigma_{\Rightarrow}^{\leftarrow}} = \frac{2\pi r_0^2}{\frac{d\sigma}{d\rho}} (1-\rho(1+a)) \left[1 - \frac{1}{(1-\rho(1-a))^2} \right]. \tag{3}$$

(r_0 is classical electron radius). The longitudinal asymmetry is maximum for $\rho = 1$, i.e $k' = k'_{max}$ (high energy scattered photon, see table 1). For a Laser with diameter d (with a negligible divergence), frequency ν, power P_L and an electron beam at intensity I_e (whose size is small with respect to the Laser beam size), the luminosity for an interacting length L, at $\alpha = 0$ crossing angle, is [9] $\mathcal{L} = \frac{I_e}{e}\frac{2}{c}\frac{P_L}{h\nu}\frac{4L}{\pi d^2}$. For polarization $P_\gamma = 100\%$ and $P_e = 50\%$, at $4GeV$, with a $0.5W$ $532\,nm$ Laser and $0.6\,cm$ interacting length, this corresponds to a 2.5 hours measurement at $\frac{\Delta P_e}{P_e} = 1\%$ (See table 1). But this Polarimeter will also be used at lower current, especially for experiments using polarized target. For $I_e = 100\,nA$, a 1% measurement is out of reach.

To increase the counting rates, one can use a high power UV Laser, emitting at higher energy such as a pulsed Excimer $80W$ (KrF), $\lambda = 248nm$, $k =$ 5eV. These types of Laser are expensive (at least 100 k$), difficult to run and not reliable. One would then need to install the Laser in an accessible room (outside the beam region) and setup a transport for the Laser light. On the other hand, with standard continuous Laser (Green Argon or InfraRed NdYag), the available power is poor (limited to $10\,W$ for Argon , and to $0.5W$ for NdYaG), but they are widely used system, cheap and very reliable, so that one can avoid the Laser beam transport problem. Their weak point being their low power one has to find amplification. An elegant solution was proposed by B. Norum et al.[1] : The Laser light is trapped in a cavity made of 2 highly reflective mirrors. In such cavity a "given" incident photon will perform a certain number of round trips inside the cavity. Setting the interaction region of the Laser with the electron beam inside this cavity gives the wanted gain. Optical Cavities fed by visible Laser light with a gain $G \simeq 10000$ are currently used for gravitational wave detection by Laser interferometry [8].

We will use a symmetric cavity consisting of two identical mirrors M_1 and M_2. Let z be the cavity axis , and the mirror radius of curvature be $R_1 = R_2 = R$. The mirror are located at $z_1 = -L/2$ and $z_2 = L/2$, using $g = 1 - \frac{L}{R}$ (with $0 \leq g^2 \leq 1$, for the cavity stability), it can be shown [2] that the allowed *transverse modes* in the cavity have a (complex) electric field given by

$$E(x,y,z)_{mn} = A\frac{d_0}{d(z)} e^{-ikz} e^{i\omega t} e^{i(m+n+1)\Psi(z)} e^{-ik\frac{x^2+y^2}{2R^2(z)}} e^{-\frac{x^2+y^2}{d^2(z)}} H_m(\sqrt{2}\frac{x}{d(z)}) H_n(\sqrt{2}\frac{y}{d(z)}), \tag{4}$$

where H_m and H_n are Hermite Polynomials. The phase $\Psi(z)$, the beam diameter $d(z)$ and the curvature radius $R(z)$ obey the evolution equations $\Psi(z) = \tan^{-1}\frac{z}{z_R}$, $d(z) = d_0\sqrt{1 + \frac{z^2}{z_R^2}}$, and $R(z) = z(1 + \frac{z_R^2}{z^2})$. The waist beam diameter d_0, the Rayleigh range z_R and the diameters d_1, d_2 of the beam on each mirror are (for a wave length $\lambda = 2\pi/k$)

$$d_0^2 = \frac{\lambda}{\pi}\frac{L}{2}\sqrt{\frac{1+g}{1-g}} \;;\quad d_0^2 = \frac{\lambda}{\pi} z_R \;;\quad d_1^2 = d_2^2 = \frac{\lambda}{\pi} L \sqrt{\frac{1}{1-g^2}} = \frac{2d_0^2}{1+g}. \tag{5}$$

For a given transverse mode (m,n), the allowed *longitudinal modes* or frequencies are given by [2] $\omega_l(mn) = 2\pi\frac{c}{2L}\left(l + \frac{1}{\pi}(m+n+1)\cos^{-1}\sqrt{g^2}\right)$, where l is a positive integer. The adaptation of the Laser modes to the cavity modes is a delicate work for which we will closely work with VIRGO specialists. To reach high cavity gain, mirrors with high reflectivity, low absorption and low scattering are used. High reflectivity is obtained using multilayer "$\lambda/4$ " mirrors. To compute the cavity gain, we introduce the reflexion and transmission coefficients of one mirror [10] $r = \frac{E_R}{E_I}$, $t = \frac{E_T}{E_I}$, where E_I, E_r and E_t are the incident, reflected and transmitted electric field. In terms of energy, the reflectivity R and the transmissivity T of this mirror are given by $R = \frac{J_R}{J_I}$, $T = \frac{J_T}{J_I}$, where (J_i) is the energy per unit area per second and is proportionnal to $|E_i|^2$: $R = |r|^2$; $T = |t|^2\frac{q_l}{q_1}$. For a Transverse Electric wave, at incident angle θ_i, q_i is given by $q_i = \sqrt{\frac{\epsilon_i}{\mu_i}}\cos\theta_i$, the indice l stands for the medium of the mirror. Let $g_{rt}(\omega)$ be the round trip gain of the gavity, so that the electric field circulating in the cavity E_{circ} is related to the incident electric field E_I by [2] $E_{circ} = it_1E_I + g_{rt}(\omega)E_{circ}$. Using α_0, the electric absorption coefficient of the medium inside the cavity, the electric field at $z = Z$ is related to field at $z = 0$ by $E(Z) = E(0)e^{-\alpha_0 Z}$, or in terms of power, $J(Z) = J(0)e^{-2\alpha_0 Z}$. This gives the intensity attenuation for a round trip , $Z = 2L$, $J(Z) = J(0)e^{-2\alpha_0(2L)} = J(0)e^{-\delta_0}$. Here $\delta_0 = 2\alpha_0(2L)$ represents the round trip absorption of the cavity medium. The mirrors themselves contribute to the absorption. Let $T_1 = |t_1|^2$ and $R_1 = |r_1|^2 = e^{-\delta_{r_1}}$ be the transmissivity and reflectivity of mirror M_1, and let A_1 (resp. D_1) be the absorption (resp. scattering) for that mirror, so that the fraction of energy loss by absorption (resp. scattering) is $A_1 = 1 - e^{-\delta_{a_1}}$ (resp. $D_1 = 1 - e^{-\delta_{d_1}}$). The energy conservation reads $1 = T_1 + R_1 + A_1 + D_1$. We then get for the round trip gain of the electric field $g_{rt}(\omega) = r_1r_2e^{-\frac{\delta_{a_1}+\delta_{d_1}+\delta_{a_2}+\delta_{d_2}+\delta_0}{2}}e^{-i2L\frac{\omega}{c}}$, and for the ratio of circulating and incident electric fields $\frac{E_{circ}}{E_I} = \frac{it_1}{1-r_1r_2e^{-\frac{\delta}{2}}e^{-i2L\frac{\omega}{c}}}$, with $\delta = \delta_{a_1} + \delta_{d_1} + \delta_0 + \delta_{a_2} + \delta_{d_2}$. For a symmetric cavity $r_1 = r_2 = r$ and $t_1 = t_2 = t$, and one gets $\frac{E_{circ}}{E_I} = \frac{it}{1-r^2e^{-\frac{\delta}{2}}e^{-i2L\frac{\omega}{c}}}$. In terms of power (with $R = r^2$ and $T = t^2$) the gain of the cavity is then :

$$G = \frac{J_{circ}}{J_I} = \frac{T}{1 - 2R\cos(\frac{\omega}{c}2L)e^{-\frac{\delta}{2}} + (Re^{-\frac{\delta}{2}})^2} = \frac{T}{(1 - Re^{-\frac{\delta}{2}})^2 + 4R\sin^2(\frac{\omega}{2c}2L)} \tag{6}$$

For frequencies $\omega_l = l\Delta\omega_{ax}, l = 1,2,3\ldots$, where $\Delta\omega_{ax} = \omega_{l+1} - \omega_l = \frac{2\pi c}{2L}$ is the axial mode interval, the cavity has maximum gain G_l, given by $G_l = \frac{T}{(1-Re^{-\frac{\delta}{2}})^2}$. Introducing the finesse $\mathcal{F}$ of the cavity, $\frac{\mathcal{F}}{\pi} = \frac{\sqrt{R}}{1-Re^{-\frac{\delta}{2}}}$, we then get for the gain $G(\omega) = G_l\frac{1}{1+\left(\frac{2\mathcal{F}}{\pi}\right)^2\sin^2\left(\pi\frac{\omega}{\Delta\omega_{ax}}\right)}$. The bandwidth of the cavity $\Delta\omega_{cav}$ is defined so that the gain G of the cavity is decreased by a factor 2 when the frequency is changed from the resonance $\omega = \omega_l$ to $\omega = \omega_l \pm \frac{\Delta\omega_{cav}}{2}$ and is related to the finesse by $\frac{\Delta\omega_{cav}}{\Delta\omega_{ax}} = \frac{2}{\pi}\sin^{-1}\left(\frac{\pi}{2\mathcal{F}}\right)$. For a high finesse cavity this equation becomes $\frac{\Delta\omega_{cav}}{\Delta\omega_{ax}} \simeq \frac{1}{\mathcal{F}}$. The maximum gain G_l at resonance can be written $G_l = \frac{\mathcal{F}}{\pi}\frac{1}{\sqrt{R}}\frac{T}{1-Re^{-\frac{\delta}{2}}}$, that , for a high finesse cavity, i.e with a high reflectivity $1 - R \simeq \delta_r \ll 1$ and with negligible scattering and mirror and medium absorption $\delta_d, \delta_a, \delta_0 \ll \delta_r$, leads to $G_l \simeq \frac{\mathcal{F}}{\pi} \simeq \frac{1}{\delta_r}\left(1 - 3\frac{\delta_a+\delta_d}{\delta_r} - \frac{\delta_0}{\delta_r}\right)$. The higher the finesse, the smaller the bandwidth of the resonance peaks and the higher the gain of the cavity at resonance.

J.M. Makowsky Laboratory at Lyon, where the VIRGO mirrors are constructed, will build and characterize our $\frac{1}{4}$ inch mirrors. The mirrors have the structure $LL(HL)_NHS$, where L (resp. H) stands for $\lambda/4$ layer of a Low (resp. Ligh) indice material, SiO_2 (resp. Ta_2O_5) with indice $n \simeq 1.47$ (resp. $n \simeq 2.1$) and S for the super polished silica substrate. To minimize scattering, the layers are made of amorphous deposit. The reflectivity of the mirror is governed by the design of the layers, we will use $(HL)_{14}$ mirrors with residual transmissivity $T = 1 - R - A - D \simeq \delta_r - (\delta_a + \delta_d) \simeq 100\,ppm$, Scattering $D \simeq \delta_d \simeq 2\,ppm$ (resp. $D \simeq 6\,ppm$) and Absorption $A \simeq \delta_a \simeq 3\,ppm$ (resp. $A \simeq 12\,ppm$) at $1064nm$ (resp. at $633nm$). The cavity medium will be vacuum to reduce absorption (e.g. $\delta_0 = 0$, for comparison $50cm$ of air has an absorption of $3\,ppm$). We thus expect a cavity gain $G \simeq 8090$ at $1064\,nm$ and $G \simeq 4000$ at $633\,nm$. For the cavity gain, a $1064\,nm$ NdYag Laser is then more efficient than a $532\,nm$ Argon Laser. To keep scattering very low, perfect surface state is mandatory. Starting with a super polished substrate with RMS roughness $0.5A$ and a peak/valley of $15A$, one can obtain after coating an RMS roughness of $0.3A$ and a peak/valley of $4A$. We also plan to test the mirror damnage by irradiation at the ORSAY synchrotron facility (LURE).

Using a $1064\,nm$ NdYaG Laser, with $0.5\,W$, and a $1\,m$ cavity at $g = -0.95$, with a gain $G = 1000$, we expect at $\alpha = 20mrad$ crossing angle ($0.8cm$ interacting lenght), $100\mu A$ beam intensity and $4\,GeV$ energy, an integrated rate of $270kHz$. If the scattered photon energy is not measured, this corresponds to a 8 minutes measurement to get $\frac{\Delta P_e}{P_e} = 1\%$ for $P_e = 50\%$. In order to increase the figure of merit of the Polarimeter, we plan to measure the energy of the scattered photon. The size and the material (Lead Glass, BGO, CsI, ...) of the photon detector are currently under study. The background will come from bremsstrahlung in the 8 meters beam pipe upstream of the polarimeter and synchrotron radiation in the first dipole of the chicane. For an Energy threshold of 5 MeV, assuming a vacuum of 10^{-8} Torr, we expect an integrated bremsstrahlung photon rate of $5kHz$ at $100\mu A$. Synchrotron photons will have characteristic energy $\epsilon_c = 4keV$ with a power, at $100\mu A$, after $1\,cm$ collimator, of $0.1W$.

We plan to commission this Polarimeter in Spring 1997.

References

[1] B. Norum, T. Welch, *CEBAF LOI 93-012.*

[2] "Lasers", A. Siegman, University Science Book, Mill Valley, California (1986).

[3] A. Denner, S. Dittmaier, *Nucl. Phys. B407 (1993) 43*

[4] C. Prescott, SLAC internal report, *SLAC TN 73 1.*

[5] D.P. Barber, et al. *NIM A329 (1993) 79.* G. Shapiro, et al. *SLAC-PUB-6261 (1993).*

[6] P. Souder et al, *CEBAF E-91-004.* E. Beise et al, *CEBAF E-91-010.* R. Lourie et al, *CEBAF E-91-011.* C. Perdrisat et al, *CEBAF PR-93-027.* Z. Meziani et al, *CEBAF PR-94-010.* C. GLashausser et al, *CEBAF E-89-033.* V. Burkert et al, *CEBAF E-89-042.* P. Ulmer et al, *CEBAF E-98-049.* J.J. Kelly et al, *CEBAF E93-013.*

[7] "Compton Polarimeter at SLAC SPEAR", *NIM 165 (1979) 177* . " Compton Polarimeter at DESY Doris", *Phys. Lett. 135B (1984) 498)* .

[8] A. Abramovici et al. , Science 256 (1992) 325. C. Bradaschia et al. , *NIM A289 (1990) 519.*

[9] D. Marchand, P. Vernin, Private Communication.

[10] M. Born, and E. Wolf, "Principles of Optics", Pergamon Press, (1975).

Experience with the Fast Polarimeter at HERA

Frank Zetsche
for the HERA Polarization Group

II. Institut für Experimentalphysik, Universität Hamburg, Germany
E-MAIL: zetsche@desy.de

Abstract. At HERA transverse and longitudinal electron polarization of 65 % has been achieved [1]. The HERA polarimeter employs the spindependent Compton scattering of laser photons off the electron beam and each minute performs a polarization measurement with statistical accuracy of typically 1–2 %. The calibration of the polarization scale and the investigation of systematic errors is done using polarization rise time measurements.

INTRODUCTION

One of the goals of the HERA physics program at DESY is the investigation of high energy interactions of polarized electrons with polarized and unpolarized nucleons. In the present phase of HERA operation studies are being done in order to increase and optimize the degree of transverse and longitudinal polarization of the electron beam [2,3,4].

The HERA electron storage ring is operated at an beam energy of around 27 GeV with typical beam currents of 20 mA in 160 bunches. In storage rings electrons become transversely polarized through the emission of synchrotron radiation which is known as the Sokolov-Ternov effect [5]. The degree of polarization which can be reached is governed by the strength of counteracting depolarizing effects which strongly depend on the parameters of the storage ring. Longitudinal polarization at an experimental interaction point in a straight section of the ring can then be achieved by the introduction of a pair of spin rotators [6].

The HERA electron polarimeter has been designed and built in order to allow for online measurements of the current degree of electron polarization. This fast information is used for interactive optimization of the polarization for which only a few dedicated runs have been scheduled. From 1995 on it will serve as a polarization monitor for the experiments HERMES, and later also H1 and ZEUS, which will make use of the polarized electron beam.

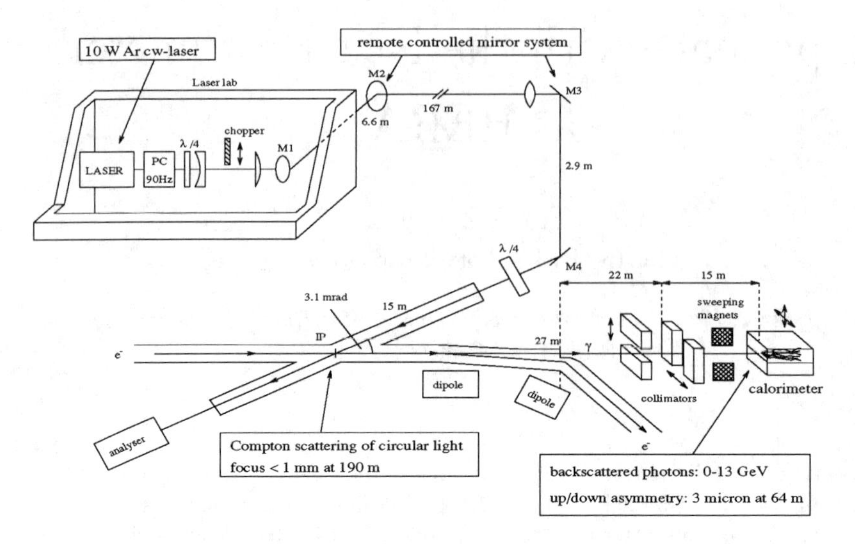

Figure 1: The setup of the HERA polarimeter showing the laser and the associated beam optics, the photon-electron interaction point and the calorimeter which detects the backscattered Compton photons.

THE POLARIMETER

The polarimeter at the HERA electron ring [7] employs the spin dependent Compton scattering of polarized laser light off the electrons. The most important components are a 10 W continuous argon ion laser and the associated laser beam optics, a calorimeter which measures energy and position of the Compton photons, and the data acquisition system. Figure 1 shows the setup of the polarimeter system in detail. The argon ion laser provides a continuous, linearly polarized laser beam at a green line of 514 nm. A Pockels cell (PC) converts the polarization state to right and left circularly polarized light, respectively, with a switching frequency of 90 Hz. The laser light is then directed onto the electron beam. Behind the interaction point its degree of circular polarization is measured with a light analyser.

It is the difference in the Compton cross sections for left and right circularly polarized laser light which is exploited for the electron polarization measurement. It results in an azimuthal asymmetry of the backscattered photons which also depends on their energy. These quantities are measured with the help of a tungsten-scintillator sandwich calorimeter which is split in an upper and a lower half. The energy asymmetry $\eta = (E_{up} - E_{down})/(E_{up} + E_{down})$ is a measure of the vertical impact coordinate y of the Compton photon on the calorimeter, whereas the energy sum measures its total energy E_γ.

ONLINE POLARIZATION VALUES

The online system [8] handles scattering rates up to 100 kHz and each minute provides a set of two-dimensional histograms containing the distributions of η and E_γ of all Compton photons recorded in the last measurement cycle for the different light polarization states. After statistical subtraction of the correctly normalized background spectra (bremsstrahlung events) the difference in the mean values of the asymmetry histograms is directly related to the transverse electron polarization P_Y,

$$\Delta\eta = \frac{\langle\eta\rangle_L - \langle\eta\rangle_R}{2} = \Delta S_3 P_Y \Pi_{\mathit{eff}}(E_\gamma)\,,$$

where $\Delta S_3 > 0.99$ is the circularity of the laser light and $\Pi_{\mathit{eff}}(E_\gamma)$ is the effective analysis power which has been determined by Monte Carlo simulations. It is basically given by the Compton cross section but it also depends on the $\eta - y$-transformation of the calorimeter and the vertical extent of the electron beam at the interaction point projected onto the face of the calorimeter. The uncertainty in the latter two is the main source of systematic errors in the polarization measurement, which in total was estimated to be 9.4 % [7]. In contrast the statistical error for a 1 minute measurement is typically 1–2 %. This means that changes of the degree of polarization can be detected quickly and reliably which is most important for the interactive optimization of polarization. However, the absolute polarization scale is less well known. A promising technique to calibrate the polarization scale more precisely is the measurement and analysis of polarization rise time curves which will be discussed in the next section.

RISE TIME MEASUREMENTS

The polarization build up due to the Sokolov-Ternov effect obeys an exponential law,

$$P_Y(t) = P_\infty(1 - e^{-t/\tau})\,. \tag{1}$$

The maximum theoretically achievable polarization, i.e. without depolarizing effects, at HERA is $P_\infty^{max} = P_{ST} = 91.6\,\%$. The ratio of the equilibrium polarization value P_∞ and the corresponding rise time τ is universal at HERA for a fixed electron energy and is given by P_{ST}/τ_{ST}, the ratio for the "ideal" values. The latter is calculable and for an electron energy of 26.7 GeV the following relation holds:

$$\frac{P_\infty}{\tau} = \frac{P_{ST}}{\tau_{ST}} = \frac{91.6\,\%}{43.2\,\mathrm{min}} = 2.1\,\%/\mathrm{min}\,. \tag{2}$$

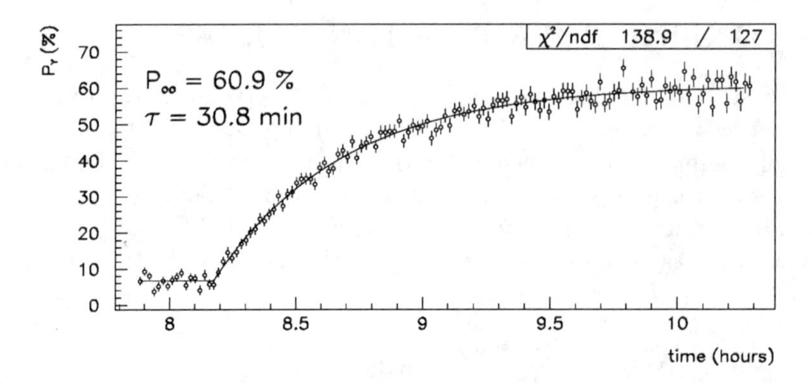

Figure 2: Measured polarization rise curve with superimposed fit according to equation 1. Displayed are the measured online polarization values versus time. Only statistical errors are shown.

With this relation the assumed polarization scale in the online calculation of the polarization values can be checked experimentally by extracting P_∞ and τ from measured polarization rise curves. The proceeding is as follows: 1) Fit equation 1 to the measured polarization rise curve thereby determining P_∞ and τ. 2) Calculate the theoretically expected equilibrium polarization $P_\infty^\tau = \tau \cdot P_{ST}/\tau_{ST}$. 3) Compare P_∞ and P_∞^τ.
Several of these rise curves have been taken after artificial depolarization of the electron beam. An example is shown in figure 2. Figure 3 displays the correlation between the measured P_∞ and the expected P_∞^τ for seven polarization rise curves which have been recorded in 1993, however mostly at different times and with different beam conditions. One observes a trend that the measured values are generally underestimated. On average the currently assumed polarization scale as given by Π_{eff} seems to be too low by 6.5 %, but the calibration measurements have a large relative spread of about 3 %.

Some remarks concerning this calibration method of the polarization scale seem to be in order: 1) Absolutely stable conditions for the electron beam and the polarimeter are essential during the polarization rise curve measurements. 2) The relatively large spread in the measurements could be due to different configurations of the storage ring during the various measurements. In the future we will take a set of polarization rise measurements after resonant depolarization with nominally the same machine parameters. This will ensure better compatibility of the measurements. 3) During routine operation of HERA at the beginning of each electron fill, i.e. about every 12 hours, a polarization rise measurement can be done. This would ensure a permanent check of the polarization scale when the HERA experiments make use of the polarized electron beam.

Figure 3: Correlation of the measured equilibrium polarization P_∞ and expected polarization value P_∞^τ as deduced from the fitted rise time τ. The dotted line displays the ideal situation $P_\infty = P_\infty^\tau$ for a perfectly calibrated polarimeter.

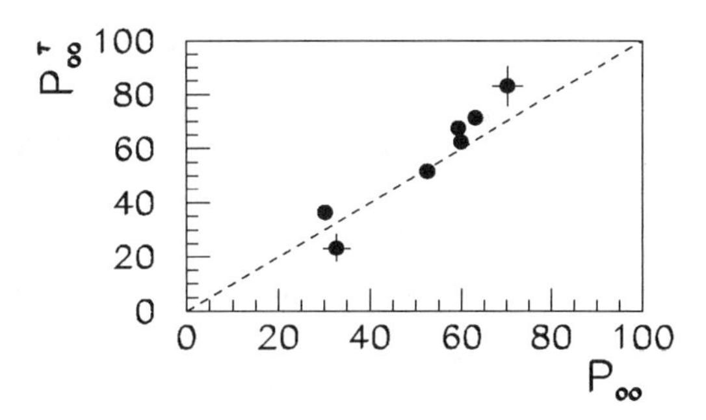

CONCLUSIONS

A fast polarimeter has been built and set up in order to measure the degree of polarization of the electron beam at HERA. This polarimeter is based on the spin dependence of the Compton scattering cross section.

Statistical errors of the polarization measurement typically amount to 1–2 % per minute depending on the electron beam current. With the help of dedicated polarization rise measurements it seems to be feasible to reduce the systematic error on the polarization scale, currently estimated to be 9.4 %, to 2–3 %. The measurements show that for 1993 data the polarization scale appears to be underestimated by about 6.5 %.

In the future the polarimeter will be used to further optimize the polarization and serve as a monitor for the HERA experiments which will make use of the polarized electron beam starting in 1995.

ACKNOWLEDGMENTS

The HERA Polarization Group would like to thank the HERA machine crew for the excellent collaboration and the DESY directorate for their strong support and encouragement.

REFERENCES

[1] D. P. Barber et al., DESY 94-174.

[2] G. Voss, these proceedings.

[3] M. Düren, these proceedings.

[4] M. Böge, these proceedings.

[5] A. A. Sokolov and I. M. Ternov, Sov. Phys. Doklady **8** 1203 (1964).

[6] J. Buon and K. Steffen, NIM **A245** 248 (1986).

[7] D. P. Barber et al., NIM **A329** 79 (1993).

[8] F. Zetsche, W. Brückner, M. Düren, IEEE Trans. Nucl. Sci., Vol. **41**, No. 1 102 (1994).

ROKK-1M is the Compton Source of the High Intensity Polarized and Tagged Gamma Beam at the VEPP-4M Collider

G.Ya.Kezerashvili, A.M.Milov, N.Yu.Muchnoi, A.N.Dubrovin, V.A.Kiselev, A.I.Naumenkov, A.N.Skrinsky, D.N.Shatilov, E.A.Simonov, V.V.Petrov, I.Ya.Protopopov.

Budker Institute of Nuclear Physics, 630090, Novosibirsk, Russia

The ROKK-1M facility (Russian abbreviation ROKK means Backward Scattered Compton Quanta) started into operation in May, 1993. This facility is the next point in the history of the similarly facilities like as ROKK-1 (1981-1985), ROKK-2 (1986-1993) [1] been constructed at BINP since 1981. The facility yielded the first results according to the experimental program that consists of the following items:

a) precise calibration of the energy scale of the Registration System of Scattered Electrons and Positrons (RSSE) of the KEDR detector for the two-photons physics program;

b) real-time measurement of the colliding beams transverse and longitudinal polarization;

c) measurement of the colliding beams energies by the method of resonance depolarization of the beam;

d) the electron-positron beams emittances measurement at the interaction point.

There are also a lot of applications for intensive outcoming high energy (30-1600) MeV γ–quanta beam produced by Backward Compton Scattering (BCS) process:

e) for the photonuclear physics experiments;

f) for different detector systems calibration;

g) for nonlinear effects of QED investigation.

The ROKK-1M facility arrangement is shown in Fig. 1. The laser system consists of two Nd:YAG lasers, argon laser "INNOVA–20" with cavity dumper and the nonlinear optical equipment. This system allows us to obtain the high power laser light in the wide spectrum range (from 1.17 eV up to 4.68 eV) in continuous and short pulse modes. The optical system transfers the laser radiation to the electron–photon interaction area. The system consists of

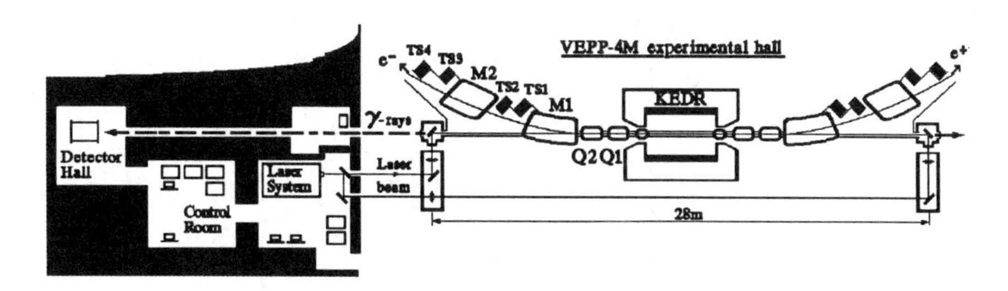

FIGURE 1. Layout of the ROKK–1M facility.

lenses and mirrors, some of them could be finally adjusted by special stepping motors. It provides less then 1 mm laser spot size at the interaction point at any place inside the VEPP–4M straight section. Pockels cell and polarimeters are used for laser light polarization control. The system conducting laser light into the VEPP–4M vacuum chamber allows to provide both 0 mrad and 15 mrad electron–photon interaction angle. The 15 mrad angle is used to make an interaction length rather short to be inside the central part of the KEDR detector, that is necessary for the RSSE calibration task. The 0 mrad angle allows to reach the highest intensity of γ–beam.

There are two approaches how to obtain monoenergetic γ–beam. The first approach is associated with the angle–energy correlation in the the BCS process and was realized for the first time at the LADONE facility at Frascati [2]. The collimated γ–spectra of the ROKK-1M is shown in Fig. 2. $\bar{N}$ is the average number of γ–quanta per one electron-photon collision. The events above the edge of the Compton spectrum (108 MeV) rise when more than one γ–quantum are detected by calorimeter simultaneously. The method leads however to the significant reduction of the γ–flux intensity because it requires the γ–beam collimation. Still it is adequate for some experimental tasks and

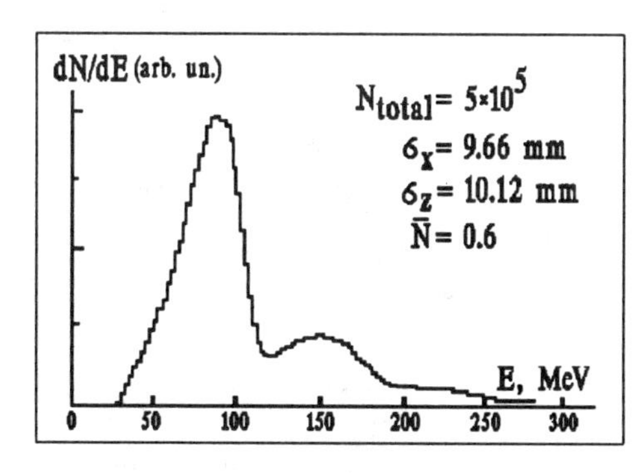

FIGURE 2. Compton spectrum after collimation.

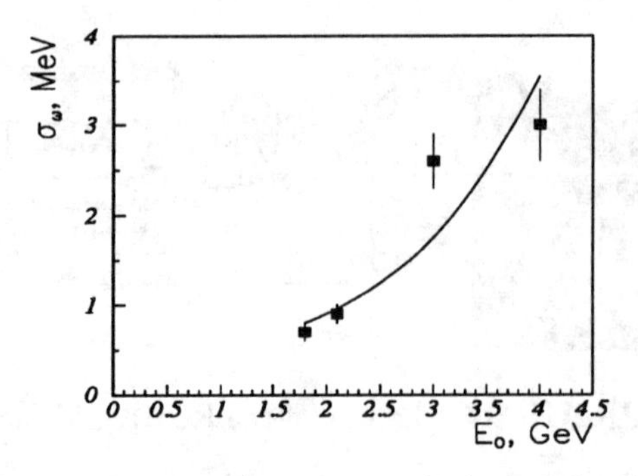

FIGURE 3. Measured energy resolution of the TS_4.

is successfully applied at the ROKK–1M facility. The second approach based on the determination of each γ–quantum energy by measuring the scattered electron momentum does not lead to the intensity loss. Moreover it possess to measure the bremsstrahlung γ–quanta energies above the Compton spectrum edge. This technique was used at the ROKK–2 facility in the BINP [3], LEGS facility in BNL [4], and at the TALADONE [2]. To demonstrate the tagging accuracy the energy resolution of the RSSE at different VEPP–4M energies obtained by the Compton spectrum edges is shown in Fig. 3. The points with error bars are the experiment, the line is the result of Monte-Carlo simulations [5]. The RSSE allows to tag the γ–quanta in $(0.4\text{–}0.98)E_0$ energy range, where E_0 is the VEPP–4M operation energy.

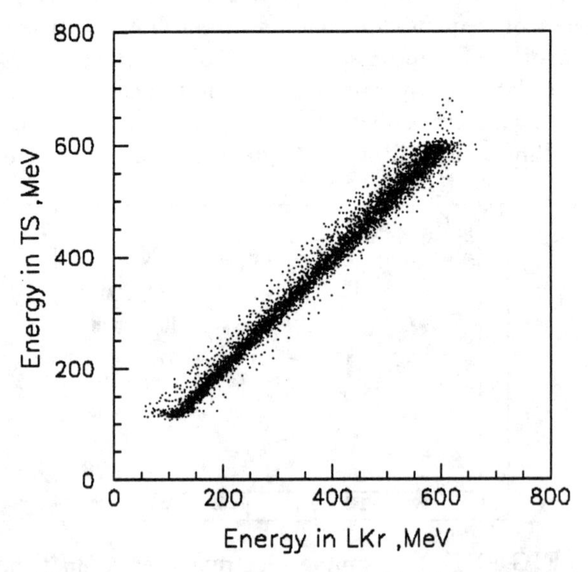

FIGURE 4. The γ–quanta energies measured by TS and LKr calorimeter.

The propotype of the Liquid Krypton Calorimeter [6] of the KEDR detector was mounted in the detector hall of the ROKK–1M facility. The γ–quanta energies measured by RSSE (vertical axis) and the calorimeter (horizontal axes) are the points shown on the two–dimensional plot in Fig. 4. As the energies of the γ–quanta are equal all the points localize near the equilibrium line $Y = X$. The RSSE has practically 100% registration efficiency. Width of the area on the plot is determined by the energy resolution of the Liquid Krypton Calorimeter. Contribution of the RSSE energy resolution in the width of the area is much smaller, and could not be deduced from this value.

Measurement of the transverse electron beam polarization in storage rings by using the BCS method allows to know the beam energy with accuracy about 0.001%. At the Budker INP the positive experience in such technique was stored at the VEPP–2 and the VEPP–4 colliders [7].

The photonuclear experiments [8] we carry out in collaboration with Istituto Nazionale di Fisica Nucleare – INFN, Italy, and Centro Brasileiro de Pesquisas Fisicas – CBPF, Brazil. The aim of this first run was to measure the total photofission cross sections for several nuclei at intermediate energies. Till now the γ–flux of the ROKK–1M is up to 10^6 photons/s. We plan to increase the γ–quanta flux up to 10^7 photons/s by development of laser controlling technique and optimization of the electron beam parameters at the electron–photon interaction area.

ACKNOWLEDGMENTS

This work was supported in part by a grant from the International Science Foundation.

REFERENCES

1. G.Ya. Kezerashvili, Proc. of the 8-th Workshop on Electromag. Interactions of Nuclei at Low and Medium Energies. Moscow, 1991, p.216.
2. L. Federici et al., Nuovo Cimento V.59B, 1980, p.247.
3. G.Ya. Kezerashvili et al., Nucl. Inst. Meth. A328, 1993, p.506.
4. D.H. Dowell et al., IEEE, NS30, (1983), p.3083.
5. V.M. Aulcnenko et al. Prep. INP 91–49, Novosibirsk, 1990.
6. V.M. Aulchenko et al. Submitted to Pa 22,17
7. P.V. Vorob'ev et al. Proc. of the 8-th Conference for Charged Particle Accelerators. Dubna, 1983, v.2, p.272.
8. G.M.Gurevich, G.Ya.Kezerashvili and V.G.Nedoresov. Proc of the 18-th Int. Symp. on Nucl. Phys. Gaussing, Germany, 1988, p.282.

RF-resonance Beam Polarimeter

Ya.S. Derbenev

Indiana University Cyclotron Facility, Bloomington, Indiana 47408

Abstract

The possibility of an RF-resonance polarimeter (RFP) for fast non-destructive measurement of beam polarization in an accelerator ring is considered. In order to accumulate the spin-dependent beam transition radiation, a passive superconducting cavity is proposed. The increase of effective voltage in the cavity (TM_{110} mode) with time, related to the free oscillating coherent spin of the beam, is calculated. The efficiency of the RFP does not decrease with particle energy and is proportional to the average beam current. Siberian snakes can be used in order to provide a sufficiently small value for the spin tune spread. Possible schemes of measurement of the accumulated voltage are presented. The noise limitations are taken into account and evaluated.

There are in the RFP dynamics different effects of the beam charge - cavity interaction, positive and negative. The negative effects are the beam noises, while the positive ones are as follows:

-the possibility to enhance the spin-dependent beam - cavity interaction, via the spin-orbit coupling in the machine focusing lattice;

-the possibility to increase, if necessary, the effective quality of the superconducting resonator, via redistribution of decrements between the TM_{110} mode and the beam coherent oscillation.

A scheme of elimination of charge effects from the measurement is proposed, if needed, which is based on use of two cavities with a spin rotator (Siberian snake) between them.

Finally, the RFP scheme is transformed to a Beam Spin Maser system, which is a spin feedback based on the superconducting cavities. This would allow one to create, to observe, and to use for polarization measurements the phenomenon of beam spontaneous coherent spin flip.

Numerical examples are given.

1. Fundamental concepts

1.1. RF-voltage accumulation rate

When a polarized beam is circulating in an accelerator ring, and the coherent spin of the beam is declined from the equilibrium direction, $\vec{n}$ (vertical or, in general case, periodical along the beam orbit), then the electromagnetic field of the beam, "observed" by a cavity located at a point of the orbit, has to have a modulation with periodicity of spin free precession in the accelerator, which is different from the periodicity's of the particle orbital motion. Apparently, this signal is very small. However, if the spin tune spread, $\Delta\nu$, enough small, the beam current, J, is sufficiently high, and the quality Q_c of the cavity, tuned in resonance with the spin free precession, is also enough high, then the beam could excite the cavity resonance mode to a measurable level while an unpolarized beam was not able to do this.

The most convenient cavity mode for the spin-cavity interaction is TM_{110} mode (see Fig. 1). The mode energy accumulation rate can be calculated starting from the Hamilton's approach driving the common coherent spin and cavity dynamics:

$$H_{int} = \sum_j \vec{W}(\vec{r}_j, \vec{p}_j)\vec{S}_j$$

where $\vec{W}$ function is taken from the BMT equation [1]

$$\dot{\vec{S}} = \vec{W} \times \vec{S},$$

$$\vec{W} \equiv -\frac{e}{mc}\left\{\left(G+\frac{1}{\gamma}\right)\vec{B}_{\perp} + \frac{1+G}{\gamma}\vec{B}_{v} + \left(G+\frac{1}{\gamma+1}\right)\vec{E}\times\frac{\vec{v}}{c}\right\},$$

where $G \approx 1.79$ for protons and $\approx 1/2\pi \cdot 137$ for electrons, and $\gamma^{-2} = 1 - v^2/c^2$.

In our case, $\vec{B}_v$ and $\vec{E}\times\frac{\vec{v}}{c}$ can be neglected.

Solving the field equations in terms of energy E_c - phase φ_c of the eigen modes, we find the effective voltage rate [2]:

$$\frac{dV}{dt} \approx 33\left(G+\frac{1}{\gamma}\right)\frac{J\hbar}{mcr_c^2}\sin\left(\frac{1.9d}{\beta r_c}\right)\cdot\xi\sin\alpha,$$

$$\left(E_c \equiv \frac{1}{2}CV^2, \quad C = r_c^2/4d\right).$$

where ξ is the polarization degree.
Note, that the bunch length must be less than cavity radius r_c:

$$\ell_b \ll r_c.$$

Remark, that the accumulation rate does not drop with particle energy, γ, and it is proportional to the average beam current, J.

1.2. Maximum accumulated voltage

The maximum voltage that can be accumulated in the resonator is defined by an effective maximum time:

$$V_{max} = \dot{V}t_{max}, \quad t_{max} = \min(\tau_{sp}, \tau_c),$$

where $\tau_{sp} = (\omega_0 \Delta\nu)^{-1}$, $\omega_0/2\pi$ is the particle's revolution frequency, and τ_c is the RF-voltage damping time due to the cavity wall resistance. It is related to the quality factor Q_c of the considered cavity mode as:

$$\tau_c = Q_c/\omega_c.$$

A typical Q_c value for TM_{110} mode of available superconducting resonators is 2×10^{10} [3]. At r_c=20 cm, τ_c value is about 5 s.

To provide τ_{sp} value compatible with τ_c, special measures for reduction of $\Delta\nu$ are required. For a bunched beam, $\Delta\nu$ is proportional to the beam emittances. It can be reduced by application of compensating sextupoles. The spin tune spread can be made especially small in rings with Siberian snakes. These possibilities should be investigated in detail separately.

1.3. RF-voltage measurement (noise criteria)

In view of the rather small value of possible accumulated voltage, one has to take into account noise limitations. There are two basic kinds of noise: thermal cavity noise and voltmeter input thermal noise.

1) The cavity noise effect is defined by the noise spectral density of the considered field mode, which is a single oscillator. When there is a state of thermodynamic equilibrium in the cavity walls, we can use a canonical formula for the oscillator energy distribution [4], which can be written for frequencies near resonance as follows:

$$\frac{dE_T}{d\omega} \approx \frac{T_c \tau_c}{1+(\omega-\omega_c)^2 \tau_c^2} \frac{1}{\pi},$$

where T_c is the cavity wall temperature (in units of energy). A voltmeter will integrate the frequencies in the interval $\Delta\omega \sim 1/t_m$, where t_m is the measurement time:

$$t_m = \max\{\tau_{sp}, \tau_c\}.$$

There are two characteristic cases:

a) $\tau_{sp} << \tau_c$, then

$$(E_c)_{max} = \frac{1}{2} C \dot{V}^2 \tau_{sp}, \quad t_m = \tau_c, \quad E_T \sim T_c$$

b) $\tau_{sp} >> \tau_c$, then

$$(E_c)_{max} = \frac{1}{2} C \dot{V}^2 \tau_c, \quad t_m = \tau_{sp}, \quad E_T \sim T_c(\tau_c/\tau_{sp}).$$

To measure the accumulated voltage with confidence, the following condition is necessary:

$$C\dot{V}^2 \tau_{sp} \min(\tau_{sp}, \tau_c) >> T_c.$$

2)The frequency bandwidth Δf of an applied RF-voltmeter should satisfy the requirement

$$2\pi t_m \Delta f >> 1;$$

on the other side, the Δf value should be small enough to eliminate the noise voltage due to the input resistance R_{in} of the voltmeter:

$$V_{in} = [2(TR_{in})\Delta f]^{1/2} << V_{max}$$

A possible voltmeter scheme is shown in Fig. 2 [5]. In the frequency region f~$10^8 - 10^9$ Hz, the characteristic R_{in} value can be about 100 Ω or less. Assume that the preamplifier channel temperature $T_{in} = 300$ K, and an f-band value 10^3 Hz; then

$$V_{in} \sim 4 \times 10^{-8} \text{ V},$$

which is much less than the thermal noise effective voltage $V_T = \sqrt{2T_c/C} \sim 10^{-6}$ V. We can conclude that, in practice, the minimum value of accumulated voltage is defined by the resonator thermal noise.

1.4. Numerical examples

Table 1 illustrates the values of basic parameters and requirements for different machines, assuming:

$V_{max}/V_T = 5$, $\xi = 1$, $\alpha = 90°$, $d = \pi\beta c/\omega_c$, $r_c = 20$ cm.

at a single resonator in a ring with $T_c = 1$ K. With N superconducting resonators, the maximum total accumulated voltage would be N times larger, and then the requirements for the $\Delta\nu$ value would be $\sqrt{N}$ times weaker.

1.5. Possible operational scheme

An operational procedure of polarization measurement could include the following steps:

1)To swing spin coherent free oscillation, use an RF-driven voltage (perhaps a different superconducting cavity), then switch off this voltage.

2)Shunt polarimeter's superconducting cavity, in order to kill an initial RF-oscillation (exited by beam charge).

3)Turn-off the shunting resistance adiabatically.

4)Wait for spin-swing of the polarimeter superconducting cavity.

5)measure the accumulated RF-voltage.

2. Beam charge effects

There is a number of beam charge different contributions to the cavity field dynamics, which have to be taken into account and reduced, if necessary.

2.1. Cavity tune shift

This is an effect of the neutral kind. It can be attributed to the definition of the TM_{110} mode frequency at the beam transverse loading and calculated in usual approach.

2.2. Renormalization of the resonator quality

Beam coherent transverse damping (deliberately arranged) can be used for the reduction of the $1\dot{1}0$ mode decrement, i.e. for increase the resonator quality, if necessary. If there is a fast beam coherent damping with decrement λ_b, then the decrement of the resonator is changed by some value $\eta\lambda_b$, where

$$\eta>0, \text{ at } \nu_c \approx k+\nu_y$$

$$\eta<0, \text{ at } \nu_c \approx k-\nu_y,$$

where ν_y is betatron tune.

A necessary precise control of the reduced τ_c^{-1} value seems to be easy to realize in this way.

2.3. Renormalization of the spin-resonator interaction

This effect takes the origin from the spin-orbital force in the machine focusing field, which creates the particle orbit modulation with spin precession frequency. It is described by the equation as follows:

$$\ddot{Q} + 2\lambda\dot{Q} + \omega_c^2 Q = Neb\langle y\rangle_{sp},$$

where $\langle y\rangle_{sp}$ is the enforced solution of the equation

$$\langle\ddot{y}\rangle_{sp} + n\omega_0^2\langle y\rangle_{sp} = -\frac{\omega_0^2}{mc}\left(G + \frac{1}{\gamma}\right)n\langle S_x\rangle(t),$$

n is the focusing field index, and b is the normalized magnetic field in the cavity. Apparently, the coupling effect increases near the spin-orbit resonances. The gain in the effective spin-resonator interaction is of about

$$k_{(sp-orb)} \sim \frac{\nu_y}{\nu_y - k \pm \nu_{sp}}.$$

This influence complicates the coherent spin measurement, although its contribution can be precisely calculated for any given energy, γ. It may be also used for a reduction of the accumulation time, if necessary.

2.4. Beam overtonal resonance noise

Particle oscillations may get in resonance with TM_{110} mode because of nonlinearities of this field, as well as because of the overtonal modulation of particle's motion due to the nonlinearities of the lattice (including beam-beam effect). It should be noted, that, in fact, one meets here the same set of resonances as for the spin motion in a ring (with snakes) in view of the spin-resonance condition. Therefore, tuning-off the spin resonances is similar to the tuning-off the orbit-resonator resonances. Estimation of dangerous high order resonances have to be made for definite situations. In practical aspects, these effects seem not to be of a dramatic meaning for the presented polarimetry concept.

2.5. Vacuum chamber-resonator beam link

A circulating beam will also transfer the vacuum chamber noise into the superconducting resonator. It has to be taken into account, in principle, due to that the chamber temperature (or temperature of the feedback elements, e.t.c.) frequently exceeds T_c. Nevertheless, and as a rule, the transferred noise must be relatively small, excluding situations of resonances between TM_{110} resonator mode and chamber eigen modes.

Note, that the resulting resonator noise spectrum has to be defined with taking into account the decrements redistribution as discussed in Sect.2.2.

2.6. Charge effects compensation

A modified sophisticated schemes of the RF-polarimeter may be used in order to avoid a complication in the polarization measurement due to the spin-orbit contribution and to reduce beam noise effect, if necessary. Fig.3 presents an example of this kind. It involves a spin rotator (Siberian snake) located between two superconducting resonators. Phase shift of value π between signals from these two resonators has to be provided to cancel charge contribution effect on the input of the resonator.

3. Beam spin maser

Next stage of the RF-polarimeter concept may be a spin feedback based on superconducting resonators. Fig.4 presents a simple principal scheme of this. If the above discussed noise limitations are fullfilled in the accumulating resonator, then the field in the resonator-kicker will be well correlated with coherent spin precession. The purpose of the feedback is to provide stability of spin coherent free precession against the spin tune spread, i.e., the condition $\Delta\omega_c >> \omega_0\Delta\nu$, where $\Delta\omega_c$ is spin tune shift by the RF-field of the kicker, should be satisfied. Then, if there would be no dissipation in the resonators, the coherent spin free precession would continue infinitely, although it may be complicated by beating as being a non-linear process. With the dissipation, the coherent spin damps to the equilibrium axis, $\vec{n}$ (periodical along the beam orbit). Since there are two possible signs of the equilibrium polarization, we can anticipate, that one of them is stable, while the another one is unstable with respect to the dissipation factor. This is the phenomenon that can be treated as beam spin maser.

Maser effect gives to the spin free precession a characteristic expression convenient for observations. The measurement procedure would become simple. Switching the feedback parameters, the initial equilibrium polarization of the beam can be made coherently unstable. After some characteristic time (a few resonator damping time) the excitation of the resonator-kicker will reach a maximum, when the polarization becomes transverse to $\vec{n}$, and then will vanish back to the noise level, with spin reverse to the stable polarization. This process can be repeated frequently, if needed. With the calibration of the involved parameters, the polarization degree can be calculated after measuring the maximum voltage of the resonator. No special RF spin kicker and hence resonator shunt would be required in here.

Note, that the noise limitation criteria have to be redefined, taking into account the tune shift, decrements redistribution and maser relaxation time.

4. Conclusion

The above consideration allows us to believe that the RF-polarimeter may become an efficient way of beam spin monitoring in high energy accelerators and storage rings.

The superconducting resonators of necessary quality value are available today. In addition, the beam transverse loading can be used to raise the resonator dynamic quality, if needed.

The RF-polarimeter principle matches well with the Siberian snakes technique, that makes the spin tune independent of energy and reduces the spin tune spread.

The RF-polarimeter efficiency grows with the beam current. At low current (low and middle energy range accelerators), the spin-resonator interaction can be enhanced using the spin-orbit coupling.

Finally, the RF-polarimeter becomes especially attractive and operationally simple with its converting to a system of a maser type using the superconducting resonator feedback.

This work was supported by a research grant from the US Department of Energy.

References

1. L.D. Landau and E.M. Lifshits, Course of Theoretical Physics, vol. 4 Quantum Electrodynamics, 2nd ed. (Pergamon, 1982).
2. Ya.S. Derbenev, Nucl. Instr. & Meth., A336 (1993) 12-15.
3. H. Padamsee, private communication.
4. L.D. Landau and E.M. Lifshits, Course of Theoretical Physics, vol. 5, Statistical Physics, 3rd ed. (Pergamon, 1980).
5. V.V.Rykalin, private communication.

Table 1
Examples
For a number N of resonators, the total V value can be N times larger, the necessary accumulation time t_{max} can be $\sqrt{N}$ times smaller, and the requirement for $\Delta\nu_{min}$ is $\sqrt{N}$ times weaker.

Machine particles	Polarized beam current[mA]	Energy γ	V [V/s]	$t_{max} = \tau_{sp}[s]$ required	$\Delta\nu_{min}$ required
IUCF Cooler Ring, p	1	1.3	5×10^{-6}	2	5×10^{-7}
FNAL, Tevatron, p	100	10^3	3×10^{-4}	0.03	10^{-4}
RHIC, p	50	250	1.5×10^{-4}	0.06	3×10^{-5}
CESR, $e^{\pm}$	100	10^4	3.3×10^{-4}	0.03	1.2×10^{-5}
HERA, $e^{\pm}$	30	5×10^4	10^{-4}	0.1	4×10^{-5}
LEP, $e^{\pm}$	3	10^5	10^{-5}	1	1.6×10^{-5}
LHC, p	100	6×10^3	3×10^{-4}	0.03	4×10^{-4}

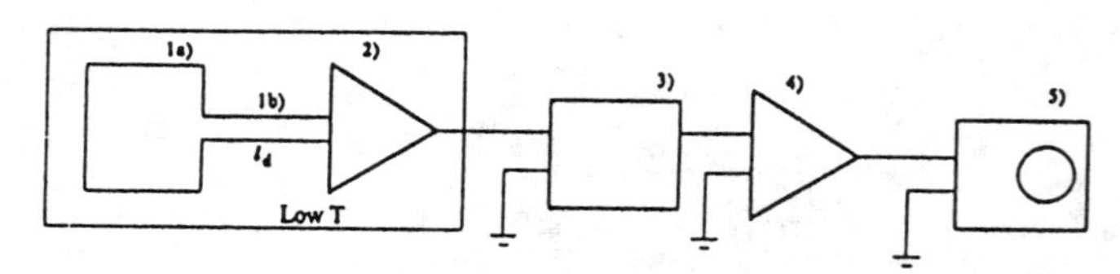

Fig. 1. Scheme of spin interaction with TM_{110} mode.

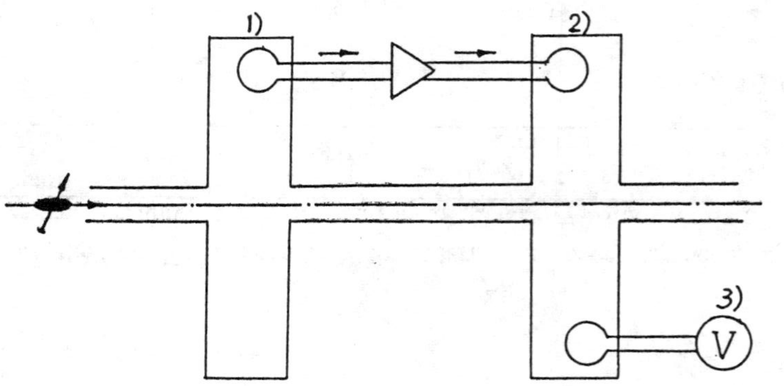

Fig. 2. RF voltage measurement scheme. 1a) A loop in the magnetic field of the superconducting RF-resonator. 1b) Waveguide (min. distance). 2) Preamplifier with narrow f-band. 3) Emitter follower. 4) Resonance circuit. 5) Scope.

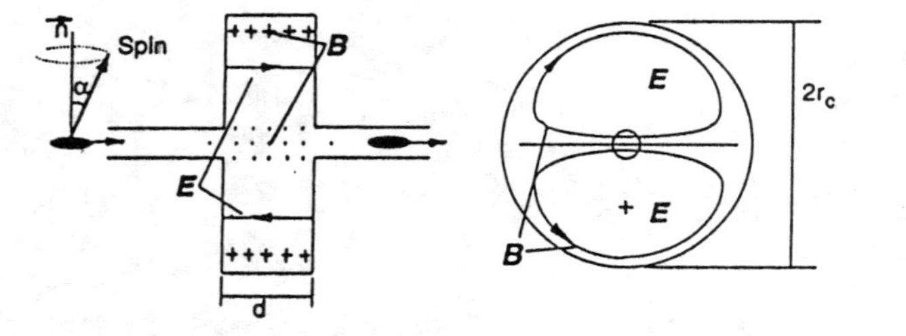

Fig. 3. Charge effect compensation scheme.
1), 3), 5) Superconducting resonators.
2) Spin rotator (Siberian snake)
4) Phase shift π circuit
6) Voltemeter

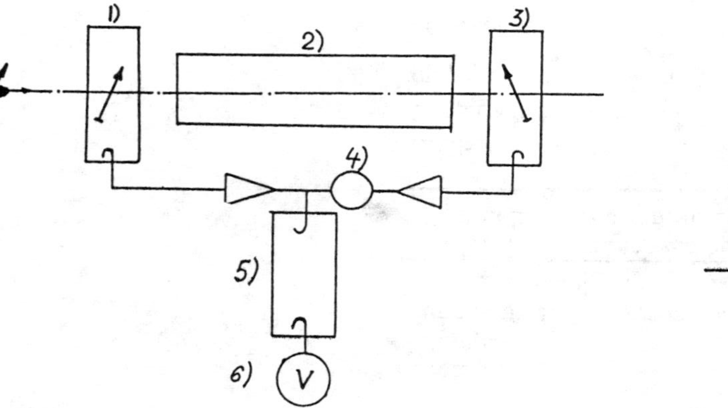

Fig 4. Spin feedback scheme
1) Superconducting resonator - accumulator
2) Superconducting resonator - kicker
3) Voltmeter

Spin Decoherence in Electron Storage Rings

D.P. Barber, M. Böge, K. Heinemann, H. Mais and G. Ripken

Deutsches Elektronen Synchrotron, DESY,
22603 Hamburg, Germany

Abstract

A simple model of spin decoherence in electron storage rings is presented and its relevance to rf spin flipping at high energy is discussed.

INTRODUCTION

For high-energy stored electron beams which are polarized by the emission of synchrotron radiation and when spin rotators provide longitudinal polarization at the interaction points for the high-energy physics experiments (1,2), it would be useful to be able to reverse the helicity either temporarily or repeatedly. This can be done by changing the rotator geometry (3) but a faster, more convenient method was already considered many years ago (4,5) and utilizes a radial rf field. It has been suggested that if flipping were repeated at the suitable intervals it would perhaps be possible to reach a periodic limit cycle for the polarization (6,7).

The rf field (or a combination of fields forming a closed bump) (4,8) would be installed at a position on the ring where the polarization were vertical and it would run in resonance with the natural spin precession frequency. Such techniques are routine at Novosibirsk (9,10) at low energy.

Flipping would involve sweeping slowly enough across resonance to ensure that the polarization vector would be tipped over adiabatically. This requires that the spins in a bunch remain tightly bundled. However, if the projections of the spins on the horizontal plane become spread out uniformly over the range $\pm\pi$ (in an appropriate coordinate system) during the sweep process, i.e. if there is complete decoherence, the polarization vector will not be flipped but instead the polarization will vanish. The decoherence results from the stochastic nature of synchrotron radiation photon emission. In proton rings, decoherence of this nature cannot occur and full spin flip is not difficult to achieve (11).

Spin flip is sometimes observed at LEP during energy calibrations (12,13) using rf fields of just a few gauss metres but the value of the polarization is much reduced and the effect is not consistently reproducible.

The fact that flip *can* be achieved suggests that the spin projections remain coherent at least for several seconds during the sweep. Thus in order to better understand the measurements it would be useful to estimate the decoherence rate. One such calculation suggests that the characteristic decoherence time is proportional to ${Q_s}^4$ and could indeed be several minutes at LEP (13).

In this article we show by a more careful treatment of the photon emission process and with the same linear "smooth ring" model for the synchrotron motion as in reference 13, that actually the spin distribution reaches *equilibrium* in a few damping times and that there need not be full decoherence. We then consider other sources of decoherence and their consequences.

EQUATIONS OF LINEARIZED ORBIT MOTION

The linearized equation of orbit motion with respect to the closed orbit in the presence of stochastic excitation and damping due to synchrotron radiation takes the form used in the SLIM program (14):

$$\frac{d}{ds}\vec{\hat{y}} = \underline{\hat{A}}\cdot\vec{\hat{y}} + \delta\underline{\hat{A}}\cdot\vec{\hat{y}} + \delta\vec{\hat{c}}\,, \tag{1}$$

where s is the distance around the ring and $\vec{\hat{y}}$ is the vector of orbit variables $(\hat{x}, \hat{p}_x, \hat{z}, \hat{p}_z, \hat{\sigma}, \hat{p}_\sigma)$. Here, $\hat{\sigma}$ is the distance to the centre of the bunch and $\hat{p}_\sigma$ is the fractional energy deviation. $\underline{\hat{A}}$ represents the 'hamiltonian' motion due to the Lorentz forces and $\delta\underline{\hat{A}}$ describes damping. Both are s–dependent 6 x 6 matrices. The vector $\delta\vec{\hat{c}} = (0,0,0,0,0,\delta c)$ accounts for the stochastic excitation in the energy variable due to photon emission (15):

$$\delta c = \sqrt{\omega}\cdot\xi(s)\,, \tag{2}$$

where in terms of the curvatures K_x and K_z, $\omega = (|K_x|^3 + |K_z|^3)\cdot C_2$ with

$$C_2 = \frac{55\cdot\sqrt{3}}{48}\cdot C_1\cdot\Lambda\cdot\gamma_0^2\,;\quad C_1 = \frac{2}{3}e^2\cdot\frac{\gamma_0^4}{E_0}\,;\quad \Lambda = \frac{\hbar}{m_0 c}\,, \tag{3}$$

and where the stochastic averages of the kicks $\xi(s)$ are

$$<\xi(s)\cdot\xi(s')> = \delta(s-s')\,;\quad <\xi(s)> = 0\,. \tag{4}$$

Thus, as is usual and sufficient (15,16,17), we take the synchrotron radiation to be a white noise process. We now decouple transverse and longitudinal motion by introducing the dispersion by means of a canonical transformation to obtain a new set of variables $(x, p_x, z, p_z, \sigma, p_\sigma)$ defined by:

$$x = \hat{x} - p_\sigma \cdot D_1 ; \tag{5}$$

$$p_x = \hat{p}_x - p_\sigma \cdot D_2 ; \tag{6}$$

$$z = \hat{z} - p_\sigma \cdot D_3 ; \tag{7}$$

$$p_z = \hat{p}_z - p_\sigma \cdot D_4 , \tag{8}$$

where the D_i are the components of the dispersion vector (18).

In terms of the variables x, p_x, z, p_z, σ, p_σ the equation of motion now takes the form:

$$\frac{d}{ds}\vec{y} = \underline{A} \cdot \vec{y} + \delta\underline{A} \cdot \vec{y} + \delta\vec{c}. \tag{9}$$

If we ignore transverse–longitudinal coupling the matrices $\underline{A}$ and $\delta\underline{A}$ have the simple block diagonal form. For example:

$$\underline{A}(s) = \begin{pmatrix} \underline{A}^{(\beta)}_{(4\times4)}(s) & \underline{0}_{(4\times2)} \\ \underline{0}_{(2\times4)} & \underline{A}^{(\sigma)}_{(2\times2)}(s) \end{pmatrix} . \tag{10}$$

The matrix $\underline{A}^{(\beta)}_{(4\times4)}(s)$ describes betatron motion in the focussing fields. $\underline{A}^{(\sigma)}_{(2\times2)}(s)$ describes the synchrotron motion. When acting alone it gives :

$$\frac{d}{ds}\sigma = -[K_x \cdot D_x + K_z \cdot D_z] \cdot p_\sigma ; \tag{11}$$

$$\frac{d}{ds}p_\sigma = h \cdot \frac{2\pi}{L} \cdot \frac{eV(s)}{E_0} \cos\varphi \cdot \sigma , \tag{12}$$

where the symbols have their usual meaning.

In this calculation we also work in the "smooth ring" approximation, considering only synchrotron motion. Thus we follow exactly the philosophy of reference 13. So the coefficients in $\underline{A}^{(\sigma)}_{(2\times2)}$ and $\delta\underline{A}^{(\sigma)}_{(2\times2)}$ are averaged over one turn (of length L) and we obtain :

$$\begin{pmatrix} \sigma' \\ p'_\sigma \end{pmatrix} = \begin{pmatrix} 0 & -\kappa \\ \Omega_s^2/\kappa & 0 \end{pmatrix} \cdot \begin{pmatrix} \sigma \\ p_\sigma \end{pmatrix} + \delta\underline{A}^{(\sigma)}_{(2\times2)} \cdot \begin{pmatrix} \sigma \\ p_\sigma \end{pmatrix} + \delta\vec{c} , \tag{13}$$

where $\delta\underline{A}^{(\sigma)}_{(2\times2)}$ and $\delta\vec{c}$ take the forms:

$$\delta\underline{A}^{(\sigma)}_{(2\times2)} \equiv \begin{pmatrix} 0 & 0 \\ 0 & -2\cdot\alpha_s/L \end{pmatrix} , \qquad \delta\vec{c} \equiv \sqrt{\bar{\omega}} \cdot \begin{pmatrix} 0 \\ \xi(s) \end{pmatrix} . \tag{14}$$

Here, α_s is the one turn synchrotron damping decrement and $\tilde{\omega}$ is the one turn averaged ω. Also, $\Omega_s = 2\pi \cdot Q_s/L$ and κ is the compaction factor.

The equilibrium covariance matrix for σ and p_σ then takes the usual value (19) viz :

$$\underline{\sigma}_2(\infty) = \begin{pmatrix} \sigma_\sigma^2 & 0 \\ 0 & \sigma_{p_\sigma}^2 \end{pmatrix}, \qquad \sigma_{p_\sigma}^2 = \frac{\tilde{\omega} \cdot L}{4 \cdot \alpha_s}, \qquad \sigma_\sigma^2 = \frac{\kappa^2}{\Omega_s^2} \cdot \sigma_{p_\sigma}^2 . \tag{15}$$

INCLUSION OF SPIN

After this recapitulation of the basis for the matrix formulation of the standard smoothed description of synchrotron motion we are in a position to introduce spin motion. The Thomas-BMT equation describing the precession of a classical spin $\vec{S}$ in electric and magnetic fields reads as (14):

$$\frac{d}{ds}\vec{S} = \vec{\Omega} \times \vec{S}, \tag{16}$$

where the precession vector $\vec{\Omega}$ is a function of the magnetic and electric fields and of the particle velocity and energy. As is usual in this context we now write $\vec{\Omega}$ as a sum of a piece $\vec{\Omega}_0$ accounting for the fields on the closed orbit and a piece $\vec{\Omega}_{osc}$ accounting for synchro-betatron motion with respect to the closed orbit.

We will assume that the ring has no vertical bends or solenoids, and that it is perfectly aligned so that there is no vertical closed orbit deviation. For this naive estimate the radial rf field will be ignored. For electrons the vertical emittance can then be taken to be zero and only motion in the horizontal plane need be considered. Spin motion will be calculated with respect to a pair of mutually orthogonal axes precessing at the rate $|\vec{\Omega}_0|$ in the horizontal plane around the vertical dipole field. The position of a horizontal spin in this frame is denoted by a phase angle ψ so that we have $\psi' = |\vec{\Omega}_{osc}|$. After averaging we then obtain $\psi' = 2\pi\nu/L \cdot p_\sigma$ where $\nu = (g-2)/2 \cdot \gamma$ is the number of spin precessions per turn, the spin tune (14). Thus ψ only couples to and is only driven by p_σ. When the spin phase ψ is included, the stochastic differential equation for the system takes the form :

$$\begin{pmatrix} \sigma' \\ p_\sigma{}' \\ \psi' \end{pmatrix} = \underbrace{\begin{pmatrix} 0 & a & 0 \\ b & 0 & 0 \\ 0 & d & 0 \end{pmatrix} \cdot \begin{pmatrix} \sigma \\ p_\sigma \\ \psi \end{pmatrix}}_{\text{Hamiltonian motion}} + \underbrace{\begin{pmatrix} 0 & 0 & 0 \\ 0 & c & 0 \\ 0 & 0 & 0 \end{pmatrix} \cdot \begin{pmatrix} \sigma \\ p_\sigma \\ \psi \end{pmatrix}}_{\text{Damping}} + \underbrace{\sqrt{\tilde{\omega}} \cdot \begin{pmatrix} 0 \\ \xi \\ 0 \end{pmatrix}}_{\text{Excitation}}, \tag{17}$$

where the constants a, b, c and d are defined as :

$$a = -\kappa \ , b = \Omega_s^2/\kappa \ , c = -2 \cdot \alpha_s/L \ , d = 2\pi\nu/L \, . \tag{18}$$

This can be rewritten in the form:

$$\vec{x}' \ = \ \underline{\mathcal{A}} \cdot \vec{x} + \delta\vec{c}_3 \, , \tag{19}$$

where

$$\vec{x} \equiv \begin{pmatrix} \sigma \\ p_\sigma \\ \psi \end{pmatrix} , \qquad \underline{\mathcal{A}} \equiv \begin{pmatrix} 0 & a & 0 \\ b & c & 0 \\ 0 & d & 0 \end{pmatrix} , \qquad \delta\vec{c}_3 \equiv \sqrt{\tilde{\omega}} \cdot \begin{pmatrix} 0 \\ \xi \\ 0 \end{pmatrix} . \tag{20}$$

This linear Langevin equation is interpreted according to the Stratanovich convention and leads to the following Fokker-Planck (20,21) equation for the distribution function $W(\sigma, p_\sigma, \psi)$:

$$\frac{\partial W}{\partial s} \ = \ -\sum_{j=1}^{3} \frac{\partial}{\partial x_j}[\mathcal{D}_j \cdot W] + \sum_{i,j=1}^{3} \frac{\partial^2}{\partial x_i \partial x_j}[\mathcal{D}_{ij} \cdot W] \, , \tag{21}$$

where

$$\mathcal{D}_j \equiv \sum_{k=1}^{3} \mathcal{A}_{jk} \cdot x_k \, , \qquad \mathcal{D}_{ij} \equiv \frac{\tilde{\omega}}{2} \cdot \delta_{ij} \cdot \delta_{i2} \qquad (i, j = 1, 2, 3) \, . \tag{22}$$

So the F-P equation has the final form :

$$\frac{\partial W}{\partial s} \ = \ -c \cdot W - a \cdot p_\sigma \cdot \frac{\partial W}{\partial \sigma} - [b \cdot \sigma + c \cdot p_\sigma] \cdot \frac{\partial W}{\partial p_\sigma} - d \cdot p_\sigma \cdot \frac{\partial W}{\partial \psi} + \frac{\tilde{\omega}}{2} \cdot \frac{\partial^2 W}{\partial {p_\sigma}^2} \, . \tag{23}$$

With this F-P equation for W we can carry out a detailed study of spin decoherence under all possible conditions just by looking for the possible solutions for $W(\sigma, p_\sigma, \psi)$ compatible with the initial conditions. For example, by starting with a delta function distribution in σ, p_σ and ψ, corresponding to a pointlike beam and a tight bundle of spin projections, the distribution function (i.e. the transition probability in this case) evolves so that the covariance matrix for the σ, p_σ and ψ is given by (20,21) :

$$\underline{\sigma}_3(s) \ = \ 2 \cdot \int_0^s ds' \, \underline{M}(s') \cdot \underline{\mathcal{D}} \cdot \underline{M}^T(s') \, , \tag{24}$$

where M is the real valued transfer matrix solving :

$$\underline{M}' = \underline{A} \cdot \underline{M}\,, \qquad \underline{M}(s=0) = \underline{1}\,. \tag{25}$$

After some initial damped oscillatory behaviour, in a few synchrotron damping times the elements of $\underline{\sigma}_3(\infty)$ reach the asymptotic values :

$$\underline{\sigma}_3(\infty) = \begin{pmatrix} \sigma_\sigma^2 & 0 & \frac{d}{a}\cdot\sigma_\sigma^2 \\ 0 & \sigma_{p_\sigma}^2 & 0 \\ \frac{d}{a}\cdot\sigma_\sigma^2 & 0 & \frac{d^2}{a^2}\cdot\sigma_\sigma^2 \end{pmatrix}. \tag{26}$$

This result follows *exactly* from the F-P equation given. Thus the σ and p_σ distributions acquire the equilibrium spreads given earlier. This is expected since in these approximations the spin has no influence on the orbital motion. However, and this is perhaps unexpected, the ψ distribution *also* reaches equilibrium (on the same time scale) with a value $\sigma_\psi = \frac{d}{a}\cdot\sigma_\sigma = \nu\sigma_{p_\sigma}/Q_s$: there is no continual decoherence in this model with these starting conditions! But of course, if σ_ψ is large enough the spins *are* effectively decoherent.

In the HERA electron ring at 27.5 GeV, ν is about 62.5, p_σ is about 10^{-3} and Q_s is about 0.06. So the asymptotic σ_{p_σ} is about 60 degrees. However, the last column of $\underline{A}$ is empty and the $\underline{\sigma}_3(s)$ is singular for all s. Then the asymptotic gaussian W function is not unique but reaches an equilibrium form depending on the initial conditions. For example to discuss decoherence according to the picture in the Introduction, one begins with gaussian distributions in σ and p_σ with their equilibrium asymptotic variances and with a delta function distribution $\delta(\psi)$ in ψ. Then the asymptotic ψ distribution has a variance of $2(\nu\sigma_{p_\sigma}/Q_s)$ which is about 120 degrees! So with those starting conditions, the spins are almost decoherent after a few damping times and the spin distribution "wraps around" on itself. This latter effect means that account must be taken of the fact that the physical domain of ψ is from $-\pi$ to $+\pi$. This can be done by solving the F-P equation with periodic boundary conditions from the beginning or, more naively, by superimposing "strips" of ψ distribution. The asymptotic variances of σ and p_σ are unique. According to our model, in machines running at one or two GeV, the asymptotic σ_{p_σ} is just a few degrees. So within this simple linear model there is no complete decoherence in such machines. Conventional wisdom suggests instead that σ_ψ should increase as $\sqrt{s}$. This is not the case as we have just seen. However, in the simpler 2 x 2 pure diffusion problem for p_σ and ψ without synchrotron oscillations the $\sqrt{s}$ growth does emerge after a few damping times. So the synchrotron motion is an essential ingredient in our calculation. Our model is much too simple to represent a realistic storage ring but it has enabled us to reconsider the calculation in reference 13.

Elaboration of the model shows that if the "smoothness" is abandoned there can be decoherence but that the rate is sensitive to the details [1]. Indeed, so far we have neglected the detailed structure of the ring and misalignments which tilt the equilibrium polarization axis and generate vertical dispersion. Horizontal and vertical betatron motion have been neglected as have the non-linear spin tune spread and skew terms caused by sextupoles. In the presence of these effects SITROS (22) predicts complete decoherence in about fifteen damping times at HERA. This shows that our simple smooth linear model was completely inadequate. Also, it appears initially that it would be impossible to obtain spin flip at HERA unless the flip could be achieved within a few damping times by applying a strong enough rf field. But in considering decoherence in isolation, a key component, the rf field itself, was ignored and partial spin flip *is* sometimes seen at LEP with small rf fields. Even if decoherence calculations of this type are relevant at HERA and LEP energies it is likely that the extracted decoherence time is very sensitive to the details included.

REFERENCES

1. Barber,D.P.,*Proc.9th Int.Symp.High Energy Spin Physics*, Bonn,Germany,1990.
2. Barber, D.P., *Proc.2nd Euro.Part.Acc.Conf,*, Nice, France, 1990.
3. Barber, D.P., et al., DESY Report 94-171 (1994). To appear in Phys.Letts. B.
4. Schwitters, R., DESY Report M-82-09 (1982).
5. Shatunov, Yu., DESY Report M-82-09 (1982).
6. Jowett, J.M. and Ruth, R.D., CERN Report LEP Th/83-26 (1983).
7. Yokoya, K., Particle Accelerators, **14**, 39 (1983).
8. Barber D.P., *DESY "Harz Seminar":February 1994.*
9. Vasserman, I.B., et al., Phys. Letts., **B198**, 302 (1987) and references therein.
10. Shatunov, Yu., These Proceedings.
11. Roser, T., These Proceedings.
12. Dehning, B., These Proceedings and references therein.
13. Koutchouk, J.P., CERN Note SL/AP-16 (1991).
14. Chao, A.W., Nucl. Inst. Meth., **180**, 29 (1981).
15. Barber, D.P., et al., DESY Report 91-146 (1991).
16. Jowett, J.M., AIP Conference Proceedings, **153**, 1987.
17. Ruggiero, F., et al., Annals of Physics **197**, 396 (1990).
18. Barber, D.P., et al., DESY Report HERA 94-02 (1994).
19. Mais, H. and Ripken,G., DESY Report 86-29 (1986).
20. Risken, H., *The Fokker-Planck Equation*, 2nd edition, Springer 1989.
21. Gardiner, G.W., *Handbook of Stochastic Methods*, 2nd edition, Springer 1985.
22. Böge, M., DESY Report 94-87 (1994).

[1]For example if a Siberian Snake is included in the otherwise smoothed ring there *is* decoherence.

Transverse and Longitudinal Electron Polarization at HERA

Michael Düren

Phys. Inst. der Univ. Erlangen-Nürnberg, D-91058 Erlangen, Germany
E-mail: dueren@vxdesy.desy.de

On behalf of the HERA Polarization Group

Abstract

Large transverse electron polarization of 60 to 70% is routinely achieved at HERA during luminosity operation. In May 1994 spin rotators were brought into operation and for the first time longitudinal electron polarization was produced in a high energy storage ring. The spin rotators caused no significant loss in the degree of polarization. Longitudinal polarization at HERA will be used by the HERMES experiment starting in 1995 and later also by the collider experiments H1 and ZEUS.

INTRODUCTION

Longitudinally polarized electrons are perfect probes in two basic fields in high energy physics: the study of the spin structure of the proton and neutron as in the HERMES experiment (1) and the study of weak interactions by the H1 and ZEUS experiments at HERA.

HERA is the world's first high energy electron storage ring to achieve longitudinal polarization (2). Although polarization has always been an integral part of the design of HERA, there were many doubts in the past concerning the strength of the depolarizing mechanisms, the degree of polarization and the functionality of the spin rotators. Only the systematic study of the response of the electron polarization to variation of the various machine parameters lead to today's situation where polarization at HERA is high, stable, reproducible and understood. The reliable and fast HERA polarimeter (3) was an essential experimental precondition and the SITROS spin tracking program (4) was the theoretical counterpart.

POLARIZATION IN HIGH ENERGY STORAGE RINGS

Electrons in storage rings become naturally transversely polarized due to the Sokolov-Ternov effect (5): the synchrotron radiation process contains a small asymmetric spin-flip amplitude that enhances the polarization state anti-parallel to the magnetic fields of the bending magnets. In the case of HERA at $E = 27.5$ GeV this process leads to an initial rise of polarization of $dP/dt|_0 = 2.5$ %/min. Polarization grows exponentially according to the formula

$$P(t) = P_\infty \cdot (1 - e^{-t/\tau}) \qquad (1)$$

with

$$\frac{1}{\tau} = \frac{1}{\tau_P} + \frac{1}{\tau_D} \quad \text{and} \quad P_\infty = P_{ST} \cdot \frac{\tau}{\tau_P}. \tag{2}$$

$P_{ST} = 92.38\%$ is the maximal polarization and $\tau_P = 37$ min is the corresponding rise time for HERA at an energy of 27.5 GeV. In the case that there are non-vertical magnetic fields (e.g. when the spin rotators are active) those numbers obtain small corrections.

The depolarization time τ_D summerizes the depolarizing effects that counteract the polarization build-up. In the case of HERA τ_D varies between a few minutes (i.e. low polarization) and about two hours in the case the machine is optimized for polarization.

High energy machines are inherently difficult to optimize for polarization for the following reason: The spin tune is $\nu_s = a\gamma = 62.5$ at HERA energies which means that the spin precession angle is 63.5 times the angular deflection as can be seen from the equation of spin motion (6):

$$\frac{d\vec{P}}{ds} = \frac{e\vec{P}}{m_e c\gamma} \times [(1 + a\gamma)B_\perp + (1 + a)B_\parallel] \tag{3}$$

with a being the gyromagnetic anomaly of the electron. Any deflection in the machine e.g. due to a small misalignment of a quadrupole, is magnified by this factor and leads to large precession angles at each turn of the electron beam. Severe depolarization occures when there are disturbances in the machine that are in resonance with the spin precession (7).

OPTIMIZATION OF TRANSVERSE POLARIZATION

The most efficient method at HERA to obtain high polarization is the empirical optimization of harmonic correction bumps (8). It functions as follows:

In an electron storage ring there is an unavoidable horizontal excitation of the orbital motion due to synchrotron radiation. On the other hand, spin motion is not very sensitive on horizontal motion and only vertical deflections precess the spin vector away from the vertical direction and thus lead to depolarization. In a realistic storage ring there is however a spin coupling between the horizontal and vertical motion. The harmonic correction scheme allows to reduce this coupling by adding additional vertical distortions that compensate the effect of the original distortions. It uses a combination of 8 vertical closed bumps at 'strategic' positions at HERA. The coupling can not be completely reduced. It is sufficient however to reduce those harmonic components which are close to the spin tune (61 to 64).

The harmonic correction bumps are introduced empirically at HERA by varing bump amplitudes and optimizing the measurend polarization values. In future the harmonic bumps will be determined directly from the measured orbit distortions without empirical optimization. Böge (9) described how the method of beam based alignment will be used to determine the orbit distortions with the required precision.

RF-DEPOLARIZATION AND ENERGY CALIBRATION

The method of resonant spin depolarization has been sucessfully tested at HERA and was used to calibrate precisely the HERA beam energy (10). A variable sweeping frequency of about half the revolution frequency was applied to a vertical dipole magnet with a strength of $B \cdot l \approx 2 \cdot 10^{-4}$ Tm. Resonant depolarization occures when the precession of the polarization vector is in resonance with the excitation frequency. Besides the main resonance also synchrotron side bands are present and have been observed. In order to distinguish side bands and mirror frequences from the main resonance, depolarization has been studied at slightly different beam energies and synchrotron tunes. Figure 1 shows several frequency scans and its effect on polarization. Spin flip has not been observed as the magnet was not strong enough to rotate the spin vector before spin diffusion occures (11).

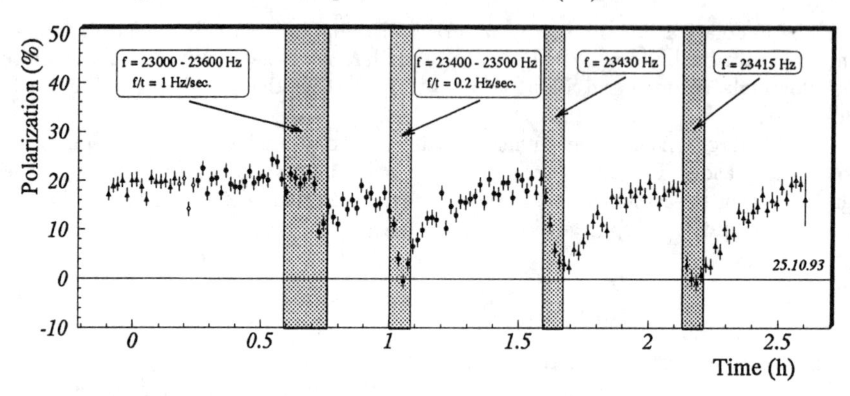

Figure 1: Resonant depolarization of the HERA beam using different frequencies and sweeping speeds.

The beam energy in an ideal storage ring is determined from the depolarization frequency f_{dep} and the revolution frequency f_{rev} according to the formula

$$E_{beam} = \left(n \pm \frac{f_{dep}}{f_{rev}}\right)\frac{m_e}{a} = (61 - f_{dep}/47317\ Hz) \cdot 440.649\ MeV \qquad (4)$$

with n being the integer part of the spin tune. The true HERA beam energy in '93 was about 33-36 MeV below the nominal value that was calculated from the dipole currents. The deviation of about one per mille is within the systematic uncertaincies of the dipole fields.

THE HERA SPIN ROTATORS

The Sokolov-Ternov effect produces vertical polarization in the arcs of an electron storage ring. As the experiments require longitudinal polarization, two spin rotators are required at each interaction region that rotate the spin into the longitudinal direction and back to the vertical direction behind the IP (figure 2). A pair of 'Mini-Rotators' was installed in winter 93/94 in HERA around the HERMES interaction region. The 'Mini-Rotators', designed by Buon and Steffen (12) consist of interleaved horizontal and vertical dipole magnets as shown in figure 3. They have a length of only 56 m and produce a vertical closed bump of about 20 cm. The horizontal bending of the rotators is an integral part of the arc. The purpose of the additional two weak dipoles H3b and H3c in figure 3 is to reduce background in the HERMES experiment. The spin precession of those magnets is taken into account for by a reduced strength of the last rotator magnet H3a.

Figure 4 illustrates how the rotators work: Small angular orbital deflections in the dipoles are magnified by the factor 63.5 according to eq. (3). As the spin

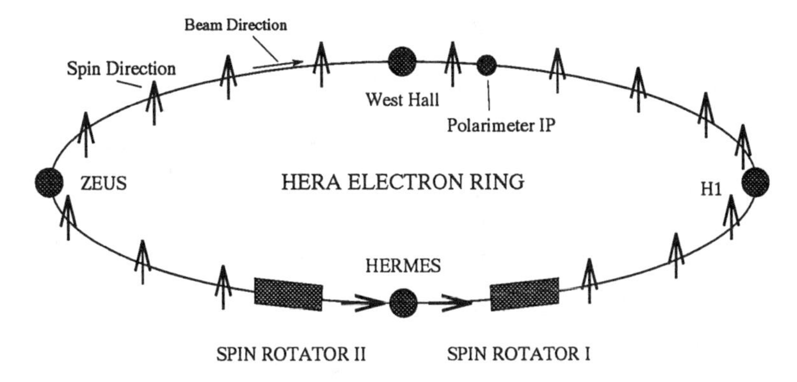

Figure 2: A sketch of the HERA electron ring showing the positions of the spin rotators.

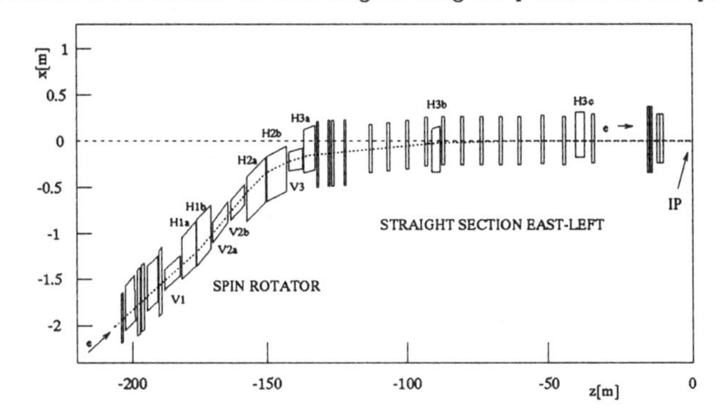

Figure 3: Horizontal projection of the layout of one 90° spin rotator. H1-3 (resp. V1-3) indicate the horizontal (resp. vertical) bending dipoles.

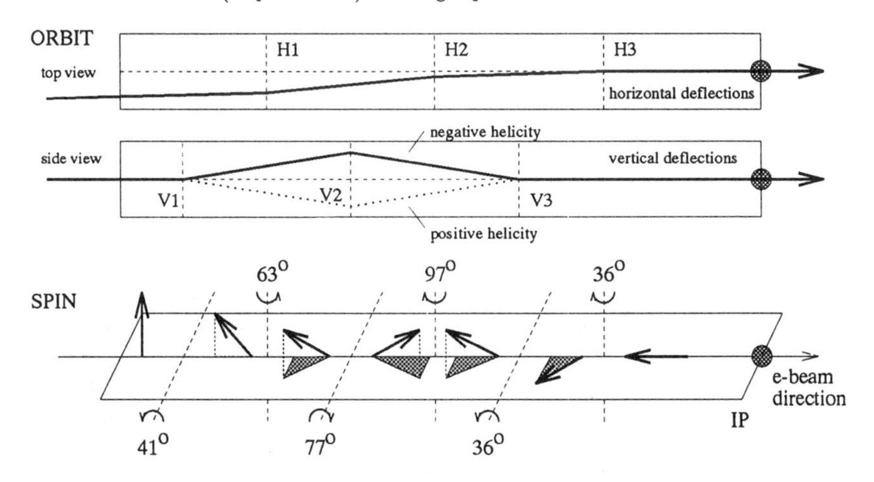

Figure 4: A functional diagram of the Mini-Rotator showing the horizontal and vertical deflections of the orbit and the corresponding spin precession angles.

rotations do not commute, horizontal and vertical bending magnets can be combined in such a way that the spin direction is turned by effectively 90° and at the same time the beam position and slope at the IP is not changed.

To invert the helicity of the electrons at the IP, the direction of the vertical bumps has to be reversed. To facilitate that the spin rotator magnets are mounted on remotely controlled jacks that move the chain of magnets without breaking the vacuum in the chambers. The rotators function over an energy range of 27 to 35 GeV. However, an energy change of greater than ±200 MeV requires manual horizontal adjustment of the rotator magnets.

LONGITUDINAL ELECTRON POLARIZATION

Spin rotators are potential sources of depolarization for several reasons: a small deviation from a 90° rotation would tilt the equilibrium spin direction and thus depolarize the beam. Therfore the magnets have to be calibrated very precisely. Secondly, as the spin is not vertical in the straight section, the large horizontal size of the beam and the horizontal betatron motion can easily lead to spin diffusion due to the vertical magnetic field in the quadrupoles. Quadrupoles inside the spin rotator could be avoided at HERA due to the short length of the 'Mini'-Rotators.

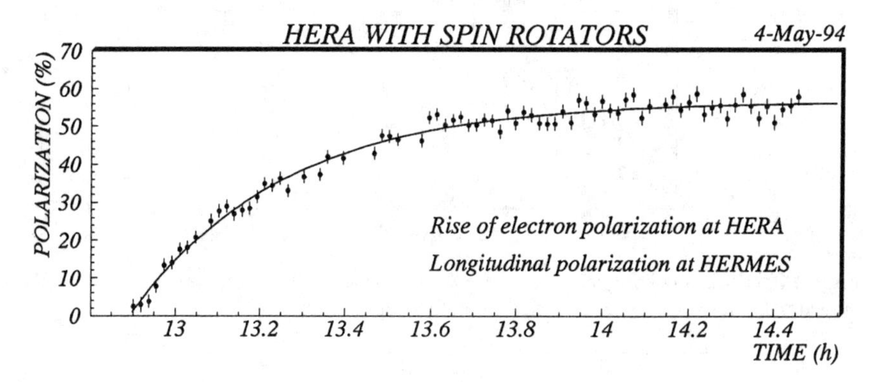

Figure 5: First observation of longitudinal polarization at HERA. Overlaid is the result of a fit of the expected exponential build-up curve with the parameters $P_\infty = 56.6 \pm 0.5\%$ and $\tau = 20.8 \pm 0.7$ min.

A special spin matching procedure was used to reduce the spin diffusion (13): the lattice in the straight section between the rotators was made horizontally 'spin transparent' and the remaining arc is vertically 'spin transparent'. A piece of lattice is described as 'spin transparent' when certain matrix elements in the linearized spin-orbit equations vanish. Then the spin precession of the individual electrons is independant of their betatron motion. Thus this piece does not contribute to spin diffusion.

On May 4th, 1994 the spin rotators were brought into operation for the first time. Due to the non-abelian behaviour of rotations the spin rotators cause a change of spin tune ($\nu_s \neq \gamma a$) which was compensated by changing the beam energy by 20 MeV. Without any fine-tuning the polarization went up to a value of about 57% at the first try (fig. 5). After a small change of the beam energy longitudinal polarization went up close to 70% and reached values which were equal to the maximal achieved polarization without spin rotators (fig. 6).

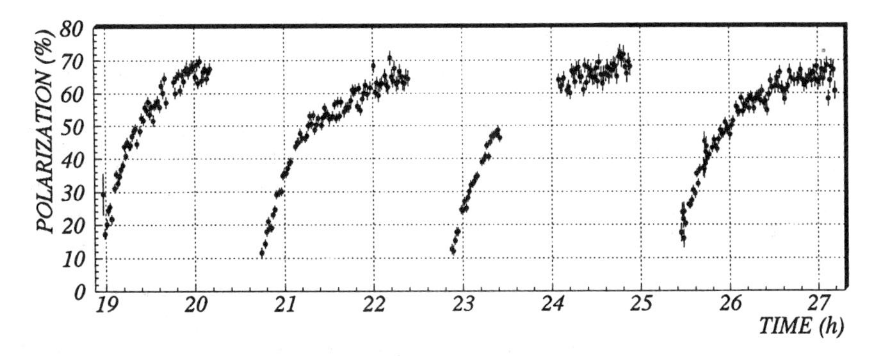

Figure 6: Recording of longitudinal polarization during HERA bake-out runs. After each injection and ramping of a new beam, polarization rises exponentially and reproduces high values of polarization of close to $P = 70\%$.

Polarization is measured with the transverse polarimeter in the HERA West area. The degree of polarization is the same all over the ring but the direction is manipulated by the rotators. There is confidence that the spin direction is longitudinal in the HERMES region within a few degrees due to the precise calibration of the rotator magnets and due to the fact that if the spin direction were tilted from the nominal direction, large polarization would not be possible due to strong depolarizing effects in the rest of the ring. HERMES plans to build a second, longitudinal polarimeter between the rotators to reduce the systematic error of the polarization measurement.

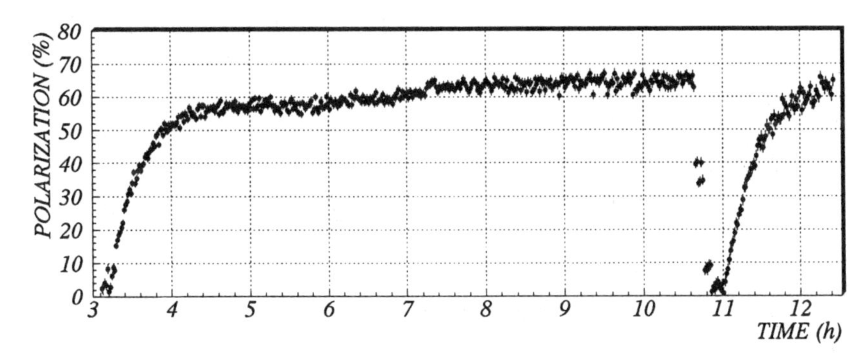

Figure 7: Polarization during a long luminosity run. Polarization rises after ramping the beam to 27.5 GeV. At 6:00 probably the beam conditions have slightly changed which lead to an increase of polarization. At 10:30 the beam was depolarized by RF-resonance to perform an energy calibration and to cross check the polarimeter calibration. At 12:30 the beam was dumped.

POLARIZATION DURING LUMINOSITY CONDITIONS

HERA switched from e^-p to e^+p operation in July '95 to solve a lifetime problem that is caused by positive dust particles which are captured by the negatively charged electron beam. At the same time the spin rotators had a vacuum problem which will be solved by replacing some vacuum chambers in Nov. '94. Therefore HERA operated in 1994 mostly with positrons and without spin rotation. High, stable (transverse) positron polarization of 60 – 70% is achieved routinely at HERA during luminosity conditions. The large solenoid fields of the H1 and ZEUS detectors are compensated locally by anti-solenoids. There is no evidence for beam-beam depolarization by the proton beam. Figure 7 shows polarization during a full luminosity fill of over 9 hours.

It should be mentioned that the achieved polarization is already now well above the design goal for the HERMES experiment of 50%. HERMES data taking will start in spring 1995. Spin rotators for the collider experiments H1 and ZEUS are ordered and will be installed at the earliest in 1996.

REFERENCES

1. HERMES Collaboration, *Technical Design Report*, DESY-PRC 93/06 (1993);
   H. Jackson, *HERMES*, contribution to this conference.
2. D.P. Barber et al., DESY report, DESY 94-171 (1994).
3. D.P. Barber et al., Nucl. Instr. Meth. **A329** (1993) 79;
   F. Zetsche, *Experience with the fast polarimeter at HERA*, contribution to this conference;
   M. Düren, *The HERA Polarimeter*, Proc. of 10th Int. Symp. of High Energy Spin Physics, Nov. 1992, Nagoya; eds. T. Hasegawa et al. (Univ. Acad. Press, Tokyo, 1993).
4. M. Böge, *Analysis of Spin Depolarizing Effects in Electron Storage Rings*, Ph.D. Thesis, Univ. of Hamburg, 1994, DESY 94-087 (1994)
5. A.A. Sokolov and I.M. Ternov, Sov. Phys. Doklady **8**, (1964) 1203.
6. V. Bargman, L. Michel, V. Telegdi, Phys. Rev. Lett. **2** (1959) 435.
7. D.P. Barber, *Theory and Observation of Electron Polarization in High Energy Storage Rings*, Proc. of 10th Int. Symp. of High Energy Spin Physics, Nov. 1992, Nagoya; eds. T.Hasegawa et al. (Univ. Acad. Press, Tokyo, 1993).
8. D.P. Barber et al., Nucl. Instr. Meth. **A338** (1994) 166.
9. M. Böge, *Optimization of spin polarization in the HERA electron ring using beam-based alignment procedures*, contribution to this conference.
10. C. Grosshauser, diploma thesis, Univ. Erlangen-Nürnberg, 1994, DESY-HERMES 94-01 (1994).
11. D.P. Barber, *Spin Depolarization in Storage Rings*, contribution to this conference.
12. J. Buon and K. Steffen, Nucl. Instr. Meth. **A245** (1986) 248.
13. D.P. Barber et al., Part. Accel. **17** (1985) 243.

Optimization of Spin Polarization in the HERA Electron Ring using Beam-Based Alignment Procedures

M. Böge and R. Brinkmann

Deutsches Elektronen-Synchrotron DESY
Notkestr. 85, D-22603 Hamburg

Abstract. The maximum degree of electron spin polarization in a real storage ring is mainly limited by the tilt of the equilibrium polarization direction $\vec{n}_0$ with respect to the direction of the main bending fields. The tilt is mainly caused by random vertical closed orbit kicks introduced by nonzero vertical offsets inside the quadrupoles. Methods for minimizing the average tilt of $\vec{n}_0$ are discussed and a correction algorithm is introduced which makes use of the known correlations between transverse offsets of quadrupoles and adjacent beam position monitors. The correlation can be established by a beam-based alignment technique. The results of first measurements are presented.

INTRODUCTION

One important prerequisite for obtaining high spin polarization in an electron storage ring is a well corrected vertical closed orbit in order to avoid depolarization due to a tilted $\vec{n}_0$-axis. Furthermore, the closed orbit kicks generate spurious vertical dispersion which also gives rise to depolarizing effects. In principle, provided that a sufficient number of correction coils and efficient closed orbit optimization algorithms are available, the nominal closed orbit as measured by the beam position monitors (BPM's) can be made very small (a few tenths of a mm (rms) in a real machine). However, due to mechanical and electronical imperfections even for a nominal orbit which reads zero in all monitors, there will remain random offsets of the beam in the quadrupoles because the monitor and the quadrupole axis do not coincide. Therefore, a tilt of the spin $\vec{n}_0$-axis remains even for an apparently perfectly corrected machine, and empirical polarization optimization procedures have to be applied (e.g. the harmonic bumps scheme [1][2]). We present a method to improve the alignment of the monitor axis with respect to the magnetic axis of the quadrupoles. That enables us to optimize polarization in a more systematic way and, if combined with empirical procedures, can eventually lead to a higher degree of polarization as well as a faster setup of the machine for optimum performance regarding polarization [3].

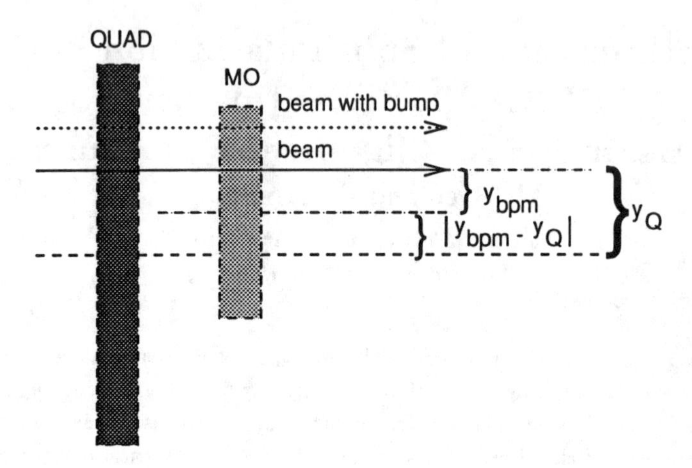

FIGURE 1: *Illustration of the beam-based alignment technique applied to the vertically focusing quadrupoles (QUAD) with adjacent monitors (MO)*

BEAM-BASED ALIGNMENT PROCEDURE

Our method is based on the well known fact that if the strength of a single quadrupole in the ring is changed, the resulting difference in the closed orbit $\Delta y(s)$ is proportional to the original offset y_Q of the beam in this quadrupole.

The equation for the resulting difference orbit is

$$\Delta y''(s) - (k(s) - \Delta k(s))\Delta y(s) = \Delta k(s) y_Q(s). \tag{1}$$

The difference orbit is thus given by the closed orbit formula for a single kick, but calculated with the perturbed optics including $\Delta k(s)$. From the measured difference orbit the kick and thus y_Q can be easily determined and compared to the nominal orbit y_{bpm} in the monitor adjacent to the quadrupole, yielding the offset between monitor and quadrupole axis. The precision of the method is very much improved by taking difference orbit data for several local beam positions y_Q varied with an orbit bump. The principle of the method is illustrated in figure 1. The error of the nominal position y_{bpm} for which the beam goes through the centre of the quadrupole is then given by the resolution of the BPM system. In the HERA electron ring (HERA-e), a difference orbit with an amplitude of 0.1 mm can be clearly resolved. This results in a resolution for the local kick of about 0.005 mrad. Since a change of a quadrupole strength of $\Delta kl = 0.03\ \text{m}^{-1}$ is possible without losing the beam, a minimum beam offset of $y_{Q,min} = 0.15$ mm can be easily detected. Taking several data points by varying the local bump, the quadrupole-to-monitor alignment can be done with a precision of about 0.05 mm.

This alignment method has been first tested in November 1992 [4] for one quadrupole circuit in HERA-e. In this case, two quadrupoles are powered in

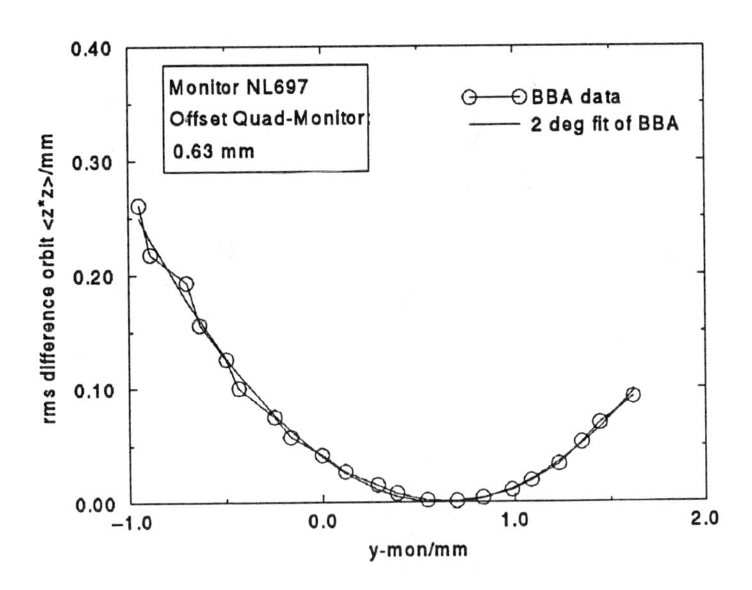

FIGURE 2: *Square of the rms value of the vertical difference orbit which is defined by the closed orbit before and after a change of the quadrupole strength by 15%, versus the nominal monitor value* y_{bpm} *for 21 different local orbit bump amplitudes. The minimum of the resulting parabola defines the position of the quadrupole axis with respect to* y_{bpm} *which is calculated to be 0.63 mm*

series (they are positioned symmetrically with respect to the interaction point East) so that the analysis of the difference orbit data is somewhat complicated in the sense that one has to take into account two kicks, without affecting the precision of the method, though. The measurement was repeated for the same quadrupole pair (and the adjacent monitors) one year later [5]. Within the limits of error (0.05 mm) this offset had not changed after one year. This means that once the alignment is established, it will be stable for a long time, probably because the mechanical imperfections remain constant since the transverse positioning of the vacuum chamber in the quadrupoles is fixed.

During the last winter shutdown switches have been installed that allow the strength of individual quadrupoles in the arcs to be varied although being powered in series with many other quadrupoles [6]. During the machine shifts in November this year an automatic alignment procedure for the vertically focussing arc quadrupoles was successfully tested, including the control of the switches, the change of the quad strengths, difference orbit measurements and the variation of local orbit bumps. Due to some hard- and software problems only a few quads could be aligned automatically. Nevertheless it was shown that the alignment of the arc quads can in principle be done in less than 24

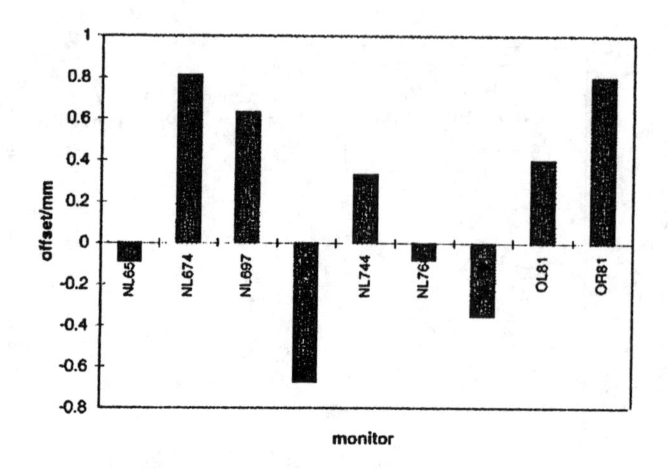

FIGURE 3: *Summary of the results of the 1994 measurements together with the data taken in 1992[4]. The resulting vertical offsets for the quadrupoles are an order of magnitude larger than the measurement error*

hours without human intervention. As an example the alignment of the monitor NL697 is shown in figure 2. The plot visualizes the square of the rms value of the vertical difference orbit which is defined by the closed orbit before and after a change of the quadrupole strength by 15%, versus the nominal monitor value y_{bpm} for 21 different local orbit bump amplitudes. The minimum of the resulting parabola defines the position of the quadrupole axis with respect to y_{bpm} which is calculated to be 0.63 mm in the presented example. Figure 3 summarizes the results of the 1994 measurements together with the data taken in 1992 [4]. The resulting vertical offsets for the quadrupoles are an order of magnitude larger than the measurement error, implying that a significant reduction of the closed orbit kicks in the ring should be possible, provided that the method is applied everywhere around the ring.

POLARIZATION OPTIMIZATION PROCEDURES

In this section the methods for improving spin polarization by making use of the beam-based alignment are discussed. The first method simply minimizes the rms closed orbit kick instead of the orbit in the monitors itself. At quadrupole positions with a BPM and a vertical correction coil nearby, the local orbit kick is given by the sum of the change of y'_{co} due to the orbit offset in the quadrupole and the kick of the corrector. We assume that this offset is known with an rms error Δy_A which is the error of the beam-based alignment procedure. With a perfect optimization algorithm one expects that eventually the rms orbit kick can be reduced to $\delta y'_{rms} = \Delta y_A \times kl$, where kl is the average quadrupole strength. However, in HERA-e only every second quadrupole (every vertically focusing quadrupole QD) has a BPM and a corrector nearby. The horizontally focusing quadrupoles (QF's) in-between can also be "beam-based aligned" [1], but the kick at those positions cannot be locally corrected. A global minimization of the rms orbit kick is still possible, though. We used the MICADO algorithm to test the method in a computer simulation for HERA-e without spin rotators with realistic tolerances. First the orbit was corrected with the standard MICADO algorithm down to an rms error of about $\Delta y_{co} = 0.5$ mm, then the rms kick optimization was applied, iterating several times with 20...50 correctors per step until no improvement of the rms closed orbit kick was observed anymore. We assume a precision of the alignment of $\Delta y_A = 0.1$ mm. For the optics of the HERA-e ring used during 1993 operation, we find by simulating with four different random seeds that the rms closed orbit kick of $62 \pm 2\mu$rad after standard orbit correction is reduced to $31 \pm 2\mu$rad with beam-based alignment and minimization of the rms kick.

What is the influence on the $\vec{n}_0$-axis tilt and thus the polarization ?

The vector $\vec{n}_0(s)$ is the periodic solution of the T-BMT equation [7]:

$$\frac{d\vec{n}_0}{ds} = \left(\vec{\Omega}_0^d + \delta\vec{\Omega}_0\right) \times \vec{n}_0 \tag{2}$$

where s is the longitudinal coordinate of the usual storage ring coordinate system represented by the unit vectors $\vec{e}_x$, $\vec{e}_y$ and $\vec{e}_s$. $\vec{\Omega}_0^d$ contains the fields on the design orbit, $\delta\vec{\Omega}_0$ the additional fields on the closed orbit.

Assuming $|\delta\vec{\Omega}_0| \ll |\vec{\Omega}_0^d|$, $\vec{n}_0$ can be decomposed into two parts:

$$\vec{n}_0 = \vec{n}_0^d + \delta\vec{n}_0. \tag{3}$$

The modulus $|\delta\vec{n}_0|(s)$ describes the tilt of $\vec{n}_0$ with respect to $\vec{n}_0^d$ mainly due to the presence of a nonzero vertical closed orbit with respect to the design orbit. $|\delta\vec{n}_0|$ is given by [2]:

$$|\delta\vec{n}_0|^2 = \frac{1}{2(1-\cos 2\pi a\gamma)}\left(\left[\int_s^{s+L}\delta\Omega_{0x}\cos\phi ds\right]^2 + \left[\int_s^{s+L}\delta\Omega_{0x}\sin\phi ds\right]^2\right) \tag{4}$$

with $\delta\Omega_{0x} = (a\gamma+1)k(s)y(s)$ assuming that the vertical deflections are mainly generated by the nonzero vertical beam offsets y_Q inside the quadrupoles before vertical orbit correction, and $\phi = a\gamma\alpha$, where α is the deflection angle in the

[1] In this case, the axis of the QF's with respect to the BPM's close to the QD's on either side of the QF is determined

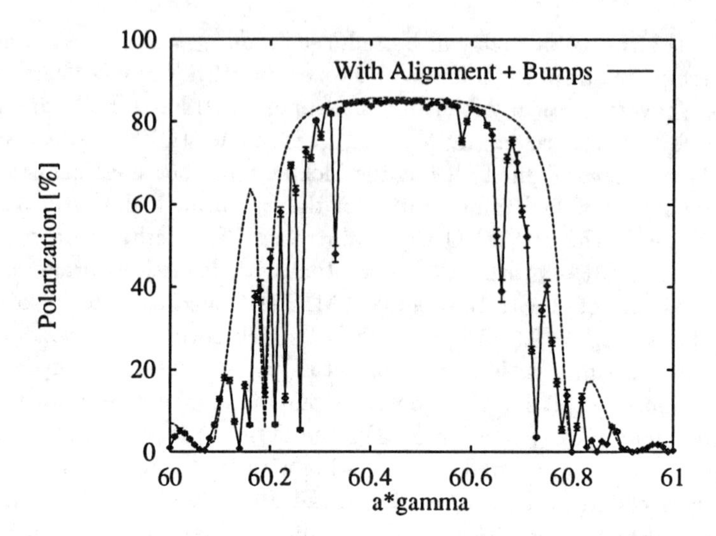

FIGURE 4: *Further empirical optimizition of polarization using the harmonic bumps scheme [1][2] after the beam-based alignment and the minimization of the rms closed orbit kick. The points with errorbars represent the result of a Monte-Carlo calculation with SITROS [8]. The smooth line represents the corresponding SITF result based on linear perturbation theory*

bending magnets. The constant a denotes the gyromagnetic anomaly $(g-2)/2$, γ the relativistic γ-factor and L the length of the ring. Thus the rms value of $|\delta\vec{n}_0|$ is approximately proportional to the rms closed orbit kick $\delta y'_{rms}$ which is reduced due to the application of the beam-based alignment procedure. The simulation leads to a reduction of the rms $\vec{n}_0$-axis tilt from 23 ± 1 mrad to 9 ± 3 mrad for a certain random seed.

Owing to the fact that the depolarization rate $1/\tau_d$ depends quadratically on the $\vec{n}_0$-axis tilt variation in linear approximation, the equilibrium polarization P which is given by [9]:

$$P = P_\infty \frac{1}{1 + (\tau_p/\tau_d)} \tag{5}$$

with $P_\infty \approx 92\%$ and a polarization buildup time $\tau_p \approx 2600$ sec at a spin tune $a\gamma = 60.5$ corresponding to a beam energy of 26.66 GeV, increases from $18 \pm 4\%$ to $60 \pm 15\%$.

The additional application of the harmonic bumps scheme leads to a further reduction of the rms tilt and an equilibrium polarization of about 85% as shown in figure 4. This has to be compared to 75% polarization which can be achieved by the empirical optimization using the bumps without prior beam-based alignment and rms kick minimization [10].

The method discussed can be improved by taking into account the spin phase advance $\phi(s)$ between the quadrupoles in order to minimize $|\delta\vec{n}_0|$ given by eq. 4.

SUMMARY

This beam-based alignment procedure is a powerful tool to get high polarization in HERA-e without time consuming empirical optimization of polarization with the harmonic bumps scheme. Using both methods the simulations indicate that polarization values of about 85% are possible compared to 75% without beam-based alignment. An automatic alignment procedure will be used next year to provide high longitudinal polarization for the HERMES experiment which starts taking data in spring 1995.

REFERENCES

[1] Barber,D.P. et al., "High Spin Polarisation at the HERA Electron Storage Ring", Nucl.Instr.Meth. A338 (1994) 166.

[2] Rossmanith,R. and Schmidt,R., "Compensation of Depolarizing Effects in Electron-Positron Storage Rings", Nucl.Instr.Meth. A236 (1985) 231.

[3] Brinkmann,R. and Böge,M., "Beam-Based Alignment and Polarization Optimization in the HERA Electron Ring", presented at the EPAC, London (1994).

[4] Brinkmann,R., in: Willeke,F. (ed.), "HERA Seminar Bad Lauterberg/Harz" (1993) 269.

[5] Gianfelice-Wendt,E. and Lomperski,M., Private communication.

[6] Bialowons,W. and Gode,W.D., Private communication.

[7] Thomas,L., Philos.Mag. 3 (1927) 1.
Bargmann,V., Michel,L. and Telegdi,V.L., Phys.Rev.Lett. 2 (1959) 435.

[8] Kewisch,J. et al., Phys.Rev.Lett. 62 (1989) 419.

[9] Montague,B.W., "Polarized Beams in High Energy Storage Rings", Phys.Rep. 113, Vol.1 (1984).

[10] Böge,M., "Analysis of Spin Depolarizing Effects in Electron Storage Rings", Ph.D. thesis, DESY 94-087 (1994).

Spin Control System for the South Holl Ring at Bates Linear Accelerator Center

S. Kovalsky, T. Zwart,* P. Ivanov and Yu. Shatunov†

*Bates Linear Accelerator Center, Middleton, MA
†Institute for Nuclear Physics, 630090, Novosibirsk, Russia

INTRODUCTION

The physical programme with longitudinal polarized electrons at SHR includes fixed target experiments on extracted beam and an internal target facility [1]. In this paper we show a possible solution to provide the longitudinal polarization in the both SHR operation modes by an implementation in the machine optics a Siberian Snake.

LATTICE MODIFICATION

A consideration of the layout of the SHR facility shows a way for the snake positioning in the ring. If we insert the snake before a electrostatic septum in the extraction straight section we have to obtain (according to SS concept [2]) the longitudinal polarization in the opposite straight section where the internal target is located. Besides that the spin of the extracted beam lies in the horizontal plane and then rotates along the momentum after 180°bend to the detector an arbitrary energy.

However a realisation of this idea meets some troubles. First of all the standard drift length between two quadrupoles (see Fig. 1) is not enough to insert a solenoid with a resonable field strength (10.5 T*m for 1 GeV) and a number of skew quads for a coupling compensation. A solution of this problem can be done probably in many ways. We developed inside the mentioned straight section a new optics that is suitable for the SS insertion and does not touch the beam extraction system. To provide this optics one needs to move two quadrupoles (LQ 41 and symmetrical to it LQ 49) to new positions. It gives two 4.81 m drifts: one is available for the snake; second can be used for many other purposes. Fig. 1 (solid lines) shows the matched behavior of the vertical β_y function along the half straight section.

This solution is sufficient for the all requirements which are needed for both SHR operation mode. There is some optical flexibility for the tune adjustment to off machine resonances in the storage mode. The optical functions near t he extraction septum are under control. The positioning of three octupoles for the beam extraction does not create an additional problem. In the section for snake insertion the β_x, β_y-functions are relatively low and smooth to avoid difficulties in the machine tuning with a strong solenoids.

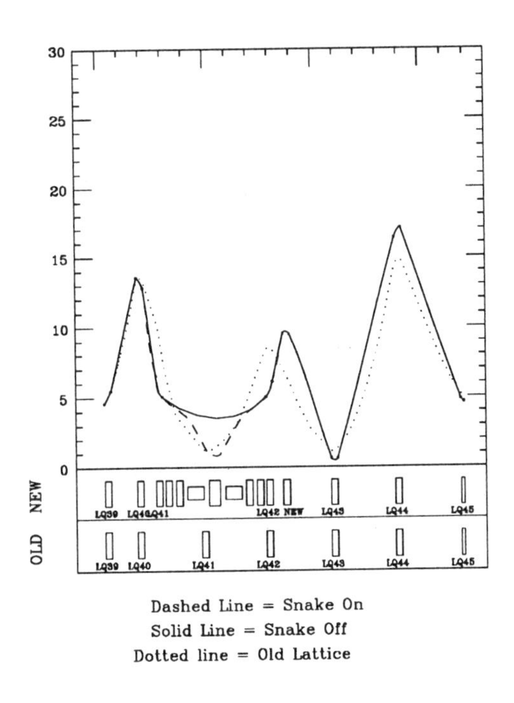

Figure 1: β_y-function on half extraction straight

SIBERIAN SNAKE SCHEME

The SS insertion will not disturb the machine optics if its transfer matrix is equivalent to the drift length physically occupied by the insertion and betatron tunes are shifted by a $m \cdot \frac{1}{2}$ (m-*integer*). This approach has been suggested in [3] and recently applied to the AmPS ring [4]. Fig.2 shows a mirrow-symmetric SS scheme that consists of two solenoids, two pairs of skew quads at both ends and two regular quads in the middle. This scheme has 4 parameters to vary its focusing: i.e. 3 quadrupole strengths and the length of the solenoid; provided that solenoid's field integral is held unchanged so that the total spin pricession angle equals exactly 180°. A reliable solution for the energy 1 GeV which satisfies all assumptions is given in Table 1.

	L (m)	G (kG/m)	B (kG)	α
q_1	0.300	61.00	0.00	45°
o	0.180	0.00	0.00	0°
q_2	0.300	-48.50	0.00	45°
o	0.175	0.00	0.00	0°
S	0.805	0.00	65.00	0°
o	0.175	0.00	0.00	0°
f	0.300	-13.97	0.00	0°
o	0.030	0.00	0.00	0°
f	0.300	-13.97	0.00	0°
o	0.175	0.00	0.00	0°
S	0.805	0.00	65.00	0°
o	0.175	0.00	0.00	0°
$-q_2$	0.300	48.50	0.00	−45°
o	0.180	0.00	0.00	0°
$-q_1$	0.300	-61.00	0.00	−45°

Note, that present solution which is very economical in number of elements and their strengths shifts the horizontal tune by integer while

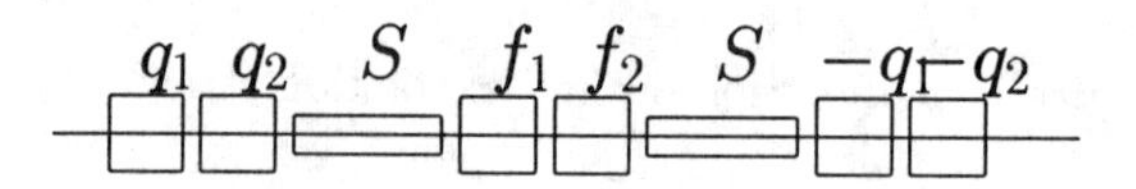

Figure 2: Siberian Snake scheme

the vertical tune will be shifted by an integer and half. An important feature of this optics is that inside the insertion the β-functions decrease considerably (Fig.1, doted lines) due to the strong solenoidal focusing dominating over the quadrupoles action. It means that aperture requirements over the insertion are relaxed.

TOLERANCES AND NONLINEARITIES

An investigation of the effects of errors in the positioning, orientation and powering shows that are not any specific requirements to the alingment and the power supply of the snake elements. A possible resudial coupling caused by a mutual mispositioning of the magnets can be easy compensated by the skew quads adjustment.

An estimation of the solenoidal edge field contribution to the lattice nonlinearities gives negligible numbers.

A real field configuration in the solenoids and quadrupole magnets must be taken into account for the actual design.

References

[1] J.Flanz, *"South Hall Ring Design Report"*, 1990.

[2] Ya.S.Derbenev, A.M.Kondratenko, A.N.Skrinsky, *Sov. Phys. Doklady*, 15, 1970, p.583.

[3] A.Zholents, V.Litvinenko, *Preprint INP 81-60*, Novosibirsk, 1980.

[4] V.V.Danilov *et al*, *Proc. of 10th Int. Symp. on High Energy Spin Physics*, 1992,p.445.

Spin Physics and Lie Algebra Technique

Eidelman Yu.I., Shatunov Yu.M., Yakimenko V.Ye.
Budker Institute of Nuclear Physics
Novosibirsk, Russia

September 8, 1994

Abstract

The application of Lie algebra technique for spin calculation is described. The calculations of both orbital and spin motions are included. The special $\mathcal{P}$- and $\mathcal{D}$-functions are introduced to formalize the determination of the orbital and spin Lie operators and to apply the systems of the analytical calculations to find them. Computer code SPINLIE based on this method ha been created to calculate the level of equilibrium polarization in colliders. The spin tune shifts due to spin resonances and rms energy shift is estimated with the help of this technique. Code SPINLIE has been used for simulation of the rms energy shifts for colliders with low (VEPP-2M), medium (VEPP-4M) and high (HERA-e) energies.

The problem of nonlinear orbital and spin dynamics can be considered as one of the most urgent ones. The reasons are as follows. First, the magnetic structure of the modern accelerators has become more rigid and therefore, nonlinear effects have turned to be more significant. Second, unfortunately there is no universal method for these tasks yet.

Different approaches are used in accelerator physics. The choice between them depends on the problem or author's preference. We have used Lie technique, but our results can be applied to any other method.

As it is known from [1],[2] the solution of the BMT equation

$$\frac{d\vec{s}}{d\theta} = [\vec{W}, \vec{s}]$$

can be found with the help of Lie operators tecnique. Here $\vec{s}$ and $\vec{W}$ are spin and its precision frequency corespondingly and θ is the azimuth over the ring. The BMT frequency $\vec{W}$ can be divided into the $\vec{W}^{(0)} + \vec{\omega}$. Here $\vec{\omega}$ is a small correction.

The solution of this equation can be written as a map for spin vector from initial azimuth θ_0 to final θ:

$$\vec{s}(\theta) = S(\theta_0, \theta)\vec{s}(\theta_0).$$

The map S is found as the expansion of the exponential Lie operators product:

$$S(\theta_0, \theta) = e^{:\vec{\mathcal{W}}^{(0)}(\theta_0,\theta)\vec{s}:} \cdot e^{:\vec{\mathcal{W}}^{(r)}(\theta_0,\theta)\vec{s}:} = S^{(0)}(\theta_0, \theta) \cdot e^{:\vec{\mathcal{W}}^{(r)}(\theta_0,\theta)\vec{s}:}. \tag{1}$$

Here $S^{(0)}(\theta_0, \theta)$ is the usual rotation matrix. It describes the spin vector rotation due to $\vec{W}^{(0)}(\theta)$ part of BMT frequency. The expression for vector $\vec{\mathcal{W}}^{(r)}(\theta_0, \theta)$ is [2] :

$$\begin{aligned} \vec{\mathcal{W}}^{(r)}(\theta_0, \theta) = \int_{\theta_0}^{\theta} d\theta' \left(S^{(0)}(\theta_0, \theta')\right)^{-1} \vec{\omega}(\theta') + \\ + \frac{1}{2} \int_{\theta_0}^{\theta} d\theta' \int_{\theta_0}^{\theta'} d\theta'' \cdot \\ \cdot \left[\left(S^{(0)}(\theta_0, \theta'')\right)^{-1} \vec{\omega}(\theta''), \left(S^{(0)}(\theta_0, \theta')\right)^{-1} \vec{\omega}(\theta') \right] + \cdots \end{aligned} \tag{2}$$

and one can find the following result for map $\mathcal{S}$:

$$S_{\alpha\beta}(\theta_0, \theta) \approx S^{(0)}_{\alpha\beta}(\theta_0, \theta) + S^{(0)}_{\alpha\delta}(\theta_0, \theta) e_{\delta\beta\gamma} \mathcal{W}^{(1)}_{\gamma i}(\theta, \theta_0) Z_i(\theta_0) + \cdots,$$

where

$$\mathcal{W}^{(1)}_{\alpha i}(\theta_0, \theta) \quad = \quad \int_{\theta_0}^{\theta} d\theta' \left(\mathcal{S}^{(0)}_{\alpha\beta}(\theta_0, \theta')\right)^{-1} \omega^{(1)}_{\beta j} \mathcal{A}_{ji}(\theta_0, \theta').$$

Hence, one can see that the rules of spin Lie operators determination include repeated integrations of different polinomials of the trigonometrical functions. As a result of these integrations the constructions, which are a division of small differences by other small differences, appear. For this reason the acqurracy of the numerical calculations decreases. Both of these problems can be solved by means of introduction of the special $\mathcal{P}$- and $\mathcal{D}$-functions.

$\mathcal{P}$-functions. What is a $\mathcal{P}$-function? It is a "single" series

$$\mathcal{P}_i(Q,s) = s^i \sum_{j=0}^{\infty} \frac{(-Q^2 s^2)^j}{(2j+i)!}.$$

It is easy to verify that the relations between the functions $\mathcal{P}_i(Q,s)$ and usual trigonometrical ones are determined by the following expressions [1]:

$$\begin{aligned}
\mathcal{P}_0(Q,s) &= \cos Qs, \\
\mathcal{P}_1(Q,s) &= \frac{\sin Qs}{Q} = s\left(\frac{\sin Qs}{Qs}\right), \\
\mathcal{P}_2(Q,s) &= \frac{1-\cos Qs}{Q^2} = s^2\left(\frac{1-\cos Qs}{(Qs)^2}\right)
\end{aligned}$$

and etc. The usage of $\mathcal{P}$- functions allows to eliminate small denominators (Q and it powers) in the right parts of these expressions. $\mathcal{D}$-functions are included in analogous way as "double" series with [3].

An increase in the accuracy of numerical calculations is achieved through the usage of explicit expansion $\mathcal{P}$- and $\mathcal{D}$-functions in the series. These series converge very quickly in cases of accelerator problems since the function's arguments are usually less than unit.

The rules for the operations with these functions allow to formalize the calculations and to apply systems of analytical calculations for Lie operators determination. Special REDUCE code was created for these calculations and analytical expressions for all orbital and spin Lie operators were obtained.

As the next step of the application of this technique it is necessary to obtain the rules of addition of spin transformations passing the collider structure. These rules were found [2] and after this were programed in FORTRAN code SPINLIE as well as formulae for all Lie operators.

What can the code SPINLIE do? Its main features are as follows:

- input language which is compatible with that of MAD;
- input data as well as output data are entered from files or interactively ;

[1] If Q is imaginary ($Q = i\tilde{Q}$) then $\mathcal{P}_0(Q,s) = \cosh \tilde{Q}s$ and etc.

- elements are described as "thick lenses" for orbital motion as well as for spin calculations;
- simulation of the standard collider rings is realized (all machine functions, emittances and etc.);
- it is possible to assign random alignment errors and field errors to elements;
- it is possible to compensate the orbit distortion (when alignment and field errors are taken into account) with the help of kickers' family. This way the simulatation of a distorbed orbit in real accelerator is achieved;
- it is possible to optimize different functions on machine parameters. As an example, the harmonic spin matching for HERA-e storage ring was simulated in such a manner [2];
- equilibrium polarization level is calculated. The second order resonances are taken into account in these calculations (contributions of all sextupole terms are uncluded);
- it is possible to plot interactively a graphic data with help of code GNUPLOT;
- the code still in process of development (the possibility to investigate spin resonances for proton colliders was added recently).

Spin tune shifts. Let us describe the application of Lie technique for estimation of the spin tune shifts into storage rings. In order to do that let us take into account only two terms of the expansion of $e^{:\mathcal{W}^{(r)}\vec{s}:}$ in the equation (1):

$$e^{:\vec{\mathcal{W}}^{(r)}\vec{s}:} =: E : + : \vec{\mathcal{W}}^{(r)}\vec{s} : + \frac{1}{2} : \vec{\mathcal{W}}^{(r)}\vec{s} :^2 .$$

One can find the following expression for the map S in this case:

$$S_{ij} = S_{ik}^{(0)}\Big[\delta_{kj} + e_{kjm}\mathcal{W}_m^{(r)} + \frac{1}{2}e_{klm}\mathcal{W}_l^{(r)}e_{mnj}\mathcal{W}_n^{(r)}\Big] =$$
$$= S_{ik}^{(0)}\Big[\delta_{kj} + e_{kjm}\mathcal{W}_m^{(r)} + \frac{1}{2}\Big(\mathcal{W}_k^{(r)}\mathcal{W}_j^{(r)} - \mid \vec{\mathcal{W}}^{(r)} \mid^2 \delta_{kj}\Big)\Big].$$

The spin tune $\nu(\theta)$ for nonequilibrium particles is determined by the trace of one turn matrix $S(\theta, \theta+2\pi)$ and after some transformations and averaging over the ring one can find the final result for spin tune shift $\Delta\nu = \nu - \nu_0$:

$$2\pi\overline{\Delta\nu} = \overline{\mathcal{W}_{\parallel}^{(r)}} + \frac{1}{4}\overline{\left|\vec{\mathcal{W}}^{(r)}\right|^2} ctg\pi\nu_0. \tag{3}$$

Here ν_0 is a spin tune for particles on the equilibrium orbit. For calculation of $\overline{\left|\vec{\mathcal{W}}^{(r)}\right|^2}$ let us introduce the spectrum [4]:

$$\left(S^{(0)}(\theta_0, \theta)\right)^{-1} \vec{\omega}(\theta) e^{i\nu_0\theta} = \frac{1}{2\pi} \sum_k \vec{w}_k e^{i\nu_k\theta + i\xi(\theta_0)}, \tag{4}$$

where $\nu_k = k + k_x\nu_x + k_y\nu_y + k_s\nu_s$ are combinations of orbital motion frequencies ν_x, ν_y, ν_s with integer coefficients k, k_x, k_y, k_s and $\vec{w}_k$ are amplitudes of these resonance harmonics. The sum $\sum_k$ means the summation over all k, k_x, k_y, k_s. After certain calculations one can find:

$$\left|\vec{\mathcal{W}}^{(r)}\right|^2 = \sum_k |\vec{w}_k|^2 \left(\frac{\sin\pi(\nu_0 - \nu_k)}{\pi(\nu_0 - \nu_k)}\right)^2 +$$

$$+ \sum_{k,k' \neq k} (\vec{w}_k \vec{w}_{k'}^*) \frac{\sin\pi(\nu_0 - \nu_k)}{\pi(\nu_0 - \nu_k)} \cdot \frac{\sin\pi(\nu_0 - \nu_{k'})}{\pi(\nu_0 - \nu_{k'})}.$$

The double sum in this expression is responsible for the overlapping of different spin resonances. Now one can finaly write the following formula after substitution of this expression into (3):

$$2\pi\overline{\Delta\nu} = \overline{\mathcal{W}_{\parallel}^{(r)}} + \frac{1}{4}\frac{1}{tg\pi\nu_0} \sum_k \overline{|\vec{w}_k|^2} \left(\frac{\sin\pi(\nu_0 - \nu_k)}{\pi(\nu_0 - \nu_k)}\right)^2 +$$

$$+\frac{1}{4}\frac{1}{tg\pi\nu_0} \sum_{k,k' \neq k} \overline{(\vec{w}_k \vec{w}_{k'}^*)} \frac{\sin\pi(\nu_0 - \nu_k)}{\pi(\nu_0 - \nu_k)} \cdot \frac{\sin\pi(\nu_0 - \nu_{k'})}{\pi(\nu_0 - \nu_{k'})}. \tag{5}$$

The important item of the last formula is the averaging of the values $|\vec{w}_k|^2$ and $(\vec{w}_k \vec{w}_{k'}^*)$ over the beam distribution. One can find a disscussion on this problem in [5].

RMS Energy shifts. As it is known, the energy shift ΔE appears due to different reasons:

- finite beam rms sizes [6] (due to quadratic nonlinearity of guide magnetic field: $s = \frac{e}{E}\frac{\partial^2 B_y}{\partial x^2}$);
- optic imperfections Δx, Δy;
- RF-cavity misalignments;
- electrical fields of vacation pumps;
- most surprisingly due to the terrestrical tides caused by the sun and moon.

The contributions of the first two items will discuss without the repetition of the derivation of the equation for $\overline{\left(\frac{\Delta E}{E}\right)}$ from [5]:

$$\begin{aligned}\frac{\overline{\Delta E}}{E} &= \frac{\overline{\Delta\gamma}}{\gamma} \approx \\ &\approx \frac{2\epsilon_x \langle s \mid f_x \mid^2 \psi\rangle + \overline{\left(\frac{\Delta\gamma}{\gamma}\right)^2} \langle s\psi^3\rangle + \langle s\psi\Delta x^2\rangle}{2 < k\psi >} + \\ &+ \frac{2\epsilon_y \langle s \mid f_y \mid^2 \psi\rangle - \langle s\psi\Delta y^2\rangle}{2 < k\psi >}.\end{aligned} \tag{6}$$

The following notations are used in this formulas: $k = \frac{eB_y}{E}$ is a ring curvature; f_x, f_y and ψ are horizontal and vertical Floke-functions and dispersion function respectively; ϵ_x and ϵ_y are horizontal and vertical beam emitances; $\overline{\left(\frac{\Delta\gamma}{\gamma}\right)^2}$ is a beam energy spread and $< \cdots >$ denotes the averaging over the ring.

Spin tune shifts with energy and all terms from equation (6) do not "destroy" the connection between spin tune and beam energy. The first of them gives the spin tune a spreading which is averaged over the beam in contrast to the last two terms, which describe the coherent energy and spin tune shifts. In Table 1 the numerical results [2] [3] of simulation with the help of the code

[2]The terms $\epsilon_y \frac{\langle s|f_y|^2\psi\rangle}{<k\psi>}$ is proportional to vertical emittance and therefore it is neglected for colliders which are listed bellow.

[3]Unfortunately we have vertical realistic distorted orbit only during simulations therefore term $\frac{\langle s\psi\Delta x^2\rangle}{2<k\psi>}$ does not present in the table. Nevertheless it can be estimated from comparision of the $\langle\Delta x^2\rangle$, $\langle\Delta y^2\rangle$ and result for $\frac{\langle s\psi\Delta y^2\rangle}{2<k\psi>}$.

SPINLIE for $\overline{\Delta\nu/\nu}$ for colliders with different electron energies are presented.

	VEPP-2M	VEPP-4M	HERA-e	LEP
ν	1.5	11	60.5	101.5
Δy_{max}, mm	1.7	1.2	3.2	-
Δy_{rms}, mm	0.5	0.8	0.8	-
$\epsilon_x \frac{<s\psi\beta_x>}{<k\psi>}$	$0.28 \cdot 10^{-4}$	$0.10 \cdot 10^{-6}$	$0.30 \cdot 10^{-5}$	$0.18 \cdot 10^{-5}$
$\left(\frac{\Delta E}{E}\right)^2 \frac{<s\psi^3>}{2<k\psi>}$	$-.54 \cdot 10^{-6}$	$0.27 \cdot 10^{-7}$	$0.15 \cdot 10^{-5}$	$0.8 \cdot 10^{-6}$
$\frac{<s\psi\Delta y^2>}{2<k\psi>}$	$-.14 \cdot 10^{-4}$	$0.14 \cdot 10^{-6}$	$-.19 \cdot 10^{-5}$	-

Table 1: $\overline{\Delta\nu/\nu}$ simulation for different colliders.

Besides, the usage of vertical bumps for the harmonic compensation of the spin-orbit coupling can produce local distorsion of the vertical orbit. 8-closed orbit bumps are used for HERA-e, as an example. The orbit was distorted inside eight regular arcs with sextupoles on this basis (the total number of arcs is 192) and inside these arcs $\Delta y \approx 1$ cm. Under these conditions the spin tune from these bumps will be six times as much as the shift which is connected with the rest of the sextupoles (the results presented in the table 1 were received without taking into consideration these bumps).

Spin chromaticity. Let us describe the usage of the Lie operators method for calculation of the level of the equilibrium polarization $\mathcal{P}$ which equals [4]:

$$\mathcal{P} = \frac{8}{5\sqrt{3}} \frac{<|\, k\,|^3 \,(n_y - \gamma\frac{\partial n_y}{\partial\gamma})>}{\left\langle \,|\, k\,|^3 \left[1 - \frac{2}{9}n_s^2 + \frac{11}{18}\left(\gamma\frac{\partial\vec{n}}{\partial\gamma}\right)^2\right]\right\rangle} \approx$$

$$\approx \frac{8}{5\sqrt{3}} \frac{<|\, k\,|^3>}{\left\langle \,|\, k\,|^3 \left[1 + \frac{11}{18}\left(\gamma\frac{\partial\vec{n}}{\partial\gamma}\right)^2\right]\right\rangle}. \tag{7}$$

Here $\vec{n}$ and $\gamma\frac{\partial\vec{n}}{\partial\gamma}$ are periodical spin and spin-orbit coupling vectors for nonequilibrium particle. Let us write an equation for periodical solution $\vec{n}$ [2] (we'll take into account only the first term in the expansion of $\vec{n}$ over powers of $\vec{\mathcal{W}}^{(r)}$):

$$\vec{n}(\theta) = \vec{n}^{(0)}(\theta) + \left[\widetilde{\vec{\mathcal{W}}^{(r)}}(\theta), \vec{n}^{(0)}(\theta)\right] + \cdots$$

where:

$$\widetilde{\vec{\mathcal{W}}^{(r)}}(\theta) = \lim_{N\to\infty} \sum_{n=0}^{N} \vec{\mathcal{W}}^{(r)}(\theta + 2\pi(n-1), \theta + 2\pi n).$$

But

$$\vec{\mathcal{W}}^{(r)}(\theta + 2\pi(n-1), \theta + 2\pi n) = \int_{\theta+2\pi(n-1)}^{\theta+2\pi n} d\theta' \Big(S^{(0)}(\theta + 2\pi(n-1), \theta')\Big)^{-1} \vec{\omega}(\theta')$$

and after using the spectrum expansion (4) and taking into account the periodicity of the phase ξ ($\xi(\theta + 2\pi) = \xi(\theta)$) one can find the following result:

$$\widetilde{\vec{\mathcal{W}}^{(r)}}(\theta) = \cdots = -\frac{e^{i\xi(\theta)}}{2\pi i} \sum_k \frac{\vec{w}_k e^{-i(\nu_0 - \nu_k)\theta}}{\nu_0 - \nu_k}.$$

Now it is possible to find the spin chromaticity vector $\gamma \frac{\partial \vec{n}}{\partial \gamma}$ as standard derivative of $\vec{n}(\theta)$ with respect to dimensionless energy γ [4]:

$$\gamma \frac{\partial \vec{n}}{\partial \gamma} = \nu_0 \frac{\partial \vec{n}}{\partial \nu_0} = \nu_0 \left[\frac{\partial \widetilde{\vec{\mathcal{W}}^{(r)}}}{\partial \nu_0} \right] \approx \frac{e^{i\xi(\theta)}}{2\pi i} \nu_0 \sum_k \frac{\vec{w}_k e^{-i(\nu_0 - \nu_k)\theta}}{(\nu_0 - \nu_k)^2}$$

and neglecting the interaction between overlaping resonances we finally find the result from [4]:

$$\left(\gamma \frac{\partial \vec{n}}{\partial \gamma}\right)^2 = \frac{\nu_0^2}{4\pi^2} \Big[\sum_k \frac{|\vec{w}_k|^2}{(\nu_0 - \nu_k)^4} + \sum_{k,k' \neq k} \frac{\vec{w}_k \vec{w}^*_{k'}}{(\nu_0 - \nu_k)^2 (\nu_0 - \nu_{k'})^2} \Big] \approx$$
$$\approx \frac{\nu_0^2}{4\pi^2} \sum_k \frac{|\vec{w}_k|^2}{(\nu_0 - \nu_k)^4}. \quad (8)$$

It is necessary to note that the same comment [5] as for formula (5) is valid in this case.

[4]It is necessary to take into account energy dependences of the average spin precession frequency as well as the orbit modulation (dependence of the strength charmonics $\vec{w}_k$ on energy). These dependences produce different powers of $\nu_0 - \nu_k$ in the denomiator. Therefore we'll preserve the high power only.

[5]It implies the averaging of the value $|\vec{w}_k|^2$ over the beam distribution when the implicit dependence $\vec{\omega}$ on orbital vector $\vec{Z}$ is taken into account.

It is interesting to make a connection between values of spin tune shift and level of the equilibrium polarization for any isolated spin resonance. It is possible due to formulas (7), (8). As an example, near integer spin resonance $\nu_0 \approx k$ after simple transformation one can find the following result:

$$\frac{\overline{\Delta\nu}}{\nu} = \frac{9}{11} \frac{\frac{8}{5\sqrt{3}} - \mathcal{P}}{\mathcal{P}} \frac{(k-\nu_0)^2}{\nu_0^3} \sin 2\pi\nu_0. \tag{9}$$

Unfortunately we only have an experimental result for the evaluation of $\overline{\Delta\nu}$, which was obtained for the collider VEPP-4 [7]. It is as follows: for $E \approx$ 5 Gev ($\nu_0 \approx 11$; it is an area near Υ-mesons family) and level polarization more then 20 % the value of the $\frac{\overline{\Delta\nu}}{\nu} \approx 2 \cdot 10^{-6}$. There is a good agreement with formula (9).

References

[1] Yu.I.Eidelman, V.Ye.Yakimenko, *The Application of Lie Method to the Spin Motion in Nonlinear Collider Fields,* Part.Accel., (1994).

[2] D.P.Barber, Yu.I.Eidelman, V.Ye.Yakimenko, *Calculating Electron Spin Polarization in Storage Rings Using Lie Algebra,* will be published.

[3] Yu.I.Eidelman, V.Ye.Yakimenko, *Calculation of the Lie Operators for Beam Transport Elements.* Preprint CERN SL/93-52 (AP), 1993.

[4] Ya.S.Derbenev, A.M.Kondratenko, A.N.Skrinsky, *Radiative Polarization at Ultra-High Energies,* Part.Accel., **9,** 247, (1979).

[5] Yu.I.Eidelman, Yu.M.Shatunov, V.Ye.Yakimenko, *Spin Tune Shifts in a Storage Rings,* submitted to NIM.

[6] A.P.Lysenko, A.A.Polunin, Yu.M.Shatunov, *Spin Frequency Spread Measurements in a Storage Ring,* Part.Accel., **18,** 215, (1986).

[7] A.N.Skrinsky, Yu.M.Shatunov, *Precision Measurements of Masses Elementary Particles Using Storage Rings with Polarized Beams,* Sov.Phys.Usp. **32** (6), June, 1989 (Usp. Fiz. Nauk **158** , 315, 1989).

A Short Spin Rotator for LEP

H. Grote

CERN, CH-1211 Geneva 23, Switzerland

Abstract. A short spin rotator for LEP compatible with the LEP-200 optics is presented. The normally vertical spin vector is rotated by 90 degrees to become horizontal at the intersection point, and is then rotated back to vertical. The spin rotator is spin matched (i.e. the polarization is not depressed by the rotator) at the Z_0 energy of 45.6 GeV. The rotator presented here is matched to intersection point five of LEP. The maximum vertical displacement of the beams (up on one side, down on the other side) is ± 0.48 m.

Introduction

The possibility of having longitudinally polarized beams in LEP has been studied in the past [5]. Since it seems impossible to maintain longitudinally polarized circulating beams in the machine continuously, all such investigations have been based on spin rotators in the straight sections. In this case, the circulating beams are polarized vertically throughout, except in the place where the spin rotator turns the spin axis horizontally, and back again to vertical. Such a rotator has to be spin-matched [1].

The calculations for this study have all been performed with the MAD program [2], [4]. They are based on the current LEP-200 optics for 1995. The constraints imposed on this design differ considerably from those of the previous proposal for a movable demonstration spin rotator (see [3]): the new rotator has to be "short", i.e. the stretch of beam pipe to be moved vertically should be as short as possible; it has to be "flat", i.e. the vertical displacement of the beam pipe has to be minimized; it is for "physics", i.e. the horizontal and vertical beam sizes at the interaction point have to be comparable to those in operation at the even intersections.

Layout

The layout of the Richter-Schwitters spin rotator around point five of LEP (IP5) can be seen in Figure 1 (only one half is shown, the interaction point is at s = 0). In total, 16 vertical bend magnets (length 5.775 m), and eight quadrupoles of type MQ (length 1.55 m) are needed in addition to the standard layout. The maximum vertical displacement of the orbit is ± 0.48 m, the orbit angle with the horizontal plane is -15.2 mrad at the intersection point.

Overall design constraints

In reality, there are three different optics layouts to be considered. All three have the elements of the insertion at the same horizontal positions; all three have to accomodate the beam dump on one side; all three have to limit their beta values to be compatible with the physical aperture of the vacuum chamber. Their main characteristics and constraints can be summarized as follows:

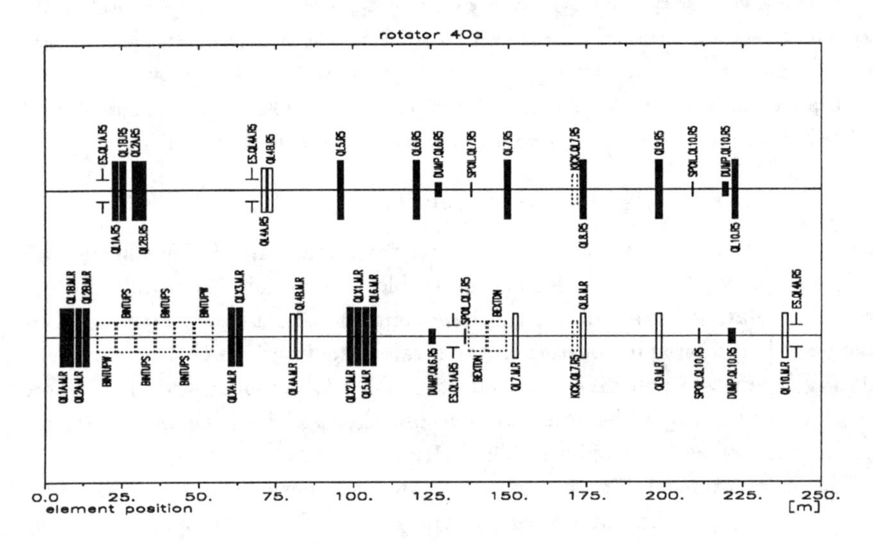

Figure 1: Layout of the standard insertion and the spin rotator(schematic top view). Dotted: vertical bend magnets; hollow: MQ quadrupoles, 1.60 m long; solid: MQA quadrupoles, 1.95 m long; "capacitors": vertical electrostatic separators

Optics A: for physics with longitudinally polarized beams at IP5. No other experiment running (beams separated at even IPs). The bend magnets are ON, the elements are displaced vertically. The constraints are: 1. fully matched for rotation, 2. tunes with high polarization at the Z_0 energy, 3. low beta values at the interaction point for high luminosity, 4. minimal increase of the vertical emittance, 5. low background at IP5.

Optics B: for injection to prepare physics with longitudinally polarized beams at IP5, and for tests with higher energy. The bend magnets are ON, the elements are displaced vertically. The constraints are: 1. good tunes for injection, 2. no reduction of the horizontal dynamic aperture, 3. vertical beam separation at IP5, 4. sufficient dipole strength for vertical deflection at higher energy, 5. minimal increase of the vertical emittance.

Optics C: for standard LEP operation. The bend magnets are OFF, the elements are NOT displaced vertically. The constraints are: 1. transparent (i.e. with respect to the original insertion, the same phase advances, and identical alpha and beta values at some insertion "border"), 2. no reduction of the horizontal dynamic aperture, 3. vertical beam separation at IP5, and at the three parasitic crossing points on either side of IP5 when operated with bunch trains.

Since optics B and C are necessary but not interesting, only the spin-rotation optics A will be discussed in more detail now.

Optics A

One starts with the standard LEP optics foreseen for 1995 with integer tunes 90 (horizontally) and 76 (vertically). When matching the optics for rotation, the vertical tune increases by about one unit. This tune of 77.2 is unfortunately bad for polarization at the required spin tune of 103.5 as can be seen from linear predictions. The neighbouring integer tune 78 is good, however, so finally the tunes $Q_x = 90.1$, $Q_y = 78.2$ were retained for polarization. For these "polarization tunes" the polarization in LEP was calculated with the help of MAD's module SITF/SODOM [6]. By carefully adjusting the betatron and synchrotron tunes in such a way that the ascending and descending higher order Q_s resonances coincide one can achieve a rather large area around 103.5 with high polarization. The curve in Figure 2 shows how the RS rotator lowers the uniform 92 % level of polarization of a machine without errors. The non-linear calculations with SODOM show the strong influence of the higher order synchrotron-tune resonances. The maximum polarization level (according to these calculations) occurs at the spin tune corresponding to the Z_0 energy. The next refinement of the predictions is then the addition of realistic errors on the positions of the vertical quadrupoles, and their subsequent correction with the MICADO algorithm to r.m.s. values of 5 cm for the vertical dispersion, and 0.3 mm for y. Finally, harmonic spin matching is applied: a Fourier analysis of the orbit as a function of the azimutal angle permits to calculate closed bumps that suppress simultaneously the two integer spin resonances at 103 and 104.

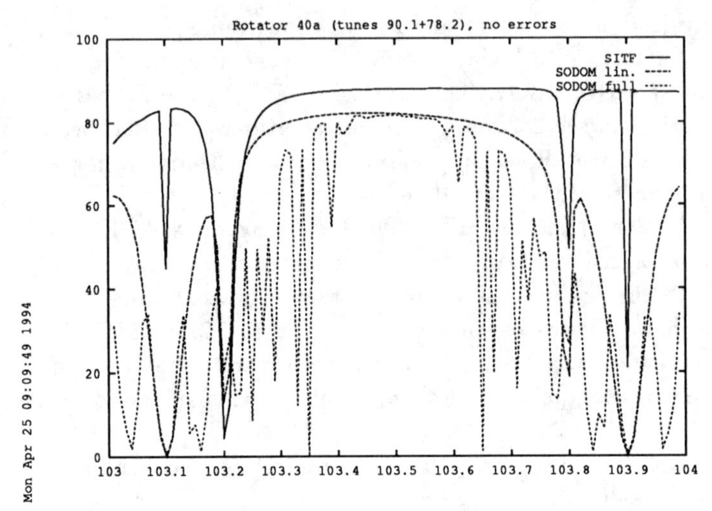

Figure 2: Prediction of polarization levels for the RS spin rotator, without errors on the vertical position of the quadrupoles.

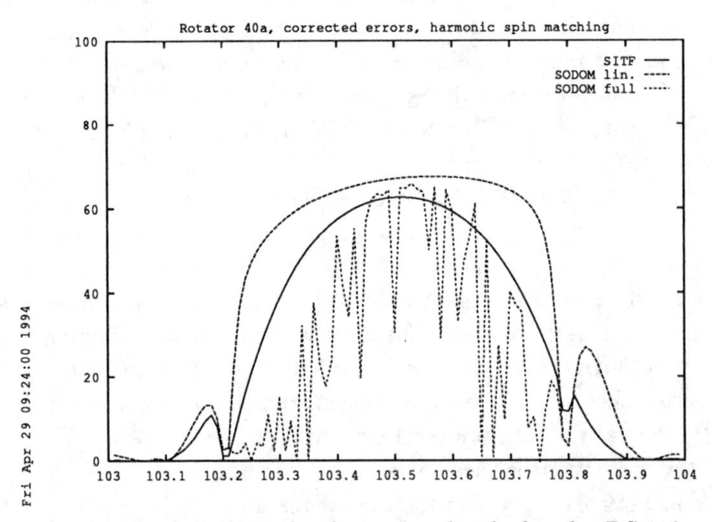

Figure 3: Prediction of polarization levels for the RS spin rotator, with corrected errors in the vertical position of the quadrupoles, and harmonic spin matching.

The resulting polarization - which therefore comprises quadrupole position errors, their correction, and harmonic spin matching - is shown in Figure 3. The polarization levels predicted here are compatible with the achievements in LEP after harmonic spin matching was applied (57% polarization). One may therefore say that one can expect a similar result with optics A. There is one caveat left, however: the simulations presented here do not comprise the beam-beam effect. From calculations with a linear beam-beam kick, and from some results with a non-linear beam-beam element one can estimate that the beam-beam effect could reduce the maximum polarization by one third, approximately.

Conclusion

The Richter-Schwitters spin rotator presented here is fully spin matched and would thus allow to rotate the spin axis from vertical to horizontal, and back again (almost) without loss of polarization. A modified optics for injection and separation at IP5 exists as well. It allows a maximum beam energy of 68.5 GeV. The identical layout of all elements in flat position is compatible with normal LEP operation, including bunch trains.

References

[1] Blondel, A., *Spin-matching conditions for the Richter-Schwitters spin rotator*, LEP-Note 603 (1988)

[2] Grote, H. and Iselin, F.C., *The MAD Program (Version 8.1) User's Reference Manual.* CERN/SL-AP/90-13 (1990)

[3] Grote, H., *A movable demonstration spin rotator for LEP*, CERN/SL-AP/91-48

[4] Grote, H., *MAD-SODOM User's Guide*, CERN/SL Note 93-40 (AP)

[5] Keil, E.and Koutchouk, J.P. (editors), *Longitudinal Polarization at LEP - A Feasibility Study -*, CERN/SL-AP/LEPC 91-11

[6] Yokoya, K., *Non-perturbative calculation of equilibrium polarization of stored electron beams*, KEK Report 92-6 (1992).

Radiation and Spin Separation of High Energy Positrons by Bent Crystal

Yu.A.Bashmakov.

P.N.Lebedev Physical Institute, Russian Academy of Science, Leninsky Prospect 53, 117924 Moscow, Russia

Abstract

The channeling of ultrahigh energy positrons both in bent and straight crystals can serve as an effective source of hard photons. The presence in the channel of the strong inhomogeneous electrical field gives rise to a number of interesting peculiarities of the motion of the channeling particles. The radiative separation of particles depending on spin orientation with respect to the plane of motion and radiative polarization can make it possible to obtain by means of bent crystals a positron beam with a degree of polarization about 10 to 20%. Because of availability of the ultrahigh energy positron beams such experiments can be carried out nowadays.

INTRODUCTION

Radiation from low mass channeled particles are of a great interest, since it offers wide prospects for creation of intense quasimonochromatic electromagnetic radiation. The problem of obtaining polarized high energy particles that attracts much attention is closely connected with the problem of radiation of relativistic particles. Development of several hundred GeV energy electron and positron beams on the largest proton synchrotrons makes it impotant to obtain polarized ultrahigh energy electrons and positrons. Propagation of charged particles in bent crystals was studied in [1], where it was suggested that bent crystals could be used for deflection of charged particles [2]. Crystal septa are used now for beam extraction from proton synchrotrons [3]. Possibilities of the control of a charged particles spin by means of bent crystals are discussed in [4,5]. The first experimental observation of magnetic momentum preccesion of channeled particles was carried out [6].

DYNAMICS OF POSITIVELY CHARGED PARTICLES UNDERGOING PLANAR CHANNELING IN A BENT CRYSTAL

In a first approximation the planar channel potential has the form

$$U(x) = 4U_0(x^2/d^2), \tag{1}$$

where U_0 is the potential magnitude at the channel boundary, d is the distance between the crystallografic planes forming the channel, x is the displacement from the channel midplane. For a given bending radius R_0 particles at an energy, which satisfies an inequality

$$\beta\varepsilon < \beta_c\varepsilon_c = (4eU_0R_0)/d, \tag{2}$$

where $\beta = v/c$, v is particle velosity, will be captured into the channel. For a particle having energy $\varepsilon \leq \varepsilon_c$ the transverse electric field E_0 on the equlibrium orbit and the displacement x_0 of this orbit of the channel midplane are

$$E_0 = \beta\varepsilon/eR_0, \; x_0 = d\varepsilon/2\varepsilon_c. \tag{3}$$

We define the equilibrium orbit as that orbit for wich the separation from the channel walls is everywhere constant. Introduce the elecrical field index in the channel as $n = (d\ln E/d\ln R)_{x=x_0} = (2R_0/d)\varepsilon_c/\varepsilon$, where practically always $n = R_0/x_0 \gg 1$. Inhomogeneity of the electrical field in the channel gives rise to particle betatron oscillations about the equilibrium orbit [7]

$$\eta = x - x_0 = a\cos(\sqrt{n+1}(l/R_0) + \phi), \tag{4}$$

where $l = \beta ct$ is the particle path length along the equilibrium orbit. Since $n \gg 1$ the period of particle oscillation does not depend on the bend radius and can be written as follows:

$$\lambda_0 = \pi d\sqrt{\varepsilon/2eU_0}. \tag{5}$$

The boundary of the stable radial motion of particles in the channel is determined by the maximum possible amplitude of betatron oscillations $a_m = d/2 - x_0$. Correspondingly the maximum angle $x'_m = (dx/dl)_m$ formed by the trajectory with the equilibrium orbit is

$$x'_m = \frac{d/2 - x_0}{R_0}\sqrt{n} = \frac{d/2 - x_0}{d/2}\theta_{ch}, \tag{6}$$

where

$$\theta_{ch} = \sqrt{\frac{2eU_0}{mc^2\gamma}} \tag{7}$$

is the critical channeling angle [8]. If the angle of a particle exceeds this value, such a particle is not captured into the channel. The patern of the radial motion can be clearly depicted in the phase-space $x, (R_0/\sqrt{n})x'$ [7]. The distance δR between the orbits corresponding to the particles of energy ε and $\varepsilon + \delta\varepsilon$ is

$$\delta R/R_0 = 1/n(\delta\varepsilon/\varepsilon). \tag{8}$$

The difference between the real and oscillatory potentials (1) results in a dependence of frequency on the amplitude for the betatron oscillation (the nonlinearity of oscillations).

RADIATION OF A PARTICLE PLANAR CHANNELING IN A BENT CRYSTALS

The characteristic of the radiation accompanyng particle channeling in bent crystals depend on the relation between the particle oscillation period λ_0 and R_0/γ, which is the length in which the radiation is formed when a particle is moving along a circle of R_0 ($\gamma = \varepsilon/mc^2 >> 1$) [7]. If $\lambda_0 \ll R_0/\gamma$ or according to (5)

$$\gamma \ll \tilde{\gamma} = (\frac{2eU_0R_0^2}{\pi^2 d^2 mc^2})^{1/3} \tag{9}$$

the radiation is quasi-undulatory [9]. The effective number of periodicity elements is $N_{eff} = R_0/\gamma\lambda_0$. If $\lambda_0 \gg R_0/\gamma$ ($\gamma \gg \tilde{\gamma}$) the radiation spectrum resembels that of synchrotron radiation with the maximum at the characteristic frequency

$$\omega_m \simeq \frac{c}{d}\frac{6eU_0}{mc^2}\gamma^2. \tag{10}$$

Relation (10) is true if the energy of the emitted photons is much less than the positron energy. That is if

$$\chi = \frac{E_c\gamma}{H_0} << 1, \tag{11}$$

where $H_o = m^2c^3/e\hbar = 4.41 \cdot 10^{13} Oe$ ($E_0 = 1.32 \cdot 10^{16} V/cm$), $E_c = (4U_0)/d$ is the maximum electric field in the channel. In silicon for 250 GeV positrons $\chi = 0.25$ and quantum mechanical corrections to synchrotron-radiation theory become significant. In oder to determine photon energy one must use instead of (10) the following relation [10]

$$\hbar\omega_m = (\varepsilon\chi)/(\frac{2}{3} + \chi). \tag{12}$$

The full radiation intensity I of a relativistic particle moving through a transverse electrical field is proportional to the mean square value of the electrical field E along the particle trajectory. If the particle accomplishes a betatron oscillation with an amplitude a relative to the equlibrium orbit one has

$$I = \frac{2e^4}{3m^2c^3}(\frac{8U_0}{d})^2\frac{{x_0}^2 + \frac{1}{2}a^2}{d^2}\gamma^2. \tag{13}$$

The radiation is accompained by a particle energy loss. For particles of small betatron oscillation amplitudes ($a^2 \ll {x_0}^2$) the law of the change in energy is

$$\varepsilon = \varepsilon_0/(3z_1 + 1)^{1/3}, \tag{14}$$

where $z_1 = I_{01}l/(\varepsilon_0 c)$; $I_{01} = 2e^4/3m^2c^3(8U_0/d)^2(dmc^2/2\varepsilon_c)^2\gamma_0^4$; $\varepsilon_0(\gamma_0)$ is the initial particle energy. For particles at large oscillation amplitudes $a \leq d/2$ one has

$$\varepsilon = \varepsilon_0/(z_2 + 1), \tag{15}$$

where $z_2 = I_{02}l/(\varepsilon_0 c)$; $I_{02} = 2e^4/3m^2c^3(8U_0/d)^2\gamma_0^2/8$; By reason of large n value the radiation friction and the adiabatic decrease of the energy due to the radiation cause antidamping of the oscillation.

SPIN EFFECTS IN PARTICLE CHANNELING IN CRYSTALS

The quantum correction to synchrotron radiation theory are determined by a parameter χ. The important manifeistation of the quantum effects is the dependence of the radiation intensity of the spin partice orientation relative to the plane of motion [11]. This effect can be used for spin separation of high energy positrons by transmission through a bent crystals [7].

The relation for the mean energy of particles with different spin orientation, having passed through a homogeneous magnetic field H is [12]

$$< \varepsilon >= \frac{\varepsilon_0}{1+z}[1 + \frac{\chi_0}{1+z}(\frac{3}{2}\zeta L + 4cL + \frac{2cz}{1+z})], \tag{16}$$

where $z = I_0 l/(\varepsilon_0 c)$; $I_0 = 2/3(e^4/m^2c^3)H^2\gamma_0^2/8$; $\chi_0 = (H/H_0)\gamma_0$; $\zeta = \pm 1$ is a spin variable, $c = 55/32\sqrt{3}$; $L = \ln(1+z)$. From (16) one has a difference in mean energy of particles with different spin orientation

$$< \varepsilon >_{\zeta=+1} - < \varepsilon >_{\zeta=-1} = \frac{3\varepsilon_0\chi_0 L}{(1+z)^2}. \tag{17}$$

The difference between $< \varepsilon >_{\zeta=+1}$ and $< \varepsilon >_{\zeta=-1}$ results in a predominance of particles at a certain spin sign.

In an inhomogeneous crystal field the process is more complicated. The results given can be used if value z_1 is such that difference between $z_1 + 1$ and $(3z_1 + 1)^{1/3}$ is small. In this case one can use the value of the electrical field on an equilibrium orbit for the magnitude of H.

Quantum effectts in synchrotron radiation give rise to the radiative polarization of lepton beams. In [4] attention was paid to the possibility of radiative self-polarization of relativistic particle in bent crystals. The electrical field on the particle trajectory during one betatron oscillation is essentially changed for particles with large amplitudes ($a \sim d/2 - x_0$). The inverse polarization time is [14]

$$T^{-1} \sim \gamma^2 \frac{1}{\lambda_0}\int_0^{\lambda_0} E_x^3 dl = \gamma^2 < |E_x|^3 >, \tag{18}$$

here the integral is taken along the particle trajectory. The degree of polarization is [14]

$$\zeta = \frac{8\int_0^{\lambda_0} E_x^3 dl}{5\sqrt{3}\int_0^{\lambda_0} |E_x|^3 dl}. \tag{19}$$

Equation (19) can also be written as

$$\zeta = \frac{9\pi}{5\sqrt{3}} \frac{x_0}{a}. \tag{20}$$

The inverse polarization time is determined by

$$T^{-1} = \frac{4 \cdot 8^2 \cdot 5}{\pi\sqrt{3}} \alpha \frac{\hbar^2}{m^5} \frac{(eU_0)^3}{d^6} a^3 \gamma^2, \tag{21}$$

where it has been taken into account that $< |E_x|^3 >= \frac{4}{3\pi}(\frac{8U_0}{d^2})^3 a^3$.

Let us consider, for example, the radiation of a positron of energy $\varepsilon = 250GeV (\gamma = 5 \cdot 10^5)$ in the channel (110) of a 2 cm-thick silicon crystal bent at a radius $R_0 = 200cm$. The characteristic parameters in this case are $\varepsilon_c = 1250GeV$, $x_0 = 0.2(d/2)$ (where $x_0^2 \ll (d/2)^2$), $x'_m = 1.2 \cdot 10^{-5}$, excceds the angle $1/\gamma$ by a factor of six. If the angular spread of the beam is $x'_0 \simeq x'_m$ efficiency of capture into the channel is $\kappa = 60\%$. The critical radiation frequences are: for particles with small oscillation amplitudes $\hbar\omega_m = 9GeV$; for particles with large amplitudes $\hbar\omega_m = 45GeV$. Particles with small oscillation amplitudes lose, due to radiation, about half ($z_1 \simeq 1$) of their energy and thereafter they can be separated on spin by means of a bending magnet. For approximately half of the particles the degree of polarization can reach about 20%. The energy of the particles at large oscillation amplitudes ($\chi = 5\chi_0 = 0.22$) decreases more than ten times ($z_2 = 15$) and the degree of polarization is $\zeta = 0.89$).

REFERENCES

1. Tsyganov, E.N., Fermilab, August 1976.
2. Vodopjanov, A.S., et al. *Pis'ma Zh. Eksp. Teor. Fiz.* **30**, pp. 474-477 (1979).
3. Akbari,H., et al., *Phys. Lett.* **B313** pp. 491-497 (1993).
4. Barishevsky, V.G., and Grubich, A.O., *Pis'ma Zh. Tekh. Fiz.* **5**, pp. 1527-1530 (1979).
5. Lyuboshitz, V.L., Preprint P2-12559, Dubna,1979.
6. Chen, D., et al. FERMILAB-Pub-92242 E761, 1992.
7. Bashmakov, Yu.A., *Rad. Eff.* **56**, pp. 55-60, (1981).
8. Lindhard, J., Danske Vld. Selsk. *Mat-fys. Medd.* **34** pp. 14-48 (1965).
9. Alferov, D.F., Bashmakov, Yu.A., and Bessonov, E.G., *Tr. Fiz. Inst. Akad. Nauk* **80**, pp. 100-125 (1975).
10. Berestesky, V.B., Lifchits, E.M., and Pitaevsky, L.P., *Reljativistskaja kvantovaja teorija.* P.I.M. "Nauka", 1968.
11. Sokolov, A.A., and Ternov, I.M., *Dokl.Akad. Nauk SSSR* **153**, pp. 1053-1056 (1963).
12. Baier, V.M., Katkov,V.M., and Strakhovenko,V.M., *Phys. Lett.* **70B**, pp. 83- 88 (1977).
13. Baier, V.M., Katkov,V.M., and Strakhovenko,V.M., *Zh. Eksp. Teor. Fiz.* **66**, pp. 81-92 (1974).
14. Derbenev, Ya.S., Kondratenko, A.M., and Skrinsky, A.N., *Particle Accelerators* **9**, pp. 247-153 (1979).

Spin Flip by RF-Field at Storage Rings with Siberian Snakes

I. Koop, Yu. Shatunov

Budker Institute for Nuclear Physics, Novosibirsk, 630090, Russia

INTRODUCTION

A measurement of spin depending effects demands as rule to reverse periodically polarizations of beams and targets. The beam polarization in experiments at linear accelerators has been controlled usually in a beam source. Adiabatic spin flip by RF-field [1] is a convinient way to change the beam polarization sign in measurements at storage rings when the beam injection is relatively seldom. This procedure does not touch practically other beam parameters. A moderate power RF- field (transverse to polarization direction) provides relatively fast flipping with small depolarization.

However in case when the spin tune is equal half integer number (a strorage rings with Siberian snake and a conventional machine on half magic energies) there are some features which will be discussed in this paper.

ADIABATIC SPIN FLIP

Let us remind an usual approach to consideration of the spin motion at storage rings [2]and its application to the spin flip.

In arbitrary machine field configuration we can introduce a travelling frame $\vec{e}_1, \vec{e}_2, \vec{n}$ so that the BMT-equation[3] for a spin vector $\vec{S}$ is simplified to:

$$\frac{d\vec{S}}{d\theta} = [\vec{W}\vec{S}], \tag{1}$$

where $\vec{W} = \nu\vec{n} + \vec{w}$ is a spin precession frequency which consist from regular rotations around fixed $\vec{n}$-axis with a frequency ν (in revolution frequency units) and small perturbation $\vec{w}$.

Considering RF-field with the frequency ν_f as the perturbation we can write for example for the longitudinal field $\vec{H}_{||} = H_{||} \cdot \vec{e_1} \cdot \cos(\nu_f\theta)$

which is excited on a short orbit section Θ_0:

$$\vec{\omega} = \frac{H_{\|} \cdot \Theta_0}{4\pi \cdot < H >}\{(e^{i\nu_f\theta} + e^{-i\nu_f\theta}) \cdot \sum_{k=-\infty}^{\infty} e^{ik\theta} \cdot \vec{e}\}, \quad (2)$$

where $< H >$ is average guiding field and $\vec{e} = \vec{e_1} + i\vec{e_2}$.

An influence of circular harmonics $k \pm \nu_f$ on the spin motion is negligible due to a weakness of the perturbation except the "resonant" harmonic ω_k for that $\epsilon = \nu - (k + \nu_f) \ll 1$. In the resonant frame that rotates around $\vec{n}$ with the frequency $k + \nu_f$ the vector $\vec{\omega_{\perp k}} = const$ and the spin motion is a slow precession around the axis $\vec{h} = \epsilon\vec{n} + \vec{\omega_{\perp k}}$ with the angular velocity $h = \sqrt{\epsilon^2 + \omega_{\perp k}^2}$. A projection $S_h = \frac{\vec{S} \cdot \vec{h}}{h} = const$ when the detuning ϵ is changed adiabaticaly [4]

$$|\dot{\epsilon}| \ll \omega_k^2 \quad (3)$$

and we can reverse the spin while the detuning goes from ∞ to $-\infty$ or *controversa*.

HALF INTEGER SPIN TUNE

In the case if the spin tune is equal to $m + \frac{1}{2}$ (m is *integer*) the spin flip by one directional RF-field is impossible because besides the resonant harmonic $k + \nu_f$ its "mirrow" partner $k + 1 - \nu_f$ contributes also essentialy to the spin motion. This harmonic rotates in the same direction and has the amplitude $\omega_{k+1} = \omega_k = \frac{H_{\|} \cdot \Theta_0}{4\pi \cdot <H>}$. So in the resonant frame both harmonics give the perturbation that oscilates with the difference frequency 2ϵ and the adiabaticity condition (3) will be violated in vicinity of the resonance.

From the first look a generation of the rotating field $\vec{H} = H_0 \cdot Re(e^{i\nu_f\theta} \cdot \vec{e^*})$ gives only harmonic $k + \nu_f$ and could be applied for the spin flip

However in this case at least one from the field components is transverse to the particle velocity. The transverse RF-field disturbs not only the spin precession but also excites particle oscillations which gives additional spin rotations while the particle travels through machine quadrupole magnets. An enhancment of the effective perturbation that can be considerable is described so called a spin response function [5] which strongly depends on machine optical parameters. An uncertainty

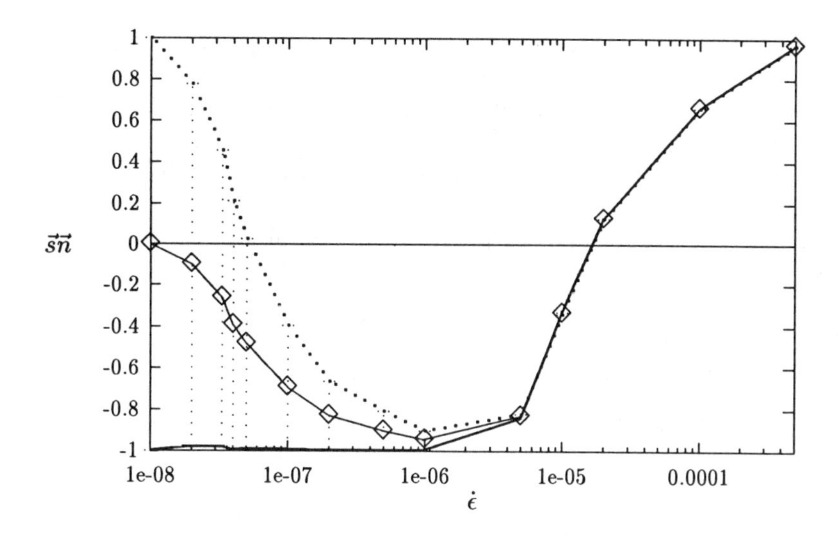

Figure 1: Resudial polarization after the resonance crossing

of the real machine optics leads to a deviation of the effective RF-field polarization from the circular one.

An influence of some elipticity of the field polarization was estimated numerically for the spin tune $\nu = \frac{1}{2}$. The Fig.1 shows the results of the spin flip simulation for different rates of the resonance crossing in the detuning range $\epsilon_0 = \pm 10^{-2}$ with the resonant circular harmonic amplitude $\omega_k = 10^{-3}$ (for real spin flip is enough $\omega_k = 10^{-4}$ and $\epsilon_0 = \pm 10^{-3}$) and 10 percent admixture of the linear field $\Delta\vec{\omega}_k \perp \vec{\omega_k}$. It is easy to see that the linear polarization presence leads (additional to (3)) to the condition:

$$(\Delta\omega_k)^2 \ll \dot{\epsilon} \ll \omega_k^2. \tag{4}$$

The resudial polarization is unpredictable (inside an interval between $S_n = -1$ and dashed line) when the crossing rate becomes too slow. This restriction comlicates actually the flipper operation and demands from time to time some calibration of the effective RF-field components. But only this way is open for the half magic energy points.

Outsides a narrow bands around these points there is a possilility for succesful spin flip by the linear RF-field. A snake mismatching from the perfect scheme can give the spin tune shift $\Delta\nu \simeq 2 \div 3 \cdot 10^{-3}$. It corresponds to snake magnet field corrections by few percents and is enough to suppress the mirrow harmonic action. On the other hand such mismatching deviates the $\vec{n}$-axis not too much and can be done only for a short flipping time.

We believe this approach provides the spin flip by the 1D RF-field at machines with Siberian snakes.

References

[1] A.A.Polunin, Yu.M.Shatunov, *Preprint INP 82-16*, Novosibirsk, (1982).

[2] Ya.S.Derbenev, A.M.Kondratenko, A.N.Skrinsky, *Soviet* JETP60 (1971) 1216.

[3] V.Bargman, L.Mishel, V.Telegdi, Phys.Rev.Lett.2 (1959) 435.

[4] M.Froisart, R.Stora, NIM, 7 (1960) 297.

[5] Ya.S.Derbenev, A.M.Kondratenko, A.N.Skrinsky, *Preprint INP 77-60*, Novosibirsk, (1977).

Differential Algebraic Description and Analysis of Spin Dynamics

Martin Berz
Department of Physics and Astronomy and
National Superconducting Cyclotron Laboratory
Michigan State University
East Lansing, MI 48824

Abstract

We present a method to compute spin maps to arbitrary order and for arbitrary field arrangements in the framework of a differential algebraic algorithm. Because of the linearity of the problem, it is possible to reduce the dimensionality to that of the orbit motion both for piecewise constant fields and for continuous fields.

It is shown how the resulting spin map can be used for tracking and how spin amplitude tune shifts can be calculated. Furthermore, we present an algorithm similar to the orbit normal form algorithm that allows the computation of the equilibrium polarization direction.

1 Introduction

The transport and acceleration of polarized particle beams requires a detailed knowledge of the nonlinear motion of the spin vector. Moreover, the motion has to be analyzed to determine quantities of interest, including the spin tune shifts and the equilibrium polarization direction $\bar{n}$. One way to study the motion is by tracking of spin orbits, which is employed in the codes SITROS [1] and SODOM [2] Tracking methods can also be used to estimate spin tune shifts and the equilibrium polarization direction.

Besides the mere tracking of spins, it is often beneficial to describe the motion of the spin in terms of a map. In a linear representation, this was first done in the program SLIM [3], and the treatment of nonlinear effects was first pursued in the code SMILE [4]. Lie Algebraic approaches were proposed by Yokoya [5] and in the code SPINLIE [6], but so far in practice it has not been possible to raise the orders beyond three. The first arbitrary order method to compute spin maps was implemented in the code VASILIE [7]; it employs DA and Lie methods to determine maps for systems that have piecewise constant fields.

In recent years, DA methods have proven useful for the computation of maps of orbit motion for arbitrary order and arbitrary arrangements of fields, including fringe fields, wigglers, and general measured fields [8, 9]. In the following, we present a method that allows the computation of spin maps for general fields to arbitrary order. We also show how the spin tune shifts and the equilibrium polarization can be computed in a DA framework.

2 The Dynamics of the Spin Matrix

The equation for the classical spin vector $\vec{s}$ of a particle in the electromagnetic field, the BMT equation, has the form [10]:

$$\frac{d\vec{s}}{dt} = \vec{w} \times \vec{s} \tag{1}$$

where

$$\vec{w} = -\frac{e}{m_0 \gamma c}\{(1+\gamma G)\vec{B} - \frac{G}{(1+\gamma)}\frac{1}{m_0^2 c^2}(\vec{p}\cdot\vec{B})\vec{p} - (G + \frac{1}{1+\gamma})\frac{1}{m_0 c}\vec{p}\times\vec{E}\}. \tag{2}$$

Here $\vec{E}$ and $\vec{B}$ are the fields, e and m_0 are the charge and the rest mass of the particle, γ is the Lorentz factor, $\vec{p}$ is the kinetic momentum of the particle, and t is the time, which plays the role of independent variable. The code COSY and others describe the motion in terms of the natural coordinates (x, y, s) in a curvilinear coordinate system with the arc length s as independent variable. In these coordinates, the spin motion equation (2) takes the form

$$\frac{d\vec{s}}{ds} = \vec{w}_s \times \vec{s}, \tag{3}$$

where $\vec{w}_s$ also accounts for both the change of variables and additional forces in the accelerated frame of reference, and the cross product form of the equations of motion is retained.

In order to compute the map of the system, we first observe that the equations of motion are linear in the spin, and thus the final transformation of the spin variables can be described in terms of a matrix that depends on the orbital quantities only. The orbital quantities themselves are unaffected by the spin motion, such that altogether the map has the form

$$\begin{pmatrix} \vec{x}_f \\ \vec{s}_f \end{pmatrix} = \begin{pmatrix} \vec{\mathcal{M}}(\vec{x}_i) \\ A(\vec{x}_i)\cdot\vec{s}_i \end{pmatrix}. \tag{4}$$

Because of the cross product form of the spin motion, the resulting matrix describes an orbit-dependent rotation in $SO(3)$, i.e. $A(\vec{x})$ satisfies $A^t(\vec{x}) \cdot A(\vec{x}) = I$ and $\det(A(\vec{x})) = 1$.

The practical computation of the spin-orbit map can be achieved in a variety of ways. Conceptually the simplest way is to interpret it as a motion in the nine variables consisting of orbit and spin described by the orbit equations (see for example [11]) as well as the spin equations (3). In this case, the DA method allows the computation of the spin-orbit map in the two conventional ways, namely via a propagation operator for the case of the s-independent fields like main fields, and via integration of the equations of motion with DA [12][9]. However, in this simplest method, the number of independent variables increases from six to nine, which particularly in higher orders entails a rather substantial increase of computational and storage requirements. This severely limits the ability to perform analysis and computation of spin motion to high orders.

Due to the special structure of the equations of motion, it is possible to rephrase the dynamics such that it is still described in terms of only the six orbital variables. For this purpose, we derive the equation of motion for the individual elements of the matrix $A(\vec{x})$. To this end, we write $\vec{s}_f = A(\vec{x})\cdot\vec{s}_i$ and insert this into the spin equation of motion (3). Comparing coefficients of $\vec{s}$, which only appears linearly, we find that the matrix $A(\vec{x})$ obeys the differential equation

$$A'(\vec{x}) = W(\vec{x}) \cdot A(\vec{x}), \tag{5}$$

where the matrix $W(\vec{x})$ is made from the vector $\vec{w}_s(\vec{x}) = (w_1, w_2, w_3)$ appearing in the particle optical BMT equation (3) via

$$W(\vec{x}) = \begin{pmatrix} 0 & -w_3 & w_2 \\ w_3 & 0 & -w_1 \\ -w_2 & w_1 & 0 \end{pmatrix}. \tag{6}$$

Integrating the equations of motion for the matrix $A(\vec{x})$ along with the orbital equations now allows the computation of the spin motion based on only six initial variables. Since orthogonal matrices have orthogonal columns and furthermore their determinant is unity and hence the orientation of a dreibein is preserved, it follows that the third column of the matrix $A(\vec{x}) = (A_1(\vec{x}), A_2(\vec{x}), A_3(\vec{x}))$ can be uniquely calculated via

$$A_3(\vec{x}) = A_1(\vec{x}) \times A_2(\vec{x}). \tag{7}$$

Thus altogether, only six additional differential equations without new independent variables are needed for the description of the dynamics. For the case of integrative solution of the equations of motion, which is necessary in the case of s-dependent elements, these equations can just be integrated in DA with any numerical integrator.

However, for the case of main fields, the explicit avoidance of the spin variables in the above way is not possible, because for reasons of computational expense, it is desirable to phrase the problem in terms of a propagator operator

$$\begin{pmatrix} \vec{x}_f \\ \vec{S}_f \end{pmatrix} = \exp(\Delta s \cdot L_{\vec{F}}) \begin{pmatrix} \vec{x} \\ \vec{S} \end{pmatrix}. \tag{8}$$

Here $L_{\vec{F}} = \vec{F} \cdot \vec{\nabla}$ is the nine dimensional vector field belonging to the spin-orbit motion. In this case, the differential vector field $L_{\vec{F}}$ describes the whole motion including that of the spin, i.e. $d/ds(\vec{x}, \vec{S}) = \vec{F}(\vec{x}, \vec{S}) = (\vec{f}(x), \vec{w} \times \vec{S})$. In particular, the operator $L_{\vec{F}}$ contains differentiation with respect to the spin variables, which requires their presence. Therefore, the original propagator is not directly applicable for the case in which the spin variables are dropped, and has to be rephrased for the new choice of variables. For this purpose, we define two spaces of functions $g(\vec{x}, \vec{s})$ on spin-orbit phase space as follows:

X: Space of functions depending only on $\vec{x}$
S: Space of linear forms in $\vec{s}$ with coefficients in X

Then we have for $g \in Z$:

$$L_{\vec{F}}\, g = (\vec{f}^t \cdot \vec{\nabla}_{\vec{x}} + (\hat{W} \cdot \vec{s})^t \cdot \vec{\nabla}_{\vec{s}})\, g = \vec{f}^t \cdot \vec{\nabla}_{\vec{x}}\, g = L_{\vec{f}}\, g, \tag{9}$$

and in particular, the action of $L_{\vec{F}}$ can be computed without using the spin variables; furthermore, since $\vec{f}$ depends only on $\vec{x}$, we have $L_{\vec{F}}\, g \in Z$. Similarly, we have for $g = |\, a_1, a_2, a_3 >= \sum_j^3 a_j \cdot s_j \in S$:

$$\begin{aligned} L_{\vec{F}}\,|\, a_1, a_2, a_3 \;> &= (\vec{f}^t \cdot \vec{\nabla}_{\vec{x}} + (\hat{W} \cdot \vec{s})^t \cdot \vec{\nabla}_{\vec{s}})\, (\sum_j^3 a_j \cdot s_j) \\ &= \sum_j (\vec{f}^t \cdot \vec{\nabla}_{\vec{x}}) a_j \cdot s_j + \sum_{j,k} s_j W_{kj} a_k \\ &= |\, L_{\vec{f}} a_1 + \sum_k W_{k1} a_k, L_{\vec{f}}\, a_2 + \sum_k W_{k2} a_k,\; L_{\vec{f}},\; a_3 + \sum_k W_{k3} a_k \end{aligned} \tag{10}$$

and in particular, the action of $L_{\vec{F}}$ can be computed without using the spin variables; furthermore, $L_{\vec{F}}\ |\ a_1, a_2, a_3 > \in S$. Thus, X and S are invariant subspaces of the operator $L_{\vec{F}}$. Furthermore, the action of nine dimensional differential operator $L_{\vec{F}}$ on S is uniquely described by (10), expressing it in terms of the six dimensional differential operator $L_{\vec{f}}$. This now allows the computation of the action of the original propagator $\exp(\Delta s L_{\vec{F}})$ on the identity in R^9, the result of which actually describes the total nine dimensional map. For the upper six lines of the identity, note that the components are in Z, and hence the repeated application of $L_{\vec{F}}$ will stay in Z; for the lower three lines, of the identity map are in S, and hence the repeated application of $L_{\vec{F}}$ will stay in S, allowing the utilization of the invariant subspaces. Since elements in either space are characterized by just six dimensional functions, $\exp(s L_{\vec{F}})$ can be computed in a merely six dimensional differential algebra.

To conclude we note that one turn maps are often made up of small pieces of maps, and it is often necessary to compute the map describing the combination of maps. Let $(\vec{M}_{1,2}, \hat{A}_{1,2})$ and $(\vec{M}_{2,3}, \hat{A}_{2,3})$ be given; then the we get

$$\begin{aligned} \vec{M}_{1,3} &= \vec{M}_{2,3} \circ \vec{M}_{1,2} \\ \hat{A}_{1,3}(\vec{z}) &= \hat{A}_{2,3}(\vec{M}_{1,2}) \cdot \hat{A}_{1,2}(\vec{z}); \end{aligned} \tag{11}$$

note the necessity of inserting $\vec{M}_{1,2}$ into $\hat{A}_{2,3}$ before composition.

3 Spin Tracking

Once the map of the system is obtained, perhaps its most direct use is for the purpose of spin tracking. This is performed iteratively with the following steps; first it is necessary to evaluate the orbit map on the current coordinates $\vec{z}_n$ to get new coordinates $\vec{z}_{n+1}$ via

$$\vec{z}_{n+1} = \vec{M}(\vec{z}_n);$$

if needed or desired, symplectification has to be performed. Next, the new new orbit coordinates are inserted into the spin matrix to get

$$\hat{A}^* = \hat{A}(z_{n+1});$$

if needed or desired, orthogonalize $\hat{A}^*$, which requires renormalization of $\vec{s}_{n+1}$. Then, multiply current spin matrix $\hat{A}^*$ with current spin coordinates $\vec{s}_n$ to get new spin coordinates $\vec{s}_{n+1}$, and display output these coordinates as needed.

Altogether, it is worthwhile to observe that the satisfaction of the symmetry of orthogonality of the spin motion is substantially easier to impose in a well defined way than that of the corresponding symplectification of the orbit variables.

4 The Determination of Spin Tune Shifts

In order to determine the spin tune shifts with the orbital amplitude, it is first necessary to bring the orbital map into normal form, for example using the DA normal form algorithm [13, 9]. As the next step, the spin matrix $\hat{A}$ is expressed in terms of orbital normal form variables.

The spin tune is then obtained as the nontrivial eigenvalue of the three by three spin matrix, which can be solved analytically. At this point it is very important that the orbital variables appear only in the role of parameters, and so it is possible to execute the algorithm for the solution of the eigenvalue problem in DA to obtain the full nonlinear behavior of the tune. This is conceptually identical to the respective algorithm to obtain orbit tune shifts depending only on parameters described in [14].

5 The DA Spin Normal Form Algorithm

The computation of the equilibrium polarization $\bar{n}(\vec{z})$ is more complicated and requires an algorithm similar to the DA orbit normal form method [13]. The vector $\bar{n}(\vec{z})$ satisfies

$$\hat{A}(\vec{z}) \cdot \bar{n}(\vec{z}) = \bar{n}(\vec{\mathcal{M}}(\vec{z})).$$

For its determination, we assume the orbital map is already in normal form and linear spin map is diagonalized. We now proceed in an iterative way.

For zeroth order, observe that $\vec{\mathcal{M}}(\vec{z}) =_0 0$, and thus the equation reads $\hat{A}_0 \cdot \bar{n}_0 = \bar{n}_0$; thus, $\bar{n}_0$ is the eigenvector to unit eigenvalue of the constant part of $\hat{A}$.

For higher orders, assume we already know $\bar{n}$ to order $m-1$ and want to determine it to order m. Assume $\vec{\mathcal{M}}(\vec{z})$ is in normal form, i.e. $\vec{\mathcal{M}}(\vec{z}) = \mathcal{R}+\mathcal{N}$. Write $\hat{A} = \hat{A}_0 + \hat{A}_{\geq 1}$, $\bar{n} = \bar{n}_{<m} + \bar{n}_m$, and obtain

$$(\hat{A}_0 + \hat{A}_{\geq 1}) \cdot (\bar{n}_{<m} + \bar{n}_m) = (\bar{n}_{<m} + \bar{n}_m) \circ (\mathcal{R} + \mathcal{N})$$

To order m, this can be rewritten as

$$\hat{A} \cdot \bar{n}_{<m} + \hat{A}_0 \cdot \vec{n}_m =_m \bar{n}_{<m} \circ (\mathcal{R} + \mathcal{N}) + \bar{n}_m \circ R, \text{ or}$$
$$\hat{A}_0 \vec{n}_m - \bar{n}_m \circ \mathcal{R} =_m \bar{n}_{<m} \circ (\mathcal{R} + \mathcal{N}) - \hat{A} \cdot \vec{n}_{<m}.$$

The right hand side has to be balanced by choosing $\bar{n}_m$ appropriately. But like in orbit case, the coefficients of $\hat{A}_0 \vec{n}_m - \vec{n}_m \circ \mathcal{R}$ differ from those of $\bar{n}_m$ only by resonance denominators, and so the task can be achieved as soon as these resonance denominators do not vanish; this requires the absence of spin amplitude tune shifts.

Acknowledgments

For stimulating my interest in spin dynamics, I am indebted to Desmond Barber and Sateesh Mane. For many other valuable discussions about the matter, I would like to thank Nina Golubeva, Volodaj Balandin and Georg Hoffstätter.

Financial support was appreciated by the US National Science Foundation and the Alfred P. Sloan Foundation.

References

[1] T. Limberg R. Rossmanith, J. Kewisch. *Phys. Rev. Lett.*, 62:62, 1989.

[2] K. Yokoya. Technical Report 92-6, KEK, 1992.

[3] A. W. Chao. *Nuclear Instruments and Methods*, 29:180, 1981.

[4] S. R. Mane. *Phys. Rev.*, A36:120, 1987.

[5] K. Yokoya. *Nuclear Instruments and Methods*, A258:149, 1987.

[6] Y. Eidelmann and V. Yakimenko. The spin motion calculation using lie method in collider magnetic field. In *Proc. 1991 IEEE Particle Accelerator Conference*, 1991.

[7] V. Balandin and N. Golubeva. *International Journal of Modern Physics A*, 2B:998, 1992.

[8] M. Berz. COSY INFINITY Version 6 reference manual. Technical Report MSUCL-869, National Superconducting Cyclotron Laboratory, Michigan State University, East Lansing, MI 48824, 1993.

[9] M. Berz. *High-Order Computation and Normal Form Analysis of Repetitive Systems, in: M. Month (Ed), Physics of Particle Accelerators*, volume AIP 249, page 456. American Institute of Physics, 1991.

[10] L. Michel V. Bargmann and V. L. Telegdi. *Phys. Rev. Lett.*, 2:435, 1959.

[11] M. Berz. Computational aspects of design and simulation: COSY INFINITY. *Nuclear Instruments and Methods*, A298:473, 1990.

[12] M. Berz. Differential algebraic description of beam dynamics to very high orders. *Particle Accelerators*, 24:109, 1989.

[13] M. Berz. Differential algebraic formulation of normal form theory. In *M. Berz, S. Martin and K. Ziegler (Eds.), Proc. Nonlinear Effects in Accelerators*, page 77. IOP Publishing, 1992.

[14] M. Berz. Direct computation and correction of chromaticities and parameter tune shifts in circular accelerators. In *Proceedings XIII International Particle Accelerator Conference, JINR D9-92-455*, pages 34–47(Vol.2), Dubna, 1992.

The Brookhaven Muon g-2 Experiment*

G. Bunce for the Muon g-2 Collaboration**
Brookhaven National Laboratory, Upton, N.Y. 11973-5000

1 Introduction

A new experiment is being mounted at BNL to measure the anomalous magnet moment of the muon to 3 parts in 10^7. In this talk I will describe the physics issues that this precision allows us to explore, the experimental method, and an interesting new device which we will use to inject muons into our muon storage ring. The device is a 1.45T non-ferrous superconducting magnet, where all fringe field is contained by a superconducting sheet.

2 Physics Issues

The gyromagnetic ratio of a particle, g, is the ratio of the magnetic field created by a revolving charge and the angular momentum of the particle from revolving mass. g-2 measures the difference between the charge distribution and the mass distribution. If the distributions are the same, whatever their size, (g-2) = 0. Also, if the particle is point-like, (g-2) = 0. For the proton, g-2 = 3.6, and the large value is due to the constituents: the proton is an extended object with charged quarks and neutral mass distributed inside. The electron and muon g-2 $\approx$ .002, and they are thought to be very small objects.

A precision measurement of g-2 gives us a means of seeing the average effect of quantum fluctuations of the particle. A charged particle emits and absorbs photons, fluctuating within the Uncertainty Principle, producing an electric field. When the particle travels through an external magnetic field, photons from the external field interact with the particle, which affects the spin and g-2. These interactions see the fluctuating particle, as well as the bare particle. The g-2 value is an average of the effects from the different states. For the electron, the difference between theory and experiment for g-2 is $(48 \pm 28) \times 10^{-12}$, in remarkable

* Work performed under the auspices of the U. S. Department of Energy and National Science Fourdation.

agreement for this precision, and this confirms our understanding of electromagnetism[1].

The original muon g-2 experiments were done to see if the muon charge and mass distributions might differ, since the muon is 200x more massive than the electron. Also, any new field carried by the muon, vs. the electron, would also be probed by g-2. From an experiment in 1961 at CERN[2],

(g-2)/2 muon	=	0.001162 ± 5
compared to the electron	=	0.001160. . .

The muon and electron g-2 values were the same at this level of precision.

The muon, because of its mass, is more sensitive than the electron to fields carried by heavier particles. This increased sensitivity is proportional to the square of the mass, and the muon is 40,000x more sensitive than the electron to contributions from the emission and absorption of heavier particles. This is the reason it is exciting to measure the muon g-2 to great precision.

A precise measurement of muon g-2 is sensitive to the emission and absorption of the very heavy particles that carry the weak force, the W and Z bosons. Table I shows the predicted contributions to the muon g-2. The W and Z contribute at a level of 4 and -2 x 10^{-6} of the muon (g-2). The present theory[3] predicts the anomaly a = (g-2)/2 to 2 x 10^{-6} and the 1979 CERN measurement[4] has a precision of 7 x 10^{-6}. To see an effect of 2 x 10^{-6}, both the theory and the experiment must be improved. The present agreement between theory and experiment is (4 ± 9) x 10^{-6} of the anomaly. Again, there is remarkable agreement at this precision, which verifies the electromagnetic contributions, and the new hadronic contribution (for the electron, this contributions is not important) where the electric field photon fluctuates to a pion pair, as shown in Table I.

The theoretical error is mainly from the estimate of the hadronic contribution. This contribution is obtained from experiment, a measurement of the cross section for e^+e^- creating $\pi^+\pi^-$. Experiment CMD2 at VEPP-2M, Novosibirsk[5] is currently measuring this, and we expect that the theoretical error in the muon anomaly will be reduced to about 0.5 x 10^{-6}.

The goal of the Brookhaven experiment is to improve the muon g-2 measurement by a factor of 20, or 3 x 10^{-7} of the anomaly. At this level of precision, and with the reduction of the theoretical error, we will be able to:

- measure the W and Z contribution
- "*see*" muon size for $\geq$ 3 x 10^{-18} cm

- "*see*" possible "*excited muons*" to 400x proton mass
- "*see*" W size for $\geq 8 \times 10^{-18}$ cm
- "*see*" the g-2 of the W boson for ≥ 0.04
- "*see*" speculated new particles where

$$\Delta \frac{g-2}{2} \frac{m_\mu^2}{m_x^2}, \quad m_x < 5000 \ \textit{proton} \quad \textit{masses} \ .$$

How do we get all of this from one number? The beauty of g-2 is that not only can we discover that there is something new over a broad range of possibilities, but, if that doesn't happen, we can show that there is nothing new for that same range of possibilities (> SSC domain), and we show that we understand the interactions of γ, W, Z.

TABLE I - Muon g-2

x represents the interaction of the photon from the external magnetic field. The underlined digits refer to the error.

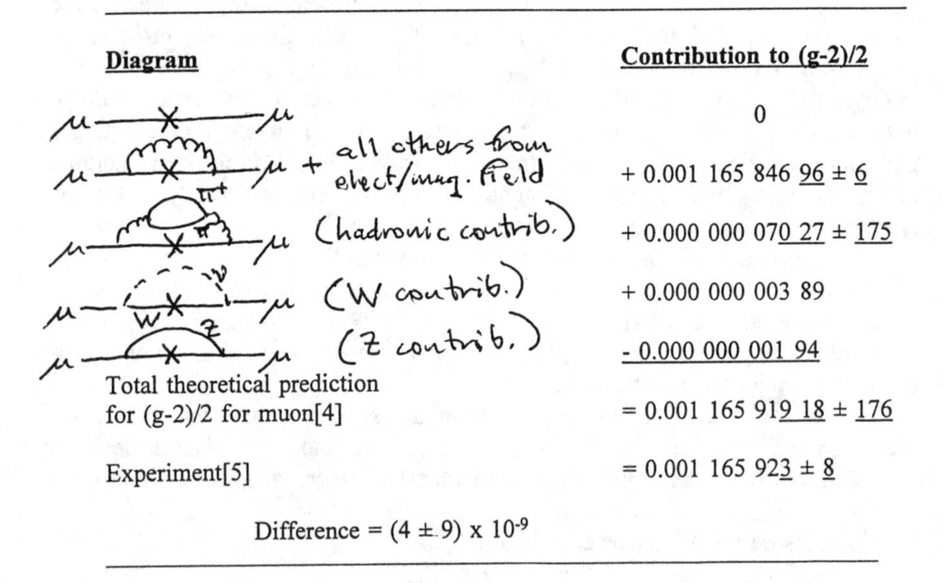

Diagram	Contribution to (g-2)/2
μ—x—μ	0
all others from elect/mag. field	+ 0.001 165 846 96 ± 6
(hadronic contrib.)	+ 0.000 000 070 27 ± 175
(W contrib.)	+ 0.000 000 003 89
(Z contrib.)	- 0.000 000 001 94
Total theoretical prediction for (g-2)/2 for muon[4]	= 0.001 165 919 18 ± 176
Experiment[5]	= 0.001 165 923 ± 8

Difference = $(4 \pm 9) \times 10^{-9}$

3 The Experiment

The muon spin precesses in a magnetic field slightly faster than the momentum:

$$\omega_s = \omega_c + \frac{g-2}{2}\frac{eB}{mc},$$

where ω_s is the spin precession frequency, ω_c is the frequency of the momentum, and B is the magnetic field. By storing muons in a precise magnetic field and observing the frequency of the spin rotation relative to the momentum (ω_s - ω_c), we are able to measure the small number g-2 directly.

The new muon g-2 experiment is constructing a muon storage ring, which will be one magnet driven by 14 meter diameter superconducting coils, with the magnetic field shaped by iron in a C-configuration, open side facing the center (Fig. 1). Muons, collected from π decay in a long beam line, and with their spins polarized in their direction of motion from the parity-violating $\pi \rightarrow \mu + \nu$ decay, are brought through a hole in the iron backleg into a field-free region near the storage ring. This field-free region is created by a 1.7 meter long superconducting injector magnet (Fig. 2) which fits in-between the poles of the storage ring magnet. The muons then cross the storage volume about a quarter of the way around the circumference, and are kicked onto the storage orbit by magnetic kicker. 3.1 GeV muons live an average of 64 microseconds, and decay to electrons (and two neutrinos). The electrons remember the spin direction of the parent muon: more energetic electrons are emitted when the muon spin points along its momentum. The electrons, with a lower energy than the muons, spiral to the inside of the ring where they hit electron calorimeters, which are distributed around the inside circumference of the storage ring vacuum chamber. The calorimeters register more high energy electrons when the spin of the muons point forward, so that one observes a rise and fall of counts in the calorimeter as the muon spins sweep around. Now, (g-2)/2 $\approx$ .001, and this is boosted by the relativistic gamma factor of 29, giving a phase advance of the spin of .034 x 360° = 12.3° per rotation in the storage ring. One turn takes 149 nanoseconds, so the frequency of the muon spins is 4.4 microseconds, which is observed in the calorimeters. This is shown in Figure 3 for the CERN experiment.

The 4.4 microsecond spin precession sits on the exponentially falling muon decay rate. The experiment must measure the spin frequency very precisely, as well as the average magnetic field seen by the muons.

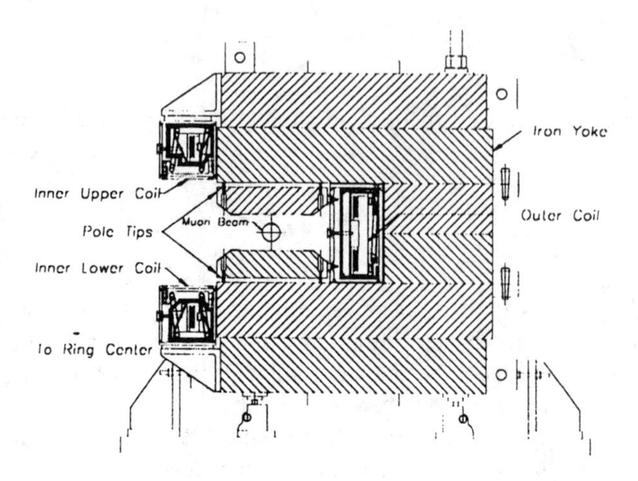

Fig. 1. A cross section view of the storage ring magnet. The muon beam will be contained in the 9 cm diameter circle shown, with the center of the storage ring 7.1 meters to the left. The coils are superconducting solenoids 13.4 and 15.1 meters in diameter.

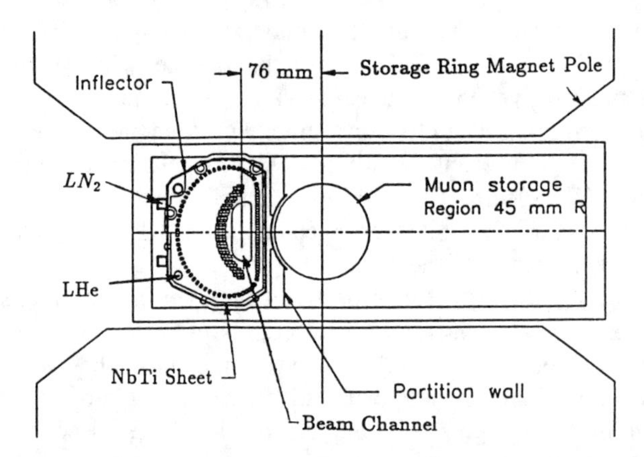

Fig. 2. Inflector and storage ring cross section at the downstream end of the inflector. The entering beam is 76mm from the center of the storage ring at this point. The inflector is a 1.5T superconducting magnet with no escaping fringe field.

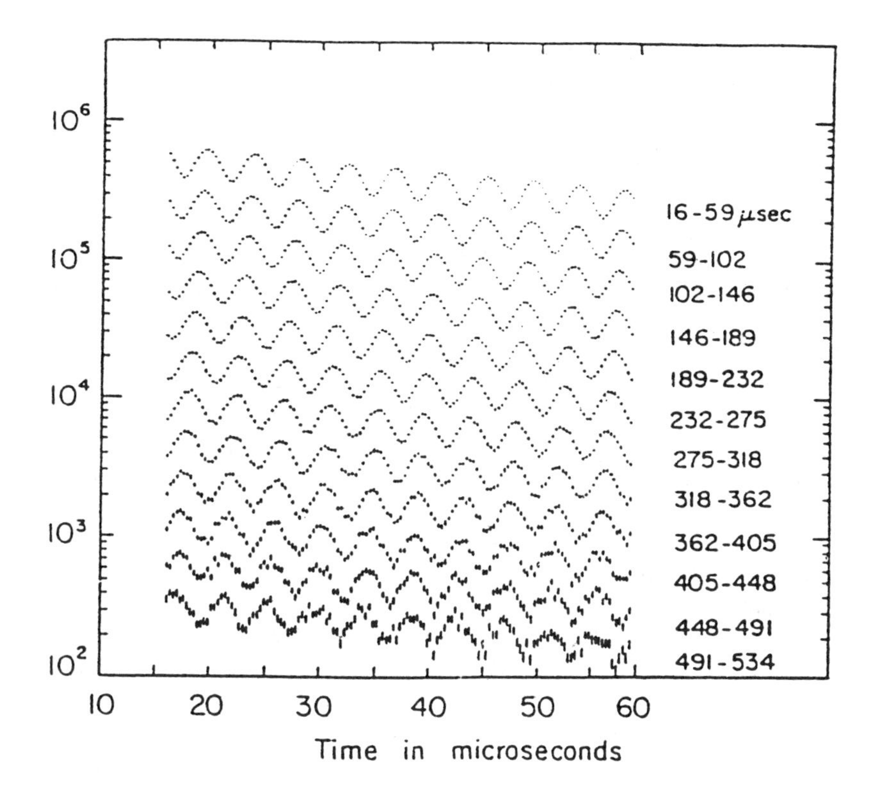

Fig. 3. Number of high energy electrons observed by the CERN III experiment[4] vs. storage time. In order to show 8 muon lifetimes of data, the plot reuses the time axis every 43 μsec.

4 Experimental Challenges

There has been considerable effort at developing tools to obtain a flat magnetic field, to 10^{-6}, and to measure the field to 10^{-7} using a trolley to carry NMR probes inside the storage ring vacuum.

A lead-scintillating fiber calorimeter system has been developed to be flat in response over the huge dynamic range--from 5 MHz at the beginning of the store to KHz later. We have developed blanking circuits for the injection time, studied the PMs and the wave form digitizer read-out for rate issues, and developed a very stable binning procedure to systematically assign times for the events.

5 The Superconducting Inflector and Shield

We have inherited the name "*inflector*" from the CERN g-2 experiment. The inflector (or injector) cancels the magnetic field near the storage ring without affecting the main storage ring field at a level of 10^{-6}. CERN's precision was less, and they used a pulsed 1.5T magnet. We were concerned with the effects of the pulsing circuit on our electronics, with remnant fields, and we needed to be able to inject each of the twelve AGS bunches at short intervals (33 msec between stores). Our solution was to construct a DC inflector and capture the return magnetic field with a cosine θ bucking coil so that the main magnetic field would not be affected. The device must fit within the 18 cm gap of the storage ring.

The device that was developed uses a double cosine θ winding (Fig. 2)[6]. A short model of the correct cross section was built, and the design works well[7]. However, there is significant field leakage at the end near the storage region and from the granularity of the conductors. A field of roughly 200 gauss was measured at the position of the center of the storage ring near the inflector. After studying shimming approaches we have adopted and tested an approach of shielding the stray field with a superconducting sheet[8].

The superconducting shield is a remarkable device. If the shield were superconducting when the main magnetic field was turned on, it would exclude this field from the inside of the inflector, and also cause very large perturbations in the storage ring field. The proposal was to only activate the shield after the main field is on. The shield then locks in the main field, not affecting it at at least 0.2 x 10^{-6} level (as measured). When the inflector magnet is then turned on, it correctly bucks out the field seen by the entering muons, and its stray field is trapped by the shield, so the field seen by the stored muons is not affected. This is described in Ref. 8 and one test of this approach is shown in Figure 4.

6 Summary

Many innovations have not been mentioned. We use the large proton flux provided by the AGS/Booster complex; we are building a beam line capable of both pion injection (used by the CERN experiment: the pion decay in the storage ring provided the kick that stores the muons) and muon injection; we will have devices to measure the muon distribution in the ring to weight the average magnetic field seen by the muons; we are building a precision NMR system with absolute calibration to better than 10^{-7}; a new electrostatic quadrupole system has been developed with a current lead configuration that dumps charge that builds up due to E x B containment; in addition to the calorimeters, we will have scintillator hodoscopes which give the muon position, reduce confusion from overlap, and provide an independent time measurement; we plan a new analysis of the calorimeter data which measures charge and is not sensitive to overlapping pulses.

The experiment requires $3x10^{10}$ muons to reach the statistical level of 3×10^{-7}, for the standard analysis. These data can be obtained in two weeks. Therefore, the ultimate sensitivity of the new experiment will depend on our control of systematic errors.

The magnet shimming will begin in early 1995, and the first run with beam will be January 1996.

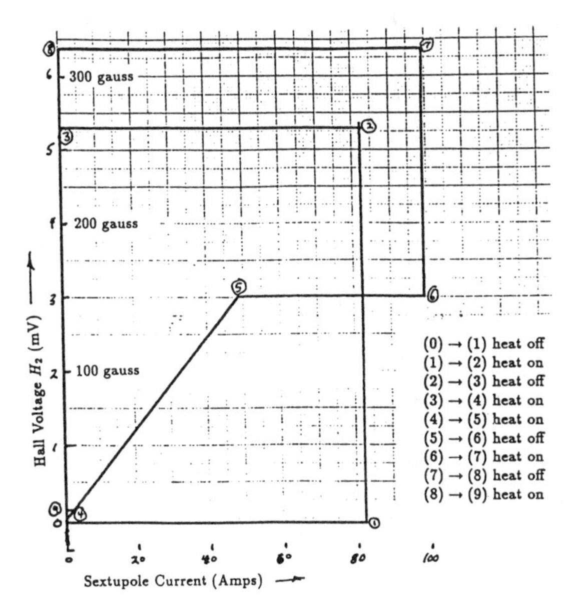

Fig. 4. An SSC sextupole was wrapped with a superconducting shield, and its field was measured outside the device. A heater was used to change the shield from superconducting (heat off) to normal (heat on) . Several cycles are shown[8].

** The g-2 Collaboration
Boston University: D. H. Brown, R. M. Carey, E. S. Hazen, F. Krienen, Zhifeng Liu, J. P. Miller, J. Ouyang, B. L. Roberts, L. R. Sulak, W. Worstell; Brookhaven National Laboratory: J. Benante, H. N. Brown, G. Bunce, J. Cullen, G. T. Danby, K. Gardner, J. Geller, H. Hseuh, J. W. Jackson, L. Jia, R. Larsen, Y. Y. Lee, R. E. Meier, W. Meng, W. Morse, C. Pai, I. Polk, A. Prodell, S. Rankowitz, J. Sandberg, Y. Semertzidis, R. Shutt, L. Snydstrup, A. Soukas, A. Stillman, T. Tallerico, P. A. Thompson, F. Toldo, K. Woodle;Budker Institute of Nuclear Physics: L. M. Barkov, D. N. Grigoriev, B. I. Khazin, E. A. Kuraev, Ya. M. Shatunov, E. Solodov; Cornell University: Y. Orlov, T. Kinoshita;Fairfield University: D. Winn; University of Heidelberg: K. Jungmann, R. Prigl, P. von Walter, G. zu Putliz; University of Illinois: D. Cronin,

P. Debevec, W. Deninger, D. Hertzog, T. Jones, K. McCormick; Lawrence Berkeley Laboratory: M. Green; Los Alamos National Laboratory: W. Lysenko; Max-Plank Inst.: U. Haeberlen; University of Minnesota: P. Cushman S. Giron, J. Kindem, D. Maxam, D. Miller, C. Timmermans; National Laboratory for High Energy Physics (KEK): K. Endo, H. Hirabayshi, S. Kurokawa, Y. Mizumachi, T. Sato, A. Yamamoto; Riken Institute: K. Ishida; University of Tokyo: M. Iwasaki, K. Nagamine, K. Nishiyama, S. Sakamoto; Yale University: S. K. Dhawan, A. A. Disco, F. J. M. Farley, X. Fei, S. Hou, V. Hughes, M. Janousch, S. Redin, Q. Xu

References

[1] R. Van Dyck, Jr., et al. Phys. Rev. Lett. 59, 26 (1987); T. Kinoshita in Quantum Electrodynamics, ed. by T. Kinoshita (World Scientific, Singapore, 1990), p. 218.

[2] G. Charpak et al., Phys. Rev. Lett. 6, 128 (1961).

[3] T. Kinoshita and W. J. Marciano, in Quantum Electrodynamics, ed. by T. Kinoshita (World Scientific, Singapore, 1990), p. 419; T. Kinoshita, Bull. Am. Phys. Soc. 39, 1051 (1994).

[4] J. Bailey et al., Nucl. Phys, B150, 1 (1979).

[5] B. I. Khazin for CMD2 collaboration, Intl. Conf. on HEP, Glasgow (1994); V. M. A. Auldrenko et al., Nucl. Inst. Meth. A252, 299 (1986).

[6] F. Krienen, D. Loomba, W. Meng, NIM A283, 5 (1989).

[7] A. Yamamoto and Y. Saito, g-2 Note 196 (1994); W. Meng and K. Woodle, g-2 Note 209 (1994).

[8] G. T. Danby, W. Meng, W. B. Sampson, K. Woodle, paper presented to 13th Int. Conf. on Magnet Technology, Victoria, Canada (1993) and g-2 Note 191 (1994); W. Meng and K. Woodle g-2 Note 210 (1994).

THE SMC MUON BEAM POLARIMETER

F. Feinstein

On behalf of the Spin Muon Collaboration (SMC)

Abstract

A muon beam polarimeter has been built for the SMC experiment at the CERN SPS, for muon energies of 100 to 200 GeV. The beam polarisation is determined by two methods, from the shape of the decay in flight positrons energy spectrum and from the asymmetry in the elastic scattering off polarised electrons of a magnetised ferromagnetic target whose magnetisation is reversed. The first method gives a typical polarisation of -0.8 with a relative systematic accuracy of 2.5%, the second one is expected to achieve a relative systematic accuracy of 2%.

1 INTRODUCTION

The SMC experiment is currently measuring the spin-dependent structure functions of the nucleon. They are extracted from the asymmetry measured in deep inelastic scattering of polarised muons off polarised protons and deuterons [1, 2]. This asymmetry is proportional to the beam polarisation. In order to match other main sources of systematics such as the knowledge of $F_2(x)$ and the target polarisation, the beam polarisation has to be known with a relative accuracy better than 5%.

Two polarimetry methods are used. The first method is described in [3] and will only be mentioned briefly. It is based on the measurement on the positron energy spectrum coming from $\mu^+ \rightarrow e^+\nu_e\overline{\nu}_\mu$ decay in flight. The shape of this spectrum depends on the parent muon polarisation. The polarisation of the CERN muon beam at 190 GeV has been measured to be $P_\mu = -0.803 \pm 0.029 \pm 0.020$. The second method uses polarised elastic muon electron scattering [4]. It is similar to polarised Møller scattering which is commonly used to measure the polarisation of electron beams. The experimental asymmetry is $A_{exp} = A_{\mu e} \times P_e \times P_\mu$ where the theoretical asymmetry $A_{\mu e}$ is known from QED calculations, the electron polarisation P_e is measured, and P_μ is the beam polarisation.

2 POLARISED MUON ELECTRON SCATTERING

2.1 Kinematics

From energy conservation, neglecting the electron mass, one has $E_\mu = E'_\mu + E_e$. One can define a dimensionless variable $y = E_e/E_\mu = 1 - E'_\mu/E_\mu$, which is the fraction of energy transferred to the knock-on electron. The maximum fraction of energy transferred is

$$Y = \frac{1}{1 + M^2/2mE_\mu}$$

where M and m are respectively the muon and the electron masses. The muon and the electron scattering angles are given by

$$\theta_\mu = \left[\frac{2m}{E_\mu}.\frac{y}{1-y} - \left(\frac{M}{E_\mu}.\frac{y}{1-y}\right)^2\right]^{1/2} \text{ and } \theta_e = \frac{1-y}{y}\,\theta_\mu$$

Figure 1.a shows θ_e and θ_μ as a function of y for a nominal muon beam energy of 190 GeV. An interesting feature of the process is that $\theta_e + \theta_\mu$ which is the opening angle between the two scattered tracks is almost constant around 5 mrad over a wide range of y.

2.2 First order cross section and asymmetry

The differential cross section for elastic muon scattering on polarised electron at first order QED at high energy can be found in [5]. Here it is expressed as a function of y:

$$\frac{d\sigma}{dy} = \frac{2\pi r_e^2 m}{E_\mu}\left(\frac{1}{y^2} - \frac{1}{y.Y} + \frac{1}{2}\right)(1 + P_\mu A_{\mu e})$$

where $2\pi r_e^2 m = 255$ μbarn.GeV, P_μ is the loongitudinal muon polarisation and $A_{\mu e}$ is the theoretical cross section asymmetry

$$A_{\mu e} = \frac{d\sigma^{\uparrow\downarrow} - d\sigma^{\uparrow\uparrow}}{d\sigma^{\uparrow\downarrow} + d\sigma^{\uparrow\uparrow}} = y\frac{1 - y/Y + y/2}{1 - y/Y + y^2/2}$$

The asymmetry for $E_\mu = 190$ GeV is shown in figure 1.b. The statistical figure of merit commonly defined for polarisation measurement $A.\sqrt{d\sigma/dy}$ is almost constant. Therefore, each bin of y will give a polarisation measurement with a comparable statistical accuracy.

2.3 Application to our measurement

The CERN muon beam intensity is around 7.10^5 μ/s which is several orders of magnitude lower than any electron beam for which Møller polarimeters are commonly used. In order to get a statistical accuracy of a few percent in a reasonable time, one needs to use an iron alloy target with a thickness up to one radiation length. In this case, the knock-on electron looses a significant fraction of its energy by radiative processes such as Bremsstrahlung on the iron nuclei, before exiting the target. Therefore, the measured energy of the electron is not relevant to evaluate the energy E_e that was transferred by the muon at the vertex and it is impossible to sign the elastic process by checking the energy balance of the charged particles of the reaction. Fortunately, the muon is much less sensitive to Bremsstrahlung which depends on the square of the mass of the particle. It is possible to measure the fraction of energy transferred to the electron by measuring the energy of the muon before and after the interaction. The scattering angles θ_e and θ_μ can also be used to check the μe scattering kinematics and reject competing processes.

3 EXPERIMENTAL SET UP

3.1 The muon beam

A secondary beam of 225 GeV/c π^+/K^+ is extracted from the hadronic shower produced by the SPS proton beam hitting a beryllium target. The spill duration is 1.8 s and the repetition time is 14.4 s. The μ^+ from π/K decay in $\mu^+\nu_\mu$, are selected at 190 GeV/c. The fraction of the momentum of the parent meson carried by the muon is the most important factor for determining the muon polarisation. Knowing others factors such as the ratio π/K and the beam optics, the polarisation can be calculated. A simulation of the beam line gives a polarisation value of -0.78 ± 0.05 [6]. With 4×10^{12} proton per spill on the Be target, one gets 4.4×10^7 muons. The momentum spread of the beam is roughly gaussian with a typical width of 5 GeV/c. Upstream of the SMC experimental hall, every muon is momentum analysed by the Beam Momentum Station (BMS) which consits of two sets of scintillator counters located upstream and downstream of a dipole (see fig. 2).

3.2 The spectrometer

Figure 2 shows a schematic of the layout. A large part of the apparatus is the same as the one described in [3]. The incoming muon is detected in two highly segmented planes of scintillator and is tracked in a set of 3 beam chambers upstream of the target. The scattered muon and the knock-on electron are tracked in 3 additional beam chambers situated downstream of the target, then are deflected in opposite directions by the MNP26, a dipole magnet providing a very homogenous field, which integral is 11.7 T.m. The electron is tracked in 3 MWPC and identified in a large lead glass calorimeter. The scattered muon is tracked by a set of 3 MWPC and a beam chamber, then identified in a scintillator hodoscope downstream of a 2 m thick iron absorber. The large MWPC installed immediately downstream of the magnet, common to both detection arms, is also a beam chamber. Characteristics of all detectors are given in Table 1.

Detector use	Type	Planes orientation $0^\circ \equiv Y$, $90^\circ \equiv Z$	Segmentation (cm) $Y \times Z$	Size (cm) $Y \times Z$
Beam hodoscope	Scintillators	$0^\circ, 90^\circ$	0.4 to 2	20×20
Beam chambers	MWPC	$0^\circ, 90^\circ, -45^\circ, 45^\circ$	1 mm pitch	19×16
Large MWPC	MWPC	$0^\circ, -37^\circ, 37^\circ, 0^\circ$	1 mm pitch	96×20
Downstream chambers	MWPC	$0^\circ, -30^\circ, 30^\circ$	2 mm pitch	50 to 100 $\times$ 30
e^- calorimeter	Lead Glass		10×10	$100 \times 30 \times 27X_0$
μ^+ hodoscope	Scintillators	0°	0.8 to 7.5	160×30

Table 1: Detectors characteristics

3.3 The polarised electron target

The target is a sheet of ferromagnetic alloy (49%Fe, 49% Co, 2% V). The ferromagnetic foil is installed in the gap of a flat magnetic circuit described in [7]. The magnetisation of the material is measured using a pickup coil wrapped around the target . To determine the foil magnetisation value, the current is inverted in the magnet and the voltage V induced in the pickup coil is integrated giving the flux change. The air flux is measured after removing the foil, and substracted. From this, one deduces the magnetic induction within the foil. One needs the magnetomechanical ratio g' to know the contribution of the angular momentum J of the electrons of the material to the magnetic induction of the foil. As no g' data exist for ternary alloys Fe-Co-V, we have taken the most accurate values of g' for 50%Fe-50%Co [8] to do a conservative estimate of $g' = 1.92 \pm 0.01$ for our alloy. One gets the polarisation in the foil plane $P = 0.0834$, and for the electron polarisation along the muon beam axis P_l (for a 25° angle of the target) the value

$$P_l = 0.0756 \pm 0.0008$$

where the guessed error stems mostly from the uncertainty on g'. At this angle, the target is around 0.4 radiation length thick.

3.4 Triggers, electronics and data acquisition

Three triggers are defined. The beam trigger is a coincidence between the two planes of the beam hodoscope. The normalisation trigger is a coincidence between the beam trigger and a random hit coming from a low rate radioactive source. It is similar to the one used by EMC [9] and NMC for absolute cross section measurement. It allows us to normalise to the same incoming μ flux, the number of events for the two electrons longitudinal polarisations. The μe scattering trigger is a coincidence between the beam trigger, a hit in the muon hodoscope and a signal in the Lead Glass array corresponding to an electromagnetic shower having deposited more than 15 GeV.

The RMH readout electronics from CERN is used for the small beam chambers, PCOS2 and PCOS3 from Le Croy for the other chambers. The hodoscopes and Lead Glass time and amplitude informations are encoded by FERA modules from Le Croy. The BMS time information is encoded by standard Le Croy TDC.

The data acquisition program is running on a VME processor. During the spills, all data from CAMAC modules, RMH and the BMS are read by the VME processor. In between the spills, the VME processor does the event building, reads the hodoscopes and Lead Glass scalers as well as information concerning the electron target. Then, it transmits the data to a microVax2 from DEC that writes it on a Exabyte tape, and to a SUN Sparcl+ workstation for online monitoring of the data.

4 MEASUREMENTS AND ANALYSIS

4.1 Experimental procedure

The target field is reversed at each SPS spill in order to change the sign of the longitudinal electron polarisation P_l. Due to the vertical field near the target, for the two different polarisations, the particles are deflected in two different parts of the apparatus which have different acceptances. This gives rise to a false asymmetry. The target was flipped every 2 hours from 25° to -25°, so that the same P_l would be obtained with a vertical field of opposite sign and therefore, a false asymmetry of opposite sign would arise. It was checked with an unpolarised target where only the false asymmetry remained, that indeed, this procedure cancelled it.

4.2 Event reconstruction

The time of flight between the beam hodoscope on one hand, the BMS, the Lead Glass calorimeter and the muon hodoscope on the other hand is measured using the TDC information. All the hits in time within ± 3 ns are accepted, the others are rejected as accidental.The information from the BMS is used to calculate the incoming muon momentum.

The beam hodoscope elements that are in time are used to build a road in the first 3 beam chambers in order to reconstruct the incoming muon track. The intersection between the incoming track and the target is used to build a road in the last 3 beam chambers in order to reconstruct the outgoing electron and muon tracks. The vertex is reconstructed as the closest point to the 3 tracks. The typical size of the vertex is $\sigma_t = 0.7$ mm and $\sigma_l = 15$ cm. Events for which the vertex is further than 50 cm from the target position, or for which more than two tracks could be reconstructed after the target, are rejected.

At this stage, one does not know which of the outgoing tracks is the muon and which is the electron. The two possibilities are tried for each outgoing track. A road is built in the downstream chambers, using the muon hodoscope element in time, and the track is propagated through the dipole magnetic field. A fit of the momentum is performed that minimises the distances between the hits in the downstream chambers on the muon arm. Then, the same procedure is applied to the same track on the electron arm using the Lead Glass element in time to build a road. The best χ^2 of the two fits allows us to determine whether the track is an electron or a muon. The other outgoing track is tried the same way on both arms. A few percent of the events are ambigous, they are indeed rejected. Once the scattered tracks momenta are calculated, the scattering angles are corrected for the deflection of the field near the target.

The fraction y of the μ^+ energy transferred to the e^-, is calculated: $y = 1 - P'_\mu/P_\mu$. The momenta P_μ and P'_μ are measured by two different spectrometers, one located upstream, the other one located downstream of the SMC target and detectors. Therefore, two effects can distort the determination of y. First, due to the energy loss of the μ^+ in the SMC target and detectors, P_μ at the polarimeter is always smaller than P_μ at the BMS. Second, the sytematic errors of the two momentum determinations do not cancel. The reconstruction of beam trigger events where the μ^+ are momentum analysed by the two spectrometers allows us to do a cross-calibration of P_μ^{BMS} versus P_μ^{MNP26}.

In summary, for each event, the time coincidence between the incoming particle and the two outgoing particles is verified, the vertex is reconstructed, the scattered muon and the knock-on electron are identified, the variables y, θ_e and θ_μ are calculated.

4.3 Analysis

The 100 cm of air included in the vertex cut ate taken into account in the calculation of the effective target polarisation.

The scattering angles allow us to check the μe scattering kinematics. The angles θ_μ^{calc} and θ_e^{calc} are calculated using y and P_μ, and compared to θ_μ^{meas} and θ_e^{meas}. The distribution of $\theta^{meas} - \theta^{calc}$ has a width of 0.2 mrd for the electrons and 0.14 mrd for the muons.

Direct pair production $\mu^+ \rightarrow \mu^+ e^+ e^-$ and to a less degree muon Bremsstrahlung with subsequent materialisation of the photon in the target are a source of background as they may produce a $\mu^+ e^-$ coincidence. For an asymmetrical pair where the e^- carries most of the pair momentum, the e^+ may not be detected and the process may fake the μe scattering kinematics. Data have been taken recently using a μ^- beam and triggering on $\mu^- e^+$ coincidences, which correspond to the $\mu^+ e^-$ background. The analysis is under way, first results show that the background is of the order of 3% of the signal. The formula for the asymmetry given in section 2 is calculated at the first order of QED. One needs to take into account radiative effects. A detailed calculation of the radiative corrections at the vertex to μe scattering has been published [10]. The application to our set-up is currently being done.

The experimental asymmetry as a function of y is fitted by the theoretical one upto to a constant with a χ^2 equal to 1.15 per degrees of freedom. On figure 3, the value of this constant is superimposed to the experimental asymmetry divided by the theoretical one. The relative precision of the constant is 3.2% obtained with about 10^{12} μ on the target. In order to yield the beam polarisation, the value of this constant will have to be corrected for radiative effects and $\mu^+ e^-$ background, and divided by the electron polarisation P_l. The final systematic error on the beam polarisation is expected to be of the order of 2%.

5 CONCLUSION

The μe scattering method is characterised by a good control of the systematics for the measurement of the beam polarisation. First, we measure an asymmetry as a function of y. In each of the 14 bins in y, the asymmetry is proportional to the theoretical one. By performing frequent target rotations, false asymmetries cancel out. We don't depend on a Monte Carlo calculation of the apparatus acceptance to extract the beam polarisation. Second, the reaction is clearly signed by its kinematics and by identifying the scattered particles.The remaining background is small and has been measured using a μ^- beam. Third, by reconstructing the vertex, we make sure that the reaction occured in the polarised target.

The radiative corrections to the theoretical asymmetry still need to be known to allow us to extract the beam polarisation.

New data taken in 1994 with thicker targets and a longer beam time exposure will result in a statistical accuracy of the order of 1.5%, matching the expected 2% systematic error.

References

[1] R. Windmolders, these proceedings

[2] A. M. Zanetti, these proceedings

[3] B. Adeva et al., NIM A **343**(1994)363

[4] P. Schüler, Proc. of the 8th International Symposium on High Energy Spin physics, Mineapolis, Sept. 12-17, 1988, American Institute of Physics (1989), 1401

[5] A. M. Bincer, Phys. Rev. **107**(1957), 1434

[6] N. Doble et al., NIM A **343**(1994), 351

[7] N. de Botton, A. Daël, J. Martino, IEEE Transactions on Magnetics **vol. 30 number 4** (July 1994), 2447

[8] G. Scott and H. Sturner, Phys. Rev. **184**(1969), 490

[9] R. P. Mount, NIM **343**(1994), 351

[10] T. V. Kukhto et al., J. Phys. G. **13**, 725

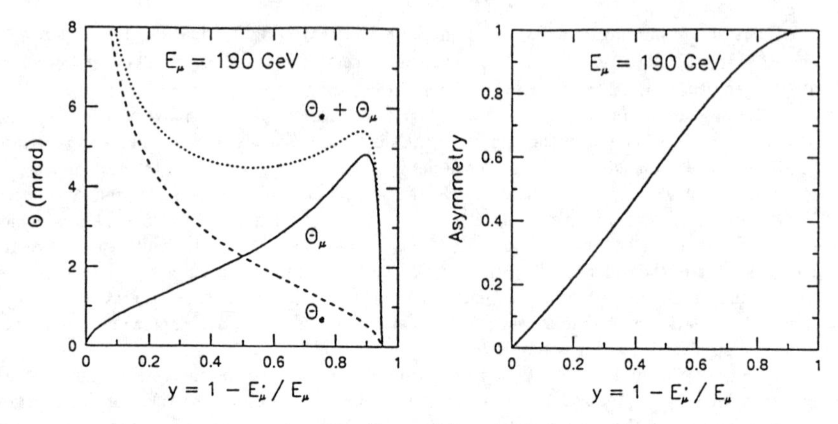

Figure 1: Scattering angles of the electron and the muon (a) and asymmetry (b) as a function of y for 190 GeV muons

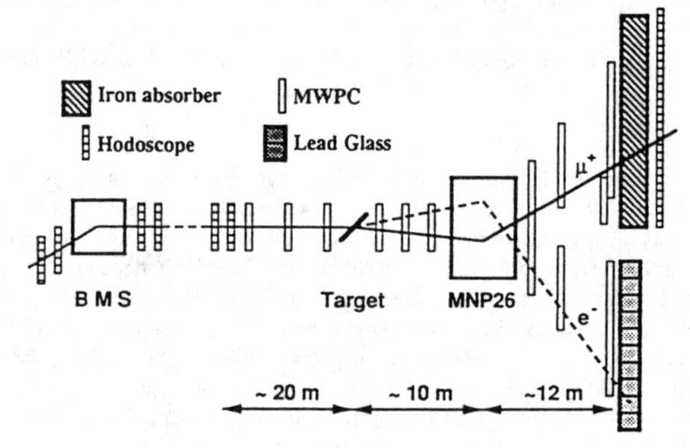

Figure 2: Layout of the SMC μe scattering polarimeter

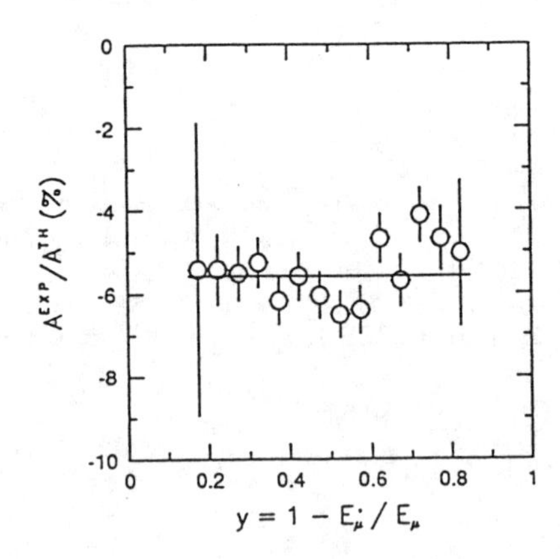

Figure 3: Experimental asymmetry divided by the theoretical asymmetry as a function of y

IV. STRONG INTERACTIONS

Strong interactions at high energy

M. Anselmino

Dipartimento di Fisica Teorica, Università di Torino and
Istituto Nazionale di Fisica Nucleare, Sezione di Torino,
Via P. Giuria 1, I-10125 Torino, Italy

Abstract Spin effects in strong interaction high energy processes are subtle phenomena which involve both short and long distance physics and test perturbative and non perturbative aspects of QCD. Moreover, depending on quantities like interferences between different amplitudes and relative phases, spin observables always test a theory at a fundamental quantum mechanical level; it is then no surprise that spin data are often difficult to accomodate within the existing models. A report is made on the main issues and contributions discussed in the parallel Session on the "Strong interactions at high energy" in this Conference.

In the parallel Session on "Strong interactions at high energy" a total of 22 talks were presented on several and various subjects, both theoretical and experimental. As usual, spin effects prove to be unexpected, rich and difficult to understand, testing the deep and basic properties of any theory. In high energy interactions the underlying fundamental theory we are testing is QCD, both in its perturbative and non perturbative aspects; the latter, at least in the energy region so far explored, still play a crucial role, as has emerged from most of the presentations.

Rather then considering the single contributions I have selected the main topics and experimental or theoretical problems discussed in the Session and will try to comment and report on these issues. For more details on the single contributions one is referred to the original talks published in these Proceedings. I apologize with some authors whose contributions will receive less attention than others; this is only due to my organization of the material and by no means implies a negative judgement on their work.

The arguments most widely discussed can be divided into:
1) *pp* elastic scattering and total cross-sections;
2) single spin inclusive asymmetries;
3) hyperon polarization;
4) other topics;
5) nucleon spin crisis.

The last subject was actually extensively treated in the parallel Session on "Nucleon spin structure functions" to which one is referred; I shall only make here a short comment to answer many questions raised at various stages during the Conference.

pp elastic scattering and total cross-sections

Surprising spin effects in large angle proton-proton elastic scattering at high energy have been known for quite a long time [1, 2]: they concern both single and double spin asymmetries, A, A_{NN}, A_{LL} and A_{SS}. Consider the C.M. scattering of two protons moving along the z-axis and let (xz) be the scattering plane; then the single spin asymmetry A, or analyzing power, is defined as

$$A \equiv \frac{d\sigma^{\uparrow} - d\sigma^{\downarrow}}{d\sigma^{\uparrow} + d\sigma^{\downarrow}} \tag{1}$$

where $d\sigma^{\uparrow(\downarrow)}$ is the cross-section for the elastic scattering of an unpolarized proton off a proton polarized perpendicularly to the scattering plane, that is along $(\uparrow)$ or opposite $(\downarrow)$ the $\hat{y}$-axis.

The double spin asymmetry A_{NN} is defined by

$$A_{NN} \equiv \frac{d\sigma^{\uparrow\uparrow} - d\sigma^{\uparrow\downarrow}}{d\sigma^{\uparrow\uparrow} + d\sigma^{\uparrow\downarrow}} \tag{2}$$

where, now, both initial protons are polarized in the direction normal (N) to the scattering plane, as explained for A.

Analogously to A_{NN}, one can define the double spin asymmetries A_{SS} and A_{LL} which only differ from A_{NN} by the direction along which the initial protons are polarized: the S direction is the $\hat{x}$ direction, always choosing (xz) as the scattering plane, and the L direction is along $\hat{z}$.

A, A_{NN}, A_{SS} and A_{LL} can be written in terms of the five independent helicity amplitudes

$$\begin{aligned}
\phi_1 &\equiv \langle + + |\phi| + + \rangle \\
\phi_2 &\equiv \langle + + |\phi| - - \rangle \\
\phi_3 &\equiv \langle + - |\phi| + - \rangle \\
\phi_4 &\equiv \langle + - |\phi| - + \rangle \\
\phi_5 &\equiv \langle + + |\phi| + - \rangle
\end{aligned} \tag{3}$$

as

$$\begin{aligned}
A &= \Sigma^{-1}\mathrm{Im}\,[\Phi_5^*(\Phi_4 - \Phi_1 - \Phi_2 - \Phi_3)] \\
A_{NN} &= \Sigma^{-1}\mathrm{Re}\,[\Phi_1\Phi_2^* - \Phi_3\Phi_4^* + 2|\Phi_5|^2] \\
A_{SS} &= \Sigma^{-1}\mathrm{Re}\,[\Phi_1\Phi_2^* + \Phi_3\Phi_4^*] \\
2A_{LL} &= \Sigma^{-1}[|\Phi_3|^2 + |\Phi_4|^2 - |\Phi_1|^2 - |\Phi_2|^2] \\
2\Sigma &= [|\Phi_1|^2 + |\Phi_2|^2 + |\Phi_3|^2 + |\Phi_4|^2 + 4|\Phi_5|^2]
\end{aligned} \tag{4}$$

According to the standard QCD description of large angle exclusive reactions $AB \to CD$ [3], assuming the dominance of collinear valence quark configurations in the proton wave function, one has the helicity conservation rule

$$\lambda_A + \lambda_B = \lambda_C + \lambda_D \tag{5}$$

which immediately implies, for pp elastic scattering, $\phi_2 = \phi_5 = 0$ and consequently, via Eqs. (4),

$$A = 0 \qquad\qquad A_{NN} = -A_{SS}\,. \tag{6}$$

The experimental data [1, 2] do not agree with such conclusions.

All this has been known for quite a long time and several models and attempts to explain the data can be found in the literature. In summary they amount to introduce non perturbative contributions which should still be important at the energies of the performed experiments; quark masses, higher order corrections or intrinsic transverse momenta, which violate the helicity conservation rule, ought to be properly taken into account. Also possible lower energy mechanisms might still interfere with the hard scattering picture and give a sizeable contribution. However, one expects that at higher energies the perturbative QCD predictions should eventually turn out to be correct* and further experimental information would be of great importance.

A rich and interesting program for measuring several spin observables in pp elastic scattering and inclusive production is in progress at UNK (Protvino) and RHIC (Brokhaven); reports on the status of the planned experiments can be found in the talks by V. Solovianov, A.M.T. Lin and P.A. Draper.

On the theoretical side it was pointed out in the talk by G. Ramsey how a careful analysis of the data on A_{NN} at the C.M. scattering angle $\theta = \pi/2$, implemented with some reasonable assumptions, might supply information on the scattering amplitudes. $\theta = \pi/2$ is an interesting region in that the data on A_{NN} present a rich structure as a function of the energy and the number of independent amplitudes is reduced to three ($\phi_5(\pi/2) = 0\,, \phi_3(\pi/2) = -\phi_5(\pi/2)$); moreover, one can assume the double-flip helicity amplitude $\phi_2 \simeq 0$.

S. Troshin has presented a model which attempts to explain spin effects via the interference between two phases of QCD, a non perturbative one, described by an effective Lagrangian which allows helicity flips, and the usual perturbative one; such interference is significant in the intermediate energy region of the existing data.

Spin effects are also expected in the measurements of $\Delta\sigma_L(pp) = \sigma_{tot}^{\rightarrow\rightarrow} - \sigma_{tot}^{\rightarrow\leftarrow}$ and $\Delta\sigma_N(pp) = \sigma_{tot}^{\uparrow\uparrow} - \sigma_{tot}^{\uparrow\downarrow}$, the differences between proton-proton total

*This need not be true for high energy *small angle* processes, where non perturbative anomalous contributions might persist and remain energy independent in polarization measurements [4]

cross-sections in pure lonfitudinal and normal spin states respectively. $\Delta\sigma_L(pp)$ should be sensitive to the gluon polarization inside the proton (Δg) and any information on Δg is of crucial importance to settle the nucleon spin crisis issue. Some preliminary results on this quantity and, for the first time, on $\Delta\sigma_L(\bar{p}p)$ were reported by D. Grosnick; however, these results are small, consistent with zero, with large errors, and do not allow yet to discriminate between different theoretical models.

Single spin inclusive asymmetries

Large single spin asymmetries have also been observed [5] in the inclusive production of pions in the collision of a high energy polarized proton beam on an unpolarized target, $p^\uparrow + p \to \pi + X$, where the proton spin is up or down with respect to the scattering plane. These asymmetries

$$A_N(x_F, p_T) = \frac{d\sigma^\uparrow - d\sigma^\downarrow}{d\sigma^\uparrow + d\sigma^\downarrow} \tag{7}$$

are found to be large for the transverse momentum of the pion in the range $(0.7 \lesssim p_T \lesssim 2)$ GeV/c and large values of x-Feynman (x_F); actually, $|A_N|$ increases with x_F. A_N also shows interesting isospin dependences ($A_N^{\pi^+} \approx -A_N^{\pi^-}$).

Such results are unexpected because a naive generalization of the QCD-factorization theorem suggests that the single spin asymmetry can be written qualitatively as:

$$A_N \sim \sum_{ab\to cd} \Delta_T G_{a/p} \otimes G_{b/p} \otimes \hat{a}_N \hat{\sigma}_{ab\to cd} \otimes D_{\pi/c} \tag{8}$$

where $G_{a/p}$ is the parton distribution function, that is the number density of partons a inside the proton, and $\Delta_T G_{a/p} = G_{a^\uparrow/p^\uparrow} - G_{a^\downarrow/p^\uparrow}$ is the difference between the number density of partons a with spin $\uparrow$ in a proton with spin $\uparrow$ and the number density of partons a with spin $\downarrow$ in a proton with spin $\uparrow$; $D_{\pi/c}$ is the number density of pions resulting from the fragmentation of parton c; $\hat{a}_N$ is the single spin asymmetry relative to the $a^\uparrow b \to cd$ elementary process and $\hat{\sigma}$ is the cross-section for such process.

The usual argument is then that the asymmetry (8) is bound to be very small because $\hat{a}_N \sim \alpha_s m_q/\sqrt{s}$ where m_q is the quark mass. This originated the widespread opinion that single spin asymmetries are essentially zero in perturbative QCD.

However, as it has been discussed by several authors [6]-[12], this conclusion need not be true because subtle spin effects might modify Eq. (8). Such modifications should take into account the parton transverse motion, higher twist contributions and possibly non perturbative effects hidden in the spin dependent distribution and fragmentation functions.

An explicit model [8] which takes into account the orbital motion of quarks inside polarized protons or antiprotons and describes meson formation via $q\bar{q}$ annihilation, has been presented by T. Meng, who advocates the use of non perturbative mechanisms to describe spin effects. In this model the different sign of $A_N^{\pi^+}$ and $A_N^{\pi^-}$ simply originates from the $SU(6)$ valence quark spin content of the protons; the results are in qualitative agreement with data and predictions are given for the single spin asymmetry in Drell-Yan processes $\bar{p}^\uparrow + p \to l^+l^- + X$.

An approach closer to perturbative QCD has been discussed by F. Murgia; application of the QCD-factorization theorem in the helicity basis (for which it has been proved) modifies Eq. (8) into [6]

$$A_N \sim \sum_{ab\to cd} I_{+-}^{a/p} \otimes G_{b/p} \otimes \hat{\sigma}_{ab\to cd} \otimes D_{\pi/c} \tag{9}$$

with

$$I_{+-}^{a/p}(x_F, \boldsymbol{k}_\perp) = \sum_h [G_{h/p}(x_F, \boldsymbol{k}_\perp) - G_{h/p}(x_F, -\boldsymbol{k}_\perp)] \tag{10}$$

where h is a helicity index. I_{+-} is a new phenomenological function of relevance for spin observables, analogous to $G_{a/p}(x_F)$ in the unpolarized case; notice that it vanishes when $k_\perp = 0$. A simple model for I_{+-} yields results in very good agreement with the data [13].

This approach has been criticized by Collins [11] on the ground that QCD time-reversal invariance should forbid a non zero value of Eq. (10); he suggests a similar effects in the parton fragmentation process rather than in the parton distribution functions. His criticism is discussed in Ref. [13].

Further theoretical work on spin asymmetries can be found in the talks by G.J. Musulmanbekov, M.P. Chavleishvili and, together with a general analysis of twist-3 single spin observables, by O. Teryaev.

W. Novak has presented a proposal to perform spin measurements at HERA by inserting a polarized target into the unpolarized proton beam. N. Saito has shown the first results on the single spin asymmetry for direct photon production in $p^\uparrow + p \to \gamma + X$ processes at Fermilab: the results are small, consistent with zero within the large errors, in agreement with the theoretical prediction of Ref. [7].

Hyperon polarization

Hyperon polarization is yet another example of puzzling single spin asymmetry in inclusive processes (mainly $p + N \to H^\uparrow + X$, with unpolarized beam and target); however, it is so well known and so much studied (experimentally) that it deserves attention in its own. Only experimental contributions on the subject have been presented during the Conference; when adding this new information to the previously available one it will become clear why no theoretical interpretation of the data has been attempted.

With some idealization and optimism the bulk of data on hyperon polarization at our disposal up to some time ago could be summarized as:
- all hyperons (with exception of Ω) produced in the proton fragmentation region are polarized perpendicularly to the production plane; antihyperons are not polarized;
- the Λ^0 polarization is negative and opposite to the Σ;
- the Λ^0 polarization is almost energy independent;
- the magnitude of the polarization increases with p_T up to $p_T \simeq 1$ GeV/c;
- the magnitude of the polarization is independent of p_T above 1 GeV/c and increases linearly with x_F;
- the polarization depends weakly on the target type (A^α dependence).

Even at such "simple" stage there was no clear theoretical explanation of the data: some semi-classical models could explain some of the features [14] and some other attempts were made which either related the Λ polarization observed in $pp \to \Lambda X$ to that observed in $\pi p \to \Lambda K$ [15] or used triple Regge models [16]. However, there is no fundamental explanation of the old set of data and it is very hard to understand how polarized strange quarks, which should be present inside polarized hyperons, can be created in the scattering of unpolarized hadrons.

The situation is now even more hopeless. These are the new data, from Fermilab E761 Collaboration, as summarized by S. Timm:
- antiparticle polarization: anti-Σ^+ and anti-Ξ^- are polarized, anti-Λ is not;
- energy dependence: Σ^+ polarization decreases with energy, Ξ^- polarization increases (in magnitude) and Λ polarization remains constant;
- p_T dependence: Σ^+ polarization decreases above $p_T \simeq 1$ GeV/c;
- x_F dependence: Σ^+ polarization increases with x_F, with a p_T dependent slope, Λ polarization increases (in magnitude) with a fixed slope and Ξ^- polarization is independent of x_F for $x_F \gtrsim 0.4$.

No kind of regular pattern emerges from the data. It is then clear how a successful description of these experimental results cannot originate from general features of the underlying dynamics, but has to rely on subtle non perturbative effects in the hadronization process for each single hyperon.

In the attempt of better understanding the basic mechanisms at work in the quark recombination process the Fermilab E800 Collaboration has measured the Ξ^- and Ω^- polarization from polarized and unpolarized neutral hyperon beams. The results, presented by K.A. Johns, are, once more, partially expected and partially unexpected.

Other topics

I'll just mention some other talks which dealt with different subjects. The polarization of the J/ψ in pion-nucleus collisions was discussed within perturbative QCD by W. Tang and some spin effects in the production and decay of D^{*+} and Λ_c^+ were studied by R. Rylko. Also a measurement of the Ω^- magnetic moment by the E800 Collaboration was shown by P.M. Border.

A. Brandenburg noticed how the angular distribution of leptons produced in $\pi^- p \to \mu^+\mu^- + X$ processes is in disagreement with the standard QCD parton model. A coherent treatment of $q\pi^-$ scattering succeeds in accounting for the data and favours the QCD sum rule pion wave function. A.P. Contagouris has presented a complete computation of higher order corrections for Drell-Yan processes with transversely polarized protons, $p^\uparrow p^\uparrow \to l^+ l^- + X$ and for $e^+e^- \to \Lambda^\uparrow \bar{\Lambda}^\uparrow$.

A simulation for the production of $W^\pm$ and Z_0 in the collision of high energy polarized protons at RHIC, with the aim of measuring the sea polarization, has been discussed by V. Rykov.

Nucleon spin crisis

The nucleon spin structure functions have been discussed in the dedicated parallel Session; however, as many questions were raised during the talks regarding the present situation with the "spin crisis", I'll briefly comment on that. A comprehensive review on the subject can be found in Ref. [17].

Recall that the measurement of $\Gamma_1^p \equiv \int_0^1 dx\ g_1^p(x)$, combined with information from hyperon β-decays, allows to obtain a value of a_0, defined by

$$\langle P, S|\bar{\psi}\gamma_\mu\gamma_5\psi|P, S\rangle = 2m_N a_0 S_\mu\,, \tag{11}$$

i.e. the expectation value between proton states (of four momentum P and covariant spin vector S) of the flavour singlet axial vector current $J^0_{5\mu} \equiv \bar{\psi}\gamma_\mu\gamma_5\psi$.

In the naive parton model (with free quarks) one has

$$a_0 = \sum_{q,\bar{q}} \Delta q \equiv \Delta\Sigma = 2S_z^q \tag{12}$$

where S_z^q is the total z-component of the spin carried by the quarks; then one expects $a_0 \simeq 1$. Instead, the EMC data [18] yield, at $\langle Q^2\rangle \simeq 10$ (GeV/c)^2,

$$A_0 = 0.06 \pm 0.12 \pm 0.17\,. \tag{13}$$

Such result, consistent with zero and certainly $\neq 1$, originated the "spin crisis" in the parton model.

However, due to the axial anomaly, we know that $J^0_{5\mu}$ is not a conserved current,

$$\partial_\mu J_5^{0\mu} = \frac{\alpha_s}{4\pi} G^a_{\mu\nu}\tilde{G}^{\mu\nu}_a\,, \tag{14}$$

so that the expectation value of $J^0_{5\mu}$, $\langle P, S|J^0_{5\mu}|P, S\rangle$, is scale dependent. Then a_0 is not forbidden to be small at $\langle Q^2\rangle \simeq 10$ (GeV/c)^2 and it might be close to 1 at smaller Q^2 values.

Moreover, this reflects into the fact that there is a large gluonic contribution to a_0, so that what we actually measure at large Q^2 is

$$a_0(Q^2) = \Delta\Sigma - 3\,\frac{\alpha_s(Q^2)}{2\pi}\Delta g(Q^2) \tag{15}$$

where Δg is the spin carried by the gluons. Notice that the gluon contribution $\alpha_s \Delta g$ does not vanish when $Q^2 \to \infty$ as the decrese of α_s is compensated by an increase of Δg. Eq. (15) tells us that $a_0 \simeq 0$ does not necessarily imply $\Delta\Sigma \simeq 0$.

To summarize this brief digression on the nucleon spin problem: a_0, being the expectation value of a non conserved current, is allowed to evolve in Q^2, both perturbatively and non perturbatively; there is both a quark and gluonic contribution to a_0 which, perturbatively, is given in Eq. (15). The large Q^2 perturbative evolution of a_0 is computed to be small ($\mathcal{O}(\alpha_s^2)$), but the $\alpha_s \Delta g$ contribution does not vanish. The small Q^2 non perturbative evolution may be large and its description requires non perturbative models. The relative amount of non perturbative and Δg contributions is still an open question; in this respect any measurements of the gluon spin would be of great importance.

Conclusions

Needless to say spin effects are fascinating, but tough to understand; in high energy hadronic interactions both the perturbative, short distance physics and the non perturbative, large distance one, are interrelated in subtle and sometime mysterious ways. In order to make successful predictions or even to explain the existing data, we need a knowledge of hadron wave functions, hadronization mechanisms, quark-gluon correlations, and so on, in addition to the usual QCD dynamics. More, detailed and precise spin data are crucial in allowing the work on spin effects to continue; it might be a long and hard work but the eventual outcome – a real understanding of the hadron structure – is certainly worth the effort.

Acknowledgements

I would like to thank the organizers of the Conference and Gerry Bunce for his work and help in the organization of the Session on "Strong interactions at high energy".

References

[1] Cameron, P.R., *et al.*, *Phys. Rev.* **D32**, 3070 (1985).

[2] Crabb, D., *et al.*, *Phys. Rev. Lett.* **41**, 1257 (1978); Crosbie, E.A., *et al.*, *Phys. Rev.* **D23**, 600 (1981); Auer, I.P., *et al.*, *Phys. Rev. Lett.* **52**, 808 (1984).

[3] See, *e.g.*, Brodsky, S.J., and Lepage, G.P., in *Perturbative Quantum Chromodynamics*, A.H. Mueller Editor, World Scientific (1989).

[4] Anselmino, M and Forte, S., *Phys. Rev. Lett.* **71**, 223 (1993).

[5] Adams, D.L., *et al., Z. Phys.* **C56**, 181 (1992); *Phys. Lett.* **B264** (1991) 462; *Phys. Lett.* **B276** (1992) 531.

[6] Sivers, D., *Phys. Rev.* **D41**, 83 (1990); *Phys. Rev.* **D43**, 261 (1991).

[7] Qiu, J., and Sterman, G., *Phys. Rev. Lett.* **67**, 2264 (1991).

[8] Boros, C., Liang, Z., and Meng, T., *Phys. Rev. Lett.* **70**, 1751 (1993); see also the talk by T. Meng in these Proceedings.

[9] Szwed, J., *Phys. Lett.* **B105**, 403 (1981).

[10] Efremov, A.V. and Teryaev, O.V., *Yad. Fiz.* **36**, 242 (1982) [*Sov. J. Nucl. Phys.* **36**, 140 (1982)].

[11] Collins, J., *Nucl. Phys.* **B396**, 161 (1993).

[12] Artru, X., Czyzewski, J., and Yabuki, H., *Single spin asymmetry in inclusive pion production, Collins effect and the string model*, preprint LYCEN/9423 and TPJU 12/94, May 1994.

[13] Anselmino, M., Boglione, M.E., and Murgia, F., *Single spin asymmetry for* $p^{\uparrow} + p \to \pi + X$ *in perturbative QCD*, preprint DFTT48/94 and INFNCA-TH-94-27, in preparation.

[14] Andersson. B., Gustafson, G., and Ingelman, G., *Phys. Lett.* **85B**, 417 (1979); De Grand, T.A., and Miettinen, H., *Phys. Rev.* **D24**, 2419 (1981).

[15] Soffer, J., and Törnqvist, N.A., *Phys. Rev. Lett.* **68**, 907 (1992).

[16] Barni, R., Preparata, G., and Ratcliffe, P.G., *Phys. Lett.* **B296**, 251 (1992).

[17] Anselmino, M., Efremov, A., and Leader, E., preprint CERN-TH/7216/94, to appear in *Phys. Rep.*

[18] Ashman, J., *et al, Nucl. Phys.* **B328**, 1 (1989).

The Discovery of Neutral Hyperon Polarization

Thomas Devlin

Department of Physics and Astronomy
Rutgers - The State University of New Jersey

Abstract

In 1975, a Wisconsin--Michigan--Rutgers collaboration working on Experiment 8 at Fermilab discovered a strong polarization signal in Λ hyperons produced by 300 GeV protons on various targets. Subsequent experiments by this and other groups demonstrated this to be true for most of the strange hyperons and anti-hyperons with lifetimes in the range of 10^{-10} sec. This talk is a personal narrative of the early stages of these experiments.

In 1970 I fled New Jersey in a dark mood looking for new directions. A focus of my work for some years had been The Princeton-Penn Accelerator (PPA), and it was being shut down. Bogdan Maglich was then at Rutgers, and it was a tense place. I called Jack Steinberger who arranged a guest appointment for me at CERN. My wife and I obtained sabbaticals, packed up our three small sons and, for a year, took refuge in Geneva.

I joined the charged-hyperon experiment under construction at the CERN PS, and worked with Jacques Lefrancois, Robert Meunier, Jean-Marc Gaillard and their colleagues. At PPA I had built a Hydrogen/Deuterium target which was perfect for this work. I had it shipped to CERN. We completed two very nice experiments,[1] and this set my compass on the heading I was to follow for fifteen years.

That same summer, a Wisconsin-Michigan group proposed work on neutral hyperons at the National Accelerator Lab - later Fermilab. During a visit by Oliver Overseth to CERN, I was given a copy of the proposal and invited to join. I accepted on the spot. As it turned out, we hit a "gold mine" of physics which exceeded all my expectations. Among the goals outlined in this proposal, which was approved as experiment number 8 (E8), were

- Production cross sections
  - for various incident beam-particles
  - for various targets (A-dependence)
  - and for various final state particles, Λ, $\overline{\Lambda}$, K_S^0 and Ξ^0
- A search for $\Xi^0 \rightarrow p\,\pi^-$, a $\Delta S = 2$ process.
- Λ-p scattering cross sections.
- **P**(Λ), the Λ production polarization.

Experiments in a Neutral Hyperon Beam

R. H. March and L. G. Pondrom*
University of Wisconsin

O. E. Overseth
University of Michigan

June 1970

Oliver Overseth

Lee Pondrom

Bob March

These particles have lifetimes ~10^{-10} sec. This proposal exploited Albert Einstein's time dilation and high energies from Bob Wilson's synchrotron to stretch out time. Quantitatively, $\tau' = \gamma\tau$, where τ and τ' are the lifetime in the rest frame and laboratory. The ratio of energy to mass is $\gamma = E/mc^2$. For our early experiments, γ was 50 to 200 and gave 10-meter mean decay lengths in our hyperon beam.

Fermilab was still under construction when I returned from Europe, and we set out to build equipment. We had test runs over the next two years at Cornell for the

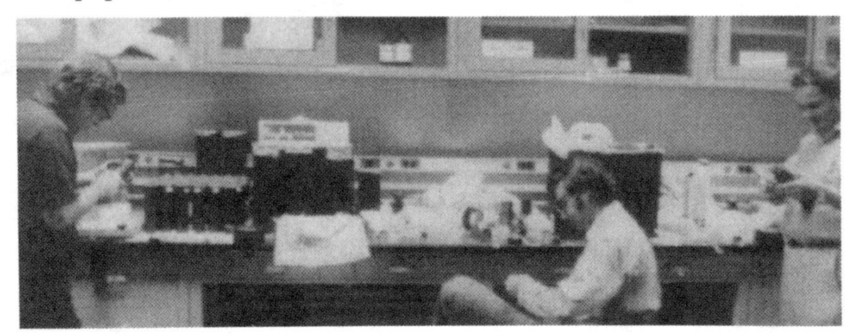

Pat Skubic **Ralph Edwards** **Bill Cooper**

lead-glass counters and at Argonne for the proportional chambers. By 1974, we were installing the apparatus in the M2 beamline at the Meson Lab, and were frantically developing event reconstruction software.

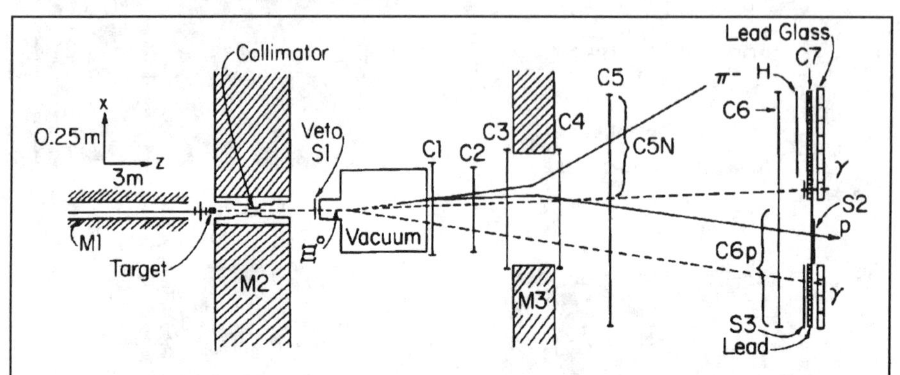

Hyperons and K_S^0 were detected by their decay products using a charged-particle momentum spectrometer and, in later runs, a lead-glass Cerenkov-counter array to measure γ-ray positions and energies. This is a plan view of the apparatus. M1 is the final magnet to control the angle of the incident proton beam relative to the collimator axis. M2 is the precession magnet. S1-S3 are scintillators; C1-C7 are proportional chambers. The event shown is $\Xi^0 \rightarrow \Lambda\pi^0$, $\Lambda \rightarrow p\pi^-$, $\pi^0 \rightarrow 2\gamma$.

The analysis programs were not yet working when the first events were collected. Oliver and I sat down with the raw wire-chamber data and decoded it by hand. We were prepared for trash, but the first three events were all good - two Λ's and a K_S. The programs soon confirmed the high yield. A typical one-hour run produced a full magnetic tape with 80,000 triggers containing 31,000 Λ, 4,200 K_S^0, 430 $\overline{\Lambda}$ and 30 Ξ^0. Subsequent trigger refinements produced yields above 90%.

Invariant-mass distributions for pπ and ππ hypotheses showed clear peaks for Λ, $\overline{\Lambda}$ and K_S^0 with good resolution and low background. Histograms of vertex

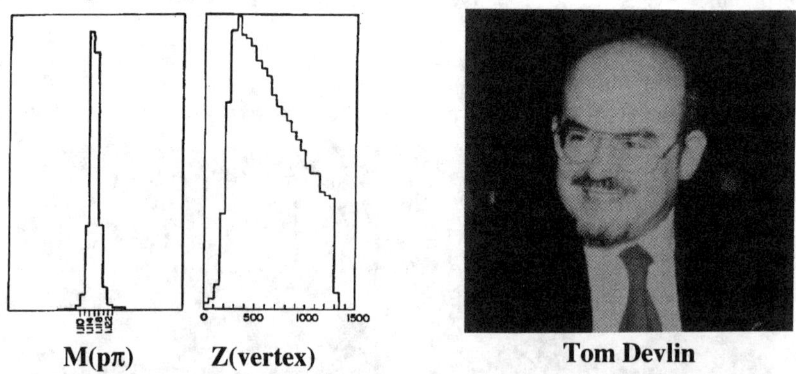

M(pπ) **Z(vertex)** **Tom Devlin**

distributions along the beam axis were completely consistent with the expected exponential decay distribution. We took a great deal of data on production cross sections for Λ, $\overline{\Lambda}$ and K_S^0 a a function of production angle by deflecting the the

300 GeV proton beam away from a straight line and then back onto the hyperon production target at angles from 0 to 9 mrad.[2,3,4] From these data we calculated distributions for Feynman-x, x_F, and transverse momentum, p_T. We used Be, Cu and Pb to study the A-dependence.[5] One striking feature of these results was the ratio of $\overline{\Lambda}$ to Λ production cross sections which was independent of production

Phil Martin

Production cross sections for Λ hyperons produced by 300 GeV protons on Beryllium at angles from 0 to 9.8 mrad.

angle, but which was a steeply falling function of x_F extrapolating to unity at $x_F=0$. In later experiments we and other groups studied production of Ξ^0, $\overline{\Xi}^0$ and the production of charged hyperons and anti-hyperons. Two reviews have described this work.[6,7]

Several experiments were also performed with a liquid hydrogen target in the hyperon beam. We measured the differential cross section[8] and final state Λ polarization[9] for Λ-p elastic scattering.

Brian Edelman

Ken Heller

Bob Handler

Marleigh Sheaff

In the summer of 1975, theorists in a conference at Argonne confidently stated that polarization effects would be negligibly small at Fermilab energies. Twenty miles away, my colleagues and I were studying a rather strong polarization signal in our earliest sample of Λ hyperons. We soon announced it at conferences, and our first paper appeared the following Spring. It reported the discovery of the polarization, the first test of parity conservation in strong interactions at Fermilab energies, and, as a by-product, a measurement of the Λ magnetic moment.[10]

Gerry Bunce

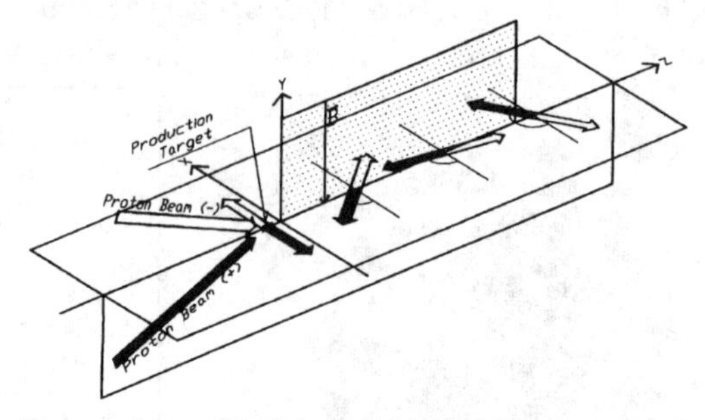

Precession with (+) and (-) Production Angle

Parity conservation in the strong production process forces the polarization at the production target to be perpendicular to the production plane. In the cartoon above, the z-axis is the axis of the straight channel through M1, the y-axis is vertically upward and x completes a right-handed coordinate system. The figure shows two runs with all conditions identical except for the direction in the y-z plane of the proton beam on the target. The "+7.2 mr"condition produced a beam with its polarization vector, **P**, in the negative x-direction, while the "-7.2 mr" condition gave the opposite polarization. All changes needed for this took place upstream of M1, a 400-ton magnetized-iron, tungsten-lined beam channel with a 4-mm-diam. aperture for the hyperon beam. The decay volume and detector were downstream, unaffected by the production angle. The polarization is perpendicular to the magnetic field and **P** will experience Larmor precession. With the field integral of 13.5 Tesla-meters downward as shown, the polarization vector precessed clockwise about the y-axis by 150°.

The weak decay, $\Lambda \rightarrow p\pi^-$, has a parity-violating asymmetry in the angular distribution of the daughter proton proportional to $(1 + \alpha P_i \cos\theta_i)$, where P_i is the component of the polarizaton along some axis i=x,y,z or helicity, $\alpha=0.642\pm0.013$ is the "parity-violating" decay parameter, and θ_i is the polar angle between the proton momentum and the ith axis. Uncorrected angular distributions below were taken with opposite production angles and show combined effects of

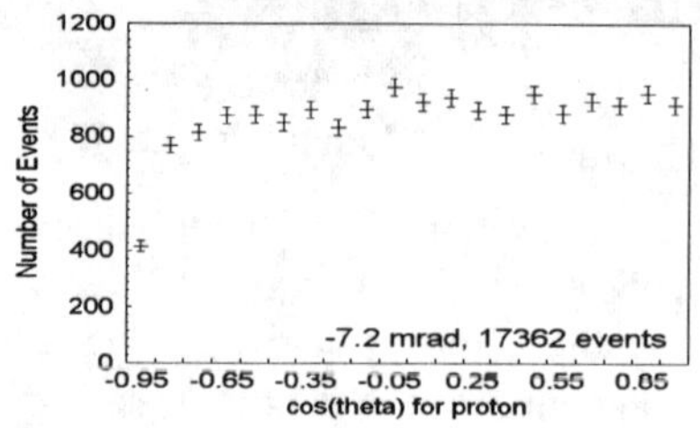

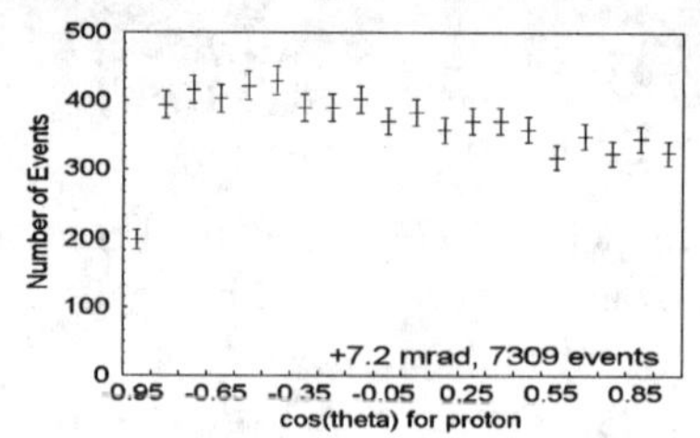

acceptance and decay asymmetry. The polarization reverses but biases do not. The bin-by-bin asymmetry $(f_+ - f_-)/(f_+ + f_-)$ plotted below shows the asymmetry with respect to the laboratory z-axis, where $f_\pm$ is the fraction of counts in a bin for ±7.2 mrad. The slope of the fitted line is αP_z. Similar analyses were done for the other axes to yield the polarization vector, **P.** The polarization was studied at a variety of production angles and as a function of Λ momentum and found to be as high as 15%. At 7.2 mrad, averaged over momentum spectrum of detected Λ's, it was about 8%.

As important as the discovery of polarization itself, was its use as a tool to measure magnetic moments of hyperons. The exploratory run which discovered

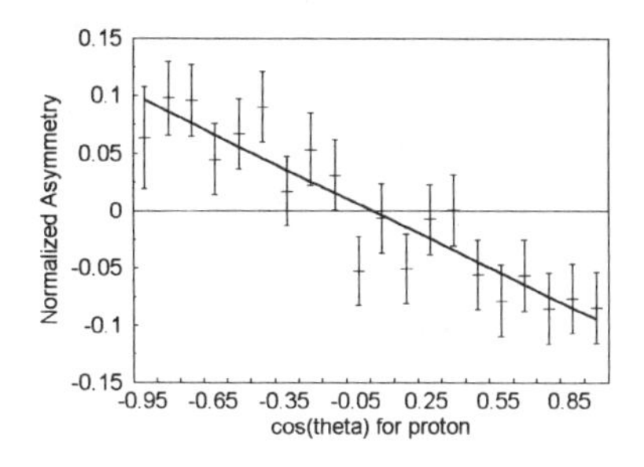

Lindsay Schachinger

the polarization gave us the Λ moment, μ(Λ), to an accuracy of 10%, about as good as all previous experiments combined. What could we do if we optimized conditions for measuring magnetic moments? My student, Lindsay Schachinger, claimed the precise measurement of the Λ magnetic moment as her thesis project, and we proposed E440, a run aimed at reducing statistical and systematic uncertainties on μ(Λ) by an order of magnitude.

The technique was simple. $\mathbf{P_o}$ at the production target was in the direction $-\mathbf{k_p}\times\mathbf{k_\Lambda}$, where $\mathbf{k_p}$ and $\mathbf{k_\Lambda}$ are the momentum vectors of the primary proton and produced Λ, respectively. $\mathbf{P_d}$ was measured downstream of M1, and ϕ_L, the precession angle is given by the difference in direction of $\mathbf{P_o}$ and $\mathbf{P_d}$. The equations of motion for

Andy Beretvas
Gordon Thomson

Ron Grobel

Russ Whitman

Larmor precession yield the relationship: $\phi_L = (4\pi\mu_\Lambda/hc)\int Bdz$, where $\mu_\Lambda = \mu(\Lambda)$ is the Λ magnetic moment, and $\int Bdz$ is the field integral. A directly useful form of this equation is: ϕ_L = (18.30 degrees/Tesla-meter)$(\mu_\Lambda/\mu_B)\int Bdz$, where μ_B is the proton Bohr magneton or nuclear magneton.

A complex choreography of runs at different magnet settings was adopted to provide redundant cancellation of systematic errors by frequently reversing the production angle, the precession field, and the spectrometer field. Our goal was to anticipate every possible criticism and ensure that our result was beyond reproach. I recall our concern about precession ambiguities in which the transformation $\phi_L \rightarrow \phi_L + n \bullet 180^o$ fits the data because the parity-allowed production polarization could be along either the positive or negative x-axis and **P** could precess an unknown number of full turns. We might resolve this by running at several values of the precession field, but an unlucky combination could still fool us. I derived the optimum choice: the maximum possible field and two rational fractions of it - even/odd and odd/odd ratios of small integers. This ruled out phony solutions for all reasonable values of the moment. Including field off and both polarities, we ran at seven fields over a range of 27 Tesla-meters and 300° in precession angle.

The asymmetry plot on the previous page shows the asymmetry along the z-axis for about 25,000 events in just two runs The full data sample of 3-million events, taken in many runs, gave us multiply redundant cross checks and a value of $\mu(\Lambda)$ = -0.6138(47) nuclear magnetons.[11]

Pete Border
Keith Krueger
Keith Thorne

Jay Dworkin (and Jason)

A few years earlier such precision for hyperon moments had been unthinkable, and our result was subjected to searching questions in talks at Fermilab and elsewhere. Here the cross checks paid off. Our results were accepted, and the value of the technique was proven. This and our other work were generally regarded as quite successful.

However, there were some disappointments. Our attempt to measure Λp and Λd total cross sections by attenuation of the neutral beam in a vacuum, hydrogen and deuterium targets was compromised by differences in charged-particle production rates in those targets. The efficiencies of the proportional chambers varied only slightly with charged particle flux, but it was enough effect to make the result unreliable. With a spin-transfer method, we measured the polarization of secondary-beam protons and found it to be zero with an accuracy of ±0.03,[12] a precise, but disappointing result which was characterized as "a failure" by one Fermilab official. Lee Pondrom and I wrote an article for Scientific American about the hyperon work which was rejected as not interesting.

Physical Review Letters, too, rejected a piece of work which they regarded as uninteresting, and I would like to describe it to you. A number of Λ's were deleted from the sample used to measure μ(Λ) because their momentum vectors, extrapolated upstream, did not point at the production target. One obvious source was production in the walls of the hyperon beam channel, but we made cuts to eliminate these. The remaining Λ's were almost certainly from $\Xi^0 \to \Lambda\pi^0$, and their decay vertex positions showed the typical parent-daughter behavior of a two-step decay process.

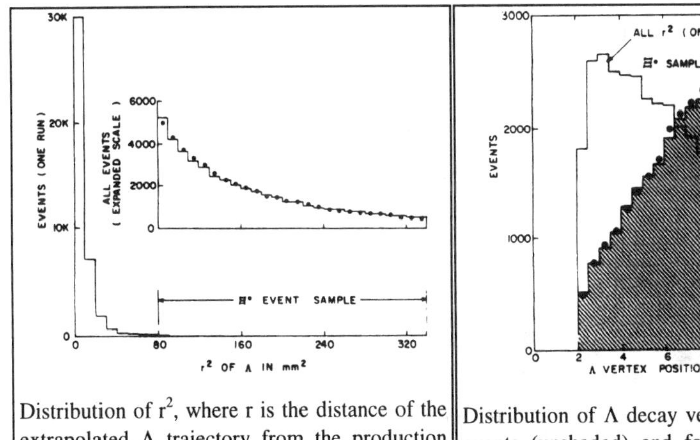

Distribution of r^2, where r is the distance of the extrapolated Λ trajectory from the production target. Beam Λ's fall in the peak near zero in the primary histogram. Daughter Λ's from Ξ^0 decay populate the long tail of the distribution above 80 mm^2.

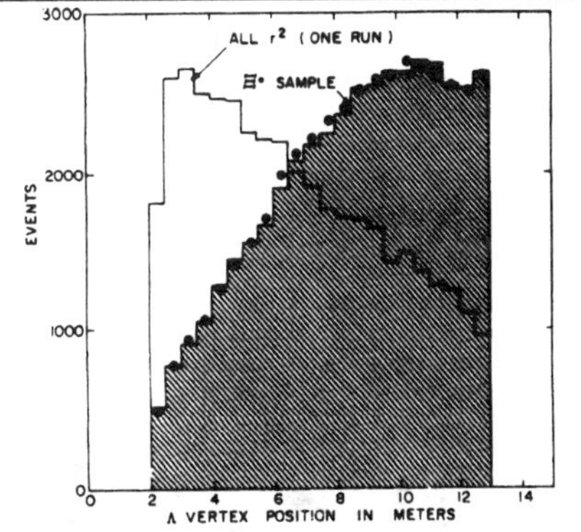

Distribution of Λ decay vertex positions for all events (unshaded) and for those with $r^2 > 80$ mm^2 (shaded). The former has the shape expected for a sample dominated by beam Λ's,while the latter behaves like a two-stage decay.

We found these daughter Λ's to be polarized, but in a direction different from that of the directly produced Λ's. When averaged over the full angular distribution of

Λ's from $\Xi^o \rightarrow \Lambda\pi^o$, the Λ "remembers" the parent Ξ polarization. This can be expressed as $\mathbf{P}_\Lambda = \mathbf{P}_\Xi(1 + 2\gamma_\Xi)/3$, where γ_Ξ is the parity-conserving decay parmater in $\Xi^o \rightarrow \Lambda\pi^o$. A study of the direction of polarization at various values of the field in M1 gave precession angles different from those of beam Λ's and showed convincingly that Ξ^o's were also polarized. The precession analysis gave us the first-ever measurement of the Ξ^o magnetic moment, $\mu(\Xi^o)$ which we published in Physics Letters[13] after it was rejected by PRL.

Tim Cox

Peter Yamin Ralph Edwards

We immediately proposed E495 to make a precise measurement of $\mu(\Xi^o)$, but this first detection of Ξ^o polarization had greater significance. It showed that the Λ polarization was not an isolated phenomenon. It was almost certain that Ξ^-, the isospin partner of Ξ^o, would be polarized. That generated a proposal for a series of charged hyperon experiments. Over the next ten years, most types of hyperons we produced, Λ, Σ^+, Σ^-, Ξ^0, Ξ^- were strongly polarized.[14,15,16,17] The Ω^- and $\overline{\Lambda}$ were exceptions.[18,19]

Les Deck Carol Wilkinson Byron Lundberg Craig Dukes

This series of measurements completed the magnetic moment matrix for the entire SU(3) baryon octet, including the one off-diagonal element.[20] They are described in our published papers. In my mind, however, each result recalls a name and a face. All these measurements are inseparable from the thesis students and other colleagues who worked on them.

Cat James Priscilla Cushman

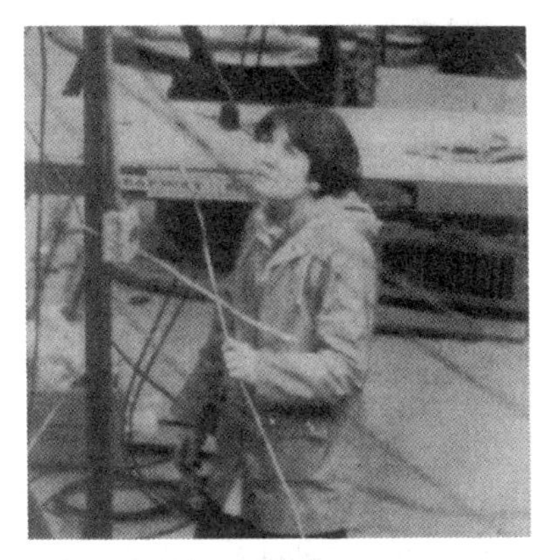

Gina Rameika

Kam-Biu Luk

Lee Pondrom and I left the E8 collaboration in the mid-1980's to do other work, but it continued with some our former students joining the leadership.

Now very precise measurements are available for all these hyperons from our work and that of some friendly competitors. In this age of very high energy colliding beams,high transverse-momentum collisions and infinite momentum frames, it convenient to downplay the importance of valence quarks in the structure of baryons. However, in the baryon rest frame, the magnetic moment data make it difficult to ignore the valence quarks. A very simple valence quark model gives a good account of the overall pattern of magnetic moments. One of the earliest, and still one of the best, arguments against the "elementary" nature of baryons is the pattern of their magnetic moments. In his talk, Lee Pondrom will review the current situation.

It is said that the E8 collaboration lasted longer than any other at Fermilab. Cooperation for such a long period is often difficult when large egos are involved. We had our share of these in E8, but, aside from brief technical disagreements,our group remained pretty amicable through this long series of runs. That is testament to good physics, although I often suspect that good cooking had something to do with it.These experiments had an elegance and simplicity which made them widely appealing and easy to describe in semi-classical terms. They were a joy to run, and they are a pleasure to recall.

---

[1] J. Bader et al., Phys. Lett. **39B**, 414 (1972); J. Badier et al., Phys. Lett. **41B**, 387 (1972).

[2] P. Skubic et al., Phys. Rev. **D18**, 3115 (1978).

[3] T. Devlin et al., Nucl. Phys. **B123**, 1 (1977).

[4] R. T. Edwards et al., Phys. Rev. **D18**, 76 (1978).

[5] K. Heller et al., Phys. Rev. **D16**, 2737 (1977).

[6] J. Lach and L. G. Pondrom, Ann. Rev. Nucl. Part. Sci. **29**, 203 (1979).

[7] L. G. Pondrom, Physics Reports **122**, 57 (1985).

[8] P. Martin, Ph.D. Thesis, U. of Wisconsin, Madison, 1979 (unpublished).

[9] B. Edelman et al., Phys. Rev. Lett. **40**, 491 (1978).

[10] G. Bunce et al., Phys. Rev. Lett. **36**, 1113 (1976).

[11] L. Schachinger et al., Phys. Rev. Lett. **41**, 1348 (1978).

[12] P. Yamin et al., Phys. Rev. **D23**, 31 (1981).

[13] G. Bunce et al., Phys. Letters **86B**, 386 (1979).

[14] C. Wilkinson et al., Phys. Rev. Lett. **58**, 855 (1987).

[15] L. Deck et al., Phys. Rev. **D28**, 1 (1983).

[16] P. T. Cox et al., Phys. Rev. Lett. **46**, 877 (1981).

[17] R. Rameika et al., Phys. Rev. Lett. **52**, 581 (1984).

[18] K.-B. Luk et al., Phys. Rev. **D38**, 19 (1988).

[19] K. Heller et al., Phys. Rev. Lett. **41**, 607 (1978).

[20] P. C. Petersen et al., Phys. Rev. Lett. **57**, 949 (1986).

Consequences of Hyperon Polarization

L. Pondrom
University of Wisconsin

1 November 1994

Abstract

Hyperon polarization has been observed almost everywhere. Unpolarized protons produce polarized hyperons. Unpolarized hyperons produce polarized hyperons. Polarized hyperons also produce other polarized hyperons. While some of these reactions make sense, others remain a puzzle. The polarization of antihyperons, which has resulted in impressive measurements of antibaryon magnetic moments, is a pleasant surprise.

1 Recent Polarization Measurements

The early experimental work on hyperon polarization by unpolarized protons has been reviewed by Heller [1], and by Pondrom [2]. The reaction $p + A \rightarrow \Lambda + X$ was the most thoroughly studied, and remains so today although thanks to recent data $p + A \rightarrow \Xi^- + X$ is a close second. Other inclusive reactions of this type where polarization has been found are $p + A \rightarrow \Xi^0$, $p + A \rightarrow \Sigma^+$, $p + A \rightarrow \Sigma^0$, and $p + A \rightarrow \Sigma^-$, while no polarization has been observed for $p + A \rightarrow \bar{\Lambda}$. Define the polarization vector in the reaction $a + b \rightarrow c + X$ as $P_c = P_0\hat{n}$, where $\hat{n} = \vec{k_a} \times \vec{k_c}/|\vec{k_a} \times \vec{k_c}|$. Write $\vec{P_\Lambda} = -f(x, p_\perp)\hat{n}$, where $x = p_{\Lambda\|}/p_{beam}$. Then the behavior of the polarization may be summarized by the following: (a) f has no strong s dependence- similar polarizations have been observed in the range $20GeV^2 < s < 3600GeV^2$; (b) $f > 0$ over the range of $(x, p_\perp)$ studied; (c) $\partial f/\partial x > 0$; (d) $\partial f/\partial p_\perp > 0$ for $p_\perp < 1.5GeV/c$, and $\partial f/\partial p_\perp = 0$ for $p_\perp > 1.5GeV/c$. For a Be target $f(x =$

$0.5, p_\perp = 1.4 GeV/c) = 0.19 \pm 0.014$[3]. The magnitude of the polarization for a lead target is about 75% of the beryllium value. Analytic forms for $f(x, p_\perp)$ are given in Refs [2] and [3]. Other hyperon polarizations may be roughly described in terms of the Λ polarization $\vec{P_\Lambda}$ by the following relations: $\vec{P}_{\Sigma^+} \sim \vec{P}_{\Sigma^-} \sim -\vec{P}_\Lambda$; and $\vec{P}_{\Xi^-} \leq \vec{P}_{\Xi^0} \sim \vec{P}_\Lambda$.

This is all ancient history. What has happened lately? Work has been done in three related areas: (a) further studies of hyperon polarization produced by an unpolarized proton beam; (b) studies of hyperon polarization by an unpolarized hyperon beam; and (c) studies of spin transfer from one polarized hyperon to another.

Duryea et al. [4] have extended the kinematic range of $\vec{P}_{\Xi^-}$. They find that although $\vec{P}_{\Xi^-}$ and $\vec{P}_\Lambda$ are similar, $|P_{\Xi^-}|$ does not exhibit the same monotonic increase with increasing x. It flattens out for $x > 0.4$. In the quark model this could be due to a difference between one quark and two quark exchanges. The cross sections for the two reactions have very different x dependences.

Woods et al. [5] have pursued polarization of Ξ^- and Ω^- all three ways mentioned above, by building an intermediate neutral beam between the proton beam and the charged hyperon beam with two production targets, the first for the neutrals and the second for the negatives. By transporting the proton beam directly to the second target they could study $p \to \Omega^-$, or by inserting both targets they could study $p \to n, \Lambda, \Xi^0 \to \Omega^-$, where in each case the final state could just as well be a Ξ^-. The polarization conditions could be varied by changing the production angles θ_1 for the $p \to neutral$ reaction and θ_2 for the $neutral \to negative$ reaction. Two combinations were used: $\theta_1 = 0$ and $\theta_2 = 1.8$ mr for an unpolarized neutral beam to produce possibly polarized hyperons; and $\theta_1 = 1.8$ mr and $\theta_2 = 0$ for a polarized neutral beam to produce possibly polarized hyperons by spin transfer. The average momentum of the negative beam was fixed at 395 GeV/c, giving $p_\perp = 0.7$ GeV/c for $\theta_2 = 1.8$ mr. Since the incident proton beam momentum was 800 GeV/c, x = 0.5 for direct production. The average x in the two stage beam depended on the energies of the neutrals, and their relative contributions to the flux of 395 GeV/c negatives. Without disentangling the various channels, it is safe to say that $x > 0.5$ regardless of which neutral produced the negative hyperon. Their results are summarized in Table 1.

The last row in the Table refers to direct production of the negative beam by 800 GeV/c protons. The value for $\vec{P}_{\Omega^-}$ in the lower right corner comes

Table 1: $\vec{P}$ for Ξ^- and Ω^-

θ_1	θ_2	$\vec{P}_{\Xi^-}$	$\vec{P}_{\Omega^-}$
0.	1.8 mr	0.0062 ± 0.0042	0.053 ± 0.012
1.8 mr	0.	-0.1172 ± 0.0062	-0.076 ± 0.021
no target	1.8 mr	-0.120 ± 0.005	-0.01 ± 0.01

Table 2: $\vec{P}_Y$ by Unpolarized Σ^-

$p_\perp$ GeV/c	$< x >$	$\vec{P}_\Lambda$	$\vec{P}_{\Sigma^+}$	$\vec{P}_{\Xi^-}$
1.0	.3	-0.022 ± 0.008	-0.031 ± 0.023	-0.090 ± 0.041
1.3	.3	-0.055 ± 0.015	-0.051 ± 0.035	-0.092 ± 0.064

from Luk et al. [6].

Work closely related to the first row of numbers in Table 1 has been reported by Adamovich et al. [7]. These authors measured the polarization of Λ, $\bar{\Lambda}$, Σ^+, and Ξ^- produced by an unpolarized 330 GeV Σ^- beam striking a Cu or C target. The π^- component of the negative beam was suppressed by a transition radiation veto detector; the pion contribution to the final state hyperons with $x \geq 0.2$ was negligible. An open geometry magnetic spectrometer detected the decay products of the final state hyperons. Representative results from these measurements, omitting the $\bar{\Lambda}$ polarizations which were all consistent with zero, are given in Table 2.

So far the data accumulated for three quark exchange processes, that is $p \to \Omega^-$ [6] and $p \to \bar{\Lambda}$ [3] [7], showed no polarization for unpolarized protons. It was natural to assume that such reactions were too complicated to give any polarization. Assumptions of this sort have proven incorrect in the past, and history has repeated itself. Two experiments have observed significant polarization in the inclusive reactions $p \to \bar{\Xi}^+$ [8], and $p \to \bar{\Sigma}^-$ [9]. These startling results are given in Table 3.

Note that the $\vec{P}_{\bar{Y}}$ polarizations have the same signs and roughly the same magnitudes as the corresponding $\vec{P}_Y$! How can the $\bar{\Sigma}$ be polarized and not the $\bar{\Lambda}$?

Table 3: $\vec{P}_{\bar{Y}}$

$\bar{Y}$	$p_\perp$ GeV/c	$< x >$	$\vec{P}_{\bar{Y}}$
$\bar{\Xi}^+$	0.76	0.39	-0.097 ± 0.015
$\bar{\Sigma}^-$	0.7	0.47	0.088 ± 0.011
$\bar{\Sigma}^-$	1.07	0.47	0.068 ± 0.011

2 Discussion of the Polarization

The simplest framework for thinking about the polarization is the constituent quark model of the baryons. This model accounts adequately, if not perfectly, for the baryon magnetic moments, which are spin and flavor dependent quantities. In this context the Λ polarization in $p \to \Lambda$ is created by acquiring a negatively polarized strange quark: $(u, u, d) \to (s \downarrow, u, d)$. The (ud) diquark in the Λ is in a singlet spin state. Promoting this $s \downarrow$ quark to a universal feature satisfactorily accounts at least for the algebraic signs of the Λ, Σ^+, Σ^-, Ξ^0, and Ξ^- polarizations.

These ideas work when applied to Table 2, hyperon polarizations from an unpolarized Σ^- beam. Thus $\Sigma^- \to \Lambda$ or $\Sigma^- \to \Sigma^+$, involving no exchange of polarized strange quarks, should show zero, or at least small polarizations, and they do. On the other hand, $\Sigma^- \to \Xi^-$ does require the exchange of a $s \downarrow$ quark, and shows larger negative polarization. The errors are not small, but there are no glaring inconsistencies. Likewise the spin exchange data, row two of Table 1, is consistent with the $s \downarrow$ quarks transferring from one hyperon to the other. How about the top row of Table 1, hyperon production by an unpolarized neutral beam? Well, the Ξ^- polarization is zero within errors. That makes sense if the dominant channel is $\Xi^0 \to \Xi^-$, which exchanges no strange quarks.

The positive value of $\vec{P}_\Omega$ unfortunately does not agree with our picture at all. Neutral production of polarized Ω^- must be either $\Lambda \to \Omega$ or $\Xi^0 \to \Omega$, since $n \to \Omega$ has no polarization. Setting aside the details of the baryon spin structure in terms of quarks, the quark exchange for $\Lambda \to \Omega$ is $(u, d) \to (s, s)$, which is the same as $p \to \Xi^0$, while $\Xi^0 \to \Omega$ is $u \to s$, the same as $p \to \Lambda$. Since each of the proton reactions leads to negatively polarized hyperons, the positive sign of the Ω^- polarization is difficult to explain. This is the first example that I am aware of which does not involve the exchange of three quarks, and yet does not work at all in the $s \downarrow$ scheme.

Is the measured sign of the Ω^- polarization above suspicion? No, because the sign of the decay parameter γ for $\Omega^- \to \Lambda + K^-$ is unknown, and assigning positive polarization to the Ω amounts to choosing $\gamma = +1$. On the other hand, the Ω polarization changes sign from unpolarized neutral production to spin exchange. So if $\gamma = -1$ is chosen, the dilemma regarding neutral production disappears, but then the spin exchange reaction does not make sense. You can't win. Besides, $\gamma = +1$ is most likely correct.

The reactions in Table 3, polarizations for $p \to \bar{Y}$, are all three quark exchanges, and $Y\bar{Y}$ pair production as well, since baryon number must be conserved. The cross section ratio $\bar{Y}/Y$ produced by protons increases with increasing strangeness [10], from about 1% for $\bar{\Lambda}/\Lambda$ to 5% for $\bar{\Xi}/\Xi$ to 20% for $\bar{\Omega}/\Omega$. It is interesting that Ω^- production with three K mesons is favored over $\bar{\Omega}\Omega$ pairs, despite the fact that the masses are about the same for the two channels. Since at the present time we have no knowledge of what exclusive final states contribute to the Ξ^- polarization, we might speculate that $\bar{\Xi}\Xi$ is a major player. Thus if Ξ is polarized, and the baryons are produced in pairs, it is plausible that $\bar{\Xi}$ is also polarized. The relative sign depends on the total spin state, and whether the pair is produced on the same or opposite sides of the proton beam, the transverse momentum being small. This idea does not work for $\bar{\Sigma}$, where the $\bar{Y}/Y$ ratio is too small, and of course the inconsistency with the zero polarization for $\bar{\Lambda}$ remains.

It may be that the antihyperon polarizations are due to an unrelated phenomenon, even though they appear to be similar to the polarizations of the corresponding hyperons. Just because there may not be a common origin does not mean that one should not try to understand what is going on. However, the positive polarization of the Ω^- from the neutral hyperon beam should make sense within the standard framework, and it does not.

3 The Magnetic Moments

One approach to hyperon polarization is to not worry about it – just enjoy it. The phenomenon has permitted the measurement of hyperon magnetic moments to unprecedented accuracy. The experimental situation a few years

Table 4: Magnetic Moments

baryon	magnetic moment (nm)
Σ^+	$2.4613 \pm 0.0034 \pm 0.0040$
$\bar{\Sigma}^-$	$-2.428 \pm 0.036 \pm 0.007$
Ξ^-	-0.6505 ± 0.0025
$\bar{\Xi}^+$	$0.657 \pm 0.028 \pm 0.020$
Ω^-	-2.024 ± 0.056

ago was reviewed by Lach[11]. There are five results of interest which have since appeared. Precision measurements of μ_{Σ^+} [12], μ_{Ξ^-} [13], μ_{Ω^-} [14]; and the antihyperon moments $\mu_{\bar{\Xi}^+}$ [8], and $\mu_{\bar{\Sigma}^-}$ [12]. These results are displayed in Table 4.

In the Table where two errors are quoted the first is statistical and the second is systematic. The new value of the Σ^+ moment settles a mild controversy between two earlier measurements. The errors on the magnetic moments are less than 1% for the Λ, the Σ^+, and the Ξ^-, with the Ξ^0, the Σ^-, and now the Ω^- not far behind. This accuracy demonstrates the power of polarized beam magnetic moment technique. The antihyperon magnetic moments are an unexpected bonus. Each agrees well in magnitude and has the opposite sign of its respective hyperon, consistent with the CPT theorem. The errors for the $\bar{\Xi}^+$ moment are smaller than the original errors on the Ξ^- moment [15] using the same technique. The simple quark model of the baryon magnetic moments predicts that $\mu_\Omega = 3 \times \mu_\Lambda = -1.893 \pm 0.012$ nm. The difference is $\Delta = 0.185 \pm 0.057$ nm, about three standard deviations, and a magnitude typical of other differences between the magnetic moments and the model.

4 Future Prospects

The strange quark magnetic moment has been fairly well studied. That leaves the charm quark and the bottom quark. One would expect $\mu_c \sim 0.67 \times |\mu_s|$, while $\mu_b \sim 0.11 \times \mu_s$. The bottom quark magnetic moment is very small, and not much is known about b baryons, so the charm quark is the more likely prospect. The analog of the Λ, the Λ_c^+, with quark content (u,d,c), has a flight path $c\tau = 60\mu$, so techniques other than precession in a conventional magnet must be used. Fortunately the spin precession in a bent channel crystal has been experimentally observed for the Σ^+, and could in principle be applied to the Λ_c [16]. We await further developments.

5 Acknowledgments

I want to thank Ken Heller and other organizers of this Conference for the opportunity to give this report. Private communications from David Woods were especially helpful.

References

[1] K. Heller, VII International Symposium on High Energy Spin Physics, Serpukhov, 1987, p 81.

[2] L. Pondrom, Phys Rep **122**, 57 (1985).

[3] B. Lundberg et al., Phys Rev D **40**, 3557 (1989).

[4] J. Duryea et al., Phys Rev Lett **67**, 1193 (1991).

[5] D. M. Woods, PhD Thesis University of Minnesota, 1995(unpublished). See also Proceedings of the Fermilab 1993 DPF Meeting, (to be published).

[6] K. B. Luk et al., Phys Rev Lett **70**, 900 (1993).

[7] M. I. Adamovich et al (WA89 Collaboration), Zeit Phys A (to be published)

[8] P.M. Ho et al., Phys Rev D **44** 3402 (1991).

[9] A. Morelos et al., Phys Rev Lett **71**, 2172 (1993).

[10] See Figure 58 of Reference [2].

[11] J. Lach, VIII International Symposium on High Energy Spin Physics, Ed K Heller, AIP Conference Proceedings No. 187, New York, 1989(p353).

[12] A. Morelos et al., Phys Rev Lett **71**, 3417 (1993).

[13] J. Duryea et al., Phys Rev Lett **68**, 768 (1992).

[14] P.M. Border, these Proceedings. See also N.B. Wallace, PhD Thesis, Univ of Minnesota, 1995 (unpublished).

[15] R. Rameika et al., Phys Rev Lett **52**, 581(1984).

[16] D. Chen et al., Phys Rev Lett **69**, 3286(1992).

Measurements of $\Delta\sigma_L(pp)$ and $\Delta\sigma_L(\bar{p}p)$ at 200 GeV/c*

D.P. Grosnick†

High Energy Physics Division,
Argonne National Laboratory, Argonne, IL 60439, USA

A measurement was made at Fermilab of the difference in the total cross sections between states with beam and target polarizations aligned antiparallel and parallel, $\Delta\sigma_L = \Delta\sigma\left(\substack{\rightarrow\\ \leftarrow}\right) - \Delta\sigma\left(\substack{\rightarrow\\ \rightarrow}\right)$, using 200-GeV/c, polarized proton and antiproton beams and a polarized proton target. This measurement explores the spin dependence of particle interactions and the constituent dynamics. A difference in the spin-dependent total cross sections has been observed in previous experiments at lower energies [1], and this experiment was the first to explore possible spin effects in $\Delta\sigma_L$ at much higher energies.

The polarized proton beam is produced from parity-nonconserving decays of the Λ^0 hyperon, where protons emitted from these decays have their spins aligned along the direction of their momenta [2]. The Λ hyperons are produced, along with other particles, when unpolarized, 800-GeV/c protons strike a beryllium target. A virtual source of protons is produced at the target from these decays and the proton polarization is correlated to the transverse distance from the target. A beam of polarized antiprotons can be produced in an analogous manner using $\overline{\Lambda^0}$ decays. The beam transport system goal was to produce no net spin precession and to preserve the correlation between the beam particle polarization and the transverse position at the virtual source. The beam transport used four sets of quadrupole magnets to focus the beam at intermediate and final focal points. An electronic, particle-tagging system was located at the intermediate focus, while the polarized target was near the final focus. A detailed description of the polarized beam is given in Ref. 3.

Because this beam is produced from decays, the polarization of each beam particle was determined through a tagging system. The momentum and polarization of each beam particle is measured using a series of overlapping scintillator hodoscopes and look-up electronics. Signals from the tagging system were then sent to the experimental trigger. Plus and minus polarization magnitudes were in bins of 10% between 25% and 64%, while those particles tagged with polarizations between -25% and +25% were assigned zero polarization. An absolute calibration of the beam polarization was performed using two polarimeters: one using the Primakoff effect [4] and the other Coulomb-nuclear interference (CNI) [5]. These measurements gave an absolute measurement of the beam polarization to about 15%.

A series of twelve spin-rotation (snake) dipole magnets rotated the particle spin direction from the transverse horizontal (S) to the longitudinal (L) direction. The particle spin direction was changed by reversing the magnet polarity, and was reversed every 10 spills, with about 5 reversals per hour. Two beam Čerenkov counters detected the pion contamination, which for the polarized proton beam was about 13%, while the polarized antiproton beam contained five times more π^-'s than $\overline{\mathrm{p}}$'s. Typical intensities for this experiment were 1.2×10^7 and 8.5×10^6 per 20-second spill for the polarized p and $\overline{\mathrm{p}}$ beams, respectively. The average beam polarization was 0.42 for polarization values between 25% and 55%.

The polarized proton target was a frozen spin target and produced a polarization of about 75% through the dynamic nuclear polarization process. The target material was pentanol, with a 13% hydrogen fraction and a target constant of 1040 ± 38 mb. The target size was a 3-cm-diameter cylinder, 20 cm in length. A ^{3}He-^{4}He dilution refrigerator was used to cool the target to about 40 mK and a superconducting solenoid, capable of a 6.5 T magnetic field, was used to align the spin direction. The spins of the target protons were reversed approximately every 24–36 hours to reduce systematic effects.

The apparatus used to measure the particle scattering consisted of several overlapping scintillator hodoscopes. Two of these, one upstream and one downstream of the snake magnets, measured the beam particle trajectory on to the polarized target. A third transmission hodoscope located 13 m downstream of the target measured the amount of scatter from the beam-target collisions. Each hodoscope could locate the hit position within a 2 mm segment size in both the vertical and horizontal directions. The amount of scattering was found by calculating the difference between the undeflected particle trajectory projected to a segment on the transmission hodoscope and the segment actually struck. A t value was calculated from $t \approx -(p\,\theta)^2 = -p^2(\Delta x^2 + \Delta y^2)/d^2$, where p is the beam particle momentum, Δx and Δy the amount of deflection in the horizontal and vertical directions, respectively, and d is the distance between the polarized target and the transmission hodoscope. This calculation was performed online using memory-lookup electronics. The number of particles scattered into each t bin for each beam polarization bin was then counted; no single events were recorded, and hence large number events could be counted rapidly. Corrections adjusted the number of counts in each t bin due to the grid geometry.

The value of $\Delta\sigma_L$ per t bin is given by: $\Delta\sigma_L = -2\,A\,\epsilon/P_B \cdot P_T$, where ϵ is the asymmetry of the number of particles transmitted for beam and target spin aligned antiparallel (+) or parallel (−), $\epsilon = (N_T^+/N_0^+ -$

$N_T^-/N_0^-)\,/\,(N_T^+/N_0^+ + N_T^-/N_0^-)$, where N_T is the number of transmitted particles and N_0 is the total number of particles, A is the target constant, P_B is the average beam polarization, and P_T is the average target polarization. The statistical accuracy scales as the inverse square root of the total number of particles, $1/\sqrt{N_0}$. Three quantities were periodically reversed to reduce systematic errors: (1) the polarization state of tht beam particle, (2) the spin rotation direction by the snake magnets, and (3) the spin direction of the protons in the polarized target. A total of eight quantities were then summed appropriately to determine the number of particles in each of the parallel and antiparallel states. From the data, the first t bin contains about 97% of the transmitted number of particles to about 3% background counts. The transmitted asymmetry can be found from the formula, $\epsilon_{Trans} = (1 + B/N)\,\epsilon_{T1} - (B/N)\,\epsilon_{Back}$, where B is the total number of background particles, N is the total number of particles measured, ϵ_{T1} is the asymmetry calculated from the first t bin, and ϵ_{Back} is the asymmetry due to the background particles. The value of ϵ_{Back} was found by making a fit of the asymmetries formed from the number of particles scattered into each t bin, excluding the first one, as a function of t. The value of ϵ_{Back} is fairly insensitive to the value of B. The preliminary result for $\Delta\sigma_L(pp)$, along with the values of ϵ_{T1}, ϵ_{Back}, and ϵ_{Trans}, are given in Table I for the polarization values between 35 and 55% and for the expanded polarization values between 25 and 55%. These same quantities are given for $\bar{p}p$ scattering in Table II. Both sets of results are consistent with zero and the errors given are statistical only. A second detector system also gave results consistent with zero.

TABLE I. $\Delta\sigma_L\,(pp)$ results for 2 different beam polarization bins.

Quantity	Beam Pol 35–55%	Beam Pol 25–55%
ϵ_{T1}	0.000000 ± 0.000006	-0.000001 ± 0.000005
ϵ_{Back}	0.000026 ± 0.000061	0.000046 ± 0.000053
ϵ_{Trans}	-0.000001 ± 0.000006	-0.000003 ± 0.000006
$\Delta\sigma_L(pp)$	$-4 \pm 40\,\mu$b	$-18 \pm 38\,\mu$b

TABLE II. $\Delta\sigma_L\,(\bar{p}p)$ results for 2 different beam polarization bins.

Quantity	Beam Pol 35–55%	Beam Pol 25–55%
ϵ_{T1}	-0.000031 ± 0.000016	-0.000025 ± 0.000014
ϵ_{Back}	0.000051 ± 0.000150	0.000074 ± 0.000131
ϵ_{Trans}	-0.000034 ± 0.000017	-0.000029 ± 0.000015
$\Delta\sigma_L(\bar{p}p)$	$-199 \pm 99\,\mu$b	$-183 \pm 94\,\mu$b

Many systematic effects such as beam motion, momentum and polarization distributions, geometrical effects, and snake magnet effects have been studied. False asymmetries such as adding + and − polarization states to produce a "fake zero" polarization and summing every other spill together, have also been calculated in an analogous manner to $\Delta\sigma_L$. As a cross check of the data, a CNI-like measurement was made using a vertically-aligned (N) beam polarization. The measured left-right asymmetry was found to be $+0.002491 \pm 0.000183$, a $13\,\sigma$ effect, while the up-down asymmetry is consistent with zero. Calculating left-right and up-down asymmetries using the longitudinally-polarized beam gives nonzero contributions, which may correspond to a S- or N-type component. Similar results were found in studies of the $\overline{\text{p}}$ beam. At present, the systematic errors have not been included in the $\Delta\sigma_L$ result, but are not expected to be dominant.

There are currently two different theoretical predictions for $\Delta\sigma_L$. One [6] of these predictions is based on conventional Regge phenomenology, where $\Delta\sigma_L\,(pp;s) \sim s^{-1.15}$, and s is the square of the energy. For $\sqrt{s} = 20\,\text{GeV}$, $\Delta\sigma_L \approx -19\,\mu\text{b}$. The other prediction [7] is based on phenomenology from jet physics, where $\Delta\sigma_L\,(pp;s)$ is divided into two parts, one from coherent hadronic dynamics and the other, $\Delta\sigma_L^{jet}$, is the contribution from constituent parton scattering. A measurement of $\Delta\sigma_L^{jet}$ can provide information on the spin-dependent gluon contribution, ΔG, since the gluon processes are dominant in this kinematic region. For a large ΔG, $\Delta\sigma_L^{jet} \approx 26\,\mu\text{b}$, and for no ΔG contribution, $\Delta\sigma_L^{jet} \approx 2\,\mu\text{b}$. Figure 1 shows the comparison of theoretical predictions and results of this experiment.

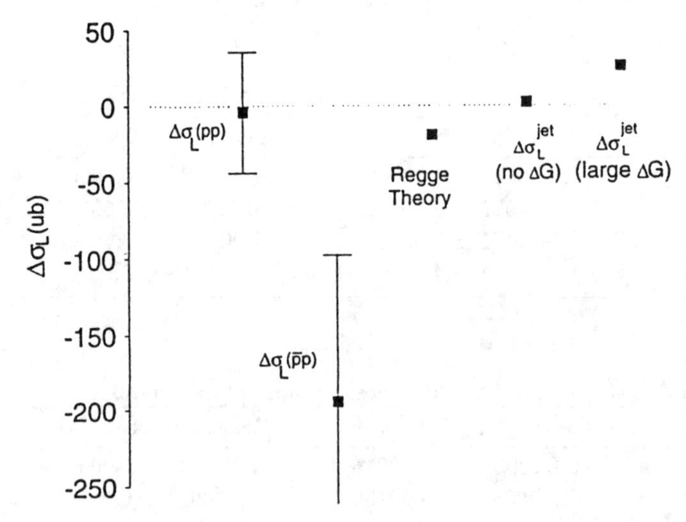

FIGURE 1. Comparison of theoretical predictions with results from this experiment.

The measurements of $\Delta\sigma_L\,(pp)$ and $\Delta\sigma_L\,(\bar{p}p)$ at 200 GeV/c are both consistent with zero, within statistical errors. The current results are not able to distinguish between theoretical models and predictions.

References

* Work supported by the U.S. Department of Energy, Division of High Energy Physics, Contract W-31-109-ENG-38.

† For the FNAL E-581/704 Collaboration: Argonne, CEN-Saclay, Fermilab, Fort Hays St., IHEP-Serpukhov, INFN-Messina, INFN-Trieste, Iowa, Kita-Kyushu, Kyoto, Kyoto Education, Kyoto-Sangyo, LAPP-Annecy, Los Alamos, Northwestern, Rice, Udine.

1. See for example, J. Bystricky *et al.*, Phys. Lett. **B142**, 130 (1984); I.P. Auer *et al.*, Phys. Rev. Lett. **62**, 2649 (1989).
2. J.W. Cronin and O.E. Overseth, Phys. Rev. **129**, 1795 (1963).
3. D.P. Grosnick *et al.*, Nucl. Instrum. Methods **A290**, 269 (1990).
4. D.C. Carey *et al.*, Phys. Rev. Lett. **64**, 357 (1990).
5. N. Akchurin *et al.*, Phys. Lett. **B229**, 299 (1989); N. Akchurin *et al.*, Phys. Rev. D **48**, 3026 (1993).
6. E. Berger *et al.*, Phys. Rev. D **17**, 2971 (1978); W. Grein and P. Kroll, Nucl. Phys. **137** 173 (1978).
7. G. Ramsey, D. Richards, and D. Sivers, Phys. Rev. D **37**, 3140 (1988); G. Ramsey and D. Sivers, Phys. Rev. D **43**, 2861 (1991).

Theoretical Aspects of Single-Spin Asymmetries Studies

S. M. Troshin, N. E. Tyurin
Institute for High Energy Physics,
142284 Protvino, Moscow Region, Russia

Abstract

We consider theoretical background for experimental measurements of single-spin asymmetries. We stress the non-perturbative QCD aspects of observed asymmetries in hadronic reactions.

The very important direction in spin studies is connected with the long–standing problem of one–spin transverse asymmetries observed in violent hadron reactions [1], [2]. It is well known fact that the experimental data manifest significant one–spin transverse asymmetries.

For example, the behavior of analyzing power in hadronic scattering is rather surprising. Indeed, we could expect significant spin effects in soft reactions where the chiral $SU(3)_L \times SU(3)_R$ symmetry of QCD Lagrangian is spontaneously broken down to $SU(3)_V$ and therefore, there is no ground for helicity conservation. However, the observed analyzing power in the region of low transferred momenta is small and decreases with energy like an inverse power of energy.

On the other side, contrary to our QCD expectations analyzing power increases with transverse momentum when we trying to explore the region of short distances. In this kinematical region we should observe helicity conservation due to chiral invariance of QCD Lagrangian. Hadron helicity conservation in hard processes is a general principle of perturbative QCD. Violation of this principle have been observed in elastic *pp*–scattering, in two-body hadronic decays of J/ψ and there are also indications for such violation in the measurements of Pauli form factor $F_2(Q^2)$.

It is evident now that new ideas and experimental data are urgently needed to study dynamics of the spin effects.

We consider possible dynamical mechanism of spin effects in elastic scattereing. In Ref. [3] we used the notions of effective chiral quark model for the description of elastic scattering at small and large angles. Different aspects of hadron dynamics were accounted in the framework of effective Lagrangian presented as a sum of three terms:

$$\mathcal{L} = \mathcal{L}_\chi + \mathcal{L}_I + \mathcal{L}_C. \quad (1)$$

$\mathcal{L}_\chi$ is the term responsible for the spontaneous chiral symmetry breaking:

$$\mathcal{L}_\chi = \bar{\psi}(i\partial^\mu \gamma_\mu - \hat{m})\psi + \mathcal{L}_4 + \mathcal{L}_6. \quad (2)$$

$\mathcal{L}_4$ is the NJL four-fermion interaction, $\mathcal{L}_6$ is the $U_A(1)$–breaking 6–quark interaction. $\mathcal{L}_\chi$ is responsible for providing constituent quark masses and for the structure of constituent quark which includes valence quark and cloud of quark-antiquark pairs [4]. $\mathcal{L}_I$ describes the interaction of constituent quarks and $\mathcal{L}_C$ — their confinement. These parts of effective interaction were taken into account at phenomenological level.

In such a model quarks appear as quasiparticles and have a complex structure. Besides its mass (consider u-quark as an example)

$$m_u = m_u^0 - g_4\langle uu\rangle - g_6\langle \bar{d}d\rangle\langle \bar{s}s\rangle \quad (3)$$

the constituent quark has a finite size. We assume that the strong interaction radius of q-quark r_q is determined by its mass: $r_q = \xi/m_q$. The common feature of the chiral models is the representation of a baryon as an inner core carring the baryonic charge and an outer condensate surrounding this core [5]. Following this picture it is natural to represent a hadron consisting of the inner region where valence quarks are located and the outer region filled with quark condensate [3]. Such a picture for the hadron structure implies that overlapping and interaction of peripheral condensates at hadron collision occurs at the first stage. In the overlapping region the condensates interact and as a result the massive quarks appear. Being released the part of hadron energy carried by the peripheral condensates goes for the generation of massive quarks. In another words nonlinear field couplings transform kinetic energy into internal energy of dressed quarks (see the arguments for this mechanism in [6] and references therein for the earlier works). Of course, the number of such quarks fluctuates. The average number of quarks should be proportional to convolution of the condensate distributions D_c^H of colliding hadrons:

$$N(s,b) \propto N(s) \cdot D_c^A \otimes D_c^B, \quad (4)$$

where the function $N(s)$ is determined by the thermodynamics of transformation of kinetic energy of interacting condenstates to the internal energy of massive quarks. To estimate the $N(s)$ it is feasible to assume that it is determined by the maximal possible energy dependence

$$N(s) \simeq \kappa \frac{(1-\langle x\rangle_q)\sqrt{s}}{m_q}, \quad (5)$$

where $\langle x \rangle_q$ is the average fraction of energy carried by valence quarks, m_q is the mass of constituent quark.

In the model [3] valence quarks located in the central part of a hadron are supposed to scatter in a quasi-independent way by the produced massive quarks at given impact parameter and by the other valence quarks. The averaged scattering amplitude of valence quark then may be represented in the form

$$\langle f_q(s,b) \rangle = [N(s,b) + N - 1] \langle V_q(b) \rangle, \tag{6}$$

where $N = N_1 + N_2$ is the total number of valence quarks in colliding hadrons, and $\langle V_q(b) \rangle$ is the averaged amplitude of single quark-quark scattering [3].

In this approach elastic scattering amplitude satisfies unitarity equation since it is constructed as a solution of the following equation [7]

$$F = U + iUDF \tag{7}$$

which is presented here in operator form. This relation allows one to satisfy unitarity provided the inequality $\mathrm{Im}U(s,b) \geq 0$ is fulfilled. The function $U(s,b)$ (generalized reaction matrix) [7] — the basic dynamical quantity of this approach — is chosen as a product of the averaged quark amplitudes

$$U(s,b) = \prod_{q=1}^{N} \langle f_q(s,b) \rangle \tag{8}$$

in accordance with assumed quasi-independent nature of valence quark scattering.

The b-dependence of function $\langle f_q \rangle$ is related to the quark formfactor behavior $\propto (\vec{q}^2 + m_q^2/\xi^2)^{-2}$ and has a simple form [3] $\langle f_q \rangle \propto \exp(-m_q b/\xi)$.

Following the lines of the above considerations, the generalized reaction matrix in the pure imaginary case can be represented in the form

$$U(s,b) = iG(N-1)^N \left[1 + \alpha \frac{\sqrt{s}}{m_q}\right]^N \exp(-Mb/\xi), \tag{9}$$

where $M = \sum_{q=1}^{N} m_q$. This expression allows one to get the scattering amplitude as a solution of Eq. 7 which reproduces the main regularities observed in elastic scattering at small and large angles and consider spin phenomena.

For that purposes system of equations for helicity amplitudes has been solved and dynamical mechanism of quark scattering with and without helicity flip has been considered.

In particular spin of constituent quark in this model comes from the orbital moment of cloud of quark-antiquark pairs while the polarization of valence current quark and the polarization of the cloud of $\bar{q}q$ pairs compensate each other, e.g. for z-component of spin it means

$$S_q = S_{qv} + S_{\{\bar{q}q\}} + \langle L_{\{\bar{q}q\}} \rangle = 1/2 - 1/2 + 1/2 = 1/2.$$

The above compensation occurs due to account of axial $U(1)_A$ anomaly in the framework of effective QCD. While considering the constituent quark as an extended object we can represent its spin as follows:

$$S_q = \langle L_{\{\bar{q}q\}} \rangle = \omega I_q,$$

where ω is the angular velocity of quark matter inside the constituent quark and I_q its moment of inertia. These notions on spin of constituent quark follows from consideration of spin in the framework of effective lagrangian approach to QCD [8], [9].

It should be noted that since spin of constituent quark is due to its orbital angular momentum the corresponding wave function should be equal to zero at $r = 0$ due to centrifugal barrier. Such picture was advocated for the proton as a whole by Ralston and Pire [10].

Quark helicity flip in the model is provided by the mechanism of quark exchange where valence quark is exchanged with the quark produced under interaction of condensates. These quarks have different helicities and therefore such mechanism can lead to helicity flip quark scattering. Helicity non-flip quark scattering has another origin resulting from optical type of interaction. The above difference of these mechanisms leads to the different energy dependence and different phases of helicity flip and non-flip quark scattering amplitudes. Helicity amplitudes at hadron level in this approach as it was already mentioned are obtained as solutions of coupled system of equations which accounts unitarity in direct channel. Analyzing power in the framework of this model does not decrease with energy and has a non–zero value at $s \to \infty$. The value of analyzing power depends on the fraction of energy carried by valence quarks k and the phase difference $\Delta(s) \propto (1-k)\sqrt{s}/m_q$ and has the following form

$$A(s,\theta) = \frac{4\sin\Delta(s)}{(1-k)N} f(\theta)\left[1 + O\left(\frac{m_q^2}{s}\right)\right], \tag{10}$$

where N is the total number of valence quarks in colliding hadrons and $f(\theta)$ is the known function of scattering angle. Asymmetry here results from interference of helicity amplitudes which occurs due to resonance type of quark helicity flip scattering and continuum type of quark helicity non-flip scattering.

Analyzing power at $\sqrt{s} = 2$ TeV and $-t = 10$ $(GeV/c)^2$ in $p_\uparrow\bar{p}$–elastic scattering is predicted to be 12% while at $-t = 5$ $(GeV/c)^2$ — $A = 7\%$. Other non–perturbative models [11], [12] also predict non–zero values for analyzing power in TeV energy range.

To summarize, the measurement of analyzing power in elastic scattering at high energies will allow

- test perturbative QCD, mechanism, get knowledge on the region of applicability of perturbative QCD, study the transition from nonperturbative to perturbative phase of QCD;

- study of hadron structure and non–perturbative effects: spontaneous breaking of chiral symmetry and confinement.

Acknowledgements

We are grateful to N. Akchurin, M. Anselmino, A. D. Krisch and J. P. Ralston for interesting discussions, one of us (S.T.) express also his gratitude to A. D. Krisch for support of the visit to this Simposium.

References

[1] A. D. Krisch, Plenary lecture given at the 9th Intern. Symp. on High Energy Spin Physics, Bonn, Germany 1990.

[2] K. Heller, Proc. of the 10th Intern. Symp. on High Energy Spin Physics, Nagoya, 1992, p. 177.

[3] S.M. Troshin and N.E. Tyurin, , Particle World **3**, No. 4, 165 (1994).

[4] K. Steininger and W. Wise, Phys. Rev. **D48** 1433 (1993).

[5] M.M. Islam, Z. Phys. C – Particles and Fields **53** 253 (1992).

[6] P. Carruthers and Minh Duong-Van, Phys. Rev. **D28** 130 (1983).

[7] A. A. Logunov, V. I. Savrin, N. E. Tyurin and O. A. Khrustalev, Teor. Mat. Fiz. **6** 157 (1971).

[8] H. Lipkin, Phys. Lett. **B230** , 135 (1989);
H. Fritzsch, Phys. Lett. **B256** , 75 (1991);
U. Ellwanger and B. Stech, Z. Phys. C **49** , 683 (1991).

[9] J. Ellis, S.J. Brodsky, M. Karliner, Phys. Lett. **B206** , 309 (1988),
J. Ellis, Y. Frishman, A. Hanany, and M. Karliner, Nucl. Phys. **B382** , 189 (1992).

[10] J. P. Ralston and B. Pire, High energy helicity violation in hard exclusive reactions of hadrons, Preprint Kansas 5-15-92, 1992.

[11] G. Preparata and J. Soffer. Phys.Lett. **86B** , 304 (1979);
C.Bourelly and J.Soffer. Phys.Rev. **D35** , 145 (1987).

[12] M. Anselmino, P.Kroll and B.Pire, Z.Phys. **C36** , 89 (1988).

Polarization and N-N Elastic Scattering Amplitudes *

Gordon P. Ramsey

Physics Department, Loyola University Chicago, Chicago, IL 60626
and
High Energy Physics Division, Argonne National Laboratory, IL 60439

Abstract

We discuss the role of polarization measurements and scattering amplitudes for elastic nucleon-nucleon processes at high energy. The relative normalization of these amplitudes involves a "leading order form factor" which we determine empirically. These amplitudes provide an economical description of a large body of existing data and make some nontrivial predictions for spin observables. In particular, we have investigated cross sections and asymmetry data at large angles (including 90° c.m.) and the fixed $| t |$, large s region, dominated by the three-gluon exchange mechanism. Our results indicate that polarization experiments can test basic QCD elements such as helicity conservation, wave function properties and the interplay between various interaction mechanisms. The significance of this analysis will be discussed in terms of specific polarization experiments, which can be performed at Fermilab and Brookhaven (RHIC, AGS).

I. Phenomenological Normalization of the Spin Amplitudes

The study of exclusive hard scattering processes provides an interesting theoretical challenge. Existing data on fixed-angle differential cross sections and spin observables at high energies suggest that predictions made using the constituent counting rules are accurate in describing exclusive hard scattering and that the data may be in a kinematic regime where hadronic amplitudes can be calculated perturbatively. There exist difficulties in the calculations (1) for $pp \to pp$ scattering which imply that there is little chance of a reliable "first principles" calculation based on QCD perturbation theory. However,

*Paper presented at the 11th International Symposium on High Energy Spin Physics, Indiana University, September, 1994. Work supported by the U.S. Department of Energy, Division of High Energy Physics, Contract W-31-109-ENG-38.

experimentally, high intensity proton beams and a clean signature for the elastic scattering process make this one of the best studied set of experimental cross sections. Spin dependent observables and the spin-averaged cross section are accurately measured over a wide range of kinematic variables (2). Thus, for a given helicity amplitude, (3) it is possible to do a phenomenological analysis where the normalization of the amplitude is fit to data at one point while its s and t dependence are extracted from the theoretical calculation. This approach leads to a large number of asymptotic predictions, but there has been considerable discussion about which of the involved mechanisms are important.

We define the observables for $NN \to NN$ elastic scattering in terms of the Jacob-Wick helicity amplitudes: $\Phi_1(s,t)$ through $\Phi_5(s,t)$ (3). Other helicity amplitudes are related to these independent ones using parity conservation, time-reversal invariance and identical particle symmetry.

The differential cross section in terms of these amplitudes is: (4)

$$\frac{d\sigma}{dt} = \frac{\pi}{2s(s-4m^2)}\Big[\mid \Phi_1 \mid^2 + \mid \Phi_2 \mid^2 + \mid \Phi_3 \mid^2 + \mid \Phi_4 \mid^2 + 4 \mid \Phi_5 \mid^2\Big], \tag{1}$$

where the total cross section is given by: $\Sigma = s(s-4m^2)\frac{d\sigma}{dt}$.

We assume that we can separate a soft, coherent Regge contribution for each independent amplitude, which dominates the observables at small t. We also assume there exists a "hard" component for each amplitude which obeys the Brodsky-Lepage factorization at large t. The coherent Regge components should be exponentially suppressed at large t, reflecting the size of the individual proton. The overall amplitudes at large t then consist primarily of the Landshoff and Quark-Interchange components (4). This separation is largely a matter of convention. The s and t dependence of the Landshoff and QIM amplitudes are specified by simple QCD calculations, and the assumption is that the proton is well described by its minimal three-quark Fock state. A consistent normalization of the theoretical amplitudes demands that the scattering is dominated by a "quark interchange" mechanism in kinematic regions where sharp structure has been observed.

Donnachie and Landhoff (5) have done a thorough phenomenological study of the differential cross sections for pp and $p\bar{p}$ at high energies. The normalization of the Landshoff 3-gluon mechanism amplitude can be reliably specified at high energy and fixed-t from elastic scattering data at the ISR collider. From the known kinematic dependence of the Landshoff amplitude, we can extrapolate its effect to the energies of the ANL data. The differential cross section at ISR energies can be pametrized as:

$$\frac{d\sigma}{dt} = \pi \frac{| L(t) |^2}{(2m_p^2 - t)^8}\Big[1 + 8 | \epsilon |^2 \frac{(-t)}{(2m_p^2 - t)^2} + \cdots\Big], \tag{2}$$

where the second term in brackets is associated with $| \Phi_5 |^2$ and other terms are suppressed by powers of $\frac{t}{s}$. Comparing eq. (2) with data at large s and with $-t$ outside the coherent Regge region is the first step in a phenomenological normalization. At large $-t$, (≥ 4 GeV^2), the three gluon exchange mechanism dominates. In this kinematic regime, there is strong experimental support for an approximately energy-independent component of the cross section which behaves as t^{-8}. Experimentally, the numerical value extracted for $| L(t) |^2$ from these data can be used to normalize this mechanism.

Note that the $| t |$ range of these high energy collider data overlaps with the $| t |$-range of the low energy large angle data where structure has been observed in the two-spin observable A_{NN} and the 90° c.m. differential cross section (6). Since the Landshoff normalization factors $| L(t) |$ are dependent only upon $-t$, we can use existing cross section data at various s for comparable $-t$ values to phenomenologically determine their behavior. We can then extrapolate the Landshoff amplitudes to the energies of the 90° c.m. data to find their effect on the behavior of the cross section and A_{NN}. We have fit elastic *pp* data from the Argonne ZGS and CERN (ISR) for various $\sqrt{s}$ ranging from 3 GeV/c to 62 GeV/c and for a common range of $-t$.

Our average fit to the the ISR data, for the differential cross section, which is relatively independent of s, is

$$\frac{d\sigma}{dt}(\mu b/GeV^2) = 1.6 \exp[-1.7 | t |]. \tag{3}$$

The ANL data (7) at 90° c.m. covers a range of $\sqrt{s}$ from 3.3 to 5.1 GeV/c. There is a marked change in structure of the cross section for $-t$ near 7 (GeV/c)^2. Our fit to the data is:

$$\begin{aligned}\frac{d\sigma}{dt}(\mu b/GeV^2) &= 6836 \exp[-1.59 | t |], \quad 3.8 \leq -t \leq 6.8\,(GeV/c)^2 \\ &= 24 \exp[-0.76 | t |], \qquad 7.3 \leq -t \leq 11.3\,(GeV/c)^2.\end{aligned} \tag{4}$$

If we now assume the ISR data are in a region dominated by the Landshoff mechanism, we can then calculate a ratio of the cross sections between the ISR and ANL data in the mid-t region:

$$\left(\frac{d\sigma}{dt}\right)_{ISR} \Big/ \left(\frac{d\sigma}{dt}\right)_{ANL} \approx (2.4 \times 10^{-4}) \exp[-0.12 | t |]. \tag{5}$$

When considering the effects of the Landshoff mechanism at 90° cm, the u-channel terms become important, but are but are correctly accounted for in our parametrization. Thus at ANL energies, the total Landshoff contribution is at most 3×10^{-4} for this range of t values. Thus, although the Landshoff contribution to the ISR cross section may be large, its effect on the ANL 90° data is down by a factor of 10^{-4}. We conclude that another mechanism must be responsible for the structure of the data in this region. Given the known s dependence of the three-gluon amplitude, we can easily rule out the possibility that the Landshoff mechanism is involved in these low energy structures.

II. Structure of the Amplitudes at 90° c.m.

There are two distinct types of phenomena which have been observed in the 90° c.m. cross section for pp elastic scattering: (1) oscillations in $s^{10}\frac{d\sigma}{dt}$ and (2) structure in A_{NN}, both as a function of energy.

It is possible to use the symmetries which occur at 90° in the cm to do a simple amplitude analysis which can extract the basic features of the sub-asymptotic mechanism responsible for the structure observed. This may have important ramifications for studies of nuclear transparency and for the study of other exclusive processes.

Since the Landshoff normalization, as determined from ISR data, rules out the possibility that this mechanism is responsible for this "low energy" behavior, it appears that we must search elsewhere for the explanation of these unusual structures. We consider that the basic symmetries of the QIM model are reflected in the large angle amplitudes and that the structures may be corrections to these amplitudes. Due to the symmetries at 90° c.m., we have the kinematic constraints that $\Phi_5 = 0$ and $\Phi_4 = -\Phi_3$. Also, the double helicity-flip amplitude, Φ_2, can be neglected at the energies where the data exist. Helicity conservation therefore implies that the observables at 90° can be understood in terms of only two independent amplitudes. For the observables we consider here, the expressions are

$$\Sigma = \frac{1}{2}\Big[\mid \Phi_1 \mid^2 + 2 \mid \Phi_3 \mid^2\Big]$$
$$A_{NN} = \frac{2 \mid \Phi_3 \mid^2}{\Big[\mid \Phi_1 \mid^2 + 2 \mid \Phi_3 \mid^2\Big]}. \tag{6}$$

We now implement the constraint that the amplitudes approach the QIM result: $\Phi_1^Q = 2\Phi_3^Q$ at 90° c.m., to write

$$\Phi_1 = 2\Phi^Q + \hat{\Phi}_1$$
$$\Phi_3 = \Phi^Q + \hat{\Phi}_3, \tag{7}$$

where Φ^Q is a smooth power-law behaved amplitude which characterizes the asymptotic observables, while $\hat{\Phi}_1$ and $\hat{\Phi}_3$ are sub-asymptotic corrections. If we write

$$\Sigma_0 \equiv \frac{1}{2} 6 \mid \Phi^Q \mid^2, \tag{8}$$

then

$$\begin{aligned} A_{NN}(\Sigma/\Sigma_0) &= \frac{1}{3}\frac{\mid \Phi_3 \mid^2}{\mid \Phi^Q \mid^2} = \frac{1}{3}\frac{\mid \Phi^Q + \hat{\Phi}_3 \mid^2}{\mid \Phi^Q \mid^2} \equiv \frac{1}{3} R_3 \\ (1 - A_{NN})\Sigma/\Sigma_0 &= \frac{1}{6}\frac{\mid \Phi_1 \mid^2}{\mid \Phi^Q \mid^2} = \frac{2}{3}\frac{\mid \Phi^Q + \frac{1}{2}\hat{\Phi}_1 \mid^2}{\mid \Phi^Q \mid^2} \equiv \frac{2}{3} R_1, \end{aligned} \tag{9}$$

where $R_1 \equiv \frac{|\Phi_1|^2}{|\Phi_1^Q|^2}$ and $R_3 \equiv \frac{|\Phi_3|^2}{|\Phi_3^Q|^2}$. The data show interference effects in both amplitudes with the structure in Φ_1 occurring at a lower energy than that in Φ_3. A plot of these amplitudes is shown in figure 1.

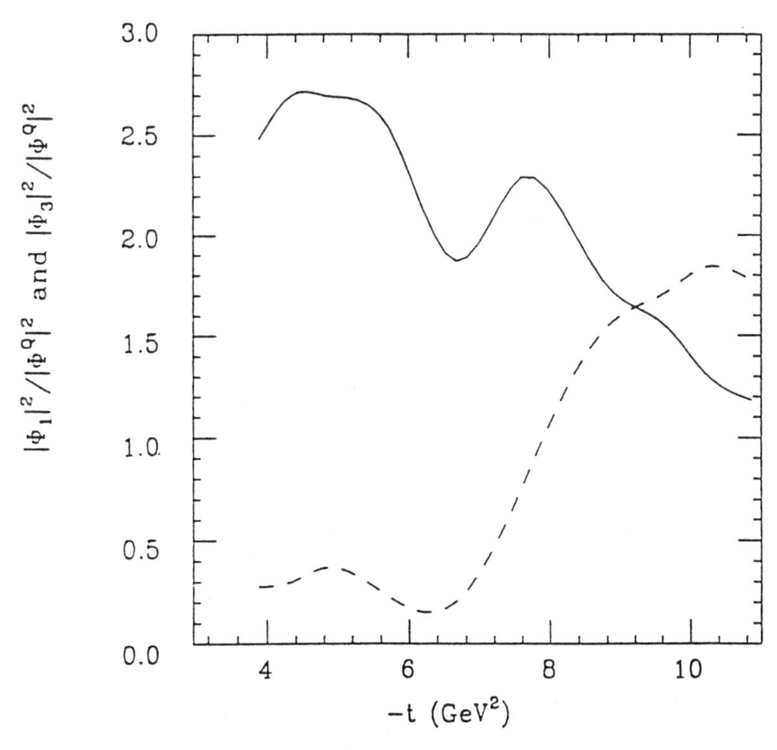

Figure 1. Amplitude ratios as a function of -t at 90 degrees c.m. The solid line represents the Φ_1 ratio and the dashed line the Φ_3 ratio.

The structure of A_{NN} for the range $4 <| t |< 7$ GeV^2 is relatively flat. The value of A_{NN} however, differs significantly from the QIM prediction of $\frac{1}{3}$. Given the average value of A_{NN} from the data, equation 9 implies that $R_1 \approx 18\ R_3$. If the QIM mechanism were dominant, we would observe that $R_1 \approx 4\ R_3$. We can conclude that the additional sub-asymptotic mechanism appears to enhance the process described by Φ_1 over that of Φ_3 significantly at these energies. The differential cross section in this region exhibits a relatively steady $(-t)^{-10}$ behavior. If we make the Ansatz that $\Sigma_0 = 3.5 \times 10^8/(-t)^{10}$, then Σ has the same t behavior and $| \Phi_1 |^2 / | \Phi_1^Q |^2 / \approx 3$. This is a measure of the relative strength of the mechanism which "interferes" with the QIM amplitudes here to cause A_{NN} to dip below the QIM prediction.

In the region $| t |> 7$ GeV^2, the slope of the cross section changes and A_{NN} simultaneously begins to rise sharply. The differential cross section exhibits more of a t^{-8} behavior in this region. The amplitude Φ_3 also becomes more significant here, since $\frac{|\Phi_1|^2}{|\Phi_3|^2} \approx 1$ for $| t |\geq 8$ GeV^2. The mechanism responsible for the change in slope of the cross section and the rapid rise in A_{NN} will then couple more strongly to Φ_3 than in the other region. This may not be the same mechanism responsible for the behavior of the data in the lower $-t$ region. Eventually, the mechanism interfering with Φ_3 dominates the QIM contributions in a way to cause A_{NN} to rise well above the QIM prediction.

We have not discussed the specific dynamical mechanisms which may be responsible for these effects, except to rule out a significant contribution from the Landshoff three-gluon exchange diagrams discussed above. Brodsky and deTeramond, (8) have discussed the structure of A_{NN} in terms of a specific dibaryon resonance, which couples to Φ_3, but not to Φ_1. The structure of these amplitudes indicates that existing data are not in a regime where "nuclear transparency" should be a feature of $pA \rightarrow pp(A-1)$. Thus it is important to push for experiments at higher energy on nuclear targets to see if nuclear transparency emerges in this region. The work of Jain and Ralston (9) implies that the subasymptotic corrections are suppressed in the nuclear environment. The point-like cross section, Σ_0 is a more useful quantity for understanding nuclear effects in their approach than is the physical cross section. An important requirement for understanding these effects is to seek new data at higher energies, such as at RHIC.

III. The Polarization Asymmetry at Large Energies

The structure of the elastic amplitudes within the framework of the constituent based hard-scattering model can be severely constrained by measurements of the elastic polarization asymmetry at large s and for $| t |\geq 4$ GeV^2.

In the traditional Regge theory approach, the polarization asymmetry vanishes at large s. In contrast, the hard-scattering approach allows for possible helicity-flip effects associated with the hadronic wave function. These effects vanish at large $| -t |$, but the helicity-flip amplitudes can share the same s dependence as the helicity conserving ones. This leads to a polarization asymmetry which is almost s-independent and hence, falls off only as a function of t.

We can write the polarization asymmetry in terms of the helicity amplitudes. Our parameterization of the helicity-flip amplitude, Φ_5, is given as a combination of the helicity-conserving amplitudes and is based on the idea of restructuring a proton wave function from scattered quarks which are approximately collinear. This approach builds in the constraint of hadronic helicity conservation so that at fixed angles, the polarization asymmetry vanishes. The $-t$ dependence is chosen to match dimensional counting rules. In this approach, the polarization asymmetry becomes independent of energy at fixed t. We can reproduce the existing polarization data very well. Details of this work will appear in a forthcoming publication. The behavior of the polarization at large energies can be summarized as:

(1) small $| t |$: There is an overall $(-t)^{\frac{1}{2}}$ factor and the polarization should fall off with energy, reflecting the coherent behavior of factorizable Regge poles.

(2) large $| t |$: Outside the coherent region, the polarization becomes asymptotically energy-independent and should behave like $(-t)^{-\frac{1}{2}}$ at large s.

Those polarization measurements which exist (10) are consistent with these predicted regularities. Further measurements of the polarization asymmetry at high energies are needed in order to test these underlying principles.

IV. Summary and Conclusions

In this paper, we have used the normalization of the Landshoff model helicity amplitudes which are taken from the phenomenological studies mentioned above and continued the amplitudes to smaller s values. There is little uncertainty in this exercise since the s dependence of the Landshoff amplitudes are well specified by the model. The continuation shows that the Landshoff amplitudes are too small to be involved in the oscillations observed in $d\sigma/dt$ at 90° c.m. or in the sharp structure observed in A_{NN}. The most natural explanation of these striking phenomena involves the interference of some asymptotic mechanism with a dominant QIM model amplitude set.

Using the symmetries in the amplitudes at 90° c.m., we have expressed the data in terms of Φ_1 and Φ_3. The data suggest a structure which interferes with the dominant QIM amplitudes. This interference occurs at a lower energy in Φ_1 than in Φ_3, but is approximately the same magnitude in each amplitude. The data disappear in a kinematic regime where there is a lot of structure and it would be interesting to continue these measurements at some higher energies. There exists a real opportunity to do these measurements at the Brookhaven AGS with the addition of a partial Siberian snake to allow for polarized beams. It will be interesting to see whether the data approach the value of $A_{NN} = \frac{1}{3}$ as predicted by the QIM model and whether the cross section oscillations fade away at higher energy.

The other measurement which can provide new insight involves the single-spin polarization asymmetry. Our simple model relates different amplitudes using a spin-flip "form factor" involving a small parameter related to SU(6) breaking. Experiments which could be performed at Fermilab and Brookhaven would further test the validity of this model in the hard scattering region: $m_p^2 \ll -t \ll s$.

Finally, within the context of our model, we have looked at the data for evidence of non-trivial behavior of the observables associated with the running of the QCD coupling or with the falloff of Sudhakov form factor. (11) We could find no evidence of these effects. This may indicate that while constituent based models can provide important insight into the structure of $pp \to pp$ amplitudes, the data are not in a kinematic regime where perturbative calculations can be attempted.

Acknowledgements

This paper is based on work done with D. Sivers. We are grateful for advice and suggestions from A. White, S. Brodsky, J. Ralston and C. Carlson.

References

1. G.R. Farrar, H. Zhang, A.A. Globlin and I.R. Zhitnitsky, Nucl. Phys. **B311**, 585 (1989); G.R. Farrar, E. Maina, and F. Neri, Phys. Rev. Lett. **53**, 28 (1984) and ibid., 742(1984).

2. A. Bohm, *et. al.*, Phys. Lett. **49B**, 491 (1974); E. Nagy, *et. al.*, Nucl. Phys. **B150**, 221 (1979). Foley, *et.al.*, Phys. Rev. Lett. **15**, 45 (1965); C.W. Akerlof, *et. al.*, Phys. Rev. **159**, 1138 (1967); R.C. Kammerud, *et. al.*, Phys. Rev. **D4**, 1309 (1971).

3. M. Jacob and G.C. Wick, Ann. Phys.**7**, 404 (1959).

4. G. P. Ramsey and D. Sivers, Phys. Rev. **D45**, 79 (1992) and Phys. Rev. **D47**, 93 (1993); G.R. Farrar, et. al. Phys. Rev. **D20**, 202 (1979); S.J. Brodsky, et.al., Phys. Rev. **D20**, 2278 (1979).

5. A. Donnachie and P.V. Landshoff., Z. Phys. C **2**, 55 (1979); Phys. Lett. **123B**, 345 (1983); Nucl. Phys. **B231**, 189 (1983); **B244**, 322 (1984); **B267**, 690 (1986).

6. A. Bohm, *et. al.*, Phys. Lett. **49B**, 491 (1974); E. Nagy, *et. al.*, Nucl. Phys. **B150**, 221 (1979).

7. C.W. Akerlof, *et. al.*, Phys. Rev. **159**, 1138 (1967); R.C. Kammerud, *et. al.*, Phys. Rev. **D4**, 1309 (1971).

8. S.J. Brodsky and G. deTeramond, Phys. Rev. Lett. **60**, 1924 (1988).

9. P. Jain and J.P. Ralston, KUHEP 93-47. Proc. of Int. Conf. on elastic and diffractive scattering, 5th Blois Conf., Providence RI, Jun 93.

10. G.W. Abshire, *et. al.*, Phys. Rev. Lett. **32**, 1261 (1974).

11. V. Sudhakov, Zh. Eksp. Theor. Phis. **30**, 87 (1956) and Sov. Phys. JETP **3**, 65 (1965); J. Botts and G. Sterman, Phys. Lett. **B224**, 201 (1989) and Nucl. Phys. **B325**, 62 (1989).

Proton-Proton Elastic Scattering Experiment at RHIC

Paul A. Draper[1]

Department of Physics, University of Texas at Arlington, Arlington, Texas 76019

Abstract. We describe an experiment to study proton-proton elastic scattering at the Relativistic Heavy Ion Collider. RHIC allows a systematic study of elastic scattering in the center of mass energy range 60-500 GeV and four-momentum transferred $|t|$ up to 1.5 $(GeV/c)^2$. There are two experimental configurations: in the small $|t|$ (Coulomb-nuclear interference) region, one can measure the total cross section, the ratio of the real to imaginary parts of the scattering amplitude, and the slope parameter simultaneously with an error of a few percent (with a special lattice tune) and at large $|t|$ (with the standard lattice), one can reach beyond the dip region. Over this entire range the same setup can be used to study elastic scattering exploiting polarized protons at RHIC to measure the difference in the total cross sections as a function of initial transverse spin states, the analyzing power, and the transverse spin correlation parameter.

The Relativistic Heavy Ion Collider (RHIC) is a natural place to perform an elastic scattering experiment, for several reasons. First, it allows study of proton-proton scattering over a wide range of center-of-mass energy, up to an unprecedented 500 GeV, a unique opportunity until the onset of the LHC physics program in the next century. Secondly, RHIC is designed to deliver transversely polarized proton beams, with polarization fractions around 70%, allowing studies of spin-orbit and spin-spin correlations in an interesting t range. Thirdly, since the details of the lattice are still in design, accommodations can be made to optimize the experimental environment for such a study. In this paper, we will describe a proposed experiment (PP2PP) to study elastic scattering at RHIC.

In the next section, we will discuss the experimental objectives for cross-section measurements in unpolarized beams, followed by a description the spin physics program for PP2PP. Finally we will describe the detector development and simulation status, and mention some milestones for future work.

CROSS-SECTION MEASUREMENTS

The proton-proton elastic differential cross-section is typically modeled as having an electromagnetic component and a hadronic component:

[1] Representing the PP2PP collaboration

$$\frac{d\sigma_{el}}{dt}=\frac{4\pi\alpha^2(\hbar c)^2G^4(t)}{t^2}+\frac{\alpha(\rho-\alpha\phi)\sigma_{tot}G^2(t)}{|t|}e^{-\frac{b|t|}{2}}+\frac{\sigma_{tot}^2(1+\rho^2)}{16\pi(\hbar c)^2}e^{-b|t|} \quad (1)$$

where the second term stems from the interference of Coulombic (first term) and hadronic (third term) amplitudes. The Coulombic term gives rise to a fast slope at very low values of t, falling as 1/t. The hadronic term falls exponentially, with b the nuclear slope parameter. The total cross-section is connected to the differential elastic cross-section at t = 0 by the optical theorem. All three parameters — the total cross-section, the hadronic slope parameter b, and the ratio ρ of real and imaginary forward scattering amplitudes — are seen to depend on the center-of-mass energy, as shown in Fig. 1. It is important to note that all experimental points above 62 GeV in the plot are from proton-antiproton collisions. The RHIC realm is indicated, the only opportunity to compare proton-proton and proton-antiproton results in this range.

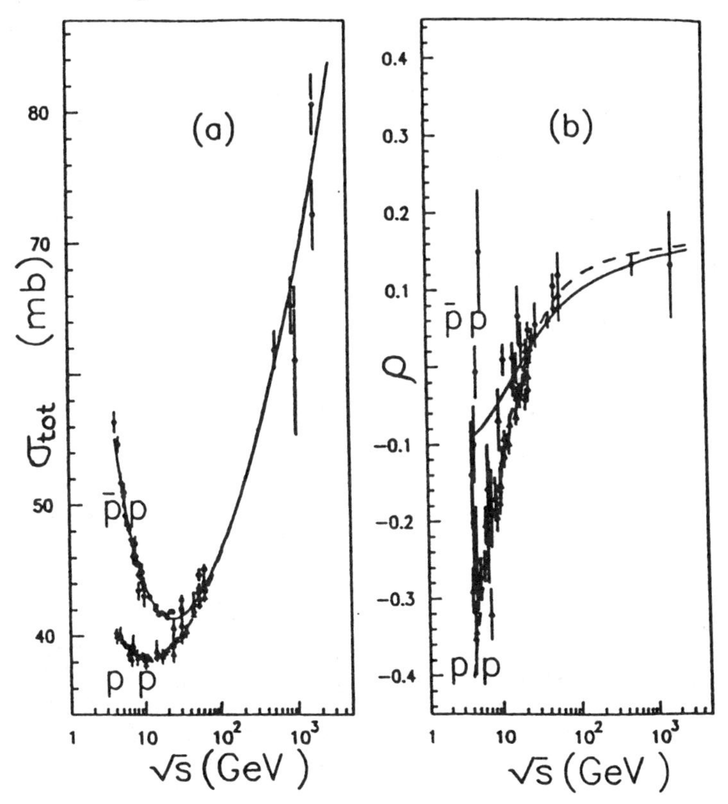

FIGURE 1. Center of mass energy dependence of elastic scattering parameters: total cross section (a), and the ρ parameter (b), for pp and p$\bar{p}$ scattering.

The interference between electromagnetic and hadronic contributions to the cross-section is maximal (the so-called CNI region) at t = 1.1 x 10^{-3} (GeV/c)^2, corresponding at beam energies of 250 GeV to a scattering angle of 80 μrad, and this sets the scale for the experimental design. Measuring below this point allows one the advantage of normalization for free, since the Coulombic cross-section is calculable, and this thereby reduces one of the primary sources of systematic error. We intend to measure scattering events over the $|t|$ range 4 x 10^{-4} to 1.5 $(GeV/c)^2$.

In the $|t|$ region around 1 $(GeV/c)^2$, there is expected to be a dip in the pp differential cross-section that is not seen in the $p\bar{p}$ case, as was observed at lower center-of-mass energies at ISR around t = 1.5 $(GeV/c)^2$. According to a calculation by Desgrolard *et al.* (1), the dip in the pp cross-section is expected to deepen with beam momentum and move slowly to lower momentum transfers. Clearly this region is of interest.

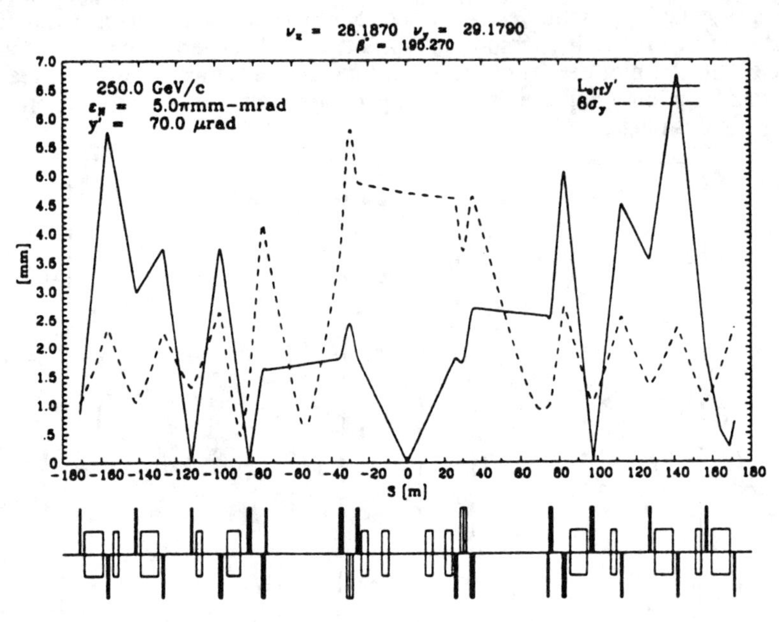

FIGURE 2. The transverse position of protons scattered at 70 μrad (solid line) and the six sigma width of the beam (dashed line) as a function of longitudinal position. The special tune parameters apply here.

The accelerator lattice has been designed with a solution allowing measurements at very small t. The approach taken is called parallel-to-point focusing, in which particles emerging from the interaction at a given scattering angle will focus to a single point at the detector downstream, independent of the transverse location of the interaction. To see this, note briefly that the tranverse deflection y of a particle scattered at a polar angle θ_y^* at the interaction point is given by

$$y = a_{11}y^* + L_{eff}\theta_y^* \tag{2}$$

where the slopes are related to the betatron function β, and the phase advance Ψ, by

$$a_{11} = \sqrt{\frac{\beta}{\beta^*}}\left[\cos\Psi + \left(\frac{\partial\beta}{\partial z}\right)^* \sin\Psi\right]$$
$$L_{eff} = \sqrt{\beta\beta^*}\sin\Psi \tag{3}$$

Clearly, parallel-to-point focussing is obtained when the phase advance is chosen such that a_{11} vanishes, and detector sensitivity is optimized by making β^* as large as possible. This experiment therefore will require a running period with a special tune at $\beta^* = 195$ m, about 20 times that of the normal operating tune. Even with this large beam spot, the luminosity is ample. In addition, the emittance at this tune is too high, and we expect to scrape the beam before data collection. Fig. 2 shows, for these running conditions, the transverse position of protons scattered at 70 μrad (well into the Coulomb-dominated scattering realm) compared to the beam envelope.

SPIN PHYSICS

One of the driving reasons for proton-proton collisions at RHIC is the chance to do spin physics, which in recent years has offered many experimental surprises not yet understood. The s-channel scattering amplitude can be written (2) as the sum of five helicity amplitudes: $\langle\uparrow\uparrow|\phi|\uparrow\uparrow\rangle$, $\langle\uparrow\uparrow|\phi|\downarrow\downarrow\rangle$, $\langle\uparrow\downarrow|\phi|\uparrow\downarrow\rangle$, $\langle\uparrow\downarrow|\phi|\downarrow\uparrow\rangle$, $\langle\uparrow\uparrow|\phi|\uparrow\downarrow\rangle$. The second and the fifth amplitudes represent spin flip terms. The conventional wisdom is that QCD chiral symmetry will cause all hadronic spin dependencies to vanish as s increases. There are experimental indications, however of a hadronic spin-flip amplitude that survives to surprisingly high energies. For example, Fidecaro *et al.* (3) measured the single spin asymmetry at fixed-target beam momentum of 200 GeV/c, and observed its sharp passage through zero at $t \approx 1.5$ $(\mathrm{GeV/c})^2$, in the vicinity of the dip in the differential cross-section.

The RHIC SPIN collaboration has proposed implementing Siberian snakes, spin rotators and polarimeters into the RHIC lattice. They estimate that the proton polarization at 250 GeV/c can be as high as 70% at a luminosity of 10^{32} cm^{-2}s^{-1}. The polarization at the PP2PP interaction point would be transverse, though the inclusion of spin rotators is being considered. The most important aspect is that we would be able to make measurements with polarized beams without change in experimental configuration.

This experiment will be able to measure the difference in the total cross-section which is directly related to the s-channel helicity amplitude:

$$\Delta\sigma_{tot} = \sigma_{tot}(\uparrow\downarrow) - \sigma_{tot}(\uparrow\uparrow) \tag{4}$$

We can also measure the singly-polarized analyzing power

$$A_N = \frac{1}{P_{1,2}(\cos\phi)} \frac{N_{\uparrow\uparrow} + N_{\downarrow\downarrow} \pm N_{\uparrow\downarrow} \mp N_{\downarrow\uparrow}}{N_{\uparrow\uparrow} + N_{\downarrow\downarrow} + N_{\uparrow\downarrow} + N_{\downarrow\uparrow}} \tag{5}$$

which is related to the spin-orbit terms in the interaction and would reveal the presence of hadronic spin-flip amplitudes. Here, $P_{1,2}$ is the polarization of either beam, ϕ is the azimuthal angle between the scattering plane and the polarization vector, and N is the number of events with the polarization of the beams as indicated. Several fixed target experiments (at relatively modest center of mass energies) have measured the analyzing power (see for example 3,4,5,6,7), but the experimental errors are rather large, and there is a unmeasured gap in the region $0.05 < |t| < 0.15$ (GeV/c)^2 , as well as the interesting behavior in the dip region mentioned above.

Finally, we can measure the double spin correlation, which probes the spin-spin interaction:

$$A_{NN} = \frac{1}{P_1 P_2(\cos\phi)^2} \frac{N_{\uparrow\uparrow} + N_{\downarrow\downarrow} - N_{\uparrow\downarrow} - N_{\downarrow\uparrow}}{N_{\uparrow\uparrow} + N_{\downarrow\downarrow} + N_{\uparrow\downarrow} + N_{\downarrow\uparrow}}. \tag{6}$$

Because of the strong dependence on the azimuthal angle, this measurement is limited primarily by how close we can approach the beam. Note that the asymmetry measurements require only the ratios of number of events, with only acceptances and measured polarizations limiting the precision of the measurement.

We are investigating ways to extend our reach to larger momentum transfers, up to around 6 (GeV/c)^2, where there is strong theoretical interest in the behavior of spin-flip amplitudes.

SIMULATIONS AND DETECTORS

Monte Carlo simulations continue to be developed to determine the experimental resolution of reconstructed scattering parameters. Key ingredients of the simulation are scattering event generation, propagation of the scattered protons through the lattice, detector simulation, track reconstruction and the fitting of measured t distributions. Some factors already incorporated in the simulation are spreads in beam momentum, vertex position, crossing angle, detector resolution and alignment offsets.

Some results of the simulation are shown in Fig. 3. The plots indicate that the experimental uncertainty in the total cross section, the hadronic slope b, and the parameter ρ, is dominated by the minimum scattering angle accepted in the detector. Also shown in the figure is how the resolution improves from a sample of 1 million events to 4 million events. The improvement in the statistical error by roughly a factor of two shows that we are driven by event statistics and not systematic errors. The assumed detector position resolution of 100 μm is an insignificant contributor to the error and is fairly easily obtained with straightforward detector techniques. We have also studied the effects of emittance growth during the store, especially for the small-$|t|$ case where beam scraping is required.

Future refinements of the experimental simulation will include inherent uncertainties in the lattice (due to magnet variations) and detector nonuniformity and cross-talk. We are also including background events, as well as the reconstruction cuts to be used in the analysis to remove them.

The detector of choice will likely be scintillating fibers or strips, chosen for their fine granularity, speed (both for timing resolution and live time), and reasonable resistance to radiation from the proximate beams. To reduce the number of channels to be read out, we are considering multiplexing schemes as suggested by the RD-17 group (8), or multi-channel phototransducers.

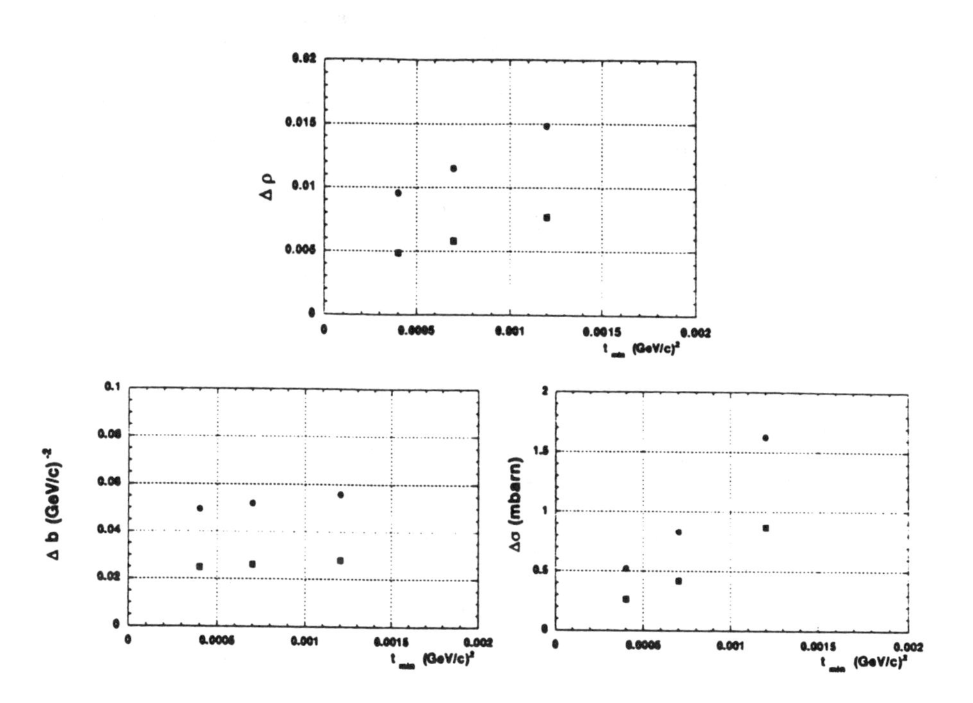

FIGURE 3. Simulation results for measurement errors in the ρ parameter (a), the hadronic slope parameter (b), and total cross section (c), as a function of minimum t reached. The lower (upper) points are for event samples of 1 million (4 million) events.

These detectors would be placed in so-called "Roman pots", devices integrated into the beampipe that allow the detectors to be moved close to the beam once the store is established. Members of the UA4/2 collaboration have agreed to contribute their Roman pots to this apparatus.

OUTLOOK

The Letter of Intent to propose an elastic scattering experiment at RHIC was submitted in June 1991. Since that time, PP2PP collaborators have done extensive simulations of the performance of the experiment, based on our understanding of the RHIC lattice design and nominal detector resolution parameters. The integration of our efforts with the RHIC accelerator design group is key to success, both for understanding the special tunes required to obtain the low-t measurement and the incorporation of detector elements in the rather limited physical space of an accelerator ring.

The proposal was given scientific approval by the Brookaven Program Advisory Committee in March 1994. An accelerator impact study, including the implications of PP2PP's requirements, is under way. Collection of physics data is expected to commence in mid-1999, and this schedule demands that major detector construction would begin in 1996. Prototype testing in beam may happen as early as spring 1995.

The special tune, with $\beta^* = 195$m and beam scraping to lower emittance, for small-$|t|$ running, will require a short (of order days) dedicated run for PP2PP. The duration is short simply because of the higher cross-section at low $|t|$. For the higher-$|t|$ running, we require the standard tune, and so can coexist with other RHIC experiments for a more extended run. We would also request running at different values of beam energy to map the dependence of the scattering on $\sqrt{s}$ over the unmeasured region.

REFERENCES

1. P. Desgrolard, M. Giffon and E. Predazzi, LYCEN 9339 (1993), unplublished.
2. M. Jacob and G.C. Wick, *Ann. Phys.* **7**, 404 (1959).
3. G. Fidecaro *et al., Phys. Lett.* **B76**, 369 (1978).
4. N. Akchurin *et al., Phys. Rev.* **B229**, 299 (1989).
5. N. Akchurin *et al., Phys. Rev.* **D48**, 3026 (1993).
6. J. H. Snyder *et al., Phys. Rev. Lett.* **41**, 781 (1978).
7. M. Corcoran *et al., Phys. Rev.* **D22**, 2624 (1980).
8. V. Agoritsas *et al.*, CERN-DRDC-93-47 (1993), unpublished.

NEPTUN-A Spectrometer for measuring the Spin Analyzing Power in p-p elastic scattering at large $P_{\perp}^2$ at 400 GeV (and 3 TeV) at UNK.[1]

Ali M. T. Lin

Randall Laboratory of Physics
The University of Michigan, Ann Arbor, MI 48109-1120

We are constructing the NEPTUN-A spectrometer for measuring the Spin Analyzing Power in $p + p_{\uparrow} \rightarrow p + p$ at $P_{\perp}^2 = 2$ to 10 $(\mathrm{GeV/c})^2$ at 400 GeV (or at 3 TeV) when the UNK accelerator in Protvino, Russia, becomes operational. The spectrometer consists of a 55 m long recoil arm with 3 horizontally bending magnets to guide the recoil protons onto a fixed 37° line. Then two vertical dipole magnets bend the protons up by 12° for momentum analysis. The momentum will be measured to an accuracy of 0.1 % using chambers. In order to accept a large solid angle, the spectrometer contains a strong-focusing pair of quadrupoles looking at the polarized proton jet target. The forward arm consists of scintillator hodoscopes for measurement of the forward vertical angle. Acceptances and event rates are calculated. The status of the spectrometer is reported.

INTRODUCTION

We had measured the one-spin Analyzing Power in p-p elastic scattering up to the large $P_{\perp}^2$ value of 7.1 $(\mathrm{GeV/c})^2$ at the incident momentum of 24 GeV/c at the Brookhaven AGS some years ago (1). As shown in Fig. 1, that analyzing power data exhibited interesting structure with a dip around 3.5 $(\mathrm{GeV/c})^2$ and then it began to rise to about 20 % at the largest measured $P_{\perp}^2$ value of 7.1 $(\mathrm{GeV/c})^2$.

Perturbative QCD predictions (2) required this analyzing power to approach zero at a large enough $P_{\perp}^2$ and at a large enough incident energy. Apparently, such a prediction is at variance with our experimental result, but one could argue that the energy is not large enough.

Our group then decided to pursue the goal of measuring the one-spin analyzing power at even larger $P_{\perp}^2$ values and at larger incident energies. A new "Mark II" polarized hydrogen gas jet (refer to the talk by V. G. Luppov of our group in these conference proceedings) is being developed for use as an

[1]Research supported by the U. S. Department of Energy

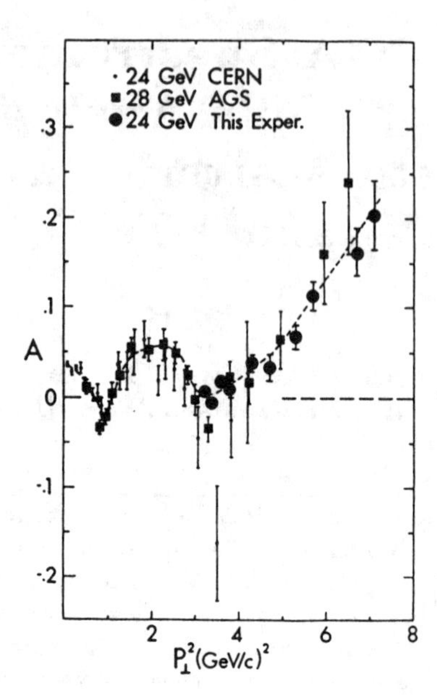

FIG. 1. Analyzing Power in p-p elastic scattering at 24–28 GeV/c.

internal target in the UNK ring. The NEPTUN-A collaboration [2] was formed for measuring the analyzing power at $P_\perp^2$ of 2–10 $(GeV/c)^2$ at 400 GeV/c and perhaps later at 3 TeV (if UNK II gets built). (For information on NEPTUN and other experiments looking at the same target, refer to the talk by V. L. Solovianov in these proceedings.)

RELEVANT KINEMATICS

The kinematics of p-p elastic scattering for $P_\perp^2$ values of 2–10 $(GeV/c)^2$ at 400 GeV/c determined the ranges of the momenta and angles of the forward and recoil protons. Table 1 shows these ranges that we are interested in measuring.

EXPERIMENTAL SET-UP

It was decided, early on, that because the forward protons have too high a momenta [> 390 (GeV/c)] and too close to the beam [< 10 mrad], its

[2]The present NEPTUN-A collaboration consists of about 50 physicists from the University of Michigan, MIT, IHEP(Protvino), and JINR(Dubna)– with spokesperson Prof. A. D. Krisch of Michigan, and Prof. V. L. Solovianov of IHEP as spokesperson for the Russian contingent.

TABLE 1. Momenta of some Forward and Recoil scattered protons at 400 GeV/c.

$P_\perp^2(GeV/c)^2$	$P_F(GeV/c)$	@	θ_F	$P_R(GeV/c)$	@	θ_R
2	398.9	@	0.20°	1.77	@	52.9°
6	396.8	@	0.35°	4.05	@	37.2°
10	394.6	@	0.46°	6.27	@	30.3°

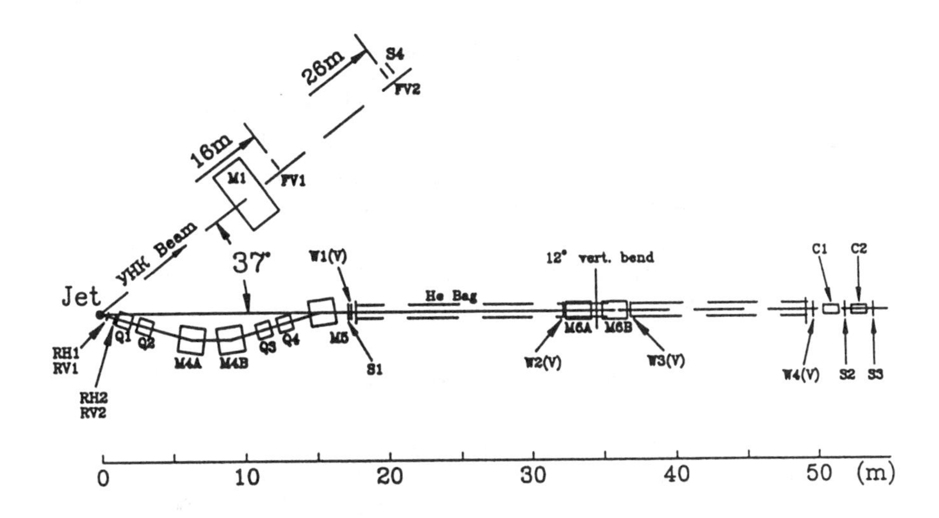

FIG. 2. Top view of the NEPTUN-A spectrometer setup.

momentum would not be measured. Because of the vicinity to the beam, and because of the unknown aspects of the beam halo, we decided to use fast small scintillation counters to detect the vertical angle of the forward scattered protons, with the horizontal angular acceptance overmatching the corresponding recoil arm acceptance.

We decided to concentrate on measuring the momentum *vector* of the recoil protons very accurately by using a 55-m long magnetic spectrometer (see Fig. 2). First, the angle of the recoil proton will be measured with vertical and horizontal solid state detectors RV1,RH1, and RV2,RH2. The about-median value of 37° was chosen for the axis of the spectrometer, with three horizontal bending magnets M4A, M4B and M5 steering the recoil protons of interest onto the axis. Two long field-free regions before and after the two vertically bending magnets M6A and M6B bending up by 12° total were then allotted for the placement of wire and proportional chambers for measuring the tracks to within 1 mm. We then expect to measure the momentum of the recoil proton to an accuracy of about 0.1 %. Note that all magnets are symmetrically placed as much as possible for simplicity and ease of surveying.

To obtain a large solid angle acceptance of the recoil protons, particularly in the vertical plane, a pair of strong-focusing quadrupoles are inserted close to the target.

For particle identification, a pair of threshold Cherenkov counters at the end of the spectrometer line will be used in classifying the higher momenta recoil particles. Lower momentum recoil particles will be identified by time-of-flight through the scintillators S1·S2·S3.

Interaction region

The fixed target interaction region is defined by the transverse beam size of (recently proposed) 5 mm vertical by 10 mm horizontal proton beam interacting with the Mark-II polarized jet which will travel down with transverse widths of 10 mm and 20 mm along beam. Thus the longitudinal interaction length is 20 mm.

Sideways, the interaction region will appear to be 15-19 mm width for the recoil arm and about 10 mm width for the forward arm.

TRANSPORT beam optics

For each $P_{\perp}^2$ point, by geometrical considerations from a (mid)point target, one obtained first the angles of bend for the central ray in each bending magnet, from which the fields were then calculated. The quadrupoles were placed with their axis on the central ray and their positions adjusted using the TRANSPORT program[3].

Applying appropriate constraints, one obtained the required quadrupole field gradients. The maximum half-widths of the beam envelope in both transverse horizontal and transverse vertical directions were then calculated from the program about every meter along the principal axis. This calculation was then repeated for a finite size target (e.g. for half-sizes of 10 mm (h) x 3 mm (v)) and the beam envelopes were plotted for both horizontal and vertical directions, as in Fig. 3. Note that the plot contained both horizontal (X - direction up) and vertical (Y-direction down) envelopes for different half-size of targets. The magnet apertures were about 20 cm bores for the quadrupoles and 20 cm gap and 40 cm width for the benders, and so most of the limiting half apertures were about 10 cm. Vertical dashed lines show roughly the positions of the wire and proportional chambers.

EVENT RATES AND BACKGROUNDS

The circumference of UNK tunnel is 21 km, so cycle time is 70 μs. Since the number of protons stored in one cycle is expected to be about 6 10^{14}, the

[3]Program obtained courtesy of Dr. David C. Carey of Fermilab.

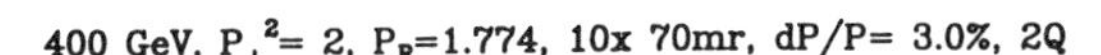

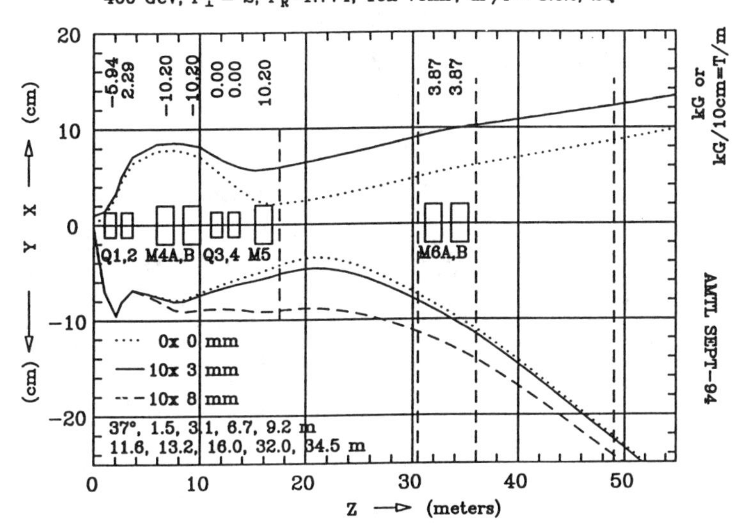

FIG. 3. TRANSPORT beam envelopes for transverse horizontal (X) and vertical (Y) directions at different longitudinal positions for some half-target sizes.

beam intensity should be 8.6 10^{18} per second. As the initial target thickness of Mark-II jet is expected to be 10^{13} cm^{-2}, the target luminosity, L is 8.6 10^{31}. The event rate for elastic scattering is obtained from the formula:

$$\text{Rate } (Events/hr) = L \quad \frac{d\sigma}{dt} \quad \Delta t \quad \frac{\Delta\phi}{2\pi} \quad \epsilon \quad 3600 \quad s/hr.$$

At $P_{\perp}^2 = 2\ (GeV/c)^2)$, with a detection efficiency $\epsilon \sim 1$, and azimuthal angular acceptance $\Delta\phi \sim \frac{140\ mrad}{sin\ 52.8^{\circ}} \simeq 0.175$, one obtains the Event Rate $= 8.6\ \ 10^{31}\ \ 42\frac{nb}{GeV^2}\ \ 0.17(GeV^2)\ \ \frac{0.175}{2\pi}\ \ 1\ \ 3600\ \ 10^{-33}\ \ \frac{cm^2}{nb} = 62\ \ events/hr.$

We expect to obtain larger luminosities from both the beam and the target in time. Because of the low event rates, one has to be careful about the backgrounds especially in the forward arm, where the counters are close to the beam line. Preliminary estimates show that the one-arm forward background rate is of order 1 kHz. The single count rates will be much more than this, although an accurate assessment is not possible without obtaining beam halo measurements.

REFERENCES

1. D. G. Crabb et al., Phys. Rev. Letters, **65**, 3241 (1990).
2. S. J. Brodsky et al., Phys. Rev. **D20**, 2278 (1979); G. R. Farrar et al., Phys. Rev. **D20**, 202 (1979).

Measurement of Single Spin Asymmetry for Direct Photon Production in *pp* Collisions at 200 GeV/c

presented by Naohito Saito, *Kyoto University*

for FNAL E704 Collaboration

D.L. Adams[a], N. Akchurin[b], N.I. Belikov[c], A. Bravar[n],[5], J. Bystricky[d], M.D. Corcoran[a], J.D. Cossairt[e], J. Cranshaw[a], A.A. Derevschikov[c], H. En'yo[f], H. Funahashi[f], Y. Goto[f], O.A. Grachov[c], D.P. Grosnick[g], D.A. Hill[g], T. Iijima[f], K. Imai[f], Y. Itow[f], K. Iwatani[h], K. W.Krueger[i], K. Kuroda[j], M. Laghai[g], F. Lehar[d], A. de Lesquen[d], D. Lopiano[g], F.C. Luehring[k],[1], T. Maki[l], S. Makino[f], A. Masaike[f], Yu.A. Matulenko[c], A.P. Meschanin[c], A. Michalowicz[j], D.H. Miller[k], K. Miyake[f], T. Nagamine[f],[2], F. Nessi-Tedaldi[a],[3], M. Nessi[a],[3], C. Nguyen[a], S. B.Nurushev[c], Y. Ohashi[g],[4], Y. Onel[b], D.I. Patalakha[c], G. Pauletta[m], A. Penzo[n], A.L. Read[e], J.B. Roberts[a], L. van Rossum[e], V.L. Rykov[c], N. Saito[f], G. Salvato[o], P. Schiavon[n], J. Skeens[a], V.L. Solovianov[c], H. Spinka[g], R. Takashima[p], F. Takeutchi[q], N. Tamura[r], N. Tanaka[s],[6], D.G. Underwood[g], A.N. Vasiliev[c], A. Villari[o], J.L. White[a], S. Yamashita[f], A. Yokosawa[g], T. Yoshida[t], and A. Zanetti[n]

[a] T.W.Bonner Nuclear Laboratory, Rice University, Houston, TX 77251, USA
[b] Department of Physics, University of Iowa, Iowa City, IA 52242, USA
[c] Institute for High Energy Physics, 142284, Protvino, Russia
[d] CEN-Saclay, F-91191 Gif-sur-Yvette, France
[e] Fermi National Accelerator Laboratory, Batavia, IL 60510, USA
[f] Department of Physics, Kyoto University, Kyoto 606, Japan
[g] Argonne National Laboratory, Argonne, IL 60439, USA
[h] Hiroshima University, Higashi-Hiroshima 724, Japan
[i] Northeastern State University, , OK 74464, USA
[j] Laboratoire de Physique des Particules, B.P.909, F-74017 Annecy-le-Vieux, France
[k] Physics Department, Northwestern University, Evanston, IL 60201, USA
[l] University of Occupational and Environmental Health, Kita-Kyushu 807, Japan
[m] University of Udine, I-33100 Udine, Italy
[n] Dipartimento di Fisica, Universita di Trieste, I-34100 Trieste, Italy
[o] Dipartimento di Fisica, Universita di Messina, I-98100 Messina, Italy
[p] Kyoto University of Education, Kyoto 612, Japan
[q] Kyoto-Sangyo University, Kyoto 612, Japan
[r] Okayama University, Okayama 700,Japan
[s] Los Alamos National Laboratory, Los Alamos, NM 87545, USA
[t] Osaka City University, Osaka 558, Japan
[1] Present address: Indiana University, Bloomington, IN 47405, USA.
[2] Present address: Bubble Chamber Physics Laboratory, Tohoku University, Sendai 980, Japan.
[3] Present address: CERN, CH-1211 Geneva 23, Switzerland.
[4] Present address: The Institute of Physical and Chemical Research, Saitama 351-01, Japan.
[5] Present address: Department of Physics, University of Iowa, Iowa City, IA 52242, USA
[6] Deceased.

Abstract. The single spin asymmetry for inclusive direct photon production has been measured using a polarized proton beam of 200 GeV/c with an unpolarized proton target at $-0.15 < x_F < 0.15$ and $2.5 < p_T < 3.1$ GeV/c at Fermilab. The measurement was done using lead glass calorimeters and the photon detectors which surround the fiducial area of the calorimeters. Background rejection has been done using the surrounding photon detectors. The single spin asymmetry, A_N, for the direct photon production is consistent with zero within experimental uncertainty.

INTRODUCTION

Measurements of longitudinal polarization asymmetries in deep inelastic lepton-nucleon scattering [1] have stimulated both experimental and theoretical works. Some of them have devoted to the understanding of the transverse spin structure of the proton. Qiu and Sterman have derived the relation between the single transverse spin asymmetry, A_N, for direct photon production and the twist-3 matrix element and suggested the substantial magnitude of A_N even at $x_F = 0$ [2]. Here the parameter, A_N, represents the left-right asymmetry for the production cross section by vertically-polarized incident protons. They pointed out that the asymmetry at large positive x_F in pp scattering with a polarized beam and an unpolarized target is dominated by quark-gluon correlations. Ji suggested that the pure gluon correlations contribute to the asymmetry at large negative x_F value [3]. To determine the strength of those correlations, experimental data are indispensable. The asymmetry, however, has never been measured yet. In this paper, we present results of the measurement of the single spin asymmetry, A_N, as well as the cross section for the direct photon production in pp collisions at 200 GeV/c.

EXPERIMENT

The experiment was carried out using the Fermilab Spin Physics Facility in 1990. The experimental apparatus is shown schematically in Figure 1. The 200 GeV/c polarized proton beam came from parity-violating decay of Λ-hyperons produced by the 800 GeV/c proton beam from the TEVATRON on a Be target [4]. The typical intensity of the polarized beam was $\approx 10^6$ protons/sec during the measurement. The polarized proton beam was divided into three phase space regions with the average polarization of 0.45, 0.0, and -0.45. About a half of the full beam phase space is substantially polarized. Twelve dipole magnets on the beam line were used as a spin-rotator. The liquid hydrogen target had a dimension of 5 cm in diameter and 100 cm in length.

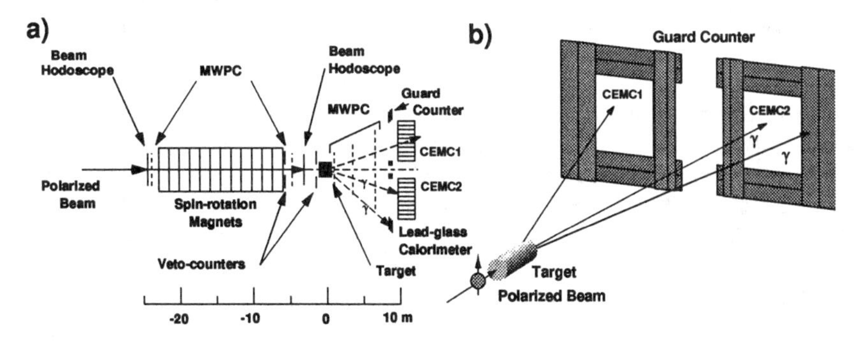

Figure 1: Schematic view of experimental apparatus.

Outgoing γ-rays were detected by Central Electro-Magnetic Calorimeters, CEMC1 and CEMC2, located to the left and to the right of the beam axis, respectively. Each calorimeter comprised 504 lead glass counters in an array of 21 columns by 24 rows. The size of the lead glass was 3.81 cm $\times$ 3.81 cm $\times$ 45 cm (whose radiation length was 2.5 cm). The front surfaces of the CEMC's were placed at 10 m from the center of the target. Each CEMC covered the polar angle of $5.5° \pm 2.2°$ and the azimuthal angle of $\pm 25°$ in the laboratory frame. The energy resolution of 3% (r.m.s.) and the spatial resolution of ± 1.5 mm (r.m.s) were obtained for 30 GeV positron beams. Multi-wire proportional chambers, MWPC's, were used for detecting charged particles, which were placed in front of the CEMC's as shown in Figure 1(a).

We installed photon detectors (the Guard Counter) to reduce the backgrounds from neutral meson decays as shown in Figure 1(b). These detectors consisted of three layers of plastic scintillators (5 mm-thick) with two layers of lead plate converters (12.5 mm-thick) inserted between them and surrounded the fiducial area of each set of lead glass calorimeters like a "picture frame". The photon-detection efficiencies were measured at Institute for Nuclear Study, University of Tokyo with tagged photons and found to be more than 98% in the energy range from 0.2 GeV to 0.9 GeV. The spatial resolution in longitudinal direction of the counter was about 9 cm (r.m.s.). Using the Guard Counter, charged particles were also identified.

ANALYSIS

Reconstruction

We have reconstructed the energy and position of showers from the energy deposits observed in each lead glass counter. The shower reconstruction has been done comparing the energy deposit to each lead glass counter with the "shower reference table" which is a table of the average energy deposition for the electromagnetic showers produced by 30 GeV positrons. The deviation of the energy deposit from the table is represented by the parameter, ρ, which is defined as $\Sigma(E_{obs} - E_{table})^2/\Sigma E_{obs}$. Here E_{obs} represents the energy deposit observed in a certain lead glass counter while E_{table} represents the values expected from the "shower reference table". The parameter, ρ, is minimized during the shower reconstruction procedure to determine the energy and the position of the shower. This parameter has been used also for the discrimination of the hadron-induced showers from electromagnetic showers. The contamination of the hadrons with the same energy as photons has been measured to be $\sim$0.1% at IHEP. The reconstruction efficiencies have been studied using both calibration data and the data simulated with GEANT 3.15 [10]. For the cut with $\rho < 0.3$ GeV, the efficiency is 95%. Further details of the shower reconstruction procedure are described elsewhere [5, 6, 7].

According to the results, the yields of π^0 and η production have been estimated. The yields give the basis of estimation of the background for direct photons. The reconstruction efficiencies are estimated using the Monte Carlo simulations with the GEANT 3.15 simulation package [10] tuned to reproduce the calibration data and found to be 80% at p_T = 2.5 GeV/c and 53% at 4.0 GeV/c. The number of the identified π^0's has been corrected with the reconstruction efficiencies as well as geometrical acceptance. The invariant cross section for the π^0 production is consistent with previous experimental data [5, 8].

Single-photon selection

We have applied the following selection criteria for the reconstructed showers to select single photons. (i)The p_T of the shower is above 2.5 GeV/c and it has no associated shower in the CEMC of the same side. (ii)The events have vertex in a target. The vertex finding efficiency was 33%. (iii)The parameter, ρ, mentioned above, is less than 0.3. This cut reduces the contamination of hadron-induced showers to about 0.2% in the p_T range of the data. (iv)The shower is isolated from charged tracks on the surface of the CEMC at least by 2.0 cm. This cut discriminates showers originating from charged particles. (v)The shower is not associated with photon-like hits in the Guard Counter of the same side. The photon-like hit is defined as the hit with correct timing measured at the third layer and with no signal at the first layer. This cut reduces the background from neutral mesons. The cut (v) with the Guard Counter kills some fraction the of direct photon events. According to the Monte Carlo simulation using PYTHIA 5.7 [9], about 30% of direct photon events are thrown away with this cut.

After the selection with those criteria, we got 473 direct photon candidates. There are still three sources of the fake events: (a) two photons from a π^0 decay merge into one shower (b) one photon from a π^0/η decay is not detected in the CEMC because the energy is lower than the threshold value (=0.5 GeV) (c) one photon from a π^0/η decay misses the CEMC and the Guard Counter. The amount of such fake events has been estimated by means of two types of Monte Carlo simulations, one is using PYTHIA 5.7 [9]+GEANT 3.15 [10] (type I) and the other is developed on the basis of the ISR experimental data [11] (type II). The type II simulation outputs 220±22 as the remaining backgrounds and the type I outputs similar numbers.

RESULTS

After the background subtraction, the direct photon cross sections are derived with correction of the cut efficiency as well as the reconstruction efficiency, the geometrical acceptance, and the vertex finding efficiency. The cross section has been obtained using all the beam phase space. The results are plotted as a function of p_T in Figure 2 a). The results of previous experiments at CERN-SPS [12] and at Fermilab [13] using the proton beam and the carbon target are found to be consistent with ours assuming a nuclear dependence of $A^{1.0}$.

Since the vertex finding efficiency was 33%, the statistics to of the single spin asymmetry was limited. To get rid of this, we have used another cut, so called "*away-side cut*" instead of the vertex cut. The *away-side cut* is the requirement to observe at least one particle with any of CEMC, the Guard Counter, or the MWPC in the region of $\cos\Delta\phi < -0.6$, where $\Delta\phi$ represents the difference of azimuthal angles between the hi-p_T photon and the other particles. The efficiency of this cut has been estimated with the two types of Monte Carlo simulations mentioned above. The type II simulation outputs 94% and the type I gives 93%. The high-p_T π^0 data ($p_T > 2.5$ GeV/c) give 93.2 ± 3.4%. This supports the simulation results on the assumption that the particle distribution in the away-side for the π^0 events is identical to that for the direct photon events.

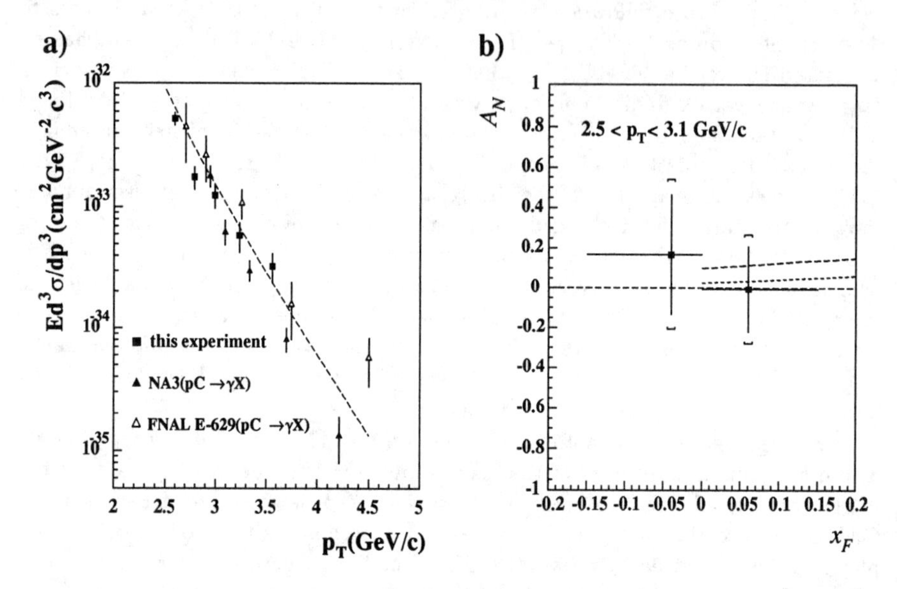

Figure 2: a)Cross section and b)single spin asymmetry for direct photon production in pp collision at 200 GeV/c .

After the selection with criteria (i)-(v) and *away-side cut* instead of vertex cut, we got 1,557 direct photon candidates. The fake events are evaluated to be 618±56. The cross section derived with this approach agrees with the one with vertex cut up to 3.1 GeV/c. It should be noted that 98% of vertex cut events satisfy the *away-side cut*. Hence we have employed the *away-side cut* for obtaining the single spin asymmetry.

The derived asymmetries are plotted as a function of x_F in Figure 2 b). Those are the ones integrated over $2.5 < p_T < 3.1$ GeV/c. The statistical errors are indicated as error bars. The systematic errors mainly come from the uncertainties in the number of π^0 and the π^0 multiplicity used for calculation of the fake event contribution. The dotted and dashed lines are the predicted asymmetries for $\sqrt{s} =$ 30 GeV and $p_T = 4.0$ GeV/c in Ref. [2] assuming two types of the strength of quark-gluon correlations correspondingly. The measured asymmetries are consistent with zero within statistical uncertainties. In spite of the difference in $\sqrt{s}$ and p_T, the data suggests that the strength of the quark-gluon correlation does not exceed their estimation.

In conclusion, the first measurement of the single spin asymmetry for direct photon production has been performed at the Fermilab Spin Physics Facility. The cross sections agree to the results of previous experiments using the proton beam of 200 GeV/c and nucleus target assuming the nuclear dependence of $A^{1.0}$. The obtained asymmetries are consistent with zero within the experimental uncertainties.

ACKNOWLEDGEMENTS

We gratefully acknowledge the assistance of the staff of Fermilab and of all the participating institutions. This research was supported by the Ministry of Education, Science and Culture in Japan, the USSR Ministry of Atomic Power and Industry, the US Department of Energy, the Commissariat a l'Energie Atomique and the Institut National de Physique Nucleaire et de Physique des Particules in France, and the Istitutodi Fisica Nucleare in Italy. One of us (N.S.) was supported by the JSPS Fellowships for Japanese Junior Scientists.

REFERENCES

1 J. Ashman *et al.*, Phys. Lett. **B206** 364 (1988); J. Ashman *et al.*, Nucl. Phys. **B328** 1 (1989); J. Ashman *et al.*, Phys. Lett. **B302** 533 (1993).

2 J. Qiu and G. Sterman, Phys. Rev. Lett. **67** 2264 (1991) and Nucl. Phys. **B378** 52 (1992).

3 Xiandong Ji, Phys. Lett. **B289** 137 (1992).

4 D.P. Grosnick *et al.*, Nucl. Inst. Meth. **A290** 269 (1990).

5 D.L. Adams *et al.*, IHEP-preprint, IFVE-94-88 (to be submitted).

6 D.L. Adams *et al.*, IHEP-preprint, IFVE-91-99 (unpublished).

7 N. Saito PhD thesis (in preparation).

8 G. Donaldson *et al.*, Phys. Lett. **B73** 375 (1978).

9 H.-U. Bengtsson and T. Sjöstrand, Computer Physics Commun. **46** 43 (1987).

10 CERN Library Long Write up W5013.

11 R. Kephart *et al.*, Phys. Rev. **D14** 2909 (1976).

12 J. Badier *et al.*, Z. Phys. **C31** 341 (1986).

13 M. MacLaughlin *et al.*, Phys. Rev. Lett. 11 971 (1983).

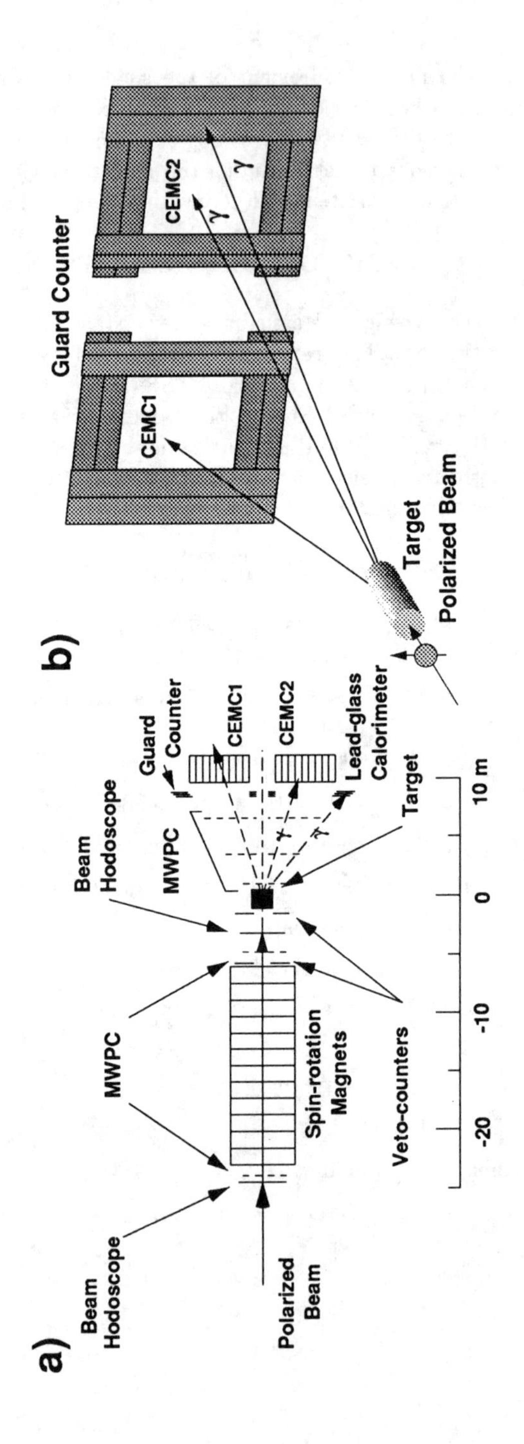

a)
Beam Hodoscope
Polarized Beam
MWPC
Spin-rotation Magnets
Veto-counters
Beam Hodoscope
MWPC
Target
Guard Counter
CEMC1
CEMC2
Lead-glass Calorimeter
-20
-10
0
10 m
b)
Guard Counter
CEMC1
CEMC2
γ
γ
Target
Polarized Beam

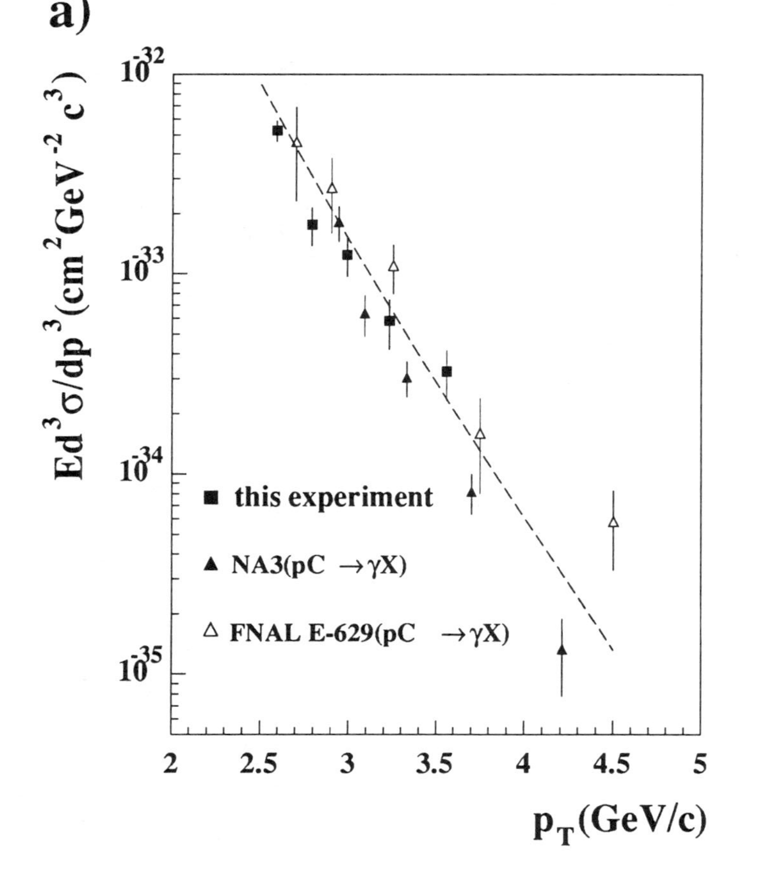
a)
$Ed^3\sigma/dp^3(cm^2GeV^{-2}c^3)$
■ this experiment
▲ NA3(pC → γX)
△ FNAL E-629(pC → γX)
p_T(GeV/c)

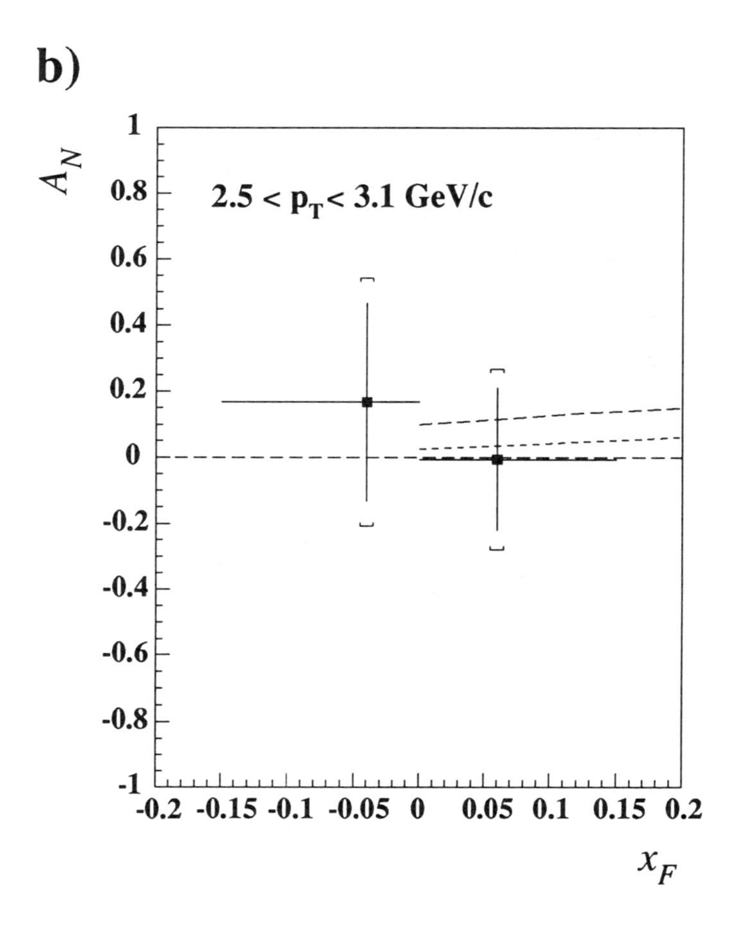
b)
A_N
$2.5 < p_T < 3.1$ GeV/c
x_F

Single Spin Asymmetries in Proton-Proton and Proton-Neutron Scattering at 820 GeV

Wolf-Dieter Nowak

DESY-Institut für Hochenergiephysik
Platanenallee 6, D-15738 Zeuthen, Germany

Abstract. The physics case is summarised for the investigation of high energy spin phenomena by placing an internal polarised target into HERA's unpolarised proton beam. Luminosity and experimental sensitivity are discussed. Estimating the physics reach of single spin asymmetries in different final states reveals a considerable physics potential, especially in testing the spin sector of perturbative QCD.

INTRODUCTION

During design and construction of the HERA proton ring unfortunately no provisions were taken to accelerate polarised particles. Nevertheless, with the now growing support there might be once polarised protons at HERA, as well. Before reaching this ultimate goal, a viable first step could be an experiment scattering HERA's unpolarised 820 GeV protons off polarised nucleons utilising a polarised internal gas target. Both proton-proton and proton-neutron spin asymmetries would be readily accessible since modern polarised gas targets can be operated with Hydrogen, Deuterium, or ^{3}He.

Experimentally single spin asymmetries are usually generated by switching the direction of the initial spin vector. In the given situation they can be measured as correlation between the polarisation of the target nucleon on one hand, and the final state polarisations and angular distributions on the other.

In the next section the anticipated experimental sensitivity is estimated for two different luminosity scenarios. Then, the physics reach of several interesting final states is discussed in some detail. The magnitude of the gluon spin might be probed with jets or possibly dimuons. It appears feasible to measure (leading) twist-3 contributions with direct photons where theoretical predictions distinguishing between proton and neutron target are of special interest. New pion data at higher transverse momenta could (dis)prove the pQCD contradicting non-zero transverse pion asymmetry data. Analysing dipions from the same jet might allow for accessing the valence quark transversity distribution.

LUMINOSITY AND SENSITIVITY

A realistic estimate for the average HERA proton beam current would be $\bar{I}_B =$ 80 mA $= 0.5 \cdot 10^{18}\ s^{-1}$ constituting half of the design current. An internal target area desity of $n_T = 10^{12}$ atoms/cm^2, as can be delivered by a standard polarised jet target, would not deteriorate the present HERA proton beam performance. This safe "low luminosity option" would have a luminosity of

$$\mathcal{L}_L = n_T \cdot \bar{I}_B = 0.5 \cdot 10^{30}\ \text{cm}^{-2} s^{-1}\ .$$

Allowing for a moderate beam life time deterioration on the 10% level the polarised internal target could presumably be operated with an area density of a few 10^{13}, say $n_T = 3 \cdot 10^{13}$ atoms/cm^2. Note that the UA6 unpolarised internal target was successfully run at a comparable density in the CERN $Sp\bar{p}S$ collider. Today's polarised H/D targets (1) and ^{3}He targets (2) with storage cells are capable of running at those densities with polarisations as high as 80% and 50%, respectively. Hence a "high luminosity option" with

$$\mathcal{L}_H = 1.5 \cdot 10^{31}\ \text{cm}^{-2} s^{-1}$$

appears feasible although still to be proven under actual HERA conditions.

To assess the physics reach of different final states a total running time of $T = 1.6 \cdot 10^7\ s$ with 100% efficiency is assumed. This corresponds to about 3 calendar years of HERA operation with 6 months per year physics running and 33% combined up–time for accelerator and experiment. Hence the integrated luminosities per year for the two discussed running scenarios are

$$\mathcal{L}_L \cdot T = 8\ pb^{-1} \quad \text{for the low luminosity option and}$$
$$\mathcal{L}_H \cdot T = 240\ pb^{-1} \quad \text{for the high luminosity option.}$$

The experimental sensitivity in the measured single spin asymmetry A is

$$\delta A = \frac{1}{p_{targ}} \cdot \frac{1}{\sqrt{N}}\ ,$$

where p_{targ} is the degree of target polarisation and $N = \mathcal{L} \cdot T \cdot C \cdot \sigma$ the total number of recorded events. Here σ is the unpolarised cross section and C the combined trigger and reconstruction efficiency. Then

$$\delta A = \frac{1}{p_{targ}} \cdot \frac{1}{\sqrt{\mathcal{L} \cdot T \cdot C}} \cdot \frac{1}{\sqrt{\sigma}}\ ,$$

and with $p_{targ} = 0.8$ and C = 50% one obtains as experimental sensitivities

$$\delta A_L = 0.6/\sqrt{\sigma\,[pb]} \quad \text{for the low luminosity option and}$$
$$\delta A_H = 0.1/\sqrt{\sigma\,[pb]} \quad \text{for the high luminosity option.}$$

PHYSICS OBJECTIVES

Probing the Gluon Spin with Inclusive Jets. At $\sqrt{s}$ = 40 GeV and p_t = 5 GeV the huge unpolarised cross section for inclusive jet production allows for sensitivities of 0.0001 [0.0006] in the high [low] luminosity option. At p_t = 10 GeV the sensitivities are still 0.003 [0.018], always meant for 1 GeV bins. Stratmann and Vogelsang (3) calculated the corresponding hard scattering cross sections for both *transverse* and *longitudinal* singly polarised proton–proton scattering including all underlying pQCD $2 \rightarrow 2$ subprocesses. The *transverse* asymmetry at p_t = 10 GeV is $A_T \simeq 12\%$, unfavourably with insignificant dependence on the polarised transverse sea. The *longitudinal* asymmetry A_L is very sensitive to the size of ΔG. Over the accessible p_t range $5 \div 10$ GeV A_L smoothly rises from 5 to 25% if ΔG = 0, whereas it stays approximately constant at 25% when assuming a very large gluon distribution.

In the given kinematical situation the c.m. backward jet will emerge under a few hundred milliradian in the laboratory system. Anticipating that the three fastest particles in the jet can be isolated a handedness (4,5) analysis is believed to measure the spin of the fragmenting parton. The supposedly process independent handedness parameter could possibly be about 0.05 (6). This, together with dilutions from the internal target polarisation [0.8] and an anticipated average parton spin [0.25], would result in a total dilution factor of about 100. Hence the 25% parton level asymmetry would be reduced to a 2.5 per mille, i.e. 0.0025 hadronic level asymmetry. This, at p_t = 5 GeV, is a 4 σ effect even in the low luminosity option, although the systematic error has to be kept on the permille level, as well.

Probing the Gluon Spin with Dimuons. Carlitz and Willey (7) calculated the *longitudinal* single spin asymmetry A_L for dimuon production in proton-proton collisions. It is non-zero if the axial vector built from the muon momenta has a longitudinal component. The relevant subprocesses are gluon Compton scattering and quark–antiquark annihilation. Using different assumptions on the total gluon spin Nadolsky (8) found this asymmetry ranging from A_L = 0.01 for ΔG = $0 \div 1$ up to A_L = 0.08 for ΔG = $5 \div 6$, although today ΔG = $3 \div 4$ might be more a realistic upper limit. Obtained for dimuon masses above 10 GeV this scenario corresponds to an unpolarised cross section below 1 pb. Even in the high luminosity option the expected experimental sensitivity is at best $\delta A \simeq 0.1$ and hence comparable to the whole asymmetry difference.

The steeply rising dimuon cross section suggests to reconsider the case at smaller dimuon masses, e.g. 4 GeV. Here the cross section of about 100 pb leads to $\delta A \simeq 0.01$ in the high luminosity option. However, a real improvement in the sensitivity to asymmetry ratio remains to be shown since in most models a smaller gluon spin is expected if probing at smaller dimuon masses.

Probing Twist-3 Matrix Elements with Direct Photons. Based upon a twist-3 parton distribution involving the correlation between quark fields and the gluonic field strength, the leading single *transverse* spin asymmetry for high p_t direct photon production was estimated by Qiu and Sterman (9). The essential subprocess is gluon Compton scattering with the gluon carrying the initial polarisation information. The hard scattering asymmetry rises to about 20% for $x_F \simeq$ - 0.8. Estimating the same matrix element differently Ehrnsperger et al. (10) reconsidered the case with special emphasis to differences between proton and neutron target. Only a small negative proton asymmetry but a rather large positive neutron asymmetry of several 10% is predicted. Additionally, Ji (11) considered the variety of relevant three–gluon–correlations. At large negative x_F the pure gluon correlations are expected to become dominant compared to quark–gluon correlations.

The parton to hadron level asymmetry dilution amounts to a factor of 5, since there is no fragmentation. Hence the above discussed neutron asymmetry requires $\delta A \leq 0.005$, i.e. it can be studied up to $p_t \simeq 4.5$ GeV in the high luminosity option. Note that the considered gaseous H/D target is a good tool to minimise systematic errors in the study of proton–neutron differences.

New Physics from Transverse Inclusive Pion Asymmetries ? Being the only significant data on single spin asymmetries at all, E704 has measured the reactions $p^{\uparrow} + p \rightarrow \pi^{0\pm} + X$ in a *transversely* polarised beam at 200 GeV (12). In strong contradiction to pQCD predictions significant non–zero asymmetries were measured. With increasing x_F the charged pion asymmetry is smoothly rising up to 40%, opposite in sign for different pion charge. Years after publication this data still constitutes the same challenge to perturbative QCD. Confirmation by an independent experiment at twice as high transverse momenta would certainly trigger new theoretical activities.

An example for alternative physics ideas on this subject is the existence of orbiting valence quarks proposed by Boros et al. (13). Within a semi–classical model the constituents of a polarised hadron are assumed to perform an orbital motion about the polarisation axis and left/right asymmetries are expected to arise from annihilations of these valence quarks. The model is able to give a fair description of the inclusive pion data as it does for dimuon data, as well.

Accessing the Valence Quark Transversity with Dipions. The measurement of two–pion correlations within the same jet when scattering *transversely* polarised hadrons off unpolarised ones is proposed by Collins et al. (14) as a tool to jointly probe the valence quark transversity distribution and the polarised fragmentation function. A twist-2 asymmetry in the latter is predicted to be accessible. The underlying pQCD subprocess is quark gluon scattering with the outgoing polarised valence quark fragmenting into a jet which is supposed to carry spin information from the transversely polarised target parton. To be analysed is the angular correlation between the axial vector built from the two pion momenta and the transverse component of the initial polarisation vector.

The experimental sensitivity was already discussed above. The spin transfer in the hard subprocess is estimated to be possibly large thus a large asymmetry in the overall process seems possible. Undoubtedly, more theoretical work is needed to arrive at numerical predictions.

CONCLUSIONS

There is good reason for optimism that very interesting and important, even completely new fundamental information on the nucleon spin could be accessed by placing an internal polarised target into the unpolarised HERA proton beam. A rather broad physics programme aiming at testing the spin sector of perturbative QCD and beyond can be based upon measurements of single spin asymmetries in several final states. Note however that for most of the potentially accessible physics a lot more theoretical predictions are necessary to seriously justify the considered experiment.

ACKNOWLEDGEMENTS

Many thanks to D.Trines and E.Steffens for valuable support on beam and target issues. Enlightening discussions with S.Manayenkov, T.Meng, M.Ryskin, A.Schäfer, J.Soffer, O.Teryaev, and W.Vogelsang are warmly acknowledged. Special thanks to A.Schäfer for critically reading the manuscript.

REFERENCES

1. K.Zapfe, contribution to this conference
2. L.Kramer, contribution to this conference
3. M.Stratmann, W.Vogelsang, *Phys. Lett.* **B295**, 277 (1992)
4. A.V.Efremov et al., *Phys. Lett.* **B284**, 394 (1992)
5. S.L.Belostotski et al., *PNPI preprint* **NP-37-1993**, 1906 (1993)
6. A.V.Efremov, contribution to this conference
7. R.D.Carlitz, R.S.Willey, *Phys. Rev.* **D45**, 2323 (1992)
8. P.M.Nadolsky, *Z. Phys.* **C62**, 109 (1994)
9. J.Qiu, G.Sterman, *Phys. Rev. Lett.* **67**, 2264 (1991)
10. B.Ehrnsperger et al., *Phys. Lett.* **B321**, 121 (1994)
11. X.Ji, *Phys. Lett.* **B289**, 137 (1992)
12. D.L.Adams et al., *Z. Phys.* **C56**, 181 (1992)
13. C.Boros et al., *Phys. Rev. Lett.* **70**, 1751 (1993)
14. J.C.Collins et al., *preprint* **PSU/TH/101** (April 1993)

Ξ^- and Ω^- POLARIZATION FROM A NEUTRAL HYPERON BEAM

K.A. Johns[a], P.M. Border[b], D.P. Ciampa[b], Y.T. Gao[c,e], G. Guglielmo[b,f], K.J. Heller[b], M.J. Longo[c], R. Rameika[d], N.B. Wallace[b], D.M. Woods[b]

[a]*Department of Physics, University of Arizona, Tucson, AZ 85721*
[b]*School of Physics and Astronomy, University of Minnesota, Minneapolis, MN 55455*
[c]*Department of Physics, University of Michigan, Ann Arbor, MI 48109*
[d]*Fermi National Accelerator Laboratory, Batavia, IL 60510*
[e]*now at Lanzhao University, Lanzhao, China*
[f]*now at University of Oklahoma, Norman, OK*

Introduction

The lack of understanding of the origin of hyperon polarization is a problem in high energy physics which remains nearly 20 years after its discovery[1]. Several phenomenological models exist[2–5] but their predictions are for the most part qualitative or semi-quantitative. Possible mechanisms for hyperon polarization vary. In the DeGrand-Miettinen model[2], polarization results from Thomas precession of the quark spins in the recombination process. In the Lund model[3], polarized $s\bar{s}$ pairs are produced by stretching of a color field and conservation of angular momentum leads to a preferred spin direction. There are also several Regge type models[4,5] in which polarization effects arise from interference between different production channels.

More recent data[6–8] taken to gain additional insight has, if anything, created additional confusion. For example, $\bar{\Lambda}$'s and Ω^-'s produced by protons are unpolarized while $\bar{\Xi}^+$'s and $\bar{\Sigma}^-$'s are produced with sizeable polarization. Additional data is clearly needed to untangle this picture. With this in mind we present recent results from Fermilab experiment E-800 on the polarization of Ξ^- and Ω^- produced from polarized and unpolarized neutral hyperon beams. Little data exists on the polarization of hyperons produced from strange baryon beams such as is presented here.

A simple model borrowing heavily from the DeGrand-Miettinen model[2] can be invoked to understand some features of the hyperon polarization data. Working with SU(6) wavefunctions and the quark recombination model, we assume a sufficient condition for hyperon polarization occurs when there are quarks in common between beam and target particles and when at least one s quark is picked from the sea in the recombination process. This simple model when applied to proton production "predicts" opposite signs for Λ, Ξ^-, and Ξ^0 polarization compared to Σ^+, Σ^-, and Σ^0 polarization. The model however says nothing about the P_T or x_F dependence of the polarization, anti-hyperon polarization, $K^- \rightarrow \Lambda$ polarization, etc.

Experimental Method

E-800 used two different methods for producing Ξ^- and Ω^-particles. The first used a polarized neutral hyperon beam and was called spin transfer mode. In this mode, 800 GeV protons were incident at $\pm$ 1.8 *mrad* on a 0.37λ Be target. The secondary beam passed through a neutral collimator and consisted of polarized Λ and Ξ^0 as well as neutrons and photons. The sign of the sweeper magnetic field was along the direction of polarization and hence no spin precession occurred. This polarized neutral hyperon beam was then incident at 0 *mrad* on a second 0.37λ Be target. The resultant tertiary beam was passed through a charged collimator (hyperon magnet with field integral of 24.36 T-m) which selected negatively charged particles. This magnet also precessed the spin of the hyperons to allow a measurement of the Ω^- magnetic moment. In spin transfer mode, one would expect that polarized Ξ^- and Ω^- would be produced via the transfer of polarized s quarks from the polarized Ξ^0 and Λ particles.

The second production method was called neutral production mode. In this case, 800 GeV protons were incident at 0 *mrad*. The secondary beam again passed through the neutral collimator but now consisted of unpolarized neutral particles including Λ and Ξ^0. This unpolarized neutral beam was then used to produce a tertiary negative beam as above however in this method was incident at a production angle of $\pm$ 1.8 *mrad*. In the neutral production mode one might expect that Ξ^- and Ω^- produced from this unpolarized neutral hyperon beam would be polarized via the same (unknown) mechanism which polarizes hyperons produced from protons (since the produced particles have quarks in common with the incident beam).

The Ξ^- and Ω^- particles and their charged daughters were tracked by a magnetic spectrometer consisting of scintillation counters, 8 planes of $100\mu m$ pitch SSD's, 12 MWPC's with 1 and $2mm$ spacing, and a spectrometer magnet with a P_T kick of 1.45 GeV/c. The decays used were $\Omega^- \rightarrow \Lambda K^-$ and $\Xi^- \rightarrow \Lambda\pi^-$ where subsequently $\Lambda \rightarrow p\pi^-$. The trigger required scintillation counter hits which selected particles exiting the collimator and MWPC hits from the two most downstream chambers which selected events with at least one positively and one negatively charged track.

The offline analysis used the $2mm$ MWPC hits to find three tracks. Next a geometric fit to a three track, two vertex topology was done followed by a kinematic fit to the proton and π^- momenta constrained to the Λ mass. The $1mm$ MPWC and SSD hits were used to improve the two vertex positions. Finally mass cuts were applied under the Ξ^- or Ω^- hypothesis. In the case of the Ω^-, a set of additional kinematic cuts were applied to remove remaining Ξ^- background.

Hyperons decay weakly through the parity violating process $H \rightarrow B\pi$, where H is the hyperon and B is the daughter baryon. The hyperon polarization is found by noting the spin direction of the daughter baryon follows the spin direction of the parent hyperon. For Ω^- and Ξ^- decays, the daughter Λ polarization follows that of the parent as

$$\vec{P}_\Lambda = \gamma_\Omega \vec{P}_\Omega \tag{1}$$

and

$$\vec{P}_\Lambda = \alpha_\Xi \hat{\Lambda} + \gamma_\Xi \vec{P}_\Xi \tag{2}$$

where $\hat{\Lambda}$ is the direction of the daughter Λ in the Ξ^- rest frame.

The Λ polarization is determined in the usual way by measuring the angular distribution

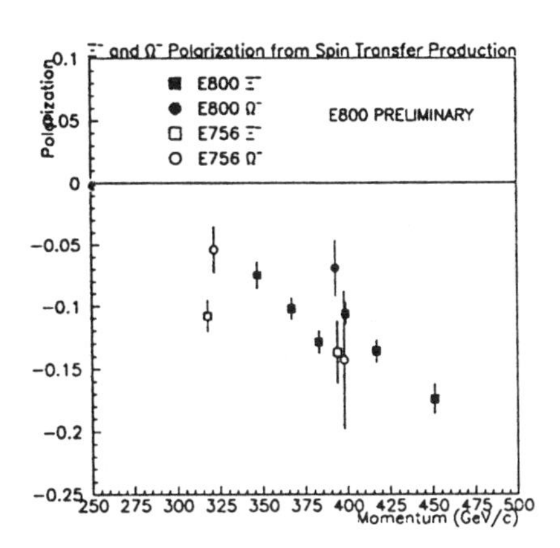

Figure 1: Ξ^- and Ω^- polarization from spin transfer production mode

of the proton in the Λ rest frame.

$$\frac{dn^{\pm}}{dcos\theta_i} = \frac{1}{2}(1 \pm \alpha_\Lambda P^i_\Lambda cos\theta_i) * \epsilon(cos\theta_i) \tag{3}$$

where $\epsilon(cos\theta_i)$ is the acceptance and the $\pm$ refer to positive and negative production angles. In practice, to account for the effect of detector acceptances on the angular distribution, a hybrid Monte Carlo technique[9] is used to determine the Λ polarization.

Results

The Ξ^- and Ω^- polarizations from the spin transfer production mode as measured by E-800 are shown in Figure 1. Also shown are earlier data from E-756. The Ξ^- polarization is large ($\approx$ 10%) and increases with momentum. The Ω^- polarization is about half as large as the Ξ^- polarization and no clear momentum dependence is observed.

These results may be described using the simple recombination model described above. Given that the polarized neutral hyperon beam contains Λ's and Ξ^0's with large, negative polarization, we expect (and observe) a large, negative polarization of Ξ^-'s via spin transfer of the polarized s quark. The same arguments can be invoked to describe the Ω^- polarization. That the Ω^- polarization is not as large as the Ξ^- polarization is difficult to understand in this model.

The Ξ^- and Ω^- polarizations from the neutral production mode are shown in Figure 2. In this mode the Ξ^- polarization is found to be consistent with zero while the Ω^- polarization is small and positive. That the Ξ^- polarization is indeed zero is seen by plotting the Ξ^- polarization components in the $\hat{x}$ and $\hat{z}$ directions. Parity conservation requires the initial polarization direction at production to be in the $\pm\hat{x}$ direction. Any real polarization would therefore be observed as a precession off this axis. In fact, the data show the precession

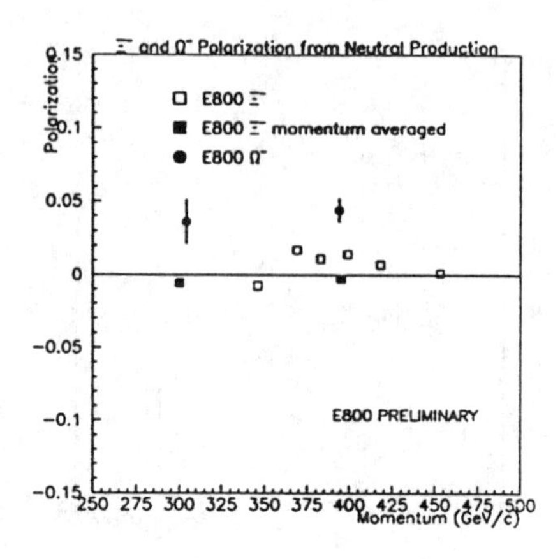

Figure 2: Ξ^- and Ω^- polarization from neutral production mode

angles to be randomly distributed. The same test may be applied to the Ω^- data. In this case, Ω^- polarization from spin transfer mode and Ω^- polarization from neutral production show the same precession angle (see Figure 3). Note that although the precession angles are the same, the directions of the initial polarization direction are opposite.

We can again turn to our simple model for interpretation of the data. Recall one of our assumed requirements for polarization was that an s quark from the sea be recombined with the fast quark(s) in common between beam and produced particle. By considering quark content, Ξ^- production from Λ's should be similar to Λ production from protons. Similarly, Ξ^- production from Ξ^0's should be similar to neutron production from protons. Making the further assumption that Ξ^- production is primarily from Ξ^0's, then we expect (and find) no Ξ^- polarization since there is no s quark picked from the sea. (Similarly one would expect neutrons to be unpolarized when produced from protons). If Ξ^- production were from Λ, one would expect a negative polarization similar to that found in $p \rightarrow \Lambda$.

For the Ω^- again assuming that most Ω^- are produced from Ξ^0's, then Ω^- polarization from $\Xi^0 \rightarrow \Omega^-$, should be similar to either $p \rightarrow \Sigma^+$ or $p \rightarrow \Lambda$. The data indicate that Ω^-'s are produced polarized similar to the former. Note that both $\Xi^0 \rightarrow \Omega^-$ and $p \rightarrow \Sigma^+$ have a diquark with identical quarks in common between beam and produced particle but in $p \rightarrow \Lambda$ the diquark consists of unlike quarks. In both cases however an s quark is produced from the sea. If Ω^- production were from Λ, one would expect a negative polarization similar to that found in $p \rightarrow \Xi^-$.

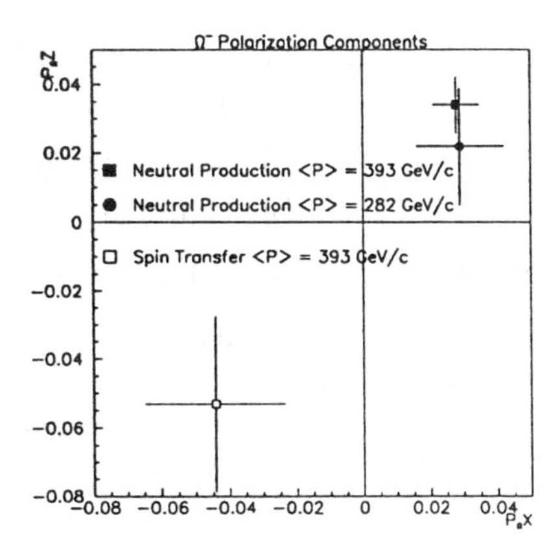

Figure 3: Ω^- polarization components from spin transfer and neutral production mode

Conclusions

E-800 has measured the polarization of Ξ^-'s and Ω^-'s produced from both polarized and unpolarized neutral hyperon beams. The negative Ξ^- and Ω^- polarizations observed in spin transfer mode are consistent with the simple quark recombination model however the small Ω^- magnitude is not understood in this context. In the neutral production mode (unpolarized neutral hyperon beam production) the Ξ^- polarization is zero and the Ω^- is small and positive. The signs of the Ξ^- and Ω^- polarizations in neutral production mode are possibly consistent with the simple quark recombination model though uncertainties in the production and polarization mechanisms weaken this argument considerably. In particular the positive Ω^- polarization is difficult to predict a priori.

A detailed understanding of hyperon polarization data continues to elude us. This problem demands additional experiments in which the quark and spin content of the beam is manipulated (such as has been pioneered by E-756 and E-800).

This work was supported by the United States Department of Energy and the National Science Foundation.

1. G. Bunce *et al.*, Phys. Rev. Lett. **36** (1976) 1113.
2. T.A. DeGrand and H.I. Miettinen, Phys. Rev. **D24** (1981) 2419.
3. B. Andersson *et al.*, Phys. Lett **85B** (1979) 417.
4. R. Barni *et al.*, Phys. Lett **296B** (1992) 251.
5. J. Soffer and N.A. Tornqvist, Phys. Rev. Lett **68** (1992) 907.
6. P.M. Ho *et al.*, Phys. Rev. Lett. **65** (1990) 1713.
7. K.B. Luk *et al.*, Phys. Rev. Lett. **70** (1993) 900.
8. A. Morelos *et al.*, Phys. Rev. Lett. **71** (1993) 2172.
9. G. Bunce, Nucl. Inst. Meth. **172** (1980) 353.

Single-Spin Asymmetries and Invariant Cross Sections of the High Transverse-Momentum Inclusive π^0 Production in 200 GeV/c pp and $\overline{p}p$ Interactions

D. L. Adams,[1] N. Akchurin,[2] N. I. Belikov,[3] A. Bravar,[14] J. Bystricky,[4] M. D. Corcoran,[1] J. D. Cossairt,[5] J. Cranshaw,[1] A. A. Derevschikov,[3] H. En'yo,[6] H. Funahashi,[6] Y. Goto,[6] O. A. Grachov,[3] D. P. Grosnick,[7] D. A. Hill,[7] K. Imai,[6] Y. Itow,[6] K. Iwatani,[8] K. W. Krueger,[9] K. Kuroda,[10] M. Laghai,[7] F. Lehar,[4] A. de Lesquen,[4] D. Lopiano,[7] F. C. Luehring,[11,*] T. Maki,[12] S. Makino,[6] A. Masaike,[6] Yu. A. Matulenko,[3] A. P. Meschanin,[3] A. Michalowicz,[10] D. H. Miller,[11] K. Miyake,[6] T. Nagamine,[6,†] F. Nessi-Tedaldi,[1,‡] M. Nessi,[1,‡] C. Nguyen,[1,§] S. B. Nurushev,[3] Y. Ohashi,[7,**] Y. Onel,[2] D. I. Patalakha,[3] G. Pauletta,[13] A. Penzo,[14] A. L. Read,[5] J. B. Roberts,[1] L. van Rossum,[5] V. L. Rykov,[3] N. Saito,[6] G. Salvato,[15] P. Schiavon,[14] J. Skeens,[1] V. L. Solovianov,[3] H. Spinka,[7] R. Takashima,[16] F. Takeutchi,[17] N. Tamura,[18] N. Tanaka,[19,††] D. G. Underwood,[7] A. N. Vasiliev,[3] J. L. White,[1,‡‡] S. Yamashita,[6] A. Yokosawa,[7] T. Yoshida,[20] and A. Zanetti[14]

(FNAL E704 Collaboration)

[1] *T.W.Bonner Nuclear Laboratory, Rice University, Houston, TX 77251*
[2] *Department of Physics, University of Iowa, Iowa City, IA 52242*
[3] *Institute for High Energy Physics, 142284, Protvino, Russia*
[4] *CEN-Saclay, F-91191 Gif-sur-Yvette, France*
[5] *Fermi National Accelerator Laboratory, Batavia, IL 60510*
[6] *Department of Physics, Kyoto University, Kyoto 606, Japan*
[7] *Argonne National Laboratory, Argonne, IL 60439*
[8] *Hiroshima University, Higashi-Hiroshima 724, Japan*
[9] *Northeastern State University, Talequah, OK 74464*
[10] *Laboratoire de Physique des Particules, B.P.909, F-74017 Annecy-le-Vieux, France*
[11] *Physics Department, Northwestern University, Evanston, IL 60201*
[12] *University of Occupational and Environmental Health, Kita-Kyushu 807, Japan*
[13] *University of Udine, I-33100 Udine, Italy*
[14] *Dipartimento di Fisica, Universita di Trieste, I-34100 Trieste, Italy*
[15] *Dipartimento di Fisica, Universita di Messina, I-98100 Messina, Italy*
[16] *Kyoto University of Education, Kyoto 612, Japan*
[17] *Kyoto-Sangyo University, Kyoto 612, Japan*
[18] *Okayama University, Okayama 700,Japan*
[19] *Los Alamos National Laboratory, Los Alamos, NM 87545, USA*
[20] *Osaka City University, Osaka 558, Japan*

Abstract

The π^0 inclusive and semi-inclusive, single-spin asymmetries have been measured using transversely-polarized, 200-GeV/c proton and antiproton beams colliding with an unpolarized hydrogen target. The measured asymmetries are consistent with a value of zero within the error bars for the kinematic regions, $-0.15 < x_F < +0.15$ and $1 < p_T < 4.5$ GeV/c. These data indicate that the higher-twist contribution in QCD to the single-spin asymmetry in inclusive π^0 production may not be as large as was previously expected. Additional evidence for such a conclusion comes from the measurement of a semi-inclusive π^0-charged particle asymmetry, where the associated charged particles are detected opposite to the π^0 azimuthal direction. This experiment also has a high-statistics measurement of the inclusive π^0 cross sections for pp and $\bar{p}p$ collisions at 200 GeV/c.

There has been a recent growing interest in the measurement of single transverse spin asymmetries, due to the discovery of a large left-right asymmetry in the inclusive production of pions at 13 and 18 GeV/c [1], 24 GeV/c [2], and 40 GeV/c [3], and the measurement of a large polarization in the production of hyperons [4]. In PQCD the expected single transverse spin effects become less than a percent. The data, however, indicate spin effects on the order of ten percent. Methods of reconciling such an apparent contradiction were proposed in several theoretical papers, for example, quark-gluon correlations [5], the color string model [6], and unsuppressed higher-twist effects [7]. These approaches appear to be very promising, but experimental data is badly needed for verification. The theoretical models also differ in their predictions for the single transverse spin asymmetry by an order of magnitude, for example, compare Refs. 8 and 9.

The single transverse spin asymmetries for inclusive and semi-inclusive π^0 production were measured in the E-704 experiment at the Fermi National Accelerator Laboratory using 200-GeV/c polarized beams of protons or antiprotons and a liquid hydrogen target. The transverse momentum range of the scattered particles was measured between 1 to 4.5 GeV/c. The beam and transverse momenta in this experiment were such that the QCD predictions ~~may be~~ applicable at the highest measured p_T values. The reliability of the

*Present address: Indiana University, Bloomington, IN 47405.
†Present address: SLAC, Stanford, CA 94305.
‡Present address: CERN, CH-1211 Geneva 23, Switzerland.
§Present address: University of Texas, Austin, TX 78712.
**Present address: The Institute of Physical and Chemical Research, Saitama 351-01, Japan.
††Deceased.
‡‡Present address: American University, Washington, D.C. 20016.

asymmetry data has been checked by reconstructing the inclusive π^0 cross sections and comparing them with published data.

The Fermilab polarized proton (antiproton) beam with 200-GeV/c momentum is produced using the party-violating decays of $\Lambda(\overline{\Lambda})$ hyperons. The beam-transport system has been designed to minimize the beam depolarization effects. The maximum polarization value of the proton (antiproton) was 65%. Further details on the polarized beam line are described in Ref. [10].

Photons from the decays of neutral mesons produced in the target were detected in two Central Electromagnetic Calorimeters, CEMC1 and CEMC2, located symmetrically to the left and to the right of the beam axis at 10 m from the target. Each calorimeter is an array of 504 lead-glass counters, stacked in 21 columns by 24 rows. The dimensions of each lead-glass block were 3.81 cm $\times$ 3.81 cm $\times$ 18 radiation lengths. Each array covered a polar angle of $(5.5 \pm 2.2)^\circ$ in the laboratory frame, where the 5.5° angle corresponds to 90° CM, and azimuthal angles of $\pm 25^\circ$ with respect to the horizontal plane containing the beam axis. A set of wire chambers, located between the target and each CEMC, was used for charged particle detection and track reconstruction.

Many events in the CEMC contained overlapping electromagnetic showers from the two photons in the π^0 decay. The details of the π^0 reconstruction program are given in Ref. 11.

A total of 2×10^7 events were recorded with incident protons and a total of 2×10^6 events recorded with incident antiprotons.

The invariant cross sections for inclusive π^0 production in pp interactions are shown in Fig. 1, where the errrors shown are statistical only. An additional normalization uncertainty was estimated to be $\pm 15\%$ and a p_T scale uncertainty was $\pm 1\%$. No significant differences were observed in a comparison of the π^0 production in pp and $\overline{p}p$ interactions.

The results from the current experiment on the invariant cross sections for the reaction $\mathrm{p} + \mathrm{p} \rightarrow \pi^0 + \mathrm{X}$ are in qualitative agreement with the results of other experiments [12] - [14]. The invariant cross section data from the NA24 experiment at CERN [13] are also shown in Fig. 1 for comparison.

Single-spin asymmetries of inclusive π^0 production in pp and $\overline{p}p$ interactions near 90° CM are presented in Fig. 2. Those asymmetries produced in $\overline{p}p$ interactions are given in Fig. 3.

Further anlysis on the entire data sample has produced different results from a previous publication [15]. The correction of a problem with the polarization decoding, which increases the overall final data sample by 30%, mostly accounts for this difference in results. In the $1 < p_T < 3$ GeV/c region, the current result is the same as that given previously, which is a zero asymmetry value. The statistics are large in this region, and the difference in the amount of data analyzed does not impact the asymmetry value. However, in the $3.1 < p_T < 4.6$ GeV/c region, where the amount of statistics is much smaller, the asymmetry values have changed to zero. The previous publication indicated a nonzero asymmetry in this region.

In addition to the events for the purely inclusive reaction, $p\uparrow + p \to \pi^0 + X$, events were also selected that contained at least one charged particle having an azimuthal angle of $\varphi = (180 \pm 30)^\circ$ relative to the produced π^0 direction. The single-spin asymmetries for inclusive π^0 production with an associated charged particle in pp interactions are given in Fig. 2. Those asymmetries produced in $\bar{p}p$ interactions are given in Fig. 3.

The asymmetry is observed to be zero for single-spin inclusive π^0 production in pp interactions in the $1 < p_T < 3\,\mathrm{GeV/c}$ region within an accuracy up to 2%. At larger p_T, the errors grow from 5% at $p_T \approx 3.3\,\mathrm{GeV/c}$ up to 15% at $p_T \approx 4.1\,\mathrm{GeV/c}$. The errors associated with the data may not completely rule out the possibility of a nonzero asymmetry, as reported previously[15].

No significant difference was observed when selected events were added with at least one additional charged particle, produced at $(180 \pm 30)^\circ$ relative to the π^0 direction. By investigating this reaction with the criteria chosen above, the hard parton-parton interactions could be extracted. A measure of the asymmetry would not be diluted by soft processes. However, the observed asymmetry was equal to zero.

The amount of data for studying $\bar{p}p$ interactions was an order of magnitude less than that for pp interactions. The asymmetry is equal to zero within the statistical accuracy for the $1 < p_T < 3\,\mathrm{GeV/c}$ region, except for two points.

Observations of a small or zero asymmetry at large p_T agree with the recent calculation of a twist-3 effect [16]. It also agrees with the model of an orbiting valence quark around a polarization axis that produces a zero asymmetry value at $x_F = 0$, but a large asymmetry at large x_F values [17].

Acknowledgements

This work was performed at the Fermi National Accelerator Laboratory, which is operated by University Research Associates, Inc., under contract DE-AC02-76CH03000 with the US Department of Energy. Work supported in part by the US Department of Energy, Division of High Energy Physics, Contracts W-31-109-ENG-38, W-7405-ENG-36, DE-AC02-76ER02289, DE-AS05-76ER05096. This research was also supported by the former USSR Ministry of Atomic Power and Industry, the Ministry of Education, Science and Culture in Japan, the Commissariat a l'Energie Atomique and the Institut de Physique Nucleaire et de Physique des Particules in France, and the Istituto di Fisica Nucleare in Italy.

References

[1] S. Saroff *et al.*, Phys. Rev. Lett. **64**, 995 (1990).
[2] J. Antille *et al.*, Phys. Lett. B **94**, 523 (1980).
[3] V. D. Apokin *et al.*, Phys. Lett. B **243**, 461 (1990).

[4] K. Heller, In *Proc. of the VII Int. Symp. on High Energy Spin Physics, Protvino (Russia), 1986*, edited by A. A. Antipova (Serpukhov, Russia, 1987), Vol. 1, p. 81.
[5] A. V. Efremov and O. V. Teryaev, Yad. Fiz. **36**, 242 (1982) [Sov. J. Nucl. Phys. **36**, 140 (1982)].
[6] M. G. Ryskin, Yad. Fiz. **48**, 1114 (1988) [Sov. J. Nucl. Phys. **48**, 708 (1988)].
[7] J. Collins, Nucl. Phys. B **396**, 161 (1993).
[8] J. Qiu and G. Sterman, Phys. Rev. Lett. **67**, 2264 (1991).
[9] A. Schäfer, L. Mankiewicz, P. Gornicki, and S. Güllenstern, Phys. Rev. D **47**, R1 (1993).
[10] D. P. Grosnick *et al.*, Nucl. Instrum. Methods **A290**, 269 (1990).
[11] D. L. Adams *et al.*, Institute for High Energy Physics, Protvino, Russia preprint number IHEP 91-99, 1991.
[12] G. Donaldson *et al.*, Phys. Lett. B **73**, 375 (1978).
[13] C. De Marzo *et al.*, Phys. Rev. D **36**, 16 (1987).
[14] F. W. Büsser *et al.*, Nucl. Phys. B **106**, 1 (1976).
[15] D. L. Adams *et al.*, Phys. Lett. B **276**, 531 (1992).
[16] A. Schäfer *et al.*, Phys. Rev. D **47**, R1 (1993).
[17] C. Boros, L. Zuo-tang, and M. Ta-chung, Phys. Rev. Lett. **70**, 1751 (1993).

Figure Captions

Fig. 1. The invariant cross sections for the reaction $p + p \rightarrow \pi^0 + X$ at 200 GeV/c at $x_F = 0$ (black points), and at 300 GeV/c at $x_F = 0$ (open points).

Fig. 2. The asymmetry parameter A_N as a function of p_T at $x_F = 0$ (a) for the purely inclusive reaction, $p\uparrow + p \rightarrow \pi^0 + X$, and (b) for the same reaction, but when at least one charged particle is also detected at an azimuthal angle within $(180 \pm 30)^\circ$ relative to the π^0.

Fig. 3. The asymmetry parameter A_N as a function of p_T at $x_F = 0$ (a) for the purely inclusive reaction $\overline{p}\uparrow + p \rightarrow \pi^0 + X$, and (b) for the same reaction, but when at least one charged particle is also detected at an azimuthal angle $(180 \pm 30)^\circ$ relative to the π^0.

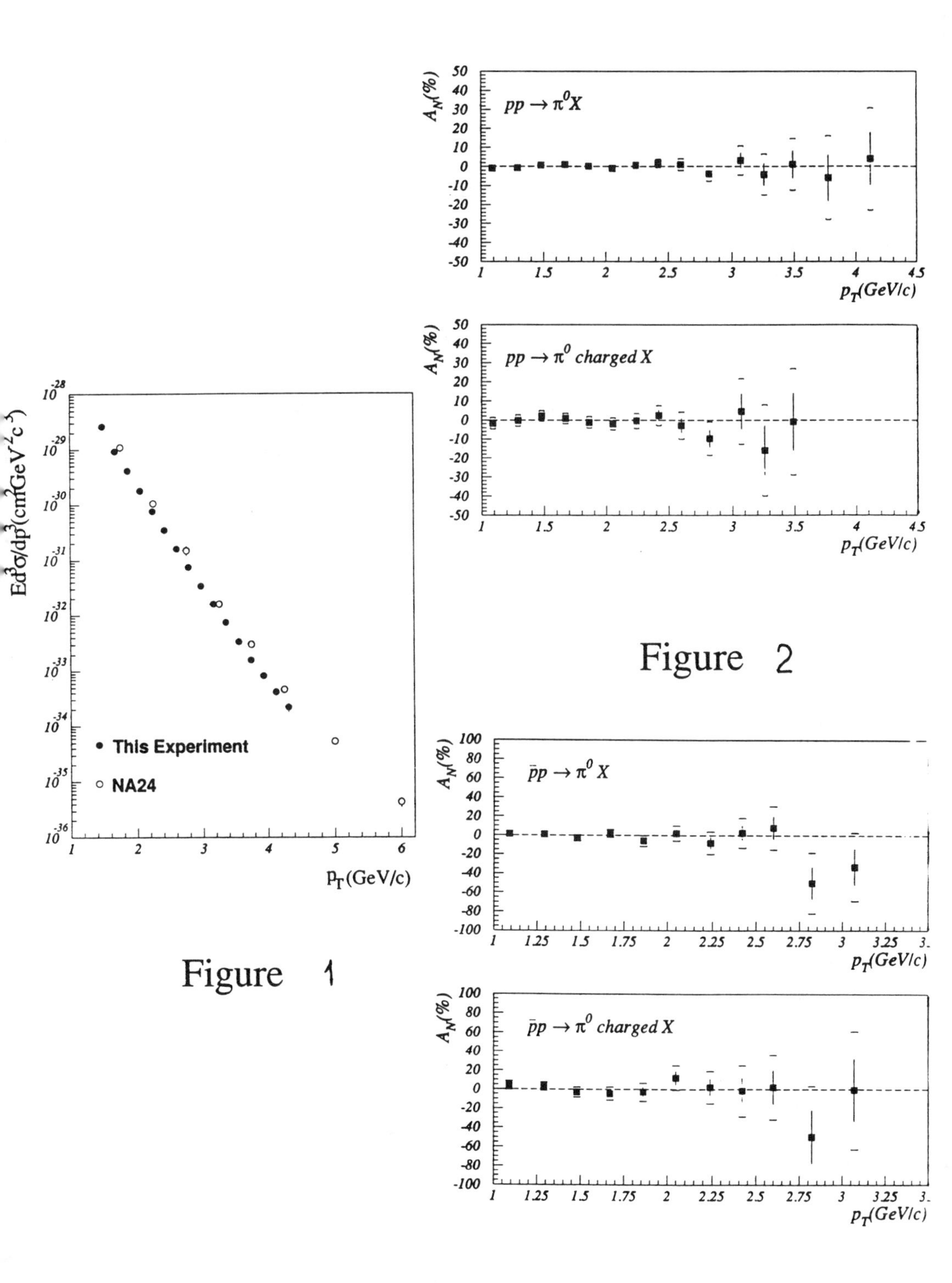

Figure 1

Figure 2

Figure 3

Modified Cascade Model of Nucleus-Nucleus Collisions

J. J. Musulmanbekov

December 27, 1994

Abstract

The modified version of cascade model of hadron - nucleus and nucleus - nucleus collisions has been adopted for simulation of inclusive pion asymmetry in polarized hadron - nucleus interactions. Pion asymmetry in single spin proton - proton collision is used as input in the model. Cascade - cascade interactions, $s-$wave resonance production and the concept of formation time of secondaries are included into the description of multiparticle production.

0.1

1 Introduction

It was found that the polarization effects (pion asymmetry and hyperon polarization in $\vec{p}p$ collisions) at high energies in the range of high x and $p_\perp$ is not negligible [1,2]. One-spin pion asymmetry A_N in $\vec{p}p \to \pi^{\pm,0}X$ processes have been studied at BNL and Fermilab and it was found that the asymmetry is small at low x and $p_\perp$ and reaches $\sim 40\%$ at high x and $p_\perp$ [5,6]. The sign of asymmetry strongly corellates with the sign of produced pion. These effects could not be explained in the frame of perturbative QCD and one has to take into account high twist corrections [3] or to use nonperturbative mechanisms [4].

These two mechanisms differ by characteristic scales at which the main contribution to the proton polarization appears.

Perturbative QCD leads to color transparency effect which is based on the idea that if small distances dominate in the process then the participating quark configurations are not absorbed in the nuclei. This results in A^1 – dependence of the cross section of these processes. Nonperturbative QCD is characterized by large distances between quarks that leads to color transparency breaking.

The nonperturbative mechanism of quark-quark interaction induced by instantons [5] have been proposed to describe the one-spin pion asymmetry in $\vec{p}p$ and $\vec{p}A$ process.. It has been shown that the asymmetry on nuclei increases with x_F and the week A-dependence of A_N has been found. The color transparency breaking has been predicted.

In this work we apply the proposed mechanism for the simulation of pion asymmetry in the $\vec{p} + A \to \pi^{\pm,0} + X$ process. We considered that pions are produced both directly and via decay of vector mesons and Δ - resonances in $S-$state. Asymmetry signs for vector mesons are given to be the same as for pions in accordance with their charges. For Δ-isobars they are the following: negative for Δ^{++}, Δ^{+} and positive for Δ^0, Δ^-. (In general case the asymmetry sign of produced intermediate particles are defined in accordance to the asymmetry sign rule [7].)

Monte-Carlo code including the concept of formation time of secondaries has been elaborated for hadron - nucleus and nucleus-nucleus interactions [8]. It enables one to simulate multiparticle production both for unpolarized and polarized beams.

Simulation of $A_N(x_F)$ on nuclei Al has been performed at AGS energy $E_L = 28$ GeV and shown in Fig.1. The black points are the experimental data for pion asymmetry in $\vec{p}p$ - collisions at $p_L = 200$ GeV. It is seen the asymmetry does not vanish with increasing atomic number A.

There are two types of explanation of large spin asymmetries of pions in inclusive hadron production by polarized protons. One of them is based on perturbative QCD, other - on nonperturbative approaches. Perturbative QCD leads to color transparency effect which is based on the idea that if small distances dominate in the process then the participating quark configurations are not absorbed in the nuclei. This results in A^1 – dependence of the cross section of these processes. Nonperturbative QCD is characterized by large distances between quarks that leads to color transparency breaking. S.Brodsky and G. De Teramon [1] put forward the idea of connection of large spin asymmetry mechanism and color transparency breaking in exclusive processes on nuclei.

So it would be convenient to use nuclear targets bombarded by polarized beams to distinguish these approaches, since nuclei can serve as useful tools for filtering high - perturbative and non - perturbative processes. In ref. [2] authors obtained analytical estimation of the influence of the nuclear matter on pion asymmetry using the approach based on non - perturbative QCD vacuum structure.

Our goal is to investigate polarization phenomena on nuclei using the cascade mechanism of multiparticle production. Traditional cascade model of hadron - nucleus and nucleus - nucleus interactions in which multiparticle production is a superposition of nucleon - nucleon and meson - nucleon interactions was in a satisfactory agreement with data at intermediate energies (up to few GeV). At higher energies it overestimates multiparticle production and thus is inconsistent for data description since the basic element of the model is the assumption that the final hadrons are created instantaneously at the point of the interaction. In the last few years various parton (multichain, string) models of hadron - nucleus and nucleus - nucleus collisions have been elaborated. In most contemporary models the multiparticle production is considered as two - step process: first, partons are created and then they are converted into the final hadrons emerged [3]. The consequence of this two - step mechanism of hadron production is that the creation of a hadron is not instantaneous but takes time, so called "formation time". So we include the formation time concept into the cascade mechanism of multiparticle production. The production of low laying resonances and clusters are taken into account. To simulate of the asymmetry of secondaries in proton - nucleus interactions we use as input the approximations of data on pion asymmetry in single polarized proton - proton interactions.

2 Formation time concepts

The inclusion of the formation time into multiparticle process in hadron – nucleus and nucleus – nucleus collisions is not trivial procedure. There are alternative ways how one can do it. The first one relates to the assumption that the fast produced particles have no possibilities for secondary interactions inside the nucleus. This assumption can be formulated using the argument based on uncertainty principle. According to uncertainty principle for creation of a hadron a proper rest frame is

$$\tau_0 = \frac{1}{m_t} = \frac{1}{(m^2 + p_t^2)^{1/2}} \tag{1}$$

In the laboratory frame this time is multiplied by Lorentz factor

$$\tau = \tau_0 \gamma = E_h / m_t^2 \tag{2}$$

where E_h - is the energy of the hadron in the laboratory. This approach is specific to the models of uncorrelated quark-parton cascades. One of well developed of these is Monte-Carlo code of J.Ranft [4]. Another way for formation time introduction is associated with string models where color exchange fields lead to the production of intermediate extended objects - strings which then emit final state hadrons. If a hadron is created inside a nucleus it can interact with nuclear matter. As regards for string interaction situation is not so clear. Two extreme possibilities can be envisaged:
(i) the string does not interact at all in nuclear matter;
(ii) the string is fully absorbed with a typical hadronic cross section.
It has been shown in papers [5,6] that an intermediate solution is needed i.e. the string is absorbed partly. In this case the formation time of emitted hadron is determined by space-time evolution of the string and depends not only upon hadron energy but on its x_F as well. Our approach, as will be seen below, embodies in a phenomenological way the peculiarities of formation time of both types of models but translates them into hadron language. The formation time is introduced by the following manner. As soon as two - hadron collision occurs, we apply the following sum rule for succeeding collisions of particles produced in forward (backward) hemisphere in c.m. frame

$$x^l \sigma^I + \sum x^m \sigma^I = \sigma^I \tag{3}$$

where x^l is Feynmam variable for the remnant of the incident particle, x^m - the same one for produced particles and σ^I is the inelastic cross section of the interacting hadrons. The first term in this identity is the cross section of the incident particle remnant (leading particle) and the second term is the sum of the cross sections of produced particles at the instant after interaction. At that moment the leading particle and other secondaries possess abnormal (reduced) cross section and could be considered as semibare and bare particles. However, this sum rule is correct only for soft collisions. We assume that all particles produced in hard collisions, which have jet - like structure, are bare particle.

The formation time when taken into account leads to the development of cross sections of these particles during their propagation inside the nucleus. We can use the exponential form of evolution of cross sections until the subsequent collision occurs

$$\sigma_2^l = \sigma_1^I(1 - (1 - x^l)e^{-\tau/(\gamma\tau_0)}) \tag{4}$$

for leading particle,

$$\sigma_2^m = \sigma_1^m - (\sigma_1^m - x^m\sigma_1^I)e^{-\tau_1/(\gamma\tau_0)} \tag{5}$$

for m - th produced particle, where σ_1^I - normal cross section of the incident particle in the first collision, σ_2^l - cross section of the remnant of the projectile (leading particle) in the second collision, σ_2^m - cross section of the m - th produced particle, σ_1^m - cross section for this type of particles in the normal state, γ — Lorentz-factor and τ_0 is adjustable parameter corresponding to the mean value of the formation time in the rest frame of the particle. Comparison of results of simulation with experimental data gives $\tau_0 = 1.4 \cdot 10^{-23} s$. One can notice that the expressions (4) and (5) are identical in the case of $\sigma_1^m = \sigma_1^I$. For $r + 1$ - th inelastic rescattering of the incident particle the cross section is defined as

$$\sigma_{r+1}^l = \sigma_1^I - \sigma_1^I(1 - x_1^l)e^{-\tau/(gamma_1\tau_0)} - \sigma_2^l(1 - x_2^l)e^{-\tau_2/(\gamma_2\tau_0)} - \dots$$
$$-\sigma_r^l(1 - x_r^l)e^{-\tau_r/(\gamma_r\tau_0)} = \sigma_1^I \prod_{i=1}^{r}(1 - (1 - x_i^l)e^{-\tau_i/(\gamma_i\ tau_0)}) \tag{6}$$

Our approach to formation time consideration is rather geometric one. The expressions (3), (4) and (5) correspond to the pattern where only overlapping parts of colliding hadrons is to evolve during formation time. The variable x characterizes the part of geometric size of Lorentz – contracted disk of the interacting hadron which is occupied by a bare or semibare secondaries at the instant of interaction. The closer x to 1, the closer the cross section of a particle to the normal one that corresponds to the case when the formation time approaches zero.

3 Formulation of the cascade model with formation time

Each of two colliding nuclei is considered as a fermi - gas of nucleons bounded in definite volume with diffuse boundary. Nucleon coordinates and their fermi moments are generated according to standard nuclear density distributions. Inelastic collision of two nuclei constitutes a superposition of baryon - baryon, meson - baryon and meson - meson elastic and inelastic interactions, which arbitrarily can be arranged into four groups.
Group C - interactions of the nucleons of the projectile nucleus with those from the target nucleus. All secondary particles produced in any group of interactions are considered as cascade particles.
Group A - interactions of the cascade particles with the nucleons of the target nucleus;
Group B - interactions of the cascade particles with the nucleons of the incident nucleus;
Group D - so called "cascade - cascade" interactions — interactions of cascade particles with each other.

The probability of any two particle interaction is expressed in cross section

$$q = \sigma_{tot}/\pi(r_0 + \lambda)^2 \tag{7}$$

where σ_{tot} - total cross section, r_0 - parameter related to the radius of strong interactions and λ – de-Broglie wave – length. Resonance cross sections in subsequent interactions are taken to be the same as for stable meson and nucleon interactions. As for meson - meson interactions their inelastic cross sections are expressed in meson - nucleon cross sections: $\sigma_{\pi\pi} = \frac{2}{3}\sigma_{\pi p}$. We did not take into consideration elastic meson – meson collisions.

Evolution of the interacting system is considered by the method proposed in paper [7]. At some instant of time t all possible interacting pairs are determined. Partners for interactions for each group (A,B,C,D) are defined; among all possible interactions that one is chosen which realizes before others, i.e. $\Delta t = min\{t_i\}$; then the positions of both nuclei and all cascade particles are moved to new ones corresponding to a new instant of time $t_i \rightarrow t_i + \Delta t$. During evolution of the system produced resonances may decay before their collisions within nuclei or with other cascade particles. It is checked both for all interactions and decay of resonances whether the Pauli principle is satisfied. Cascade stage of particle generation is completed when all cascade particles have left both nuclei or have absorbed by them.

The number and total charge of remaining nucleons in each nucleus specifies mass and charge number of the residual nucleus. Excitation energy of each residual nucleus is determined by the energy of absorbed particles and "holes" originated during intranuclear cascade process. Momentum and angular momentum of the residual nucleus is evaluated by momentum conservation law sequentially followed for each intranuclear interaction. We describe the behavior of excited residual nuclei by using the evaporation model [8].

4 Hadron - hadron event generator

According to the bremsstrahlung analogy the multiparticle production in hadron - hadron interactions is considered as a result of emission and subsequent decay of intermediate clusters [9]. Monte - Carlo simulation of exclusive events is performed in several steps.

i) On the first step initial c.m. energy portion available for production of secondaries is evaluated

$$W = \sum E_i = k\sqrt{s} \tag{8}$$

where E_i is the energy of i – th particle (excluding leading particles), k is inelasticity. Fluctuation of the inelasticity from event to event leads to the distribution $P(k)$. There is no elaborated theoretical method for calculation of $P(k)$. It has been shown in Ref. [10] that one may approximate the inelasticity distribution with beta distribution

$$P(k,s) = k^{a-1}(1-k)^{b-1}/B(a,b) \tag{9}$$

$$B(a,b) = \Gamma(a)\Gamma(b)/\Gamma(a,b) \tag{10}$$

$$< k(s) >= a/(a+b) \tag{11}$$

where $\Gamma(a), \Gamma(b)$ and $\Gamma(a,b)-$ gamma functions; s - dependence of $P(k,s)$ and $< k(s) >$ is enclosed in parameters a and b. Up to the ISR energies one can neglect this s - dependence.

ii) Considerable amount of kinetic energy of colliding hadrons is converted into the exitation energy (effective mass) of produced clusters. For simulation of cluster mass we adopted the result of Chou Kuang-Chao et al. [11], the idea of which was that the statistical nature of cluster production lead to mass spectra

$$P(M) = cM \exp(-bM) \tag{12}$$

Parameters c and b are determined by the following normalization conditions.

$$\int P(M)dM = 1 \tag{13}$$

$$\int M \cdot P(M)dM = \langle M \rangle \tag{14}$$

If M less than 1 GeV we put $M = m_\pi, m_\rho, m_K$ in correspondence with their relative weights.

One of the assumption of our approach – dependence of the average mass of cluster $< M >$ on collision energy. The reason of this dependence arise from the necessity to reproduce the following features of multiparticle production: growth of the central rapidity density, growth of the mean transverse momentum, increasing with collision energy yield of so called "minijets" [12]. We used the following expression for this dependence

$$\langle M \rangle = a_1 + a_2 \cdot s^\nu \tag{15}$$

where a_1, a_2 and ν are adjustable parameters. Comparison with experimental data gives for the values of these parameters: $a_1 = 0.03, a_2 = 0.09, \nu = 0.25$.

iii) On the third step the energy W is distributed between secondary particles which kinematical characteristics are generated in correspondence to a cylindrical phase space model. Parameters of cylindrical phase space model are adjusted by comparing the results of simulation of pion – nucleon and nucleon – nucleon interactions with data in wide range of collision energy. The remaining part of c.m. energy $(1-k)\sqrt{s}$ is distributed between remnants of interacting particles (so called leading particles) according to energy – momentum conservation lows

$$\overline{P}_I + \overline{P}_{II} = \sum \overline{P}_i \tag{16}$$

$$E_I + E_{II} = (1-k)\sqrt{s} \tag{17}$$

where $\overline{P}_i$ is a momentum of i-th produced particle, $\overline{P}_I$, $\overline{P}_{II}$ and E_I, E_{II}, are momenta and energies of leading particles. Colliding nucleons (mesons) can transform into nucleons (stable mesons) and s – wave resonances (Δ – isobars and ρ, ω – mesons). For transition probabilities we use the results of OPE – model [13].

. iiii) We assume that there is a definite possibility for cluster to decay on final particles in jet - like manner. Such assumption is motivated by the fact of increasing with collision energy fraction of events with minijets. For more detailed consideration of cluster decay one must include definite mechanism such as cascade branching. However, we restrict ourselves by phenomenological approach and apply for cluster decay cylindric phase space model with parameters derived from hadronproduction data in e^+e^-– annihilation processes provided

$$M_{cl} = \sqrt{s}_{e^+e^-} \tag{18}$$

The direction of such minijet axis is taken randomly to be isotropic in c.m.s of the decaying cluster.

5 Simulation of left - right asymmetry of secondaries

In the case of unpolarized hadron - hadron interactions seconaries are emitted isotropically in azimuthal angles, i.e. inclusive cross section of particle production doesn't depend on azimuthal angle. To include this dependence for interactions of polarized hadron with unpolarized one we apply the following expression

$$\sigma(\theta, \varphi) = \sigma(\theta) \cdot [1 - A(\theta) P_B \cos(\varphi)] \tag{19}$$

where θ is zenith angle, φ–azimuthal angle, $A(\theta)$–analyzing power and P_B–polarization of beam. For simulation of asymmetry of secondaries by this formula we used experimental data on x–dependence of asymmetry of inclusive pion production in polarized proton - proton interaction obtained in experiment E704. Signs of asymmetry for intermediate states (vector mesons and isobars) were evaluated according to nonperturbative approach [2] and given in the following table

Table 1

Type	Sign	Type	Sign	Type	Sign
π^+	$+$	ρ^+	$+$	Δ^{++}	$-$
π^0	$+$	ρ^0	$+$	Δ^+	$-$
π^-	$-$	ρ^-	$-$	Δ^0	$+$
		ω	$+$	Δ^-	$+$

x_F-dependence of asymmetry for resonances is taken the same as for pions with corresponding sign of asymmetry.

The collisions of polarizied proton with target nucleon inside the nucleus can be elastic or inelastic. In elastic collisions rescattered proton conserve its polarization. In the first inelastic collision produced particles are emitted asymmetrically, according to formula (12). In subsiquent collisions of secondaries with target nucleons produced particles are emitted isotropically in azimuthal angles in c.m.s. of interacting particles. Thus, the resulting asymmetry of particles produced on nucleus is defined by the asymmetry obtained in the first inelastic interaction and smeared in subsequent rescatterings inside nucleus.

We run the model in two modes:

i) all pions are produced directly without any intermediate state;

ii) some pions are produced via intermediate states: clasters, $s-$wave resonances which decay sequentially into final pions.

In Fig.1 one can see that inclusion of intrmediate states results in the behaviour of $x-$dependence of asymmetry which differs significantly from that one in the case of direct produced pions. So the inverstigation of the influence of intermediate states on final asymmetry of pions could help us in understanding of the nature of this phenomena.

6 Summary and conclusions

We evaluated one-single inclusive pion asymmetry in proton-nucleus interactions using Monte-Carlo code for hadron-nucleus and nucleus-nucleus collisions. The code is based on modified cascade model which includes $s-$wave resonance production and the concept of formation time (length).

We have elaborated modified cascade model of hadron - nucleus and nucleus - nucleus interactions for description of high energy data. It uses traditional hadron language for binary interactions. The main features of the model are the following:
A. Process of cascading takes place both in the target nucleus and the projectile one.
B. Cascade - cascade interactions have been taken into account.
C. All the interactions are ordered in time.
D. Formation time which phenomenologically characterizes the evolution of both interacting nucleons and produced mesons is included into the process of cascading inside both colliding nuclei.
E. Hadron - hadron event generator is based on cylindrical phase space model. Production of ρ,ω and Δ resonances have been taken into account.

The model is in a good agreement with experimental data for nucleus - nucleus interactions in the energy range $0.1 - -200$ GeV. An estimation of formation time of secondaries gives $(1.4 \pm 0.2) \cdot 10^{-23}$ s.

Table 2

E (A GeV)	3.6	14.6	60	200
N_{ev}^{exp}	1760	385	372	503
N_{ev}^{mod}	1000	1000	1000	1000
$< n_s >^{exp}$	10.6 ± 0.2	21.2 ± 1.1	40.6 ± 2.2	58.1 ± 2.8
$< n_s >^{mod}$	10.6 ± 0.4	21.1 ± 0.5	39.3 ± 0.5	55.5 ± 0.5
$< n_g >^{exp}$	6.5 ± 0.2	4.9 ± 0.3	5.7 ± 0.4	4.3 ± 0.2
$< n_g >^{mod}$	6.6 ± 0.4	5.6 ± 0.4	5.8 ± 0.4	5.6 ± 0.3
$< n_b >^{exp}$	4.7 ± 0.1			4.0 ± 0.2
$< n_b >^{mod}$	5.0 ± 0.2	4.9 ± 0.2	4.8 ± 0.2	5.0 ± 0.2

References

[1] S.J.Brodsky and G.De Teramon, Phys. Rev. Lett. 60 (1988) 1924.

[2] N.I.Kochelev and M.I.Tokarev, Phys. Lett. B 309 (1993) 416.

[3] A. Bialas, preprint Jagellonian University PJU-3/91.

[4] J.Rauft, Z.Phys.C-Particles and Fields 43 (1989) 439.

[5] A. Bialas and Gyalassy, Nucl.Phys. B291 (1987) 793.

[6] B. Z. Kopeliovich, L. I. Lapidus, Proc. of the VI Balaton Conf. on Nucl. Phys., p. 73, Balatonfured, Hungary, 1983; B. Z. Kopeliovich, F. Niedermayer, Sov. J. Nucl. Phys., 42 (1985) 504.

[7] V. S. Barashenkov, F. G. Geregi, J. J. Musulmanbekov, Yad. Fiz., 39 (1984) 1133.

[8] V. S. Barashenkov, V. D. Toneev, Interactions of high energy particles and nuclei with nuclei, Moscow.

[9] S.Pokorsky and L.van Hove, Nucl. Phys. B 86 (1975) 243; G.J.Musulmanbekov, Preprint JINR, E2-88-809.

[10] G. N. Fowler, R. M. Weiner and G. Wilk, Phys. Rev. Lett., 55 (1985) 173.

[11] Chou Kuang-chao, Liu-Lian-sou and Meng Ta-chung, Phys. Rev., D 28 (1983) 1080.

[12] G.F.Alner et al., Phys. Lett., B 160 (1985) 199.

[13] K. G. Boreskov, A. A. Grigorian, A. B.Kaidalov, Yad. Fiz., 24 (1976) 789.\\

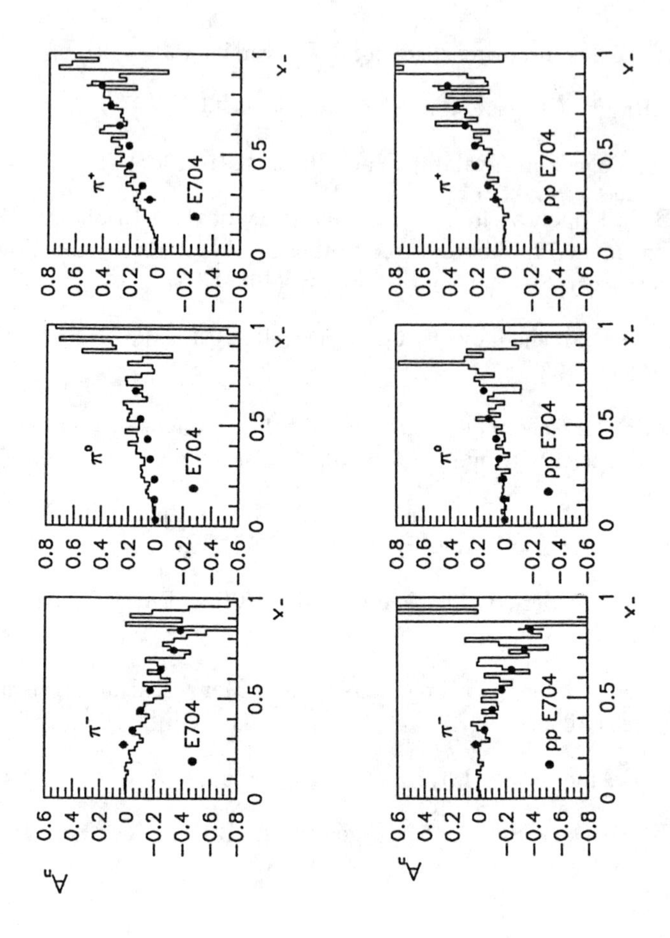

A_N
π⁻
π⁰
π⁺
E704
pp E704
x_F

Polarization of Charmonium in πN Collisions*

Wai-Keung Tang†
Stanford Linear Accelerator Center, Stanford University, Stanford, CA 94309

Abstract. Measurements of the polarization of J/ψ produced in pion-nucleus collisions are in disagreement with leading twist QCD prediction where J/ψ is observed to have negligible polarization whereas theory predicts substantial polarization. We argue that this discrepancy cannot be due to poorly known structure functions nor the relative production rates of J/ψ and χ_J. The disagreement between theory and experiment suggests important higher twist corrections, as has earlier been surmised from the anomalous non-factorized nuclear A-dependence of the J/ψ cross section.

INTRODUCTION

One of the most sensitive tests of the QCD mechanisms for the production of heavy quarkonium is the polarization of the J/ψ in hadron collisions. In fact, there are serious disagreements between leading twist QCD prediction [2] and experimental data [3] on the production cross section of 'direct' J/ψ and χ_1. We would like to advocate that polarization of J/ψ provides strong constraints on the production mechanisms of J/ψ and thus can pinpoint the origin of these disagreements.

In this paper we will present results on the theoretical calculation of the polarization of J/ψ in πN collisions. The detailed analysis will be published in a later paper[1]. We found that the polarization of J/ψ provides important constraints on the nature of the production mechanisms and urge that polarization measurement of J/ψ should be included in the design of future charm production experiment.

PRODUCTION RATES OF ψ AND χ_J STATES

In leading twist QCD, the production of the J/ψ at low transverse momentum occurs both 'directly' from the gluon fusion subprocess $gg \to J/\psi + g$ and indirectly via the production and decay of χ_1 and χ_2 states. These states have sizable decay branching fractions $\chi_{1,2} \to J/\psi + \gamma$ of 27% and 13%, respectively.

In this model, we assume that the non-perturbative physics, which is described by the wave function at the origin in cases of production of J/ψ and ψ', is separable from the perturbative hard subprocess, *i.e.*, factorization holds.

*Talk presented at The Eleventh International Symposium on High Energy Spin Physics and the Eighth International Symposium on Polarization Phenomena in Nuclear Physics, Indiana University, Bloomington, In., September 15-22, 1994.
†Work supported by Department of Energy contract DE–AC03–76SF00515.

As the wave function at the origin can be related to the leptonic decay amplitude, the ratio of ψ' to direct J/ψ production can be expressed in terms of the ratio of their leptonic decay width. More precisely, taking into account of the phase space factor,

$$\frac{\sigma(\psi')}{\sigma_{dir}(J/\psi)} \simeq \frac{\Gamma(\psi' \to e^+e^-)}{\Gamma(J/\psi \to e^+e^-)} \frac{M_{J/\psi}^3}{M_{\psi'}^3} \simeq 0.24 \pm 0.03 \tag{1}$$

where $\sigma_{dir}(J/\psi)$ is the cross section for direct production of the J/ψ. The ratio (1) should hold for all beams and targets, independently of the size of the higher twist corrections in producing the point-like $c\bar{c}$ state. The energy should be large enough for the bound state to form outside the target. The available data is indeed compatible with (1). In particular, the E705 values [4] with different projectiles are all consistent with 0.24.

The anomalous nuclear target A-dependence observed for the J/ψ is also seen for the ψ' [9], so that the ratio (1) is indeed independent of A. Therefore, at high energies, the quarkonium bound state forms long after the production of the $c\bar{c}$ pair and the formation process is well described by the non-relativistic wavefunction at the origin.

In leading twist and to leading order in α_s, J/ψ production can be computed from the convolution of hard subprocess cross section $gg \to J/\psi g$, $gg \to \chi_j$, *etc.*, with the parton distribution functions in the beam and target. Higher order corrections in α_s, and relativistic corrections to the charmonium bound states, are unlikely to change our qualitative conclusions at moderate x_F. Contributions from direct J/ψ production, as well as from indirect production via χ_1 and χ_2 decays, will be included. Due to the small branching fraction $\chi_0 \to J/\psi + \gamma$ of 0.7%, the contribution from χ_0 to J/ψ production is expected (and observed) to be negligible. Decays from the radially excited 2^3S_1 state, $\psi' \to J/\psi + X$, contribute to the total J/ψ rate at the few per cent level and will be ignored here.

In Table 1 we compare the χ_2 production cross section, and the relative rates of direct J/ψ and χ_1 production, with the data of E705 and WA11 on $\pi^- N$ collisions at $E_{lab} = 300$ GeV and 185 GeV [4]. The χ_2 production rate in QCD agrees with the data within a 'K-factor' of order 2 to 3. This is within the theoretical uncertainties arising from the J/ψ and χ wavefunctions, higher order corrections, structure functions, and the renormalization scale. A similar factor is found between the lowest-order QCD calculation and the data on lepton pair production [10] On the other hand, Table 1 shows a considerable discrepancy between the calculated and measured relative production rates of direct J/ψ and χ_1, compared to χ_2 production. A *priori* we would expect the K-factors to be roughly similar for all three processes. We conclude that leading twist QCD appears to be in conflict with the data on direct J/ψ and χ_1 production. Although in Table 1 we have only compared our calculation

	$\sigma(\chi_2)$ [nb]	$\sigma_{dir}(J/\psi)/\sigma(\chi_2)$	$\sigma(\chi_1)/\sigma(\chi_2)$
Experiment	$188 \pm 30 \pm 21$	$0.54 \pm 0.11 \pm 0.10$	$0.70 \pm 0.15 \pm 0.12$
Theory	72	0.19	0.069

Table 1: Production cross sections for χ_1, χ_2 and directly produced J/ψ in $\pi^- N$ collisions. The data from Ref. [4, 5] include measurements at 185 and 300 GeV. The theoretical calculation is at 300 GeV.

with the E705 and WA11 $\pi^- N$ data, this comparison is representative of the overall situation (for a recent comprehensive review see [6]).

POLARIZATION OF THE J/ψ

The polarization of the J/ψ is determined by the angular distribution of its decay muons in the J/ψ rest frame. The angular distribution of massless muons, integrated over the azimuthal angle, has the form

$$\frac{d\sigma}{d\cos\theta} \propto 1 + \lambda\cos^2\theta \tag{2}$$

where we take θ to be the angle between the μ^+ and the projectile direction (*i.e.*, we use the Gottfried–Jackson frame). The parameter λ can be calculated from the $c\bar{c}$ production amplitude and the electric dipole approximation of radiative χ decays.

In Fig. 1a we show the predicted value (solid curve) of the parameter λ of Eq. (2) in the GJ-frame as a function of x_F, separately for the direct J/ψ and the $\chi_{1,2} \to J/\psi + \gamma$ processes. Direct J/ψ production gives $\lambda \simeq 0.25$, whereas the production via χ_1 and χ_2 result in $\lambda \simeq -0.15$ and 1 respectively. Smearing of the beam parton's transverse momentum distribution by a Gaussian function $\exp[-(k_\perp/500 \text{ MeV})^2]$ (dashed curve) has no significant effect in λ except for the production via χ_2 which brings λ down to $\lambda \simeq 0.85$. The $\lambda(x_F)$-distribution obtained when both the direct and indirect J/ψ production processes are taken into account is shown in Fig. 1b and is compared with the Chicago–Iowa–Princeton [7] and E537 data [8] for 252 GeV πW collisions and 150 GeV $\pi^- W$ collisions respectively. Our QCD calculation gives $\lambda \simeq 0.5$ for $x_F \lesssim 0.6$, significantly different from the measured value $\lambda \simeq 0$.

The discrepancies between the calculated and measured values of λ is one further indication that the standard leading twist processes considered here are not adequate for explaining charmonium production. The J/ψ polarization is particularly sensitive to the production mechanisms and allows us to make further conclusions on the origin of the disagreements, including the above

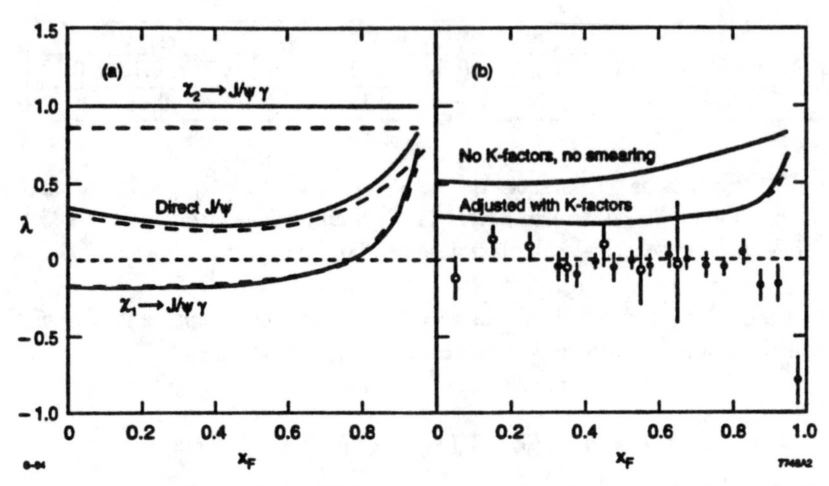

Figure 1: CIP(•) and E537 (o) data compared with theoretical prediction.

discrepancies in the relative production cross sections of J/ψ, χ_1 and χ_2. If these discrepancies arise from an incorrect relative normalization of the various subprocess contributions (*e.g.*, due to higher order effects), then we would expect the J/ψ polarization to agree with data when the relative rates of the subprocesses are adjusted according to the measured cross sections of direct J/ψ, χ_1 and χ_2 production The lower curve in Fig. 1b shows the effect of multiplying the partial J/ψ cross sections with the required K-factors. The smearing effect is insignificant as shown by the dashed curve. The λ parameter is still predicted incorrectly over most of the x_F range.

A similar conclusion is reached (within somewhat larger experimental errors) if we compare our calculated value for the polarization of direct J/ψ production, shown in Fig. 1a, with the measured value of λ for ψ' production. In analogy to Eq. (1), the ψ' polarization data should agree with the polarization of directly produced J/ψ's, regardless of the production mechanism. Based on the angular distribution of the muons from $\psi' \to \mu^+\mu^-$ decays in 253 GeV $\pi^- W$ collisions, Ref. [11] quotes $\lambda_{\psi'} = 0.02 \pm 0.14$ for $x_F > 0.25$, appreciably smaller than our QCD values for direct J/ψ's in Fig. 1a.

DISCUSSION

We have seen that the J/ψ and χ_1 hadroproduction cross sections in leading twist QCD are at considerable variance with the data, whereas the χ_2 cross section agrees with measurements within a reasonable K-factor of 2 to 3. On the other hand, the relative rate of ψ' and direct J/ψ production (Eq. 1), which at high energies should be independent of the production mechanism, is in agreement with experiment. It is therefore improbable that the treatment of the $c\bar{c}$ binding should require large corrections.

In a leading twist description, an incorrect normalization of the charmonium production cross sections can arise from large higher order corrections or uncertainties in the parton distributions[6]. Taking into account that the normalization may be wrong by as much as a factor of 10 and that even such a $K-$factor does not explain the polarization data of J/ψ, a more likely explanation may be that there are important higher-twist contributions to the production of the J/ψ and χ_1 as suggested in large x_F case [12, 13].

Further theoretical work is needed to establish that the data on direct J/ψ and χ_1 production indeed can be described from higher twist mechanisms. Experimentally, it is important to check whether the J/ψ's produced indirectly via χ_2 decay are transversely polarized. This would show that χ_2 production is dominantly leading twist, as we have argued. Thus, the polarization of J/ψ production from different channels provides a very sensitive discriminant of different production mechanisms.

REFERENCES

1. M. Vänttinen, P. Hoyer, S. J. Brodsky and W.-K. Tang, preprint SLAC-PUB-6637 and HU-TFT-94-29.
2. V. Barger and A. D. Martin, *Phys. Rev.* **D31**, 1051 (1985).
3. A. G. Clark, *et al.*, *Nucl. Phys.* **B142**, 29 (1978); R806: C. Kourkoumelis *et al.*, *Phys. Lett.* **B81**, 405 (1979); WA11: Y. Lemoigne, *et al.*, *Phys. Lett.* **B113**, 509 (1982); E673: S. R. Hahn, *et al.*, *Phys. Rev.* **D30**, 671 (1984); D. A. Bauer, *et al.*, *Phys. Rev. Lett.* **54**, 753 (1985); F. Binon, *et al.*, *Nucl. Phys.* **B239**, 311 (1984).
4. E705: L. Antoniazzi, *et al.*, *Phys. Rev. Lett.* **70**, 383 (1993)
5. E705: L. Antoniazzi, *et al.*, *Phys. Rev.* **D46**, 4828 (1992)
6. G. A. Schuler, preprint CERN-TH.7170/94.
7. C. Biino, *et al.*, *Phys. Rev. Lett.* **58**, 2523 (1987).
8. E537: C. Akerlof, *et al.*, *Phys. Rev.* **D48**, 5067 (1993).
9. E772: D. M. Alde, *et al.*, *Phys. Rev. Lett.* **66**, 133 (1991).
10. J. Badier, *et al.*, *Z. Phys.* **C18**, 281 (1983). J. S. Conway, *et al.*, *Phys. Rev.* **D39**, 92 (1989).
11. J. G. Heinrich, *et al.*, *Phys. Rev.* **D44**, 1909 (1991).
12. S. J. Brodsky, P. Hoyer, A. H. Mueller and W.-K. Tang, *Nucl. Phys.* **B369**, 519 (1992).
13. P. Hoyer, M. Vänttinen and U. Sukhatme, *Phys. Lett.* **B246**, 217 (1990).

Single Spin Asymmetry In Inclusive Pion Production

M. Anselmino*, M.E. Boglione* and F. Murgia†

**Dipartimento di Fisica Teorica, Università di Torino and Istituto Nazionale di Fisica Nucleare, Sezione di Torino, Via P. Giuria 1, I-10125 Torino, Italy*
†Istituto Nazionale di Fisica Nucleare, Sezione di Cagliari, Via Ada Negri 18, I-09127 Cagliari, Italy

Abstract. It is shown how the single spin asymmetry observed in inclusive pion production is related, in the helicity basis, to the imaginary part of the product of two different distribution amplitudes, rather than to the usual quark and gluon distribution functions; there is then no reason why it should be zero even in massless perturbative QCD, provided the quark intrinsic motion is taken into account. A simple model is constructed which reproduces the main features of the data.

Spin physics in large p_T inclusive hadronic processes has unique features; not only it probes the internal structure of hadrons, but, as spin dependent observables involve delicate interference effects among different amplitudes, it tests the theory at a much deeper level than unpolarized processes.

We consider here the single spin asymmetry in inclusive pion production in $p-p$ collisions, $p^\uparrow + p \to \pi + X$. Let the two protons move along the $\hat{z}$-axis in their c.m. frame and $\hat{x}$-$\hat{z}$ be the scattering plane. The proton moving in the $+\hat{z}$ direction is polarized transversely to the scattering plane, *i.e.* along ($\uparrow$) or opposite ($\downarrow$) the $\hat{y}$-axis. The single spin asymmetry A_N is then defined by:

$$A_N(x_F, p_T) = \frac{d\sigma^\uparrow - d\sigma^\downarrow}{d\sigma^\uparrow + d\sigma^\downarrow} \tag{1}$$

where $d\sigma$ is the differential cross section and $\uparrow$, $\downarrow$ refer to the proton spin directions; we denote by p_L and p_T the c.m. longitudinal and transverse pion momentum respectively; $x_F = 2p_L/\sqrt{s}$ is the Feynman variable and $\sqrt{s}/2$ is the c.m. energy of each incident proton.

Several experimental results are available on A_N (for a list of references see [1]); the E704 Collaboration has produced the most recent, high energy ones ($\sqrt{s}/2 \simeq 10$ GeV). Two sets of measurements are relevant to our analysis:

i) $A_N(x_F, p_T)$ for $p^\uparrow + p \to \pi^\pm, \pi^0 + X$, *vs.* x_F in the p_T range $0.7 \le p_T \le 2.0$ GeV/c [2, 3]; these data show intriguing x_F dependence (Fig. 1).

ii) $A_N(x_F, p_T)$ for $p^\uparrow + p \to \pi^0 + X$, as a function of p_T (up to $p_T \simeq 4$ GeV/c), in the central region ($|x_F| \le 0.1$) [4]; in this case no p_T dependence seems to be observed and $A_N \simeq 0$ in the whole p_T range (notice that this updates and corrects some previous results of the same collaboration [5]).

A naive generalization of the QCD-factorization theorem suggests that the single spin asymmetry can be written qualitatively as:

$$A_N \sim \sum_{ab \to cd} \Delta_T G_{a/p} \otimes G_{b/p} \otimes \hat{a}_N \hat{\sigma}_{ab \to cd} \otimes D_{\pi/c} \tag{2}$$

where $G_{a/p}$ is the parton distribution function, that is the number density of partons a inside the proton, and $\Delta_T G_{a/p} = G_{a\uparrow/p\uparrow} - G_{a\downarrow/p\uparrow}$ is the difference between the number density of partons a with spin $\uparrow$ in a proton with spin $\uparrow$ and the number density of partons a with spin $\downarrow$ in a proton with spin $\uparrow$; $D_{\pi/c}$ is the number density of pions resulting from the fragmentation of parton c; $\hat{a}_N$ is the single spin asymmetry relative to the $a^\uparrow b \to cd$ elementary process and $\hat{\sigma}$ is the cross-section for such process.

The usual argument is then that the asymmetry (2) is bound to be very small because $\hat{a}_N \sim \alpha_s m_q/\sqrt{s}$ where m_q is the quark mass. This originated the widespread opinion that single spin asymmetries are essentially zero in perturbative QCD.

However, it has become increasingly clear in the last years that such conclusion need not be true because subtle spin effects might modify Eq. (2). Such modifications should take into account the parton transverse motion, higher twist contributions and possibly non perturbative effects hidden in the spin dependent distribution and fragmentation functions. Several models have been proposed which differ in practice by which part of A_N, Eq. (2), is responsible for these effects: $\Delta_T G_{a/p}$ [6, 7, 8], $\hat{\sigma}$ [9, 10] or $D_{\pi/c}$ [11, 12].

We briefly discuss here a reformulation of Eq. (2) in the helicity basis, which is more suitable for applying the factorization theorem and which allows to formulate a model for the spin dependence of the quark distributions [1]. Our approach is reminiscent of that of Ref. [6].

In the helicity basis the differential cross-section for the inclusive process $p_1(\lambda_1) + p_2(\lambda_2) \to \pi + X(\lambda_X)$ can be written in terms of helicity amplitudes as:

$$d\sigma \sim \sum_{X,\lambda_X} \sum_{\lambda_1,\lambda_1',\lambda_2,\lambda_2'} M_{\lambda_X;\lambda_1,\lambda_2}\, \rho_{\lambda_1,\lambda_2;\lambda_1',\lambda_2'}(p_1,p_2)\, M^*_{\lambda_X;\lambda_1',\lambda_2'} \tag{3}$$

where the sum over X includes also a phase space integral for the undetected particles and the matrix ρ is the helicity density matrix describing the polarization state of the initial protons. In our case p_2 is unpolarized, while p_1 is transversely polarized along $\pm\hat{y}$ direction, so that Eq. (1) becomes

$$A_N = 2\,\frac{\sum_{X,\lambda_X,\lambda_2} \mathrm{Im}[M_{\lambda_X;+,\lambda_2} M^*_{\lambda_X;-,\lambda_2}]}{\sum_{X,\lambda_X,\lambda_1,\lambda_2} |M_{\lambda_X;\lambda_1,\lambda_2}|^2}\,. \tag{4}$$

Eq. (4) shows how a non zero value of A_N implies non zero interference effects between two amplitudes which only differ by one helicity index; its denominator, instead, proportional to $d\sigma^\uparrow + d\sigma^\downarrow = 2\,d\sigma^{unp}$, only depends on moduli squared of amplitudes and can be written in the parton model as:

$$d\sigma^{unp} \sim \sum_{abcd} \int dx_a dx_b\, \frac{1}{x_c}\, G_{a/p}(x_a)\, G_{b/p}(x_b)\, \frac{d\hat{\sigma}}{d\hat{t}}(ab \to cd)\, D_{\pi/c}(x_c)\,. \tag{5}$$

In order to express the numerator of Eq. (4) in terms of parton interactions we have to define $\mathcal{G}^{a/h}_{\lambda_{X_h},\lambda_a;\lambda_h}(x_a, \boldsymbol{k}_{\perp a})$ as the helicity distribution amplitude for the process $h(\lambda_h) \to a(\lambda_a) + X_h(\lambda_{X_h})$, where $\boldsymbol{k}_{\perp_a}$ is the transverse momentum of the parton a inside the hadron h; these amplitudes are related to the unpolarized partonic distribution function by:

$$G_{a/p}(x_a) = \sum_{X_p,\lambda_{X_p}} \int d\boldsymbol{k}_{\perp a} \left\{ |\mathcal{G}^{a/p}_{\lambda_{X_p},+;+}(x_a, \boldsymbol{k}_{\perp a})|^2 + |\mathcal{G}^{a/p}_{\lambda_{X_p},-;+}(x_a, \boldsymbol{k}_{\perp a})|^2 \right\}. \tag{6}$$

By applying the same steps which lead to the partonic expression of $d\sigma^{unp}$ we get for the numerator of A_N an expression similar to Eq. (5), with $G_{a/p}(x_a)$ replaced by $\int d\boldsymbol{k}_{\perp a}\, I^{a/p}_{+-}(x_a, \boldsymbol{k}_{\perp a})$, where

$$I^{a/p}_{+-}(x_a, \boldsymbol{k}_{\perp a}) \equiv \sum_{X_p,\lambda_{X_p}} \mathrm{Im}[\mathcal{G}^{a/p}_{\lambda_{X_p},+;+}(x_a, \boldsymbol{k}_{\perp a})\, \mathcal{G}^{a/p*}_{\lambda_{X_p},+;-}(x_a, \boldsymbol{k}_{\perp a})]\,. \tag{7}$$

Notice that $I^{a/p}_{+-}(x_a, \boldsymbol{k}_{\perp a})$ has to vanish for $\boldsymbol{k}_{\perp a} = 0$, as required by helicity conservation in the forward direction; moreover, since $I^{a/p}_{+-}(x_a, \boldsymbol{k}_{\perp a})$ is an odd function of $\boldsymbol{k}_{\perp a}$[1], we must keep into account $\boldsymbol{k}_{\perp a}$ effects also in the partonic cross sections, otherwise we are left with $\int d\boldsymbol{k}_{\perp a}\; I^{a/p}_{+-}(x_a, \boldsymbol{k}_{\perp a}) = 0$. Then

$$\int d\boldsymbol{k}_{\perp a} I^{a/p}_{+-}(x_a, \boldsymbol{k}_{\perp a}) \frac{d\tilde{\sigma}}{d\tilde{t}}(\boldsymbol{k}_{\perp a}) = \int_{(k_{\perp a})_x > 0} d\boldsymbol{k}_{\perp a} I^{a/p}_{+-}(x_a, \boldsymbol{k}_{\perp a}) \left[\frac{d\tilde{\sigma}}{d\tilde{t}}(+\boldsymbol{k}_{\perp a}) - \frac{d\tilde{\sigma}}{d\tilde{t}}(-\boldsymbol{k}_{\perp a}) \right] \tag{8}$$

where $d\tilde{\sigma}/d\tilde{t}$ means that now the partonic cross section includes $\boldsymbol{k}_{\perp a}$ effects.

To give numerical estimates we need a model for the non perturbative functions $I^{a/p}_{+-}(x, \boldsymbol{k}_\perp)$; these non diagonal distribution functions play for spin

[1] This is more easily seen if we observe that our $I^{a/p}_{+-}(x_a, \boldsymbol{k}_{\perp a})$ equals $\Delta^N G_{a/p^\uparrow}(x_a, \boldsymbol{k}_{\perp a}) = \sum_{\lambda_a} \{G_{a(\lambda_a)/p^\uparrow}(x_a, \boldsymbol{k}_{\perp a}) - G_{a(\lambda_a)/p^\uparrow}(x_a, -\boldsymbol{k}_{\perp a})\}$ defined by Sivers [6].

observables the same rôle plaid by the usual diagonal distribution functions $G_{a/p}$ in unpolarized cross-sections. We parameterize their x dependence with simple power behaviours. The dependence on $\boldsymbol{k}_\perp$ is treated, at this stage, in a simplified way: we replace the integral in Eq. (8) by the value of the integrand at some average $k_{\perp a} = \langle \boldsymbol{k}^2_{\perp a}\rangle^{1/2}$. That is we set

$$\int d\boldsymbol{k}_{\perp a}\, I^{a/p}_{+-}(x_a, \boldsymbol{k}_{\perp a})\, \frac{d\tilde{\sigma}}{d\tilde{t}}(\boldsymbol{k}_{\perp a}) = \frac{\hat{k}_\perp}{M_h} N_a x_a^{\alpha_a}(1-x_a)^{\beta_a}\, \frac{d\tilde{\sigma}}{d\tilde{t}}(k_{\perp a}) \tag{9}$$

where M_h is some hadronic mass scale, of the order of 1 GeV, and we assume $\hat{k}_\perp \simeq 0.5$ GeV/c. N_a can be taken from the usual distribution functions [1].

Our final expression for the single spin asymmetry A_N is then

$$A_N = \frac{\sum_{abcd}\int dx_a dx_b \frac{1}{x_c} I^{a/p}_{+-}(x_a, k_{\perp_a}) G^{b/p}(x_b)\left[\frac{d\tilde{\sigma}}{d\tilde{t}}(k_{\perp_a}) - \frac{d\tilde{\sigma}}{d\tilde{t}}(-k_{\perp_a})\right] D_{\pi/c}(x_c)}{2\sum_{abcd}\int dx_a dx_b \frac{1}{x_c} G^{a/p}(x_a) G^{b/p}(x_b) \frac{d\hat{\sigma}}{d\hat{t}}(ab \to cd) D_{\pi/c}(x_c)}\,. \tag{10}$$

In Eq. (10) we take into account, at lowest perturbative QCD order, all possible elementary interactions involving quarks and gluons. According to $SU(6)$ proton wave functions we take $I^{u/p}_{+-} > 0$ for u quarks and $I^{d/p}_{+-} < 0$ for d quarks (see footnote after Eq. (7)); the sign of $I^{a/p}_{+-}$ for the other partonic contributions is less relevant, and for the moment we assume all these contributions to be positive. However, a more careful analysis is in progress [1]. The unpolarized distribution and fragmentation functions are taken from Ref. [13, 14].

In Fig. 1 we compare our results, at $p_T = 2$ GeV/c, with the experimental data. Most contributions come from $qg \to qg$ and $gg \to gg$ processes and the parameters α and β of Eq. (9) yielding these results are $\alpha_u = \alpha_d = \alpha_g \simeq -0.6$, $\beta_u = \beta_d = 2.5$ and $\beta_g = 3.5$. We also find that, at $x_F = 0$, $A_N \simeq 0$, independently of p_T, in agreement with the most recent data [4].

Our results clearly show how a careful treatment of spin observables and the inclusion of intrinsic $k_\perp$ effects can yield sizeable values of single spin asymmetries in hadronic inclusive pion production, via perturbative QCD dynamics, contrary to widespread belief.

The off-diagonal distribution functions $I^{a/p}_{+-}$ introduced in Eq. (7) contain all the relevant non perturbative information; similarly to the parton distribution functions in unpolarized processes, they cannot be computed, but have to be taken from experiment. This is essentially what we have done here, resulting in reasonable expressions for $I^{a/p}_{+-}$; further discussions can be found in Ref. [1]. Once the non diagonal distribution functions have been obtained from one set of experiments, they can be used to make genuine perturbative QCD predictions for other spin observables like single spin asymmetries in $\pi + p^\uparrow \to \pi + X$ and $p^\uparrow + p \to \gamma + X$. Some experimental results are already available and more are soon expected.

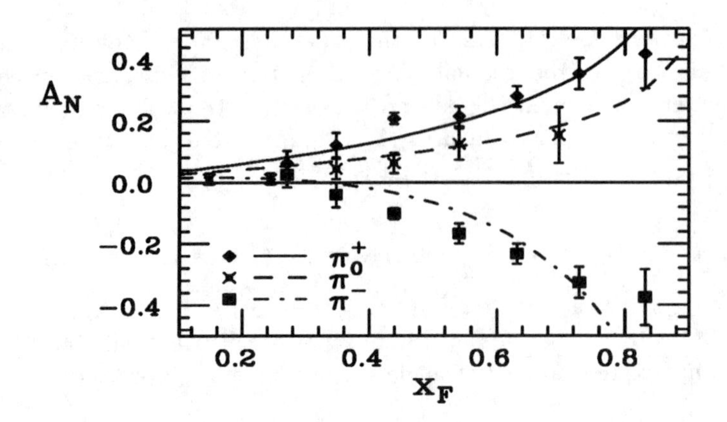

Figure 1: Single spin asymmetry for π^+, π^0, π^-, *vs.* x_F at $p_T = 2$ GeV/c, from Eq. (10), compared to experimental results [2, 3] (see text for details).

References

[1] Anselmino, M., Boglione, M.E., and Murgia, F., *Single spin asymmetry for* $p^\uparrow + p \to \pi + X$ *in perturbative QCD*, preprint DFTT48/94 and INFNCA-TH-94-27, October 1994.

[2] Adams, D.L., *et al.*, *Z. Phys.* **C56**, 181 (1992).

[3] Adams, D.L., *et al.*, *Phys. Lett.* **B264**, 462 (1991).

[4] Nurushev, S.B., *Single spin asymmetries and invariant cross-sections of high* p_T *inclusive* π^0 *production in* $\bar{p} - p$ *interactions*, these Proceedings.

[5] Adams, D.L., *et al.*, *Phys. Lett.* **B276**, 531 (1992).

[6] Sivers, D., *Phys. Rev.* **D41**, 83 (1990); *Phys. Rev.* **D43**, 261 (1991).

[7] Qiu, J., and Sterman, G., *Phys. Rev. Lett.* **67**, 2264 (1991).

[8] Boros, C., Liang, Z., and Meng, T., *Phys. Rev. Lett.* **70**, 1751 (1993); see also the talk by T. Meng in these Proceedings.

[9] Szwed, J., *Phys. Lett.* **B105**, 403 (1981).

[10] Efremov, A.V. and Teryaev, O.V., *Yad. Fiz.* **36**, 242 (1982) [*Sov. J. Nucl. Phys.* **36**, 140 (1982)].

[11] Collins, J., *Nucl. Phys.* **B396**, 161 (1993).

[12] Artru, X., Czyzewski, J., and Yabuki, H., *Single spin asymmetry in inclusive pion production, Collins effect and the string model*, preprint LYCEN/9423 and TPJU 12/94, May 1994.

[13] Brodsky, S.J., Burkardt, M., and Schmidt, E., *Perturbative QCD constraints on the shape of polarized quark and gluon distributions*, preprint SLAC-PUB-6087, January 1994.

[14] Field, R.D., *Applications of perturbative QCD*, New York: Addison Wesley, 1989, ch. 3.

Inclusive meson and lepton-pair production in single-spin hadron-hadron collisions [1]

Meng Ta-chung

Institut für Theoretische Physik, Freie Universität Berlin, 14195 Berlin, Germany [2]

Abstract. The following are pointed out in this summary: The left-right asymmetries observed in such production processes can be understood in terms of a relativistic quark model which describes the magnetic moments of the baryons. The existing data strongly suggest the existence of orbital motion of valence quarks in polarized protons and antiprotons. Quantitative predictions for future experiments can be made. Details are given in Refs. 1-4.

This is a brief summary of my talk. The results presented here are obtained in collaboration with C. Boros and Z. Liang. Part of these results are published [1,2], others are available as preprints [3,4].

High-energy single-spin hadron-hadron collisions, in which either the projectile or the target is transversely polarized with respect to the scattering plane, are of particular interest for a number of reasons: (a) Conceptually, this kind of collision processes is one of the simplest — if not *the* simplest — in which polarization phenomena in hadron-hadron collisions can be studied experimentally and theoretically. (b) Such experiments have already been performed. Data [5-8] are now available not only for proton-proton elastic scattering $p+p(\uparrow) \rightarrow p+p$ where $p(\uparrow)$ indicates the polarized proton target [5], but also for inclusive production processes with negatively charged pion beam and polarized proton target $\pi^-+p(\uparrow) \rightarrow (\pi^0 \text{ or } \eta)+X$ [6] and for inclusive production processes using polarized proton and antiproton beams: $p(\uparrow)+p \rightarrow (\pi^0, \eta^0, \pi^+, \pi^-$ or $e^+e^-) + X$, $\bar{p}(\uparrow) + p \rightarrow (\pi^0, \pi^+, \pi^-) + X$ [7,8]. Here π, η, e^+e^- stand for pion, η-meson and electron-positron pair respectively. Further experiments at higher energies are planned (See in particular the talks given by Krisch, Nowak and Nurushev at this symposium). (c) While in deep-inelastic lepton-nucleon experiments, in which both the lepton and the nucleon are longitudinally polarized, comparison between experimental results and theoretical ideas can be

[1]Supported in part by Deutsche Forschungsgemeinschaft (DFG:Me 470/7-1)
[2]E-mail address: meng@spin.physik.fu-berlin.de

made only after data-extrapolation and sum-rule-evaluation, it is possible to compare data and theory in single-spin hadron-hadron collision experiments without such extrapolations and/or evaluations. (d) Results obtained in the single-spin hadron-hadron collisions experiments mentioned in (b) are in *profound disagreement* with the conventional quark-parton-QCD picture [9]. In particular, the inclusive meson production experiments at 200 GeV show not only a significantly non-zero left-right asymmetry in the projectile fragmentation region, but also a striking flavor-dependence!

What do these experimental results, especially the above-mentioned striking features observed in inclusive meson-production processes, tell us? To answer this, it is useful to recall:

First, inclusive particle production processes in high-energy hadron-hadron collisions with unpolarized projectiles and targets have been extensively studied already in the 1960's and 1970's, and the following are known since then: Particles observed in the kinematical region $x_F \geq 0.4$ (x_F is the Feynman-x) are predominantly fragments of the projectile. Various experimental results – including the well-known leading particle effect – show that valence quarks of the projectile play the dominating role. Taken together with the fact that mesons are quark-antiquark boundstates, it is not difficult to imagine that part of the meson observed in this kinematical region – especially these with relatively large transverse momentum ($p_\perp > 0.7$ GeV/c) are due to *direct-formation (fusion)* of the valence quarks and suitable antiquarks from the sea.

Second, the observed left-right asymmetry implies that the transverse motion of the produced mesons is asymmetric with respect to the polarization axis. Hence, once we accept that these mesons are due to the valence quarks of the transversely polarized projectile, we are immediately led to the question: Can valence quarks perform transverse motion with respect to the polarization axis in general, and at the moment of the quark-antiquark fusion in particular? This question should be answered in the affirmative. The reason is: Having in mind that valence quarks are bounded spin-1/2 objects with relative small masses (compared to the hadron mass), it is reasonable to consider such a quark as Dirac-particle in an effective confining potential (caused by the other constituents in the hadron). This implies, among other things, that only the *total* but not the *orbital* angular momentum can be used to characterize the state of such a quark, because the latter is not a good quantum number! In particular, the ground state wave function for the valence quark is characterized, beside flavor f and energy ε, by its *total angular momentum* $j = 1/2, m = \pm 1/2$ and its parity $P = +1$. From the solution of the corresponding Dirac equation one can explicitly see that orbital motion is always involved — also when these valence quarks are in their ground states! Furthermore, one also sees that the effective orbital motion is counter clockwise with respect to the polarization axis.

Third, baryon wave functions can be obtain as usual by requiring that they should be completely antisymmetric in color and thus (according to Pauli's principle) symmetric in flavor, spin and space. In fact they can be obtained by replacing the Pauli-spinors in the static model [10] by the corresponding Dirac spinors mentioned above. Using these wave functions, the magnetic moments of the baryons can be readily calculated [11,12]. It turns out that the result is exactly *the same* as that in the static-model; and this result is *independent* of the explicit expression of the confining potential. The same proton wave function is used to determine the polarization of the valence quarks in a polarized proton: Among the two u-valence quarks, on the average, 5/3 is polarized in the same, 1/3 is polarized in the opposite direction as the proton. The corresponding chance for the d-valence quark is 1/3 in the same and 2/3 is polarized in the opposite direction as the proton. Hence, there is asymmetry in valence quark polarization and such asymmetry is *flavor-dependent!*

Fourth, consistent with the fact that hadrons are spatially extended objects, within the range of which the constituents (quarks and gluons) interact with one another through color forces and only color singlet can leave color fields, a significant "surface effect" is expected to exist in such production processes. To be more precise, when one of the colliding hadrons is transversely polarized (up or down), only color-singlet $q\bar{q}$ systems directly formed near the front surface can acquire extra momenta due to the orbital motion of the valence quarks. This is because *only such valence quarks do not have enough time to become randomly distributed before they meet a suitable antiquark.* As a consequence, they either go left or go right, depending on the (transverse) polarization of their parent valence quarks.

A number of direct consequences can be deduced from the proposed picture, and they can be tested experimentally: (A) In the projectile-fragmentation region of the above mentioned inclusive meson production processes $p(\uparrow) + p(0) \rightarrow \pi^+(\pi^-, \pi^0 \text{ or } \eta) + X$, in which the valence quarks of the upwards polarized projectile proton contribute, the produced π^+, π^0 and η go left, while π^- go right. (B) By using transversely polarized antiproton-beam instead of proton-beam, one should see the following: While π^0 and η behave in the same way as that in the proton beam case, π^+ and π^- behave differently: They interchange their roles! (C) In the corresponding production processes using pseudoscalar meson beams – irrespective of what kind of target is used and whether the target is polarized – there should be *no* left-right asymmetry *in the projectile fragmentation region.* (D) The asymmetry of the produced mesons is expected to be more significant for large x_F in the fragmentation region of the transversely polarized projectile. (E) Not only mesons but also lepton-pairs in such experiments are expected to exhibit left-right asymmetry.

The qualitative features mentioned in (A), (B), (C), (D) and (E) agree well with experiments. The associations mentioned in (B), (C) and (E) have been predicted [11,12] before the corresponding data [6,7,8] were available.

Encouraged by the good agreement between the experimental findings [6,7,8] and the qualitative features an attempt has been made to describe the data *quantitatively.* The method and the results can be summarized as follows. In $p(\uparrow) + p(0) \rightarrow (q\bar{q}) + X$ at a given c.m.-system energy $\sqrt{s}$, where $(q\bar{q})$ stands for a color-singlet quark-antiquark system which either appears as a meson (through fusion) or becomes a lepton pair (through annihilation) with invariant mass Q, the left-right asymmetry $A_N \equiv A_N(x_F, Q|s)$ at x_F is $[N(x_F,Q|s,\uparrow) - N(x_F,Q|s,\downarrow)]/[N(x_F,Q|s,\uparrow) + N(x_F,Q|s\downarrow)]$ where $N(x_F,Q|s,i),(i=\uparrow,\downarrow)$, is the normalized number density of the $q\bar{q}$ system observed in a given kinematical region. Note that the denominator of this ratio is nothing else but $2N(x_F,Q|s)$, two times the spin-averaged number density in the corresponding reaction with *unpolarized* beams. The numerator is proportional to the difference of the above-mentioned number densities of the $q\bar{q}$ systems directly formed by the valence quarks of the transversely polarized projectiles and sea antiquarks of the target. Such number densities can be obtained by considering the corresponding products of the momentum fraction distributions of the quarks/antiquarks. In other words, these number-densities are products of the usual spin-dependent valence quark distributions $u_v^{\pm}(x^P;Q^2|tr), d_v^{\pm}(x^P;Q^2|tr)$ and the sea antiquark distributions $\bar{u}_s(x^T,Q^2) = \bar{d}_s(x^T,Q^2) \equiv \bar{q}_s(x^T;Q^2)$. Here, u/d indicates the flavor; v/s refers to valence/sea; x^P and x^T are the momentum fractions with P/T standing for projectile/target; $+/-$ indicates the polarization of the valence quark with respect to the spin-direction of the transversely (tr) polarized projectile proton. There is one unknown constant C, namely the one which we use to parametrize the degree of the surface effect.

I do not show the corresponding formulae in this summary, but just mention that A_N for π^+,π^- and π^0 are directly proportional to $\Delta u_v, \Delta d_v$ and $\Delta u_v + \Delta d_v$ respectively, where $\Delta u_v \equiv u_v^+ - u_v^-$ etc. Note that the baryon wave functions, which describe the magnetic moments very well, dictate the following: For protons, the integrals over x of $u_v^{\pm}(x|s,tr)$ and $d^{\pm}(x|s,tr)$ between 0 and 1 should be 5/3, 1/3, 1/3, and 2/3 respectively. This implies that the corresponding integrals of their differences $\Delta u_v(x|s,tr)$ and $\Delta d_v(x|s,tr)$ should be 4/3 and $-1/3$ respectively. This is where one can explicitly see why and how A_N depends on flavor! The corresponding expression for $p(\uparrow)+p \rightarrow \ell\bar{\ell}+X$ for lepton-pair $\ell\bar{\ell}$ of invariant mass $Q, A_N^{\ell\bar{\ell}}(x_F,Q|s)$ can be written down in a similar manner. Here, due to the fact that lepton-pair in such processes are produced by the Drell-Yan mechanism, also the denominator can be explicitly expressed by the quark momentum distributions.

The question which we then asked is: What can we do, before empirically determined Δu_v and Δd_v (or $u_v^{\pm}$ and $d_v^{\pm}$) are available? Since, on the average, the probability of finding a u-valence quark with the same polarization as the polarized proton is $\frac{5}{6}$, while that of finding such a d-valence quark is $\frac{1}{3}$, the simplest possibility to satisfy these conditions is the following ansatz: $u_v^{\pm}(x, Q^2|tr)$ are $(\frac{5}{6}, \frac{1}{6})$ of $u_v(x, Q^2)$, and $d_v^{\pm}(x, Q^2|tr)$ are $(\frac{1}{3}, \frac{2}{3})$ of $d_v(x, Q^2)$, respectively, where $u_v(x, Q^2)$ and $d_v(x, Q^2)$ are the corresponding spin-averaged valence-quark distribution functions. By inserting this ansatz into the corresponding equation and by noting that $N(x, \pi^{\pm}|s)$ and $u_v(x, Q^2), d_v(x, Q^2)$ and $\bar{q}_s(x, Q^2)$ are empirically known, A_N^{π} for π^+, π^-, and π^0 as well as for $\bar{\ell}\ell$ have been evaluated. The obtained results [1,2,11,12] are in good agreement with the experimental findings.[7,8] In the same approximation and with the same parameter, quantitative predictions have been made [3] for the production of lepton-pairs (The possible influences of perturbative QCD corrections for these processes are discussed in [4]) and for the production of different kinds of mesons in reactions with different projectile-target combinations at various energies. A more detailed review of the present approach will be given elsewhere [13].

References

[1] C. Boros, Z. Liang, and T. Meng , Phys.Rev.Lett.**70**,1751 (1993).

[2] Z. Liang and T. Meng, Phys.Rev. **D49**,3759 (1994).

[3] C. Boros, Z. Liang and T. Meng, FU Berlin preprint FUB-HEP/94-7 (1994).

[4] C. Boros and T. Meng, FU Berlin preprint FUB-HEP/94-8 (1994).

[5] P.R. Cameron et al, Phys.Rev.**D32**,3070 (1995) and the papers cited therein.

[6] V.D. Apokin et al., Phys.Lett.**B234**,461 (1990); "X_F-dependence of the asymmetry in inclusive π^0 and η^0 production in beam fragmentaion region", Serpuhkov-Preprint (1991).

[7] FNAL E581/704 Collaboration, D.L. Adams et al., Phys.Lett.**B261**,201 (1991); **B264**,462 (1991), and **B276**,531 (1992);
A. Yokosawa in Proc. of the 10th Intern. Symp. on High Energy Spin Phys., Nagoya, Japan (1992); S.B. Nurushev, dito.

[8] A. Yokosawa, D. Patalakha, private communications (1993).

[9] G. Kane, J. Pumplin, and W. Repko, Phys.Rev.Lett. **41**, 1689 (1978).

[10] See, for example, J. Franklin, in Proc. of the 8th Inter. Symp. on High Energy Spin Phys., Minneapolis, USA 1988, ed. K. Heller (AIP Conf. Proc. No.187), AIP, N.Y. 1989, p. 298 and p. 384.

[11] Z. Liang and T. Meng, Z. Phys. **A344**, 171 (1992).

[12] T. Meng, in Proc. of the 4th Workship on High Energy Spin Physics, Protvino, Russia, Sept. 1992, ed. S.B. Nurushev, p.121.

[13] T. Meng, "Nucleon structure and single-spin reactions", FU Berlin preprint (in preparation).

Higher Twist Effects in the Drell-Yan Process *

A. Brandenburg, S.J. Brodsky, V.V. Khoze, D. Müller

Stanford Linear Accelator Center, Stanford University, Stanford, California 94309

Abstract. We calculate the angular distribution of the lepton produced in the Drell-Yan reaction taking into account pion bound state effects. We work in the kinematic region where one of the pion constituents goes far off shell, which allows us to treat the bound state problem pertubatively. We show that the angular distribution is very sensitive to the shape of the pion distribution amplitude. The model we discuss fits the data if we choose a two-humped pion distribution amplitude suggested by QCD sum rules.

In the Drell-Yan process

$$h_1 + h_2 \to V + X, \qquad h_{1,2} \text{ hadrons}, \; V = \gamma^*, Z, W^{\pm} \qquad (1),$$

(for reviews cf. [1]), the polarization of the V reveals itself in the angular distribution of the decay products, e.g. leptons in $\gamma^* \to \ell^+\ell^-$. A measurement of the angular distribution should provide a clean test of the underlying QCD production mechanism since uncertainties due to imprecise knowledge of parton distribution functions and of the value of α_s have practically no effect. Surprisingly, however, the QCD improved parton model result [2,3] *fails* to describe the data [4-6]. Thus one is forced to think about modifications of the standard approach.

The basic assumptions made within the framework of the parton model may be summarized as follows:
(*i*) Initial state partons are uncorrelated in spin and color.
(*ii*) Hadronic cross sections are obtained by convoluting the partonic cross sections with process independent parton distribution functions.

In ref. [7] it was shown that the discrepancy between data and parton model prediction may be resolved if spin correlations of the initial state partons are present. Such correlations could be induced by the nontrivial structure of the QCD vacuum, thus violating assumption (*i*) above. Here we will concentrate on modifying assumption (*ii*) by taking into account hadronic bound state effects. Our approach is explained in detail in [8] where also further references can be found.

The angular distribution of the μ^+ in

$$\pi^- + N \to \gamma^* + X \to \mu^+ + \mu^- + X \qquad (2)$$

* Talk presented by Arnd Brandenburg at the Spin '94 conference in Bloomington, Indiana, September 15th– September 22nd 1994

may be parameterized in general as follows:

$$\frac{1}{\sigma}\frac{d\sigma}{d\Omega} \sim 1 + \lambda \cos^2\theta + \mu \sin 2\theta \cos\phi + \frac{\nu}{2}\sin^2\theta \cos 2\phi. \tag{3}$$

Here θ and ϕ are angles defined in the muon pair rest frame and λ, μ and ν are angle–independent coefficients. The naive parton model (Drell-Yan picture [9]) views the production of the virtual photon γ^* in (2) as originating from the annihilation of two uncorrelated constituent quarks, resulting in an angular distribution of the form $1+\cos^2\theta$. Higher order partonic processes generate nonzero Q_T through real gluon emission and absorption. In [3] the angular distribution at fixed transverse momentum was computed by resumming soft gluons to the leading double logarithmic accuracy. The deviations from the $1+\cos^2\theta$ behavior were found to be less than 5% in the range $0 < Q_T < 3$ GeV [3]. However, the NA-10 measurements from CERN [4] and the Chicago-Iowa-Princeton collaboration [5,6] show a quite different behavior. In the limit where the momentum fraction x of one of the pion constituents is very close to 1 and for moderate transverse momenta of the muon pair, the value of λ turns strongly negative [6], consistent with a $\sin^2\theta$ distribution. Furthermore, the data [4-6] is observed to have a strong azimuthal modulation (nonzero μ and ν in (3)), an effect which is missing in standard QCD. The Lam–Tung sum rule [2], $1-\lambda-2\nu=0$, which follows from the approach used in [3] is also badly violated by the experimental data.

One way to go beyond the standard treatment is to take into account the pion bound state effects [10,11]. We want to treat the bound state problem perturbatively; thus we will restrict ourselves to a specific kinematic region in which the momentum fraction x of one of the pion constituents is large, $x > 0.5$. In the large x region the off-shell nature of the annihilating quark from the projectile is crucial and the dominant subprocess is $\pi^- q \to \mu^+\mu^- q$. We resolve the pion by a single hard gluon exchange [12]. The main contribution to reaction (2) then comes from the diagrams of Fig. 1a,b [10,11].

We see from diagram 1a that the $\bar{u}$ quark propagator is far off-shell, $p_{\bar{u}}^2 = -Q_T^2/(1-x_{\bar{u}})$. The second diagram is required by gauge invariance.

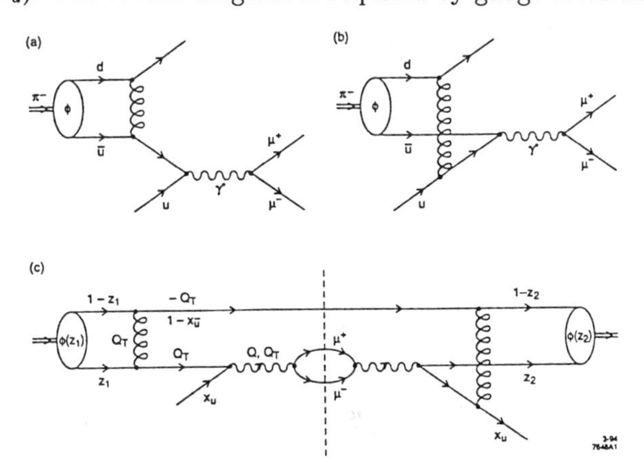

FIG. 1: Diagrams (a) and (b) give the leading contribution to the amplitude of reaction (4). Diagram (c) gives a typical (one out of four) contribution to the cross section (6).

The leading contribution to the amplitude M for the reaction

$$u + \pi^- \to \gamma^* + X \to \mu^+ + \mu^- + X \tag{4}$$

is obtained [12] by convoluting the partonic amplitude $T(u + \bar{u}d \to \gamma^* + d \to \mu^+ + \mu^- + d)$ with the pion distribution amplitude $\phi(z, \tilde{Q}^2)$,

$$M = \int_0^1 dz\ \phi(z, \tilde{Q}^2)\ T, \tag{5}$$

where $\tilde{Q}^2 \sim Q_T^2/(1-x)$ is the cutoff for the integration over soft momenta in the definition of ϕ.

For the hadronic differential cross section we have

$$\frac{Q^2 d\sigma(\pi^- N \to \mu^+\mu^- X)}{dQ^2 dQ_T^2 dx_L d\Omega} = \frac{1}{(2\pi)^4}\frac{1}{64}\int_0^1 dx_u G_{u/N}(x_u) \int_0^1 dx_{\bar{u}} \frac{x_{\bar{u}}}{1 - x_{\bar{u}} + Q_T^2/Q^2}|M|^2$$

$$\delta(x_L - x_{\bar{u}} + x_u - Q_T^2 s^{-1}(1-x_{\bar{u}})^{-1})\ \delta(Q^2 - s x_u x_{\bar{u}} + Q_T^2(1-x_{\bar{u}})^{-1}) + \{u \to \bar{d}, \bar{u} \to d\}. \tag{6}$$

Here Q^μ is the four-momentum of γ^* in the hadronic center of mass system, $x_{u(\bar{u})}$ is the light-cone momentum fraction of the $u(\bar{u})$ quark and $G_{u/N}$ is the parton distribution function of the nucleon. The longitudinal momentum fraction of the photon is defined as $x_L = 2Q_L/\sqrt{s}$. In Fig. 1c we show a typical contribution to the hadronic cross section.

We note that no primordial or intrinsic transverse momenta have been introduced. The single gluon exchange is the only source of Q_T in the model discussed. We also neglected the quark masses and the mass of the projectile which are small compared to $\tilde{Q}$.

In analogy to eq. (3) we parameterize the angular distribution as follows,

$$\frac{Q^2 d\sigma}{dQ^2 dQ_T^2 dx_L d\Omega}\left(\frac{Q^2 d\sigma}{dQ^2 dQ_T^2 dx_L}\right)^{-1} = \frac{3}{4\pi}\frac{1}{\lambda+3}(1+\lambda\cos^2\theta+\mu\sin 2\theta\cos\phi+\frac{\nu}{2}\sin^2\theta\cos 2\phi) \tag{7}$$

where the angular distribution coefficients λ, μ and ν are now functions of the kinematic variables x_L, Q_T^2/Q^2 and Q^2/s.

We work in the Gottfried-Jackson frame where the $\hat{z}$ axis is taken to be the pion direction in the muon pair rest frame and the $\hat{y}$ axis is orthogonal to the $\pi^- N$ plane. The structure of our result for λ (and likewise for μ and ν) is given by

$$\lambda = N^{-1}\int_0^1 dz_1 \frac{\phi(z_1, \tilde{Q}^2)}{z_1(z_1 + \tilde{x} - 1 + i\epsilon)}\int_0^1 dz_2 \frac{\phi(z_2, \tilde{Q}^2)}{z_2(z_2 + \tilde{x} - 1 - i\epsilon)}\{z_1 z_2 l_2 + (z_1 + z_2) l_1 + l_0\},$$

$$\tilde{x} \equiv \frac{x_{\bar{u}}}{1 + Q_T^2/Q^2} \tag{8}$$

The factors $1/z$ in eq. (8) come from the gluon propagators and the factors $1/(z + \tilde{x} - 1 \pm i\epsilon)$ arise from the quark propagator of Fig. 1b. The coefficients l_i $(i = 0, 1, 2)$ depend only on $\tilde{x}$ and Q_T^2/Q^2. The explicit results for λ, μ and ν are given in [8].

We note that the internal quark line of Fig. 1b can go on-shell. The amplitude M of equation (5), however, is always regular due to the z-integration for realistic choices of $\phi(z,\tilde{Q}^2)$ as can be read off from (8). The fact that the internal line goes on-shell does not cause a Sudakov suppression since our diagrams are the lowest order contribution of an *inclusive* process. In other words gluon emission to the final state will occur in the higher order corrections. Only when $x_{\bar{u}}$ approaches unity, where gluon emission is prohibited by kinematics, the Sudakov suppression will arise.

Our model and the parton model are not complementary, but rather different approximations to the Drell-Yan process. The diagrams of Fig. 1a,b give the *whole* leading order contribution in the specific kinematic region of large enough $x_{\bar{u}}$, $x_{\bar{u}} > 0.5$ [11].

Now we can present our final results for λ, μ and ν for different choices of the pion distribution amplitude $\phi(z,\tilde{Q}^2)$. We find in general that the values of μ and ν are very sensitive to the choice of $\phi(z,\tilde{Q}^2)$ which we always take to be positive, symmetric, i.e. $\phi(z,\tilde{Q}^2) = \phi(1-z,\tilde{Q}^2)$, and normalized, $\int_0^1 dz\, \phi(z,\tilde{Q}^2) = 1$. Thus we will not restrict ourselves to the simplest case of $\phi(z) \sim \delta(z-1/2)$ considered in [11].

In Fig. 2 we plot λ, μ, ν and $2\nu - (1-\lambda)$ versus $x_{\bar{u}}$ for $\sqrt{Q_T^2/Q^2} = 0.25$ for different choices of $\phi(z,\tilde{Q}^2)$ together with the data of Ref. [5]. For the two-humped distribution amplitude we have chosen the evolution parameter $\tilde{Q}^2$ to be effectively $\sim 4\ \mathrm{GeV}^2$. The solid line is the result for the two-humped $\phi(z)$ where powers of $(Q_T^2/Q^2)^{n/2}$ were dropped for $n \geq 3$ in the results for λ, μ and ν. We note that corrections to our model may induce such terms, thus the difference between the dashed and the solid lines should be viewed as the uncertainty of our predictions. We also show the data points of Ref. [5] averaged in the intervals $4.05 < \sqrt{Q^2} < 8.55$ GeV and $0 < \sqrt{Q_T^2} < 5$ GeV. In Fig. 3 the same quantities are shown versus $\sqrt{Q_T^2}$ for $x_{\bar{u}} = 0.6$ and $\sqrt{Q^2} = 6$ GeV.

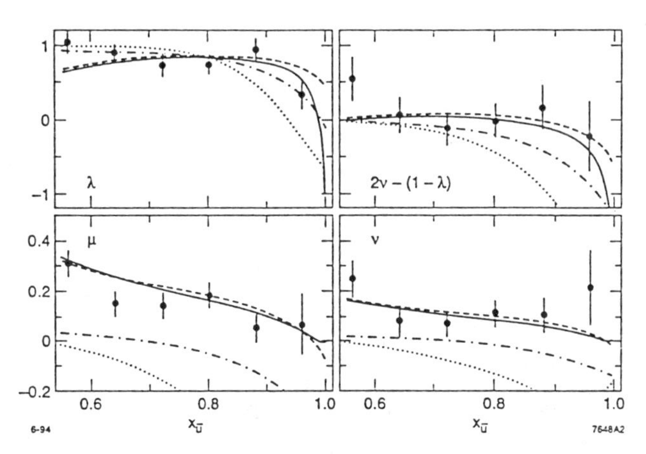

FIG. 2: The angular distribution coefficients λ, μ and ν and the Lam–Tung combination, $2\nu - (1-\lambda)$, in the Gottfried-Jackson frame, vs $x_{\bar{u}}$ for $\sqrt{Q_T^2/Q^2} = 0.25$. The dotted line corresponds to $\phi(z) = \delta(z-1/2)$, the dashed-dotted line corresponds to the asymptotic $\phi(z) = 6z(1-z)$ and the dashed line shows the results for the two humped distribution amplitude, $\phi(z) = 26z(1-z)(1-50/13\, z(1-z))$. The solid line is the result for the two-humped $\phi(z)$ where powers of $(Q_T^2/Q^2)^{n/2}$ were dropped for $n \geq 3$ in the results for λ, μ and ν. The data points (averaged as explained in the text) are taken from Ref. [5] .

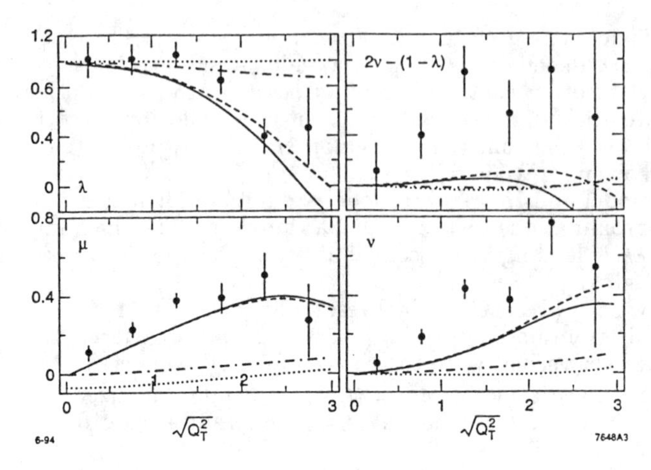

FIG. 3: The same quantities as in Fig. 2 vs $\sqrt{Q_T^2}$ for $x_{\bar{u}} = 0.6$ and $\sqrt{Q^2} = 6$ GeV.

The data points in this case are averaged over intervals $4.05 < \sqrt{Q^2} < 8.55$ GeV and $0.2 < x_{\bar{u}} < 1$ and taken from Ref. [5].

Our analysis shows that the broad, two-humped, distribution amplitude for the pion which was obtained within the context of QCD sum rules [13] can account for the main features of the data. In contrast, narrow momentum distributions, characteristic of weak hadronic binding, predict the wrong sign for the observed azimuthal angular coefficients μ and ν.

Acknowledgments

A.B. would like to thank the Max Kade Foundation for support. This work was supported by the Department of Energy, contract DE-AC03-76SF00515.

References

[1] K. Freudenreich, *Int. J. Mod. Phys.* **A 5** (1990) 3643 ; Yu.L. Dokshitzer, D.I. Dyakonov and S.I. Troyan, *Phys. Rep.* **58** (1980) 269; Proceedings of the Workshop on Drell–Yan Processes, Fermilab, Batavia, 1982

[2] C.S. Lam and W.K. Tung, *Phys. Rev.* **D 21** (1980) 2712

[3] P. Chiappetta and M. Le Bellac, *Z. Phys.* **C 32** (1986) 521

[4] NA10 Collab. S. Falciano et al., *Z. Phys.* **C 31** (1986) 513; NA10 Collab. M. Guanziroli et al., *Z. Phys.* **C 37** (1988) 545

[5] J.S. Conway et al., *Phys. Rev.* **D 39** (1989) 92

[6] J.G. Heinrich et al., *Phys. Rev.* **D 44** (1991) 44

[7] A. Brandenburg, O. Nachtmann and E. Mirkes, *Z. Phys.* **C 60** (1993) 697

[8] A. Brandenburg, S.J. Brodsky, V.V. Khoze, D. Müller, *Phys. Rev. Lett.* **73** (1994) 939

[9] S.D. Drell and T.M. Yan, *Phys. Rev. Lett.* **25** (1970) 316

[10] E.L. Berger and S.J. Brodsky, *Phys. Rev. Lett.* **42** (1979) 940; S.J. Brodsky, E.L. Berger and G.P. Lepage in Proceedings of the Workshop on Drell–Yan Processes, Fermilab, Batavia, 1982, p. 187

[11] E.L. Berger, *Z. Phys.* **C 4** (1980) 289

[12] G.P. Lepage and S.J. Brodsky, *Phys. Rev.* **D 22** (1980) 2157

[13] V.L. Chernyak and A.R. Zhitnitsky, *Phys. Rep.* **112** (1984) 173

Higher Order Corrections to Processes with Transverse Polarization

A.P. Contogouris,[a,b] B. Kamal,[a] O. Korakianitis,[b] F. Lebessis[c] and Z. Merebashvili[a]

a. Department of Physics, McGill University, Montreal H3A 2T8, Canada
b. Nuclear and Particle Physics, University of Athens, Athens 15771, Greece
c. Institute of Nuclear Physics, NRCPS Democritos, Athens 15310, Greece

Abstract. QCD higher order corrections (HOC) are determined for two processes involving transversely polarized particles: (i) $p_\uparrow p_\uparrow \to l^- l^+ + X$ and (ii) $e^- e^+ \to \Lambda_\uparrow \overline{\Lambda}_\uparrow(\text{back-to-back}) + X$. K-factors well exceeding unity are found. The dominance of HOC by soft, collinear and virtual gluons is discussed. For process (ii), for the two cases that the spin vectors of Λ, $\overline{\Lambda}$ are: (a) on the production plane, (b) perpendicular to it, cross sections of opposite sign are found. This feature of the Born term persists with HOC and is independent of the size and shape of the fragmentation functions or other aspects of the calculation.

Processes with transversely polarized hadrons and the related transversity quark distributions $\Delta_T F_{q/h}(x, Q)$ have been of much interest. The most appropriate reaction to study them is Drell-Yan lepton pair production[1,2]

$$A_\uparrow + B_\uparrow \to l^- l^+ + X \tag{1}$$

and with $A = B =$ proton, experiments are planned at RHIC.

Another reaction of interest is the time-reversed[3]

$$e^- e^+ \to A_\uparrow + B_\uparrow + X \tag{2}$$

with e.g. $A = \Lambda$ and $B = \overline{\Lambda}$, which gives information on transversity fragmentation functions $\Delta_T D_{\Lambda/q}$. In reaction (2) as well as in (1), chiral symmetry is spontaneously broken.

At lowest QCD order (Born) both reactions, in particular (1), have been studied. Here we present higher order (one-loop) corrections to them.

QCD CORRECTIONS TO $p_\uparrow p_\uparrow \to l^- l^+ + X$

The suprocess to consider is

$$q(p_1, s_1) + \bar{q}(p_2, s_2) \to V^* + [g] \to l^-(p_3) + l^+(p_4) + [g] \tag{3}$$

where p_i $(i = 1, \ldots, 4)$ are 4-momenta, s_1 and s_2 transverse spin 4-vectors and $V = \gamma$ or Z. Let $M^2 = (p_3 + p_4)^2$ and $\hat{\theta}_3$, φ_3 specifying the direction of l^- in the c.m. of q and $\bar{q}$. We are interested in

$$\Delta_T \frac{d\hat{\sigma}}{dM^2 d\hat{\Omega}_3} \equiv \frac{1}{2} \left\{ \frac{d\hat{\sigma}(s_1, s_2)}{dM^2 d\hat{\Omega}_3} - \frac{d\hat{\sigma}(s_1, -s_2)}{dM^2 d\hat{\Omega}_3} \right\} \tag{4}$$

The HOC come from graphs with gluon loops (vertex and self-energy) and gluon Bremsstrahlung (Brems); these involve singularities, and to regulate them we work in $n = 4 - 2\varepsilon$ dimensions. Dealing, however, with polarized quarks implies the presence of the Dirac matrix γ_5, so that a proper scheme must be used. We choose dimensional reduction[4], where momenta are in $n \neq 4$ dimensions and everything else in $n = 4$. An important question is the satisfaction of the Ward identities, here one of the QED-type. We find that to satisfy it we must add a vertex counterterm; as a result, a term $\sim s_1 \cdot s_2 \delta(1 - \hat{\tau})$ is eliminated $(\hat{\tau} \equiv M^2/\hat{s})$.

The cross section for $p_\uparrow p_\uparrow \to l^- l^+ + X$ has the form

$$\Delta_T \frac{d\sigma}{dM^2 d\varphi_3 d\eta_3} = \sum_q \int \frac{dx_a}{x_a} \frac{dx_b}{x_b} \Delta_T F_{q/p}(x_a) \Delta_T F_{\bar{q}/p}(x_b) k(x_a, x_b, \eta_3)$$

$$\times \left\{ \Delta_T \frac{d\hat{\sigma}_B}{dM^2 d\hat{\Omega}_3} \delta(1 - \hat{\tau}) + \frac{\alpha_s}{\pi} \Delta_T f \, \theta(1 - \hat{\tau}) \right\} \tag{5}$$

where η_3 is the rapidity of l^- in the c.m. of $p - p$, $d\hat{\sigma}_B / dM^2 d\hat{\Omega}_3$ stands for the Born term, $\Delta_T f$ for the HOC and $k(x_a, x_b, \eta_3)$ is a known kinematic factor.[4]

For the valence quark distributions we take $\Delta_T F_{q/p}(x, Q_0) = \Delta_L F_{q/p}(x, Q_0)$ as in nonrelativistic quark models and in fair accord with the MIT bag model. For the sea we use $(Q_0 = 2$ GeV$)$

$$\Delta_T F_{\bar{q}/p}(x, Q_0) = -0.24 x^{0.1} (1 - x)^{9.5};$$

this satisfies $|\Delta_T F_{\bar{q}/p}(x, Q)| < F_{\bar{q}/p}(x, Q)$, Regge requirements at $x \sim 0$ etc. Its size, however, is rather arbitrary, and RHIC experiments should provide valuable information.

Denoting $\Delta_T d\sigma / dM^2 d\varphi_3 = \sigma$ the η_3 integrated cross section, Fig. 1(K) presents the K-factor $K = (\sigma_{\text{Born}} + \sigma_{\text{HOC}})/\sigma_{\text{Born}}$. The peaks and dips are due to the Z-pole, γ-Z interference and the presence of more than one q in (5). Now the point to stress is that, for all s, K exceeds 1, i.e. the HOC *enhance* the Born contributions.[5] The K-factors for the cross sections differential in η_3 are very similar.

Regarding magnitudes, at $\sqrt{s} = 500$ GeV, at the Z-peak we predict $\sigma \simeq 10^{-4}$pb/GeV^2; and with integrated luminosity 800 pb^{-1}, in the range $80 < M_{l^- l^+} < 100$ GeV: ~ 250 events/rad.

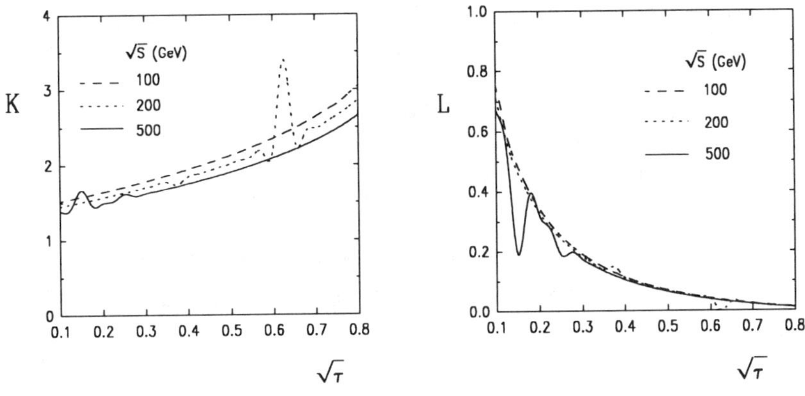

FIGURE 1.

Finally, we ask to what extent HOC are dominated by soft, collinear and virtual gluons. First, in general, the structure of HOC is:[4]

$$\Delta_T f = a\delta(1-\hat{\tau}) + b\frac{1}{(1-\hat{\tau})_+} + c\left(\frac{\ln(1-\hat{\tau})}{1-\hat{\tau}}\right)_+ + d + e\ln(1-\hat{\tau}) + \cdots + h\frac{\ln^2(\hat{\tau})}{1-\hat{\tau}} \tag{6}$$

where the terms d, e, ..., h contain no distributions. The first 3 terms originate from soft, collinear and virtual gluons. Denoting by σ_S the contribution of these 3 terms to σ_{HOC} and by σ_H $(= \sigma_{HOC} - \sigma_S)$ that of the rest, Fig. 1(L) presents the ratio $L \equiv \sigma_H/\sigma_S$. Clearly, as $\sqrt{\tau}$ increases $\rightarrow 1$ (kinematic endpoint), on the average, L decreases fast. The L-factors for the cross sections differential in η_3 are very similar.

PROCESS $e^-e^+ \rightarrow \Lambda_\uparrow + \overline{\Lambda}_\uparrow + X$ AND ITS HIGHER ORDER CORRECTIONS

For the reaction (2) the most interesting case is $A = \Lambda$, $B = \overline{\Lambda}$ (Λ: polarization self-analyser). In general, the contributing subprocess is

$$e^-(p_1) + e^+(p_2) \rightarrow V^*(Q) \rightarrow q(p_3, s_3) + \bar{q}(p_4, s_4) + [g] \tag{7}$$

where s_3, s_4 are transverse spin 4-vectors and $V = Z$ or γ. We consider the process at LEP1, at $\sqrt{s} = M_Z$, so γ^* is negligible. Also we consider Λ, $\overline{\Lambda}$ in opposite hemispheres (back-to-back[3]), so they do not originate from the same parton.

Here the proper variables are

$$z \equiv 2p_3 \cdot Q/s, \qquad \xi \equiv p_3 \cdot p_4/p_3 \cdot Q\ ; \tag{8}$$

in the c.m. of e^-e^+: $z = 2|\mathbf{p_3}|/\sqrt{s}$ and for q, $\bar{q}$ in opposite directions: $\xi = 2|\mathbf{p_4}|/\sqrt{s}$. In this frame defining by $\hat{\theta}_3$, φ_3 the direction of $\mathbf{p_3}$, we first determine:

$$\Delta_T \frac{d\hat{\sigma}}{dz d\xi d\hat{\Omega}_3} = \frac{1}{2}\left\{\frac{d\hat{\sigma}(s_3, s_4)}{dz d\xi d\hat{\Omega}_3} - \frac{d\hat{\sigma}(s_3, -s_4)}{dz d\xi d\hat{\Omega}_3}\right\} \tag{9}$$

The coupling to the fermion f $(=l,q)$ is $-ie\gamma_\mu(\alpha_f - \beta_f\gamma_5)$, and we subsequently use:

$$\Pi(s) \equiv \alpha^2(\alpha_l^2 + \beta_l^2)(\alpha_q^2 - \beta_q^2)s[(s - M_Z^2)^2 + (s\Gamma_Z/M_Z)^2]^{-1};$$

Γ_Z = width of Z. Also define $t \equiv (p_3 - p_1)^2$, $s' \equiv (p_3 + p_4)^2$, $u' \equiv (p_4 - p_1)^2$, and

$$T \equiv (2s'(p_1 \cdot s_3)(p_1 \cdot s_4) + tu'(s_3 \cdot s_4))/s^2. \tag{10}$$

The Born contribution is then:

$$\Delta_T d\hat{\sigma}/dz d\xi d\hat{\Omega}_3 = (3/4)\Pi(s)T_B\delta(1-z)\delta(1-\xi) \tag{11}$$

where $T_B = T$ with $s' = s$ and $u' = -(s+t)$. Now, in the c.m. of e^-e^+ take the production plane to be defined by Λ and consider the cases that the spin vectors $\mathbf{s_3}$, $\mathbf{s_4}$ are: (a) on the production plane, (b) perpendicular to it. Then

$$T_B = \pm(1/4)\sin^2\hat{\theta}_3$$

where the sign + corresponds to (a) and − to (b). This change of sign is an interesting overall result and is further discussed below.

The HOC are determined from graphs with gluon loops and Brems similar to those of $p_\uparrow p_\uparrow \to l^-l^+ + X$. Again, to regulate the singularities we use dimensional reduction and, to satisfy the Ward identity we must add a counterterm; this eliminates a term $\sim s_3 \cdot s_4\delta(1-z)\delta(1-\xi)$.[6] Then the finite HOC is

$$\begin{aligned}\Delta_T \frac{d\hat{\sigma}_{HOC}}{dz d\xi d\hat{\Omega}_3} &= \frac{\alpha_s}{\pi}\Pi(s)[\frac{1}{2}(\pi^2 - 7)T_B\delta(1-z)\delta(1-\xi) + \{f_1(z)\delta(1-\xi)\\ &+ f_2(\xi)\delta(1-z) + \frac{1}{(1-z)_+(1-\xi)_+}\}T - \frac{1}{2}z\xi^2(s_3 \cdot s_4)\\ &+ \{\frac{1}{z}\frac{1}{(1-z)_+}\delta(1-\xi) + \frac{1}{\xi}\frac{1}{(1-\xi)_+}\delta(1-z) + \frac{3}{2}\delta(1-z)\delta(1-\xi)\}T\ln\frac{s}{M^2}]\end{aligned} \tag{12}$$

where M = factorization scale and

$$f_1(z) \equiv \left(\frac{\ln(1-z)}{1-z}\right)_+ + 2\frac{\ln(z)}{1-z}, \quad f_2(\xi) \equiv \left(\frac{\ln(1-\xi)}{1-\xi}\right)_+ + \frac{\ln(\xi)}{1-\xi} \tag{13}$$

We may integrate (11) and (12) over $\hat{\theta}_3$ and obtain $\Delta_T d\hat{\sigma}(z,\xi)/dz d\xi d\varphi_3$. Then the cross section for $e^-e^+ \to \Lambda_\uparrow\bar{\Lambda}_\uparrow + X$ is

$$\Delta_T \frac{d\sigma}{dz d\xi d\varphi_3} = \int_z^1 \frac{dz'}{z'}\int_\xi^1 \frac{d\xi'}{\xi'}\Delta_T\mathcal{D}_{\Lambda/q}(\frac{z}{z'})\Delta_T\mathcal{D}_{\bar{\Lambda}/\bar{q}}(\frac{\xi}{\xi'})\Delta_T\frac{d\hat{\sigma}(z',\xi')}{dz' d\xi' d\varphi_3}; \tag{14}$$

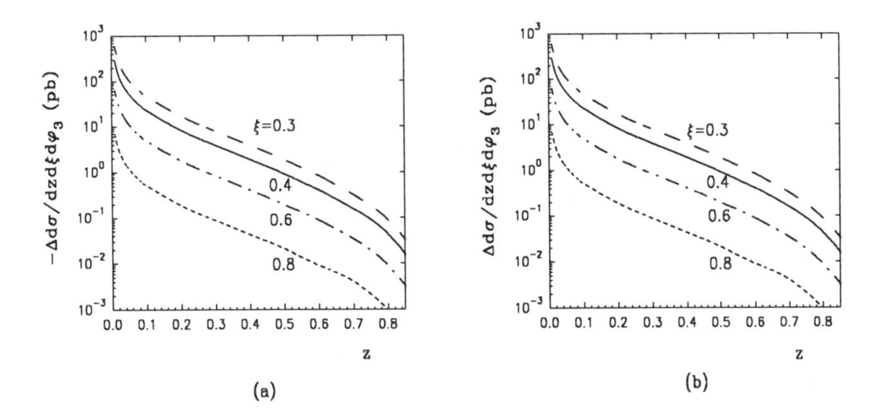

FIGURE 2.

here $z \equiv 2p_\Lambda \cdot Q/s$, $\xi \equiv p_\Lambda \cdot p_{\overline{\Lambda}}/p_\Lambda \cdot Q$ and z', ξ' are given by (8).

In (14), $\Delta_T \mathcal{D}_{\Lambda/q}$ is the transversity fragmentation function for $q \to \Lambda$. From data on Z^0 hadronic decays[7] we estimate the unpolarized $\mathcal{D}_{\Lambda/q}(z) \simeq 0.286(1-z)^{2.45}/z$, $q = u, d, s$. Then, in view of $|\Delta_T \mathcal{D}_{\Lambda/q}(z)| \leq \mathcal{D}_{\Lambda/q}(z)$ and assuming, as usual, that the polarization of Λ comes entirely from the s-quark, we simply take

$$\Delta_T \mathcal{D}_{\Lambda/s}(z) = 0.2(1-z)^3/z, \quad \Delta_T \mathcal{D}_{\Lambda/u}(z) = \Delta_T \mathcal{D}_{\Lambda/d}(z) = 0 \qquad (15)$$

We stress that the size of $\Delta_T \mathcal{D}_{\Lambda/s}$ is arbitrary, and experiment can give valuable information.

Fig. 2 presents the Born + HOC cross sections. We first stress the *opposite sign* for the cases (a) and (b). Since $\Delta_T \mathcal{D}_{\overline{\Lambda}/\bar{q}} = \Delta_T \mathcal{D}_{\Lambda/q}$, this is *independent* of the size and shape of $\Delta_T \mathcal{D}_{\Lambda/s}$, or of whether $\Delta_T \mathcal{D}_{\Lambda/u}$, $\Delta_T \mathcal{D}_{\Lambda/d}$ are nonzero. Moreover, it holds for all z and ξ (Fig. 2). Thus, this feature, already remarked at the Born level and persisting with HOC, can be tested even with a small number of events.

Fig. 3(K) presents our K-factors for case (a); for case (b) they are almost identical.[6] Now we remark that for all z, ξ: HOC enhance the Born cross sections ($K > 1$). Also, K-factors are insensitive to the size and shape of $\Delta_T \mathcal{D}_{\Lambda/q}$.

Finally, with σ_S and σ_H defined as in $p_\uparrow p_\uparrow \to l^- l^+ + X$, Fig. 3(L) presents $L = \sigma_H/\sigma_S$ for case (a); for (b) L are of opposite sign and almost equal in magnitude. Again, as we approach the kinematic endpoints (z, $\xi \to 1$), $|L|$ decrease fast, i.e. HOC are dominated more and more by soft, collinear and virtual gluon contributions. This is in common with unpolarized and longitudinally polarized processes.[8]

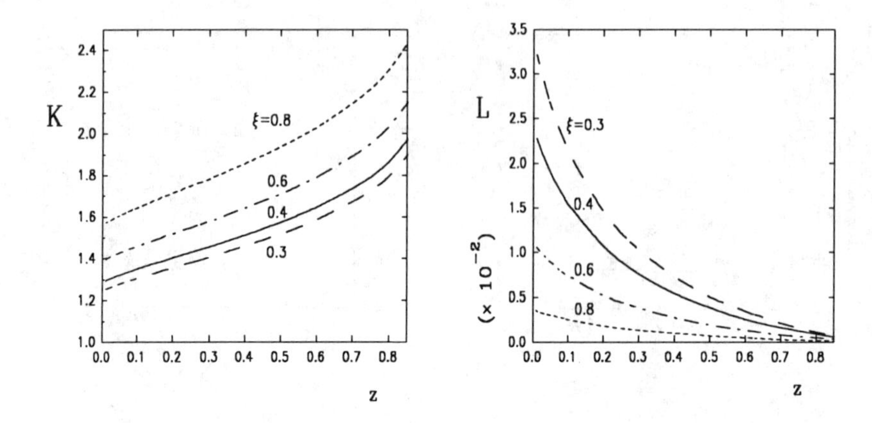

FIGURE 3.

REFERENCES

1. R. Jaffe and X. Ji, *Phys. Rev. Lett.* **67**, 552 (1991) and *Nucl. Phys.* **B375**, 527 (1992); J. Ralston and D. Soper, *ibid* **B152**, 109 (1979); X. Artru and M. Mekhfi, *Z. Phys.* **C55**, 409 (1992); G. Bunce et al, *Part. World* **3**, 1 (1992).
2. RHIC Spin Collaboration, Updated Proposal, September 1993.
3. X. Artru, *Proceed. of the 28th Rencontre de Moriond, Les Arcs*, 1993.
4. A.P. Contogouris, B. Kamal and Z. Merebashvili, *Phys. Lett.* **B** (in press) and references therein.
5. W. Vogelsang and A. Weber, *Phys. Rev.* **D48**, 2073 (1993) also study HOC for $p_\uparrow p_\uparrow \to l^- l^+ + X$ with a non-dimensional regularization (assigning a mass to the gluon); also they consider only $V = \gamma$ and low $M_{l^-l^+}$. As far as we can tell, their K-factor is similar to ours.[4]
6. A.P. Contogouris, O. Korakianitis, F. Lebessis and Z. Merebashvili, to be published.
7. P. Abreu et al (DELPHI Collaboration), *Phys. Lett.* **B318**, 249 (1993).
8. A.P. Contogouris, B. Kamal, Z. Merebashvili and F.V. Tkachov, *Phys. Rev.* **D48**, 4092 (1993).

On The Single and Double Twist−3 Asymmetries

O.V. Teryaev [a,b]

a. Department of Physics, McGill University, Montreal H3A 2T8, Canada
b. Bogoliubov Laboratory of Theoretical Physics,
Joint Institute for Nuclear Research, Dubna 141980, Russia

Abstract. The contributions of various twist −3 partonic subprocesses to single asymmetries of dileptons, pions and jets production at large p_T are computed. The quark-gluon correlations, describing the long-distance part of these processes, are related to the spin-dependent deep inelastic scattering structure functions. This is done by the set of sum rules. One of those sum rules allows to relate the valence contributions to g_1 and g_2 structure functions.

The single transverse spin asymmetries are known to be one of the most subtle effects in QCD. They should be proportional to mass scale (just because massless particle is always polarized longitudinally), and the only scale in "naive" perturbative QCD is that of the current quark mass. For the light quarks it is about three order of magnitude less than typical transverse momenta which are appearing in the denominator to produce the dimensionless asymmetry. The additional suppression comes from the fact, that single asymmetries are related to the antisymmetric part of the density matrix. Due to its hermiticity, the imaginary part of scattering amplitude is relevant. As a result, the spin-dependent contribution to the hard scattering cross section starts at the one-loop level only. At the same time, the Born graph provides a leading approximation to the spin-averaged cross section and the asymmetry is proportional to α_s.

Such a suppression was definitely a main reason for the statement [1], that large single asymmetries contradict either QCD or its applicability. Although this point of view is still popular, the more accurate application of perturbative QCD, including twist −3 effects, results in a completely different picture. The twist −3 quark–gluon correlations give rise to the QCD single transverse asymmetries suppressed neither by the quark mass (it is substituted by the hadronic one M) nor by α_s(see e.g. [2, 3] and refs.therein). The physical interpretation is as follows. The collective gluon field of the polarized hadron in which quark is propagating, provides the latter by the mass of order that of the hadron. This field is also the source of the phase shift and the loop integration in the short-distance subprocess is no more required.

Below given is the expression [4] for the left-right asymmetry of the lepton pair with the high transverse momentum P_T and arbitrary mass Q. The pair is produced by the high energy real photon in the inclusive Compton process on the transverse polarized nucleon:

$$
\begin{aligned}
A_{\gamma\gamma^*} &= \frac{x(1-y)}{x_F(1-y-x_F)[(x(1-y)-yx_F)^2+x^2x_F^2]}\frac{1}{f(x)}\frac{2Mp_T}{s} \\
&\times \; [(b_A(0,x)-b_V(0,x))(x_F(1-y)^2-\frac{y}{x}x_F^2(2(1-y)+x_F))) \\
&+ \; (b_A(0,x)-\frac{y}{x}b_V(0,x))2\frac{y}{x}x_F^3 \\
&+ \; (1-y)[(b_A(y,x)+b_V(y,x))\frac{y}{x}(1-y-x_F)^2 \\
&- \; (b_A(y,x)-b_V(y,x))x_F^2]].
\end{aligned}
$$

Here $x_F=-u/s, x=-t/(s+u), m_T^2=ut/s, y=Q^2/(Q^2-u)$, b_A, b_V and f are the quark-gluon correlations and "ordinary" quark distribution, respectively. Note that the t is the squared photon–lepton pair momentum transfer, while s and u are expressed through the polarized hadron (not quark) momentum. Studying the production of the pairs of the different mass, transverse momentum and rapidity, one may probe the correlations for the different fractions of the proton momentum carried by quark and gluon. In particular, the pole in $b_V(x,y)$ at $y\to x$ results in the infinite rise of asymmetry near the edge of the phase space. This seems to contradict the existence of "gluonic poles" [3], which, in principle, may provide the dominant contributions to single asymmetries.

The single asymmetry of dilepton production is non-zero for the longitudinally polarized particles, provided the angular integration in the lepton c.m. frame is not performed ([5]). It is significantly enhanced if the dilepton is produced by the J/ψ decay ([6]). However, the mentioned twist-3 mechanism does not contribute in the case of the longitudinal polarization, because the correlations are reduced by use of Ward identities to the ordinary parton distributions and no imaginary phases are produced.

This is not the case for the higher twist contribution in the pion-proton scattering. The imaginary phase, whose origin is in fact similar to the inclusive case, was discovered recently [7]. The resulting π^2 term contribute significantly to the lepton angular distribution. The result is very sensitive to the shape of the pion wave function.

It is necessary to stress, that if such an imaginary phase is really present, it should provide a single spin asymmetry (which is zero otherwise) for the pion scattering on the longitudinally polarized proton. Such an asymmetry would be a natural "partonometer" for pion wave function.

The non-existence of the gluonic poles leads to the significant numerical difference between the single asymmetries of the gluon

$$A_{gg} = \frac{b_A(0,x) - b_V(0,x)}{f(x)} \frac{Mp_T}{m_T^2} \times \frac{(C_F - C_A/2)(1-x_F)}{C_F(1+x_F^2)} \frac{((x_F^4+1)C_A/2 - C_F x_F(1-x_F)^2)}{(x_F(C_F - C_A/2) - (1+x_F^2)C_F/2)},$$

and quark

$$A_{gq} = \frac{b_A(0,x) - b_V(0,x)}{f(x)} \frac{Mp_T}{m_T^2} \times \frac{1}{C_F(C_F(x_F^2(1-x_F) + x_F^4/2) + C_A/2(2 - 4x_F + 3x_F^2 - x_F^3))} \times \Big(C_F^2(x_F^2(x_F - 1)) - C_F C_A/2(x_F^3 - x_F^2 - x_F) + C_A^2/4(x_F^5 - 5x_F^4 + 11x_F^3 - 14x_F^2 + 9x_F - 4)\Big)$$

production on the transverse polarized nucleon by the gluon from the unpolarized one. Namely, the latter is several times larger[8]. One may expect, that the pion single asymmetry is due to the pions resulting from the quark (not gluon!) fragmentation. As the result, the π^+ meson asymmetry is related to the correlation of gluon and u-quark, while π^- meson asymmetry is related to the correlation of the gluon and d- quark; π^0 can be produced by u- and d-quarks with equal probability.

Making use of sum rules relating the correlations to the "ordinary" quark distributions[9] (we shall discuss these sum rule in some details below) one can easily get

$$A_{\pi^+} \sim \frac{\Delta u}{u}, \quad A_{\pi^-} \sim \frac{\Delta d}{d}, \quad A_{\pi^0} \sim \frac{\Delta u + \Delta d}{u + d}.$$

Note that mirror asymmetries for A_{π^+} and A_{π^-} and $A_{\pi^0} = 1/3 A_{\pi^+}$ can be obtained if $u = 2d$ and $\Delta u = -2\Delta d$, what is not far from experimental observation.

While the short-distance contributions to single asymmetries are calculable in perturbative QCD, the quark-gluon correlations should be extracted from the experimental data. As it is hardly possible yet, the sum rules [9] for the correlations are of special importance. The quark equation of motion for each flavour allows to relate them to the "transverse" spin-dependent quark distribution

$$\frac{1}{2\pi} \int dx dy (b_A(y,x)[\sigma(x) + \sigma(y)] - [\sigma(x) - \sigma(y)] b_V(y,x)) = 2 \int dx x c_T^A(x). \quad (1)$$

Here $\sigma(x)$ is an arbitrary test function. The independence on the choice of the vector n, fixing gauge and transverse direction results in the relation:

$$\frac{1}{2\pi}\int dxdy(b_A(y,x)\frac{\sigma(x)-\sigma(y)}{x-y} = \int dx(c_L^A(x) - c_T^A(x)), \tag{2}$$

where $c_L^A(x)$ is the most familiar distribution of the longitudinally polarized quarks. All the integrals in (1,2) are performed in the regions $|x, y, x-y| \leq 1$. For ordinary parton distributions positive arguments correspond to the quarks and the negative ones – to the antiquarks. In the Born approximation the structure functions g_1 and g_2 are:

$$g_1(x) + g_2(x) = \frac{1}{2}\sum_f e_(^2 c_T(x) + c_T(-x)) \tag{3}$$

$$g_1(x) = \frac{1}{2}\sum_f e_(^2 c_L(x) + c_L(-x))$$

Combining (3) with (1,2) and choosing the test functions to be x^n one immediately get [9]:

$$\int_0^1 x^n(\frac{n}{n+1}g_1(x) + g_2(x))dx = \tag{4}$$

$$-\frac{1}{\pi(n+1)}\int_{|x_1,x_2,x_1-x_2|\leq 1} dx_1dx_2\sum_f e_f^2[\frac{n}{2}b_V(x_1,x2)(x_1^{n-1} - x_2^{n-1}) +$$

$$b_A(x_1,x_2)\phi_n(x_1,x_2)],\quad n = 0,2...$$

The analog of the sum rule for $n = 2$ was used for numerical estimation of the single asymmetry of direct photon production [10]. One should note that only the gluonic pole contribution was considered, which is zero in our approach.

The sum rules[9] provide an important restrictions for double spin asymmetries, in particular, for the deep inelastic scattering structure functions g_1, g_2. As the sum rules were derived in the case of the arbitrary hard process, they are valid also for the "wrong" odd moments of g_1 and g_2, which can never appear in the framework of the operators product expansion. Physically, they correspond to the valence quarks contributions to these structure functions. The $n = 1$ moment is of special interest, because the correlations are cancelled out at all:

$$\int_0^1 dxx(g_1^{val}(x) + 2g_2^{val}(x)) = 0 \tag{5}$$

It is interesting to write down this sum rule for proton and neutron. While proton g_1 structure function exceeds that of neutron, QCD sum rules[11] predict that $\int_0^1 dx x^2 g_2(x)$ for the neutron is larger, than for the proton. To match these results with (5) the strong sea contribution to g_2 is required. Otherwise, one may expect either the bad convergence of the QCD sum rules or the large radiative (and/or power) corrections to (5).

The recent experimental results of SMC collaboration [12], indicate (with large errors) that transverse asymmetry, g_2 structure function and its twist−3 part are small in the proton case.

It seems necessary to stress, that if qualitative result (the small transverse asymmetry) would become exact enough in order to prove that $c_T \sim g_1(x) + g_2(x)$ is also "small" (say, much less than g_1), this would mean the large $g_2(x)$ and its twist−3 part. While the large g_2 is a consequence of large g_1 and small $g_1 + g_2$, the large twist −3 contribution is required due to the violation of Wandzura-Wilczek sum rule (this is just (4) with r.h.s. neglected), requiring large $g_1 + g_2$ at small x. One should note also that there are the additional complications for studying g_2 in the small x region. Namely, the kinematical suppression of the transverse asymmetries, mentioned at the very beginning, is proportional to x. The present SMC result is to the large extent the kinematical effect and the much better accuracy is required to probe QCD at large distances.

REFERENCES

1. G.L.Kane, J.P.Pumplin and W.Repko, Phys. Rev. Lett. **41** (1978) 689.
2. A.V.Efremov and O.V.Teryaev, Phys. Lett.**150B** (1985) 383.
3. J.Qiu, G.Sterman, Nucl. Phys.**B378** (1992) 52.
4. V.M.Korotkiyan, O.V.Teryaev, JINR Preprint E2-94-200.
5. B.Pire, J.P.Ralston, Phys. Rev.**D28** (1983) 260; R.D.Carlitz, R.S.Willey, Phys. Rev.**D45** (1992) 2323.
6. D.I.Kazakov, D.Ross, F.Renzoni, O.V.Teryaev, in preparation.
7. A.Brandenburg, S.J.Brodsky, V.V.Khoze and D.Muller, Phys. Rev. Lett. **73** (1994) 939.
8. A.V.Efremov, V.M.Korotkiyan and O.V.Teryaev, ICTP Preprint IC/94/169.
9. A.V.Efremov and O.V.Teryaev, Phys. Lett.**200B** (1988) 363.
10. B.Ehrnsperger, A.Shäfer, W.Greiner and L.Mankiewicz, Phys. Lett.**321B** (1994) 121.
11. I.I. Balitsky, V.M.Braun and A.V.Kolesnichenko, Phys. Lett.**242B** (1990) 245.
12. D.Adams *et al.* (The Spin Muon Collaboration), Phys. Lett.**336B** (1994) 125.

Drell-Yan Pairs, $W^{\pm}$ and Z^0 Event Rates and Background at RHIC

A. A. Derevschikov[1], V. L. Rykov[1,2], K. E. Shestermanov[1] and A. Yokosawa[3]

[1] *Experimental Physics Division, Institute of High Energy Physics, Protvino, Moscow District 142284, Russia*
[2] *Department of Physics and Astronomy, Wayne State University, Detroit, Michigan 48202, USA*
[3] *High Energy Physics Division, Argonne National Laboratory, Argonne, Illinois 60439, USA*

September, 1994

Abstract. The estimates for the Drell-Yan pairs, $W^{\pm}$ and Z^0 acceptances and event rates in the STAR and PHENIX detectors at RHIC are presented. The background to $W^{\pm}$ in STAR evaluated. The results were obtained by Monte-Carlo simulations with the *PYTHIA/JETSET* and *GEANT* programs.

Introduction

The measurements of Drell-Yan lepton pairs, $W^{\pm}$ and Z^0 production in polarized proton collisions appear to be the only way to determine polarization properties of the sea quarks. Moreover, the intermediate bosons are produced due to the parity violating mechanism providing an unique opportunity to study a parity violation effects in the hard hadron interactions. All these phenomena have been discussed in details elsewhere (see, for example, (1–4)), and a quintessence of these discussions is reflected in the Proposal on Spin Physics at RHIC (5).

In this report we provide the event rate estimations (6) for the processes above for the two major detectors, STAR and PHENIX. We also present the first results of the background study for the $W^{\pm}$ detection in STAR (7). All data were obtained by Monte-Carlo simulations with the *PYTHIA/JETSET* and *GEANT V3.15* programs. The *EHLQ1* (8) set of proton structure functions has been used. The parameters for *PYTHIA* were tuned up in order to produce results in a reasonable agreement with the available experimental data from *E772, ISR, UA1, UA2* and *CDF* (see (6) and referencies therein).

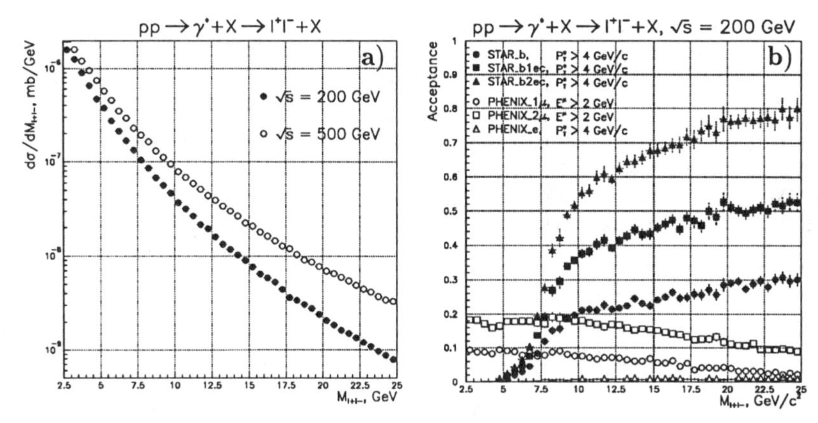

Figure 1: Differential Drell-Yan pairs production cross section in pp collisions (**a**) and STAR and PHENIX acceptances (**b**).

Acceptances and Event Rates

The differential cross sections for Drell-Yan pairs production at $\sqrt{s} = 200$ and $500\,GeV$ are shown in Fig. 1a. For $W^{\pm}$ and Z^0, it has been assumed that only lepton decay modes are detected, i.e. one high P_T lepton, for $W^{\pm} \rightarrow l^{\pm}\nu$, and two high P_T leptons, for $Z^0 \rightarrow l^+l^-$, where $l^{\pm}$ are either electron or muon. *PYTHIA V5.6* provides the following estimates for $W^{\pm}$ and Z^0 production cross sections in pp interactions at $\sqrt{s} = 500\,GeV$:

$$\sigma \cdot B(pp \rightarrow W^+ + X \rightarrow l^+\nu_l + X) = \mathbf{120} \text{ pb};$$
$$\sigma \cdot B(pp \rightarrow W^- + X \rightarrow l^-\overline{\nu}_l + X) = \mathbf{43} \text{ pb};$$
$$\sigma \cdot B(pp \rightarrow Z^0 + X \rightarrow l^+l^- + X) = \mathbf{10} \text{ pb}.$$

The integrated luminosity, during the exposition time of $4 \cdot 10^6$ seconds (100 days, 50% efficiency), has been taken as:
- At $\sqrt{s} = 500\,GeV$, $\int L\,dt = 800\,pb^{-1}$ $(L = 2 \cdot 10^{32}\,cm^{-2} \cdot sec^{-1})$;
- At $\sqrt{s} = 200\,GeV$, $\int L\,dt = 320\,pb^{-1}$ $(L = 8 \cdot 10^{31}\,cm^{-2} \cdot sec^{-1})$.

The STAR and PHENIX setups had been described elsewere (9,10). Here we considered six geometries for the detection single leptons and lepton pairs:
STAR_b: STAR detector with the Barrel Electromagnetic Calorimeter (EMC) of the acceptance (fiducial): $-0.95 < \eta < 0.95$ and 2π coverage in φ;
STAR_b1ec: STAR detector with the Barrel + one End Cup EMC of the acceptance (fiducial): $-0.95 < \eta < 1.9$ and 2π coverage in φ;
STAR_b2ec: STAR detector with the Barrel + two End Cup EMC of the acceptance (fiducial): $-1.9 < \eta < 1.9$ and 2π coverage in φ;
PHENIX_1μ: PHENIX one-arm muon detector of the summary acceptance (geometry): $1.1 < \eta < 2.5$ and 2π coverage in φ;
PHENIX_2μ: PHENIX two-arm muon detector of the summary acceptance (geometry): $1.1 < \eta < 2.5$ plus $-2.5 < \eta < -1.1$ with 2π coverage in φ;

Table 1: Drell-Yan pairs event rates, integrated over mass intervals $M_1 < M_{l^+l^-} < M_2$, at $\sqrt{s} = 200\,GeV$ for $\int L\,dt = 320\,pb^{-1}$; Cuts: $P_T^e > 4\,GeV/c$ or $E_\mu > 2\,GeV$.

$M_1 - M_2$, GeV/c^2	2–5	5–9	9–12	12–15	15–20	20–25	2–25
STAR_b	550	11,500	7,200	2,900	1,800	600	24,500
STAR_b1ec	800	19,000	13,000	5,400	3,400	1,100	43,000
STAR_b2ec	1,000	27,000	19,000	8,100	5,100	1,700	62,000
PHENIX_1μ	95,000	19,400	2,600	850	360	60	118,000
PHENIX_2μ	190,000	41,000	6,300	2,100	1,000	200	240,000
PHENIX_e	–	450	260	110	70	25	915

Table 2: The same as Table 1, but at $\sqrt{s} = 500\,GeV$ for $\int L\,dt = 800\,pb^{-1}$; Cuts: $P_T^e > 5\ GeV/c$ or $E_\mu > 2\,GeV$.

$M_1 - M_2$, GeV/c^2	2–5	5–9	9–12	12–15	15–20	20–25	2–25
STAR_b	500	12,000	21,000	12,300	9,000	3,800	59,000
STAR_b1ec	600	21,500	37,000	23,500	17,500	7,300	107,000
STAR_b2ec	660	31,000	53,000	35,000	27,000	11,300	107,000
PHENIX_1μ	235,000	67,500	14,000	6,200	4,100	1,400	328,000
PHENIX_2μ	470,000	140,000	30,000	14,000	9,500	3,500	670,000
PHENIX_e	–	600	900	430	340	140	2,400

PHENIX_e: PHENIX EMC, consisting of two separated segments of the acceptance (geometry): $-0.35 < \eta < 0.35$ with the coverage in φ: $-112.5^o < \varphi < -22.5^o$ and $22.5^o < \varphi < 112.5^o$.

The STAR and PHENIX acceptances to the Drell-Yan pairs are shown in Fig. 1b. Besides geometries, the thresholds to the lepton energies E_l and/or transverse momenta P_T, which are supposed to be applied at the lowest trigger levels, have also been taken into account. The expected event rates are presented in Tables 1 and 2.

One can observe, PHENIX_μ covers well the low pair mass region, while STAR is more sensitive to the higher mass di-electrons. The sensitivities, in terms of the "hardness" of the primary hard-colliding proton constituents producing detected lepton pair, are shown in Fig. 2. PHENIX_μ acquires better muon pairs, originated from the hard interaction of quark with high x_q but of rather soft antiquark, while STAR has a better sensitivity to the antiquark sea with the higher $x_{\bar{q}}$.

The expected event rates for $W^\pm$ and Z^0 are shown in Table 3. Apparently, Z^0 is almost unreachable in PHENIX due to its low acceptance to the high mass dilepton pairs[1]. High background level in the forward and backward directions might also make difficult to detect $W^\pm \to \mu\nu$ signal with the PHENIX_μ setups.

[1] J. Moss pointed out that the requirement $P_T^\mu > 20\,GeV/c$ might not be necessary for background suppression in PHENIX_μ. Without P_T^μ-cut the event rate for PHENIX_1μ remains unchanged compared to the Table 3, but for PHENIX_2μ it increases from 310 to 700 events.

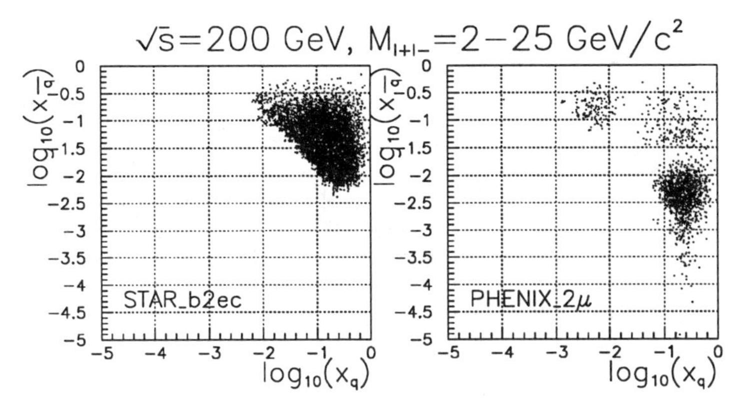

Figure 2: The x_q vs $x_{\overline{q}}$ plot of the primary constituents, producing lepton pairs within STAR_b2ec and PHENIX_2μ acceptances.

Background for $W^{\pm}$ in STAR

Any event with a high, local energy deposition in the EMC, matching a high-P_T charged particle, and observed in the tracking system, may be considered a potential candidate for a $W^{\pm}$ decay. Other sources can also provide such a signature, however, for example: Z^0 decays with one missing electron; electrons from π^0 Dalitz decays; misidentified high-P_T charged hadrons as electrons; overlapping in the EMC γ-quanta from π^0 decays with charged high P_T hadrons.

For the STAR barrel EMC the background from $Z^0 \rightarrow e^+e^-$, with one electron missing, is expected to be quite low: ~2.5% for W^+, and about 10% for W^-. With the end cup(s), this background will be even lower due to a better detection efficiency to electron pairs.

The P_T spectrum of $e^{\pm}$ originated from $\pi^0 \rightarrow \gamma\, e^+e^-$ Dalitz decays drops rapidly when P_T increases (2,7). As a result, with the only requirement being $P_T^e > 20 - 25\ GeV/c$, the background from Dalitz pairs will be reduced to about the same level or even lower than from Z^0 decays.

Table 3: $W^{\pm}$ and Z^0 event rates at $\sqrt{s} = 500\ GeV$ for $\int L\ dt = 800\ pb^{-1}$; Cuts: $P_T^{e,\mu} > 20\,GeV/c$.

	STAR_b	STAR_b1ec	STAR_b2ec	PHENIX_1μ	PHENIX_2μ	PHENIX_e
W^+	64,600	71,500	78,400	4,650	9,300	14,900
W^-	15,000	20,600	26,200	5,050	10,100	2,600
Z^0	2,700	4,200	6,200	25	310	120

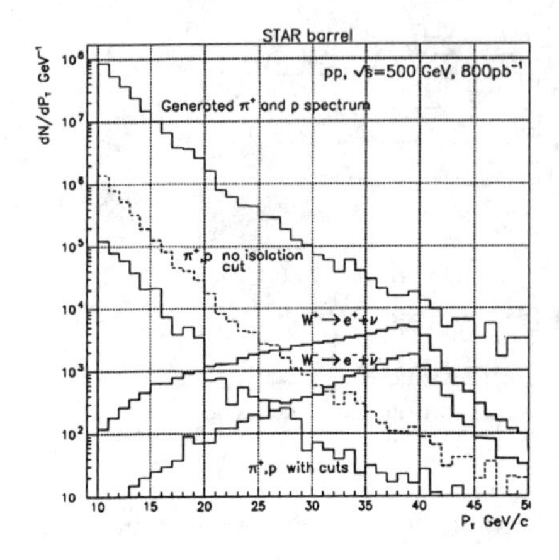

Figure 3: Charged hadron background vs P_T before and after applying rejection criteria along with $e^{\pm}$ from $W^{\pm}$ decays.

The contamination of the $W^{\pm}$ event samples by high-P_T charged hadrons misidentified as electrons is expected to be the most serious source of background. A number of the STAR detector features can be used to separate charged hadrons from electrons in order to get a necessary rejection power of $\sim 10^3$.

At the lowest trigger level a threshold ~20 - 25 GeV/c should be applied to the $P_T^{e^{\pm}}$. Since hadrons mostly deposit only a fraction of their energy in the EM calorimeter, a single hadron will effectively be "seen" in the EMC as a particle with a lower P_T than it actually is. Thus, due to the rapid drop of the hadron P_T spectrum when P_T increases, it effectively provides a hadron P_T spectrum, measured in the EMC, lying well below the actual one. The effective hadron suppression power of this mechanism varies from about 50 to 150 in the P_T region of 10 – 50 GeV/c. Some other selection criteria, common for the practice of other collider experiments, also have been studied (isolation cut; limit to the shower width, defined with the fine-graned Shower Maximum Detector placed at the depth $5 \cdot X_0$ in the EMC).

The results of simulations are shown in Fig. 3, providing that the W^+ signal can be extracted in STAR_b at the background level of ~3-7%, while the detection efficiency, due to applying cuts, drops by ~15-25%. Background to W^- can be rejected to the level of ~10-30%, depending on the selection criteria.

Conclusion

The estimated event rates prove the possibility for STAR and PHENIX to carry out the substantial spin physics with Drell-Yan pairs, $W^{\pm}$ and Z^0 at the statistical error level of a few percents, provided the polarization of colliding protons is ~50-70%. The background to the $W^{\pm}$ signal in STAR is expected to be at the acceptable level.

Acknowledgments

We are pleased to express our appreciation to G. Bunce, D. Grosnick, S. Heppellmann, Yo. Makdisi, S. Nurushev, J. Soffer, H. Spinka, M. Tannenbaum, A. Vasiliev and D. Underwood for the useful and encouraging discussions.

The work of A.A.D. and V.L.R. was supported in part by the U.S. Department of Energy, Division of Basic Energy Sciences under Contract no. DE-AC02-76CH00016 and also (V.L.R.) by the U.S. Department of Energy, Grant DE-FG0292ER40713. The work of K.E.S. and A.Y. was supported in part by the U.S. Department of Energy, Division of High Energy Physics, Contract W-31-109-ENG-38.

References

1. Bourrely, C. and Soffer, J., *Parton Distributions and Parity-Violating Asymmetries in $W^{\pm}$ and Z^0 Production at RHIC*, CPT-93/P.2865, CNRS Luminy (France).
2. Tannenbaum, M. J., *Polarized Protons at RHIC*, Proc. of the Polarized Collider Workshop, University Park, PA (1990), AIP conf. proc. No 223, AIP, New York (1991), p. 201.
3. Doncheski, M. A., et al., *Hadronic W production and the Gottfried Sum Rule*, Preprint MAD/PH/744, June 1993.
4. Cheng, H.-Y. and Lai, S.-N., *Phys. Rev.*, **D41**, 91 (1991);
   Jaffe, R. L. and Ji, X., *Phys. Rev. Lett.*, **67**, 552 (1991);
   Ji, X., *Phys. Lett.*, **B284**, 137 (1992).
5. RHIC Spin Collaboration, *Proposal on Spin Physics Using the RHIC Polarized Collider*, August 14, 1992; *Proposal Update*, September, 1993, and references therein.
6. Derevschikov, A. A. and Rykov, V. L., *Notes on the Drell-Yan Pairs, Z^0 and $W^{\pm}$ in STAR and PHENIX at RHIC*, Internal RSC Report RSC-BNL/IHEP-4, August 1992 (unpublished).
7. Rykov, V. L. and Shestermanov, K. E., *$W^{\pm}$ and Z^0 Event Rates and Background Estimates for the STAR Detector at RHIC in pp Collisions*, ANL-HEP-TR-93-89, Argonne, 1993 (unpublished).
8. Eichten, E., et al., *Rev. Mod. Phys.*, **56** 579 (1984) and *Erratum* **58**, 1065 (1986).
9. The STAR collaboration, *Conceptual Design Report for the Solenoidal Tracker at RHIC (STAR)*, June 15, 1992.
10. The PHENIX collaboration, *PHENIX Conceptual Design Report*, January 29, 1993.

A PRECISION MEASUREMENT OF THE Ω^- MAGNETIC MOMENT

N.B. Wallace[a], P.M. Border[a], D.P. Ciampa[a], Y.T. Gao[b,e], G. Guglielmo[a,f], K.J. Heller[a], K.A. Johns[c], M.J. Longo[b], R. Rameika[d], D.M. Woods[a]

[a]*School of Physics and Astronomy, University of Minnesota, Minneapolis, MN 55455*

[b]*Department of Physics, University of Michigan, Ann Arbor, MI 48109*

[c]*Department of Physics, University of Arizona, Tucson, AZ 85721*

[d]*Fermi National Accelerator Laboratory, Batavia, IL 60510*

[e]*now at Lanzhao University, Lanzhao, China*

[f]*now at University of Oklahoma, Norman, Oklahoma*

The structure of baryons can be probed at long range by measuring their magnetic moments. The particularly simple valence quark structure (three strange quarks with their spins aligned) of the Ω^- should make a precise measurement its magnetic moment a useful test of models of baryon structure. The only previous measurement of the Ω^- magnetic moment(Ref. 1), to a precision of 10%, could not clearly differentiate between these models.

Polarized Ω^-s used for this measurement were produced using two different techniques: the spin transfer technique from a polarized neutral beam (PNB), which was used in the previous Ω^- magnetic moment measurement(Ref. 2), and a new technique that used an unpolarized neutral beam (UNB)(Ref. 3). In both cases a neutral beam containing Λ and Ξ^0 hyperons, as well as γs, neutrons, and K^0s, was produced by an 800-GeV/c proton beam in the inclusive reaction p+Be $\rightarrow$ (neutral particle) + X. In the unpolarized neutral beam mode, the protons struck an upstream target at 0 mrad. The resulting particles passed through a collimator embedded in a sweeping magnet with a 1.8 T field. This neutral beam was then targeted at vertical production angles of $\pm$ 1.8 mrad on a second Be target to produce Ω^-s primarily by the reaction (Λ,Ξ^0) + Be $\rightarrow \Omega^-$ + X. The polarized neutral beam was produced by targeting the proton beam at vertical targeting angles of $\pm$ 1.8 mrad producing polarized Ξ^0 and Λ(Refs. 4,5). Since the sweeping magnet field was perpendicular to the production plane, the spins of the neutral particles were not precessed as they passed through the channel. The Ω^-s were then produced by targeting the polarized neutral beam at 0 mrad. Table 1 shows the average polarizations for each of these modes. The Ω^- yield per incident proton for unpolarized neutral beam production was roughly three times that for polarized neutral beam production.

The Ω^- production target (Be, $5.14\times5.28\times147$ mm^3) was located 55 cm upstream of the spin-precession/momentum selection magnet. The magnet was

Production method	Precession field integral (T·m)	Sample size 10^4 events	P_{Ω^-}
Unpolarized neutral beam	-24.36±0.26	16.7	0.044±0.008
Unpolarized neutral beam	-17.48±0.17	5.02	0.036±0.015
Polarized neutral beam	-24.36±0.26	1.83	-0.069±0.023

Table 1: The sample sizes and average polarizations measured for the three Ω^- samples used in this analysis. The initial polarization is in the $\pm\hat{x}$ direction in a right-handed coordinate system defined by the Ω^- momentum direction ($\hat{z}$) and the vertical ($\hat{y}$).

7.315 m long with a field in the $-\hat{y}$ (where $\hat{y}$ is the vertical) direction. The magnet was fitted with a curved brass-tungsten channel(Ref. 6), with a total bend of 18.7 mrad in the x-z (where $\hat{z}$ is along the Parent particle momentum and $\hat{x} = \hat{y} \times \hat{z}$) plane and a defining aperture of 5.08 mm$\times$ 5.08 mm. The curved channel selected negatively charged particles with a momentum range of 300 to 550 GeV/c when the magnet was operating at a field of 3.33 T. The field integral was measured using a Hall probe(Ref. 1) and checked by measuring the Ξ^- magnetic moment. It was found to be accurate to better than 1%.

The parent Ω^- and its charged daughters for the decay $\Omega^- \to \Lambda K^-$ and $\Lambda \to p\pi^-$ were tracked by a spectrometer consisting of 8 planes of silicon microstrip detectors with 100 μm pitch, 12 multiwire proportional chambers with 1 and 2 mm wire spacing, and an analyzing magnet consisting of two dipole magnets which gave a deflection of 1.45 GeV/c to the daughter p, π^-, and K^-(Ref 7). Signals from scintillation counters and wire chambers were used to form a trigger that required at least one positively charged and one negatively charged track.

This trigger produced a reasonably unbiased data sample which contained 3.4% Ξ^- and 0.035% Ω^- with a spectrometer live time of 70%. Approximately 1.35×10^9 triggers were processed by a multi-pass offline reconstruction program which fit the three-track, two-vertex topology with an overall efficiency of 97%(Ref. 3,6). Selected events were required to fit the topology of the parent/daughter hyperon decay with the parent hyperon pointing back to within 8 mm of the center of the target in x and to within 9 mm in y. Ω^- candidates were also required to have a Λ-K^- invariant mass between 1657 and 1687 MeV/c^2. The $\Xi^- \to \Lambda\pi^-$ decays reconstructed under the ΛK^- hypothesis which satisfy that mass criterion occupy a small range of decay angles in the Ω^- rest frame. All events in this range of decay angles were removed from the data sample which reduced the Ξ^- background to the 0.6% level(Ref. 7). The predominant remaining background, $\Omega^- \to \Xi^0\pi^-$ decays, was at the 2.4% level. The Λ-K^- invariant mass distribution for the final Ω^- sample is shown in Figure 1.

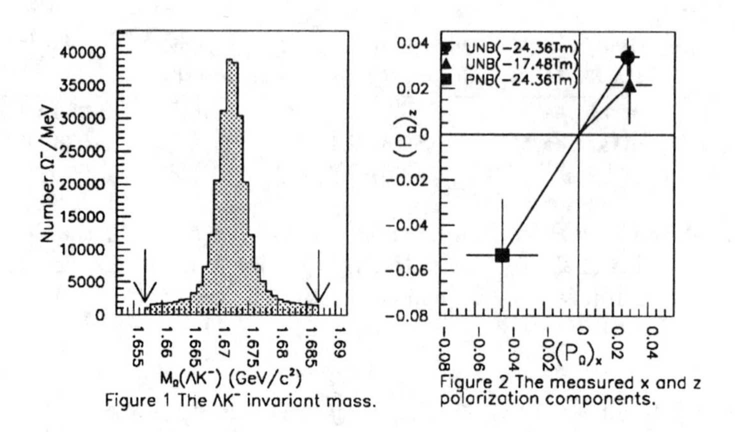

Figure 1 The ΛK⁻ invariant mass.

Figure 2 The measured x and z polarization components.

Assuming $\gamma_{\Omega^-} = +1$, the vector polarization of the Ω^-, $\vec{P}_\Omega$, is related to the daughter Λ polarization, $\vec{P}_\Lambda$, by $\vec{P}_\Omega = \vec{P}_\Lambda$(Ref. 7). The Λ polarization was determined by measuring the decay asymmetry of the proton in the Λ rest frame(Ref. 9). This measured asymmetry was corrected for acceptance by a a hybrid Monte Carlo analysis(Ref. 9). The resulting proton distributions were then fit to a linear function in $\cos\theta$ for each component of the asymmetry.
For a hyperon H with spin J (in units of $\hbar$) passing through a precession field perpendicular to the initial polarization, the precession angle relative to its momentum direction is given by

$$\Phi = \frac{e}{\beta m_H c}\left(\frac{m_H \mu_H}{2J m_p \mu_N} + 1\right)\int Bdl, \tag{1}$$

where Φ is the precession angle in radians, e is the magnitude of the electron charge, β = v/c, m_p is the mass of the proton, μ_N is the nuclear magneton, m_H and μ_H are the hyperon's mass and magnetic moment, and $\int Bdl$ is the precession field integral in T·m. Because parity is conserved in strong interactions, the initial polarization direction must be perpendicular to the production plane. In this case the measured polarization components are given by $P_x = P_0\cos\Phi$ and $P_z = P_0\sin\Phi$, where $\hat{z}$ is parallel to the Ω^- momentum, $\hat{y}$ is the vertical, $\hat{x} = \hat{y} \times \hat{z}$, and P_0 is the initial hyperon polarization at the target. The precession angle is given by $\Phi = \tan^{-1}\left(\frac{P_z}{P_x}\right) + n\pi$ where n is an integer. The x and z components of the polarization signals, shown in Figure 2, were significantly different from zero, while the y component for each sample was consistent with zero. Table 2 gives the precession angles and magnetic moments for the three samples for this experiment as well as the two samples from the

Neutral beam type	Precession field integral (T·m)	Φ (radians)	μ_Ω (nuclear magnetons)
Unpolarized neutral beam	−24.36±0.26	0.88±0.17	−2.023±0.065
Unpolarized neutral beam	−17.48±0.17	0.65±0.43	−2.03 ±0.23
Polarized neutral beam	−24.36±0.26	0.88±0.32	−2.02 ±0.12
Polarized neutral beam(Ref. 1)	−19.53±0.19	0.58±0.42	−1.96 ±0.20
Polarized neutral beam(Ref. 1)	−14.77±0.14	0.34±0.46	−1.90 ±0.29

Table 2: The precession angles and magnetic moments measured for the three samples used in this analysis and the two samples from the previous measurement of μ_Ω.

previous experiment(Ref. 1).

Using the three data samples of this experiment we found μ_Ω by minimizing,

$$\chi^2 = \sum_{ij} \left(\frac{P_{x_{ij}} - P_{0_{ij}} cos\Phi_j}{\sigma_{x_{ij}}} \right)^2 + \left(\frac{P_{z_{ij}} - P_{0_{ij}} sin\Phi_j}{\sigma_{z_{ij}}} \right)^2 \tag{2}$$

with Φ_j given by Eq. 1. $P_{0_{ij}}$ is the initial polarization which depends on production method. $P_{x_{ij}}$ and $P_{z_{ij}}$ are the measured x and z polarization components, and $\sigma^2_{x_{ij}}$ and $\sigma^2_{z_{ij}}$ include uncertainties from $P_{x_{ij}}$ and $P_{z_{ij}}$. The subscript i indicates the production method, and the two precession field values are represented by the sum over j. The observed polarization at the target depends on the field integral since the momentum spectrum of the particles entering the spectrometer changes with the field value. Minimizing the χ^2 from Eq. 2 gave $\mu_{\Omega^-} = (-2.024 \pm 0.056)\mu_N$ with a χ^2 of 1×10^{-3} for two degrees of freedom. The uncertainty for the combined result is given by the variation in the magnetic moment which changes the χ^2 by one. This uncertainty is equal to the uncertainty obtained when the statistical uncertainties of the three sample are combined by standard methods. The momentum averaged magnetic moments for the three samples as well as μ_Ω vs momentum for the largest data sample are shown compared to the constrained fit result in Figure 3.

The Ω^- magnetic moments measured independently for all the samples including those of the previous measurement agree to within their measurement uncertainties. To remove the ambiguity of the precession angle due to rotations by an additional $n\pi$ a linear fit for Φ as a function of the precession field, constrained to include $\Phi = 0$ for zero field, was made. The best fit for the points shown in Figure 4 was for $n = 0$ with a $\chi^2 = 0.3$ for four degrees of freedom; the next best fit had $\chi^2 = 10$ for $n = 1$.

Using the spin precession technique on polarized samples of Ω^- produced by both polarized and unpolarized neutral beams, we have measured the magnetic

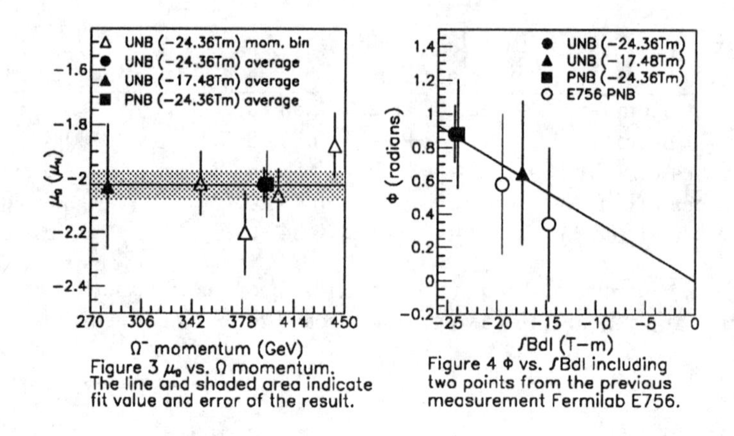

Figure 3 μ_Ω vs. Ω momentum. The line and shaded area indicate fit value and error of the result.

Figure 4 Φ vs. $\int Bdl$ including two points from the previous measurement Fermilab E756.

moment of the Ω^- to be $\mu_{\Omega^-} = (-2.024 \pm 0.056)\mu_N$ which is in agreement with the previous measurement of $\mu_{\Omega^-} = (-1.94 \pm 0.17 \pm 0.14)\mu_N$. Combining our result with the previous measurement gives a world average of $\mu_{\Omega^-} = (-2.019 \pm 0.054)\mu_N$ including the systematic uncertainty of the previous measurement. The line shown in Figure 4 corresponds to this value of the magnetic moment. This measurement disagrees with the static quark model value of $-1.84\mu_N$ at the 3σ level and all other models known to the authors which sucessfully predict other magnetic moments. It is hoped that this measurement will provide a stringent test for future models of baryon structure.

This work was supported by the U.S. Department of Energy and the National Science Foundation. We would like to thank G. Allan, A. Ayala-Mercado, E. Berman, V. DeCarlo, D. Fein, M. Groblewski-Higgins, J. Jallian-Marian, E. James, R. McGriff, L. Morris, A. Nguyen, T. Tynan, and the staff of Fermi National Laboratory.

1. H. T. Diehl *et al.*, Phys. Rev. Lett. **67**, 804 (1991).
2. K. B. Luk *et al.*, Phys. Rev. Lett. **70**, 900 (1993).
3. D. M. Woods, Ph.D. Thesis, University of Minnesota, 1995.
4. G. Bunce *et al.*, Phys. Rev. Lett. **36**, 1113 (1976).
5. P. T. Cox *et al.*, Phys. Rev. Lett. **51**, 2025 (1983).
6. G. M. Guglielmo, Ph.D. Thesis, University of Minnesota, 1994.
7. N. B. Wallace, Ph.D. Thesis, University of Minnesota, 1995.
8. K. B. Luk *et al.*, Phys. Rev. D **38**, 19 (1988).
9. G. Bunce *et al.*, Nucl. Inst. Meth. **172**, 553 (1980).
10. Particle Data Group, Phys. Rev. D **50**, 1771 (1994).

Spin Effects in Hadroproduction of Charmed Hadrons

Robert Ryłko[1]
Department of Physics, Brunel University, Uxbrigde UB8 3PH, United Kingdom.

Abstract. We review the results on spin effects in the inclusive production of $D^{*+}(2010)$ and Λ_c^+ based on the CERN NA32 data. The spin alignment parameter of $D^{*+}(2010)$ meson is measured to be $\eta = 0.10^{+0.12}_{-0.11} \pm 0.01$. The angular distribution of decay products in the Gottfried-Jackson frame is consistent with the quark model assignment for Λ_c^+ spin. For Λ_c^+ $p_T >$ 1.1 GeV/c, the product of the Λ_c^+ transverse polarization and the kaon type asymmetry for $\Lambda_c^+ \to pK^-\pi^+$ is measured to be $\alpha_K P_{\Lambda_c} = -0.65^{+0.22}_{-0.18}$ (anti-Basel). The Λ^0 polarization in the same experiment is measured to be $-0.28 \pm 0.09 \pm 0.02$ (Basel), for Λ^0 $p_T >$ 1 GeV/c.

INTRODUCTION

Currently, in the charm sector, there are two particles[2] with nonzero spin for studying spin effects via exclusive decay channels with reasonable statistics. These are $D^{*+}(2010)$ and Λ_c^+, with quark model assignments $J^P = 1^-$ and $J^P = \frac{1}{2}^+$, respectively. In both cases the spin measurements are lacking[3] and studies of spin effects with these particles assume their quark model spin assignment.

This paper reviews the results of spin effects in the inclusive hadroproduction of $D^{*+}(2010)$ and Λ_c^+. The results are based on data from the CERN NA32 fixed Cu target experiment, using the 230 GeV/c π^- beam, described elsewhere [Ref.(1)]. In addition, other spin effects involving charmed hadrons are listed.

[1]Previously at *Queen Mary & Westfield College, University of London, Mile End Rd., London E1 4NS, United Kingdom.*

[2]Throughout the paper a charmed particle symbol stands for particle and antiparticle, i.e. any reference to a specific state implies the charge-conjugate state as well.

[3]In this paper we review the results on the measurement of the Λ_c^+ spin with the use of the $\Lambda_c^+ \to pK^-\pi^+$ channel.

THE $D^{*+}(2010)$ MESON [Ref.(2)]

The $D^{*+}(2010)$ sample consisted of 127 events. Using the same selection criteria 20 wrong sign events were found. The angular distribution for the $D^{*+}(2010) \to D^0\pi^+_{ext}$ decay is given by

$$I(x) = \frac{1}{2}[1 + d_2 P_2(x)] \equiv \frac{1}{2} + \eta P_2(x) , \tag{1}$$

where $P_2(x)$ is the second Legendre polynomial, $\eta = (3\rho_{00} - 1)/2$ is the spin alignment parameter, and the variable $x \equiv \cos\theta_{hel}$ stands for the cosine of the angle between the D^{*+} direction in the laboratory frame and the π^+_{ext} direction in the D^{*+} rest frame.

The results of the maximum-likelihood fit, with the background subtracted, for the whole sample is $\eta = 0.10^{+0.12}_{-0.11} \pm 0.01$ [Ref.(2)]. In four equally populated subsamples, differing in x_F and p_T cuts, the spin alignment parameter η is consistent with zero.

The relative abundance of vector mesons $P_V^\sigma = V/(P+V)$, where P and V stands for the cross sections for the production of D^+ and $D^{*+}(2010)$ mesons, may be calculated using η within the statistical assumption (namely, for the spin dependent hadronization of the c quark combining with the sea antiquark to form a meson [Ref.(3)]). The P_V calculated within this picture

$$P_V^\eta = \frac{3}{4(1-\eta)} , \tag{2}$$

should be equal to that from the cross section measurements, i.e. $P_V^\sigma = P_V^\eta$.

This simple picture works for a number of different production processes for π^0/ρ^0 and K_s^0/K^{*0} systems (see Ref.(4) for references). The only other measurements of the spin alignment parameter for the $D^{*+}(2010)$ come from e^+e^- collisions [Refs.(4), (5)]. For charm mesons the (e^+e^-) average, recalculated using the latest charm branching fractions (see Ref.(2) for details), is $P_V^\sigma = 0.70 \pm 0.07$, which is consistent with $P_V^\eta = 0.77 \pm 0.02 \pm 0.01$ [Ref.(4)]. Hence the e^+e^- results for charm mesons agree well with the statistical approach.

The measured spin alignment $\eta = 0.10^{+0.12}_{-0.11} \pm 0.01$ gives $P_V^\eta = 0.83^{+0.11}_{-0.10} \pm 0.01$, which is inconsistent with the directly measured values of P_V^σ. The NA32 $\sigma(D^+)$ and $\sigma(D^{*+})$ cross sections yield $P_V^\sigma = 0.47 \pm 0.11$, and similarly using the same branching fractions, one obtains $P_V^\sigma = 0.46$ with a comparable error from the recent measurements of the E653 collaboration in π^-*Emuls.* interactions at 600 GeV/c [Ref.(6)].

There is a definite discrepancy between P_V^η and P_V^σ for hadroproduction. A possible explanation is that contrary to the other processes, the hadroproduction of the $D^{*+}(2010)$ proceeds mainly via the decays of higher D^{**} mesons.

THE Λ_c^+ BARYON

The Λ_c^+ is the lowest mass charmed baryon and decays weakly. A natural decay channel to study production/decay spin effects for Λ_c^+ is the two-body channel $\Lambda_c^+ \to \Lambda^0\pi^+$. There are interesting results from studies of the decay spin effects, namely the weak-cascade decay of an unpolarized $\Lambda_c^+ \to \Lambda^0\pi^+ \to (p\pi^-)\pi^+$. A negative asymmetry $\alpha_{\Lambda_c}^{NL}$ of $-1.0^{+0.4}_{-0.0}$ and -0.96 ± 0.42 has been observed in this decay by CLEO and ARGUS [Ref.(7)], respectively. A similar cascade was used to measure the decay asymmetry parameter in the semileptonic decay of $\Lambda_c^+ \to \Lambda^0 l^+\nu_l$, giving $\alpha_{\Lambda_c}^{SL}$ equal to $-0.89^{+0.17+0.09}_{-0.11-0.05}$ and -0.91 ± 0.49 from CLEO and ARGUS [Ref.(8)], respectively. The branching ratios for the above channels are few times smaller than for $\Lambda_c^+ \to pK^-\pi^+$ (being of order of 4% only). Besides that, the $\Lambda_c^+ \to pK^-\pi^+$ channel has a much better experimental signature than the above mentioned channels.

In the rest frame of the decaying particle, the angular orientation of the back-to-back two-body decay is specified by one direction only. For the three-body decay, the momenta of the final particles are bound to the decay plane, thus the three-body angular orientation is fixed by two directions that specify the decay plane and the rotation within it. These two directions may be chosen as: the normal to the decay plane and the direction of a final particle, or just the two directions of final particles. Thus, the resulting angular decay distributions have different decay coefficients, depending on the angular parameterization of the final state.

For the spin measurement, and the polarization & decay asymmetry studies the practically background free sample of 121 $\Lambda_c^+ \to pK^-\pi^+$ events with positive x_F of Λ_c^+ was used (see Refs.(9), (10)).

Testing Λ_c^+ Spin Hypothesis [Ref.(9)]

The $\cos\theta$ distribution, i.e. the angular decay distribution integrated over the decay plane rotation angle and the azimuthal angle of the first direction ϕ, has $(2J+1)$ parameters (d_l moments, see Ref.(9)) for the spin J. Choosing the Z direction of the reference frame to be the beam direction (the Gottfried-Jackson frame) their number is reduced by parity conservation (the d_l moments vanish for odd l). Thus, this frame is suitable for spin measurements with limited data samples, especially for higher spin hypotheses. On the other hand, among the different angular parameterizations of the final state, the one using the normal to the decay plane as the first direction (i.e. θ_n^{GJ} and ϕ_n^{GJ}, the polar and the azimuthal angle of the normal to the decay plane) appears to have a moderate acceptance dependence.

The experimental distribution of $\cos\theta_n^{GJ}$ was fit with the corresponding distribution for the Λ_c^+ spin J=1/2, J=3/2 and J=5/2 hypotheses [Ref.(9)]. The results of the maximum-likelihood fits for higher nonvanishing moments are consistent with zero, both for the whole sample and for the subsample of

$p_T(\Lambda_c^+) > 0.7$ GeV/c. It should be recalled here that the range of d_2 values for $J = 3/2$ is [-1,1] and for both d_2 and d_4, in the case of $J = 5/2$, is even wider. The integrated distribution in the azimuthal angle ϕ_n^{GJ} is essentially flat, both for the total sample and the $p_T > 0.7$ GeV/c subsample. Its coefficients $S_m^{\pm}$, which do not vanish identically for $J = 1/2$ or $J = 3/2$, are consistent with zero, within the statistical error of 0.13 for the whole sample and of 0.19 for the $p_T > 0.7$ GeV/c subsample.

Λ_c^+ Polarization and Decay Asymmetry [Ref.(10)]

The Λ_c^+ polarization was studied with respect to the normal to the production plane, within the anti-Basel convention $\vec{p}_{\Lambda_c} \times \vec{p}_{beam}$ used in Ref.(10). Assuming the J=1/2 spin assignment for the Λ_c^+, the integrated distribution of the polar angle of the first direction reads

$$I(\cos\theta_i^T) = \frac{1}{2}[1 + d_1^i \cos\theta_i^T] \equiv \frac{1}{2}[1 + \alpha_i P_{\Lambda_c} \cos\theta_i^T], \qquad (3)$$

where P_{Λ_c} is the Λ_c^+ polarization with respect to the normal to the *production* plane and α_i are the decay asymmetry parameters depending on the choice of the angular parameterization of the final state. The θ_n^T, θ_p^T, θ_π^T and θ_K^T refer, respectively, to the angle between the normal to the *production* plane and: the normal to the *decay* plane, the proton, the pion and the kaon direction.

The results of maximum-likelihood fits of the d_1^p and d_1^π are consistent with zero, regardless of p_T. The d_1^n is consistent with zero within two standard deviations for Λ_c^+ $p_T >$ 1.1 GeV/c. For the distribution of $\cos\theta_K^T$ and $p_T >$ 1.1 GeV/c, the product of the Λ_c^+ polarization and the kaon type $\Lambda_c^+ \to pK^-\pi^+$ asymmetry is measured to be $\alpha_K P_{\Lambda_c} = -0.65^{+0.22}_{-0.18}$. The systematic error estimated by varying the acceptance within one standard deviation is of order of 0.02, for high p_T fits. Assuming the suggestion in Ref.(11) for the $\Lambda_c^+ \to pK^-\pi^+$ decay and the heavy quark effective theory predictions and first results [Refs.(7), (8)], the kaon type decay asymmetry should be negative, thus the Λ_c^+ polarization should be positive (negative) within the anti-Basel (Basel) convention and of order of 0.6-0.8.

Comparison of Λ_c^+ and Λ^0 Polarization [Ref.(12)]

There is a different production mechanism for Λ_c^+ and Λ^0. For positive x_F, the measured N(Λ_c^+)/N(Λ_c^-)=0.99±0.16, while N(Λ^0)/N($\overline{\Lambda}^0$)=1.27±0.03 [Refs.(1), (12)]. Thus, the Λ_c^+ is expected to be oppositely polarized to the Λ_c^- and, neglecting the CP violation in the decay, the angular decay distributions look the same for both. This is not the case with the hyperon, where the hadronization effects have more impact on the light s quark. Nevertheless, it is interesting to compare the polarization of the two baryons in the same production process.

The hyperon polarization was studied with respect to the normal to the production plane, with the more widely used Basel convention $\vec{p}_{beam} \times \vec{p}_{\Lambda}$, for a very clean sample of 27217 Λ hyperons produced in 230 GeV/c π^-Cu interactions [Ref.(12)]. For negative x_F there is basically low p_T data, thus the measured polarization is consistent with zero. For x_F>0 and p_T>1 GeV/c we find P(Λ^0)=-0.28±0.09(stat.)±0.02(syst.).

CONCLUSIONS

We have shown that interesting spin effects arise for inclusive $D^{*+}(2010)$ and Λ_c^+ hadroproduction. The spin effects in $D^{*+}(2010)$ hadroproduction do not support the statistical picture of spin dependent c quark hadronization, possibly indicating the importance of higher resonance contribution to the $D^{*+}(2010)$ hadroproduction. The $\Lambda_c^+ \to pK^-\pi^+$ decay channel has been shown to be useful for the Λ_c^+ spin measurements and the polarization & decay asymmetry measurements. Evidence has been shown for the high p_T Λ_c^+ transverse polarization and the kaon type $\Lambda_c^+ \to pK^-\pi^+$ asymmetry.

ACKNOWLEDGMENTS

I would like to thank the Royal Society of London and the organizers of the SPIN'94 conference for financial support.

REFERENCES

(1). ACCMOR coll., Barlag S., *et al.*, *Phys. Lett.* **B218**(1989)374, **B247**(1990)113, **B278**(1992)480.
(2). Rybicki K., Ryłko R., *Acta Phys. Pol.* **B24**(1993)1049.
(3). Donoghue J.F., *Phys. Rev.* **D19**(1979)2806.
(4). CLEO coll., Kubota Y., *et al.*, *Phys. Rev.* **D44**(1991)593.
(5). HRS coll., Abachi S., *et al.*, *Phys. Lett.* **B199**(1987)585; TPC/Two-Gamma coll., Aihara H., *et al.*, *Phys. Rev.* **D43**(1991)29.
(6). E653 coll., Kodama K., *et al.*, *Phys. Lett.* **284**(1992)461.
(7). CLEO coll., Avery P., *et al.*, *Phys. Rev. Lett.* **65**(1990)2842; ARGUS coll., Albrecht H., *et al.*, *Phys. Lett.* **B274**(1992)239.
(8). CLEO coll., Bergfeld T., *et al.*, *Phys. Lett.* **B323**(1994)219; ARGUS coll., Albrecht H., *et al.*, *Phys. Lett.* **B326**(1994)320.
(9). Jeżabek M., Rybicki K., Ryłko R., *Acta Phys. Pol.* **B23**(1992)771.
(10). Jeżabek M., Rybicki K., Ryłko R., *Phys. Lett.* **B286**(1992)175.
(11). Bjorken J.D., *Phys. Rev.* **D40**(1989)1513.
(12). ACCMOR coll., Barlag S., *et al.*, *Phys. Lett.* **B325**(1994)531.

Spin effects for high energy binary processes at large angles

M.P. Chavleishvili

Bogoliubov Laboratory of Theoretical Physics,
Joint Institute for Nuclear Research, Dubna 141980, Russia

Abstract. We consider spin asymmetries for two concrete reactions (proton-proton elastic scattering and proton antiproton annihilation into photino and antiphotino) on the basis of general formalism for description of any binary processes with particles of arbitrary spins and masses. In a formalism provides an analysis where spin-kinematics is fully taken into account and clearly separated from dynamics. In this "dynamic amplitude" approach obligatory kinematic factors arise in the expressions of observables. These spin structures for high energies give a small parameter that orders the contributions of helicity amplitudes to observables. This results in hierarchy of amplitude contributions, connections between asymmetry parameters and even numerical values for some of them.

In recent years the attention of many physicists has been paid to the "spin crisis", which is connected to European Muon Collaboration results. The crisis consists in that the spin of the proton is difficult to explain in terms of naive quark constituent models [1,2].

There exists another less sensational and less discussed contradiction between the perturbative QCD and experiment. This is connected with proton-proton scattering at high energies and large fixed angles. This is just the region where PQCD must work. But one can say that "the naive PQCD" fails here.

"Serious challenge to PQCD predictions" [3] for exclusive scattering is the observed behavior of the normal spin-spin correlation asymmetry A_{nn} measured in large momentum transfer proton-proton elastic scattering. At $p_{lab} = 11.75 GeV/c$ and $\theta_{cm} = \pi/2$, Ann rises to 0.6, corresponding to four times more probability for protons to scatter plane and parallel, rather than normal and opposite [4,5].

The point is that PQCD yields a "helicity conservation rule" [6-8] which gives in the lowest orders of perturbation theory a zero value for polarization and the value 1/3 for the asymmetry parameter A_{nn}. It is difficult to expect for changing (and improving) the situation by calculating an enormous number of diagrams in higher order. (It seems that helicity conservation will remain in any finite order of a perturbative expansion.)

Based on the symmetry principles we can separate the kinematics and dynamics of the binary processes [9-12]. In this approach we can define the dynamic amplitudes which are free of kinematic singularities, have the same dimensions and are connected with observable quantities very simply. At extremely high energies with fixed angles observables are expressed via dynamic amplitudes with definite kinematical factors. These kinematical factors in the considered region in fact are small parameters in different powers. This gives us something like perturbation theory: the hierarchy of contributions in observables ordered (sorted) by small parameter. In the first approximation such "a kinematic hierarchy" gives us definite relations between observables. For proton-proton scattering we will have relations between asymmetry parameters and even numerical values for them. For the asymmetry parameter A_{nn} it seems that its observed value just goes up to the obtained asymptotic number.

For proton-proton scattering we have five independent amplitudes. The standard choise of them is the following:

$$f_1(s,t) = f_{1/2,1/2;1/2,1/2}(s,t), \quad f_2(s,t) = f_{1/2,1/2;-1/2,-1/2}(s,t),$$
$$f_3(s,t) = f_{1/2,-1/2;1/2,-1/2}(s,t), \quad f_4(s,t) = f_{1/2,-1/2;-1/2,1/2}(s,t),$$
$$f_5(s,t) = f_{1/2,1/2;1/2,-1/2}(s,t). \tag{1}$$

For three asymmetry parameters (initial state correlation parameters) A_{nn}, A_{ss} and A_{ll} we have [13]:

$$\frac{d\sigma}{dt}A_{nn} = Re[f_1 f_2^* - f_3 f_4^* - 2 \mid f_5 \mid^2], \tag{2}$$

$$\frac{d\sigma}{dt}A_{ss} = Re[f_1 f_2^* + f_3 f_4^*], \tag{3}$$

$$\frac{d\sigma}{dt}A_{ll} = -\frac{1}{2}\{\mid f_1 \mid^2 + \mid f_2 \mid^2 - \mid f_3 \mid^2 - \mid f_4 \mid^2\}. \tag{4}$$

In these equations

$$\frac{d\sigma}{dt} = \frac{1}{2}\{\mid f_1 \mid^2 + \mid f_2 \mid^2 + \mid f_3 \mid^2 + \mid f_4 \mid^2 + 4 \mid f_5 \mid^2\}. \tag{5}$$

After separation of kinematic factors we have the rest of information in so called dynamic amplitudes. For elastic scattering of equal mass spin-J particles we can determine dynamic amplitudes by the following equation:

$$f_{\lambda_3\lambda_4,\lambda_1,\lambda_2}(s,t) = \left(\frac{\sqrt{-t}}{m}\right)^{|\lambda-\mu|}\left(\frac{\sqrt{s-4m^2+st}}{m}\right)^{|\lambda+\mu|}$$
$$\left(\frac{s-4m^2}{m^2}\right)^{-2J}\left(\frac{\sqrt{s}}{m}\right)^{K} D_{\lambda_3\lambda_4,\lambda_1,\lambda_2}(s,t). \tag{6}$$

where

$$K = \frac{1-(-1)^{\lambda_1+\lambda_2+\lambda_3+\lambda_4}}{2}. \quad (7)$$

As the connection between the helicity and dynamic amplitudes is one-to-one, every helicity amplitude for elastic scattering is expressed in terms of one dynamic amplitude. Hence it follows that all attractive features of the helicity amplitudes, clear physical meaning, simple relations with observables, and equal dimensions, are also inherent in the dynamic amplitudes. The formalism of dynamic amplitudes is simple for low spins and remains such also for higher spins.

Asymmetry parameters are expressed via dynamic amplitudes by the relations:

$$\frac{d\sigma}{dt}A_{nn} = Re[(\frac{m}{\sqrt{s}})^4 D_1 D_2^* - \frac{1}{4}\sin^2\theta D_3 D_4^* - \frac{1}{2}(\frac{\sqrt{s}}{m})^2 \sin^2\theta \mid D_5 \mid^2]. \quad (8)$$

$$\frac{d\sigma}{dt}A_{ss} = Re[(\frac{m}{\sqrt{s}})^4 D_1 D_2^* + \frac{1}{4}\sin^2\theta D_3 D_4^*]. \quad (9)$$

$$\frac{d\sigma}{dt}A_{ll} = -\frac{1}{2}\{(\frac{m}{\sqrt{s}})^4[\mid D_1 \mid^2 + \mid D_2 \mid^2] - \cos^4\frac{\theta}{2} \mid D_3 \mid^2 - \sin^4\frac{\theta}{2} \mid D_4 \mid^2\}. \quad (10)$$

In the high-energy large-fixed-angle region

$$\frac{m}{\sqrt{s}} \ll 1 \quad (11)$$

and the helicity amplitudes are splitted into three classes in the order of smallness determined by the kinematic factors. So we obtain in terms of dynamic amplitudes:

$$\begin{aligned} &\frac{\sqrt{s}}{2m}\sin\theta D_{1/2,1/2;1/2,-1/2} \gg \\ &\gg \cos^2\frac{\theta}{2} D_{1/2,-1/2;1/2,-1/2)} \sim \sin^2\frac{\theta}{2} D_{1/2,-1/2;-1/2,1/2} \gg \\ &\gg \left(\frac{m}{\sqrt{s}}\right)^2 D_{1/2,1/2;1/2,1/2} \sim \left(\frac{m}{\sqrt{s}}\right)^2 D_{1/2,1/2;-1/2,1/2}. \end{aligned} \quad (12)$$

here "$a \gg b$" means that the contribution of b is suppressed relative to the contribution of a in the observables.

For proton-proton scattering at $\theta_{c.m} = 90°$ we have from $s-u$ crossing symmetry that

$$f_{1/2,1/2;1/2,-1/2}(90°) = 0, \quad (13)$$

$$f_{1/2,-1/2;1/2,-1/2}(90°) = -f_{1/2,-1/2;-1/2,1/2}(90°). \quad (14)$$

Taking into account the dominating amplitudes we get in fact two type of results.

A."Hierarchy relations". Because the number of the dominating amplitudes is smaller than that of independent amplitudes (describing all the observable quantities), the hierarchy scenario contains the smaller number of amplitudes describing the same number of observables. So we must have the relations between the observables, hierarchy relations or hierarchy sum rules.

$$A_{nn} = A_{ll} = -A_{ss}. \tag{15}$$

B.Numerical values for asymmetry parameters.

$$A_{nn} = 1, A_{ll} = 1, A_{ss} = -1. \tag{16}$$

For pp scattering at $\theta = 90^o$ by definition there is the following exact relation between asymmetry parameters:

$$A_{nn} - A_{ll} - A_{ss} = 1. \tag{17}$$

Ouer results are in agreement with this relation.

Reaction proton + antiproton $\rightarrow$ photyno + antiphotyno is interesting because at new asselerators we will have chance to discover particles predicted by theories based on supersymmetry, in particular, photino.

Dynamic amplitudes for inelastic reaction with massive particles of masses $m+m \rightarrow M+M$ are expressed by binary combinations of helicity amplitudes (kinematic factors are considered as functions of the variables s and θ) [12]:

$$D^{s,\pm}_{\lambda_3\lambda_4,\lambda_1,\lambda_2} = \left(\frac{\sqrt{s}}{m+M}\right)^a \left(\frac{\sqrt{s-4m^2}}{m+M}\right)^b \left(\frac{\sqrt{s-4M^2}}{m+M}\right)^c$$
$$\left\{\left(sin\frac{\theta}{2}\right)^{-|\lambda-\mu|}\left(cos\frac{\theta}{2}\right)^{-|\lambda+\mu|} F_{\lambda_3\lambda_4,\lambda_1,\lambda_2}\right.$$
$$\left.\pm\left(sin\frac{\theta}{2}\right)^{-|\lambda+\mu|}\left(cos\frac{\theta}{2}\right)^{-|\lambda-\mu|} F_{\lambda_3\lambda_4,-\lambda_1,-\lambda_2}\right\}. \tag{18}$$

We have helicity amplitudes

$$F_1 = F_{1/2,1/2;1/2,1/2}, \quad F_2 = F_{1/2,1/2;-1/2,-1/2}, \quad F_3 = F_{1/2,-1/2;1/2,-1/2},$$
$$F_4 = F_{1/2,-1/2;-1/2,1/2}, \quad F_5 = F_{1/2,1/2;1/2,-1/2}, \quad F_6 = F_{1/2,-1/2;1/2,1/2}. \tag{19}$$

For asymmetry parameters we have:

$$\frac{d\sigma}{dt}A_{nn} = Re[F_1F_2^* - F_3F_4^* - F_5F_5^* - F_6F_6^*], \tag{20}$$

$$\frac{d\sigma}{dt}A_{ss} = Re[F_1F_2^* + F_3F_4^* + F_5F_5^* - F_6F_6^*], \tag{21}$$

$$\frac{d\sigma}{dt}A_{ll} = \frac{1}{2}\{| F_1 |^2 + | F_2 |^2 - | F_3 |^2 - | F_4 |^2 +2 | F_5 |^2 -2 | F_6 |^2\}. \quad (22)$$

In these equations

$$\frac{d\sigma}{dt} = \frac{1}{2}\{| F_1 |^2 + | F_2 |^2 + | F_3 |^2 + | F_4 |^2 +2 | F_5 |^2 +2 | F_6 |^2\}. \quad (23)$$

Using the explicit form of connection between helicity and dynamic amplitudes and smallnest of kinematic factor appeared we have "kinematic hierarchy" for the discussed reaction. Helicity amplitudes are splitted into three classes in the order of smallness determined by the kinematic factors.

$$F_5, F_6 \gg F_3, F_4 \gg F_1, F_2, \quad (24)$$

If we take into account only the dominating amplitudes, we get for asymmetries (for some other reasons, for example, for the $s-u$ symmetry and at the definite angles there can be additional restrictions, which have not been taken into account here).

So, for this reaction we also have have two tipe of oresults.

1."Hierarchy relation:" $A_{ll} = A_{ss}$.

2.Numerical value for asymmetry parameter: $A_{nn} = -1$.

REFERENCES

1.Proceedings of the 10th International Symposium "High Energy Spin Physics", Nagoya, 1992.

2.Proceedings of the International Conference "Polarization Dynamics in Nuclear and Particle Physics", Eds. A.Barut, R.Raczka, Trieste, 1992.

3.S.J.Brodsky, In Proceedings of the 10th International Symposium "High Energy Spin Physics", Nagoya, 1992; Preprint SLAC-PUB-6068, Stanford, 1993.

4.E.A.Crosbie et al., *Phys. Rev.*, **D23** (1981) 600.

5.A.D.Krish, *Phys.Rev. Lett.*, **63** (1989) 1137.

6.S.J.Brodsky, G.P.Lapage, *Phys. Rev.*, **D24** (1981) 2848.

7.S.J.Brodsky, C.E.Carlson, H.J.Lipkin, *Phys. Rev.*, **D20** (1979) 2278.

8.G.R.Farrar, S.Gottlieb, D.Sivers, G.H.Thomas, *Phys. Rev.*, **D20** (1979) 202.

9.M.P.Chavleishvili, *Ludwig-Maximilian University Preprint LMU-02/93*, Munich, 1993.

10.M.P.Chavleishvili, *Ludwig-Maximilian University Preprint LMU-03/93*, Munich, 1993.

11.M.P.Chavleishvili, Symmetry, Wigner functions and particle reactions, Plenary report at the III International Wigner Symposium, Oxford, 1993. (In print); Preprint JINR, E2-94-22, Dubna, 1994.

12.M.P.Chavleishvili, Symmetry and spin in binary processes. (Review). Particles & Nuclei, Dubna. (In print).

13.C.Bourrely, E.Leader, J.Soffer, *Phys. Reports*, **59** (1980) 96.

V. POLARIZED TARGETS AND SOURCES

Present status of polarized solid targets

W. Meyer

Physikalisches Institut der Universität Bonn, Germany

Abstract. The current situation of the polarized solid targets for their use in particle beams is briefly reviewed. This includes some general remarks on the latest polarized target experiments, a brief description of the dynamic nuclear polarization (DNP) principle and a review of the recent developments in the field of polarized target materials and techniques.

INTRODUCTION

The study of polarization phenomena in particle physics using polarized targets began with the development of polarized solid proton targets in the early 1960's. Since then enormous progress in the field of target technology has been made. The advances fall into two categories:

(I) increase in operating values of the magnetic field B/temperature T, which range from ~ 2 Tesla/Kelvin in 1960 to ~ 50 Tesla/Kelvin in 1990

(II) discovery of suitable electronic spin systems in materials of high free proton (deuteron) content

The majority of these targets are used in quite an appreciable number of particle experiments in order to study spin effects up to the highest energies. These experiments are performed with primary beams such as protons, photons and electrons as well as with secondary beams such as pions, kaons and muons. Recent surveys of the polarization phenomena and results are given in the proceedings of several spin physics conferences [1, 2].
The recent experiments on the deep inelastic scattering of polarized leptons from polarized protons and neutrons have raised interesting questions on the spin structure of the nucleon [3]. At intermediate energies, from some hundred MeV to several GeV, the understanding of the saturation of the color forces is incomplete. Precise and extensive experimental data are needed for a deeper understanding of the confinement region with respect to theory. Spin physics is a powerful tool for exploring the nucleon or nuclei dynamics, especially when there is already a large body of experimental data measuring the unpolarized 'Spin

Averaged' cross sections. Hence, many of the most interesting and important features of either the reaction mechanism (final state interaction, meson exchange currents) or the properties of the target nucleus (wave function, form factors) can only be discovered by directly measuring spin dependences using polarized targets and beams.
Important technical advances over the last years make such studies feasible:
- the availability of continuous beams of electrons and high quality proton beams
- the reliability of polarized beams and polarized targets or polarimeters

In the following the general scheme of the polarized solid targets is briefly explained. Then some recent work is discussed, which includes (I) polarized proton and deuteron targets for spin structure experiments, (II) new materials and (III) frozen spin techniques. The use of polarized targets as spin filters [4] as dark matter detector [5] and in biological structure research [6] are only mentioned here. Detailed information about these applications and on all other present research work will be given in the proceedings of the 7th Workshop on Polarized Target Material and Techniques.

Polarized Solid Targets

The starting point of any general discussion of polarized targets is the magnetic moment of the particle that is of interest - in the case of particle experiments - the proton or the deuteron (neutron). A polarized target can be assumed to be an ensemble of such particles placed in a high magnetic field and cooled to a low temperature. Unfortunately the magnetic moment of the proton is small, and that of the deuteron is even smaller. Consequently the polarization obtained in this way is very low. A simple calculation, using the Boltzmann equation gives 0.5 % for protons and 0.1 % for deuterons in a magnetic field of 2.5 Tesla and at a temperature of 0.5 Kelvin. Obviously, these polarization values are not very useful for experiments. However, the technique of dynamic nuclear polarization - developed in 1953 for metals (Overhauser effect) and in 1958 for solid insulators (solid effect) - allows very high nuclear polarization that can be obtained.
A simplified description of the DNP process can be given as follows. A suitable solid target material with a high concentration of H or D atoms is doped with paramagnetic radicals which provide unpaired electron spins. Since the magnetic moment of the electron is very much larger than that of the nucleon, the electron polarization is very high (99 % at 2.5 T and 0.5 K). The dipole-dipole interaction between the nucleon and electron spins leads to hyperfine splitting. The application of microwaves at a frequency very close to the electron spin resonance frequency induces a double spin flip of an electron-nucleus pair followed by a relaxation back flip of the electron spin. The DNP works because the electron spins relax much faster than the nuclcon spins, which results in a greatly

enhanced nucleon polarization. The nucleon polarization can be directed either parallel or antiparallel to the applied magnetic field by using slightly different values of frequency. No other parameter which could influence the experiment must be changed. This is a very important feature of the DNP, as systematic errors are reduced to a very low level. It has turned out that the DNP is a very practical technique and it is used in almost all polarized targets.
The main problem with DNP is finding a suitable combination of hydrogenous material and paramagnetic radicals (electrons). Suitable means that the relaxation time of the electron spins is small (msec.) and that of the nucleons is long (min.). The ideal target material, hydrogen, is at low temperature in the para-state with spin zero and hence unpolarizable.
Due to the improvements in cryogenics (dilution refrigerators) and magnet technology (superconducting magnets) protons can be polarized to 100 % and deuterons to more than 50 % (see later). Standard target materials for the particle experiments are alcohols (butanol, d-butanol) or ammonia (NH_3, ND_3). Target volumes, generally determined by the experiment requirements, vary between 1 cm^3 and some litres.
A schematic diagram of the main components of a polarized target is shown in figure 1.

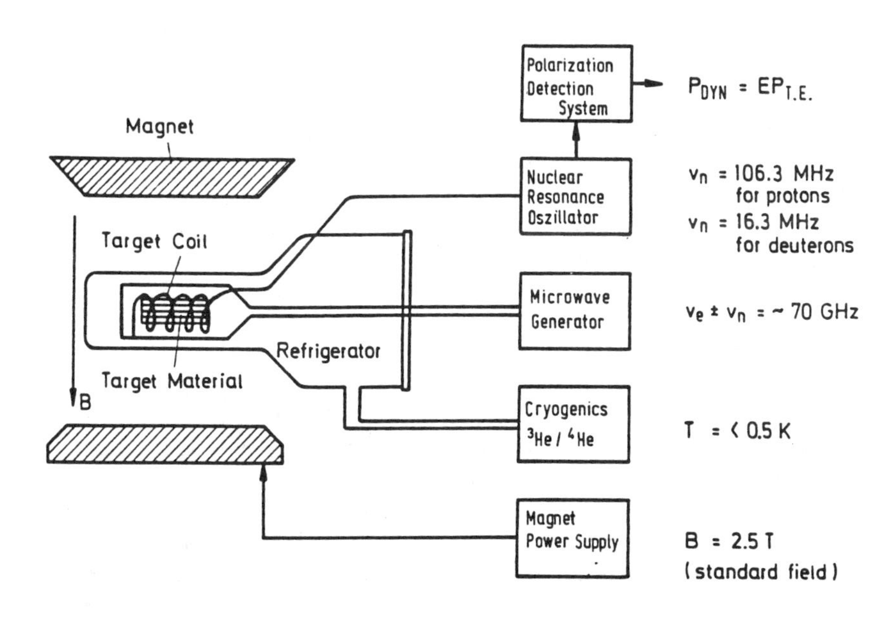

FIGURE 1. Schematic diagram of the main components of a polarized solid target

Polarized Proton and Deuteron Targets for Spin Structure Experiments

The development of the polarized target material ammonia in the early 1980's has brought improved and new polarization measurements in deep inelastic scattering. Due to its higher polarizable nucleon content compared to the standard target material butanol, NH_3 was used by the EMC-collaboration at CERN. The exploration of a wider kinematic region, using muons of higher energy, indicated that quarks carry only little of the proton spin. This unexpected result demanded - besides immense theoretical work - new experimental activities at CERN and SLAC. These activities will be later supported by experiments at HERA, where different polarization techniques for the beam as well as for the targets are going to be used. In the meantime polarized solid deuteron (neutron) targets with a high polarization degree are available, which enable the test of the Bjorken sum rule using both electrons and muons.
Due to the different beam conditions at CERN and SLAC, the polarized target builders have to react on these requirements. At the SMC-measurements sufficient luminosities can be obtained using a huge target (~ 2.5 litres target volume) to overcome the relatively small number of muons in the beam [7]. In order to measure the required asymmetries $\sigma^{\rightleftarrows}$ and $\sigma^{\rightrightarrows}$ the target is divided into two sections, each oppositely polarized as muons are only available in one helicity state.
The discovery of a large gain in the dynamic nuclear polarization after the application of a microwave frequency modulation (FM) in the large SMC-target [8] was very important. In the glassy deuterated butanol material doped with a paramagnetic EDBA-chromium (V) complex the maximum polarization increased by about 100 % with FM and the polarization build-up time decreased by a factor of almost two. Up to now there is no complete understanding of this behaviour and future work has to be done. Nevertheless, in the latest experiment at CERN a deuteron polarization of more than 50 % has been reached with FM and this was a key point for their statistical accuracy obtained in the deuteron spin structure measurements up to now.
High statistics experiments in this field are performed with a polarized electron beam at SLAC [3]. However, polarized target experiments with intense electron beams suffer from the depolarization effects due to radiation damage and beam heating. Therefore NH_3 or ND_3 have to be used because their depolarization caused by radiation damage is much less compared to that of the alcohol materials [9]. As already known, the ND_3 target preparation for the DNP by irradiation is a little bit tricky. Highest deuteron polarization values can only be achieved by means of an in situ "low temperature" irradiation [10]. For this doses of 4×10^{14} particles/cm^2 are needed, which can be delivered by intense electron beams in a reasonable amount of time (< 1 h).
Detailed studies showed that polarized solid targets with ^4He cooling at about 1

K can accept electron beam intensities of more than 50 nA with only a small (< 10 %) temperature dependent depolarization [11]. In addition, this depolarization effect can be minimized by scanning the electron spot size (some millimeters in diameter) over the target front face.
Due to the above mentioned depolarizing effects, an average proton polarization of 65 % and 25 % average deuteron polarization could be obtained during the SLAC-experiments. Typical electron beam intensities were 40 - 60 nA [12].
The results of the spin structure experiments have indicated that the systematic error due to the present accuracy of the polarization measurement is a limiting factor. The polarization is measured by the nuclear magnetic resonance method (series Q-meter). The system is calibrated using the calculable polarization at thermal equilibrium of the nucleon spins with the solid lattice at a known temperature in a known magnetic field. Continuous progress was achieved all over the years and the polarization accuracy is now better than ± 2.5 % for the protons and < ± 4.5 % for the deuterons [12, 13, 14, 15]. Although the series Q-meter technique has reached the status of a well explored technology [16], it can still be improved and developed for specific applications. A challenge might be the measurement of the polarization of ^{14}N in ammonia.

New Target Materials

Since the first experiments with polarized solid targets, there has been a strong need for more pure target materials. A first major breakthrough occured at the end of the sixties by the development of the alcohol materials, which are still used in many experiments.
The paramagnetic radicals in such target materials are introduced by chemical admixtures. For the first time promising DNP results in thin polymer foils and tubes with the chemical composition $(CX_2)_n$, [X = ^1H, ^2D, ^{19}F], doped with the stable nitroxyl radical TEMPO, are reported.
The radical can be easily introduced via a diffusion process [17]. Proton polarizations of about 70 % were measured in thin polymer tubes [18]. New is the fact that the samples (12 - 100 μ thick foils and tubes) can be handled at room temperature for at least a few hours, which provides obviously practical advantages. Especially the sample handling during the loading into the refrigerator is facilitated, which must be done under liquid nitrogen in case of all other polarized target materials. This could be extremely helpful for experiments, in which tiny and geometrically well-shaped samples are required. However, the polarization resistance against radiation damage has been explored.
In experiments with high intensity beams NH_3 and ND_3 are the standard target materials. They have been well investigated in the early eighties and their very good polarization resistance against radiation damage is mainly a result of radiation induced radicals.

Radiation doping has also been employed in different compounds of lithium hydrides and lithium deuterides for more than 15 years. Although very high DNP have been observed with maximum polarizations of 95 % (H) and 80 % (^{7}Li) and 71 % in ^{6}LiD, the long polarization build-up times has been a problem. For a summary of the work of the different SACLAY groups see [19]. Present work was done by the Bonn group with the following intention [20]:

1) finding a physically satisfactory explanation of the dependence of the polarization behaviour on the preirradiation conditions (especially the temperature)
2) on the basis of this knowledge shortening of the long polarization build-up times
3) investigations in the polarization behaviour of the materials under an intense particle beam in order to test ^{6}LiD as a possible polarized deuteron material in high current scattering experiments

With the extremely helpful tool of EPR studies some new aspects could be demonstrated in order to explain the preparation requirements for an efficient DNP process in the lithium hybrides. Adequate paramagnetic centres can be created in the material by means of irradiation with "high energy" electrons at temperatures allowing the formed point defects to coagulate to larger paramagnetic structures. These structures possess a sufficiently strong coupling of their magnetic moment to the lattice phonons.
With more than 50 % and 22 % very high deuteron polarizations could be obtained in ^{6}LiD at a magnetic field of 5 T in a ^{3}He/^{4}He-dilution refrigerator (T $\approx$ 200 mK) and in a ^{4}He-evaporation cryostat (T $\approx$ 1 K), respectively. No significant loss in the polarizability was observed during the irradiation of the lithium hydrides under the conditions of an electron scattering experiment up to an accumulated radiation dose of 10^{16} e/cm^2, a result even better than the radiation resistance of ammonia. In addition, subsequent low temperature irradiation shortens the polarization build-up time by a factor 5 to 6 (all lithium compounds) and increases the maximum polarization by about 25 % (^{7}LiH). Similar effects have been measured in recent experiments at SACLAY [21]. The main idea consists in the introduction of crystalline displacements caused by the particle beam which should be stable at temperatures at 1 K and below.
For further improvement of the polarization characteristics of the lithium hydrides (maximum polarization and polarization build-up time) other coloration methods should be tested. Their gain of proper paramagnetic impurities should be controlled with in situ EPR measurements under the DNP conditions.
A very promising polarized solid target material is HD which has many important properties. For example, it is possible to polarize either the H or the D, or both. All of the free protons in solid HD are polarizable and these represent half of all protons, while all the deuterons (and thus all the neutrons) are polarizable.

More than 25 years ago, it was suggested that solid HD could be used as a polarized target [22]. Attempts to polarize HD by the DNP mechanism were not very successful [23]. Recent efforts are concentrated on the "brute force" polarization [24], which works with every polarized target material. In the case of HD the extreme long polarization build-up time can be shortened. The spin-lattice relaxation time for pure HD is essentially infinite. However, by condensing the HD gas molecules with small concentrations ($\approx 10^{-4}$) of the J = 1 species of ortho-H_2 and para-D_2, the H and D spins in HD are coupled to the lattice via these J = 1 impurities. This decreases the initial relaxation time to the order of the day, and permits one to obtain equilibrium polarization of the H and D in the HD molecule in a reasonable time. These J = 1 impurities at low temperatures (< 4.2 K) convert to the magnetically inert para-H_2 and ortho-D_2 (J = 0 species) with time contants of 6.5 days and 18.2 days, respectively. Therefore, by aging the targets at low temperatures and at high fields for the order of a month, the relaxation times increase sufficiently for use as a practical target as the J = 1 dopants convert to the J = 0 state. Once polarized, these solid targets retain their polarization at a temperature of 1.5 K with a field of 0.4 T for the order of a week.
Up to now a deuteron polarization of 38 % has been obtained after 6 days in a target volume of 0.015 mole. Further improvements can be expected which includes an increase of the target volume, a better fast passage efficiency and the operation at a lower temperature (5 mK) and in a higher field (17 T). The successful application in particle experiments depends on controlling relaxation rates and implementing cold-transfer operations [25].

Frozen Spin Targets

For dynamically polarized targets a magnetic field in the range of 2.5 - 5 Tesla is required to achieve sizeable polarizations. In addition the field must be uniform to about one part in 10^4 over the whole target volume as the electron spin resonance linewidth is relatively narrow. Therefore special magnets are needed which should also allow maximum experimental access for the beam and the outgoing particles.
All early targets used a C-type electromagnet combined with horizontal refrigerators. With this type of target the polarization direction can be only oriented perpendicular to the scattering plane of the produced particles. For experiments which require other polarization orientations superconducting magnets must be used. With a split pair configuration all three possible orientations of the polarization direction with respect to the scattering plane can be obtained by rotating the coils. The superconducting magnets are normally operated in conjunction with a vertical refrigerator. Nevertheless, the experimental access is limited by the finite dimensions of the coils.

This limitation is serious if experiments with small cross sections or low intensity beams, e. g. tagged photon beams are performed. To obtain a reasonable counting rate a large solid angle has to be covered simultaneously. This can be achieved with a frozen spin target. Its operation depends on the experimental fact that the nucleon relaxation time T_n is a very steep inverse function of the temperature. T_n characterizes the polarization decay, after the polarizing mechanism (microwaves) is switched off. Typical values for T_n are minutes at a temperature of 1 K and days at 100 mK. The principle of the frozen spin operation mode is to polarize the material at e. g. 2.5 T and around 300 mK. After reaching the maximum polarization the microwave power is turned off. Then the spin is "frozen in" and the target can be placed into a "holding" field which can be much lower than the polarizing magnet field. An appropriate setting of the "holding magnets" allow the target to be polarized in any arbitary direction combined with a good experimental access.
Up to now targets, e. g. at SACLAY and Bonn, have operated with external superconducting magnets providing the holding field of the order of 0.5 T in the target area. These magnets are large and have a strong fringe field. Therefore such frozen spin targets cannot be operated in combination with 4 π detectors, where detector components are placed close around the target refrigerator.
A new frozen spin target built by the PSI group [26], where the dilution refrigerator has been incorporated in a cryostat in which also the superconducting holding coil system has been integrated. This system consists of a split pair magnet (1.2 T) and a saddle coil magnet (1.1 T), providing the vertical and horizontal holding fields, respectively. A linear combination of these fields allows any quantization direction in space. The opening angle for the particles is only limited by the coil packages above and below the target volume and by two pillows of the magnet support. A similar technique is used at TRIUMF [27].
This limitation can be avoided by a new type of small, thin superconducting holding coil, developed by the Bonn polarized target group [28]. The coil (200 μ thick) has been wound on the inner cooling shield of the refrigerator in the target area. It is cooled by conduction to less than 1.5 K, thus obtaining a magnetic field of 0.35 T. This scheme allows a substantial reduction of the magnetic field affecting the detector components: at a distance of 20 mm axial and radial form the centre of the coil the field strength diminishes to a value of about 10 G. On the other hand this solenoid version of the internal holding coil allows for the first time to measure the polarization of the nucleons during the frozen spin mode with the standard NMR-technique.
Experiments studying spin effects with frozen spin targets suffer from systematic uncertainties due to the infrequent reversal of the target polarization. Normally this reversal is done by the time consuming process of the DNP at low temperatures. Fast polarization reversal by adiabatic fast passage with high efficiencies in a variety of polarized target materials was demonstrated by the PSI group [29]. Good results with efficiencies of δP as high as - 90 % were obtained

in slowly polarizing materials (low concentration of paramagnetic centres in spin-1/2 systems) and in systems with large inhomogeneous quadrupole broadenings (spin-1 system in deuterated butanol).
Nowadays experiments at intermediate energies with frozen spin targets are performed at ELSA, PSI, SACLAY and TRIUMF. However, new techniques like the construction of tiny internal holding coils will extend their applications. A new variety of experiments has already been proposed and will be started in the future at facilities like CEBAF, COSY and MAMI. In such experiments a maximum beam intensity of 10^7 - 10^8 particles/s $\cdot$ cm^2 - given by the cooling power of the ^{3}He - ^{4}He dilution refrigerators and not at least by the Kapitza resistance - can be accepted. The prize which has to be paid for the possibility of the integration of a frozen spin target into a e. g. 4 π detection system is that parts or the whole detector must be movable.

CONCLUSION

For more than thirty years polarized solid targets have been a powerful tool in particle physics to study spin effects. The latest results in deep inelastic scattering experiments with polarized leptons on polarized protons and deuterons (neutrons) demonstrate its importance. There has been a steady progress in the field of polarized target instrumentation. Especially the latest developments in the frozen spin target technology (internal tiny superconducting holding coil, fast adiabatic passage) and progress in the preparation of highly polarized HD-targets make improved and new experiments possible. Such experiments are essential to reveal the complex structure of the nucleon in the confinement region. It seems that the mature polarized solid target technology can even be more developed.

REFERENCES

1. Eds. K.-H. Althoff, W. Meyer, *Proceedings of the 9th International Symposium on High Energy Spin Physics*, Springer Verlag, 1991
2. Eds. T. Hazegawa, N. Horikawa, A. Masaike, S. Sawada, *Proceedings of the 10th International Symposium on High Energy Spin Physics*, Nagoya, 1992, Universal Academy Press, Tokyo, 1993
3. J. Ashman et al., *Phys. Lett. B 206*, 364 (1988)
   J. Ashman et al., *Nucl. Physis. B 328*, 1 (1989)
   B. Adeva et al., *Phys. Lett. B 302*, 533 (1993)
   P. L. Anthony et. al., *Phys. Rev. Lett 71*, 959 (1993)
   D. Adams et al., *Phys. Lett. B 329*, 399 (1994)
4. P. P. J. Delheij et al., *Proceedings of the 7th Workshop on Polarized Target Materials and Techniques*, Bad Honnef, Germany 1994
   Eds. H. Dutz, W. Meyer, *Nucl. Instr. and Meth. A356* (1995)
5. Y. K. Semertzidis et al., ibid

6. H. B. Stuhrmann et al., ibid
7. J. Kyynäräinen et al., ibid
8. Yu. F. Kisselev et al., ibid
9. W. Meyer et al. *Nucl. Instr. and Meth. 215*, 65 (1983)
K.-H. Althoff et al., *Proceedings of the 4th International Workshop on Polarized Target Materials and Techniques*, Bonn, 1984, p. 23
10. W. Meyer et al., *Nucl. Instr. and Meth. 227*, 35 (1984)
11. A. Thomas et al., *Proceedings of the 7th Workshop on Polarized Target Materials and Techniques*, Bad Honnef, Germany, 1994
Eds. H. Dutz, W. Meyer, *Nucl. Instr. and Meth. A356* (1995)
12. D. Crabb et al., ibid
13. D. Krämer et al., ibid
14. Y. Semertzidis et al., ibid
15. G. Reicherz et al., ibid
16. T. O. Niinikoski et al., ibid
17. E. I. Bunyatowa, ibid
18. B. van den Brandt et al., ibid
19. P. Chaumette et al., *Proceedings of the 8th International Symposium on High Energy Spin Physics*, Minneapolis, 1988, Conf. Proc. No. 187, Vol. 2, p. 1275
20. St. Goertz et al, 7th Workshop on Polarized Target Material and Techniques, Bad Honnef, Germany 1994, Eds. H. Dutz, W. Meyer, *Nucl. Instr. and Meth. A356* (1995)
21. J. Ball, private communication
22. A. Honig, *Phys. Rev. Lett. 19*, 1009 (1967)
23. J. C. Solem, *Nucl. Instr. and Meth. 117*, 477 (1974)
24. A. Honig et al., 7th Workshop on Polarized Target Materials and Techniques, Bad Honnef, Germany 1994, Eds. H. Dutz, W. Meyer, *Nucl. Instr. and Meth. A356* (1995)
25. M. Rigney et al., ibid
26. B. van den Brandt et al., ibid
27. P. P. J. Delheij et al., ibid
28. H. Dutz et al., ibid
29. P. Hautle et al., ibid

Report on the Madison Workshop on Polarized Ion Sources and Polarized Gas Targets

L. W. Anderson

Department of Physics, University of Wisconsin, Madison, WI 53706

Abstract. A brief report on the current status of polarized ion sources and polarized gas targets as reported at the Workshop on Polarized Ion Sources and Polarized Gas Target is given.

I. INTRODUCTION

The Workshop on Polarized Ion Sources and Polarized Gas targets was held in Madison, Wisconsin, from 23 May to 27 May 1993. It is one of a series of on going workshop held at two to three year intervals. It was attended by about 100 scientists with attendees from almost all the world's major laboratories where spin polarized physics is studied. The principal topics discussed at the workshop include the following:

- Polarized H or D Targets
  - A. Atomic Beam H or D Targets
  - B. Low Temperature Polarized H or D Targets
  - C. Spin Exchange Optically Pumped Polarized H or D Targets
- Optically Pumped ^{3}He Targets
- Polarized H or D Ion Sources
  - A. Atomic Beam Polarized H or D Ion Sources
  - B. Optically Polarized H or D Ion Sources
- Other Polarized Ion Source or Polarized Target Topics
- Machine Target Interactions

In this paper a short survey of the material discussed at the workshop on each of the above topics is presented. The proceedings of this workshop have been published. For additional details on any of the above topics one is referred to the proceedings of the workshop - Polarized Ion Sources and Polarized Gas Targets (AIP Conference Proceedings 293), Edited by L. W. Anderson and W. Haeberli, American Institute of Physics, New York, 1994.

II. POLARIZED H OR D TARGETS

There are several ways to make a polarized H or D atoms for a polarized target. Some methods of obtaining nuclear spin polarized H or D atoms are the following: (i) In the atomic beam method an atomic beam passes through a strong field 6-pole magnet which selects the electron spin up states. In order to obtain nuclear spin polarization one must use rf transitions after the 6-pole magnet. If one desires to select a single hyperfine state one must use both rf transitions and a second 6-pole magnet. (ii) In the ultra-cold method very cold electron spin polarized H or D atoms are prepared in the high magnetic field of a solenoid ($kT<<u_eB$ where u_e is the electronic magnetic moment of the H or D atoms). The electron spin polarized atoms are then ejected by the fringing magnetic field of the solenoid. For this method one must also use rf transitions to obtain nuclear polarization. (iii) In the optical pumping method nuclear spin polarized H or D atoms can be prepared via spin exchange optical pumping. (iv) Finally nuclear spin polarized HD can be prepared by brute force ($kT<<u_nB$ where u_e is the nuclear magnetic moment of the H or D atoms) and then evaporate the solid to make a gaseous target.

The polarized atoms once formed can be used in a storage ring in two ways. One can insert the polarized atoms into a storage cell located in the ring, or one can form a focal spot of an atomic beam at the location of the ring. The storage cell has the advantage of higher density than a focussed beam but has the disadvantage of requiring cell walls that can interact with the circulating electron or ion beam in the storage ring.

A. Atomic Beam H or D Polarized Targets

There were six papers on atomic beam H or D polarized targets presented at the workshop. Zapfe reported on the HERMES-FILTEX. This internal target utilizes a polarized atomic beam injected into a storage cell. The HERMES-FILTEX target density is 10^{14} atoms/cm^2 with a polarization equal to 90% of the maximum possible value. Zapfe reported the demonstration use of the target as a spin filter for the polarization of circulating ion beams. Stock reported on the optimization of the HERMES-FILTEX atomic beam source. Roberts et al. reported on the atomic beam source for use in the polarized H or D target for the IUCF cooler ring. They reported atomic intensities as high as 6.7×10^{16} atoms/s (two spin states) into a 1 cm id 13 cm long compression tube 26 cm beyond the last 6-pole magnet. Ross et al. reported target polarization higher than 0.70 (single spin state) for the storage cell for the IUCF cooler ring. Price reported measurements of the deuterium tensor polarization for gas storage targets both with Teflon wall coatings and uncoated Al walls as functions of the temperature and at low and high magnetic fields. With Teflon coated walls the nuclear polarization is about 0.95 of the maximum value. With uncoated Al walls Price reported the remarkable result that in a high magnetic field the nuclear polarization is about 0.7 of the maximum possible value whereas in a low magnetic field the nuclear polarization is only about 0.3 of the maximum value. These results may be useful for situations where Teflon coatings can not be used.

Boyd and Kubischta discussed simulations of gas dynamics in atomic beams using monte-carlo techniques. Gas dynamic simulations may be important since the intensity limitations for an atomic beam are not completely understood. Van den Brand described polarized internal targets for electronuclear experiments.

B. Low Temperature Polarized H or D Targets

Raymond reported tests of a prototype ultra-cold spin polarizing H jet (0.3K) ejected from the fringing field of a high magnetic field solenoid. The ejected H atoms are focussed into a compression tube with a 6-pole magnet. The flux of H atoms is $3x10^{15}$ atoms/s, which leads to a density of $3x10^{11}$ atoms/cm^3. They hope ultimately to obtain a density of H atoms of $5x10^{12}$ for use in the experiment NEPTUN-A at UNK. Kaufman analyzed the transport of ultra-cold spin polarized H atoms, and Luppov et al. discussed the use of a He coated quasi-parabolic mirror to focus a beam of ultra cold spin polarized H atoms.

Honig et al. described a brute force method (B=13T and T=0.02 K) to polarize solid HD with the possibility of evaporating the solid HD to produce a polarized gas target. The proton polarization is over 0.5 for the above magnetic field and temperature, but the deuteron polarization is low. The deuteron polarization can be increased by driving a forbidden NMR transition by adiabatic fast passage thereby transferring some of the proton polarization to the deuteron. The spin lattice relaxation time for HD at 1.5 K and in a magnetic field of 0.1 T is about one day so that the HD can be produced and then the material stored passively until it is to be used.

C. Spin Exchange Optically Pumped Polarized H or D Targets

Poelker et al. reported on the Argonne spin exchange optically pumped H or D target. Their results indicate that the H or D targets are approaching spin exchange equilibrium and are nuclear polarized without the use of rf transitions. They have a flow rate of 1.6×10^{18} atoms/s, 75% dissociation, and they infer a nuclear polarization of 0.51 under the assumption of spin exchange equilibrium. They use a Teflon lined transport tube to eliminate K contamination. Jones et al. discussed a proposed method to measure the D tensor polarization of the Argonne target. Walker and Anderson analyzed the effects of spin exchange collisions on a spin exchange optically pumped polarized target. They have shown that even in a high magnetic field spin exchange hydrogen-hydrogen collisions transfer angular momentum between the electron spin and the nuclear spin driving the system toward spin exchange equilibrium. For a hydrogen target in a low field spin exchange collisions do not alter the nuclear spin polarization, but in a high field they do change the nuclear polarization. For a deuterium target both the vector and tensor nuclear polarizations are affected by spin exchange collisions. The tensor polarization of D may be severely degraded. In a separate paper Anderson and Walker have shown that the prospects for producing a highly polarized hydrogen target without rf transitions by spin exchange optical pumping is good. Martin et al. reported experimental results on the optical pumping of a K vapor with linearly polarized light incident at right angles to the magnetic field. Frolov et al. reported on experiments with an optically pumped D target at VEPP.

III. POLARIZED ^{3}He TARGETS

There are two methods for producing optically pumped ^{3}He targets. In the first method one optically pumps the metastable 2^3S level of ^{3}He and utilizes metastability exchange to produce nuclear polarized ground level ^{3}He. In the second method one optically pumps Rb and utilizes Rb-^{3}He spin exchange collisions to produce nuclear polarized ^{3}He. The first method is carried out at low ^{3}He pressures (~1 Torr), and the second method is carried out at high ^{3}He pressures. The rates of polarization of ^{3}He atoms are similar for the two methods are similar. In the Rb spin exchange method for polarizing ^{3}He, even though the Rb density is high, radiation trapping is avoided by adding N_2 to the gas which quenches the excited Rb atoms.

Leduc et al. described the nuclear polarization of ^{3}He via metastability exchange optical pumping. In this method the metastable 2^3S level of ^{3}He is optically pumped with high power LNA lasers operating at the wave length corresponding to the 2^3S→2^3P transition. They also described methods for the measurement of the ^{3}He nuclear polarization. Milner also discussed a nuclear polarized ^{3}He target using metastability exchange optical pumping. The target had a nuclear polarization of 0.5 and a flow rate of 1×10^{17} atoms/s. A cryogenically cooled thin walled storage cell is used to enhance the target density. The target has been operated for a year. Heil presented a talk on the development of a high density ^{3}He target using metastability exchange optical pumping to polarize the ^{3}He at a pressure of about 1 Torr followed by compression of the polarized ^{3}He. A target operating at a pressure after compression of 1 bar is installed on an experiment, and a target for operation at a compressed pressure after compression of 13 bars is being tested.

Middleton et al. described the SLAC high density polarized ^{3}He target. This target uses spin exchange collisions between ^{3}He and optically pumped Rb. The target operates with a density of 2.3×10^{20} atoms/cm^3 and with a polarization of 0.3-0.4. Cells with relaxation times of several days were developed in order to achieve these polarizations. Häusser et al. reported on the TRIUMF spin exchange optically pumped ^{3}He target. This target operates at pressures of about 12 bars, a polarization of 0.6 and with a volume that can be as large as 70 cm^3. Hersman discussed a plan for installing a ^{3}He target at CEBAF.

IV. POLARIZED H OR D ION SOURCES

There are in use currently two different types of ion source to produce beams of polarized H or D ions, the atomic beam ion source and the optically pumped ion source. In the atomic beam ion source H or D atoms are polarized passing an H or D atomic beam through a 6-pole magnet followed by rf transitions to produce nuclear polarized H or D atoms. The nuclear polarized H or D atoms can be ionized in two different ways. They can be partially converted into H^+ or D^+ ions in an electron impact collision ionizer (such as an ECR ionizer). If one desires the positive ions can be partially converted into negative ions by two change exchange collisions in an alkali vapor to make nuclear polarized H^- or D^- ions. Alternatively the nuclear polarized $H°$ or $D°$ atoms can be partially converted into nuclear polarized H^- or D^- ions directly in a charge changing collision such as $H^0 + Cs \rightarrow H^- + Cs^+$ or $H^0 + D^- \rightarrow H^- + D^0$. The optically pumped ion source a H^+ ion beam is extracted from an ECR ion source and is passed through an optically pumped alkali vapor where the beam is partially converted by a charge exchange collision into fast electron spin polarized H^0 atoms. The electron spin polarization of the fast H^0 atoms is then transferred into nuclear spin polarization. Finally the fast nuclear spin polarized H^0 atoms are partially converted into nuclear spin polarized H^- ions by charge changing collisions in a second alkali vapor target.

The question arose as to how much current is desirable from a polarized ion source. Haeberli provided what seemed to all the participants a reasonable answer. He felt that one should strive to develop polarized ion sources that had currents as large as can be accepted by accelerators or storage rings. If such sources are available then one can run any experiment with polarized or unpolarized beams without the loss of intensity.

In addition to the construction of polarized ion sources the measurement of the polarization of the ion beams is of interest. Crosson et al. described a polarimeter constructed using the spin filter from an old Lamb shift ion source. This polarimeter is based on a three level interaction in the n=2 states of hydrogen. The polarimeter enables one to make rapid measurements of the hyperfine state populations. It works for either hydrogen or deuterium.

A. Atomic Beam Polarized H or D Ion Sources

Schmelzbach described ECR ionizer and analyzed the operating experience at PSI with ECR ionizers. Derenchuk discussed the IUCF high intensity polarized ion source. The PSI source has an atomic beam flux of 5×10^{16} atoms/s, and it utilizes an ECR ionizer. After mass analysis a polarized H^+ or D^+ ion beam of 245 uA with a normalized emittance of 0.78 π mm mrad was obtained. Clausnitzer et al. described the conversion of the Giessen Lamb shift polarized ion source into an atomic beam polarized ion source. Clausnitzer et al. discussed the design of the Balzers' polarized ion source. Lemaire investigated the limitations of an atomic beam polarized ion source produced by the electron beam ionizer. Okamura et al. reported on the RIKEN polarized ion source, and Maehata et al. reported on a compact D polarized ion source under construction at Kyushu. Eversheim discussed the status of the polarized ion source for COSY-JÜLICH, which is a source of the colliding beam type ionizer, and provides polarized H^- or D^- ions. Paetz gen. Schieck et al. described the status of the fast Cs° beam for the COSY-JÜLICH ion source. Belov et al. presented results on an atomic beam polarized ion source with a deuterium plasma colliding beam ionizer. They obtained a peak current of about 150uA for the pulsed polarized H^- current with an atomic beam flux of $2x10^{17}$ atoms/s and with a peak D^- ion current of 1.2 mA. The ion source has a 100μs pulse duration and a 5 Hz repetition rate.

B. Optically Pumped Polarized H or D Ion Sources

Mori presented a review paper on the present status of the optically pumped polarized ion source (OPPIS) for polarized H^- ions. The various OPPIS's that have been constructed can operate in either dc or pulsed mode. Polarized H^- ion currents between 20uA and 400uA have been achieved with polarizations between 0.65 and 0.80 and with normalized emittances near 1.0xπ mm mrad. Mori also reported experiments on a novel method for the production of polarized D^- ions with either vector or tensor polarization. The method had been suggested by Clegg. Prior to this work by Mori only polarized H^- ions had been produce by the OPPIS. Mori also outlined several methods for possible future significant improvements of the OPPIS with the H or D polarization obtained with the use of spin exchange collisions rather than charge exchange collisions. Zelenski and Kokhanovski reported an experimental study of spin exchange polarization transfer in H-Rb collisions. In a separate paper Zelenski described a proposal for a pulsed OPPIS which can produce currents of 1.2 mA of H^- ion current with a polarization of 0.8 and a normalized emittance of 1.5 π mm mrad. Levy et al. discussed the status of the TRIUMF OPPIS. This polarized ion source produces 20 uA DC H^- ion current with a polarization of 0.8 and a normalized emittance of 1.0 π mm mrad. The laser system can produce spin flips at a repetition rate of 200 Hz. In another paper Levy et al. described the measurement of spin correlated current modulation in OPPIS. They have developed techniques that permit one to keep the spin correlated current modulation below 10^{-5}. The small spin correlated current modulation is a major achievement and will be useful for parity violation experiments in strong interactions. York et al. discussed the performance of the LAMPF OPPIS. Their source can be operated at 50 uA of H^- ion current on target with a polarization of 0.56, at 25 uA with a polarization of 0.65, or at 2 uA with a polarization of 0.77. The source was very reliable operating well over 0.95 of the time. Swenson et al. described polarization diagnostics for the LAMPF ion source including a low energy ^{6}Li polarimeter for the measurement of the nuclear polarization and Faraday rotation experiment for the measurement of the density and polarization of the optically pumped alkali vapor charge exchange target. At the round table discussion there was a general optimism that optically pumped ion sources can be improved to yield substantially more current than is currently available from polarized ion sources.

V. OTHER OPTICAL PUMPING RESULTS

There were three contributors to the using optical pumping for interesting experiments that did not fit into previously discussed subjects. In the first of these, Tanaka et al. discussed the development of an optically pumped polarized ion source for heavy ions with the immediate goal of producing polarized $^{3}He^{+}$ ions. Myers et al. reported on the status of the optically pumped $^{6,7}Li^{-}$ ion source at Florida State University for use with a tandem-linac accelerator. This ion source produces currents greater than 50 particle nA on target. Finally Voytas et al. reported the polarization by optical pumping of ^{21}Na. The ^{21}Na is produced in a high pressure ^{20}Ne cell by the ^{20}Ne(d,n)^{21}Na reaction. Polarizations of 0.63 have been achieved for the ^{21}Na.

VI. MACHINE TARGET INTERACTIONS

Finally there was a session on the very important subject of machine target interactions. A study of storage cell induced background at NIKHEF was reported by de Jager. Sowinski discussed the interactions between an internal polarized ^{3}He target and the polarized proton beam at the IUCF Cooler Ring. He demonstrated that very low backgrounds are possible. In order to obtain the very low backgrounds careful design of the target is essential. The optimum luminosity of polarized gas targets in storage ring was studied by V. Przewoski. In particular the influence on the ion beam lifetime in the storage ring due to a restriction, such as a storage cell, to the size of the ion beam was studied. Kinney reported on the depolarization induced by the interaction between the magnetic fields induced by intense electron pulses and the polarized H or D in a storage cell. He found that a strong magnetic field can ameliorate the depolarization and that care must be taken to avoid values of the holding field at which atomic transitions are equal to integral multiples of the circulating resonance frequency.

VII. CONCLUSION

The workshop on Polarized Ion Sources and Polarized Gas Targets had papers on most of the subjects of importance in the field. In addition there were stimulating round table discussions on various subjects. The workshop was supported in part by grants from the National Science Foundation and the International Committee on High Energy Spin Physics.

LARGE, ACCESSIBLE, HIGHLY POLARIZED FROZEN-SPIN SOLID HD TARGETS*

A. Honig[1], Q. Fan[1], X. Wei[1], M. Rigney[2+], A. M. Sandorfi[3] and C. S. Whisnant[4]

[1]*Physics Dept., Syracuse University, Syracuse, NY 13244*
[2]*IN2P3, IPN Orsay, France*
[3]*LEGS Project, Brookhaven National Laboratory, Upton, NY 11973*
[4]*Physics Dept., University of S. Carolina, Columbia, SC 29208*

ABSTRACT

New advances in frozen-spin polarized solid HD make practical the use of large targets of independently polarizable H and D for experiments in fusion, photonuclear and particle physics. Long polarization-retention times of about a year at 1.5K and 7.5 Tesla allow multiple target production and long term storage in economical cryostats, for off-the-shelf availability of polarized HD material and targets. Polarization-retention times at sub-Tesla fields, which exceed a week, provide a fair match between polarization-production time and target-utilization time in moderate fields, for weakly ionizing beam fluxes in which radiation damage to the target is small. Because of the convenient 1.5-4K frozen-spin temperature range, cold-transfer (4K) portable apparatuses, which have already been constructed and used, can move the targets readily from polarization-production systems to storage or in-beam cryostats with negligible polarization loss. This permits total separation of polarization-production and target-utilization apparatuses, and the possibility of assembling very large targets.

INTRODUCTION

Solid HD polarized targets have favorable nuclear properties, which include highest free proton ratio compared with currently operational targets, independent polarization and polarization reversal of proton and deuteron, allowing a polarized neutron target, and high polarization values, comparable with the best among targets in present use. These targets were first proposed more than 25 years ago[1], and were among the earliest of the frozen-spin type. Unlike most

---

* Work supported by NSF, through the Univ. of S. Carolina, and by the U. S. Dept. of Energy, under contract #DE-AC02-76-CH00016.

+ Presently working at Syracuse University.

current frozen-spin targets, where the target is dynamically polarized via electron-nucleon coupling at T~1K, followed by reduction of T to ~100mK to increase the passive polarization retention time, these HD targets are prepared at low mK temperatures, but after an aging period, long polarization retention occurs at 1.5-4K temperatures. Because of this accessible frozen-spin region, complete separation of polarization production and utilization apparatuses can be effected through use of recently developed portable cold-transfer (4K) apparatuses[2], which extract and insert polarized targets between polarization-production systems, storage systems, and in-beam cryostats without appreciable loss of polarization. Thus, in addition to the excellent nuclear properties, the HD targets also provide important configurational advantages.

The high polarizations in the HD are attained at very low mK temperatures and at high magnetic fields through thermal equilibrium with a cold "lattice", after which the spin-lattice relaxation mechanism is "switched off" through a process of conversion of ortho-H_2 and para-D_2 impurities (which have rotational angular momentum and thus act as relaxation catalysts during polarization production) to respectively para-H_2 and ortho-D_2 (which have no rotational angular momentum and cannot facilitate relaxation of the H and D in the HD molecules). The advantages have already been noted, but a clear disadvantage is that the aging process is a long one, since the conversion rates[3] of ortho-H_2 and para-D_2 in HD are 0.16 d^{-1} and 0.055 d^{-1}, respectively. Typically, about 40 days aging in the polarization-production process is required to achieve sufficiently long relaxation (depolarization) times at 1.5K operating conditions, which in conjunction with multiple target production match polarized target production and utilization times. These times are also sensitive to the value of the surrounding magnetic field which the experiment may require. Prior to this past year, relaxation times longer than several days had never been obtained, and it was problematical if times of the order of months were feasible with any amount of aging, since the possibility that HD might have an intrinsic relaxation rate which would cut in at some point could not be excluded. During this past year, spurred by the potentially very advantageous application of these targets to photonuclear experiments with polarized gammas obtained from back scattering of laser photons from an electron synchrotron beam[4], an effort to obtain very long relaxation times was undertaken and the successful results were reported this past summer[5]. In effect it was found that at typical experimental usage fields of about 1 Tesla, an approximate match of polarization-production time to polarization utilization time of a target could be obtained if 4 targets were simultaneously polarized, and were employed one at a time for about a week while the remaining polarized targets were stored in a 1.5K cryostat in a magnetic field of about 7.5T or larger, where their relaxation times exceed a year. With such long polarization retention times in storage, one effectively has off-the-shelf polarized HD targets. In the photonuclear experiments contemplated as thc first nuclear target application[4], the gamma beam flux is less than 10^7 photons/cm^2s, and the relaxation time degradation over a 40 day period due to radiation damage is negligible. For stronger ionizing beams, radiation damage[6] could result in a mismatch between the polarization-production and usage times, unless faster polarization-production methods and/or use of very large beam-rastered targets are developed. There exist several interesting ideas on how to accomplish both of these objectives. In this paper, we briefly review previously described HD polarization methods[7], give both attained polarization results as well as projected results with a new apparatus which will soon be in service, review the cold-transfer methodology[2], and present new results which indicate increased longevity of the polarization in the usage mode compared with the polarization retention times reported only a few months ago[5].

REVIEW OF POLARIZATION METHOD

Only a brief account is provided here, since the solid HD polarization method has been described in considerable detail in previous publications[5,7]. The basic polarization results from thermally equilibrating the nuclear spin systems at mK temperatures and high magnetic fields. For small targets, such as inertial confinement fusion target shells, or larger targets with an abundance of cooling wires in contact with the HD, our refrigerator can cool the HD to 9 mK, which together with the 13 Tesla magnet in our present system, yields polarization limits derived from thermal equilibrium for H and D of 91% and 29%, respectively. P^H, the polarization value of the protons, is satisfactory, but the thermal equilibrium value for P^D is insufficient for many applications. This is overcome by employing a nucleon-nucleon dynamic polarization technique[8] in which 'forbidden' transitions among intermolecular dipolar-coupled D and H spin states are driven by radio-frequency energy through an adiabatic rapid passage, which reverses populations of the coupled states and results in polarization transfer from H to D. This is carried out at low field (about 0.03T) for reduced 'forbiddenness'. At 100% efficiency, P^D would attain 2/3 of the original P^H, while the latter would become -1/3 of its original value. The H can subsequently be repolarized, with the D relatively unaffected since we usually choose D thermal relaxation to be very weak. Thus far, only 50% efficiency has been achieved with this protocol, resulting in a transfer of 1/3 of the original P^H to P^D. However, the polarization transfer cycle is repeatable, and D polarizations appreciably above the thermal equilibrium limit have been achieved. The other important consideration is the thermal polarization of the H. The relaxation time, $T_1{}^H$, must be reasonably short (a few days, for example) at the polarizing conditions, and subsequently at the end of the polarization process, it must be very long at target operating conditions. $T_1{}^H$ is known to be controlled by the concentration, $c_1{}^H$, of ortho-H_2 impurities, which are metastably in the molecular J=1 rotation state due to symmetry requirements on the homonuclear molecules, and in HD at low temperatures have a temperature-independent decay rate to the J=0 rotational state of $0.16d^{-1}$. The ortho-H_2 nuclear spins relax rapidly through spin-rotation and rotation-lattice mechanisms, and since they precess at the same frequency as the H nuclei on the HD molecules, they provide through cross-relaxation a source of relaxation of the HD, whose relaxation rate in the low ($<10^{-3}$) c_1^H regime decreases monotonically, but not necessarily linearly, with decreasing c_1^H. Similar arguments apply to para-D_2 molecules, with the D thermal relaxation rate in HD dependent on c_1^D (and weakly on c_1^H, as well). However, as noted above, one often keeps c_1^D intentionally very low to curtail its thermal relaxation rate, obtaining a P^D higher than its thermal equilibrium value from dynamic polarization transfer from P^H. There are a variety of polarization options, but they all have common elements. The initial c_1^H is almost always between 3 and 4×10^{-4}, with a corresponding T_1^H initial value at 1.5K and 0.3T of the order of 10 seconds and an initial T_1^H at polarizing conditions, near 15 mK and 15T, of a few days. The initial c_1^D varies from less than 10^{-4} to about 5×10^{-4}, depending upon the P^H and P^D retention times desired at the end of the polarization production process. After polarization by a combination of thermal relaxation and polarization transfer cycles, the HD is aged for a period between about 15 days, if only P^D is desired, as in a fusion experiment, to up to 50 days, if very long P^H and P^D retention times are desired, as for the targets we are now preparing for photonuclear experiments[4].

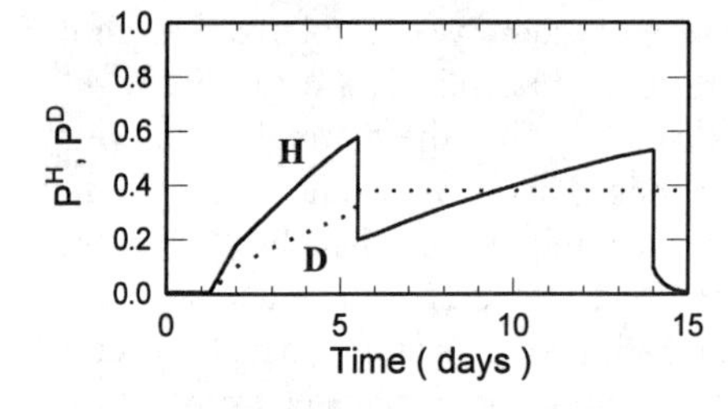

Fig.1a. Actual experimental time-line of polarization process in a 12 Tesla field and at a temperature of 9mK, for solid HD when only D polarization is to be preserved at 1.5K and moderate holding magnetic field. At day 5.5, polarization transfer from H to D via "forbidden" rf-induced transitions takes place.

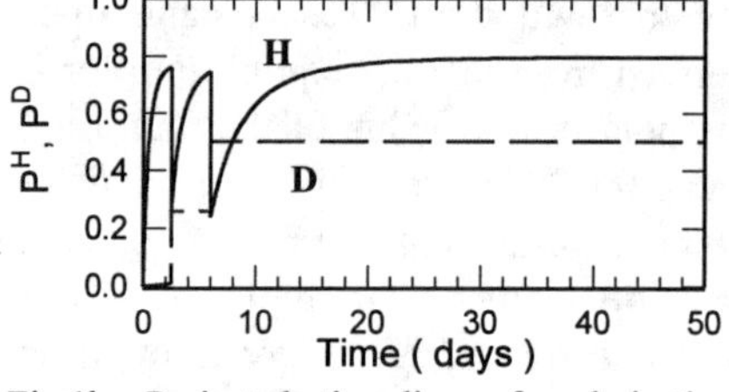

Fig.1b. Projected time-line of polarization process in a 20 Tesla field and at temperatures decreasing approximately linearly in time from 20mK at t=0 to 12mK at t=10 days, for solid HD when both D and H polarizations are to be preserved at 1.5K and moderate holding magnetic field. At days 2.5 and 6, polarization transfer from H to D via "forbidden" rf-induced transitions takes place.

In Fig. 1a, we illustrate a polarization cycle for an HD sample of about 0.014 moles, in which only P^D is of interest[9], using a starting c_1^H of approximately 3×10^{-4} and a starting c_1^D of about 4×10^{-4}. These c_1's decay exponentially at the conversion rates given earlier, resulting in continuously increasing T_1's. At day 1, the temperature is lowered to 10 mK and the field raised to 12 Tesla. Both D and

H polarizations grow at their respective relaxation rates. At day 5, the field is lowered to 0.03T, the adiabatic rapid passage is executed in about 2 minutes, and the field is then restored to its high value for H repolarization (even though not of interest) and aging, to insure an appreciable P^D retention time after the polarization. For this sample, the final relaxation time of the P^D at 1.5K and in a 7 Tesla field was over 8 hours, sufficient for a single fusion shot experiment, but not for a polarized D target beam experiment. The final D polarization was 38%, but by small changes in the starting c_1's and using the full 13 Tesla field our magnet can provide, a P^D of 55% should be attainable. The 10 mK HD temperature during the early polarization buildup was possible despite the heat of conversion generated in the sample by the c_1^H and c_1^D decays (2.6μW and 0.46μW respectively per milli-mole), since we used an abundance of copper cooling wires, more than would be permissible in a nuclear physics target experiment where the unpolarized cooling wire nucleons would dilute the benefits of the polarized H and D nucleons. In Fig. 1b, a projected polarized target preparation in a 20T magnet system is illustrated. The assumed sample is still 0.014 moles, but only a limited number of gold-coated aluminum cooling wires, whose total nucleon content is <15% of the number of protons or deuterons in the target, is employed. Here, the temperature of the HD exceeds the near-base temperature of the dilution refrigerator mixing chamber by about 10mK, as has been demonstrated by direct measurements with NMR thermometry. We start with c_1^H of about 4×10^{-4} and $c_1^D<10^{-4}$, leading to virtually no thermal relaxation of the D. At days 2 and 5, rf induced polarization transfer builds up the D polarization at the expense of the P^H. Only two cycles are feasible because of the increasing T_1^H as c_1^H decreases through conversion. Higher initial c_1^H is not a cure for this since it results in too short a T_1^H at the polarization-transfer field-temperature conditions. The narrow range of initial c_1^H for which this method works effectively is a feature of these targets. After the extended 50 day aging period shown in Fig. 1b, the relaxation times at 1.5K are very long, ranging from several days to over a year, depending on the magnetic field at which the target is kept. The polarization-retention times are at the heart of this polarized target enterprise. They have been separately determined, and are discussed in the next section. Before we turn to that, we present Table I which gives expected D and H polarizations under various assumptions. The various rows listed in the tables correspond to different possible scenarios of target improvement. The top two rows of each of the tables reflect currently established performances, except for larger magnetic fields than we have thus far used, for which the modest scaling should be quite safe. The other rows give expected polarizations for specific improvements which we believe are achievable. These are respectively lower Kapitza resistance using improved cooling wire geometry, improved efficiency of adiabatic rapid passage from the present value of 50% to 75% (even 100% cannot be excluded), and careful adjustment of c_1^D to establish some thermal P^D before

Table I **A. Only $^{\uparrow}$D Polarization of Interest**

B (Tesla)	T (mK)	Forbid. Transit. Effic. (%)	$P^D_{thermal}$ (1st cycle) (%)	P^D (%)	P^H (%)	T_1^D(day) at 1.5K		T_1^H(day) at 1.5K		Polariz. Produc. Time (day)	
						1.0T	8.0T	1.0T	8.0T		
20	20-15	50	18	65		0.5	1			15	a
17	20-15	50	14	58		0.5	1			15	a
20	15-12	50	24	75		0.5	1			15	a
17	15-12	50	21	70		0.5	1			15	a
20	20-15	75	18	78		0.5	1			15	a
17	20-15	75	14	72		0.5	1			15	a
20	20-15	50	18	66		10	100			50	b
17	20-15	50	14	58		10	100			50	b

B. Only $^{\uparrow}$H Polarizations of Interest

B (Tesla)	T (mK)	Forbid. Transit. Effic. (%)	$P^D_{thermal}$ (1st cycle) (%)	P^D (%)	P^H (%)	T_1^D(day) at 1.5K		T_1^H(day) at 1.5K		Polariz. Produc. Time (day)	
						1.0T	8.0T	1.0T	8.0T		
20	20-15	50			84			5	>300	50	c
17	20-15	50			78			5	>300	50	c
20	15-12	50			90			5	>300	50	
17	15-12	50			87			5	>300	50	

C. $^{\uparrow}$D and $^{\uparrow}$H Polarizations Retained

B (Tesla)	T (mK)	Forbid. Transit. Effic. (%)	$P^D_{thermal}$ (1st cycle) (%)	P^D (%)	P^H (%)	T_1^D(day) at 1.5K		T_1^H(day) at 1.5K		Polariz. Produc. Time (day)	
						1.0T	8.0T	1.0T	8.0T		
20	20-15	50	0	50	79	15	>500	5	>300	50	c
17	20-15	50	0	46	73	15	>500	5	>300	50	c
20	15-12	50	0	57	85	15	>500	5	>300	50	d
17	15-12	50	0	52	83	15	>500	5	>300	50	d
20	20-15	75	0	70	79	15	>500	5	>300	50	e
17	20-15	75	0	64	73	15	>500	5	>300	50	e
20	20-15	50	10-15	60-65	80	15	>500	5	>300	50	f
17	20-15	50	8-12	54-58	73	15	>500	5	>300	50	f

a. For short retention of P^D.
b. For long retention of P^D.
c. Under demonstrated feasibility conditions, in Janis Vari-Temp experimental cryostat.
d. Assuming improvement in conduction wire geometry and reduced interface thermal resistance, thereby lowering target temperature.
e. Assuming increase in r.f. induced "forbidden" transition efficiency.
f. Assuming P^D growth via relaxation in first cycle, to between 1/2 and 3/4 equilibrium value.

Note: Each improvement is considered to act separately in these Tables. For a combination of improvements, the resultant polarization is higher, but not a simple sum of the individual polarization improvements.

the first dynamic polarization transfer, without sacrificing significant polarization-retention-time at the end of the preparation. We now turn to the recent study of the relaxation times under potential usage conditions.

POLARIZATION RETENTION TIMES

We have seen how the key to efficient use of polarized solid HD for nuclear physics target experiments depends upon matching the production time to the target utilization time, and that these times are of the order of month. The early results of a feasibility study to investigate whether such long relaxation times are indeed attainable were reported in Ref. 5. In Figs. 2a and 2b, we extract some of those results, illustrating respectively growths of T_1^D and T_1^H at 1.5K and

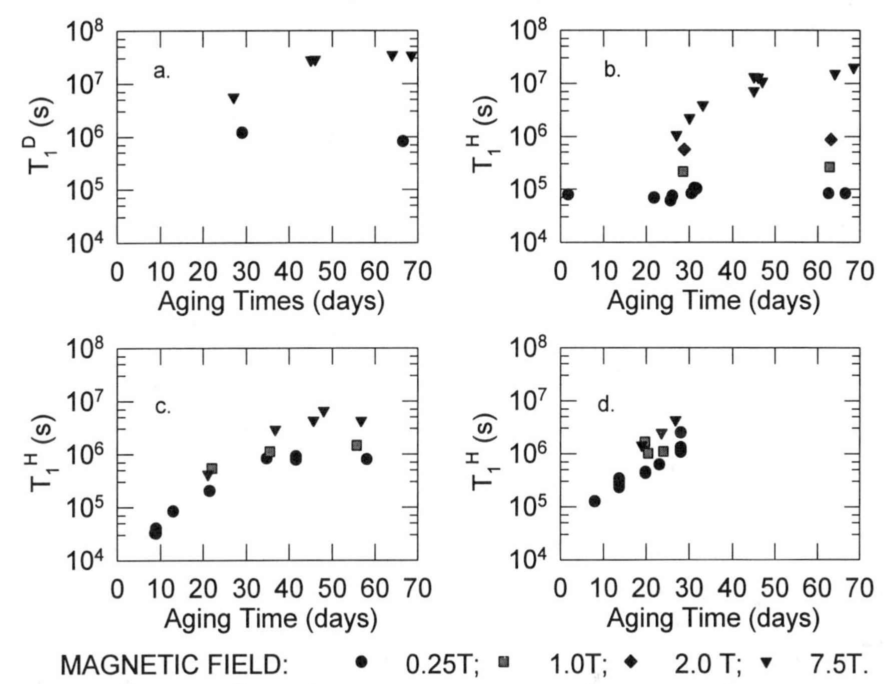

Fig. 2. Growth of T_1^D and T_1^H under continuous aging: a) T_1^D growth, in Janis Vari-Temp cryostat at 1.5K. b) T_1^H growth, in Janis Vari-Temp cryostat at 1.5K. c) T_1^H growth, in dilution refrigerator at 1.5K. d) T_1^H growth, in dilution refrigerator at 0.45K. (See text for description of HD target environments for a, b, and for c, d)

at several values of the magnetic field, from a starting c_1^H of about 4×10^{-4}, under continuous aging in a 1.5-4.2K environment. In this case, the test apparatus was a Janis Vari-Temp cryostat, and our condensed HD sample was about 1.5 cm long and located at the bottom of a 4 mm i.d. Pyrex glass tube, with no cooling wires.

The T_1's reach limiting values where they do not increase with further aging. These values, even though unlikely to be due to an intrinsic HD relaxation mechanism[10] because of the T_1 dependence on T and B, are still sufficiently long for the initially contemplated photonuclear experiments. In Figs. 2c and 2d, the growth of T_1^H is shown for comparable aging of the same sample in the dilution refrigerator. In this case, the principal differences in sample environment are the presence of cooling wires, and the isolation of the solid target from its vapor column and from background infrared radiation. T_1^H at 7.5T is somewhat reduced from its values in Fig. 2b, possibly due to the wires, but very gratifyingly, T_1^H at the sub-Tesla fields is considerably increased. At these fields, there is also a temperature dependence, absent in the limiting state for the Vari-Temp cryostat experiments (Figs. 2a and 2b), which opens the possibility for considerably improved performance with a liquid ^{3}He evaporation-cooled in-beam cryostat. Details of the relaxation studies, which extend down to about 20 mK, will be published elsewhere.

TARGET CRYOSTAT CONSIDERATIONS

Target mounting in the dilution refrigerator is shown in Fig. 3. Similar mountings will be used in the storage and in-beam cryostats. Basic cold-transfer technology and engagement-disengagement procedures have already been implemented[2] in connection with polarized D inertial confinement fusion investigations. Adaptation to these larger targets is straight forward., and preliminary designs have been presented[11] recently. A large holding-field magnet in the 77K shroud can be provided by a copper-wound solenoid. However, a bismuth-based HTC superconducting tape[12] could eliminate the several watts power dissipation expected from the copper coil.

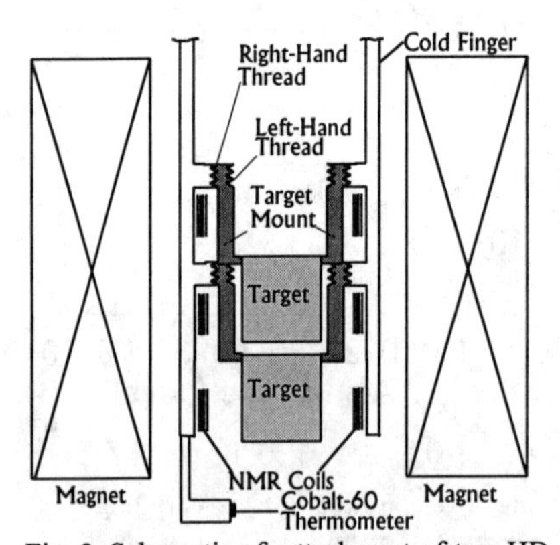

Fig. 3. Schematic of attachment of two HD targets (4 will actually be used) to cold finger of dilution refrigerator. Left-right hand threaded target holders allow engagement and disengagement from cold finger, using appropriately matched threads on cold-transfer apparatus (not shown).

The in-beam cryostat will have provision for polarization reversal, using both adiabatic fast passage of the allowed transitions for H, and field rotation combined with the latter for D, since the efficiency of allowed adiabatic rapid passage for H is about 98%, while for D it is only about 85%, although possibly amenable to improvement.

SUMMARY

Frozen-spin HD polarized targets operating between 0.4 and 4K, used with cold-transfer (4K) techniques, provide great configurational flexibility. Their long depolarization times under target usage conditions assure reasonable match between polarization production and usage times, for weakly ionizing beam fluxes, and the very long relaxation times at fields above 7T (~1 yr.) provide an economical storage mode and "off-the-shelf" availability.

REFERENCES

1. A. Honig, Phys. Rev. Lett. 19 (1967) 1009.
2. N. Alexander, J. Barden, Q. Fan and A. Honig, Rev. Sci. Instrum. 62 (1991) 2729.
3. H. Mano, PhD thesis, Syracuse University, 1978 (unpublished).
4. LEGS-Spin Collaboration, C. S. Whisnant *et al.*, "Using SPHICE: A Strongly Polarized Hydrogen and Deuterium Ice Target", paper presented at this conference.
5. A. Honig, Q. Fan and X. Wei, A. M. Sandorfi and C. S. Whisnant, "New Investigations of Polarized Solid HD Targets", paper presented at the 7th International Workshop on Polarized Target Materials and Techniques, Bad Honnef, Germany, June 20-22, 1994. To be published in Nucl. Inst. Meth.
6. H. Mano and A. Honig, Nucl. Inst. Meth. 124 (1975)1.
7. A. Honig, N. Alexander, and S. Yucel, Workshop on Muon-Catalyzed Fusion, Sanibel Island, FL., May 2-6, 1988. eds. S. E. Jones, J. Rafelski and H. J. Monkhorst. AIP Conf. Proc. No. 181 (1989) 199.
8. A. Honig and H Mano, Phys. Rev. B14 (1976) 1858.
9. A. Honig, N. Alexander, Q. Fan, X. Wei, Y. Y. Yu, Workshop on Polarized Ion Sources and Polarized Targets, Univ. of Wisconsin, Madison, May 23-27, 1993; eds: L. W. Anderson and Willy Haeberli. AIP Conf. Proc. No. 213 (1994) 50.
10. Mechanisms which can bring about the saturated values of the T_1's include wall relaxation, modified rotational energy levels at surfaces and walls, vapor phase interaction, small amounts of dissolved oxygen, defects in the HD solid, and others.
11. M. Rigney *et al.* "Solid HD Polarized Target: Conceptual Design and Transport and In-Beam Cryostat", paper presented at the 7th International Workshop on Polarized Target Materials and Techniques, Bad Honnef, Germany, June 20-22, 1994. To be published in Nucl. Inst. Meth.
12. A workable solenoid, based on multifilament BSCCO2223 tape, is available from IGC Advanced Superconductors, Waterbury, CT.

Measurements of the Spin Structure of the Nucleon Using SPHICE: A Strongly Polarized Hydrogen and Deuterium Ice Target*

LEGS-Spin Collaboration

D. Babusci[2], M. Blecher[8], M. Breuer[4], A. Caracappa[1], C. Commeaux[4], J.-P. Didelez[4], Q. Fan[7], G. Giordano[2], K. Hicks[3], S. Hoblit[9,1] P. Hoffmann-Rothe[4], A. Honig[7], O. C. Kistner[1], M. Khandaker[8,1], Z. Li[10], M. A. Lucas[6], G. Matone[2], L. Miceli[1,6], B. M. Preedom[6], M. Rigney[4], A. M. Sandorfi[1], C. Schaerf[5], C. E. Thorn[1], X. Wei[7], and C. S. Whisnant[6]

[1]*Brookhaven National Laboratory, Upton, NY 11973*
[2]*Laboratori Nationali di Frascati-INFN, Rome, Italy*
[3]*Ohio University, Athens, OH 45701*
[4]*IN2P3, IPN Orsay, 91406 Orsay-Cedex, France*
[5]*Universita di Roma Tor Vergata and INFN Sezione di Roma 2, Rome, Italy*
[6]*University of South Carolina, Columbia, SC 29208*
[7]*Syracuse University, Syracuse, NY 13244*
[8]*Virginia Polytechnic Institute and State University, Blacksburg, VA 24061*
[9]*University of Virginia, Charlottesville, VA 22901*
[10]*Christopher Newport University and CEBAF, Newport News, VA 23606*

The partition of the total nucleon spin into the intrinsic and orbital components of its constituents is crucial to models of hadron structure. The experimental tests of a model of the spin structure come from measurements of helicity amplitudes and these have received considerable attention in recent years. Through the spin-dependent polarizability of the nucleon, γ, and the Gerasimov-Drell-Hearn (GDH) sum rule integrals, the helicity amplitudes reflect the underlying spin structure of the nucleon. The combination of the LEGS photon beam, the SPHICE frozen spin HD target and the large solid angle SASY (Spin ASYmmetry) detector array is uniquely suited for measurements that will make significant impacts on these quantities.

Using circularly polarized photons with a longitudinally polarized HD target, the total cross sections for one- and two-pion production from threshold to 470 MeV will be measured. From this data a direct measure of the spin-dependent sum rules can be constructed for both the proton and neutron. Estimates of the GDH sum rule for both the proton and neutron using current pion multipole analyses are 40% different from the sum rule predictions. This discrepancy is almost entirely due to the Vector-Scalar (VS) component. In sharp contrast, the VS contribution to g is well reproduced by 1-loop chrial perturbation theory calculations. Direct measurement of spin observables is required to resolve this discrepancy.

*Supported by DOE contract DE-AC02-76CH00016 and grants from NSF, INFN and IN2P3.

THE SPIN STRUCTURE OF THE NUCLEON

At low photon energies, the spin-dependent part of the forward Compton scattering of real photons from a spin $\frac{1}{2}$ particle, $f_2(\omega)$, can be expanded in powers of the energy to give

$$f_2(\omega) = \frac{1}{4\pi}\left(-\frac{2\pi\alpha}{M^2}\kappa^2\right)\omega + (\gamma)\omega^3 + O(\omega^5) \quad (1)$$

The first term containing the anomalous magnetic moment of the nucleon, κ, is the low-energy limit due to Low[1] and Gell-Mann and Goldberger[2]. In 1966, Gerasimov[3] and, independently Drell and Hearn[4] found the sum rule

$$GDH = \int_{\omega_0}^{\infty} \frac{\sigma_{1/2} - \sigma_{3/2}}{\omega} d\omega = -\frac{2\pi^2\alpha}{M^2}\kappa^2 \quad (2)$$

relating the integral of the total spin-dependent cross sections measured with the photon and nucleon polarizations parallel, $\sigma_{3/2}$, and anti-parallel, $\sigma_{1/2}$, to this term.

By analogy with the corresponding term in the spin-independent part of the amplitude, γ, is called the spin-dependent polarizability. The nucleon polarizabilities also obey sum rules[5] that are constructed from the helicity dependent cross sections. In particular, the spin-dependent polarizability is given by

$$\gamma = \frac{1}{4\pi^2} \int_{\omega_0}^{\infty} \frac{\sigma_{1/2} - \sigma_{3/2}}{\omega^3} d\omega \quad (3)$$

The sum rule for γ has two very important features. First, the denominator in the integral contains the third power of the photon energy. This means a rapid convergence, limiting the range of energies that must be spanned and a reduction of the experimental uncertainties in its evaluation. Secondly, the polarizabilities are exactly calculable in chiral perturbation theory (χPT) from the charged pion mass, the axial vector coupling constant and weak pion decay constant[5]. As yet, there is no direct verification of this prediction for this fundamental nucleon property.

The first two coefficients in the low energy expansion of $f_2(\omega)$ depend on $\sigma_{1/2} - \sigma_{3/2}$ and thus, both reflect the underlying spin structure of the nucleon. A valid description of this structure must account for both sum rules.

MULTIPOLE ESTIMATES OF THE SUM RULES

In absence of direct observation of the helicity-dependent cross sections, it is possible to estimate the spin-dependent polarizability and the GDH sum rule integrals using existing multipole analyses constructed from (mostly unpolarized) measurements of the different charge channels in meson production. Sandorfi, Whisnant and Khandaker[6] have recently made such estimates using the FA93 N(γ,π) multipoles of Workman and Arndt[7]. To do this, an isospin decomposition of the integrals is made, producing three components: the isovector (VV), isoscalar (SS) and mixed (VS) terms.

Assuming that single-π production is the dominant process, the $\frac{3}{2}$ and $\frac{1}{2}$ helicity total cross sections for the proton and neutron are computed. Following the earlier estimates of Karliner[8], the known $\pi\pi$N/πN branching ratios of N* resonances have been used to estimate 2π photo-production.

Sandorfi *et al.* find that the multipole estimates γ for the proton and neutron are reasonably close to the relativistic 1-loop χPT calculations of Bernard *et al*[5]. In particular, the $VS = \frac{1}{2}(proton - neutron)$ term is within 8% of the calculated value. Although corrections for the Δ are large at the 1-loop level, they are the same for both proton and neutron and do not affect the VS term. However, the multipole estimates for both the proton and the neutron, are 40% different from the full GDH sum rule predictions. This discrepancy is almost entirely due to the VS contribution which differs in magnitude from the magnetic-moment value by a factor of 4, and is of the opposite sign. Almost $\frac{2}{3}$ of the GDH and 90% of γ integrals are saturated in integrating up to 500 MeV because of their respective energy weighting and the dominant effect of the Δ. This is the energy region containing the greatest concentration of published measurements and, thus, is precisely where multipole analyses would be expected to be the most reliable. Unless there is considerable missing strength at high energy, there are only two other possibilities: Either (a) *both* the 2-loop corrections to γ are large *and* the existing multipoles are wrong, *or* (b) modifications to the GDH sum rule are needed to fully describe the isospin structure of the nucleon. Direct measurement of the helicity dependent cross sections is required to resolve this question.

MEASURMENTS USING THE SPHICE TARGET AT LEGS

SPHICE, a **S**trongly **P**olarized **H**ydrogen and Deuterium **ICE** target, represents a new technology utilizing molecular HD in the solid phase[9]. This target offers several advantages for such an experiment. Because the only polarized species in this target are hydrogen and deuterium and there is only 15% unpolarized material (by weight), much smaller backgrounds are expected for many reactions, making it possible to measure not only asymmetries but absolute cross sections. At present there are no absolute cross section data taken with polarized targets. SPHICE will provide the first opportunity, and the ability to make proton and neutron measurements simultaneously minimizes systematic uncertainties. This is crucial since the key physics questions require a determination of helicity cross sections for the proton-neutron difference. The dilution factors are quite advantageous: $\frac{1}{2}$ of all the protons are usefully polarized and all the deuteron (neutrons) are polarized. Since the polarizing facility is not in-beam, the in-beam dewar is simplified, making large solid angle available. The proton polarization (80%) and deuteron polarization (50%) are comparable to that achieved with conventional materials. Using rf techniques, the H and D spins can be oriented independently allowing all beam/target spin combinations to be explored, again having a major impact on systematic errors.

A double polarization experiment is characterized by a quality factor of $(P_\gamma f_t)^2$, where P_γ is the γ-ray polarization and f_t is the target dilution factor. The laser-backscattered beams available at LEGS produce circular polarizations of ≈100%, while ≈50% can be achieved with a bremsstrahlung beam. The backscattered beam thus represents an improvement of a factor of 4 over a bremsstrahlung beam. The resulting enhancement of the backscattered-beam/SPHICE-target compared to the bremsstrahlung/ammonia-target combination is an order of magnitude for experiments on protons and over a factor of 40 for neutrons. These represent dramatic potential gains in data quality.

SASY, the **S**pin **ASY**mmetry array, a high efficiency, large solid angle detector for measuring total reaction cross sections, is being constructed for these experiments[11]. This detector will provide complete determination of angle, energy, and particle identity for all reactions induced by photons on hydrogen and deuterium for incident γ-ray energies up to 471 MeV (the maximum at LEGS). The ability to completely identify all reaction channels will allow angular distributions to be measured separately for each individual channel. Such measurements will provide important new constraints on photoproduction amplitudes.

At the center of the detector, the SPHICE target will be surrounded by cylindrical wire chambers inside a pair of superconducting Helmholtz coils with a diameter of 32 cm. These coils will produce a uniform 1.3 Tesla holding field at the target. This field is also used to determine the sign of charged pions. The wire chambers will cover scattering angles from 40° to 140°. A plastic scintillator box surrounding these chambers will serve as a trigger, as well as providing an energy-loss measurement. Around this entire assembly will be the Crystal Box[12], containing 432 NaI(Tl) crystals providing good efficiency for g-ray detection.

At the open front end (small scattering angles) of the Crystal Box, a stack of large area detectors will cover the range from about 30° to 10°. The first layer consists of three planes of multi-wire proportional chambers. At a distance of 140 cm from the target an array of 160 x 160 x 20 cm plastic scintillator bars will detect protons, pions, and neutrons. Behind the bars a wall of 136 lead glass blocks will determine photon energies and angles. At backward angles, the opening in the Crystal Box will be sampled by 12 phoswich detectors.

To estimate the yields from reactions on $\vec{H}$ that can be anticipated in these measurements (or yields from neutrons in $\vec{D}$), we use the typical LEGS flux of 10^5 s^{-1} in a 5 MeV tagging interval of photon energy. The density of the solid HD is 0.15 g/cc and the cell length is 7.0 cm. This represents a free-proton thickness of 0.35 g/cm^2. Assuming SASY covers of 80% of 4π and a data acquisition live-time of 80 %, 1% statistical accuracy can be achieved in 10 days for a reaction channel with a total cross section of 1 μb. The dominant reaction channel in the LEGS energy range, single-π production, has cross sections that are 2 orders of magnitude larger. The accuracy of these measurements will clearly be dominated by systematic uncertainties.

The FA93 amplitudes have been used to estimate the polarization-difference cross section $\sigma_{1/2} - \sigma_{3/2}$ that appears in the integrand of both γ and the GDH sum rules. Five days of running will reduce these errors below 1% at all energies. To measure all target/beam spin combinations, target full and empty with the four LEGS operating modes required to span the full energy range will take two months for measurements on $\vec{H}$. The corresponding time required for measurements on the neutron is five months, due to the lower deuteron polarization.

1) F. E. Low, *Phys. Rev.* **96**, 1428-1432 (1954).
2) M. Gell-Mann and M. L. Goldberger, *Phys. Rev.* **96**, 1433-1438 (1954).
3) S.B. Gerasimov, *Sov. J. Nucl. Phys*. **2**, 430-433 (1966).
4) S. D. Drell and A.C. Hearn, *Phys. Rev. Lett.* **16**, 908-911 (1966).
5) V. Bernard, N. Kaiser, J. Kambor and Ulf-G. Meiβner, *Nucl. Phys.* **B388**, 315-345 (1992).
6) A. M. Sandorfi, C. S. Whisnant and M. Khandaker, BNL int. rep. 60616, submitted to Phys. Rev. **D**.
7) R. L. Workman and R. A. Arndt, private comm.; the **S**cattering **A**nalysis **I**nteractive **D**ial-in (SAID) program, available by TELNET to VTINTE
8) I. Karliner, *Phys. Rev.* **D7**, 2717-2723 (1973).
9) A. Honig, Q. Fan, X. Wei, A. M. Sandorfi and C. S. Whisnant, "New Investigations of Polarized, Solid HD targets", presented at the 7th Workshop on Polarized Target Materials and Techniques, Bad Honnef, Germany, June 20-22, 1994. To be published in *Nucl. Inst. Meth.*
10) S. L. Wilson, *et. al, Nucl. Inst. Meth*. **A264**, 263 (1988).
11)"Measurements of the Spin Structure of the Nucleon.", BNL int. rep., 1994.

A Statically Polarized Solid ^{3}He Target

D. G. Haase*, C. D. Keith*, C. R. Gould*, P. R. Huffman+, N. R. Roberson+, M. L. Seely*, and W. S. Wilburn+

**North Carolina State University, Raleigh, North Carolina 27695-8202 and the Triangle Universities Nuclear Laboratory, Durham, North Carolina 27708-0308*
+Duke University, Durham, North Carolina 27708-0308 and the Triangle Universities Nuclear Laboratory, Durham, North Carolina 27708-0308

Abstract. We describe the construction and operation of a large (0.4 mole) solid ^{3}He target polarized to 37% by the "brute force" technique at 12 mK in a 7 Tesla magnetic field. The target has been employed at the Triangle Universities Nuclear Laboratory to study by neutron transmission measurements the excited states of the ^{4}He nucleus. Such a target has particular advantages for experiments using neutrons or other beams which produce little target heating.

POLARIZED CONDENSED ^{3}He

^{3}He has been successfully polarized in the gas phase by the technique of optical pumping(1). Such a target can operate in a relatively low magnetic field and near room temperature. The diminished thickness of the target is a drawback for some experiments, as is the necessity of large lasers to provide the pumping photons. A solid ^{3}He target can be made much thicker and is statically polarized - the polarization can be calculated from the temperature and the local magnetic field. For temperatures above 5 mK the polarization of solid ^{3}He (bcc crystal structure) is given by

$$P = \tanh\left\{\frac{1}{kT}\left(\Theta P + K P^3 + \mu B_{ext}\right)\right\}$$

where the magnetic moment of the ^{3}He nucleus and the external magnetic field are μ and B_{ext} respectively, and Θ and K are constants describing the spin exchange interactions between neighboring nuclei. At low temperatures these interactions lead to transitions from a nuclear paramagnet to antiferromagnetic or weakly ferromagnetic phases.

Because of its large zero-point motion, ^{3}He does not solidify under its own vapor pressure but must be pressurized to greater than 45 bars to be a solid at T = 0 K. At this pressure the density of the solid is more than 0.124 gm/cm^3, about 900 times greater than the density of a 1 atmosphere polarized gas target. This increase in density makes possible experiments that are inaccessible to polarized gas targets. On the other hand the brute force polarized target has the disadvantages of the large external magnetic field, provided by a superconducting magnet, and the low temperature which requires that beam heating effects be kept below 1 microwatt on the target.

THE TUNL POLARIZED SOLID ^{3}HE TARGET

The ^{3}He container is constructed of beryllium copper and OFHC copper to maximize thermal conduction to the solid ^{3}He. It is known that the spin-dependent part of the n-Cu interaction is small. The container walls that the neutrons traverse are 1.27 mm thick. The ^{3}He target itself is in the shape of a rectangular parallelepiped with a cross-sectional area 38.1 x 14.0 mm and 21.6 mm thick. The rectangular shape of the target maximizes the thickness of solid ^{3}He. The sample space is filled with 3 μm silver powder packed to 19% of the density of bulk silver. The powder assures homogeneous cooling of the solid, which by itself has a low thermal conductivity.

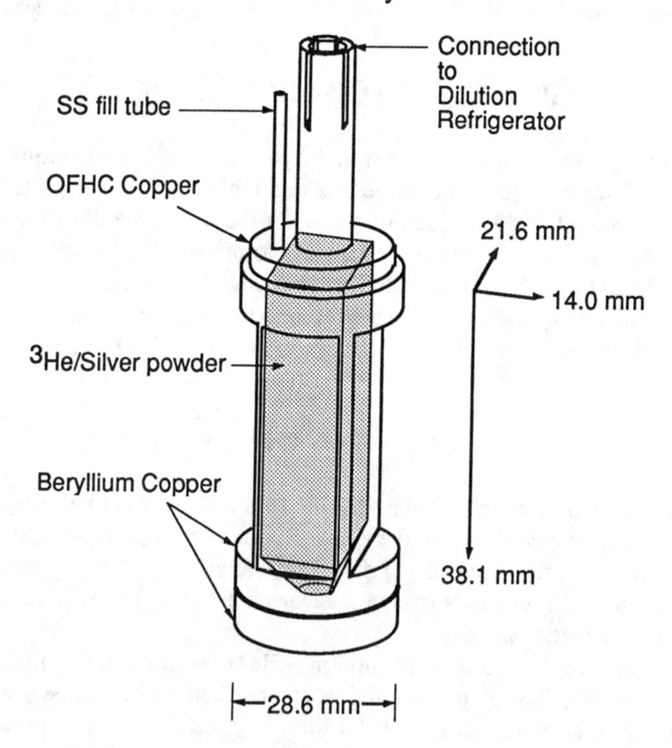

FIGURE 1. Schematic of TUNL solid ^{3}He target cell . The sample volume is indicated by the shaded portion and the interior dimensions are shown to the right.

The solid ^3He target was cooled in the TUNL spin-spin cryostat which uses a commercial dilution refrigerator (SHE Model 430) having a cooling power of about 1.5 microwatts at 10 mK. The magnetic field is supplied by a 7 T split-coil superconducting magnet (American Magnetics) having a homogeneity of 0.1% in the region of the target. This cryostat has been previously used for brute force polarization of several nuclear targets (2). The temperature of the refrigerator and the target are monitored by nuclear orientation and ^3He melting curve thermometers(3). The ^3He solid was grown from the liquid at an initial pressure of 45.5 bar by the blocked capillary technique. The target density was determined by reference to the published ^3He melting curve. The target could be cooled to less than 20 mK ten hours after solidification and reached 12 to 13 mK in 36 hours as shown in Figure 2.

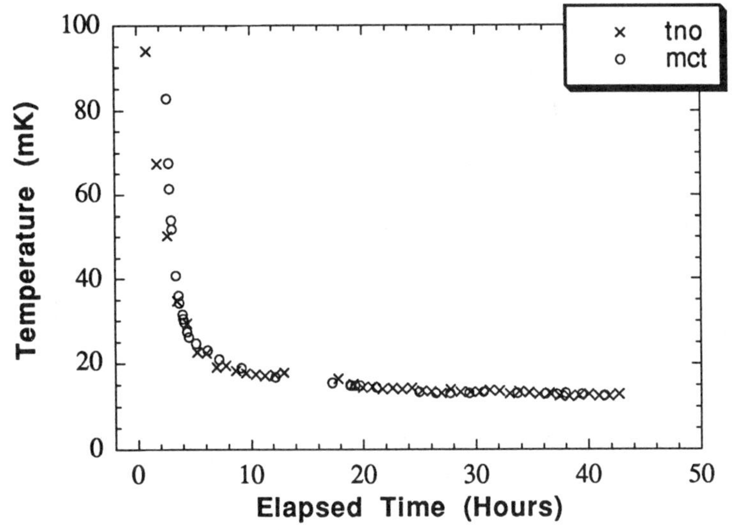

FIGURE 2. Cooling of the solid ^3He polarized target. The temperature is measured simultaneously by nuclear orientation (tno) and ^3He melting curve thermometers (mct). At 20 mK the nuclear polarization is 24% and at 10 mK it is 42%.

The TUNL target has been used to investigate the excited states of the ^4He nucleus through neutron transmission measurements.(4) The incident beam consisted of MeV neutrons polarized perpendicular to their momentum. The ^3He target was polarized transverse to the beam momentum. The transmitted neutrons were detected at 0° by liquid scintillators. The spin dependent asymmetry in the transmission was determined by flipping the neutron spin every 0.1 seconds. The asymmetry ε is

$$\varepsilon = \frac{N^+ - N^-}{N^+ + N^-} \approx -\frac{1}{2} x_3 P_3 P_n \Delta\sigma_T$$

where P_3 and P_n are the target and beam polarizations, x_3 is the target thickness and $\Delta\sigma_T$ the spin-dependent cross section: the difference in total cross section between parallel and antiparallel spin geometries,

$$\Delta\sigma_T = \sigma_{tot}(\uparrow\uparrow) - \sigma_{tot}(\uparrow\downarrow)$$

In Figure 3 we display data for transmission asymmetries measured with a polarized solid target, a polarized liquid target, and an unpolarized liquid target. At low temperatures liquid ^{3}He is Fermi-Dirac paramagnet. The polarization at 7 T is 3% and relatively temperature independent. The slope of the line is $2\Delta\sigma_T$, demonstrating that the asymmetry itself can be used to determine target polarization.

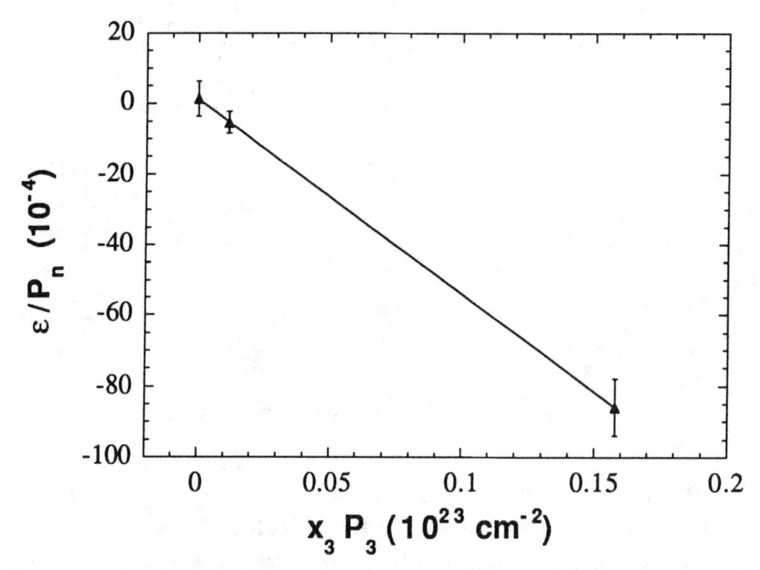

FIGURE 3. Transmission asymmetries for 3.65 MeV neutrons. The ratio of transmission asymmetry to beam polarization (ε/P_n) is plotted versus the product of the ^{3}He thickness and polarization. The data points are shown for (left to right) warm liquid, cold liquid and cold solid ^{3}He targets.

FURTHER DEVELOPMENT

There are several directions for further development of such a target. Because ^{3}He is highly compressible the density of the target can be increased significantly by growing the solid at a higher pressure. Solid ^{3}He can be grown in the hcp crystal structure by the application of 105 bars. Because the hcp solid is a simple paramagnet, without antiferromagnetic interactions, nuclear polarization is greater than bcc ^{3}He at a given B and T. The hcp solid reaches P = 50% at 10 mK and 7 T, in comparison with 42% for the bcc solid in the same conditions. Because the polarization is dependent on the ratio of B_{ext}/T, it is also possible to achieve the same polarizations at lower magnetic fields using a more powerful dilution refrigerator or by direct cooling through the Pomeranchuk effect (5). Either of these methods is increasingly sensitive to problems of beam heating. The Pomeranchuk effect has certain advantages because it can be used to polarize a ^{3}He target more quickly than the present technique, and the target can be rapidly repolarized. An attractive option which may be useful in certain situations is the polarization of solid ^{3}He in the very low temperature ($T \leq 0.9$ mK) weakly ferromagnetic phase. In this phase polarizations in excess of 60% may be achieved at very low magnetic fields --- $B_{ext} \leq 0.6$ T.

ACKNOWLEDGEMENTS

This research has been supported, in part by U.S. DOE Grants No. DE-FG05-91-ER40619 and DE-FG05-88ER40441.

REFERENCES

1. R. G. Milner, *Nuclear Physics*, **A508**, 599c (1990).
2. D. G. Haase, C. R. Gould, and L. W. Seagondollar, *Nucl. Instrum. and Meth.*, **A243**, 305 (1986).
3. C. D. Keith, C. R. Gould, D. G. Haase, N. R. Roberson, W. Tornow, and W. S. Wilburn, *Hyperfine Interactions* **75**, 525 (1992).
4. C. D. Keith, C. R. Gould, D. G. Haase, N. R. Roberson, W. Tornow, G. M. Hale, H. M. Hoffmann and H. Postma, *Phys. Rev.* **C50**, 237 (1994).
5. R. T. Johnson and J. C. Wheatley, *J. of Low Temp. Phys.* **2**, 423 (1970).

A Neutron Beam Polarizer for Study of Parity Violation in Neutron-Nucleus Interactions

S. I. Penttilä[(1)], J. D. Bowman[(1)], P. P. J. Delheij[(2)], C. M. Frankle[(1)], D. G. Haase[(3)], H. Postma[(4)], S. J. Seestrom[(1)], and Yi-Fen Yen[(1)]

(The TRIPLE Collaboration)

[(1)] *Los Alamos National Laboratory, Los Alamos, New Mexico 87545*
[(2)] *TRIUMF, Vancouver, British Columbia, V6T2A3, Canada*
[(3)] *North Carolina State University, Raleigh, North Carolina 27695 and Triangle Universities Nuclear Laboratory, Durham, North Carolina 27708*
[(4)] *University of Technology, P.O. Box 5046, 2600 GA, Delft, The Netherlands*

Abstract. A dynamically-polarized proton target operating at 5 Tesla and 1 K has been built to polarize an epithermal neutron beam for studies of parity violation in compound-nuclear resonances. Nearly 0.9 proton polarization was obtained in an electron-beam irradiated ammonia target. This was used to produce a neutron beam polarization of 0.7 at epithermal energies. The combination of the polarized proton target and the LANSCE spallation neutron source produces the most intense pulsed polarized epithermal neutron beam in the world. The neutron-beam polarizer is described and methods to determine neutron beam polarization are presented.

INTRODUCTION

Development of neutron spallation sources has created a new capability to study fundamental symmetries such as parity and time-reversal violation, in compound-nuclear resonances. These experiments require a high-intensity, polarized low-energy neutron beam with a relatively short pulse length. In this paper we describe the polarized neutron-beam facility at the Los Alamos Neutron Scattering Center (LANSCE) built by the TRIPLE collaboration for parity-violation (PV) studies in neutron-nucleus interactions.

The large spin dependence of the n-p scattering cross section has been used to polarize low-energy neutron beams by transmitting neutrons through material containing polarized protons. This method was first used at the Joint Institute for Nuclear Research, Dubna, in the 1960's (1). Since, significant progress has been made in building targets for spin experiments in nuclear and high-energy physics (2). Recently, almost 100% proton polarization was reported using the dynamic nuclear polarization method at 1 K and 5 Tesla with electron beam irradiated NH_3 material (3). The energy independence of the spin-spin n-p cross section makes a cryogenic polarized proton filter a useful neutron beam polarizer in the energy interval from 0.1 eV up to 50 keV.

The TRIPLE cryogenic neutron-spin filter produces longitudinally polarized epithermal neutrons for PV experiments. The experiment and preliminary results are discussed by Yen *et al.* (4) in these Proceedings. The requirements for the neutron-spin filter include high proton polarization, resistance to radiation damage from γ-rays, small beam attenuation, and operational simplicity and reliability.

Polarized ^{3}He can be also used to polarize low-energy neutrons. Recent developments in optically polarized ^{3}He targets, especially the availability of low-cost and high-power GaAlAs diode laser arrays, have made the polarized ^{3}He cells an alternative for producing polarized low-energy neutron beams. At present, 70% polarization has been achieved with ^{3}He gas in pressures up to 10 atmospheres. Use of optically polarized ^{3}He has been discussed in a number of papers in the Conference (5). The total absorption cross section, almost the same as the capture cross section, for the n-^{3}He reaction is about 980 b at 0.7 eV. The absorption cross section results from a strong subthreshold *s*-wave resonance and falls off inversely with the neutron velocity (6). This energy dependence of the cross section limits a practical neutron-spin filter below 10 eV. In our PV experiments the energy range of interest is from 0.1 eV to 1 keV.

The polarization of a neutron beam transmitted through a longitudinally polarized proton target with polarization of f_H is given by

$$f_n = \tanh(f_H n \sigma_P t), \tag{1}$$

where n is the number density of the protons per cm^3 in the target, σ_P is the polarization cross section, and t is the thickness of the filtering material. The polarization cross section is $\sigma_P = (\sigma_{\rightleftarrows} - \sigma_{\rightrightarrows})/2$, where the arrows indicate the direction of the neutron spin with respect to the direction of the proton polarization. In the energy region of 1 eV to several keV, these cross sections are nearly constant, $\sigma_{\rightleftarrows}$ is 37.2 b and $\sigma_{\rightrightarrows}$ is 3.7 b, giving σ_P = 16.7 b (7,8). Neutrons with spin direction opposite to the polarization of the filter will be scattered. Neutrons with parallel spin direction will be attenuated by part of the triplet n-p cross section and scattering from the other background nuclei present in the material. The transmission of the beam is given by

$$T = T_o \cosh(f_H n \sigma_P t), \tag{2}$$

where

$$T_o = \exp\left(-\sum_i n_i \sigma_0 t\right). \tag{3}$$

The sum is over all the types of nuclei present in the material. The unpolarized total cross section for the n-p reaction is σ_0 = 20.5 b and is σ_0 ~ 10 b for n-^{14}N scattering (6). Figure 1 presents the neutron beam polarization and transmission as a function of proton polarization in NH_3 for three different thicknesses of the polarizing material. The transmissions do not include contributions from liquid ^{4}He or other cryostat materials in the path of the neutron beam.

PV effects are proportional to f_n and the measurement time required to obtain a given statistical accuracy is inversely proportional to the figure-of-merit (FOM) of the filter, If_n^2, where I represents the beam intensity and f_n depends on the proton polarization. The FOM is optimized using the thickness of the polarizing material and the cross-sectional area of the filter.

THE POLARIZED PROTON FILTER

A layout of the TRIPLE apparatus for PV experiments is schematically shown in Fig. 2 and is described in detail in Ref. (9). Intense epithermal neutron pulses are produced by impinging 800-MeV proton pulses with a FWHM of 125 ns at the

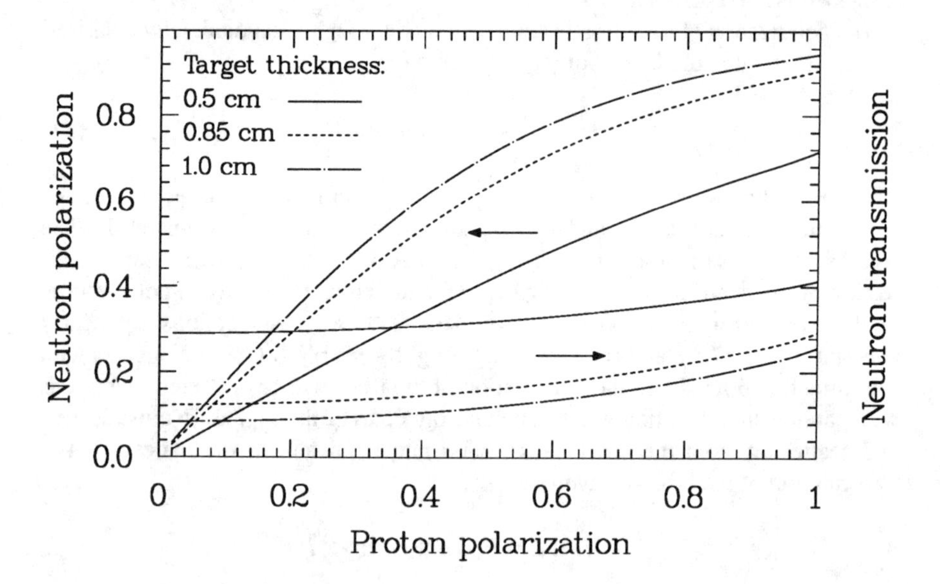

FIGURE 1. Neutron beam polarization and transmission as a function of proton polarization in NH_3 for the effective polarizing thicknesses of 0.5, 0.85, and 1.0 cm.

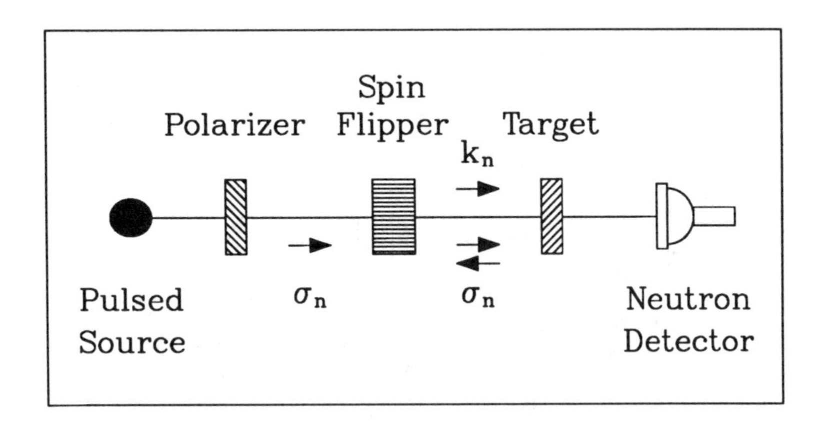

FIGURE 2. TRIPLE apparatus for PV measurements.

rate of 20 Hz on a tungsten target. A water moderator is used to decrease the neutron energy to the epithermal range. After a collimation, the beam is polarized by the spin filter and then the neutron helicity state is controlled by the spin flipper (9,10). The flipper consists of a system of longitudinal and transverse magnetic fields and is effective over a neutron energy range of E_{max}/E_{min} ~ 1000. The systematic errors of the experiment were reduced by changing the neutron helicity state every 10 seconds and the direction of the proton polarization every few days.

In the neutron-spin filter, which is a polarized proton target, protons are polarized dynamically by microwave pumping at 1 K and 5 Tesla. Figure 3 shows schematically the proton filter. The system consists of a 5-Tesla split-coil superconducting magnet. The field homogeneity, as measured in the LANSCE target room in the presence of iron shielding, was $1.3 \cdot 10^{-4}$ over a volume of 8 cm in diameter and 2 cm in length. The magnet was manufactured with an undistorted homogeneity of 1 part in 10^5 (11).

The polarizing material is NH_3 in the form of 1–2-mm grains. Paramagnetic polarizing centers were produced by irradiating the material with 30-MeV electrons with the dose of $2\text{–}4 \cdot 10^{16}$ e^-/cm^2 and rate of 0.5–1 $\mu A/cm^2$. The NH_3 filter is a cylindrical disk 80 mm in diameter and 13 mm in length cooled to 1 K in a pumped ^4He bath. A cooling power capacity of 2 W at 1.05 K is obtained with a pumping speed of 8200 m^3/h for ^4He gas. This cooling capacity is needed to remove heat caused by the microwave pumping, which with NH_3 was 20 mW/cm^3. We have also tested 1-Butanol material doped with EHBA (12) to a level of $3 \cdot 10^{19}$ $spins/cm^3$. With this sample, a microwave power of 45 mW/cm^3 was required.

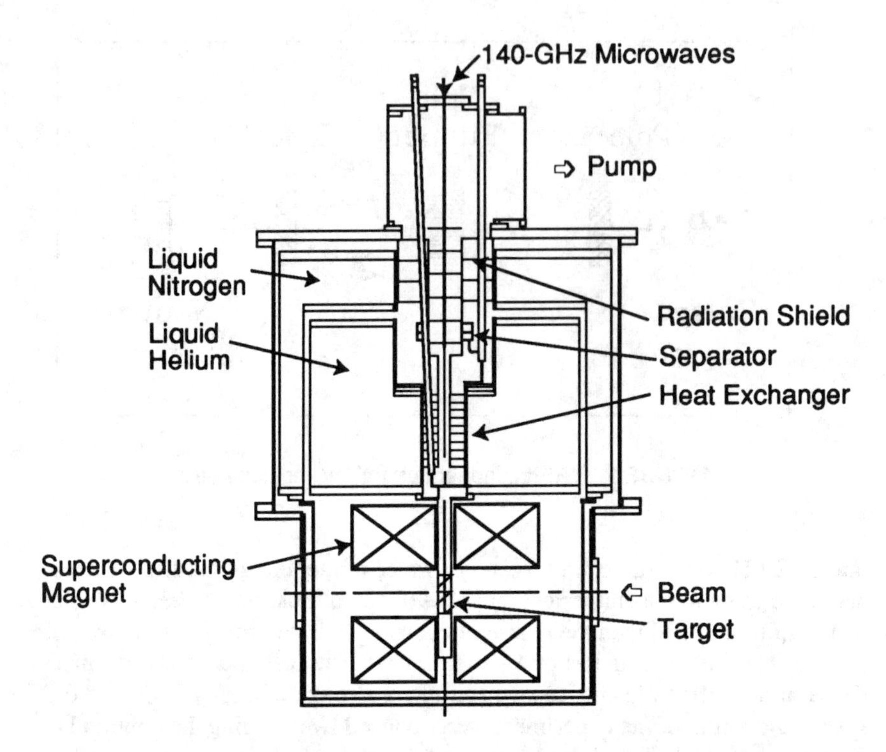

FIGURE 3. Schematic of the TRIPLE longitudinally-polarized proton filter.

PERFORMANCE OF THE POLARIZER

Proton polarization was measured with NMR using the Liverpool-type detector system (13). Typical proton polarizations with NH_3 during the experiments were +0.85 and –0.90. The neutron beam polarizations were +0.73 and –0.67, respectively. The neutron beam polarization measurement is described below. From Fig. 1, one can estimate that the effective thickness of the NH_3 was less than 0.8 cm, which was the design value. Also, the comparison of the proton polarization values measured with NMR and the beam polarization values indicate that there was a linearity problem with NMR (13). The time constant for the build-up of the proton polarization is about 30 min.

A sample composed of 1-Butanol beads was also polarized in the same cryostat. The density of polarizing centers was $6 \cdot 10^{19}$ spins/cm^3. Proton polarization of ±0.60 was achieved in a sample of 8 cm^3. In a sample of 65 cm^3, polarizations were +0.24/ –0.34. The electron spin density in this sample was $3 \cdot 10^{19}$ spins/cm^3.

POLARIZATION MEASUREMENT OF THE NEUTRON BEAM

Three different techniques were used to determine the neutron beam polarization.

(a) The proton polarization of the polarizer was measured with NMR and then the neutron polarization was calculated from Eq. (1) by using the known thickness of the filter, *t*. The thickness of the filter material was determined with the attenuation of 0.662-MeV gamma rays from a ^{137}Cs source. As a result, the filling factor was obtained to be 0.63. NMR is not sensitive to inhomogeneities in the polarization or in the thickness of the material, which can lead to an inhomogeneous beam polarization. However, an NMR measurement is accurate, it does not disturb the neutron beam and a NMR measurement takes less than 30 seconds. Therefore, NMR was used in the experiment as a continuous neutron-beam polarization monitor.

(b) Neutron beam polarization could be determined from the absolute value of the transmission of the neutron beam by using Eq. (2), where the factor $f_H n \sigma_p t$ was extracted and then substituted in Eq. (1) (7,14). A better way to measure the beam polarization is to determine the transmission ratio of the polarized and unpolarized filter. According to Eq. (2),

$$f_n = \sqrt{1 - \frac{T_0^2}{T_{\text{pol}}^2}}, \tag{4}$$

where T_0 (T_{pol}) is the transmission through the unpolarized (polarized) filter. With this method, one does not need to know the thickness of the material, proton polarization, or cross sections. In an hour we can obtain a beam polarization with the accuracy of 2%.

(c) The parity-violating (PV) longitudinal asymmetry P of the 0.734-eV p-wave resonance in ^{139}La has been determined to the accuracy of $P = (9.55 \pm 0.35)\%$ (14). To measure the neutron polarization produced by the spin filter at 0.734 eV, a measurement of the PV longitudinal asymmetry was performed using the method described in Ref. (14). The measured neutron asymmetry of the 0.734-eV p-wave resonance is proportional to $f_n P$, thus allowing f_n to be determined. The upper panel of Fig. 4 shows a TOF spectrum of ^{139}La around 0.7 eV. The lower panel shows the PV asymmetry of the 0.734-eV resonance after a 20-min run (32000 proton pulses). A 5% neutron polarization measurement can be achieved in two hours. A problem of this method is that the beam polarization is measured only at one energy, at the lower part of our energy region of interest. Therefore, the energy dependence of the n-p cross section has to be known, as well as the effect of the spin flipper on the beam polarization as a function of the neutron energy (9,10).

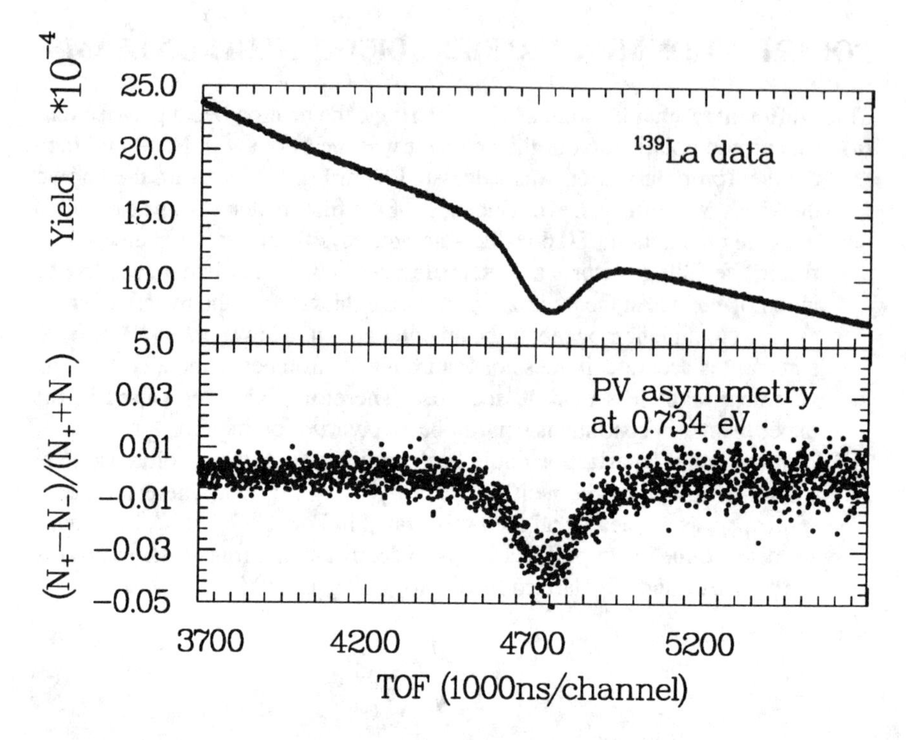

FIGURE 4. Top: The ^{139}La TOF spectrum in the vicinity of 0.7 eV. Bottom: The parity-violating asymmetry of 0.734-eV *p*-wave resonance in ^{139}La.

ACKNOWLEDGMENTS

This work was supported in part by the U.S. Department of Energy, Office of High Energy and Nuclear Physics, under grant Nos. DE-FG05-88-ER40441 and DE-FG05-91-ER40691.

REFERENCES

1. Draghicescu, D., Lushchikov, V. I., Nikolenko, V. G., Taran, Yu. V., and Shapiro, F. L., *Phys. Lett.* 12, 334 (1964).
2. Jeffries, C. D., "History of the Development of Polarized Targets," in *Proceedings of the 9th International Symposium on High Energy Spin Physics*, eds. Althoff, K.-H. and Meyer, W., Springer-Verlag, 1991, Vol. 1, pp. 3–19.
3. Crabb, D. G., "Polarization Studies with Radiation Doped Ammonia at 5 T and 1 K," in *Proceedings of the 9th International Symposium on High Energy Spin Physics*, eds. Althoff, K.-H. and Meyer, W., Springer-Verlag, 1991, Vol. 2, pp. 289–300.
4. Yen, Yi-Fen *et al.*, "Study of Parity and Time-Reversal Violation in Neutron-Nucleus Interactions," in these Proceedings.

5. Cummings, W. *et al.*, "Application of High-Power GaAlAs Diode Laser Arrays for Optically Pumped Spin Exchanged Polarized ^{3}He Targets," in these Proceedings; Coulter, K. P. *et al.*, "Advances in Alkali Spin-Exchange Pumped ^{3}He Targets," in these Proceedings; Thomson, A. K. *et al.*, "A Polarized ^{3}He Based Neutron Spin Filter at NIST," in these Proceedings.
6. ENDF/B-IV Data File for ^{1}H, ^{3}He, and ^{14}N (National Nuclear Data Center, Brookhaven National Laboratory, Upton, New York, April 1989), evaluated by Hale, G. and Young, P.
7. Lushchikov, V. I., Taran, Yu. V., and Shapiro, F. L., *Yad. Fiz.* **10**, 1178 (1969).
8. Hoshizaki, N. and Masaike, A., *Japanese J. Appl. Phys.* **25**, L244 (1986).
9. Roberson, N. R. *et al.*, *Nucl. Instrum. Methods A* **326**, 549 (1993).
10. Bowman, J. D. *et al.*, "A Spin-Reversal System for Epithermal Neutrons," to be published in *Nucl. Instrum. Methods.*
11. Magnet and refrigerator were fabricated by Oxford Instruments Ltd., England.
12. Krumpolc, M. and Rôcek, J., *J. Am. Chem. Soc.* **101**, 3206 (1979).
13. Court, G., "Review of Non-linear Corrections in CW Q-meter Target Polarization Measurements," in these Proceedings.
14. Yuan, V. W. *et al.*, *Phys. Rev. C* **44**, 2187 (1991).

^{13}C Cross-Relaxation Measurements

S. Bültmann[1], G. Baum[1], C. M. Dulya[2], N. Hayashi[6], A. Kishi[6], Y. Kisselev[5], D. Krämer[1], J. Kyynäräinen[3], J.-M. Le Goff[4], A. Magnon[4], T. Niinikoski[3] and Y. K. Semertzidis[3]

[1] *Universität Bielefeld, Fakultät für Physik, D-33615 Bielefeld, Germany*[a]
[2] *University of California, Dept of Physics, Los Angeles, 90024 CA, USA*[b]
[3] *CERN, CH-1211 Genève 23, Switzerland*
[4] *DAPNIA, C. E. Saclay, F-91191 Gif-sur-Yvette, France*
[5] *JINR, Laboratory of Super High Energy, Dubna, Russia*
[6] *Nagoya University, Dept of Physics, Furo-Cho, Chikusa-Ku, 464 Nagoya, Japan*[c]

Abstract. The time constants τ of the spin coupling between protons and ^{13}C-nuclei in the SMC butanol target was measured as a function of the applied microwave power. A clear decrease of τ in the presence of a microwave field was found. This has to be related to the cross-relaxation between the different spin species. The application for cold dark matter detection is outlined.

INTRODUCTION

The cross-relaxation studies of ^{13}C-nuclei to be described were undertaken with the large polarized target employed by the Spin Muon Collaboration at the CERN experiment NA 47 [3,4]. A proton target material was used composed of 90.5% $C_4H_{10}O$ and 5.0% H_2O with 4.5% EHBA-Cr(V), a paramagnetic dopant. The used target cell had a length of 60 cm and a diameter of 5 cm corresponding to a volume of 1180 cm^3. The target contained $N(^1\mathrm{H}) = 93$ mol of polarizable protons and $N(^{13}\mathrm{C}) = 0.4$ mol of ^{13}C-nuclei. Their Larmor frequencies at 2.5 Tesla are 106.5 MHz and 26.78 MHz, respectively. The polarization of protons and ^{13}C-nuclei was measured by five longitudinally

[a]Supported by Bundesministerium für Forschung und Technologie
[b]Supported by the Department of Energy
[c]Supported by Ishida Foundation, Mitsubishi Foundation and Monbusho International Science Research Program

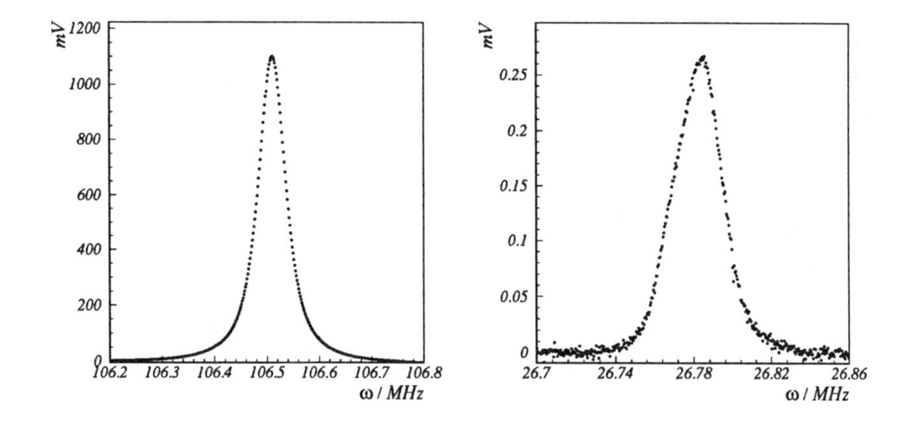

Figure 1: NMR signals of protons (left) and ^{13}C-nuclei (right). The sweep width
and number of sweeps are
^{1}H : $\Delta\omega = 600$ kHz, 400 sweeps, ^{13}C : $\Delta\omega = 160$ kHz, 20000 sweeps.

aligned NMR coils. Four of them were calibrated for the measurement of the proton signals, while one was tuned for the measurement of the ^{13}C signal. The later was not calibrated by measuring the natural ^{13}C TE signal.
The polarization is proportional to the absorptive part of the resonance signal measured by a series Q-meter [2]. It reads

$$P \propto \frac{1}{\gamma^2 N I} \cdot \Gamma, \qquad (1)$$

where γ denotes the gyromagnetic ratio, I the spin and Γ the integral of the absorption lineshape over the frequency range (sweep width) $\Delta\omega$, centered around the Larmor frequency.
The target was polarized in a 2.5 Tesla magnetic field by the dynamic nuclear polarization (DNP) process using EIO tubes oscillating at around 69 GHz. In figure 1 the resonance signals of protons (left) and ^{13}C-nuclei after several hours of polarizing by DNP are shown. For the integrals of the lines we measured $\Gamma(^1\mathrm{H}) = 82.5$ V·kHz and $\Gamma(^{13}\mathrm{C}) = 8$ mV·kHz, respectively. From the integral of the proton signal we get the polarization $P(^1\mathrm{H}) = 0.81$ using calibration at 1 K. The sweep width $\Delta\omega$ and number of sweeps are different; their values are given in the caption.

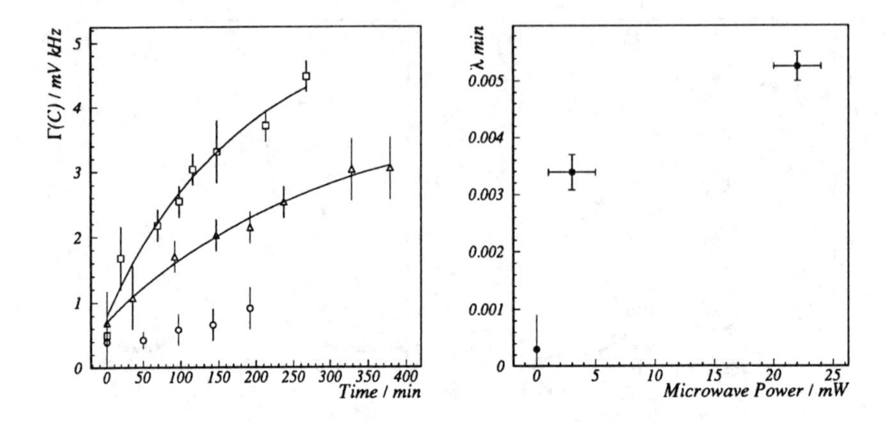

Figure 2: $\Gamma(^{13}C)$ vs. time at different levels of applied microwave power (left) (22 mW □, 3 mW △ and 0 ○). The time constants τ of the fits are given in the text and displayed as $\lambda = \tau^{-1}$ vs. microwave power (right).

MEASUREMENT OF RELAXATION

After polarizing the target with DNP, the carbon-13 polarization was destroyed by saturating the ^{13}C spins at their Larmor frequency. The output power of the RF synthesizer used was not sufficient to destroy the polarization completely. The data points at time zero in figure 2 (left) show the remaining $\Gamma(^{13}C)$. These were taken as an offset for the cross-relaxation measurements. Without microwave field the polarization is growing very slowly with a time constant τ of the order of $3{\cdot}10^3$ minutes. This is shown in figure 2 (left) together with data points from two measurements in a microwave field of different strength. The power was at 3 mW (△) and 22 mW (□). The build-up time constants of the indicated exponential fits are τ(3 mW) = 295 min and τ(22 mW) = 190 min. These results are also shown in figure 2 (right). A sudden increase of $1/\tau$ due to the applied microwaves is visible already at a very low power level. This has to be related to cross-coupling between the two spin species which also should lead to a decrease of the proton polarization. Due to the small amount of ^{13}C-nuclei in the sample, this decrease could not be detected.

AXION SEARCH

One candidate for solving the dark matter problem of the Universe is the axion, a boson which was proposed first by Weinberg and Wilczek as a consequence of the strong CP problem [7,8]. An overview of experiments searching for galactic dark matter can be found in [1,5] and the references cited therein. The fact that axions may couple directly to fermions could enable their detection in a large polarized target. The usage of such a target seems feasible, because the de Broglie wavelength of the axions, due to the virial velocity of the sun, is in the range of ten to hundred meters. An axion mass in the range of $5{\cdot}10^{-3}$ to $5{\cdot}10^{-5}$ eV corresponds to a frequency in the microwave range. If the electron paramagnetic resonance (EPR) coincides with the axion frequency, cross-relaxation between different nuclear spin species can be induced.
A practical approach would consist of the measurement of the polarization build-up time constant τ of depolarized spins at different fields. As there is nearly no increase in polarization without a microwave field present, a sharp decrease of τ at the resonance frequency of the axions should be observed. The range of investigation is determined by the accessible magnetic field. Our magnet is limited to 2.5 Tesla, while the other boundary is given by the faster increasing relaxation of the nuclei at magnetic fields below 0.25 Tesla. This corresponds to a frequency range of roughly 7 GHz to 70 GHz, a mass range which has not yet been covered by other experiments.
The results of our preliminary feasibility studies were described above. Measurements of the cross-relaxation between protons and deuterons in a deuterated target with a larger content of protons are foreseen for this year. With the predicted axion absorption rate around $3.5 \cdot 10^{-3}$ $\sec^{-1}$ [6], we might be able to reach the solar limit [1].

CONCLUSIONS

Cross-relaxation between the spins of protons and ^{13}C-nuclei was observed as a build-up of non-equilibrium polarization of the ^{13}C-nuclei. A further decrease of the polarization build-up time as a consequence of stronger spin coupling due to an applied microwave field was observed. The relevance for the possible detection of cold dark matter in a large polarized target was pointed out.

REFERENCES

1. R. Cameron et al., Phys. Rev. D **47**, 3707 (1993).
2. G. R. Court et al., Nucl. Instr. Meth. **A324**, 433 (1993).
3. D. Krämer et al., these proceedings.
4. J. Kyynäräinen et al., these proceedings.
5. Y. K. Semertzidis et al., Phys. Rev. Lett. **64**, 2988 (1990).
6. Y. K. Semertzidis et al., proceedings of the 7th *Workshop on Polarized Target Materials and Techniques*, Bad Honnef 1994, to be published in Nucl. Instr. Meth
7. S. Weinberg, Phys. Rev. Lett. **40**, 223 (1978).
8. F. Wilczek, Phys. Rev. Lett. **40**, 279 (1978).

Target with a Frozen Nuclear Polarization for Experiments at Low Energies

N.S. Borisov[1], V.N. Matafonov[1], A.B. Neganov[1], Yu.A. Plis[1], O.N. Shchevelev[1], Yu.A. Usov[1], I. Jánský[2], M. Rotter[2], B. Sedlák[2], I. Wilhelm[2], G.M. Gurevich[3], A.A. Lukhanin[4], J. Jelínek[5], A. Srnka[5], L. Skrbek[6]

[1] *Joint Institute for Nuclear Research, Laboratory of Nuclear Problems, 141980 Dubna, Moscow Region, Russia*
[2] *Faculty of Mathematics and Physics, Charles University, V Holešovičkách 2, 180 00 Prague 8, Czech Republic*
[3] *Institute for Nuclear Research, Russian Academy of Sciences, 60th October Anniversary prospect 7A, 117312 Moscow, Russia*
[4] *Kharkov Institute of Physics and Technology, Academicheskaya str. 1, 310108 Kharkov, Ukraine*
[5] *Institute of Physical Instrumentation, Academy of Sciences of the Czech Republic, Královopolská 147, 612 64 Brno, Czech Republic*
[6] *Institute of Physics, Academy of Sciences of the Czech Republic, 250 68 Řež, Czech Republic*

Abstract. The short history of the development of frozen spin polarized targets at the Laboratory of Nuclear Problems JINR is given. The latest development is the target with a frozen spin polarization of protons in 1,2- propanediol with a paramagnetic Cr^V impurity, intended for polarization parameter studies in np-scattering at approximately 15 MeV neutron energy. The target of cylindrical shape of 2 cm diameter and 6 cm long with an initial polarization of $95 \pm 3\%$ obtainable by the dynamic polarization technique is placed at a temperature about 20 mK in a magnetic field of 0.37 T generated by a magnetic system, which provides a large aperture for scattered particles. The relaxation time for the spin polarization is about 1000 hours.

Introduction

A polarized solid-state target is one of the most important elements in the technique of polarization experiments. Achieving of very low stationary temperatures in 1966 [1] at the Laboratory of Nuclear Problems JINR (Dubna) gave rise to the idea of using a radically new cooling technique based on dissolving

^{3}He in ^{4}He to create "frozen" spin polarized target. In 1975 at LNP the experiments on measurement of C_{nn}^{pp} in the elastic pp-scattering were conducted with frozen spin proton polarized target[2] . The frozen spin proton and deuteron polarized targets have been developed at the LNP JINR[3–7], and for 20 years they have been successfully used in experiments with accelerators at IHEP (Protvino), LNP (Dubna), SPINP (Gatchina) and CU (Prague). In Table 1 the parameters of the frozen spin polarized targets created at the LNP JINR are given.

In particular, we have developed a version of the target intended for polarization parameter studies in np-scattering using 15 MeV neutron beam produced by the Van de Graaf accelerator of the Charles University Nuclear Center in Prague. The target is a complex including a stationary cryostat with a dilution refrigerator, a movable magnetic system providing a "warm" field and consisting of a superconducting solenoid and a superconducting split pair magnet with a large aperture, and electronic equipment for providing a dynamic polarization and NMR signal detection. The detailed description of the target is given elsewhere[7], here we discuss the most essential points.

The dilution refrigerator

The vertical dilution refrigerator with a horizontal tailpiece is placed stationary in a neutron beam. The low-temperature system consists of three autonomous matchable units: the cryostat, a ^{3}He pumping, pre-cooling and condensing unit, and ^{3}He/^{4}He dilution step.

The cryostat is provided with a 17 l nitrogen vessel ensuring 24 h of work without refilling. A liquid helium stock of about 19 l ensures dilution refrigerator work for 30 h without refilling.

^{3}He circulation in the dynamic polatization mode ($\approx$ 10 mmole/s) and during an experimental run ($\approx$ 2 mmole/s) is produced by a set of pumps WS-2000, WS-250 (Leybold) and H-2030 (Alcatel). ^{4}He pumping from the condensing bath is produced by pumps WS-500 and H-2060.

A special multichannel automatic ohmmeter[8] is used for temperature measurements of various target units with the help of 15 carbon resistors of Speer, Allen-Breadley and TVO types.

Magnetic system

The target magnetic system consists of two superconducting magnets with a "warm" field, intended respectively for the dynamic polarization mode and experimental runs in the neutron beam.

The polarizing magnet is a solenoid which is being put on the horizontal tailpicce of the refrigerator in the dynamic polarization mode, producing in the target volume a 2.7 T horizontal field with an uniformity better than 10^{-4}.

Table 1: Parameters of the frozen polarized targets LNP JINR

Year of first publ.	Volume (cm^3)	Material	Maximum polarization (%)	Magnetic field (T) dynamic/ frozen	Accelerator (place)	Authors and references
1976	15	$C_3H_6(OH)_2$ 1,2-propanediol with Cr^V ($1.8\times \times10^{20}cm^{-3}$)	$P_\pm=$ $=98\pm2$	2.69/ 2.69	Dubna, Gatchina (in use)	N.S.Borisov, E.I.Bunyatova, Yu.F.Kiselev, V.N.Matafonov, B.S.Neganov, Yu.A.Usov[3]
1980	60	$C_3H_6(OH)_2$ with Cr^V ($1.8^{+0.1}_{-0.2}\times \times10^{20}cm^{-3}$)	$P_\pm=$ $=87\pm3$	2.06/ 0.45	Protvino (in use)	N.S.Borisov, E.I.Bunyatova, A.G.Volodin, M.Yu.Liburg, V.N.Matafonov, A.B.Neganov, B.S.Neganov, Yu.A.Usov[4]
1985	60	$(CD_2OD)_2$ deuterated ethanediol with Cr^V	$P_\pm=$ $=37\pm3$	2.06/ 0.45	Protvino (in use)	N.S.Borisov, E.I.Bunyatova, M.Yu.Liburg, V.N.Matafonov, A.B.Neganov, Yu.A.Usov[5]
1992	120	only refrigeratir and magnet are available		2.5	Protvino (is planned)	N.S.Borisov, V.V.Kulikov, A.B.Neganov, Yu.A.Usov[6]
1994	20	$C_3H_6(OH)_2$ with Cr^V ($1.5\times \times10^{20}cm^{-3}$)	$P_+=$ $=93\pm3$ $P_-=$ $=98^{+2}_{-3}$	2.7/ 0.37	Prague (in use)	this paper and (ref.[7])

The liquid helium stock in the vertical part of the cryostat is about 4 l which allows solenoid work for 12 h. The solenoid produces a field of 3 T at a current of 30 A.

The holding magnet is a superconducting split pair magnet with a vertical field placed stationary at the beam line. Its cryostat is divided into two parts connected rigidly by two tubes, through which the helium vessel of the lower coil communicates with the upper main helium vessel. Apertures for scattered particles are ±50° in the verical plane and about 360° in the horizontal plane (neglecting the connecting tubes). 16 l liquid helium stock over the upper coil ensures magnet work for about 32 h.

After achievement of a high nuclear polarization, using the dynamic technique, in a highly uniform 2.7 T field the solenoid current is decreased until a field of about 0.25 T is achieved. After that the field of the holding magnet is increased up to the same value. Then the solenoid field is decreased to zero, and the solenoid is moved firstly along the beam axis and further in a transverse direction to remove it from the vicinity of the target.

To increase the holding field a soft steel core is used to be introduced into the gap. As a result the magnetic field increases to 0.37 T . The rate of the field manipulations is limited mostly by the permissible heat load of the refrigerator whose temperature was held within 0.05-0.1 K with a polarization loss of no more than 2%. A complete transition from the pumping mode to the experimental run takes usually about one hour.

Dynamic Polarization and Target Polarization Measurement

1,2-propanediol with a paramagnetic Cr^V impurity having a spin concentration of 1.5×10^{20} cm^{-3} (it was kindly provided by E.I. Bunyatova [9]) was used as a target substance.

The target polarization measurement is carried out using a Q-meter with an automatic phase control of resonance frequency of an input series-tuned circuit similar to one described in ref.[10]. To connect the circuit to the Q-meter, a specially prepared coaxial cable with a low heat conductivity was used. An electric length of the whole cable equals $\lambda/2$. A gated analog integrator is used to average NMR signals. The total error of the polarization determination does not exceed 3%.

After measurement of the thermal signal the refrigerator is put into a large thermal load mode for irradiation of the target with microwave power. An ATT diode generator with an output power of 200 mW at a frequency about 75 GHz is used for the dynamic building up of polarization. The thermal load of the refrigerator in this mode equals 30-40 mW at a temperature 0.3-0.4 K in the mixing chamber and a ^{3}He circulation rate about 10 mmole/s. The time demanded to achieve 80% polarization is about one hour. Maximum values of the polarization obtainable at a longer building up time equal 93% and 98% for positive and negative polarization respectively.

After the measurement of polarization is obtained whole target complex is put into work mode in the neutron beam. The target temperature in the working conditions is about 20 mK at a circulation rate of 2 mmole/s. A duration of one continuous run at a given sign of a polarization equals 10-12 hours. Polarization degradation during this period is insignificant since the nuclear spin relaxation time in a magnetic field of 0.37 T is about 1000 hours.

Acknowledgments

The authors are expressing their gratitude to Profs V.P. Dzhelepov and Yu.M. Kazarinov, Drs N.A. Russakovich and S. Šafrata for their help and support with the work; Dr. B.S. Neganov for stimulating discussions and help; and M.Yu. Liburg and E.I. Bunyatova for their assistance in carrying out the experiments. The authors are also grateful to R.L. Khamidulin, V.G. Kolomiets, M. Trhlík and F. Trenčnský for their help during the installation of the facility.

References

1. Neganov,B.S., Borisov,N.S., Liburg,M.Yu., *JETP* 50, 1445-1457 (1966)
2. Borisov, N.S., Glonti, L.N., Kazarinov, M.Yu., Kazarinov, Yu.M., Kiselev, Yu.F., Kiselev, V.S., Matafonov, V.N., Macharashvili, G.G., Neganov, B.S., Strakhota, I., Trofimov, V.N., Usov, Yu.A., Khachaturov, B.A., *JETP* 72, 405-410 (1977); Dubna: JINR Preprint, 1976, P1-9912.
3. Borisov, N.S., Bunyatova, E.I., Kiselev, Yu.F., Matafonov, V.N., Neganov, B.S., and Usov, Yu.A., *Prib. tekhn. eksp.* 2, 32-40 (1978); Dubna: JINR Preprints, 1976, 13-10253; 13-10257.
4. Borisov, N.S., Bunyatova, E.I., Volodin, A.G., Liburg, M.Yu., Matafonov, V.N., Neganov, A.B., Neganov, B.S., Usov, Yu.A., Proton Polarized Frozen Target for High Energy Particles Secondary Beams, Dubna: JINR Communication, 1980, 1-80-98.
5. Borisov, N.S., Bunyatova, E.I., Liburg, M.Yu., Matafonov, V.N., Neganov, A.B., Usov, Yu.A., *J. Phys. E: Sci. Instrum.* 21, 1179-1182 (1988); Dubna: JINR Preprint, 1985, P1-85-292.
6. Borisov, N.S., Kulikov, V.V., Neganov, A.B., Usov, Yu.A., Dubna: JINR Preprint, 1992,P8-92-238; *Cryogenics* 33, 738-741 (1993).
7. Borisov, N.S., Matafonov, V.N., Neganov, A.B., Plis, Yu.A., Shchevelev, O.N., Usov, Yu.A., Jánský, I., Rotter, M., Sedlák, B., Wilhelm, I., Gurevich, G.M., Lukhanin, A.A., Jelínek, J., Srnka, A., Skrbek, L., *Nuclear Instruments and Methods in Physical Research* A345, 421-428 (1994).
8. Neganov, A.B., Multichannel Automatic Device for Measuring Ultralow Temperatures, Dubna: JINR Communication, 1985, 8-85-291.
9. Bunyatova, E.I., Galimov, R.M., Luchkina, S.A., The Research with an ESR Method of the Stable Cr^V Complex in the Different Solvents, Dubna: JINR Communication, 1982, 12-82-732.
10. Kiselev, Yu.F. and Matafonov, V.N., *Prib. tekhn. eksp.* 5, 55-60 (1977).

Polarized Target for Nucleon-Nucleon Experiments at Saturne II

J. Ball[1], B. Benda[2], P. Chaumette[2], M. Combet[1], J. Derégel[2], G. Durand[2], C. Gaudron[2], Z. Janout[3,4], T.E. Kasprzyk[5], B.A. Khachaturov[3], F. Lehar[2], A. de Lesquen[2], V.N. Matafonov[3], J.-L. Sans[1] and Yu.A. Usov[3]

[1] *Laboratoire National SATURNE, CNRS/IN2P3, CEA/DSM and*
[2] *CEA/DAPNIA, CE-Saclay, 91191 Gif-sur-Yvette Cedex, France*
[3] *Laboratory of Nuclear Problems, JINR, Dubna Moscow Region 141980, Russia*
[4] *Present address: Faculty of Nuclear Sciences and Physical Engineering, Czech Technical University, Břehová 7, 11519 Prague 1, Czech Republic*
[5] *ANL-HEP, 9700 South Cass Ave., Argonne, IL 60439, USA*

Abstract. Continuous improvements of SATURNE polarized target resulted in a flexible and reliable facility for spin physics. For polarized neutron target, two cartridges loaded with ^6LiD and ^6LiH are set in the refrigerator and can be quickly inserted in the beam. The polarized proton target is a 70 cm^3 cartridge loaded with Pentanol-2, a promising material according to the results obtained. Angular distribution as a function of a kinematically conjugate angle and coplanarity in nucleon-nucleon scattering is shown for different targets.

1. POLARIZED TARGET SET-UP

The frozen spin polarized target, set at Saturne, on the Nucleon Nucleon experiment in 1980 has been evoluating for 14 years to enable convenient measurements of spin observables. We are going to describe the last improvements that occured since (1).

The $^3He/^4He$ dilution refrigerator is adapted to carry two types of targets according to the choice of the polarized nucleon. For polarized proton target (PPT), the mixing chamber contains a 70 cm^3 parallelelipedic Voltalef cartridge filled with Pentanol beads chemically doped with EHBA-Cr(V). For polarized deuteron target (PDT), two 14 cm^3 cylindrical cartridges of 2 cm diameter are set in the mixing chamber. The vertical gap between the two cartridges is 1 cm. By the means of a mechanical translation system linked to the outer top part of the refrigerator with vertical bellows adapted on the pumping tubes, the upper or the lower target can be put at the beam level. One of the cartridge is filled with ^{6}LiD and the other with ^{6}LiH in order to discriminate scattering on D and ^{6}Li.

The cooling power of the refrigerator is 300 mW at 300 mK and frozen spin temperature is 40 mK. Polarizing flows are 29 mmoles/sec ^{4}He and 23 mmoles/sec ^{3}He, as for frozen spin mode they are 7 mmoles/sec and 4 mmoles/sec respectively. Temperature is measured by 3 Ruthenium oxide resistors (RO-600 Scientific Instruments Inc.), placed above and below the target. One of the resistors was calibrated by Lake Shore and was used to calibrate the two others. The choice of these probes was determined by their insensivity to magnetic fields, namely 1.4% at 2.5 T. Polarizing field is 2.5 T produced by polarising solenoid with an homogeneity of 10^{-4} in the 70 cm^3 volume of the large target. Stability of power supply has been fixed to 10^{-6}. Vertical and horizontal holding magnets keep the target in a 0.33 T field (1).

The target polarization is measured classically by NMR. For the large 70 cm^3 cartridge, two single turn rectangular coils 40 mm wide and 49 mm high are placed at each end of the target. They are holding mechanically the cartridge as well as measuring the polarization. For the twin cylindrical target set there is one coil per target. Each of these coils of 21 mm in diameter consists of 5 turns embedded in Voltalef in order to assure good geometrical stability. The connection on each coil can be done on the number of turns required by the tuning of the system depending on the resonating frequency of the materials investigated.

The NMR measurement system consists on a resonating LCR circuit connected to "Liverpool type" Q-meters (Ultra Physics Ltd.) through semi-rigid line. The RF frequency sweeping is produced by a synthesizer (Marconi 2040). The data are processed through an acquisition card (National Instruments ATMIO16X) set in a PC (HP vectra 486). The computer is also used to generate the polarization measument program, to pilot the synthesizer through GPIB bus and the polarizing magnet power supply. The software used for the whole programming and signal treatment from data is LabView/PC (National Instruments).

2. PENTANOL-2 POLARIZATION

For PPT, alcohols with a good ratio of free protons on bound protons have lead to the now obvious choice of Butanol or Pentanol. These two products are equivalent towards this ratio, but the advantage of pentanol is that its melting temperature is 195 K which is higher than for butanol where it is 183 K. Moreover their glass transition temperatures are 170 K and 162 K,

TABLE 1. Pentanol-2 polarization and relaxation time

Sign of the polarization	P(H) after 1 hour	Nuclear relaxation time (days)
positive	84%	25
negative	87%	21

respectively (2). This remark is important for target operators as the loading of the cartridge in the refrigerator at liquid nitrogen temperature is often a period of short warming up of the sample through rinsing of the Helium circuit and injection of warm $^3He/^4He$ gas mixture. If warming up reaches glass transition temperature, there will be a significant reduction of polarization and relaxation time. Recent works on Pentanol-2 (3), led us to use this product on experiment 225 at SATURNE II. Pentanol-2 ($CH_3CH_2CH_2CHOHCH_3$) differs from Pentanol-1 ($CH_3CH_2CH_2CH_2CH_2OH$) by isomeric structure, increasing the stability of the molecule. On the other hand, Pentanol-2 is strictly amorphous, independantly of the cooling rate, which enables the sample to be safer towards devitrification processes.

Polarization and relaxation time results are displayed in TABLE 1. Conditions were 2.5 T polarizing field, 42 mK frozen spin temperature and 0.33 T holding field. The target while in holding mode was constantly submitted to a proton beam the intensity of which being less or equal to 10^8 particles/burst. Due to physics requests, target polarization had to be reversed every day and time spent for building up was strictly confined to one hour.

3. POLARIZATION OF ^{6}Li PRODUCTS

^6LiH as well as ^6LiD are crystalline solids. To be able to polarize them one has to go through a well defined routine (4) the main steps of which we are going to explain here. The raw products are delivered as small stones which have first to be crushed to calibrated grains in nitrogen atmosphere to prevent any oxidation. The size of each grain is about 2 mm. To create vacancies in the crystal net frame which will act as paramagnetic centers, the grains are irradiated at low temperature in an inert atmosphere by electron beam. After irradiation, the samples are kept in liquid nitrogen until there are transfered to the dilution refrigerator for polarization.

Fig.1 shows polarization building up of three ^6LiD samples irradiated at the same temperature (185 K), but with different electron doses. The polarizing field was 2.5 T. Results are summarized in TABLE 2, the nuclear relaxation time was measured after 20 hours under a 0.33 T holding field and 50 mK temperature.

TABLE 2. Deuteron polarization towards dose

Sample	Electron dose per cm^2	P(D) after 1 hour	P(D) after 7 hours	Nuclear relaxation (days)
1	3×10^{17}	30%	42%	29
2	1×10^{17}	22%	40%	66
3	3×10^{16}	7%	30%	70

The choice of the sample is directly connected to the physics experiment conditions. If target polarizations have to be reversed often, sample 1 would fit and sample 2 would be better for long running periods without any change on polarization direction.

In physics off-line analysis, measured data from scattering of polarized beam on polarized target are processed through a continuous range of cuts. One of the first steps is the study of angular correlation and coplanarity. It consists mainly to work out the differences between the measured and calculated angles of scattered and recoil tracks of particles $\Delta\theta_{CM} = \theta_{calc} - \theta_{meas}$ and $\Delta\phi = \phi_{calc} - \phi_{meas}$, for the experimental set up (5). These kinematics cuts select elastic (or quasielastic) events on a "background piedestal", which has to be subtracted. In Fig.2 we compare the $\Delta\theta_{CM}$ distributions in np scattering at 0.88 GeV for the Pentanol-1 PPT, ^6LiD and ^7LiH targets. The shape of the distribution for ^6LiD differs from those of the two other targets, background being uneasy to estimate. This feature was one of the reasons for using also a ^6LiH target in order to work out correctly the ^6Li contribution and enable background normalization.

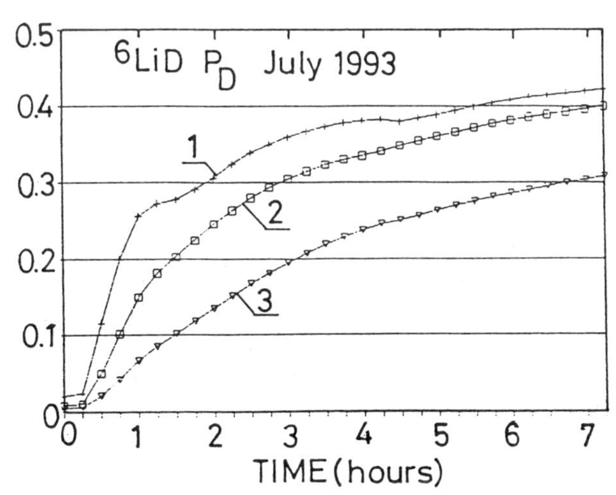

FIGURE 1. Deuteron polarization building up of three ^6LiD samples (see TABLE 2).

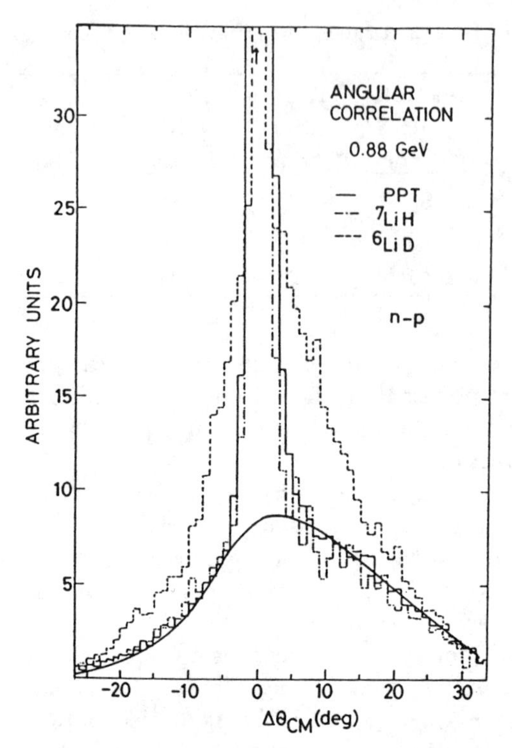

FIGURE 2. Comparison of the $\Delta\theta_{CM}$ distribution in *np* scattering at 0.88 GeV for Pentanol-1 PPT (solid line), ^6LiD (dashed line) and ^7LiH target (dot-dashed line). The solid curve is provided by a Monte Carlo calculation for the carbon background.

REFERENCES

(1) Bernard, R. et al., *Nucl.Instrum.Methods,* **A249**(1986), pp. 176-184

(2) Takala, S, and Niinikoski, T.O., Measurements of glass properties and density of hydrocarbon mixtures of interest in polarized targets, in *Proceedings of the 9th International Symposium on High Energy Spin-Physics, Bonn, 6-15 September 1990,*Vol.2: Workshops, (1991), pp. 347-352

(3) Sahling, S., Low temperature thermal properties of Pentanol-2 - A perspective polarized target material, in *Proceedings of the 9th International Symposium on High Energy Spin-Physics, Bonn, 6-15 September 1990,*Vol.2: Workshops, (1991), pp. 353-357

(4) Durand, G. et al., Progress Report on polarizable Lithium hydrides, in *Proceedings of the 10th International Symposium on High Energy Spin-Physics, Nagoya, November 1992,*(1993), pp. 355-364

(5) Ball, J. et al., *Nucl.Instrum.Methods,* **A327**(1993), pp. 308-318

The New SMC Dilution Refrigerator

J. M. Kyynäräinen[a], M.P. Berglund[b], and T.O. Niinikoski[a]

[a] *CERN, 1211 Geneva 23, Switzerland*
[b] *Low Temperature Laboratory and Institute of Particle Physics Technology Helsinki University of Technology, 02150 Espoo, Finland*

Abstract. The dilution refrigerator for the SMC twin solid polarized target at CERN is by far the largest one in operation. The 2.5 liter, 1.5 m long target is loaded horizontally into the cold mixing chamber. A cooling power of 400 mW at 300 mK temperature is available for dynamic nuclear polarization. Proton and deuteron targets are polarized to ±94% and ±47%, respectively. The base temperature of 30 mK enables rotation of the polarization vector without losses.

INTRODUCTION

The experiment NA47 of the Spin Muon Collaboration (SMC) at CERN SPS determines the spin dependent structure functions of the nucleons by scattering polarized muons from solid polarized proton and deuteron targets (1,2). The target material is either normal or deuterated butanol frozen into glassy beads. The nuclear spins are polarized with dynamic nuclear polarization (DNP), based on microwave saturation of impurity electron spins near their paramagnetic resonance in a 2.5 T longitudinal field.

In our experiment the same beam penetrates oppositely polarized target halves to minimize the effect of beam flux variations. The spin directions are frequently reversed to cancel residual false asymmetries. This is done by rotating the magnetic field at 0.5 T.

Dilution refrigeration is essential because of the possibility to effectively "freeze" the spins as the spin-lattice relaxation becomes very slow at low temperatures and thereby allows the field rotation procedure with minor loss of polarization.

CONSTRUCTION OF THE REFRIGERATOR

The geometry of the refrigerator is dictated by the requirement to provide minimum amount of additional material in the muon beam path and by the handling of the target material below 100 K. In Fig. 1 the refrigerator and the magnets (3) are shown with the target in place.

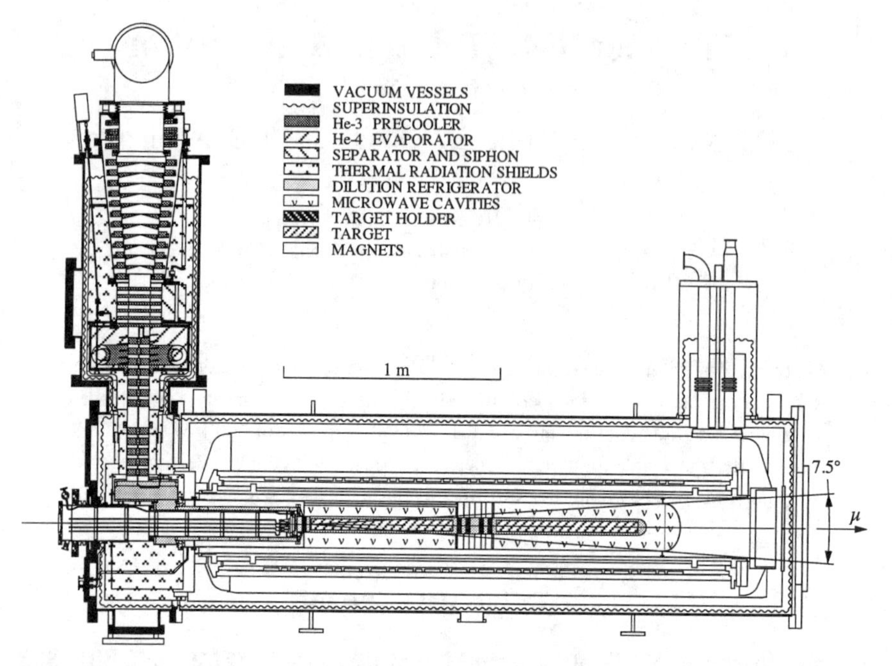

FIGURE 1. The longitudinal cross-section of the dilution refrigerator, shown together with the magnets. The muon beam enters the cryostat from the left.

The Precooler, the Evaporator and the Still

Compared with refrigerator (4) of the predecessor experiment NA2 of EMC, our refrigerator has a considerably more elaborate system for precooling. A schematic flow diagram of the refrigerator, showing the arrangement of the heat exchangers, is shown in Fig. 2. ^{4}He vapour at atmospheric pressure is taken from a phase separator to precool the incoming ^{3}He in a parallel tube counterflow heat exchanger (No. 1 in Fig. 2). This is also in thermal contact with the low-pressure ^{3}He and ^{4}He in a 4-flow heat exchanger (No. 2). Fin type exchangers have been chosen to give reasonable pressure drop with high gas flow at low pressure.

The evaporator volume is 27 liters, which gives ample space for a ^{3}He condenser and provides good tolerance to sudden heat load variation without risk of drying up or overflowing. A copper grid type heat exchanger (No. 3) is used to cool the incoming ^{3}He and the ^{4}He liquid from the separator. This reduces the ^{4}He boil-off rate by up to 30%. During initial cooldown the heat exchanger can be bypassed. The condenser (No. 4) is made of copper tubes with inner surface area of 0.35 m^2. Between the condenser and the Joule-Thompson needle valve a fin/grid type heat exchanger is placed in the still pumping tube (No. 5).

The still has two main volumes, the lower one containing a tubular heat exchanger for incoming ^{3}He and the upper one a heater and a level gauge. The heat exchanger (No. 6), made of copper tubing, has surface areas of 0.12 m^2 and

0.23 m^2 for the ^{3}He-rich and the dilute phases, respectively. The still heater is made of a stainless steel strip, placed vertically and having a surface area of 0.58 m^2. At the maximum design flow of 500 mmol/s about 16 W of heating power is needed, corresponding to a power density less than 3 mW/cm^2 to avoid film boiling. No special precautions were taken to suppress the superfluid film creep into the still pumping tube, as the estimated film flow rate is much smaller than the lowest practical ^{3}He flow rate.

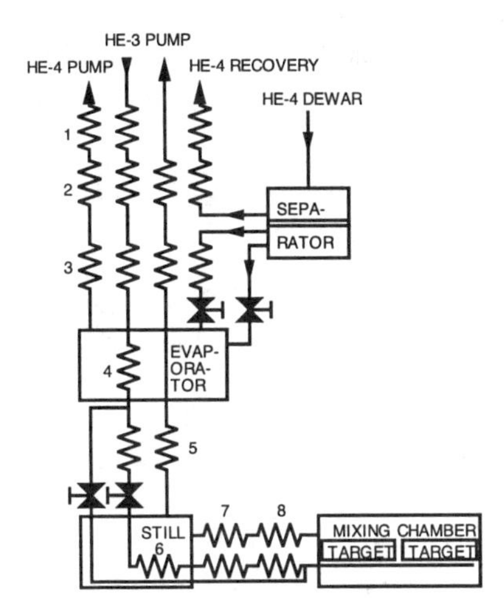

FIGURE 2. The arrangement of the heat exchangers in the refrigerator.

The Main Heat Exchanger

Between the still and the mixing chamber a tubular (No. 7) and a sintered copper (No. 8) heat exchanger are placed into a helicoidal groove formed by glassfibre-epoxy spacers. The groove defines also the dilute phase flow channel. A machined stainless steel tube with tight fitting to the spacers seals off the flow channel. The length of the flow channel along the centerline is 1.7 m and the free cross-sectional area is about 11.5 cm^2. The latter was calculated using data from previous smaller dilution refrigerators (5).

The tubular heat exchanger is used to condense vapour bubbles which may have formed in the flow at the expansion needle valve and to cool the helium below the phase separation temperature before entering the narrow channels of the sintered heat exchanger. The unit is made of flattened stainless steel tubes and has a total inner surface area of 0.1 m^2.

The sintered heat exchanger consists of 12 elements with dimensions of 12 x 4 cm^2 arranged in two parallel streams. To prevent cold plug formation due to the increasing viscocity of ^{3}He below 0.5 K, the flow in the streams is crossed at several points. The base material is 0.2 mm Cu(P) foil. The elements consist of two sintered plates which were first bent to fit into the flow channel and then electron beam welded together. The average thickness of the layer of the nominally 18 μm grain size sinter is 0.75 mm, yielding 375 g of sinter and a geometrical surface area of 12 m^2 on both the concentrated and the dilute phase streams. The heat transfer is enhanced by grooves on the sinter surface to increase the interfacial area with fluid flow and to produce turbulence in it.

The calculation of the required surface area was based on the principles of optimized design of Ref. 6. As the refrigerator has to provide high cooling powers at high temperatures, the effective surface area cannot be increased by increasing the sinter thickness, as the transverse heat conduction in the helium and in the sinter limits the heat transfer. Instead, one has to rely on increasing the interfacial surface area between the sinter and the fluid streams.

The Mixing Chamber, the Target Holder and the Microwave Cavity

The mixing chamber, with a length of 1500 mm and a diameter of 70 mm, is made of glassfibre reinforced epoxy with 0.6 mm wall thickness to ensure sufficient rigidity and to withstand slight overpressures in the case of a pump failure.

The target holder slides into the horizontal access tube and is sealed with a cold indium seal at the still back flange. A stainless steel vacuum chamber provides thermal isolation and a plastic part confines and supports the material. The vacuum chamber has two 0.1 mm stainless steel windows for the beam access and 6 aluminium foil thermal shields, and it provides thermal anchors for the coaxial lines for NMR and for the instrumentation wires.

The plastic part is mostly made of Kevlar-epoxy composite for rigidity and reduced thermal contraction. The target material is located in two cells with a length of 650 mm and a diameter of 50 mm, separated by 200 mm. The net volume of the target is 2x800 cm^3. Good heat transfer is assured by making the cells of polyester net with 60% open area. The containers weigh only 30 g each.

To ensure uniform temperature, ^{3}He is fed into the mixing chamber through 40 holes in a CuNi tube which is fixed to the target holder. A spring-loaded conical connector couples the outlet of the heat exchanger to the CuNi tube when the target holder is in place.

The mixing chamber is surrounded by a cylindrical microwave cavity of 210 mm diameter, made of copper. The cavity is divided axially in two compartments by placing graphite foil coated copper baffles and copper reflectors in the center. Inside the mixing chamber isolation was obtained using graphite coated Nomex honeycomb absorbers and fine copper mesh reflectors, designed to ensure free diffusion and convection in the dilute solution. The microwave isolation was measured to be 20-30 dB in the 69-70 GHz band with empty cavities; this prevents the microwave power from leaking excessively from one side of the cavity to the other. The cavity is cooled to 3 K by ^{4}He flow controlled by a cold needle valve. The two wave guides feeding the cavity enter the main vacuum through FEP windows and continue with circular sections made of silvered thin-walled CuNi tubing and further with rectangular ones. FEP windows close the guides before the coupling slots to the cavity, isolating its vacuum from the main vacuum and preventing the loss of ^{3}He in case of mixing chamber rupture.

PUMPS

We use the same pumping system for ^{3}He as the EMC experiment with 8 Root's blowers in series, giving a volume speed of 13500 m^3/h. This is sufficient to lower the pressure of the still to 0.2 mbar at the flow rate of 30 mmol/s and 1.2 mbar at 200 mmol/s. The pumps are connected to the refrigerator via a 20 m long line, 320 mm in diameter. At a flow rate of 200 mmol/s the pressure drop is about 0.4 mbar across both the 320 mm line and the heat exchangers above the still.

Charcoal traps at room temperature and at 77 K are used to filter out impurities in the ^{3}He return flow, in addition to the zeolite filters in the pumps.

The ^{4}He evaporator is pumped by a Root's blower followed by a rotary pump with a nominal speed of 2000 m^3/h, maintaining the evaporator at 1.5 K. The cavity cooling line is connected directly to the rotary pump.

CONTROL, DATA ACQUISITION AND SAFETY SYSTEMS

Temperature measurement below about 10 K is based on carbon and RuO resistors, read by 4-wire AC resistance bridges. Two carbon resistors are dedicated to monitor the microwave power at the target; other resistors are partly shielded against the microwave field. Higher temperatures are measured using Si diodes. High-accuracy pressure measurements, e.g. that of the ^{3}He vapour pressure during NMR signal calibration, are done using capacitive pressure gauges. The evaporator and still levels are monitored by capacitance bridges.

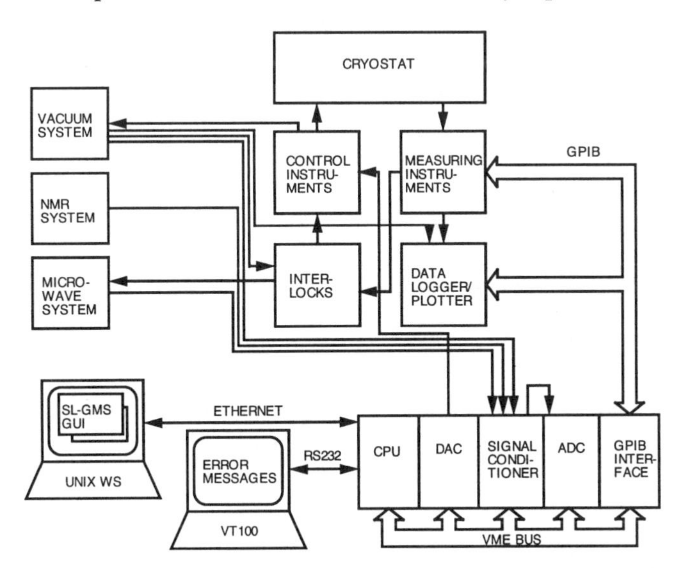

FIGURE 3. Block diagram of the cryogenic control, data acquisition and safety system.

A graphical user interface, based on a commercial software package SL-GMS, is running in a Unix workstation and X terminals. The control programs are running in VME processors. More than 100 cryogenic parameters are logged by the programs which also generate alarms. Most readout instruments are read via a GPIB bus, either directly or using a 32-channel data logger/plotter (see Fig. 3).

The controlled parameters include the needle valve for evaporator filling, the separator and microwave cavity flow rates and the still heater. The latter is to be controlled by the microwave power to maximize the cooling power.

A PLC-based interlock system, powered by an uninterruptible supply, protects the target against loss of polarization and of ^{3}He.

PERFORMANCE

The cooldown of the magnet from room temperature to 4 K takes about one week. The dilution refrigerator is then cooled in 8 hours to 77K, the cells of the target holder are filled with the material in a special LN_2 bath and the target holder is pushed into the precooled mixing chamber. Purging, final cooldown and condensing the ^{4}He-^{3}He mixture (8700 litres STP) takes less than 8 hours.

During DNP the temperature of the helium mixture decreases slowly from about 350 mK to 200 mK as the optimum microwave power is reduced with increasing polarization. The maximum polarizations obtained in the proton and deuteron targets are ±94% and ±47%, respectively. About 95% of the maximum polarization is reached in 12 hours of DNP.

The target is cooled down below 100 mK by turning the microwave power off 0.5-1 hours before the field rotation. The ultimate temperature is reached only after several hours of running without microwaves. This suggests that the microwaves heat the isolator and that the thermal contact is sufficiently poor to the mixing chamber to prevent a rapid cooldown. However, practically no loss of polarization takes place during the field rotation.

The cooling power in the mixing chamber with optimum ^{3}He circulation is shown in Fig. 4. The temperature was measured at the outlet of the dilute phase from the mixing chamber. Residual heat leak of about 1 mW to the mixing chamber is mainly from this end, and the temperature there is 20-30 mK higher than in the downstrean end where the lowest temperature of 30 mK is obtained.

The ^{3}He flow rate has practical minimum and maximum values of 27 and 350 mmol/s with the ^{4}He contamination of about 25% which is a consequence of the rather high still temperature of 0.95-1 K.

The liquid ^{4}He consumption of the refrigerator varies between 15 and 40 l/h depending on the ^{3}He flow rate. This compares favourably with the maximum of 70 l/h of the EMC dilution refrigerator, and gives a safety margin to the helium liquifier with a maximum capacity of 100 l/h and which supplies also the magnets.

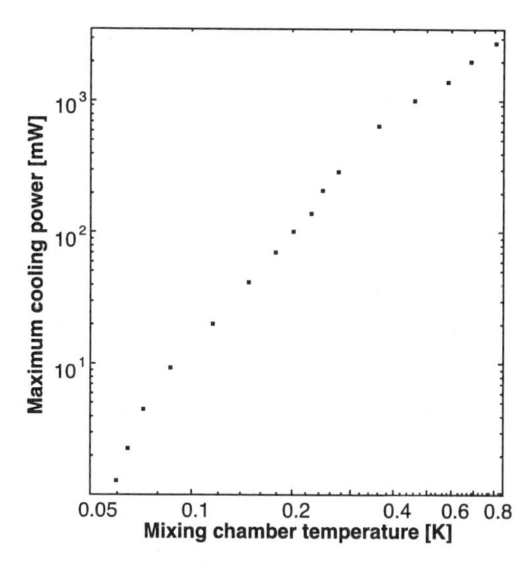

FIGURE 4. The maximum cooling power vs. mixing chamber temperature.

CONCLUSION

Dilution refrigeration has been shown to be a very suitable method to reach high polarizations in a large target and to manipulate the polarization vector. The practical limit of the target size is set by the cooling power of the refrigerator, determined by the available pumping speed of ^{3}He. Several innovations have improved the reliability considerably, e.g. the glassfibre-epoxy mixing chamber, vacuum brazed heat exchangers and the use of demountable metal seals at superfluid ^{4}He temperatures. Computer control and graphical user interface enable nonexperts to handle the refrigerator on routine basis.

ACKNOWLEDGEMENTS

We wish to thank G. Bonnefond, M. Bron, R. van Danzig, B. Feral, D. Geiss, H. Herbert, J. Homma, M. Houlmann, A. Isomäki, J. Kaasinen, S. Kaivola, J.-C. Labbe, Y. Lefevere, L. Naumann, A. Staude, B. Trincat, and S. Utriainen for their help in machining, assembling and testing the parts of the refrigerator.

REFERENCES

1. B. Adeva et al., *Phys. Lett. B* **302**, 533 (1993).
2. D. Adams et al., *Phys. Lett. B* **329**, 399 (1994).
3. A. Daël et al., *IEEE Trans. on Magnetics* **28**, 560 (1992).
4. T. O. Niinikoski, *Nucl. Instr. and Meth.* **192**, 151 (1982).
5. T.O. Niinikoski, *Nucl. Instr. and Meth.* **97**, 95 (1971).
6. T.O. Niinikoski, "Dilution refrigeration: new concepts", in *Proceedings of the Sixth International Cryogenics Engineering Conference*, 1976, pp. 102-111.

First Test of the CHAOS Polarized Proton Target.

P.P.J. Delheij, G. Mantel and I. Sekachev.
TRIUMF, 4004 Wesbrook Mall, Vancouver, B.C. V6T 2A3, Canada.

Abstract

The CHAOS polarized proton target (CPPT) is a new dynamically polarized target for the recently commissioned Canadian High Acceptance Orbit Spectrometer at TRIUMF. During the first test with a sample, a polarization of 0.70 was obtained in 1 hour and a decay time of 63 hours was measured at a temperature of 60 mK. The target must travel 2 m through a field free region between the polarizing magnet and the spectrometer magnet. Therefore, a superconducting magnet, giving a field of 0.13 T on the target, is incorporated in the cryostat shell surrounding the dilution refrigerator. The magnetic field in the spectrometer will be lowered to 0.5 T. These conditions require the operation in frozen spin mode. Particle access and exit is by two windows spanning 320 degrees in the scattering plane. Vertically the opening angle is ± 20 degrees. Particles with an energy down to 30 MeV must be observed which requires very thin windows. A major goal is the determination of the πN σ-term from low energy pion scattering. Then the strange sea quark content of the proton can be derived.

A new dynamically polarized target has been build for the CHAOS spectrometer at TRIUMF, which started operation recently. The spectrometer has a very large opening angle of 360 degrees in the horizontal plane and ± 20 degrees in the vertical direction. The target provides an opening angle of 320 degrees because the two window support posts extend each over only 20 degrees in angular range. In the vertical direction the full 40 degrees is supported. An important measurement is the determination of the analyzing power for elastic scattering of pions with an incident energy as low as 30 MeV. Extrapolation of the scattering amplitudes to zero energy and the application of chiral perturbation theory will set limits for the strange sea quark content of the proton.

The magnetic field in the spectrometer is only 1.4 T. Furthermore, the homogeneity is not sufficient for the process of dynamical polarization. Therefore, the polarized target must be transported over a distance of 2 m between the polarizing position in a solenoid with high homogeneity, which is located on top of the spectrometer as is shown in Fig. 1, and the scattering position in the spectrometer. To preserve the polarization a magnetic field must be applied during the transport. Therefore, a superconducting magnet, giving a minimum magnetic field of 0.13 T on the target, is incorporated in the cryostat shell. Due to the space limitations the magnet is very thin. It is wound from multifilament NbTi wire (purchased from Supercon) with a diameter of 225 μm. The magnet consists of 12 layers with 215 windings each. The magnet can carry 28.9 A, which is close to the design value of 30 A.*

*We appreciate the discussion with R. Gehring from the University of Bonn regarding the design of the current leads.

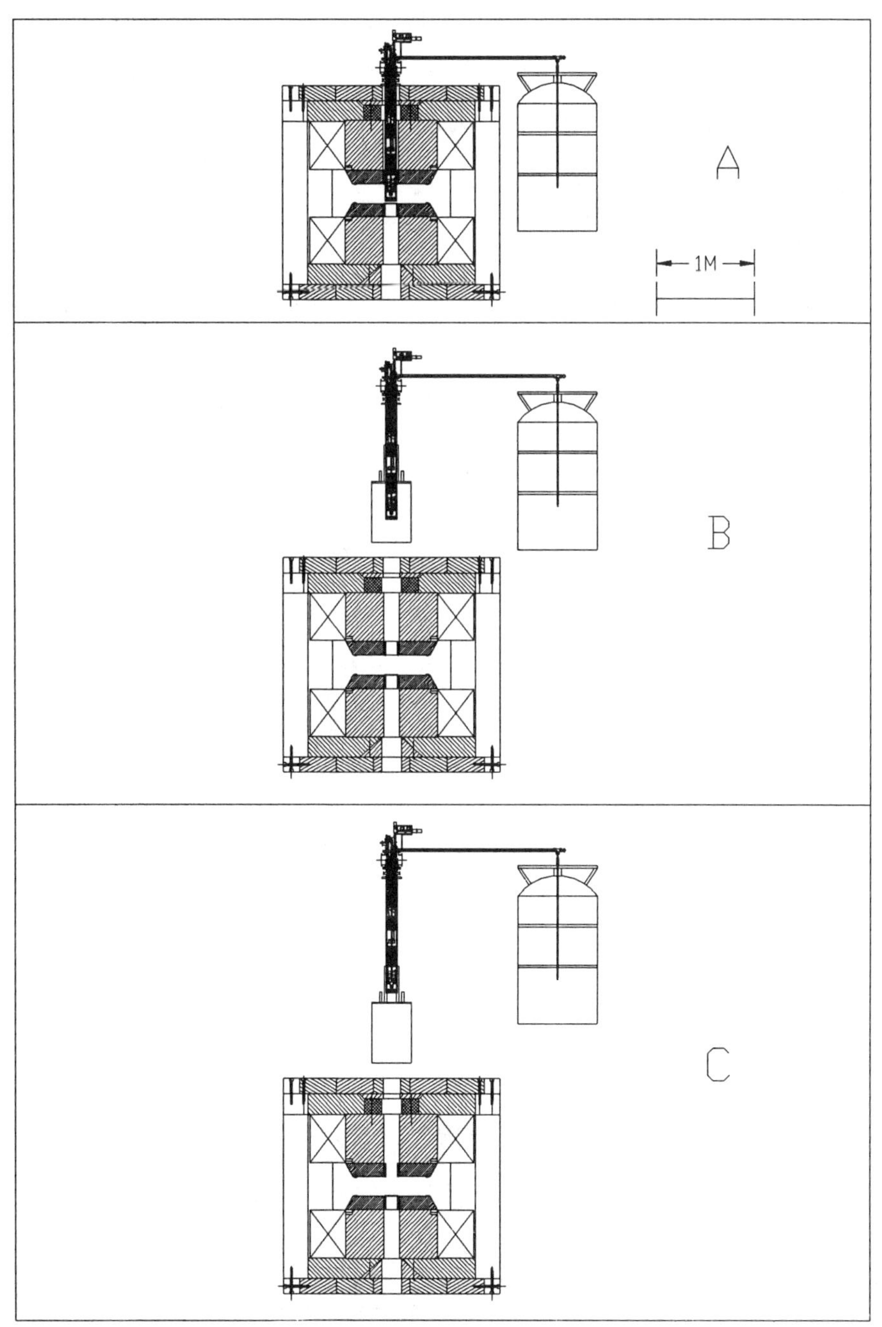

Figure. 1. Three positions of the target: a) The scattering position on the beamline, b) the target inserted in the polarizing magnet and c) the highest target point for repositioning the polarizing magnet.

The design of the setup followed the scheme that we succesfully developed previously(1). The refrigerator consists of a tubular counter flow heat exchanger, ^{4}He pots at a temperature of 4.2 K and 1.3 K respectively, tubular still heat exchangers, and a dilution stage mounted on a phenolic plug. In comparison with the previous refrigerator the number of sintered copper heat exchangers was increased from two to six. The NMR system is a constant current, series resonant Q-meter setup. Phase sensitive detection is applied to the difference of the output from the signal arm and a compensating arm. The tuning capacitor in the present setup is mounted outside the cryostat. The microwaves are generated with a tunable IMPATT source. A new pumping system has been assembled around a Balzers WKP-4000 Roots pump. Also a new control system was build based on Motorola 68040 processor in VME. Over ethernet any X-window terminal can control the setup. Remotely, pumps can be turned on or off, solenoid valves opened or closed, settings of mass flow controllers changed, heater output modified etc. Also all sensors, like thermometers, pressure transducers, flow meters etc. are read into the computer.

The target dimensions (Fig. 2.) are 30 mm wide, 25 mm high and 5 mm thick. It is a solid slab of 1-butanol with 5% water and a concentration of 5 x 10^{19} EHBA molecules per cm^3. Because of the low incident pion energies special care is taken to minimize the thickness of the walls. The sample is contained in a copper cup with walls that are 17 μm thick. There is 0.7 mm spacing between the sample cup and the inner vacuum wall of the cryostat. Alignment pins ensure that the target enters its rectangular cryostat chamber as the refrigerator is lowered. Two copper thermal shields with a thickness of 17 μm surround the target. The window in the outer wall of the cryostat consists again of 30 μm thick havar.

During the first test with a sample a polarization of 0.70 was obtained in 1 hour. Next a relaxation time of 63 hours was measured at a temperature of 60 mK in a magnetic field of 0.5 T. This is shorter than expected when the results from our previous frozen spin target at a temperature 55 mK in a magnetic field of 0.22 T are scaled(2) to the present temperature and magnetic field. Under those circumstances typical decay times were 200 hours. This reduction in relaxation time could be attributed to the different treatment of the sample material. Previously, beads were formed by dripping the solution into liquid nitrogen. In the present setup the solution is loaded in the copper cup. Then it is slowly immersed in liquid nitrogen as the refrigerator is lowered into the cryostat. A change of the loading procedure might improve this result.

With this setup we also continue our study of the influence of the microwave frequency modulation. At CERN a large increase in the deuterium polarization was obtained(3). In the spring of 1993 we observed an increase of the polarizing rate for protons by a factor 4 at high polarization when the modulation was turned on as Fig. 3 illustrates. However, these studies were hampered by the limited reproducibility of our IMPATT sources, with or without the application of the modulation. With the new setup a more systematic study of this process should be feasible.

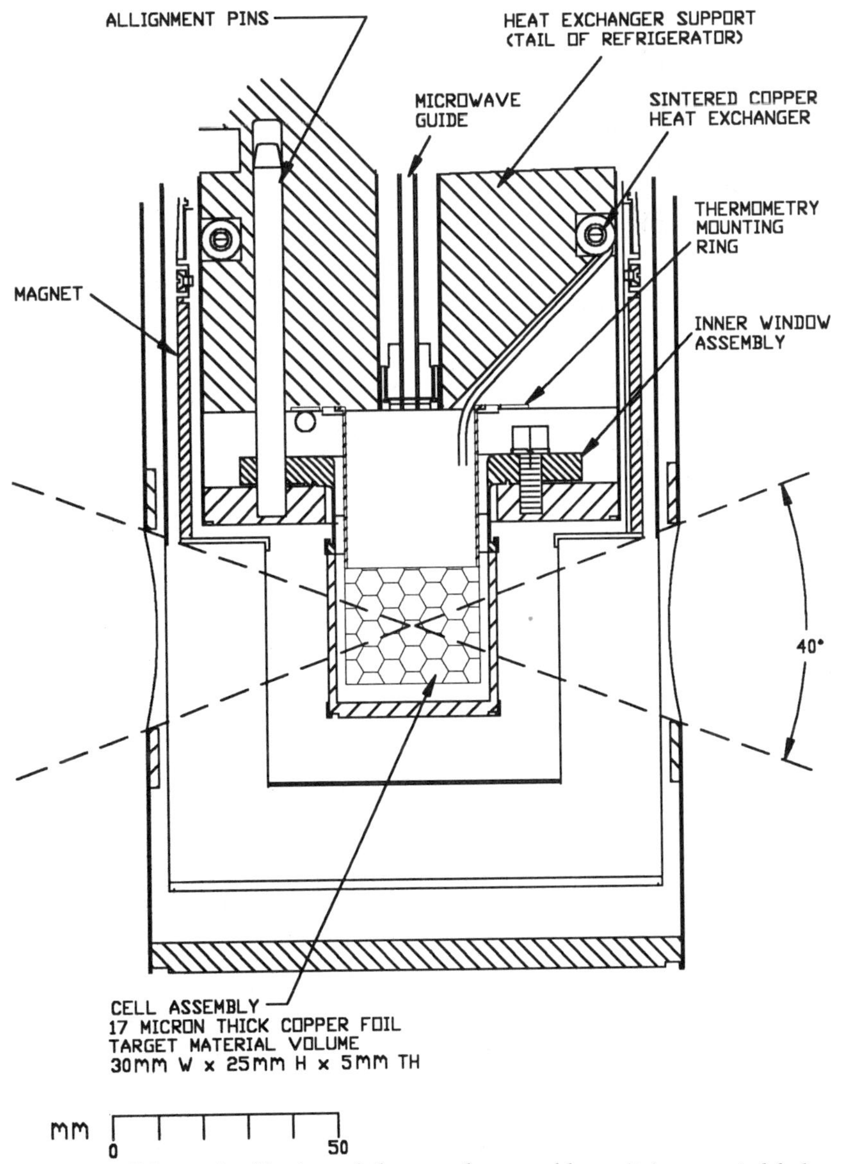

Figure. 2. Schematic sideview of the sample assembly as it is mounted below the dilution refrigerator.

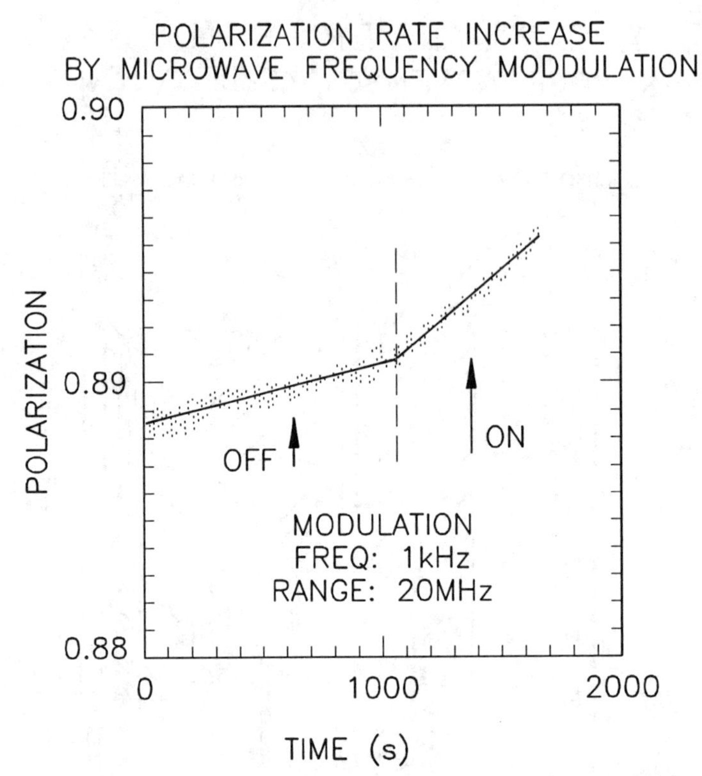

Figure. 3. The increase of the proton polarizing rate after turning on the microwave frequency modulation.

REFERENCES

1. Delheij, P.P.J., Healey, D.C. and Wait, G., Nucl. Instrum. Meth. **A264** (1988) 186.
2. Delheij, P.P.J., Healey, D.C. and Wait, G., Proc. Sixth Int. Symp. on Polarization Phenomena in Nucl. Phys., Osaka 1985, J. Phys. Soc. Jpn. **55** (1986) 1090.
3. Kisselev, Y., 7th Workshop on Polarized Target Materials and Techniques, Bad Honnef 1994, to be published In Nucl. Instr. Meth.

Proton Polarization at Room Temperature

M. Daigo[a], N. Hirota[c], M. Iinuma[b], A. Masaike[b], I. Shake[b], H. M. Shimizu[d], Y. Takahashi[b], R. Takizawa[b], M. Terazima[c], and T. Yabuzaki[b]
(presented by Masataka Iinuma)

[a] Premedical Course, Wakayama Medical College, Wakayama 649-63, Japan
[b] Department of Physics, Kyoto University, Kyoto 606, Japan
[c] Department of Chemistry, Kyoto University, Kyoto 606, Japan
[d] National Laboratory for High Energy Physics (KEK), Ibaraki-ken 305, Japan

Abstract

We have polarized protons in naphthalene doped with pentacene at higher temperature($\geq$77K) and lower magnetic field($\sim$3kG) than those for ordinary polarized proton targets. Pentacene molecules have been excited with a laser beam. Protons in naphthalene have been polarized dynamically on the intermediate state of pentacene. We obtain about 13% polarization at liquid nitrogen temperature in about 3kG with a N_2-laser of about 150mW. It has been found to be also possible to obtain high polarization at room temperature.

1 Introduction

The polarized proton target is an important tool in particle and nuclear physics. An organic material containing paramagnetic impurity [1,2,3,4] is used as a target material in many experiments. Usually, dynamic nuclear polarization method (DNP)[5] is used for obtaining the high polarization of protons. The electron spin polarization of the impurities is transferred to the proton spin by means of microwave irradiation in a high magnetic field($\geq$ 2.5T) at a low temperature ($\leq$ 1K). Recently, the group of Leiden University developed a new DNP method to polarize the proton by using a laser at room temperature in a low magnetic field [6]. They obtained the proton polarization of about 0.5%. The method will be applicable to real polarized proton targets, if higher polarization is obtained. We report on the study of obtaining proton polarization for a useful polarized target in a low magnetic field at high temperature based on this method.

2 Experiment

The polarization is transferred from electrons to protons on the excited state in an organic crystal by laser excitation. Our sample is a single crystal of naphthalene doped with pentacene (0.01mol%). The pentacene in the ground singlet (S=0) state (Fig.1) is excited to higher singlet states with a laser. The spin-orbital interaction causes the transition from the singlet excited state to the intermediate triplet (S=1) state. It is essential for this experiment that the electron spin in the intermediate triplet state is spontaneously aligned [7]. It is independent of the temperature and the strength of the magnetic field. The population difference between two levels among the three sublevels is transferred to protons in pentacene and naphthalene by means of microwave irradiation during the life time of this intermediate state. After electrons decay to the ground state, the proton spin is kept polarized for long time, since there is no spin-spin interaction between electrons and protons. This cycle is repeated. As the result, we can obtain the high polarization of protons. Our

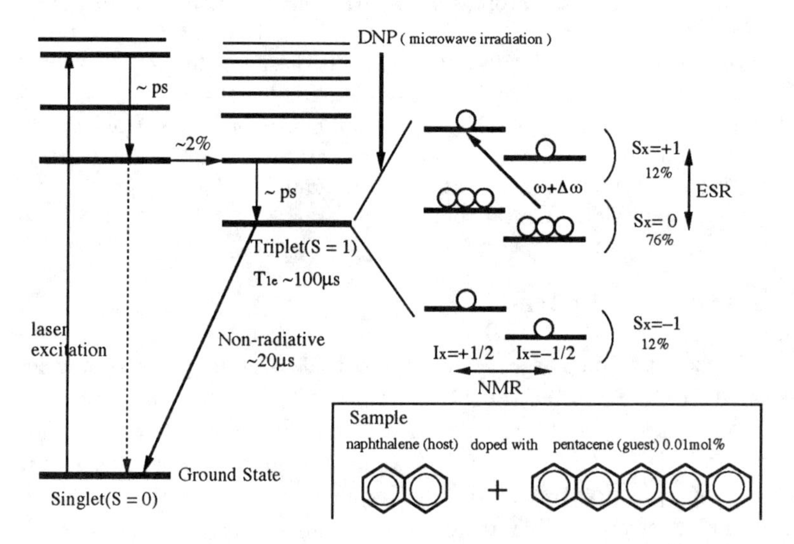

Fig.1: Energy Levels of pentacene

major experimental apparatus consists of an electron spin resonance (ESR) system, a nuclear magnetic resonance (NMR) system, and a laser. Our sample is placed inside the cylindrical cavity of which resonance frequency is about 9.3GHz. We applied magnetic field of about 3kG. The typical size of our sample is 3mm×5mm×2mm. We can confirm the alignment of electrons in the triplet state by detecting the ESR signal. The proton polarization is derived from the ratio of the size of NMR signal after DNP to the one in the thermal equilibrium. The radio frequency of NMR is 12.95MHz. We used two types of lasers, one is N_2-laser ($\lambda = 337$nm), the other is the solid state laser which we call TFR laser ($\lambda = 523$nm).

3 Experimental results

Our experimental results at room temperature are presented in Table 1. Since the relaxation time of protons in naphthalene without laser excitation is 40min.[6], the sample was irradiated by the laser beam for 40min. For obtaining higher polarization, we tried to excite pentacene in our sample with higher power of a laser. However, the polarization did not become so large. Furthermore, the polarization decreased when the laser increased so much. To solve this problem, we measured the laser power dependence of the proton relaxation time around the room temperature. The closed circle in Fig.2 shows the result. The relaxation time of the proton in naphthalene became much shorter with the higher laser power. We conclude that this behavior is caused by the slight heating of the crystal by laser excitation, since the relaxation time of protons in naphthalene molecule strongly depends on the temperature

laser	wavelength	R.R.	power	DNP time	pol.
N_2-laser	337nm	70Hz	24mW	40min.	0.08%
TFR-laser	523nm	1kHz	70mW	40min.	0.13%

Table 1: The results at room temperature R.R. : repetition rate

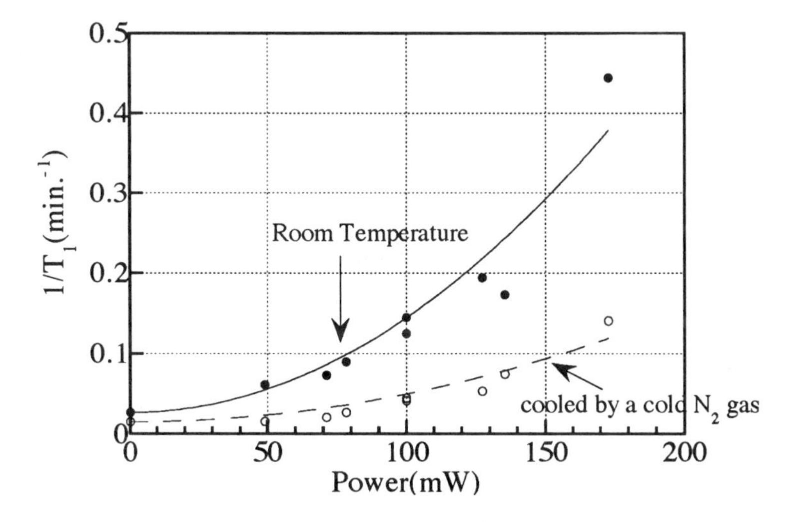

Fig.2: The laser power dependence of the relaxation time

near the room temperature[8]. To avoid this effect, we tried to cool down our sample by cold nitrogen gas. The opened circle in Fig.2 shows the laser power dependence of the relaxation time. The heating effect by laser excitation was reduced in this cooled sample. We also measured the relaxation time of the proton at liquid nitrogen temperature. It was about 1000min. even in the presence of N_2-laser with 150mW. Then, we performed the DNP experiment at liquid nitrogen temperature by using N_2-laser of 150mW. The Fig.3 shows the result. The polarization reached about 13% at liquid nitrogen temperature. We expect that the polarization more than 50% is achieved by using

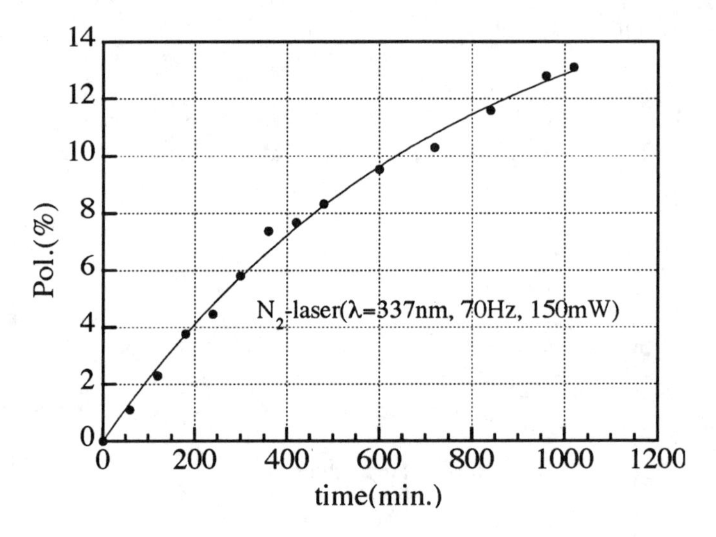

Fig.3: DNP at liquid nitrogen temperature

the higher power laser with $\lambda \sim 600$nm which is the most efficient wavelength for the excitation of pentacene. We have started to study other host crystals which do not strongly depend on the temperature. It is also possible to obtain high proton polarization in such a crystal at room temperature.

References

[1] D. Hill, J. B. Ketterson, R. C. Miller, A. Moretti, R. C. Niemann, L. R. Windmiller, A. Yokosawa, and C. F. Hwang, Phys. Rev. Lett. **A23** (1969) 460

[2] S. Mango, O. Runolfsson, and M. Borghini, Nucl. Instrum. & Methods **72** (1970) 45

[3] A. Masaike, H. Glattli, J. Ezratty and A. Malinovski, Phys. Lett. **30A** (1969) 63

[4] W. de Boer, Nucl. Instrum. & Methods **107** (1973) 99

[5] A. Abragam, and M. Goldman, Nuclear magnetism:order and disorder (Clarendon Press, Oxford, 1982)

[6] A. Henstra, T-S. Lin, J. Schumidt and W. Th. Wenckbach, Chem. Phys. Letters **165** (1990) 6

[7] D. J. Sloop, H. -L. Yu, T. -S. Lin, and S. I. Weissman, J. Chem. Phys. **75** (1981) 3746

[8] O. Lauer, D. Stehlik, and K. H. Hausser, J. Magn. Resonance **6** (1972) 524

A New Movable Polarized Target at Dubna

N.G. Anischenko[1], N.A. Bazhanov[3], B. Benda[5], N.S. Borisov[2], Yu.T. Borzunov[1], V.F. Burinov[2], G. Durand[5], A.P. Dzyubak[6], A.V. Gevchuk[6],L.B. Golovanov[1], G.M. Gurevich[4], A.I. Kovalev[3], A.B. Lazarev[2], Ph. Leconte[5], F. Lehar[5], A. de Lesquen[5], A.A. Lukhanin[6], V.N. Matafonov[2], E.A. Matyushevsky[1], S. Mironov[1], A.B. Neganov[2], N.M. Piskunov[1], Yu.A. Plis[2], S.N. Shilov[2], Yu.A. Shishov[1], P.V. Sorokin[6], V.V. Teterin[2], S. Topalov[4], V.Yu. Trautman[3], A.P. Tsvinev[1], Yu.A. Usov[2].

1.Laboratory of High Energies, JINR, Dubna, Moscow Region 141980 Russia

2.Laboratory of Nuclear Problems, JINR, Dubna, Moscow Region 141980 Russia

3.St Petersburg Nuclear Physics Institute, 188350 Gatchina, Russia

4.INR RAN, 60th Anniversary Prospect, 7A Moscow 117312, Russia

5.CEA DAPNIA, CE Saclay, 91191 Gif-sur-Yvette Cedex, France

6.KPTI, Akademicheskaya str.,1, 310108 Kharkov, Ukraine

Several experiments in Dubna were performed during last ten years using only polarized deuteron beam up to 9 GeV/c. The main goal of these experiments was devoted to study the deuteron wave function at the highest internal momentum [1-3].

Investigation of polarization phenomena gives more powerful information when polarized beam and polarized target are used simultaneously.

The first suggestion to use a polarized proton target (PPT) at Dubna was given by F. Lehar several years ago. But the existing situation at JINR did not give possibility to build such a target because its cost is about 1M \$. A new opportunity has arisen two years ago when the PPT , used previously at FERMILAB, has become available. Prof. A. Baldin (LHE, Dubna) supported the program of researches with the PPT and polarized deuteron and neutron beams.

This PPT (20 cm long and 3 cm diameter) was built at Saclay and Argonne in 1985-1988, mounted and tested at FERMILAB in 1988-1989 [4]. Despite the rich physical information obtained during five experiments on 200 GeV proton and antiproton polarized beams, this target was used only two months. The PPT remained unused during 1989-1992 and in December 1992 FNAL ordered to disassemble it. That was made in April 1993.

After the decision of Prof. J. Haissinski (DAPNIA, Saclay) and Prof. A. Yokosawa (ANL, Argonne) to use this equipment in JINR, a collaboration has been established in order to reconstruct this PPT. The LNP decided to participate in this collaboration. This target was originally designed for a set of experiments at a unique place at Fermilab. Since it is expected that several experiments at different beam lines will be performed. The collaboration decided to reconstruct it as a "MOVABLE" polarized target (MPT), which could be easily transported from one beam line to another one and to get this unit operated in Russia.

Experiments in Dubna will also require transverse polarization. For this purpose, transversal holding coils must be added to the target. Therefore, the target must be reassembled and upgraded.

Thus the MPT concept is following:
All the elements positioned close to the beam line are placed on two separated decks: the magnet deck and the dilution refrigerator deck, both can be easily translated relatively the frame (Fig. 1). Magnet deck is equipped with the polarizing magnet, its power supply, ^4He dewar and holding coils (in future). Dilution refrigerator deck includes ^3He pumping system, NMR system, microwave system, dilution refrigerator and two 4He dewars. Control room includes ^4He control unit, ^3He control unit, ^3He purification system, remote control of the magnet, NMR and microwave systems. Those decks and units may be transported without disassembly. Very small connection effort is required between those elements.

The status in September 1994 is following:
a) All the necessary parts of equipment for completion of the MPT in frozen spin mode with longitudinal polarization are available in Dubna for laboratory tests.
b) A main frame was reconstructed. Distance from the beam to ground was adjusted to 170 cm.
c) Polarizing magnet was installed and tested to 6.5 T, homogeneity is better than $2 \cdot 10^{-4}$.
d) New basic version of NMR was tested.
e) The surrounding for the dilution refrigerator is under assembly.
f) The pumping system of a new design is under assembly.
g) A beam line is being prepared for receiving the MPT.

The limited goal of the 1994 is obtaining good polarization in the target cell. This does not require the holding coils, since no transverse polarization is needed. After the test of polarization, the target will be placed on the beam line and tested again.

The further tasks are following:
Building a new polarizing magnet.

- Building a set of holding coils.
- Integrating the target on beam line and managing its use for physics.

All the functional elements of the MPT may be stayed in a sea - container without disassembly and transported. This will give the possibility to shorten the time for preparation to experimental run to 2-3 months. The reconstruction of the MPT will open new possibilities for the physicists from the member states of JINR and new states of the former Soviet Union and will also be attractive for the scientists from EC and other countries. An exhaustive experimental program may be determined using Dubna polarized beams and the MPT.

Three experiments have already been proposed:

- Measurements of the difference in the transmission cross sections of polarized neutrons through the PPT when beam and target polarization directions are parallel or antiparallel. The measurements will be done at several energies up to 3.6 GeV, and in two different spin orientations with respect to the beam direction: transversal and longitudinal [5]. Former measurements were already performed, below 1.1 GeV only, in Villingen (Switzerland), Los Alamos (USA) and Saclay (France). They showed unexpected behaviors, which have to be understood in doing measurements at higher energies.

- Measurements of the spin effects in the backward elastic scattering of transversally polarized deuterons on transversally polarized protons. This experiment was proposed by JINR groups and by the group from Kharkov Institute. The aim is to clarify the exact knowledge of the deuteron wave function which shows a disagreement with classical models [6].

- The third experiment concerns φ meson production with polarized deuteron or proton beams and with the PPT. Recent experiments with antiprotons at LEAR show that the ratio of φ to ω meson production is by a factor 50 - 100 times higher than expected from the OZI rule predictions. A strong dependence of the φ to ω ratio on quantum numbers of the initial state (e.g. on a total spin of the nucleon-nucleon system) was found. In ref. [7] it was suggested that this effect is connected with the polarized strange sea in the nucleon. The crucial experiments for tests of this model consist in measurement of $pp \Rightarrow pp + \varphi$ or $dp \Rightarrow^3 He + \varphi$ reactions using polarized proton or deuteron beams and a proton polarized target. An intrinsic nucleon strangeness will manifests itself in an asymmetry of the φ production for parallel or antiparallel polarizations of initial particles. This asymmetry will be independent on the space orientation of the particle spins.

The International Scientific Users Committee which will hold at Dubna in January 1995 will evaluate the relative interest of proposed experiments and will define priorities in order to share the use of the MPT. Reconstruction of the MPT is supported in part by the INTAS grant No. 93-3315.

REFERENCES

[1] L.S. Azhgirey, et al., in: Proc. of 14th International IUPAP Conference on Few Body Problems in Physics, May 26-31, 1994,Williamsburg, Virginia, USA, p.18 and p.22.

[2] B. Kuehn, et al., *Phys. Lett.*, **B334** (1994), 298-303.

[3] A.A. Nomofilov, et al., *Phys. Lett.*, **B325** (1994), 327-332.

[4] J. Chaumette, J. Deregel, H. Desportes, G. Durand, J. Fabre, L. van Rossum, D. Hill, Proc. Workshop on Polarized Sources and Targets, SIN-Montana, Jan. 1986, Helvetica Physica Acta Vol. 59, 1986, 786-787.

[5] J. Ball, et al., in: Proc. of International Workshop "DUBNA DEUTERON 91", Dubna, 11-13 June 1991, Dubna Preprint E2-92-25,1992, p. 12-21

[6] I.M. Sitnik, V.P. Ladygin, and M.P. Rekalo, Preprint JINR, E-1-94-23, 1994,

[7] J. Ellis, M. Karliner, D. Kharzeev, M. Sapozhnikov, Preprint CERN-TH/7326-94.

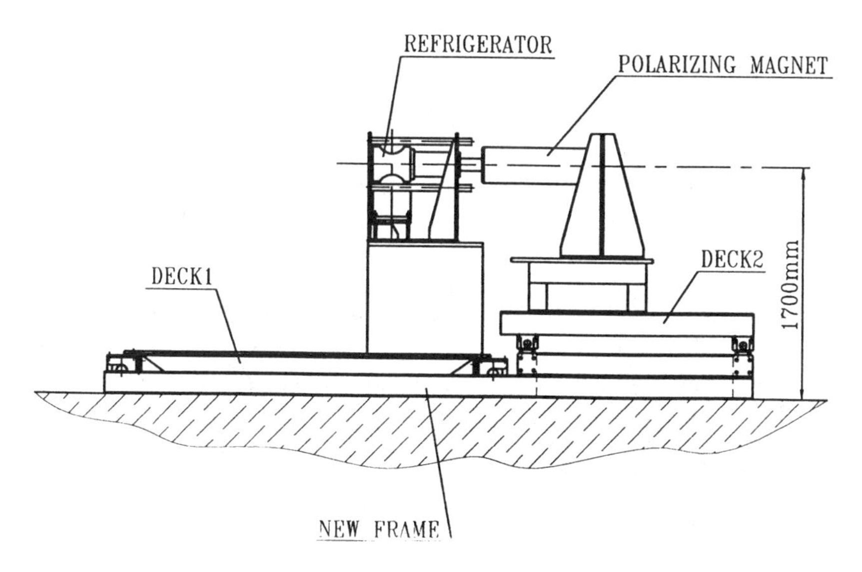

Figure 1. The MPT layout.

Operation of Polarized $^{15}NH_3$ and $^{15}ND_3$ Targets in a High Intensity Electron Beam: SLAC-E143 Target Report

T. Averett–Dept. of Physics, University of Virginia[1]

Abstract

Several radiation doped targets have recently been used in a polarized target experiment (E143) at the Stanford Linear Accelerator Center using a 29GeV electron beam ($5.0 \times 10^{11} e^-/sec$). The targets used were $^{15}NH_3$ and $^{15}ND_3$ that were pre-irradiated under liquid argon by electrons with energies of 35-350 MeV. Microwaves were used to dynamically polarize the nucleons in a 5 T superconducting magnet. Temperatures of ~1.0 K were achieved by means of a high cooling power ^{4}He evaporation refrigerator. After in-situ 1.0 K irradiation by 29 GeV electrons, polarizations up to ~75% and ~40% with beam on target were achieved for protons and deuterons respectively. Detailed polarization results, radiation damage and annealing characteristics, and general target performance will be presented here.

INTRODUCTION

An experiment (E143) was recently performed at SLAC to measure the longitudinal and transverse spin structure functions of both the proton and neutron using a high intensity polarized electron beam ($5.0 \times 10^{11} e^-/sec$) and a dynamically polarized nuclear target. Several spin structure measurements (EMC[1], E142[2], SMC[3, 4]) had been previously performed, but this was the first high statistics measurement of both the proton and deuteron, and the first to measure both the longitudinal and transverse structure functions. The goal of the target group was to construct and operate a polarized target

[1] On behalf of the E143 target group.

capable of maintaining high polarizations at these high beam intensities.

TARGET SYSTEM

The target system is shown in Fig.1. The target was of the dynamically polarized type. For this we used a 5.1 Tesla superconducting magnet from Oxford Instruments. It was constructed in a split Helmholtz pair arrangement allowing beam access from two orthogonal directions. The field direction was in the horizontal plane and the entire target system could be rotated with respect to the incident beam to allow for both longitudinal and transverse spin measurements.

The target material itself was $^{15}NH_3$ for protons and $^{15}ND_3$ for deuterons. Ammonia was chosen for its high polarizability and superior radiation resistance. The material was in the form of 2mm slow frozen granules and was pre-irradiated under liquid argon with 30-350 MeV electron beams. The typical dose was $10^{17}e^-/cm^2$. ^{15}N was used instead of ^{14}N in the ammonia since it does not contain any unpaired neutrons.

To drive the nuclear spins to the desired state requires the use of microwave radiation at the appropriate frequency ($\sim$ 138GHz at 4.87 Tesla). We used a 20 Watt EIO tube produced by Varian Canada. Power was monitored by a thermistor and the frequency was monitored using an EIP frequency counter. In addition, we had the capability to frequency modulate the microwaves about the central frequency by sweeping the cathode voltage on the EIO tube.

The largest obstacle to maintaining high polarization is overcoming the heating of the target material by the beam and the microwaves ($\sim 1.3Watts$). For this purpose we used a high cooling power ^4He evaporation refrigerator with a $12,000m^3/hr$ roots pumping assembly.

The polarization measurements were made using an NMR system similar to the one used by SMC at CERN[3, 4]. The system consisted of 5 Liverpool Q-meters, a PTS frequency synthesizer, and a STAC interface for readout and control of the system. All polarization measurements and general target monitoring was done via GPIB interface on Macintosh computers running LabView software. The polarization was measured on a continuous basis.

The target material was held in a movable target insert. This insert contained $^{15}ND_3$, $^{15}NH_3$, and empty target cells as well as a carbon target for calibration measurements. Any target could be placed into the beam region on demand. The insert also contained all instrumentation directly related to the the target material such as the heater and thermocouples for annealing, NMR coils, and temperature monitoring devices. Temperatures were monitored at the target region primarily by ^3He and ^4He manometers and various resistive devices.

TARGET PERFORMANCE

NMR Calibration

To measure the target polarization requires calibration of the NMR system. The system was calibrated in the usual way by measuring the area under the thermal equilibrium (TE) signal at a temperature of $1.6K$. Boltzmann statistics allows us to calculate the value of the polarization and this calibration is then used to calculate the enhanced polarization from the area under its NMR signal. For protons, the TE signal is large and the error in the calibration is dominated by systematic effects. For deuterons, the TE signal is very small and the statistical error is significant. For this experiment, our polarization errors are 2.5% and 4.0% relative for the protons and deuterons respectively.

Polarization Characteristics

Typical target polarization curves can be seen in Figs. 2 and 3. With beam on the target, maximum polarizations of 75% and 40% were reached for protons and deuterons respectively. In the case of the deuteron material, in–situ 1K irradiation increased the maximum polarization by a factor of 3 from its pre-beam value of 13%. This was consistent with what had been previously seen in $^{14}ND_3$ targets[5].

The deuteron polarization rate was increased by a factor of 2 by modulating the microwave frequency about its central value. Typically we modulated the central frequency by ±22MHz at a rate of 1.0kHz. No effect was seen for the proton material.

It is also worth noting that the deuteron material was located nearest the microwaves in the target insert with the proton material directly below. The result of this configuration was that the proton material was starved for microwave radiation. When the proton material was placed in the direct path of the microwaves, polarizations of greater than 90% could be achieved very quickly (with no beam on target). For the future it would be advantageous to couple microwaves directly into each target cell to increase the maximum polarization.

Radiation Damage

It is well known that the target material becomes radiation damaged due to the beam flux. The effect is a slow reduction in polarization. This is seen in Figs. 2 and 3. The proton material was typically run for 1×10^{16} incident electrons and then annealed at 80K for 5 minutes to recover from the damage. The deuteron material was typically run for 2×10^{16} incident electrons. For the deuteron material, the maximum polarization increased slightly after each anneal allowing us to eventually achieve polarizations of $\sim 40\%$. For the proton material, there was no increase or decrease seen in the maximum polarization after annealing. Figs. 4 and 5 show the polarization and anneal cycle for the life of a typical target.

Polarization of ^{15}N and Residual Protons

Because of the unpaired proton in ^{15}N, the nitrogen becomes polarized as the protons polarize in the $^{15}NH_3$ material. For the $^{15}ND_3$ material, both the nitrogen and residual protons in the material become polarized as the deuteron becomes polarized. Since these polarized species contribute to the measured asymmetry, their polarizations must be measured and corrected for. The equal spin temperature hypothesis gives a relationship between the polarization of any two species present in a material by assuming the spins of the species are all in equilibrium at a specific spin temperature. The polarizations of all of these materials was measured and was not consistent with the equal spin temperature hypothesis.

SUMMARY

We have successfully built and operated a dynamically polarized proton and deuteron target to measure both the longitudinal and transverse spin structure functions of the nucleon. This target operated at polarizations of 75% and 40% for protons and deuterons respectively at a beam intensity of $5 \times 10^{11} e^{-}/sec$. In-situ irradiation boosted the deuteron polarization by a factor of 3 over the pre-irradiated value of 13%. In addition, frequency modulating the microwaves increased the deuteron polarization rate by a factor of 2. The proton polarization was not affected by frequency modulating the microwaves. As was seen in previous experiments, ammonia is able to withstand high radiation doses with regular target annealing. In terms of polarization and radiation damage, nitrogen-15 ammonia seemed to behave in the same manner as nitrogen-14. For the future we expect to increase the proton polarization by directly coupling the microwaves into the target cavity.

References

[1] J. Ashman et al., *Nuc. Phys.* **B328**, 1 (1989).

[2] P.L. Anthony et al., *Phys. Rev. Lett.* **71**, 959 (1993).

[3] B. Adeva et al., *Phys. Lett.* **B302**, 553 (1993).

[4] B. Adeva et al., *Phys. Lett.* **B329**, 399 (1994).

[5] B. Boden et al., *Particles and Fields* **49**, 175 (1991).

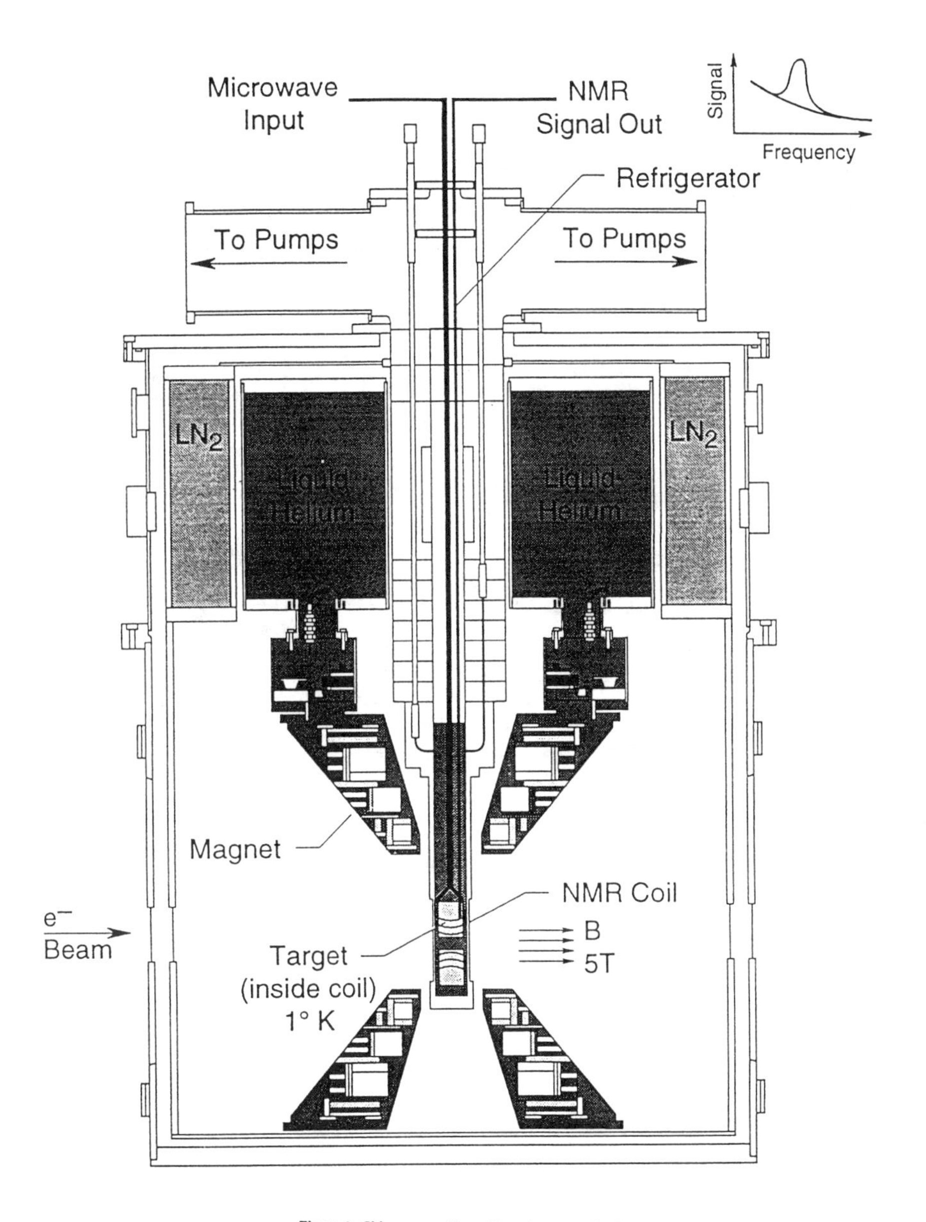

Figure 1: Side cross section of target system showing magnet coils, refrigerator, and target insert.

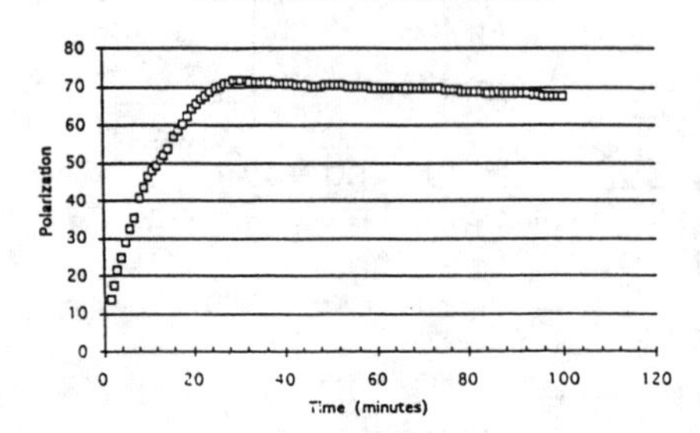

Figure 2: Proton polarization versus time showing slow decrease due to radiation damage.

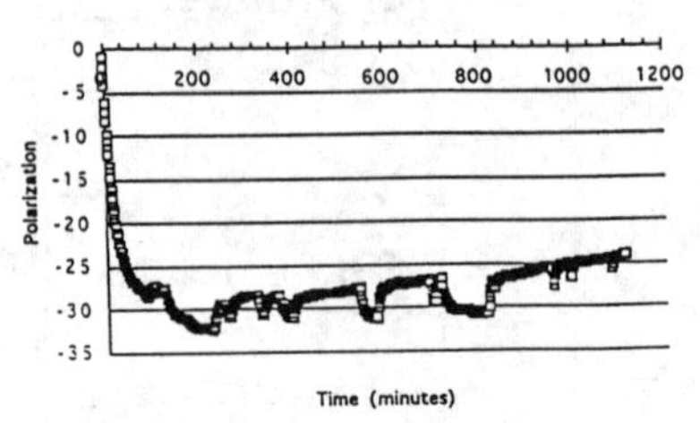

Figure 3: Deuteron polarization versus time showing slow decrease due to radiation damage. The bumps in the data occur when the beam is momentarily off.

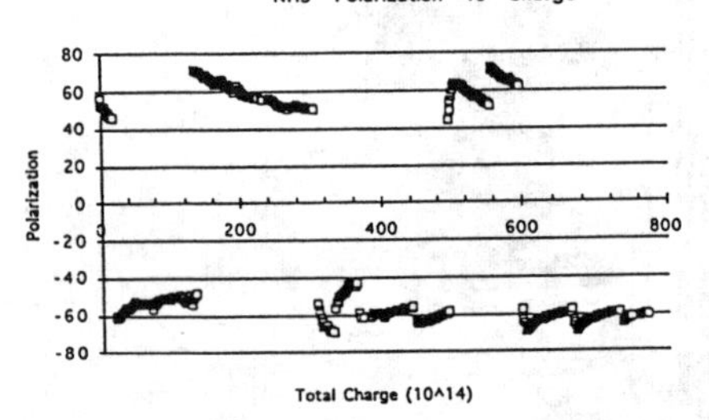

Figure 4: Proton polarization versus charge. Note that the polarization is recovered after annealing.

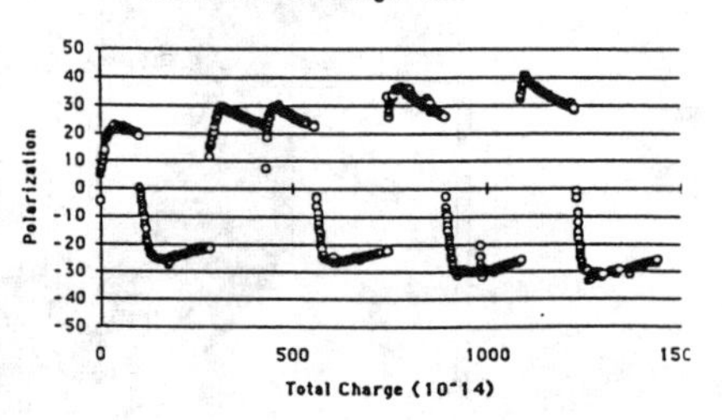

Figure 5: Deuteron polarization versus charge. Note the increase in maximum polarization after each anneal.

A new frozen spin target for the measurement of the Gerasimov-Drell-Hearn sum rule

H. Dutz, Ch. Bradtke, R. Gehring, S. Goertz, W. Meyer, M. Plückthun, G. Reicherz, A. Thomas, H. Werth

Physikalisches Institut der Universität Bonn
Nussallee 12
53115 Bonn
Germany

Abstract For the measurement of the Gerasimov-Drell-Hearn sum rule with real photons at the Mainz accelerator facility MAMI [1] a new horizontal polarized frozen spin target is under construction by the Bonn polarized target group. The horizontal $^3He/^4He$-dilution refrigerator includes an internal superconducting holding coil to maintain the polarization in the frozen spin mode longitudinal to the incoming γ-beam.

1 Introduction

To measure the polarized cross section difference of the Gerasimov-Drell-Hearn sum rule :

$$\int_0^\infty d\nu \frac{\sigma_{3/2} - \sigma_{1/2}}{\nu} = \frac{2\pi^2\alpha}{m^2}\kappa^2$$

a polarized solid state target is needed, whereby the spins of the nucleons in the target material are oriented parallel or antiparallel to the incoming polarized γ-beam. The γ-beam will be a tagged beam with a proposed intensity of $\dot{N}_\gamma = 5 \times 10^5 sec^{-1}$. Under this beam condition, the polarized target has to be a solid-sate target with its high polarized nucleon density ($\sim 10^{23}/cm^3$). As the Gerasimov-Drell-Hearn sum rule measurement requires the particle detection over a large range of scattering angles, the concept of the so called frozen spin target has to be used. To fulfill the special demands of the particle detection the target has to be designed as a horizontal frozen spin target including a $^3He/^4He$ dilution refrigerator and a horizontal superconducting polarizing magnet. A suitable detector for the first measurement in Mainz at MAMI will be the DAPHNE - detector of the Saclay/Pavia/Mainz - collaboration [1]. Because of the kinematics and the compact size of the detector, the beam pipe has to be implemented in the $^3He/^4He$ dilution refrigerator in the backward of the detector. This concept allows a large acceptance for the particle detection. To maintain the polarization during the 'frozen spin mode' a 'holding coil' which provides a longitudinal (horizontal) magnetic field is required. Here we will report about the design of the refrigerator and the magnet system as well as the pumping performance.

2 The dilution refrigerator

The horizontal ^{3}He/^{4}He dilution refrigerator represents the central part of the frozen spin target. The outer dimensions of the refrigerator are limited by the inner diameter of the detector tubes and the warm bore of the horizontal polarizing magnet. Taking this conditions into account, the front part of the cryostat has been designed with a diameter of 95 mm and the back part of 204 mm. The total length will be 2215 mm. A schematic drawing of the ^{3}He/^{4}He dilution refrigerator is shown in fig. 1. Since the refrigerator has to fit into the detector set-up the cryostat will be installed longitudinal to the beam axis. Thus the beamline is the central part of the refrigerator. The inner diameter of the tube varries between 78 mm in the back and 47 mm in the front part. Around the beam line tube, the cryogenic components of the refrigerator are placed. The tube will be evacuated and has to fulfill the demand of an internal isolation vacuum and is tightened against the mixing chamber and the ^{3}He circuit with a cold indium seal. To load the refrigerator the target cylinder, containing NMR coils with attached cables, microwave guide and the temperature sensors will be inserted into the refrigerator along the horizontal axis of the beam-line. On the outside the refrigerator is enclosed by an isolation vacuum. To reduce the thermal heat load given by thermal radiation the system is surrounded by two cooling shields.
The incoming ^{3}He will be precooled in a series of counter current heat exchangers by ^{4}He. The refrigerator is equipped with a separator to cool the ^{3}He stream in a coaxial heat exchanger down to about 4 K. The liquefaction of the ^{3}He gas will be accomplished in an evaporator at an operational temperature of 1.6 K. After passing a final heat exchanger the 'still-temperature' of 0.8 K will be obtained, with maximal recovery of the enthalpy of the outcoming ^{3}He-gas from the still.
The dilution unit consisting of the still, heat exchanger and mixing chamber is designed as a tube in tube system to reduce the heat load of the refrigerator. The heat exchanger will be build as a helic cooper tube with cooper sintered walls to increase the surface and the thermal conductivity of the tube. This type was former used in various ^{3}He/^{4}He dilution refrigerators [2]. The designed minimum temperature of the refrigerator is of the order of 50 mK.
For the dynamic nuclear polarization process a horizontal superconducting 6.5 Tesla magnet of the Saclay polarized target group will be used [3]. The magnet will be movable in the direction of the beam axis. In the polarization mode the magnet encloses the target. After switching into the frozen spin mode, the

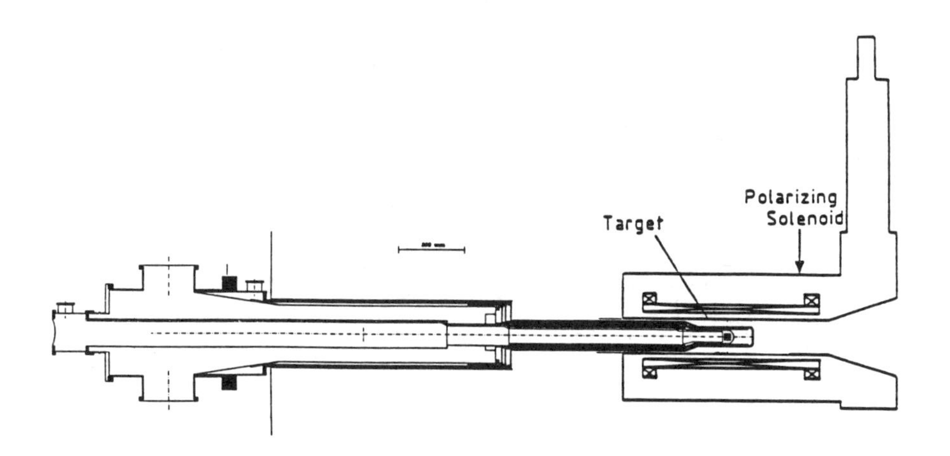

Figure 1: Cut view of the horizontal $^3He/^4He$ dilution refrigerator and the polarizing magnet

magnet will be moved out of the target position. Then the detector than can be placed in the data taking position around the refrigerator.

3 The internal superconducting holding coil

The polarization in the frozen spin mode will be maintained by a small thin superconducting coil, which is integrated in the dilution refrigerator around the target material. The coil is designed as a two layer winded solenoid. The compact inner field with a weak outer fringe field is one of the characteristics of a solenoid. By this scheme, which is different from operational frozen spin targets, a minimum field influence on the detector components is achieved.
To minimize the affect of the additional material for the outgoing particles we limited the thickness of the coil of about 500 μm. This thickness is divided in 300 μm of the cu - carrier and 2 layers of 120 μm NbTi - wire. The length of the solenoid is 120 mm, the diameter 44 mm. The maximum field of 0.35 Tesla in the target area is generated at a current of 17 A. The homogeneity of the central field of the magnet allows an online polarization measurement via NMR during the frozen spin mode. This will considerable lower the systematic error using a frozen spin target. A magnet of the same characteristics has been tested successfully in the existing Bonn frozen spin target [4]. The mixing chamber, the cold seal and the internal superconducting holding coil is schematically shown in fig. 2.

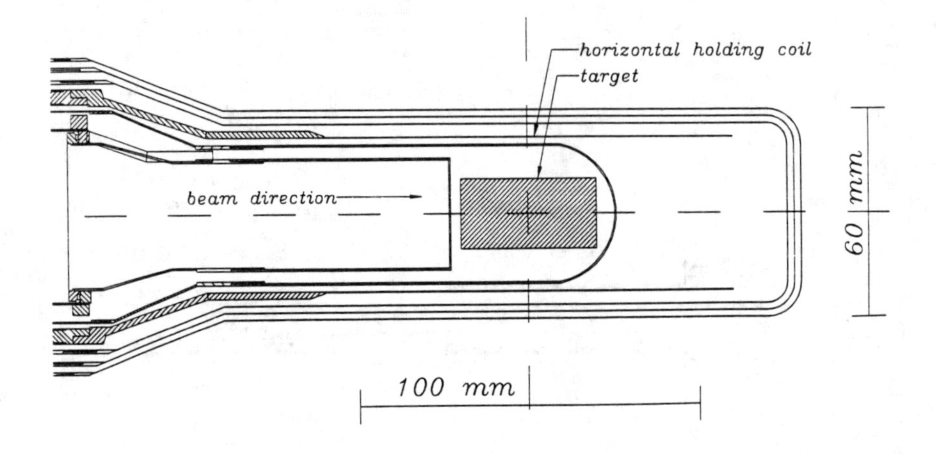

Figure 2: Cut view of the mixing chamber of the $^3He/^4He$ dilution refrigerator. The target material is encosed by an internal superconducting 'holding magnet'.

4 The pumping system

The pumping system is designed for a maximum ^{3}He circulation rate of $\dot{n}_3 = 30$ mmole sec^{-1}. It consists of a combination of oil-free compressing roots blowers Balzers WKP 4000, WKP 1000, WGK 500 and WGK 250, which are equipped with canned motors.

References

[1] J. Ahrends et al., Proposal to the PAC MAMI Mainz (1992).
G. Anton et al., Proposal to the PAC ELSA Bonn (1992).

[2] T.O. Niinikoski, Cryogenics 11 (1971) 232.

[3] P. Chaumette et al., Ad. in Cryogenic Engineering 35 (1990) 1067.

[4] H. Dutz et al., NIM A 356, (1995).

The SMC Polarized Target
— Systems and Operations —

Dirk Krämer

Universität Bielefeld, Fakultät für Physik, D-33615 Bielefeld, Germany[a]

(On behalf of the Spin Muon Collaboration)

Abstract. The SMC polarized target with all its functional subsystems, namely the dilution refrigerator, the superconducting magnet, the NMR, and the microwave system, is described. The performance of the used proton and deuteron target material is outlined.

1 Introduction

The SMC experiment measures the spin dependent structure functions g_1 of the nucleons by deep inelastic scattering of longitudinally polarized muons off polarized proton and deuteron targets [1, 2, 3]. The related questions addressed with these measurements are the Bjorken and the Ellis - Jaffe sum rules and the spin structure of the nucleons. The primary observable is the counting rate asymmetry of the parallel and antiparallel polarized beam and target which is linked to the structure functions via the virtual photon asymmetry A_1.
The SMC apparatus consists of three main parts: The *spectrometer* for vertex reconstruction, muon identification, and momentum determination, the *polarimeter* [4], which measures the beam polarization either by the decay method or via Møller scattering on a polarized iron foil, and the *polarized target*, providing a huge amount of spin aligned nucleons in two cells of opposite polarization direction.
The advantage of this *twin target* concept is the substantially decreased systematic error due to acceptance and flux variations. The common muon flux on both target halves together with frequent polarization reversals cancels to a large extent the effects of the absolute acceptances in each cell, so that the main cause of false asymmetries arises from the *time dependence* of the acceptance ratio between upstream and downstream target.

[a]Supported by Bundesministerium für Forschung und Technologie

2 System Overview

The purpose of a polarized solid state target is to provide a given amount of target material with a large fraction of polarizable nucleons in a state of utmost high polarization. Further demands are the homogeneity and the accurate measurement of the nucleon polarization, as well as the possibility to reverse the spin orientation. The last claim leads to the most common used principle for polarized targets, the so called Dynamic Nuclear Polarization (DNP), which allows the polarization reversal by simply adjusting the microwave frequency without changing the magnetic field.
In general a polarized target system can be subdivided in several functional subsystems:

— **Cryogenics:** Low temperatures ($T \leq 1\,\mathrm{K}$), mostly provided by a dilution refrigerator, are needed to obtain high electron population of the top Zeeman level in the paramagnetic radicals of the target, i. e. a electron polarization close to 100 %.

— **Magnetic Field:** In addition, a high magnetic field of the desired spin direction leads to the appropriate Zeeman splitting of the electron and the nucleon energy levels. It should be mentioned that also the small thermal equilibrium (TE) signal, which is essential for the calibration, is in first order a linear function of B/T, thus influencing the accuracy of the polarization measurement.

— **NMR:** The nucleon polarization is usually derived by nuclear magnetic resonance techniques (NMR), which link the measurement of the RF - susceptibility to the value of the polarization.

— **Microwaves:** The DNP process involves irradiation by microwaves to transfer the high electron polarization via simultaneous e^-/p spin flips to the nucleon spin system.

— **Material:** Requirements to be fulfilled for a suitable target material are a high proton (deuteron) content, a short polarization build-up time, and a reproducible manufacture. Further requirements depend on the particular particle experiment.

The following sections will describe each of these items with respect to the SMC polarized target.

2.1 The Dilution Refrigerator

The SMC employs the world's largest dilution refrigerator cooling a total amount of target material of almost 2.5 l, shared between two cells of 65 cm in length and 5 cm in diameter. A detailed description of the SMC refrigerator can be found in Ref. [5] in these proceedings. It is operated either in the DNP mode with a high ^{3}He flow and large cooling power ($\sim$ 150 mmol/s, 0.4 W) around 250 mK while irradiating the target with microwaves, or in the frozen spin mode below 100 mK during field reversals to preserve the polarization.

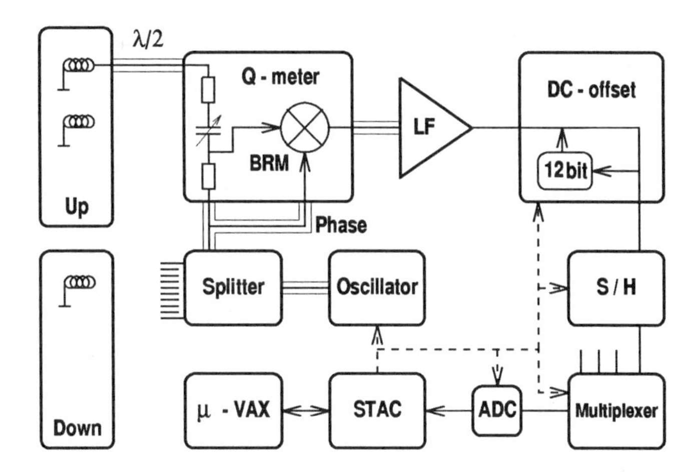

Figure 1: *Schematic drawing of one of the 10 NMR circuits.*

2.2 The Magnet System

The superconducting magnet system consists of a main solenoid with a longitudinal field of 2.5 T aligned with the beam axis, a perpendicular 'holding' dipole of 0.5 T used for field reversals and for a recent measurement of g_2^p with transverse target polarization [6]. 16 trim coils provide an excellent field homogeneity of $\pm 3.5 \cdot 10^{-5}$ over the target volume. In addition, the trim coils have been used to suppress the superradiance effect[7], which destroys the negative polarization while lowering the field, by creating a field gradient throughout the target.
The solenoid has a bore of 26.5 cm in which the target can be loaded while the refrigerator has already been cooled to liquid nitrogen temperatures. The free

opening angle for scattered particles is $\pm 7.5°$.
Both the magnets as well as the cryogenics are controlled via GPIB interfaces by a VME system connected to a SPARC station [8]. The control includes automatical field reversals in a time interval of 5 h.

2.3 The NMR System

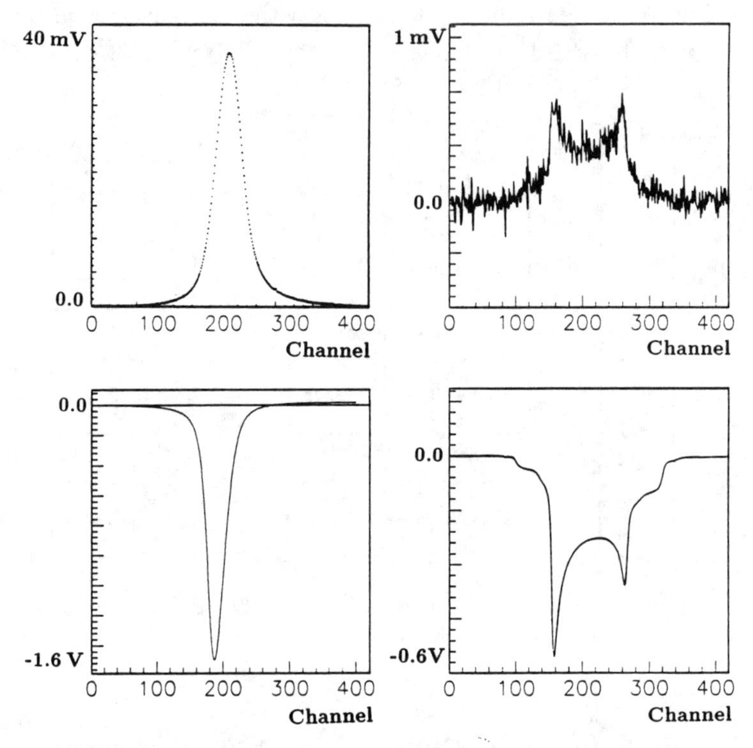

Figure 2: *NMR signals of deuterons and protons at 16.35 and 106.5 MHz, resp., with scan width of 500 and 600 kHz. TE signals (*$P_d = 0.05\%, P_p = 0.25\%$*) at 1 K averaged 2000 times are drawn above. The lower plots show enhanced signals (200 sweeps) with* $P_d = -49\%$ *and* $P_p = -80\%$*, resp.. Note the difference in dynamic scale!*

Fig. 1 shows the principal layout of the NMR circuit which is used to determine the nucleon polarization [9, 10]. Accounting for the large dimension of the target cells, the polarization signal is picked up by 10 series Q-meters simultaneously. The system is based on the commercially available *Liverpool*

NMR electronics [11], which provides real part detection of the output voltage by so-called BRM's[2]. The tuned Q-meters are connected to the probes inside the cryostat via coaxial cables of half-integer wavelength. An RF synthesizer sweeps the frequency in 400 steps across the nucleon's Larmor resonance. After RF and LF amplification the DC offset of the signal is subtracted automatically by a 12-bit ADC-DAC combination. A sample-and-hold amplifier stores the signal before feeding it through a 4-to-1 multiplexer into a 16-bit ADC. The digitized signal is averaged and stored in a STAC[3] computer, which also controls the various hardware of the circuit. Finally, the signals are send to a μVAX computer which hosts the user interface. Fig. 2 shows typical examples of TE and enhanced signals for deuterons and protons after subtraction of the Q-meter baselines, which are obtained outside the Larmor resonance. Data and circuits have been carefully analyzed and corrected to reach the final accuracy of $\pm 5\,\%$ for the deuteron and $\pm 3\,\%$ for the proton polarization [9, 12].

2.4 The Microwave System

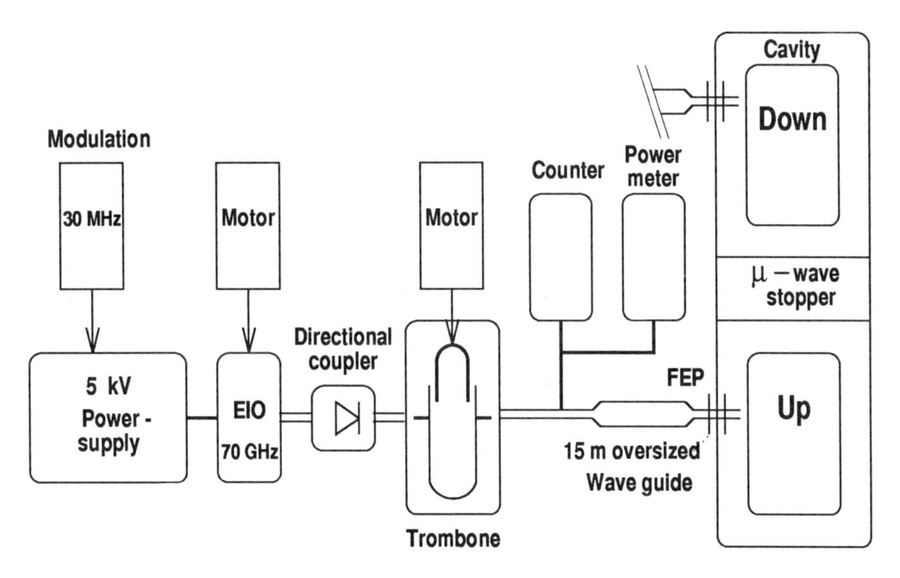

Figure 3: *The SMC microwave system.*

[2]Balanced Ring Modulators
[3]Stand Alone Camac module

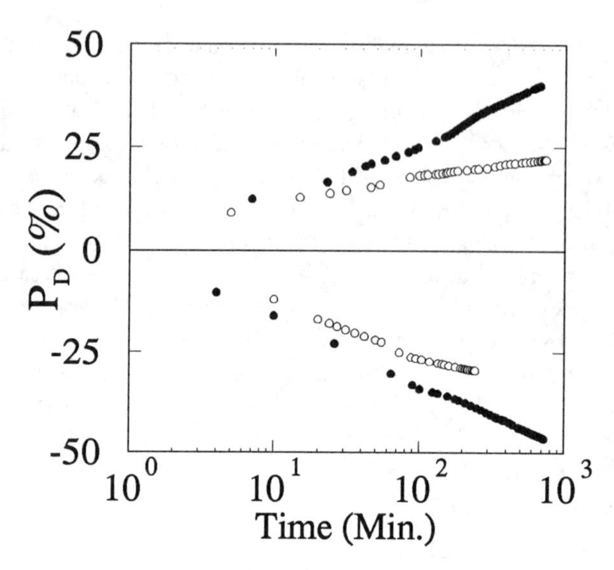

Figure 4: *Enhancement of the deuteron polarization* with *μwave modulation of 30 MHz.*

Due to the desired opposite spin directions in the target halves, two independent microwave systems, based on 20 W EIO's[4], operate around 70 GHz with 0.5 GHz spacing (Fig. 3). Both targets are located in a common multimode cavity, separated by a graphite coated honeycomb microwave absorber of 30 dB attenuation. A motor driven 'trombone-like' structure provides adjustable power inside the cavity. The frequency can be tuned either coarse by changing the cavity of the EIO or fine by adjusting the cathode voltage of the HV power supply, which in addition is used to modulate the microwaves in a band of 30 MHz just outside the EPR line at a rate of 1 kHz. This leads in the case of deuterons to a dramatic increase of the final polarization values by a factor of 1.7 and almost to a doubling of the build-up speed as indicated in Fig. 4 [13, 14].

2.5 The Target Material

1-Butanol doped with $7 \cdot 10^{19}\ \frac{\text{spins}}{\text{ml}}$ chromium (V) paramagnetic radicals has been produced in form of spherical beads of ⌀ 1.8 mm for both proton and deuteron targets used in previous SMC experiments [15]. The characteristic

[4]Extended Interaction Oscillators

build-up times $\tau = P_{max}(1 - 1/e)$ lay in the order of 2 to 3 h, but the highest polarizations were achieved only after several days of DNP. For protons the maximum polarizations reached so far are +93/-94 %, while the average value throughout the '93 data run was 86 %. In the case of deuterons, the maximum polarizations are +46/-52 %, obtained during the '94 data taking, which are the highest values ever reached at 2.5 T.

3 Conclusion

Summarizing, the SMC polarized target has turned out to be a very reliable system, showing a reasonable performance in terms of stability and high nucleon polarization. Further improvements, mostly concerning the accuracy of the polarization measurement, are in progress for the 1995 physics run.

References

[1] B. Adeva et al. (SMC), Phys. Lett. B 302 (1993) 533.

[2] B. Adams et al. (SMC), Phys. Lett. B 329 (1994) 399.

[3] R. Windmolders, these proceedings.

[4] F. Feinstein, these proceedings.

[5] J. Kyynäräinen et al., these proceedings.

[6] B. Adams et al., (SMC) Phys. Lett. underline B 336 (1994) 125.

[7] L. A. Reichertz et al., Nucl. Inst. Meth. A 340 (1994) 278.

[8] J.-M. Le Goff et al. to be publ. in Nucl. Inst. Meth. (1994).

[9] SMC, B. Adeva et al., Nucl. Inst. Meth. A **349** (1994), p. 334.

[10] N. Hayashi et al., to be publ. in Nucl. Inst. Meth. (1994).

[11] G. Court et al., Nucl. Inst. Meth. A 324 (1993) 433.

[12] D. Krämer et al., to be publ. in Nucl. Inst. Meth. (1994).

[13] Y. Kisselev et al., to be publ. in Nucl. Inst. Meth. (1994).

[14] B. Adams et al., to be submit. to Phys. Let. A.

[15] S. Bültmann et al., to be publ. in Nucl. Inst. Meth. (1994).

Review of Non-Linear Corrections in CW Q-Meter Target Polarisation Measurements

G.R.Court and M.A.Houlden

Department of Physics - Liverpool University
Liverpool L69 3BX UK

ABSTRACT

A review is given of the circuit effects which can cause systematic errors when a constant current Q-meter, with real part signal output, is used to measure NMR signals to determine the nucleon polarisation in solid state polarised targets. With proton type signals significant errors can occur when the constant current assumption is not satisfied due of the use of a too low input impedance voltmeter to measure the circuit voltage output. With deuteron type signals the major source of error is generally the drift with time of the values of certain critical circuit components. It is shown that these errors can be reduced by a large factor when an optimised method of subtracting the background signal associated with the NMR signal is used.

INTRODUCTION

A number of experiments are in progress or have recently been and completed [1,2] which use solid state polarised nucleon targets and which require that the overall systematic uncertainty in the measurement of the target polarisation is in the region of 3% or less. The target polarisation is generally measured by the technique of NMR using a constant current Q-meter system [3]. In this talk we review the possible origin of a range of intrinsic non-linear effects which can arise in this type of Q-meter system and which can have a significant effect at these levels of precision. We suggest ways in which these effects can be minimised and methods of calculating corrections. The discussion will involve only the type of Q-meter in which the real part of the output signal is measured, as is described in reference [3].

THE STANDARD CIRCUIT

The polarised nucleons have a complex susceptibility which is a function of applied frequency (ω) and is described by the equation

$$\chi(\omega) = \chi'(\omega) - i\,\chi''(\omega) \qquad (1)$$

The polarisation is given by the equation

$$P = K \int_{o}^{\infty} \chi'' (\omega)\, d\omega \tag{2}$$

where K is a calibration constant.

For the nucleons of interest χ'' has a finite value over a limited range of frequencies and for a simple spin 1/2 nucleon has the general form shown in fig 1. where ω_0 is the nucleon resonance frequency. The polarisation is proportional to the area under the χ'' curve.

In the standard Q-meter circuit (fig 2) the target material is sampled using a coil of inductance L_0 which is operated in series resonance with capacitor C adjusted to be in resonance at frequency ω_0. The inductor which must of necessity be in the low temperature region of the target is connected to the rest of the circuit via a cable which is $n\lambda/_2$ long at ω_0 .

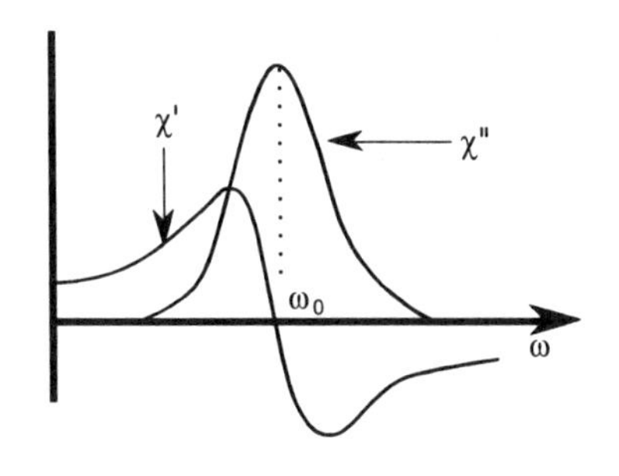

FIGURE 1. Real and imaginary parts of a proton type signal

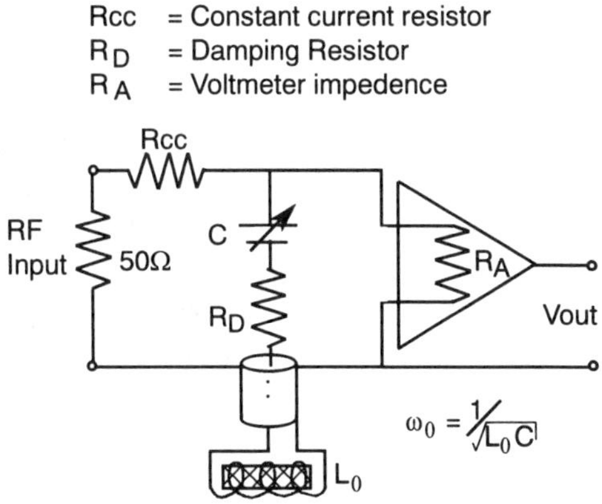

Rcc = Constant current resistor
R_D = Damping Resistor
R_A = Voltmeter impedence

Rcc
RF Input
50Ω
C
R_D
R_A
Vout
$\omega_0 = 1/\sqrt{L_0 C}$
L_0

FIGURE 2. Q-meter circuit

The change in impedance of the inductor due to the complex susceptibility of the target material is given by

$$L(\omega) = L_0 \,[1 + 4\pi\eta\, \chi(\omega)] \tag{3}$$

where η is the filling factor of the coil.

The change in impedance as a function of ω is determined by feeding the resonant circuit with a constant current and measuring the real part of the voltage that appears across the circuit when the frequency is scanned. The factor η is difficult or impossible to calculate precisely so the absolute calibration is obtained by measuring the signal obtained when the target material has a known polarisation (thermal equilibrium signal calibration). The value of this calibration polarisation, and hence the corresponding signal size, is always very much smaller than the operating polarisation of the target material.

The impedance of the resonant circuit and cable combination varies as the frequency is scanned, so the NMR signals appear on a curved background (the Q curve). The modulation generated by this signal (M) is defined as $\Delta V/V$ as shown in fig 3. This Q-curve is generally measured and stored with the magnetic field shifted so that the nucleons are not at resonance. It can then be subtracted from the output signal to obtain the area under the resonance curve.

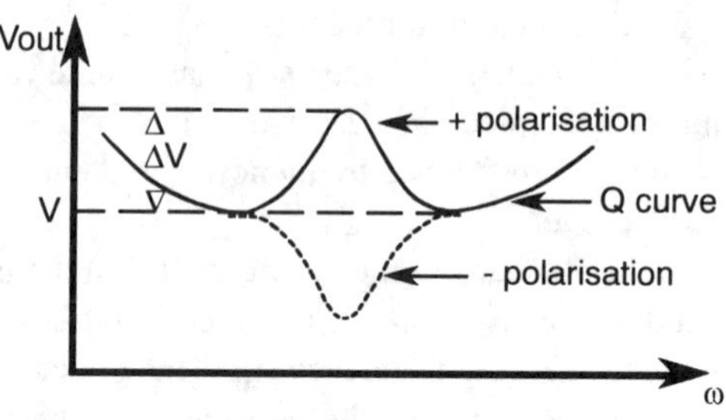

FIGURE 3. Q-meter output

COMPUTER SIMULATION

A computer simulation has been set up to calculate the transfer function for the standard circuit with theoretical proton and deuteron line shapes as input.

Proton lineshape.

This calculation is described in detail in ref [3]. A summary of the relevant results is given here.

(a) Line shape errors.

A symmetric input signal does not produce a symmetric output signal. This change in shape arises because the imaged impedance of the coil, as seen through the cable, is not symmetric. It is higher on the high side of the signal than it is on the low side. This is a small effect for proton type signals as $\Delta\omega$ is small, being typically of the order of 0.5% The effect is independent of M for small M and has a negligible effect on the measurement of polarisation because the associated errors cancel when the signals are symmetric.

(b) Dispersive signal interaction.

In general the dispersive signal χ' is not zero at the ends of the frequency scan. This dispersive signal interacts with the circuit components, particularly the cable, to produce an output signal from the real part voltmeter. As a consequence, a finite error signal is generated at the ends of the frequency scan. This effect is independent of M for small M and in general has a very small effect on the accuracy of the polarisation measurement [3].

(c) Constant current assumption.

The resistor Rcc in the circuit in fig 2 must have a high value compared with the total resonant circuit impedance Z_{tot}. Then if the voltmeter impedance R_A is also large compared with Z_{tot} the current through the sampling coil will be constant, as is required. However, in practical circuits R_A is generally made rather low in order to maximise the noise performance of the input stage of the voltmeter. In this situation the current through the coil will not be constant when M is large due to the change in current split between the two parallel paths in the circuit. This leads to a non-linearity error which has an opposite sign for positive and negative polarisations. With proton signals, which typically have maximum M values in the range 0.1 to 0.5, these errors can be important, particularly as they generate a systematic uncertainty in the asymmetry measurement in the experiment. It is possible to calculate correction factors to take account of these effects if all the circuit parameters are precisely known [3].

Deuteron lineshape.

In this case effects (a) and (b), the line shape errors and dispersion effects, become relatively more important as the relative frequency scan range is much larger than for the proton case, typically of the order of 3%. Effect (c) is generally not significant because M values are always very small.

(a) Non physical input signal

A very simple non-physical input signal example can be used to illustrate the problems. The output signal obtained when two identical proton type signals set at a spacing of 0.007 ω_0 are used as input is shown in fig 4. This clearly shows an asymmetry in peak signal heights of approximately 2.5% as well as non-zero signals at the edges of the scan range. These effects do not depend on the signal size or the value of M.

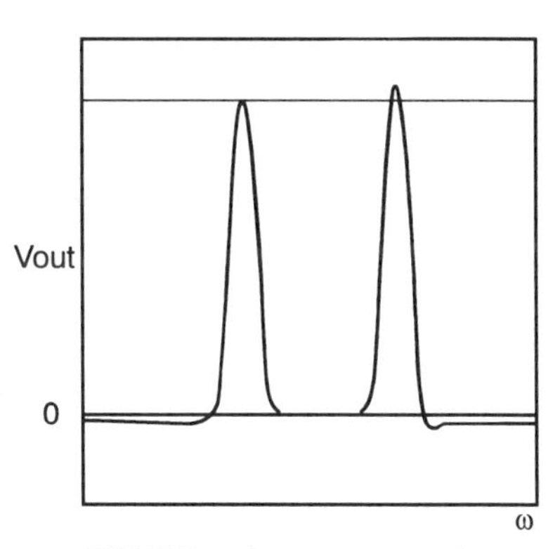

FIGURE 4. Q-meter output for non physical test signal.

(b) Simulated deuteron signal

A set of calculations have been made using a simulated deuteron signal. The absorptive and dispersive parts of this signal, calculated as in ref. [3], are shown in fig 5. The shape corresponds to a deuteron thermal equilibrium signal as the heights of the two peaks are identical.

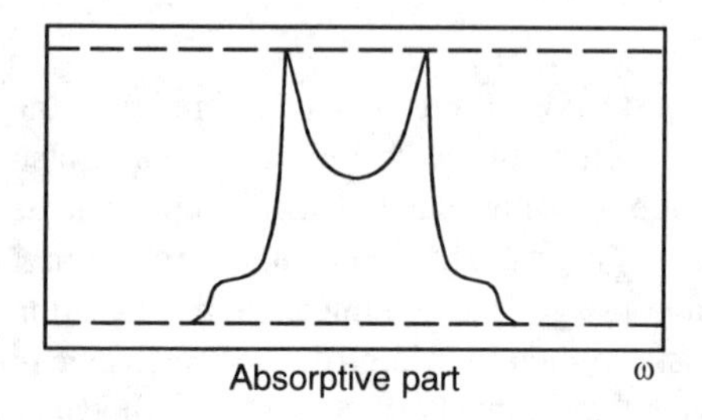

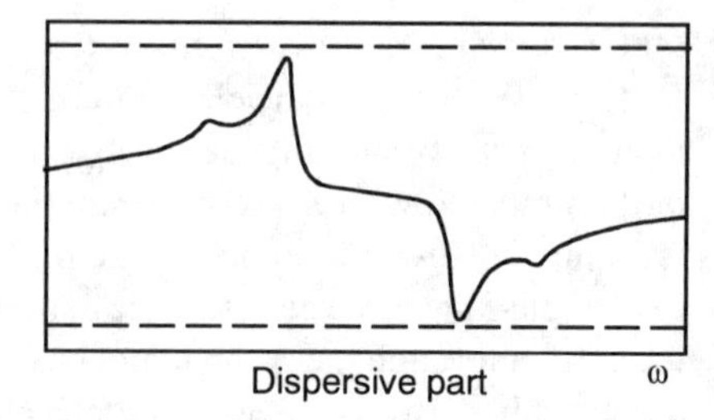

FIGURE 5. Simulated deuteron signal

The circuit parameters used in the calculation were f_0 = 16.3Mhz, $\Delta f = \pm 250$ KHz, $L_0 = 0.4\mu H$, $R_D = 6.8\Omega$, r(coil) = 0.2Ω, $R_A = 50\Omega$, and cable = $\lambda/2$.

The output signal with the Q curve subtracted is shown in fig 6. This shows the expected small peak height asymmetry and relatively large negative signal at the edges of the frequency scan. The signal regions labelled A,B and C have areas which are in the ratio 2.3:100.0:3.7.

(c) Signal backgrounds

These results suggest that the simple subtraction of the Q curve background does not give the correct value for the integral of $\chi''(\omega)$. This is true because there is a residual background due to the dispersion signal effects. This residual background must be zero when $\omega = \pm\infty$ and when $\omega = \omega_0$ because $\chi' = 0$ at these points. It follows that it will have a cusp shape for a proton type signal. The shape is much more complicated for the deuteron signal. A calculated shape, with a zero offset added for clarity, is included in fig 6. The residual background has been observed experimentally [1, 2]. Its effect can only be seen in the wings of the signal and a parabola can generally be fitted to these regions.

The question now arises, what effect do these background uncertainties have on the precision of the polarisation measurement? We have made calculations with three different methods of background determination which can be used experimentally.

(i) A simple Q-curve subtraction ignoring the sign of the signal area.

(ii) A Q-curve subtraction followed by a linear fit to the wings of the signal.

(iii) A Q-curve subtraction followed by a parabolic fit to the wings of the signal.

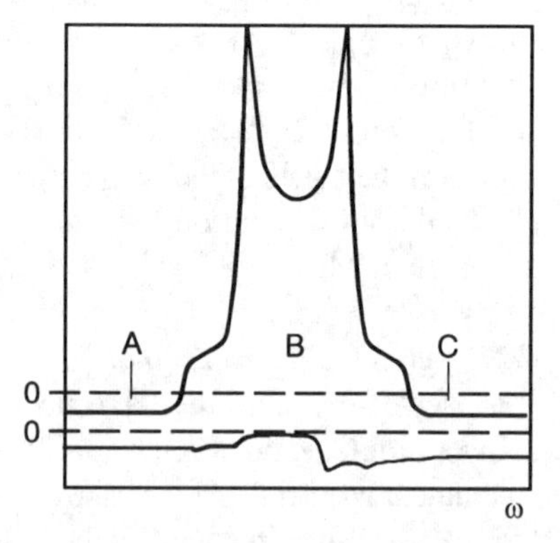

FIGURE 6. Q-meter output for deuteron signal showing signal background.

All these methods give a negligible negative/positive asymmetry error and an error in the absolute magnitude of the polarisation of less than 1% provided that the same technique is used for both the calibration signal and the operational signal.

These errors do not increase significantly when account is taken of the changes in the relative heights of the two peaks in the signal which occurs when the deuterons are highly polarised. It was originally thought that the intrinsic peak height asymmetry generated by the Q-meter circuit would result in a negative/positive asymmetry error in this situation. However, these effects cancel to a high degree when the value of the polarisation is calculated, at least for the values of circuit parameters used in these calculations.

Methods (ii) and (iii) have an important experimental advantage. This arises because in a practical situation the Q-curve background must be measured with no signal present. This involves changing the target magnetic field by a small amount so that the NMR resonance line is not present on the Q-meter output signal. This process takes a finite period of time to complete. It has been observed both experimentally and theoretically that the shape of the Q-curve is very sensitive to changes in circuit parameters ,and in particular to changes in the cable propagation constant β [4]. The shape of the Q curve therefore generally changes with time and this can lead to very large errors in the absolute value of the polarisation when method (i) is used. Both methods (ii) and (ii) give much smaller errors in these circumstances. Fig. 7 shows the effect of small changes in the propagation constant and attenuation constant of the cable on signals which have the same Q-curve background subtraction. A parabolic background has been fitted in each case and the maximum change in area is 1.4%

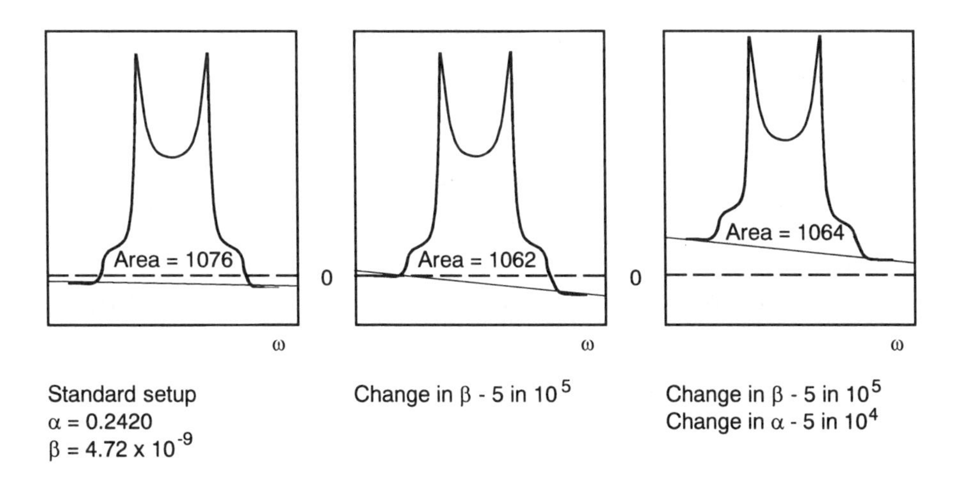

FIGURE 7 Output signal with fitted parabolic background.

CONCLUSIONS

(a) Proton signals

Relatively large non-linear effects which generate a negative/positive asymmetry in the polarisation measurement can occur if the voltmeter input impedance is comparable with the total resonant circuit impedance and the signal modulation levels are high. Corrections can be calculated if the circuit parameters are accurately known.

(b) Deuteron signals

(i) Non linear corrections are generally not significant.
(ii) The use of a simple Q-curve subtraction technique can lead to large errors in the measurement of the absolute polarisation when circuit parameters drift.
(iii) The errors due to (ii) can be reduced to a very low level by using a linear or parabolic fit to the wings of the NMR signal to determine the signal background.
(iv) The Q-meter circuit introduces an intrinsic line shape asymmetry. This does not introduce significant errors into the measurement of polarisation when the system is calibrated using a thermal equilibrium signal. It can lead to large errors when the peak height method of determining the deuteron polarisation is used.

(c) General

(i) The signal background subtraction technique which uses a stored experimentally measured Q-curve has a number of limitations. Full background subtraction using a calculated function fitted to the wings of the signal may be a better alternative.
(ii) The cable which links the sampling coil to the room temperature circuits is a major source of non-linear effects and technical problems. The development of a system which is less sensitive to cable parameters would be a major advance.

REFERENCES

[1] Measurement of deuteron polarisation in a large target
SMC collaboration CERN-PPE/94 - 54 1994

[2] The Virginia/Basel/SLAC polarised target - D. G. Crabb and D. B. Day
To be published in Proc. 7th workshop on Polarised Target Materials and Techniques Bad Honnef June 20 - 22 1994

[3] G. R. Court, D. W. Gifford, P. Harrison, W. G. Heyes and M. A. Houlden
NIM A 324 (1993) 433-44

[4] S. K. Dhawan IEEE Trans Nucl. Sci. NS-39 (5) 1992

STATUS OF THE HYDROGEN AND DEUTERIUM ATOMIC BEAM POLARIZED TARGET FOR NEPTUN EXPERIMENT

N.I. Balandikov, V.P. Ershov, V.V. Fimushkin,
M.V. Kulikov, Yu.K. Pilipenko, V.B. Shutov
Joint Institute for Nuclear Research,
Dubna, Moscow Region, RU-141980, Russia

ABSTRACT
NEPTUN - NEPTUN-A is a polarized experiment at Accelerating and Storage Complex (UNK, IHEP) [1] with two internal targets. Status of the atomic beam polarized target [2] that is being developed at the Joint Institute for Nuclear Research, Dubna is presented.

The general layout of the target and interaction region is given in Fig.1. The principles of the target operation were described in details in [3].

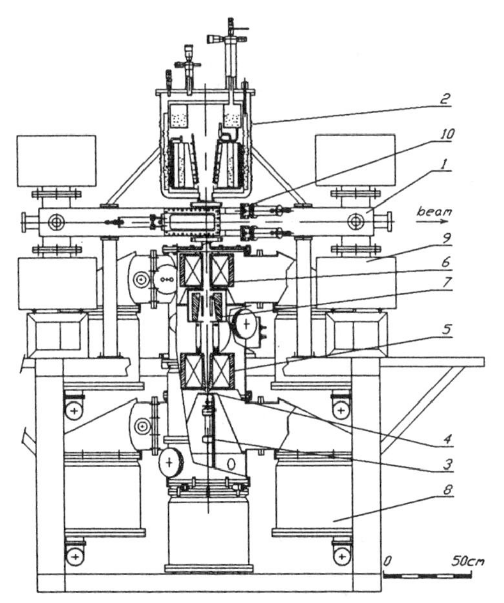

Figure 1. General layout of the target and interaction region.
1 - interaction box, 2 - cryocatcher, 3 - RF dissociator, 4 - gas forming system, 5,6 - sextupole magnets, 7 - RF transition cell, 8 - turbomolecular pump, 9 - ion pump, 10 - vacuum valve.

The beam goes from the down to the interaction box (1) and being caught by the cryocatcher(2). The DC target consists of the RF dissociator(3) with the cold (40K) nozzle, the gas forming system(4), two water cooled sextupole magnets(5,6), the radio frequency transition cell(7) and the pumping system.

The geometry of the sextupole magnets (lengths and apertures) for the most probable velocity of atoms (1000 m/s) was optimized by computer calculations. The pole tips are made of permendur. Geometrical parameters of the produced sextupole magnets are presented in the table.

	1st sextupole		2nd sextupole
	(1st section)	(2nd section)	
Length	75mm	120mm	200mm
Aperture	15/28mm	28/34mm	34/34mm
Yoke (ϕ)	350/300mm		350/300mm

The RF cells will provide 2-4 transition for hydrogen atoms and 1-4, (2-6)+(3-5) or (1-4)+(3-5) transitions for deuterium. Sign changing of the polarization will be realized by reversing the magnetic field in the interaction region. High and low holding fields are planned.

The gas load of the first stage will be about $1mbar*l/s$. With an effective pumping speed of 2500 l/s an equilibrium pressure around $4*10^{-4}$ mbar in the first pumping stage can be obtained. With 5000 l/s effective pumping speed in the second stage an equilibrium pressure of $< 8*10^{-6}$ mbar will be maintained. The pressure inside the first sextupole chamber will be lower than $4*10^{-7}$ mbar. The pumping speed of this chamber is 5000 l/s. Vacuum in the second sextupole chamber will be lower than $2*10^{-7}$ mbar. This chamber will be pumped by the 3K cryopump. Its design is very similar with the cryocatcher one and shown on the Fig.2.

3K cryopump consists of two liquid helium vessels at 4.2 K (P_{He} = 1atm) and at 2.6 K (P_{He} = 0.25atm). The total area of cryosurfaces is $1.5m^2$. Estimated frozen hydrogen capacity is 500 l (STP), a pumping speed is about 20 m^3/s and an extrem partial hydrogen pressure is $< 8*10^{-9}$ mbar. Time of the continuous operation of the 3K cryopump is about 300 hours because the main part of hydrogen is pumped by the turbomolecular pumps.

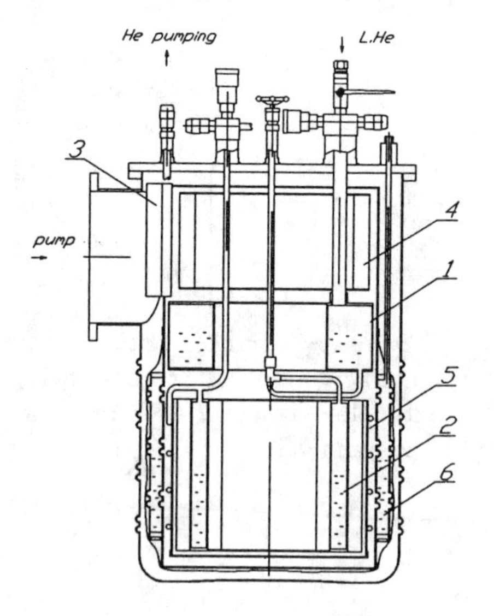

Figure 2. 3K cryopump.
1 - 4K He vessel, 2 - 3K He fin vessel, 3 - Nitrogen baffles, 4 - He baffles, 5 - 3K shield panel, 6 - L. Nitrogen vessel.

A sublimation of the gas from the cryopump and the cryocatcher is needed once time in a week.

The strong vacuum requirements (10^{-9}mbar) around the interaction point and complexity with organizing of sideways ports to spectrometers predefines a design of the interaction box. Main peculiarities of the box are separate pumping system with four ion pumps(9), flat builded in the interaction box body vacuum valves(10) and pulsed coils of the holding field.

The target is assembled on a test bench for preliminary tests. The cryocatcher and cryopump are fabricated. The interaction box is submitted to the JINR machine shop.

References

1. V.I. Balbekov et al. - In: Proc. of XII Intern. Conf. on High Energy Accel., Batavia, 1983, p. 40.
2. V.P. Ershov et al. - In: Proc. of IX Intern. Symp. on High Energy Spin Physics, Bonn, 1991, v.2, p.160.
3. V.P. Ershov et al. - In Proc. of the Workshop on Polarized Gas Targets for Storage Rings, Heidelberg, 1991, p. 40-44.

THE MOVABLE POLARIZED TARGET (STATUS OF THE RECONSTRUCTION)

Yu. A. Plis (Laboratory of Nuclear Problems, Joint Institute for Nuclear Research, 141980, Dubna, Russia)

Collaboration: Belarus, France, Japan, Netherland, Russia, Switzerland, Ukraine, USA

The polarized proton target (PPT), used previously at FERMILAB, has become available. A collaboration has been set up in order to reconstruct this PPT as a movable polarized target (MPT), which could be easily transported from one beam line to another and to get this unit operated in Russia.

The Saclay-Argonne frozen PPT is 20 cm long and 3 cm diameter. The experiment E-704 was closed at the end of 1992, the French owned parts have been shipped back to Saclay and then in March 1994 from Saclay to Dubna. The available ANL-HEP parts, stored in a sea-container at Argonne, were shipped to Dubna in May 1994.

Experiments in Dubna will also require transverse polarization. For this purpose, transversal holding coils must be added to the set-up. Therefore, the target must be reassembled and upgraded. It needs also to be made transportable from one experimental area to another by bonding firmly every parts on separate functional decks.

This is the MPT concept:

1. All the elements positioned close to the beam line (target cell, magnets, power supplies, dewars etc.) are set on two decks which can be moved as blocks to set the apparatus in and out of the beam for easy maintenance.
2. The pumps will be mounted on a separate deck for easy access during beam operation.
3. The contol room will be installed inside a trailer. It will include a remote control for the entire operation of the target.

The collaborating laboratories propose to assemble the target at Dubna during 1994, in a close co-operation between their various experts. Installation on the Synchrophasotron- Nuclotron complex will start in 1995. Tests of the target will then demonstrate its operability. The whole project is divided into 5 distinct parts. Each of them is well defined and will be terminated by an operating status and clear commissioning tests.

1) Obtaining good polarization level in the target cell.

This requires to assemble all the available elements of the target on decks, provide the missing elements, use the existing polarizing magnet. An operating vacuum system and an operating dilution refrigerator with 4He and 3He pumping system must be built. A full control of the target consists in an interlock system, microwave system and NMR system. This item does not require the holding coils, since no transverse polarization is needed. The final test will consist of polarization and temperature measurements.

2) Installation on the beam line.

After the test of polarization, the target will be placed on the beam line and tested again.This requires power and fluid supplies, cooling and pumping capability, magnetic field for polarization, interlock system, microwave system and NMR system. Final test is operation of the target in physics experiment conditions.

3) Building a new polarizing magnet.

Drawing of the present polarizing magnet, built in Saclay, are available. Final test is field mapping.

4) Building a set of holding coils.

The exact configuration was decided. Final test is field mapping.

5) Integration the target on beam line
and managing its use for physics.

Prepare the integration of the whole system together with the particle detection system at the experimental area. Provide alignment procedure. Provide power and fluid supplies. Final test is full operation of the target, including spin rotation.

Within the 1993 INTAS allocation the participating laboratories have decided to support only the task 1) in 1994

There are some problems, which we have in the course of the reconstruction of the target.

1) All the elements are being made so that the target enters into sea-container in blocks for transportation to new place. This will give the possibility to shorten the time for the preparation to experimental run.

2) Different heights of the beam line in FERMILAB and Laboratory of High Energy demand a new frame.

3) Pumping group for 3He and 4He circulation are made anew.

4) Different standards for mechanics and electronics in Europe and USA create some difficulties.

The status in September 1994 is following:

a) All the necessary parts of equipment for completion of the MPT in frozen spin mode with a longitudinal polarization are available in Dubna.
b) Necessary funds and manpower for test of the MPT in 1994 are available.
c) Basis and frame are reconstructed.
d) Polarizing magnet was tested.
e) NMR new basic version was tested.
f) The dilution refrigerator are under assembly.
g) A beam line is preparing for receiving the MPT.

This project has been selected by INTAS for financial support
(Reference number: INTAS-93-3315)

LONG TIME EXPERIENCE OF POLARIZED DEUTERON SOURCE POLARIS AT THE JINR SYNCHROPHASOTRON

V.P.Ershov, V.V. Fimushkin, M.V. Kulikov,
Yu.K. Pilipenko, V.B. Shutov, A.I. Valevich
Joint Institute for Nuclear Research,
Dubna, Moscow Region, RU-141980, Russia

ABSTRACT
Since 1981 the JINR 4.5 GeV/nucleon synchrophasotron accelerates a polarized deuteron beam. A cryogenic source of a polarized deuterons POLARIS and some experimental equipment used for this purpose are briefly described.

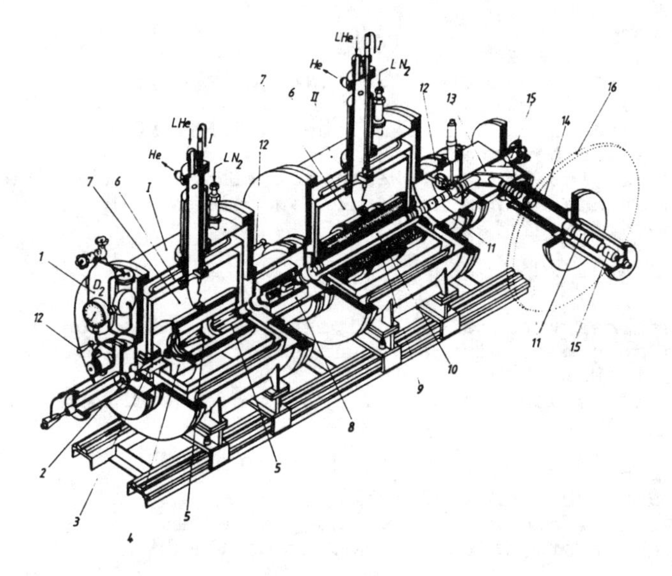

Figure 1. A general view of the polarized deuteron source POLARIS.
I - polarized atomic source, II - ionizer. 1 - deuterium volume, 2 - electromagnetic gas valve, 3 - dissociator, 4 - nozzle chamber, 5 - sextupole magnet, 6 - nitrogen shield, 7 - helium cryostat, 8 - RF cell, 9 - SC solenoid, 10 - electron optics, 11 - ion optics, 12 - vacuum gate , 13 - electrostatic mirror, 14 - solenoid of the spin-precessor, 15 - Faraday cup, 16 - position of the preaccelerator flange.

The deuterium atomic beam polarized source POLARIS, Fig.1. (more detail see [1,2,3,4,5]) has the following features:

-magnetic fields are set up by superconducting magnets operating in a persistent current state,

-vacuum in the source is supplied due to gas condensation on the LHe temperature surfaces of the cryostats,

-cooling of the dissociator, nozzle and skimmer is produced by a thermal contact with the cryostats,

-source is very compact and requires power only for the RF and control systems. The set-up POLARIS consists of two modules a cryogenic source of polarized deuterium atoms and a cryogenic high magnetic field (2.6 T) Penning ionizer. A generation of nuclear polarization of an electron polarized atomic beam is based on RF-transitions interchanging the population of hyperfine substates. Two RF cells provide 1-4, 3-6 transitions for vector polarization and 3-5, 2-6 transitions for tensor polarization.

The essential stabilization of the Penning ionizer operation has been reached under the optimal tuning of the quantity of feeding gas using a fast electromagnetic valve. Magnetic shields and correcting coils for RF cells allow to decrease the scattering magnetic field of the SC ionizer solenoid in the RF transition region and as result to increase the efficiency of the nuclear polarization. The energy of the deuteron beam is about 3 keV at the output of the source while the current is about 200 μA.

At the JINR accelerator the source POLARIS is installed on a 750 kV terminal. Information exchange is performed by a fiber glass optic system. A special controller placed on the high voltage terminal gives a possibility of the information exchange between controller and console computer.This one has allowed to carry out an effective monitoring of the Penning ionizer observing the current signals at the electrodes of the electron optics. A typical operating pulse of the Penning ionizer on a console display is shown on Fig.2

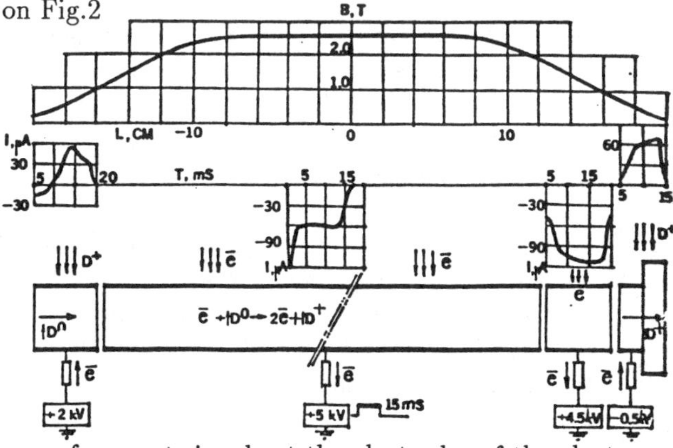

Figure 2. A diagram of current signals at the electrodes of the electron optics.

The source can stable run for a month without a sublimation of the condensed gas. The module variant of the set-up allows to accomplish a replacement of the RF cell for vector or tensor polarization mode, a HF acid washing of the dissociator without warming of the cryostats.

A week need of the cryogenic liquids is 200 l of the LHe and 300 l of the LN_2. A long time experience of the source POLARIS allows to contend that the cryogenic version of the set-up was justified.

The acceleration of the polarized deuterons is approximately 15% of the synchrophasotron running time. Experience of the accelerator runs allows to use only one polarimeter for polarization measurements on the extracted deuteron beam. Low energy polarimeter behind the linac and internal beam polarimeter inside the ring of the synchrophasotron are not used now.

The results of the polarization measurements of the accelerated beam ($P_z^+ = 0.54 \pm 0.01$, $P_z^- = -0.57 \pm 0.01$, $P_{zz}^+ = 0.76 \pm 0.02$, $P_{zz}^- = -0.79 \pm 0.02$.) proof the absence of depolarization during the acceleration.

In the future it is planned to increase the intensity of the polarized deuteron beam up to 2 mA by using a pulsed H^+ plasma ionizer (Fig.3) developed at the INR (Moscow) by A.S. Belov.

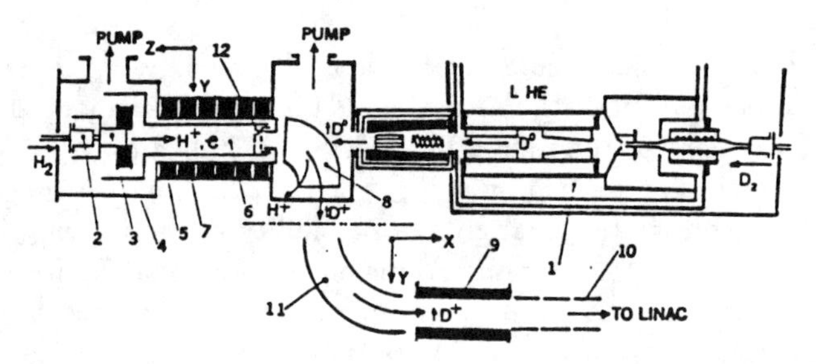

Figure 3. A principal diagram of the POLARIS charge exchange ionizer.
1 - cryogenic atomic source, 2 - electromagnetic H_2 gas valve, 3 - hydrogen plasma source, 4 - magnet, 5 - solenoid, 6 - ionization volume, 7 - shield, 8 - deflecting magnet, 9 - solenoid of the spin- precessor, 10 - electrostatic lens, 11 - spherical mirror, 12 - accelerating and beam-forming ion-optical system.

Polarized deuterons are produced by charge exchange between polarized deuterium atoms and ions of hydrogen plasma $D^o_\uparrow + H^+ = D^+_\uparrow + H^o$ and ionization of D by plasma electrons. The energy of the deuteron beam will be 15 keV at the emittance 0.2 π $cm * mrad$. The fabrication of the new ionizer was finished and the assembling on a test bench is completed.

References

1. Anischenko N.G. et al. - In: Proc. 5th Int.Symp. on High Energy Spin Physics, Brookhaven, 1982 (AIP Conf. Proc. N95,N.Y., 1983,p.445)
2. Belushkina A.A. et al. - In: High Energy Physics with Polarized Beams and Polarized Targets,Argonne,1978 (AIP Conf.Proc. N51, N.Y., 1979 p.351).
3. Belushkina A.A. et al. - In: High Energy Physics with Polarized Beams and Polarized Targets, Basle, 1981, p.429.
4. Anischenko N.G. et al. - In: Proc. 6th Int. Symp. on High Energy Spin Physics, Marseille, 1984 (Jorn. De Phys.,Colloque C2, Supplement an n° 2,Tome 46, 1985, p.C2-703)
5. Ershov V.P. et al. JINR Communication E13-90-331, Dubna,1990.

EXPERIMENTS AT ULTRACOLD SOURCE PROTOTYPE FOR POLARIZED ATOMIC HYDROGEN BEAM

V.P.Ershov, V.V.Fimushkin, G.I.Gai, M.V.Kulikov, L.V.Kutuzova, A.V.Levkovich, Yu.K.Pilipenko, V.B.Shutov, A.I.Valevich.
Joint Institute for Nuclear Research,
Dubna, Moscow Region, RU-141980, Russia

ABSTRACT
Experimental investigations presented here continue a study of an ultracold beam extraction of polarized hydrogen atoms by the back side gradient separation method [1,2,3]. This study is carried out at JINR to get a high intensity atomic beam polarized jet source.

Main ideas of an ultracold source were discussed in [4,5,6]. The experimental setup ATOM-H was described earlier [2]. It was grade up to study extraction of the ultracold polarized beam (Fig.1).

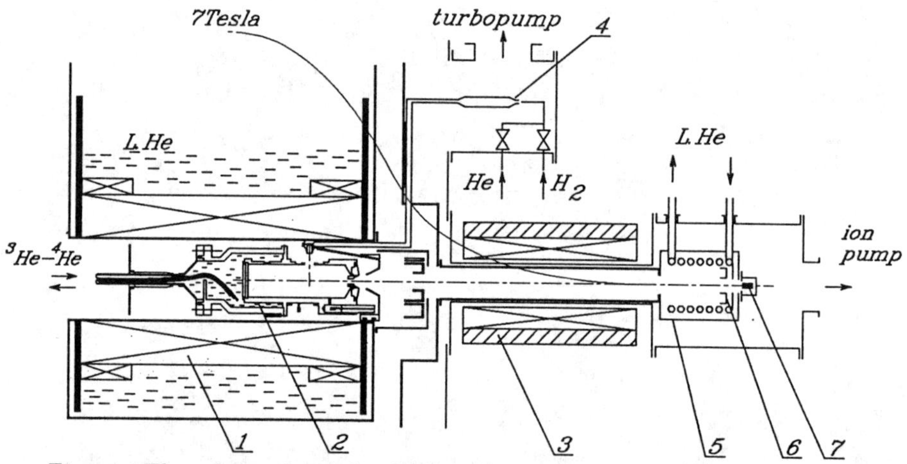

*Figure 1.*The schematic of the ATOM-H source.
1 - S.C. solenoid, 2 - ultracold source cell, 3 - sextupole magnet, 4 - dissociator, 5 - L. nitrogen shield, 6 - thermal detector, 7 - ion gauge.

To reduce a cell temperature, a ^{3}He pumping cryostat [7] was changed for 10 mW dilution refrigerator. A mixing chamber of the refrigerator and the low temperature cell were designed as a single block (Fig.2). The cell has a high thermoconductivity and it is cooled down to 0.15 - 0.3 K. The cell is located in a magnetic field gradient of the superconducting solenoid so that the magnetic fields at the entrance and exit hole positions are about 80% and 50% of the maximum field (7T) respectively.

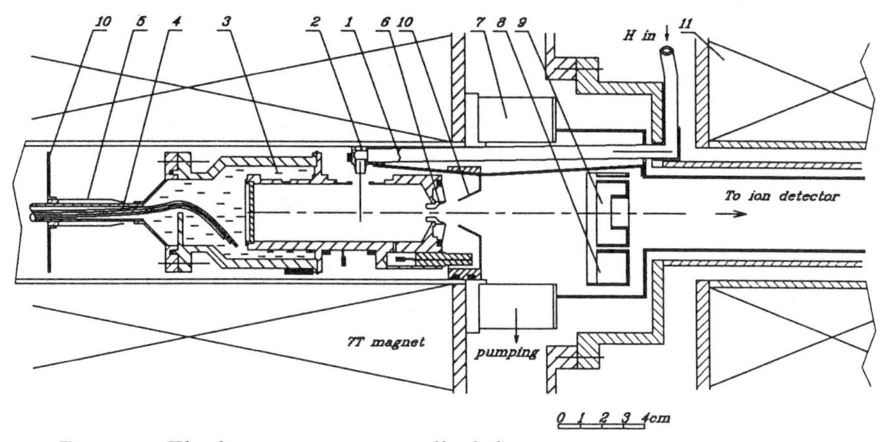

Figure 2. The low temperature cell of the source.
1,2 - teflon pipe and accommodator, 3 - mixing chamber and cell, 4 - final ^{3}He heatexchanger, 5 - film burner, 6 - 4 mm cell exit, 7 - baffles, 8 - peripheral detector, 9 - central detector, 10 - He differential pressure diaphragm, 11 - sextupole magnet.

Atomic hydrogen is produced by an RF discharge in a small size (7 cm^3) dissociator. To reduce recombination of atoms, atomic hydrogen is cooled down in a teflon pipe and an accommodator to 8 - 10 K. A cold atomic hydrogen jet is injected into the low temperature cell (0.25 K) through a vacuum gap (Fig.2). After thermalization ultracold H$\uparrow$ atoms pushed out of the cell are accelerated by the magnetic field gradient up to velocities corresponding to $\sim$5 K and form a polarized beam.

To study properties of the beam and an efficiency of atoms separation at the back side gradient of the S.C. magnet additional sextupole and thermal and ion detectors were installed.

To prevent a recombination of the hydrogen atoms the cell surface is covered by a superfluid helium film. The film thickness depends on a setup vacuum pressure. To make the S.F. film more stable the He differential pressure diaphragms (10) around the cell (Fig.2) were installed. At a typical pressure of 1.5*10^{-7} mbar the film thickness measured by the bolometer is about 3 nm (Fig.3).

Due to insufficient refrigerator cooling capacity the setup operates pulsewise. Each cycle starts every ten seconds with an injection of helium (4*10^{17} mol/pulse) to restore the cell film. Hydrogen and RF discharge pulses (t = 300 ms) follow in four seconds. The hydrogen flow rate is 4.5*10^{17} mol/pulse and the flow of atoms is estimated as 6*10^{17} at/pulse.

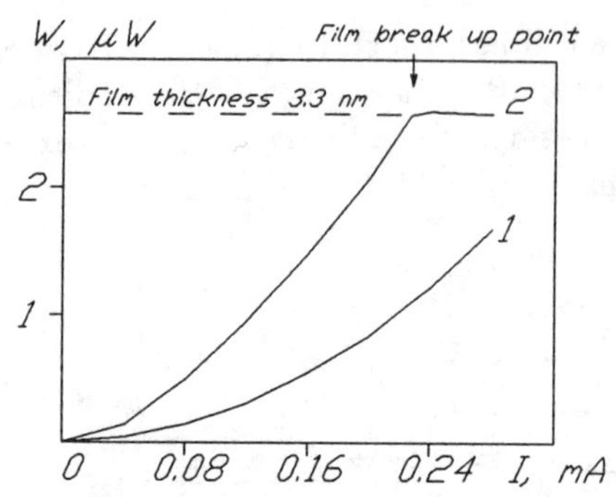

Figure 3. The film thickness measurements by the bolometer.
1 - no S.F. film, 2 - with S.F. film.

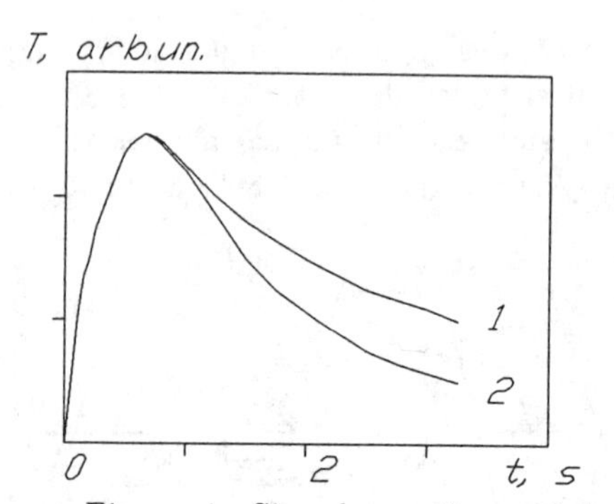

Figure 4. Signals on the peripheral detector.
1 - atomic beam pulse, 2 - electrical amplitude calibrating pulse.

A part of atoms entering the cell is recombined and their heat can be measured by the calorimetric method. The same idea is used to measure the beam flux on thermal detectors. There are two ring detectors which are placed near the cell (central detector $\phi 28/15$ mm, peripheral one $\phi 64/30$ mm) and final detector after the sextupole ($\phi 42/12$ mm). A typical atomic beam pulse and an electrical calibrating signal are shown on Fig.4. It is seen the beam pulse area is larger then calibrating one and the some correction is needed.

Main experimental results are summarized in the table:

$N_{at\ cell}$	$N_{at\ c.det}$	$N_{at\ p.det}$	$N_{at\ sum.}$	Comments
5.4	1.2	2.2	8.8	no sf.film and magn.field
8.6%	1.9%	3.5%	14%	% of total N_{at}
7.9	4.1	8.2	20.2	sf.film, no magn. field
12.5%	6.5%	13%	32%	% of a total N_{at}
17	4.1	6.9	28	sf.film,7T magn.field
121.4%	29.3%	49.3%	-	% of a $N_{at} = 28/2$ H↑

All numbers are quantity of atoms at the calibration $N_{at}/10^{16}$ at/pulse. For this case total atomic beam intensity is: $N_{at\ total} = 2.0 * 10^{17}$ at/pulse.

To increase the input atomic hydrogen flow the 100 mW dilution refrigerator has been designed and its fabrication is mainly finished.

This work was supported by the grant 93-02-3780 from the Russian Fund of Fundamental Research.

References

1. M.Mertig, A.V.Levkovich, V.G.Luppov, Yu.K.Pilipenko In: Proc. of the 9-th Int.

Symp. on High Energy Spin Physics, Bonn, Springer-Verlag, V.2, (1991), p.164-167.
2. V.P.Ershov et al. In: Proc. Workshop on Polarized Gas Targets for Storage Rings, Heidelberg, Max-Planck-Institut, (1991), p.68-74.
3. M.Mertig et al. In: Proc. Workshop on Polarized Gas Targets for Storage Rings, Heidelberg, Max-Planck-Institut (1991), p.87-89.
4. T.O. Niinikovski In: Proc. Int. Symp. on High Energy Physics with Polarized Beams and Polarized Targets, Lausanne (Birkhauser EXS-38, Basel, 1981), p.191.
5. D.Kleppner and T.J.Greytak In: Proc. of the 5-th Int. Symp. on High Energy Spin Physics, Brookhaven, Ed. by G.M. Brunce, AIP Conf. Proc. N 95, p.546 (1983).
6. T.Roser et al., Nucl. Instrum. Methods A301e (1991) p. 42-46.
7. V.G.Luppov et al. In: JINR Rapid Communications N 5 [31]-88, Dubna, 1988, p.21.

PULSED CRYOGENIC POLARIZED ATOMIC HYDROGEN JET SOURCE

V.P. Ershov, V.V. Fimushkin, M.V. Kulikov,
A.V. Levkovich, Yu.K. Pilipenko, V.B. Shutov
Joint Institute for Nuclear Research,
Dubna, Moscow Region, RU-141980, Russia

ABSTRACT
Design and test results of a pulsed cryogenic polarized atomic hydrogen jet source are presented. The source uses cryopanels and turbomolecular pumps (3*500 l/s). The source has three superconducting sextupole magnets. The beam measurements are produced by two 4.7K thermal detectors.

A development of the cryogenic polarized atomic hydrogen beam source was based on a large experience of the cryogenic polarized deuteron source [1,2]. The main problem is a large hydrogen vapour pressure (about 10^{-6} mbar) over cryopanels. For deuterium a saturation vapour pressure is significantly lower.

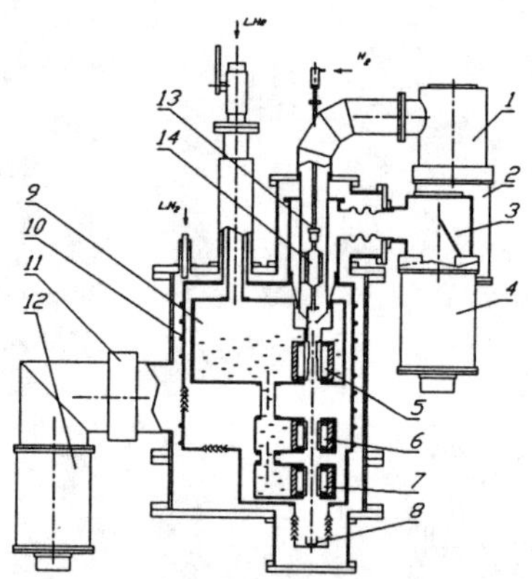

Figure 1. Schematic of the source. 1,3,11 - valves, 2,4,12 - turbomolecular pumps (500 l/s), 5,6,7 - superconducting sextupole magnets, 8 - thermal detector, 9 - helium cryostat, 10 - nitrogen shield, 13 - electromagnetic gas valve, 14 - dissociator.

Atomic hydrogen is produced by a high frequency discharge of 80 MHz in 6 cm^3 pyrex volume. The discharge power is about 700 W. The dissociator tube is tightly fitted into a teflon block having a thermal contact with a liquid N_2 shield. After the dissociator a gas flow is passed through the teflon pipe (l=50 mm, ϕ 5 mm) and the nozzle (l=20 mm, ϕ 3mm, T=30 K). The atomic beam under molecular flow conditions is formed by the skimmer (ϕ 4.5mm) and the collimator (ϕ 10mm). The distance between the nozzle and the skimmer is 11 mm.

Spatial spin separation of the atomic beam is realized in gradient fields of three superconducting sextupole magnets placed in the liquid He cryostats. The pole tips are closely inserted into the cryostat pipe. The aperture of the pole tips increases from ϕ 10mm to ϕ 28mm in the first magnet. The other magnets have the apertures of ϕ 28mm. The length of each magnet is 70mm. A magnetic field on the tips is 1T.

Two low heat capacity thermal detectors are used for measurements of the atomic beam intensity. The detectors are made as copper foil cups with fins and carbon resistors as thermometers. The detectors temperature is about of 4.7 K. A sensitivity of the detectors is about 10^{14} atoms/sec. There are electrical heaters on the detectors for calibration measurement. There is a diaphragm in front of the first detector. The second detector is placed in the diaphragm shadow and used for a background measurement. A movable beam stopper is placed between the first and the second magnets to interrupt the beam for a background vacuum measurement.

A molecular hydrogen and a RF power pulses start simultaneously with a frequency of 0.1 Hz. The pulses duration is 150 ms. A molecular hydrogen throughput is about 0.6 mbar*l/sec.

Typical observed heat loads on the detectors due to recombination of the beam atoms correspond to the atomic beam intensities of $2 * 10^{15}$ atoms/sec and 10^{15} atoms/sec for the first and the second detectors respectively. A magnets system focussing factor of three for the first detector was measured.

Vacuum measurements show that both mechanical (turbomolecular) and cryopumping have essential influence on the vacuum conditions in the second pumping stage (see Fig.2). The results indicate unsatisfactory vacuum conditions along the beam path. Runs with improved pumping system are planned in the nearest future.

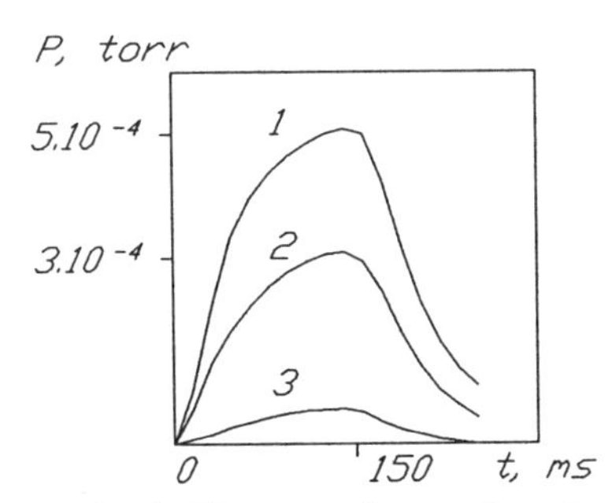

Figure 2. The pressure pulse in the second pumping stage for the operation modes: 1 - T=300 K, mechanical pumping; 2 - T=4.2 K, cryopumping only; 3 - T=4.2 K, cryo and mechanical pumping.

References

1. V.P. Ershov et al. JINR Communication E13-90-331.
2. V.P. Ershov et al. In: Proc. of Workshop on Polarized Proton Ion Sources and Polarized Gas Jets,1993.

VI. POLARIZED ELECTRON SOURCES

REVIEW OF THE SLAC AND LES HOUCHES WORKSHOPS

Charles Y. Prescott
Stanford Linear Accelerator Center
Stanford University, Stanford CA 94309

I. Introduction. Polarized Electron Source workshops have been held at varying intervals, beginning in 1983 when Charles Sinclair convened the first at SLAC. Since that time, three workshops were held in conjunction with the International Spin Symposia and two at other occasions. The increasing importance of polarized electron beams at accelerators has stimulated interest in these workshops. Two workshops have been held since the last International Spin Symposium in Nagoya. In 1993, a workshop was held at SLAC, and in 1994 at Les Houches, a polarized electron beam session was held as part of a polarized beam and targets workshop. This report summarizes highlights from the latter two workshops.

History of Polarized Electron Sources Workshops

Year	Place	Convenor
1983	SLAC	C. Sinclair
1988	Minneapolis	C. Sinclair
1990	Bonn	E. Reichert
1992	Nagoya	T. Nakanishi
1993	SLAC	J. Clendenin(SLAC)
		T. Maruyama (SLAC)
		D. Schultz (SLAC)
1994	Les Houches	M. Leduc(ENS Paris)
		E. Steffen(MPI Heidelberg)

The workshops have experienced a growing attendance. These workshops report on progress in all aspects of polarized electron beams and support the polarization progress at labs like SLAC, MIT-Bates, and Mainz, where experiments are running, and at labs such as CEBAF, NIKHEF, and KEK, where future experiments are likely to be run. Interest in polarized electrons for a future linear collider is strong. The frequency of one workshop per year should continue into the future as long as this interest in experiments with polarized electron beams remains strong.

II. The SLAC Workshop on Photocathodes for Polarized ElectronSources for Accelerators (September 8-10, 1993). This workshop was organized by J. Clendenin, T. Maruyama, and D. Schultz, and supported by SLAC. The focus of this workshop was on photocathodes for use in polarized electron sources at accelerators. The reports presented in this workshop have been combined into a proceedings consisting of copies of transparencies from each talk [1]. The workshop contributions can be divided into two categories, accelerators and cathode studies. Listed below are the laboratories and institutions represented, the speakers, and their topics. The workshop had 43 participants representing 22 institutions and laboratories

Accelerators	Speaker	Subject
Mainz	K. Aulenbacher	MAMI operating experiences
SLAC	S. Ecklund	NLC and Linear Colliders
SLAC	L. Klaisner	SLC experiences

MIT-BATES	M. Farklendeh	Bates operating experiences
Los Alamos	R. Sheffield	AFEL Accelerator
CEBAF	C. Sinclair	Emittances of GaAs sources
KEK	M. Yoshioka	JLC-ATF source

Cathode Studies

Accelerators	**Speaker**	**Subject**
Bielefeld	G. Baum	Strained GaAs
Ferrara	V. Guidi	Timed resolved emission
Ecole Polytechnique	G. Lampel	Layered structures
SLAC	R. Mair	Photoluminescence studies
St. Petersburg	Y. Mamaev	InGaAs,AlGaInAs, GaAsP
SLAC	T. Maruyama	Strained GaAs
SLAC	G. Mulhollan	Low energy test facility
Nagoya	T. Nakanishi	Superlattice structures
Spire Corp.	S. Vernon	MOCVD process
Heidelberg	S. Zwickler	Energy analysis of electrons
Novosibirsk	A. Terekov	InGaAsP on GaAs
SLAC	H. Tang	Charge limits
Mainz	E. Reichert	Thermal studies
ETH Zürich	H Seigmann	High pulsed laser power studies
NIST	D. Pierce	Studies of surface magnetism
SLAC	G. Mulhollan	Low energy test facility
Stanford	W. Spicer	Theory

Three of the accelerator labs, Mainz, SLAC, and MIT-Bates reported on operational aspects of polarized beams. The remaining laboratory presentations reported on R&D activities aimed at possible future operations. A partial list of accelerator parameters is given below. It should be noted that parameters listed here are a snapshot given at the time of the workshop, and that they may change and can be expected to be different at future times.

Parameter	**Mainz MAMI**	**SLAC SLC**	**MIT-Bates**	**CEBAF**	**KEK JLC ATF**	**Los Alamos AFEL**
Pulsed/cw	cw	pulsed	pulsed	cw	pulsed	pulsed
Rate (Hz)		120	600		150	100
Gun type	triode	diode	diode	diode	diode	rf
Voltage (KeV)	100	120	380	100	100	10^3
Pulse length		2 ns	15 μs		< 1 ns	< 20 ps
Q/pulse		8×10^{10}	2×10^{12}		1.5×10^{10}	6×10^{10}
Cathode	GaAsP	strn GaAs	GaAs	GaAs	GaAs	multi-alkali
QE(%)	0.9	0.9	6.5			< 0.1
ϵ(π mm mrad)	0.32	20	0.1	0.6		3
Polarization (%)	40	80	40	—	—	0
QE lifetime (hrs)	20	100–500	200			3
Load lock?	no	yes	no	no	no	no
Laser type	dye	ti-sapp.	ti-sapp.e			

III. The Les Houches Workshop on Polarized Beams and Targets; Sensitive Tools for the Study of Solids, Nuclei, and Particles (June 7-10, 1994). This *Scientific Network on "Polarized Beams and Targets"* workshop was organized under the auspices of the European Community "Human Capital and Mobility." The organizers were Michele Leduc (ENS Paris) and Erhard Steffens (MPI Heidelberg). Sixty three participants representing 36 institutions attended the workshop.

One day-long session was devoted to polarized electron sources and semiconductor photocathodes. Other topics at the workshop included lasers, optical pumping of gas targets, ^{3}He targets, and selected topics in polarized neutrons and atomic physics.

The talks covering the electron sources and semiconductor photocathodes were combined into one day and summarized in one evening session. Below are listed the presentations from those sessions relating to electron sources and photocathodes.

The talks are summarized in a collection of abstracts available from the organizers [2].

Institution	**Speaker**	**Subject**
Palaiseau	G. Lampel	The Physics of the Semiconductor Spin-Polarized Electron Source
ETH Zürich	F. Meier	Emission of Polarized Electrons from Photocathodes using Continuous and Pulsed Light Sources
SLAC	G. Mulhollan	Polarized Electron Beams at SLAC
Mainz	S. Plützer	Photoemission of Spin-Polarized Electrons from Strained $In_{1-x}Ga_xAs_{1-y}P_y$ and $GaAs_{1-y}P_y$
Mainz	H. Fischer	XPS Studies of NEA-Cathode Fatigue
Palaiseau	D. M. Campbell	High Polarization and High Yield from Strained GaAs Photoelectric Sources
Orsay	S. Essabaa	Recent Developments and Latest Results of the Orsay Polarized Electron Source

IV. Summary. Gallium arsenide cathodes and layered structures of strained gallium arsenide and gallium arsenide superlattices have been shown to provide high polarization and high currents in photoemission by polarized laser beams. At accelerator facilities with running experiments, these sources have proven to operate with high reliability. Pulsed currents of up to 10 amps in 2 nanoseconds have been used at SLAC for the SLC. Polarizations up to 85% have been achieved for experiments. Mainz has operated photoemission sources with continuous currents, but in some conditions, the cathodes have shown a loss of quantum efficiency at the spot where the laser hits the photocathode when steady currents in excess of 150 μA are drawn. Advances in laser technology, vacuum technology and cathodes materials continue to make the electron source devices better and easier to fabricate and use.

Helium afterglow sources at Rice University and Orsay offer an alternative to photoemission sources for high average currents ($\geq 100\mu$A) with polarization in the range of 60-80%.

1. Proceedings of the Workshop on Photocathodes for Polarized Electron Sources for Accelerators, organized by J. Clendenin et al., SLAC-432 (1993).
2. Abstracts of the Workshop on Polarized Beams and Targets; Sensitive Tools for the Study of Solids Nuclei, and Particles, organized by M. Leduc (ENS Paris) and E. Steffens (MPI Heidelberg), Les Houches (1994).

Recent Progress on Cathode Development from Nagoya and KEK

T. Nakanishi, H. Aoyagi,[1] S. Nakamura,[2] S. Okumi, C. Suzuki, C. Takahashi, Y. Tanimoto, M. Tawada, K. Togawa, and M. Tsubata

Department of Physics, Nagoya University, Nagoya 464, Japan

Y. Kurihara, T. Omori, Y. Takeuchi, and M. Yoshioka

KEK, National Laboratory for High Energy Physics, Tsukuba 305, Japan

T. Kato, and T. Saka

New Research Laboratory, Daido Steel Co.Ltd., Nagoya 457, Japan

T. Baba, and M. Mizuta

Fundamental Research Laboratory, NEC Corp., Tsukuba 305, Japan

Abstract. Recently the high electron spin polarization (ESP) photocathodes which have also high quantum efficiencies (QE) have been developed in Japan. These improvements were achieved using new techniques of a resonant absorption and a modulated doping, which are applied to a strained GaAs grown on GaAsP and an AlGaAs-GaAs superlattice, respectively. The former photocathode provides the QE of 1.0 % with the ESP of 85% at the laser wavelength of 866 nm. The latter one does the QE of 0.5% with the ESP of 71% at 757 nm. Using a new type of strained-layer superlattice cathode, the ESP of 83% which is higher than that of our previous superlattices was observed. These development of photocathodes not only with a high ESP but also with a high QE are discussed.

INTRODUCTION

The highly polarized electron beams are required for the future linear colliders, such as JLC (Japan Linear Collider) and NLC (Next Linear Collider), as well as for the nuclear physics experiments. Linear colliders are ideal for the acceleration of

---

[1] Presented by H. Aoyagi

[2] Present address: Physics Institute, Bonn University

longitudinally polarized electrons, since there are no drastic depolarization mechanisms which lie on synchrotrons. JLC Polarized Electron Source group has been developing photocathodes to obtain both high ESP and high QE. The breakthroughs against the theoretical limitation of the 50% ESP for a bulk GaAs were achieved in 1991. The ESP of 71% was observed with a AlGaAs-GaAs superlattice by KEK/Nagoya/NEC [1], 71% with a strained InGaAs grown on GaAs by SLAC/Wisconsin/Berkeley [2], and 86% with a strained GaAs grown on GaAsP by Nagoya/Osaka Pref./Toyota Tech. Inst./Daido [3], respectively. Their works were done independently. In these kinds of semiconductors, the degeneracy between heavy-hole and light-hole bands is removed at Γ-point to allow a single excitation only from heavy-hole bands.

The AlGaAs-GaAs superlattices had been tested systematically by KEK/Nagoya/NEC [4]. The variable parameters are total thicknesses, acceptor concentrations, and surface characters. We observed that the thinner total thickness of active layers and the lower doping acceptor concentration result in the higher ESP. The highest ESP of 76% among these superlattices were obtained.

The systematic study on a strained GaAs grown on GaAsP had been made experimentally by Nagoya/Osaka Pref./Toyota Tech.Inst./Daido [5] and SLAC/Wisconsin [6], independently. We found that the maximum ESPs for each sample with different parameters, such an active layer thickness and a phosphorus fraction, strongly depend on the strain induced in the active layer of GaAs. As a typical result, using a 100 nm-thick strained GaAs grown on GaAsP, the ESP of about 80% and the QE of ~0.2% are achieved at current SLAC/SLC94 running.

For high peak current operations, an amount of charge from a photocathode is limited by the NEA surface property instead of by the space charge limit of the gun. This phenomena which is called "cathode charge limit" was observed at SLAC [7]. JLC requires the higher charge than SLC in a macro-bunch with 90 micro-bunches separated by 1.4 ns, typically, so this phenomena should be well studied by further experiments.

QE IMPROVEMENTS

The above two types of semiconductors, such a AlGaAs-GaAs superlattice and a strained GaAs grown on GaAsP, have big advantages in ESP, however QE is still not sufficient compared with a bulk GaAs. This situation is brought by the single state excitation only from the heavy-hole band near the band-edge to avoid electrons excited from the light hole band, and by the thinner active layer than that of a bulk GaAs to avoid the strain relaxation and the depolarization. Therefore we have developed the following two new techniques to improve QE without lowering ESP.

Resonant Absorption Strained GaAs

The number of absorbed photons inside photocathode material can be enhanced, using a Fabry-Perot optical cavity which is formed by a couple of mirrors; a quarter-wave distributed Bragg reflector (DBR) and a surface boundary of GaAs and vacuum (Fig.1) (Daido/Nagoya/Sophia/Osaka Pref. [8]). The DBR which is a stack of alternating quarter-wave layers with high and low refractive indexes is inserted between GaAsP layer and GaAs substrate. It is designed to have the highest reflectivity at the laser wavelength of the maximum ESP. To satisfy the condition of resonance in a Fabry-Perot cavity, a cavity length L must be given by $2 \cdot n \cdot L = m \cdot \lambda_R$, where n is a refractive index, m is an integer number, and λ_R is a wavelength which corresponds to the maximum ESP.

We applied this technique to a strained GaAs grown on GaAsP. The parameters are selected for the first sample as followings. The 30 pairs of $Al_{0.1}Ga_{0.9}As$ and $Al_{0.6}Ga_{0.4}As$ layers form a DBR. The refractive indexes are 3.52 and 3.20, respectively. The thicknesses of each layer are 60.2 nm and 66.5 nm, respectively. The thicknesses of GaAs layer and $GaAs_{0.82}P_{0.18}$ layers are set to 140 nm and 2000 nm, respectively, which are the same as one of the normal strained GaAs grown on GaAsP.

The QE and ESP have been measured as shown in Figure 2. The maximum ESP of 85% is the same level as the normal strained GaAs, where the QE is enhanced up to 1.0% at 866 nm. From this data, the technique of a resonance absorption had been verified to be effective for QE improvements preserving the high ESP.

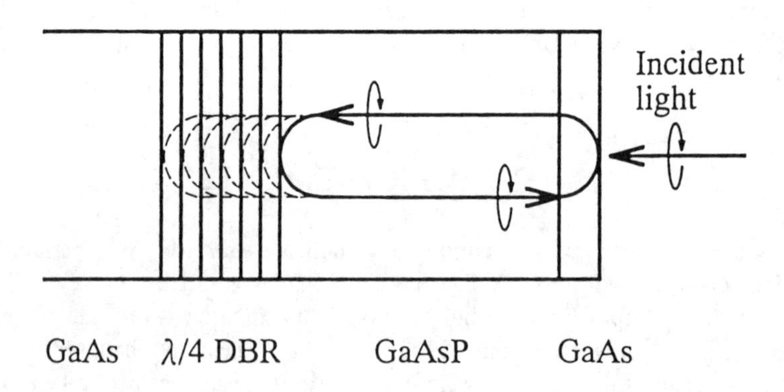

FIGURE 1. The principle of the Fabry-Perot optical cavity with DBR. The absorption of incident light is enhanced between the DBR and the surface of GaAs.

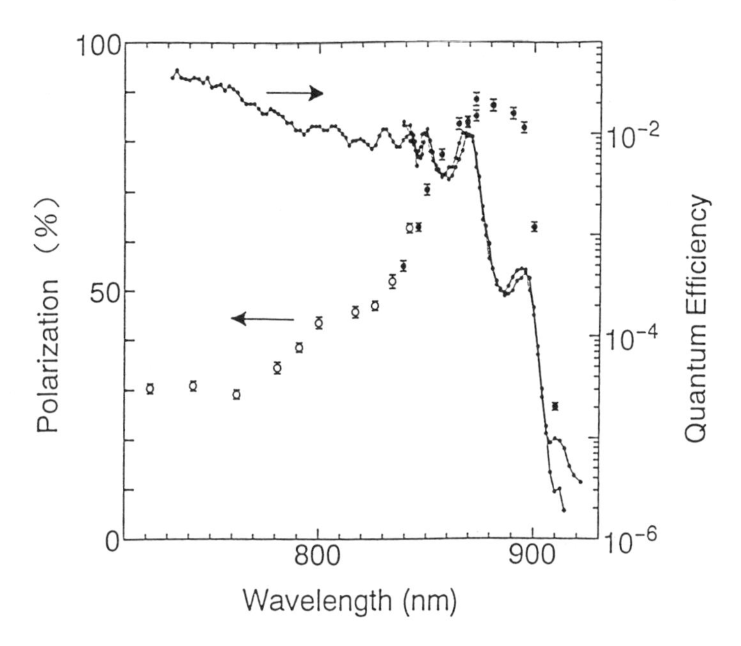

FIGURE 2. Experimental results of the ESP and the QE by the strained GaAs grown on GaAsP with DBR, measured as a function of laser wavelength.

Modulated Doping Superlattice

In general, the heavily p-doped material is used for the NEA cathode, while the doping may cause the depolarization due to electron-impurity scattering. We have expected that the higher doping of the surface region and the lower doping level of the rest in one sample lead to both the higher QE and the higher ESP. Therefore the technique of modulated doping had been applied to superlattice at KEK/Nagoya/NEC [9]. The sample (a) and the sample (b) are compared here. The sample (a) and (b) have the doping level of 4×10^{19}/cc and 5×10^{18}/cc in the surface region, respectively, and the same level of 5×10^{17}/cc in the rest. These cathodes had been tested with a bias of -4 kV. The QE of 0.5% was obtained with sample (a) at 757 nm, while the QE of 0.03% had been obtained with sample (b) at 764 nm. On the other hand, there is no significant difference in the maximum ESPs (71% and 75%, respectively) as shown in Figure 3.

Using the sample (a), the total extracted charge and the QE were measured at SLAC/SLC polarized electron gun (KEK/Nagoya/NEC/SLAC [10]). The total charge of 2.3×10^{11} electrons in 2.5 ns pulse at 757 nm was extracted. The cathode was biased to -120 kV and its laser-irradiated diameter was 20 mm. The QE was 2.0% at 752 nm and at -120 kV. Total charge dependence on the laser pulse energy is shown in Figure 4. This test of a modulated doping superlattice makes it clear that the charge can be extracted as much as the level of space charge limit even at -120 kV. Thus, this technique seems to be useful especially for the future linear colliders.

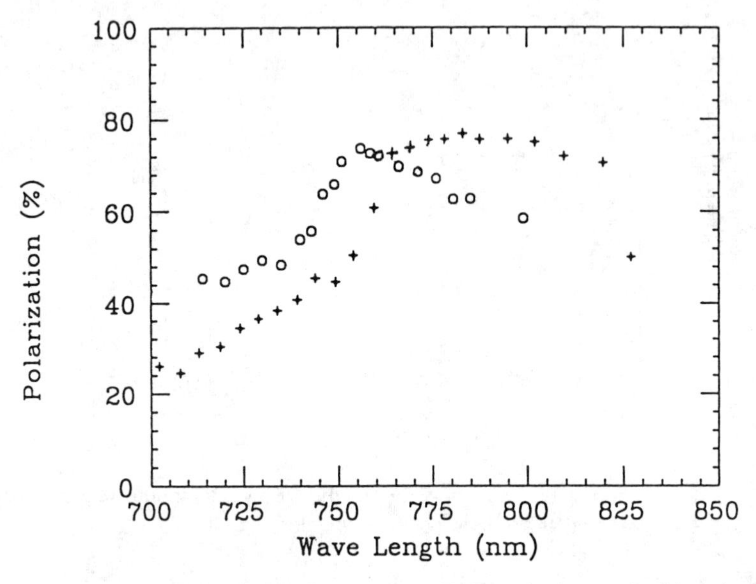

FIGURE 3. The experimental results of the ESP by both sample (a) and sample (b) which are plotted with (o) and (+), respectively.

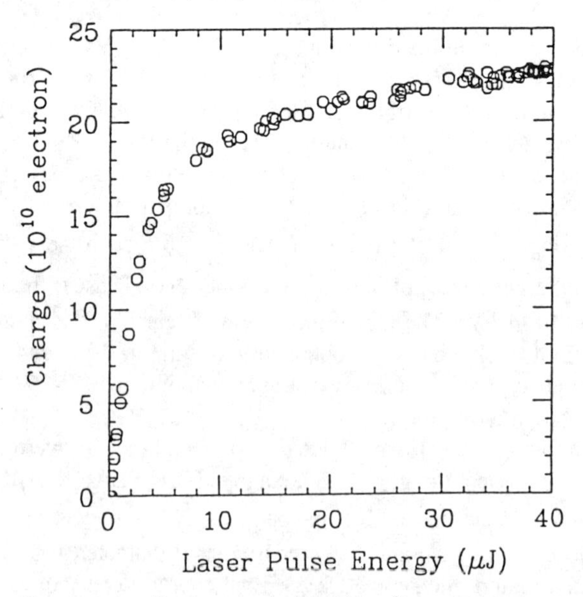

FIGURE 4. Charge saturation curve as a function of laser pulse energy by sample (a).

STRAINED-LAYER SUPERLATTICE

The higher-order-mixing between heavy-hole bands and light-hole bands seems to be a main reason of insufficient ESPs in our previous AlGaAs-GaAs superlattices. Therefore we tried to test a strained-layer superlattice (SLS).

An InGaAs-GaAs strained-layer superlattice grown on a GaAs substrate was chosen for our first test (KEK / Nagoya / NEC [11]). The lattice constant of

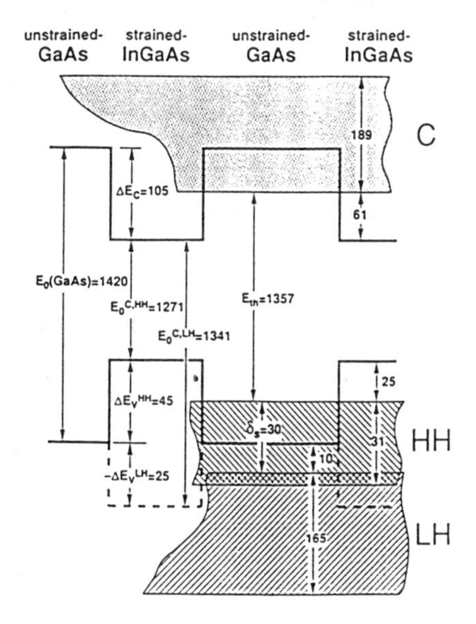

FIGURE 5. The band structure of the $In_{0.15}Ga_{0.85}As$-GaAs strained-layer superlattice.

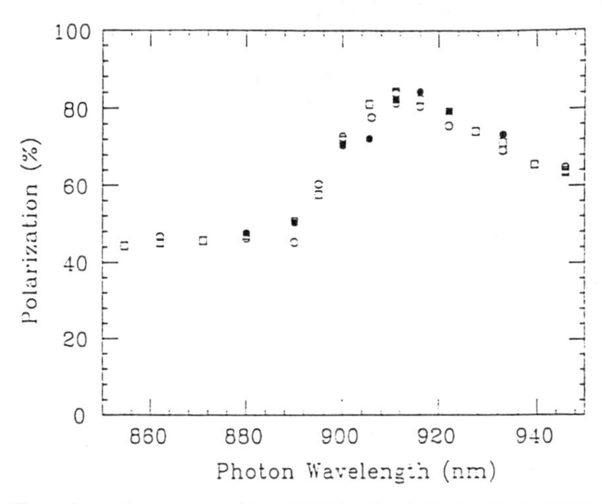

FIGURE 6. Experimental results of the ESP by the InGaAs-GaAs SLS.

$In_{0.15}Ga_{0.85}As$ is bigger than that of GaAs by an amount of ~1%, and the energy band gap of $In_{0.15}Ga_{0.85}As$(1.27 eV) is smaller than that of GaAs(1.42 eV). The heavy-hole bands of InGaAs layer and GaAs layers act as quantum wells and barrier layers, respectively (Fig.5). (Note that the GaAs layers are wells in previous AlGaAs-GaAs superlattices.) In this configuration, the tensile strain is applied only to InGaAs layers, and GaAs layers are supposed to be free of strain.

The maximum ESP of 83% was observed at the laser wavelength between 911 nm and 916 nm (Fig.6), and the typical QE was 0.015% at 911 nm. The ESP of superlattice cathodes have been improved significantly by this strained-layer superlattice. But the QE at the wavelength of the maximum ESP is not high enough. The QE improvement of the strained-layer superlattice has been investigated at present.

In conclusion, as the recent progress on the QE improvements, we have demonstrated the usefulness of a strained GaAs with DBR and a modulated doping superlattice. Using the former cathode, we observed the QE of 1.0% and the ESP of 85% at λ=866 nm. Using the latter one, the QE of 0.5% at the maximum ESP of 71% (λ=757 nm) was observed by a -4 kV biased gun, and the total charge of 2.3×10^{11} electrons in 2.5 ns pulse (λ=757 nm) was extracted from a 20 mm diameter photocathode by a -120 kV biased SLAC/SLC gun. As the progress on the ESP improvement, using a strained-layer superlattice, the ESP of 83% was recently observed with the QE of 0.02 % (λ=911 nm). The further progress seems to be possible because there are still many adjustable parameters.

REFERENCES

[1] T. Omori et al., Phys.Rev.Lett. 67(1991)3294
[2] T. Maruyama et al., Phys.Rev.Lett. 66(1991)2376
[3] T. Nakanishi et al., Phys.Lett.A 158(1991)345
[4] Y Kurihara et al., Nucl.Instrum. Methods, A313(1992)393
[5] H .Aoyagi et al., Phys.Lett.A 167(1992)415
T. Saka et al., J.Cry.Growth 124(1992)346
[6] T. Maruyama et al., Phys.Rev.B46(1992)4261
[7] M. Woods et al., J.Appl.Phys. 73(1993)8531
[8] T. Saka et al., Jpn.J.Appl.Phys. 32(1993)1837
[9] T. Omori et al., Int.J.Mod.Phys.A (proc.Suppl.)2A(1993)157
[10] This work was done by the collaboration between JLC PES group and SLAC. Y. Kurihara et al., to be published in Jpn.J.Appl.Phys., KEK Preprint 94-59/ DPNU-94-27/ SLAC-PUB-6530
[11] T. Omori et al., to be published in Jpn.J.Appl.Phys., KEK Preprint 94-39/ DPNU-94-15

Recent Progress on Cathode Development from SLAC/Wisconsin*

Takashi Maruyama and Edward L. Garwin

Stanford Linear Accelerator Center, Stanford University,
Stanford, California 94309

Robin A. Mair and Richard Prepost

Department of Physics, University of Wisconsin,
Madison, Wisconsin 53706

Abstract. Strained GaAs photocathodes are used for SLAC high energy physics experiments producing high intensity electron beams with up to 85% polarization. We describe three distinctive features of cathode seen in the last two years of usage, and two techniques to investigate strained cathodes.

INTRODUCTION

Polarized electron sources based on negative-electron-affinity GaAs have been widely used. Although this type of source has many advantages, one major drawback has been a 50% polarization limit due to the valence-band degeneracy of the heavy-hole and light-hole bands of GaAs. After much effort devoted to achieving higher than 50% polarization by removing this degeneracy, a SLAC-Wisconsin-Berkeley collaboration demonstrated a significant polarization enhancement with strained InGaAs in January 1991.[1] Subsequently, high polarization was also achieved with strained GaAs.[2] Since GaAs is a better photoemitter than InGaAs due to its larger band gap energy, the cathode developments that followed these pioneering works focused on strained GaAs structures.[3][4]

By growing a GaAs layer on $GaAs_{1-x}P_x$, strain is introduced in the GaAs layer due to lattice-mismatch. Two important parameters in the strained GaAs are: 1) the phosphorus fraction (x) in $GaAs_{1-x}P_x$, and 2) the thickness (t) of the active GaAs layer. A higher phosphorus fraction will produce a larger lattice-mismatch and correspondingly larger strain in the GaAs layer, for typical values of x = 0.25 – 0.30. While thicker active GaAs layer will

* Work supported in part by the Department of Energy contracts: DE-AC03-76SF 00515 (SLAC) and DE-AC02-76ER00881 (Wisconsin)

provide a higher quantum efficiency, there is a critical thickness beyond which a significant strain relaxation takes place. The critical thickness depends on the phosphorus fraction, and the thickness is typically 0.1 – 0.3 μm.

Strained GaAs cathodes were used for the first time for high energy experiments during the 1993 SLC run at SLAC producing about 62% polarization at the SLD interaction point. Since then, two more strained GaAs cathodes have been used for the proton/neutron spin structure function experiment (E143) and the 1994 SLC/SLD run, producing average polarizations of 84% (E143) and 80%(SLD), respectively.

Although strained GaAs cathodes have been used successfully for high energy experiments, there are characteristics which are not fully understood and require further investigation. The following sections describe techniques based on photoluminescence and X-ray diffraction which have been developed to meet these needs.

CATHODE CHARACTERISTICS OBSERVED DURING E143 AND SLC

During the last two years of using strained GaAs cathodes for the SLAC polarized electron source, three distinctive features have been observed.

1) Quantum efficiency dependence of polarization - Fig. 1 shows the electron spin-polarization and cathode quantum efficiency as a function of time during the E143 experiment. The spin-polarization was measured by the double-arm Moller polarimeter of the E143 experiment, and the quantum efficiency was measured at 833 nm at the polarized electron source. The beam polarization increases as the quantum efficiency drops. This behavior has been known for many years, but the effect is dramatic for strained cathodes due to the large value of the polarization.

2) Charge limit - When a GaAs photocathode is excited by an intense light, the maximum photoemitted charge is limited by properties of the cathode itself, and is proportional to the cathode quantum efficiency. This effect was first observed with a bulk GaAs cathode in 1991.[5] Since the quantum efficiency of strained cathodes was typically 1/10 of bulk GaAs, the charge limit was a real concern for actual usage of strained cathodes. The charge limit for strained GaAs has turned out to be much higher than that for bulk GaAs with the same quantum efficiency.[6] Presently the charge limit does not hamper the physics program at SLAC, but further investigation is required in order to understand the mechanism, particularly for application to future linear colliders.

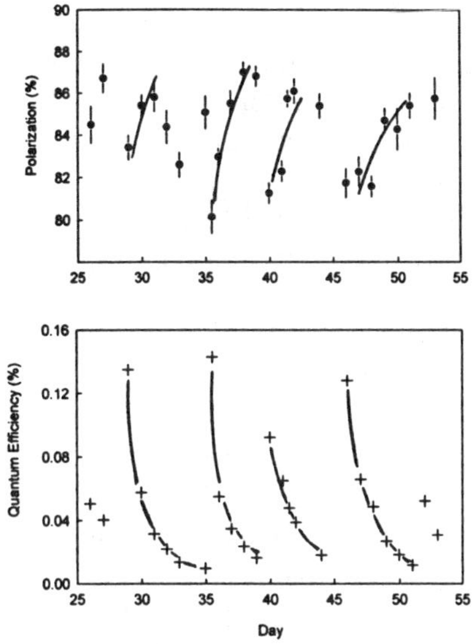

Figure 1: Polarization and cathode quantum efficiency as a function of time.

3) Quantum efficiency asymmetry - During the early commissioning period of the E143, a large helicity-dependent asymmetry was observed in the beam intensity. It was determined that this asymmetry was caused by a small linear polarization contamination in the excitation laser due to an optics problem. This incident prompted a further investigation of cathode quantum efficiency with linearly polarized light. Fig. 2 shows the quantum efficiency as a function of the azimuthal angle of electric field of the incident linearly polarized light.[7] Quantum efficiency asymmetries as large as 20% have been observed. The asymmetry does not arise as long as the light is 100% circularly polarized, but even a 0.1% linear polarization contamination can produce a helicity dependent intensity asymmetry of the order of 10^{-4}, too large for the physics asymmetry of 10^{-5} expected in parity violation experiments.

Figure 2: QE asymmetry

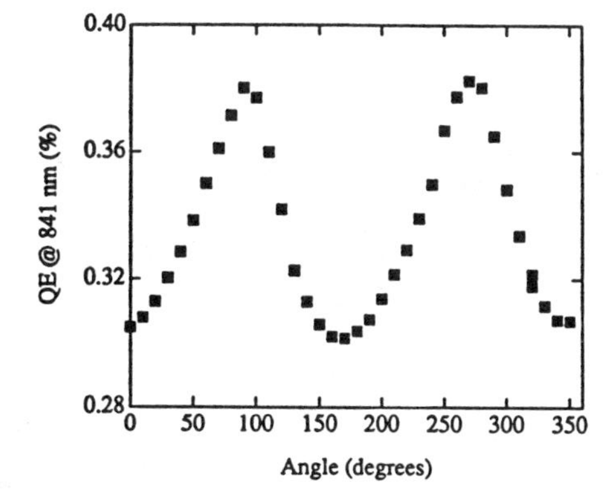

PHOTOLUMINESCENCE

Photoluminescence measurements detect the recombination photons resulting from electrons recombining with a hole in the valence band. The wavelength of the resulting recombination radiation measures the band gap energy for the particular band-to-band transition that has been excited. When a short wavelength excitation (such as HeNe laser) is used so that the light is completely absorbed in the $GaAs_{1-x}P_x$ layer, two recombination lines are observed, one from the $GaAs_{1-x}P_x$ layer and the other from the strained GaAs. Since the relationship between band gap energy and phosphorus fraction of $GaAs_{1-x}P_x$ is well known, the observed luminescence from the $GaAs_{1-x}P_x$ layer can be used to measure the phosphorus fraction. This technique is used to measure the variation of the phosphorus fraction over the 5 cm diameter of the strained GaAs wafers. The variation is typically less than 2%, and the performance of photocathodes cut from these wafers is expected to be similar. However, one wafer from a different vendor showed as much as an 8% variation. Since the strain relaxation is critically dependent on the phosphorus fraction, a variation as large as 8% would limit the usable area on the wafer.

Photoluminescence is also a convenient and powerful method for characterizing the quality of the strain introduced into epitaxial structures. Strain introduced in epitaxial GaAs increases the band gap energy. Since the band gap energy is directly related to the luminescence wavelength, the shift in the luminescence wavelength is a measure of the strain in the GaAs layer. Fig. 3 shows photoluminescence spectra for samples with different strain. As the strain is increased, the peak wavelength shifts towards the shorter wavelength. Fig. 4 shows the correlation between the electron polarization measured by Mott scattering[3] and the energy shift measured by the photoluminescence technique. Since the excitation light can be focused to 100 μm, local strain variations can be readily explored as well.

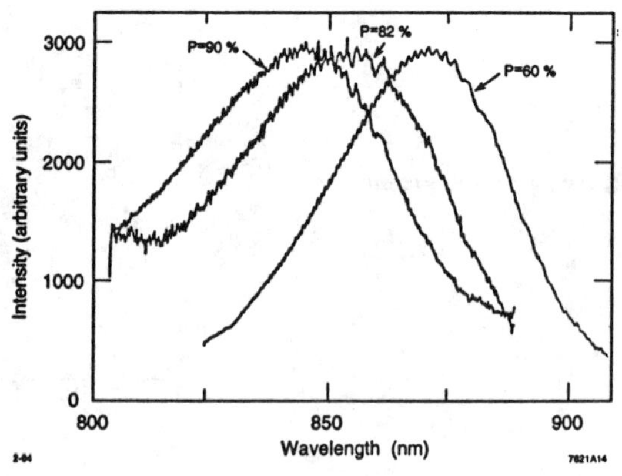

Figure 3: Photoluminescence spectra from GaAs samples with different strain.

Figure 4: **Electron polarization and band gap energy shift**

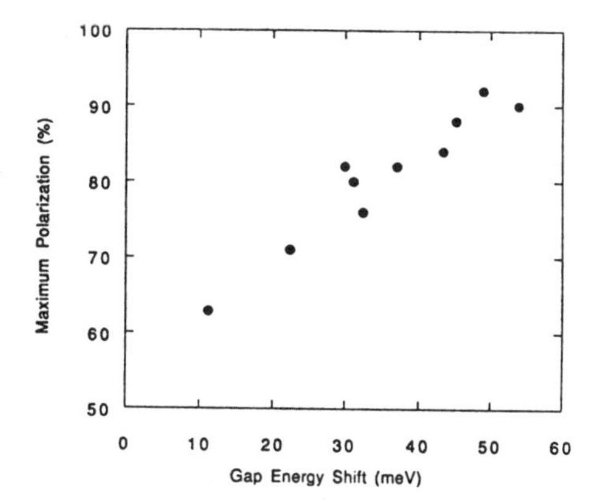

When the excitation light is circularly polarized, the luminescence is also circularly polarized as a result of the recombination of spin-polarized electrons. Measurements of luminescence polarization with near band edge excitation provide information similar to the photoemission polarization. However, with near band edge excitation the buffer $GaAs_{1-x}P_x$ layer is transparent, and the light is absorbed in the GaAs substrate producing luminescence signals much larger than the signals from the epitaxial GaAs layer. To substantially reduce the substrate luminescence, a hole of about one millimeter radius is etched into the substrate layer, removing almost all of the substrate material. A non-reflective coating is then applied to the inside of the hole to reduce reflective light. Fig. 5 shows the photoluminescence polarization measured at 850 nm as a function of excitation wavelength for a sample with (t,x) = (0.1 μm, 0.28). The polarization enhancement is well correlated to the electron polarization measured by Mott scattering.[3] The luminescence polarization (P_{PL}) is related to the electron polarization (P_e) by $P_{PL} = H\ \tau_s/(\tau_s+\tau_e)\ P_e$, where H is the hole coupling factor, τ_s the spin relaxation time, and τ_e the electron lifetime. The luminescence polarization can in principle be used to study the spin depolarization mechanisms.

Figure 5: Photoluminescence polarization (solid circle) electron polarization (open circle).

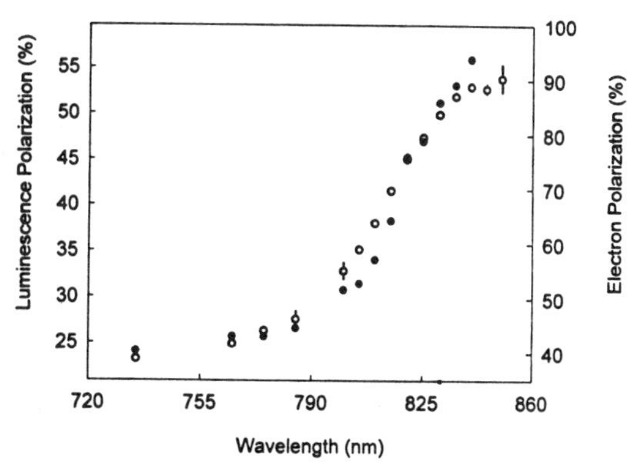

X-RAY DIFFRACTION

X-ray Bragg diffraction is a well established technique and can be used to measure strains as small as 10^{-4} in epitaxial structures. Since strained GaAs has a tetragonal structure, two lattice constants perpendicular ($a_{\perp}$) and parallel ($a_{\parallel}$) to the interface need to be determined. The lattice spacing d is related to the Bragg angle θ_B for layers with Miller indeces (h, k, l) as follows,

$$\frac{1}{d} = \frac{2sin\theta_B}{\lambda} = \sqrt{\frac{h^2 + k^2}{a_{\parallel}^2} + \frac{l^2}{a_{\perp}^2}},$$

where λ is the X-ray wavelength. Since symmetric diffraction such as (004) measures only $a_{\perp}$, at least one additional asymmetric diffraction is needed to measure both constants.

To investigate the quantum efficiency asymmetry with linearly polarized light as mentioned above, asymmetric diffractions (113) and (115) were used to measure strain asymmetries along the [110] and $[1\bar{1}0]$ directions. Fig. 6 shows the (113) diffractions along the two orthogonal directions for a sample with (t,x) = (0.3 μm, 0.24). The strained GaAs peaks appear at different angles, clearly demonstrating the strain asymmetry. The measured in-plane strains are $\epsilon_{\parallel}([110])= -6.5\times10^{-3}$, and $\epsilon_{\parallel}([1\bar{1}0])= -4.5\times10^{-3}$, indicating that the strained GaAs layer is not tetragonal but in fact orthorombic. The quantum efficiency asymmetry can thus be related to the strain asymmetry. More systematic studies are in progress.

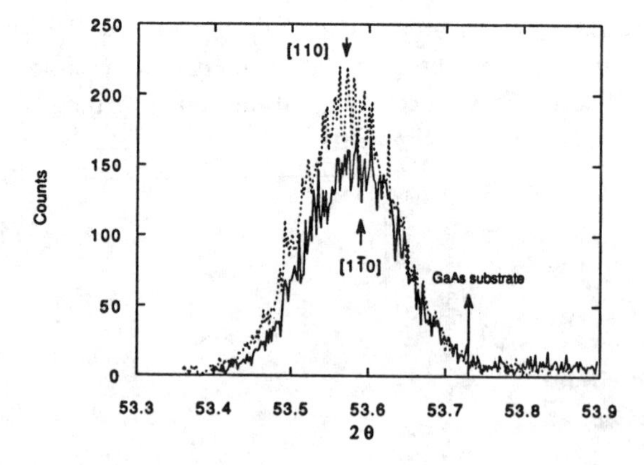

Figure 6: (113) X-ray diffraction spectra along [110] and $[1\bar{1}0]$.

SUMMARY

During the last two years at SLAC, strained GaAs cathodes have been used successfully for high energy physics experiments, producing up to 85% electron polarization. Cathode diagnostic measurements are still needed to fully characterize the cathodes, and techniques based on photoluminescence and X-ray diffraction have been developed to meet this need.

REFERENCES

1. T. Maruyama, E. L. Garwin, R. Prepost, G. H. Zapalac, J. S. Smith, and J. D. Walker, Phys. Rev. Lett. 66, 2376 (1991).

2. T. Nakanishi *et al.*, Phys. Lett. A158, 345 (1991).

3. T. Maruyama, E. L. Garwin, R. Prepost, and G. H. Zapalac, Phys. Rev. B46, 4261 (1992).

4. H. Aoyagi *et al.*, Phys. Lett. A167, 415 (1992).

5. M. Woods *et al.*, J. of Appl. Phys. 73, 8531 (1993).

6. H. Tang, in *Proceedings of the Workshop on Photocathodes for Polarized Electron Sources for Accelerators*, p. 344, SLAC-432, 1994.

7. G. Mulhollan (SLAC), private communication.

Latest results of the Orsay polarized electron source

S. Cohen, J. Arianer, S. Essabaa, R. Frascaria and O. Zerhouni

Institut de Physique Nucléaire, 91406 Orsay Cedex, France

Abstract. The latest results as well as the operating principle and experimental arrangement of the Orsay polarized electron source are presented. The '94 performance, which has revealed 82%-70% electron polarization for currents up to 30 μA, 60% for 100 μA and a maximum quality factor $IP^2 \approx 3.9\ 10^{-5}$ Amp, is discussed. Finally we comment up-on our future development.

DESCRIPTION OF THE SOURCE

The Orsay polarized electron source resembles in many respects the one first proposed and realized by Rice University (1). Our concern apart of obtaining an appreciable electron polarization at an extended and more useful current range was its preparation for proper installation to an accelerator environment. Our present results together with the operational simplicity and reliability of the source show that these tasks have already been largely accomplished.

The polarized electron production of this type of source is based on the spin-preserving Penning reaction involving optically aligned He triplet metastable atoms :

$$He(2^3S_1)\uparrow\uparrow + CO_2\uparrow\downarrow \rightarrow He(1^1S_0)\uparrow\downarrow + CO_2^+\uparrow + e^-\uparrow \qquad (1)$$

The CO_2 molecule is chosen due to its large chemi-ionization cross section. It is also a very efficient electron cooler to thermal energies which are desirable in order to avoid diffusion and to optimize the emittance of the electron beam formed.

Our set-up is schematically presented in figure 1a. A purified He gas flows through a Laval nozzle placed in a microwave cavity. The discharge preferentially excites He atoms to 2^3S metastable level. The metastable density, calculated by absorption measurements, is 10^9 - 10^{10} cm^{-3} for pressures ranging from 0.05 to 0.15 mbar.

Excited and ground state neutrals as well as electrons and ionic species expand to a ≈30 cm straight pyrex tube. A 50 MHz RF coil is used at this point to eliminate at least partially these charged particles. The 1000 l/sec Roots blower evacuating the system results in an average atomic flow velocity of ≈100 m/sec.

The Pyrex tube is connected to a metallic chamber where the optical pumping and the Penning reaction take place. Two laser beams, one circularly and the other linearly polarized, delivered by the same Kr-lamp-pumped LNA laser (4 Watts, 2.5 GHz linewidth) (2), pump optically the He metastables according to the scheme of fig. 2. The atoms are prepared to a selected m_j=+1 or -1 state depending on the helicity of the circularly polarized beam. We have measured experimentally (3) the atomic polarization, and we found it to be ≈92% for pressures up to 0.15 mbar (current >130 μA).

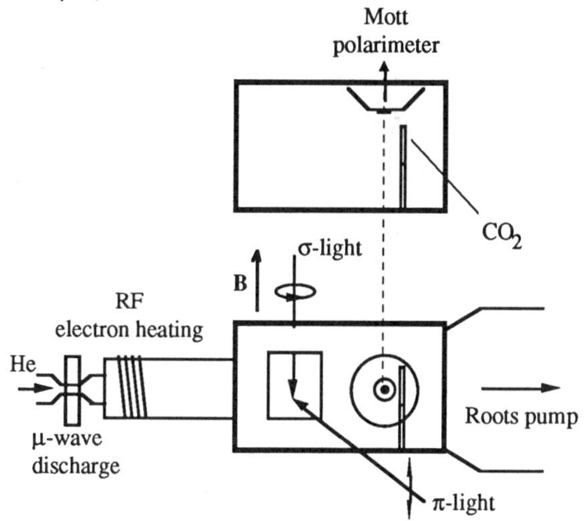

FIGURE 1. Schematic view of the source.

We make use of the He $2^3S_1 \rightarrow 2^3P_0$ (D_0) transition (fig. 2) instead of the $2^3S_1 \rightarrow 2^3P_1$ (D_1) one, prefered by the Rice University group. Such a way depolarization problems caused by J-mixing between the 2^3P_1 and 2^3P_2 levels at relatively high pressures (1) are avoided. It has been experimentally demonstrated (4) that our scheme leads to almost 100% atomic polarization in a discharge cell. The principal benefit of its use however stems from the fact that the demand for ultranarrow laser linewidth is relaxed resulting in considerable simplicity of operation.

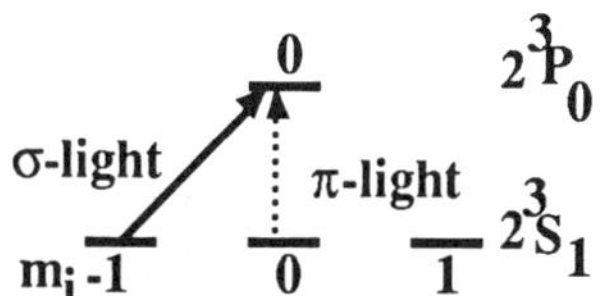

FIGURE 2. Optical pumping scheme using He $2^3S_1 \rightarrow 2^3P_0$ (D_0) transition.

Three pairs of Helmholtz coils cancel the earth's magnetic field and provide a 2.5 Gauss highly homogeneous one ($\Delta B/B \approx 10^{-3}$) which defines the quantization axis and ensures efficient pumping. The quantization axis is perpendicular to the He beam and the electron extraction direction.

About 10 cm downstream the pumping region, CO_2 molecules are injected through a perforated copper ring and interact with the aligned atoms. CO_2 pressure is 0.01-0.1 times the He one. The produced electrons are extracted from the chamber through a small hole (1-2.5 mm diameter) at the center of a metallic button mounted on a plexiglass cone. Extraction is facilitated by biasing the chamber at a negative potential V_c with respect to ground ($-300 \leq V_c \leq 0$ Volts) and the button at a positive V_b with respect to the chamber ($30 \leq V_b \leq 0$ Volts). An electrostatic transport system directs the electron beam to a Mott polarimeter for polarization measurement.

RESULTS AND DISCUSSION

In the course of 1993 (5), we were able to reproduce the results obtained by the Rice University group. However at our latest measurements using a straight tube, the RF field and the chamber grounded or biased at a voltage lower than -200 Volts the performance of the source has considerably improved. Two sets of polarization (P) vs current (I) results recorded lately with the chamber at ground and at -100 Volts are shown in Figure 3. Summarizing the data, we now obtain 82%-75% polarization for 0.1-3 μA, 75%-70% for 3-30 μA, ≈65% for ≈60 μA and ≈60% for ≈100 μA. A probably more useful criterion of the source's performance is the quality factor curve (IP^2 vs I) which is shown on Figure 4.

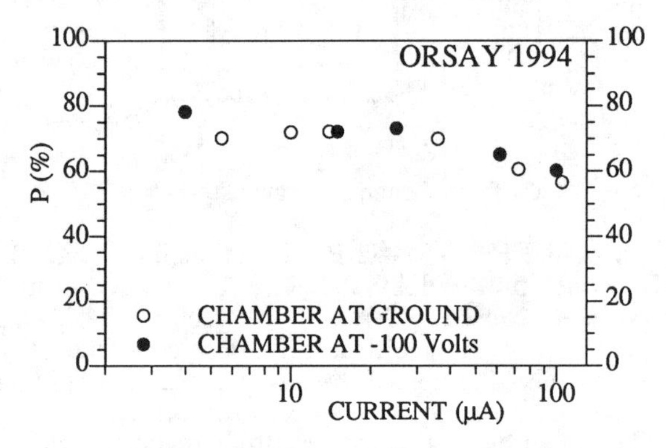

FIGURE 3. The Orsay polarized electron source 1994 performance with the interaction chamber grounded and biased at -100 Volts.

The RF field has no observable effect on the electron polarization at low pressure (<0.09 mbar, currents <60 μA). On the contrary its effect is dramatic for pressures above this value. The polarization is almost doubled with respect to the value measured without it. The phenomena dictating this behaviour have not yet been understood.

It has to be emphasized that it is a necessity to operate the source with a grounded chamber in order to measure the polarization at pressures at which electrical breakdown occurs otherwise. It is also implied by safety considerations at an accelerator environment. For the moment the chamber is connected to the Roots pump via an isolating PVC tube while a grounded chamber would allow its removal. Currently we study the last remaining problem for this modification to be realized, namely vibration compensation.

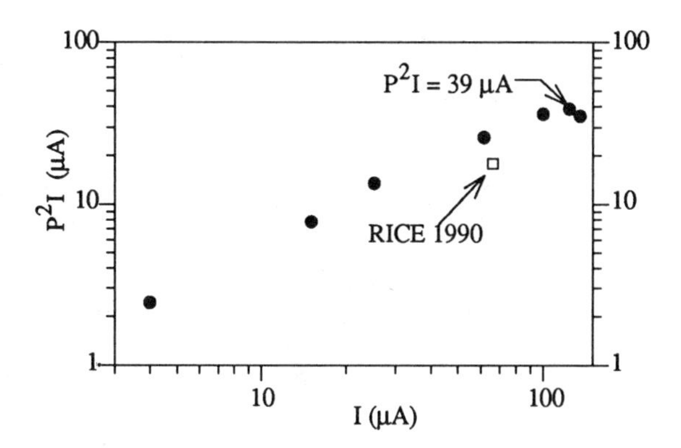

FIGURE 4. The Quality Factor curve of the 1994 Orsay polarized electron source performance. Together is shown the maximum IP^2 value obtained by the Rice group (ref. 1).

It is worth discussing now possible depolarization mechanisms at high currents (and hence pressures). One could safely neglect radiation trapping effects associated to the optical pumping process as experimental evidence shows (3). The artificial decrease of the Roots pumping speed while all the other parameters kept unaltered, demonstrated that we can also exclude the increasing number of neutrals. On the contrary the increasing number of unpolarized electrons originating from the microwave discharge seems to be a good depolarizing candidate. Parasitic current measurements point indeed to this direction. The RF field while efficient is not able to remove the totality of parasitic electrons. Furthermore since they are expected to be more energetic than the Penning ionization produced electrons they may contribute to electron multiplication processes under certain voltage conditions.

It is well known that equation (1) describes only one of the possible paths of the He* + CO_2 reaction. Other products like CO, CO^+,O and O^+ exist. However, new rather recent experimental studies (6) revealed the production of O_2^+ (which may create O_2 by neutralization), scaling as the square of the CO_2 density (three body collision). O_2 has a triplet ground state and thus spin exchanging e + O_2 collisions would depolarize the electrons. Of course the above process is important at high CO_2 pressures but considering that higher CO_2 density is necessary to produce the highest available currents this depolarization mechanism seems quite probable. Even more it is also supported by preliminary measurements.

PERSPECTIVES AND FURTHER DEVELOPMENT

Our future experiments are directed towards a more complete beam characterization and further increase of the polarization obtained at high currents.

Currently we performe energy distribution studies of the electron beam and detailed emittance measurements are about to follow. We expect to verify the emittance reported by the Rice group (7) (normalized $\varepsilon_n \approx 0.16\pi$ mrad mm) already indicated by our spot diameter measurements along the beam.

It is also our concern to improve the routine qualities of the source which they will lead to its even greater simplicity of operation and to prepare the study of all the necessary development for the device's coupling to an accelerator.

A number of diagnostics as well as theoretical modeling are devoted to a better understanding of the physical processes involved prior of extraction. Part of these are the use of a new variable frequency RF circuit to clarify the ununderstandable behaviour mentioned above and more detailed measurements with different CO_2 density at constant currents.

Finally we discuss the most important modification to the actual configuration of the source namely its operation with the magnetic field parallel to the extraction direction. This is because extraction (and transport) of low energy electrons perpendicularly to the direction of the magnetic field is inconvenient. The goal would be to extract a higher percentage of produced electrons for a given pressure and thus to operate the source at the trouble-free low pressure regime. Electron current measurements have been already performed and verified this expectation. The polarization however can not be measured with the present arrangement because in this case the extracted electrons are longitudinally polarized. Their spin will be rotated by an electrostatic deviator and the polarization will be measured using a vertical scattering plane. Our present Mott polarimeter is not flexible enough to be rotated. For this reason a new one has been constructed and it will be installed in the very near future.

ACKNOWLEDGMENTS

It is a pleasure for the authors to acknowledge the exellent technical assistance of Pierre Julou.

REFERENCES

1. Rutherford G. H., Ratliff J. M., Lynn J. G., Dunning F. B. and Walters G. K., Rev. Sci. Instrum., **61**, 1460-1463, (1990).
2. Aminoff C. G., Larat C., Leduc M. and Laloe F., Rev. Phys. Appl. **24**, 827-832, (1989); Aminoff C. G., Essabaa S., Brissaud I. and Arianer J., Optics Commun. **86**, 99-105, (1991).
3. Essabaa S., Schearer L. D., Arianer J., Brissaud I., Humblot H. and Zerhouni O., Nuclear Instr. and Methods A , **344**, 315-318, (1994).
4. Schearer L. D. and Tin P., J. Appl. Phys. **68**, 943-949,(1990).
5. Arianer J., Brissaud I., Essabaa S., Humblot H. and Zerhouni O., Nuclear Instr. and Methods A , **337**, 1-2, (1993).
6. Le Nadan A., Le Coz G. and Tuffin F., J. Phys.(Paris) **50**, 387-395, (1989).
7. Gray L. G., Giberson K. W., Cheng C., Keiffer R. S., Dunning F. B. and Walters G. K., Rev. Sci. Instrum., **54**, 271-274, (1983).

Prospects for a Polarized Electron Source for Next Generation Linear Colliders based on a SLC-Type Gun*

H. Tang, J.E. Clendenin, J.C. Frisch, G.A. Mulhollan, D.C. Schultz, and K. Witte

Stanford Linear Accelerator Center, Stanford University, Stanford, CA 94309

Abstract. The successful operation of a GaAs-based polarized electron source utilizing a DC high voltage gun for the SLC program at SLAC has raised the prospects for a similar source for next generation linear colliders (NGLC). A major challenge in meeting the NGLC requirements is to produce $>10^{12}$ electrons per macrobunch from the cathode. The physics issues that are involved in limiting charge extraction from a GaAs-type cathode and the prospects of realizing a NGLC polarized electron source based on an SLC-type gun will be discussed.

INTRODUCTION

The SLC has been operating with a polarized electron source since the spring of 1992 [1]. The source produces two 2-ns electron bunches of about 62 ns apart and with an intensity of up to 8×10^{10} electrons/bunch at a repetition rate of 120 Hz. The first bunch of polarized electrons, after being bunched and accelerated but with its intensity significantly reduced (by about 50%) due to beam losses, is used to collide with a similar bunch of positrons at the interaction point (IP) with a center of mass energy of about 91 GeV to produce Z° bosons. The second electron bunch, which does not need to be polarized, is used to generate the positron bunch. The polarized electron source uses a DC high voltage gun operated at 120 kV across its Pierce diode structure [2]. The gun uses a thin strained GaAs(100) photocathode, whose surface is treated with cesium and NF_3 to have negative electron affinity (NEA). Coupled with a Nd:YAG pumped Ti:Sapphire laser system that is tuned to the appropriate wavelengths and that has the desired pulse structure, the photocathode gun produces the desired beam for the SLC injector. The performance of the polarized source in all aspects, including reliability, beam polarization and intensity, has been excellent [1].

The successful operation of the SLC polarized source has raised the prospects for a similar polarized source for the next generation of linear colliders (NGLC) [3]. In order to overcome the inherent low repetition rate in a linear collider and also to maximize its energy efficiency, an NGLC requires the use of high-charge macrobunches consisting of about 100 microbunches for the purpose of increased luminosity [3]. The total charge in a single macrobunch is expected to be close to

* Work supported by Department of Energy contract DE-AC03-76SF00515.

1×10^{12} electrons at the IP, or at least about 1.5×10^{12} electrons at the source, which is more than an order of magnitude beyond the present SLC requirement. Although this large amount of charge is distributed over the width of the macrobunch, about 130 ns, and hence the instantaneous current at the source may actually be lower than that for the SLC, the large quantity of the charge in each macrobunch presents the major challenge for developing a NGLC polarized source due to inherent charge-limiting physics that operates in an NEA GaAs photocathode. In this paper, we will discuss the prospects of realizing such a source based on an existing SLC gun from charge considerations only. Other important issues, e.g., laser issues, are beyond the scope of this paper.

CATHODE PHYSICS AND DISCUSSION

The difficulty of extracting a large number of polarized electrons from a GaAs photocathode within a short period is due to the charge limit (CL) effect, which was discovered during the course of polarized source research and development at SLAC [4]. The CL effect dictates that the maximum extractable charge from a GaAs cathode is determined by its intrinsic and surface properties (specifically, doping density and surface NEA condition) and is generally lower than the space charge limit of the gun. In practice, the NEA condition is characterized by the cathode's quantum efficiency (QE), defined as the number of photoemitted electrons normalized by the number of incident photons.

The data shown in Figures 1 and 2, obtained from two 300 nm strained GaAs(100) cathodes of different doping densities with 2-ns excitation laser pulses, illustrate the following important characteristics of the CL effect: (i) once the photoemission is charge limited, further increase in the laser pulse energy leads to a monotonic decrease in the emitted charge; (ii) for charge limited emission, the charge pulse peaks at an earlier time and shrinks in width due to suppressed emission in the later part of the pulse; (iii) there is a significant inter-bunch effect, i.e., the CL for a later coming bunch is further decreased by the presence of another at an earlier time; and (iv) the time constant characterizing the decay of the inter-bunch effect depends critically on the doping density.
A simple picture to understand the CL effect is that electrons excited earlier suppress the emission of subsequently excited electrons within a time scale defined by the decay of the inter-bunch effect. The suppression is most likely caused by an increase in the surface work function due to accumulation of excited electrons at the surface. It happens because only a fraction of the excited electrons reaching the surface may ultimately escape whereas the majority of them will be trapped at the surface. This mechanism is known as the photovoltaic effect, based on which certain aspects of the CL effect have already been numerically simulated by Herrera-Gomez and Spicer [5]. The data shown in Figure 1 is for a relatively low QE condition so as to emphasize the CL effect. Under the best QE condition, this cathode is capable of producing 10×10^{10} electrons in a 2-ns pulse, approaching what the space charge limit permits. As the cathode's QE decays with time, the CL also decreases almost proportionally, and the inter-bunch effect becomes more pronounced [4]. While these results are derived from 2-ns pulse studies, they can be used to help us analyze the feasibility of generating the long NGLC macrobunches.

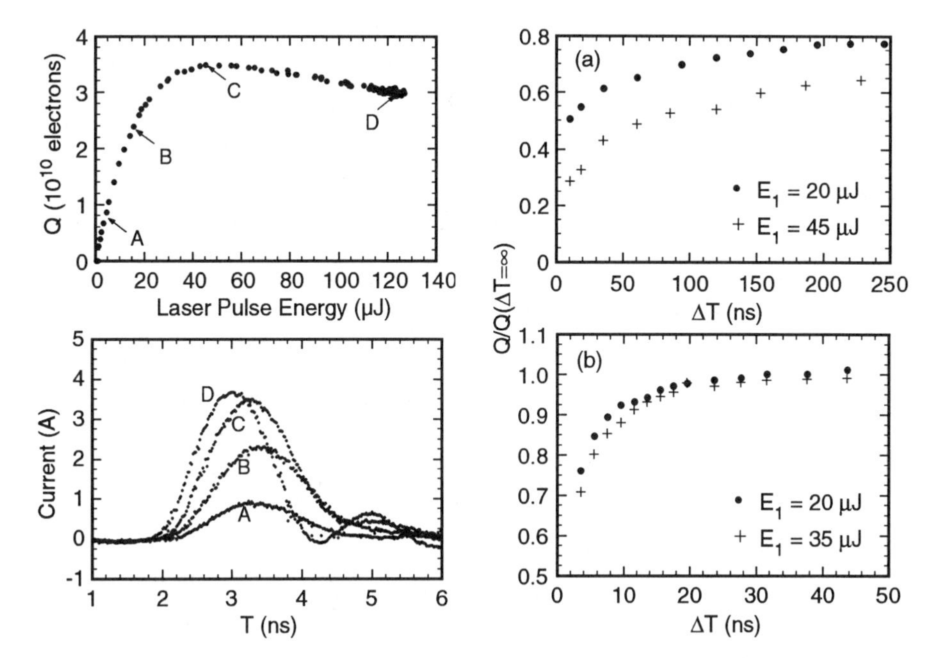

Figure 1: Photoemitted charge versus laser pulse energy (upper panel) and representative charge pulse shapes at various laser pulse energies (lower panel) for a 300 nm 5×10^{18} cm^{-3} Zn-doped strained GaAs cathode. The laser wavelength is 865 nm and pulse width is about 1.8 ns.

Figure 2: Normalized maximum charge in bunch 2 versus time separation from bunch 1 for (a) a 5×10^{18} cm^{-3} and (b) a 2×10^{19} cm^{-3} Zn-doped 300 nm strained GaAs cathode with bunch 1 just into CL (•) and deep into CL (+). The laser wavelengths for the two bunches are 775 nm and 865 nm.

We will consider the case of using a long, continuous laser pulse to generate an NGLC macrobunch of polarized electrons with a GaAs photocathode based SLC gun [6]. In this scheme, the microbunch structure is created with a bunching system which would admit only about 50% of the charge in the unchopped macrobunch from the gun. Thus, the total charge in the continuous macrobunch from the gun must be around 3×10^{12} electrons, or nearly 4 A in current over the duration of the pulse. For comparison, the highest operating current out of an SLC polarized gun for the SLC high energy physics program was about 6.5 A, while the space charge limited current is just about twice as much.

A major difference between generating 2-ns and >100-ns charge pulses from a GaAs cathode is that in the former case the cathode may be operated below its CL, as for the 1994 SLC run, while in the latter case the cathode will probably have to be always operated in its CL regime. This is because in the long pulse mode the number of excited electrons accumulated at the cathode's surface will probably inevitably exceed the necessary level for subsequent photoemission to become charge limited. Therefore, the CL effect plays an even more important role in an NGLC polarized source. In particular, the inter-bunch effect of the CL phenomenon is the most important limiting factor for generating the high-current long charge pulses. This is because it is essentially *all* of the previously excited electrons *over* a time duration characteristic of the decay of the inter-bunch effect that contribute to suppressing the emission of subsequently excited electrons. It becomes immediately clear that a short decay time for the inter-bunch effect is very desirable. As it takes a certain time for the CL effect to fully build up and reach a steady state, there will be a transient period at the beginning of the charge pulse during which the intensity will vary with time and consequently may be unusable for the accelerator. The time scale of such a transient period should be comparable to the inter-bunch effect decay time. In what follows we shall discuss semi-quantitatively the possibility of sustaining a 4 A photoemission out of an SLC gun following the initial transient period.

Let us assume that a cathode has been prepared to have an excellent NEA condition such that an SLC gun equipped with such a cathode is capable of generating a maximum of 1.4×10^{11} electrons in a 2-ns pulse, or about 11 A of current. This is a realistic assumption since even better performances have been achieved at SLAC. We shall show that the response of the cathode to long intense laser pulses may be inferred from its 2-ns single- and two-bunch CL behaviors.

Take the two cases in Figure 2 for example. For the 5×10^{18} cm^{-3} doped sample (Figure 2(a)), the inter-bunch effect decay time is on the order of 100 ns. In order to produce the desired current of 4 A, the laser intensity during the long pulse needs to be, say, about 50% of that of bunch 1 in the upper branch of the data in Figure 2(a), or 50 kW (= 10 μJ/2 ns) at 775 nm [7]. At this laser intensity, the charge emitted in a 2-ns pulse should be about 70% of the maximum, i.e., about 1×10^{11} electrons or 8 A. Within one decay time of 100 ns, the total number of excited electrons would be a factor of 25 more than that during the excitation of bunch 1. According to Figure 2(a), as the total number of excited electrons excited during bunch 1 is increased by about a factor of 2, the charge in bunch 2 is suppressed from about 50% to about 25% of its single-bunch value at a bunch separation of 10 ns. For a 25-fold increase in bunch 1 excitation, the charge in bunch 2 is expected to be suppressed to <10% of its single-bunch value, i.e., the current in bunch 2 would be below 1 A. Clearly, as this current is well below the desired 4 A level, an SLC gun using such a GaAs cathode would fail to produce the desired high intensity macrobunch for an NGLC.

For the 2×10^{19} cm^{-3} doped sample (Figure 2(b)), the inter-bunch effect decay time is on the order of 10 ns, one order of magnitude shorter than for the 5×10^{18} cm^{-3} doped sample. All of the above quantitative analysis remains valid except that the 25-fold increase is now reduced to a mere 2.5-fold increase. According to Figure 2(b), the charge intensity in bunch 2 at a time separation of 4 ns from bunch 1 is reduced to about 70% of its single-bunch value, or about 6 A. Instead of being 4 ns apart, however, photoemission must proceed continuously in the long macrobunch. This would certainly reduce the steady emission current to below the estimated 6 A, but probably would still exceed the desired 4 A level. Thus, it appears that an SLC gun using such a cathode might be adequate for use as an NGLC polarized gun.

Of course, the above discussion is solely based on the available data which are obtained from 2-ns pulse studies. It may be regarded as a generalization of the CL effect from short to long time scales. While we believe that our discussion is quantitatively reasonable, it is possible that new unknown phenomena may arise in going from 2-ns pulses to >100-ns pulses which may complicate our analysis. Therefore, the ultimate check of its validity is only possible through experiments. A Q-switched flash-lamp pumped Ti:Sapphire laser system is now under development at SLAC to produce the desired long, high intensity laser pulses. We expect to conduct experiments to study the CL effects on 100-ns time scales toward the end of 1994.

OUTLOOK

The CL effect plays a very important role for the development of an NGLC polarized source. Our analysis suggests that high doping strained GaAs cathodes hold the best promise for a possible high-polarization electron source for an NGLC. Optimizing a cathode's NEA condition, including the use of larger band gap materials, should always be helpful as it reduces the inter-bunch effect as well as raises the CL. Use of high doping, however, noticeably compromises the polarization obtainable from a strained GaAs cathode due to increased band tailing and other effects. A possible approach to take advantage of the high current capability of high doping cathodes while retaining the high polarization characteristic of low doping cathodes is to use a modulated doping scheme, i.e., high doping in a thin (50 – 100 Å) surface layer and low doping in the rest of the active layer. Such a doping scheme has been tested in a superlattice cathode which indeed yielded encouraging results [8]. From the standpoint of charge considerations, it is fair to state that the present SLC guns are viable candidates for NGLC polarized sources.

REFERENCES

1. D.C. Schultz, *et al.*, "Polarized source performance in 1992 for SLC-SLD", Proc. of *the 10th Intern. Symp. on High Energy Spin Phys.*, Nagoya, Japan, 1992, p. 833; D.C. Schultz, *et al.*, "The polarized electron source of the Stanford Linear Accelerator Center", SLAC-PUB-6606, 1994, presented at *the 17th Intern. Linear Accel. Conf.*, Tsukuba, Japan, 1994.
2. R.K. Alley, *et al.*, SLAC-PUB-6489, to be published.
3. Specifically, NLC, JLC, and SBLC, see e.g. R.H. Siemann, "Linear colliders: the last ten years and the next ten years", Proc. of *the Sessler Symp. on Beam Physics*, Berkeley, 1993.
4. For details, see M. Woods, *et al.*, J. Appl. Phys. **73**, 8531 (1993); H. Tang, *et al.*, "Experimental studies of the charge limit phenomenon in GaAs photocathodes", Proc. of *the Workshop on Photocathodes for Polarized Electron Sources for Accelerators*, Stanford, CA, 1993, SLAC-432 (Rev), 1994, p. 344.
5. A. Herrera-Gomez and W.E. Spicer, "Physics of high intensity nanosecond electron source", Proc. of *1993 SPIE Intern. Symp. on Optics, Imaging and Instrum.*, San Diego, 1993, p. 51.
6. The most feasible approach to generate a stable NGLC macrobunch is probably to use a long laser pulse to generate a long charge pulse, and then to create the desired microbunch structure using a bunching system.
7. The wavelength of the laser pulse used for bunch 1 is not important here. If the wavelength used for bunch 1 is to be the same as that for bunch 2 – a case perhaps more appropriate for the discussion on long pulses – the laser pulse energy for bunch 1 simply needs to be larger by about a factor of 3 to drive the photoemission into CL. Also, the exact laser intensity in the long pulse is hard to predict. 50% appears to be a reasonable assumption.
8. Y. Kurihara, *et al.*, "A high polarization and high quantum efficiency photocathodes using a GaAs-AlGaAs superlattice", SLAC-PUB-6530, submitted to Japn. J. Appl. Phys.

VII. ELECTROWEAK INTERACTIONS

Electroweak Coupling Measurements from Polarized Bhabha Scattering at SLD

The SLD Collaboration*
Stanford Linear Accelerator Center
Stanford University, Stanford, California 94309

Represented by

Kevin T. Pitts†
Department of Physics, University of Oregon
Eugene, OR 97403

Abstract. The cross section for Bhabha scattering ($e^+e^- \rightarrow e^+e^-$) with polarized electrons at the center of mass energy of the Z^0 resonance has been measured with the SLD experiment at the SLAC Linear Collider (SLC) during the 1992 and 1993 runs. The first measurement of the left-right asymmetry in Bhabha scattering ($A_{LR}^{e^+e^-}(\theta)$) is presented. From $A_{LR}^{e^+e^-}(\theta)$ the effective weak mixing angle is measured to be $sin^2\theta_W^{\rm eff} = 0.2245 \pm 0.0049 \pm 0.0010$. When combined with the measurement of A_{LR}, the effective electron couplings are measured to be $v_e = -0.0414 \pm 0.0020$ and $a_e = -0.4977 \pm 0.0045$.

The SLD Collaboration has recently performed the most precise single measurement of the effective electroweak mixing angle, $sin^2\theta_W^{\rm eff}$, by measuring the left-right cross section asymmetry (A_{LR}) in Z boson production at the Z^0 resonance [1]. The left-right cross section asymmetry is a measure of the initial state electron coupling to the Z^0, which allows all visible fermion final states to be included in the measurement. For simplicity, the e^+e^- final state (Bhabha scattering) is omitted in the A_{LR} measurement due to the dilution of the asymmetry from the large QED contribution of the t-channel photon exchange. Here, two new results are presented: the first measurement of the left-right cross section asymmetry in polarized Bhabha scattering ($A_{LR}^{e^+e^-}(|cos\theta|)$), and measurements of the effective electron coupling parameters based on a combined analysis of the A_{LR} measurement [1] and the Bhabha cross section and angular distributions. The vector coupling measurement is the most precise yet presented [2].

In the Standard Model, measuring the left-right asymmetry yields a value for the quantity A_e, a measure of the degree of parity violation in the neutral current, since:

$$A_{LR} = A_e = \frac{2v_e a_e}{{v_e}^2 + {a_e}^2} = \frac{2[1 - 4sin^2\theta_W^{\rm eff}]}{1 + [1 - 4sin^2\theta_W^{\rm eff}]^2}, \tag{1}$$

†current address: Fermilab, P.O. Box 500, Batavia, IL 60510

Table 1: Number of accepted events for the 1992 run. ($<\mathcal{P}_e>= 22.4\%$)

region	left-handed	right-handed	$A_{LR}^{e^+e^-}(raw)$
$0.0 < cos\theta_{CM} < 0.70$	157	137	0.068 ± 0.058
$0.70 < cos\theta_{CM} < 0.94$	208	205	0.0073 ± 0.049
$0.94 < cos\theta_{CM} < 0.98$	305	318	-0.021 ± 0.040
$0.998 < cos\theta_{CM} < 0.9994$	12,395	12,353	0.0017 ± 0.0064

Table 2: Number of accepted events for the 1993 run. ($<\mathcal{P}_e>= 63.0\%$)

region	left-handed	right-handed	$A_{LR}^{e^+e^-}(raw)$
$0.0 < cos\theta_{CM} < 0.70$	864	702	0.103 ± 0.0253
$0.70 < cos\theta_{CM} < 0.94$	1,039	946	0.047 ± 0.022
$0.94 < cos\theta_{CM} < 0.98$	1,566	1,479	0.029 ± 0.018
$0.998 < cos\theta_{CM} < 0.9996$	93,727	94,319	-0.0032 ± 0.0023

where the effective electroweak mixing parameter is defined as $sin^2\theta_W^{\rm eff} = \frac{1}{4}(1-v_e/a_e)$, and v_e and a_e are the effective vector and axial vector electroweak coupling parameters of the electron. The partial width for Z^0 decaying into e^+e^- is dependent on the coupling parameters:

$$\Gamma_{ee} = \frac{G_F M_Z^3}{6\sqrt{2}\pi}({v_e}^2 + {a_e}^2)(1+\delta_e), \tag{2}$$

where $\delta_e = \frac{3\alpha}{4\pi}$ is the correction for final state radiation. G_F is the Fermi coupling constant and M_Z is the Z^0 boson mass. By measuring A_e and Γ_{ee}, the above equations can be utilized to extract v_e and a_e.

Event selection is calorimetry-based and makes use of the distinct topology of the e^+e^- final state. The efficiency and contamination for the wide angle events are calculated from Monte Carlo simulations. Corrections are applied as a function of scattering angle to account for angle-dependent changes in response.

Tables I and II show the number of events accepted, by beam helicity, for the 1992 and 1993 SLC runs. The raw asymmetry is defined as:

$$\tilde{A}_{LR}^{e^+e^-}(\theta) = <\mathcal{P}_e> \mathrm{A}_{\mathrm{LR}}^{\mathrm{e^+e^-}}(\theta) \doteq (\mathrm{N_L} - \mathrm{N_R})/(\mathrm{N_L} + \mathrm{N_R}),$$

where $N_L(N_R)$ is the number of events tagged with a left-(right-) handed electron beam as a function of the $|cos\theta|$, where θ is the center-of-mass scattering angle for the e^+e^- system after initial state radiation. Aside from the charge ambiguity which is unresolved by the calorimeter measurement, the center-of-mass scattering angle is derived trivially from the measured electron and positron laboratory scattering angles. The angular regions in the table are chosen to emphasize the different regimes of the $e^+e^- \to e^+e^-$ distribution: for $|cos\theta| < 0.7$ the s-channel Z^0 decay dominates; from 0.7 to 0.94 the s-channel Z^0 decay, the t-channel photon exchange and the interference between those two interactions all contribute; for $|cos\theta| > 0.94$, the t-channel photon exchange dominates. The region of $0.998 < |cos\theta| < 0.9996$ is that which is covered by the small angle silicon/tungsten luminosity monitor (LUM). The expected asymmetry $(A_{LR}^{e^+e^-}(\theta))$ is largest at $cos\theta = 0$, and may be approximately written as $A_{LR}^{e^+e^-}(\theta) = A_e(1 - f_t(|cos\theta|))$, where $f_t(|cos\theta|)$ represents the t-channel contribution. For the region $|cos\theta| < 0.7$, $< f_t > \simeq 0.12$. The expected asymmetry falls to very small values ($\sim 10^{-4}$) in the small angle region where the t-channel photon exchange dominates.

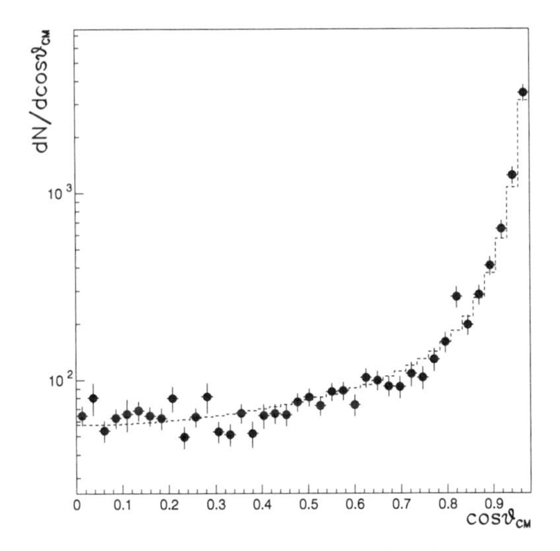

Figure 1: Differential angular distribution for $e^+e^- \to e^+e^-$. The points are the corrected data, the dashed line is the fit.

To extract Γ_{ee} and A_e, the data are fit to the differential e^+e^- cross section using the maximum likelihood method. Two programs are used to calculate the differential e^+e^- cross section: EXPOSTAR [4] and, as a cross check, DMIBA [5]. The EXPOSTAR program calculates the differential cross sections within the framework of the Standard Model. The DMIBA program calculates the differential e^+e^- cross section in a model independent manner. To extract the maximal amount of information from the differential polarized Bhabha scattering distribution, the fit is performed over the entire angular region accepted by the liquid argon calorimeter (LAC), where $|cos\theta| < 0.98$. No t-channel subtraction is performed. All ten lowest order terms in the cross section are included in the fit: the four pure s-channel and t-channel terms for photon and Z^0 exchange, and the six interference terms [6]. The fit also includes initial state radiation. Since the measurement is calorimetric it is insensitive to final state radiation.

The partial width Γ_{ee} is extracted from the data in two ways: (1) using the full fit to the differential cross section for $|cos\theta| \leq 0.98$, and (2) measuring the cross section in the central region ($|cos\theta| < 0.6$) where the systematic errors are smaller, yielding a more precise measurement. For the fits we use $M_Z = 91.187$ GeV/c^2 and $\Gamma_Z = 2.489$ GeV/c^2 [7]. Figure 1 shows the fit to the full $e^+e^- \rightarrow e^+e^-$ distribution, which yields $\Gamma_{ee} = 83.14 \pm 1.03$ (stat) ± 1.95 (sys) MeV. The 2.4% systematic error is dominated (2.1%) by the uncertainty in the efficiency correction factors in the angular region $0.6 < |cos\theta| < 0.98$, where the LAC response is difficult to model due to materials from interior detector elements [3].

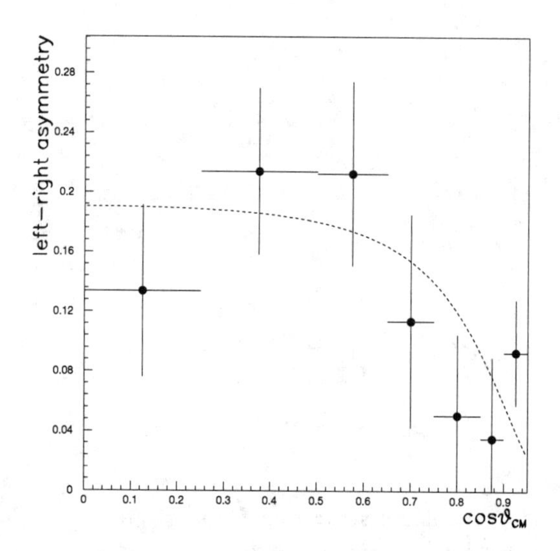

Figure 2: Left-right asymmetry, $\tilde{A}_{LR}^{e^+e^-}(|cos\theta|)$ for polarized $e^+e^- \rightarrow e^+e^-$. The points are the correctd data, the dashed curve is the fit.

A more precise determination of Γ_{ee} was performed using only the central region of the LAC ($|cos\theta| < 0.6$) and the small angle region in the LUM [8]. The program MIBA [9] is then used to calculate Γ_{ee} based on the total measured cross section within the defined fiducial region. From this method, we find:

$$\Gamma_{ee} = 82.89 \pm 1.20 \text{ (stat)} \pm 0.89 \text{ (sys) MeV}.$$

The loss in statistical precision of the limited fiducial region is more than compensated by the improvement in the systematic error. The 1.1% systematic error is dominated by the accuracy of the detector simulation (0.74%) and the uncertainty in the absolute luminosity (0.52%).

To extract A_e from the Bhabha events, the right- and left-handed differential $e^+e^- \rightarrow e^+e^-$ cross sections are fit directly for v_e and a_e using EXPOSTAR. This yields

$$A_e = 0.202 \pm 0.038 \text{ (stat)} \pm 0.008 \text{ (sys)}.$$

Figure 2 shows the measured left-right cross section asymmetry for $e^+e^- \rightarrow e^+e^-$ ($A_{LR}^{e^+e^-}(|cos\theta|)$) compared to the fit. The measurement of A_e is limited by the statistical uncertainty. The 3.8% systematic is dominated by a 3.2% uncertainty in the angle-dependent response correction factors. The polarization uncertainty contributes 1.7% and asymmetry factors from the SLC contribute 0.06% as discussed in Refs. [1] and [3].

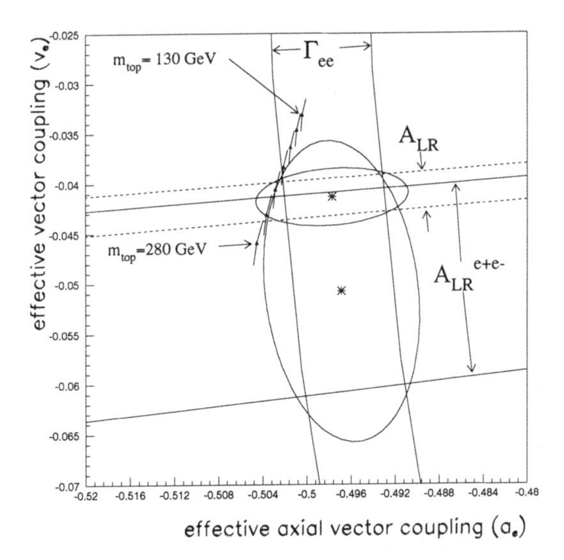

Figure 3: One standard deviation (68%) contour in the a_e, v_e plane. The large ellipse is for $e^+e^- \rightarrow e^+e^-$, the smaller ellipse includes the measurement of A_{LR}. The triangles indicate the Standard Model calculation as a function of the mass of the top quark and the Higgs boson.

The results for Γ_{ee} and A_e from above may now be used in equations 1 and 2 to extract the effective vector and axial vector couplings to the Z^0: $v_e = -0.0507 \pm 0.0096$ (stat) ± 0.0020 (sys), $a_e = -0.4968 \pm 0.0039$ (stat) $\pm$ 0.0027 (sys), where lower energy e^+e^- annihilation data have been utilized to assign $|v_e| < |a_e|$, and $\nu_e e$ scattering data have been utilized to establish $v_e < 0$

and $a_e < 0$ [10]. Figure 3 shows the one standard deviation (68%) contour for these electron vector and axial vector coupling measurements. Most of the sensitivity to the electron vector coupling and, hence, $sin^2\theta_W^{\rm eff}$ arises from the measurement of A_e, while the sensitivity to the axial vector coupling arises from Γ_{ee}. Also shown are standard model calculations using the program ZFITTER [11].

The effective electroweak mixing angle represented by these vector and axial vector couplings is:

$$sin^2\theta_W^{\rm eff} = 0.2245 \pm 0.0049\ (\rm stat) \pm 0.0010\ (\rm sys).$$

We reiterate that this measurement derives strictly from the Bhabha events.

The SLD Collaboration has published a more precise measurement of A_e from the left-right cross section asymmetry (A_{LR}) measurement [1]. Combining the Bhabha results with the SLD measurement of A_{LR} gives:

$$v_e = -0.0414 \pm 0.0020 \qquad a_e = -0.4977 \pm 0.0045,$$

the most precise measurement of the electron vector coupling to the Z^0 published to date. The v_e, a_e contour including the A_{LR} measurement is also shown in Figure 3, demonstrating the increased sensitivity in v_e from A_{LR}.

We thank the personnel of the SLAC accelerator department and the technical staffs of our collaborating institutions for their outstanding efforts on our behalf.

References

[1] SLD Collaboration, K. Abe *et al.*, Phys. Rev. Lett. **73**, 25 (1994).
[2] SLD Collaboration, K. Abe *et al.*, SLAC-PUB-6605, August 1994.
[3] K.T. Pitts, Ph.D. Thesis, University of Oregon, SLAC Report 446 (1994).
[4] D. Levinthal, F. Bird, R.G. Stuart and B.W. Lynn, Z. Phys. C **53**, 617 (1992).
[5] P. Comas and M. Martinez, Z. Phys. C **58**, 15 (1993).
[6] M. Greco, Nucl. Phys. **B177**, 97 (1986).
[7] The LEP Collaborations, Report No. CERN-PPE-93-157, August, 1993.
[8] J.M. Yamartino, Ph.D. Thesis, MIT, SLAC Report 426, February 1994.
[9] M. Martinez and R. Miquel, Z. Phys. C **53**, 115 (1992).
[10] S.L. Wu, Phys. Rep. **107**, 59 (1984).
[11] D. Bardin *et al.*, Report No. CERN-TH-6443-92, May 1992.

This work was supported by Department of Energy contracts: DE-FG02-91ER40676 (BU), DE-FG03-92ER40701 (CIT), DE-FG03-91ER40618 (UCSB), DE-FG02-91ER40672 (Colorado), DE-FG02-91ER40677 (Illinois), DE-FG02-91ER40661 (Indiana), DE-AC03-76SF00098 (LBL), DE-FG02-92ER40715 (Massachusetts), DE-AC02-76ER03069 (MIT), DE-FG06-85ER40224 (Oregon), DE-AC03-76SF00515 (SLAC), DE-FG05-91ER40627 (Tennessee), DE-AC02-76ER00881 (Wisconsin), DE-FG02-92ER40704 (Yale); National Science Foundation grants: PHY-91-13428 (UCSC), PHY-89-21320 (Columbia), PHY-92-04239 (Cincinnati), PHY-88-17930 (Rutgers), PHY-88-19316 (Vanderbilt), PHY-92-03212 (Washington); the UK Science and Engineering Research Council (Brunel and RAL); the Istituto Nazionale di Fisica Nucleare of Italy (Bologna, Ferrara, Frascati, Pisa, Padova, Perugia); the Natural Sciences and Engineering Research Council of Canada (British Columbia, Victoria, TRIUMF); and the Japan-US Cooperative Research Project on High Energy Physics (KEK, Nagoya, Tohoku).

*The SLD Collaboration

K. Abe,[28] I. Abt,[14] T. Akagi,[26] W.W. Ash,[26] D. Aston,[26] N. Bacchetta,[21] K.G. Baird,[24] C. Baltay,[32] H.R. Band,[31] M.B. Barakat,[32] G. Baranko,[10] O. Bardon,[16] T. Barklow,[26] A.O. Bazarko,[11] R. Ben-David,[32] A.C. Benvenuti,[2] T. Bienz,[26] G.M. Bilei,[22] D. Bisello,[21] G. Blaylock,[7] J.R. Bogart,[26] T. Bolton,[11] G.R. Bower,[26] J.E. Brau,[20] M. Breidenbach,[26] W.M. Bugg,[27] D. Burke,[26] T.H. Burnett,[30] P.N. Burrows,[16] W. Busza,[16] A. Calcaterra,[13] D.O. Caldwell,[6] D. Calloway,[26] B. Camanzi,[12] M. Carpinelli,[23] R. Cassell,[26] R. Castaldi,[23] A. Castro,[21] M. Cavalli-Sforza,[7] E. Church,[30] H.O. Cohn,[27] J.A. Coller,[3] V. Cook,[30] R. Cotton,[4] R.F. Cowan,[16] D.G. Coyne,[7] A. D'Oliveira,[8] C.J.S. Damerell,[25] S. Dasu,[26] R. De Sangro,[13] P. De Simone,[13] R. Dell'Orso,[23] M. Dima,[9] P.Y.C. Du,[27] R. Dubois,[26] B.I. Eisenstein,[14] R. Elia,[26] D. Falciai,[22] C. Fan,[10] M.J. Fero,[16] R. Frey,[20] K. Furuno,[20] T. Gillman,[25] G. Gladding,[14] S. Gonzalez,[16] G.D. Hallewell,[26] E.L. Hart,[27] Y. Hasegawa,[28] S. Hedges,[4] S.S. Hertzbach,[17] M.D. Hildreth,[26] J. Huber,[20] M.E. Huffer,[26] E.W. Hughes,[26] H. Hwang,[20] Y. Iwasaki,[28] P. Jacques,[24] J. Jaros,[26] A.S. Johnson,[3] J.R. Johnson,[31] R.A. Johnson,[8] T. Junk,[26] R. Kajikawa,[19] M. Kalelkar,[24] I. Karliner,[14] H. Kawahara,[26] H.W. Kendall,[16] M.E. King,[26] R. King,[26] R.R. Kofler,[17] N.M. Krishna,[10] R.S. Kroeger,[18] J.F. Labs,[26] M. Langston,[20] A. Lath,[16] J.A. Lauber,[10] D.W.G. Leith,[26] X. Liu,[7] M. Loreti,[21] A. Lu,[6] H.L. Lynch,[26] J. Ma,[30] G. Mancinelli,[22] S. Manly,[32] G. Mantovani,[22] T.W. Markiewicz,[26] T. Maruyama,[26] R. Massetti,[22] H. Masuda,[26] E. Mazzucato,[12] A.K. McKemey,[4] B.T. Meadows,[8] R. Messner,[26] P.M. Mockett,[30] K.C. Moffeit,[26] B. Mours,[26] G. Müller,[26] D. Muller,[26] T. Nagamine,[26] U. Nauenberg,[10] H. Neal,[26] M. Nussbaum,[8] Y. Ohnishi,[19] L.S. Osborne,[16] R.S. Panvini,[29] H. Park,[20] T.J. Pavel,[26] I. Peruzzi,[13] L. Pescara,[21] M. Piccolo,[13] L. Piemontese,[12] E. Pieroni,[23] K.T. Pitts,[20] R.J. Plano,[24] R. Prepost,[31] C.Y. Prescott,[26] G.D. Punkar,[26] J. Quigley,[16] B.N. Ratcliff,[26] T.W. Reeves,[29] P.E. Rensing,[26] L.S. Rochester,[26] J.E. Rothberg,[30] P.C. Rowson,[11] J.J. Russell,[26] O.H. Saxton,[26] T. Schalk,[7] R.H. Schindler,[26] U. Schneekloth,[16] B.A. Schumm,[15] A. Seiden,[7] S. Sen,[32] M.H. Shaevitz,[11] J.T. Shank,[3] G. Shapiro,[15] S.L. Shapiro,[26] D.J. Sherden,[26] C. Simopoulos,[26] H.J. Simpson,[26] N.B. Sinev,[20] S.R. Smith,[26] J.A. Snyder,[32] M.D. Sokoloff,[8] P. Stamer,[24] H. Steiner,[15] R. Steiner,[1] M.G. Strauss,[17] D. Su,[26] F. Suekane,[28] A. Sugiyama,[19] S. Suzuki,[19] M. Swartz,[26] A. Szumilo,[30] T. Takahashi,[26] F.E. Taylor,[16] A. Tolstykh,[26] E. Torrence,[16] J.D. Turk,[32] T. Usher,[26] J. Va'vra,[26] C. Vannini,[23] E. Vella,[26] J.P. Venuti,[29] P.G. Verdini,[23] S.R. Wagner,[26] A.P. Waite,[26] S.J. Watts,[4] A.W. Weidemann,[27] J.S. Whitaker,[3] S.L. White,[27] F.J. Wickens,[25] D.A. Williams,[7] D.C. Williams,[16] S.H. Williams,[26] S. Willocq,[32] R.J. Wilson,[9] W.J. Wisniewski,[5] M. Woods,[26] G.B. Word,[24] J. Wyss,[21] R.K. Yamamoto,[16] J.M. Yamartino,[16] S.J. Yellin,[6] C.C. Young,[26] H. Yuta,[28] G. Zapalac,[31] R.W. Zdarko,[26] C. Zeitlin,[20] and J. Zhou[20]

[1]*Adelphi University, Garden City, NY 11530,* [2]*INFN Sezione di Bologna, I-40126 Bologna, Italy*
[3]*Boston University, Boston, MA 02215,* [4]*Brunel University, Uxbridge, Middlesex UB8 3PH, UK*
[5]*California Institute of Technology, Pasadena, CA 91125*
[6]*University of California at Santa Barbara, Santa Barbara, CA 93106*
[7]*University of California at Santa Cruz, Santa Cruz, CA 95064*
[8]*University of Cincinnati, Cincinnati, OH 45221,* [9]*Colorado State University, Fort Collins, CO 80523*
[10]*University of Colorado, Boulder, CO 80309,* [11]*Columbia University, New York, NY 10027*
[12]*INFN Sezione di Ferrara and Università di Ferrara, I-44100 Ferrara, Italy*
[13]*INFN Lab. Nazionali di Frascati, I-00044 Frascati, Italy,* [14]*University of Illinois, Urbana, IL 61801*
[15]*Lawrence Berkeley Laboratory, University of California, Berkeley, CA 94720*
[16]*Massachusetts Institute of Technology, Cambridge, MA 02139*
[17]*University of Massachusetts, Amherst, MA 01003*
[18]*University of Mississippi, University, MS 38677*
[19]*Nagoya University, Chikusa-ku, Nagoya 464 Japan,* [20]*University of Oregon, Eugene, OR 97403*
[21]*INFN Sezione di Padova and Università di Padova, I-35100 Padova, Italy*
[22]*INFN Sezione di Perugia and Università di Perugia, I-06100 Perugia, Italy*
[23]*INFN Sezione di Pisa and Università di Pisa, I-56100 Pisa, Italy*
[24]*Rutgers University, Piscataway, NJ 08855*
[25]*Rutherford Appleton Laboratory, Chilton, Didcot, Oxon OX11 0QX UK*
[26]*Stanford Linear Accelerator Center, Stanford University, Stanford, CA 94309*
[27]*University of Tennessee, Knoxville, TN 37996,* [28]*Tohoku University, Sendai 980 Japan*
[29]*Vanderbilt University, Nashville, TN 37235,* [30]*University of Washington, Seattle, WA 98195*
[31]*University of Wisconsin, Madison, WI 53706,* [32]*Yale University, New Haven, CN 06511*

PRECISE DETERMINATION OF THE WEAK MIXING ANGLE FROM A MEASUREMENT OF A_{LR} IN $(e^+e^- \to Z^0)$*

M. Woods

Stanford Linear Accelerator Center
Stanford University, Stanford, CA 94309

Representing

The SLD Collaboration

Abstract. In the 1993 SLC/SLD run, the SLD recorded 50,000 Z events produced by the collision of longitudinally polarized electrons on unpolarized positrons at a center-of-mass energy of 91.26 GeV. The luminosity-weighted average polarization of the SLC electron beam was (63.0±1.1)%. We measure the left-right cross-section asymmetry in Z boson production, A_{LR}, to be 0.1628±0.0071(stat.)±0.0028(syst.) which determines the effective weak mixing angle to be $\sin^2\theta_W^{\rm eff} = 0.2292 \pm 0.0009({\rm stat.}) \pm 0.0004({\rm syst.})$.[1]

1. Introduction

The left-right asymmetry is defined as $A_{LR}^0 \equiv (\sigma_L - \sigma_R)\,/\,(\sigma_L + \sigma_R)$, where σ_L and σ_R are the e^+e^- production cross sections for Z bosons at the Z pole energy with left-handed and right-handed electrons, respectively. The Standard Model predicts that this quantity depends upon the vector (v_e) and axial-vector (a_e) couplings of the Z boson to the electron current,

$$A_{LR}^0 = \frac{2v_e a_e}{v_e^2 + a_e^2} = \frac{2\left[1 - 4\sin^2\theta_W^{\rm eff}\right]}{1 + \left[1 - 4\sin^2\theta_W^{\rm eff}\right]^2}, \qquad (1)$$

where the effective electroweak mixing parameter is defined[2] as $\sin^2\theta_W^{\rm eff} \equiv (1 - v_e/a_e)/4$.

Using the SLD detector, we count the number (N_L, N_R) of hadronic and $\tau^+\tau^-$ decays of the Z boson for each of the two longitudinal polarization states (L,R) of the electron beam. The electron beam polarization is measured precisely with a Compton polarimeter. From these measurements we determine the left-right asymmetry,

$$A_{LR}(\langle E_{cm}\rangle) = \frac{1}{\langle \mathcal{P}_e^{lum}\rangle} \cdot \frac{N_L - N_R}{N_L + N_R} \qquad (2)$$

where $\langle E_{cm}\rangle$ is the mean luminosity-weighted collision energy, and $\langle \mathcal{P}_e^{lum}\rangle$ is the mean luminosity-weighted polarization. This measurement does not require knowledge of the absolute luminosity, detector acceptance, or detector efficiency.

---

* Work supported in part by the Department of Energy, contract DE-AC03-76SF00515

2. Beam Polarization

Polarized electrons are produced by photoemission from a GaAs cathode; SLC operation with a polarized electron beam is illustrated in Figure 1.[3] The longitudinal electron beam polarization ($\mathcal{P}_e$) at the IP is measured by a Compton polarimeter.[4] This polarimeter detects Compton-scattered electrons from the collision of the longitudinally polarized electron beam with a circularly polarized photon beam; the photon beam is produced from a pulsed Nd:YAG laser operating at 532 nm. After the Compton Interaction Point (CIP), the electrons passes through a dipole spectrometer; a nine-channel Cherenkov detector then measures electrons in the range 17 to 30 GeV.

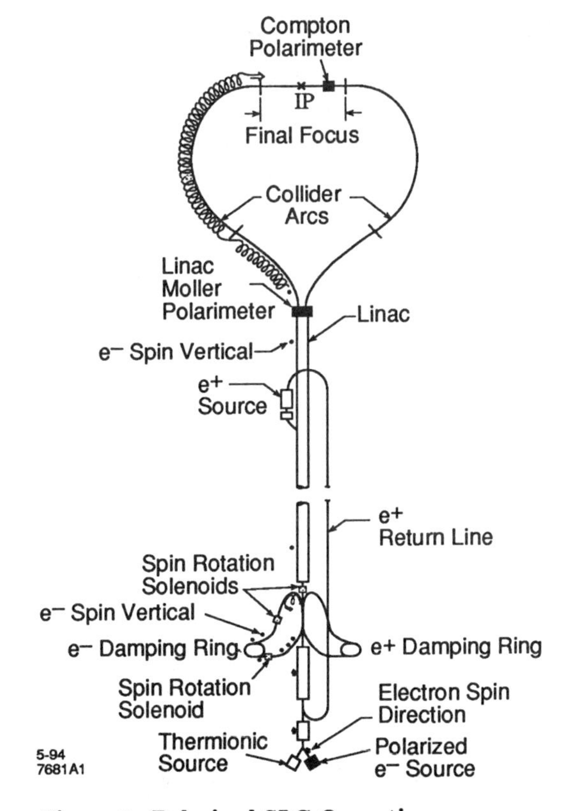

Figure 1: Polarized SLC Operation

The counting rates in each Cherenkov channel are measured for parallel and anti-parallel combinations of the photon and electron beam helicities. The asymmetry formed from these rates is given by

$$A(E) = \frac{R(\rightarrow\rightarrow) - R(\rightarrow\leftarrow)}{R(\rightarrow\rightarrow) + R(\rightarrow\leftarrow)} = \mathcal{P}_e \mathcal{P}_\gamma A_C(E)$$

where $\mathcal{P}_\gamma$ is the circular polarization of the laser beam at the CIP, and $A_C(E)$ is the

Compton asymmetry function. Measurements of $\mathcal{P}_\gamma$ are made before and after the CIP. By monitoring and correcting for small phase shifts in the laser transport line, we are able to achieve $\mathcal{P}_\gamma = (99 \pm 1)\%$. $A_C(E)$ and the unpolarized Compton cross-section are shown in Figure 2. The Compton spectrum is characterized by a kinematic edge at 17.4 GeV, corresponding to an 180° backscatter in the center of mass, and the zero-asymmetry point at 25.2 GeV. $A_C(E)$ is modified from the theoretical asymmetry function [5] by detector resolution effects. This effect is about 1% for the Cherenkov channel at the Compton edge. Detector position scans are used to locate precisely the Compton edge. The position of the zero-asymmetry point is then used to fit for the spectrometer dipole bend strength. Once the detector energy scale is calibrated, each Cherenkov channel provides an independent measurement of $\mathcal{P}_e$. The Compton edge is in channel 7, and we use this channel to determine precisely $\mathcal{P}_e$. The asymmetry spectrum observed in channels 1-6 is used as a cross-check; deviations of the measured asymmetry spectrum from the modeled one are reflected in the inter-channel consistency systematic error. Figure 3 shows the good agreement achieved between the measured and simulated Compton asymmetry spectrum.

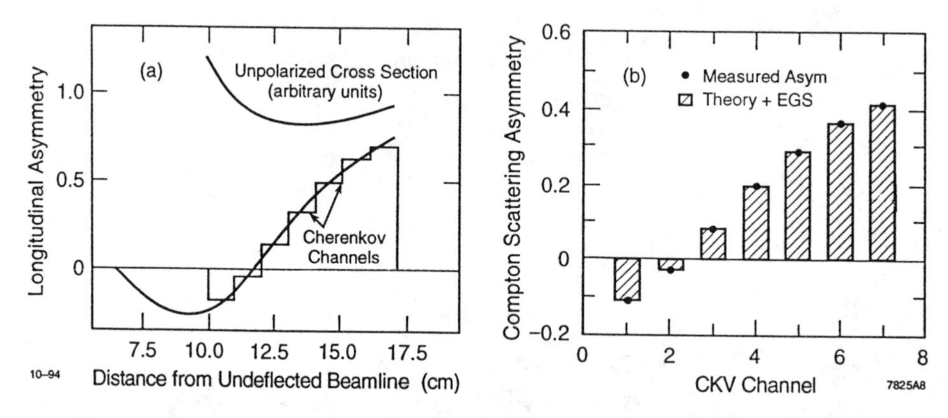

Figure 2: Compton cross-section and asymmetry **Figure 3: Measured and simulated Compton asymmetry**

Polarimeter data are acquired continually during the operation of the SLC. The measured beam polarization is typically 61-64%. The absolute statistical precision attained in a 3 minute interval is typically $\delta\mathcal{P}_e = 1.0\%$. Averaged over the 1993 run, we find the mean beam polarization to be $\langle\mathcal{P}_e\rangle = (61.9 \pm 0.8)\%$. The systematic uncertainties that affect the polarization measurement are summarized in Table 1.

The Compton polarimeter measures $\mathcal{P}_e$ which can differ slightly from $\mathcal{P}_e^{lum}$. The main contribution to this difference arises from a chromatic effect. The electron beam is not monochromatic, but has an energy distribution, $N(E)$, which is characterized by a narrow core ($\Delta E/E < 0.2\%$) and a low-energy tail extending to $\Delta E/E \simeq -1\%$ defined by collimators at the end of the linac. The luminosity and beam polarization at the IP also have a dependence on energy given by $L(E)$ and $P(E)$. For the 1993 running, the energy dependence of $L(E)$ resulted from the small vertical spotsize and the resulting limitation on the luminosity due to third order chromatic aberrations in the final focus. $P(E)$ results from the effective number of spin rotations in the plane of the SLC Arc, which is measured to be 17.9 and is proportional to the energy. The three energy distributions $N(E), L(E), P(E)$ yield the beam polarization, $\mathcal{P}_e$ (ie. $P(E)$ weighted by by $N(E)$), and the luminosity-weighted beam polarization, $\mathcal{P}_e^{lum}$ (ie. $P(E)$ weighted by $N(E) \cdot L(E)$). These are related by $\mathcal{P}_e^{lum} \equiv \mathcal{P}_e(1+\xi)$, which defines the parameter ξ.

Off-energy electrons have reduced longitudinal polarization at the IP due to spin precession in the arc. They contribute less to the luminosity than on-energy electrons because they do not focus to a small spot at the IP; however they contribute the same as on-energy electrons to the Compton measurement of the beam polarization. Thus, $\mathcal{P}_e^{lum}$ can be greater than $\mathcal{P}_e$ due to this chromatic effect. $\mathcal{P}_e^{lum}$ is constrained, however, to be less than the polarization in the linac, $\mathcal{P}_e^{lin}$, since no spin precession occurs before the SLC Arc. Hence,

$$\mathcal{P}_e < \mathcal{P}_e^{lum} < \mathcal{P}_e^{lin} \tag{3}$$

From measurements of $N(E)$ and $P(E)$ and from studies with a low-energy spread beam (core energy distribution rms less than 0.1% and no low-energy tail), we determine $(\mathcal{P}_e^{lin} - \mathcal{P}_e) < 4.7\%$ relative. From measurements of $N(E)$ and $P(E)$ and from calculations of the worst-case distribution for $L(E)$, we determine $(\mathcal{P}_e^{lin} - \mathcal{P}_e^{lum}) > 1.4\%$ relative. Using equation (3), these two results constrain ξ to be in the range from (0-3.3)%. Thus, we find $\xi = (1.7 \pm 1.1)\%$. We correct the Compton measurement of $\langle \mathcal{P}_e \rangle$ for this, and we find the luminosity-weighted polarization for the 1993 run to be $\langle \mathcal{P}_e^{lum} \rangle = (63.0 \pm 1.1)\%$.

The experiments described above to address the chromatic effect allow a determination of the beam polarization in the linac. We find this to be $\mathcal{P}_e^{lin} = (65.7 \pm 0.9)\%$ from measurements made on two separate occasions. This can be compared directly to measurements made by a diagnostic Moller polarimeter located at the end of the linac. This polarimeter analyzes the rate asymmetry in elastic scattering of the polarized electron beam from polarized electrons in a magnetized iron foil. After corrections for atomic momentum effects in the Moller target (the Levchuk effect) [6] we find the Moller measurements to give $\mathcal{P}_e^{lin} = (66 \pm 3)\%$, in good agreement with the Compton measurement.

Following the 1993 SLC/SLD run, the photocathode used for that run was taken to a test beamline with a newly commissioned Mott polarimeter. This polarimeter analyzes the rate asymmetry in elastic scattering of a polarized electron beam from nuclei in an uranium target. This polarimeter was built at UC Irvine and was calibrated there against another Mott polarimeter.[7] The SLAC Mott polarimeter measured the 1993 SLC photocathode to give a beam polarization of $(64 \pm 2)\%$, providing another cross-check on the Compton measurement.

3. Z° Event Selection

The e^+e^- collisions are measured by the SLD detector which has been described elsewhere.[8] The trigger relies on a combination of calorimeter and tracking information, while the event selection is entirely based on the liquid argon calorimeter.[1] We estimate that the combined efficiency of the trigger and selection criteria is (93±1)% for hadronic Z decays. Less than 1% of the sample consists of tau pairs. Because muon pair events deposit only small energy in the calorimeter, they are not included in the sample. The residual background in the sample is due primarily to beam-related backgrounds and to e^+e^- final state events. We use our data and a Monte Carlo simulation to estimate the background fraction due to these sources to be $(0.23 \pm 0.10)\%$. The background fraction due to cosmic rays and two-photon processes is (0.02±0.01)%.

4. Measurement of A_{LR}

Applying the selection criteria, we find 27,225 (N_L) of the events were produced with the left-polarized electron beam and 22,167 (N_R) were produced with the right-polarized beam. The measured left-right cross section asymmetry for Z production is

$$A_m \equiv (N_L - N_R)/(N_L + N_R) = 0.1024 \pm 0.0045.$$

To determine A_{LR} from this, we use Equation (2) modified by some small correction terms,

$$A_{LR}(\langle E_{cm}\rangle) = \frac{A_m}{\langle \mathcal{P}_e^{lum}\rangle} + \frac{1}{\langle \mathcal{P}_e^{lum}\rangle}\left[f_b(A_m - A_b) - A_{\mathcal{L}} + A_m^2 A_{\mathcal{P}} - E_{cm}\frac{\sigma'(E_{cm})}{\sigma(E_{cm})}A_E - A_\varepsilon\right] \quad (4)$$

where f_b is the background fraction; $\sigma(E)$ is the unpolarized Z cross section at energy E; $\sigma'(E)$ is the derivative of the cross section with respect to E; A_b, $A_{\mathcal{L}}$, $A_{\mathcal{P}}$, A_E, and A_ε are the left-right asymmetries of the residual background, the integrated luminosity, the beam polarization, the center-of-mass energy, and the product of detector acceptance and efficiency, respectively.

The corrections defined in square brackets in equation (4) are found to be small. Of these corrections, the most significant one is that due to background contamination. The correction for this is moderated by a non-zero left-right background asymmetry ($A_b = 0.031 \pm 0.010$) arising from e^+e^- final states which remain in the sample. Backgrounds give a net fractional correction to A_{LR} of $(+0.17 \pm 0.07)\%$. Including all the corrections due to backgrounds and left-right asymmetries in luminosity, polarization, energy and efficiency gives a net correction to A_{LR} of $(+0.10 \pm 0.08)\%$ of the uncorrected value.

Using equation (4), we find the left-right asymmetry to be

$$A_{LR}(91.26\text{ GeV}) = 0.1628 \pm 0.0071(\text{stat.}) \pm 0.0028(\text{syst.}).$$

The contributions to the systematic error are summarized in Table 1. Correcting this result to account for photon exchange and for electroweak interference which arises from the deviation of the effective e^+e^- center-of-mass energy from the Z-pole energy (including the effect of initial-state radiation), we find the effective weak mixing angle to be

$$\sin^2\theta_W^{\rm eff} = 0.2292 \pm 0.0009(\text{stat.}) \pm 0.0004(\text{syst.}).$$

Table 1: Systematic uncertainties for the A_{LR} measurement

Systematic Uncertainty	$\delta\mathcal{P}_e/\mathcal{P}_e$ (%)	$\delta A_{LR}/A_{LR}$ (%)
Laser Polarization	1.0	
Detector Calibration	0.4	
Detector Linearity	0.6	
Interchannel Consistency	0.5	
Electronic Noise	0.2	
Total Polarimeter Uncertainty	1.3	1.3
Chromaticity Correction (ξ)		1.1
Corrections in Equation (4)		0.1
Total Systematic Uncertainty		1.7

5. Conclusions

We note that this is the most precise single determination of $\sin^2\theta_W^{\rm eff}$ yet performed. Combining this value of $\sin^2\theta_W^{\rm eff}$ with our previous measurement at $E_{CM} = 91.55$ GeV, we obtain the value, $\sin^2\theta_W^{\rm eff} = 0.2294 \pm 0.0010$. This result can be compared to the determination of $\sin^2\theta_W^{\rm eff}$ from measurements of unpolarized asymmetries at the Z^0 resonance performed by the LEP collaborations (Aleph, Delphi, L3, and OPAL). The LEP collaborations combine roughly 30 individual measurements of quark and lepton forward-backward asymmetries and of final state τ-polarization, to give a LEP global average of $\sin^2\theta_W^{\rm eff} = 0.2321 \pm 0.0004$.[9] The LEP and SLD results differ by 2.5 standard deviations.[10]

The SLD experiment is currently engaged in a physics run which will end in March 1994. By the end of this run, SLD expects to achieve a precision of 0.0005 for $\sin^2\theta_W^{\rm eff}$.

REFERENCES

1. K. Abe et. al., *Phys. Rev. Lett.* **73**, 25 (1994).
2. We follow the convention used by the LEP Collaborations in *Phys. Lett.* **B276**, 247 (1992).
3. For details of SLC operation with a polarized electron beam, see M. Woods, *'Polarization at SLAC'*, SLAC-PUB-6694, November 1994; presented at this conference.
4. M.J. Fero *et al.*, SLAC-PUB-6423; in preparation.
5. See S.B. Gunst and L.A. Page, *Phys. Rev.* **92**, 970 (1953).
6. L.G. Levchuk, *Nucl. Inst. Meth.* **A345**, 496 (1994); and M. Swartz *et al.*, SLAC-PUB-6467, in preparation.
7. The calibration of the UC Irvine Mott polarimeter is described in H. Hopster, D.L. Abraham, *Rev. Sci. Instrum.* **59**, 49 (1988).
8. The SLD Design Report, SLAC Report 273, 1984.
9. The LEP results are taken from The LEP Electroweak Working Group, LEPEWWG-94-02, July 1994; presented at the International Conference on High Energy Physics, Glasgow, Scotland, July 1994.
10. For recent reviews of precision electroweak physics results from LEP and SLC, see M. Swartz, SLAC-PUB-6384, November 1993; and A. Blondel, CERN-PPE/94-133, August 1994.

Measurement of the τ Polarization with the L3 Detector

Oscar Adriani

I.N.F.N. Firenze, Largo Fermi 2, 50125 Firenze, ITALY

Abstract. With a data sample of 86 000 $Z \to \tau^+\tau^-(\gamma)$ events collected by the L3 detector we have measured the polarization of τ-leptons as a function of the production polar angle. We obtain for the ratio of vector to axial-vector weak neutral couplings for electrons $g_{\mathrm{Ve}}/g_{\mathrm{Ae}} = 0.0791 \pm 0.0099(\mathrm{stat}) \pm 0.0025(\mathrm{syst})$ and taus $g_{\mathrm{V}\tau}/g_{\mathrm{A}\tau} = 0.0752 \pm 0.0063(\mathrm{stat}) \pm 0.0045(\mathrm{syst})$ consistent with the hypothesis of e-τ universality. Assuming universality of the e-τ neutral current we determine the effective electroweak mixing angle to be $\sin^2\theta_{\mathrm{w}}^{\mathrm{eff}} = 0.2309 \pm 0.0016$.

INTRODUCTION

In the reaction $e^+e^- \to Z \to \tau^+\tau^-$ even with unpolarized beams at $\sqrt{s} \approx M_Z$ the final state τ leptons are polarized. This polarization is due to the different couplings of left and right-handed leptons to the Z boson. The τ polarization, $\mathcal{P}_\tau(\cos\theta)$, is defined as the asymmetry in the production cross section of τ^- leptons with positive helicity ($h = +1/2$) and negative helicity ($h = -1/2$)

$$\mathcal{P}_\tau(\cos\theta) \equiv \frac{\sigma(h = +1/2) - \sigma(h = -1/2)}{\sigma(h = +1/2) + \sigma(h = -1/2)}, \tag{1}$$

where θ is the angle between the e^- beam and τ^- flight direction. In the improved Born approximation $\mathcal{P}_\tau(\cos\theta)$ at the Z pole is given by [1]:

$$\mathcal{P}_\tau(\cos\theta) = -\frac{\mathcal{A}_\tau + 2\mathcal{A}_e \cos\theta/(1+\cos^2\theta)}{1 + 2\mathcal{A}_\tau\mathcal{A}_e\cos\theta/(1+\cos^2\theta)}. \tag{2}$$

The quantities $\mathcal{A}_\ell$ ($\ell = e, \tau$) are defined as $\mathcal{A}_\ell \equiv 2g_{V\ell}g_{A\ell}/(g_{V\ell}^2 + g_{A\ell}^2)$, where $g_{V\ell}$ and $g_{A\ell}$ denote the effective vector and axial-vector coupling constants.

The measurement of $\mathcal{P}_\tau(\cos\theta)$ yields both $\mathcal{A}_\tau$ and $\mathcal{A}_e$, thus making it possible to check whether the e and τ couplings to the Z are equal, as required by the lepton universality hypothesis. In the framework of the Standard Model the lepton couplings are equal and we use the average of $\mathcal{A}_\tau$ and $\mathcal{A}_e$ to determine the effective electroweak mixing angle through $g_{V\ell}/g_{A\ell} = 1 - 4\sin^2\theta_w^{eff}$.

The following 1-prong τ decay channels have been used in this analysis:

$$\begin{aligned}
\tau^- &\to e^-\overline{\nu}_e\nu_\tau \\
\tau^- &\to \mu^-\overline{\nu}_\mu\nu_\tau \\
\tau^- &\to \pi^-(K^-)\nu_\tau \\
\tau^- &\to \rho^-\nu_\tau \to \pi^-\pi^0\nu_\tau \\
\tau^- &\to a_1^-\nu_\tau \to \pi^-\pi^0\pi^0\nu_\tau
\end{aligned}$$

where the charge conjugate decays are implied here and throughout this paper.

The results presented below are based on a data sample of 86 000 τ pairs from a total integrated luminosity of 74 pb^{-1} collected with the L3 detector [2] in the center of mass energy range $88.2 < \sqrt{s} < 94.3$ GeV in the 1990-1993 running periods.

SELECTION OF τ DECAYS

Selection of $\tau^- \to e^-\overline{\nu}_e\nu_\tau$ and $\tau^- \to \mu^-\overline{\nu}_\mu\nu_\tau$ decays is similar to that described in reference [3]. The selection efficiency are calculated using Monte Carlo simulation of $Z \to \tau^+\tau^-(\gamma)$ including full simulation of the L3 detector response. For the $\tau^- \to e^-\overline{\nu}_e\nu_\tau$ channel the efficiency is estimated to be 76% inside the fiducial region $|\cos\theta| < 0.7$ and is independent of electron energy above 8 GeV; the total background is 4.3%. The selection efficiency for $\tau^- \to \mu^-\overline{\nu}_\mu\nu_\tau$ is 70% inside the fiducial region $|\cos\theta| < 0.8$ and is independent of the muon momentum above 4 GeV; the background contribution is 5.2%.

The selection of hadronic τ decays relies mainly on the particle identification in each hemisphere, which is based upon the topological properties of the energy deposition in the electromagnetic and hadron calorimeters. The hemispheres with identified electron or muon candidates are first rejected. Then an algorithm [3] for finding overlapping neutral energy clusters in the vicinity of hadronic shower in the electromagnetic calorimeter is applied in order to determine the number of neutral clusters and their energies. The invariant mass of each neutral cluster is estimated by fitting its transverse profile with the sum of two electromagnetic shower shapes. A single neutral cluster forms a π^0 candidate if its energy exceeds 1 GeV and its transverse energy profile is consistent with an electromagnetic profile or its invariant mass is within 50 MeV of the π^0 mass. Two distinct neutral clusters form a π^0 candidate if their invariant mass is within 40 MeV of the π^0 mass.

The $\tau^- \to \pi^-(K^-)\nu_\tau$ selection admits no π^0 candidates and no neutral clusters with energy greater than 0.5 GeV. The efficiency of $\tau^- \to \pi^-(K^-)\nu_\tau$

selection is 72% in the barrel ($|\cos\theta| < 0.7$) and 64% in the endcap ($0.82 < |\cos\theta| < 0.94$) region. The efficiency is relatively independent of the pion energy above 5 GeV. The total background is 15.1% in the barrel and 32.5% in the endcap.

To select a $\tau^- \rightarrow a_1^- \nu_\tau$ decay two π^0 candidates are required in the hemisphere. If the two π^0 candidates each consist of a single neutral cluster, then the invariant mass of these two neutral clusters must be incompatible with the mass of a π^0. The selected decays are next subjected to a neural network selection in order to further reduce background from $\tau^- \rightarrow \rho^- \nu_\tau$ and $\tau^- \rightarrow \pi^-\pi^0\pi^0\pi^0\nu_\tau$ events. The final selection efficiency is 33% in the fiducial volume $|\cos\theta_{\rm thrust}| < 0.7$, with a total background of 28%.

To select a $\tau^- \rightarrow \rho^- \nu_\tau$ decay exactly one π^0 candidate is required in the hemisphere. The invariant mass of $\pi^-\pi^0$ system must be in the range 0.45-1.20 GeV. The efficiency of the selection is 70% in the barrel and 51% in the endcap. The total background is 10.9% in the barrel and 17.3% in the endcap. Fig. 1 show the invariant mass spectra of π^0 and $\pi^-\pi^0$ respectively for the selected sample of $\tau^- \rightarrow \rho^- \nu_\tau$ events.

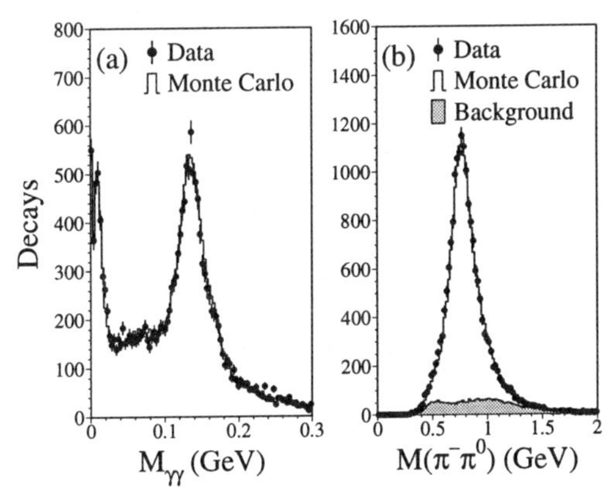

FIGURE 1: a) Mass of $\gamma\gamma$ pairs for selected $\tau^- \rightarrow \rho^- \nu_\tau$ candidates showing a clear π^0 peak. b) Mass spectrum of $\pi^-\pi^0$ of $\tau^- \rightarrow \rho^- \nu_\tau$ candidates in the range $0 < M_{\pi^-\pi^0} < 2$GeV.

MEASUREMENT OF THE POLARIZATION

The goal of this measurement is the determination of the ratios of vector to axial-vector weak neutral couplings for electrons and taus. For each τ decay channel and in each angular bin, $\mathcal{P}_\tau$ is measured by obtaining the linear combination of the $h = +1/2$, $h = -1/2$ Monte Carlo and the non-τ background distributions of proper kinematical quantities which best fits the data. We use a binned maximum likelihood function which properly accounts for the finite statistics both in the data and in the Monte Carlo. For electrons, muons and pions the kinematical variable which is used is simply the energy of the charged particle. To fit $\tau^- \rightarrow a_1^- \nu_\tau$ decays we use the optimal variable fitting following the strategy of references [4, 5] for the construction of the polarization sensitive variable ω_{a_1}. For the analysis of the decay mode $\tau^- \rightarrow \rho^- \nu_\tau$ we fit a 10×15 matrix in the parameter space of $\cos\theta^*_\rho$ and $\cos\psi^*_\rho$, which are respectively the angle between the flight line of the τ^- and the ρ^- in the τ^- rest frame, and the angle between the flight line of the ρ^- and the π^- in the ρ^- rest frame [6].

Ref. [6] contains a detailed discussion of the systematic uncertainties present in this measurement; here we want only remark that the main sources of systematic errors are the selection procedure, background estimation, calibration, charge confusion and theoretical uncertainties.

In order to fit $\mathcal{A}_\tau$ and $\mathcal{A}_e$ using Eqn. 2 we combine bin-by-bin all the individual $\mathcal{P}_\tau$ measurements taking into account the corrections due to initial and final state radiation, γ-exchange and γZ-interference that must be first applied to the data. The results obtained in this way are

$$\begin{aligned} \mathcal{A}_\tau &= 0.150 \pm 0.013 \pm 0.009 \\ \mathcal{A}_e &= 0.157 \pm 0.020 \pm 0.005. \end{aligned}$$

The first error is statistical and the second is systematic. Fitting Eqn. 2 to the corrected data with the assumption $\mathcal{A}_\tau = \mathcal{A}_e$ we obtain

$$\mathcal{A}_{e-\tau} = 0.152 \pm 0.011 \pm 0.007.$$

The corrected $\mathcal{P}_\tau$ points together with the best fit curves are shown in Fig. 2.

CONCLUSIONS

From the measurement of $\mathcal{A}_\tau$ and $\mathcal{A}_e$ we derive the ratio of vector to axial-vector weak neutral couplings for electrons and taus to be

$$\begin{aligned} g_{Ve}/g_{Ae} &= 0.0791 \pm 0.0099 \pm 0.0025 \\ g_{V\tau}/g_{A\tau} &= 0.0752 \pm 0.0063 \pm 0.0045. \end{aligned}$$

This measurement supports the hypothesis of $e - \tau$ universality of the weak neutral current. Assuming lepton universality we derive the ratio of vector to axial-vector weak neutral couplings for leptons and the effective electroweak mixing angle:

$$\begin{aligned} g_V/g_A &= 0.0763 \pm 0.0054 \pm 0.0033 \\ \sin^2\theta_w^{eff} &= 0.2309 \pm 0.0016. \end{aligned}$$

This is consistent with other L3 measurements of $\sin^2\theta_w^{eff}$ extracted from the study of the Z lineshape and forward-backward charge asymmetries in the processes $Z \rightarrow e^+e^-(\gamma)$, $Z \rightarrow \mu^+\mu^-(\gamma)$, $Z \rightarrow \tau^+\tau^-(\gamma)$ and $Z \rightarrow b\bar{b}$ [7, 8].

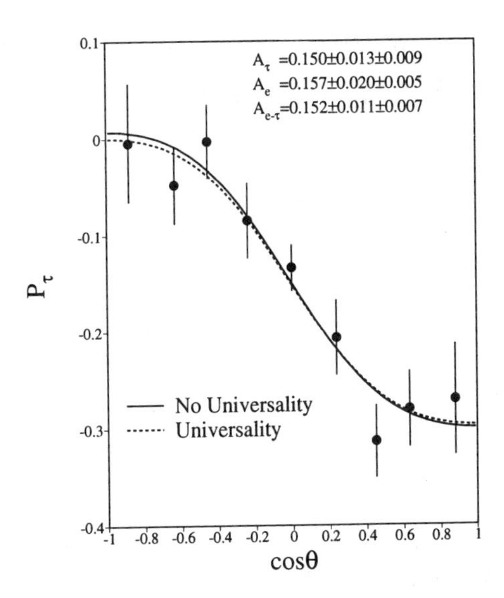

FIGURE 2: The measured dependence $\mathcal{P}_\tau(\cos\theta)$ for all channels combined (dots). The solid and dashed lines represent the fits using Eqn. 2 with and without assumption of lepton universality.

REFERENCES

[1] S. Jadach, Z. Was *et al.* in "Z Physics at LEP1", CERN Report CERN-89-08, (CERN, Geneva, 1989) Vol. 1, p. 235.

[2] L3 Collaboration, B. Adeva *et al.*, *Nucl. Instr. and Meth.* **A289** (1990) 35.

[3] L3 Collaboration, O. Adriani *et al.*, *Phys. Lett.* **B294** (1992) 466.

[4] P. Privitera, *Phys. Lett.* **B308** (1993) 163.

[5] M. Davier *et al.*, *Phys. Lett.* **B306** (1993) 411.

[6] L3 Collaboration, M. Acciarri *et al.*, CERN–PPE/94–145.

[7] L3 Collaboration, M. Acciarri *et al.*, *Z. Phys.* **C62** (1994) 551.

[8] L3 Collaboration, M. Acciarri *et al.*, Preprint CERN-PPE 94-89, June 1994.

Study of Reaction $e^+e^- \to \gamma\gamma(\gamma)$ at the Z^0 Pole With The L3 Detector

Manat Maolinbay

Swiss Federal Institute of Technology
CH-8093, Zürich, Switzerland

Abstract. We have measured the total and the differential cross-sections of the process $e^+e^- \to \gamma\gamma(\gamma)$ at the Z^0 pole region. The kinematics and the absolute rate of higher order process $e^+e^- \to \gamma\gamma\gamma$ are also measured. For the analysis we have used 1.7×10^6 Z^0 events recorded with the L3 detector in 1991 – 1993. Good agreements between QED predictions and the measurements are observed for both processes. At 95% confidence level(CL), we set lower limits on the mass of an excited electron m_{e^*} >154 GeV, the cutoff parameters $\Lambda_+ > 155$ GeV and $\Lambda_- > 152$ GeV; and the cutoff parameter for a non-pointlike coupling $\Lambda_6 > 623$ GeV. We searched for rare Z^0 decays with photonic signatures in the final state. The 95% CL limits on the branching ratios are set to: $\mathrm{Br}(Z^0 \to \pi^0\gamma/\gamma\gamma) < 5.2 \times 10^{-5}$, $\mathrm{Br}(Z^0 \to \eta\gamma) < 7.2 \times 10^{-5}$ and $\mathrm{Br}(Z^0 \to \gamma\gamma\gamma) < 6.3 \times 10^{-6}$.

1. INTRODUCTION

At LEP energies around the Z^0 pole, the reaction $e^+e^- \to \gamma\gamma(\gamma)$ remains to be the only clean QED process which is not affected by the Z^0 decays, thus can be used to test QED. A deviation from QED can be looked for via a search for an excited electron[1,2] or by fitting form factors related to finite size of a composite electron[3]. For both cases it is possible to express differential cross-sections by adding a deviation term δ_{new} to the known QED cross-section $(d\sigma/d\Omega)_{QED}$:

$$(d\sigma/d\Omega) = (d\sigma/d\Omega)_{QED}\,(1 + \delta_{new})$$

where $\delta_{new} = \pm s^2/2\ (1/\Lambda_\pm^4)(1-\cos^2\theta)$ for the excited electron with mass m_{e^*} and cut-off parameters Λ_+ and Λ_-; $\delta_{new} = s^2/\alpha(1/\Lambda_6^4)(1-\cos^2\theta)$ for the non-pointlike electron with cut-off parameter Λ_6 which is related to the size of a non-pointlike electron. θ is the angle of emitted photons with respect to the beam axis and $\sqrt{s}$ is the energy of center-of-mass(E_{cm}). The measurement of $(d\sigma/d\Omega)$ allow to set lower limits on δ_{new} consequently on m_{e^*}, Λ_+, Λ_- and Λ_6.

The decay of $Z^0 \to \gamma\gamma$ is forbbiden[4]. However the rare decays of $Z^0 \to \pi^0\gamma$ and $Z^0 \to \eta\gamma$ [5] would have the same experimental signatures as the

QED process when π^0 or η decays into photons or neutrals. The measurement of total cross-section as a function of E_{cm} can set limits on these branching ratios. The rare process $Z^0 \to \gamma\gamma\gamma$ has a branching ratio of 7×10^{-10} [6] in the Standard Model, which is too small to be observed. However if Z^0 is indeed a composite particle, it could decay to $\gamma\gamma\gamma$ through its charged constituents and the branching ratio could be as high as 10^{-4} [7,8]. Furthermore the photons coming from this decay mode could be seperated from the QED events through their distinct topology. The measurement of event rate of $e^+e^- \to \gamma\gamma\gamma$ can set limit on branching ratio of $Z^0 \to \gamma\gamma\gamma$.

2. ANALYSIS

The L3 detector is described in detail in ref.[9]. The description of the event selections can be found in ref.[10]. The selection efficiencies for the QED events are calculated by Monte-Carlo(MC)[11] simulation and reconstruction where geometrical acceptance, photon conversions and the status of the detector are taken into account. The averaged overall selection efficiency, in the fiducial volume of the detector defined by $|cos\theta| < 0.97$, is found to be $53\% \pm 1\%$. Fig.1(a), (b) and (c) show the comparisons between the data and the MC for the energies of the three most energetic clusters normalized to the beam energy. Fig.1(d) shows the collinearity angle ζ of the two most energetic clusters. The measurement of the differential cross-section as a function of $|cos\theta|$, the QED Born level calculation and radiative corrections are shown in fig.2(a) and (b), where in (b) the data points and the cross-section are normalized to the Born approximation. The numerical values of the measurement are listed in table 1. The comparison of the data points with the QED expectation leads to $\chi^2 = 11$ for 16 degrees of freedom indicating a good agreement for the differential cross-section. To set lower limits on the parameters in δ_{new} we use unbinned maximum likelihood methods [12]. The likelihood function is chosen as:

$$L(m_{e^*}, \Lambda_\pm, \Lambda_6) = \frac{1}{\sqrt{2\pi\sigma^2}} \exp\left(\frac{-(N_{exp} - N_{theo})^2}{2\sigma^2}\right) \prod_{i=1}^{N_{exp}} P(\theta_i; m_{e^*}, \Lambda_\pm, \Lambda_6)$$

where N_{exp} is the total number of observed events, N_{theo} is the total number of expected events, $P(\theta_i; m_{e^*}, \Lambda_\pm, \Lambda_6)$ is the event probability density depending on parameters m_{e^*}, $\Lambda_\pm$, Λ_6 and the polar angle θ_i of the event. σ is the error including statistical and systematic error of 1%. At 95% CL we found $m_{e^*} >$ 154 GeV, $\Lambda_+ >$ 155 GeV, $\Lambda_- >$ 152 GeV and $\Lambda_6 >$ 623 GeV. Fig.3 shows the ratio of measured to radiative-corrected QED differential cross section. The dashed, dotted and dashed-dotted curves illustrate the deviations from QED caused by Λ_+, Λ_- and Λ_6 with the values obtained above.

$\lvert\cos\theta\rvert$	$N_{\gamma\gamma(\gamma)}$ (1991-1993)	$\left(\frac{d\sigma_{meas}}{d\Omega_{\gamma\gamma(\gamma)}}\right)$ (pb/sr) (1991-1993)	$\left(\frac{d\sigma_{QED}}{d\Omega_{\gamma\gamma(\gamma)}}\right)$ (pb/sr)
0.027	37	1.8 ± 0.3	1.76
0.082	47	2.3 ± 0.3	2.29
0.135	60	2.8 ± 0.4	2.68
0.190	51	2.5 ± 0.3	2.98
0.244	61	3.4 ± 0.4	3.21
0.299	61	3.3 ± 0.4	3.43
0.353	72	4.0 ± 0.5	3.64
0.408	64	3.5 ± 0.4	3.89
0.463	87	4.3 ± 0.5	4.20
0.517	88	4.3 ± 0.5	4.61
0.572	107	5.3 ± 0.5	5.17
0.627	132	6.5 ± 0.6	5.95
0.681	122	6.5 ± 0.6	7.07
0.844	222	15.5 ± 1.0	16.01
0.890	301	23.7 ± 1.5	23.89
0.926	267	39.7 ± 2.4	37.58
0.958	103	82.2 ± 8.0	68.68

Table 1: First column gives the center-of-gravity values of the bins. $N_{\gamma\gamma(\gamma)}$ is the number of events. The 3rd and 4th columns give measured and predicted cross-sections.

$\sqrt{s}$ (GeV)	$\mathcal{L}_{int}$ (pb^{-1}) (1991-1993)	$N_{\gamma\gamma(\gamma)}$ (1991-1993)	σ_{meas} (pb)	σ_{QED} (pb)
88.5	0.67	13	64.9 ± 18.0	59.02
89.4	9.00	260	59.6 ± 3.7	57.84
90.3	0.70	20	51.9 ± 11.6	56.69
91.2	43.16	1303	56.0 ± 1.5	55.58
92.0	0.64	26	74.1 ± 14.5	54.62
93.0	9.18	244	51.8 ± 3.3	53.44
93.7	0.78	16	37.7 ± 9.4	52.65

Table 2: The first column is E_{cm}, second the integrated luminosities, third the number of events and the fourth and fifth are the measured and the predicted total cross-sections.

A further deviation from the QED could come from the decays of $Z^0 \to \gamma\gamma$, $\pi^0\gamma$ or $\eta\gamma$. The total cross section as a function of E_{cm} is described by a Breit-Wigner ansatz:

$$\sigma(s) \;=\; \frac{12\pi}{m_Z^2}\,\frac{\Gamma_{ee}\,\Gamma_X}{\Gamma_Z^2}\,\frac{s\Gamma_Z^2}{(s-m_Z^2)^2\;+\;(s\Gamma_Z/m_Z)^2}$$

where Γ_X is the width of the decay mode under consideration, Γ_{ee} is the electronic width of the Z^0, Γ_Z is the total Z^0 width and m_Z is the mass of Z^0. The selection efficiencies for these decays are estimated by MC events with angular distribution of $(1+\cos\theta)$. This leads to same efficiency of $(73\pm1)\%$ for $Z^0 \to \gamma\gamma$ and $Z^0 \to \pi^0\gamma$ and $(52\pm1)\%$ for $Z^0 \to \eta\gamma$. To calculate the limits on these decay widths Γ_X we optimized the likelihood function:

$$L(\Gamma_X) \;=\; \prod_{i=1}^{7} P(N_i, N_{exp}(\Gamma_X))$$

where P_i is the Poisson distribution function at an energy bin 'i', N_{exp} is the measured number of events and N_{theor} is the expected number of events from QED plus the $Z^0 \to X$. At 95% CL the limits are set to: $Br(Z^0 \to \pi^0\gamma) < 5.2\times10^{-5}$ and $Br(Z^0 \to \eta\gamma) < 7.2\times10^{-5}$. The measurement of the total cross-section as a function of E_{cm} together with the limits on the above decay modes are shown in fig.4(a). The measurement is also listed in table 2 with the radiative corrected QED predictions, the integrated luminosities and the number of events at each energy.

Fig.5 illustrates the kinematics of reaction $e^+e^- \to \gamma\gamma\gamma$, where (a) and (b) show the angular and energy distributions of the least energetic photons for data and QED MC. Fig.5(c) shows invariant mass of all photon combinations, from which no significant mass clustering can be seen. The measurement of total cross-section of $e^+e^- \to \gamma\gamma\gamma$ as a function of E_{cm} is shown in fig.4(b). For $Z^0 \to \gamma\gamma\gamma$ decays the 95% CL upper limit is set to $Br(Z^0 \to \gamma\gamma\gamma) < 6.3\times10^{-6}$ or $\Gamma(Z^0 \to \gamma\gamma\gamma) < 16$ KeV. The enhanced resonant line-shape(dashed line) in fig.4(b) shows this limit.

We conclude that at Z^0 pole region all observed variables of reactions $e^+e^- \to \gamma\gamma(\gamma)$ and $e^+e^- \to \gamma\gamma\gamma$ are well described by QED and the Standard Model and no deviations are found.

References

[1] F.E. Low, Phys. Rev. Lett. **14** (1965) 238; R. P. Feynman, Phys Rev. Lett. **74** (1948) 939; F. M. Renard, Phys Lett. **116B** (1982) 264; S. Drell, Ann. Phys. (N.Y.) **4** (1958) 75.

[2] A. Litke, Harvard Univ., Ph.D Thesis (1970) unpublished.
[3] O. J. P. Eboli et al.,Phys. Lett. **271B** (1991) 274.
[4] C. N. Yang, Phys. Rev. **77** (1950) 242.
[5] M. Jacob,T. T. Wu, Phys. Lett. **232B** (1989) 529; G. B. West, Mod. Phys. Lett. **5A** No. 27 (1990) 2281; S. Ghosh, D. Chatterjee, Mod. Phys. Lett. **5A** No. 19 (1990) 1493.
[6] See "Rare Decays", I.J. van der Bij and E.W.N. Glover (Conveners), in Z Physics at LEP 1, CERN Yellow Book, 1989, edt. by G.Altarelli, R. Kleiss and C. Verzegnassi.
[7] F. M. Renard, Phys. Lett. **116B** (1982) 269; M. Bardadin-Otwinowska, CERN-PPE/92-6 (1992).
[8] M. Baillargeon, F. Boudjema,"New Physics Through Final State Photons", to be published, CERN-Yellow-Book "Photon Radiation from Quarks".
[9] L3 Collab., B. Adeva et al., Nucl. Instr. and Meth. **A289** (1990) 35.
[10] L3 Collab., B. Adeva et al., Phys. Lett. **250B** (1990) 199; L3 Collab., O. Adriani et al., Phys. Lett. **288B** (1992) 404; DELPHI Collab., P.Abreu et al., Phys. Lett. **268B** (1991) 296; OPAL Collab., M. Z. Akrawy et al., Phys. Lett. **257B** (1991) 531.
[11] F.A. Berends and R. Kleiss, Nucl. Phys. **186B** (1981) 22.
[12] W. T. Eadie et al., Statistical Methods in Experimental Physics, North Holland (1971) p.268.

Figure Captions

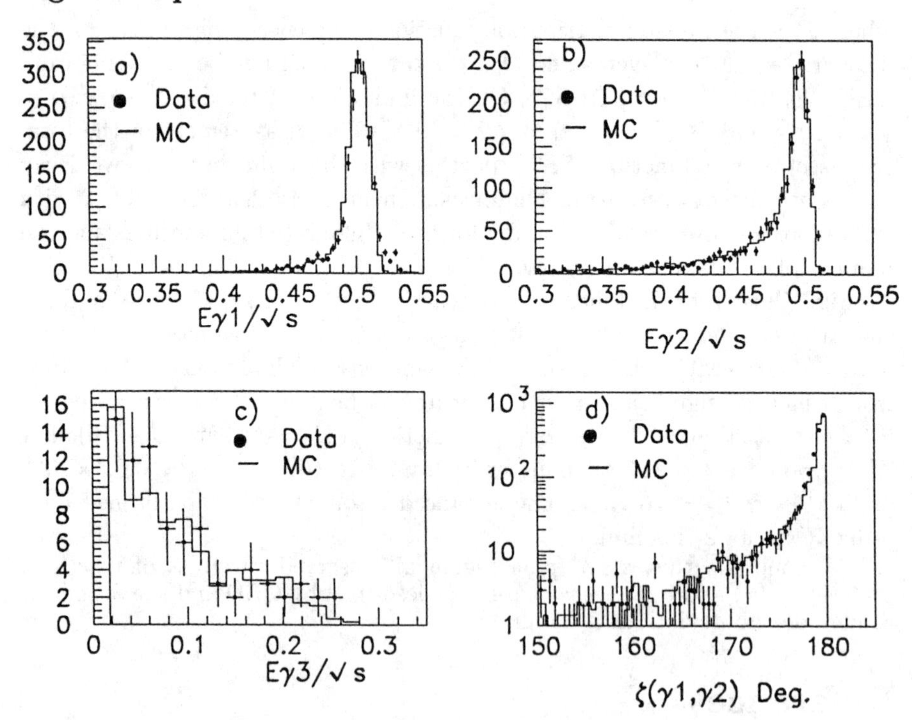

Figure 1: The energy of the most(a), second(b) and third(c) energetic photons divided by the beam energy. The collinearity angle ζ between the two most energetic photons(d).

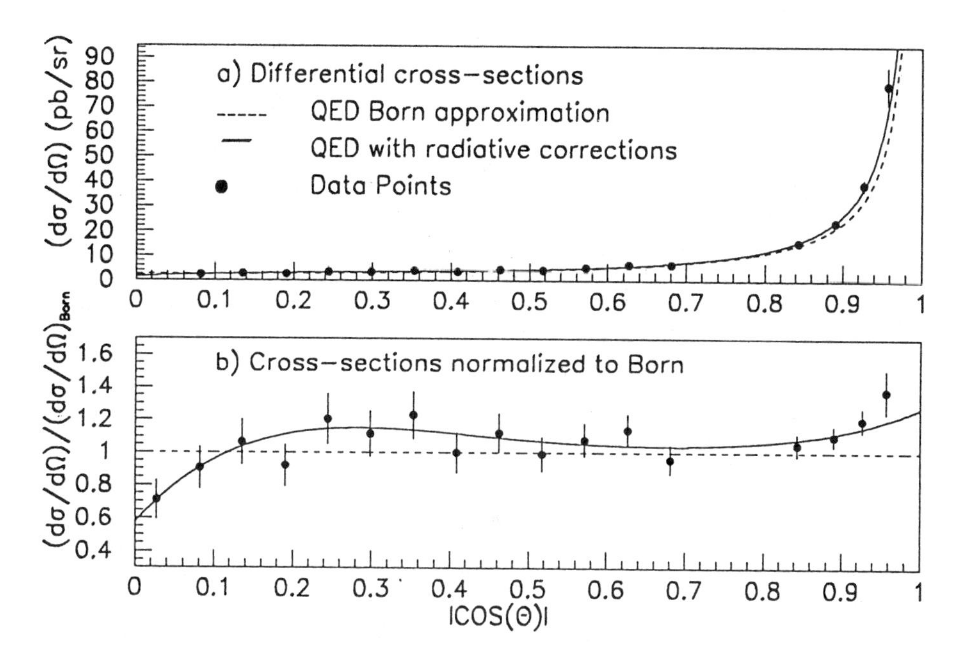

Figure 2: Differential cross-section for $e^+e^- \rightarrow \gamma\gamma(\gamma)$. (a). The QED Born approximations(dashed line), the QED with radiative corrections(solid line) and the measured points (solid dots). (b). Differential cross sections in fig.(a) normalized to the Born approximation.

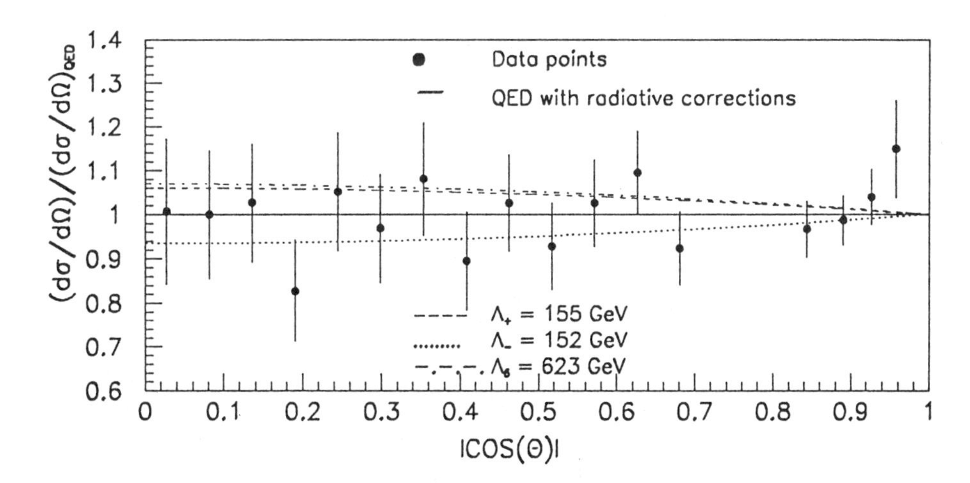

Figure 3: Differential cross-section normalized to the radiatively corrected QED cross-section(line), the measured points (solid dots), the deviation by Λ_+(dashed line), by Λ_-(dotted line), and by Λ_6 (dashed and dotted line).

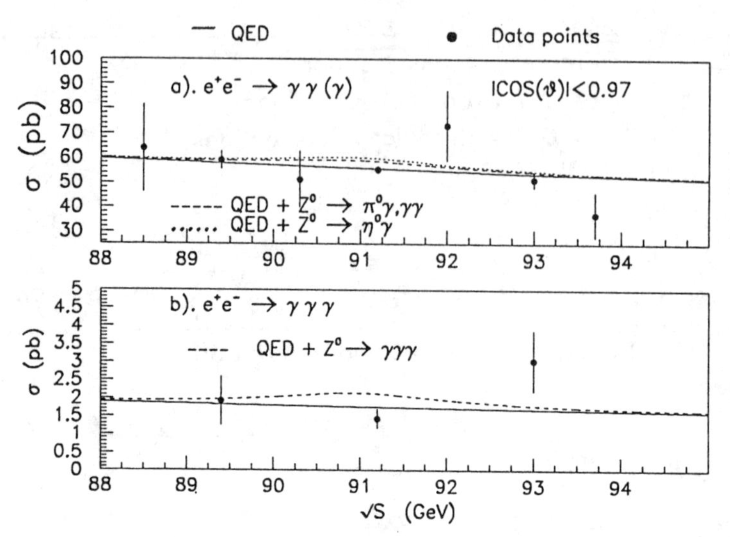

Figure 4: Integrated cross-sections as functions of E_{cm}. (a). $e^+e^- \to \gamma\gamma(\gamma)$: the radiatively corrected QED cross-section (solid line), the measured points (solid dots), the contribution from $Z^0 \to \pi^0\gamma/\gamma\gamma$(dashed line) or $Z^0 \to \eta\gamma$(dotted line). (b). $e^+e^- \to \gamma\gamma\gamma$: the QED cross-section (solid line), the measured points (solid dots), and the contribution from $Z^0 \to \gamma\gamma\gamma$.

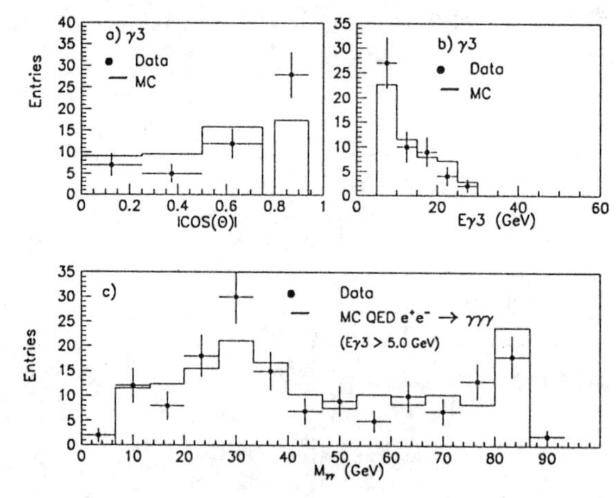

Figure 5: Comparisons of data(solid dots) and QED MC(solid lines) for reaction $e^+e^- \to \gamma\gamma\gamma$: (a). The angular distributions of the least energetic photons; (b). The energy distributions of the least energetic photons; (c). The invariant mass of all photon combinations.

Polarization in Top Pair Production and Decay near Threshold *

R. Harlander[a], M. Jeżabek[a,b], J.H. Kühn[a] and T. Teubner[a]
[a] *Institut für Theoretische Teilchenphysik, D-76128 Karlsruhe, Germany*
[b] *Institute of Nuclear Physics, Kawiory 26a, PL-30055 Cracow, Poland*

Abstract

Theoretical results are presented for top quarks produced in annihilation of polarized electrons on positrons. Polarization studies for $t\bar{t}$ pairs near threshold are free from hadronization ambiguities. This is due to the short lifetime of the top quark. Semileptonic decays are discussed as well as their applications in studying polarization dependent processes involving top quarks. The Green function formalism is applied to $t\bar{t}$ production at future e^+e^- colliders with polarized beams. Lippmann–Schwinger equation is solved numerically for the QCD chromostatic potential given by the two-loop formula at large momentum transfers and Richardson ansatz at intermediate and small ones. The polarization dependent momentum distributions of top quarks and their decay products are calculated.

1 Introduction

The top quark is the heaviest fermion of the Standard Model. Its large mass allows to probe deeply into the QCD potential for nonrelativistic $t\bar{t}$ system produced near energy threshold. Such a system will provide a unique opportunity for a variety of novel QCD studies. The lifetime of the top quark is shorter than the formation time of top mesons and toponium resonances. Therefore top decays intercept the process of hadronization at an early stage and practically eliminate associated nonperturbative effects.

The analysis of polarized top quarks and their decays has recently attracted considerable attention, see [1, 2] and references cited therein. The reason is that this analysis will result in determination of the top quark coupling to the W and Z bosons either confirming the predictions of the Standard Model

*Presented by M. Jeżabek.

or providing clues for physics beyond. The latter possibility is particularly intriguing because m_t plays an exceptional role in the fermion mass spectrum.

The polarization fourvector s^μ of the top quark can be determined from the angular-energy distributions of the charged leptons in semileptonic t decays. In the t quark rest frame this distribution is in Born approximation the product of the energy and the angular distributions[3]:

$$\frac{\mathrm{d}^2\Gamma}{\mathrm{d}E_\ell\,\mathrm{d}\cos\theta} = \frac{1}{2}\left[1 + S\cos\theta\right]\frac{\mathrm{d}\Gamma}{\mathrm{d}E_\ell} \tag{1}$$

where $s^\mu = (0, \vec{s})$, $S = |\,\vec{s}\,|$ and θ is the angle between $\vec{s}$ and the direction of the charged lepton. QCD corrections essentially do not spoil factorization[4]. Thus, the polarization analyzing power of the charged lepton energy-angular distribution remains maximal. There is no factorization for the neutrino energy-angular distribution which is therefore less sensitive to the polarization of the decaying top quark. On the other hand it has been shown [5] that the angular-energy distribution of neutrinos from the polarized top quark decay will allow for a particularly sensitive test of the V-A structure of the weak charged current.

A number of mechanisms have been suggested that will lead to polarized top quarks. However, studies at a linear electron-positron collider are particularly clean for precision tests. Moreover, close to threshold and with longitudinally polarized electrons one can study decays of polarized top quarks under particularly convenient conditions: large event rates, well identified rest frame of the top quark, and large degree of polarization. At the same time, thanks to the spectacular success of the polarization program at SLC [6], the longitudinal polarization of the electron beam will be an obvious option for a future linear collider[1]. In this article some results are presented of a recent calculation [9] of top quark polarization for the reaction $e^+e^- \to t\bar{t}$ in the threshold region. In Sect.2 we discuss the dependence of the top quark polarization on the longitudinal polarizations of the beams. Due to restricted phase space the amplitude is dominantly S wave and the electron and positron polarizations are directly transferred to the top quark. For a quantitative study this simple picture has to be extended and the modifications originating from $S-P$ wave interference should be taken into account. In Sect.3 all these corrections are calculated from numerical solutions of Lippmann-Schwinger equations.

[1] Another proposed and closely related facility is a photon linear collider. At such a machine the high energy photon beams can be generated via Compton scattering of laser light on electrons accelerated in the linac. The threshold behaviour of the reaction $\gamma\gamma \to t\bar{t}$ has been reviewed in [7] and the top quark polarization has been recently considered in [8].

2 Top quark polarization

We adopt the conventions of ref.[10] and describe the longitudinal polarization of the e^+e^- system in its center-of-mass frame as a function of the variable

$$\chi = \frac{P_{e^+} - P_{e^-}}{1 - P_{e^+}P_{e^-}} \tag{2}$$

where $P_{e^\pm}$ denote the polarizations of $e^\pm$ with respect to the directions of e^+ and e^- beams, respectively[2]. In the absence of phases from final state interaction, which can be induced by higher orders in α_s and will be considered elsewhere [9], the top quark polarization is in the production plane. $P_\parallel$ and $P_\perp$ denote the longitudinal and the transverse components of top polarization vector (in its rest frame) with respect to the electron beam. The angle ϑ denotes the angle between e^- and the top quark. In the threshold region the top quark is nonrelativistic (with velocity $\beta = p/m_t \sim \alpha_s$) and the kinetic energy of the $t\bar{t}$ system $E = \sqrt{s} - 2m_t$ is of the order $\mathcal{O}(\beta^2)$. Retaining only the terms up to $\mathcal{O}(\beta)$ one derives the following expressions for the components of the polarization vector

$$P_\parallel = C_\parallel^0(\chi) + C_\parallel^1(\chi)\Phi(E)\cos\vartheta \tag{3}$$
$$P_\perp = C_\perp(\chi)\Phi(E)\sin\vartheta \tag{4}$$

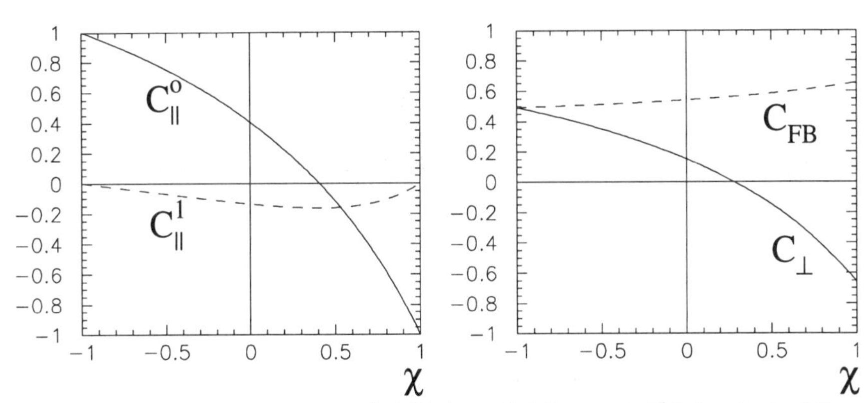

Figure 1: Coefficient functions: a) $C_\parallel^0(\chi)$ – solid line and $C_\parallel^1(\chi)$ – dashed line, b) $C_\perp(\chi)$ – solid line and $C_{FB}(\chi)$ – dashed line.

The coefficients $C_\parallel^0(\chi)$, $C_\parallel^1(\chi)$ and $C_\perp(\chi)$ depend on the polarization χ, the electroweak coupling constants, the Z mass and the center-of-mass energy

[2]It is conceivable that for a future linear e^+e^- collider $P_{e^+} = 0$, $P_{e^-} \neq 0$ and then $\chi = -P_{e^-}$.

$\sqrt{s} \approx 2m_t$. They are plotted in Fig.1 for $m_t = 174$ GeV. $C_{\parallel}^0(\chi)$ and $C_{\parallel}^1(\chi)$ are shown in Fig.1a as the solid and the dashed lines, respectively, and $C_{\perp}(\chi)$ as the solid line in Fig.1b.
The function $\Phi(E)$ describes the complicated dynamics of the $t\bar{t}$ system near threshold. In particular it includes effects of the would-be toponium resonances and Coulomb enhancement. Nevertheless, it is possible to calculate this function using the Green function method. The same function $\Phi(E)$ also governs the forward-backward asymmetry in $e^+e^- \to t\bar{t}$

$$A_{FB} = C_{FB}(\chi)\Phi(E) \tag{5}$$

where C_{FB} is shown as the dashed line in Fig.1b. Eqs.(3) and (4) extend the results of [11] into the threshold region.

3 Lippmann-Schwinger equations

The Green function method has become a standard tool for studying e^+e^- annihilation in the threshold region [12, 13, 14, 15]. We follow the momentum space approach of [15] and solve the Lippmann-Schwinger equations numerically for the S-wave and P-wave Green functions

$$G(p,E) = G_0(p,E) + G_0(p,E)\int \frac{d^3q}{(2\pi)^3}\tilde{V}(|\vec{p}-\vec{q}|)\,G(q,E) \tag{6}$$

$$F(p,E) = G_0(p,E) + G_0(p,E)\int \frac{d^3q}{(2\pi)^3}\frac{\vec{p}\cdot\vec{q}}{p^2}\tilde{V}(|\vec{p}-\vec{q}|)\,F(q,E) \tag{7}$$

where $p = |\vec{p}|$ is the momentum of the top quark in $t\bar{t}$ rest frame

$$G_0(p,E) = \left(E - p^2/m_t + i\Gamma_t\right)^{-1} \tag{8}$$

Γ_t denotes the top width and $\tilde{V}(p)$ is the QCD potential in momentum space; see [15, 9] for details. The function $\Phi(E)$ is related to $G(p,E)$ and $F(p,E)$:

$$\Phi(E) = \frac{\left(1-\frac{4\alpha_s}{3\pi}\right)\frac{1}{m_t}\int_0^{p_m} dp\, p^3 \mathcal{R}e\,(G\,F^*)}{\left(1-\frac{8\alpha_s}{3\pi}\right)\int_0^{p_m} dp\, p^2\,|G|^2} \tag{9}$$

where p_m has been introduced in order to cut off a logarithmic divergence of the numerator. The denominator remains finite for $p_m \to \infty$. In experimental analyses the contributions of very large intrinsic momenta will be automatically suppressed by separation of $t\bar{t}$ events from the background. In our calculation we use $p_m = m_t/3$. The function $\Phi(E)$ is plotted in Fig.2a for $m_t = 174$ GeV. The QCD potential depends on $\alpha_s(m_Z)$ and our results have been obtained for $\alpha_s = 0.12$. For a comparison in Fig.2b the annihilation cross section $\sigma(e^+e^- \to t\bar{t})$ is shown in units of the annihilation cross section $\sigma(e^+e^- \to \mu^+\mu^-)$.

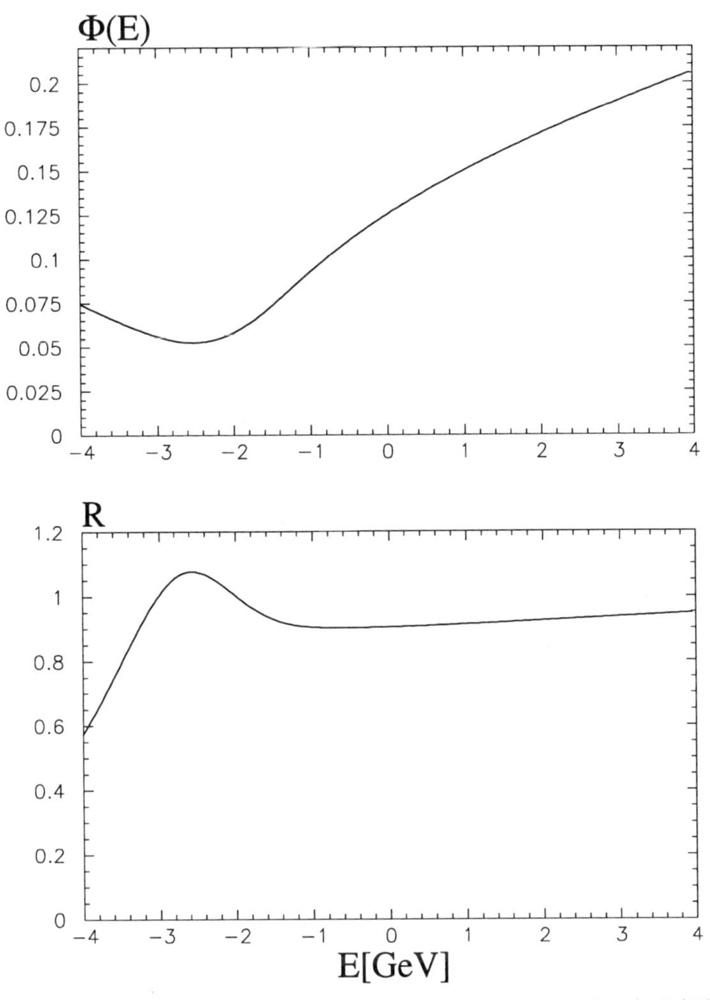

Figure 2: Energy dependence in the threshold region of: a) $\Phi(E)$ and b) $R = \sigma(e^+e^- \to t\bar{t})/\sigma(e^+e^- \to \mu^+\mu^-)$ for $m_t = 174$ GeV and $\alpha_s(m_Z) = 0.12$.

Acknowledgments

MJ would like to thank the Stefan Batory Foundation for a travel grant and the Local Organizing Committee for an additional support which enabled his participation in SPIN'94. This work was supported in part by KBN grant 2P30225206 and by DFG contract 436POL173193S.

References

[1] J.H. Kühn, "Top Quark at a Linear Collider", in *Physics and Experiments with Linear e^+e^- Colliders*, eds. F.A. Harris et al., (World Scientific, Singapore, 1993), p.72.

[2] M. Jeżabek, *Top Quark Physics*, in proceedings of Zeuthen workshop *Physics at LEP 200 and Beyond*, to appear in Nucl.Phys. B(1994) Suppl.; Karlsruhe preprint TTP94-09.

[3] M. Jeżabek and J.H. Kühn, Nucl. Phys. B320 (1989) 20.

[4] A. Czarnecki, M. Jeżabek and J.H. Kühn, Nucl.Phys. B351 (1991) 70;
A. Czarnecki and M. Jeżabek, Nucl.Phys. B427 (1994) 3.

[5] M. Jeżabek and J.H. Kühn, Phys.Lett. B329 (1994) 317.

[6] M. Woods, *Polarization at SLAC*, in these proceedings.

[7] P.M. Zerwas (ed.), e^+e^- Collisions at 500 GeV: The Physics Potential, DESY Orange Reports DESY 92-123A, DESY 92-123B and DESY 93-123C.

[8] V.S. Fadin, V.A. Khoze and M.I. Kotsky, Z. Phys. C64 (1994) 45.

[9] R. Harlander, M. Jeżabek, J.H. Kühn and T. Teubner, to be published.

[10] G. Alexander et al. (eds.), *Polarization at LEP*, CERN 88-06, Geneva 1988.

[11] J.H. Kühn, A. Reiter and P.M. Zerwas, Nucl. Phys. B272 (1986) 560.

[12] V.S. Fadin and V.A. Khoze, JETP Lett. 46 (1987) 525; Sov. J. Nucl. Phys. 48 (1988) 309.

[13] J.M. Strassler and M.E. Peskin, Phys. Rev D43 (1991) 1500.

[14] Y. Sumino, K. Fujii, K. Hagiwara, H. Murayama and C.-K. Ng, Phys. Rev. D47 (1993) 56; H. Murayama and Y. Sumino, Phys. Rev. D47 (1993) 82.

[15] M. Jeżabek, J.H. Kühn and T. Teubner, Z. Phys. C56 (1992) 653;
M. Jeżabek and T. Teubner, Z. Phys. C59 (1993) 669.

Neutrino Physics with KARMEN

W. Kretschmer[a]
KARMEN Collaboration[b]

[a] *Universität Erlangen-Nürnberg, D-91058 Erlangen, Germany*

Abstract

The KARMEN experiment at the pulsed neutron facility ISIS is investigating neutrino properties and interactions by measuring neutrino-induced charged current (CC) $^{12}C(\nu_e, e^-)^{12}N$ reactions and, for the first time, neutral current (NC) $^{12}C(\nu, \nu')^{12}C^*$ excitations of nuclei. In this contribution we present cross section results for these reactions in the energy range of beam dump neutrinos $E_\nu \leq 52.8 MeV$ and discuss first results for $\bar{\nu}_\mu \rightarrow \bar{\nu}_e$ neutrino oscillations in the appearance mode.

The KARMEN Experiment

The KARMEN experiment is being performend since 1990 at the spallation neutron facility ISIS at the Rutherford Appleton Laboratory in Chilton, UK. An 800 MeV proton beam with an average current of 200 μA is stopped in a TaD_2O beam dump thus producing neutrons and pions. These pions are quickly stopped still within the target zone, so that only a small fraction less than 10^{-3} will decay in flight. Negative pions at rest are captured and finally absorbed in nuclei, whereas in the successive decays $\pi^+ \rightarrow \mu^+ + \nu_\mu$ ($\tau = 26ns$) and $\mu^+ \rightarrow e^+ + \nu_e + \bar{\nu}_\mu$ ($\tau = 2.2\mu s$) monoenergetic ν_μ (29.8 MeV) and ν_e, $\bar{\nu}_\mu$ of energies up to 52.8 MeV are generated with exactly the same flux. The unique time structure of two 100 ns wide proton bunches 330 ns apart (repetition rate 50 Hz) allows a clear separation of ν_μ- and ν_e+ $\bar{\nu}_\mu$- induced reactions. Neutrinos are detected in a high resolution 56 t liquid scintillator

---

[b]KARMEN Collaboration: B. Armbruster, G. Drexlin, V. Eberhard, C. Eichner, K. Eitel, H. Gemmeke, W. Grandegger, D. Hunkel, T. Jannakos, M. Kleifges, J. Kleinfeller, P. Plischke, J. Rapp, J. Weber, J. Wochele, J. Wolf, S. Wölfle, B. Zeitnitz: *Kernforschungszentrum Karlsruhe* and *Universität Karlsruhe.* D. Blaser, B. A. Bodmann, A. Dirschbacher, M. Ferstl, E. Finckh, M. Hehle, J. Hößl, P. Jünger, F. Schilling, H. Schmidt, O. Stumm: *Universität Erlangen-Nürnberg.* R. Maschuw: *Universität Bonn.* J. A. Edgington, B. Seligmann: *Queen Mary and Westfield College, London.* A. C. Dodd: *Rutherford Appleton Laboratory* (now at *University of Guelph, Canada*). N. E. Booth: *Oxford University.*

calorimeter (effective distance: 17.6 m from the target) which consists entirely of hydrocarbons and thus serves as a massive life target for the investigation of neutrino interactions with ^{12}C. The scintillator volume is segmented by totally reflecting double lucite sheets into 512 independent modules with an energy resolution of $\sigma(E)/E = 11.5\%/\sqrt{E(MeV)}$ and a time resolution of 1 ns [1]. Bulk shielding is provided by a 6000 t steel blockhouse, cosmic muons are effectively rejected by active and passive shields [2] in combination with the small duty factor of 10^{-5} for ν_μ and 2.5×10^{-4} for ν_e and $\bar{\nu}_\mu$.

Exclusive CC Reaction $^{12}\mathrm{C}\,(\,\nu_e\,,\mathrm{e}^-\,)\,^{12}\mathrm{N}_{\mathrm{g.s.}}$

The charged current transition from $^{12}C(0^+,0)$ to the $^{12}N_{gs}(1^+,1)$ involves both a spin flip and an isospin flip which reduces the rather complex structure of weak hadronic currents to a predominant isovector-axialvector term. The signature of this reaction is a spatially correlated delayed coincidence of a prompt electron in the ν_e time window within 0.5 to 9 μs after beam on target and a positron from the subsequent β-decay of $^{12}\mathrm{N}_{\mathrm{g.s.}}$ in the beam pause within 0.5 to 36 ms. The positron from the ^{12}N-decay uniquely identifies ν-induced transitions to its ground state since excited states are proton unstable. After software cuts on time, energy and position [3] and a pretrigger veto to the sequence of any cosmic ray activity in a 20 μs interval preceeding the prompt or delayed event a total of 351 $\pm$ 20 correlated events with complete signature and a signal to background ratio of 27:1 remain. The energy and time distribution of the prompt and sequential events are in good agreement with expectations [3]. The flux averaged cross section for the exclusive CC reaction $^{12}\mathrm{C}\,(\,\nu_e\,,\mathrm{e}^-\,)\,^{12}\mathrm{N}_{\mathrm{g.s.}}$, derived from the number of neutrino events including efficiency factors for the applied time-, energy- and position- cuts as well as for dead time due to pretrigger vetoing is

$$< \sigma_{CC}(\nu_e) >= [9.0 \pm 0.5_{stat.} \pm 0.8_{sys.}] \times 10^{-42} cm^2.$$

This result is in good agreement with recent theoretical predictions [4, 5, 6] ranging from $(8.0-9.4)\times10^{-42}cm^2$ indicating that neutrino induced transitions between specific nuclear states can be predicted reliably. The good calorimetric properties of the KARMEN detector also allowed the first measurement of the energy dependence of the $^{12}\mathrm{C}\,(\,\nu_e\,,\mathrm{e}^-\,)\,^{12}\mathrm{N}_{\mathrm{g.s.}}$ cross section. The precise measurement of the electron kinetic energy enables the extraction of the isovector axialvector form factor $F_A(q^2)$ for ^{12}C at low momentum transfer $q^2 \leq 0.008\ GeV^2$. Due to the limited q^2-range KARMEN is mainly sensitive to the rms radius R_A of the weak axial charge of ^{12}C. Assuming a dipole form $F_A(q^2) = F_A(0)[1 - \frac{1}{12}R_A^2q^2]^{-2}$ a value of $R_A = (3.8^{+1.4}_{-1.8})\ fm$ has been deduced from a maximum likelihood fit to the observed electron spectrum [7], which agrees within the errors with the rms charge radius of ^{12}C R_m=2.478 fm determined in electron scattering.

Neutral and Charged Current Nuclear Excitation

The neutrino induced neutral current excitation of nuclei has been observed by KARMEN in the reaction $^{12}C(\nu,\nu')^{12}C^*$ (1^+1;15.1 MeV) for the first time [8]. The signature of this reaction is a localized event of 15 MeV from the γ-ray deexcitation of the 15.1 MeV level to the ground state. Since the statistics for ν_μ-induced reactions is still poor, results are only presented for (ν_e+ $\bar{\nu}_\mu$)-induced reactions in the muon decay time window. The energy distribution of the single prong events is shown in fig 1a with a residual background from a pre-beam analysis subtracted. The prominent peak structure around 15 MeV is attributed to the $^{12}C(\nu,\nu')^{12}C^*$ (1^+1;15.1 MeV) reaction. The

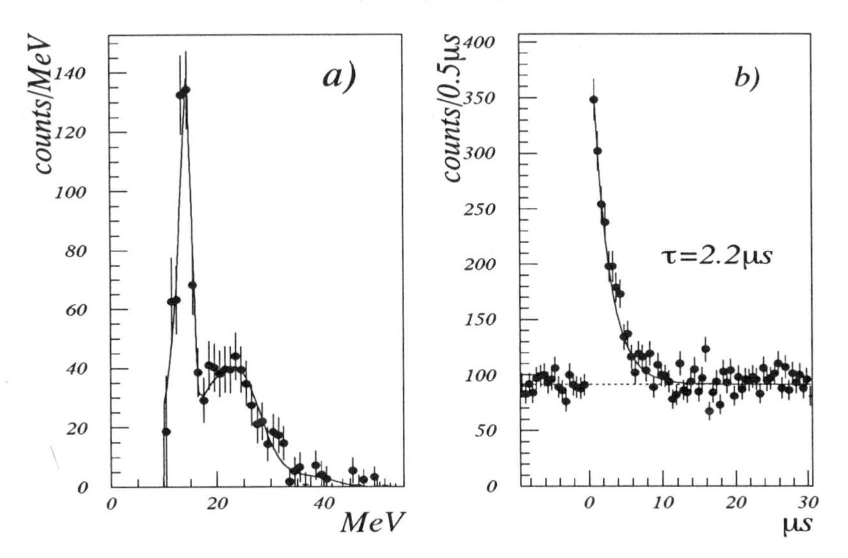

Figure 1: Visible energy and time distribution of ν-induced single prong events in the μ decay time window. (a) energy distribution compared to expectation from MC simulations; (b) time distribution with expected μ decay curve.

resulting flux averaged NC cross section for both neutrino flavours amounts to

$$< \sigma_{NC}(\nu_e + \bar{\nu}_\mu) >= [10.6 \pm 0.9_{stat.} \pm 0.9_{sys.}] \times 10^{-42} cm^2$$

which is again in good agreement with theoretical predictions [4, 5, 6] of $(9.9 - 11.5) \times 10^{-42} cm^2$. The time distribution of single prong events above

10 MeV (fig 1b) is consistent with 2.2 μs decay time and clearly demonstrates the neutrino induced nature of the observed signals. The broad distribution of events with energies above 17 MeV (fig 1a), which have the same time dependence, has been assigned mainly to the inclusive CC reactions $^{12}C(\nu_e, e^-)^{12}N$ and is well reproduced by Monte Carlo simulations including small contributions from ν-e scattering and the $^{13}C(\nu_e, e^-)^{13}N$ reaction. After subtraction of the ground state transition determined in the exclusive CC measurement, the flux averaged cross section for the CC transitions $^{12}C(\nu_e, e^-)^{12}N^*$ has been determined via a maximum likelihood fit:

$$< \sigma_{CC}(\nu_e) >= [6.3 \pm 0.8_{stat.} \pm 1.1_{sys.}] \times 10^{-42} cm^2$$

This result is in contradiction to a shell modell prediction of $3.7 \times 10^{-42} cm^2$ by Donelly [5], but it agrees well with a CRPA calculation of $6.3 \times 10^{-42} cm^2$ by Kolbe et al. [6] based on the Bonn potential for the description of the residual particle hole interaction.

Search for $\bar{\nu}_\mu \rightarrow \bar{\nu}_e$ neutrino flavour oscillations

One of the most important open problems in particle physics is the determination of the neutrino rest mass. With the pulse structure of the spallation source ISIS and the fact, that there is only a negligible number of $\bar{\nu}_e$ generated (contamination rate 8×10^{-4}), the KARMEN experiment allows a high sensitivity search for $\bar{\nu}_\mu \rightarrow \bar{\nu}_e$ neutrino oscillations in the appearance mode, which is a precision test of the Standard Model of weak interaction at low energies. Evidence for the oscillation $\bar{\nu}_\mu \rightarrow \bar{\nu}_e$ is the statistically significant detection of $\bar{\nu}_e$ using the inverse β-decay reaction on the free protons of the scintillator.

$$\bar{\nu}_\mu \xrightarrow{osc} \bar{\nu}_e \quad ; \quad \bar{\nu}_e + p \rightarrow n + e^+$$

The reaction is identified by a prompt positron with an energy ranging up to 51 MeV and spatially correlated delayed gamma rays from the neutron capture reaction Gd(n,γ) with a sum energy of 8 MeV and an mean capture time of 107 μs due to thermalisation and diffusion effects. For this purpose Gd_2O_3 paper with approximately 7.4 mg/cm^2 Gd is placed between the double lucite sheets of the module boundaries.

Performing a maximum likelihood analysis on both the energy (20 - 50 MeV) and time distribution of the 31 detected possible oscillation candidates we find a contribution of 25.5 – 29.6 background events for different Δm^2 and a contribution of 3.3 events from the exclusive CC reaction. Thus we have no evidence for $\bar{\nu}_\mu \rightarrow \bar{\nu}_e$ oscillations. For large Δm^2 we exclude any more than 5.6 oscillation events at a 90% confidence limit and from a comparison to the expected 1840 events for full oscillation we derive an upper limit for the oscillation probability of

$$P(\bar{\nu}_\mu \rightarrow \bar{\nu}_e) < 3.1 \times 10^{-3} (90\% C.L.).$$

Fig 2 shows the excluded region compared to other experiments [9].

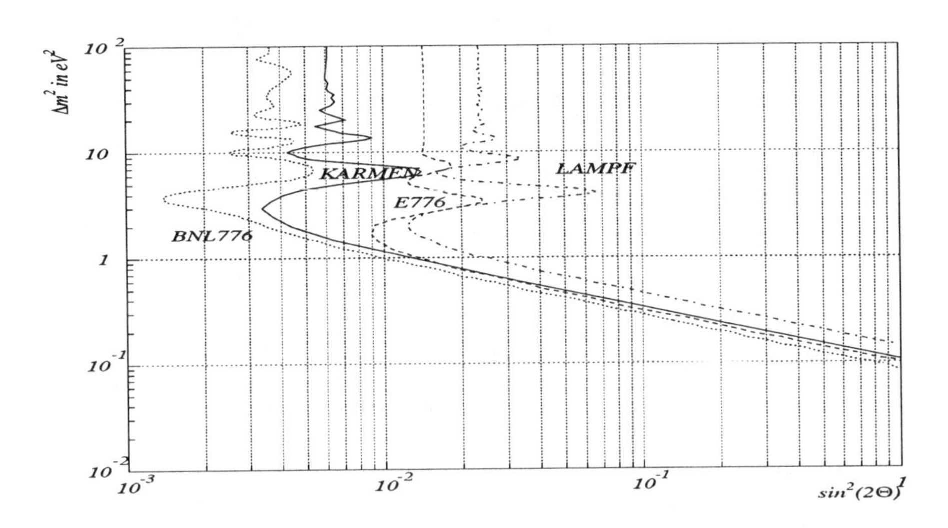

Figure 2: Limits for Δm^2 and $\sin^2 2\Theta$ (90% C.L.) from the KARMEN experiment for the $\bar{\nu}_\mu \rightarrow \bar{\nu}_e$ oscillation compared to other experiments.

This work was supported by BMFT(FRG) and SERC(UK).

References

[1] G. Drexlin et al, *Nucl. Instr. Meth.* A289(1990)490

[2] B. Bodmann et al, *Nucl. Instr. Meth.* A286(1990)214

[3] B. E. Bodmann et al, *Phys. Lett.* B332(1994)251

[4] S. L. Mintz and M. Pourkaviani, *Phys. Rev.* C40(1989)2458

[5] T. W. Donelly, *Phys. Lett.* B43(1973)93 and *private communication*

[6] E. Kolbe et al, *Phys. Rev.* C49(1994)1122; *Nucl. Phys.* A540(1992)599

[7] B. A. Bodmann et al, *Phys. Lett.* B (1994) accepted

[8] B. Bodmann et al, *Phys. Lett.* B267(1991)321

[9] F. Boehm, P. Vogel, *Physics of Massive Neutrinos*, Cambridge University Press (1992) p. 116

Measuring the Muon Polarization in $K^+ \to \pi^0 \mu^+ \nu$ with the E246 detector at KEK

P. Gumplinger *

TRIUMF, Vancouver, B.C., Canada V6T 2A3

Abstract

In the decay, $K^+ \to \pi^0 \mu^+ \nu$, a value for the T-violating triple correlation $P_T \equiv \vec{s}_{\mu^+} \cdot (\vec{p}_{\mu^+} \times \vec{p}_{\pi^0}) / \mid \vec{p}_{\mu^+} \times \vec{p}_{\pi^0} \mid$, of order 10^{-3}, would signal new physics beyond the standard model. An experimental search at this level will be carried out at KEK. It is essential to be able to detect such a small T-odd transverse component of the muon polarization in the presence of a dominant T-even component in the plane defined by the muon and pion momenta, and also to eliminate any systematic errors associated with the apparatus.

1 Introduction

The transverse muon polarization (P_T) in the decay, $K^+ \to \pi^0 \mu^+ \nu$ ($K^+_{\mu 3}$) [1], is that component of the muon polarization which is normal to the plane defined by the three decay particle momenta. The expression for P_T

$$P_T \equiv \frac{\vec{s}_{\mu^+} \cdot (\vec{p}_{\mu^+} \times \vec{p}_{\pi^0})}{|\vec{p}_{\mu^+} \times \vec{p}_{\pi^0}|}, \tag{1}$$

is a triple correlation which changes sign under the time-reversal operation. Consequently, a measured non-zero value of P_T would signal T-violation.

To the extent that the CPT theorem is valid, T-violation must exist at the same level as CP-violation. However, we do not have to invoke CPT-invariance to argue that there is enough experimental evidence for T-violation. It can be shown that under very general assumptions, the scalar product

*representing the KEK-PS E246 collaboration - Japan, Russia, Korea, Canada, USA

$$\zeta \equiv < K_S^0 \mid K_L^0 > = \mid \zeta \mid e^{i\Phi_\zeta} \simeq \frac{\Gamma_S(\pi^+\pi^-)\eta_{+-} + \Gamma_S(\pi^0\pi^0)\eta_{00} + \ldots}{\frac{1}{2}(\Gamma_L + \Gamma_S) + i(m_L - m_S)} \quad (2)$$

must be real if CPT holds, independent of T-invariance; and it must be purely imaginary if T holds, independent of CPT-invariance [2]. The measured phases ($\Phi_{+-} \simeq \Phi_{00} \simeq 45^o$) of the CP-violating parameters η_{+-} and η_{00} in the decay of the neutral kaon into two pions

$$\eta_{2\pi} \equiv \frac{< 2\pi \mid H_{wk} \mid K_L^0 >}{< 2\pi \mid H_{wk} \mid K_S^0 >} \quad (3)$$

and the remarkable fact that $\Delta m = (m_L - m_S) \simeq \frac{1}{2}\Gamma_S \gg \frac{1}{2}\Gamma_L$, where m and Γ are the mass and the decay-width of the kaons, lead to $\Phi_\zeta \simeq 0^o$ and give further evidence for CPT-invariance as opposed to T-invariance.

Among various models proposed to explain observed CP-violation phenomena in K^0 decay, the most appealing is the standard model with an imaginary phase in the Cabbibo-Kobayashi-Maskawa (CKM) matrix elements. Although two experiments were carried out at FNAL and CERN to verify the non-zero CKM phase, no definite statement about the model can be drawn from their results. Furthermore, it is speculated that this standard model CP violation may not be large enough to explain the other observed CP violating phenomenon, the baryon asymmetry in the universe (BAU). It is therefore worthwhile to undertake a search for other sources of CP violation.

The measurement of P_T in $K_{\mu 3}^+$ decay has two striking virtues as a background-free search for new CP violation mechanisms. First, the fake T-odd effect that could be introduced by final-state electromagnetic interactions (FSI) has been calculated to be negligibly small, O(10^{-6}), since only one charged particle exists in the final state [3]. Second, P_T in $K_{\mu 3}^+$ decay has no contributions from the CKM phase in the minimal standard model at the tree level, and higher-order contributions are negligibly small. This implies that the observation of a non-zero P_T value would be a definite signature of physics beyond the standard model. Also, there have been general discussions [4,5] that neither an effective vector (V) nor an axial-vector (A) interaction (such as in the minimal standard model and the left-right symmetric model) introduce P_T, but that only a scalar (S) or a psuedoscalar (P) interaction gives a non-zero P_T. Therefore, P_T can be regarded as being one of the best observables to examine the extended Higgs sector and the scalar leptoquark model.

P_T is related to the two structure form factors in the $K_{\mu 3}^+$ hadronic current, $f_+(q^2)$ and $f_-(q^2)$. T-invariance places a constraint that the relative phases

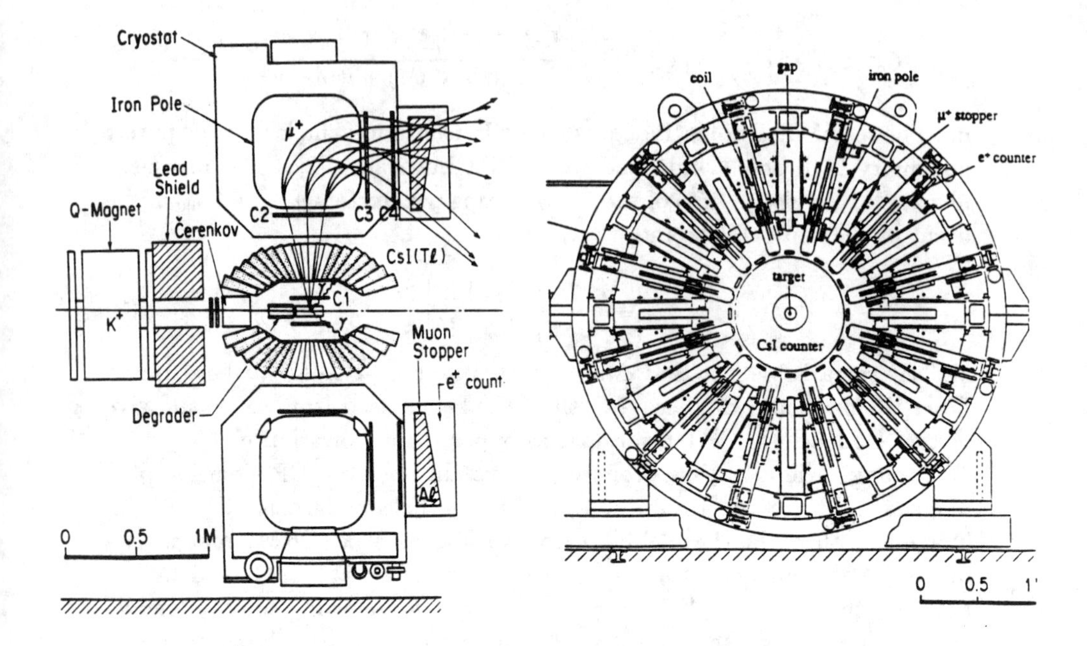

Figure 1: Schematic side and end views of the E246 detector

of f_+ and f_- are the same (or opposite); in other words, the parameter $\xi(q^2) = f_-(q^2)/f_+(q^2)$ should be a real number [2]. P_T is given by [6]

$$P_T \cong \text{Im}\xi \left(\frac{m_\mu}{m_K}\right) \frac{|\vec{p}_\mu|}{[E_\mu + |\vec{p}_\mu|\vec{n}_\mu \cdot \vec{n}_\nu - m_\mu^2/m_K]} \tag{4}$$

2 KEK E246 Experiment

A new experiment, E246, is being prepared at the KEK 12-GeV proton synchrotron (KEK-PS) to search for P_T in $K_{\mu 3}^+$ decays with high sensitivity. In contrast to previous experiments which used in-flight K^+ decays, K^+ decays at rest will be employed in this experiment, providing several advantages: Foremost, it allows the kinematics of the decay to be precisely determined. This, together with the high degree of symmetry in the detector will reduce systematic errors, as will be described later. In fact, the design of the experiment is motivated by the attempt to cancel all systematic errors in first order.

Schematic side and end views of the E246 detector being constructed are shown in Fig.1. Incident K^+s are slowed down in the degrader and stopped in a target located at the center of the detector. The K^+ stopping target consists of an array of $\sim$ 200 plastic-scintillating fibers with square cross section. After exiting the target, the muons from K^+ decay enter a spectrometer built around a superconducting toroidal magnet, where their momenta are analyzed. The magnet has 12 identical magnet-gaps with accurate 30° rotational symmetry. Four sets of wire chambers track the muon. One multiwire proportional chamber is placed at the entrance and two at the exit of each of the 12 magnet gaps. The fourth chamber is a cylindrical drift chamber surrounding the target.

The photons from the decay of the π^0s are detected with a highly-segmented photon detector, consisting of 768 thallium-doped CsI crystals. The crystals form an array surrounding the K^+ stopping target, with each element pointing to the target center. The assembly covers 75% of 4π and has a beam hole upstream to allow the K^+s into the target, a similar opening downstream to permit the instrumentation of the inner detector and 12 slits for the μ^+s to enter the magnetic spectrometer. A photon directional resolution of about 3 degrees and a reconstructed π^0 mass resolution of 5 MeV/c^2 are expected based on Monte Carlo calculations.

The μ^+s exiting the spectrometer are stopped in a muon polarimeter installed at the exit of each magnet gap. Counters on either side of the muon stopping region, detect e^+s in $\mu^+ \rightarrow e^+\nu\overline{\nu}$ decay ($\tau_\mu = 2.2$ μs). The muon polarization is deduced from the asymmetric distribution of the detected positrons, which are emitted preferentially along the μ^+ spin direction. The direction of P_T is parallel to the fringing magnetic field of the toroidal magnet, which will hold P_T against any disturbances, while precessing the in-plane polarization component.

3 Sensitivity to P_T

The narrow azimuthal acceptance of each magnet gap to muons from the target, when combined with the requirement that the π^0 momentum be co-linear with the detector axis, will select $K^+_{\mu 3}$ events whose decay plane is radial. For these events, P_T is directed azimuthally in a *screw-sense* around the detector axis, as shown in Fig.2 but with an opposite sense of rotation for forward and backward going π^0s.

P_T will then manifest itself as a difference in the e^+ counts between the counters located at the clockwise (*cw*) and counter-clockwise (*ccw*) sides of each muon stopper. By summing the *cw*- and *ccw*- counts of all the 12 magnet gaps, P_T is given by

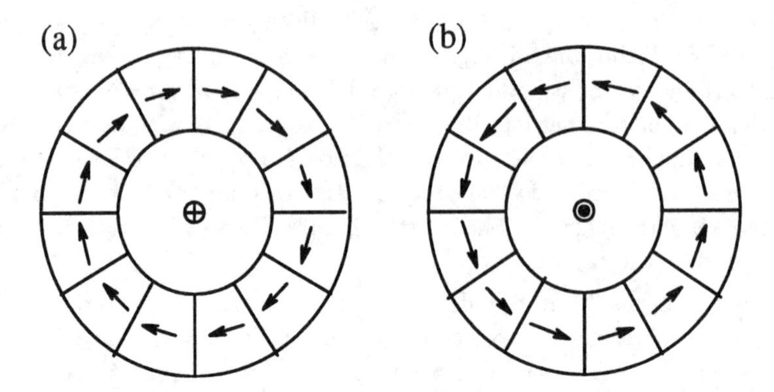

Figure 2: Direction of P_T in a schematic end view of the detector with the 12 sectors. (a) is for $K_{\mu 3}$ events with forward-going π^0s and (b) is for those with backward-going π^0s.

$$\frac{\sum_{i=1}^{12} N_i(cw)}{\sum_{i=1}^{12} N_i(ccw)} \cong 1 \pm 2AP_T, \tag{5}$$

where $N_i(cw)$ and $N_i(ccw)$ are the total number of e^+ counts in the *cw*- and *ccw*- counters as seen from the muon stopper in the *i*th magnet gap and A is the effective analyzing power of the polarization measurement. Since in the toroidal geometry, the same e^+ detector acts as the *cw* counter in one magnet-gap, as well as the *ccw* counter in the neighboring magnet-gap, any fake asymmetries resulting from differences in e^+ counter efficiencies thus introduced, would have opposite signs in these two magnet gaps and cancel in first order.

A double ratio can be formed by combining results for the two groups of events, with the π^0 going either forward or backward along the detector axis.

$$\frac{[\sum_{i=1}^{12} N_i(cw)/\sum_{i=1}^{12} N_i(ccw)]_{fwd}}{[\sum_{i=1}^{12} N_i(cw)/\sum_{i=1}^{12} N_i(ccw)]_{bwd}} \cong 1 + 4AP_T. \tag{6}$$

By taking such a double ratio, we can obtain an enhancement factor of two in the asymmetry measurement as well as a remarkable cancelation of spurious asymmetries which may have the same sign for both samples. In addition, by using the $K_{\mu 3}^+$ events with the π^0s moving transverse to the detector axis, we will be able to examine the zero-asymmetry level and to further reduce any spurious asymmetries. For those events, the decay plane is set transverse to the detector axis and the in-plane μ^+ polarization can be measured. Since the amount of in-plane polarization of the *cw* and *ccw* directions are equal, a null asymmetry should be detected. This fact can be used to monitor and correct for any systematics during the data collection time.

To reduce the background from other major decay modes, only those $K^+_{\mu 3}$ events whose μ^+ momentum is above 100 MeV/c and below the $K^+ \to \pi^+\pi^0$ ($K_{\pi 2}$) peak (of 205 MeV/c) will be taken. It covers about 65% of the total $K^+_{\mu 3}$ spectrum. The major background process in this kinematic region comes from in-flight π^+ decays in $K_{\pi 2}$ decay. This can be discriminated by cuts on the angle between μ^+ and π^0 and on the total π^0 energy. It should be noted that most background processes do not introduce any spurious asymmetry, but rather dilute the asymmetric effect of P_T by reducing the effective analysing power.

Using the estimated μ^+ acceptance of 5% in the spectrometer and a reconstructed π^0 acceptance of 20% together with a detection efficiency of $\mu^+ \to e^+\nu\bar{\nu}$ decay of about 20%, the net acceptance of the $K_{\mu 3}$ events is estimated to be 2×10^{-3}. Assuming an average K^+ stopping rate of about 100 k/sec and a 10^7-sec running time, we should obtain a total of 65 M $K^+_{\mu 3}$ events whose decay plane is essentially radial. The statistical error on P_T can be estimated from

$$\triangle P_T \sim \frac{1}{A} \cdot \sqrt{\frac{1}{< \sum_{i=1}^{12} N_i >}}, \tag{7}$$

where A is the analyzing power, which is expected to be about 0.25. From a total of $< \sum_{i=1}^{12} N_i > \sim$ 65 M events, we estimate $\triangle P_T \sim 5 \times 10^{-4}$. Using the average value of $P_T/\mathrm{Im}\xi \sim 0.25$, estimated from Monte-Carlo calculations, a sensitivity to $\mathrm{Im}\xi$ of about 2×10^{-3} is expected. Extensive Monte Carlo simulations indicate that remaining higher order systematic errors are small compared to the statistical sensitivity.

References

[1] Sakurai, J.J., Phys. Rev. **109**, 980 (1958)

[2] Lee, T.D., Particle Physics and Introduction to Field Theory, Harwood Academic Publishers, 1981, ch.13, pp 313-314 and ch.14, pp 355-359.

[3] Zhitnitskii, A.R., Sov. J. Nucl. Phys. **31**, 529 (1980)

[4] Leurer, M., Phys. Rev. Lett. **62**, 1967 (1989)

[5] Castoldi, P., Frère, J.-M., and Kane, G.L., Phys. Rev, **D39**, 2633 (1989)

[6] Bélanger, G., and Geng, C.Q., Phys. Rev. **D44**, 2789 (1991)

A PRECISE MEASUREMENT OF THE NON-LEPTONIC WEAK DECAY PARAMETERS α AND ϕ IN THE SPIN 3/2 DECAY $\Omega^- \to \Lambda^0 + K^-$

D.P. Ciampa[a], P.M. Border[a], Y.T. Gao[b,e], G. Guglielmo[a,f], K.J. Heller[a], K.A. Johns[c], M.J. Longo[b], R. Rameika[d], N.B.Wallace[a], D.M. Woods[a]

[a] *School of Physics and Astronomy, University of Minnesota, Minneapolis, MN 55455*
[b] *Department of Physics, University of Michigan, Ann Arbor, MI 48109*
[c] *Department of Physics, University of Arizona, Tucson, AZ 85721*
[d] *Fermi National Accelerator Laboratory, Batavia, IL 60510*
[e] *now at Lanzhao University, Lanzhao, China*
[f] *now at University of Oklahoma, Norman, Oklahoma*

To get a physical feeling for the non-leptonic weak decay parameters α_Ω and $\Phi_\Omega = tan^{-1}(\beta/\gamma)$, and to understand how they manifest themselves in the decay of the spin 3/2 baryon $\Omega^- \to \Lambda^0 + K^-$, we first consider the more familiar (and topologically identical) case of the spin 1/2 hyperon decay sequence $\Xi^- \to \Lambda^0 + \pi^-$, $\Lambda^0 \to \pi^- + p$. The polarization of the daughter baryon ($\vec{P}_\Lambda$) is related to the polarization of the parent ($\vec{P}_\Xi$) as:

$$\vec{P}_\Lambda = \frac{(\alpha_\Xi + \hat{\Lambda} \cdot \vec{P}_\Xi)\hat{\Lambda} + \beta_\Xi(\vec{P}_\Xi \times \hat{\Lambda}) + \gamma_\Xi(\hat{\Lambda} \times \vec{P}_\Xi) \times \hat{\Lambda}}{1 + \alpha_\Xi \hat{\Lambda} \cdot \vec{P}_\Xi} \tag{1}$$

where $\hat{\Lambda}$ is a unit vector defining the momentum direction of the daughter Λ in the Ξ rest frame. $\hat{\Lambda}$ and the vector cross-products that appear in the expression are mutually orthogonal and can be used to construct a very natural coordinate system known as the helicity axes.

$$\hat{X} = \frac{\vec{P}_\Xi \times \hat{\Lambda}}{\left|\vec{P}_\Xi \times \hat{\Lambda}\right|}, \qquad \hat{Y} = \frac{\hat{\Lambda} \times \left(\vec{P}_\Xi \times \hat{\Lambda}\right)}{\left|\hat{\Lambda} \times \left(\vec{P}_\Xi \times \hat{\Lambda}\right)\right|}, \qquad \hat{Z} = \hat{\Lambda} \tag{2}$$

The parameters β and γ thus provide information about the strength of the daughter Λ's polarization as projected onto the $\hat{X}$ and $\hat{Y}$ helicity axes. Note that if the parent Ξ is unpolarized, α_Ξ is seen as the helicity of the Λ (i.e., from (1), with $\vec{P}_\Xi = 0$, $\vec{P} = \alpha_\Xi \hat{\Lambda}$).

The decay parameters α_Ξ, β_Ξ, and γ_Ξ also appear in the expressions for the distributions of the proton (from the Λ decay) as seen in the Λ rest frame. These expressions reduce to their simplest form when calculated with respect to the helicity axes.

$$I(\hat{X} \cdot \hat{p}) = \frac{1}{2}(1 - \gamma_\Xi[\frac{\pi}{4}\alpha_\Lambda P_\Xi \hat{Y} \cdot \hat{p}]) \tag{3}$$

$$I(\hat{Y}\cdot\hat{p}) = \frac{1}{2}(1+\beta_{\Xi}[\frac{\pi}{4}\alpha_{\Lambda}P_{\Xi}\hat{Y}\cdot\hat{p}])$$
$$I(\hat{\Lambda}\cdot\hat{p}) = \frac{1}{2}(1+\alpha_{\Xi}[\alpha_{\Lambda}\hat{\Lambda}\cdot\hat{p}])$$

where $\hat{p}$ is the momentum direction of the proton in the rest frame of the parent Λ. Each of the decay parameters is a measure of the asymmetry of its associated distribution and indicates the strength of the parity violation taking place in the particular weak decay. The corresponding expressions in the spin-3/2 case of $\Omega^- \to \Lambda^0 + K^-$, $\Lambda^0 \to \pi^- + p$ are (Ref. 1):

$$\begin{aligned} I(\hat{X}\cdot\hat{p}) &= \frac{1}{2}(1-\gamma_{\Omega}[\frac{3\pi}{10}\alpha_{\Lambda}\hat{X}\cdot\hat{p}(P_{\Omega}-\frac{5}{16}\sqrt{\frac{7}{5}}t_{30}])) \\ I(\hat{Y}\cdot\hat{p}) &= \frac{1}{2}(1+\beta_{\Omega}[\frac{3\pi}{10}\alpha_{\Lambda}\hat{Y}\cdot\hat{p}(P_{\Omega}-\frac{5}{16}\sqrt{\frac{7}{5}}t_{30}])) \\ I(\hat{\Lambda}\cdot\hat{p}) &= \frac{1}{2}(1+\alpha_{\Omega}[\alpha_{\Lambda}\hat{\Lambda}\cdot\hat{p}]) \end{aligned} \tag{4}$$

The most striking difference between the spin-1/2 case and the spin-3/2 is that the asymmetry parameters β and γ are now tangled up with a tensor polarization term t_{30}. Thankfully, this term cancels in the calculation of ϕ_{Ω}. To see this, first note that the dot products of the momentum direction of the proton in the Λ rest frame with each of the helicity axes are just cosines, and so the distributions plotted as a function of these cosines should be linear with slope

$$\begin{aligned} \hat{X}: &\quad -\gamma_{\Omega} \quad [\frac{3\pi}{10}\alpha_{\Lambda}(P_{\Omega}-\frac{5}{16}\sqrt{\frac{7}{5}}t_{30})] \\ \hat{Y}: &\quad \beta_{\Omega} \quad [\frac{3\pi}{10}\alpha_{\Lambda}(P_{\Omega}-\frac{5}{16}\sqrt{\frac{7}{5}}t_{30})] \\ \hat{\Lambda}: &\quad \alpha_{\Omega} \quad [\alpha_{\Lambda}] \end{aligned} \tag{5}$$

In calculating the ϕ parameter, we use only the *ratio* of the slopes involving β and γ, which is independent of t_{30}. This ratio is also independent of the magnitude of the polarization $\vec{P}_{\Omega}$, though the direction of $\vec{P}_{\Omega}$ is necessary in constructing the helicity axes. Since each distribution is a linear function of a particular $cos\theta$, our job as experimentalists is easy: we measure the distribution of the proton in the Λ rest frame (eg. Fig 1), find its slope (properly corrected for the acceptance), and extract α_{Ω} and ϕ_{Ω}.

We can estimate the magnitude of the asymmetry parameters by expressing them in terms of the p-wave (B_p) and d-wave (B_d) amplitudes as

$$\alpha = \frac{2Re(B_p{}^*B_d)}{(|B_p|^2+|B_d|^2)},\quad \beta = \frac{2Im(B_p{}^*B_d)}{(|B_p|^2+|B_d|^2)},\quad \gamma = \frac{|B_p|^2-|B_d|^2}{(|B_p|^2+|B_d|^2)} \tag{6}$$

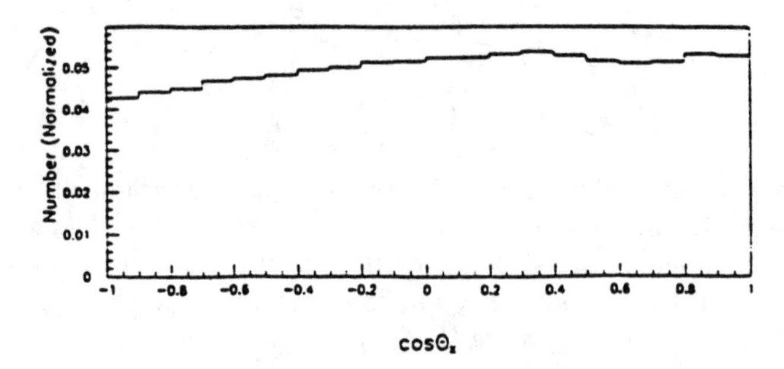

Figure 1: Distribution of protons in the Λ rest frame with respect to the X-helicity axis

with

$$\alpha^2 + \beta^2 + \gamma^2 = 1 \tag{7}$$

Since B_d is suppressed with respect to B_p by a factor of $(M_{\Xi^0} - M_\Lambda)/(M_{\Xi^0} + M_\Lambda) \sim 0.08$ (Ref. 2), we would expect that $\alpha_\Omega \sim 0$, $\beta_\Omega \sim 0$, and $\gamma_\Omega \sim 1$, so that $\phi_\Omega \sim 0$.

The E800 spectrometer at Fermilab was a simple particle tracking device consisting of (Fig. 2): 12 Multi-Wire Proportional Chambers (MWPCs) shown as C1-C12 in the figure, 8 Silicon Strip Detectors (SSDs), 4 scintillation counters (S1,S2, and veto counters V1 and V2), and 2 analysis magnets (PC4AN1 and PC4AN2). Polarized Ω^-s (and Ξ^-s) were produced upstream of the spectrometer using both polarized and unpolarized neutral beams (Ref. 3) in conjunction with two magnets (PC3SW and PC3ANA) and two targets (TGT1 and TGT2). The hardware trigger ensured that the parent particles accepted were traversing the zero-line of the spectrometer and that their decay products evinced the characteristic V-topology associated with the decay of the Λ hyperons. This loose trigger configuration allowed us to write 1.35×10^9 events to tape, although only $\sim 3\%$ of these proved to be good three-track events. Of these three-track events, most were Ξ^-s with an admixture of about 1% Ω^-s.

To extract a small Ω signal buried amidst the deluge of Ξ background, we employed several kinematic selection criteria in addition to the set of cuts used to garner the good three-track events. In reconstructing the events, the mass of the parent particle was calculated twice: once assuming the event was an $\Omega^- \rightarrow \Lambda^0 + K^-$ and again assuming it was a $\Xi^- \rightarrow \Lambda^0 + \pi^-$. Some Ξ^-s reconstructed both as good Ξ^-s and as good Ω^-s and could be eliminated from the Ω sample by requiring that $cos\theta_K > 0.775$, where θ_K is the angle made by the daughter kaon in the Ω rest frame with respect to the $\hat{z}$ axis of the spectrometer. Moreover, a subset of Ξs which decayed in the charged particle collimator and whose decay products were bent in the fringe field of the

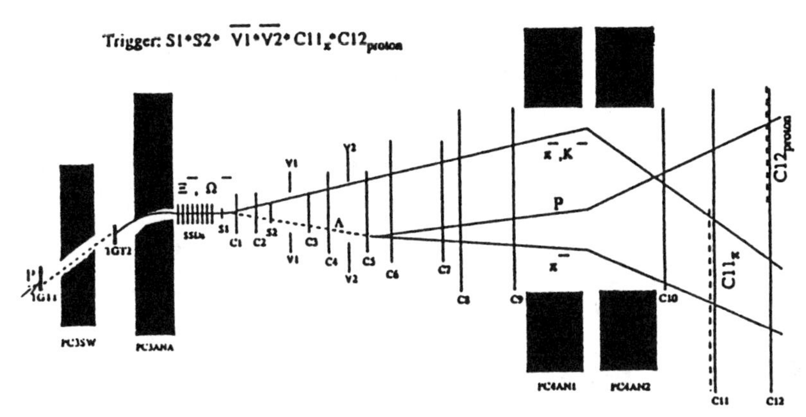

Figure 2: The E800 spectrometer

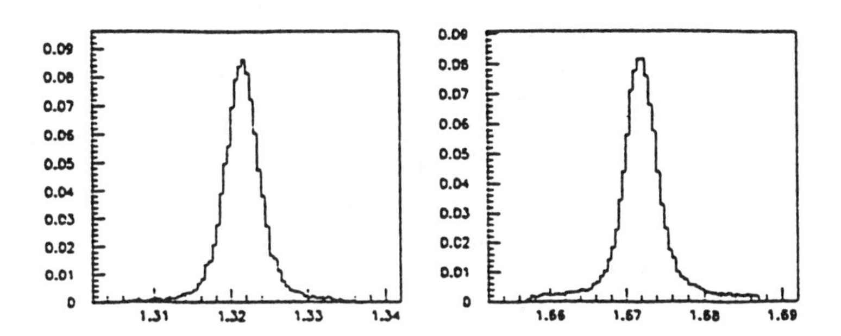

Figure 3: The Cascade (left) and Omega (right) mass peaks (scale in Gev.)

the PC3ANA magnet also reconstructed as good Ω^- candidates. Imposing a second kinematic cut, $\cos\theta_K > (\mid 0.008125 \times \phi_K \mid -1.8125)$ (where ϕ_K is the azimuthal angle associated with θ_K) expunged these events from the Ω^- event sample. Monte carlo studies indicated that these cuts removed only $\sim 5\%$ of the Ωs while reducing the background by 99.9%. The final data sample used for this analysis contained 252×10^3 $\Omega^- \rightarrow \Lambda^0 + K^-$ events. Figure 3 shows the cleanliness of the cascade and omega mass peaks.

The most critical element in the measurement of α_Ω and ϕ_Ω was the accuracy with which we could reconstruct the decay angle in the Λ rest frame (i.e., $cos\theta$). The bin size used in the analysis was 0.1 (20 bins from -1.0 to +1.0). We fed monte carlo data into the reconstruction algorithms to determine the percentage of reconstructed events that were within a bin width of the known monte carlo value; the results indicated that 99.8% were reconstruced into the bins from which they came.

To correct for the non-uniform acceptance of the spectrometer and reconstruction programs, we used a hybrid monte carlo (Ref. 4) which used all of the

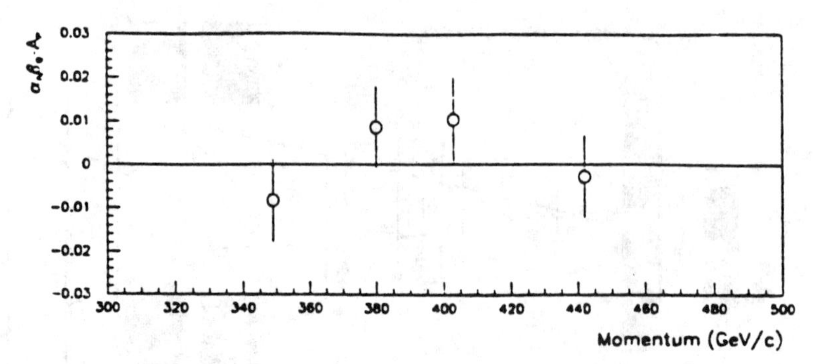

Figure 4: A systematic study of the asymmetry involving β_Ω vs momentum

characteristics of the real events except those associated with $cos\theta$, which was generated randomly. In essence, the hybrid monte carlo allowed us to require that every good event could have had any value of $cos\theta$ and still have been accepted by the spectrometer and reconstruction.

Data collection at opposite production angles at the target allowed us to use a bias cancellation technique to help eliminate any systematic effects. For the α_Ω measurement, we studied the data as a function of momentum, uncertainty in the bias measurement, run type, time and selection criteria to estimate the magnitude of the systematics. For ϕ_Ω, the polarization direction was also studied. In all cases, the systematic errors were negligible when compared with the statistical errors. A sample is shown in Figure 4.

The final answers are:

$$\alpha_\Lambda \alpha_\Omega = 0.0126 \pm 0.0042 \tag{8}$$

$$\alpha_\Omega = 0.0196 \pm 0.0066$$

$$\phi_\Omega = -3.4^\circ \pm 10.3^\circ \tag{9}$$

E800's measurement of α_Ω is almost four times more precise than the previous world average value of -0.026 ± 0.026 (Ref. 5) and shows this parameter to be inconsistent with zero. This is the first measurement of the parameter ϕ_Ω.

This work was supported by the U.S. Department of Energy and the National Science Foundation.

1. G. M. Guglielmo, Ph.D. Thesis, University of Minnesota, (1994).
2. M. Suzuki, Prog. of Theor. Phys., **32** 1:138, (1964).
3. K. Johns, these proceedings.
4. G. Bunce *et al.*, Nucl. Inst.. Meth. **172**,553 (1980).
5. Particle Data Group, Phys. Rev. D **50**, 1781 (1994).

A New Fermilab Experiment to Search for Direct CP Violation in Hyperon Decays

Edmond C. Dukes[1]
High Energy Physics Laboratory
University of Virginia
Charlottesville, Virginia 22901

ABSTRACT

The theoretical and experimental status of *CP* violation in hyperons is briefly reviewed. The present experimental limit for the difference between the decay parameter α in Λ and $\overline{\Lambda}$ decays is about 10^{-1}, far above the standard model predictions of approximately 10^{-4}. A new dedicated Fermilab effort — the HyperCP experiment (E871) — which intends to measure the difference in α between both Ξ^- and $\overline{\Xi}^+$ and Λ and $\overline{\Lambda}$ to better than 10^{-4} is discussed. The experiment is scheduled to run in 1996. A positive result would be the first observation of *CP* violation outside of the decays of the K_L and unambiguous evidence of direct *CP* violation.

INTRODUCTION

Despite 30 years of intense experimental effort, our understanding of *CP* violation is quite incomplete [2]. Experimentally it still remains a small peculiarity unique to the decays of the K_L. The standard model — which although not explaining the origin of *CP* violation, accommodates it quite nicely — tells us that *CP* violation should be evident elsewhere; in the decays of hyperons, for example. The standard model also tells us that *CP* violation should be evident in the $|\Delta S| = 1$ decays of baryons — so-called direct *CP* violation — and not only in the $|\Delta S| = 2$ weak transitions responsible for kaon mixing. To date, however, no experiment has seen *CP* violation outside of the decay of the K_L and the evidence for direct *CP* violation remains inconclusive [3]. We describe here a new Fermilab experiment which seeks to address both of these important issues: the search for direct *CP* violation and the search for *CP* violation outside of the kaon system.

THEORETICAL EXPECTATIONS

It has long been known that *CP* violation should manifest itself in the decays of hyperons [4]. Observables include the differences in the partial de-

[1]Representing the E871 collaboration [1].

cay rates and differences in the angular distributions of the decay daughters between hyperon and antihyperon:

$$\Delta = \frac{\Gamma - \overline{\Gamma}}{\Gamma + \overline{\Gamma}}, \quad A = \frac{\alpha + \overline{\alpha}}{\alpha - \overline{\alpha}}, \quad B = \frac{\beta + \overline{\beta}}{\beta - \overline{\beta}}. \tag{1}$$

The overlined quantities refer to the antihyperons and

$$\Gamma \propto |S|^2 + |P|^2, \quad \alpha = \frac{2\mathrm{Re}(S^* P)}{|S|^2 + |P|^2}, \quad \beta = \frac{2\mathrm{Im}(S^* P)}{|S|^2 + |P|^2}, \tag{2}$$

where S and P refer to the final state angular momentum amplitudes. *CP* invariance implies that $\Gamma = \overline{\Gamma}$, $\alpha = -\overline{\alpha}$, and $\beta = -\overline{\beta}$. Nonzero values for both A and B result from the interference of S-and P-wave amplitudes whereas a nonzero value of Δ results from the interference of $|\Delta \mathrm{I}| = 1/2$ and $|\Delta \mathrm{I}| = 3/2$ amplitudes.

Model independent calculations of the above observables [5] indicate that for Ξ and Λ decays B should exhibit the largest asymmetry, with A being approximately an order of magnitude smaller, and Δ yet another order of magnitude smaller[2]. Measuring B is prohibitively difficult because the daughter polarization from a parent with precisely known polarization must be measured. Measuring partial decays rates to the required precision appears to be even more difficult. The asymmetry A is the most experimentally accessible and can be determined if the parent hyperon and antihyperon polarizations are known.

Exact calculations of *CP* violation in hyperon decays are difficult and it can be said that only order of magnitude estimates exist. For example, Donoghue [6] predicts, in a standard model calculation, asymmetries for A which range from $-(0.3 \to 4.0) \times 10^{-4}$ for Λ hyperons and $-(0.4 \to 4.8) \times 10^{-4}$ for Ξ^- hyperons. Theories with no $|\Delta \mathrm{S}| = 1$ *CP*-odd effects, such as the superweak model [7] and models with a very heavy neutral Higgs, predict no *CP* asymmetries.

PRESENT EXPERIMENTAL LIMITS

A dedicated hyperon *CP* violation experiment has yet to be done. The present experimental limits on *CP* violation in hyperon decays come from three experiments comparing the α parameters in Λ and $\overline{\Lambda}$ decays. These are the R608 experiment at the ISR ($A_\Lambda = 0.02 \pm 0.14$ [8]), the DM2 experiment at the Orsay DCI e^+e^- colliding ring ($A_\Lambda = 0.01 \pm 0.10$ [9]), and the PS185 experiment at LEAR ($A_\Lambda = -0.07 \pm 0.09$ [10]). In the R608 experiment the Λ and $\overline{\Lambda}$ are separately produced in pp and $\overline{\mathrm{p}}$p reactions and hence their *CP* measurement depends on the assumption that the Λ and $\overline{\Lambda}$ polarizations are

[2]For Ξ^- decays Δ should be identically zero since there is only one isospin channel.

the same. In the other two experiments $\Lambda\overline{\Lambda}$ pairs are exclusively produced: at Orsay through the process $e^+e^- \rightarrow J/\Psi \rightarrow \Lambda\overline{\Lambda}$, and at LEAR in the threshold reaction $p\overline{p} \rightarrow \Lambda\overline{\Lambda}$. In the DM2 experiment neither the Λ or $\overline{\Lambda}$ is produced polarized and hence only the product $\alpha_\Lambda\overline{\alpha}_\Lambda$ is measured. Their *CP* measurement requires knowledge of α_Λ from other experiments. In the PS185 experiment, the assumption of C-parity invariance in the strong interaction leads to equal polarizations of Λ and $\overline{\Lambda}$ and hence the polarizations do not need to be known.

These results are several orders of magnitude away from standard model predictions, and only the PS185 experimental technique appears viable at the level of the theoretical predictions. Getting sufficient statistics is a problem and would require a higher luminosity LEAR (LEAR-2 or SuperLEAR) [11]. Proposals to carry out such an experiment at CERN and a similar effort at Fermilab [12] have not been approved.

THE HYPERCP EXPERIMENT

A new Fermilab experiment — the HyperCP experiment (E871) [13] — intends to measure both A_Λ and A_Ξ to a sensitivity of better than 10^{-4}. It has recently been approved for the 1996 fixed target run. In this experiment the hyperon and antihyperon are not produced simultaneously, and hence they are no longer required by symmetry arguments to share the same polarization, as is the case in threshold $p\overline{p} \rightarrow\Lambda\overline{\Lambda}$ production.

The HyperCP experiment, however, takes advantage of the fact that Ξ^- and $\overline{\Xi}^+$ hyperons with *zero* polarization can be made by requiring that they be produced at zero degrees, where parity conservation in the strong interaction requires that the polarization be identically zero. A Λ from the weak decay of an unpolarized Ξ ($\rightarrow \Lambda\pi^+$) is found in a pure helicity state with a polarization magnitude given by the parent Ξ alpha parameter: $P_\Lambda = \alpha_\Xi$. Hence, if *CP* is a good symmetry in Ξ decays, then Λ and $\overline{\Lambda}$ have equal and opposite polarizations. The decay distributions of the proton and antiproton in the frame in which the Λ polarization defines the polar axis — the Λ helicity frame (see Fig. 1) — are given by:

$$\frac{dN}{d(\cos\theta)} = \frac{N_0}{2}(1+\alpha P\cos\theta) = \frac{N_0}{2}(1+\alpha_\Lambda\alpha_\Xi\cos\theta) = \frac{N_0}{2}(1+\overline{\alpha}_\Lambda\overline{\alpha}_\Xi\cos\theta). \quad (3)$$

If *CP* symmetry is good in *both* Ξ and Λ decays then $\overline{\alpha}_\Xi = -\alpha_\Xi$ and $\overline{\alpha}_\Lambda = -\alpha_\Lambda$ and the decay distributions of the proton and antiproton are identical, as is every other kinematic parameter in the Ξ decays. Hence reversing the polarity of the spectrometer magnet results in a *CP* invariant apparatus.

It is evident from the above equation that differences between the slopes of the two $\cos\theta$ distributions can be due to *CP* violation in either the Ξ or Λ

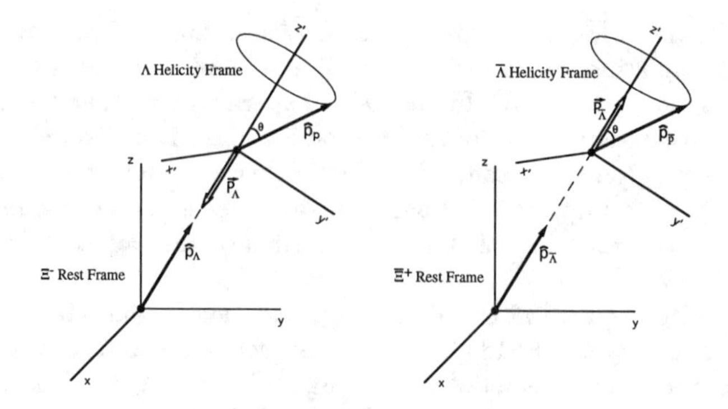

Figure 1: Frames used in the Λ and $\overline{\Lambda}$ polarization analysis.

decay — the experiment is by necessity sensitive to *CP* violation in both:

$$\mathcal{A} = \frac{\alpha_\Lambda \alpha_\Xi - \overline{\alpha}_\Lambda \overline{\alpha}_\Xi}{\alpha_\Lambda \alpha_\Xi + \overline{\alpha}_\Lambda \overline{\alpha}_\Xi} = A_\Lambda + A_\Xi. \tag{4}$$

Most theoretical calculations predict that the Ξ and Λ asymmetries should have the same sign. If a nonzero asymmetry is measured the experiment cannot distinguish which hyperon, if not both, is the source.

The design of the HyperCP experimental apparatus is based on 15 years of experience in doing hyperon physics at Fermilab, and in particular, the experience gathered in E756 [14]. A plan view of the HyperCP experimental apparatus is shown in Fig. 2 below. The apparatus has been kept simple in order to facilitate its understanding at the level of the expected asymmetry. It consists of a target immediately followed by a collimator — with a 4.88 μsr solid angle acceptance — embedded in a 6 m long dipole magnet. Following a vacuum decay region is a conventional magnetic spectrometer employing high-rate, narrow pitch wire chambers. The spectrometer analyzing magnet has sufficient strength to insure that the proton from the Λ decay is always bent to one side of the spectrometer and that the two pions from the Ξ and Λ decays are always bent to the opposite side, and that both are well separated from the charged beam exiting the collimator. A simple, yet selective trigger is formed by requiring the coincidence at the rear of the spectrometer of charged particles on either side of the spectrometer. A scintillator hodoscope (on the pion side) and a hadronic calorimeter (on the proton side) are used to detect the charged particles. The calorimeter is needed to make the trigger "blind" to muons. The main trigger background is interactions of the charged particle beam exiting the collimator with material in the spectrometer.

The mean momentum acceptance of the magnetic channel is 150 GeV/c, at

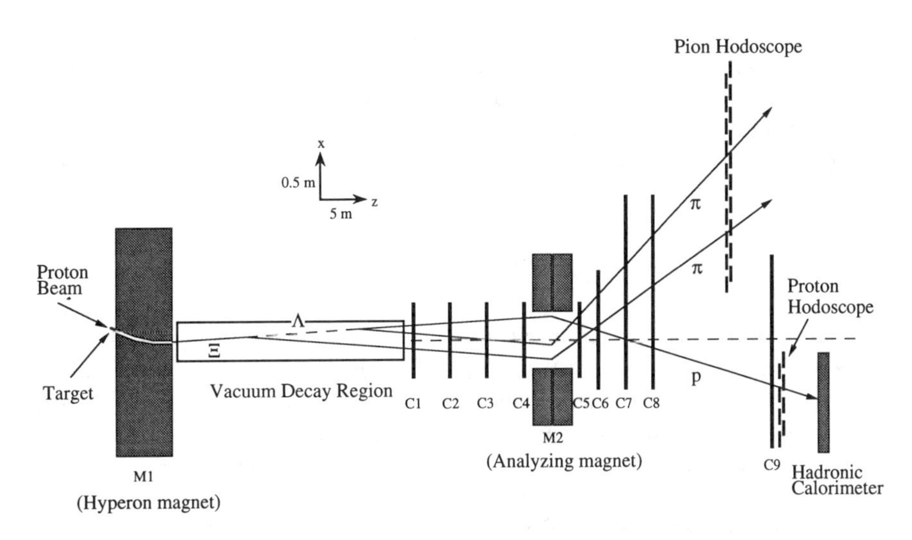

Figure 2: Plan view of the HyperCP apparatus.

which the Ξ to charged particle ratio is approximately maximal. Yields have been estimated using a Monte Carlo which has been tuned to the E756 data. With 10^{10} protons/s incident on a 21% λ_I Be target, it is estimated that 2,500 (1,400) Ξ^- ($\overline{\Xi}^+$) events pass all kinematic and reconstruction cuts. In a 200 day run with a 50% duty factor approximately 6×10^9 Ξ^- and $\overline{\Xi}^+$ events will be accumulated resulting in a sensitivity in $\mathcal{A} = A_\Xi + A_\Lambda$ of 0.8×10^{-4}. The trigger rate is estimated to be 30 KHz per spill second for a sustained data logging rate of approximately 4.5 Mbyte/s.

Sources of systematic error are: 1) acceptance differences; 2) differences in the Ξ^- and $\overline{\Xi}^+$ polarizations; 3) differences in the p and $\overline{\text{p}}$ as well as the π^+ and π^- interaction cross sections with material in the spectrometer; and 4) different backgrounds under the Λ and $\overline{\Lambda}$ and the Ξ^- and $\overline{\Xi}^+$ mass peaks. Acceptance differences will be minimized by the use of redundant chambers and by changing from Ξ^- to $\overline{\Xi}^+$ running as often as possible.

Investigations of systematic errors using Monte Carlo simulations and analysis of the E756 Ξ^- data indicate that false asymmetries should contribute to the asymmetry at a level less than 10^{-4}. It should be emphasized that the analysis method naturally minimizes potential biases. This is because as the Λ direction in the Ξ rest frame changes from event to event, so too does the direction of the Λ polarization and hence the analysis frame in which the proton $\cos\theta$ is measured. As a consequence, acceptance differences localized to a particular part of the apparatus do *not* map to a particular part of the proton or antiproton $\cos\theta$ distribution and the proton $\cos\theta$ acceptance in the Λ helicity frame is quite flat. Indeed, in the limit of uniform Λ acceptance, any correlation at all would vanish. That this is the case has been confirmed by measurement of $\alpha_\Lambda \alpha_\Xi$ from pairs of E756 Ξ^- data sets with manifestly different acceptances and polarizations. With no attempt to correct for acceptance differences, no difference in $\alpha_\Xi \alpha_\Lambda$ is seen at level of 10^{-3}, the limit of the statistical precision.

CONCLUSIONS

The ultimate statistical sensitivity that can be achieved by the HyperCP experimental technique is not yet limited by Ξ^- or $\overline{\Xi}^+$ yields, but rather by rate and lifetime limitations of wire chambers. New technologies, such as gas microstrip wire chambers, promise much higher rates than conventional wire chambers, although problems such as chamber lifetime remain. Eventually, at high enough rates, the number of random charged tracks exiting the collimator in the same RF bucket as the Ξ events will limit the Ξ yield. An estimate of the ultimate systematic limit to the sensitivity would require more *hubris* than this author possess and more space than is allowed here. With suitable resources, however, there appears to be no fundamental reason why sensitivities of 10^{-5} cannot be attained in the future.

REFERENCES

[1] The E871 collaboration is: J. Antos, Y.C. Chen, C.N. Chiou, C. Ho, A. Sumarokov, P.K. Teng, M. Botlo, G. Abrams, C. Ballagh, H. Bingham, D. Chapman, G. Gidal, P.M. Ho, K.B. Luk, J. Lys, C. James, J. Volk, L. Pinsky, R. Burnstein, D. Kaplan, L. Lederman, H. Rubin, M. Jenkins, K. Clark, E.C. Dukes, K. Nelson, D. Pocanic, T. Alexopoulos, A. Erwin, and M. Thompson.

[2] J.F. Donoghue, B.R. Holstein, and G. Valencia, Inter. Jour. of Mod. Phys. **A2** (1987) 319.

[3] B. Winstein and L. Wolfenstein, Rev. Mod. Phys. **65** (1993) 1113.

[4] A. Pais, Phys. Rev. Lett. **3** (1959) 242.

[5] J.F. Donoghue, X.-G. He, and S. Pakvasa, Phys. Rev. **D34** (1986) 833.

[6] J.F. Donoghue, Third Conference on the Intersections between Particle and Nuclear Physics, Rockport, ME, 14–19 May, 1988.

[7] L. Wolfenstein, Phys. Rev. Lett. **13** (1964) 562.

[8] P. Chauvat et al., Phys. Lett. **B163** (1985) 273.

[9] M.H. Tixier et al., Phys. Lett. **B212** (1988) 523.

[10] P.D. Barnes et al., Phys. Lett. **B199** (1987) 147.

[11] N.H. Hamann, Inst. Phys. Conf. Ser. No 124 (1992) 211.

[12] S.Y. Hsueh, *Search for direct CP violation in* $\overline{p}+p \rightarrow \overline{\Lambda}+\Lambda \rightarrow \overline{p}\pi^+ + p\pi^-$, Proposal to Fermilab, January 2, 1992.

[13] J. Antos et al., *Search for CP Violation in the Decays of* $\Xi^-/\overline{\Xi}^+$ *and* $\Lambda/\overline{\Lambda}$ *Hyperons*, Fermilab Proposal P-871 (Revised Version), March 26, 1994.

[14] P.M. Ho et al., Phys. Rev. **D44** (1991) 3402.

STUDY OF POLARIZED $^5_\Lambda$He WEAK DECAY VIA (π^+, K^+) REACTION ON ^{6}Li

H. Noumi,[a] S. Ajimura,[b] H. Bhang,[c] H. Ejiri,[b] T. Hasegawa,[d] O. Hashimoto,[d] R. Hazama,[b] T. Inomata,[b] M. Ishikawa,[b] T. Kishimoto,[b] K. Kume,[b] K. Maeda,[e] K. Manabe,[a] T. Nagae,[d] T. Nakano,[b] J. Okusu,[b] H. Park,[c] M. Sekimoto,[d] T. Shibata,[d] N. Shinkai,[b] T. Takahashi,[f] and M. Youn,[c]

[a] *National Laboratory for High Energy Physics, Tsukuba, Ibaraki 305, Japan*
[b] *Department of Physics, Osaka University, Toyonaka, Osaka 560, Japan*
[c] *Department of Physics, Souel National University, Souel, Korea*
[d] *Institute for Nuclear Study, University of Tokyo, Tanashi, Tokyo 188, Japan*
[e] *Department of Physics, Tohoku University, Sendai, Miyagi 980, Japan*
[f] *Department of Physics, Kyoto University, Kyoto 606, Japan*

Abstract. We carried out an experiment to study asymmetric weak decays of polarized $^5_\Lambda$He produced via (π^+, K^+) reaction on ^{6}Li. The experiment aims mainly at investigation of the ΛN weak interaction through the hypernuclear weak decays. We introduce the experiment and report current status of analysis.

I. INTRODUCTION

There are two predominant hadronic processes, mesonic and nonmesonic ones, in Λ-hypernuclear weak decays. Mesonic process can be described by a lambda decay in free space ($\Lambda \rightarrow p\pi^-$ and $\Lambda \rightarrow n\pi^0$). Nonmesonic process is characterized by a lambda decay involving a neighboring nucleon ($\Lambda p \rightarrow pn$ and $\Lambda n \rightarrow nn$). One can learn ΛN interactions through hypernuclear weak decays. In particular, hypernuclear nonmesonic process provides only a good opportunity to investigate lambda-nucleon (ΛN) weak interaction since the transition can take place due to Λ bound in nucleus. It is one of the biggest interests in hypernuclear physics to clearify the ΛN weak interaction.

There are three observables available for the study of the hypernuclear weak decays, life time (τ=$1/\Gamma_{tot}$), partial decay rates (Γ_{π^-}, Γ_{π^0}, Γ_p, Γ_n), and asymmetry parameter (a_1). Here the Γ_{π^-}, Γ_{π^0}, Γ_p, and Γ_n denote decay rates in π^-- and π^0-stimulated mesonic decays, and proton- and neutron-induced nonmesonic decays, respectively. The Γ_{tot} is total decay rate.

In early 1960's Block and Dalitz analyzed limited data of decay rates from emulsion experiments in $^4_\Lambda$H and $^4_\Lambda$He in order to understand the structure of the $\Lambda N \rightarrow NN$ transition classifying it into 6 amplitudes as seen in Table 1.[1] According to the classification the ratio Γ_n/Γ_p provides isospin dependence of the transition, *i.e.*, Γ_n/Γ_p=$2\Gamma(I$=$1)/[\Gamma(I$=$1)+\Gamma(I$=$0)]$. Recent years a complete set of weak decay rates (τ, Γ_{π^-}, Γ_p, and Γ_n) in $^5_\Lambda$He were measured at

TABLE 1: Six amplitudes in the $\Lambda N \rightarrow NN$ transition are presented.

Transition			Matrix Element	Rate	Final Isospin	Parity Change
$^1S_0$	$\rightarrow$	$^1S_0$	a	a^2	1	no
		$^3P_0$	$(b/2)(\sigma_1 - \sigma_2)q$	b^2	1	yes
$^3S_1$	$\rightarrow$	$^3S_1$	c	c^2	0	no
		$^3D_1$	$(\sqrt{2}/4)dS_{12}(q)$	d^2	0	no
		$^1P_1$	$(\sqrt{3}/2)e(\sigma_1 - \sigma_2)q$	e^2	0	yes
		$^3P_1$	$(\sqrt{6}/4)f(\sigma_1 + \sigma_2)q$	f^2	1	yes

BNL,[2] where the $\Gamma_{nm}(=\Gamma_p+\Gamma_n)$ and Γ_n/Γ_p were discussed. The measurement gave the Γ_n comparable to the Γ_p, while theory predicted the Γ_n much smaller than the other. Meson exchange model favors tensor force ($^3S_1 \rightarrow {}^3D_1$) which dominates the transition to the final state of isospin equal to 0, thus resulting the Γ_n is suppressed.[3] Heavier meson exchanges can be considered significantly as a short range effect since the $\Lambda N \rightarrow NN$ involves large momentum transfer of ~0.4 GeV/c. It improves the ratio,[4] however there is still a large discrepancy between experiment and theory. It is one of big problems to be solved in hypernuclear physics.

Nonmesonic decay protons from polarized hypernuclei show asymmetric angular distribution. It is described as $W(\theta) \propto 1+a_1P_\Lambda\cos\theta$, where P_Λ is the Λ-spin polarzation. The asymmetry a_1 is caused by interference of the parity violating amplitude with the conserving one. The a_1 is written as $2\sqrt{3}(\sqrt{2}c+d)f/[a^2+b^2+3(c^2+d^2+e^2+f^2)]$. Therefore it provides information complementary to the decay rates to study the ΛN weak interaction. No measurement of the asymmetry parameter has been available until recently. We have succeeded in observing the large asymmetry for the first time in polarized $^{11}_{\Lambda}$B and $^{12}_{\Lambda}$C nonmesonic decays.[5] The result was $a_1 < -0.6$ in about 40 % precision. A recent theoretical calculation based on meson exchange model agrees with the measured asymmetry.[6]

Until now we have no theoretical calculation to reproduce both of the measured weak decay rates and asymmetry. While, experiment should be improved in precision in order to discriminate the theoretical models. New measurement with better precision has been being required.

Recently we carried out an experiment of polarized $^5_\Lambda$He weak decays at KEK (PS E278).[7] The $^5_\Lambda$He was produced exclusively by gating the $^6_\Lambda$Li ground state produced by the (π^+, K^+p) reaction on ^{6}Li since it is unstable against a proton emission (0.6 MeV). The $^5_\Lambda$He is advantageous in the followings. The $^6\mathrm{Li}(\pi^+, K^+p)^5_\Lambda\mathrm{He}$ reaction gives large polarization of P_Λ=0.26~0.4 with large cross section of $d\sigma/d\Omega$=7~10 μb/sr at θ_K=~15°.[8] This improves a factor of 4~10 larger than the previous experiment in terms of $P_\Lambda^2 d\sigma/d\Omega$. The polarization of $^5_\Lambda$He can be determined experimentally by measuring the asymmetry of pionic decay because the asymmetry parameter of the $^5_\Lambda$He mesonic decay is almost equal to that of the Λ decay in free space.[9] In the previous study polarization was estimated by DWIA calculations.[10] Only relative s wave of initial ΛN system is concerned in $^5_\Lambda$He. This simplifies the comparison with theoretical calculations. Attenuation of the asymmetry due to the nuclear effect such as fermi motion and internuclear cascade is estimated to be 6 %

which is about 3 times smaller than that of the ${}^{12}_{\Lambda}$C case.

Technically, there is a difficulty to measure the Γ_n which causes sizable errors on the results. Then the E278 experiment measured the Γ_{π^-} and Γ_{π^0} precisely as well as the Γ_p since the Γ_{nm} and Γ_n can be obtained by $(\Gamma_{tot}-\Gamma_m)$ and $(\Gamma_{nm}-\Gamma_p)$, respectively. The present experiment is expected to give almost a complete set of decay rates in addition to the asymmetry parameter with better precision although the analysis is under way at present.

II. EXPERIMENTAL INSTRUMENTS

The E278 experiment carried out at the KEK 12-GeV PS K6 beam line. The beam line provides about 3×10^6 pions at 1.05 GeV/c every 4 seconds with the duty factor of 50 %. Pion momentum was analyzed by the last QQDQQ magnets of the K6 beam line in 0.1 % resolution. In the present experiment the Superconducting Kaon Spectrometer (SKS) is used to measure scattered K^+ sitting at the end of the beam line. It has an angular acceptance of θ_K=$-16\sim+16$ degrees (100 msr) and a momentum resolution of 0.1 %. The large angular acceptance realized simultaneous measurements of positive and negative polarizations with respect to the reaction plane. It eliminates spurious asymmetry due to instrumental asymmetry. This spectrometer system has achieved a resolution of 2 MeV in the ${}^{12}_{\Lambda}$C excitation energy spectrum.[11] This is quite enough to separate the ${}^{6}_{\Lambda}$Li ground state from the Λ-unbound region, which is essential to measure the polarization of ${}^{5}_{\Lambda}$He without contamination of escape Λ decays.

Decay counter system used is illustrated in Fig.1. Two sets of detector system are placed symmetrically above and below the target. Each system consists of a Si strip detector (SSD), an MWPC, a range shower counter (RSC), 36 NaI detectors, and hodoscopes (DH and CTC). The SSD is an energy loss (ΔE) detector for charged particles. The SSD and MWPC are used as a tracking device. The RSC has 32 layers of 0.2 mm lead and 1 mm plastic sheets in order to detect total energy and range for charged particles. It acts as an EM shower calorimeter with combining NaI's for γ rays from π^0. The decay counter systems are sensitive to nonmesonic protons and mesonic negative and neutral pions. The DH and CTC are thin plastic scintillator hodoscopes located before and behind the RSC, which give additional information on charged particle hit position.

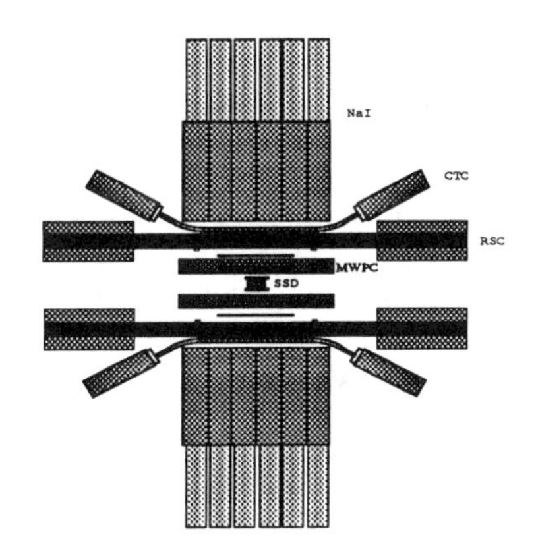

FIGURE 1. The decay counter system is shown.

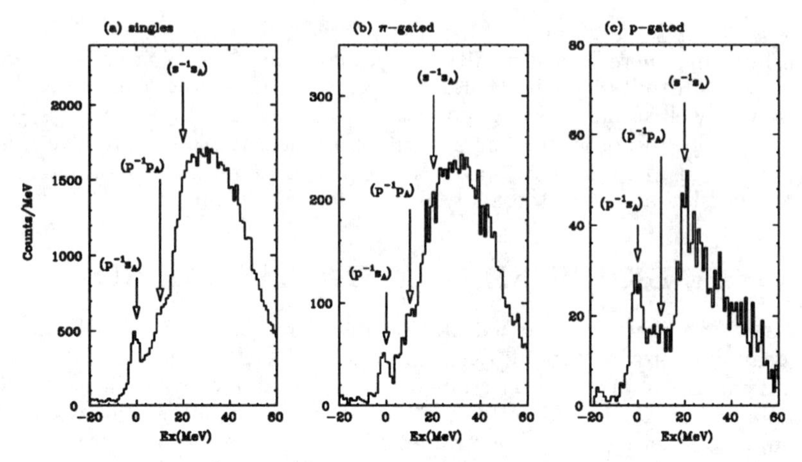

FIGURE 2. The ${}^{6}_{\Lambda}$Li excitation energy spectra are displayed. (preliminary)

III. STATUS OF ANALYSIS

The ^{6}Li$(\pi^{+}, K^{+})X$ reaction is characterized by three particle-hole states, namely $p_n^{-1}s_\Lambda$, $p_n^{-1}p_\Lambda$, and $s_n^{-1}s_\Lambda$. The $p_n^{-1}s_\Lambda$ corresponds to the ground state lying 4.5 MeV below the Λ-escape threshold. The ground and the $(s_n^{-1}s_\Lambda)$ states form ${}^{5}_{\Lambda}$He emitting a proton eventually. The other state is unstable against a Λ emission. Fig.2 displays a preliminary result on the ${}^{6}_{\Lambda}$Li excitation energy spectra obtained: a singles (inclusive) spectrum and those in coincidence with π^{-} and proton decays. The ${}^{6}_{\Lambda}$Li ground state at around E_x=0 MeV is observed clearly. The $(s_n^{-1}s_\Lambda)$ state is enhanced in the p-gated spectrum, as seen in previous (K^{-}, π^{-}) experiment at BNL.[2] It is noted that the π^{-}-gated spectrum shows better quality than that obtained in the BNL experiment. The ground state can be seen separately from Λ-unbound region raising up at E_x=4.5 MeV in all spectra. It demonstrates that the polarization of ${}^{5}_{\Lambda}$He can be measured without contamination of escape Λ.

A prospect of the present experimental study is as follows.

Motoba *et al.* calculated Γ_{π^-} and Γ_{π^0} in ${}^{5}_{\Lambda}$He, where the YNG and ORG interactions are employed.[9] These quantities depend upon the Λ wave function in ${}^{5}_{\Lambda}$He. We can justify the models. The theory predicts Γ_{π^0}=0.55$\times\Gamma_{\pi^-}$ almost independent of the model. This is due mainly to the ΔI=1/2 rule. Validity of this assumption can be tested since we measured the π^{0}-mesonic decay.

To extract nonmesonic weak decay rates careful estimation of the nuclear effects, such as fermi motion of a nucleon (hyperon) and mean free path of a stimulated nucleon, is necessary. These effects affect the energy spectrum of observed nucleons much. Reversely the spectrum shape tells information on what is going on in the decay. Recently multinucleon-induced nonmesonic process is proposed to solve the discrepancy between measurement and calculation on the decay rates.[12] If it is correct, the final nucleon energy spectrum is expected to be characterized in low energy region. This can be investigated since the SSD-ΔE counter makes sensitive energy region lower.

Ramos *et al.* calculated the asymmetry parameter (a_1) based on meson exchange model with relativistic nuclear treatment.[6] They gave a_1=−0.43 in ${}^5_\Lambda$He. Quark model calculation has been being applied to the $\Lambda N \to NN$ transition amplitude.[13] In particular, they studied contributions of ΔI=1/2 and 3/2 amplitudes, and also of one pion exchange in comparison. Each gave different a_1. The present experiment can test those models both with the asymmetry parameter and decay rates.

IV. SUMMARY

We carried out the E278 experiment to study asymmetric weak decay of polarized ${}^5_\Lambda$He hypernuclei produced via (π^+, K^+) reaction on ^{6}Li. Preliminary results of ${}^6_\Lambda$Li excitation energy spectra are presented. They show better quality than those observed in the BNL experiment. Analysis is now in progress. We will see the asymmetry and the decay rates with better precision soon. These are expected to test the validity of various model calculations. Once angular correlation methods with polarized hypernuclei are established, one can proceed studies of electromagnetic properties of hypernucleus, for example, its magnetic moment. It is very sensitive to baryon modification in nuclear medium.

ACKNOWLEDGEMENTS

The authors thank professors K. Nakai, T. Ohshima, M. Takasaki, T. Sato, J. Chiba, T. K. Ohska, M. Kihara and the crew of the KEK PS for their supports to carry out the experiment. They are grateful to the staff of the KEK cryogenic group for the maintenance of SKS.

REFERENCES

1. M. M. Block and R. H. Dalitz, Phys. Rev. Lett. **11**, 96-100(1963).
2. J. J. Szymanski *et al.*, Phys. Rev. Lett. **C43**, 849-862(1991).
3. K. Takeuchi, H. Takaki, and H. Bandō, Prog. Theor. Phys. **73**, 841-844(1985).
4. J. F. Dubach, Nucl. Phys. **A450**, 71c-84c(1986).
5. S. Ajimura *et al.*, Phys. Lett. **B282**, 293-298(1992).
6. A. Ramos, E. van Meijgaard, C. Bennhold, and B. K. Jennings, Nucl. Phys. **A544**, 703-730(1992).
7. T. Kishimoto *et al.*, KEK-PS Proposal E278 (1992), unpublished.
8. T. Motoba and K. Itonaga, private communication, 1993.
9. T. Motoba, K. Itonaga, and H. Bandō, Nucl. Phys. **489**, 683-715(1988).
10. K. Itonaga *et al.*, Proc. 5th Int. Symp. on Meson and Light Nuclei, Prague, 1991, and private communication, 1991.
11. T. Nagae *et al.*, Proc. of PANIC XIII, Perugia, Italy, 1993.
12. M. Ericson, Nucl. Phys. **A547**, 127c-132c(1992); W.M. Alberico, A. De Pace, M. Ericson, and A. Molinari, Phys. Lett. **B256**, 134-140(1991).
13. T. Inoue, S. Takeuchi, and M. Oka, Nucl. Phys. **A577**, 281c-286c(1994).

VIII. NUCLEON SPIN STRUCTURE

Nucleon Spin Structure

Rüdiger Voss

PPE Division, CERN, CH-1211 Geneva 23, Switzerland

Abstract. Recent results are reviewed on the spin-dependent structure functions of neutrons and protons, tests of the Bjorken and Ellis-Jaffe sum rules, and on the spin content of nucleons, with emphasis on recent experimental data.

In 1988, the EMC Collaboration discovered (1) that the spin-dependent structure function of the proton violated the Ellis-Jaffe sum rule (2), and that the quarks appear to carry only a small fraction of the total spin of the proton. This result has triggered intense theoretical and experimental activities, which have resulted in a plethora of papers and in a new generation of experiments to study the internal spin structure of the nucleon.

POLARIZED LEPTON-NUCLEON SCATTERING

Until now, polarized deep inelastic scattering experiments have focused on measuring the cross-section asymmetry of the scattering of longitudinally polarized beams on longitudinally polarized targets:

$$A_{\parallel} = \frac{\sigma^{\uparrow\downarrow} - \sigma^{\uparrow\uparrow}}{\sigma^{\uparrow\downarrow} + \sigma^{\uparrow\uparrow}}, \tag{1}$$

which is related to the virtual photon-nucleon asymmetries A_1 and A_2 by

$$A_{\parallel} = |P_b P_t| f D \left[A_1 + \frac{2(1-y)}{2-y} \gamma A_2 \right] \tag{2}$$

where P_b and P_t are the beam and target polarizations, respectively, the dilution factor f is the fraction of polarized nucleons in the target material, and D and γ are kinematic factors. A_1 is an asymmetry and A_2 a ratio of virtual photon absorption cross-sections. They are related to the spin-dependent structure function g_1 by

$$g_1 = \frac{F_2}{2x(1+R)(1+\gamma^2)}(A_1 + \gamma A_2) \quad (3)$$

where F_2 and R are spin-independent structure functions. A_1 and A_2 are constrained by positivity relations to $|A_1| \leq 1$ and $|A_2| \leq \sqrt{R}$; A_2 is therefore expected to give only a small contribution to $A_{\parallel}$ and g_1, and the relation $g_1 \approx A_1 F_2 / 2x \approx A_1 F_1$ holds to a good approximation. The structure function g_1 has a simple interpretation in the Quark-Parton Model where

$$g_1(x) = \tfrac{1}{2}\sum_i e_i^2 \delta q_i(x), \quad (4)$$

$$\delta q(x) = q^+(x) + \overline{q}^+(x) - q^-(x) - \overline{q}^-(x). \quad (5)$$

The distribution functions $q^+(\overline{q}^+)$ and $q^-(\overline{q}^-)$ refer to quarks (antiquarks) with spin parallel and antiparallel to the nucleon spin, respectively.

RECENT EXPERIMENTAL RESULTS

New programmes to measure the spin-dependent structure functions of nucleons have been launched at CERN, DESY, and SLAC. The SMC experiment (NA47) at CERN and the E-143 experiment at SLAC have presented results from deep inelastic scattering of muons and electrons, respectively, on protons and neutrons (Table 1). These experiments use cryogenic solid state targets. SLAC E-143 has used a ^{3}He target; in this nucleus, the two proton spins are antiparallel most of the time and it behaves, for a measurement of spin-dependent cross sections, similar to a free neutron target (3). The HERMES experiment at DESY, which employs the internal electron beam of the HERA ep-collider and a storage cell target, will start data-taking in 1995.

As an example, recent results on the virtual photon asymmetry A_1 for the proton are compared in Fig. 1. This figure illustrates both the good agreement of SLAC and CERN results and the complementarity of electron and muon

Table 1. Recent measurements of spin-dependent structure functions

Experiment	Beam	Beam energy (GeV)	Target	Reference
SMC	μ^+	100	C_4D_9OD	(4)
SMC	μ^+	190	C_4D_9OH	(5,6)
E-142	e^-	≤ 25	^3He	(7)
E-143	e^-	≤ 29	NH_3	(8)
E-143	e^-	≤ 29	ND_3	(8)

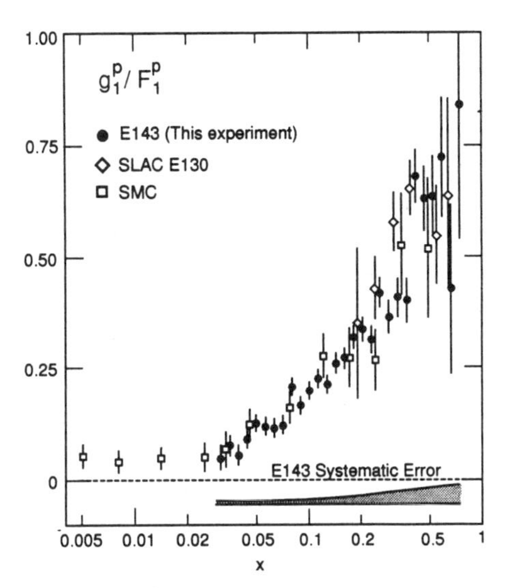

Figure 1. The ratio $g_1^p / F_1^p \approx A_1^p$ measured by the SMC and E-143 experiments. Data from the earlier SLAC E-130 experiment are also shown.

experiments. The electron beam experiments excel in statistical accuracy thanks to their superior beam intensity, whereas the muon experiments can explore a larger kinematic range owing to their higher beam energy, an important advantage when measuring moments and testing sum rule predictions.

From Cross-Section Asymmetries to Sum Rule Tests

The evaluation of moments of the spin-dependent structure functions from cross-section asymmetries relies on experimental data from other experiments, and on theoretical guidance. None of the experiments discussed here measures the unpolarized structure functions F_2 and R; to evaluate g_1 from A_1 (Equ. (3)), the "NMC fit" is usually employed, which is a QCD-inspired phenomenological parametrization of high-statistics proton and deuterium data from electron and muon scattering at $x > 0.007$ (9).

The experimental situation is less satisfactory for R where present analyses rely on the "SLAC parametrization" of electron, muon, and neutrino data (10). This parametrization is limited to $x > 0.1$ and the accuracy poor compared to the F_2 fit; however, R is small and the net sensitivity of g_1 to uncertainties on R is weak (Equ. (3)).

Until recently, no experimental data existed on the A_2 asymmetry which also enters into the determination of g_1 (cf. Equ. (3)). For most of the data published so far, $A_2 = 0$ was assumed and an allowance made for a systematic uncertainty which corresponds to the limit $|A_2| \leq \sqrt{R}$. A first measurement of A_2 from muon scattering on a transversely polarized target has recently been published by SMC (Fig. 2) (11). More and much more accurate data on A_2 are expected soon from the E-143 experiment (8).

The Q^2 dependence of A_1 and g_1 is an important issue in the interpretation of the data; in the experimental results, the average Q^2 rises approximately linearly with x due to kinematics and the acceptance behaviour of the spectrometers, whereas moments of structure functions need to be evaluated at constant Q^2 to allow for meaningful comparisons to sum rule predictions. Experimentally, this problem is not yet settled conclusively. The present data exhibit no Q^2 dependence of A_1 within the errors (12), in agreement with theoretical expectations that such a dependence should be weak (13). It is therefore customary to assume $\mathrm{d}A_1/\mathrm{d}Q^2=0$, such that the Q^2 dependence of g_1 is controlled by F_2 and R (Equ. (3)).

Finally, g_1 must be extrapolated to $x=0$ and $x=1$ to evaluate its first moment

$$\Gamma_1=\int_0^0 g_1(x)\,\mathrm{d}x. \tag{6}$$

The extrapolation $x\to 1$ is not critical and gives only a small contribution to the integral, whereas the extrapolation $x\to 0$ is, to some extent, controversial. So-called Regge-inspired models predict that $g_1\propto x^{\alpha}$ with $0\le\alpha\le 0.5$ at small x, where it is not entirely clear over how large an x range this relation holds (14,15). Both E-143 and SMC assume $g_1=const$ and fit $const$ to their low x data, whereas E-142 assumes $A_1^n\propto x^{1.2}$. Neither assumption is based on hard theory, and both have been questioned (16,17). The recent SMC proton data (5,6) do indeed suggest a rise at small x which is, however, far from being significant. It should be kept in mind that this also is an issue which is not yet settled conclusively – all experiments allow for a more or less generous systematic uncertainty on the

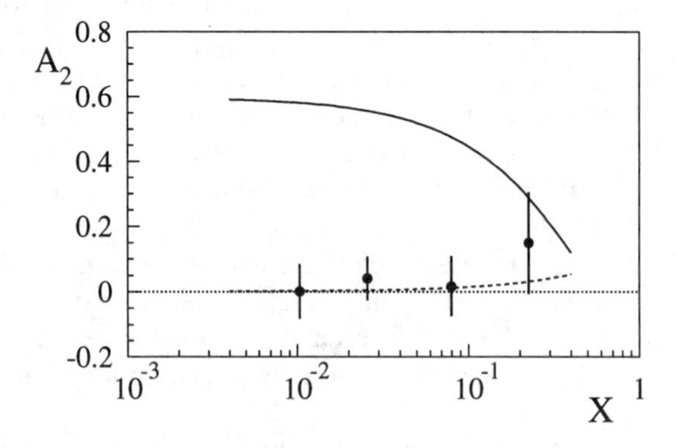

Figure 2. The asymmetry A_2^p measured by SMC, shown at the average Q^2 of each x bin. Only statistical errors are shown. The solid line represents the limit $|A_2|\le\sqrt{R}$, assuming the SLAC parametrization of R (10).

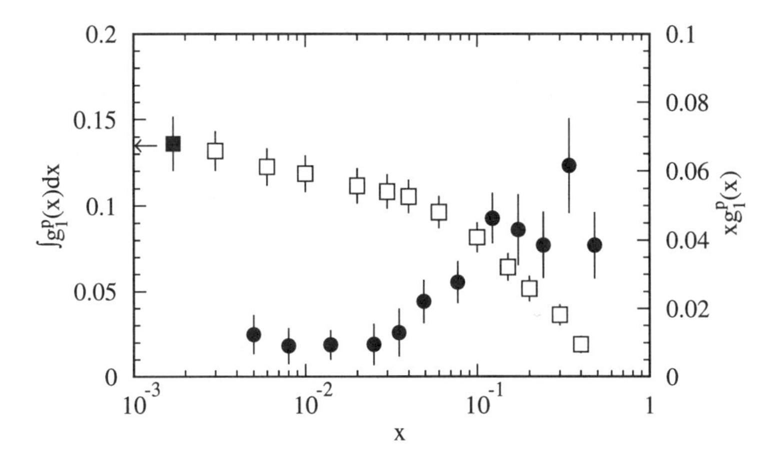

Figure 3. SMC data on g_1. The solid circles and the right-hand axis show the structure function $xg_1^p(x)$ at $Q^2 = 10\ \mathrm{GeV}^2$. The open boxes and the left-hand axis show the integral $\int_{x_m}^{1} g_1^p(x)\,dx$, where the lower integration limit x_m is the value of x at the lower edge of each x bin. Only statistical errors are shown. The solid square shows Γ_1^p, with statistical and systematic errors combined in quadrature.

unmeasured contribution to the first moment, but they all assume that g_1 is well-behaved at small x, and we should beware of surprises. As an example, the recent g_1^p data from SMC and their extrapolation to $x = 0$ are shown in Fig. 3.

The Ellis-Jaffe Sum Rules and the Spin Content of the Nucleon

Ellis and Jaffe have derived sum rules which relate the first moments of g_1^p and g_1^n to the coupling constants F and D of the axial currents between states of the baryon octet (2). E-143 and SMC have found that these sum rules are violated, both for the proton and the neutron data, and have thus confirmed the earlier EMC result for the proton. E-142 had found their neutron data to be in agreement with the sum rule prediction (7); however, in a more recent analysis, this agreement appears to be fortuitous and due to the neglect of perturbative QCD corrections which are substantial at the low Q^2 of these data (18).

The Ellis-Jaffe sum rules were derived under the assumption that the strange sea is unpolarized and does not contribute to the spin of the nucleon; in Quark-Parton Model (QPM) language, the most straightforward interpretation of a violation of these sum rules is that only a fraction of the total nucleon spin is carried by quarks and that the strange sea *is* polarized.

Numerical results on tests of the Ellis-Jaffe sum rules and the quark contributions to the nucleon spin from the individual experiments can be found in Refs. (4–8). The sum rules are subject to higher order QCD corrections. Perturbative corrections have been computed up to $O(\alpha_s^3)$ (19); $O(\alpha_s^4)$ corrections have also been estimated (20). Non-perturbative ("higher-twist") corrections have been computed by several authors; the results differ substantially from one another – even in sign – but these corrections appear to be small and have so far been mostly ignored. The perturbative corrections are, however, sizeable and it is important to take them into account when comparing data from experiments at different Q^2. In a recent combined analysis, Ellis and Karliner (21) concluded that the total quark contribution to the nucleon spin is (Fig. 4)

$$\Delta\Sigma = \Delta u + \Delta d + \Delta s = 0.33 \pm 0.04$$

at $Q^2 = 10\ \mathrm{GeV}^2$, where Δu etc. are the first moments of δu etc. of Equ. (5). The corresponding result for the strange quark contribution to the nucleon spin is

$$\Delta s = -0.10 \pm 0.03.$$

The Bjorken Sum Rule

Back in 1966, Bjorken derived a sum rule for the difference of the spin-dependent proton and neutron structure functions from current algebra and flavour SU(2) symmetry (22). This sum rule was later recognised as a cornerstone of the QPM and is today understood to be a firm prediction of QCD (23). Including higher order QCD corrections, it reads

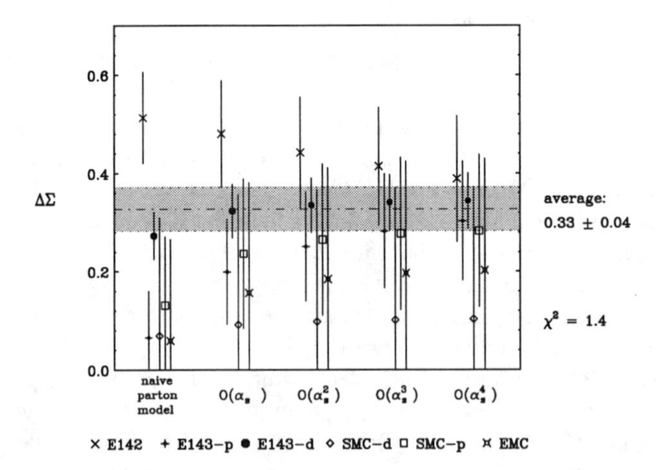

Figure 4. The total quark contribution to the nucleon spin at $Q^2 = 10\ \mathrm{GeV}^2$ from the SLAC and CERN experiments, shown as a function of the perturbative higher order QCD corrections. The average refers to the rightmost data points (from Ref. 21).

$$\Gamma_1^p(Q^2) - \Gamma_1^n(Q^2) = \frac{1}{6}\left|\frac{g_A}{g_V}\right|\left(1 - \frac{\alpha_s(Q^2)}{\pi} - .....\right) \tag{7}$$

where g_A and g_V are the axial-vector and vector constants measured in nuclear β decay. Perturbative QCD corrections have also been computed/estimated up to the 4th order in α_s (19,20).

The advent of neutron and deuteron data has allowed to test this sum rule eventually, more than 25 years after it was first derived. For numerical results, the reader is referred again to the original publications (4-8); there is now a consensus that the Bjorken sum rule is experimentally verified within one standard deviation, at the 10% level. This is illustrated in Fig. 5 where recent experimental data are compared to the Bjorken prediction in the $\Gamma_1^p - \Gamma_1^n$ plane.

Ellis and Karliner have argued that the Bjorken sum rule should be taken for granted, in which case it can be employed to determine the strong coupling constant (21). They find $\alpha_s(Q^2 = 2.5\ \text{GeV}^2) = 0.375^{+0.062}_{-0.081}$, corresponding to

$$\alpha_s(Q^2 = M_Z^2) = 0.122^{+0.005}_{-0.009}.$$

This result is in very good agreement with the present world average $\alpha_s(M_Z^2) = 0.117 \pm 0.005$ (24), with remarkably competitive errors.

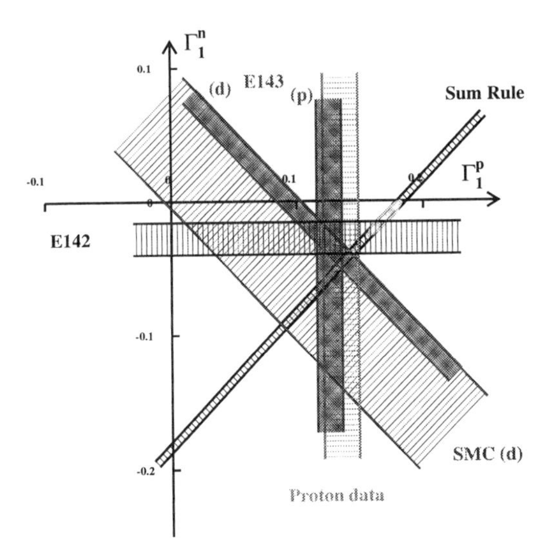

Figure 5. First moments of the spin-dependent structure functions g_1 of proton, neutron, and deuteron, evaluated at $Q^2 = 5\ \text{GeV}^2$ and compared to the prediction of the Bjorken sum rule. The finite width of the band representing the sum rule prediction is due to the large uncertainty on the strong coupling constant at small Q^2.

OUTLOOK

Nucleon spin physics is flourishing, experimentally and theoretically. SMC will continue during 1995, SLAC experiments E-154 and E-155 will repeat E-142 and E-143, respectively, with increased accuracy and beam energy (50 GeV), and HERMES will also start in 1995 (25). All experiments combined will improve substantially the accuracy of the present data. The study of semi-inclusive final states in SMC and HERMES will provide information on flavour-separated spin-dependent quark distributions. Finally, RHIC will open up a new frontier in high-energy nucleon spin physics (26).

Lack of space does unfortunately not allow to review here the many interesting theoretical developments reported at this conference and the reader is again referred to the original contributions to these proceedings.

REFERENCES

1. EMC, Ashman, J., et al., *Phys. Lett.* **B206**, 364–370 (1988);
   EMC, Ashman, J., et al., *Nucl. Phys.* **B328**, 1–35 (1989).
2. Ellis, J., and Jaffe, R.L., *Phys. Rev.* **D9**, 1444–1446 (1974); **D10**, 1669–1670 (1974).
3. Milner, R., these proceedings.
4. SMC, Adeva, B., et al., *Phys. Lett.* **B302**, 533–539 (1993).
5. SMC, Adams, D., et al., *Phys. Lett.* **B329**, 399–406 (1994).
6. Windmolders, R., these proceedings; Zanetti, A.-M., these proceedings.
7. SLAC E-142, Anthony, P.L., et al., *Phys. Rev. Lett.* **71**, 959–962 (1993).
8. E-143, Abe, K., et al., SLAC-PUB-6508 (1994);
   Borel, H., these proceedings; Day, D., these proceedings; Feltham, A., these proceedings.
9. NMC, Amaudruz, P., et al., *Phys. Lett.* **B295**, 159–168 (1992).
10. Whitlow, L.W., et al., *Phys. Lett.* **B250**, 193–198 (1990).
11. SMC, Adams, D., et al., *Phys. Lett.* **B336**, 125–130 (1994).
12. See e.g. SMC, Adeva, B., et al., *Phys. Lett.* **B320**, 400–406 (1994).
13. See e.g. Altarelli, G., Nason, P., and Ridolfi, G., *Phys. Lett.* **B320**, 152–158 (1994).
14. Heimann, R.L., *Nucl. Phys.* **B64**, 429–463 (1973).
15. Ellis, J., and Karliner, M., *Phys. Lett.* **B213**, 73–80 (1988).
16. Bass, S.D., and Landshoff, P.V., Cavendish preprint HEP 94/4 (DAMTP 94/50).
17. Close, F.E., and Roberts, R.G., Rutherford preprint RAL-94-071.
18. Ellis, J., and Karliner, M., *Phys. Lett.* **B313**, 131–140 (1993).
19. Larin, S.A., Tkachov, F.V., and Vermaseren, J.A.M., *Phys. Rev. Lett.* **66**, 862–863 (1991);
   Larin, S.A., and Vermaseren, J.A.M., *Phys. Lett.* **B259**, 345–352 (1991);
   Larin, S.A., CERN-TH-7208/94 (hep/ph 9403383).
20. Kataev, A.L., and Starshenko, V., CERN-TH-7198/94 (hep-ph/9405294).
21. Ellis, J., and Karliner, M., *Phys. Lett.* **B341**, 397 (1995).
22. Bjorken, J.D., *Phys. Rev.* **148**, 1467–1478 (1966); *Phys. Rev.* **D1**, 1376–1379 (1991).
23. Kodaira, J., et al., *Phys. Rev.* **D20**, 627 (1979);
   Kodaira, J., et al., *Nucl. Phys..* **B159**, 99–124 (1979);
   Kodaira, J., *Nucl. Phys.* **B165**, 129–140 (1980).
24. Particle Data Group, Montanet, L., et al., Review of Particle Properties, *Phys. Rev.* **D50,** 1173–1826 (1994).
25. Jackson, H., these proceedings.
26. Soffer, J., these proceedings; Ozaki, S., these proceedings; Yokosawa, A., these proccedings.

Nucleon Spin Structure Functions from the Spin Muon Collaboration

R.Windmolders

Mons University, 7000 Mons, Belgium

(on behalf of the S.M.C.)

Abstract. The spin muon collaboration (SMC) has studied polarized deep inelastic scattering of muons on protons and deuterons using a high energy muon beam at CERN. We present here a summary of the results obtained from data taken in the last two years.

INTRODUCTION

The interest in nucleon spin structure functions has been considerably enhanced by the earlier EMC result showing that the first moment of the spin structure function of the proton was smaller than expected. The interpretation of this result in the quark-parton model leads to the conclusion that the spins of the quarks contribute relatively little to the nucleon spin. The SMC experiment is part of the effort undertaken to improve the understanding of nucleon spin by remeasuring the spin structure function of the proton with a good precision over a wide kinematic range and by measuring for the first time the spin structure function of the neutron. These combined measurements provide the possibility to test the Bjorken sum rule, which relates the first moments of the spin structure functions to the coupling constant of β decay and is considered as a fundamental test of QCD.

In this report we describe the analysis of the SMC data collected in 1992-1993. In the first section we discuss the experimental aspects of the evaluation of spin asymmetries. In the second one we present the spin structure functions g_1^p and g_1^n and compare their integrals with the predictions given by the Bjorken and Ellis-Jaffe sum rules. In the next section we determine the second spin structure function of the proton from a data sample taken with transverse target polarization. The last section is devoted to the analysis of semi-inclusive spin asymmetries for positive and negative hadrons with the aim to separate the valence and sea quark contributions to the structure function.

SPIN ASYMMETRIES IN THE SMC EXPERIMENT

The differential cross section $d^2\sigma/dx\, dQ^2$ measured in unpolarized deep-inelastic scattering experiments is generally expressed in terms of two spin averaged structure functions $F_1(x, Q^2)$ and $F_2(x, Q^2)$. The scaling variable x is defined by $x = Q^2/2M\nu$ where $(-Q^2)$ and $\nu = E - E'$ are the square of the mass and the energy of the virtual photon respectively. When beam and target are polarized,two different cross sections are obtained for parallel and anti-parallel spin orientations and their difference can be written in a similar way in terms of two spin dependent structure functions $g_1(x, Q^2)$ and $g_2(x, Q^2)$.

Polarized deep inelastic experiments measure the spin asymmetry

$$A_{\parallel}(x, Q^2) = \frac{d\sigma^{\uparrow\downarrow} - d\sigma^{\uparrow\uparrow}}{d\sigma^{\uparrow\downarrow} + d\sigma^{\uparrow\uparrow}} \,. \tag{1}$$

The cross section $d\sigma^{\uparrow\uparrow}(d\sigma^{\uparrow\downarrow})$ corresponds to $d^2\sigma/dx\, dQ^2$ for parallel (antiparallel) beam and target polarization. The spin asymmetry $A_{\parallel}$ is directly related to the virtual photon asymmetries A_1 and A_2 by

$$A_{\parallel} = D(A_1 + \eta A_2) \tag{2}$$

where

$$A_1 = \frac{\sigma_{1/2} - \sigma_{3/2}}{\sigma_{1/2} + \sigma_{3/2}}\,, \qquad A_2 = \frac{2\sigma_{TL}}{\sigma_{1/2} + \sigma_{3/2}}\,. \tag{3}$$

Here $\sigma_{1/2}$ and $\sigma_{3/2}$ are the virtual photon-nucleon cross sections for total helicity 1/2 and 3/2 and D the depolarization factor of the virtual photon. The kinematical factor η is small at high energy and A_2 is bound by the positivity condition $|A_2| \leq \sqrt{R}$. Therefore the second term on the right hand-side of eq. (2) can generally be dropped and this relation reduces to

$$A_{\parallel} = DA_1 = D\frac{g_1}{F_1} \tag{4}$$

or

$$g_1 = \frac{A_{\parallel}}{D}\frac{F_2}{2x(1+R)}. \tag{5}$$

The measured asymmetries combined with the values of F_2 and R thus provide a direct evaluation of the spin structure function g_1.

The SMC has measured spin asymmetries on hydrogen and on deuterium using large polarized targets exposed to a high energy muon beam at the CERN SPS. The scattered muons and hadrons produced in the interaction are detected in a spectrometer where their momenta are measured and their trajectories reconstructed up to the interaction vertex. A calorimeter is used to separate hadrons from electrons. The beam polarization is determined by measuring the positron energy spectrum in the decay $\mu^+ \rightarrow e^+ \nu_e \overline{\nu}_\mu$.

Two different polarized targets have been used. For the runs on deuterium which took place in 1992,the old EMC target was filled with deuterated butanol and polarized to about 35 %. For the runs on hydrogen, a new 50 % longer target was installed and the average polarization was 86 %. Both targets are divided in two parts which are always polarized in opposite directions. The polarization is reversed at regular intervals several times a day by inverting the direction of the magnetic field,and once per week by dynamic nuclear polarization (DNP).

A summary of the running conditions is given in table 1.

Table 1: The SMC data (1992-1993)

	Deuteron runs (1992)	Proton runs (1993)
BEAM (μ^+)		
E(GeV)	100	190
intensity	$4 \times 10^7 \mu$pp	$4 \times 10^7 \mu$pp
$\langle P \rangle$	-0.82 ± 0.06	-0.80 ± 0.04
TARGET		
material	C_4D_9OD	C_4H_9OH
Dil.fact. (f)	$\sim$ 0.19	$\sim$ 0.12
configuration	2×40 cm	2×60 cm
Field........		
solenoid	2.5 T	2.5 T
dipole	0.2 T	0.5 T
$\langle P \rangle$	0.35 ± 0.05	0.86 ± 0.03
Reversals....		
field rot.	every 8 hrs.	every 5 hrs.
DNP	every week	every week
KINEM.RANGE		
Q^2	$1 - 30\,\mathrm{GeV}^2$	$1 - 60\,\mathrm{GeV}^2$
x	$0.006 - 0.6$	$0.003 - 0.7$
Nr.EVENTS	3.2×10^6	4.4×10^6

The energy of the incident muons was 100 GeV for the runs on deuterium and 190 GeV for the runs on hydrogen. The polarized target and the polarimeter are described in other communications in these proceedings [1,2]. A general description of the experimental set-up and the running conditions has been presented at the previous spin conference [3]. The asymmetries (2) are calculated from the ratio of the number of muons scattered in the two parts of the target before and after a polarization reversal. In this way the incident flux and the number of target nucleons cancel out so that only the ratio of the acceptances of the two parts of the target has to be kept constant [3].

The resulting asymmetries [4,5] are shown as a function of the scaling variable x in figs.(1a) and (1b) for the proton and the deuteron respectively.

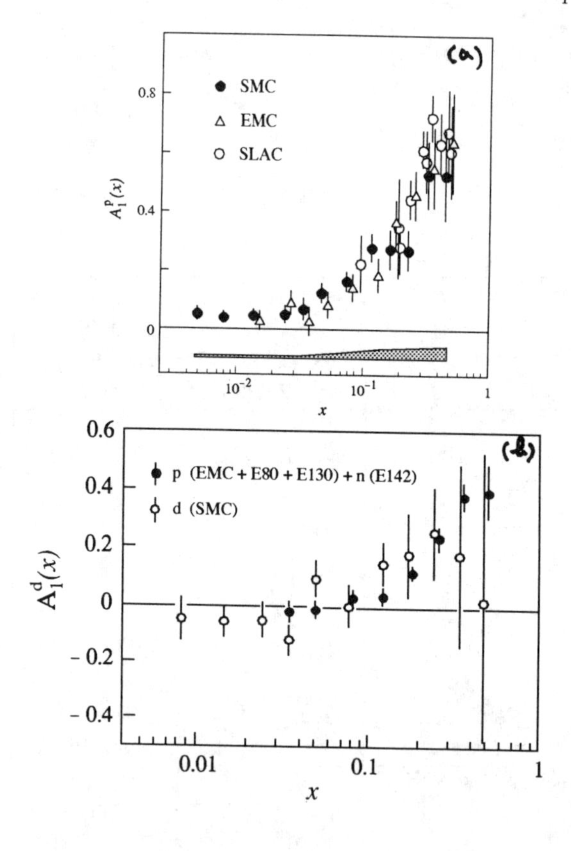

Fig.1: The virtual-photon cross section asymmetries on protons (a) and deuterons (b) as a function of the scaling variable x. Only statistical errors are given for each point.The shaded area shows the systematic error.

For comparison we show in the same figures the proton asymmetries measured by the EMC [6] and deuteron asymmetries computed from the E142 data on neutrons [7] combined with averaged proton data. In both cases the agreement with the SMC asymmetries is excellent.

The average Q^2 for the different x intervals increases from 1.5 GeV^2 in the lowest x interval to about 50 GeV^2 (15 GeV^2) for the largest x values in the proton (deuteron) data. The average Q^2 of the samples are 10 GeV^2 and 4.6 GeV^2 for the proton and the deuteron data respectively. At fixed x, we do not observe any systematic Q^2 dependance of the asymmetries (fig.2). This is in agreement with a recent QCD calculation which has shown that the expected Q^2 dependance is well beyond the sensitivity of the present data [8].

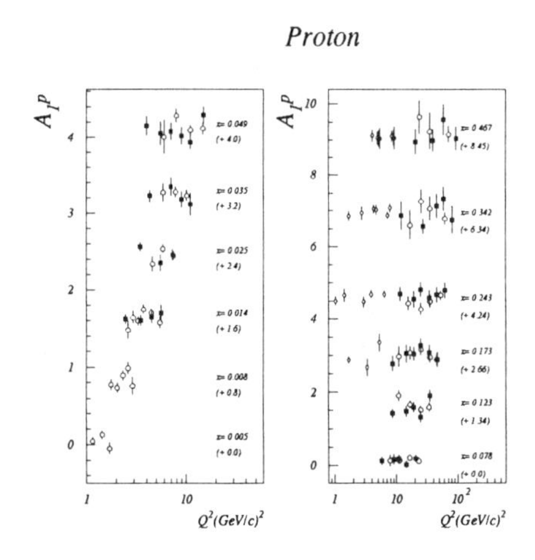

Fig.2: A_p^1 as a function of Q^2 for different values of x.The open circles are the SMC results;the squares represent the EMC data from ref.[6].

THE SPIN STRUCTURE FUNCTIONS $g_1^{p,d}(x)$.

The spin structure functions have been derived from the measured asymmetries using eq.[5] with F_2 taken from the NMC parametrisation [9] and R from a fit to the SLAC and other high energy data [10]. All data points have been converted to the average Q^2 of the data sample assuming that g_1 and F_2 have the same Q^2 dependence. The resulting values are shown in fig. 3. The same figure shows $\int_{x_{min}}^{1} g_1(x)dx$ as a function of x_{min} and the estimated values of the first moments

$$\Gamma_1^{p,d} = \int_0^1 g_1^{p,d}(x)dx \tag{6}$$

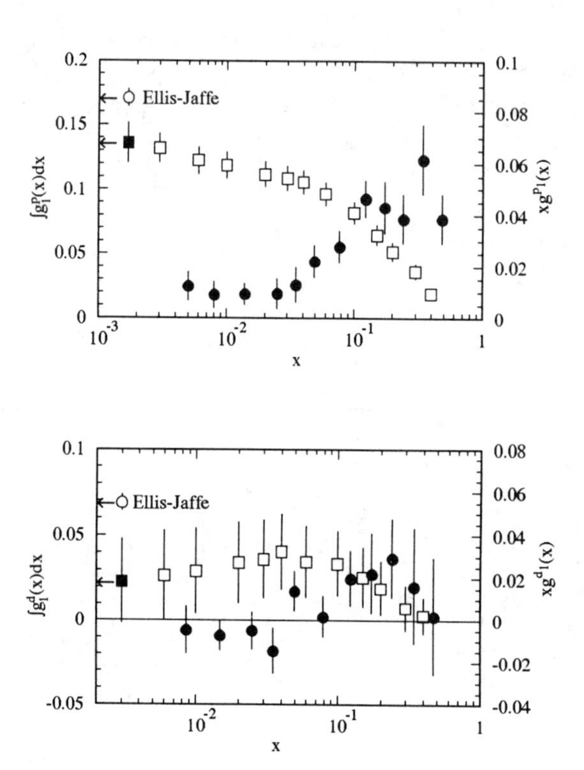

Fig.3: The structure functions xg_1^p and xg_1^d as a function of x at the average Q^2 of the data (solid circles). The open squares show $\int_{xm}^{1} g_1(x)dx$ where xm is the lower edge of each bin. The full square indicates the extrapolation to $xm = 0$.

For the extrapolation from x_{min} to zero a constant value of g_1 has been assumed, in agreement with the expected Regge behaviour. Although the low x data points for protons may suggest some increase of g_1 , this effect remains non-significant with the accuracy of the present data. For this reason we are unable to discriminate between the various forms of extrapolation which have been proposed recently [11, 12].

The following values are obtained for the first moments [13,4]:

$$\Gamma_1^p + \Gamma_1^n = 0.049 \pm 0.044(stat.) \pm 0.032(syst.) \qquad (Q^2 = 4.6GeV^2) \qquad (7)$$

$$\Gamma_1^p = 0.136 \pm 0.011(stat.) \pm 0.011(syst.) \qquad (Q^2 = 10GeV^2) \qquad (8)$$

The first value is derived from Γ_1^d after correcting for the probability of the deuteron to be in a D state ($P_D = 0.058$).

The various sources of systematic error are listed in table 2. It may be noted that the error due to the neglect of the A_2 term in eq.(2) is strongly reduced for the proton data due to the measurement of this quantity which will be discussed in the next section.

Table 2: Contributions to the systematic error on Γ_1. For the items marked by a (*) sign,the error is proportionnal to the value of Γ_1.

origin	deuteron data	proton data
Pbeam(*)	0.0015	0.0057
Ptarget(*)	0.0010	0.0039
F_2(*)	0.0012	0.0052
R	0.0005	0.0018
A2	0.0041	0.0017
f	0.0010	0.0034
Acc.var.	0.0130	0.0030
Rad.corr.(*)	0.0009	0.0023
Mom.meas.	0.0005	0.0020
Resolution	0.0008	0.0010
Proton bgd.	0.0005	—
Extrap.low x	0.0030	0.0040
Extrap.high x	0.0040	0.0007
Total	0.0148	0.0113

The Bjorken sum rule [14] predicts that the difference of the first moments Γ_1^p and Γ_1^n is related to the coupling constant g_A of the β decay by

$$\Gamma_1^p - \Gamma_1^n = \frac{1}{6}|g_A|C_{NS} \qquad (9)$$

where C_{NS} is a QCD correction factor which has been evaluated up to the order $(\alpha_s/\pi)^3$ [15].

For this test the first moments have been computed at a common Q^2 of 5 GeV^2. The result is shown in fig.4 together with the theoretical predictions obtained for two "extreme" values of α_s. In both cases very good agreement is observed. The same conclusions hold for the E143 results at Q^2= 3 GeV^2

[16]. The predictions of the Bjorken sum rule are thus fully supported by all existing data. Since this sum rule is a fundamental test of QCD it has been suggested to use the values of $\Gamma_1^p - \Gamma_1^n$ in order to determine α_s. Although the experimental errors are still quite large, it has been shown that a global fit yields a rather accurate value of $\alpha_s(M_Z^2)$ [17].

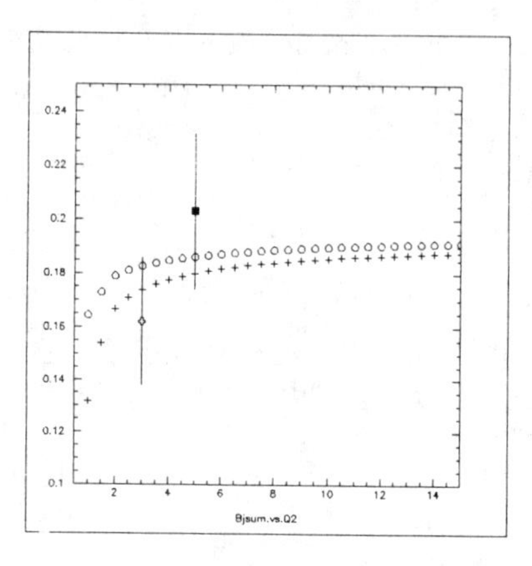

Fig.4: Values of $\Gamma_1^p - \Gamma_1^n$ for the SMC data (full square) and for SLAC experiment E143 (open cross). The predictions of the Bjorken sum rule are shown as a function of Q^2 for $\alpha_s = 0.114$ (circles) and $\alpha_s = 0.121$ (+ signs)

The values expected for Γ_1^p and Γ_1^n are derived from the formula

$$\Gamma_1^{p,n} = C_{NS}(\pm\frac{1}{12}(F+D) + \frac{1}{36}(3F-D)) + C_{SI}\frac{\Delta\Sigma}{9} \tag{10}$$

where $\Delta\Sigma = \Delta u + \Delta d + \Delta s$ and C_{SI} is the singlet QCD correction factor,which has been calculated to the order $(\alpha_s/\pi)^2$ [18]. In the Ellis-Jaffe sum rule [19],the strange sea is assumed to be unpolarized ($\Delta s = 0$) and F and D are taken from hyperon decay assuming SU_3 flavor symmetry ($F+D = 1.2573 \pm 0.0028$ and $F/D = 0.575 \pm 0.024$)[20]. The predictions are shown in fig.5 as a function of Q^2 for the two values of $\alpha_s(M_Z^2)$ used before. The experimental results are systematically below the predicted values. In the case of the SMC proton data at $Q^2 = 10$ GeV^2 the discrepancy is larger than 2 standard deviations.

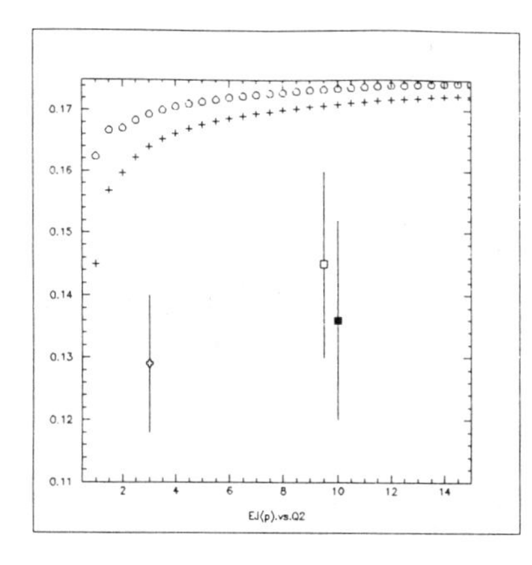

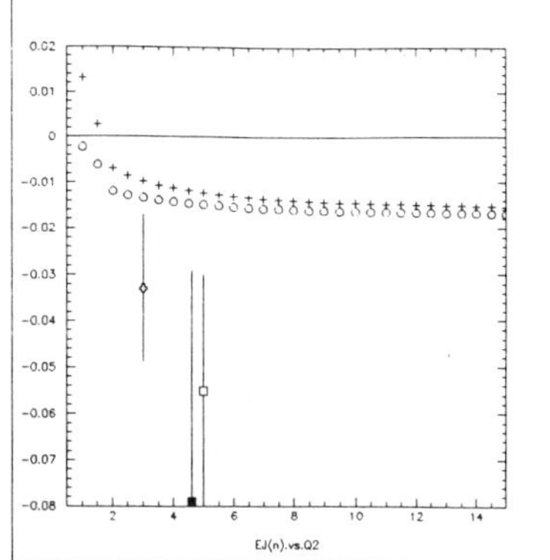

Fig.5: Values of Γ_1^p (a) and Γ_1^n (b) for the SMC data (full squares) , for the SMC data combined with earlier results (open squares) and for SLAC experiment E143 (open crosses). The predictions of the Ellis-Jaffe sum rules are shown as a function of Q^2 for $\alpha_s = 0.114$ (circles) and $\alpha_s = 0.121$ (+ signs).

Inserting the measured value of Γ_1^p in formula [10] and solving for $\Delta\Sigma$ we obtain :

$$\Delta\Sigma = 0.22 \pm 0.11(stat.) \pm 0.11(syst.) \tag{11}$$

to be compared with $\Delta\Sigma = 0.58$ as expected from the Ellis-Jaffe sum rule. With the assumed values of F and D, this result implies

$$\Delta s = -0.12 \pm 0.04 \pm 0.04 \tag{12}$$

TRANSVERSE SPIN ASYMMETRIES.

The neglect of A_2 in formula (2) and its replacement by the upper limit imposed by the positivity condition induces a relatively large contribution to the systematic errror on Γ_1. In order to reduce this effect for the proton data,the SMC has directly measured A_2 in a dedicated run with transverse polarization.

When the target is polarized in a direction transverse to the longitudinally polarized beam,the measured asymmetry is proportional to $\cos(\phi)$ where ϕ is the angle between the direction of the target polarization and the lepton scattering plane. The transverse asymmetry

$$A_\perp = \frac{1}{cos(\phi)} \frac{d\sigma^{\downarrow\rightarrow} - d\sigma^{\uparrow\rightarrow}}{d\sigma^{\downarrow\rightarrow} + d\sigma^{\uparrow\rightarrow}} \tag{13}$$

is independent of ϕ and related to the virtual photon asymmetries A_1 and A_2 by

$$A_\perp = D'(A_2 + \eta' A_1) \tag{14}$$

where D' and η' are kinematic factors. In contrast with formula (2),A_2 is here the dominant factor and can thus be directly evaluated from the measured values of $A_\perp$.

The measurements have been performed at an incident muon energy of 100 GeV [21]. The two target cells were initially polarized longitudinally by DNP and kept in frozen spin mode. The proton spins were rotated to the vertical direction by applying the dipole field and reducing the longitudinal field to zero. An average transverse polarization of 0.80 ± 0.04 was obtained. The transverse field was always applied in the same direction to avoid possible changes in acceptance and the polarization was reversed 10 times by DNP during the 17 days of running.

As for the longitudinal configuration,physical asymmetries are derived from data taken before and after a polarization reversal.However,instead of using events originating from the two target cells, the calculation is based on the number of events recorded at azimuthal angles ϕ and $\phi + \pi$. The data from the two target cells are treated independently. The corresponding values of $A_\perp$ are shown in fig.6 as a function of x. They are compatible and consistent with zero over the full range of x.

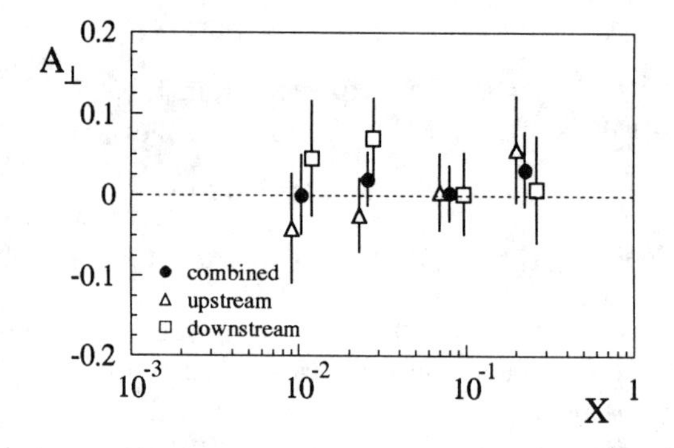

Fig.6: The transverse cross-section asymmetry $A_\perp$ as a function of x for interactions in the upstream and downstream targets and for the combined data. The error bars represent the statistical errors.

The derived values of A_2 are shown in fig.7. For $x < 0.15$ they are well below the positivity limit $\sqrt{R}$.

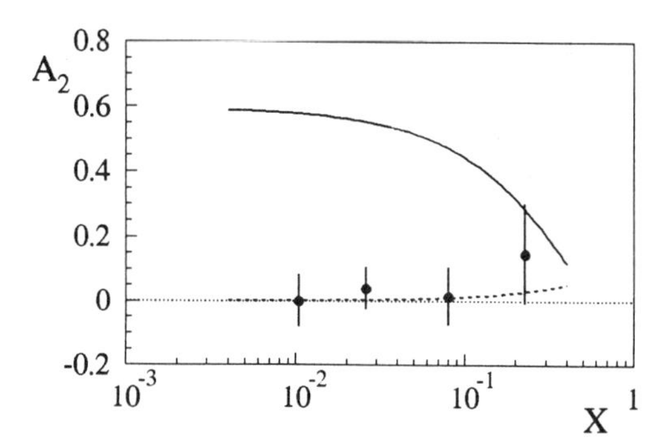

Fig.7: The asymmetry A_2 as a function of x. The solid line shows $\sqrt{R}$ from the SLAC parametrization [10], and the dashed line the results obtained with $\overline{g_2} = 0$ in Eq. (16). The error bars represent statistical errors only.

The spin structure function g_2 can be derived from A_2 by the relation

$$g_2 = -g_1 + \frac{\sqrt{Q^2}}{2Mx} A_2 F_1. \tag{15}$$

This function contains a term directly calculable from g_1 and a higher twist term $\overline{g_2}$ which is so far unknown:

$$g_2 = (-g_1 + \int_x^1 g_1(t) \frac{\mathrm{dt}}{\mathrm{t}}) + \overline{g_2} \tag{16}$$

Although the present data do not provide an accurate determination of g_2, it is observed that the experimental values of A_2 are in very good agreement with those expected under the assumption $\overline{g_2} = 0$ (fig.7). Therefore the present data do not show any evidence for the presence of a large higher twist contribution in $g_2(x)$.

HADRON ASYMMETRIES

Spin asymmetries of positive and negative hadrons produced in lepton-nucleon collisions provide complementary information with respect to the lepton asymmetries [22]. In the quark-parton model their study may lead for instance to a discrimination between the contributions of different types of quarks to the spin structure function g_1.

The SMC spectrometer has good acceptance for forward produced hadrons with momentum larger than a few GeV.Electrons are rejected from the hadron sample by measuring the energy deposited in the electro-magnetic part of the calorimeter. The spectra of positive and negative hadrons have been determined for interactions on protons and deuterons and integrated over the relative hadron energy $z = E_{had}/\nu$ for $z > 0.2$. The four corresponding semi-inclusive asymmetries A_1 are shown in fig.8 as a function of the scaling variable x. The quoted systematic errors include contributions due to the uncertainty in the hadron selection and to secundary interactions.

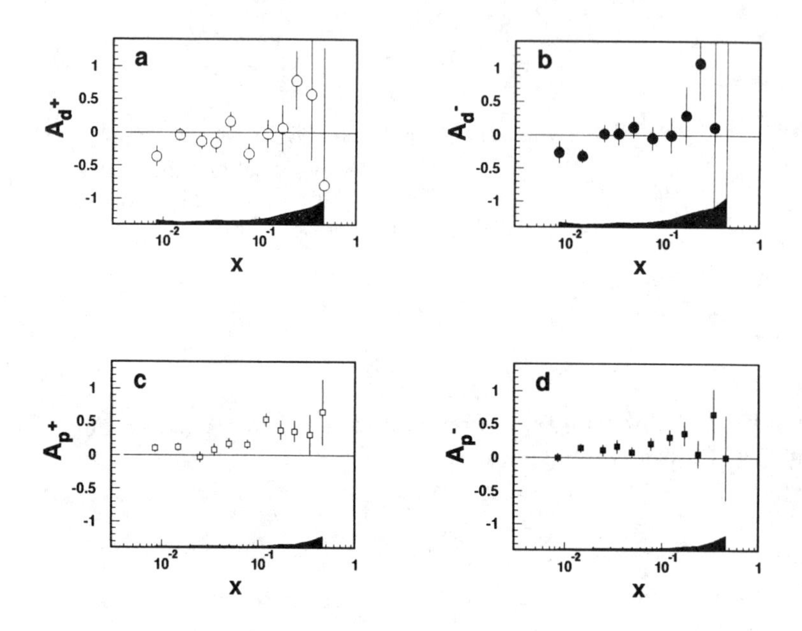

Fig.8: Semi-inclusive asymmetries of spin dependent cross sections for production of positive hadrons on deuterons (a),negative hadrons on deuterons (b),positive hadrons on protons (c) and negative hadrons on protons (d). The errors associated to each point are statistical;the shaded areas represent the systematic errors.

In each x interval,each semi-inclusive asymmetry $A_1^{(k)}$is a linear combinations of the various quark polarizations $\tilde{\Delta} q_i(x)$:

$$A_1^{(k)} = c_1^{(k)}\tilde{\Delta}u_v + c_2^{(k)}\tilde{\Delta}d_v + c_3^{(k)}\tilde{\Delta}q_s + c_4^{(k)}\tilde{\Delta}s \tag{17}$$

Here u_v,d_v,q_s and s represent the densities of the valence u and d quarks,of the non-strange sea quarks ($q_s = \overline{u} = \overline{d}$) and the strange quarks ($s = \overline{s}$) respectively. The quark polarizations $\tilde{\Delta}q$ are the differences between quark densities with spin parallel and antiparallel to the nucleon spin. The coefficients $c_i^{(k)}$ depend on unpolarized quark densities,for which the parametrisation [23] was assumed, and on ratios of fragmentation functions, which were derived from the EMC results [24]. We have checked that the values obtained for the valence quarks are consistent with the spin structure functions g_1^p and g_1^n derived from the muon asymmetries,i.e.that they are within errors compatible with the relation

$$g_1^p(x) - g_1^n(x) = \frac{1}{6}(\tilde{\Delta}u_v(x) - \tilde{\Delta}d_v(x)) \tag{18}$$

We have then complemented the set of eqs.(17) by the two relations corresponding to the muon asymmetries on protons and deuterons and solved the system by minimum χ^2. The results show that the asymmetries are almost insensitive to $\tilde{\Delta}s$ which cannot be determined from the present data. Positive (negative) values are obtained for $\tilde{\Delta}u_v$ ($\tilde{\Delta}d_v$) while the polarization of the non-strange sea quarks is found to be small and compatible with zero for all values of x. Integrating the polarizations over x and summing the valence and sea contributions we obtain:

$$\Delta u = 0.93 \pm 0.26(stat.) \qquad \Delta d = -0.54 \pm 0.25(stat.) \tag{19}$$

The large positive contribution of the u quarks to the spin structure function and the negative contribution of the d quark are well known results since the EMC experiment [6]. They are generally obtained from the integral of the spin structure function g_1 and from results on hyperon decay. Here we confirm these results by a different method using only muo-production data.

CONCLUSIONS

The SMC has measured the spin structure function g_1 of the proton and the deuteron in high energy muon scattering experiments covering a large range of $x(0.003-0.7)$.For $x < 0.03$ these data are the only existing ones, except for the older EMC results on the proton. The values of $\Gamma_1^p - \Gamma_1^n$ are in good agreement with the Bjorken sum rule, which is tested with a precision corresponding to about 15 % of its value. The individual values of Γ_1^p and Γ_1^n are both below the predictions of the Ellis-Jaffe sum rules. For the proton, the discrepancy is larger than 2 standard deviations. In the quark-parton model, the SMC results on $\Gamma_1^{p,n}$ imply that the sum of the quark contributions to the nucleon spin $\Delta\Sigma$ is equal to 0.22 ± 0.15.

A study of spin asymmetries in a configuration where the target polarization is perpendicular to the beam direction shows that the values of A_2 are much smaller than their positivity limit $\sqrt{R}$. These data do not show evidence for large higher twist contributions to the spin structure function g_2 of the proton.

The analysis of semi-inclusive hadron asymmetries confirms the large positive and negative contributions to the nucleon spin by u and d quarks respectively,while the contribution of the non-strange sea quarks is found to be small.

REFERENCES

1. D. Krämer et al., these proceedings.
2. F. Feinstein et al., these proceedings.
3. G. K. Mallot, Proc.10th Int.Symp. on High Energy Spin Physics, Nov.1992,Nagoya,Japan, pp.103-113.
4. SMC,D.Adams et al.,Phys.Lett. **B329** (1994) 399.
5. SMC,B.Adeva et al.,Phys.Lett. **B320** (1994) 400.
6. EMC,J.Ashman et al., Nucl. Phys. **B328** (1989) 1.
7. SLAC E142, D.L.Anthony et al.,Phys.Rev.Lett. **71** (1993) 959.
8. G.Altarelli et al.,Phys. Lett. **B320** (1994) 152.
9. NMC,P.Amaudruz et al.,Phys.Lett. **B295** (1992) 159.
10. L.W.Whitlow et al., Phys.Lett. **B250** (1990) 193.
11. S.D.Bass and P.V.Landshoff, "The small x behaviour of g_1", Cavendish preprint HEP 94/4 and DAMTP 94/50.
12. F.E.Close and R.G.Roberts,"The spin dependence of the diffractive processes and implications for the small x behaviour of g_1 and the spin content of the nucleon",RAL report,RAL-94-071 (July 1994).
13.SMC,B.Adeva et al., Phys.Lett. **B302** (1993) 533.
14.J.D.Bjorken,Phys.Rev. **148** (1966) 1467;Phys.Rev.**D1**(1970) 1376.
15.S.A.Larin and J.A.M.Vermaseren,Phys.Lett.**B259** (1991) 345.
16.E143,J.Mc Carthy,Int.Conf. on High Energy Physics,Glasgow,July 1994.
17.J.Ellis and M.Karliner,"Determination of α_s and the Nucleon Spin Decomposition using Recent Polarized Structure Function Data", CERN-TH-7324/94 and TAUP-2178-94.
18.S.A.Larin,Phys.Lett.**B334** (1994) 192.
19.J.Ellis and R.L.Jaffe,Phys.Rev. **D9** (1974) 1444;**D10**(1974)1669 .
20.Z.Dziembowski and J.Franklin,Nucl.Part.Phys. **17**(1991) 213.
21.SMC,D.Adams et al.,Phys.Lett. **B336** (1994) 125.
22.L.L.Frankfurt et al.,Phys.Lett. **B230** (1989) 141.
23.A.D.Martin,W.J.Stirling and R.G.Roberts,Phys.Lett. **B306** (1993) 145.
24.EMC,M.Arneodo et al.,Nucl.Phys. **B321** (1989) 541.
25.SMC,D.Adams et al, "Polarisation of valence and sea quarks in the nucleon from semi-inclusive spin asymmetries" (to be published).

Spin Structure Results from E143

Donal B. Day*
Institute of Nuclear and Particle Physics
University of Virginia
Charlottesville, VA 22901

Abstract

We present the spin structure functions $g_1^p(x)$ and the preliminary results for $g_1^d(x)$ from SLAC experiment E143 at 29.1 GeV along with a brief description of the experiment. These new data are consistent with the Bjorken sum rule; however they do not support the Ellis–Jaffe sum rules. These new data on the proton indicated that the spin fraction carried by the quarks is 0.29 ± 0.10 consistent with other experiments at CERN and SLAC.

Introduction

Measurements of the spin–dependent structure functions of the nucleon have received great attention since the the publication of the EMC results [1] in 1988 which indicated that the fractional spin of the proton carried by the quarks was small. In fact, since their results were consistent with zero (their results, $\Sigma^p = 0.11 \pm 0.17$ were also consistent with 28%), a "spin crisis" developed, at least among the more excitable. Since 1988 there have been further measurements at CERN [2, 3, 4] and SLAC [7]. I will report on the latest results from E143 which ran at SLAC for three months starting in November 1993 and measured the spin–dependent structure functions from polarized protons and polarized deuterons.

*For the E143 Collaboration: The American University, Institut für Physik der Universität Basel, 3LPC IN2P3/CNRS, University Blaise Pascal, CEBAF, DAPNIA-Service de Physique Nucleaire Centre d'Etudes de Saclay, Lawrence Livermore National Laboratory, University of Massachusetts, University of Michigan, Naval Postgraduate School, Old Dominion University, University of Pennsylvania, Stanford Linear Accelerator Center, Stanford University, Temple University, Tohoku University, University of Virginia, University of Wisconsin, R. Arnold and O. Rondon, spokesmen.

The original motivation for polarized electron–polarized nucleon experiments derives from the early days of deep inelastic scattering. In a naive quark picture in which the nucleon is made of of three spin $\frac{1}{2}$ constituents a polarized virtual photon can only scatter from a quark whose helicity is opposite to the virtual photon polarization. This would lead to a large positive asymmetry $A_1 = \frac{\sigma_a - \sigma_p}{\sigma_a - \sigma_p}$ where $a = \uparrow\downarrow, b = \uparrow\uparrow$. Formally, the virtual photon asymmetries are

$$A_1 = \frac{\sigma^T_{1/2} - \sigma^T_{3/2}}{\sigma^T_{1/2} + \sigma^T_{3/2}} = \frac{g_1(x,Q^2) - \gamma^2 g_2(x,Q^2)}{F_1(x,Q^2)} \quad \text{and}$$

$$A_2 = \frac{2\sigma^{TL}}{\sigma^T_{1/2} + \sigma^T_{3/2}} = \frac{\gamma(g_1(x,Q^2) + g_2(x,Q^2))}{F_1(x,Q^2)}.$$

$\sigma^T_{1/2(3/2)}$ is the virtual photon–nucleon scattering amplitude when the virtual photon spin is (anti)aligned with the nucleon spin. σ^{TL} is the interference between longitudinal and transverse amplitudes. For the evaluation of the fundamental sum rules we must establish $g_1(x)$. If $g_2 = 0$ or $\gamma = \frac{Q^2}{\nu^2} = 0$ (high energy), then $A_1 \approx g_1/F_1$. EMC neglected g_2 (which is OK at high energy) while this experiment measured it. It should be noted that there is a constraint on $A_2, \mid A_2 \mid < \sqrt{R}$, and we can expect g_2 to be small. The earliest spin structure measurements at SLAC [5, 6] were consistent with this naive picture. It was the CERN experiment [1] that exposed how much more was needed to be learned.

Experimental Determination of g_1, g_2

We do not have a source of 100% longitudinally polarized virtual photons but only the partially polarized virtual photons from the scattering of polarized electrons from the target. Consequently the relationship between the fundamental asymmetries and the experimental asymmetries is not a clean as shown above. Nonetheless it is transparent. The measured physics asymmetries are $A_\parallel$ and $A_\perp$, where $A_\parallel$ designates the asymmetry when the beam and target polarizations are parallel and $A_\perp$ when the target polarization is perpendicular to the beam polarization,given by

$$\begin{aligned} A_\parallel &= \frac{\sigma^{\uparrow\downarrow} - \sigma^{\uparrow\uparrow}}{\sigma^{\uparrow\downarrow} + \sigma^{\uparrow\uparrow}} \\ &= f\left[g_1(x,Q^2)[E + E'\cos\theta] - \frac{Q^2}{\nu} g_2(x,Q^2)\right] \quad \text{and} \\ A_\perp &= \frac{\sigma^{\downarrow\leftarrow} - \sigma^{\uparrow\leftarrow}}{\sigma^{\downarrow\leftarrow} + \sigma^{\uparrow\leftarrow}} \end{aligned}$$

$$= fE'\sin(\theta)\left[g_1(x,Q^2)+\frac{2E}{\nu}g_2(x,Q^2)\right]$$

where

$$f=\frac{\sigma_{Mott}}{\sigma}\frac{2\tan^2(\theta/2)}{M\nu}=\frac{1}{F_1(x,Q^2)}\frac{1}{\nu}\frac{1-\epsilon}{1+\epsilon R(x,Q^2)}.$$

The spin structure functions g_1 and g_2 written in terms of $A_\parallel$ and $A_\perp$ are:

$$g_1(x,Q^2)=\frac{F_1(x,Q^2)}{D'}\left[A_\parallel+\tan(\frac{\theta}{2})A_\perp\right] \text{ and } \quad (1)$$

$$g_2(x,Q^2) = \frac{F_1(x,Q^2)}{D'}\frac{y}{2\sin(\theta)}\times \left[\frac{E+E'\cos(\theta)}{E'}A_\perp-\sin(\theta)A_\parallel\right].$$

$g_{1,2}(x,Q^2)$ depend on other quantities, most of which are extracted from the information about the events but also on quantities (R and F_2) which have been measured in other experiments. Note that the angle term in both expressions is small and that g_1 and g_2 are essentially determined by measuring $A_\parallel$ and $A_\perp$ respectively. The other quantities are:

$$\begin{aligned}
D' &= \frac{(1-\epsilon)(2-y)}{y\left[1+\epsilon R(x,Q^2)\right]}\\
R(x,Q^2) &= \frac{\sigma_L}{\sigma_T}=\frac{(1+\gamma^2)F_2}{2xF_1}-1\\
y &= \frac{E-E'}{E}=\frac{\nu}{E}\\
\epsilon &= \frac{1}{1+2\left(1+\frac{\nu^2}{Q^2}\right)}\tan^2(\frac{\theta}{2})\\
\gamma &= \frac{Q^2}{\nu^2}.
\end{aligned}$$

g_1 and g_2 can be written in terms of A_1 and A_2 [8, 9].

Sum Rules

In 1966, Bjorken [10] derived, using only current algebra and isospin symmetry, an important relationship between spin–dependent deep inelastic scattering and the weak coupling constant found in β decay. His derivation, while consistent with QCD, predated it and is considered a benchmark of our understanding of high–energy physics. Strictly true at infinite momentum transfer the Bjorken sum rule relates the difference between the proton and neutron spin structure functions to the nucleon beta decay coupling constants and reads

$$\Gamma_1^p - \Gamma_1^n = \int (g_1^p(x) - g_1^n(x))dx = \frac{1}{6}\frac{g_A}{g_V} \quad (Q^2 = \infty)$$

where g_A and g_V are nucleon axial–vector and vector couplings determined from β decay and $g_A/g_V = 1.2573 \pm 0.0038$ [11]. Experiments are not done at infinite momentum transfer and this result must be corrected for the finite momentum transfer. These non–singlet corrections to third order in α_s are [12]

$$C_{NS} = 1 - \frac{\alpha_2(Q^2)}{\pi} - 3.58\left(\frac{\alpha_s(Q^2)}{\pi}\right)^2 - 20.22\left(\frac{\alpha_s(Q^2)}{\pi}\right)^3,$$

where $\alpha_s(Q^2)$ is the strong coupling constant. Evaluated at $Q^2 = 3$ (GeV/c)^2 with $\alpha_s(Q^2 = 3) = 0.35 \pm 0.05$ and $n_f = 3$, the BSR prediction becomes

$$\Gamma_1^p - \Gamma_1^n = \frac{1}{6}\frac{g_A}{g_V}c_1 = 0.171 \pm 0.008.$$

Less rigourous individual sum rules for the proton and neutron have been derived by Ellis and Jaffe [14] using two critical assumptions. First, that SU(3) flavor symmetry is valid and second, that the strange sea quarks in the nucleon are unpolarized. These are, including the QCD corrections evaluated at $Q^2 = 3(\text{GeV/c})^2$, for the proton and neutron,

$$\Gamma_1^p = \frac{c_1(F+D)}{12} + \frac{(c_1 + 4c_0)(3F - D)}{36} = 0.160 \pm 0.006$$

$$\Gamma_1^n = -\frac{c_1(F+D)}{12} + \frac{(c_1 + 4c_0)(3F - D)}{36} = -0.011 \pm 0.006$$

when using the weak hyperon decay constants $F/D = 0.575 \pm 0.016$ and $g_A/g_V = F + D$ and the singlet QCD corrections [13],

$$c_0 = 1 - \frac{\alpha_s(Q^2)}{3\pi} - 0.55\left(\frac{\alpha_s(Q^2)}{\pi}\right)^2.$$

Quark Model Interpretation

Knowledge of $A_\parallel$ and $A_\perp$ (or equivalently A_1 and A_2) and the unpolarized structure functions F_2 and R determines the spin structure function g_1 (See Eq. 1.) In the quark parton model $g_1(x)$ has a simple interpretation as the charge–weighted difference between momentum distibutions for quarks with helicities aligned parallel (↑) and antiparallel (↓) to the nucleon:

$$g_1(x) = \frac{1}{2}\sum_i e_i^2\left[q_i^\uparrow(x) - q_i^\downarrow(x)\right] \equiv \sum_i e_i^2 \Delta q_i(x)$$

where e_i^2 is the charge of the quark flavor i, and $q_i^{\uparrow(\downarrow)}(x)$ are the quark plus antiquark momentum distributions. Thus the integrals over x of the structure functions, $g_1(x)$ and $g_2(x)$, can be related directly to the spin carried by the different quark flavors (i) integrated over all x, $\Delta q_i = \int_0^\infty q_i(x)dx$.

$$\Gamma_1^p = \frac{1}{18}\left[4\Delta u + \Delta d + \Delta s\right], \; \Gamma_1^n = \frac{1}{18}\left[4\Delta d + \Delta u + \Delta s\right]$$

and the total spin carried by the quarks is $\sum = \Delta u + \Delta d + \Delta s$ From neutron decay $n \to pe^- \bar{\nu}$ we have

$$\Delta u - \Delta d = \left(\frac{g_A}{g_V}\right) = 1.2573 \pm 0.003.$$

and from hyperon decay, $\Sigma^- \to ne^- \bar{\nu}$, assuming $SU(3)$ flavor symmetry, we have another equation,

$$\Delta s - \Delta d = D - F = 0.340 \pm 0.017.$$

Using Γ_1^p together with the two other measurements, (g_A/g_V, and F/D) along with the appropriate QCD corrections we can solve for $\Delta u, \Delta d, \text{and} \Delta s$. For example including QCD corrections, we find

$$\sum = \left(\frac{9}{c_0}\right)\left[\Gamma_1^p - \left(\frac{3F+D)}{18}\right)c_1\right]$$

Similar expressions hold for Γ_1^n and Γ_1^d. If the Bjorken Sum Rule is correct, we should get the same $\sum$ from all experiments. Individual contributions can be found:

$$\Delta u = \frac{\Sigma}{3} + 0.725 \qquad \Delta d = \frac{\Sigma}{3} - 0.532 \qquad \Delta s = \frac{\Sigma}{3} - 0.193.$$

Experiment E143

Unlike the experiments at CERN [1, 2, 3] which use low intensity muon beams, E143 had the benefit of a very large data rate such that the statistical errors were 2% for the proton and 5% for the deuteron in a three month run. Statistical precison is only useful if a major effort is made to push down the systematic errors to the point where they approach the statistical ones. This battle was fought on many fronts. The spectrometer acceptances and efficiencies were measured and two independent measurements of the beam polarization [15] were made throughout the experiment. These two measurements agreed with each other within 2%. Pion contamination of the data sample was measured as were any pion asymmetries. False asymmetries were eliminated by random flipping the beam helicity on a pulse by pulse basis and by reversing the target polarization and the target magnetic field direction. The beam profile and location was measured on every pulse. Both the field and polarization were reversed often so that any false asymmetries due to one or the other would be recognized. We show in Fig. 1 that the asymmetry we measured was independent of the four different combinations of field or polarization direction.

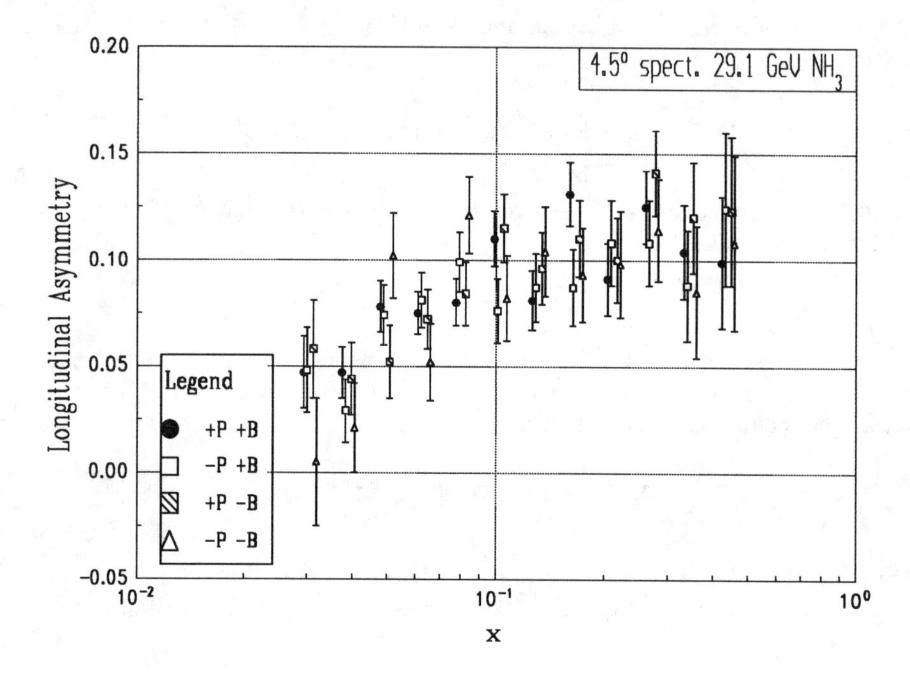

Figure 1. $A_{\parallel}$ in the 4.5° spectrometer at 29.1 GeV from NH_3 for four different combinations of target field (B) and polarization (P) directions.

Experimental Details

I will dedicate only a brief section to the experimental details. The reader should refer to the contributions to these proceedings by Maruyama [16] on the polarized electron source, by Band [15] on the electron polarimeter and by Averett [17] on the target. A summary of the experimental parameters of SLAC experiment E143 is given in the accompanying table.

The longitudinally polarized electron beam was produced by photoemission

The E143 Experiment	
Electron Beam	Energy: 9.7 - 29.1 GeV Electrons from a strained GaAs photocathode illuminated by a flash-lamp pumped TI-sapphire laser Polarization: $\pm 0.85 \pm 0.03$ Polarization randomly reversed from pulse-to-pulse
Polarized target	Proton, deuteron targets Composition: Ammonia granules $^{15}NH_3$ and $^{15}ND_3$ Polarized via DNP at 5T and 1 K Longitudinal and transverse polarization Polarization: Proton 0.55 – 0.80 Deuteron: 0.20 – 0.40 Polarization reversal every 8 – 12 hours
Kinematic Coverage	$0.029 < x < 0.8; Q^2 > 1.0 (GeV/c)^2$ $< Q^2 >= 3.0 (GeV/c)^2$

Table 1: Experimental Parameters of E143 at SLAC

from a strained-lattice GaAs crystal illuminated by a flash-lamp-pumped Ti-sapphire laser [18]. Beam pulses were typically 2 μsec long, contained 2–4$\times 10^9$ electrons, and were delivered at a rate of 120 Hz. The helicity was selected randomly on a pulse-to-pulse basis.

The beam current was measured for each beam pulse by two independent toroid systems with an uncertainty of $< 1\%$. A steering feedback system kept the average angle and position of the beam at the polarized target essentially constant.

The polarized target assembly contained permeable target cells filled with granules of $^{15}NH_3$ and $^{15}ND_3$ and immersed in a vessel filled with liquid He, maintained at 1 K using a high-power evaporation refrigerator. A superconducting Helmholtz coil provided a uniform field of 4.8 T. The ammonia granules were pre-irradiated [19] with 30 to 350 MeV electron beams to create a dilute assembly of paramagnetic atoms. During the experiment, they were exposed to 138 GHz microwaves to drive the hyperfine transition which aligns the nucleon spins. This technique of dynamic nuclear polarization produced proton polarizations of 65 to 80% in 10 to 20 minutes. Maximum deuteron polarizations of 35–40% were achieved after 40 minutes. The polarization then slowly decreased due to radiation damage: after eight to twelve hours of exposure to the incident electron beam the polarization had dropped to 50 to 55% (NH_3) and 20 to 25% (ND_3.) Most of the radiation damage was repaired by annealing the target at about 80 K. The electron beam was rastered over the 4.9 cm^2 front surface of the target to uniformly distribute beam heating and radiation damage. The target polarization direction was usually reversed after each anneal by adjusting the microwave frequency. Also, the direction of the magnetic field was reversed several times during the experiment (see Fig. 1). The target polarization P_t was measured using a series LCR resonant circuit and Q-meter detector [20]. The inductance was supplied by an NMR coil embedded in the ammonia granules, calibrated by measuring the thermal-equilibrium (TE) signal near 1.6 K with beam and microwaves off.

Scattered electrons with energy E' between 6 and 25 GeV were detected in two independent magnetic spectrometers [21] (first used in experiment E142 [7]) positioned at angles of 4.5° and 7° with respect to the incident beam. Electrons were distinguished from a background of pions in each spectrometer using two threshold gas Čerenkov counters and a 24-radiation-length shower-counter array composed of 200 lead-glass blocks. Seven planes of plastic scintillator hodoscopes were used to measure particle momenta and scattering angles. Fig. 2 gives an example of the electron identification using tracking and the shower counter energy.

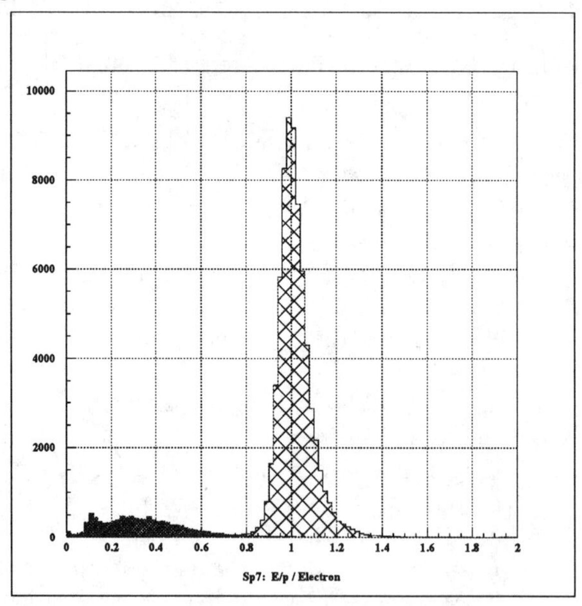

Figure 2: An example of the electron identification in the 7 degree spectrometer. Energy in the shower counter array is divided by the reconstructed momentum p. Electrons are centered around 1 and pions between 0 and 0.8.

Physics Asymmetries

The experimental asymmetries are constructed by tagging the electrons per incident beam charge in each x and Q^2 bin by the direction of the beam helicity such that the quantities N_L and N_R were the electron scattering rates for left (L) and right (R) beam helicity after being corrected for pair symmetric contributions and electronics deadtime. The longitudinal ($A_\parallel$) and transverse ($A_\perp$) experimental asymmetry could be constucted as

$$A_\parallel(\text{or})A_\perp = \left(\frac{N_L - N_R}{N_L + N_R}\right)\frac{C_N}{f P_b P_t} + A_{RC}.$$

C_N is a correction for the polarized nitrogen in the target. Nitrogen is polarized along with the proton or deuteron in the molecule. The relation between the nitrogen polarization and the proton and deuteron polarization was measured after the experiment [17]. ^{15}N has only an unpaired proton whose polarization can be related to the ^{15}N polarization through Clebsch–Gordon coefficients. This factor was approximately 2% for NH_3. For ND_3 it included a correction for hydrogen contamination in the target.

The dilution factor f is an x dependent quantity which accounts for the fact that much of the target is material that is unpolarized. The makeup of the target (besides polarizable protons and deuterons) included ^{15}N, ^{4}He, Al, Cu, and Ti in varying amounts. The dilution factor itself is calculated from the known nucleon structure functions while accounting for the EMC effect. The dilution factor f varied with x between 0.13 and 0.17 for NH_3 and between 0.22 and 0.25 for ND_3.

The beam polarization, P_b, was measured throughout the experiment by measuring the Möller asymmetry from thin ferromagnetic foils (49% Fe, 49% Co, 2% Va) magnetized by a Helmholtz coil. Two independent measurements were made. One was single arm Möller similar to that used in E142 [7]. The second was a double–arm Möller which detected both the incident and the scattered electron in coincidence. P_b ranged from 0.82 to 0.86.

The target polarization P_t was determined from continous NMR measurements from a coil inbeded in the target material. Average target polarization for the runs on the NH_3 target varied from 0.55 to 0.80; for ND_3, from 0.20 to 0.40.

The internal radiative corrections for both $A_\parallel$ and $A_\perp$ were evaluated using the formulae of Kukhto and Shumeiko [22]. The cross section components of the asymmetry were "externally radiated" according to Tsai [23] to form the "fully radiated" asymmetry corrections A_{RC}. The corrections varied slowly with x and changed $A_\parallel$ by typically $< 2\%$.

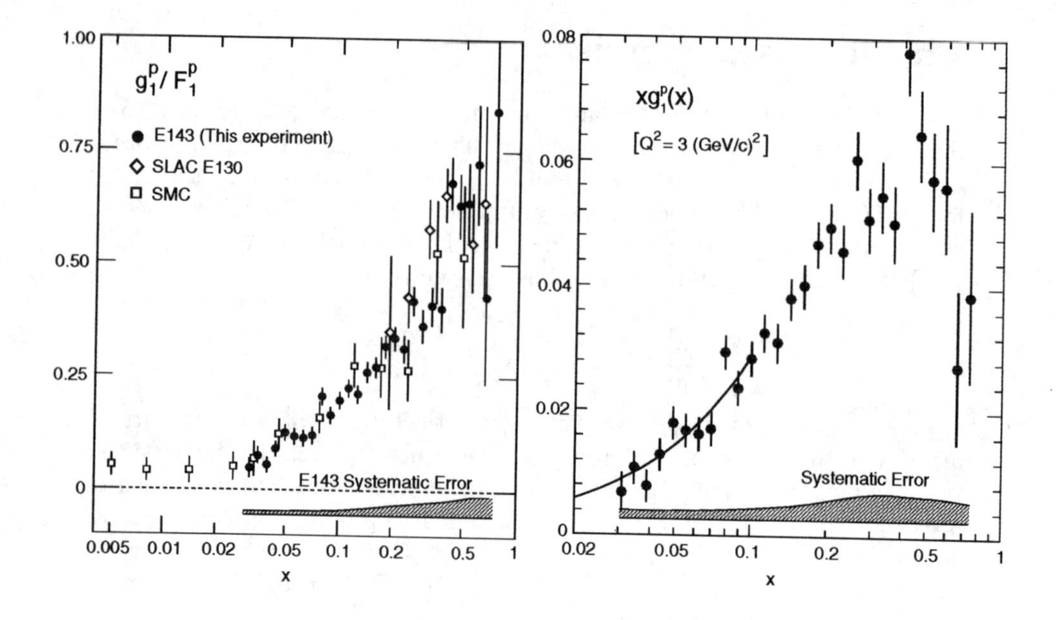

Figure 3: g_1^p/F_1^p world Proton data (left) and $xg_1^p(x)$ at $Q^2 = 3.0(\text{GeV/c})^2$ (right).

Data and Spin Fractions

The world data set for g_1^p/F_1^p is shown against $\log x$ in Fig. 3 (left) with the E143 shown in closed symbols with statistical errors. The inset shows the systematic errors. The E143 data [24] are shown as xg_1^p in on the right hand side of Fiq. 3.

To calculate the integral over all x we wish to obtain the value of the $g_1(x)dx$ at $Q^2 = 3$ GeV^2 which presents two alternative methods of evaluating g_1. We can assume g_1/F_1 independent of Q^2, and multiply by F_1 from models at $Q^2 = 3$ (GeV/c)^2 to get g_1. Or we can assume A_1 *and* A_2 independent of Q^2 and use $g_1 = F_1 \frac{A_1+\gamma A_2}{1+\gamma^2}$. To get F_1 we use the SLAC global fit to R [25] and either the NMC [26] or the SLAC [27] global fits to F_2. We have chosen the first alternative, g_1/F_1 independent of Q^2, have taken g_1 to be constant at low x, have used the NMC fit for F_2 and SLAC's fit for R, all at $Q^2 = 3(\text{GeV/c})^2$. Similar analysis of the preliminary deuteron data shown in Fig. 4 has been performed and the Table 2 gives the contributions from the measured region as well as the unmeasured regions at low and high x.

The result for the proton is well below Ellis-Jaffe prediction of $\Gamma_1^p = 0.160 \pm 0.006$ but agrees with SMC where $\Gamma_1^p = 0.131 \pm 0.015$ evaluated at $Q^2 = 10$

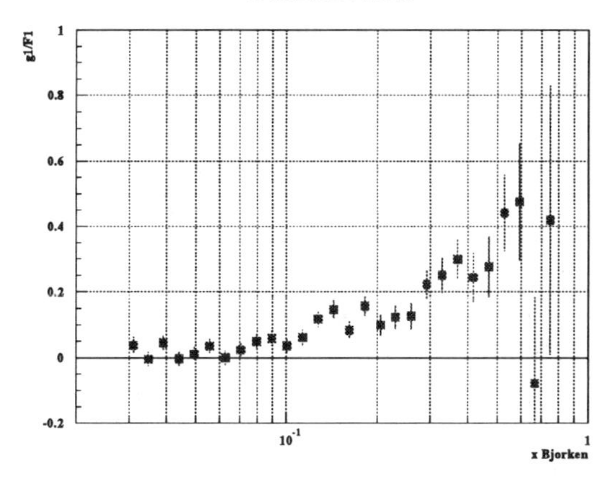

Figure 4: Preliminary deuteron data g_1^D from this experiment (solid squares) and data from Ref. [3].

x region	$\int(g_1^p(x)dx$	$\int(g_1^d(x)dx$
$0.029 < x < 0.8$ (measured)	$0.120 \pm 0.004 \pm 0.09$	$0.041 \pm 0.004 \pm 0.003$
$0.8 < x < 1$ (extrapolation)	0.001 ± 0.001	0.000 ± 0.001
$0 < x < 0.029$ (extrapolation)	0.008 ± 0.005	0.002 ± 0.001
TOTAL	$0.129 \pm 0.004 \pm 0.010$	$0.043 \pm 0.004 \pm 0.004$

Table 2: Contributions to the integral $\Gamma_1^{p,d}$.

(GeV/c)^2. Using $\alpha_s = 0.35 \pm 0.05$, $F/D = 0.575 \pm 0.016$, we find $\sum^p = 0.29 \pm 0.10$ and $\Delta s = -0.10 \pm 0.04$.

The deuteron integral is related to the proton and neutron integrals with a correction for the D–state of the deuteron wavefunction where

$$\Gamma_1^d = \frac{1}{2}(\Gamma_1^p + \Gamma_1^n)(1 - \frac{3}{2}\omega_d)$$

For $Q^2 = 3$ (GeV/c)^2 we find $\Gamma_1^D = 0.043 \pm 0.004 \pm 0.004$ with the contributions shown in Table 2. Again this is well below Ellis-Jaffe prediction of $\Gamma_1^d = 0.068 \pm 0.004$. However it agrees with the results from SMC $\Gamma_1^d = 0.023 \pm 0.025$ at $Q^2 = 5$ (GeV/c)^2. We find

$$\sum = 0.35 \pm 0.065 \text{ and } \Delta s = -0.08 \pm 0.025.$$

Preliminary results for Γ_1^n are available. We find that $\Gamma_1^n = 2\Gamma_1^d/(1-\frac{3}{2}\omega_d) - \Gamma_1^p = -0.035 \pm 0.009 \pm 0.011$ where $\omega_D = 0.058$. This is well below Ellis-Jaffe prediction of $\Gamma_1^n = -0.011 \pm 0.005$. However it agrees with E142 which found $\Gamma_1^n = -0.022 \pm 0.011$ at $Q^2 = 2$ (GeV/c)^2. Using $\alpha_s = 0.35 \pm 0.05$, $F/D = 0.575 \pm 0.016$, we find (still preliminary)

$$\sum = 0.38 \pm 0.15 \quad \text{and} \quad \Delta s = -0.07 \pm 0.05.$$

We can summarize these results in the table below.

	Proton	Deuteron	Neutron
g_1 integral	$0.129 \pm .004 \pm .010$	$.043 \pm .004 \pm .008$	$-.035 \pm .0096 \pm .011$
Δu	$.821 \pm .034$	$.842 \pm .022$	$.853 \pm .050$
Δd	$-.437 \pm .035$	$-.416 \pm .022$	$-.404 \pm .050$
Δs	$-.098 \pm .037$	$-.077 \pm .025$	$-.065 \pm .052$
Σ	$.287 \pm .104$	$.349 \pm .065$	$.384 \pm .151$

Table 3: Summary of the spin fractions carried by the different quark flavors for the proton, deuteron and the neutron.

Conclusions and Future

From these new data we can test the Bjorken sum rule which predicts at $Q^2 = 3(\text{GeV/c})^2$ $\Gamma_1^p - \Gamma_1^n = 0.171 \pm 0.008$. Using the E143 proton and neutron data and the extrapolation based on the Q^2 independence of g_1/F_1 we obtain $\Gamma^{p-n} = 0.160 \pm 0.011 \pm 0.017$. Using the extrapolation based on the Q^2 independence of A_1 and A_2 we obtain $\Gamma^{p-n} = 0.154 \pm 0.011 \pm 0.017$. Both methods are compatible with the theory. We note that the QCD corrections [28] are very important in this agreement. Finally combining the E142 Γ_1^n with our Γ_1^p we obtain $\Gamma_1^p - \Gamma_1^n = 0.149 \pm 0.014$, consistent with the others within the errors.

These new data at 29.1 GeV indicate that g_1/F_1 appears independent of Q^2 at all x, that the $g_1^d > 0$ at low x (for $x > 0.03$), and finally the Ellis-Jaffe sum rules are violated while the Bjorken sum rule is validated. From these and other experiments one can argue that (with the critical assumption of $SU(3)$ flavor symmetry) the strange sea is negatively polarized ($\Delta s \sim -0.1$) and only about a third of the total spin of the nucleon is carried by the quarks ($\Sigma \sim 0.3$).

Over the next several months the remaining data from E143 will be analyzed. This includes data at 9.7 and 16.2 GeV which should provide insight into the Q^2 dependence of the structure functions. Data on g_2 at 29.1 GeV will be analyzed. In the next several years there will be two more experiments at SLAC, utilizing 50 GeV electrons. One will use a polarized ^{3}He target as a polarized neutron target and the other the solid polarized targets of NH_3 and ND_3. The SMC experiment continues to take data [4] and the Hermes collaboration which will have polarized proton, deuteron and ^{3}He targets will begin taking data this year [29].

References

[1] EMC, J. Ashman *et al.,* Phys. Lett. **B206**, 364 (1988); Nucl. Phys. **B328**, 1 (1989).

[2] SMC, D. Adams *et al.,* Phys. Lett. **B329**, 399 (1994).

[3] SMC, B. Adeva *et al.,* Phys. Lett. **B302**, 533 (1993).

[4] R. Windmolders, contributions to these Proceedings, SPIN94.

[5] SLAC E80, M. J. Alguard *et al.,* Phys. Rev. Lett. **37**, 1261 (1976); **41**, 70 (1978).

[6] SLAC E130, G. Baum *et al.,* Phys. Rev. Lett. **51**, 1135 (1983).

[7] SLAC E142, P. L. Anthony *et al.,* Phys. Rev. Lett. **71**, 959 (1993).

[8] T. Pussieux and R. Windmolders, SMC Note 93/16, May 1993.

[9] P.E. Bosted, E143 Technical Note , 1993.

[10] J. D. Bjorken, Phys. Rev. **148**, 1467 (1966); Phys. Rev. D **1**, 1376 (1970).

[11] Particle Data Group, Phys. Rev. **D**45 S1 (1992).

[12] S. A. Larin, Phys. Lett. **B334**, 192 (1994).

[13] S. A. Larin and J. A. M. Vermaseren, Phys. Lett. **B259**, 345 (1991) and references therein.

[14] J. Ellis and R. Jaffe, Phys. Rev. D **9**, 1444 (1974); D **10**, 1669 (1974).

[15] H. Band, Contribution to these Proceedings, SPIN94, 1994.

[16] T. Maruyama, Contribution to these Proceedings, SPIN94, 1994.

[17] T. Averett, Contribution to these Proceedings, SPIN94, 1994.

[18] T. Maruyama, E. L. Garwin, R. Prepost, G. H. Zapalac, Phys. Rev. B **46**, 4261 (1992); R. Alley *et al.*, Report No. SLAC–PUB–6489 (1994).

[19] D. G. Crabb *et al.*, Phys. Rev. Lett. **64**, 2627 (1990); W. Meyer *et al.*, Nucl. Instrum. Meth. **215**, 65 (1983).

[20] G. R. Court *et al.*, Nucl. Instrum. Meth. **A324**, 433 (1993).

[21] G. G. Petratos *et al.*, Report No. SLAC–PUB–5678 (1991).

[22] T. V. Kukhto and N. M. Shumeiko, Nucl. Phys. **B219**, 412 (1983); I. V. Akusevich and N. M. Shumeiko, J. Phys. G **20**, 513 (1994).

[23] Y. S. Tsai, Report No. SLAC–PUB–848, (1971); Y. S. Tsai, Rev. Mod. Phys. **46**, 815 (1974).

[24] Tables corresponding to Fig. 1 and Fig. 2 are available in SLAC E143, K. Abe *et al.*, Report No. SLAC–PUB–6508 (1994).

[25] L. W. Whitlow *et al.*, Phys. Lett. **B250**, 193 (1990).

[26] NMC, P. Amaudruz *et al.*, Phys. Lett. **B295**, 159 (1992).

[27] L. W. Whitlow *et al.*, Phys. Lett. **B282**, 475 (1992).

[28] J. Ellis and M. Karliner, preprint CERN TH 7324/94 and TAUP 2178-94.

[29] H. Jackson, Contribution to these Proceedings, SPIN94, 1994.

Prospects for HERMES-Spin Structure Studies at HERA

H. E. Jackson

Physics Division, Argonne National Laboratory, Argonne, IL 60439-4843

Abstract. HERMES, HERA Measurement of Spin, is a second generation experiment to study the spin structure of the nucleon by using polarized internal gas targets in the HERA 35-GeV electron storage ring. Scattered electrons and coincident hadrons will be detected in an open geometry spectrometer which will include particle identification. Measurements are planned for each of the inclusive structure functions, g_1,(x), g_2(x), b_1(x) and Δ(x), as well as the study of semi-inclusive pion and kaon asymmetries. Targets of hydrogen, deuterium and ^{3}He will be studied. The accuracy of data for the inclusive structure functions will equal or exceed that of current experiments. The semi-inclusive asymmetries will provide a unique and sensitive probe of the flavor dependence of quark helicity distributions and properties of the quark sea. Monte Carlo simulations of HERMES data for experiment asymmetries and polarized structure functions are discussed.

INTRODUCTION

Recent studies of the spin structure functions of the nucleon have yielded very interesting, if unexpected, results quite different from naive expectations. The data from these experiments (1,2) have been interpreted as showing that the quarks in the nucleon carry only ~1/3 of the spin of the nucleon, i.e. about 50% of the value which results from a careful full relativistic treatment (3) of the nucleon spin structure. These data also violate at the 3-σ level the Ellis-Jaffe sum rule (4) for $\int_0^1 dx g_1^{p(n)}(x)$, and suggest that the strange sea in the nucleon is polarized. A large body of theoretical speculation on the spin structure of the nucleon, based on the data from these experiments, has developed. It is evident that detailed and comprehensive experiments will be necessary to provide a precise description of the spin structure of the nucleon. HERMES, a study of spin structure at HERA, is such an experiment. The participants and collaborating institutions are listed in Appendix A. Spin dependent deep inelastic scattering from the proton and the neutron will be measured by using the longitudinally polarized electron beam of the HERA storage ring at beam energies in excess of 27.5 GeV. Polarized internal gas targets of hydrogen, deuterium, and ^{3}He will scatter the circulating electron beam which is polarized to a level greater than 50% by the Sokolov-Ternov effect (5). Scattered electrons and associated hadrons will be detected in a large acceptance open geometry spectrometer. The x dependence of the proton and neutron spin-dependent structure functions will be measured to high precision. The determination of their integrals over x will test the Bjorken Sum rule to high precision. In addition, in HERMES, measurements will be possible of leading hadrons from spin-dependent deep inelastic scattering. This will provide important additional information on the contribution of the valence and sea quarks to the spin asymmetries. HERMES will provide a new experimental frontier in deep inelastic scattering and an unprecedented capability to study spin-dependent effects in electron scattering from the nucleon and the few-nucleon systems.

PHYSICS OF SPIN ASYMMETRIES

The basic features of the spin structure of the nucleon are characterized by the spin structure functions $g_1(x)$ and $g_2(x)$. Experimentally, these structure functions describe the difference in the deep-inelastic electron scattering cross sections for opposite target polarizations according to the equation

$$\frac{d^3(\sigma(\alpha)-\sigma(\alpha+\pi))}{dx\,dy\,d\sigma} = \frac{e^4}{4\pi^2 Q^2}\left[\cos\alpha\left(\left[1-\frac{y}{2}-\frac{y^2}{4}\gamma^2\right]g_1(x,Q^2)-\frac{y}{2}\gamma^2 g_2(x,Q^2)\right)\right.$$

$$\left.-\sin\alpha\cos\phi\sqrt{\gamma^2\left(1-y-\frac{y^2}{4}\gamma^2\right)}\left(\frac{y}{2}g_1(x,Q^2)+g_2(x,Q^2)\right)\right] \qquad (1)$$

where, as indicated in Fig. 1, α is the angle between the beam momentum and the target polarization, ϕ is the angle between the plane defined by the beam direction and target polarization, and the scattering plane defined by the directions of the incident and scattered lepton. Q^2 is the negative square of the invariant mass of the virtual photon, $x = Q^2/2M\nu$ and $y = \nu/E$ are the Bjorken scaling variables, E and E' are the energies of the incident and the scattered lepton, and M is the nucleon mass. The quantity γ is defined by $\gamma = \sqrt{Q^2/\nu}$. The structure functions can be extracted from the polarization asymmetry measured for parallel and anti-parallel target-beam spins, $A_{||}(x)$ and the transverse asymmetry, $A_\perp(x)$, which corresponds to target polarization perpendicular to the beam. These measured asymmetries are related to the virtual photon asymmetries, $A_1 = (\sigma_{1/2}-\sigma_{3/2})/(\sigma_{1/2}+\sigma_{3/2})$ and $A_2 = \sigma_{TL}/\sigma_T$, by the relations

$$A_{||} = D\cdot(A_1+\eta\cdot A_2) \qquad (2)$$

$$A_\perp = d\cdot(A_2-\xi\cdot A_1). \qquad (3)$$

where the kinematic factors are defined by

$$D = \frac{y(2-y)}{y^2+2(1-y)(1+R)} \qquad (4)$$

$$\eta = \frac{2\gamma(1-y)}{(2-y)} \qquad (5)$$

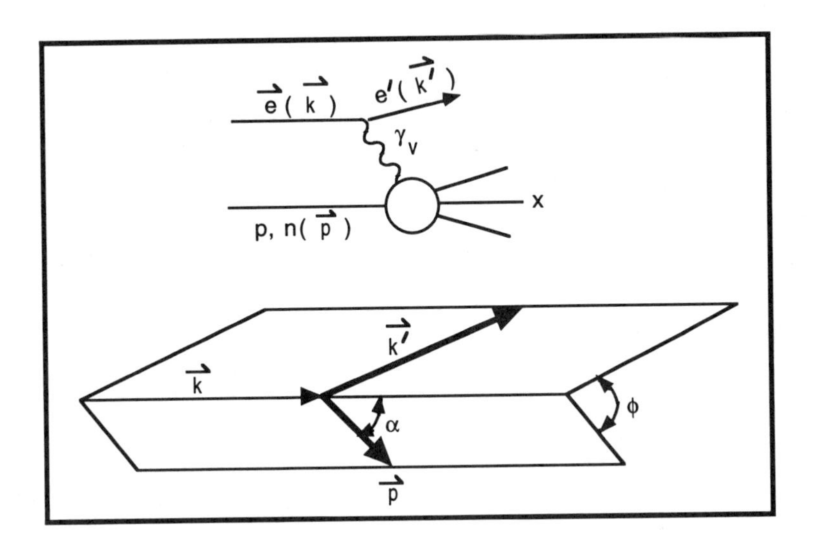

FIGURE 1. Kinematics for polarized deep-inelastic electron scattering. $\vec{p}$ is the target polarization.

$$d = D\sqrt{\frac{2\varepsilon}{1+\varepsilon}} \tag{6}$$

$$\xi = \eta\frac{1+\varepsilon}{2\varepsilon} \tag{7}$$

with $\varepsilon = (1-y)/(1-y+y^2/2)$ being the degree of transverse polarization of the virtual photon. The asymmetries A_1 and A_2 can be expressed in terms of g_1 and g_2 as

$$A_1 = \frac{g_1 - \gamma^2 g_2}{F_1};$$

$$A_2 = \frac{\gamma(g_1 + g_2)}{F_1} \tag{8}$$

Since γ, η, and R are usually small, it is customary to neglect the contribution of A_2 to the $A_{||}$ and A_1 to $A_\perp$ with the result that to good approximation

$$A_{||} \approx \frac{g_1}{F_1}\cdot D \qquad A_\perp \approx \frac{\gamma(g_1 + g_2)}{F_1}\cdot d \tag{9}$$

The experimentally measured electron asymmetries include the effects of beam and target polarizations, P_b and P_t, as well as any target dilution, f, i.e. $A_{exp} = P_b P_t f A_e$. In the Quark-Parton Model $g_1(x) = 1/2\sum e_i^2(q_i^{\uparrow}(x)-q_i^{\downarrow}(x))$, so that the net helicity distribution summed over quark flavor can be estimated from $A_{||}$ with the aid of the known unpolarized structure function, $F_1(x)$. The integral of $g_1(x)$ is of great importance (3) since it provides an independent combination of the net helicities, Δu, Δd, and Δs in the proton. The value of this integral, when combined with data on neutron and hyperon β-decay, can be used to infer values of Δu, Δd, and Δs. This is the treatment which has been used in previous experiments to conclude that the spin carried by the quarks in the nucleon is anomolously small. In HERMES it will be possible to measure A_2 with sufficient precision to avoid the approximation made in Eq. (9).

Because of the ability to identify hadrons, HERMES will be able to provide unambiguous identification of pions and later kaons and their associated spin asymmetries. These data can provide important information on the flavor dependence of quark helicity distributions in the nucleon, and be useful in separating the contributions of the valence and sea quarks to the total nucleon polarization. For polarized pion production, the Quark-Parton model gives (16) up to a constant factor

$$N_{\uparrow\downarrow}^{\pi^+} \sim \frac{4}{9}u_+(x)D_u^{\pi^+}(z)+\frac{4}{9}\bar{u}_+(x)D_{\bar{u}}^{\pi^+}(z)+\frac{1}{9}d_+(x)D_{\bar{d}}^{\pi^+}(z)+\frac{1}{9}d_+(x)D_{\bar{d}}^{\pi^+}(z)$$

$$+\frac{1}{9}S_+(x)D_s^{\pi^+}(z)+\frac{1}{9}\bar{S}_+(x)D_{\bar{s}}^{\pi^+}\ . \qquad (10)$$

where $q_+(q_-)$ is the distribution function of a quark q with helicity parallel (antiparallel) to the proton helicity and $D_q^h(z)$ is the fragmentation function of a quark into a hadron h with energy $E^h = z{\cdot}\nu$, with similar form for other pion charges and helicities. Factorization and helicity independence of the fragmentation function is assumed. An example of a useful hadron asymmetry is that of the net pion charge

$$A_R^{\pi}(x)=\frac{N_{\uparrow}^{\pi^+-\pi^-}(x)-N_{\downarrow}^{\pi^+-\pi^-}(x)}{N_{\downarrow}^{\pi^+-\pi^-}(x)+N_{\downarrow}^{\pi^+-\pi^-}(x)}\ . \qquad (11)$$

This quantity should be independent of sea quark effects, with the result that measurements of this quantity for the proton and the deuteron can be used to deduce $\Delta u_v(x)$ and $\Delta d_v(x)$ in the proton.

EXPERIMENTAL DETAILS

The internal polarized gas target for HERMES is shown schematically in Fig. 2. It consists of an open-ended thin walled storage cell through which the circulating electron beam of the HERA accelerator passes. The cell is filled with polarized

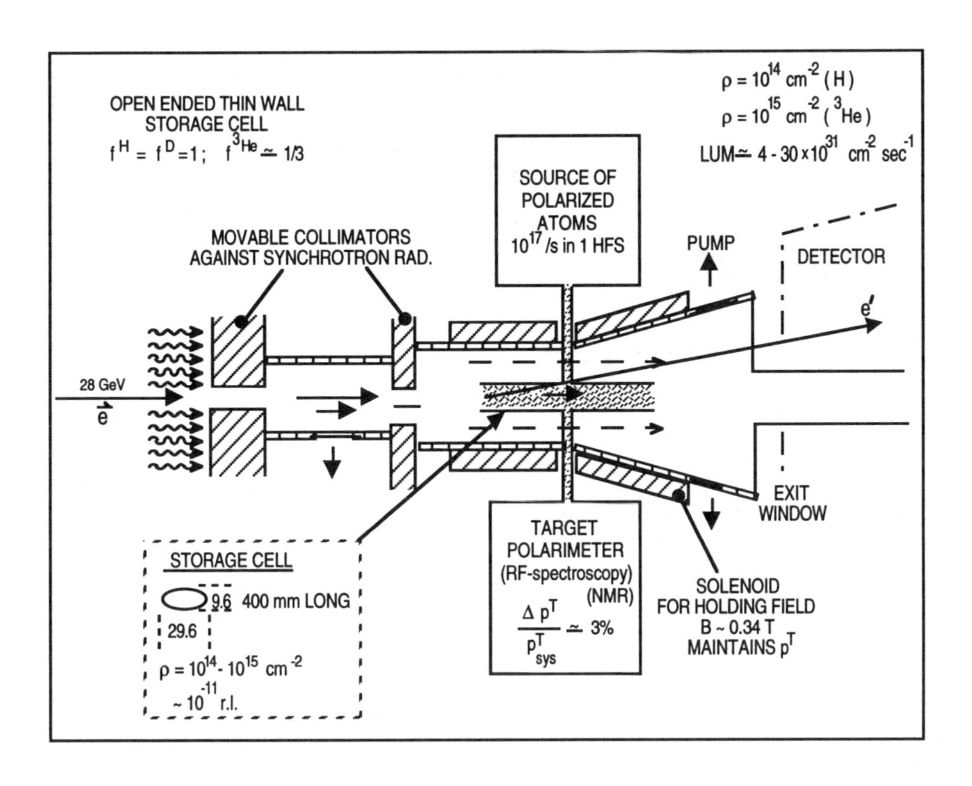

FIGURE 2. Schematic view of HERMES internal polarized gas target.

protons, deuterons, or ^{3}He from an atomic beam source as shown or from an optically pumped ^{3}He source. A solenoidal holding field provides a quantization axis for the target polarization. The target densities will be about *10^{14} atoms/cm^2* for hydrogen and deuterium and *10^{15} atoms/cm^2* for ^{3}He. Luminosities will be in the range of *4-30 10^{31} nucleons cm^{-2}sec^{-1}*. The gases are pure so that there is no target dilution. The extremely thin target density minimizes external radiative corrections.

The experiment will be located in the east Hall of the HERA accelerator with the interaction area reconfigured so that the proton beam bypasses the HERMES target. The beam is polarized transverse to the beam direction by the Sokolov-Ternov effect. In a perfect machine it will reach 92%. Spin rotators located at the entrance and exit of the hall (7) will precess the spin direction from vertical to longitudinal at the target position and following the target back to the vertical position. The polarization of the beam with the rotators in place has been measured for both positrons and electrons. The results of a typical measurement are shown in Fig. 3. Longitudinal polarizations of about 60% are stable and reproducible. The polarization time is about 20 minutes. As a result of the ease with which the direction of the target polarization can be changed, it will be possible to study

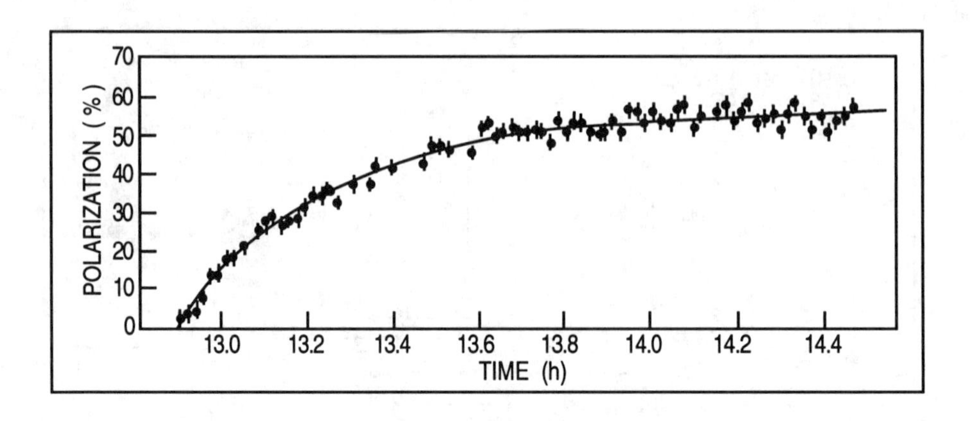

FIGURE 3. Rise of electron polarization at HERA with spin rotators in place. Polarization is longitudinal at the HERMES target.

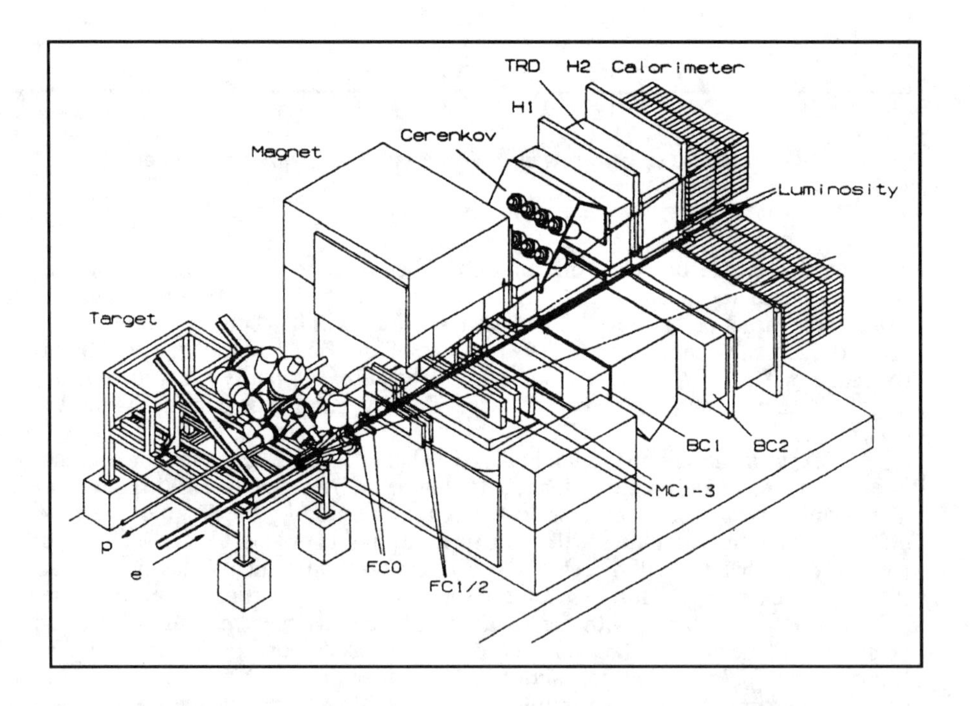

FIGURE 4. Schematic view of the HERMES magnetic spectrometer.

TABLE 1. Projected sum rule uncertainties.

Target	Polarization	Quantity to be measured
H	$\rightarrow$	$g_1^p(x)$
D	$\rightarrow$	$g_1^p(x)$, $g_1^n(x)$, $\int\left(g_1^p - g_1^n\right)dx$, $b_1(x)$
^{3}He	$\rightarrow$	$g_1^n(x)$, $\int\left(g_1^p - g_1^n\right)dx$
H	$\perp$	$g_2^p(x)$
D	$\perp$	$g_2^p(x)$, $\Delta(x)$
^{3}He	$\perp$	$g_2^n(x)$

parallel and perpendicular polarization asymmetries for all targets. In addition to measurements of g_1 and g_2, and the Bjorken sum rule, estimates will be made of the higher order structure functions b_1 (9) and Δ (10) which are relevant to the deuteron. A summary of the measurements planned is shown in Table 1.

The HERMES spectrometer, shown in Fig. 4, is an open geometry system. A magnet with a bending strength of 1.3 T-m will be used to measure particle momenta. The magnet is divided into 2 symmetric parts by a horizontal flux plate through which both the electron and proton beam pass. A fly's eye Pb glass calorimeter will provide a level one trigger. A system of segmented hodoscopes with road structure will be part of the trigger. A transition radiation detector will provide strong discrimination against pions. Tracking chambers before, in the magnetic field, and behind the magnet will be used to measure particle trajectories and momenta. For identification of hadrons in coincidence with the scattered lepton, a pair of threshold gas Cerenkov counters will be used. The radiator, tetrafluoromethane, will have Cerenkov light thresholds of 4.4 GeV for pions, 15.8 GeV for kaons, and 30.0 GeV for protons. Each unit will be divided into 20 cells viewed by individual photomultipliers. The kinematic region which will be covered by the spectrometer for a beam energy of 28 GeV is shown in Fig. 5. A minimum momentum transfer of 1 $(\mathrm{GeV/c})^2$ will allow interpretation of the data in terms of the quark-parton model. The invariant mass cut of $W > 2$ GeV will eliminate the resonance region. A cut on the energy loss at $y < 0.85$ will eliminate the region of large radiative corrections. The spectrometer angular acceptance will range from 40 mr to 220 mr.

PROJECTED PERFORMANCE

The precision of measurements of structure functions, statistical plus systematic errors in HERMES, should be at least as good or perhaps somewhat improved than any of other experiments in progress or planned. This point is illustrated by Fig. 6

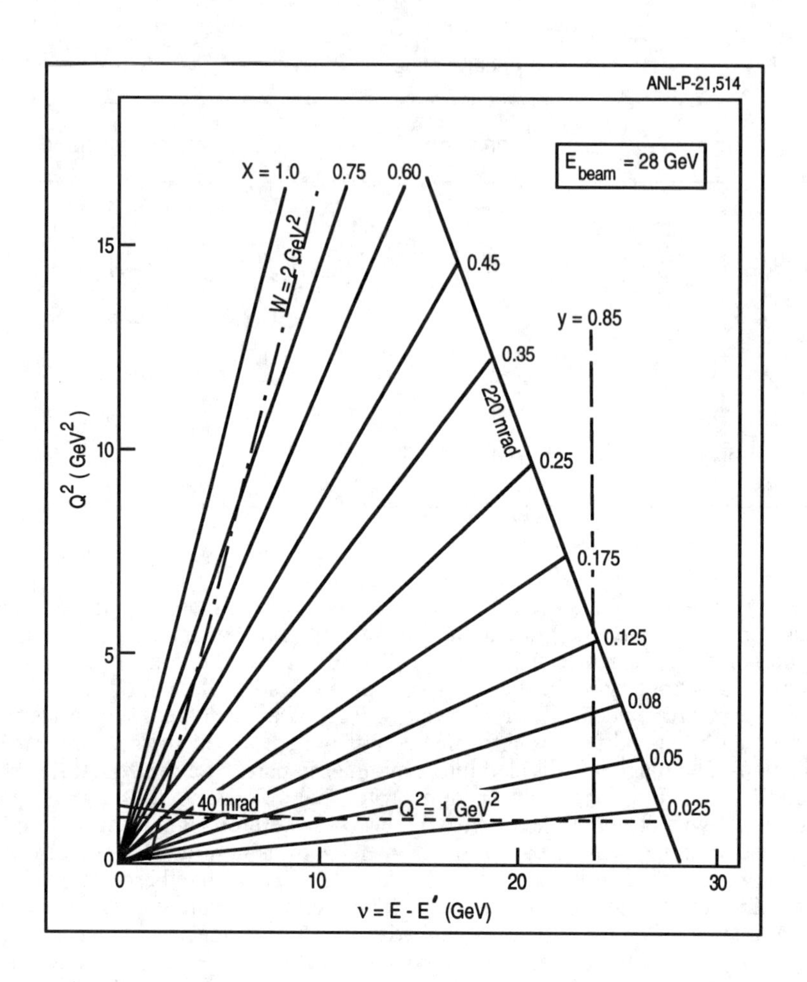

FIGURE 5. Kinematic plane in HERMES for an electron beam energy of 28 GeV.

where recent results for g_1^n are plotted together with projected data for a 600 hour run in HERMES at the expected luminosity. Projected uncertainties in the measured values of $\int g_1^p$ and $\int g_1^n$ as well as the Bjorken sum rule are presented in Table 2. HERMES should be competitive with the best experiments.

A knowledge of the structure function $g_2(x)$ is of interest theoretically as a measure of higher twist effects and because its value is required to make an unambiguous determination of $g_1(x)$ from the longitudinal asymmetry. It will be estimated in HERMES from measurements of $A_2(x)$ in a target transverse asymmetry (see Eq. (3)). In Fig. 7 the statistical uncertainties are shown for the extracted $g_2^p(x)$ together with curves for the leading twist contribution and a twist-3 contribution in a specific model. Also shown is the positivity limit for $A_2 = \sqrt{R}$. The data will constrain the x-dependence of $g_2(x)$ rather well. This will also

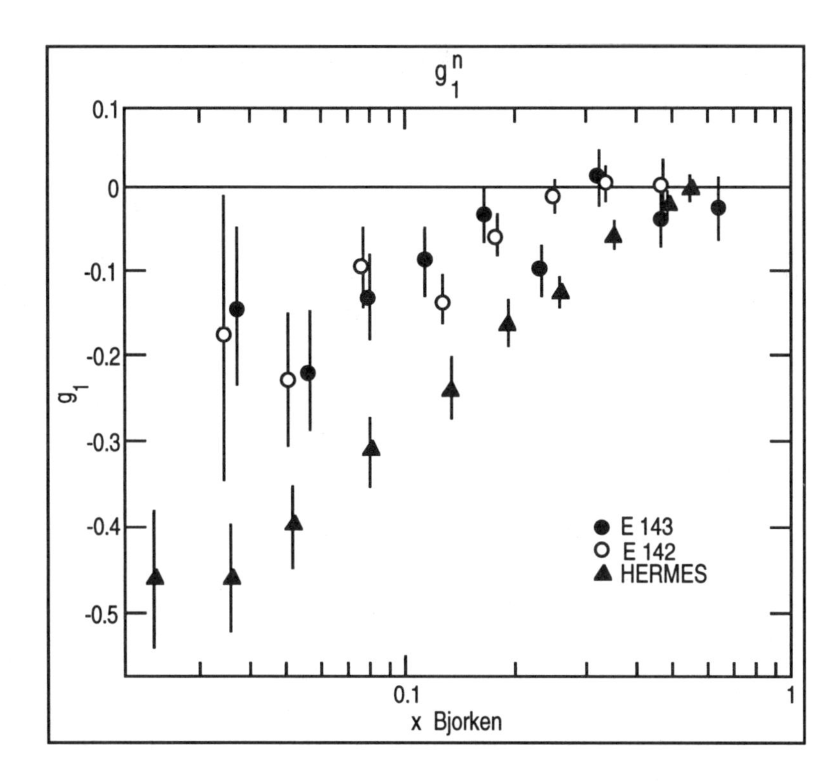

FIGURE 6. The neutron spin structure function $g_1^n(x)$ projected for a 300-hour run in HERMES. Also shown are data from recent measurements at SLAC (see Ref. (2)). The projected data are displaced from that of the SLAC experiments because of the model used in their calculation.

provide a test of the Wandzura-Wilsek relation (11) and possibly a separation of the twist-3 contribution arising from quark-gluon correlations.

In the case of the deuteron, which is a spin-1 target, there are two new structure functions, $b_1(x)$ and $\Delta(x)$ which describe the additional physics of the spin one system. $b_1(x)$ has a structure similar to a tensor polarization and can be measured in HERMES by scattering an unpolarized beam from a deuterium target arranged in each of the three possible substates $m_I = +1,0,-1$. $\Delta(x)$ is measured by scattering an unpolarized electron beam from a polarized deuterium target in the $m_I = 0$ state with the target polarization perpendicular the beam direction and measuring the ϕ distribution of the scattered electrons. It is sensitive to the gluon components in the deuteron.

TABLE 2. Projected Sum Rule Uncertainties. Values for the Bjorken Sum Rule $\int_0^1 \left[g_1^p(x) - g_1^n(x) \right] dx$ are tabulated in the last column.

	$\int g_1^p$	$\int g_1^n$	$\int g_1^p - g_1^n$
Measured Value	0.13	-0.03	0.15
E 142 (^3He)		±0.011 (36%)	±0.016 (10%)
E 143 (NH_3,NO_3)	±0.011 (8%)	±0.014 (46%)	±0.025 (17%)
SMC (C_4H_9OH, C_4D_9OD)	±0.016 (12%)	±0.043 (143%)	±0.051 (34%)
HERMES (H,D)	±0.007 (5%)	±0.009 (30%)	±0.015 (10%)
HERMES (H,^3He)		±0.007 (23%)	±0.010 (7%)

SEMI-INCLUSIVE ASYMMETRIES

With particle identification, HERMES will be able to provide unambiguous identification of pions, and later kaons, and measurements of their associated asymmetries. These data can provide important information on the flavor dependence of quark helicity distributions in the nucleon, and be useful in separating the contributions of the valence and sea quarks to the total quark polarization. Frankfurt and coworkers (12) have calculated the asymmetry in the net pion charge, Eq. (11), for the proton and the deuteron in the quark parton model. They find

$$A_p^{\pi}(x) = \frac{4\Delta u^v(x) - \Delta d^v(x)}{4u^v(x) - d^v(x)},$$

$$A_d^{\pi}(x) = \frac{\Delta u^v(x) + \Delta d^v(x)}{u^v(x) + d^v(x)}. \tag{12}$$

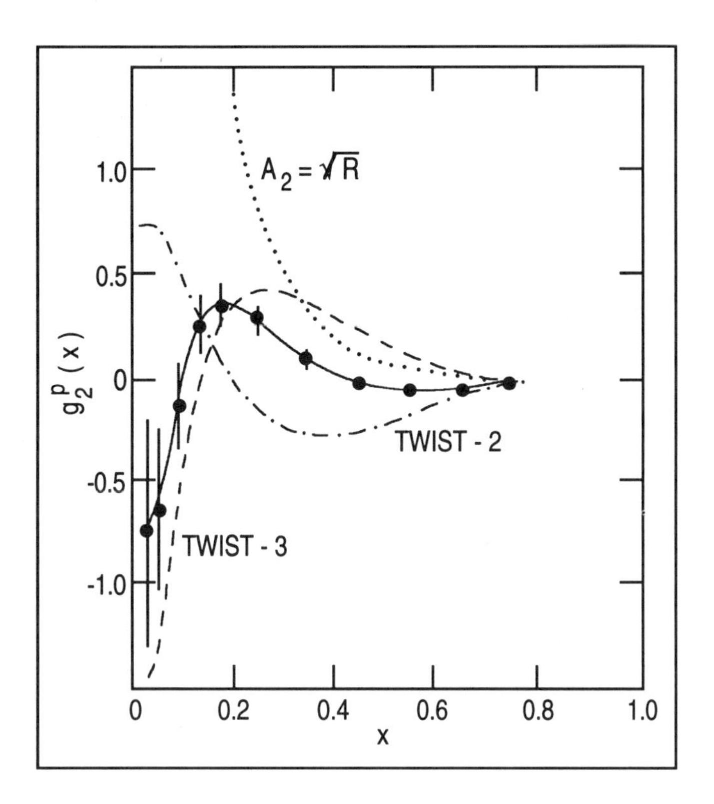

FIGURE 7. $g_2^p(x)$ as projected for an 800-hour run by HERMES.

From these two symmetries one can deduce both $\Delta u_v(x)$ and $\Delta d_v(x)$. They also note that one can use these results together with the data for the Bjorken sum rule to estimate the isospin symmetry of the polarization of the sea through the relationship

$$6\int_0^1 \left[g_1^p(x) - g_1^n(x)\right]dx = \int_0^1 \left[\Delta u^v(x) - \Delta d^v(x) + 2\left(\Delta\bar{u}(x) - \Delta\bar{d}(x)\right)\right]dx \qquad (13)$$

Monte Carlo simulations of these asymmetries have been carried out by Veltri *et al.* (13) Their results are presented in Fig. 8. The curve of Ross and Roberts (14) describes the result expected for a model with a polarized sea and glue. The dashed curve corresponds to a parton distribution due to Schäfer (15) in which the sea is unpolarized. The data corresponds to a standard HERMES run. In the region below $x = 0.3$ one should be able to distinguish between the two models. The integrals of $\Delta u_v(x)$ and $\Delta d_v(x)$ can be estimated from this data to about 15-20%.

After the first year of operation the HERMES threshold Cerenkov counters will be converted to Ring-imaging Cerenkov counters (RICH's). The current design is based on a C_4F_{10} gas radiator. The projected ring separation which has been

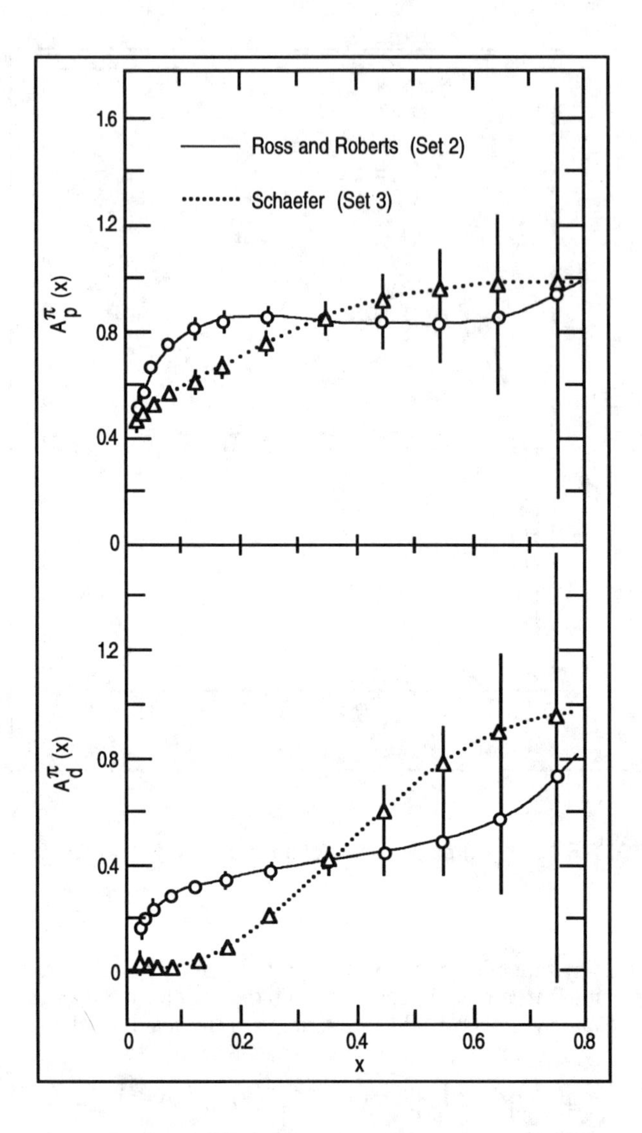

FIGURE 8. A_p^π and A_d^π for two sets of parton distributions used in the HERMES Monte Carlo simulation. The error bars indicate the projected statistical accuracy for a 800-hour run.

calculated for the converted system is shown in Fig. 9. The system will provide good separation of pions, kaons, and protons above 8 GeV up to the highest hadron energies of the experiment. K^- asymmetries will be of particular interest because of the "all sea $s\bar{u}$" character of the K^-. For a proton target, the asymmetry in K^- production

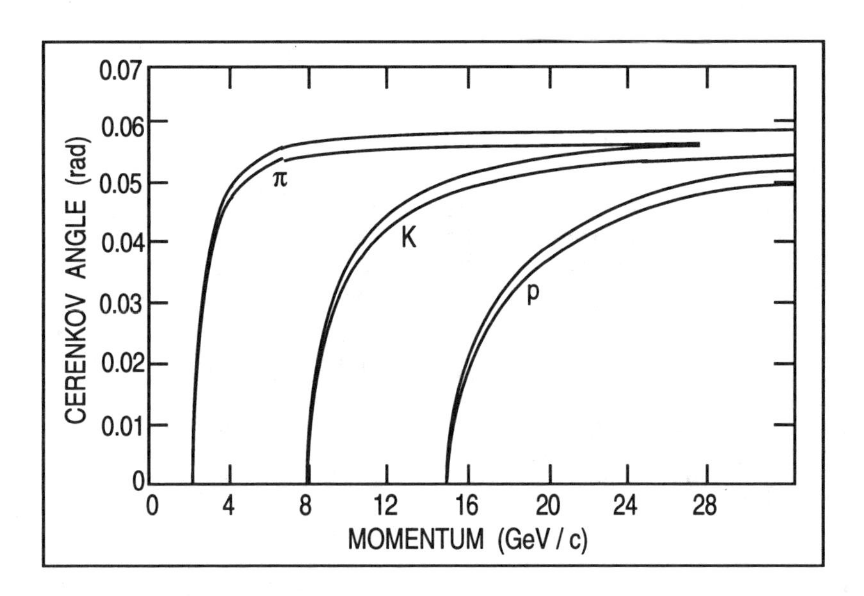

FIGURE 9. Anticipated 3σ π/K/p separation with a RICH counter using a 1m long C_4F_{10} radiator. The uncertainty in the Cerenkov angle is dominated by the granularity of the detector.

$$A_p^{K^-} = \frac{K^-_{\uparrow\downarrow} - K^-_{\downarrow\downarrow}}{K^-_{\uparrow\downarrow} + K^-_{\downarrow\downarrow}} \tag{14}$$

can be related (16) to the $q\bar{q}$ polarized sea distribution through the relation

$$A_p^{K^-} = \frac{4\Delta u^v + \Delta d^v + \left[(6+4\eta)\Delta\bar{q} + \left(\frac{\eta}{R}+1\right)\Delta s\right]}{4u^v + d^v + \left[(6+4\eta)\bar{q} + \left(\frac{\eta}{R}+1\right)s\right]} \tag{15}$$

where $\Delta\bar{q}$ indicates the polarized sea with $\Delta u_s = \Delta\bar{u}_s = \Delta d_s = \Delta\bar{d}_s$, Δs is the polarized strange quark distribution, u_v, d_v, $\bar{q}$, and s are the unpolarized counterparts. η is ratio of favored to unfavored fragmentation functions

$$D_{\bar{u}}^{K^-}(z) = \eta(z) D_u^{K^-}(z) \tag{16}$$

$$\eta(z) \approx \frac{1+z}{1-z} \tag{17}$$

In Eq. (15), R gives the suppression of $\overline{u}$ quark fragmentation due to the strange quark mass

$$D_{\overline{u}}^{K^-}(z) = R \cdot D_s^{K^-}(z); \qquad R \approx 0.3\,. \tag{18}$$

Monte Carlo simulations suggest that the statistical precision of a measurement of a K^- spin asymmetry in HERMES would be sufficient to detect a polarized sea quark distribution provided its polarization is strongly negative, in spite of a substantial dilution of the asymmetry by the valence quark contribution evident in Eq. (15).

SCHEDULE AND ACKNOWLEDGMENTS

The installation and testing of the spin rotators in the East Hall has been completed. A test experiment to explore backgrounds and beam-target interactions has been online during the 1994 running period. The spectrometer and target will be installed in the HERA East Hall during the winter shutdown of 1994/95 and data taking will begin in the spring of 1995.

This work is supported in part by the U.S. Department of Energy, Nuclear Physics Division, under contract W-31-109-ENG-38.

Appendix A. HERMES Collaboration

P. Green, G. Greeniaus, P. Kitching
University of Alberta, Canada
J.-O. Hansen, H. E. Jackson, C. E. Jones, N. C. R. Makins,
T. G. O'Neill, D. Potterveld
Argonne National Laboratory, Argonne, IL USA
E. J. Beise, B. W. Filippone, A. Lung, W. Korsch, R. D. McKeown, M. Pitt
W. K. Kellogg Laboratory, Caltech, Pasadena, CA USA
E. Kinney, R. Ristinen, J. Z. Williams
University of Colorado, Boulder, CO USA
W. Beckhusen, B. Grabowski, Y. Holler, K. Sinram, G. Wöbke, K. Zapfe
DESY, Hamburg, Germany
H. Böttcher, W.-D. Nowak, H. Roloff, A. Schwind
DESY-IfH, Zeuthen, Germany
M. Düren, F. Neunreither, K. Rith, J. Stenger, F. Stock, W. Wander
University of Erlangen-Nürnberg, Germany
D. H. Beck, R. J. Holt, R. Laszewski, C. N. Papanicolas, S. E. Williamson
University of Illinois, Urbana, IL USA
E. Cisbani, S. Frullani, F. Garibaldi, M. Jodice, G. M. Uriuoli
INFN Sezione Sanita, Rome, Italy
A. Feschenko, V. Lysyakov, I. Savin
JINR, Dubna, Russia
N. Bianchi, G. P. Capitani, V. Muccifora, E. de Sanctis, P. Rossi,
P. Levi Sandri, E. Poli, A. R. Roelon-Cora
LNF, Frascati, Italy

G. R. Court, R. Gamet, P. Hayman, T. Jones, S. Kiourkos, J. Stewart
University of Liverpool, England
G. Rosner, T. Walcher
University of Mainz, Germany
D. Fick, F. Rathmann
University of Marburg, Germany
R. Ent, J. Kelsey, L. Kramer, K. Lee, R. Milner, R. Redwine, N. Simicevic
Laboratory of Nuclear Science, MIT, Cambridge, MA USA
B. Braun, G. Graw, P. Schiemenz
University of München, Germany
G. Burleson, G. Kyle, B. Park, M.-Z. Wang
New Mexico State University, Las Cruces, NM USA
J. F. J. van den Brand (Spokesman), M. Doetz, K. de Jager, P. de Witt Huberts
F. Hartjes, B. Kaan, F. Udo
NIKHEF, Amsterdam, The Netherlands
S. Barrow, W. Lorenzon, M. Spangos
University of Pennsylvania, Philadelphia, PA
S. Belostotski, Y. Kiselev, S. Manaenkov, Y. Naryshkin, V. Nelyubin, V. Vikrov
St. Petersburg Nuclear Physics Institute, Catchina
Commonwealth of Independent States
O. Hiroshi, Y. Sakemi, T.-A. Shibata
Tokyo University, Japan
B. Cummings, P. J. Delheij, O. Häusser, R. Henderson, R. Kaiser,
M. Kueckes, C. A. Miller, R. Openshaw, A. Trudel,
M. Vetterli (Deputy Spokesman), R. Woloshyn
Simon Fraser University, TRIUMF, Vancouver, Canada
H. J. Bulten, W. Haeberli, T. Wise, Z. L. Zhou
University of Wisconsin, Madison, WI USA
S. Taroyan, H. Vartapetian, H. Voskanian
Yerevan Physics Institute, Armenia

REFERENCES

1. Ashman, J., *et al.* Phys. Lett. **B206**, 364 (1988); Nucl. Phys. **B238**, 1 (1989); Adeva, B. *et al.*, Phys. Lett. **B302**, 533 (1993).
2. Anthony, P. L., Phys. Lett. **71**, 959 (1993); also see Day, D., previous talk, this conference.
3. Jaffee, R. L., and Manohar, A., Nucl. Phys. **B337,** 509 (1990).
4. Ellis, J., and Jaffee, R. J., Phys. Rev. **D9,** 1444 (1974).
5. Sokolov, A. A., and Ternov, I. M., Sov. Phys. Doklady **8**, 1203 (1964).
6. Frankfurt, L. L., *et al.*, Phys. Lett. **B230**, 141 (1989).
7. Buon, J., and Steffen, K., Nucl. Instrum. Methods **A245**, 248 (1986).
8. Bjorken, J. D., Phys. Rev. **148**, 1457 (1966); **D1**, 1367 (1970).
9. Hoodbhoy, P., Jaffe, R. L., and Manohar, A., Nucl. Phys. **B312**, 571 (1989); Jaffe, R. L., and Manohar, A., Nucl. Phys. **B321**, 343 (1989).
10. Jaffe, R. L., and Manohar, Aneesh, Phys. Lett. **B223,** 218 (1989).
11. Wandzura, W., and Wilczek, F., Phys. Lett. **B172**, 195 (1977); Shuriak, E. V., and Vainshtein, A. F., Nucl. Phys. **B201**, 142 (1982).
12. op. cit. 6.

13. Mankiewicz, L., Schäfer, A., and Veltri, M., Proceedings of the Workshop "Physics at HERA", Vol. **3**, October (1991).
14. Ross, G. G., and Roberts, R. G., "The Glum Contribution to Polarized Nucleon Structure Functions", RAL-90-062 (1990).
15. Schäfer, A., Phys. Lett. **B209**, 175 (1988).
16. Close, F. E., and Milner, R. G., Phys. Rev. **D44,** 3691 (1991).

MEASURING POLARIZED PARTON DISTRIBUTIONS FROM ELECTROWEAK INTERACTIONS

Jacques SOFFER

Centre de Physique Théorique
CNRS - Luminy, Case 907 - F–13288 Marseille Cedex 9 - France

Abstract. We will try to review different methods based on electroweak interactions which can be used to extract polarized parton distributions. Our basic knowledge on parton distributions, coming from deep inelastic scattering in leptoproduction, will be reexamined and we will present a simple construction for unpolarized and polarized structure functions in terms of Fermi-Dirac distributions. We will also emphasize the importance of longitudinally and transversely polarized proton-proton and proton-neutron collisions at high energies for improving the determination of polarized quark distributions, in particular by means of gauge bosons and lepton-pair production.

1. INTRODUCTION

The spin structure of the nucleon is not yet fully understood and in spite of some recent measurements in polarized deep inelastic scattering, several puzzling questions remain unanswered. The spin dependent structure functions for proton and neutron are now more accurately determined in the low Q^2 region, but the validity or breakdown of the corresponding first moments sum rules are still the subject of many theoretical speculations. According to the standard interpretation of the data, it seems that only one third of the nucleon spin is carried by the quarks, a small fraction indeed, when compared to what one would naively expect. The amount of the proton spin carried by the sea quarks and the antiquarks is not firmly established and we still don't know what is the role of the axial anomaly and how much the gluon participates in the nucleon spin. These are some the reasons why polarized parton distributions are still so important and the purpose of this lecture is to review different methods involving electroweak interactions, which allow to measure these fundamental physical quantities.

The outline of the paper is as follows. In section 2 we will present a construction for the neutron and proton unpolarized structure functions in terms of Fermi-Dirac distributions by means of a very small number of parameters. By making some reasonable and simple assumptions, one can relate unpolarized and polarized quark distributions and predict the proton and neutron spin dependent structure functions. In section 3 this set of polarized quark distributions will be used to study various helicity asymmetries for $W^{\pm}, Z$ and dilepton production in pp and pn collisions with longitudinally polarized protons at RHIC. Section 4 will be devoted to the case of transversely polarized protons, transverse spin asymmetries and in particular, we will stress the relevance of lepton-pair and Z production to determine quark transversity distributions.

2. DEEP INELASTIC SCATTERING

We will start by studing deep inelastic scattering and we will make use of some simple observations and advocate the Pauli exclusion principle, to construct a reliable set of quark, antiquark and gluon distributions.

Many years ago Feynman and Field made the conjecture[1] that the quark sea in the proton may not be flavor symmetric, more precisely $\bar{d} > \bar{u}$, as a consequence of Pauli principle which favors $d\bar{d}$ pairs with respect to $u\bar{u}$ pairs because of the presence of two valence u quarks and only one valence d quark in the proton. This idea was confirmed by the results of the NMC experiment[2] on the measurement of proton and neutron unpolarized structure functions, $F_2(x)$. It yields a fair evidence for a defect in the Gottfried sum rule[3] and one finds

$$I_G = \int_0^1 \frac{dx}{x}[F_2^p(x) - F_2^n(x)] = 0.235 \pm 0.026 \tag{1}$$

instead of the value 1/3 predicted with a flavor symmetric sea, since we have in fact

$$I_G = 1/3(u + \bar{u} - d - \bar{d}) = 1/3 + 2/3(\bar{u} - \bar{d}). \tag{2}$$

A crucial role of Pauli principle may also be advocated to understand the well known dominance of u over d quarks at high x,[4] which explains the rapid decrease of the ratio $F_2^n(x)/F_2^p(x)$ in this region. Let us denote by $q^{\uparrow}(q^{\downarrow})$, u or d quarks with helicity parallel (antiparallel) to the proton helicity. The double helicity asymmetry measured in polarized muon (electron) - polarized proton deep inelastic scattering allows the determination of the quantity $A_1^p(x)$ which increases towards one for high x,[5,6] suggesting that in this region $u^{\uparrow}$ dominates over $u^{\downarrow}$, *a fortiori* dominates over $d^{\uparrow}$ and $d^{\downarrow}$, and we will see now, how it is possible to make these considerations more quantitative. Indeed at $Q^2 = 0$ the first moments of the valence quarks are related to the values of the axial couplings

$$u^{\uparrow}_{\text{val}} = 1 + F, \quad u^{\downarrow}_{\text{val}} = 1 - F, \quad d^{\uparrow}_{\text{val}} = \frac{1 + F - D}{2}, \quad d^{\downarrow}_{\text{val}} = \frac{1 - F + D}{2}, \tag{3}$$

so by taking $F = 1/2$ and $D = 3/4$ (rather near to the quoted values[7] 0.461 ± 0.014 and 0.798 ± 0.013) one has $u^{\uparrow}_{\rm val} = 3/2$ and $u^{\downarrow}_{\rm val} = 1/2$ which is at the center of the rather narrow range $(d^{\uparrow}_{\rm val}, d^{\downarrow}_{\rm val}) = (3/8, 5/8)$. The abundance of each of these four valence quark species, denoted by $p_{\rm val}$, is given by eq. (3) and we assume that the distributions at high Q^2 "keep a memory" of the properties of the valence quarks, which is reasonable since for $x > 0.2$ the sea is rather small. So we may write for the parton distributions

$$p(x) = F(x, p_{\rm val}) \tag{4}$$

where F is an increasing function of $p_{\rm val}$. The fact that the dominant distribution at high x is just the one corresponding to the highest value of $p_{\rm val}$, gives the correlation *abundance - shape* suggested by Pauli principle, so we expect broader shapes for more abundant partons. If $F(x, p_{\rm val})$ is a smooth function of $p_{\rm val}$, its value at the center of a narrow range is given, to a good approximation, by half the sum of the values at the extrema, which then implies[8]

$$u^{\downarrow}_{\rm val}(x) = 1/2 d_{\rm val}(x). \tag{5}$$

This leads to

$$\Delta u_{\rm val}(x) \equiv u^{\uparrow}_{\rm val}(x) - u^{\downarrow}_{\rm val}(x) = u_{\rm val}(x) - d_{\rm val}(x) \tag{6}$$

and, in order to generalize this relation to the whole u quark distribution, we assume that eq. (6) should also hold for quark sea and antiquark distributions, so we have

$$\Delta u_{sea}(x) = \Delta \bar{u}(x) = \bar{u}(x) - \bar{d}(x)\ . \tag{7}$$

Moreover as a natural consequence of eq. (3), we will assume

$$\Delta d_{\rm val}(x) = (F - D) d_{\rm val}(x)\ . \tag{8}$$

Finally we will suppose that the d sea quarks (and antiquarks) and the strange quarks (and antiquarks) are not polarized i.e.

$$\Delta d_{\rm sea}(x) = \Delta \bar{d}(x) = \Delta s(x) = \Delta \bar{s}(x) = 0\ . \tag{9}$$

Clearly the above simple relations (6)-(9) are enough for fixing the determination of the spin dependent structure functions $xg_1^{p,n}(x, Q^2)$, in terms of the spin average quark parton distributions. We now proceed to present our approach for constructing the nucleon structure functions $F_2^{p,n}(x, Q^2)$, $xF_3^{\nu N}(x, Q^2)$, etc... in terms of Fermi-Dirac distributions which is motivated by the importance of the Pauli exclusion principle, as we stressed above. Let us consider u quarks and antiquarks only, and let us assume that at fixed $Q^2, u_{\rm val}^{\uparrow}(x)$, $u_{\rm val}^{\downarrow}(x)$, $\bar{u}^{\uparrow}(x)$ and $\bar{u}^{\downarrow}(x)$ are expressed in terms of Fermi-Dirac distributions, in the scaling variable x, of the form

$$xp(x) = a_p x^{b_p} / (exp((x - \tilde{x}(p))/\bar{x}) + 1) \ . \qquad (10)$$

Here $\tilde{x}(p)$ plays the role of the "thermodynamical potential" for the fermionic parton p and $\bar{x}$ is the "temperature" which is a universal constant. Since valence quarks and sea quarks have very different x dependences, we expect $0 < b_p < 1$ for $u_{\rm val}^{\uparrow,\downarrow}(x)$ and $b_p < 0$ for $\bar{u}^{\uparrow,\downarrow}(x)$. Moreover $\tilde{x}(p)$ is a constant for $u_{\rm val}^{\uparrow,\downarrow}(x)$, whereas for $\bar{u}^{\uparrow,\downarrow}(x)$, it has a smooth x dependence. This might reflects, the fact that parton distributions contain two phases, a gas contributing to the non singlet part with a constant potential and a liquid, which prevails at low x, contributing only to the singlet part with a potential slowly varying in x, that we take linear in $\sqrt{x}$. In addition, in a statistical model of the nucleon[9], we expect quarks and antiquarks to have opposite potentials, consequently the gluon, which produces $q\bar{q}$ pairs, will have a zero potential. Moreover, since in the process $G \to q_{\rm sea} + \bar{q}$, $q_{\rm sea}$ and $\bar{q}$ have opposite helicities, we expect the potentials for $u_{\rm sea}^{\uparrow}$ (or $\bar{u}^{\uparrow}$) and $\bar{u}^{\downarrow}$ (or $u_{\rm sea}^{\downarrow}$) to be opposite. So we take

$$\tilde{x}(\bar{u}^{\uparrow}) = -\tilde{x}(\bar{u}^{\downarrow}) = x_0 + x_1\sqrt{x} \ . \qquad (11)$$

The d quarks and antiquarks are obtained by using eqs. (5) and (7) and concerning the strange quarks, we take in accordance with the data[10] $s(x) = \bar{s}(x) = (\bar{u}(x) + \bar{d}(x))/4$. Finally for the gluon distribution, for the sake of consistency, we take a Bose-Einstein expression given by

$$xG(x) = a_G x^{b_G} / (exp(x/\bar{x}) - 1) \qquad (12)$$

with the same temperature $\bar{x}$ and a vanishing potential, as we discussed above. Since it is reasonable to assume that for very small x, $xG(x)$ has the same dependence as $x\bar{q}(x)$, we will take $b_G = 1 + \bar{b}$, where $\bar{b}$ is b_p for the antiquarks. So, except fo the overall normalization a_G, $xG(x)$ has no free parameter. All the distributions considered so far depend upon *eight* free parameters[1] which have been determined by using the most recent NMC data[2] on $F_2^p(x)$ and $F_2^n(x)$ at $Q^2 = 4GeV^2$ together with the most accurate neutrino data from CCFR[10,12] on $xF_3^{\nu N}(x)$ and the antiquark distribution $x\bar{q}(x)$ [10].

As an example of the results of our fit, $xF_3^{\nu N}(x)$ is presented in Fig. 1. As shown in ref.[11] the description of the data is very satisfactory, taking

[1] To identify them, see ref.[11] where their values are also given.

into account the fact that we only have *eight* free parameters and this certainly speaks for Fermi-Dirac distributions. Note that we find $I_G = 0.228$ in beautiful agreement with eq. (1). The steady rise of $x\bar{q}(x)$ at small x leads to a rise of F_2^p which is consistent[11] with the first results from Hera.

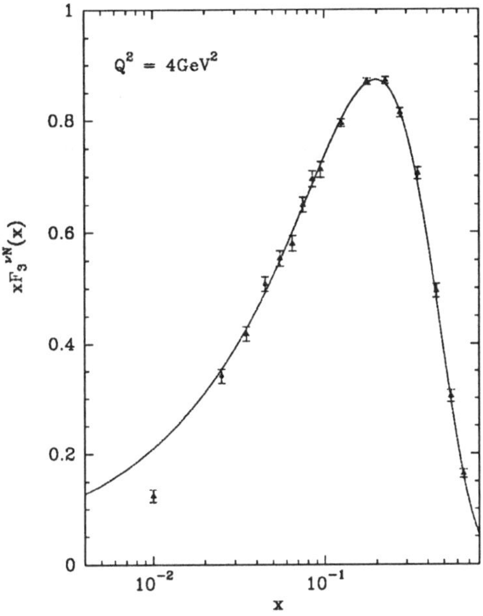

Fig.1 - The structure function $xF_3^{\nu N}(x)$ versus x. Data are from ref.[12] at $Q^2 = 3GeV^2$ and the solid line is the result of our fit from ref.[11].

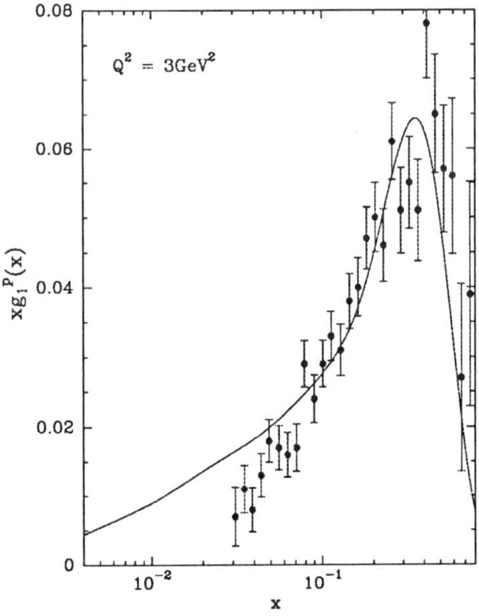

Fig.2 - $xg_1^p(x)$ at $< Q^2 >= 3GeV^2$ versus x. Data are from ref.[6] and solid line is our prediction.

Let us now turn to the polarized structure functions $xg_1^{p,n}(x, Q^2)$ which will allow to test our simple relations (6)-(9). We show in Fig. 2 our prediction together with the recent proton data from SLAC[6] at $Q^2 = 3GeV^2$ and we find

$$I_p = \int_0^1 g_1^p(x)dx = 0.138 \tag{13}$$

which is consistent with the evaluation of ref.[6], $I_p = 0.127 \pm 0.004$ (*stat.*) ± 0.010 (*syst.*). It is well below the Ellis-Jaffe sum rule[14] prediction of 0.160 ± 0.006 and we interpret it as being due to a large negative contribution of Δu_{sea} and $\Delta \bar{u}$ in the small x region (see eq. (7)).

Concerning the neutron polarized structure function $xg_1^n(x)$ we show in Fig. 3 a comparison of the SLAC data[13] at $Q^2 = 2GeV^2$ with our theoretical calculations. The dashed line corresponds to the case where d quarks are assumed to be unpolarized and it clearly disagrees with the data. However by including the d valence quark polarization according to eq. (8), we obtain the solid line in perfect agreement with the data and we find for $Q^2 = 2GeV^2$

$$I_n = \int_0^1 g_1^n(x)dx = -0.020 \ . \tag{14}$$

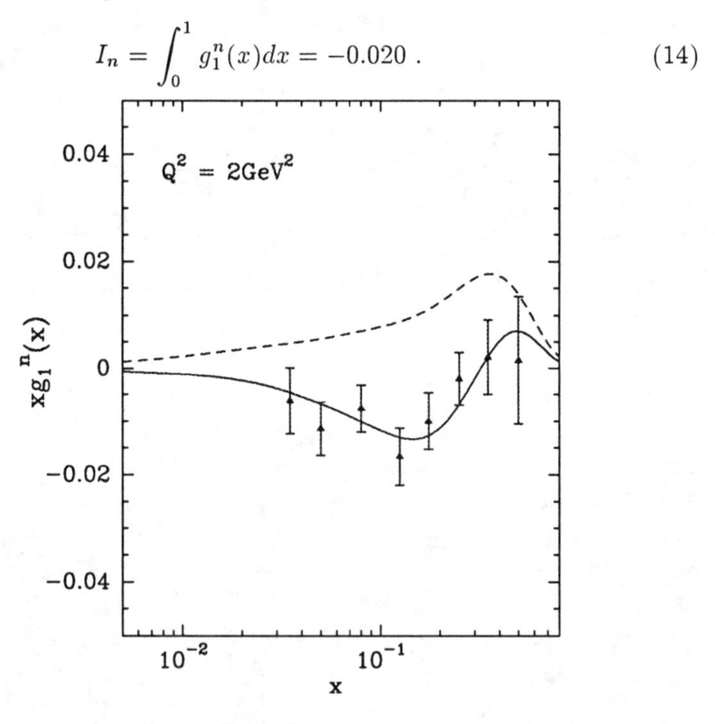

Fig.3 - $xg_1^n(x)$ at $< Q^2 >= 2GeV^2$ versus x. Data are from ref.[13] together with our predictions at $Q^2 = 2GeV^2$ from ref.[11] (Dashed line is the contribution of $\Delta u(x)$ and $\Delta \bar{u}(x)$ only and solid line contains, in addition, the contribution of $\Delta d_{\mathrm{val}}(x)$).

Finally our results for the parton polarizations $\Delta q(x)/q(x)$ at $Q^2 = 4GeV^2$ are shown in Fig. 4 and we will now discuss how polarized pp and pn collisions at high energies can provide an independent determination of these polarized distributions.

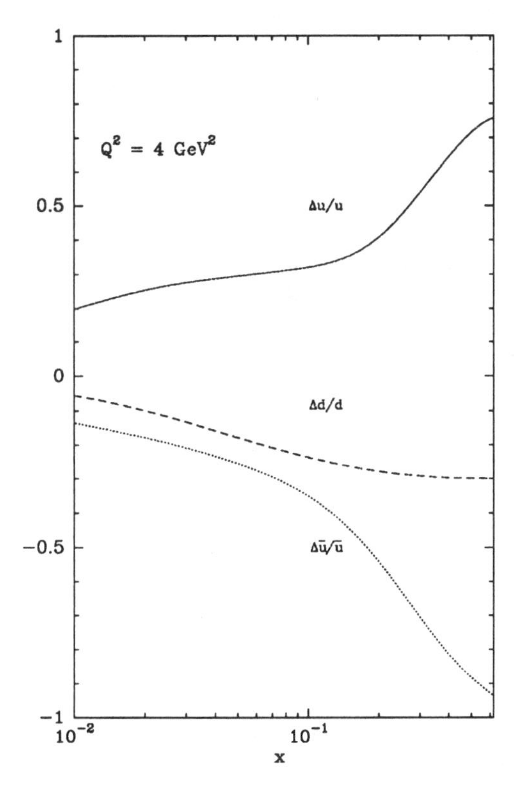

Fig.4 - Parton polarizations $\Delta q(x)/q(x)$ versus x at $Q^2 = 4GeV^2$ obtained from eqs.(6)-(9).

3. HELICITY ASYMMETRIES AT RHIC

A Relativistic Heavy Ion Collider (RHIC) is now under construction at Brookhaven National Laboratory and, already more than three years ago, it was realized that one should propose a very exciting physics programme[15], provided this machine could be ever used as a polarized pp collider. Of course all these considerations relie on the foreseen key parameters of this new facility, i.e. a luminosity up to $2.10^{32} cm^{-2} sec^{-1}$ and an energy of $50 - 250$ GeV per beam with a polarization of about 70%. Since then, the RHIC Spin Collaboration (RSC) has produced a letter of intent[16] and has undertaken several serious studies in various areas which have led to a proposal[17] which has now been fully approved.

A very copious production of $W^{\pm}$ and Z bosons[18] is expected at RHIC, because in three months running, the integrated luminosity at $\sqrt{s} = 500GeV$ will be $800pb^{-1}$. In a recent article[19], it was shown that unpolarized cross section in pp and pn collisions allow an independent test of the flavor asymmetry of the light sea quarks mentioned above i.e. $\bar{d}(x) > \bar{u}(x)$, but here we will only recall some of the results obtained for the various spin-dependent observables.

3.1 Parity-violating asymmetries A_L, A_{LL}^{PV} and $\bar{A}_{LL}^{PV}$

Since RHIC is planned to be used as a polarized pp collider, let us now investigate what we can learn from the measurement of the helicity asymmetries and in particular from parity-violating asymmetries which involve the electroweak Standard Model couplings. In principle one can consider three parity-violating asymmetries defined as

$$A_L = \frac{\sigma_- - \sigma_+}{\sigma_- + \sigma_+} \quad , \quad A_{LL}^{PV} = \frac{\sigma_{--} - \sigma_{++}}{\sigma_{--} + \sigma_{++}} \quad , \quad \overline{A}_{LL}^{PV} = \frac{\sigma_{-+} - \sigma_{+-}}{\sigma_{-+} + \sigma_{+-}} \tag{15}$$

where $\sigma_{h_1 h_2}$ is the cross section where the initial protons have helicities h_1 and h_2 and σ_h is the case where only one of the proton beam is polarized. Clearly if parity is conserved $\sigma_{h_1 h_2} = \sigma_{-h_1 -h_2}$, $\sigma_h = \sigma_{-h}$ so all these asymmetries vanish. Because of the axial vector couplings this is not the case in the Standard Model and these helicity asymmetries will be expressed in terms of the parton helicity asymmetries $\Delta f(x, Q^2)$ which are known at $Q^2 = 4GeV^2$ from the above analysis. In the Standard Model the W is a purely left handed current and A_L, in W^+ production, reads simply

$$A_L(y) = \frac{\Delta u\,(x_a, M_W^2)\,\overline{d}\,(x_b, M_W^2) - (u \leftrightarrow \overline{d})}{u\,(x_a, M_W^2)\overline{d}\,(x_b, M_W^2) + (u \leftrightarrow \overline{d})} \tag{16}$$

assuming the proton a is polarized, and a similar expression can be written for the case of pn collisions[19].

Turning to the double helicity asymmetries A_{LL}^{PV} and $\bar{A}_{LL}^{PV}$, for pp collisions they are explicitely given in ref.[20] and one finds that A_{LL}^{PV} is symmetric in y whereas $\overline{A}_{LL}^{PV}$ is antisymmetric, so we have

$$A_{LL}^{PV}(y) = A_{LL}^{PV}(-y) \quad , \quad \overline{A}_{LL}^{PV}(y) = -\overline{A}_{LL}^{PV}(-y). \tag{17}$$

A priori these helicity asymmetries are three independent observables, but if one makes the reasonable assumption $\Delta u \Delta \overline{d} << u\overline{d}$ for all x, one gets the two following relations

$$A_{LL}^{PV}(y) = A_L(y) + A_L(-y) \tag{18}$$

$$\overline{A}_{LL}^{PV}(y) = A_L(y) - A_L(-y). \tag{19}$$

So within this approximation, the double spin asymmetries A_{LL}^{PV} and $\overline{A}_{LL}^{PV}$ do not contain any additional information than that contained in the single spin asymmetry $A_L(y)$ and in particular at $y = 0$, where the cross section is maximum, $A_{LL}^{PV} = 2A_L$ and $\overline{A}_{LL}^{PV} = 0$. In order to calculate these asymmetries we have used the above model for the various parton helicity asymmetries Δu_v, Δu_s, $\Delta \overline{u}$, Δd_v, Δd_s, $\Delta \overline{d}$ evaluated at $Q^2 = M_W^2$.

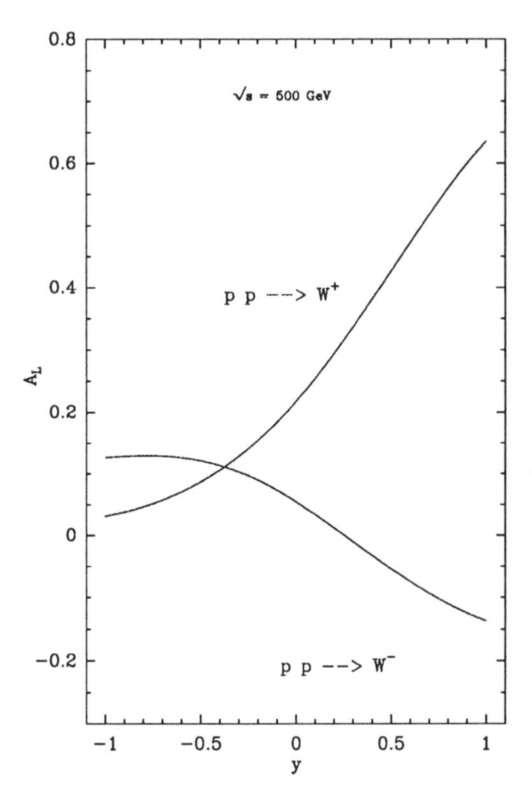

Fig.5 Parity-violating helicity asymmetry A_L versus y for W^+ and W^- production in pp collisions at $\sqrt{s} = 500\ GeV$.

The results of our calculations for $W^\pm$ production are presented in Fig.5 for A_L with this choice of polarized parton distributions. The general trend of A_L can be understood as follows: from eq.(16) and a similar expression for

W^- production by permuting u and d, one sees that for $y = 0$

$$A_L^{W^+} = \frac{1}{2}\left(\frac{\Delta u}{u} - \frac{\Delta \bar{d}}{\bar{d}}\right) \quad \text{and} \quad A_L^{W^-} = \frac{1}{2}\left(\frac{\Delta d}{d} - \frac{\Delta \bar{u}}{\bar{u}}\right) \tag{20}$$

evaluated at $x = M_W/\sqrt{s}$, for $y = -1$ one has

$$A_L^{W^+} \sim -\frac{\Delta \bar{d}}{\bar{d}} \quad \text{and} \quad A_L^{W^-} \sim -\frac{\Delta \bar{u}}{\bar{u}} \tag{21}$$

evaluated at $x = 0.059$ and for $y = +1$ one has

$$A_L^{W^+} \sim \frac{\Delta u}{u} \quad \text{and} \quad A_L^{W^-} \sim \frac{\Delta d}{d} \tag{22}$$

evaluated at $x = 0.435$.

So the region $y \simeq +1$ is controled by valence quark polarizations while $y \simeq -1$ is very sensitive to the sea quark polarizations. Similar calculations have been done for pn collisions[19] and also for Z production.

Finally by using eqs.(18) and (19), one obtains from Fig.5 the following rather good estimates i.e. $0.40 \leq A_{LL}^{PV} \leq 0.60$ for W^+ production and $\bar{A}_{LL}^{PV} \simeq 0$ for W^- production in pp collisions.

3.2 Parity-conserving asymmetries A_{LL}

In pp collisions where both proton beams are polarized, there is another observable which is very sensitive to antiquark polarizations, that is the parity-conserving double helicity asymmetry A_{LL} defined as

$$A_{LL} = \frac{\sigma_{++} + \sigma_{--} - \sigma_{+-} - \sigma_{-+}}{\sigma_{++} + \sigma_{--} + \sigma_{+-} + \sigma_{-+}} \, . \tag{23}$$

This asymmetry, in W^+ production, reads simply

$$A_{LL}(y) = -\frac{\Delta u\,(x_a,\ M_W^2)\,\Delta \bar{d}\,(x_b,\ M_W^2) + (u \leftrightarrow \bar{d})}{u\,(x_a,\ M_W^2)\,\bar{d}\,(x_b,\ M_W^2) + (u \leftrightarrow \bar{d})} \, . \tag{24}$$

For W^- production quark flavors are interchanged. It is clear that $A_{LL}(y) = A_{LL}(-y)$ and that $A_{LL} \equiv 0$ if the antiquarks are not polarized, i.e. $\Delta\bar{u}(x) = \Delta\bar{d}(x) \equiv 0$. Similarly for Z production we find

$$A_{LL}(y) = -\frac{\sum_{i=u,d} \left(a_i^2 + b_i^2\right) \left[\Delta q_i\,(x_a, M_Z^2)\,\Delta \bar{q}_i\,(x_b, M_Z^2) + (x_a \leftrightarrow x_b)\right]}{\sum_{i=u,d} \left(a_i^2 + b_i^2\right) \left[q_i\,(x_a, M_Z^2)\,\bar{q}_i\,(x_b, M_Z^2) + (x_a \leftrightarrow x_b)\right]} \tag{25}$$

which will vanish for unpolarized antiquarks. We show in Fig. 6 our predictions for the three cases at $\sqrt{s} = 500GeV$. Clearly as a consequence of eq. (9) $A_{LL} \equiv 0$ for W^+ production but if $\Delta\bar{d}(x) \neq 0$, it would be non-zero and of opposite sign to $\Delta\bar{d}(x)$. For W^- production for $y = 0$ we get $A_{LL} = -\frac{\Delta d}{d}\frac{\Delta\bar{u}}{\bar{u}}$ evaluated at $x = M_W/\sqrt{s}$ which gives around -10% and from the trend of the d and $\bar{u}$ polarizations similar to that shown in Fig. 4, we also expect A_{LL} to be almost constant for $-1 < y < +1$. For Z production as a consequence of eq. (9) $A_{LL} \simeq +10\%$ and does not depend on the d quark polarization.

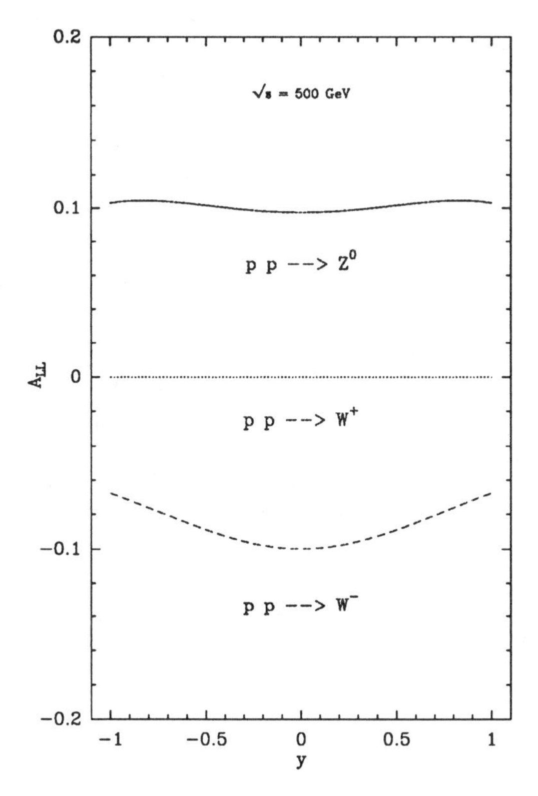

Fig.6 - Parity-conserving double helicity asymmetry A_{LL} versus y for $W^{\pm}$ and Z production in pp collisions at $\sqrt{s} = 500\ GeV$.

Finally let us consider lepton-pair production and in this case the expression for $A_{LL}(y)$ follows from eq.(25) where a_i is replaced by e_i, the electric charge of q_i, and $b_i = 0$. We have calculated A_{LL} at $\sqrt{s} = 100GeV$ which seems more appropriate to the acceptance of the detectors at RHIC and the results are shown in Fig.7. We observe that A_{LL} increases for an increasing lepton-pair mass M and of course for $\Delta\bar{u} = 0$ we would have $A_{LL} = 0$.

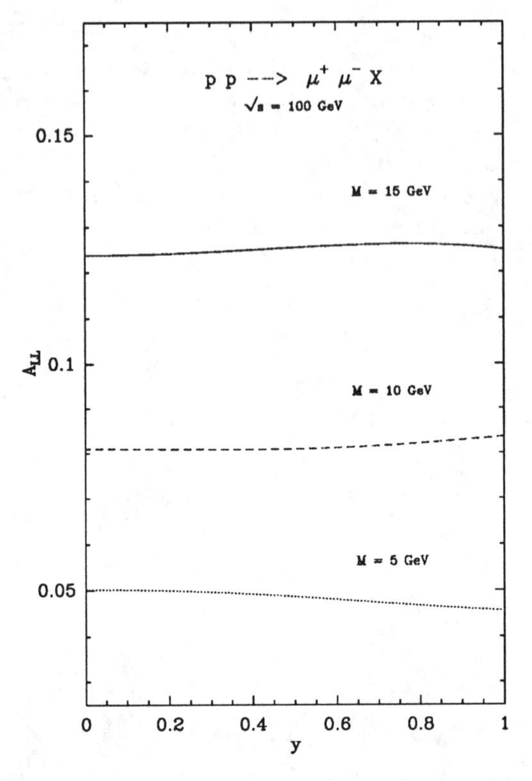

Fig.7 - Parity-conserving double helicity asymmetry A_{LL} versus y for dilepton production at $\sqrt{s} = 100GeV$ and different values of the lepton-pair mass.

4. DOUBLE SPIN TRANSVERSE ASYMMETRIES A_{TT}

So far we have considered collisions involving only longitudinally polarized proton beams, but of course at RHIC, transversely polarized protons will be available as well[17]. This new possibility is extremely appealing because of recent progress in understanding transverse spin effects in QCD, both at leading twist[21] and higher twist levels[22]. For the case of the nucleon's helicity, its distribution among the various quarks and antiquarks can be obtained in polarized deep inelastic scattering from the measurement of the structure function $g_1(x)$ mentioned above. However this is not possible for the *transversity* distribution $h_1(x)$ which describes the state of a quark (antiquark) in a transversely polarized nucleon. The reason is that $h_1(x)$, which measures the correlation between right-handed and left-handed quarks, decouples from deep inelastic scattering. Indeed like $g_1(x)$, $h_1(x)$ is leading - twist and it can be measured in Drell-Yan lepton-pair production with both initial proton beams transversely polarized[21]. Other possibilities have been suggested[23] but in the framework of this lecture, we will envisage also a practical way to determine $h_1(x)$, by using gauge boson production in pp collisions with protons transversely polarized. Let us consider the double spin transverse asymmetry defined as

$$A_{TT} = \frac{\sigma_{\uparrow\uparrow} - \sigma_{\uparrow\downarrow}}{\sigma_{\uparrow\uparrow} + \sigma_{\uparrow\downarrow}} \tag{26}$$

where $\sigma_{\uparrow\uparrow}(\sigma_{\uparrow\downarrow})$ denotes the cross section with the two initial protons transversely polarized in the same (opposite) direction. Assuming that the underlying parton subprocess is quark-antiquark annihilation, we easily find for Z production

$$A_{TT} = \frac{\sum\limits_{i=u,d} (b_i^2 - a_i^2) \left[h_1^{q_i}(x_a) h_1^{\bar{q}_i}(x_b) + (x_a \leftrightarrow x_b)\right]}{\sum\limits_{i=u,d} (a_i^2 + b_i^2) \left[q_i(x_a)\bar{q}_i(x_b) + (x_a \leftrightarrow x_b)\right]} \,. \tag{27}$$

This result generalizes the case of lepton-pair production[21] through an off shell photon $\gamma^\star$ and corresponding to $b_i = 0$ and $a_i = e_i$, as mentioned above. For $W^\pm$ production, which is pure left-handed and therefore does not allow right-left interference, we expect $A_{TT} = 0$, since in this case $a_i^2 = b_i^2$. This result is worth checking experimentally.

So far there is no experimental data on these distributions $h_1^q(x)$ (or $h_1^{\bar{q}}(x)$), but there are some attempts to calculate them either in the framework of the MIT bag model[21] or by means of QCD sum rules[24]. However the use of positivity yields to derive a model-independent constraint on $h_1^q(x)$ which restricts substantially the domain of allowed values[25]. Indeed one has obtained

$$q(x) + \Delta q(x) \geq 2|h_1^q(x)|. \tag{28}$$

which is much less trivial than

$$q(x) \geq |h_1^q(x)|, \tag{29}$$

as proposed earlier in ref.[21].

In the MIT bag model, let us recall that these distributions read[21]

$$q = f^2 + g^2,\ \Delta q = f^2 - 1/3g^2 \ \text{ and } \ h_1^q = f^2 + 1/3g^2 \tag{30}$$

and they saturate (28). In this case, we observe that $h_1^q(x) \geq \Delta q(x)$ but this situation cannot be very general because of eq.(28). As an example let us assume $h_1^q(x) = 2\Delta q(x)$. Such a relation cannot hold for all x and we see that eq.(28), in particular if $\Delta q(x) > 0$, implies $q(x) \geq 3\Delta q(x)$. This is certainly not satisfied for all x by the present determination of the u quark helicity distribution, in particular for large x where $A_1^p(x)$ is large[5,6]. The simplifying assumption $h_1^q(x) = \Delta q(x)$, based on the non-relativistic quark model, which has been used in some recent calculations[19,23] is also not acceptable for all x values if $\Delta q(x) < 0$ because of eq.(28). To illustrate the practical use of eq.(28), let us consider eqs.(6) and (7). It is then possible to obtain the allowed range of values for $h_1^u(x)$, namely

$$u(x) - \frac{1}{2}d(x) \geq |h_1^u(x)| \tag{31}$$

which is shown in Fig.8. In this case, we have checked that for $x > 0.5$, both the results of the MIT bag model[21] and the QCD sum rule[24] violate this positivity bound, combined with low Q^2 data. A similar calculation can be done for the d quarks to get the allowed region for $h_1^d(x)$ [25].

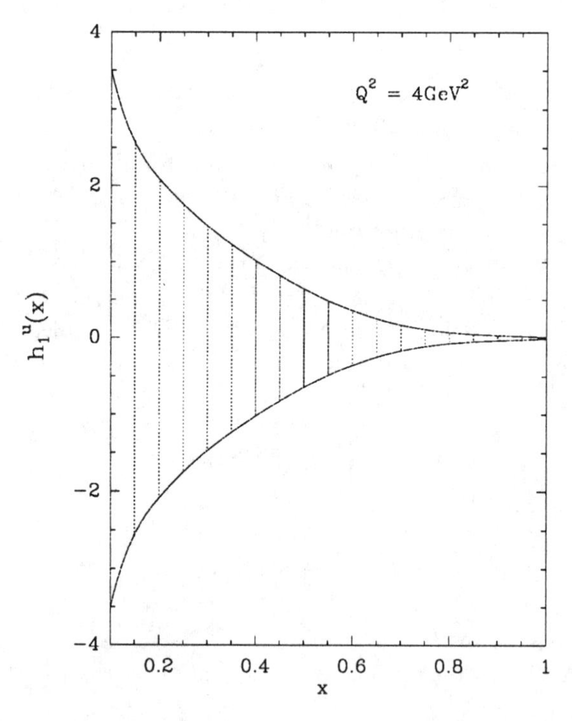

Fig.8 The striped area represents the domain allowed for $h_1^u(x)$ using eq.(31).

Finally we show in Fig.9 the results of our calculation for A_{TT} in the case of Z production by assuming the equality sign in eq.(28) and $h_1^u(x) > 0$,

$h_1^{\bar{u}}(x) < 0$. Clearly this prediction is only a guide for a future experiment at RHIC which will lead to the actual determination of $h_1(x)$.

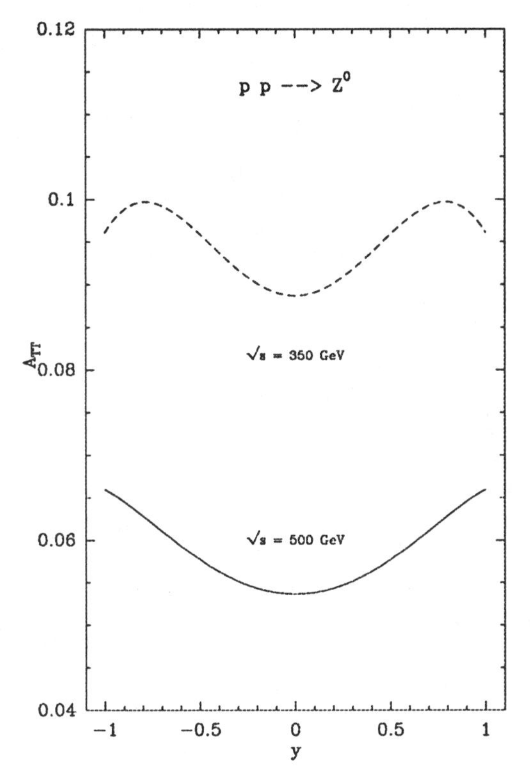

Fig.9 Double spin transverse asymmetry A_{TT} versus y for Z production in pp collisions at $\sqrt{s} = 350$ and $500\ GeV$.

ACKNOWLEDGMENTS

I am glad to thank the organizers of this Symposium, in particular Profs. K.J. Heller and J.M. Cameron, for such a pleasant and stimulating atmosphere. I also thank C. Bourrely for useful discussions during the preparation of this lecture.

REFERENCES

1. R.D. Field and R.P. Feynman, Phys. Rev. **D15** (1977) 2590.
2. M. Arneodo *et al.*, (New Muon Collaboration), Phys. Rev. **D50** (1994) R1 and references therein.
3. K. Gottfried, Phys. Rev. Lett. **18** (1967) 1174.
4. T. Sloan, G. Smadja and R. Voss, Phys. Rev. **162** (1988) 45.

5. J. Ashman *et al.*, (European Muon Collaboration), Phys. Lett. **B206** (1988) 364; Nucl. Phys. **B328** (1989) 1; D. Adams *et al.*, (Spin Muon Collaboration), Phys. Lett. **B329** (1994) 399.
6. K. Abe *et al.*, (E143 Collaboration) preprint SLAC-PUB 6508 (august 1994). See also D. Day, these proceedings.
7. Z. Dziembowski and J. Franklin, J. Phys. G: Nucl. Part. Phys. **17** (1991) 213.
8. F. Buccella and J. Soffer, Mod. Phys. Lett. A **8** (1993) 225.
9. See for example J. Cleymans, I. Dadic and J. Joubert, preprint BI-TP 92-45 (july 1993) and references therein.
10. C. Foudas *et al.*, Phys. Rev. Lett. **64** (1990) 1207; S.R. Mishra *et al.*, Phys. Rev. Lett. **68** (1992) 3499; S.A. Rabinowitz *et al.*, Phys. Rev. Lett. **70** (1993) 134.
11. C. Bourrely and J. Soffer, preprint CPT-94/P.3032.
12. P.Z. Quintas *et al.*, (CCFR Collaboration) Phys. Rev. Lett. **71** (1993) 1307; W.C. Leung *et al.*, (CCFR Collaboration) Phys. Lett. **B317** (1993) 655.
13. P.L. Anthony *et al.*, (E142 Collaboration), Phys. Rev. Lett. **71** (1993) 959.
14. J. Ellis and R.L. Jaffe, Phys. Rev. **D9** (1974) 1444.
15. C. Bourrely, J.Ph. Guillet and J. Soffer, Nucl. Phys. **B 361** (1991) 72. Proceeding of the Polarized Collider Workshop, University Park PA (1990), Eds J. Collins, S. Heppelmann and R.W. Robinett, AIP Conf. Proc. N° 223 AIP, New York (1991).
16. G. Bunce et al. Particle World, vol. 3 (1992) 1.
17. Proposal on Spin Physics using the RHIC Polarized Collider, R5 (14 august 1992), approved october 1993. See also Y. Makdisi, these proceedings.
18. V.L. Rykov, these proceedings.
19. C. Bourrely and J. Soffer, Nucl. Phys. **B314** (1993) 132.
20. C. Bourrely and J. Soffer, Phys. Lett. **B423** (1994) 329.
21. J. Ralston and D.E. Soper, Nucl. Phys. **B152** (1979) 109 ; J.L. Cortes, B. Pire and J.P. Ralston, Z. Phys. **C55** (1992) 409 ; R. Jaffe and X. Ji, Phys. Rev. Lett. **67** (1991) 552 ; R. Jaffe and X. Ji, Nucl. Phys. **B375** (1992) 527 ; X. Ji, Nucl. Phys. **B402** (1993) 217.
22. J. Qiu and G. Sterman, Nucl. Phys. **B 378** (1992) 52 ; R.L. Jaffe and X. Ji, Phys. Rev. **D 43** (1991) 724.
23. X. Ji, Phys. Lett. **B234** (1992) 137 ; R.L. Jaffe and X. Ji, Phys. Rev. Lett. **71** (1993) 2547.
24. B.L. Ioffe and A. Khodjamarian, Preprint University of München LMU-01-94.
25. J. Soffer, Preprint CPT-94/P.3059.

Spin dependent structure function $g_1(x)$ of the proton and the deuteron from the SMC experiment

the Spin Muon Collaboration (SMC)

Anna Maria Zanetti

Istituto Nazionale di Fisica Nucleare
Area di Ricerca - I34012 Trieste, Italy

Abstract. The measurements of the spin-dependent structure function g_1 of the proton and, briefly, of the deuteron in polarized deep inelastic scattering are presented. Implications for our understanding of the internal spin structure of the nucleon are discussed. Data are in disagreement with the Ellis-Jaffe sum rule at a two standard deviation level. The Bjorken sum rule is confirmed to within 10% of its theoretical value (one standard deviation level).

Introduction

In the past three years, big improvements have been achieved in the experimental knowledge of the spin dependent structure functions.
The SMC experiment has played an essential role: the first measurement of g_1 of the deuteron (g_1^d) has been performed in 1992 and g_1 of the proton (g_1^p) has been measured in 1993 with better statistical and systematic accuracy compared with previous experiments [1].

In the single photon exchange approximation and in the scaling limit, the difference in cross sections for deep inelastic scattering of a lepton polarized antiparallel and parallel to a longitudinally polarized nucleon can be expressed in terms of the spin dependent structure functions g_1 and g_2, where the contribution of the part involving g_2 is negligible. The structure function g_1 allows us to have an insight of the internal spin structure of the nucleon and, in the framework of the Quark Parton Model (QPM), has a simple interpretation similar to the one of the structure function F_1:

$$g_1(x) = \frac{1}{2}\sum_i e_i^2 \left[q_i^+(x) - q_i^-(x)\right], \quad F_1(x) = \frac{1}{2}\sum_i e_i^2 \left[q_i^+(x) + q_i^-(x)\right] \quad (1)$$

where $q_i^{\pm}$ is the distribution function of the quark of flavour i, charge e_i and spin parallel (+) or antiparallel (−) to the nucleon spin, and x is the Bjorken scaling variable.

The structure function g_1 is extracted from the measured lepton-nucleon asymmetry A, given by

$$A = (\sigma^{\uparrow\downarrow} - \sigma^{\uparrow\uparrow})/(\sigma^{\uparrow\downarrow} + \sigma^{\uparrow\uparrow}) = D(A_1 + \eta A_2) \tag{2}$$

The depolarization factor D and the coefficient η depend on the event kinematics. D depends also on R, the ratio of the longitudinal and transverse photon-nucleon absorption cross sections. The asymmetry A_2 arises from the interference between transverse and longitudinal virtual photon polarizations. A_1 is the virtual-photon nucleon asymmetry, related to g_1 by

$$A_1 = \frac{\sigma_{1/2} - \sigma_{3/2}}{\sigma_{1/2} + \sigma_{3/2}} \simeq \frac{g_1}{F_1} \tag{3}$$

where 1/2 and 3/2 are the total spin projections in the direction of the virtual photon.

Computing the first moment $\Gamma_1 = \int_0^1 g_1(x)dx$, the Ellis and Jaffe and the Bjorken sum rules can be tested.

In the late 1960's, Bjorken derived from current algebra a sum rule [2] which related the difference of the first moments of the proton and the neutron to the nucleon axial vector and vector coupling constants for β decay, g_A and g_V. It is a rigorous prediction of QCD. A verification of the validity of this sum rule is an important test of QCD and of its fundaments.

Within the QPM, Ellis and Jaffe [3] have derived sum rules for the proton and for the neutron separately, under the assumptions that the strange sea is unpolarized and that SU(3) symmetry is valid for the baryon octet decays. These rules relate the first moments of g_1 to the SU(3) coupling constants measured from hyperon decay, F and D, making predictions for Γ_1^n and Γ_1^p.

The SMC experiment

The main components of the SMC experiments are a polarized muon beam, a polarized target, a spectrometer to detect the incoming and the scattered muon, and a beam polarimeter. The muon beam polarimeter and the polarized target have been described in detail in other talks at this conference [4]. The polarization of the incoming muon beam was measured to be $P_\mu = -0.803 \pm 0.029 \pm 0.020$ at an energy of 190 GeV [5]. Here, and in the following, the first error is statistical and the second is systematic. The target consists of two cells with opposite longitudinal polarizations, allowing us to record data from both polarization simultaneously. The polarizations in both target halves are frequently reversed in regular time intervals to minimize systematic uncertainties.

For each recorded scattering event, the momentum and direction of the incident and scattered muon are measured. These two tracks determine the interaction vertex with an average resolution of 3 cm in the direction of the beam and 0.3 mm in the transverse plane. This permits a good identification of the events originating from the upstream or downstream target cell.

Asymmetry extraction

Experimentally, one measures the number N of deep inelastic events as a function of x, Q^2 in the two target cells. This can be expressed as

$$N_{u(d)} = n_{u(d)} \Phi a_{u(d)} \sigma_0 (1 - f P_\mu P_{Tu(d)} A) \tag{4}$$

where the subscripts u and d refer to the upstream and downstream target cells, n is the number of target nucleons, Φ the beam flux, a the apparatus acceptance, σ_0 the unpolarized cross section, f is the dilution factor, i.e. the fraction of the event yield from the polarized protons (for g_1^p) or from the polarized deuterons (for g_1^d) in the target material, and P_μ and $P_{Tu(d)}$ are the beam and target polarizations. The asymmetry A is extracted from combinations of data sets taken before (N, a) and after (N', a') a polarization reversal:

$$\frac{N'_u N_d}{N_u N'_d} = \frac{a'_u a_d}{a_u a'_d}(1 + 4 f P_\mu P_T A) \tag{5}$$

It is important to notice that Φ, n and the absolute value of the spectrometer acceptances cancel in the determination of A. We made the experimentally verified assumption that the ratio $r = a_u/a_d$ remains constant within the typical period of time between two polarization reversals. The asymmetry can then be extracted, without any knowledge about acceptances. A time dependence of r leads to a false asymmetry of $\Delta A_1^p = \frac{1}{4fP_\mu P_T D}\frac{\Delta r}{r}$. For instance, in the case of the measurement of g_1^p we have $\Delta r/r < 7{\times}10^{-4}$, corresponding to a false asymmetry $\Delta A_1^p < 7 \times 10^{-3}$. These effects are included as a systematic error.

Measurement of $g_1^p(x)$

The spin-dependent structure function g_1 of the proton has been measured in deep inelastic scattering of polarized muons of 190 GeV off polarized protons, in the kinematic range $0.003 < x < 0.7$ and $1\,\mathrm{GeV}^2 < Q^2 < 60\,\mathrm{GeV}^2$ [6]. The target consists of two cell, each 60 cm long and 5 cm in diameter, separated by 30 cm and with opposite longitudinal polarizations. The target material is butanol with a dilution factor f equal to 0.12. On average the target polarization is $P_T = 0.86$, measured with a relative accuracy of 3%. The spin directions were reversed every five hours.

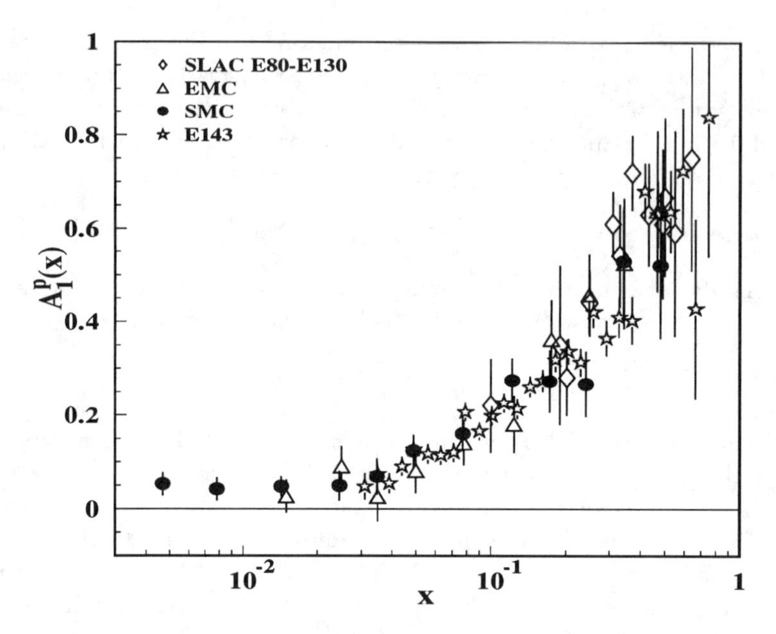

Figure 1: The virtual-photon proton cross section asymmetry A_1^p as a function of x. Only statistical errors are shown with the data points.

The virtual-photon proton asymmetry A_1^p is computed using eq. 2. A_2^p has been measured by SMC in a dedicated experiment where the protons in the target have been transversely polarized and was found to be compatible with zero within a statistical uncertainty of 0.20 [7]. Furthermore the coefficient η is small in the kinematic range covered by our experiment. The term ηA_2^p is then neglected and its possible effect included in the systematic error.

The measured values of A_1^p for each x bin are shown in fig. 1. The mean Q^2 of the data points varies with increasing x from 1.3 to 58 GeV^2. Also shown are the recent data from the E143 SLAC experiment [8]. SMC and E143 data are consistent in the region of overlap. In the higher x region ($x > 3 \times 10^{-2}$) the E143 data points provide a nice high statistic measurement while SMC is unique in its capability of measuring the low x region.

The spin dependent structure function g_1^p is evaluated from the average asymmetry A_1^p in each x bin, according to

$$g_1^p(x, Q^2) = A_1^p(x, Q^2) F_1^p(x, Q^2) = \frac{A_1^p(x, Q^2) F_2^p(x, Q^2)}{2x[1 + R(x, Q^2)]}. \qquad (6)$$

$F_2^p(x, Q^2)$ is the unpolarized structure function, taken from the NMC parametrization [9]; the uncertainty on F_2 is typically 3% to 5%. The lowest x bin is outside the kinematic region covered by the NMC data, but their parametri-

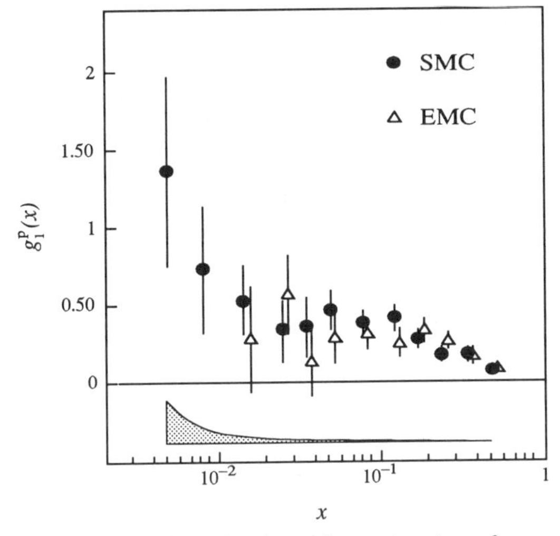

Figure 2: The spin dependent structure function $g_1^p(x)$. Error bars represent the statistical uncertainty. The size of the systematic errors for the SMC data is indicated by the shaded area.

Source of the error	$\Delta\Gamma_1^p$
Beam polarization	0.0057
Uncertainty on F_2	0.0052
Extrapolation at low x	0.0040
Target polarization	0.0039
Dilution factor	0.0034
Acceptance variation Δr	0.0030
Radiative corrections	0.0023
Neglect of A_2	0.0017
Momentum measurement	0.0020
Uncertainty on R	0.0018
Kinematic resolution	0.0010
Extrapolation at high x	0.0007
Total systematic error	0.0113
Statistics	0.0114

Table I. Contributions to the error on Γ_1^p

zation extrapolates smoothly to the HERA data, and the corresponding uncertainty on F_2 is estimated to be 15%. R is taken from a global fit of the SLAC data [10]. The results for $g_1^p(x)$ at the average Q^2 of each x bin are shown in fig.2. g_1 shows a tendency to increase at low x but we do not consider this trend significant enough to call into question the validity of the Regge behavior, that is usually expected to describe g_1 at small x values. Some theoretical work have proposed a more singular behaviour for g_1 near $x = 0$ [11].

To evaluate the integral $\int g_1^p(x, Q_0^2)dx$ at a fixed Q^2, g_1^p in each bin is recalculated at $Q_0^2 = 10\,\mathrm{GeV}^2$, using Eq. 6, under the assumption that $A_1(x, Q^2)$ is independent of Q^2. This assumption is consistent with our data and with recent theoretical calculations that predict a small Q^2 dependence of A_1, not measurable within the present experimental accuracy[12].

The integral over the measured x range is $\int_{0.003}^{0.7} g_1^p(x, Q_0^2)dx = 0.131 \pm 0.011 \pm 0.011$. The contributions to the systematic error are detailed in Table I.

To estimate the integral for $x > 0.7$, we take $A_1^p = 0.7 \pm 0.3$ for $0.7 < x < 1.0$, which is consistent with the bound $A_1 < 1$, and also with the result from perturbative QCD $A_1 \to 1$ as $x \to 1$; we compute 0.0015 ± 0.0007. The contribution to the integral from the unmeasured region $x < 0.003$ was evaluated assuming a Regge-type dependence $g_1^p(x) =$ *constant*, that we fit to our two lowest x data points. We obtain $\int_0^{0.003} g_1^p(x)dx = 0.004 \pm 0.004$.

The result (fig. 3-left) for the first moment of $g_1^p(x)$ at $Q_0^2 = 10\,\mathrm{GeV}^2$ is $\Gamma_1^p(Q_0^2) = 0.136 \pm 0.011 \pm 0.011$.

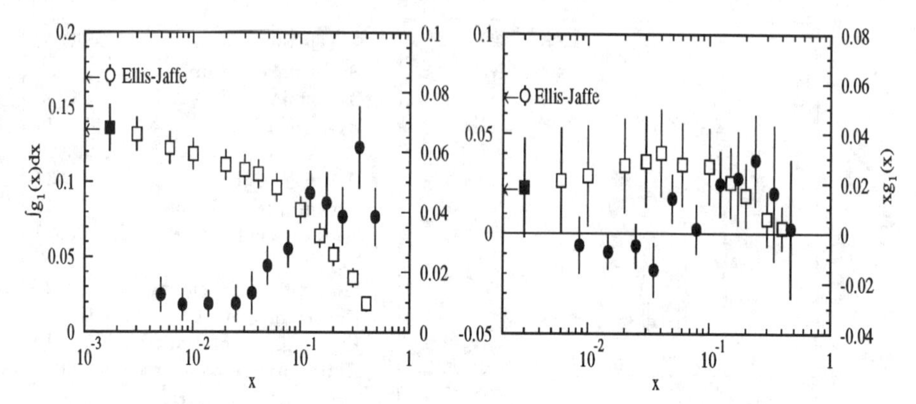

Figure 3: The left plot refers to g_1^p while the right one refers to g_1^d. The solid circles (right-hand axis) show the structure function xg_1 as a function of x, at $Q_0^2 = 10\,GeV^2$. The open boxes (left-hand axis) show $\int_{x_m}^1 g_1(x)dx$, where x_m is the value of x at the lower edge of each bin. Only statistical errors are shown. The solid square is our result $\int_0^1 g_1(x)dx$, with statistical and systematic errors combined in quadrature. Also shown is the theoretical prediction by Ellis and Jaffe.

Combining our data with those of E80/E130 and EMC, we get

$$\Gamma_1^p(Q_0^2) = \int_0^1 g_1^p(x, Q_0^2)dx = 0.142 \pm 0.008 \pm 0.011 \quad \textit{(World proton)} \qquad (7)$$

For a finite Q^2, QCD corrections to the Ellis–Jaffe sum rule are needed and are computed up to third (for the non singlet part) and second (for the singlet part) order in α_S [13]. The theoretical prediction is then $\Gamma_1^p = 0.171 \pm 0.006$. Our measurement is two standard deviations below this value.

Measurement of $g_1^d(x)$

The first measurement of g_1 of the deuteron was performed in the deep inelastic scattering of polarized muons of 100 GeV energy off polarized deuterons, in the kinematic range $0.006 < x < 0.6$, 1 GeV$^2 < Q^2 <$ 30 GeV^2.

The apparatus is basically the same as for the measurement of g_1^p apart from the polarized target. In this case, the two target cells are 40 cm long and 5 cm in diameter and are separated by 20 cm. The longitudinal polarizations were reversed every eight hours for most of the data. The target material is deuterated butanol with a dilution factor $f \simeq 0.19$. The average deuteron polarization is $P_T = 0.35$, measured with a relative accuracy of 5%.

The data analysis is very similar to the one of the measurement of g_1^p and is described in detail in ref. [14]. The results for xg_1^d and the integral of g_1^d are shown in fig. 3-right. The first moment of $g_1^d(x)$ at $Q_0^2 = 4.6\,\mathrm{GeV}^2$ is:

$$\Gamma_1^d(Q_0^2) = \int_0^1 g_1^d(x, Q_0^2)dx = 0.023 \pm 0.020 \pm 0.015. \tag{8}$$

Using the relation $\Gamma_1^p + \Gamma_1^n \simeq 2\Gamma_1^d/(1 - 1.5\omega_D)$, where ω_D is the probability of the deuteron to be in a D-state, we compute $\Gamma_1^p + \Gamma_1^n = 0.049 \pm 0.044 \pm 0.032$. This value is two standard deviations below the Ellis–Jaffe prediction that, including QCD corrections, is $\Gamma_1^p + \Gamma_1^n = 0.152 \pm 0.011$.

Spin content of the nucleon

In QPM and assuming SU(3) flavour symmetry in the baryon octet decays, the first moment Γ_1 can be expressed in terms of F, D and $\Delta\Sigma = \Delta u + \Delta d + \Delta s$, the sum of the quark spin contributions to the nucleon spin ($\Delta q = \int_0^1 (q_i^+(x) - q_i^-(x))dx$ for $q = u, d, s$). From our measurement of Γ_1^d we obtained $\Delta\Sigma = 0.09 \pm 0.25$ and $\Delta s = -0.16 \pm 0.08$. From Γ_1^p we get $\Delta\Sigma = 0.22 \pm 0.14$ and $\Delta s = -0.12 \pm 0.06$.

QCD corrections are important for a correct interpretation of the obtained values of $\Delta\Sigma$. When comparing the results from the various experiments, taking this effect into account allows a consistent picture to be obtained with a value of $\Delta\Sigma \simeq 0.30$ and $\Delta s \simeq -0.10$ with relative errors of the order of $20 - 30\%$ [15].

Test of the Bjorken sum rule

We now turn to a test of the Bjorken sum rule, using all available proton, neutron and deuteron data (the recent preliminary E143 data [16] are not included). This test is made at $Q^2 = 5\,\mathrm{GeV}^2$ in order to avoid a large Q^2 evolution of the SLAC-E142 neutron data, which have an average $Q^2 = 2\,\mathrm{GeV}^2$. A fit to Γ_1^p (Eq. 7), Γ_1^n [17] and Γ_1^d [14] yields $\Gamma_1^p - \Gamma_1^n = 0.163 \pm 0.017$ where statistical and systematic errors are combined in quadrature. Using the available deuteron and proton data to replace the extrapolation on the neutron data, as discussed in Ref. [18], one obtains $\Gamma_1^n = -0.069 \pm 0.025$ and $\Gamma_1^p - \Gamma_1^n = 0.204 \pm 0.029$ with a larger error due to the limited statistics in the deuteron experiment. The theoretical prediction, including perturbative QCD corrections up to third order in α_s, gives $\Gamma_1^p - \Gamma_1^n = 0.185 \pm 0.004$.

All the experimental results on Γ_1 are summarized in fig.4 where the prediction of the Bjorken sum rum is shown as a 45° band. The SMC measurement on Γ_1^d is a constraint on $\Gamma_1^p + \Gamma_1^n$. The experimental data overlap in a region of the Γ_1^p Γ_1^n plane showing agreement with the prediction of the Bjorken sum rule.

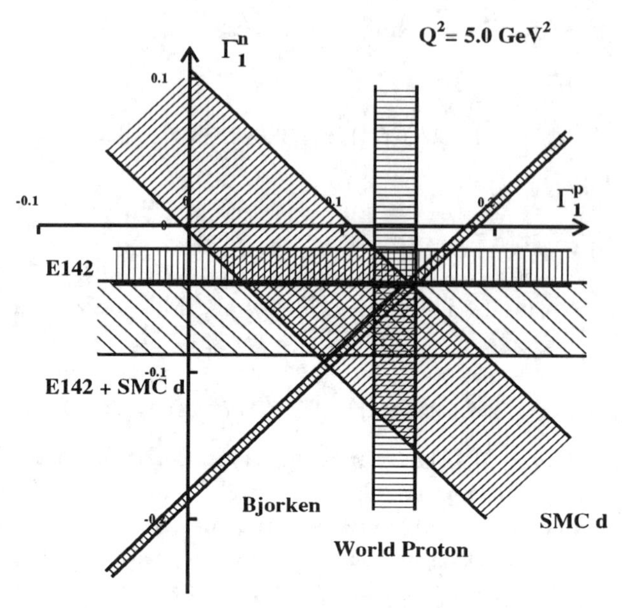

Figure 4: Summary of all the measurements on Γ_1 compared with the Bjorken sum rule in the Γ_1^p Γ_1^n plane. The shaded areas represent the uncertainty.

References

[1] SLAC E-80, M.J. Alguard et al., Phys. Rev. Lett. 37 (1976) 1261; ibid. 41 (1978) 70; SLAC E-130 G. Baum et al., Phys. Rev. Lett. 51 (1983) 1135; EMC, J. Ashman et al., Phys. Lett. B206 (1988) 364.

[2] J.D. Bjorken, Phys. Rev. 148 (1966) 1467; Phys. Rev. D1 (1970) 1376.

[3] J. Ellis and R.L. Jaffe, Phys. Rev. D9 (1974) 1444; D10 (1974) 1669.

[4] S. Bültmann, F. Feinstein, D. Kramer, J. Kyynäräinen, these Proceedings.

[5] SMC, B. Adeva et al., Nucl. Instrum. Methods A343 (1994) 363.

[6] SMC, B. Adams et al., Phys. Lett. B329 (1994) 399.

[7] SMC, B. Adams et al., Phys. Lett. B336 (1994) 125.

[8] E143, Abe et al., SLAC-PUB-6508 (August 1994).

[9] NMC, P. Amaudruz et al., Phys. Lett. B295 (1992) 159 and CERN-PPE/92-124 (July 1992); Errata Oct. 26, (1992) and Apr. 19, (1993).

[10] L.W. Whitlow et al., Phys. Lett. B250 (1990) 193.

[11] S.D. Bass and P.V. Landshoff, DAMTP 94/50 (1994); F.E. Close and R.G. Roberts, RAL-94-071 (1994)

[12] G. Altarelli, P. Nason and G. Ridolfi, Phys. Lett. B320 (1994) 152.

[13] S.A. Larin et al.,CERN-TH-7208/94 and references therein.

[14] SMC, B. Adeva et al., Phys. Lett. B302 (1993) 533.

[15] J. Ellis and M. Karliner, CERN-TH-7324/94.

[16] D. Day, A. Feltham, H. Borel, these Proceedings.

[17] SLAC E142, D.L. Anthony et al., Phys. Rev. Lett. 71 (1993) 959.

[18] SMC, B. Adeva et al., Phys. Lett. B320 (1994) 400

E143 Measurement of the Deuteron Spin Structure Function g_1^d

A. Feltham[1]

University of Basel
Klingelbergstr. 82 4056 Basel Switzerland

Abstract. The experimental aspects and results of the recent SLAC E143 measurement of the deuteron spin structure function g_1^d are discussed. These results are combined with the proton spin structure function, g_1^p, obtained concurrently in the same experiment, to provide tests of the predictions of sum rules. The results are found to be in excellent agreement with the predictions of Bjorken while a significant disagreement is observed for Ellis-Jaffe.

INTRODUCTION

Recently the SLAC E143 collaboration has completed a series of measurements which determine the deep inelastic spin structure functions $g_1(x)$, as a function of the Bjorken x-variable, for both proton and deuterium targets using polarized electron beams of 9, 16 and 29 GeV. The experiment measured the polarization dependent cross-sections for a longitudinally polarized electron beam interacting with both longitudinally and transversely polarized ammonia targets ($\sigma^{\uparrow(\downarrow)\uparrow}$, $\sigma^{\uparrow(\downarrow)\leftarrow}$, respectively). From these quantities the following experimental asymmetries were obtained:

$$A_{\|} = \frac{\sigma^{\downarrow\uparrow} - \sigma^{\uparrow\uparrow}}{\sigma^{\downarrow\uparrow} + \sigma^{\uparrow\uparrow}} \text{ and } A_{\perp} = \frac{\sigma^{\downarrow\leftarrow} - \sigma^{\uparrow\leftarrow}}{\sigma^{\downarrow\leftarrow} + \sigma^{\uparrow\leftarrow}}$$

The structure function g_1 is related to these asymmetries in the following manner:

$$g_1(x,Q^2) = \frac{F_1(x,Q^2)}{D'}\left[A_{\|} + \tan(\frac{\theta}{2})A_{\perp}\right] \tag{1}$$

with the kinematic definitions: $D' = \frac{(1-\epsilon)(2-y)}{y[1+\epsilon R(x,Q^2)]}$, $R(x,Q^2) = \frac{(1+\gamma^2)F_2(x,Q^2)}{2xF_1(x,Q^2)} - 1$, $y = (E-E')/E = \nu/E$, $\epsilon = 1/[1+2(1+\frac{\nu^2}{Q^2})\tan^2(\frac{\theta}{2})]$ and θ, E', E are the laboratory angle and energy of the scattered electron and beam energy, respectively. $F_1(x,Q^2)$ and $F_2(x,Q^2)$ are the unpolarized structure functions.

[1]for the SLAC E143 Collaboration

Nucleon spin structure functions, when evaluated in the limit of Bjorken scaling ($Q^2 \to \infty$), are important for testing several sum rules which describe our understanding of QCD and the quark structure of nucleons. In particular, the spin structure functions for both the proton, $g_1^p(x)$, and the neutron, $g_1^n(x)$, are required to test the Bjorken sum rule [1],

$$\int_0^1 [g_1^p(x) - g_1^n(x)]dx = \frac{1}{6}(F+D) \quad , \tag{2}$$

where F and D are the hadronic weak decay coupling constants. The verification or violation of this sum rule serves as an important check of our understanding of the role of QCD in the Quark-Parton model. The individual nucleon spin structure functions can be employed to test the Ellis-Jaffe sum rule [2]. This sum rule relates the integrals of the individual structure functions to F, D, and to information pertaining to the spin content of the quarks present inside the nucleon [3]:

$$\int_0^1 g_1^p(x)dx = \frac{1}{18}(9F - D + 6\Delta s) \tag{3}$$

$$\int_0^1 g_1^n(x)dx = \frac{1}{18}(6F - 4D + 6\Delta s) \tag{4}$$

where $\Delta\Sigma = \Delta u + \Delta d + \Delta s = 3F - D + 3\Delta s$, and $\Delta q = q^\uparrow - q^\downarrow$ ($q \in u, d, s$) for quark spins (anti-)parallel ($\downarrow$)$\uparrow$ to the nucleon momentum. The work done by EMC and SMC [4] to evaluate the proton Ellis-Jaffe sum rule (Eq. 3) resulted in an unexpected interpretation of the role of quarks in the spin composition of the nucleon. This has increased interest in testing the Ellis-Jaffe sum rule for the neutron (Eq. 4) to verify this interpretation. A third "Ellis-Jaffe" sum rule can be obtained from the sum of equations 3 and 4.

$$\int_0^1 [g_1^p(x) + g_1^n(x)]dx = \frac{1}{18}(15F - 5D + 12\Delta s) \tag{5}$$

This allows one to study the nucleon spin composition directly from a polarized deuteron target ($g_1^d \sim \frac{1}{2}[g_1^p + g_1^n]$).

The integrands of equations 2,4, and 5 can be expressed directly as functions of E143's measured proton and deuteron structure functions g_1^p and g_1^d as shown in Table 1. A detailed error evaluation of the expressions in Table 1 allows the cancellation of correlated errors common to both the proton and deuteron measurements (ie. beam polarization error).

The following discussion focuses on the E143 measurement of the deuteron structure function $g_1^d(x)$ for a beam energy of 29 GeV. In addition to its application in equation 5, interest in g_1^d stems from the information it contains of the neutron spin structure function g_1^n (Eq. 4). The results for $g_1^p(x)$ are discussed elsewhere [5].

An ideal deuteron target consists of a static proton and neutron whose spins are projected parallel to the deuteron spin. This is a reasonable starting point given the small deuteron binding energy and the fact that the deuteron magnetic moment is close to the sum of the proton and neutron magnetic moments ($\mu_d \sim \mu_p + \mu_n$). Simple

Table 1: Functions of the measured quantities $g_1^d(x)$ and $g_1^p(x)$ whose integrals can be directly related to predictions of the Bjorken and Ellis-Jaffe sum rules. $\frac{1}{(1-\frac{3}{2}\omega_D)}$ is the deuteron D-state correction.

Sum Rule	Expression
Bjorken	$g_1^p - g_1^n = 2g_1^p - \frac{2g_1^d}{(1-\frac{3}{2}\omega_d)}$
Ellis-Jaffe	$g_1^n = \frac{2g_1^d}{(1-\frac{3}{2}\omega_D)} - g_1^p$
Ellis-Jaffe	$g_1^p + g_1^n = \frac{2g_1^d}{(1-\frac{3}{2}\omega_D)}$

corrections must be applied for the fact that the deuteron is known to occupy, with a small probability ω_D, an angular momentum D-state. However, other corrections to the ideal deuteron are less straightforward to deal with: i) the Fermi motion of nucleons inside deuteron which alters the measurement of kinematic quantities such as x, and changes measured observables; ii) the possible scattering from mesons responsible for nuclear binding (ie. π, ρ, etc) and other non-nucleonic components (ie. Δ, N^*, etc); and iii) the modification of free nucleon structure functions in the nuclear environment. To correct for these effects, one must rely on input from various theoretical models. One can also get an idea of the magnitude of these effects in a theory independent manner, by studying g_1^n using different nuclear targets, as, for example, $^3\vec{He}$, measured in SLAC E142 [6].

EXPERIMENT

The polarized electron beam is produced by photoemission from a strained GaAs crystal. In this experiment, electron bunches 2 μsec in duration were emitted 120 times per second. The number of particles in each bunch was varied between 1 and 4×10^9 electrons, an intensity which allowed the accumulation of high statistics in short running time. The charge of each bunch was measured to an accuracy better than 1% by using two independent toroids. The charge asymmetry between both polarization states was found to be less that 0.1%. The sign of the beam polarization was chosen randomly on a pulse by pulse basis to reduce possible false asymmetries in the measurement. The beam was scanned grid-wise over the polarized ammonia target to avoid localized heating and depolarization effects therein.

The asymmetries $A_{\|}$ and $A_{\perp}$ are determined from several measured quantities:

$$A_{\|}(A_{\perp}) = \frac{(\Delta - C_P)}{fP_bP_t}C_N + A_{RC} \quad (6)$$

The raw asymmetries, Δ, were obtained directly from the count rates in the spectrometers. The "two-bounce" design [7] of the spectrometer vastly reduced the acceptance for the considerable uncharged background such as, for example, photons produced by bremsstrahlung of the intense electron beam passing through the main target. Each spectrometer arm included two gas Čerenkov detectors for the

fast trigger and π/e separation, two sets of hodoscopes for position information, and an array of lead-glass detectors for energy measurement. The count rates for each polarization state (L, and R) were normalized by the charge in each spill, $Q_{L(R)}$, and the livetime, $l_{L(R)}$, and then corrected for contributions due to charge symmetric background processes, $B_{L(R)}$:

$$\Delta = \frac{(\frac{L}{Q_L l_L} - B_L) - (\frac{R}{Q_R l_R} - B_R)}{(\frac{L}{Q_L l_L} - B_L) + (\frac{R}{Q_R l_R} - B_R)}$$

The background rates $B_{L(R)}$ were measured by reversing the spectrometer polarity and counting e^+ events.

The beam polarization, P_b, was measured about 30 m upstream of the main target, using a Møller polarimeter. Its coincidence detectors yielded high statistics results with an absolute error of ± 0.02 and a typical value of about 0.85 .

The polarized target used in this experiment has been described by Averett [8] elsewhere in these proceedings. The polarization, P_t, for the g_1^d measurements was determined with a relative error of 4%, and its value found in the range 0.2 to 0.35 .

The dilution factor, f, is defined as the ratio of the number of events scattered into the spectrometer due to polarized deuterium (D) to events from all possible scattering sources (for example nitrogen, N, and helium, He). It represents the dilution of the asymmetry of interest due to scattering off of an unpolarized background. Radiative corrections, U_i, must be applied to the cross-sections, σ_i, employed in the expression for f.

$$f = \frac{\#D\sigma_d}{\#D\sigma_d + \#N\sigma_N + \#\mathrm{He}\sigma_{\mathrm{He}} + \#\mathrm{Other}\sigma_{\mathrm{Other}}} \times \frac{U_d}{U_{All}}$$

To determine the amount of each material present in the target, three independent measurements were compared: first, the careful weighing of all target materials, second, a comparison of spectrometer rates for empty and full targets, and third, x-ray absorption measurements performed on the target, following a precision determination of the x-ray absorption coefficient for ammonia. All were in good agreement. The typical value for f was in the range 0.22 to 0.25 with a relative uncertainty of about 3%. In the case of the deuterated ammonia target, the presence of both polarized nitrogen and protons [8] (isotopic impurities of the deuterium) meant that the background contributed an asymmetry to the measurement (C_N and C_P respectively in Eq. 6). The combined size of this calculable correction was ~4%.

Lastly, the results were also corrected for radiative losses of the electron beam with the target. Both internal [9] and external [10] processes were convoluted to obtain a radiative corrected asymmetry, A_{rad}, which, when subtracted from the Born asymmetry, A_{Born}, provides the correction term, $A_{RC} = A_{Born} - A_{rad}$. The error contribution from A_{RC} is dominated by the model uncertainty of the Born asymmetries used, especially in the resonance region (low Q^2), as such events radiate into the spectrometer acceptance. This uncertainty dominates the overall measurement systematic error, particularly at low x.

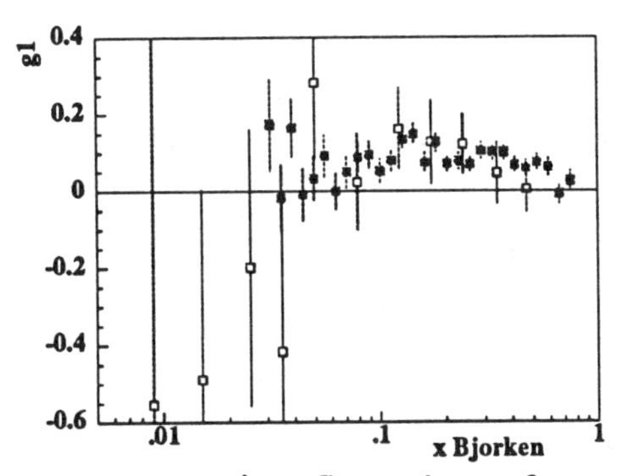

Figure 1a): Comparison of preliminary E143 (solid □) and SMC (Ref. [13]) (open □) $g_1^d(x)$ data.

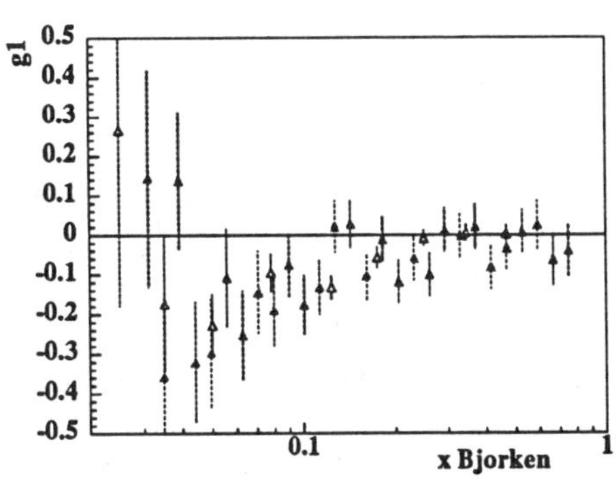

Figure 1b): Comparison of preliminary E143 (solid △) and E142 (Ref. [6]) (open △) $g_1^n(x)$ data.

EVALUATION OF INTEGRALS

Data were obtained over the kinematic range $0.029 \leq x \leq 0.8$ and 1.3 GeV^2/c^2 $\leq Q^2 \leq 10$ GeV^2/c^2. In order to evaluate the data at a common value of Q^2, it was assumed that the ratio g_1/F_1, as taken from equation 1, is independent of Q^2. This is justified by the data since the values of g_1/F_1 obtained from the overlap region of both spectrometers ($.08 \leq x \leq .5$) are in excellent agreement, although the average Q^2 seen by each spectrometer differs by a factor of two. To obtain $g_1(x, Q^2)$ at a fixed value of Q^2, the $\frac{g_1}{F_1}(x)$ data (each x-bin was averaged over all Q^2) was multiplied by $F_1(x, Q^2 = 3$ GeV/c) where 3 GeV/c is the average Q^2 of the data set. $F_1(x, Q^2)$ was obtained from parameterizations of $R(x, Q^2)$ [11] and $F_2(x, Q^2)$ [12]. The preliminary results for $g_1^d(x)$ are compared with the published SMC results [13] in Fig. 1a). The E143 results, although in good agreement, show a significant improvement in statistical errors over the earlier results. In Fig. 1b), the preliminary E143 determination of $g_1^n(x)$ is compared with the published E142 results [6]. Again the results are in good agreement, suggesting that neither determination of g_1^n is significantly influenced by the nuclear environment.

In order to integrate the functions of Table 1, three regions are considered for each integrand. First, for the region $0.029 \leq x \leq 0.80$, the integral and its error are

Table 2: Systematic errors (absolute) from various integral regions.

Quantity	$\int_0^1 g_1^d$	$\int_0^1 g_1^n$	$\int_0^1 g_1^p - g_1^n$
Beam Polarization	.001	.001	.004
Target Polarization	.002	.005	.007
dilution factor	.001	.005	.008
radiative corrections	.003	.006	.007
cross-section (F_2, R)	.001	.003	.007
low x extrap.	.001	.005	.006
high x extrap.	.001	.000	.001
total	.004	.011	.017

Table 3: Contributions from various integral regions. (errors:±stat.±sys.)

Region	$\int_0^1 g_1^d$	$\int_0^1 g_1^n$	$\int_0^1 g_1^p - g_1^n$
data	0.041±.004±.004	-0.028±.009±.010	0.146±.011±.015
low x extrap.	0.002±.000±.001	-0.004±.001±.005	0.013±.001±.006
high x extrap.	0.000±.000±.001	0.000±.000±.000	0.001±.000±.001
total	0.043±.004±.004	-0.032±.009±.011	0.160±.011±.016

taken directly from the data. Second, the low x extrapolation for $0.0 \leq x \leq 0.029$ assumes $g_1(x)$ =constant. This assumption is consistent with Regge Theory [14]. The data for $x < 0.1$ were averaged to determine this constant. Finally, in the extrapolation to high x, $0.80 \leq x \leq 1.0$, $g_1(x)$ assumes the form $(1-x)^3$, which follows from quark counting rules [15]. The parameters of this fit were obtained by fitting the data in the region $x > 0.6$.

The error of the extrapolations include both the statistical and systematic errors of the data used in the fit. In addition, a contribution is included which describes the extreme range of possible integration areas obtained when different x-limits for the fits are used. The error estimations for the extrapolations, as well as the various experimental contributions to the systematic errors in the data region, are summarized in Table 2. The integral contributions from the various regions are listed in Table 3.

In order to compare the results of the integrals with the predictions of their respective sum rules, one must account for the fact that the experiment has been performed at finite Q^2. In the results presented below, the perturbative QCD corrections of Kataev [16] have been applied. These corrections include the terms of second and third order in $\frac{\alpha_s}{\pi}$ for the flavour singlet and non-singlet terms respectively, and include estimates of the next higher order terms for each.

In Table 4 the predictions for the Bjorken and Ellis-Jaffe sum rules, evaluated at Q^2 =3 GeV/c, are compared with the experimentally determined integrals obtained in this work. One can see that the experimental results are in good agreement with the predictions of the Bjorken sum rule. The E143 data alone provide a test of this sum rule to a precision of 14%. For the Ellis-Jaffe sum rule the experimental results are not in agreement with the predictions of theory, in particular for the deuteron data, where the disagreement is greater that 3.5 standard deviations.

One can further employ equations 4 and 5 to extract the quark contribution to

Table 4: Sum rule predictions and preliminary results for integral and $\Delta\Sigma$ and Δs. The deuteron integral value has been corrected for the D-state contribution. The proton results are taken from reference [5].

Sum Rule	Prediction	E143 Results	Target	$\Delta\Sigma$	Δs
Bjorken	0.167±.011	0.160±.020			
Ellis-Jaffe	0.069±.004	0.046±.006	deuteron	0.348±.070	-0.077±.025
Ellis-Jaffe	-0.015±.005	-0.032±.014	neutron	0.399±.158	-0.060±.054
Ellis-Jaffe	0.152±.007	0.129±.010	proton	0.29±.10	-0.10±.04

the spin of the nucleon. These results are also given in Table 4 along with values obtained from the E143 proton measurement [5]. One can see here that the results from each measurement are in good agreement. In addition, these results are in agreement with the published SMC evaluation [4] of $\Delta\Sigma$ and Δs. Further all results are far away from the naive predictions of Ellis-Jaffe sum rule: $\Delta\Sigma \sim 0.6$ and $\Delta s = 0$. Of these results, the E143 deuteron data provide the best single measurement of Δs and $\Delta\Sigma$.

CONCLUSIONS

The SLAC E143 collaboration has performed a precision measurement of the spin structure function $g_1(x)$ for the deuteron. The data from this experiment are complementary to similar data for the proton, $g_1^p(x)$, obtained in the same experiment. From these measurements integrals of the quantities $g_1^p(x) - g_1^n(x)$, $g_1^p(x)+g_1^n(x)$ and $g_1^n(x)$ can be obtained, allowing tests of the Bjorken and Ellis-Jaffe sum rules. The results presented here are in good agreement with the Bjorken sum rule, providing a test at the 14% level of this fundamental prediction. The results disagree significantly with predictions of the Ellis-Jaffe sum rule, where in the case of the deuteron a difference of more than 3.5 standard deviations is observed.

REFERENCES

[1] J.D. Bjorken. *Phys. Rev.*, D1:1376, 1970.

[2] J. Ellis and R.L. Jaffe. *Phys. Rev.*, D9:1444, 1974.

[3] F.E. Close and R.G. Roberts. *Phys. Lett.*, B316:165, 1993.

[4] J. Ashman et al. *Nucl. Phys.*, B328:1, 1989; D. Adams et al. *Phys. Lett.*, B329:399, 1994.

[5] K. Abe et al. *SLAC-PUB-6508*, 1994.

[6] P. Anthony et al. *Phys. Rev. Lett.*, 71:959, 1993.

[7] G.G. Petratos et al. *SLAC-PUB-5678*, 1991.

[8] T. Averett. *these proceedings*, 1994.

[9] T.V. Kukhto and N.M. Shumeiko. *Nucl. Phys.*, B219:412, 1983.

[10] Y.S. Tsai. *Rev. Mod. Phys.*, 46:815, 1974.

[11] L.W. Whitlow et al. *Phys. Lett.*, B250:193, 1990.

[12] P. Amaudruz et al. *Phys. Lett.*, B295:159, 1992.

[13] B. Adeva et al. *Phys. Lett.*, B302:533, 1993.

[14] J. Ellis and M. Karliner. *Phys. Lett.*, B213:73, 1988.

[15] S.J. Brodsky, M. Burkardt, and I. Schmidt. *SLAC-PUB-6087*, 1994.

[16] A.L. Kataev. *CERN-TH.7333/94*, 1994.

How Gluons Spin in the Proton

John P. Ralston

Department of Physics and Astronomy
and
Kansas Institute for Theoretical and Computational Science
University of Kansas, Lawrence, KS 66045

Abstract. Measurable correlations in QCD depend on quantum numbers such as the spin of the hadronic state involved. I create an effective 3+1 dimensional local gauge theory for the gluons in the proton which depends explicitly on the spin. The theory is similar to a normal gauge theory but with the addition of a term which significantly tames the infrared behavior. The theory is related to the adiabatic approximation for the slow modes in the theory. Incorporating the Berry phase there is a new canonical symmetry which is crucial for the model's novel gauge invariance. The result also can be viewed as a coupling of the proton's spin to the gluons' topological current. Asymptotic freedom remains a property; I present gauge invariant zeroeth order dispersion relations and discuss propagation in the new theory as compared to conventional perturbative QCD.

Imagine enlarging a proton at rest to a convenient beach–ball size, about a meter in diameter. We would see an immensely complicated swarm of partons, now known to be highly correlated with the direction $\vec{s}$ of the proton's spin. Can we write down an effective theory for parts of this system without solving all of QCD?

Effective theories are "derived", or approximated, by integrating away a set of configurations that are not of interest. However, there is a severe shortage of terms that could be candidates for the infrared sector of QCD. The gluon sector, in particular, is almost totally constrained by symmetry. But if this is true, can symmetry help us to write down a local effective theory for the spin dependence of gluons?

Such a theory ideally should have several standard features: renormalizability (a practical necessity to avoid continual re–tuning of the ultraviolet with the infrared dynamics); gauge invariance; acceptable Lorentz, and rotational symmetries; asymptotic freedom, and explicit dependence on the proton's spin. Locality of an effective Hamiltonian may be an acceptable approximation in the sense of a derivative expansion; quantum mechanics will supply non–locality when the theory is solved. It turns out that there is one way to make a theory with these requirements, and it hinges on what is meant by the "spin". In a microscopic theory the spin is an operator made from the angular momentum tensor. However, an effective theory is something made to generate correlation function *in a particular state or set of states* – in this case the spin 1/2 baryons. The spin $\vec{s}$ is then a "background" vector and is represented by a set of c–number parameters taken from certain matrix elements. Connected gluon correlation functions of the form

$$< ps \mid F^{\mu v}(x) \ldots F^{\lambda\sigma}(x') \mid ps >= T^{\mu v \ldots \lambda\sigma} + S^{\mu\nu \ldots \lambda\sigma\alpha} s_{\alpha}$$

are either independent of the parameters s^{μ} or dependent linearly on them (because spinors $u(ps)\bar{u}(ps)$ dictate the LSZ reduction of the proton state). We seek, then, some generating functional depending on $\vec{s}$ to summarize such correlations.

In the discussion below I will present a complete dynamical scheme based on emphasizing some slowly varying degrees of freedom in the proton. The resulting "pure gluon" theory is possible to anticipate. The effective Lagrangian $\mathcal{L}_{eff}$ must be

$$\mathcal{L}_{eff} = -Tr[F^{\mu\nu}F_{\mu\nu}] + \frac{m}{4} s_{\mu} Tr[\varepsilon^{\mu\nu\alpha\beta} A_{v} F_{\alpha\beta} - \frac{2}{3} A_{\nu} A_{\alpha} A_{\beta}] + Tr[J_{\mu} A^{\mu}] \; , \quad (1)$$

where J^{μ} is the effective current of non–relativistic matter fields. This theory is the usual one, with the additional coupling of the proton's spin to the gluons' topological current K^{μ}– the second term in Eq. (1) being $ms^{\mu}K_{\mu}$, where $\partial_{\mu}K^{\mu} = Tr[F\tilde{F}]$. Here m is a scale (an inverse length) of hadronic origin; s^{μ} is normalized to $s^{\mu}s_{\mu} = -1$. In some gauges the topological current can be related to the gluon spin current, so Eq. (1) is just a "spin–spin" interaction. This observation, and previous work (1) showing that the polarized distribution of quarks can be related to the polarized gluon distribution, strongly suggests that the road to Eq. (1) must involve the axial anomaly. However, Eq. (1) also represents the only possible local coupling to s^{μ} which is "soft" in the ultraviolet (super–renormalizable) and gauge invariant. In the infrared, the topological current term dominates because it is a dimension–3 operator, leading to extraordinarily interesting effects.

Relation to the Adiabatic Approximation

Given that QCD is a theory of all phenomena of the strong interactions we should not hope to solve it. A rough approximation, i.e. a local theory of some matrix elements of selected states, is a much more modest goal. Different physical pictures of the proton have some things consistently in common. The proton seems to be made of three *slowly varying* constituent quarks. In the large–N limit, the proton is *slowly* rotating. This indicates considerable importance of slowly varying modes in the state.

To investigate this I have turned to the adiabatic approximation (2), an elementary technique that seems not to have been studied much in gauge theories. The proton in its rest frame is described by a Schroedinger wave functional $\Phi_{\Lambda,S}$, where Λ is a cutoff separating "fast" from "slow". The effective Schroedinger equation is $H_{\Lambda,S}\Phi_{\Lambda,s} = E\Phi_{\Lambda S}$, with E an observable independent of Λ but the effective Hamiltonian H_Λ dependent on Λ and s. This is a "one–scale" separation; more complicated schemes are of course conceivable. Denote the fast degrees of freedom by q and their momenta by p; the slow degrees of freedom will be Q with conjugate momenta P. The full wave functional Ψ_s is the overlap at fixed time of the abstract state $/E, s >$ onto field eigenstates:$\Psi_s(q; Q) =< q; Q/E, s >$.

Let $\zeta_s(q; Q)$ be the wave functional for the fast modes q solved as if Q were time-independent classical parameters, a background field. The Born-Oppenheimer(2,3) ansatz for the full wave functional is $\Psi_s(q; Q)=\zeta_s(q; Q)\Phi_s(Q)$. The effective Hamiltonian operator H_Q for the slow variables Q is given by taking the expectation value of the full Hamiltonian $H(q, p; Q, P)$ in the fast variable state ζ_s and integrating over the fast variables: $H_Q(Q, P) = (\zeta_s(q; Q)/H(q, p; Q, P)/\zeta_s(q; Q))$. Here the $(\ldots)$ bracket means the functional integral $\int d[q]$ at fixed time. The Hamiltonian functional of the slow variables is then $H_Q(\Phi_s(Q), \Phi_s^*(Q)) = \{\Phi_s(Q)/H_Q(Q, P_Q)/\Phi_s(Q)\}$ where the bracket $\{\ldots\}$ means the remaining fixed-time functional integrals, namely $\int d[Q]$.

Exploiting a Separate Symmetry

At first sight the Born–Oppenheimer approach is problematic in a gauge theory. Physically we know that there are slow modes ($Q's$) of the gauge fields, denoted by $\mathbf{A}$, with canonical momenta $\mathbf{\Pi_A}$. These coordinates and their Hamiltonian H_A are intrinsically gauge dependent. Set $A^0 = 0$ to remove time–dependent gauge transforms. Let us observe that the definition of "slowly varying modes" is still a convention, due to space-dependent gauge transformations. Closely related are the well–known technical nightmares of ultraviolet regularization, whose arbitrary conventions are quite difficult to reconcile with gauge invariance. Then the renormalized wave functional $\Phi_{\Lambda s}$ of the slow modes is not supposed to be gauge invariant. Yet, in $A^0 = 0$ gauge the full wave functional, hopefully well-enough described by the ansatz, must be gauge invariant according to Gauss' law (4). How is this to be arranged?

The answer I propose hinges on the Berry phase (5,6). Recall that the wave functional of any designated slow modes develops a phase, generated via the adiabatic connection $\mathbf{\Gamma_s(A)} = (\zeta_s(q; \mathbf{A})/i\delta/\delta\mathbf{A}/\zeta_s(q; \mathbf{A}))$. The effects of the adiabatic connection reappear in the slow sector when one takes derivatives using $\mathbf{\Pi_A} = -i\delta/\delta\mathbf{A}$; the rule is $H_A(\mathbf{\Pi_A}, \mathbf{A}) \rightarrow H_A(\mathbf{\Pi_A} - \mathbf{\Gamma_s(A)}, \mathbf{A})$. The slow modes thus get "gauged". This is the generalization of the result from quantum chemistry of Mead and Truhlar (3), whose work predates Berry's. In addition to the shift of the momenta, H_A also generally contains invariant new functionals of the slow coordinates (7), denoted $W(\mathbf{A})$, corresponding to "molecular potentials".

Consider now the question of maintaining gauge invariance (8); I use an Abelian gauge theory as an illustration. The basic Hamiltonian of the photon sector becomes $H = 1/2 \int d^3x[(-i\delta/\delta\mathbf{A} - \mathbf{\Gamma_s}(\mathbf{A}))^2 + \mathbf{B}^2(x) + \mathbf{j}\cdot\mathbf{A} + W(\mathbf{A})]$, where $\mathbf{B}(x) = \nabla \times \mathbf{A}$ is the magnetic field and $\mathbf{j}$ is the electromagnetic current. This looks bad because $\mathbf{\Gamma_s}(\mathbf{A})$ is not gauge invariant and also depends on the arbitrary separation $\Psi = \zeta\Phi$. However, through its very ambiguity, we have gained an enormous symmetry. Consider certain transformations $V(\Theta)$, which can be called ***para–gauge*** transformations of the generalized first and second kinds:

$$\begin{aligned} \Phi_s(A) &\to V(\Theta)\Phi_s(A) = exp(i\Theta(A))\Phi_s(A); \\ \Gamma_s(A) &\to \Gamma_s(A) + \delta\Theta(A)/\delta A. \end{aligned} \tag{2}$$

where $\Theta(\mathbf{A})$ is some functional of $\mathbf{A}$. (I call these para–gauge transformations because they are both beyond ordinary gauge transformations and also act in concert with them.) Under these transformations the Abelian gauge-covariant functional derivative $-i\delta/\delta\mathbf{A} - \mathbf{\Gamma_s}(\mathbf{A})$ is invariant. This symmetry was a manifest property of the Born-Oppenheimer ansatz, $\Phi_s\zeta \to \Phi_s \exp(i\Theta(\mathbf{A})\exp(-i\Theta(\mathbf{A}))\zeta_s$, before the integration over the fast variables. Naturally, the same symmetry has emerged after integrating.

Note the following observations: first, the full exact wave function is strictly gauge invariant, but the separate factors Φ_s and ζ_s cannot be so. Second, the result of the approximation after the fast variables have been eliminated has a gigantic local para-gauge symmetry (Eq. (2)), which was never a part of the original theory. Can the system be invariant under the joint action of both transformations? If so, the effects of the real gauge transformations must be compensated by the para–gauge transformations (8).

Let the standard gauge transformation be represented by an operator $U(\theta)$; $U(\theta)AU^{-1}(\Theta) = \mathbf{A} + \nabla\theta$. Suppose there is a subgroup of the transformations made jointly by the $U(\theta)$ and $V(\Theta)$ operators under which Φ_s is invariant. Then ζ_s and Γ_s must also be invariant. That is,

$$\mathbf{\Gamma_s}(\mathbf{A}) \to \mathbf{\Gamma_s}(\mathbf{A} + \nabla\theta) + \delta\Theta(\mathbf{A})/\delta\mathbf{A} = \mathbf{\Gamma_s}(\mathbf{A}).$$

It follows that the adiabatic connection's invariant curvature $\beta^{ij} = \partial\Gamma^i_s(\mathbf{A})/\partial A^j - \partial\Gamma^j_s(\mathbf{A})/\partial A^i$ satisfies $\beta_{ij}(\mathbf{A}) = \beta_{ij}(\mathbf{A} + \nabla\theta)$: *a sufficient condition is that the invariant curvature β_{ij} be gauge invariant.* A second condition comes from the definition of $\mathbf{\Gamma_s}(\mathbf{A})$, assuming that is space-translationally invariant. Taking the space divergence gives $\nabla \cdot \mathbf{\Gamma_s}(\mathbf{A}) = (\zeta_s(q;\mathbf{A})/\nabla \cdot i\delta/\delta\mathbf{A}/\zeta_s(q;\mathbf{A}))$, which is the expectation value of $\nabla \cdot \mathbf{E}$, where $\mathbf{E}$ is the electric field. Then, $\nabla \cdot \mathbf{\Gamma_s}(\mathbf{A})$ *is also gauge invariant.* (In a non–Abelian theory replace "invariant" by "covariant".) Furthermore, the two conditions are compatible: if $\mathbf{\Gamma_s}$ is divergence-free under the space derivative, then adding a functional gradient $\delta\Theta(\mathbf{A})/\delta\mathbf{A}$ preserves it as divergence free. Generally, we have sufficient conditions for maintaining gauge invariance of the approximation if we require invariance under $aU + bV$, where a and b are constants.

The Simplest Example. Suppose we set $W(A) \to 0$ for simplicity. The adiabatic magnetic field is the interesting part; for a self–consistent infrared theory, we seek behavior that is slowly varying in both $\vec{A}$ and $\vec{x}$ spaces. The simplest possible choice (Abelian theory) is

$$\beta^{ij} = 2m\varepsilon^{ijk} s^k$$

where s^k are the proton's spin components in the rest frame. This will give us "adiabatically magnetized" gluons coupled to the spin. In the non–Abelian case, multiply by τ_a, a color generator. We should think of this as an ansatz for the most slowly varying terms in a derivative expansion. Note that we must have m a constant with dimensions of mass; the dependence of m on Λ is unknown.

The corresponding connection is $\vec{\Gamma}_s = m\vec{s} \times \vec{A}$, in one choice of (adiabatic) gauge. Now since the theory is the original one but para–gauged by $\vec{\Pi}_A \to \vec{\Pi}'_A = \vec{\Pi}_A + m\vec{s} \times \vec{A}$, we can work backwards to find the Lagrangian such that $\vec{\Pi}'_A = \partial L/(\partial \vec{A}/\partial t)$. This produces the coupling to the topological current, Eq. (1). Due to its fishy way of becoming gauge invariant with a phase, it is not a fundamental theory, but it is what the adiabatic approximation gives.

Gauge Invariance. The symmetries of the theory are easier to see if we use the Lagrangian formulation. Unlike the usual fundamental theory, the para–gauge one is invariant up to a phase shift on the wave functional. It follows that the Lagrangian is invariant up to a surface term, from integration by parts. For example, the covariant Abelian version of the same theory is

$$\mathcal{L} = -\frac{1}{4} F_{\mu\nu} F^{\mu\nu} + \frac{s^\mu}{4m} \varepsilon_{\mu\nu\alpha\beta} A^\gamma F^{\alpha\beta} \,. \tag{3}$$

Under $A^\nu \to A^\nu + \partial^\nu \vartheta$, this is invariant under integration by parts and using the Bianchi identity. The same goes for the non–Abelian case, although then there are "small" gauge transforms (connected to the identity) and "large" ones associates with a winding number. The small transformations can be treated with integration by parts. Coupling to a spacelike vector s^μ is just right to eliminate the potential problem of the large transformations, since there is a global gauge (e.g., $A^0 = 0$) where the offending triple–A term vanishes. Now, provided the parameters s^μ are transformed properly when the state they refer to is transformed, the action from Eq. (1) should be both gauge and Lorentz invariant.

Perturbation Theory. An immediately interesting question is whether perturbation theory in the strong coupling can be used effectively in the new theory. For the lowest order approximation we treat the new terms depending on spin to all orders in the coupling $1/m$ at order g^0. Fortunately this can be done, because the Abelian theory is a quadratic one which can be solved.

Some care is required to get gauge invariant results. This is most readily seen by the Gauss's law constraint in the effective theory

$$D \cdot E \to \vec{\nabla} \cdot \vec{E} = m\vec{s} \cdot \vec{B} \tag{4}$$

This means that some gauge choices, such as Coulomb gauge, will be inconsistent even in the Abelian case. The correct procedure is to solve the Gauss law constraint to eliminate dependent modes, and re–insert those constraints in the

Green functions. Before any gauge choice is made, the Gauss law constraint at order g^0 is

$$-i\omega\vec{k}\cdot\vec{A}_k+\vec{k}^2A^0_k+im\vec{s}\cdot\vec{k}\times\vec{A}_k=0 \tag{5}$$

where $A^\mu(x,t)=\int d^3kA^\mu_k\exp(i\vec{k}\cdot\vec{x}-i\omega(k)t)$. It is convenient to set $\vec{k}\cdot\vec{A}_k=0$ and eliminate A^0 in terms of the other degrees of freedom, leading to the constrained wave equation $\Delta^{ij}A^j_k=0$, where

$$\Delta^{ij}=\delta^{ia}_T\Big\{(\vec{k}^2-\omega^2)\delta^{ab}+i\omega m\varepsilon^{abc}s_c+m^2\varepsilon^{alm}\varepsilon^{bpq}s^\ell\hat{k}^m s^p\hat{k}^q\Big\}\delta^{bi}_T \tag{6}$$

with $\delta^{ij}_T=\delta^{ij}-\hat{k}^i\hat{k}^j, \hat{k}^i=\vec{k}/\mid\vec{k}\mid$. The transverse part of vectors will be denoted by subscript T; for example $A^i_T=\delta^{ij}_TA_j$ The poles in the propagator are found by the eigenvalue relation

$$\begin{aligned}&\omega^2-k^2-m^2s^2_T\ \mp\ \lambda_\pm=0\ ;\\&\lambda_\pm=m^2s^2_T/2\pm(m^2/2)\sqrt{s^4_T+4\omega^2(1-s^2_T)/m^2}\ ,\end{aligned} \tag{7}$$

with polarization eigenvectors

$$\varepsilon^i_\pm=s^i_T\pm(i\omega m/\lambda_\pm)\varepsilon^{iab}s^a_T\hat{k}^b. \tag{8}$$

The dispersion relation for the frequency coming from the eigenvalue equation (Eq. (7)) turns out to be:

$$\omega^2_\pm=k^2+m^2/2\pm(m^2/2)\sqrt{1+4(\vec{k}\cdot\vec{s})^2/m^2}\ . \tag{9}$$

Remarkably, there are no tachyons. From these relations, which are the result of considerable algebra, the determined person can work out the covariant propagator which is rather complicated.

By examining Eqs.(8,9) it is easy to show that in the ultraviolet limit $\omega>>m$, $k>>m$, the propagator becomes independent of m and "standard". The divergent one–loop corrections to the theory are explicitly independent of m (as suggested by power–counting), indicating that perturbation theory at this order is asymptotically free. Power counting would suggest this remains true in higher orders, but that question is open. Perturbation theory in powers of m/ω appears smooth, indicating that the m–dependent terms are "higher twist"; however, perturbative branch cuts at all $0<m/\omega<\infty$ (see below) may possibly upset this.

A Mass Gap. The dispersion relation (Eq. (9)) reveals one mode with a gap and another mode with no gap. It is tempting to call the upper mode a "massive gluon", but more careful examination shows that is is not a true mass, but rather a result of the "magnetization" of the gluons by the adiabatic background field. Because of adiabatic magnetization, the field cannot drift away freely from the origin (in field space), but generally orbits with a finite frequency and energy like a Landau level.

The gapless mode is unusual because it is concave upward (Fig. 1). Such a dispersion relation causes a new kind of instability when the 3–gluon or 4–gluon vertex is turned on. Consider a gluon propagating on the lower branch with frequency and wave numbers (ω_0, k_0). The phase velocity ω_0/k_0 exceeds the phase velocity ω_1/k_1 of all other gluons carrying frequency $\omega_1 < \omega_0$ and $k_1 < k_0$. The larger phase velocity for the first gluon means that it will spontaneously decay by sonic gluonic booms (Cerenkov radiation) into lower frequency on–shell gluons. The lower frequency gluons, in turn, decay to yet lower frequency states and so on until finally reaching zero frequency. The "tree approximation" to the theory thus has no propagating colored states, except possibly a state with an infinite number of zero momentum particles. A color singlet is protected from this decay because the color monopole moment vanishes.

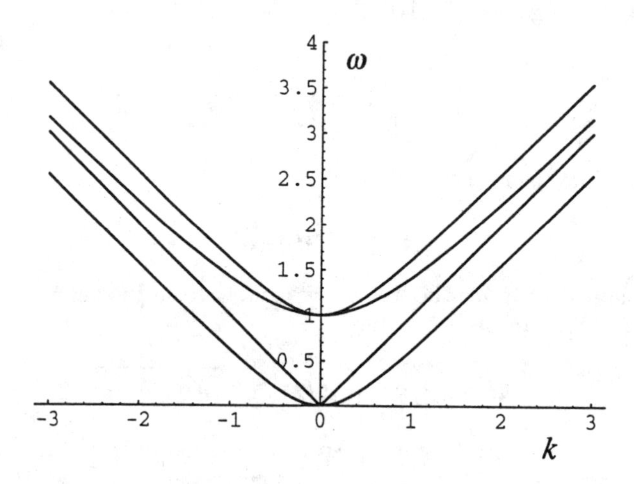

FIGURE 1. Zeroeth order dispersion is real valued, but colored monopole propagation is damped. Curves indicate range of dispersion relations as the angle between $\vec{k}$ and $\vec{s}$ is varied.

This is an interesting hint at confinement, but one must be cautious because the theory needs more investigation. It will be interesting to explore the alternatives to standard perturbation theory using the new propagators and higher order vertices from $s_\mu K^\mu$. Due to the unusual singularity the propagator at second order in perturbation theory seem to develop a gap, although this preliminary result needs confirmation. The calculations (in progress) are straightforward but novel and rather involved due to the presence of vector s^μ. When the technical issues are resolved, one can evaluate the gluon distribution and other correlations in this model (8). One can also consider reformulating the idea in null–plane quantization directly.

ACKNOWLEDGEMENTS

This research was supported in part by DOE Grant Number 85ER40214 and the *Kansas Institute for Theoretical and Computational Science.*

REFERENCES

1. Efremov, A.V. and Teryaev, O.V., *JINR Report* E2–88-278 (1988); Altarelli, G. and Ross, G.G., *Phys. Lett.* **B212**, 391 (1988); Carlitz, R.D., Collins, J.C. and Mueller, A.H., *Phys. Lett.* **214B**, 229 (1988).
2. Messiah, A., *Quantum Mechanics*, (North Holland, Amsterdam 1962) Vol. 2.
3. Mead, C.A. and Truhlar, D.G., *J. Chem. Phys.* **70**, 2284 (1979); Aitchison, I.J.R., *Physica Scripta* **T23**, 12,(1988).
4. For early work in the functional Schroedinger representation see Schwinger, J., *Nuovo Cimento* **30**, 278 (1963); Loos, H., *Phys. Rev.* **188**, 2342 (1969). For a more recent exposition, see Cheng H., and Tsai, E.-R., *Chinese J. Phys.* **25**, 95 (1987).
5. Berry, M.V., *Proc. R. Soc. London Ser.* **A292**, 45 (1984).
6. Wu, Y.S. and Zee, A., *Nuc. Phys.* **B 258**, 157 (1985); Niemi, A., and Semenoff, G., *Phys. Rev. Lett.* **55**, 927 (1985).
7. Zygelman, B., *Phys. Lett.* **A 125**, 476 (1987); Ralston, J. P., *Phys. Rev.* **A40**, 4872 (1989); *ibid* **A40**, 5400 (1989).
8. Ralston, J. P., Kansas Preprint (1994); in preparation.

Quark Spin and Quark Orbital Angular Momentum Content of the Proton

Bing An Li, Department of Physics and Astronomy,
University of Kentucky, Lexington, KY 40506, USA

Abstract

In a quark-meson theory it is found that the smallness of quark spin content and strong strange quark polarization can be understood by the anomaly and large N_c expansion very well. It has been proved in this theory that the proton spin is mostly carried by quark orbital angular momentum.

The quark spin contents of the proton are revealed from the EMC measurements[1]

$$\Delta u + \Delta d + \Delta s = 0.120 \pm 0.094 \pm 0.138, \Delta u = +0.782 \pm 0.032 \pm 0.046,$$
$$\Delta d = -0.472 \pm 0.032 \pm 0.046, \ \Delta s = -0.190 \pm 0.032 \pm 0.046. \quad (1)$$

The quark spin content of the proton is related to the matrix element of flavor-singlet axial-vector current between protons. This is a nonperturbative dynamical problem. In ref.[3], the expression of this flavor-singlet axial-vector current has been derived in a meson theory in terms of momentum expansion. It is well known that chiral symmetry is one of important nature of QCD when the masses of u, d, and s quarks are ignored. The second important fact is the pion form factor is dominant by ρ meson pole, therefore in quantum field theory pion field can be treated as point-like field. Based on these two points the $SU(3)_L \times SU(3)_R$ symmetric lagrangian has been constructed

$$\mathcal{L} = \bar{\psi}\{i\partial\!\!\!/ - mu(x)\}\psi, \quad u(x) = \exp\{i\gamma_5 \sum_{a=1}^{8} \lambda^a \phi^a(x)\}, \quad (2)$$

where ψ is the quark field, ϕ^a is the pseudoscalar field and m is a parameter. The flavor-independent axial-vector current is defined as

$$A_\mu(x) = \langle\bar{\psi}(x)\gamma_\mu\gamma_5\psi(x)\rangle = -iTr\{\gamma_\mu\gamma_5 S_F(x,y)\}|_{y\to x}, \tag{3}$$

The propagator $S_F(x,y)$ satisfies following equation

$$\{i\partial\!\!\!/ - mu(x)\}S_F(x,y) = \delta^4(x-y). \tag{4}$$

At low energy this equation can be solved by using the derivative expansion

$$S_F(x,p) = S_F^{(0)}(x,p)\sum_{n=0}^{\infty}\{(-i\partial\!\!\!/_x)S_F^{(0)}(x,p)\}^n,$$
$$S_F^{(0)} = -\frac{p\!\!\!/ - m\hat{u}}{p^2-m^2}, \quad \hat{u} = \exp\{-i\gamma_5\sum_a\lambda^a\phi^a\}. \tag{5}$$

In this effective theory there are two expansions: large N_c expansion and derivative expansion which is taken as the low energy approximation. The method using quark propagator with background fields to find quark current can be verified by using this method to derived the baryon number current.

$$B_\mu(x) = \frac{1}{3}\langle\bar{\psi}(x)\gamma_\mu\psi(x)\rangle = -iTr\{\gamma_\mu S_F(x,y)\}|_{y\to x}, \tag{6}$$

The leading term of B_μ has been found at the third order in derivatives

$$B^\mu = \frac{N_c}{24\pi^2}\frac{1}{3}\varepsilon^{\mu\nu\alpha\beta}Tr\{\partial_\nu UU^\dagger\partial_\alpha UU^\dagger\partial_\beta UU^\dagger\} \tag{7}$$

where $U = exp\{i\sum_{a=1}^{8}\lambda^a\phi_a(x)\}$. This is exact the topological current derived from Wess-Zumino term. Unlike the baryon number current, at the third order in derivatives there is no flavor-singlet axial-vector current. The leading terms of this current are at the fifth order in derivatives

$$A^\mu(x) = \frac{N_c}{15m^2(4\pi)^2}\varepsilon^{\mu\nu\rho\sigma}Tr\{L_\nu L_\rho L_\lambda\partial^\lambda L_\sigma + R_\nu R_\rho R_\lambda\partial^\lambda R_\sigma\}, \tag{8}$$

where $L_\nu = \partial_\nu UU^\dagger$, $R_\nu = \partial_\nu U^\dagger U$. The matrix element of the spatial component A^j is the quark spin content. Due to the antisymmetric tensor there must be a time derivative in A^j. In this model nucleon emerges as a soliton which is a slow rotating object. The time derivative is at order of $O(1/N_c)$. Therefore A^j is at $O(1)$ which is smaller than all other spatial components of the flavor-dependent currents by one order of magnitude in the large N_c expansion. At the low energy the higher order derivatives make less contribution. A^j is at the fifth order in derivatives. Based on these two factors we expect a small quark spin content of the proton.

In this meson theory, proton emerges as a soliton. Under the hedgehog ansatz in the case of $SU(3)$ symmetry there is

$$U = V(t)\begin{pmatrix} e^{iF(r)\hat{r}\cdot\vec{\tau}} & 0 \\ 0 & 1 \end{pmatrix} V^{\dagger}(t). \tag{9}$$

where $V(t)$ is an 3×3 matrix of the collective coordinates. The total angular momentum of proton can be defined

$$J_i = -\varepsilon_{ijk}\int d^3x \frac{\partial\mathcal{L}}{\partial\pi_l\partial x_j}\frac{\partial\pi_l}{\partial x_k}, \tag{10}$$

where $\mathcal{L}$ is the lagrangian of meson fields which is obtained by integrate out the quark fields from eq.(2). Use the ansatz (9) we obtain

$$J_i = -ia^2 TrV^{-1}\dot{V}\tau_i, \quad [J_i, J_j] = i\varepsilon_{ijk}J_k. \tag{11}$$

where a^2 is the moment of inertia. Therefore $J_i\{i = 1,2,3\}$ are the spin operators of the proton. Substituting the ansatz (9) into eq.(8) we have

$$A^j(x) = \frac{N_c}{15\pi^2 m^2}\{\frac{s^2}{r^2}F'^2(\delta_{ij} - \hat{x}_i\hat{x}_j) + \frac{s^4}{r^4}\hat{x}_i\hat{x}_j\}(-i)TrV^{\dagger}\dot{V}\tau_i \tag{12}$$

$s = \sin F(r)$. From eq.(12) the quark spin content of the proton is found

$$\Delta u + \Delta d + \Delta s = \frac{2N_c}{45\pi m^2 a^2}\int_0^\infty dr\{2s^2F'^2 + \frac{s^4}{r^2}\}, \tag{13}$$

The equation satisfied by $F(r)$ is derived by minimizing the soliton mass[4]

$$\begin{aligned}&(\frac{\tilde{r}^2}{4} + 2s^2 + \frac{9e^4F_\pi^2}{10\pi^2m^2}\frac{s^4}{\tilde{r}^2})F'' + \frac{1}{2}\tilde{r}F' - (\frac{1}{4} - F'^2 + \frac{s^2}{\tilde{r}^2})\sin 2F \\ &-\frac{9e^4F_\pi^2}{10\pi^4m^2\tilde{r}^2}(s^2\sin 2FF'^2 + \frac{2}{\tilde{r}^3}s^4F') - \frac{\mu^2}{e^2F_\pi^2}\tilde{r}^2 s = 0.\end{aligned} \tag{14}$$

where $\tilde{r} = eF_\pi r$ and $\mu^2 = (m_1^2 - \frac{2}{3}\Delta m^2)/e^2F_\pi^2$, $m_1^2 = \frac{1}{8}(3m_\pi^2 + 4m_K^2 + m_\eta^2)$ and $\Delta m^2 = m_K^2 - m_\pi^2$. Choosing $e = 6$ and $m = -1.6$GeV we obtain $g_A = 1.26$, g_A is the β decay constant and proton mass is 1260MeV. The quark spin content(13) is determined to be 0.015. As a matter of fact, the smallness of the quark spin content is very insensitive to the values of e and m. The numerical calculation shows that in a large range of $4 < e < 6$ and $20 < |m| < 6000 MeV$ the quark spin content, as we expected, is always small and consistent with data.

We can also calculate Δu, Δd, and Δs individually. Another two flavor-dependent axial-vector currents A^3_μ and A^8_μ can defined in the same way. They are at order of $O(N_c)$. To the leading order there is

$$\frac{\Delta u - \Delta d}{\Delta u + \Delta d - 2\Delta s} = \frac{\langle p\uparrow |\bar{\psi}\gamma^3\gamma_5\tau_3\psi|p\uparrow\rangle}{\sqrt{3}\langle p\uparrow |\bar{\psi}\gamma^3\gamma_5\lambda_8\psi|p\uparrow\rangle} = \frac{\langle p\uparrow |TrV^\dagger\tau_3 V\tau_3|p\uparrow\rangle}{\sqrt{3}\langle p\uparrow |TrV^\dagger\lambda_8 V\tau_3|p\uparrow\rangle} = \frac{7}{3}. \tag{15}$$

The calculation of eq.(15) is performed in terms of the wave function of proton presented in ref.[4]. From eq.(15) the ratio of F and D values is determined to be $F/D = 0.56$. The deviation from the current value of $F/D = 0.631 \pm 0.018$ is about ten percent. As discussed above, in large N_c expansion $\Delta u + \Delta d - 2\Delta s$ is at order of $O(N_c)$ while $\Delta u + \Delta d + \Delta s$ is $O(1)$. Therefore, a nonzero strange quark spin content of the proton is expected. The Bjorken sum rule can be derived in this theory

$$\Delta u - \Delta d = \langle p\uparrow | \int A^3_3(x)\, d^3x|p\uparrow\rangle = g_A = 1.26. \tag{16}$$

Using eq.(15) we obtain

$$\Delta u + \Delta d - 2\Delta s = 0.54. \tag{17}$$

From eqs.(16,17) the u, d and s spin contents are determined

$$\Delta u = 0.72, \quad \Delta d = -0.54, \quad \Delta s = -0.18. \tag{18}$$

They are consistent with data(1) within the experimental errors.

In this theory anomaly and large N_c expansion provide the explanation of smallness of the quark spin content and strange quark polarization. Due to the structure of Wess-Zumino term A_μ is identical zero at the third order in derivatives. Therefore the quark spin content is at higher order in both large N_c and momentum expansions. The smallness of quark spin content is revealed from these facts. As mentioned above the strange quark polarization is due to the nature of A_μ and A^8_μ in large N_c expansion.

In order to understand what physics effects take the responsibility of proton's spin, quark orbital angular momentum of proton has been studied in this theory. The quark orbital angular momentum of proton can be defined

$$L_i = -i\varepsilon_{ijk} < \psi^*(x)x_j\frac{\partial}{\partial x_k}\psi(x) >= \frac{-i}{(2\pi)^4}\varepsilon_{ijk}x_j\int d^4p p^k Tr\gamma^0 S_F(x,p). \tag{19}$$

Using eq.(5) the quark orbital angular momentum of proton can be calculated. At the second order of derivatives we obtain

$$\int L_i(x)d^3x = \frac{4m^2N_c}{(4\pi)^2}\frac{D}{4}\Gamma(2-\frac{D}{2})\int\frac{2}{3}sin^2F(r)d^3x\frac{1}{a^2}J_i$$
$$+\frac{4}{3(4\pi)^2}\int d^3x\{sin2F(F''+\frac{2}{r}F'-\frac{sin2F}{r^2}+sin^2FF'^2)\}\frac{1}{a^2}J_i. \quad (20)$$

In eq.(20) there is a divergence, therefore renormalization is needed.

The effective lagrangian of meson fields can be obtained by integrated out the quark fields in eq.(2).

$$\mathcal{L} = \frac{m^2N_c}{(4\pi)^2}\frac{D}{4}\Gamma(2-\frac{D}{2})Tr\partial_\mu U\partial^\mu U^\dagger + \frac{N_c}{6(4\pi)^2}Tr\partial_{\mu\nu}U\partial^{\mu\nu}U^\dagger$$
$$-\frac{N_c}{12(4\pi)^2}Tr(\partial_\mu U\partial^\mu U^\dagger\partial_\nu U\partial^\nu U^\dagger + \partial_\mu U^\dagger\partial^\mu U\partial_\nu U^\dagger\partial^\nu U - \partial_\mu U\partial_\nu U^\dagger\partial^\mu U\partial^\nu U^\dagger). (21)$$

Comparing with non-linear σ model the renormalization of the divergence of the first term of eq.(21) takes following form

$$\frac{m^2N_c}{(4\pi)^2}\frac{D}{4}\Gamma(2-\frac{D}{2}) = \frac{F_\pi^2}{16}. \quad (22)$$

The moment of inertia a^2 can be calculated by using the lagrangian(21)

$$a^2 = \int d^3x\{\frac{F_\pi^2}{6}sin^2F + \frac{4}{3(4\pi)^2}[sin^2FF'^2 + sin2F(F''+\frac{2}{r}F'-\frac{sin2F}{r^2}]\}. \quad (23)$$

Comparing eqs.(20) and (23) we obtain

$$\int L_i(x)d^3x = J_i. \quad (24)$$

Therefore, up to the fourth order of derivatives the total spin of proton is carried by the quark orbital angular momentum.

References

[1] Ashman,J. et al. (EMC collaboration), Phys.Lett.**206B** ,364 (1988) and Nucl. Phys. **B328**,1(1989).

[2] Li, Bing An, Phys. Lett., **B282**(1992)435.

[3] Kanazawa, Akira, Prog. of Theor. Phys., **77**,1240(1987).

Spin Structure Functions At SLAC

Hervé Borel
DAPNIA/SPhN, CEN Saclay
91191 Gif-sur-Yvette Cedex, France

Representing the SLAC E143 collaboration

Abstract. Preliminary results on proton and deuteron from E143 polarized deep inelastic scattering experiment at SLAC are presented at 29 GeV, in the $0.029<x<0.8$ domain. Comparison with other experiments and with predictions are made and emphasis is laid on Q^2 corrections.

The recent few years have seen new results in polarized deep inelastic scattering. Results on proton, deuteron and ^{3}He from SLAC E142 [1] and E143 [2] and CERN SMC [3] experiments come in addition to the first data of E80 [4], E130 [5] and EMC [6]. The first motivation of these experiments is to test the Björken sum rule, derived by Björken [7] from current exchange, and which also turns to be a rigourous prediction of QCD. The second motivation comes from the signification of g_1 in the quark parton model relating g_1 to the difference of probability to find a quark with a spin parallel or antiparallel to the spin of the proton. The measurement of the integral of g_1 over the Björken variable x between 0 and 1 allows a determination of the fraction of spin of the nucleon carried by the quarks (i.e : valence, sea quarks and antiquarks). Last, Ellis and Jaffe [13] have calculated sum rules separatly for proton and neutron, assuming SU(3) symmetry and an unpolarized strange sea. Theoretical work has been also accomplished specially on the Q^2 corrections [8] to apply when comparing theory and experiment where the finite Q^2 of the last one must be taken into account. It seems so appropriate to make a kind of status of what we know from the different experiments. I will first remind the way to get g_1, then briefly the principle and caracteristics of the SLAC experiments. The diverse experimental results will be compared. A special part will be devoted to the Q^2 corrections leading to a confrontation between results and theory.

In deep inelastic scattering (DIS), an electron interact with a quark in a nucleon by an exchange of a virtual photon of energy ν and four-momentum Q^2. In the polarized case where both beam and target are polarized, two spin structure functions G_1 and G_2 are added to the unpolarized one F_1 and F_2. The difference of inclusive DIS cross sections when beam and target spins are parallel and antiparallel is a combination of G_1 and G_2:

$$\frac{d^3\sigma}{d\Omega dE'}(\uparrow\downarrow - \uparrow\uparrow) = \frac{4\alpha^2 E'}{EQ^2}\{(E + E'cos\theta)MG_1(\nu, Q^2) - Q^2G_2(\nu, Q^2)\}$$

A similar combination holds when beam and target spins are perpendicular and we flip the spin of the beam :

$$\frac{d^3\sigma}{d\Omega dE'}(\downarrow\leftarrow - \uparrow\leftarrow) = \frac{4\alpha^2 E'}{EQ^2}E'sin\theta\{MG_1(\nu, Q^2) + 2EG_2(\nu, Q^2)\}$$

In the general case these structure functions depend on the transferring energy ν and momentum Q^2; in the scaling limit when $\nu \to \infty$ and $Q^2 \to \infty$ they only depend on the so-called Björken variable $x = \frac{Q^2}{2M\nu}$ (M is the nucleon mass) :

$$M^2\nu G_1(\nu, Q^2) \to g_1(x) \quad and \quad M\nu^2 G_2(\nu, Q^2) \to g_2(x)$$

In order to get g_1, we measure the two electron-nucleon asymmetries

$$A_{\|} = \frac{d\sigma^{\uparrow\downarrow} - d\sigma^{\uparrow\uparrow}}{d\sigma^{\uparrow\downarrow} + d\sigma^{\uparrow\uparrow}} \quad and \quad A_{\perp} = \frac{d\sigma^{\uparrow\rightarrow} - d\sigma^{\uparrow\leftarrow}}{d\sigma^{\uparrow\rightarrow} + d\sigma^{\uparrow\leftarrow}}$$

An other equivalent way is to look at the level of the virtual photon and to speak of virtual photon-nucleon asymmetries

$$A_1 = \frac{1}{(1+\eta\xi)}[\frac{A_{\|}}{D} + \frac{A_{\perp}}{d}] \quad and \quad A_2 = \frac{1}{(1+\eta\xi)}[\frac{A_{\|}}{D} - \eta\frac{A_{\perp}}{d}]$$

where D,d,η,ξ are kinematical factors.
The nucleon spin structure function g_1 is then obtained from $A_{\|}$, $A\perp$ and the unpolarized structure function F_1 :

$$g_1(x, Q^2) = \frac{F_1(x, Q^2)}{D'}[A_{\|} + tan(\frac{\theta}{2})A_{\perp}]$$

From A_1 and A_2, one has

$$g_1(x, Q^2) = \frac{F_1(x, Q^2)}{1+\gamma^2}[A_1 + \gamma A_2]$$

D' and γ are kinematical factors.
The terms involving $A_{\perp}$ or A_2 are small and $g_1 \approx F_1 \times A_1$; this is especially valid in the case of SMC kinematics.

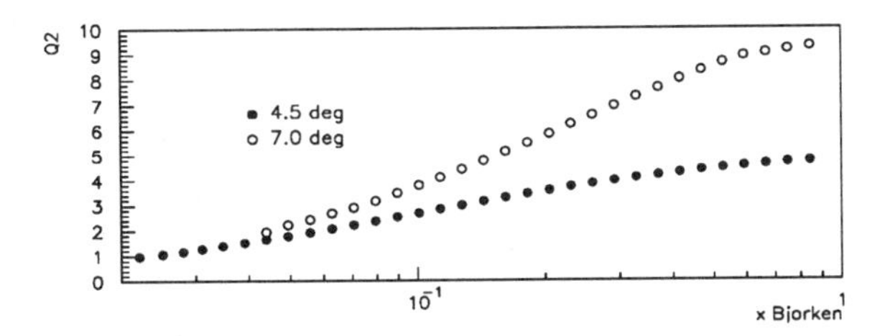

Figure 1: E143 Kinematical region in x and Q^2.

The SLAC experiments E142 and E143 have very close principles and caracteristics . The scattered electrons from a longitudinally polarized electron beam [10] off a polarized target are detected in two independent spectrometers, located at 4.5^o and 7° with respect to the direction of the beam. These two spectrometers allow for more statistics and to cover different kinematical region in x and Q^2 (figure 1).
E143 has measured g_1 and g_2 for the proton and deuteron, using polarized ammonia targets (NH_3 and ND_3) [11], in the range $0.029 < x < 0.8$ and $1 < Q^2 < 10$, with an average $< Q^2 > = 3\ (GeV/c)^2$ [2] (a detailed presentation of the experiment is presented in the contributions from D. Day and, specially for the deuteron, from A. Feltham). In fact, each x bin has a different Q^2 and we need to evolve it at the average Q^2 of the experiment. Two ways of evolving can be considered: a) one is to assume g_1/F_1 independent of Q^2, and multiply by F_1 coming from fit at the average Q^2 of the experiment, to get g_1
b) the other is to assume A_1 *and* A_2 independent of Q^2 and use

$$g_1(x,Q^2) = F_1(x,<Q^2>)\frac{[A_1+\gamma(x,<Q^2>)A_2]}{(1+\gamma(x,<Q^2>)^2)}$$

where $\gamma^2 = \frac{Q^2}{\nu^2}$
Instead of F_1, the second spin averaged nucleon structure function F_2 , coming from NMC fit, and $R = \sigma_L/\sigma_T$ the ratio of longitudinal to transverse virtual photon cross sections, coming from a SLAC global fit, are used:

$$F_1(x,Q^2) = (1+\gamma^2)\frac{F_2(x,Q^2)}{2x(1+R(x,Q^2))}$$

The first assumption (g_1/F_1 independent of Q^2) is made to get E143 results. The figure 2 compares g_1/F_1 for the two spectrometers, in the deuteron case, and does not show any Q^2 dependence within the error bars.

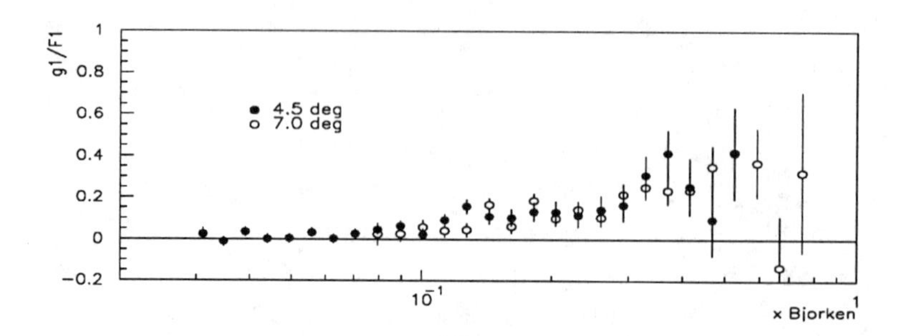

Figure 2: g_1/F_1 for the 4.5 and 7° spectrometers, in the deuteron case.

The value of the integral $\Gamma_1 = \int_0^1 g_1(x)dx$ at $Q^2 = 3$ $(GeV/c)^2$ for the proton and the deuteron is shown in table 1. An extrapolation from x = 0.8 to x = 1 is done assuming g_1 is proportional to $(1 - x)^3$ at high x . The extrapolation to x = 0 is more model dependent, and could be large if g_1 were to increase strongly at low x. Consistent with Regge theory [12], SMC, EMC and E143 data are well fit, for $x < x_{max} = 0.1$, by a constant value of g_1. The systematic error is estimated by varying x_{max} for the fit from 0.03 (for which only CERN data contribute) to 0.12 (for which E143 data dominate). The total integral $\Gamma_1^p = 0.127 \pm 0.004 \pm 0.010$ is in good agreement with the value from SMC, $\Gamma_1^p = 0.136 \pm 0.011 \pm 0.011$, obtained at $Q^2 = 10$ $(GeV/c)^2$. The deuteron result $\Gamma_1^d = 0.042 \pm 0.003 \pm 0.004$ also agrees with SMC result $\Gamma_1^d = 0.023 \pm 0.020 \pm 0.015$ at $Q^2 = 5$ $(GeV/c)^2$.

Table 1: E143 results of $\int g_1(x)dx$ for proton and deuteron at $Q^2 = 3$ $(GeV/c)^2$.

x region	$\int g_1^p(x)dx$ (preliminary)	$\int g_1^d(x)dx$ (preliminary)
measured $0.029 < x < 0.8$	$0.120 \pm 0.004 \pm 0.008$	$0.040 \pm 0.003 \pm 0.004$
extrapolation $0.8 < x < 1$	0.001 ± 0.001	0.000 ± 0.001
extrapolation $0 < x < 0.029$	0.006 ± 0.006	0.002 ± 0.001
Total	$0.127 \pm 0.004 \pm 0.010$	$0.042 \pm 0.003 \pm 0.004$

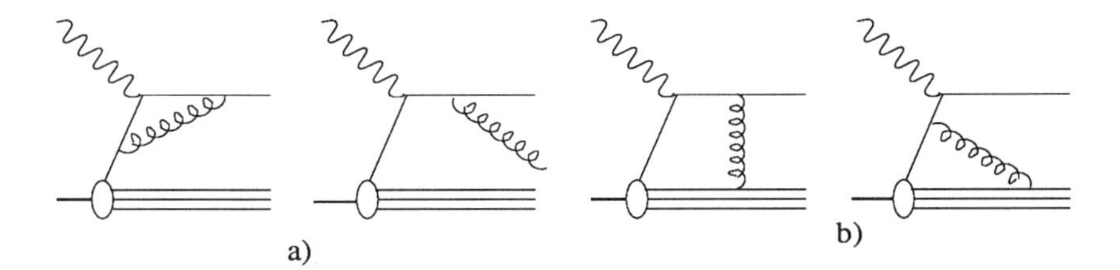

Figure 3: Q^2-dependence to deep inelastic from a) QCD radiative corrections b) higher twist effects.

The comparison of experimental results with theory must take into account the Q^2 corrections. They can be presented in two categories (figure 3): a) the first comes from QCD radiative corrections, developing in serie of the strong coupling constant α_s and giving a logarithmic dependence. This contribution is due to the radiation of gluons from the struck quark . b) the second (higher twists) has a power-dependence and comes *mainly* from an exchange of gluons between the struck quark and the spectator quarks [14] [15]. The α_s corrections only will be presented in this paper. They have been calculated by Larin et al [8] to the third order for the Björken sum rule which has only a singlet contribution.

$$\Gamma_1^p - \Gamma_1^n = \frac{1}{6}\frac{g_a}{g_v}\left[1 - \frac{\alpha_s(Q^2)}{\pi} - 3.58\left(\frac{\alpha_s(Q^2)}{\pi}\right)^2 - 20.22\left(\frac{\alpha_s(Q^2)}{\pi}\right)^3\right]$$

The Ellis-Jaffe sum rule is formed of a singlet and a non-singlet part; the non-singlet part contains the SU(3) a_3 and a_8 matrices related to the SU(3) coupling constants F and D; a_3 is known from neutron β-decay results while the hyperon decay experiments derive a_8 assuming SU(3) symmetry. The fraction of nucleon spin carried by the quark $\Delta\Sigma$ is included in the singlet part. Larin [9] has calculated the $\alpha_s(Q^2)$ corrections to the non-singlet part to the second order.

$$\begin{aligned}\Gamma_1^{p(n)} &= \Gamma_1^{NS} + \Gamma_1^S \\ &= \frac{1}{12}\left(+(-)a_3 + \frac{a_8}{3}\right)\left[1 - \frac{\alpha_s(Q^2)}{\pi} - 3.58\left(\frac{\alpha_s(Q^2)}{\pi}\right)^2 - 20.22\left(\frac{\alpha_s(Q^2)}{\pi}\right)^3\right] \\ &+\frac{1}{9}\left[1 - 0.33\left(\frac{\alpha_s(Q^2)}{\pi}\right) - 0.55\left(\frac{\alpha_s(Q^2)}{\pi}\right)^2\right]\Delta\Sigma\end{aligned}$$

$$
\begin{aligned}
\textit{where}\ \ a_3 &= \frac{g_a}{g_v} = F + D = (\Delta u - \Delta d) \qquad \textit{neutron decay} \\
a_8 &= 3F - D = (\Delta u + \Delta d - 2\Delta s) \ \ \textit{hyperon decay} \\
\Delta\Sigma &= \Delta u + \Delta d + \Delta s \qquad \textit{with}\ \ \Delta u = u^{\uparrow} - u^{\downarrow}
\end{aligned}
$$

Figure 4 shows E143 proton and deuteron results at Q^2=3(GeV/c)^2 and the corresponding predictions of Ellis-Jaffe sum rule for different orders of $\alpha_s(Q^2)$ corrections. Even up to the third order, experimental results are well below the Ellis-Jaffe predictions.The same kind of comparison for the Björken sum rule in figure 5 leads to a good agreement between data and theory when QCD radiative corrections are included. Table 2 sum up E143 experimental results and predictions. The fraction of spin $\Delta\Sigma$ carried by the quarks for each experiment has been calculated for figure 6 from the value of Γ_1 at the average Q^2 of the considered experiment, then applying Q^2 corrections [8] at this average Q^2, and finally using neutron and hyperon results [16]. The data trend to converge to an amount of the order of 30% when different orders of correction are applied. We notice that deuteron result is less sensitive to Q^2 corrections and give smaller error bars on $\Delta\Sigma$.

In conclusion, SLAC and CERN results are compatible but are below Ellis-Jaffe sum rule. They agree with the Björken sum rule when applying Q^2 corrections. Last, the amount of spin carried by the quark is of the order of 30% and the strange sea is negatively polarized at about -10%.

Table 2: E143 results of $\int g_1(x)dx$ for proton and deuteron compared with predictions, both at Q^2 = 3 (GeV/c)^2, using α_s (Q^2=3(GeV/c)^2)=0.35 ± 0.05 [17].

Q^2 = 3 (GeV/c)^2	Γ_1^{exp} E143	Γ_1^{theory} O(α_s^3)
proton	0.127 ± 0.004 ± 0.010 (preliminary)	0.160 ± 0.006
deuteron	0.042 ± 0.003 ± 0.004 (preliminary)	0.068± 0.004

Q^2 = 3 (GeV/c)^2	$(\Gamma_1^p - \Gamma_1^n)^{exp}$ E143	BJ^{theory} O (α_s^3)
E143 proton and deuteron	0.162 ± 0.024 (preliminary)	0.171 ± 0.008
E143 proton and E142 neutron	0.149 ± 0.014 (preliminary)	0.171 ± 0.008

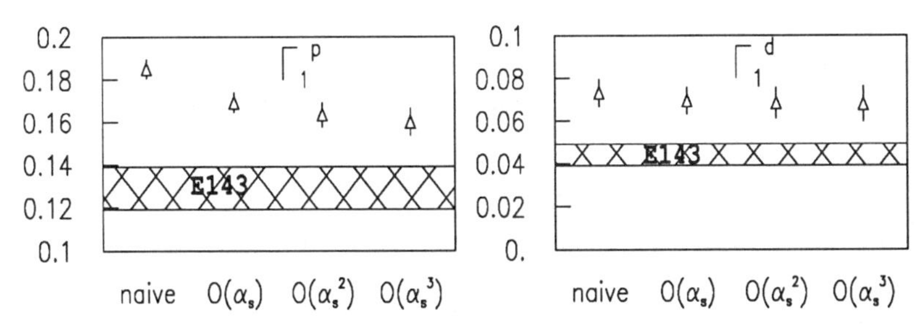

Figure 4: Comparison of Γ_1 from E143 proton and deuteron data at Q^2=3 $(GeV/c)^2$, with Ellis-Jaffe prediction ($\triangle$) for different orders of α_s corrections, using α_s (Q^2=3$(GeV/c)^2$)=0.35 $\pm$ 0.05 [17], F+D = 1.2573 $\pm$ 0.0028 and F/D = 0.575 $\pm$ 0.016 [16].

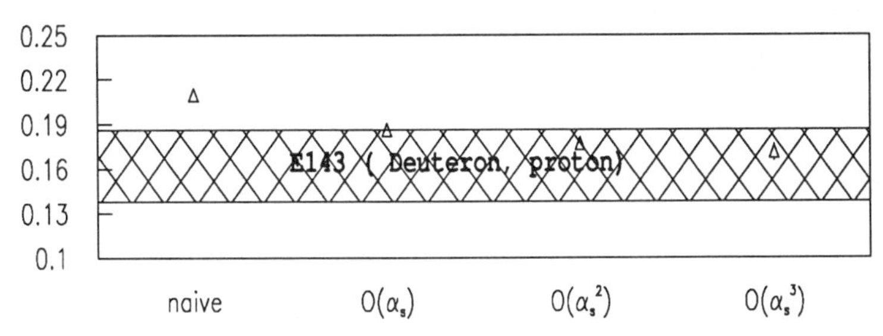

Figure 5: Comparison of Γ_1^p - Γ_1^n from E143 proton and deuteron data at Q^2=3 $(GeV/c)^2$, with Björken prediction ($\triangle$) for different orders of α_s corrections, using α_s (Q^2=3$(GeV/c)^2$)=0.35 $\pm$ 0.05 [17], F+D = 1.2573 $\pm$ 0.0028 and F/D = 0.575 $\pm$ 0.016 [16].

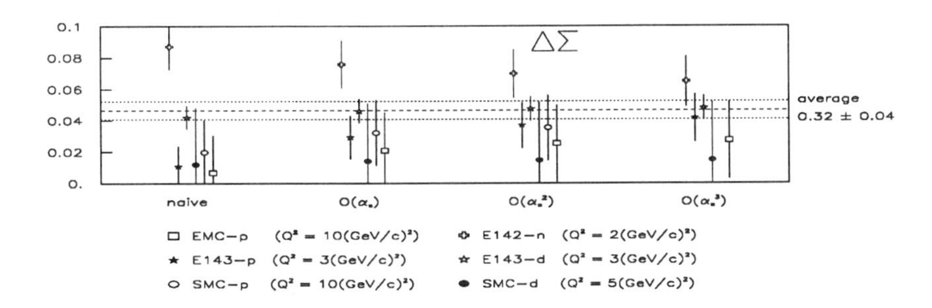

Figure 6: Proton spin fraction carried by quarks as a function of the order of α_s.

References

[1] Anthony, P. L. *et al*, Phys. Rev. Lett. **71**, 959 (1993).

[2] Abe, K. *et al*, submitted to Physical Review Letters.

[3] Adeva, B.*et al*, Phys. Lett. **B302**, 533 (1993); Adams, D.*et al*, Phys. Lett. **B329**, 399 (1994).

[4] Alguard, M. J., Phys. Rev. Lett. **37**, 1261 (1976); **41**, 70 (1978).

[5] Baum, G. *et al*, Phys. Rev. Lett. **51**, 1135 (1983).

[6] Ashman, J. *et al*, Phys. Lett. **B206**, 364 (1988); Nucl. Phys. **B328**, 1 (1989).

[7] Björken, J. D., Phys. Rev. **148**, 1467 (1966); Phys. Rev. **D1**, 1376 (1970).

[8] Larin, S. A. and Vermaseren, J. A., Phys. Lett. **B259**, 345 (1991) and references therein.

[9] Larin, S. A. and Vermaseren, J. A., Phys. Lett. **B334**, 192 (1994).

[10] Maruyama, T., Garwin, E. L., Prepost, R., Zapalac, G. H., Phys. Rev. **B46**, 4261 (1992); Alley, R. *et al*, SLAC-PUB-6489 (1994). See also contribution from Maruyama, T. .

[11] Crabb, D. G. *et al*, Phys. Rev. Lett. **64**, 2627 (1990); Meyer, W. *et al*, Nucl. Inst. and Method. **215**, 65 (1983). See also contribution from Averett, T. .

[12] Heimann, R. L., Nucl. Phys. **B64**, 429 (1973).

[13] Ellis, J. and Jaffe, R., Phys. Rev. **D9**, 1444 (1974); **D10**, 1669 (1974).

[14] Ji, X., Talk at the International Symposium on "The Spin Structure of the Nucleon" given at Yale University, January 1994.

[15] Balitsky, I. and Braun, V., and Kolesnichenko, A. V., Phys. Lett. **B242**, 245 (1990); **B318**, 648 (1993)

[16] Close, F. E. and Roberts, R. G., Phys. Lett. **B316**, 165 (1993); Ellis, J. and Karliner, M., Phys. Lett. **B313**, 131 (1993).

[17] Schmelling, M. and St. Denis, R. D., CERN/PPE93-193; Nerison, S., CERN-TH.7188/94.

The Polarized Gluon Distribution and its Determination in $\vec{p}\vec{p} \to \gamma(\text{large-}p_T) + X$

A.P. Contogouris,[a,b] B. Kamal[a] and Z. Merebashvili[a]

a. Department of Physics, McGill University, Montreal H3A 2T8, Canada
b. Nuclear and Particle Physics, University of Athens, Athens 15771, Greece

Abstract. Higher order (next-to-leading) corrections to $\vec{p}\vec{p} \to \gamma + X$ at large p_T are computed and K-factors exceeding 1 are found. The cross sections peak at pseudorapidity $\eta \simeq 1$. Asymmetries are also determined, and at $\eta \simeq 1$ are shown to be very sensitive to the size of the polarized gluon distribution. Various schemes defining the Dirac matrix γ_5 in $n \neq 4$ dimensions are also discussed.

THE POLARIZED GLUON DISTRIBUTION

The EMC experiment on $\vec{\mu}\vec{p} \to \mu + X$ brought the polarized gluon distribution $\Delta F_{g/p}$ to the center of interest in Spin Physics: Within perturbative QCD, a basic explanation of the results was a sizeable contribution from the subprocess $\vec{\gamma}^*\vec{g} \to q\bar{q}$ and therefore a large $\Delta F_{g/p}$. Since then the size and form of $\Delta F_{g/p}$ remain an important question.

Let us begin by recalling the theoretical situation regarding the form of $\Delta F_{g/p}(x, Q)$ as contrasted with that of the unpolarized $F_{g/p}(x, Q)$.

First, at $x \approx 0$ one invokes arguments based on Regge theory: At $Q^2 = Q_0^2$ = a few GeV^2, the form of the momentum distribution of the parton a is

$$F_{a/p}(x, Q_0) \sim x^{1-\alpha(0)}, \tag{1}$$

where $\alpha(0)$ is the intercept of the leading Regge exchange for the process $\gamma^*\text{a} \to \gamma^*\text{a}$.[1] For a = unpolarized gluon the proper Regge exchange is the Pomeron, and if its intercept is $\alpha_P(0) = 1$, then $F_{g/p}(x, Q_0) \sim$ const. For a = polarized quark, the leading exchange is generally believed to be a low-lying trajectory (possibly a_1) with $\alpha(0) \sim 0$. Thus $\Delta F_{q/p}(x, Q_0) \sim x$, and this is consistent with EMC and SMC data. Now, the behaviour of $\Delta F_{g/p}(x, Q_0)$ is a matter of great controversy; a rather general belief is that $\Delta F_{g/p}(x, Q_0)/F_{g/p}(x, Q_0) \to 0$, and we accept

$$\Delta F_{g/p}(x, Q_0) \to 0, \qquad x \to 0 \tag{2}$$

For the behaviour at large x ($\simeq 1$), recall that $\Delta F_{g/p} \equiv G_{++} - G_{+-}$ and $F_{g/p} \equiv G_{++} + G_{+-}$, where G_{++} (G_{+-}) is the distribution of gluons of positive

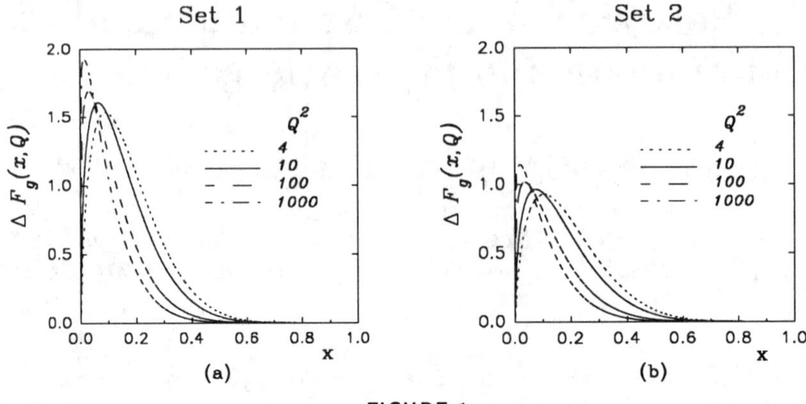

FIGURE 1.

(negative) helicity in a proton of positive helicity. At $x \sim 1$, $G_{+-} \ll G_{++}$, so that

$$\Delta F_{g/p}(x, Q_0) \simeq F_{g/p}(x, Q_0) \sim (1-x)^n, \qquad n \approx 5 \tag{3}$$

The above x behaviour and the Q - dependence of $\Delta F_{g/p}$ is shown in Fig. 1, which presents two typical cases of $\Delta F_{g/p}$: With $Q_0 = 2$ GeV, the first corresponds to $\Gamma(Q_0) \equiv \int_0^1 \frac{dx}{x} \Delta F_{g/p}(x, Q_0) = 5$ ($\Delta F_{g/p}$ large, input of set 1 of Ref. 2) and the second to $\Gamma(Q_0) = 2.8$ ($\Delta F_{g/p}$ moderate, set 2). Thus, as a result of (2) and (3), $\Delta F_{g/p}$ is believed to peak at some $x \neq 0$, in contrast to $F_{g/p}$ that always has its maximum at $x = 0$.

DIRECT γ PRODUCTION AS A PROBE OF $\Delta F_{g/p}$ AND THE NEED FOR HOC

To probe the size and form of the polarized gluon distribution, one needs reactions dominated by subprocesses with gluons in the initial state. One of the best is $\vec{p}\vec{p} \to \gamma(\text{large-}p_T) + X$, which is dominated by $\vec{g}\vec{q} \to \gamma q$, and experiments are planned at RHIC, at c.m. energy $\sqrt{s} = 100 \sim 500$ GeV.

The reaction was first studied at the lowest order of α_s (Born level).[3–5] An interesting feature has been that the inclusive cross section $E\Delta d\sigma/d^3p$ versus c.m. photon pseudorapidity η peaks at $\eta \approx 1$ (not $\eta = 0$);[4,5] see the Born cross section of Fig. 3(a) at $\sqrt{s} = 100$, $p_T = 6$ GeV (dashed line). This is due to the form of the $\vec{g}\vec{q} \to \gamma q$ cross section, which gives dominant contributions from regions where it increases with $|\eta|$. At the same time, the cross section is multiplied by parton distributions, decreasing at large $|\eta|$ - hence the peak at $\eta \neq 0$.

However, higher order corrections (HOC), due to gluon loops and gluon Bremsstrahlung (Brems), are generally large (comparable to the Born), and in QCD their study is essential. Some of the reasons are the following:
(i) If HOC are large and of opposite sign to the Born, the predicted cross sections will be small and the experiments will be inconclusive.
(ii) HOC may change the shape of the Born cross sections.
(iii) In QCD theory there are several ambiguities, like the choice of scale Q^2, of the renormalization and factorization schemes, etc... (see also below). If higher orders are included in the subprocess cross sections, in the running coupling $\alpha_s(Q^2)$ and in the Q^2-dependence of the parton distributions, then in general, the ambiguities in the prediction of the physical cross sections are significantly reduced; the reason is that then the ambiguities affect even higher orders of α_s (in $\vec{p}\vec{p} \to \gamma + X$ they affect $\mathcal{O}(\alpha_s^3)$).

For $\vec{p}\vec{p} \to \gamma + X$, complete $\mathcal{O}(\alpha_s^2)$ HOC have been determined by two groups.[6,7] We present part of the results of the first and comment of those of the second.

K-FACTORS, CROSS SECTIONS AND ASYMMETRIES

Regarding $\vec{p}\vec{p} \to \gamma + X$, we are interested in the following difference of inclusive cross sections:

$$E\frac{\Delta d\sigma}{d^3p}(s, p_T, \eta) = \frac{1}{2}\left\{E\frac{d\sigma}{d^3p}(p(+)p(+) \to \gamma X) - E\frac{d\sigma}{d^3p}(p(+)p(-) \to \gamma X)\right\} \tag{4}$$

where + and − refer to proton helicities. A complete calculation of HOC involves several subprocesses.[6,7] Taking as example the dominant $\vec{g}\vec{q} \to \gamma q$, its contribution has the form

$$E\frac{\Delta d\sigma}{d^3p} = \frac{\alpha_s}{\pi}\int \frac{dx_a}{x_a}\frac{dx_b}{x_b}\Delta F_g(x_a)\Delta F_q(x_b)\{\Delta B\delta(1+\frac{\hat{t}+\hat{u}}{\hat{s}}) + \frac{\alpha_s}{\pi}\Delta f\theta(1+\frac{\hat{t}+\hat{u}}{\hat{s}})\} \tag{5}$$

where ΔB stands for the Born term, Δf for the HOC and $\hat{s}, \hat{t}, \hat{u}$ are subprocess variables. Gluon loops and Brems contribute to Δf, each with terms involving singularities. To regularize them, we work in $n = 4 - 2\varepsilon$ dimensions. Then there are singularities of the types $\sim 1/\varepsilon^2$ (due to gluons being both soft and collinear) and $\sim 1/\varepsilon$ (soft or collinear or ultraviolet nature). The former and part of the latter cancel in the sum: loops + Brems (Bloch-Nordsieck mechanism); the remaining are eliminated by further adding proper counter-terms of known (universal) form.

$\Delta F_{g/p}$ and $\Delta F_{q/p}$ are obtained by fitting data on $\vec{l}\vec{p} \to l + X$ ($l = e$ or μ), the Bjorken sum rule etc. Subsequently, we present results using set 1 of Ref. 2. In Fig. 3(b) we also present results using set 2.

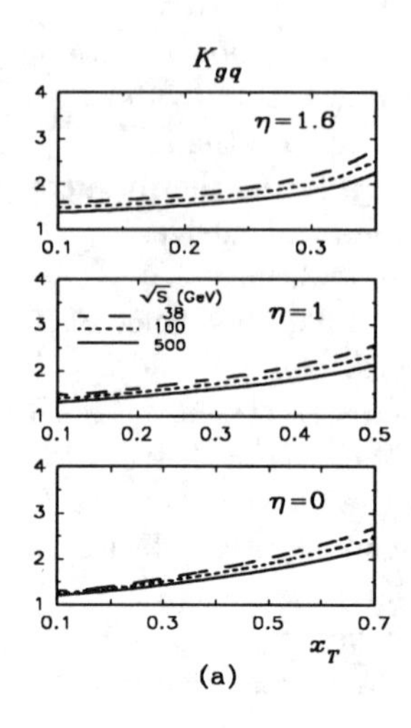

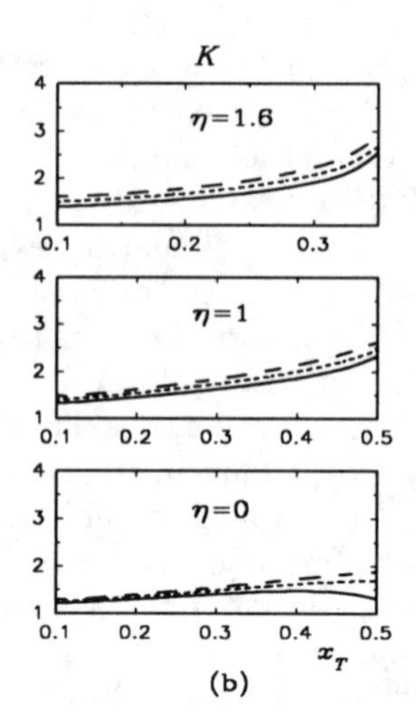

FIGURE 2.

Denoting by $\sigma_B(gq)$ and $\sigma_{HO}(gq)$ the contributions to (5) of ΔB and Δf from $\vec{g}\vec{q} \to \gamma q$, Fig. 2 shows first the K-factor

$$K \equiv \frac{\sigma_B(gq) + \sigma_{HO}(gq)}{\sigma_B(gq)} \tag{6}$$

at $\sqrt{s} = 38$, 100 and 500 GeV (dashed, dotted and solid lines) and various pseudorapidities. Then it shows the K-factor corresponding to contributions of all the subprocesses.[6,7] Notice the following features:

(i) Always the K-factors exceed 1, i.e. the HOC *enhance* the Born cross sections.

(ii) The subprocess $\vec{g}\vec{q} \to \gamma q$ dominates, and this controls the HOC to a great extent. Only at $\eta = 0$ and large x_T is there an appreciable difference between K_{gq} and K, due to a non-negligible contribution from the subprocess $\vec{q}\vec{q} \to qq\gamma$[6,7] resulting from the relative hardness of $\Delta F_{q/p}$.

(iii) In a wide range of x_T, K does not much differ from a constant.

Next, Fig. 3(a) presents total (Born+HOC) cross sections $E\Delta d\sigma/d^3p$; also, as mentioned, it presents the Born $E\Delta d\sigma/d^3p$ for $\sqrt{s} = 100$ GeV and $p_T = 6$ GeV. The point to notice is that the peak at $\eta \approx 1$ *does persist* after inclusion of HOC (in fact it is enhanced). This is due to $K \approx$ const (feature (iii)).

Finally, Fig. 3(b) presents the asymmetry

$$A(s, p_T, \eta) \equiv \frac{E\Delta d\sigma(s, p_T, \eta)/d^3p}{Ed\sigma(s, p_T, \eta)/d^3p} \tag{7}$$

for set 1 (solid lines) and set 2 (dashed).[2] The unpolarized cross sections also include HOC and have been calculated with the set S-$\overline{\text{MS}}$ of Ref. 8 (fit to all

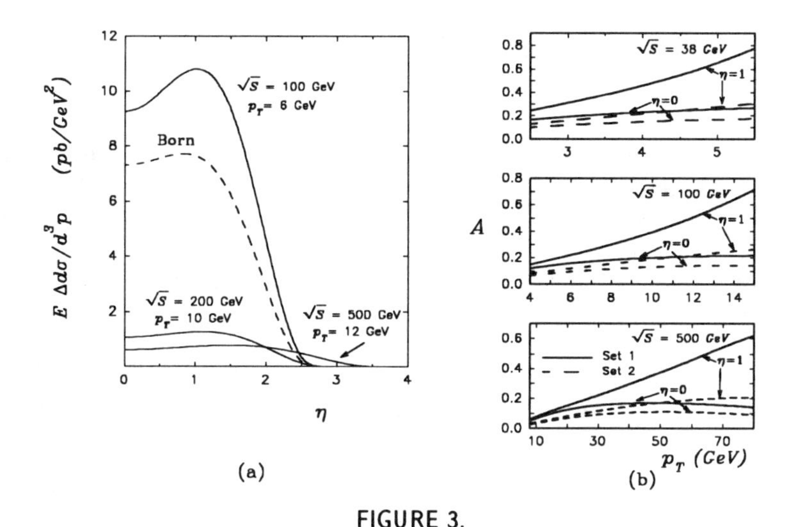

FIGURE 3.

DIS data, Q^2 evolution based on higher order split functions). Clearly the asymmetries are sizeable and, in particular at $\eta \approx 1$, are quite sensitive to the size of $\Delta F_{g/p}$.

Very similar to those of Figs 2,3 are the results of Ref. 7, in spite of the fact that, not only are the HOC determined by a different procedure, but also different $\Delta F_{g/p}$ and $\Delta F_{q/p}$ have been used. The difference in the procedure and the reason of the similarity of results is explained below.

SCHEMES FOR DETERMINING HOC IN POLARIZED PROCESSES

Since the determination of HOC involves intermediate steps with singular quantities, a regularization procedure is necessary, and most convenient is a dimensional one (working in $n \neq 4$ dimensions). Calculations, however, with polarized partons involve projection operators like $\not{p}(1 \pm \gamma_5)$. Now, $\gamma_5 (= i\gamma^0\gamma^1\gamma^2\gamma^3)$ is defined in 4 dimensions, thus a scheme for extending it to $n \neq 4$ is necessary.

One scheme (HVBM), first proposed by t'Hooft and Veltman,[9] is characterized by the following commutation relations:

$$\{\gamma_5, \gamma^\mu\} = 0 \quad \mu = 0,1,2,3; \qquad [\gamma_5, \gamma^\mu] = 0 \quad \text{other } \mu \tag{8}$$

Clearly this scheme treats differently the first 4 and the remaining $n-4$ dimensions, so intermediate expressions are non-covariant and, until recently, calculation of HOC of our type was almost impossible.

Thus a simpler scheme was developed[10] based on

$$\{\gamma_5, \gamma^\mu\} = 0 \quad \text{all } n \text{ dimensions;} \quad \gamma_5^2 = 1. \tag{9}$$

Ref. 6 uses this scheme. It also uses a scheme calculating traces in 4 dimensions, but loop and phase-space integrals in $n \neq 4$ (dimensional reduction).

More recently a routine became available[11] facilitating calculations in the HVBM scheme; this has been used in Ref. 7. Now, the reason for the similarity of the results is that, at least for the dominant subprocess $\vec{g}\vec{q} \to \gamma q$, in the expression of Δf (Eq.(5)), in spite of the presence of a very large number of terms, only a few of them differ.[12]

At present, the Q^2-dependence of $\Delta F_{a/p}(x, Q^2)$ is determined via lowest order split functions. We expect that when higher order ones become available, use of different schemes will change physical quantities even less.

REFERENCES

1. B.L. Ioffe, V.A. Khoze and L.N. Lipatov, *Hard Processes, Vol1, Phenomenology / Quark-Parton Model* (North Holland, 1984).
2. G. Altarelli and W. Stirling, *Particle World* **1**, 40 (1989).
3. E. Berger and J. Qiu, *Phys. Rev.* **D40**, 778 (1989); H.-Y. Cheng and S.N. Lai, *ibid* **41**, 91 (1990); S. Gupta et al, *Z. Phys.* **C42**, 493 (1989).
4. A.P. Contogouris, S. Papadopoulos and B. Kamal, *Nucl. Phys.* (Proc. Suppl.) **B23**, 119 (1991).
5. G. Bunce et al, *Particle World* **3**, 1 (1992).
6. A.P. Contogouris, B. Kamal, Z. Merebashvili and F.V. Tkachov, *Phys. Rev.* **D48**, 4092 (1993); *Phys. Lett.* **B304** 329 (1993).
7. L.E. Gordon and W. Vogelsang, *Phys. Rev.* **D48** 3136 (1993); *ibid* **D49** 170 (1994).
8. J. Morfin and W.K. Tung, *Z. Phys.* **C52**, 13 (1991).
9. G. 't Hooft and M. Veltman, *Nucl. Phys.* **B44**, 189 (1972); P. Breitenlohner and D. Maison, *Commun. Math. Phys.* **52**, 11 (1977).
10. A.P. Contogouris, S. Papadopoulos and F.V. Tkachov, *Phys. Rev.* **D46**, 2846 (1992).
11. M. Jamin and M.E. Lautenbacher, *Tracer (Version 1.1)*, TU München report TUM-T31-20/91.
12. A.P. Contogouris, B. Kamal, Z. Merebashvili, S. Papadopoulos and F.V. Tkachov, *Proceed. of Intern. Workshop on Spin Physics, Protvino, Russia*, September 1993 (to appear).

Jet Handedness Correlation in Hadronic Z^0-Decay

A. Efremov*and L. Tkatchev
Joint Institute for Nuclear Research, Dubna, 141980 Russia

Abstract

Abstract. A short introduction to handedness is given. A specific of its observation in $Z^0 \to$ 2-jet decay and some preliminary experimental results are reviewed. A special attention is drawn to a puzzling phenomena of the jet handedness correlation in $Z^0 \to$2-jet events. The sign of the correlation contradicts ether to charge conjugation of the two jets or to factorization of q and $\bar{q}$ fragmentation into hadrons. A theoretical significance of the effect is discussed.

1 Introduction

Recently a new property of the multiparticle parton fragmentation function was proposed [1] for experimental measurement of partons polarization – its asymmetry with respect to pseudoscalar variable

$$X = \frac{(\vec{k}_1 \times \vec{k}_2) \cdot \vec{j}}{|\vec{k}_1|\;|\vec{k}_2|} \tag{1}$$

in Lab frame of reference. (It was named the longitudinal *handedness.*) Here $\vec{k}_1$, $\vec{k}_2$ are momenta of two particles of jet, selected according to some criteria.The vector $\vec{j}$ is a unit vector in jet direction defined by thrust axis or by total jet momenta [1].

In this paper we present some theory of jet handedness, specifics of its search in e^+e^--annihilation and some result of the search is reviewed. The main aim however is to present preliminary experimental result on the longitudinal

*Partially supported by International Science Foundation under the Grant RFE000 and Russian Foundation for Fundamental Investigation under the Grant 93-02-3811.

[1]Similarly one can define two transverse component of the handedness using two unite transverse vectors instead of $\vec{j}$. So the handedness is in fact a pseudovector similar to polarization.

Instead of jet axis one can use a unit vector in direction of momenta of a triple of particles.

jet handedness correlation in 2-jet events in e^+e^- annihilation in the region of Z^0−peak and theoretical discussion of puzzling features observed.

The asymmetry with respect to X is interesting due to the following reason. The dependence on the pseudoscalar X can appear only in pair with another pseudoscalar. The only one characterizing the two-particle fragmentation of an object (quark, gluon or resonance) is a longitudinal polarization P. So measurement of the asymmetry could give an information about polarization P.

Indeed, let a probability of a right (left) handed quark (with helicity $h = \pm 1$) to fragment into a right handed jet ($X > 0$) or (assuming the P-invariance of fragmentation) a probability of left (right) handed quark to fragment into left handed jet ($X < 0$) be

$$\frac{N_R}{n^q_\pm} = \frac{N_L}{n^q_\mp} = \frac{1}{2}(1 \pm \alpha^q) \tag{2}$$

where $n^q_\pm$ and $N_{R,L}$ are numbers of right or left handed quarks and jets correspondingly. Then for jet handedness of definite quark (antiquark) flavor one can obtain ¿from (2)

$$H^{q,\bar{q}} = \frac{N_R - N_L}{N} = \alpha^{q,\bar{q}} P_{q,\bar{q}} \tag{3}$$

where $P_q = (n^q_+ - n^q_-)/N$ is a quark polarization and $N = N_R + N_L = n^q_+ + n^q_-$. So the knowledge of the *analyzing power* α allows to measure the quark polarization.

The value of α naturally depends on kinematical cuts implied in selecting the pair of particles (e.g. particle rapidity and transverce momenta y and k_T, rapidity difference Δy, pair invariant mass m_{12} etc.) and on the chosen criteria on particle "1" in (1). Concerning the selection of particle "1" one can discriminate between a charge independent criteria Z (e.g. the particle "1" is the leading one in a pair, i.e. $|y_1| > |y_2|$) and a charge dependent criteria Q (e.g. the particle "1" is positive one in (+−)-pair). So the handedness H and the analyzing power α could acquire the label Z or Q depending on chosen criteria on particle "1".

Now turn to some features of α. Charge conjugation transforms quarks into antiquarks with the same helicities and negative particle of the pair into positive one. So it does not change the handedness of jet in the criteria Z but change it for opposite in the criteria Q. As a consequence one have [1, 2]

$$\alpha^{\bar{q}}_Z = \alpha^q_Z \text{ and } \alpha^{\bar{q}}_Q = -\alpha^q_Q \; . \tag{4}$$

Another relation follows from $SU(2)$ flavor symmetry which transforms u-quark into d-quark and if the (+−)-pair arc chosen as pion pair the handedness of jets does not change under $u \leftrightarrow d$ transformation in the criteria Z but changes for the opposite for the criteria Q, i.e

$$\alpha_Z^u = \alpha_Z^d \text{ and } \alpha_Q^u = -\alpha_Q^d . \quad (5)$$

Notice however that $SU(2)$ invariance and the relation (5) could be broken for heavy flavors.

The theoretical estimates for the value of jet handedness are rather uncertain. A few very general statements however could give useful indications for the search. The handedness just as a polarization is an interference phenomena [1]. So it is most probable when a pair of particles in a resonance region interferes with a nonresonant background. Since in parton fragmentation we have to deal mostly with pions, the most prominent resonances are in a region of 1 GeV in invariant mass of the pair (e.g. in the region of the $\rho-$resonance). One can expect also that the most leading particles are the most informative about parton spin state (just as about its charge or flavor) and that the handedness will be more pronounced for large k_T where the variable X is large due to two-particle fragmentation function have to be linear in P and thus linear in PX.

Concerning the value of the handedness one can state that the commonly used QCD Monte-Carlo models like JETSET or HERWIG deal with probabilities rather than with amplitudes and so does not contain any interference phenomena like the handedness. The lowest order perturbative QCD diagrams give an effect proportional to squared quark mass while one loop calculation [4] results in a small value of analizing power $\alpha \sim \alpha_s(k_T/M_{jet})^2 z$, where M_{jet} is a jet mass, k_T is a transverse momentum and z is a fraction of longitudinal momentum of the pair. This could mean that partons transmit their helicity to hadrons at a very late nonperturbative stage of fragmentation. All this makes the problem of estimation of the handedness rather uncertain. Simplest estimations of α using an effective Feynmann diagrams of pion interference in fragmentation $q \to \pi^+ \pi^- q'$ produced via ρ-decay and produced successively give the value of few percent [3]. A similar estimation was obtained in the model of jet handedness origin recently proposed by M.Ryskin in Ref. [5] according to which the handedness arise due to turning of q and $\bar{q}$ produced in breaking of string in longitudinal chromo-magnetic field from chromo-magnetic moments of initial q and $\bar{q}$. Such value of α being experimentally confirmed in a process with known quark polarization allows to hope for applications of the handedness (3) for measuring a quark polarization in some other process with quark production.

2 Handedness in e^+e^--annihilation

The e^+e^--annihilation in the region of the Z^0-peak seems at the first sight as one of the best places for searching the handedness of quark jets and measuring the analyzing power α. This is due to the fact that the quarks from Z^0-decay are strongly polarized as a result of the interference of vector and axial couplings. In the Standard Model quark polarization are $P_u = -0.67$, $P_d = -0.93$ with production ratio $\sigma_u/\sigma_d = 0.78$ and opposite sign polarization for the antiquarks. If one do not distinguish between quark and antiquark jet one can easily find from (4) that $H_Z^{\bar{q}} = -H_Z^q$ and the total handedness cancels to zero. However for charge criteria $H_Q^{\bar{q}} = H_Q^q$ and the handedness for q and $\bar{q}$ add to each other.

With no distinction between quark flavor one has also

$$H_Q^{e^+e^-} = \sum_q w_q \alpha_Q^q P_q \quad \text{and} \quad H_Z^{e^+e^-} = 0 \; , \tag{6}$$

where the probabilities w's consist of a flavor rate production and of probability of the flavor to fragment into a pair obeying the applied cuts[2],

$$w_q = \frac{\sigma_q w_{cut}^q}{\sum_q \sigma_q w_{cut}^q} \; . \tag{7}$$

It is clear from (5) that u and d terms in (6) could be of different signs and some cancellation are possible.

It could be a reason that only a rather small value of the handedness was observed experimentally [6] in e^+e^--annihilation via Z^0. The best value

$$H_Q^{e^+e^-} = 1.19 \pm 0.48\% \tag{8}$$

was seen for leading $(++-)$ and $(--+)$ pion triples with the total longitudinal momenta $k_L = (k_1 + k_2 + k_3)_L \geq 5\ GeV/c$ in the ρ-resonance region of invariant mass of $(+-)$-pairs $0.62 < m_{13} < m_{12} < 0.92\ GeV/c^2$ while charge independent (Z) criteria gives zero value $H_Z = -0.02 \pm 0.5\%$ as it should be due to cancellation of handedness of quark and antiquark jets. This agrees with SLD observation [7] $H_Q < 2.0\%$.

As for the value of analyzing power α it should be find using a general expression (6) determined by probabilities w_{cut}^q. E.g. in the most optimistic case when only light u and d quarks are dominated in fragmenting into ρ-resonance (i.e. $w_b \approx w_c \approx w_s \approx 0$) the average quark polarization $\overline{P} = (\sigma_d P_d - \sigma_u P_u)/(\sigma_d + \sigma_u) = -0.23$ and with the value (8) $\bar{\alpha} \approx -5 \pm 2\%$.

[2]The latter could be calculated using a Monte-Carlo generated events with the same cuts.

This cancellation in H was a motivation of turning to the search for handedness correlation in 2-jet events

$$C = \frac{\Delta N_{xx}}{N_{xx}^{tot}} = \frac{N_{RL} + N_{LR} - N_{RR} - N_{LL}}{N_{RL} + N_{LR} + N_{RR} + N_{LL}} \ . \tag{9}$$

Since at the production level $e^+e^- \to q\bar{q}$ the helicities of quark and antiquark are always correlated, i.e. $n_{++}^{q\bar{q}} = n_{--}^{q\bar{q}} = 0$, one can write using (2)

$$N_{RR} = n_{+-}^{q\bar{q}} \cdot \frac{1}{4}(1 + \alpha^q)\,(1 - \alpha^{\bar{q}}) + n_{-+}^{q\bar{q}} \cdot \frac{1}{4}(1 - \alpha^q)\,(1 + \alpha^{\bar{q}})$$

and similar expressions for N_{LL}, N_{RL} and N_{LR}. An important assumption used here is that each quark in Z° decay fragments independently of its partner. Substituting this into the correlation (9) and making a sum over the quark flavors one obtains[3]

$$C = \sum_q \bar{w}_q \alpha^q \alpha^{\bar{q}}. \tag{10}$$

The probabilities $\bar{w}$'s here have to contain probabilities of both quark and antiquark to fragment into desired pairs in addition to the flavor production rate $\bar{w}^q = (w_{cut}^q)^2\sigma_q / [\sum_q (w_{cut}^q)^2 \sigma_q]$. Using the relation (4) one can find for different criteria

$$C_Q = -\sum_q \bar{w}_q (\alpha_Q^q)^2 \ \text{ and } \ C_Z = \sum_q \bar{w}_q (\alpha_Z^q)^2 \ . \tag{11}$$

So, the correlations are sign definite and no cancellation is expected. Moreover it has to be *negative* in charge criteria Q and *positive* in no charge criteria Z.

3 Experimental observation of correlation

The first observation of handedness correlation using the DELPHI data collection was reported at Moriond-94 workshop [8].

An initial statistics of about 1 MZ° hadronic events selected by a standard cuts [9] were used. According to JADE method with jet resolution parameter $Y_{cut} = 0.08$ a number of jets for each event was determined. Only 2-jet events were remained for the following analysis. In addition, acollinearity of two jets less than 15° was implied.

The unit vector $\vec{t}$ along thrust axis was taken as a jet axis vector. The jet axis $\vec{j}$ was chosen as $\pm\vec{t}$ depending on sign of rapidity of the pair. In each event a nonintersecting pairs of pions were selected satisfying one-particle and two-particle cuts:

[3] In more general case of quark helicity correlation $c_{q\bar{q}} = (n_{+-} + n_{-+} - n_{++} - n_{--})/n$ the r.h.s. of the expression should be multiplied by this number.

i. The rapidity with respect to the thrust axis $|y_t| > Y_{min} > 1$ to be in a leading (assumed the most informative) group of particles.

ii. The transverse momenta $k_T > k_T^{min} > 0.5\ GeV/c$ – an average k_T in jet – to get rid of low k_T hadrons created by hadronization of soft gluons.

iii. The difference in rapidity of pions in pair $|\Delta y_t| < \Delta Y_{max}$ to select correlated pions created mostly from the same breaking of the $q\bar{q}$ string.

iv. The invariant mass of the pair $m_{12} \leq M_{max}^{pair} \leq 1\ GeV/c^2$ to be in resonance region.

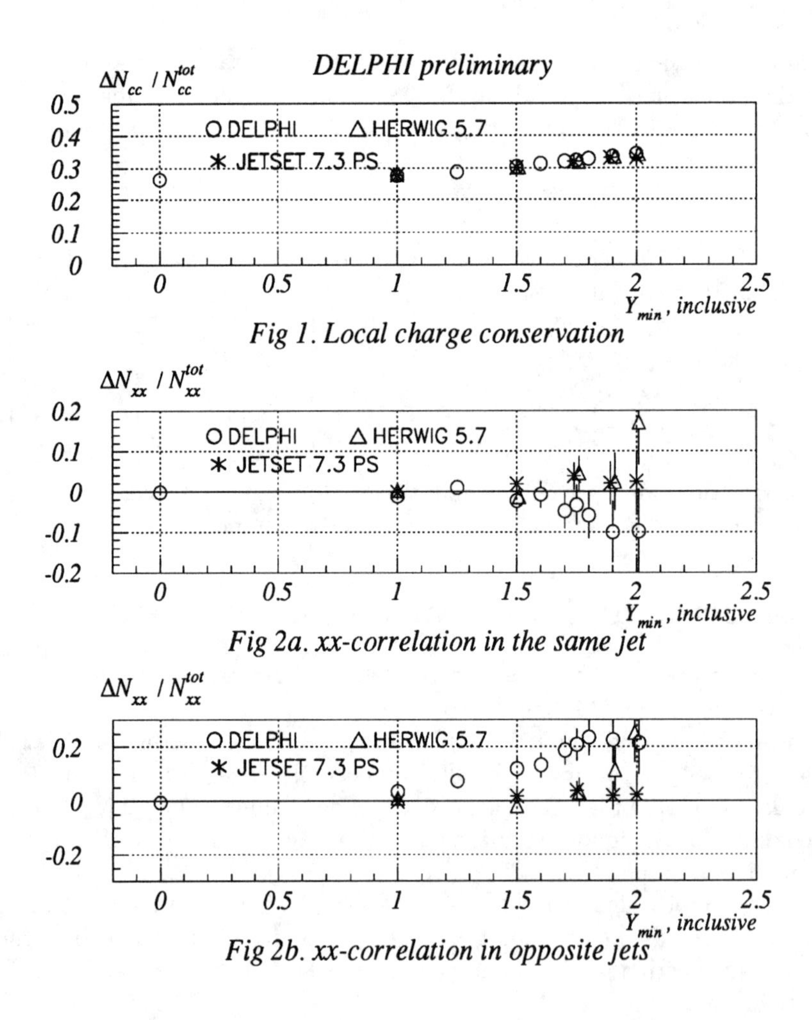

Fig 1. Local charge conservation

Fig 2a. xx-correlation in the same jet

Fig 2b. xx-correlation in opposite jets

v. The absolute value of X defined by (1) is greater then 0.01. For each given track among different pairs which satisfy the above cuts only the pair

DELPHI preliminary

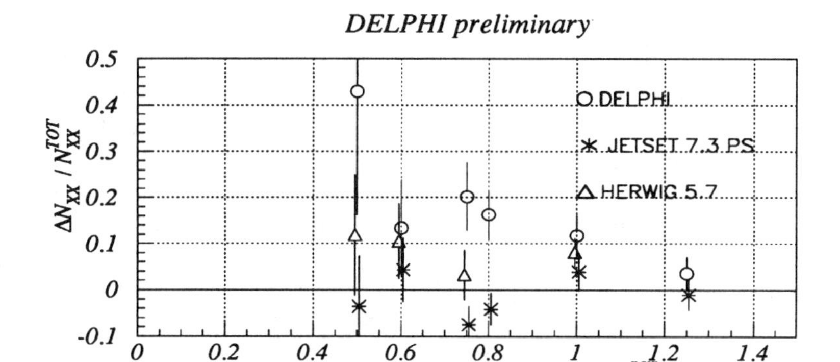

Fig 3. Mass dependence of xx-correlation for opposite jets

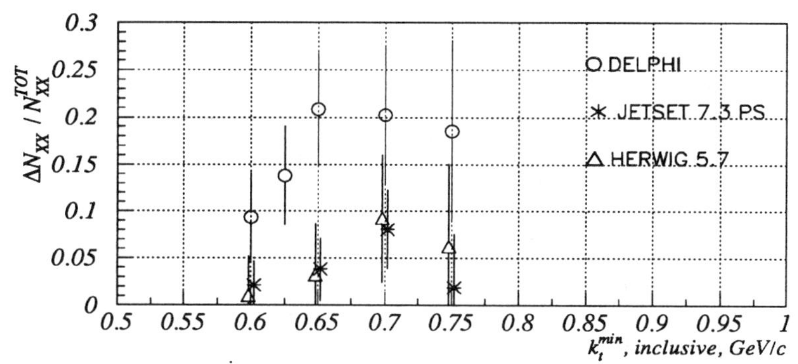

Fig 4. k_t^{min}-dependence of xx-correlation for opposite jets

with largest value of $|X|$ was selected. (The last cut is due to limited momentum resolution of the DELPHI apparatus and off-line analysis procedure [10].)

The handedness correlation (9) of two pairs in the same events both in the same and in opposite jets was investigated for the charge and for the rapidity criteria for particle "1". For the former case only $(+-)$-pairs were taken into account. For the sake of control approximately the same number (about 1 MZ^0) of Monte Carlo events produced by JETSET7.3 PS and by HERWIG5.7 generators were used with the same cuts for hadronic 2-jet events and pair selection.

In the Fig. 1-2 Y_{min} dependence of $\Delta N_{cc}/N_{cc}^{tot}$ and $\Delta N_{xx}/N_{xx}^{tot}$ is presented where ΔN_{cc} is a number of neutral and double charged pairs difference. The Fig. 1 shows correlation of the charges in the same pair. It demonstrates a dominance of neutral pairs over charged ones as a result of local charge conservation in string breaking. The JETSET and HERWIG Monte Carlo programs reproduce this behavior. This serves as an internal check of analysis procedure. The other checks were done by comparing the distributions before and after cuts for the total momenta of all charged particles, energy, charged multiplicity, lepton multiplicity etc. All the distributions well correspond to each other except the energy where the cuts result in a shift about 5 GeV between this two distributions. The shift is well reproduced by the Monte-Carlo distributions also.

The Y_{min} dependence of the C_Q-correlation for two selected pairs in jets is shown on (Figs. 2a, 2b) at the following cuts: $M_{max}^{pair} = 0.75\ GeV/c^2$, $\Delta Y_{max} \leq 1$, $k_t^{max} = 0.65\ GeV/c$, $|X| \geq 0.01$. For the opposite jets an increasing positive C_Q correlation was observed in the region of $1 < Y_{min} < 2$ (Fig. 2b). It shows that left handed pair in one jet prefers a right handed pair in opposite jet. Maximal value of the correlation about $24 \pm 5\%$ is obtained at $Y_{min} = 1.75$. Some indication to a negative correlation can be seen for pairs in the same jet (Fig. 2a). Neither JETSET7.3 PS nor HERWIG 5.7 Monte Carlo programs show this effect at the whole domain of the given cuts variation. No correlation was also seen for the pairs from different events with the same cuts including the back-to-back acollinearity cut of the two jets. This should convince that it is not an apparatus effect.

The mass cut dependence M_{max}^{pair} of the C_Q correlation for opposite jets was investigated also (Fig.3). It is difficult to say something about the effect in the region below 0.5 GeV because of decrease of statistics. The most definitely the correlation is seen at the ρ-meson mass region and clearly decreases for the higher masses. Such behavior confirms the above mentioned theoretical expectation that for handedness phenomena manifestation one needs an interference of the different amplitudes.

As a function of k_T^{min} the C_Q correlation disappeared for small $k_T^{min} < 0.6\ GeV/c$ and seems saturated at $k_T^{min} > 0.65\ GeV/c$. (Fig.4)

Figs. 1-4 present inclusive dependence from cut parameters. It is interesting to analyze the correlation differentially for an independent set of data. It was done in different ways of separation of data. In Table 1 C_Q-correlation is presented in different rapidity intervals for selected tracks. (Here $\Delta N_{xx}^{(1)}$ and $\Delta N_{xx}^{(2)}$ corresponds to correlation values obtained with different versions of DELANA - DELPHI off-line analysis program.) In all intervals the effect has the same positive sign but statistics is not enough to see some rapidity dependence. The same conclusion can be drawn from results given in Table 2 for correlation in different intervals of angles between thrust and beam axis ($q\bar{q}$-axis scattering angle). In the first two lines of Table 3 C_Q-correlation are presented for 92- and 91+93-data separately. They show the same positive sign and agrees with each other inside of error bars. In the next lines of Table 3 the results of selection with different "visible volumes", i.e. with different

Table 1: C_Q-correlation in different rapidity intervals

Y_{min}	Y_{max}	$\Delta N^{(1)}_{xx}/N^{tot}_{xx}$	$\Delta N^{(2)}_{xx}/N^{tot}_{xx}$	JETSET7.3 PS
1.00	1.25	0.05±0.20	-0.14±0.22	-0.30±0.24
1.25	1.50	0.33±0.25	0.28±0.20	0.25±0.25
1.50	1.75	0.30±0.25	0.63±0.25	-0.16±0.23
1.75	2.00	0.50±0.50	0.60±0.45	0.09±0.30
2.00	3.00	0.25±0.10	0.16±0.10	0.04±0.11

Table 2: C_Q-correlation in different intervals of thrust and beam angles

$\cos\Theta_{min}$	$\cos\Theta_{max}$	$\Delta N^{(1)}_{xx}/N^{tot}_{xx}$	$\Delta N^{(2)}_{xx}/N^{tot}_{xx}$	JETSET7.3 PS
0.00	0.25	0.12±0.15	-0.15±0.17	0.13±0.14
0.25	0.50	0.30±0.12	0.23±0.12	0.21±0.12
0.50	0.75	0.14±0.11	0.14±0.10	0.0±0.10
0.75	1.00	0.23±0.11	0.08±0.10	-0.14±0.14

cuts for polar angles of tracks and thrust axis are given. It was noticed that Θ_{th} distribution of selected events is more pronounced in the region of the inefficient zones between barrel and end cup detectors of DELPHI than before cuts. To investigate an effect of this zones the "visible volume" was shrunk at 10° from each side what results in decrease of statistics but not in change of the effect as it is seen from Table 3.

The C_Z handedness correlation with rapidity ordering ($|y_1| > |y_2|$) was also investigated. Both neutral and double charge pairs were taken into account with the same cuts. Results are presented in Table 4. The correlation in opposite jet is also positive and has the same behavior of the cut parameters but about two times smaller in magnitude. The maximal value obtained is about $10 \pm 4\%$ at approximately the same cut parameters.

4 Discussion

So, a clear evidence for the jet handedness correlations were found. The puzzling thing however is that the C_Q correlation of selected pairs ¿from opposite jets has a sign which contradicts to predicted one according to (11) based on the standard parton picture. It includes the helicity correlation of in Z^0-decay $c_{q\bar{q}} = 1$, independence of q and $\bar{q}$ fragmentation into the pair and charge conjugation of the two jets. The question now is which of the statement is

Table 3: C_Q-correlation for 92-, 91+93-data and for different selection angles of tracks and thrust axis

Data selection	$\Delta N^{(1)}_{xx}/N^{tot}_{xx}$
92-data	0.12±0.08
91+93-data	0.28±0.05
$25° < \Theta_{tr} < 155°$ $40° < \Theta_{th} < 140°$	0.18±0.05
$35° < \Theta_{tr} < 145°$ $50° < \Theta_{th} < 130°$	0.24±0.07

Table 4: C_Z-correlation for neutral and double charged pairs

Charge of pairs	0-0	0-2	2-2	total
$\Delta N^{(1)}_{xx}/N^{tot}_{xx}$	0.07±0.06	0.13±0.06	0.09±0.11	0.10±0.04

broken?

Same sign quark helicity contribution (negative $c_{q\bar{q}}$) first seems suppressed by m_q/M_{Z^0} factor and second would gives a negative C_Z correlation in contradiction with observation.

Break of factorization due to a high twist contribution seems unreliable since the opposite jet correlation should decrease with increase of rapidity interval between pairs because of decrease of wave functions overlapping region. Indeed, from a simpleminded dimensional argument one can see that

$$C_Q \propto \frac{\epsilon_{tk_1k_2}\epsilon_{tk'_1k'_2}}{(k_1k'_1)\,(k_2k'_2)+(k_1k'_2)\,(k_2k'_1)} \approx \frac{\sin\phi\sin\phi'}{\cosh^2(y-y')} < e^{-4Y_{min}} \qquad (12)$$

where ϕ and ϕ' are an azimuthal angle between transverse momenta of particles in pairs. In contrast to this the observed correlation increases with Y_{min}.

Concerning the charge conjugation it is really hardly seen in selected events. E.g. charge correlation of leading particles in the pairs was $C_{ch} = (N_{opp} - N_{same})/N \approx -9.7 \pm 6.0\%$ though similar picture was seen in the MC-generated events, $C^{MC}_{ch} = -3.1 \pm 4.4\%$, where with no doubt one have deal with $q\ \bar{q}$-jets.

The model [5] predicts negative sign correlation for pairs in the *same* jet what seems supported by observation (Fig. 2a). For the *opposite* jet handedness correlation it also gives a normal (negative) sign since the chromomagnetic moments q and $\bar{q}$ are opposite to each other. To produce the observed positive sign one needs an *universal* longitudinal chromo-magnetic field in the color tube between q and $\bar{q}$ as if quarks were chromo-magnetic monopoles rather than dipoles. One can now only guess what is a true physical reason for the field. It is clear however that there is no one in QED or Perturbative QCD. It is a question wither it could arise as a non-perturbative (topological? vacuum?) effect. Aside of the nature of the field is rather obscure it is hardly connected with spins of quarks. In fact, only a small value of handedness itself, $H \approx 2 \pm 1\%$ was observed with the same cuts. That would be natural if the direction of the field be occasional from one event to another.

In conclusion, a very nontrivial effect in handedness correlation seems observed which have no simple explanation in present theory. It is of special interest to study jet handedness correlations in the other LEP experimental data and also in data for smaller energy e^+e^--collider.

We would like to thank G. Altarelli, B. Kopeliovich, A. Olshevski, R. Jaffe, M. Ryskin, T. Sjöstrand, O. Teryaev, N. Törnqvist and T.T. Wu for fruitful discussions.

References

[1] Efremov A., Mankiewicz L., Törnqvist N., *Phys. Lett.* **B284**, 394 (1992).

[2] Belostotski S., Manayenkov S., Ryskin M., Prepr. PIYAF-1906, 1993.

[3] Efremov A.V., "A model for jet handedness estimation". To be published.

[4] Belitsky A.V., Efremov A.V., *Communication JINR*, E2-94-71, 1994.

[5] Ryskin M.G., *Phys.Lett.* **B319**, 346 (1993).

[6] Efremov A.V., Potashnikova I.K., Tkatchev L.G., Vertogradov L.S., DELPHI Collab., DELPHI 94-11 PHYS 355. 31 January 1994.

[7] SLD-Collab., "A Search for Jet Handedness in Hadronic Z^0-Decay", Preprint SLAC-PUB-6550, July 1994.

[8] Efremov A.V., Potashnikova I.K., Tkatchev L.G., "Search for Jet Handedness Correlation in Hadronic Z-decays", presented at Rancontre de Moriond, Meribel, 1994.

[9] DELPHI Collab., Aarnio P. et al., *Phys.Lett.* **B240**, 271 (1990).

[10] DELPHI Collaboration, *NIM* **A303**, 233 (1991).

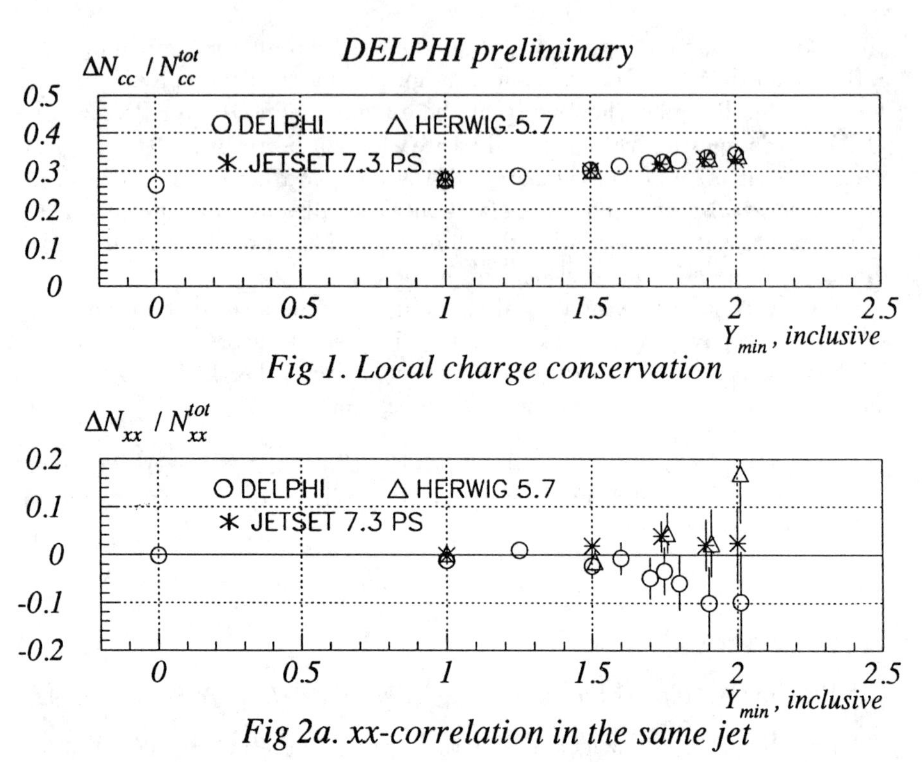

Fig 1. Local charge conservation

Fig 2a. xx-correlation in the same jet

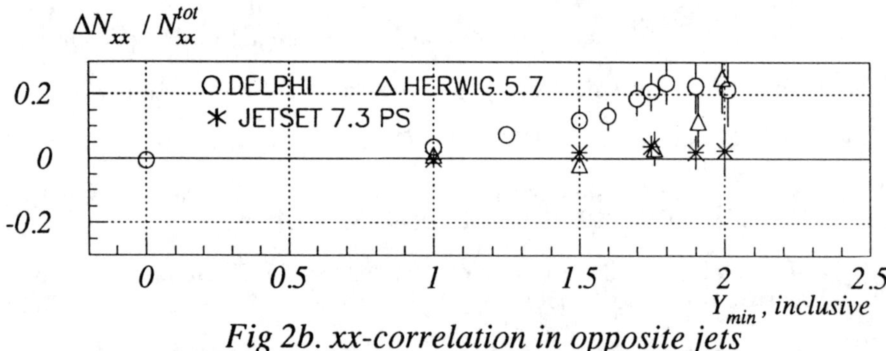

Fig 2b. xx-correlation in opposite jets

DELPHI preliminary

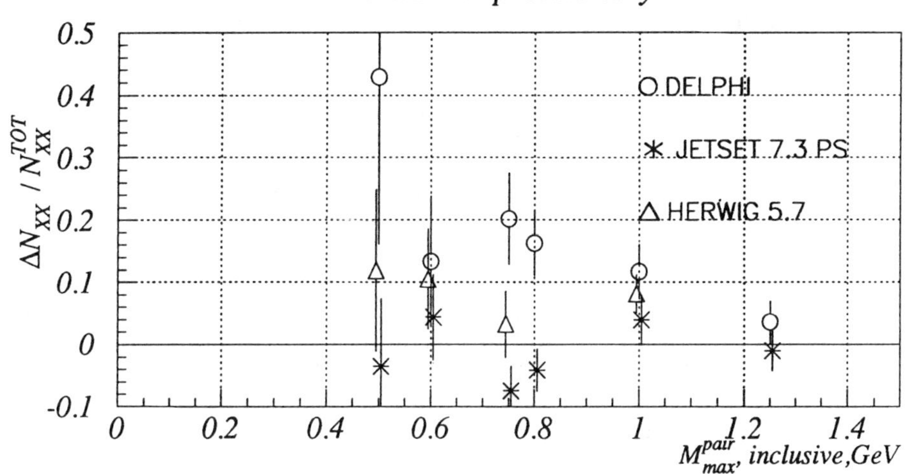

Fig 3. Mass dependence of xx-correlation for opposite jets

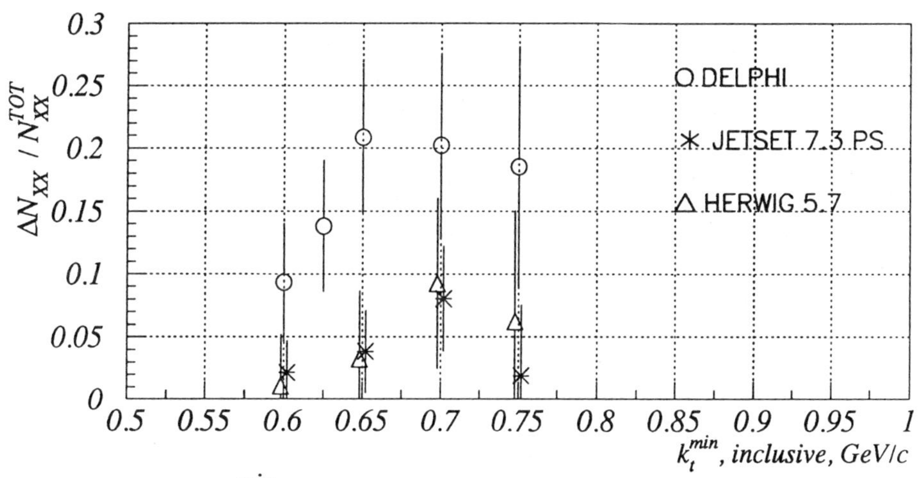

Fig 4. k_t^{min}-dependence of xx-correlation for opposite jets

INSTANTON - INDUCED HELICITY AND FLAVOR ASYMMETRIES IN THE LIGHT QUARK SEA OF THE NUCLEON

A. E. DOROKHOV
Joint Institute for Nuclear Research
Bogoliubov Theoretical Laboratory
Dubna, Moscow region, 141980, Russia

ABSTRACT

Specific helicity and flavor structure of zero modes of quarks in instanton field allows simultaneously to explain the breaking of both Ellis-Jaffe and Gottfried sum rules. From the analysis of the recent CCFR Collaboration data for the structure function $xF_3(x, Q^2)$ of the deep inelastic neutrino - nucleon scattering we conclude that Gross - Llewellyn sum rule is probably violated by the QCD vacuum polarization effects.

Experimental Tests of Parton Sum Rules

The deep inelastic lepton - nucleon scattering processes (DIS) occurring at small distances characterize the internal structure of the elementary particles. In the past few years new experimental data with high precision and in large kinematic region has become available.

This is primarily a result of the SLAC - EMC - SMC - SLAC measurements [1,2,3,4] of helicities of the charged constituents of the proton and neutron. The EMC data analysis [2] has resulted in a striking conclusion: the sum of the helicities of the quarks inside a proton, $\Delta\Sigma$, was found to be extremely small, ($\Delta\Sigma << 1$), and the Ellis-Jaffe (EJ) sum rule (SR) [5] is strongly violated.

Some time later, from the polarization experiments on neutron - contained targets it turns out that even Bjorken (Bj) SR [8] may be not that strong [4]. Then, it was concluded from NMC data analysis of the unpolarized structure function of nucleon $F_2(x, Q^2)$ [6] that $u-$ quark sea in the proton is suppressed w.r.t. $d-$ quark sea, that is the Gottfried (G) SR [7] is violated, too. All this is in dramatical contradiction with the expectation of the naive parton model where all these sum rules are fulfilled.

Recently the next - to - next - to - leading order QCD analysis of the most precise data for the neutrino - nucleon DIS structure function $xF_3(x, Q^2)$ measured by the CCFR Collaboration at the FERMILAB collider [9] has been performed [10]. This analysis results in the estimation of the Gross - Llewellyn Smith (GLS) SR [12] in the wide region of squared momentum transfer, Q^2, $2\ GeV^2 < Q^2 < 500\ GeV^2$, and reveals at the level of the statistical experimental errors the effect of deviation from perturbative QCD prediction [11].

The deficiency, $\Delta GLS \equiv GLS_{QCD} - GLS_{exp}$, at $Q^2 = 10\ GeV^2$ with four active flavors and the value of QCD parameter $\Lambda^{(4)}_{\overline{MS}} = 213 \pm 31\ MeV$ is equal to: $\Delta GLS = 0.180 \pm 0.107(stat)$ and decreases only logariphmically with the squared momentum transfer over all experimental accessible region up to $500\ GeV^2$.

The Axial Anomaly and the Nucleon Structure

In this talk I want to argue [13] that the observed inconsistency in parton sum rules are a manifestation of nonperturbative structure of the QCD vacuum. Within the framework of this approach the breaking mechanism of QCD SR is connected with a mixture of sea quarks with large transverse momentum in the nucleon wave function. This quark sea results from scattering valence quark off nonperturbative vacuum fluctuation, instanton.

This interaction in the limit of small size instanton is defined by the effective 't Hooft Lagrangian:

$$\mathcal{L}^{inst}(x) = (2n_c k^2)\Re \det(q_R q_L) \tag{1}$$

with the anomaly equation $\partial_\mu j^5_\mu = -2N_f(2n_c k^2)\Im \det(q_R q_L)$, where $n_c \approx 0.8 \cdot 10^{-3}$ is effective instanton density, and $k = \frac{4\pi\rho_C^3}{3}\frac{\pi}{(m_*\rho_C)}$ is effective instanton - quark coupling. The value $\rho_c \approx 1 \div 2\ GeV^{-1}$ defines the constituent quark radius, and $m_* = m - 2/3\rho_c^2 < 0\ |\bar{Q}Q|0 >$ is the effective quark mass in the physical vacuum.

It's important that instanton induced interaction (1) changes the chirality of a quark by the value $\Delta Q = -2N_f$ and acts only for differently flavoured quarks. From this it immediately follows that **sea quarks have negative helicity** and screen the helicity of a valence quark on which they are produced. On instanton the sea pair in the state with *Right chirality* is created and on the anti-instanton the quark pair with *Left chirality* is appeared.

Another thing is that on u-(d-) quark only $d\bar{d} - (u\bar{u}-)$ and $s\bar{s}-$ quark sea is possible. Therefore there is **more $d-$ sea quarks in the proton**. As a result it turns out that in the framework of the instanton mechanism the spin and flavor structure of nucleon quark sea is strongly correlated with the spin-flavor of the valence nucleon wave function.

Thus, specific helicity and flavor structure of quark zero mode interaction in the instanton field allows us simultaneously to explain the breaking of both Ellis-Jaffe SR related to significant breaking of quark helicity conservation and the Gottried SR caused by the violation of the $SU_f(2)$- symmetry of quark sea.

I should note that a perturbative quark-gluon vertex does not flip the helicity neither does feel the flavor of the valence quark. Thus, within the perturbative QCD it is not possible, in principle, to explain the experimentally observed significant violation of both sum rules.

From the vertex (1) we obtain the anomalous contributions to different PSR and axial charges [14]:

Gottfried sum rule

$$\Delta S_G = \bar{d} - \bar{u} = a;$$

flavour triplet and octet axial constants

$$\Delta g_A^3 = \Delta u - \Delta d = -10/3a_s; \quad \Delta g_A^8 = \Delta u + \Delta d - 2\Delta s = -4a + 2a_s;$$

flavour singlet axial constant

$$\Delta\Sigma_{inst} \equiv \Delta g_A^0 = \Delta u + \Delta d + \Delta s = -4a - 4a_s; \tag{2}$$

Ellis - Jaffe sum rules for proton and neutron

$$\Delta S_{EJ}^p = \sum_p e_q^2 \Delta q/2 = -(5a + 6a_s)/9;$$
$$\Delta S_{EJ}^n = \sum_n e_q^2 \Delta q/2 = -(5a + a_s)/9; \tag{3}$$

Bjorken sum rule

$$\Delta S_{Bj} = \Delta S_{EJ}^p - \Delta S_{EJ}^n = -5a_s/9 = \Delta g_A^3/6$$

where a ($a_s \leq a/2$) is the probability to create nonstrange (stgrange) sea quark pair in instanton field. If we attribute all $SU_f(2)$ asymmetry of sea measured by the NMC [6]

$$\int_0^1 dx\, [\bar{d}(x) - \bar{u}(x)] = 0.140 \pm 0.024, \tag{4}$$

to the instanton contribution then we obtain the value for the coupling $a = 0.140 \pm 0.024$.

To estimate the (nonanomalous) valence quark contribution we shall use the quark model where relativistic effect reduces the helicity of quarks with respect to the nonrelativistic quark model result and take as a conservative estimation the value

$$\Delta u_v + \Delta d_v = 0.8 \pm 0.15 \tag{5}$$

From (2), (4), (5) we obtain the final result for the singlet axial charge of the proton:

$$\Delta\Sigma = -0.04 \pm 0.30. \tag{6}$$

Our theoretical error is the sum of idefinetness in experimental number for GSR for sea quarks and relativistic effects for valence quarks. Physically this compenstion for the helicity of initial quark means a transformation of the valence quark spin momentum into the angular momentum of quark pair (in O^{++} state) created by instanton.

It should be stressed that in spite of that the instanton induced interaction contributes to g_A^3 it does not violate the Bjorken sum rule and violation of GSR is strongly correlated with contribution of axial anomaly to EJSR.*

Deficiency of The Gross - Llewellyn Smith Sum Rule

In the framework of the naive parton model the GLSsr for the proton structure function $F_3(x,Q^2)$ corresponds to the conservation of the baryon number, B, [15]

$$\frac{1}{3}\int_0^1 [(u(x,Q^2)-\bar{u}(x,Q^2))+(d(x,Q^2)-\bar{d}(x,Q^2))]\,dx = B(1-\frac{\alpha_s(Q^2)}{\pi}). \quad (7)$$

The baryon charge operator in the quark model is defined by

$$\hat{B}=\frac{1}{6}\int_0^1 (\{u^+(\vec{x}),u(\vec{x})\}_+ + \{d^+(\vec{x}),d(\vec{x})\}_+)d\vec{x} \quad (8)$$

and the baryon number is related to the low - energy spin - averaged matrix element of the isoscalar vector current $J_\mu(\vec{x})=\bar{u}\gamma_\mu u+\bar{d}\gamma_\mu d$ over the proton state:

$$<p|J_\mu(0)|p>=12p_\mu B. \quad (9)$$

If the proton state $|p_0>$ contained only free quarks, then the matrix element would provide $B=1$ exactly. The index 0 of $|p_0>$ means that a proton (and quarks) is considered over perturbative QCD vacuum with zero contribution of Dirac sea quarks to the matrix element: $<p_0|\hat{B}^{sea}|p_0>=0$.

However, the physical proton is immersed in the strong interacting medium and the phenomena of the confinement and of the spontaneous breaking of the chiral symmetry occur. As it has been shown by Skyrme and Witten [16,17] this highly nonlinear QCD vacuum can carry its own baryon number:

$$B^{Skyrme}=\frac{1}{24\pi^2}\epsilon_{0\mu\lambda\rho}\int Tr\{R_\mu R_\lambda R_\rho\}d\vec{x}, \quad (10)$$

where $R_\mu=(\partial_\mu U)U^+$ with $U^+U=1$ is constructed from effective bosonic fields. It is the effect of the fermion - boson transmutation. In the Skyrme model the chiral soliton baryon charge (10) is fully compensated for by the negative baryon charge induced by sea quarks.

*In the papers [13,14] the model of the sea quark distributions induced by instanton interaction has been developed.

Later Rho, Goldhaber and Brown [18] and Goldstone and Jaffe [19] have suggested that the baryon number (9) of the proton surrounded by the nontrivial (Skyrme) vacuum could be distributed between the normal (canonical) quark contribution, $B^{valence}$, and the part anomalously induced by the vacuum polarization, B^{sea}:

$$B = B^{valence} + B^{sea}. \tag{11}$$

The latter is related to the influence of the regularization procedure on the symmetry properties of the theory and is of pure quantum origin. The classical Skyrme field serves as a tool to define this procedure. Within the chiral bag model for the physical proton state $|p>$ the valence and sea polarization parts of the baryon number are equal to:

$$< p|\hat{B}^{valence}|p >= 1, \qquad < p|\hat{B}^{sea}|p >= -B^{Skyrme}, \tag{12}$$

correspondingly. We can write the following sum rule:

$$B^{valence} + B^{sea} + B^{Skyrme} = 1, \tag{13}$$

with $B^{sea} = -B^{Skyrme}$ (by definition) and B^{Skyrme} is invisible in DIS since it characterizes the property of the background vacuum field.

In the framework of the chiral bag model [20] when a massless Dirac quark field is confined to a finite region of space by means of a chiral boundary condition parametrized by a chiral angle Θ characterizing a leakage of the baryon charge, the anomalous baryon number of the vacuum is equal to [19]

$$B^{Skyrme}(\Theta) = -\frac{1}{\pi}(\Theta - \frac{1}{2}\sin 2\Theta), \qquad -\frac{\pi}{2} < \Theta < \frac{\pi}{2}, \tag{14}$$

$$B^{Skyrme}(\Theta + \pi) = B^{Skyrme}(\Theta), \qquad \text{outside the interval} \quad [-\frac{\pi}{2}, \frac{\pi}{2}].$$

The chiral boundary condition

$$-i\vec{\gamma}\cdot\hat{n}\; q|_s = \exp[i\gamma_5\Theta(\vec{\tau}\cdot\hat{n} + 1)]\; q|_s, \tag{15}$$

where $\hat{n}$ is the outward normal to the surface, is due to specific condition of the confinement of quarks in the closed region.

We have the following picture of the baryon charge leakage induced by the background field [19]. The chiral angle Θ varies from zero at very large value of bag radius, $R >> f_\pi^{-1}$ to $-\pi$ as bag radius goes to zero. It corresponds to change of the baryon charge carried by Dirac sea quarks from zero at chiral angle $\Theta = 0$ (large R) to -1 at $\Theta = -\pi$ (small R). When Θ pass $-\pi/2$ the occupied positive quark mode transit sharply into a negative - charge level and the baryon charge of the Dirac sea changes by -1 (14).

Now we can relate the deficiency of the GLSsr with the anomalous vacuum baryon number [22]

$$-\frac{1}{\pi}(\Theta - \frac{1}{2}\sin 2\Theta) = 0.060 \pm 0.036 \tag{16}$$

and then estimate the value of the chiral angle:

$$\Theta = -\frac{\pi}{4}\left(0.86 \begin{array}{c} +0.16 \\ -0.23 \end{array}\right). \tag{17}$$

The numbers (16) and (17) correspond to an upper bound of the effect.

The origin of the anomaly in the singlet vector current results from the low energy QCD box anomaly: $\omega \rightarrow \pi^+\pi^0\pi^-$ [23]. At the same time the isovector chiral flow through the surface controlled by the boundary condition (15) is zero due to the equal number of left - and right - handed chiral quarks. Thus, the Adler SR remains valid.

Finally, the question arises: does the explanation suggested have a particular signal in deep inelastic scattering, e.g. x or Q^2 dependence? The answer is positive [24] because of the structure function $F_3^{\nu p}(x, Q^2)$ defined by vector - axial vector correlator being specific. It's well known [15] that in this channel the Regge singularities have negative $C-$ parity, $C = -1$. They are the $\omega-$ meson exchange with intercept $\alpha_\omega \approx 1/2$ and the Odderon, $C-$ odd partner of the Pomeron, with high intercept $\alpha_O \geq 1$: $F_3(x) = a_\omega x^{-\alpha_\omega} + a_o x^{-\alpha_O}$. The first one is an exchange related to the momentum distribution of valence quarks. The second singularity is due to $C-$ odd vacuum exchange and determines the $x-$ dependence of sea quarks. In the high energy elastic hadron-hadron interactions the cross sections of particle and anti-particle would be different if the Odderon trajectory existed [25]. We hope to clarify the connection between the Dirac sea and Odderon singularity contributions to GLSsr in future.

Thus we can interpret the possible violation of the Gross - Llewellyn Smith sum rule observed by CCFR Collaboration in neutrino - nucleon DIS in wide Q^2 interval as a hint at a large polarization effect in the nonperturbative QCD vacuum surrounding the hadron [22]. We also stress that the peculiar interaction of the constituents induced by instantons is also responsible for large helicity and flavor asymmetry of the sea quarks in the proton wave function and the sea quark distribution functions. These and related questions are currently under investigation.

Acknowledgements

I am thankful to A.V. Sidorov and A.L. Kataev for the stimulating discussions and informing me about the CCFR Collaboration results, B.V. Struminsky for clarifying the role of the Odderon in DIS and P.N. Bogolubov, D. Broadhurst, S.B. Gerasimov, B.L. Ioffe, N.I. Kochelev, E.A. Kuraev, A.W. Thomas and O.V. Teryaev for discussions.

1. SLAC - Yale Collaboration, G.Baum et al. *Phys. Rev. Lett.* **51** (1983) 1135.

2. The EMC Collaboration, J.Ashman et al. *Nucl. Phys.* **B238** (1989) 1.
3. The SMC Collaboration, B.Adeva et al. *Phys. Lett.* **B302** (1993) 533; *Phys. Lett.* **B329** (1994) 399.
4. The E142 Collaboration, P.L.Anthony et al. *Phys. Lett. Lett.* **71** (1993) 959.
5. J. Ellis, R. L. Jaffe *Phys. Rev.***D9** (1974) 1444
6. P Amaurdruz et al. *Phys. Rev. Lett.* **66** (1991) 2712.
7. K. Gottfried *Phys. Rev. Lett.***18** (1967) 1154
8. J.D. Bjorken *Phys. Rev.* **148** (1966) 1467; *ibid* **D1** (1971) 1376.
9. CCFR Collab. *Phys. Lett.* **B317** 665 (1993); *Phys. Rev. Lett.* **71** 1307 (1993).
10. A.L. Kataev, A.V. Sidorov *Phys. Lett.* **B331** 179 (1994).
11. See discussion on all available in the literature information on this subject in: A.L. Kataev, A.V. Sidorov *CERN preprint* CERN-TH.7235/94 , May, (1994).
12. D.J. Gross, C.H. Llewellyn Smith *Nucl. Phys.* **B14** 337 (1969).
13. A.E.Dorokhov, N.I.Kochelev, Yu.A.Zubov *Int. Journ. Mod. Phys.* **A8** (1993) 603. A. E. Dorokhov, N. I. Kochelev *Phys. Lett.* **B304** 167 (1993).
14. A. E. Dorokhov, in Proc. of *Quantum Field Theoretcal Aspects of High Energy Physics*, Kyffhauser, Germany, 1993, p.33; *Adelaide preprint* **ADP-93-224/T141** (1993).
15. B.L. Ioffe, V.A. Khoze, L.N. Lipatov *Hard Processes* North Holland, Amsterdam, 1984.
16. T.H.R. Skyrme *Proc. Roy. Soc.* London, **A260** 127 (1961); *Nucl. Phys.* **31** 556 (1962).
17. E. Witten *Nucl. Phys.* **B223** 422, 433 (1983).
18. M. Rho, A.S. Goldhaber, G.E. Brown *Phys. Rev. Lett.* **51** 747 (1983).
19. J. Goldstone, R.L. Jaffe *Phys. Rev. Lett.* **51** 1518 (1983).
20. C.G. Callen, R.F. Dashen, D.J. Gross *Phys. Rev.* **D19** 1826 (1979); G.E. Brown, M. Rho, V. Vento *Phys. Lett.* **84B** 383 (1979); R.L. Jaffe, in *Pointlike Structure Inside and Outside the Nucleon*, Proceedings of the 1979 Erice Summer School "Ettore Majorana" edited by A. Zichichi (Plenum, New York, 1981); A.W. Thomas, S. Theberge, G.A. Miller *Phys. Rev.* **D24** 216 (1981).
21. S.J. Brodsky, J. Ellis, M. Karliner *Phys. Lett.* **B206** 309 (1988); S. Forte, *Phys. Lett.* **B224** 189 (1989).
22. A. E. Dorokhov, *Zh. Eksp. Teor. Fiz. Pis'ma Red.* **60** 80 (1994).
23. E. Witten *Nucl. Phys.* **B223** 422, 433 (1983).
24. B.V. Struminsky *Inst. Theor. Phys. (Kiev) preprint* ITP-93-29P (1993); *Sov. J. Nucl. Phys.* (to be published).
25. A. Donnachie, P.V. Landshoff *Nucl. Phys.* **B266** (1986) 690.

SEA AND GLUON SPIN STRUCTURE FUNCTION MEASUREMENTS AT RHIC*

A. Yokosawa

High Energy Physics Division
Argonne National Laboratory, Argonne, Illinois 60439

Abstract. The first polarized collider where we collide 250-GeV/c beams of 70% polarized protons at high luminosity is under construction. This will allow a determination of the nucleon spin-dependent structure functions over a large range in x and a collection of sufficient W and Z events to investigate extremely interesting spin-related phenomena.

INTRODUCTION

We start out with discussing the status of the RHIC polarized collider and associated experiments.

In August 1991, a Partial Snake experiment at AGS was approved by the Brookhaven Program Advisory Committee (PAC). Then in August 1992 the Relativistic Heavy Ion Collider (RHIC) Spin Collaboration (RSC), STAR/Spin PHENIX/Spin proposals were submitted to the PAC. In October 1993, full approval of the proposals on spin physics using the RHIC polarized collider was granted by the Brookhaven National Laboratory PAC. In April this year, successful polarized proton acceleration in the AGS with a partial snake was achieved.

Proposed spin experiments at RHIC are to explore QCD in a new way and allow us to make the first direct measurement on one of the proton structure functions.

High Energy Spin Physics at RHIC

We present physics issues and expected event rates for various reactions (the integrated luminosities used are shown in Appendix I).

The RHIC polarized collider allows us to explore QCD in a new way and opens challenging opportunities for QCD studies.

The proton spin consists of:

$$1/2\, \Sigma\, \Delta q + \Delta G + \langle Iz \rangle = 1/2$$

Our proposed measurements are to determine sea-quark polarization, $\Delta \bar{u}$ (x) and $\Delta \bar{d}$ (x), and gluon polarization $\Delta G(x)$. We are also to determine one of the fundamental proton structure functions (f_1, g_1, h_1).

For these measurements, two detectors, STAR and PHENIX, will be simultaneously used, and their functions are complementary. Expected event rates given in this paper are for the detector STAR.

Let us consider the hadronic reaction, pp $\rightarrow$ (hadron or gauge boson) + X. When both initial protons are longitudinally polarized, we measure an observable A_{LL} defined as:

$$A_{LL} = (1/P^2)\,(N^{++} - N^{+-})/(N^{++} + N^{+-})\,.$$

If one QCD subprocess is dominant:

$$A_{LL} \sim P_a \cdot P_b \cdot \hat{a}_{LL} \quad (a + b \rightarrow c + d)\,,$$

where $\hat{a}_{LL}$ in various reactions are shown in Appendix II.[1]

1. Measurements with Barrel EMC and Shower Maximum Detector

The proposed barrel EM calorimeter is a lead-scintillator sampling calorimeter. It is located inside the aluminum coil of the STAR solenoid and covers $|\eta| \leq 1.0$ and 2π in azimuth, thus matching the acceptance for full TPC tracking. At $\eta \sim 0$, the amount of material in front of the EMC is ~ 0.5 radiation lengths (X_o). The inner radius is 2.20 meters, and the overall length is 6.20 meters.

A detector with fine spatial resolution will be placed at a depth of approximately 5 X_o, near the location of the maximum number of shower electrons for photons of 3-5 GeV, to allow for the detection of direct photons. It will reject background photons emanating from decaying π° mesons having $p_T \leq 20$ GeV/c by examining the transverse shower profile at this depth. The radial space allotted for this device is 25 mm. (Reference - EMC CDR[2])

1.1 Jet Production at 200 GeV

Several QCD subprocesses contribute to the cross section for jet production:

a) gluon-gluon scattering at low p_T,
b) gluon-quark scattering at medium p_T (above ~ 20 GeV/c, and
c) quark-quark elastic scattering at p_T.

At low p_T:

$$A_{LL} = [\Delta G(x_1)/G(x_1)] \times [\Delta G(x_2)/G(x_2)] \times \hat{a}_{LL} \quad (gg \rightarrow gg).$$

The $\hat{a}_{LL}$ is expected to be large, $\hat{a}_{LL} = 0.8$ at 90°.

1.2 Di-Jet Production at 200 GeV

The advantage over the single jet is the kinematic constraint on the momentum fractions, x_i, of the two partons.

Jet + Jet Events		
p_T	$\vert\eta\vert$	N_{pair}
≥ 10	≤ 0.3	$1 \cdot 10^8$
≥ 20	≤ 0.3	$3 \cdot 10^6$

1.3 Direct-γ Production at 200 GeV

Direct photons are produced through the $q\bar{q}$ annihilation subprocess and the quark-gluon Compton subprocess, ($qg \rightarrow \gamma q$). The Compton process is the dominant one in pp interactions. Then,

$$A_{LL} = [\Delta u\,(x_1)/u(x_1)] \cdot [\Delta G\,(x_2)/G(x_2)] \cdot \hat{a}_{LL}\quad (qg \rightarrow \gamma q),$$

where $\Delta u(x)/u(x) = (u_+ (x) - u\_(x)]/[u+(x) + u\_(x)]$ Δu (x) being the helicity distribution of the quark, $\Delta G(x)/G(x) = [G_+(x) - G_-(x)]/[G_+ (x) + G\_ (x)]$, where $\Delta G(x)$ is the helicity distribution carried by gluon fields.

For example, $\Delta u\,(x)/u\,(x) \approx 0.4$ at $x_q = 0.2$ (from EMC - SMC), $\hat{a}_{LL} = 0.6$ at 90° scattering. Then we have $A_{LL} = 0.2 \times \Delta G/G$, and $\delta(\Delta G/G) = 5 \times \delta\, A_{LL}$.

The estimated δA_{LL} at $\sqrt{s} = 200$ GeV for $p_T = 10$ to 20 GeV, $\Delta y \pm 1$,

$$\delta A_{LL} \sim \pm 0.006$$

$$\delta(\Delta G/G) \sim \pm 0.03$$

1.4 Direct-γ+ Jet

For the p_T acceptance of 10 to 20 GeV, x_1 and x_2 vary from 0.1 to 0.2 at $\sqrt{s}$ = 200 GeV. The expected number of events is 9,000 corresponding to

$$\delta A_{LL} \sim \pm 0.03$$

1.5 $W^{\pm}$ and Z° Production at 500 GeV

a. Parity-Violating Asymmetry. The observable A_L (PV) is defined as, $A_L = (N^- - N^+)/(N^- + N^+)$, where -(+) are minus (plus) helicity. For W^+,

$$A_L = \frac{\Delta u(x_1)\bar{d}(x_2) - (u \leftrightarrow \bar{d})}{u(x_1)\bar{d}(x_2) + \bar{d}(x_1)u(x2)}.$$

When the helicities of both beams are the same, we define another observable as:

$$A_{LL}^{PV}(y) = \frac{\left[\Delta u(x_1)\bar{d}(x_2) - \Delta\bar{d}(x_2)u(x_1)\right] - (u \leftrightarrow \bar{d})}{\left[u(x_1)\bar{d}(x_2) - \Delta u(x_1)\Delta\bar{d}(x_2)\right] + (u \leftrightarrow \bar{d})}.$$

For y = 0, $A_L^{W^+}$ = 1/2 ($\Delta u/u$ - $\Delta\bar{d}/d$) and $A_L^{W^-}$ = 1/2 ($\Delta d/d$ - $\Delta\bar{u}/u$). The results of predictions[3] for $W^\pm$ production are shown in Fig. 1 with polarized sea quarks[4] and $\Delta\bar{u}$ = $\Delta\bar{d}$. In the case of W- production, one observes a drastic difference between the cases $\Delta\bar{u} \neq 0$ and $\Delta\bar{u} = 0$.

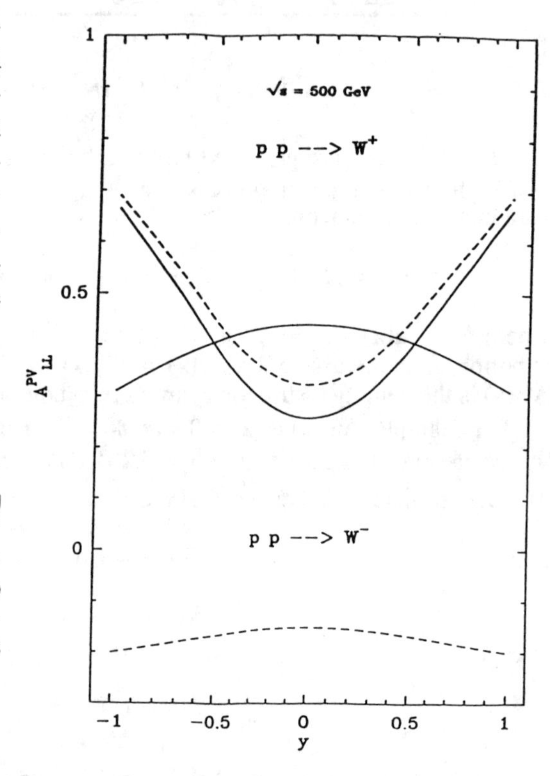

FIGURE 1. The parity violating asymmetry A_{LL}^{PV} vs. y for W+ and W- production at $\sqrt{s}$ = 500 GeV. Solid lines correspond to non-zero sea quark polarizations where as dashed lines correspond to $\Delta\bar{u} = \Delta\bar{d} = 0$.

b. Parity-Conserving Asymmetry.

$$\text{For } W^+,\ A_{LL} \sim \frac{\Delta u(x_1)\Delta\bar{d}(x_2)}{u(x_1)\bar{d}(x_2)}.$$

A similar expression is for W- production by permuting u and d.

1.6 Measurements of $h_1(x)$ in Z° Production
(Quark Tranversity Distribution in Polarized Proton)

A complete quark-parton model of the nucleon requires three quark distributions and quark spin density matrix is given as:

$$P(x) = 1/2\ [f_1(x) + g_1(x)\ \vec{s}_{//} \cdot \vec{\sigma} + h_1(x)\ \vec{s}_{\perp} \cdot \vec{\sigma}]\ ,$$

where $f_1(x)$ is related to the longitudinal momentum distribution of quarks in the nucleon, $g_1(x)$ is related to the helicity distribution in a polarized nucleon, and $h_1(x)$ is related to the correlation between left-handed and right-handed quarks.[5]

$h_1(x)$ can be determined by measuring the transverse spin correlation A_{TT} (A_{NN}) in Z° production.

In terms of $h_1(x)$, A_{TT} (A_{NN}) is given as:[6,7]

$$A_{TT}(A_{NN}) = \hat{a}_{TT} \frac{\sum_i (a_i^2 - b_i^2) h_{1i}(x_1) \bar{h}_{1i}(x_2)}{\sum_i (a_i^2 + b_i^2) f_{1i}(x_1) \bar{f}_{1i}(x_2)},$$

where a_i and b_i are the vector and axial couplings of the Z° to the quark of flavor i, $\hat{a}_{TT}$ is the partonic double-spin asymmetry, $\hat{a}_{TT} = 1$ at the vicinity of $\theta_{c.m.} = \pi/2$ and $\phi_{c.m.} = 0$.

The statistical error will be $\Delta A_{TT} = (1/P^2) \cdot 0.025$. At a first approximation $h_1 = g_1$, then $|A_{TT} / \hat{a}_{TT}| \sim |A_{LL}|$.

2. Measurements with Barrel, One Endcap, and Shower Maximum Detector

An endcap calorimeter is to be placed inside the iron pole pieces. The endcap increases solid angle and acceptance and allows better measurements of the following reactions.

2.1 Detecting the Direct-γ and the "Away-Side" Jet

In order to measure $\Delta G(x)$, the gluon spin structure function, both the direct-γ and the "away-side" jet must be detected in coincidence so that the kinematics of the incoming partons can be calculated.

The Compton and annihilation subprocesses both involve $2 \rightarrow 2$ scatterings. The incoming partons are assumed to have fractions x_1 and x_2 of the beam momentum and collide colinearly. Then x_1 and x_2 are given in terms of pseudorapidity as:

$$x_1 \simeq \left(2p_T / \sqrt{s}\right)\left(e^{\eta_1} + e^{\eta_2}\right)/2,\ x_2 \simeq \left(2p_T / \sqrt{s}\right)\left(e^{-\eta_1} + e^{-\eta_2}\right)/2.$$

Figure 2 shows the x coverage for $x_T = 0.1$ as functions of the direct-γ and jet pseudorapidities.

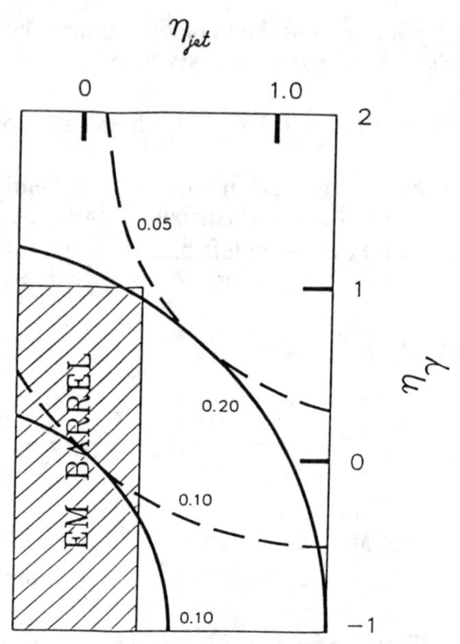

FIGURE 2. Plots of x_1 and x_2 as a function of the direct-g and jet pseudorapidities. The solid lines represent contours for x_1 and the dashed lines are contours for x_2.

2.2 *W's and Z Production at 500 GeV*

W^+ 74,000 events
W^- 22,000 events
For W's the detected electron (or positron) has $p_T > 20$ GeV/c
Z 5,300 events
For Z at least one of the detected electrons has $p_T > 20$ GeV/c

In particular, many Z events are needed for a reasonable measurement of $h_1(x)$.

REFERENCES

1) Bourrely, C., Guillet, J. Ph., and Soffer, J., *Nucl. Phys.* **B361**, 72 (1991); Bourrely, C. and Soffer, J., Renard, F. M., and Taxil, P., *Phys. Rep.* **177**, 319 (1989).
2) "The Electromagnetic Calorimeter for the Solenoidal Tracker at RHIC," PUB-5380 (1993).
3) Bourrely, C., and Soffer, J., *Phys. Lett.* **B314**, 132 (1993).
4) Martin, A. D., Sterling, W. J., and Roberts, R. G., Preprint RAL-92-021 (1992).
5) Ralston, J., and Soper, D. E., *Nucl. Phys.* **B152** 109, (1979); Cortes, J., Pire, B., and Ralston, J., *Z. of Phys.* (Sep. 1992); Jaffee, R., and Ji, X., *Phys. Rev. Lett.* **67**, 552 (1991).
6) Ji, X., *Phys. Lett.* **B284**, 137 (1992).
7) Soffer, J., to be published.

APPENDIX I

The integrated luminosities used are:

$$\int Ldt = 8 \cdot 10^{38}\ cm^{-2} \text{ at } 500\ GeV = 800\ pb^{-1},$$

and

$$\int Ldt = 3.2 \cdot 10^{38}\ cm^{-2} \text{ at } 200\ GeV = 320\ pb^{-1},$$

which means

100 days of running ($4 \cdot 10^6$ sec. 50% efficiency).

APPENDIX II

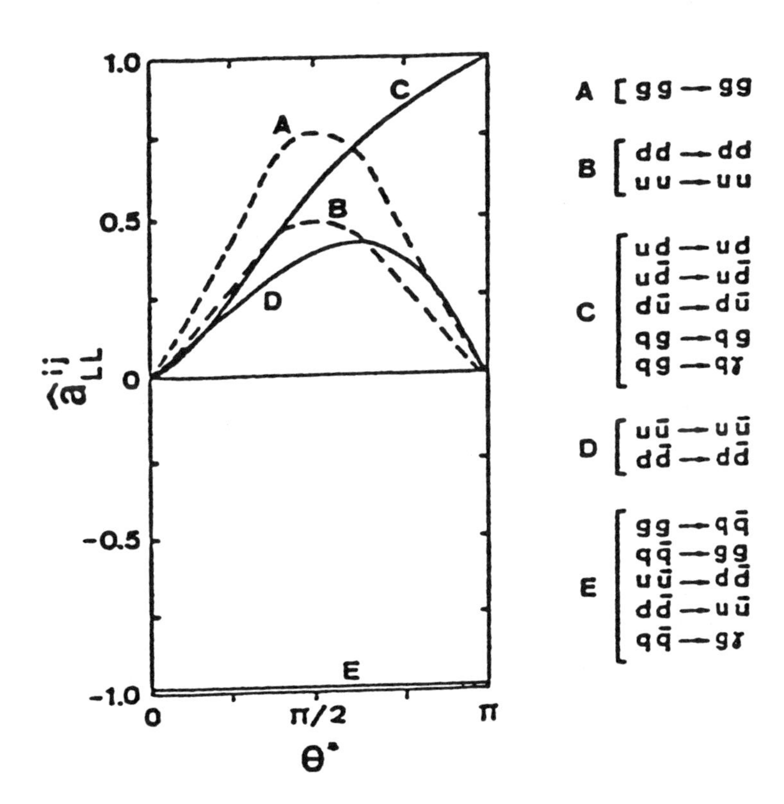

* Work supported by the U.S. Department of Energy, Division of High Energy Physics, Contract W-31-109-ENG-38.

APPENDICES

Eighth International Symposium on Polarizaton Phenomena in Nuclear Physics

INTERNATIONAL ADVISORY COMMITTEE

J. M. Cameron (chair)	IUCF	R. D. McKeown	Caltech
J. Arvieux	Saturne	W. T. H. van Oers	Manitoba
A. S. Belov	INR Moscow	H. Ohnuma	Tokyo Inst. Tech.
L. S. Cardman	CEBAF	J.-M. Richard	CERN
C. R. Gould	North Carolina State	H. Sakai	Tokyo
R. J. Holt	Illinois	F. D. Santos	Lisboa
R. L. Jaffe	MIT	M. Simonius	Zürich
M. Kondo	RCNP	J. Speth	KFA Jülich
C. Lechanoine-Leluc	Geneva	E. Steffens	Heidelberg
V. M. Lobashov	Moscow	K. Yagi	Tsukuba
J. B, McClelland	LANL		

ORGANIZING COMMITTEE

S. E. Vigdor (chair)	IUCF	L. D. Knutson	Wisconsin
T. B. Clegg	North Carolina	J. T. Londergan	Indiana
W. W. Jacobs	IUCF	S. W. Wissink	IUCF

LOCAL ARRANGEMENTS COMMITTEE

J. M. Cameron (chair)	A. D. Bacher	P. Schwandt
E. J. Stephenson (co-chair)	R. D. Bent	J. Sowinski
	B. B. Brabson	

Eleventh International Symposium on High Energy Spin Physics

INTERNATIONAL COMMITTEE

A. D. Krisch (chair)	Michigan	K. J. Heller	Minnesota
C. Y. Prescott (chair-elect)	SLAC	V. W. Hughes	Yale
D. P. Barber	DESY	D. Kleppner	MIT
O. Chamberlain	Berkeley	A. Masaike	Kyoto
E. D. Courant	Brookhaven	P. W. Schmor	TRIUMF
G. R. Court	Liverpool	A. N. Skrinsky	Novosibirsk
A. V. Efremov	Dubna	V. Soergel	Heidelberg
G. Fidecaro	CERN	J. Soffer	Marseille
W. Haeberli	Wisconsin	L. D. Soloviev	Protvino

ORGANIZING COMMITTEE

K. J. Heller (chair)	Minnesota	R. A. Phelps	Michigan
S. J. Brodsky	SLAC	R. Prepost	Wisconsin
A. W. Chao	SLAC	R. A. Rameika	Fermilab
D. G. Crabb	Virginia	J. B. Roberts	Rice
A. R. Dzierba	Indiana	T. Roser	Brookhaven
R. L. Jaffe	MIT	H. M. Steiner	Berkeley
S. Y. Lee	IUCF	D. G. Underwood	Argonne
D. B. Lichtenberg	Indiana		

SPIN'94

Conference Schedule and Plenary Session Talks

start time	talk length	author, institution title

FRIDAY MORNING, SEPTEMBER 16
[Alumni Hall]

OPENING

8:30	15	*George Walker, Indiana University Vice-President* Opening remarks

NUCLEAR: Quarks to Hadrons to Nuclear Matter Chair: B.D. Serot, Indiana

8:45	30	*A.W. Thomas, Adelaide* Non-perturbative aspects of the spin structure functions of nucleons and nuclei
9:25	30	*T.N. Taddeucci, Los Alamos* The nuclear spin-isospin response to quasifree nucleon scattering
10:05	20	**COFFEE BREAK**

HIGH ENERGY Chair: A. Masaike, Tokyo

10:25	30	*S.E. Vigdor, Indiana* LISS: Planning for spin physics with multi-GeV nucleon beams at IUCF
11:05	30	*T. Devlin, Rutgers* Discovery of hyperon polarization at Fermilab
11:45	30	*S. Nurushev, Protvino* Summary of SPIN'93 International Workshop at Protvino

FRIDAY AFTERNOON, SEPTEMBER 16

PARALLEL SESSIONS

	NUCLEAR	HIGH ENERGY
14:00	Hadron Form Factors I [Frangipani]	Lepton Accelerators, Beams and Polarimeters I [Maple]
14:00	Polarized Sources and Targets I [Oak]	Polarized Solid Targets I [Sassafras]
14:00	Low Energy Nuclear Reactions I [Persimmon]	Hadron Accelerators and Beams I [Whittenberger]

15:45	30	**COFFEE BREAK**

16:15	Hadron Form Factors II [Frangipani]	Strong Interactions at High Energy I [Walnut]
16:15	Polarized Sources and Targets II [Oak]	Polarized Electron Sources [Maple]
16:15	Low Energy Nuclear Reactions II [Persimmon]	Hadron Accelerators and Beams II [Whittenberger]

SATURDAY MORNING, SEPTEMBER 17
[Alumni Hall]

HIGH ENERGY Chair: A. Efremov, Dubna

8:30 30 *M. Placidi, CERN*
Polarization at LEP

9:10 30 *L. Pondrom, Wisconsin*
Consequences of hyperon polarization

9:50 25 **COFFEE BREAK**

NUCLEAR: Symmetry Tests in Complex Systems Chair: M. Simonius, Zurich

10:15 30 *B. Heckel, U. Washington*
Limit on the electric dipole moment of the ^{199}Hg atom

10:55 30 *J. Deutsch, Louvain*
Search for incomplete parity violation in leptonic and semi-leptonic processes: status and perspectives

11:35 30 *Y.-F. Yen, Los Alamos*
Study of parity and time-reversal violation in neutron-nucleus interactions

SATURDAY AFTERNOON, SEPTEMBER 17
[Alumni Hall]

HIGH ENERGY/NUCLEAR PLENARY SESSION: Nucleon Spin Structure Functions Chair: V. Hughes, Yale

14:00 30 *R. Windmolders, CERN*
SMC results for nucleon spin structure functions

14:40 30 *D. Day, Virginia*
E143 results from SLAC

15:20 30 *H. Jackson, Argonne*
HERMES

16:00 30 *J. Soffer, Marseille*
Use of weak interactions to measure spin distributions

16:45–17:00 **Buses leave for IUCF reception and tour**

17:00–20:00 **IUCF reception and tour**

SUNDAY MORNING, SEPTEMBER 18

9:00 **Buses depart from Union for all-day excursion and Belle of Louisville cruise**

MONDAY MORNING, SEPTEMBER 19

[Alumni Hall]

NUCLEAR: Polarization Effects in Strange Systems Chair: H. Fearing, TRIUMF

8:30 30 *H. Ejiri, Osaka*
Polarization and weak decays of hypernuclei

9:10 30 *K. Kilian, Jülich*
Polarization effects in $\bar{p}p \rightarrow \bar{\Lambda}\Lambda$

9:50 25 **COFFEE BREAK**

HIGH ENERGY Chair: G. Fidecaro, CERN

10:15 30 *M. Woods, SLAC*
Polarization at SLAC

10:55 30 *G. Voss, DESY*
Polarization at HERA

11:35 30 *I. Ternov, Moscow State*
The spin of relativistic nuclei in external magnetic fields

MONDAY AFTERNOON, SEPTEMBER 19

NUCLEAR PLENARY SESSION: Few Body Studies Chair: C. Elster, Ohio [Whittenberger]

14:00 30 *A. Sandorfi, Brookhaven*
Few-body photodisintegration with polarized photons

14:40 30 *R. Milner, MIT*
The spin structure of ^{3}He

15:20 20 *COFFEE BREAK*

15:40 30 *W. Tornow, Duke*
New results in nucleon-nucleon scattering at low energies

16:20 30 *H. Spinka, Argonne*
New results in nucleon-nucleon scattering at intermediate energies

HIGH ENERGY PARALLEL SESSIONS

14:00 Strong Interactions at High Energy II [Walnut]

14:00 Electroweak Interactions I [Frangipani]

14:00 Solid Polarized Targets II [Maple]

14:00 Hadron Accelerators and Beams III [Oak]

15:45 30 **COFFEE BREAK**

16:15 Solid Polarized Targets III [Maple]
16:15 Nucleon Spin Structure Functions I [Frangipani]
16:15 Hadron Accelerators and Beams IV [Oak]

20:00–22:00 **CONCERT and RECEPTION** [Waldron Arts Center]

TUESDAY MORNING, SEPTEMBER 20

[Alumni Hall]

HIGH ENERGY Chair: E. Courant, Brookhaven/Michigan

8:30 30 *L.W. Anderson, Wisconsin*
Report on the Madison workshop on sources and targets

9:05 30 *W. Meyer, Bonn*
RAPPORTEUR: Solid polarized targets

9:40 30 *S. Ozaki, Brookhaven*
Spin at RHIC

10:15 20 **COFFEE BREAK**

NUCLEAR: Nucleon Form Factors Chair: R. Jaffe, MIT

10:35 30 *E. Reichert, Mainz*
Determination of the neutron electric form factor in quasielastic collisions of polarized electrons with ^{3}He and ^{2}H

11:15 30 *I. Stancu, Riverside*
Measurements of the proton's axial form factor via neutrino-proton scattering

11:55 20 *E. Beise, Maryland*
RAPPORTEUR: Hadron form factors

12:15 20 *D. Fick, Marburg*
RAPPORTEUR: Low energy nuclear reactions

TUESDAY AFTERNOON, SEPTEMBER 20

PARALLEL SESSIONS

	NUCLEAR	HIGH ENERGY
14:00	Symmetries I [Walnut]	Lepton Accelerators, Beams and Polarimeters II [Maple]
14:00	Few-Body Systems I [Persimmon]	Nucleon Spin Structure Functions II [Whittenberger]
14:00	Intermediate-Energy Hadron-Induced Reactions I [Frangipani]	Strong Interactions at High Energy III [Oak]

15:45 30 **COFFEE BREAK**

	Nuclear	High Energy
16:15	Symmetries II [Walnut]	Lepton Accelerators, Beams and Polarimeters III [Maple]
16:15	Few-Body Systems II [Persimmon]	Strong Interactions at High Energy IV [Oak]
16:15	Intermediate-Energy Electromagnetic Interactions I [Frangipani]	Electroweak Interactions II [Whittenberger]

18:30–20:00 **RECEPTION** [Solarium]

20:00 **BANQUET** [Alumni Hall]

WEDNESDAY MORNING, SEPTEMBER 21

[Alumni Hall]

NUCLEAR Chair: H.-O. Meyer, Indiana

8:30 30 *W. Haeberli, Wisconsin*
Proton-proton interactions with polarized internal targets in storage rings

9:10 20 *A. Eiró, Lisbon*
RAPPORTEUR: Few Body Systems

9:30 20 *S. Page, Manitoba*
RAPPORTEUR: Symmetries

9:50 25 **COFFEE BREAK**

HIGH ENERGY Chair: W. Happer, Princeton

10:15 30 *M. Anselmino, Torino*
RAPPORTEUR: Strong interactions at high energy

10:55 30 *H. Steiner, Berkeley*
RAPPORTEUR: Electroweak interactions

11:35 30 *T. Roser, Brookhaven*
RAPPORTEUR: Hadron beams and accelerators

WEDNESDAY AFTERNOON, SEPTEMBER 21

PARALLEL SESSIONS

	NUCLEAR	HIGH ENERGY
14:00	Intermediate-Energy Hadron-Induced Reactions II [Frangipani]	Lepton Accelerators, Beams and Polarimeters IV [Maple]
14:00	Intermediate-Energy Electromagnetic Interactions II [Oak]	Nucleon Spin Structure Functions III [Whittenberger]
14:00	Polarized Sources and Targets III [Persimmon]	

15:45–17:00 **COFFEE + POSTER SESSION**

17:30 **PRE-OPERA BUFFET DINNER** [Tudor Room]

18:45 **OPERA DRESS REHEARSAL** [Musical Arts Center]

THURSDAY MORNING, SEPTEMBER 22
[Alumni Hall]

HIGH ENERGY Chair: L. Soloviev, Protvino

8:30 30 *D. Barber, DESY*
RAPPORTEUR: Polarized beams and polarimeters at lepton accelerators

9:05 30 *R. Voss, CERN*
RAPPORTEUR: Hadron Structure

NUCLEAR Chair: J. Arvieux, Saclay

9:35 20 *Y. Mori, KEK*
RAPPORTEUR: Polarized sources and gaseous targets

9:55 20 *M. Garçon, Saclay*
RAPPORTEUR: Intermediate energy electromagnetic interactions

10:15 20 *H. Sakai, Tokyo*
RAPPORTEUR: Intermediate energy hadron-induced reactions

10:35 25 **COFFEE BREAK**

CONCLUDING TALKS Chair: J. Arvieux, Saclay

HIGH ENERGY

11:00 45 *A. Krisch, Michigan*
Highlights and future prospects

NUCLEAR

11:45 45 *J. Moss, Los Alamos*
Spin physics in the next decade

12:30 **END OF CONFERENCE**

SPIN'94

Parallel Sessions

start time	talk length	author, institution title

FRIDAY, SEPTEMBER 16

NUCLEAR

HADRON FORM FACTORS I Chair: C. Horowitz, Indiana [Frangipani Room]

2:00	20	*J. Mitchell, CEBAF* Measurement of the neutron electric form factor with a polarized deuterium target at CEBAF
2:25	20	*A. Lai, Kent State* A new neutron polarimeter for measurements of G_E^n from the $d(\vec{e},e'\vec{n})p$ reaction
2:50	20	*J.O. Hansen, MIT* Measurement of G_E^n with a polarized ^{3}He target
3:15	20	*M. Pitt, Caltech* The SAMPLE experiment: parity violating electron scattering from the proton and deuteron

POLARIZED SOURCES AND TARGETS I: Ion Sources Chair: P. Schmor, Manitoba [Oak Room]

2:00	15	*A. Belov, Russian Academy of Sciences* A study of the source of polarized negative hydrogen ions with a deuterium plasma ionizer
2:18	15	*A.N. Zelenski, TRIUMF* Spin exchange polarization studies at the TRIUMF OPPIS
2:36	15	*M. Kinsho, KEK* A dual-optically-pumped polarized negative deuterium ion source
2:54	15	*V. Derenchuk, IUCF* IUCF high intensity polarized ion source operation
3:12	15	*A.J. Mendez, TUNL* Investigation of ion beam extraction system performance for the TUNL polarized ion source
3:30	15	*P.D. Eversheim, Bonn* The polarized ion source for COSY

LOW ENERGY NUCLEAR REACTIONS I: Light ions Chair: R.C. Johnson, Surrey [Persimmon Room]

2:00	15	*H. Toyokawa, Osaka* Cross section and iT_{11} for the ^{12}C(d,p) reaction and coupling between the $^{3}S_{1}$ states
2:18	15	*L. Zetta, Milano* Spectroscopy of ^{88}Y by means of the ^{91}Zr(p,α)^{88}Y reaction at 22 MeV
2:36	15	*M. Tanifuji, Hosei* Analyzing power formulae for nuclear reactions in the low energy limit, application to ^{2}H(d,p)^{3}H reactions
2:54	15	*A. Plavko, St. Petersburg Technical* Inelastic scattering of polarized protons at low energies from various medium and light-weight nuclei
3:12	15	*Y. Iseri, Chiba* Singlet state contributions to deuteron elastic scattering
3:30	15	*M.A. Al-Ohali, Duke* A dispersive optical model for the neutron-nucleus interaction from −80 to +80 MeV for nuclei in the mass region $27 \leq A \leq 32$

HIGH ENERGY

LEPTON ACCELERATORS, BEAMS AND POLARIMETERS I: Lepton Polarimeters Chair: M. Woods, SLAC [Maple Room]

2:00	15	*H.R. Band, Wisconsin* Möller polarimeters for SLAC
2:20	15	*A. Afanasev, CEBAF* Atomic electron binding effects for Möller polarimetry at electron beams
2:40	15	*C. Cavata, Saclay* The Compton polarimeter for CEBAF
3:00	15	*F. Zetsche, DESY/Hamburg* Experience with the fast polarimeter at HERA
3:20	15	*G. Kezerashvili, Novosibirsk* The fast polarimeter for the VEPP–4M collider
3:40	15	*Y. Derbenev, IUCF/Michigan* RF-resonance beam polarimeter

POLARIZED SOLID TARGETS I Chair: G. Court, Liverpool [Sassafras Room]

2:00	25	*A. Honig, Syracuse* Large, accessible, highly polarized frozen-spin solid HD targets
2:30	15	*S. Whisnant, South Carolina* SPHICE: a strongly polarized H and D target
2:50	15	*D. Haase, North Carolina State* A statically polarized solid ^{3}He target

3:10 20 *S. Penttila, Los Alamos*
LAMPF neutron spin filter

3:35 15 *S. Bultmann, Bielefeld*
^{13}C cross relaxation measurements

3:55 15 *I. Plis, Dubna*
Target with frozen nuclear polarization for experiments at low energies

HADRON ACCELERATORS AND BEAMS I: Polarized beam preservation and manipulation Chair: H. Sato, KEK [Whittenberger Auditorium]

2:00 20 *E.D. Courant, Brookhaven/Michigan*
Prospects of RHIC spin and Tevatron spin projects

2:25 10 *D.D. Caussyn, Michigan*
First partial Siberian snake test during acceleration

2:40 10 *H. Huang, Indiana/Brookhaven*
Preservation of proton polarization by a partial Siberian snake

2:55 10 *T. Toyama, KEK*
Prospect for polarized beam acceleration at the KEK PS

3:10 10 *F. Rathmann, Marburg*
Polarizing a stored, cooled proton beam by spin-dependent interaction with a polarized hydrogen gas target

3:25 20 *H.-O. Meyer, Indiana*
Towards longitudinal beam polarization in the IUCF Cooler
Beam depolarization with an internal target
Effect of a polarized hydrogen target on the polarization of a stored proton beam

3:45 30 **COFFEE BREAK**

NUCLEAR

HADRON FORM FACTORS II Chair: W. Turchinetz, MIT [Frangipani Room]

4:15 20 *D. Beck, U. Illinois*
Measurement of the g_0 proton form factor by parity-violating electron scattering at CEBAF

4:40 20 *S. Pate, MIT*
More details on HERMES

5:05 20 *A. Yu. Umnikov, Alberta*
The deuteron spin structure function in a Bethe-Salpeter approach and the extraction of the neutron spin-dependent structure function

5:30 20 *B. Vuaridel, Geneva*
Transverse polarization in deep inelastic scattering

POLARIZED SOURCES AND TARGETS II: Gaseous targets Chair: R.J. Holt, Illinois [Oak Room]

4:15 15 *F. Stock, Heidelberg*
The HERMES-FILTEX target source for polarized hydrogen and deuterium

4:33 15 *T. Wise, Wisconsin*
Spin-polarized internal gas target for hydrogen and deuterium in the IUCF Cooler Ring

4:51 15 *D.K. Toporkov, Novosibirsk*
Laser-driven internal polarized deuterium target for the VEPP-3 electron storage ring

5:09 15 *K. Zapfe, Munich/DESY*
High density polarized hydrogen gas target for storage rings

5:27 15 *L. Kramer, MIT*
An internal polarized ^{3}He target for electron storage rings

5:45 15 *V.G. Luppov, Michigan*
Status of the Michigan-MIT ultra cold polarized hydrogen jet target

LOW ENERGY NUCLEAR REACTIONS II: Heavy Ions Chair: K. Kemper, Florida State [Persimmon Room]

4:15 15 *E.L. Reber, Florida State*
Analyzing powers for elastic and inelastic scattering of polarized ^{6}Li from ^{12}C at 30 MeV

4:33 15 *R.C. Johnson, Surrey*
Semi-classical analysis of scattering of deformed heavy-ions below the Coulomb barrier

4:51 15 *R.P. Ward, Birmingham*
Polarized ^{6}Li studies at the Nuclear Structure Facility, Daresbury

5:09 15 *J.A. Christley, Surrey*
Fusion of a polarized projectile with a polarized target

5:27 15 *K. Matsuta, Osaka*
Spin polarization of ^{23}Mg in ^{24}Mg+Au, Cu and Al collisions at 91 A MeV

5:45 15 *H. Sakai, Tokyo*
Spin-flip probability via the ^{26}Mg(^{3}He,tγ)^{26}Al reaction

HIGH ENERGY

STRONG INTERACTIONS AT HIGH ENERGY I: Elastic Scattering at Small and Large Angles Chair: M. Anselmino, Torino [Walnut Room]

4:15 15 *D. Grosnick, Argonne*
Measurement of $\Delta\sigma_L$(pp) and $\Delta\sigma_L$($\bar{p}$p) at 200 GeV/c

4:35 15 *S. Troshin, Protvino*
Theoretical aspects of single-spin studies

4:55	15	*G.P. Ramsey, Loyola* Polarization and N–N elastic scattering amplitudes
5:15	15	*P. Draper, Texas at Arlington* Proton-proton elastic scattering experiment at RHIC
5:35	15	*V. Solovianov, Protvino* Experiment NEPTUN – physics and status
5:55	15	*A.M.T. Lin, Michigan* NEPTUN – a spectrometer for measuring the spin analyzing power in p–p elastic scattering at large $p_\perp^2$ at 400 GeV (and 3 TeV) at UNK

POLARIZED ELECTRON SOURCES Chair: R. Prepost, Wisconsin
[Maple Room]

4:15	20	*C. Prescott, SLAC* Review of the SLAC and Les Houches Workshops
4:40	20	*H. Aoyagi, Nagoya* Recent progress on cathode development from Nagoya and KEK
5:05	20	*T. Maruyama, SLAC* Recent progress on cathode development from SLAC/Wisconsin
5:30	15	*S. Cohen, Orsay* Results from the Orsay polarized electron source
5:50	15	*H. Tang, SLAC* Prospects for a DC-gun based polarized electron source for the next generation of linear colliders

HADRON ACCELERATORS AND BEAMS II Chair: H. Sato, KEK
[Whittenberger Auditorium]

4:15	15	*D. Underwood, Argonne* Polarimeters for high energy polarized beams
4:35	10	*R.A. Phelps, Michigan* Spin flipping a stored polarized proton beam
4:50	10	*Y. Derbenev, IUCF/Michigan* A concept for Stern-Gerlach polarization in storage rings
5:05	10	*H. Okamura, RIKEN* Technique for rotating the spin direction at RIKEN
5:20	10	*S.B. Nurushev, Protvino* Extraction and transformation of polarization at RHIC at 250 GeV/c
5:35	10	*V.I. Ptitsin, Novosibirsk* Helical spin rotators and snakes

MONDAY, SEPTEMBER 19

HIGH ENERGY

STRONG INTERACTIONS AT HIGH ENERGY II: Single Spin Asymmetry
Chair: G. Bunce, Brookhaven [Walnut Room]

2:00 15 *N. Saito, Kyoto*
Measurement of single spin asymmetry for direct photon production in pp collisions at 200 GeV/c

2:20 15 *W. Nowak, DESY*
Single spin asymmetries in proton-proton and proton-neutron scattering at 820 GeV

2:40 15 *S. Timm, Carnegie-Mellon*
Energy, p_T, and x_F dependence of the polarization of Σ^+ hyperons produced by 800 GeV/c protons

3:00 15 *K.A. Johns, Arizona*
Ξ and Ω^- polarizations produced by high energy polarized and unpolarized neutral beams

3:20 15 *S.B. Nurushev, Protvino*
Single spin asymmetries and invariant cross sections of high p_T inclusive π^0 production in $\bar{p}$–p interactions

3:40 15 *G.J. Musulmanbekov, Dubna*
Simulation of one-spin meson asymmetry in $\vec{p}$–p and $\vec{p}$–A collisions at high energy

ELECTROWEAK INTERACTIONS I Chair: H. Steiner, Berkeley [Frangipani Room]

2:00 20 *K.T. Pitts, Fermilab*
Electroweak coupling measurements from polarized Bhabha scattering at the Z resonance

2:25 20 *M. Woods, SLAC*
Precise measurement of the left-right cross section asymmetry in Z production by e^+e^- collisions

2:50 15 *O. Adriani, INFN Firenze*
Measurement of the tau polarization with the L3 detector

3:10 15 *M. Maolinbay, Zürich*
Study of the reaction $e^+e^- \rightarrow \gamma\gamma(\gamma)$ at the Z^0 pole with the L3 detector at LEP

3:30 15 *M. Jezabek, Krakow*
Polarization in top pair production and decay near threshold

SOLID POLARIZED TARGETS II Chair: S. Pentilla, Los Alamos [Maple Room]

2:00 15 *J. Ball, Saclay*
Polarized target for nucleon-nucleon experiments at Saturne

2:20	20	*J. Kyynarainen, CERN* The new SMC dilution refrigerator
2:45	15	*P. Delheij, TRIUMF* The CHAOS polarized target
3:05	15	*S. Goertz, Bonn* Irradiated lithium hydrides as polarized target materials
3:25	15	*M. Iinuma, Kyoto* Proton polarization at room temperature
3:45	15	*N. Piskunov, Dubna* The new Dubna movable target

HADRON ACCELERATORS AND BEAMS III: Polarimeters Chair: L. Ratner, Brookhaven/Michigan [Oak Room]

2:00	10	*C.D. Roper, TUNL* Static and electromagnetic field requirements for a Lamb-shift spin filter polarimeter
2:10	10	*A.J. Mendez, TUNL* Modeling the hyperfine state selectivity of a short Lamb-shift spin-filter polarimeter
2:20	10	*V. Derenchuk, IUCF* The IUCF/TUNL spin filter polarimeter
2:30	10	*Discussion of first 3 papers*
2:40	10	*R. Gilman, CEBAF/Rutgers* Physics with a focal plane polarimeter for Hall A at CEBAF
2:50	10	*S.M. Bowyer, IUCF* A calibration of the K600 focal plane polarimeter
3:00	10	*M. Yosoi, Kyoto* Focal plane polarimeter for the GRAND RAIDEN at RCNP
3:10	10	*H. Sakai, Tokyo* Facility for the (p,n) polarization transfer measurement at RCNP
3:20	10	*Y. Tagishi, Tsukuba* A proton polarimeter using a liquid helium target
3:30	10	*B. Braun, Münich* The Breit-Rabi polarimeter for polarized internal H/D targets
3:40	15	*Discussion of above papers*

3:45 30 **COFFEE BREAK**

HIGH ENERGY

SOLID POLARIZED TARGETS III Chair: D. Crabb, Virginia [Maple Room]

4:15	25	*T. Averett, Virginia* Operation of a polarized target with ammonia in a high intensity electron beam

4:45 25 *H. Dutz, Bonn*
The Bonn polarized target facility

5:15 20 *D. Kramer, Bielefeld*
The SMC polarized target – systems and operations

5:40 20 *G. Court, Liverpool*
Review of non-linear corrections in CW Q-meter target polarization measurements

NUCLEON SPIN STRUCTURE FUNCTIONS I Chair: R. Voss, CERN [Frangipani Room]

4:15 20 *A. Zanetti, Trieste*
More details on SMC experiments

4:40 20 *A. Feltham, Basel*
More details on SLAC E–143

5:05 20 *J. Ralston, Kansas*
How gluons spin in the proton

5:30 20 *B. Li, Kentucky*
Quark spin and quark angular momentum contents of the proton

HADRON ACCELERATORS AND BEAMS IV: Polarimeters Chair: L. Ratner, Brookhaven/Michigan [Oak Room]

4:15 10 *E.J. Ludwig, North Carolina*
The use of the ^{3}He(d,p)^{4}He reaction for low-energy polarimetry

4:25 10 *S. Ishida, Tokyo*
Construction of the deuteron polarimeter at RIKEN

4:35 10 *E. Tomasi-Gustafsson, Saturne*
A deuteron vector and tensor polarimeter up to 2 GeV

4:45 20 *Discussion of first 3 papers*

5:05 10 *S. Nurushev, Protvino*
Proton polarization determination by elastic p–e scattering

5:15 10 *T. Wakasa, Tokyo*
Effective analyzing powers of NPOL at 295 and 384 MeV

5:25 10 *J.W. Watson, Kent*
The Kent State "2π" neutron polarimeter

5:35 15 *Discussion of above papers*

TUESDAY, SEPTEMBER 20

NUCLEAR

SYMMETRIES I: Parity Violation Chair: P.D. Eversheim
[Walnut Room]

2:00 18 *J. Birchall, TRIUMF*
Parity violation in p–p scattering at TRIUMF

2:22 15 *M. Shmatikov, Moscow*
Theoretical overview of parity violation in p-p scattering

2:40 16 *Y. Masuda, KEK*
Neutron spin rotation and P and T violations

3:00 16 *M. Leuschner, IUCF*
Parity nonconservation in ^{207}Pb

3:20 12 *O. Yilmaz, Indiana*
Relativistic effects on parity violation in nuclei

3:35 12 *K. Kimura, Nagasaki*
Search for parity nonconservation in the compound nuclear reaction via an isobaric analog resonance of ^{90}Zr + p

FEW BODY SYSTEMS I: Two Baryons Chair: H. von Geramb, Hamburg
[Persimmon Room]

2:00 15 *R.T. Braun, TUNL*
Neutron-proton analyzing power at 12 MeV and the charged πNN coupling constant

2:18 15 *S.M. Bowyer, IUCF*
A measurement of the spin transfer observable D_{NN} for p+p elastic scattering at T_p=200 MeV

2:36 15 *C.A. Davis, TRIUMF*
Zero-crossing angle of the np analyzing power below 300 MeV

2:54 15 *B. Vuaridel, Geneva*
Spin observables in neutron-proton elastic scattering

3:12 15 *Y.D. Kim, KEK/Tsukuba*
Hyperon-nucleon scattering experiment at KEK

3:30 15 *P. Heimberg, Northwestern*
Differential cross section and analyzing power of $p(\vec{p},\pi^+)d$ near threshold

INTERMEDIATE ENERGY HADRON-INDUCED REACTIONS I
Chair: W.G. Love, Georgia [Frangipani Room]

2:00 12 *A. Tamii, Kyoto*
Test measurement of spin rotation parameters in proton elastic scattering from ^{58}Ni at E_p=300 MeV

2:15 12 *J. Liu, IUCF*
Fragmentation of "stretched" 6^- strength in $^{28}Si(\vec{p},\vec{p}')^{28}Si$

2:30	12	*S.P. Wells, IUCF* Simultaneous measurement of $(\vec{p},\vec{p}')$ and $(\vec{p},p'\gamma)$ observables for the 15.11 MeV, 1^+, T=1 state in ^{12}C at 200 MeV
2:45	12	*D.A. Cooper, Ohio State* A study of the Fermi (0^+) transition in $^{14}C(p,n)^{14}N$ at 495 MeV
3:00	12	*T. Wakasa, Tokyo* Measurement of the polarization transfer $D_{NN}(0°)$ for (p,n) reactions at 300 MeV
3:15	12	*C. Djalali, South Carolina* Isoscalar spin strength in ^{12}C and ^{40}Ca
3:30	12	*H. Okamura, Tokyo* Tensor analyzing power of the $^{12}C(d,^2He)^{12}B$ reaction at 270 MeV

HIGH ENERGY

LEPTON ACCELERATORS, BEAMS AND POLARIMETERS II

Chair: M. Minty, SLAC [Maple Room]

2:00	20	*D. Barber, DESY* Spin decoherence in electron storage rings
2:25	20	*M. Düren, Erlangen/DESY* Transverse and longitudinal electron polarization at HERA
2:50	20	*M. Böge, DESY/Hamburg* Optimization of spin polarization in the HERA electron ring using beam-based alignment procedures
3:15	20	*Y. Shatunov, Novisibirsk* Spin control system for the South Hall Ring at Bates Linear Accelerator Center
3:40	20	*Y. Eidelman, Novosibirsk* Developments in the computer code SPINLIE

NUCLEON SPIN STRUCTURE FUNCTIONS II

Chair: R. Voss, CERN [Whittenberger Auditorium]

2:00	20	*N. Kochelev, Dubna* Vacuum QCD and new information on nucleon structure functions
2:25	20	*H. Borel, Saclay* More details on SLAC E–143
2:50	20	*X. Ji, MIT* Using PQCD to probe non-perturbative QCD
3:15	20	*J. Qiu, Iowa* Twist-3 and structure functions
3:40	20	*B. Kamal, McGill* Direct γ production to measure the gluon structure function

STRONG INTERACTIONS AT HIGH ENERGY III: Inclusive processes, QCD, and other topics Chair: M. Anselmino, Torino [Oak Room]

2:00	15	*W. Tang, SLAC* Polarization as a probe to the production mechanism of charmonium in π–N collisions
2:20	15	*F. Murgia, Cagliari* Single spin asymmetry in inclusive pion production
2:40	15	*T. Meng, Berlin* Inclusive meson and lepton-pair production in single-spin hadron-hadron collisions
3:00	15	*A. Brandenburg, SLAC* Angular distributions in the Drell-Yan process: a closer look at higher twist effects
3:20	15	*A.P. Contagouris, McGill/Athens* Higher-order QCD corrections to processes with longitudinally and transversely polarized particles
3:40	15	*O. Teryaev, Dubna* On the twist–3 single and double asymmetries

3:45 30 **COFFEE BREAK**

NUCLEAR

SYMMETRIES II Chair: H.E. Conzett, Berkeley [Walnut Room]

4:15	20	*W.T.H. van Oers, TRIUMF* Measurement of charge symmetry breaking in np elastic scattering at 350 MeV
4:40	20	*G.A. Miller, U. Washington* Theoretical overview of charge symmetry violation
5:05	16	*P.R. Huffman, Duke* An experiment to test P-even time reversal invariance with MeV neutrons
5:25	16	*P.D. Eversheim, Bonn* Test of time reversal invariance in proton-deuteron scattering
5:45	12	*B.M.K. Nefkens, UCLA* Polarization and tests of discrete symmetries

FEW BODY SYSTEMS II: Three and four nucleons Chair: F.D. Santos, Lisboa [Persimmon Room]

4:15	15	*L.D. Knutson, Wisconsin* Polarization observables for p–d breakup and the nuclear three-body force
4:33	15	*I.M. Sitnik, Dubna* Status and future of polarization phenomena investigations in backward elastic deuteron-proton scattering

4:51 15 *Z. Ayer, North Carolina*
Determination of the asymptotic D- to S-state ratio for the triton and ^{3}He via ($\vec{d}$,t) and ($\vec{d}$,^{3}He) reactions

5:09 15 *M. Miller, Wisconsin*
Measurement of spin observables in quasielastic scattering of polarized protons from polarized ^{3}He: $^3\vec{He}(\vec{p},2p)$, $^3\vec{He}(\vec{p},pn)$ and $^3\vec{He}(\vec{p},pd)$

5:27 15 *W.J. Cummings, TRIUMF*
Elastic π^+ scattering on polarized ^{3}He

5:45 15 *W. Kretschmer, Erlangen*
Polarization transfer in p–d scattering at 22.7 MeV

INTERMEDIATE ENERGY ELECTROMAGNETIC INTERACTIONS I
Chair: K. de Jager, NIKHEF [Frangipani Room]

4:15 30 *J.M. Laget, Saclay*
Spin physics with an intense CW electron beam in the 15-30 GeV range

4:50 15 *H. Dutz, Bonn*
Target asymmetry measurements of $\gamma p \rightarrow \pi^+ n$ and $\gamma p \rightarrow \pi^0 p$ with PHOENICS and ELSA

5:10 20 *B. Saghai, Saclay*
Pseudoscalar meson photoproduction on the proton

5:35 15 *K.H. Hicks, Ohio*
Exclusive pion and two proton photoproduction from ^{16}O using polarized photons

HIGH ENERGY

LEPTON ACCELERATORS, BEAMS AND POLARIMETERS III
Chair: M. Minty, SLAC [Maple Room]

4:15 15 *R. Assmann, CERN*
Deterministic harmonic spin matching in LEP

4:35 15 *B. Dehning, CERN*
Energy calibration with resonant depolarization at LEP in 1993

4:55 15 *H. Grote, CERN*
A Richter-Schwitters test spin rotator for LEP

5:15 15 *Y. Bashmakov, RAS*
Radiation and spin separation of high energy positrons by a bent crystal

5:35 15 *Y. Shatunov, Novisibirsk*
Spin flip by rf field at storage rings with Siberian snakes

STRONG INTERACTIONS AT HIGH ENERGY IV: Inclusive processes, QCD, and other topics Chair: G. Bunce, Brookhaven [Oak Room]

4:15 15 *V. Rykov, Wayne State and IHEP, Protvino*
Study of W and Z production processes at RHIC

4:35	15	*P.M. Border, Minnesota* A precise mesurement of the Ω^- magnetic moment
4:55	15	*R. Rylko, London* Spin effects in production of charmed hadrons
5:15	15	*M. Chavleishvili, Dubna* Spin phenomena in two-body processes

ELECTROWEAK INTERACTIONS II Chair: C. Prescott, SLAC
[Whittenberger Auditorium]

4:15	15	*W. Kretschmer, Erlangen* Neutrino physics with KARMEN
4:35	15	*P. Gumplinger, British Columbia* Measuring the muon polarization in $K \rightarrow \pi\mu\nu$ with the E-246 detector at KEK
4:55	15	*D. Ciampa, Minnesota* A precise measurement of non-leptonic decay parameters α and β/γ for $\Omega^- \rightarrow \Lambda K^-$ decays
5:15	15	*E.C. Dukes, Virginia* A new experiment to search for CP violation in hyperon decays
5:35	15	*H. Noumi, KEK* Study of polarized ${}^{5}_{\Lambda}He$ weak decay via the (π^+, K^+) reaction on ${}^{6}Li$

WEDNESDAY, SEPTEMBER 21

NUCLEAR

INTERMEDIATE ENERGY HADRON-INDUCED REACTIONS II
Chair: N.S. Chant [Frangipani Room]

2:00	12	*C.M. Edwards, Minnesota* Study of the ${}^{3,4}He(p,n)$ reactions at T_p=100 and 200 MeV
2:15	12	*D.S. Carman, IUCF* Quasifree (p,Np) reaction studies from ${}^{2}H$ and ${}^{12}C$ at 200 MeV
2:30	12	*T. Noro, RCNP* Exclusive measurement of $s_{1/2}$ proton knockout reaction
2:45	12	*W.G. Love, Georgia* Scattering of polarized protons from polarized targets
3:00	12	*D. Prout, Ohio State* Charge exchange spin observable measurements on Pb at 795 MeV in the giant resonance region
3:15	12	*W.W. Jacobs, IUCF* Polarization observables in $\vec{p}p \rightarrow pK^+\vec{Y}$ reactions at 2.9 GeV

INTERMEDIATE ENERGY ELECTROMAGNETIC INTERACTIONS II
Chair: C. Glashausser, Rutgers [Oak Room]

2:00	30	*S. Nanda, CEBAF* CEBAF spin physics
2:35	15	*S. Popov, Novosibirsk* Status of the t_{20} electron–deuteron scattering experiment at VEPP-3
2:53	15	*S. Kox, Grenoble* The new deuteron polarimeter POLDER and the t_{20} experiment at CEBAF
3:11	15	*A. Afanasev, CEBAF* Induced nucleon polarization in the d(e,e′N) reaction near threshold
3:29	15	*E. Passchier, NIKHEF* Proton knockout from tensor-polarized deuterium

POLARIZED SOURCES AND TARGETS III Chair: K. Zapfe, DESY [Persimmon Room]

2:00	15	*K. Hatanaka, RCNP* High intensity polarized ion source at RCNP
2:20	15	*W.J. Cummings, TRIUMF* Application of high power GaAlAs diode laser arrays for optically pumped spin exchange polarized ^{3}He targets
2:40	15	*K.P. Coulter, Michigan* Advances in alkali spin-exchange pumped ^{3}He target technology
3:00	15	*C.J. Horowitz, Indiana* Polarized stored beams by interaction with polarized electrons
3:20	15	*H. Spinka, Argonne* A possible method to produce a polarized antiproton beam at intermediate energies

HIGH ENERGY

LEPTON ACCELERATORS, BEAMS, AND POLARIMETERS IV
Chair: M. Minty, SLAC [Maple Room]

2:00	15	*M. Berz, Michigan State* Description and normal form analysis of spin dynamics using differential algebra
2:20	15	*R. Lieu, U.C. Berkeley* Synchrotron radiation: inverse Compton effect
2:40	15	*Y. Derbenev, IUCF/Michigan* RF-intrinsic spin flipper
3:00	15	*G. Bunce, Brookhaven* The new muon $g-2$ experiment at BNL
3:20	15	*F. Feinstein, Saclay* Muon polarimeters at SMC

NUCLEON SPIN STRUCTURE FUNCTIONS III Chair: R. Voss, CERN
[Whittenberger Auditorium]

2:00	20	*L. Tkatchev, Dubna* Search for jet handedness in hadronic Z^0 decays at DELPHI
2:25	20	*A. Efremov, Dubna* Puzzling correlation of handedness in $Z \to 2$ jet decay
2:50	20	*A. Dorokhov, Moscow* Instanton-induced helicity and flavor asymmetries in the light quark sea of the nucleon
3:15	20	*A. Yokosawa, Argonne* RHIC valence, sea, and gluon spin structure function measurements

SPIN'94 PARTICIPANTS LIST

High Energy

Klaus Ackerstaff
(DESY) Hamburg
MEA/Hermes
Notkestr.85
22603 Hamburg
GERMANY
ack@dxhrb3.desy.de
(tel.) 49-40-8998-4480
(fax) 49-40-8998-4013

Oscar Adriani
INFN, Firenze
Largo E. Fermi 2
I-50125
ITALY
adriani@cernvm.cern.ch
(tel.) 39-55-2298141
(fax) 39-55-229330

Andrei Afanasev
CEBAF
12000 Jefferson Ave
Newport News, VA 23606
USA
afanas@cebaf.gov
(tel.) 1-804-249-7011
(fax) 1-804-249-7363

Nural Akchurin
University of Iowa
Dept. of Physics & Astronomy
Iowa City, IA 52242
USA
akchurin@iaquark.physics.uiowa.edu
(tel.) 1-319-335-1941
(fax) 1-319-335-1753

Luydmila Alexeyeva
University of Michigan
High Energy Spin Physics
1239 Kipke Pike Dr. Suite 2341
Ann Arbor, MI 48109-1010
USA
blinov@mich.physics.lsa.umich.edu
(tel.) 1-313-763-9026
(fax) 1-313-763-9027

Antonio Amorim
Universidade de Lisboa
Centro de Fisica Nuclear
Av. Gama Pinto 2
1699-Lisboa Codex
PORTUGAL
faamorim@skull.cc.fc.ul.pt
(tel.) 1-351-1-7950790
(fax) 1-351-1-7956289

L. Wilmer Anderson
University of Wisconsin
Dept. of Physics
1150 University Ave.
Madison, WI 53706
USA
lwanderson@uwnuc0.physics.wisc.edu
(tel.) 1-608-262-8962
(fax) 1-608-262-3598

Mauro Anselmino
Universita di Torino
Dept. of Theoretical Physics
Via P. GIURIA 1
10125 Torino
ITALY
anselmino@to.infn.it(tel.) 39-11-6707227
(fax) 39-11-6707214

Hideki Aoyagi
Nagoya University
Department of Physics
Furo-cho, Chikusa-ku
Nagoya 464
JAPAN
aoyagi@kekvax.kek.jp
(tel.) 81-52-789-2894
(fax) 81-52-789-2903

Jacques Arvieux
Laboratoire National Saturne
C.E. Saclay
F-91191 Gif-sur-Yvette
Cedex
FRANCE
arvieux@frcpn11.in2p3.fr
(tel.) 33-1-69082203
(fax) 33-1-69082970

Ralph Assmann
Kreuzbergstr. 46
D-53127 Bonn
GERMANY
(tel.) 49-228-25-32-74

Ulrich Atzrott
Universitut Tubingen
Physikalisches Institut
Auf der Morgenstelle 14
D-72076 Tubingen
GERMANY
atzrott@pit.physik.uni-teubingen.de
(tel.) 49-7071-293432
(fax) 49-7071-296296

Todd Averett
University of Virginia
Department of Physics
Charlottesville, VA 22901
USA
averett@barney.phys.virginia.edu
(tel.) 1-804-982-2054

Jacques Ball
Laboratoire National SATURNE
LNS
Dir. 91191 Gif-sur-Yvette
Cedex
FRANCE
ball@frcpn11.in2p3.fr
(tel.) 33-1-69088719
(fax) 33-1-69082970

Henry R. Band
SLAC
Bin 94, P.O. Box 4349
Stanford, CA 94309
USA
hrb@stanford.edu
(tel.) 1-415-926-2655
(fax) 1-415-926-2923

Desmond Barber
Deutsches Elektronen Synch. (DESY)
Notkestrasse 85
Postfach 22603 Hamburg
GERMANY
mpybar@dsyibm.desy.de
(tel.) 49-040-8998-3035
(fax) 49-040-8998-3282

Yurij Bashmakov
Lebedev Physics Institute
Russian Academy of Sciences
Leninsky Prospect 53
117924, Moscow
RUSSIA
holbsh@dsyibm.desy.de
(tel.) 7-095-135-05-77
(fax) 7-095-135-78-80

Guenter G. Baum
Universitut Bielefeld
Facultut fur Physik
Universit tstrasse 25
D-33615 Bielefeld
GERMANY
baum@physf.uni-bielefeld.de
(tel.) 49-0521-1065383
(fax) 49-0521-1062959

Douglas Beck
University of Illinois
Nuclear Physics Laboratory
23 E. Stadium Dr.
Champaign, IL 61820
USA
beck@uinpla.npl.uiuc.edu
(tel.) 1-217-244-7994
(fax) 1-217-333-1215

Alexander S. Belov
Inst. for Nuclear Research
Russian Academy of Sciences
prospect 60th Oct.
Revolution Anniversary, 7A
117312, Moscow, RUSSIA
belov@inr.msk.su
(tel.) 7-095-334-09-62
(fax) 7-095-135-22-68

Jim Birchall
University of Manitoba
Department of Physics
Winnipeg
Manitoba R3T 2N2
CANADA
birchall@physics.umanitoba.ca
(tel.) 1-204-474-6205
(fax) 1-204-269-8489

Les Bland
Indiana Univ. Cyclotron Fac.
2401 Milo B. Sampson Lane
Bloomington, IN 47408
USA
bland@iucf.indiana.edu
(tel.) 1-812-855-6051
(fax) 1-812-855-6645

Michael Boge
DESY/MPY
85 Notkestrasse
22603 Hamburg
GERMANY
mpyboe@hp-cluster.desy.de
(tel.) 49-040-8998-2052
(fax) 49-040-8998-4305

Pete M. Border
University of Minnesota
Department of Physics
116 Church St. SE
Minneapolis, MN 55455
USA
border@mnhep4.hep.umn.edu
(tel.) 1-612-624-1020
(fax) 1-612-624-4578

Herve Borel
C. E. Saclay
DAPNIA/SPhN
Orme des Merisiers
F91191 Gif-sur-Yvette
FRANCE
borel@phnx7.saclay.cea.fr
(tel.) 33-1-69-08-75-09
(fax) 33-1-69-08-75-84

Bennet B. Brabson
Indiana University
Dept. of Physics
Bloomington, IN 47405
USA
brabson@ind.physics.indiana.edu
(tel.) 1-812-855-3881
(fax) 1-812-855-0440

Arnd Brandenburg
SLAC
P.O. Box 4349
Mail Stop 81
Stanford, CA 94309
USA
arnd@slac.stanford.edu
(tel.) 1-415-926-4433
(fax) 1-415-926-2525

Stephen Bueltmann
CERN
CH-1211
Geneve 23
SWITZERLAND
stephen@na47sun05.cern.ch
(tel.) 41-22-767-3453
(fax) 41-22-785-0672

Gerry Bunce
BNL
AGS Department, Bldg. 911B
Upton, NY 11973
USA
bunce@bnldag.ags.bnl.gov
(tel.) 1-516-282-4771
(fax) 1-516-282-5954

John M. Cameron
Indiana University Cyclotron Facility
2401 Milo B. Sampson Lane
Bloomington, In 47408
USA
cameron@iucf.indiana.edu
(tel.) 1-812-855-9365
(fax) 1-812-855-6645

Peter Cameron
BNL
Bldg. 830
Upton, NY 11973
USA
cameron@bnlux1.bnl.gov
(tel.) 1-516-282-7657
(fax) 1-516-282-3079

David Caussyn
University of Michigan
High Energy Spin Physics
1239 Kipke Drive #2321
Ann Arbor, MI 48109
USA
caussyn@mich1.physics.lsa.umich.edu
(tel.) 1-313-763-9033
(fax) 1-313-763-9027

Christian Cavata
DAPNIA/SPhN CEN Saclay
F-91191 Gif-sur-Yvette
Cedex
FRANCE
cavata@phnx7.saclay.cea.fr
(tel.) 33-1-6908-3237
(fax) 33-1-6908-7584

Alex W. Chao
Stanford Linear Accel. Center
Stanford, CA 94305
USA
achao@scs.slac.stanford.edu

Michael P. Chavleishvili
Joint Inst. for Nuclear Rsch.
141980, Dubna
Moscow region
RUSSIA
chavlei@theor.jinrc.dubna.su
(tel.) 7-09621-6-2156
(fax) 7-09621-6-5084

Chungming Chu
University of Michigan
Randall Laboratory of Physics
Ann Arbor, Mi 48109-1120
USA
chu@mich.physics.lsa.umich.edu
(tel.) 1-313-764-5113
(fax) 1-313-763-9027

David Ciampa
University of Minnesota
Department of Physics
Tate Hall
116 Church Street SE
Minneapolis, MN 55455
USA
ciampa@mnhep1.hep.umn.edu
(tel.) 1-612-624-1020
(fax) 1-612-624-4578

Thomas B. Clegg
University of North Carolina
Dept. of Physics & Astronomy
Phillips Hall
Chapel Hill, NC 27599-3255
USA
clegg@tunl.duke.edu
(tel.) 1-919-962-2079
(fax) 1-919-962-0480

Mario Conte
Dipartmento di Fisica
INFN Sezione di Genova
Via Dodecaneso 33
16146 Genova
ITALY
contem@genova.infn.it
(tel.) 39-10-353208
(fax) 39-10-313358

Andreas P. Contogouris
McGill University
Dept. of Physics
Montreal, P.Q.
CANADA H3A 2T8
secretariat@hep.physics.mcgill.ca
(tel.) 1-514-398-6520
(fax) 1-514-398-3733

Homer E. Conzett
Lawrence Berkeley Laboratory
One Cyclotron Road
Berkeley, CA 94720
USA
heconzett@lbl.gov
(tel.) 1-510-486-7813
(fax) 1-510-486-7983

Daniel A. Cooper
Ohio State University
Van de Graaf Laboratory
1302 Kinnear Road
Columbus, OH 43212
USA
dcooper@mps.ohio-state.edu
(tel.) 1-614-292-4775
(fax) 1-614-292-4833

Ernest D. Courant
Brookhaven National Laboratory
Building 1005
Upton, NY 11973-5000
USA
courant@bnlcl1.bnl.gov
(tel.) 1-516-282-4609
(fax) 1-516-282-5729

G.R. Court
University of Liverpool
Dept. of Physics
P.O. Box 147
Liverpool L69 BX
ENGLAND
(tel.) 44-51-794-3383
(fax) 44-51-794-3444

Donald G. Crabb
University of Virginia
Dept. of Physics
McCormick Road
Charlottesville, VA 22901
USA
dgc3q@virginia.edu
(tel.) 1-804-924-6790
(fax) 1-804-924-4576

Don Crandell
University of Michigan
2068 Randall
Ann Arbor, MI 48109-1120
USA
crandell@mich.physics.lsa.umich.edu
(tel.) 1-313-763-8161
(fax) 1-313-763-9027

William J. Cummings
TRIUMF
4004 Wesbrook Mall
Vancouver, B.C., V6T 2A3
CANADA
cummings@erich.triumf.ca
(tel.) 1-604-222-1047
(fax) 1-604-222-1074

Daniel Dale
University of Kentucky
Dept. of Physics
Lexington, KY 40506
USA
dale@zeppo.pa.uky.edu

Mario Conte
Dipartmento di Fisica
INFN Sezione di Genova
Via Dodecaneso 33
16146 Genova
ITALY
contem@genova.infn.it
(tel.) 39-10-353208
(fax) 39-10-313358

Andreas P. Contogouris
McGill University
Dept. of Physics
Montreal, P.Q.
CANADA H3A 2T8
secretariat@hep.physics.mcgill.ca
(tel.) 1-514-398-6520
(fax) 1-514-398-3733

Homer E. Conzett
Lawrence Berkeley Laboratory
One Cyclotron Road
Berkeley, CA 94720
USA
heconzett@lbl.gov
(tel.) 1-510-486-7813
(fax) 1-510-486-7983

Daniel A. Cooper
Ohio State University
Van de Graaf Laboratory
1302 Kinnear Road
Columbus, OH 43212
USA
dcooper@mps.ohio-state.edu
(tel.) 1-614-292-4775
(fax) 1-614-292-4833

Ernest D. Courant
Brookhaven National Laboratory
Building 1005
Upton, NY 11973-5000
USA
courant@bnlcl1.bnl.gov
(tel.) 1-516-282-4609
(fax) 1-516-282-5729

G.R. Court
University of Liverpool
Dept. of Physics
P.O. Box 147
Liverpool L69 BX
ENGLAND
(tel.) 44-51-794-3383
(fax) 44-51-794-3444

Donald G. Crabb
University of Virginia
Dept. of Physics
McCormick Road
Charlottesville, VA 22901
USA
dgc3q@virginia.edu
(tel.) 1-804-924-6790
(fax) 1-804-924-4576

Don Crandell
University of Michigan
2068 Randall
Ann Arbor, MI 48109-1120
USA
crandell@mich.physics.lsa.umich.edu
(tel.) 1-313-763-8161
(fax) 1-313-763-9027

William J. Cummings
TRIUMF
4004 Wesbrook Mall
Vancouver, B.C., V6T 2A3
CANADA
cummings@erich.triumf.ca
(tel.) 1-604-222-1047
(fax) 1-604-222-1074

Daniel Dale
University of Kentucky
Dept. of Physics
Lexington, KY 40506
USA
dale@zeppo.pa.uky.edu

Charles A. Davis
TRIUMF
4004 Wesbrook Mall
Vancouver, B.C.
CANADA V6T 2A3
cymru@erich.triumf.ca
(tel.) 1-604-222-1047
(fax) 1-604-222-1074

Donal Day
University of Virginia
Department of Physics
McCormick Road
Charlottesville, VA 22901
USA
dbd@virginia.edu
(tel.) 1-804-924-6566
(fax) 1-804-924-4576

Bernd Dehning
CERN
CH-1211
Geneva 23
SWITZERLAND
dehning@vxcern.cern.ch
(tel.) 41-22-767-4783
(fax) 41-22-782-2850

Kees de Jager
NIKHEF-K
P.O. Box 41882
1009 DB Amsterdam
THE NETHERLANDS
kees@paramount.nikhefk.nikhef.nl
(tel.) 31-20-5922143
(fax) 31-20-5922165

Paul Delheij
TRIUMF
4004 Wesbrook Mall
Vancouver, B.C., V6T 2A3
CANADA
delh@triumf.ca
(tel.) 1-604-222-1047
(fax) 1-604-222-1074

Yaroslav Derbenev
Nuclear Eng. Dept.
University of Michigan
Ann Arbor, MI 48109
USA
(tel.) 1-313-764-5290
(fax) 1-313-763-4540

Jules Deutsch
Univ. Catholique de Louvain
Inst. de Physique
Chemin du Cyclotron 2
Louvain-la-Neuve, B1348
BELGIUM
deutsch@fynu.ucl.ac.be
(tel.) 32-10-473273
(fax) 32-10-452183

Tom Devlin
Rutgers University
Dept. of Physics
Box 849
Piscataway, NJ 08855-0849
USA
devlin@ruthep.rutgers.edu
(tel.) 1-908-932-4848
(fax) 1-908-932-4343

William De Zarn
Indiana University Cyclotron Facility
2401 Milo B. Sampson Lane
Bloomington, IN 47408
USA
dezarnwa@iucf.indiana.edu
(tel.) 1-812-855-3613
(fax) 1-812-855-6645

Chaden Djalali
University of S. Carolina
Dept. of Physics
Columbia, SC 29208
USA
djalali@nuc002.psc.scarolina.edu
(tel.) 1-803-777-4318
(fax) 1-803-777-3065

Alexander Dorokhov
Joint Inst. for Nuclear Rsch.
Bogoliubov Laboratory of
Theoretical Physics
114980 Moscow Region, Dubna
RUSSIA
dorokhov@thsunl.jinr.dubna.su
(tel.) 7-096-2162730
(fax) 7-096-2165084

Paul Draper
Univ. of Texas at Arlington
High Energy Physics
241 Science Hall
502 Yates Street
Arlington, TX 76019
USA
draper@utahep.uta.edu
(tel.) 1-817-273-2817
(fax) 1-817-273-2824

Michael Duren
DESY
HERMES Collaboration
Notkestrasse 85
D-22603 Hamburg
GERMANY
dueren@vxdsya.desy.de
(tel.) 49-40-89982089
(fax) 49-40-89983438

Edmond C. Dukes
University of Virginia
Dept. of Physics
McCormick Road
Charlottesville, VA 22901
USA
dukes@uvahep.phys.virginia.edu
(tel.) 1-804-982-5364
(fax) 1-804-982-5375

Fraser Duncan
University of Maryland
Department of Physics
College Park, MD 20742
USA
duncan@enp.umd.edu
(tel.) 1-301-405-6105
(fax) 1-301-314-9525

Hartmut Dutz
Universit t Bonn
Physikalisches Institut
Nussallee 12
D-53115 Bonn
GERMANY
dutz@pibl.physik.uni-bonn.de
(tel.) 49-228-733610
(fax) 49-228-737869

Carla M. Edwards
c/o University of Minnesota Group
Mail Stop H846
Los Alamos National Laboratory
Los Alamos, NM 87545
USA
carla@stpaul.lampf.lanl.gov
(tel.) 1-505-665-7695
(fax) 1-505-665-7920

Anatoli Efremov
Joint Inst. for Nuclear Rsch.
Laboratory of Theoretical Physics
Dubna
Moscow Region
141980 RUSSIA
efremov@thsunl.jinr.dubna.su
(tel.) 7-09621-65678
(fax) 7-09621-65084

Hiroyasu Ejiri
Osaka University
Research Center for Nuclear Physics
10-1 Mihogaoka, Ibaraki
Osaka 567
JAPAN
ejiri@rcnpvx.rcnp.osaka-u.ac.jp
(tel.) 81-68798929
(fax) 81-68798899

Yury I. Eidelman
Budker Inst. of Nuclear Physics
Prospekt Lavrentieva 11
Novosibirsk 630090
RUSSIA
eidelyur@tulip.inp.nsk.su
(tel.) 7-383-2359-975
(fax) 7-383-2352-163

Dieter Eversheim
Universit t Bonn
Institut fur Strahlen und Kernphysik
Nussallee 14-16
D-53115 Bonn
GERMANY
evershei@servax.iskp.uni-bonn.de
(tel.) 49-228-735299
(fax) 49-228-733728

Willie R. Falk
University of Manitoba
Dept. of Physics
Winnipeg
CANADA R3T 2N2
falk@physics.umanitoba.ca
(tel.) 1-204-474-9856
(fax) 1-204-269-8489

Harold W. Fearing
TRIUMF
4004 Wesbrook Mall
Vancouver BC V6T 2A3
CANADA
fearing@triumf.ca
(tel.) 1-604-222-1047
(fax) 1-604-222-1074

Fabrice Feinstein
C.E.N. SACLAY
S.P.P./DAPNIA
F-91191 Gif-sur-Yvette
FRANCE
feinstei@frcpnll.in2p3.fr
(tel.) 33-1-6908-8533
(fax) 33-1-6908-6428

Andrew Feltham
Inst. f r Physik der Universit t Basel
Experimental Kernphysik
Klingelbergstrasse 82
4056 Basel
SWITZERLAND
feltham@urz.unibas.ch
(tel.) 41-61-267-3728
(fax) 41-61-267-3784

Dieter Fick
Philipps Universitut-Marburg
Fachbereich Physik
D35032 Marburg
GERMANY
fick@mvl3a.physik.uni-marburg.de
(tel.) 49-06421-282017
(fax) 49-06421-287033

Giuseppe Fidecaro
CERN
CH1211 Geneva 23
SWITZERLAND
fde@cernvm.cern.ch
(tel.) 41-22-767-2686
(fax) 41-22-783-0672

Michael Finger
Technical University Prague
Faculty of Nuclear Sciences
and Physical Engineering
Brehov 7, Praha-1
CZECH REPUBLIC
finger@vxcern.cern.ch

Michel Garcon
DAPNIA/SPhN, Bat. 703
CEN-Saclay
F-91191 Gif-sur-Yvette
Cedex
FRANCE
garcon@phnx7.saclay.cea.fr
(tel.) 33-1-6908-8623
(fax) 33-1-6908-7584

Susan V. Gardner
Indiana University
Nuclear Theory Center
2401 Milo B. Sampson Lane
Bloomington, IN 47408
USA
gardner@iucf.indiana.edu
(tel.) 1-812-855-9365
(fax) 1-812-855-6645

Shalev Gilad
MIT
77 Massachusetts Avenue
Bldg. 26-449
Cambridge, MA 02139
USA
gilad@pierre.mit.edu
(tel.) 1-617-253-7785
(fax) 1-617-258-5440

Ron Gilman
Rutgers University
Department of Physics and Astronomy
P.O. Box 849
Piscataway, NJ 08855-0849
USA
gilman@ruthep.rutgers.edu
(tel.) 1-908-445-5489
(fax) 1-908-445-4343

Svetlana Gladycheva
University of Michigan
High Energy Physics
1239 Kipke Drive, Suite 2341
Ann Arbor, MI 48109-1010
USA
anferov@mich.physics.lsa.umich.edu
(tel.) 1-313-763-9026
(fax) 1-313-763-9027

Charles Glashausser
Rutgers University
Serin Physics Laboratory
Box 849
Piscataway, NJ 08854,USA
glashausser@ruthep.rutgers.edu
(tel.) 1-908-445-2526
(fax) 1-908-445-4343

Charles Goodman
Indiana University Cyclotron Facility
2401 Milo B. Sampson Lane
Bloomington, IN 47408
USA
goodman@iucf.indiana.edu
(tel.) 1-812-855-9365
(fax) 1-812-855-6645

David Grosnick
Argonne National Labratory
HEP362
9700 South Cass Ave.
Argonne, IL 60439
USA
dpg@hep.anl.gov
(tel.) 1-708-252-7529
(fax) 1-708-252-5782

Hans Grote
Route de Meyrin
CERN
1211 Geneva 23
SWITZERLAND
USA
hansg@cernvm.cern.ch
(tel.) 41-22-7674961
(fax) 41-22-7830552

P. Gumplinger
University of British Columbia
TRIUMF
4004 Wesbrook Mall
Vancouver, BC V6T 2A3
CANADA
gum@erich.triumf.ca
(tel.) 1-604-222-1047
(fax) 1-604-222-1074

Willy Haeberli
University of Wisconsin
Dept. of Physics
1150 University Avenue
Madison, WI 53706
USA
whaeberli@uwnuc0.physics.wisc.edu
(tel.) 1-608-262-0009
(fax) 1-608-262-3598

William Happer
Princeton University
Department of Physics
Princeton, NJ 08544
USA
happer@pupgg.princeton.edu
(tel.) 1-609-258-4382
(fax) 1-609-258-2496

Takeo Hasegawa
Miyazaki University
Faculty of Engineering
1-1 Gakuen-Kibanadai-Nishi-
Miyazaki-shi, 889-21
JAPAN
hastake@kekvax.kek.jp
(tel.) 81-0985-58-2811X4321
(fax) 81-0985-58-1647

Kichiji Hatanaka
Osaka University
RCNP
10-1 Mihogaoka, Ibaraki
Osaka 567
JAPAN
hatanaka@rcnpvx.rcnp.osaka-u.ac.jp
(tel.) 81-6-879-8934
(fax) 81-6-879-8899

Peter Heimberg
Northwestern University
Department of Physics & Astronomy
Evanston, IL 60208
USA
heimberg@nuhepb.phys.nwu.edu
(tel.) 1-708-491-8607
(fax) 1-708-491-9982

Kenneth J. Heller
University of Minnesota
School of Physics & Astronomy
16 Church Street SE
Minneapolis, MN 55455
USA
heller@mnhep.hep.umn.edu
(tel.) 1-612-624-7314
(fax) 1-612-624-4578

Bill Hersman
University of New Hampshire
Dept. of Physics
Durham, NH 03824
USA
(tel.) 1-603-862-3512
(fax) 1-603-862-2998

K. H. Hicks
Dept. of Physics
Ohio University
Athens, OH 45701
USA
khicks@ohiou.edu
(tel.) 1-614-593-1981
(fax) 1-614-593-1436

Gregory C. Hillhouse
University of the Western Cape
Dept. of Physics
Private Bag X17
Bellville 7535
SOUTH AFRICA
hillhouse@nacdh4.nac.ac.za
(tel.) 27-021-959-2556
(fax) 27-021-959-2266

Yorck Holler
DESY, Dpt. MEA
Postfach
D-22603 Hamburg
GERMANY
meahol@dsyibm.desy.de
(tel.) 49-40-8998-3743
(fax) 49-40-8998-3438

Haixin Huang
Brookhaven National Laboratory
AGS Department
Building 911B
Upton, NY 11973
USA
huang@bnldag.bnl.gov
(tel.) 1-516-282-5446
(fax) 1-516-282-5954

Vernon W. Hughes
Yale University
Dept. of Physics, 465 J. W. Gibbs
P.O. Box 208121
New Haven, CT 06520-8121
USA
hughes@yalph2.physics.yale.edu
(tel.) 1-203-432-3819
(fax) 1-203-432-3804

Masataka Iinuma
Kyoto University
Department of Physics
Kitashirakawa
Kyoto, 606-1
JAPAN
iinuma%kytax1.dnet@kekux.kek.jp
(tel.) 81-757533871
(fax) 81-757115175

Yasunori Iseri
Chiba-Keizai College
Todoroki-cho, 4-3-30
Inage-ku
Chiba, 263
JAPAN
(tel.) 81-43-255-3451
(fax) 81-43-252-6050

Harold E. Jackson, Jr.
Argonne National Laboratory
Bldg. 203 B-237
Argonne, IL 60439
USA
hal@anl.gov
(tel.) 1-708-252-4013
(fax) 1-708-252-3903

William W. Jacobs
Indiana University Cyclotron Facility
2401 Milo B. Sampson Lane
Bloomington, IN 47408
USA
jacobs@iucf.indiana.edu
(tel.) 1-812-855-8873
(fax) 1-812-339-6645

Robert L. Jaffe
6-311 MIT
77 Massachusetts Avenue
Cambridge, MA 02139
USA
jaffe@mitlns.mit.edu
(tel.) 1-617-253-4858
(fax) 1-617-253-8674

Brajesh K. Jain
Bhabha Atomic Research Centre
Nuclear Physics Division, BARC
Bombay 400 085
INDIA
bkjain@magnum.barctl.ernet.in
(tel.) 91-22-555-6071
(fax) 91-22-556-0750

Marek Jezabek
Institute of Nuclear Physics
High Energy Physics Department
ul. Kawiory 26a
PL-30055 Cracow
POLAND
jezabek@vsk01.ifj.edu.pl
(tel.) 48-12-333366x46
(fax) 48-12-333884

Kenneth A. Johns
University of Arizona
Department of Physics
Tuscon, AZ 85721
USA
johns@uazhep.physics.arizona.edu
(tel.) 1-602-621-6791
(fax) 1-602-621-4721

Mark Jones
College of William and Mary
Department of Physics
Williamsburg, VA 23187
USA
jones@wmspin.physics.wm.edu
(tel.) 1-804-221-3491
(fax) 1-804-221-3540

Basim Kamal
McGill University
Physics Department
Montreal, Quebec, H3A 2T8
CANADA
cxbk@musica.mcgill.ca
(tel.) 1-514-398-6502

G. Ya. Kezerashvili
Budker Institute of Nuclear Physics
630090 Novosibirsk
RUSSIA
guramkez@inp.nsk.su
(tel.) 7-3832-359420
(fax) 7-3832-352163

Kurt Kilian
Forschungszentrum Julich GmbH
Institut f r Kernphysik
Briefpost: 52425 Julich
Fracht/Paketpost: 52428 Julich
GERMANY
kph001@djukfall
(tel.) 49-2461-61-5943
(fax) 49-2461-61-3930

Yeongduk Kim
KEK
Dept. of Physics
1-1 Oho Tsukuba-shi
Ibaraki 305
JAPAN
ydkim@kekvax.kek.jp
(tel.) 81-298-64-5426
(fax) 81-298-64-7831

Kikuo Kimura
Nagasaki Inst. of Applied Sci.
536 Aba-machi
Nagasaki 851-01
JAPAN
kimura@nias.ac.jp
(tel.) 81-958-39-3111
(fax) 81-958-30-1126

Michikazu Kinsho
National Lab for High En.Phys.
(KEK), 1-1 Oho-machi
Tsukuba-shi
Ibaraki 305
JAPAN
kinsho@kekvax.kek.jp
(tel.) 81-298-64-5215
(fax) 81-298-64-3182

Nikolai Kochelev
Joint Inst. for Nuclear Rsch.
Laboratory of High Energy
Head Post Office P.O. Box 79
101000 Moscow
RUSSIA
kochelev@main1.jinr.dubna.su
(fax) 7-095-975-2381

Alexander Komives
Indiana University Cyclotron Facility
2401 Milo B. Sampson Ln.
Bloomington, IN 47408
USA
komives@iucf.indiana.edu
(tel.) 1-812-855-9365
(fax) 1-812-855-6645

Dirk Kramer
CERN
PPE Division
1211 Geneva 23
SWITZERLAND
kraemer@uxnhd.cern.ch
(tel.) 41-22-785-6428

Laird Kramer
MIT-Lab for Nuclear Science
Building 26-533
Cambridge, MA 02139
USA
kramer@mitlns.mit.edu
(tel.) 1-612-253-3761
(fax) 1-612-258-5440

Wolfgang Kretschmer
Universitut Erlangen-Nurnberg
Physikalisches Institut
Erwin-Rommel-Str.1
D-91058 Erlangen
GERMANY
pi4kret@pkvxl.physik.uni-erlangen.de
(tel.) 49-9131-857075
(fax) 49-9131-15249

Alan D. Krisch
University of Michigan
Randall Lab of Physics
Ann Arbor, MI 48109-1120
USA
krisch@umiphys
(tel.) 1-313-936-1027
(fax) 1-313-936-0794

Ronald Kunne
Laboratoire National Saturne
CE Saclay
91191 Gif-sur-Yvette
Cedex
FRANCE
kunne@frcpnll.in2p3.fr
(tel.) 31-1-6908-3358
(fax) 31-1-6908-2970

Jukka Kyynarainen
CERN
Building 892, 1-D18
CH-1211
Geneva 23
SWITZERLAND
jukka@na47sun05.cern.ch
(tel.) 41-22-7672942
(fax) 41-22-7850672

J. M. Laget
CEA (French Atomic Energy Com.)
Dapnia/SPhn
CE. Saclay, F91191-Gif-sur-Yvette
Cedex
FRANCE
laget@phnx7.saclay.cea.fr
(tel.) 33-1-6928-7554
(fax) 33-1-6928-7584

Anzhi Lai
CEBAF
Mail Stop 28F
12000 Jefferson Avenue
Newport News, VA 23606
USA
lai@cebaf.gov
(tel.) 1-804-249-7310

Shyh-Yuan Lee
Indiana University
Dept. of Physics
Swain Hall W. 117
Bloomington, IN 47405
USA
lee@iucf.indiana.edu
(tel.) 1-812-855-1247
(fax) 1-812-855-6645

C.D. Philip Levy
TRIUMF
4004 Wesbrook Mall
Vancouver B.C.
CANADA V6T 2A3
levy@triumf.ca
(tel.) 1-604-222-1047
(fax) 1-604-222-1074

Bing An Li
University of Kentucky
Dept. of Physics and Astronomy
Lexington, KY 40506
USA
li@ukcc
(tel.) 1-606-257-1486
(fax) 1-606-323-2846

R. Lieu
Center for EUV Astrophysics
2150 Kittredge Street
Berkeley, CA 94720
USA
lieu@cea.berkeley.edu
(tel.) 1-510-642-4224

Ali M.T. Lin
University of Michigan
Dept. of Physics
Ann Arbor, MI 48109-1120
USA
lin@umiphys
(tel.) 1-313-763-9033
(fax) 1-313-763-9026

Jian Liu
Indiana University Cylcotron Facility
2401 Milo B. Sampson Lane
Bloomington, IN 47408
USA
jiliu@iucf.indiana.edu
(tel.) 1-812-855-9365
(fax) 1-812-855-6645

Alfredo U. Luccio
Brookhaven National Laboratory
AGS Dept. Bldg. 911B
Upton, NY 11973-5000
USA
luccio@bnldag.ags.bnl.gov
(tel.) 1-516-282-7699
(fax) 1-516-282-5954

Vladimir Luppov
University of Michigan
High Energy Spin Physics
1239 Kipke Drive
Ann Arbor, MI 48109-1120
USA
luppov@michl.physics.lsa.umich.edu
(tel.) 1-313-764-5111
(fax) 1-313-763-9027

Malcolm MacFarlane
Indiana University
Nuclear Theory Center
2401 Milo B. Sampson Lane
Bloomington, IN 47408
USA
macfarlane@iucf.indiana.edu
(tel.) 1-812-855-2953
(fax) 1-812-855-6645

Yousef I. Makdisi
Brookhaven National Laboratory
RHIC Project, Bldg. 510C
P.O. Box 5000
Upton, NY 11973-5000
USA
makdisi@bnldag.ags.bnl.gov
(tel.) 1-516-282-4932
(fax) 1-516-282-2532

Gerhard K. Mallot
CERN PPE
CH-1211 Geneve, 23
SWITZERLAND
gkm@na47sun05.cern.ch
(tel.) 49-22-767-6422
(fax) 49-22-785-0672

Manat Maolinbay
Swiss Federal Institute of Technology
Institute for High En. Physics
ETH-H ngg
CH-8093 Zurich
SWITZERLAND
manat%czhethla.bitnet@cearn.cern.ch
(tel.) 41-1-6332028
(fax) 41-1-3720534

Takashi Maruyama
SLAC
P.O. Box 4349
Stanford, CA 94309
USA
tvm@slacvm.slac.stanford.edu
(tel.) 1-415-926-3398
(fax) 1-415-926-2923

Akira Masaike
Kyoto University
Dept. of Physics
Sakyo-ku
Kyoto 606
JAPAN
masaike@kytvax.scphys.kyoto-u.ac.jp
(tel.) 81-75-7533859
(fax) 81-75-7115175

Yasuhiro Masuda
KEK
1-1 Oho
Tsukuba-shi
Ibaraki-ken 305
JAPAN
masuda@kekvax.kek.jp
(tel.) 81-298-641171
(fax) 81-298-643202

Yasuyuki Matsuda
Kyoto University

Ta-chung Meng
Institut fur Theoretische Physik
Freie Universit t Berlin
Arnimallee 14
14195 Berlin
GERMANY
meng@spin:physik.fu-berlin.de.
(tel.) 49-30-8383031
(fax) 49-30-8383741

Hans-Otto Meyer
Indiana University
Dept. of Physics
Swain Hall West
Bloomington, IN 47405
USA
meyer@iucf.indiana.edu
(tel.) 1-812-855-9365
(fax) 1-812-855-6645

Werner Meyer
Physikalisches Institut-Bonn
Nussallee 12
D-53115 Bonn
GERMANY
meyer@pibl.physik.uni-bonn.de
(tel.) 49-228-732230
(fax) 49-228-737869

Gerald A. Miller
University of Washington
Dept. of Physics, FM-15
Seattle, WA 98195
USA
miller@alpher.npl.washington.edu
(tel.) 1-206-543-2995
(fax) 1-206-685-0635

Michael A. Miller
University of Wisconsin
1150 University Avenue
Madison, WI 53706
USA
miller@uwnucl.physics.wisc.edu
(tel.) 1-608-262-3091
(fax) 1-608-262-3598

Richard G. Milner
MIT
26-447, Dept. of Physics
Cambridge, MA 02139
USA
milner@mitlns.mit.edu
(tel.) 1-617-258-5439
(fax) 1-617-258-5440

Michiko Minty
SLAC
MS-26
P.O. Box 4349
Stanford, CA 94309
USA
minty@slac.stanford.edu
(tel.) 1-415-926-3650
(fax) 1-415-926-4999

Yoshiharu Mori
National Lab for High En. Phys
(KEK), Oho 1-1
Tsukuba-shi
Ibaraki-ken 305
JAPANmoriy@kekvax.kek.ac.jp
(tel.) 81-298-64-1171
(fax) 81-298-64-3182

Toshiyuki Morii
Kobe University
Faculty of Human Development
Tsurukabuto, Nada
Kobe 657
JAPAN
morii@cphys.cla.kobe-u.ac.jp
(tel.) 81-78-803-0917
(fax) 81-78-803-0831

Francesco Murgia
Istituto Nazionale Di Fisica Nucleare
Sezione Di Cagliari
Via Ada Negri 18
I-09127 Cagliari
ITALY
murgia@vaxca.ca.infn.it
(tel.) 39-70-670834
(fax) 39-70-657823

Genis Musulmanbekov
Joint Inst. for Nuc. Research
LCTA
Head Post Office P.O. Box 79
101000 Moscow
RUSSIA
genis@vsdl28.jinr.dubna.su
(fax) 7-096-21-6-5145

Sergey I. Nagorny
Kharkov Inst. of Phys. & Tech.
National Science Center,
Theoretic Division
Academicheskaya St., 1
Kharkov 310108
UKRAINE
kfti%kfti.kharkov.ua@relay.ussr.eu.net
(tel.) 7-057-235-6024
(fax) 7-057-235-1738

Hermann Nann
Indiana University Cyclotron Facility
2401 Milo B. Sampson Lane
Bloomington, IN 47408
USA
nann@iucf.indiana.edu
(tel.) 1-812-855-2884
(fax) 1-812-855-6645

Bernard Nefkens
UCLA
Dept. of Physics
405 Hilgard Avenue
Los Angeles, CA 90024
USA
bnefkens@uclapp.physics.ucla.edu
(tel.) 1-310-825-4970
(fax) 1-310-206-4397

Edwin Norbeck
University of Iowa
Dept. of Physics and Astronomy
Iowa City, IA 52242
USA
norbeck@iowa.physics.uiowa.edu
(tel.) 1-319-335-0903
(fax) 1-319-335-1753

Wolf-Dieter Nowak
DESY-IfH Zeuthen
D-15735 Zeuthen
GERMANY
wdn@znher2.ifh.de
(tel.) 49-33-762-77349
(fax) 49-33-762-77330

Sandibek B. Nurushev
Inst. for High Energy Physics
142284 Protvino
Moscow Region
RUSSIA
nurushev@mx.ihep.su
(tel.) 7-095-230-3228
(fax) 7-095-230-2337

Teamour Nurushev
University of Michigan
1239 Kipke Drive
2341 High Energy Physics
Ann Arbor, MI 48109
USA
nurushev@mich.physics.lsa.umich.edu
(tel.) 1-313-763-8161
(fax) 1-313-763-9027

Hideaki Otsu
University of Tokyo
Dept. of Physics
7-3-1, Hongo, Bunkyo-ku
Tokyo 113
JAPAN
otsu@rikvax.riken.go.jp
(tel.) 81-48-462-1111x4131
(fax) 81-48-461-5301

H. Paetz gen Schieck
Universit t K ln
Institut f r Kernphysik
Zuelpicher Strasse 77
D-50937 K ln
GERMANY
schieck@lucie.ikp.uni-koeln.de
(tel.) 49-221-470-3620
(fax) 49-221-470-5168

Shelley Page
University of Manitoba
c/o TRIUMF
4004 Wesbrook Mall
Vancouver, BC V6T 2A3
CANADA
page@erich.triumf.ca
(tel.) 1-604-222-1047
(fax) 1-604-222-1074

Paul V. Pancella
Western Michigan University
Dept. of Physics
Kalamazoo, MI 49008-5151
USA
pancella@wmich.edu
(tel.) 1-616-387-4962

Erik Passchier
NIKHEF-K
Postbus 41882
1009 DB Amsterdam
THE NETHERLANDS
erik@nikhefk.nikhef.nl
(tel.) 31-20-5922147
(fax) 31-20-5922165

D. I. Patalakha
Instit. of High Energy Physics
Protvino 142284
Moscow Region
RUSSIA

Stephen Pate
MIT
Building 26-405
77 Massachusetts Avenue
Cambridge, MA 02139
USA
pate@marie.mit.edu
(tel.) 1-617-253-4868
(fax) 1-617-258-5440

Seppo I. Penttila
Los Alamos National Laboratory
P-11, MS-H846
Los Alamos, NM 87545
USA
penttila@lampf.lanl.gov
(tel.) 1-505-665-0641
(fax) 1-505-665-7920

Aldo Penzo
INFN- Trieste
Dipartmento di Fisica
Via A. Valerio 2
Trieste
ITALY
penzo@vxcern.cern.ch
(tel.) 39-40-676-3385
(fax) 39-40-676-3350

Cristiana Peroni
Universita' di Torino & INFN
Istituto di Fisica
Via P. Giuria, 1
10125 Torino
ITALY
peroni@torino.infn.it
(tel.) 39-11-6707336
(fax) 39-11-6699579

Richard A. Phelps
University of Michigan
Randall Laboratory of Physics
Ann Arbor, MI 48109-1120
USA
phelps@michl.physics.lsa.umich.edu
(tel.) 1-313-764-5110
(fax) 1-313-936-0794

Yuri K. Pilipenko
Joint Institute for Nuc. Rsch.
LHE JINR
141980 Dubna
Moscow Region
RUSSIA
pilyuk@lhe02.jinr.dubna.su
(tel.) 7-096-2165044
(fax) 7-095-9752381

Gualtiero Pisent
University of Padova
Dept. of Physics
Via Marzolo 8
35131 Padova
ITALY
pisent@vstp04.pd.infn.it
(tel.) 39-49-831752
(fax) 39-49-844245

Nikolai Piskunov
Joint Institute for Nuc. Rsch.
LHE JINR
Dubna
Moscow Region
141980 RUSSIA
piskunov@lhe06.jinr.dubna.su
(tel.) 7-096-2163023
(fax) 7-096-2165889

Massimo Placidi
CERN
CH-1211
Geneva 23
SWITZERLAND
massimo@cernvm.cern.ch
(tel.) 41-22-767-6638
(fax) 41-22-782-2850

Iouri A. Plis
Joint Inst. for Nuclear Rsch.
Laboratory of Nuclear Problems
141980 Dubna
Moscow Region
RUSSIA
plis@mainl.jinr.dubna.su
(tel.) 7-096-2162757
(fax) 7-096-2166666

Lee G. Pondrom
University of Wisconsin
Physics Department
Madison, WI 53706
USA

Stanislav G. Popov
Budker Inst. of Nuc. Physics
Prosp. Lavrent'eva 11
630090 Novosibirsk
RUSSIA
stas@nikhefk.nikhef.nl
(tel.) 7-383-235-9714
(fax) 7-383-235-2163

Richard Prepost
University of Wisconsin
Dept. of Physics
1150 University Avenue
Madison, WI 53706
USA
prepost@wishep.physics.wisc.edu
(tel.) 1-608-262-4905
(fax) 1-608-263-0800

Charles Y. Prescott
Stanford Linear Acc. Center
MS 78, P.O. Box 4349
Stanford, CA 94309
USA
prescott@slac.stanford.edu
(tel.) 1-415-926-2856
(fax) 1-415-926-3587

Scott Price
University of Michigan
Randall Laboratory of Physics
Ann Arbor, MI 48109-1010
USA
price@michl.physics.lsa.umich.edu
(tel.) 1-313-764-5114
(fax) 1-313-763-9027

David Prout
Ohio State University
1302 Kinnear Road
Columbus, OH 43212
USA
prout@ohstpy.mps.ohio-state.edu
(tel.) 1-614-292-4775
(fax) 1-614-292-4833

Vadim Ptitsin
Budker Inst. of Nuclear Phys.
630090, pr. Lavrentyeva, 11
Novosibirsk
RUSSIA
ptitsin@inp.nsk.su
(tel.) 7-35-95-55

Modesto Pusterla
Padova University
Dept. of Physics
Via Marzolo
Padova 35131
ITALY
pusterla@padova.infn.it
(tel.) 39-49-831767
(fax) 39-49-844245

John Ralston
University of Kansas
Dept. of Physics & Astronomy
Lawrence, KS 66045
USA
ralston@kuphsx.phsx.ukans.edu
(tel.) 1-913-864-4626
(fax) 1-913-864-5262

Gordon P. Ramsey
Loyola University of Chicago
Dept. of Physics
6525 N. Sheridan
Chicago, IL 60626
USA
gpr@hep.anl.gov
(tel.) 1-312-508-3540
(fax) 1-312-508-3534

Frank Rathmann
University of Wisconsin
Nuclear Physics
1150 University Avenue
Madison, WI 53706-1390
USA
rathmann@uwnuc0.physics.wisc.edu
(tel.) 1-608-262-6555
(fax) 1-608-262-3598

Lazarus G. Ratner
Two Canterbury Ct.
E. Setauket, NY 11733
USA
ratner@bnldag.ags.bnl.gov
(tel.) 1-516-473-6585
(fax) 1-516-282-5954

Richard Raymond
University of Michigan
Dept. of Physics
Ann Arbor, MI 48109
USA
raymond@mich.physics.lsa.umich.edu
(tel.) 1-313-764-5113
(fax) 1-313-763-9027

Erwin Reichert
Institut fur Physik
der Johannes Gutenberg Universitut
D-55099 Mainz
GERMANY
reichert@vipmza.physik.uni-mainz.de
(tel.) 49-6131-392729
(fax) 49-6131-392991

Jabus B. Roberts
Rice University
Bonner Nuclear Laboratory
Herman Brown Hall, MS 315
Houston, TX 77251-1892
USA
roberts@physics.rice.edu
(tel.) 1-713-285-5941
(fax) 1-713-285-5215

Renato Roncaglia
Indiana University
Nuclear Theory Center
2401 Milo B. Sampson Lane
Bloomington, IN 47408
USA
renato@iucf.indiana.edu
(tel.) 1-812-855-9365
(fax) 1-812-855-6645

Chris Roper
Triangle Universities Nuclear Laboratory
Duke University
Box 90308
Durham, NC 27708-0308
USA
roper@tunl.duke.edu
(tel.) 1-919-660-2635
(fax) 1-919-660-2634

Thomas Roser
Brookhaven National Laboratory
Bldg. 911B, AGS Dept.
Upton, NY 11973-5000
USA
roser@bnl.gov
(tel.) 1-516-282-7084
(fax) 1-516-282-5954

Vladimir Rykov
Wayne State University
Department of Physics and Astronomy
666 W. Hancock
Detroit, MI 48201
USA

Robert Rylko
University of London
Queen Mary & Westfield College
Mile End Road
London El 4NS
ENGLAND
rylko@vl.ph.qmw.ac.uk
(tel.) 44-71-9755555x4003
(fax) 44-81-9819465

Naohito Saito
Kyoto University
Department of Physics
Kitashirakawa
Kyoto 606-01
JAPAN
saito@kytax.scphys.kyoto-u.ac.jp
(tel.) 81-75-753-3842
(fax) 81-75-711-5175

Takeji Sakae
Kyushu University
6-10-1, Hakozaki, Higashi-ku
Fukuoka 812
JAPAN
sakaetne@mbox.nc.kyushu-u.ac.jp
(tel.) 092-641-1101X5821
(fax) 092-641-7098

Hide Sakai
University of Tokyo
Dept. of Physics, Fac. of Science
Hongo 7-3-1, Bunkyo
Tokyo 113
JAPAN
sakai@tkyvax.phys.s.u-tokyo.ac.jp
(tel.) 81-3-5689-7343
(fax) 81-3-3811-0960

Naohito Saito
Kyoto University
Department of Physics
Kitashirakawa
Kyoto 606-01
JAPAN
saito@kytax.scphys.kyoto-u.ac.jp
(tel.) 81-75-753-3842
(fax) 81-75-711-5175

Hikaru Sato
KEK
1-1 Oho
Tsukuba-shi
305, JAPAN
satoh@kekvax.kek.jp
(tel.) 81-298-64-5272
(fax) 81-298-64-3182

Pierre A. Schmelzbach
Paul Scherrer Institute
F1, Accelerator Division
CH-5232
Villigen-PSI
SWITZERLAND
schmelzbach@cvax.psi.ch
(tel.) 41-1-992111
(fax) 41-1-993383

Bill Schmitt
Indiana University Cyclotron Facility
2401 Milo B. Sampson Lane
Bloomington, IN 47408
USA
schmitt@iucf.indiana.edu
(tel.) 1-812-855-9365
(fax) 1-812-855-6645

Paul W. Schmor
TRIUMF
4004 Wesbrook Mall
Vancouver, B.C.
V6T 2A3 CANADA
schmor@triumf.ca
(tel.) 1-604-222-1047
(fax) 1-604-222-1074

Benjamin Shahbazian
Joint Inst. for Nuclear Rsch.
Dubna
Moscow Region 141980
RUSSIA
shahbazi@lhe22.jinr.dubna.su
(tel.) 7-096-216-2985
(fax) 7-095-975-2381

Yuri Shatunov
Russian Academy of Science
Budker Institute for Nuclear Physics
77 Lavrentiev Str.
630090 Novosibirsk
RUSSIA
shatunov@inp.nsk.su
(tel.) 7-383-235-9743
(fax) 7-383-235-2163

Michael Shmatikov
Kurchatov Institute
Russian Research Center
B. Chermushkinskaya 25
117259 Moscow
RUSSIA
msh@ofpnp.kiae.su
(tel.) 7-095-196-7736
(fax) 7-095-123-6584

Markus Simonius
Institute for Particle Physics
ETH-Hoenggerberg
CH-8093 Z rich,
SWITZERLAND
simonius@imp.phys.ethz.ch
(tel.) 41-1-633-2038
(fax) 41-1-633-1067

Klaus Sinram
DESY, Dpt. MEA
Notkestr. 85, Postfach
D22603 Hamburg
GERMANY
measin@dsyibm.desy.de
(tel.) 49-40-8998-3714
(fax) 49-40-8998-3438

Igor M. Sitnik
Joint Inst. for Nuc. Research
Laboratory for High Energy
141980 Dubna
RUSSIA
sitnik@main1.jinr.dubna.su
(tel.) 7-096-2163023

Dennis Sivers
Portland Physics Institute
4780 SW Macadam #101
Portland, OR 97201
USA
sivers@anlhep
(tel.) 1-503-223-2680
(fax) 1-503-223-2750

Todd B. Smith
University of Michigan
4063 Randall Lab
Ann Arbor, MI 48109
USA
smith@mich.physics.lsa.umich.edu
(tel.) 1-313-763-5981
(fax) 1-313-763-9694

W. Mike Snow
Indiana University Cyclotron Facility
2401 Milo B. Sampson Lane
Bloomington, IN 47408
USA
snow@iucf.indiana.edu
(tel.) 1-812-855-7914
(fax) 1-812-855-6645

Jacques F. Soffer
Centre de Physique Theorique
CNRS Luminy Case 907
13288 Marseille
Cedex 09
FRANCE
soffer@frcpn11
(tel.) 33-91-26-95-46
(fax) 33-91-26-95-53

Vladimir L. Solovianov
Inst. for High Energy Physics
142284 Protvino
Moscow Region
RUSSIA
solovianov@mx.ihep.su
(tel.) 7-095-277-5858
(fax) 7-095-230-2337

L.D. Soloviev
Inst. for High Energy Physics
142284 Protvino
Serpukhov
Moscow
RUSSIA

Harold Spinka
Argonne National Laboratory
High Energy Physics Division
Building 362, 9700 S. Cass Avenue
Argonne, IL 60439
USA
hms@hep.anl.gov
(tel.) 1-708-252-6317
(fax) 1-708-252-5782

L. Stancu
U. of California at Riverside
Physics Department
Riverside, CA 92521
USA
stancu@nue.lampf.lanl.gov
(tel.) 1-505-667-0559

Herbert Steiner
Univ. of California at Berkeley
Lawrence Berkeley Laboratory
Dept. of Physics
Berkeley, CA 94720
USA
steiner@lbl.gov
(tel.) 1-510-642-3316
(fax) 1-510-643-8497

Edward J. Stephenson
Indiana University Cyclotron Facility
2401 Milo B. Sampson Lane
Bloomington, IN 47408
USA
stephenson@iucf.indiana.edu
(tel.) 1-812-855-9365
(fax) 1-812-855-6645

James Stewart
University of Liverpool
Department of Physics
P.O. Box 147
Liverpool, L69 3BX
UNITED KINGDOM
stewart@ia.ph.liv.ac.uk
(tel.) 44-51-794-3413
(fax) 44-51-794-3444

Leonid Strunov
JINR
Head Post Office
PO Box 79
101000 Moscow
RUSSIA
strunov@sunhe.jinr.dubna.su
(tel.) 7-09621-62885
(fax) 7-095-9752381

Evan R. Sugarbaker
Ohio State University
Dept. of Physics
174 West 18th Avenue
Columbus, OH 43210-1106
USA
sugarbak@mps.ohio-state.edu
(tel.) 1-614-292-4775
(fax) 1-614-292-4833

John Szymanski
Indiana University Cyclotron Facility
2401 Milo B. Sampson Lane
Bloomington, IN 47408
USA
szymanski@iucf.indiana.edu
(tel.) 1-812-855-2882
(fax) 1-812-855-6645

Mauro Taiuti
Istituto Nazionale di
Fisica Nucleare (INFN)
Via Dodecanneso 33
I-16100, Genova
ITALY
taiuti@genova.infn.it
(tel.) 39-10-3536336
(fax) 39-10-313358

Atsushi Tamii
Kyoto University
Department of Physics
Kitashirakawa Oiwaki-cho
Sakyo-ku, Kyoto, 606
JAPAN
tamii@kytvsl.scphys.kyoto-u.ac.jp
(tel.) 81-075-753-3866
(fax) 81-075-753-3887

Huan Tang
Stanford Linear Accelerator Center
MS66, P.O. Box 4349
Stanford, CA 94040
USA
tang@slac.stanford.edu
(tel.) 1-415-926-2487
(fax) 1-415-926-2407

Wai-keung Tang
SLAC
Mailstop 81, Stanford University
P.O. Box 4349
Stanford, CA 94309
USA
waitang@slacvm.slac.stanford.edu
(tel.) 1-415-925-4430
(fax) 1-415-926-2525

Makoto Tanifuji
Hosei University
Dept. of Physics
2-17-1, Fujimi, Chiyoda
Tokyo 102
JAPAN
(tel.) 81-03-3264-9431
(fax) 81-03-3264-9326

Michael J. Tannenbaum
Brookhaven National Laboratory
Dept. of Physics, 510C
P.O. Box 5000
Upton, NY 11973-5000
USA
sapin@bnldag.ags.bnl.gov
(tel.) 1-516-282-3722
(fax) 1-516-282-3253

Igor M. Ternov
Moscow State University
Faculty of Physics
Foreign Affairs Office
Moscow 119899
RUSSIA
asl@phys.msu.su
(tel.) 7-095-9393046
(fax) 7-095-9328820

Oleg Teryaev
Joint Inst. for Nuc. Research
Laboratory of Theoretical Physics
Dubna
Moscow Region
RU-141980 RUSSIA
teryaev@thsunl.jinr.dubna.su
(tel.) 7-096-21-62313
(fax) 7-096-21-65084

Anthony W. Thomas
University of Adelaide
Dept. of Physics
P.O. Box 498
Adelaide SA 5005
AUSTRALIA
athomas@physics.adelaide.edu.au
(tel.) 61-8-303-5113
(fax) 61-8-303-4380

Alan K. Thompson
National Inst. of Standards & Technology
Bldg. 235/Mail Stop A106
Gaithersburg, MD 20899
USA
akt@rrdstrad.nist.gov
(tel.) 1-301-975-4666
(fax) 1-301-921-9847

Steven Timm
Fermilab
MS221 E781
P. O. Box 500
Batavia, IL 60510
USA
timm@fnal.fnal.gov
(tel.) 1-708-840-4873
(fax) 1-708-840-4343

Leonid G. Tkatchev
CERN
CH-1211
Geneva 23
SWITZERLAND
tkatchev@vxcern.cern.ch
(tel.) 41-22-767-4870
(fax) 41-22-782-3084

Dmitri Toporkov
Sib. Div. of the Academy of Science
Budker Institute of Nuclear Physics
630090 Novosibirsk
RUSSIA
tdm@inp.nsk.su
(tel.) 7-3832-359910
(fax) 7-3832-352163

Takeshi Toyama
Nat. Lab. for High En. Physics (KEK)
1-1 Oho-machi
Tsukuba-shi
Ibaraki-ken, 305
JAPAN
toyama@kekvax.kek.jp
(tel.) 81-298-64-5277
(fax) 81-298-64-3182

Hidenori Toyokawa
Research Center for Nuclear Physics
Osaka University
10-1, Mihogaoka, Ibaraki
Osaka 567
JAPAN
toyokawa@rcnpvx.rcnp.osaka-u.ac.jp
(tel.) 81-6-879-8939
(fax) 81-6-879-8899

Sergei M. Troshin
Institute for High En. Physics
142284 Protvino
Moscow Region
RUSSIA
troshin@mx.ihep.su
(tel.) 7-095-289-2732
(fax) 7-095-230-2337

Tamotsu Ueda
Ehime University
Dept. of Physics, Faculty of Science
Bunkyo 2-5
Ehime 790
JAPAN
(tel.) 81-899-247111x3541
(fax) 81-899-232545

Alex Umnikov
University of Alberta
Theoretical Physics Institute
Dept. of Physics, Edmonton
Alberta T6G 2J1
CANADA
umnikov@phys.ualberta.ca
(tel.) 1-403-492-5575
(fax) 1-403-492-3408

David Underwood
Argonne National Laboratory
Bldg. 362
Argonne, IL 60439
USA
dgn@hep.anl.gov
(tel.) 1-708-252-6305
(fax) 1-708-252-5782

Jacques Van DeWiele
Universite Paris-sud
Institut de Physique Nucleaire
BP n½1
91406 Orsay, Cedex
FRANCE
vandewi@ipncls.in2p3.fr
(tel.) 33-1-6941-7328
(fax) 33-1-6941-6470

Willem T. H. Van Oers
University of Manitoba, TRIUMF
4004 Wesbrook Mall
Vancouver BC V6T 2A3
CANADA
vanoers@triumf.ca
(tel.) 1-604-222-1047
(fax) 1-604-222-1074

Ludo R. Vanneste
University of Leuven
IKS
Celestijneniaan, 200D
3001 Leuven
BELGIUM
lv@iks.kuleuven.ac.be
(tel.) 32-16-201015
(fax) 32-16-291959

Sergei Varzar
University of Michigan
High Energy Spin Physics
1139 Kipke Dr. Suite 2341
Ann Arbor, MI 48109-1010
USA
blinov@mich.cern.ch
(tel.) 1-313-763-9026
(fax) 1-313-763-9027

Nikolaos Vodinas
NIKHEF
Postbus 41882
1009 DB Amsterdam
THE NETHERLANDS
vodinas@nikhefk.nikhef.nl
(tel.) 31-20-5922089
(fax) 31-20-5922165

Heinrich V. von Geramb
Universit t Hamburg
Theoretische Kernphysik
Luruper Chaussee 149
D-22761 Hamburg
GERMANY
i04ger@dsyibm.desy.de
(tel.) 49-040-8998-2131
(fax) 49-040-8998-2143

Barbara von Przewoski
Indiana University Cyclotron Facility
2401 Milo B. Sampson Ln.
Bloomington, IN 47405
USA
przewoski@iucf.indiana.edu
(tel.) 1-812-855-9365
(fax) 1-812-855-6645

G. Voss
DESY
Deutsches Elektronen-Sychroton
Notkestrasse 85
D-22603 Hamburg
GERMANY

Rudiger Voss
CERN
PPE Division
CH-1211
Geneva, 23
SWITZERLAND
rvoss@cernvm.cern.ch
(tel.) 41-22-767-6447
(fax) 41-22-785-0672

Bertrand Vuaridel
University of Geneva
DPNC
24, Quai Ernest Ansermet
1211 Geneve 4
SWITZERLAND
vuaridel@sc2a.unige.ch
(tel.) 41-22-7026213
(fax) 41-22-7812192

Noah B. Wallace
University of Minnesota
School of Physics & Astronomy
116 Church St. S.E.
Minneapolis, MN 55455
USA
wallace@mnhepj.hep.umn.edu
(tel.) 1-612-624-4557
(fax) 1-612-624-4578

Richard Walter
DUKE
Dept. of Physics
Durham, NC 27708-0305
USA
walter@tunl.duke.edu
(tel.) 1-919-660-2629
(fax) 1-919-660-2634

Glen Warren
MIT
Rm 26-648
77 Massachusetts Avenue
Cambridge, MA 02139
USA
gwarren@mitlns.mit.edu
(tel.) 1-617-253-7977
(fax) 1-617-258-5440

John W. Watson
Kent State University
Dept. of Physics
Kent, OH 44242
USA
watson@ksuvxd.kent.edu
(tel.) 1-216-672-2771
(fax) 1-216-672-2938

Susan Wheeler
University of Michigan
1239 Kipke Drive
Ann Arbor, MI 48109-1010
USA
(tel.) 1-313-936-1027
(fax) 1-313-763-9027

C. Steven Whisnant
University of South Carolina
Department of Physics & Astronomy
Columbia, SC 29208
USA
whisnant@nuc003.psc.scarolina.edu
(tel.) 1-803-777-9025
(fax) 1-803-777-3065

Roland H. Windmolders
Universite de L'Etat a Mons
Faculte des Sciences
19 Avenue Maistriau
B-7000 Mons
BELGIUM
ndm@cernvm.cern.ch
(tel.) 32-65-37-33-88
(fax) 33-65-37-30-54

Victor K. Wong
University of Michigan
Office of the Provost
University of Michigan at Flint
Flint, MI 48502
USA
vkw@umich.edu
(tel.) 1-810-762-3177
(fax) 1-812-762-3178

David Woods
University of Minnesota
School of Physics and Astronomy
116 Church St. S.E.
Minneapolis, MN 55455
USA
woods@mnhep1.hep.umn.edu
(tel.) 1-612-624-4557
(fax) 1-612-624-4578

Michael Woods
SLAC
Stanford Linear Accelerator Center
P.O. Box 4349
Stanford, CA 94309
USA
mwoods@slac.stanford.edu
(tel.) 1-415-926-3609

Vitaly Yakimenko
Budker Inst. for Nuc. Physics
Prospekt Lavrenteva 11
Novosibirsk, 630090
RUSSIA
yakimenk@tulip.inp.nsk.su
(tel.) 7-383-2-359-975
(fax) 7-383-2-352-163

Teruya Yamanishi
Kobe University
Graduate School of Science & Technology
1-1 Rokkodai-machi, Nada-ku
Kobe-shi 657
JAPAN
yamanisi@cphys.cla.kobe-u.ac.jp
(tel.) 81-78-881-1212
(fax) 81-78-803-0831

Yi-Fen Yen
Los Alamos National Laboratory
MS H846
Los Alamos, NM 87545
USA
yen@lampf.lanl.gov
(tel.) 1-505-665-8322
(fax) 1-505-665-7920

Masaru Yosoi
Kyoto University
Department of Physics
Oiwake-cho, Kitashirakawa, Sakyo-ku
Kyoto 606
JAPAN
yosoi@kytvsl.scphys.kyoto-u.ac.jp
(tel.) 81-075-753-3832
(fax) 81-075-753-3887

Anna Marie Zanetti
INFN - Area di Ricerca
Palazzina L3
Padriciano 99
34012 Trieste
ITALY
zanetti@trieste.infn.it
(tel.) 39-40-3756227
(fax) 39-40-3756258

Kirsten Zapfe
DESY/HERMES Collaboration
Notkestrasse 85
D-2260 Hamburg, 3
GERMANY
zapfe@vxdesy.desy.de
(tel.) 49-40-8998-3743
(fax) 49-40-8998-3438

Anatoli Zelenski
TRIUMF
4004 Wesbrook Mall
Vancouver B.C., V6T 2A3
CANADA
zelenski@erich.triumf.ca
(tel.) 1-604-222-1047
(fax) 1-604-222-1074

Frank Zetsche
University of Hamburg
DESY-F1, Notkestr. 85
22603 Hamburg
GERMANY
zetsche@vxdesy.desy.de
(tel.) 49-40-89982339
(fax) 49-40-89983092

Pawel Zupranski
Soltan Inst. for Nuc. Studies
Hoza 69, 00-68 Warsaw, 9
POLAND
zupran@apollo.fuw.edu.pl
(tel.) 48-2-6213829
(fax) 48-2-6213829

AUTHOR INDEX

D

E

F

G

H

I

J

K

L

M

N

O

T

U

V

W

Y

Z

AIP Conference Proceedings

		L.C. Number	ISBN
No. 108	The Time Projection Chamber (TRIUMF, Vancouver, 1983)	83-83445	0-88318-307-2
No. 109	Random Walks and Their Applications in the Physical and Biological Sciences (NBS/La Jolla Institute, 1982)	84-70208	0-88318-308-0
No. 110	Hadron Substructure in Nuclear Physics (Indiana University, 1983)	84-70165	0-88318-309-9
No. 111	Production and Neutralization of Negative Ions and Beams (3rd Int'l Symposium) (Brookhaven, NY, 1983)	84-70379	0-88318-310-2
No. 112	Particles and Fields – 1983 (APS/DPF, Blacksburg, VA)	84-70378	0-88318-311-0
No. 113	Experimental Meson Spectroscopy – 1983 (7th International Conference, Brookhaven, NY)	84-70910	0-88318-312-9
No. 114	Low Energy Tests of Conservation Laws in Particle Physics (Blacksburg, VA, 1983)	84-71157	0-88318-313-7
No. 115	High Energy Transients in Astrophysics (Santa Cruz, CA, 1983)	84-71205	0-88318-314-5
No. 116	Problems in Unification and Supergravity (La Jolla Institute, 1983)	84-71246	0-88318-315-3
No. 117	Polarized Proton Ion Sources (TRIUMF, Vancouver, 1983)	84-71235	0-88318-316-1
No. 118	Free Electron Generation of Extreme Ultraviolet Coherent Radiation (Brookhaven/OSA, 1983)	84-71539	0-88318-317-X
No. 119	Laser Techniques in the Extreme Ultraviolet (OSA, Boulder, CO, 1984)	84-72128	0-88318-318-8
No. 120	Optical Effects in Amorphous Semiconductors (Snowbird, UT, 1984)	84-72419	0-88318-319-6
No. 121	High Energy e^+e^- Interactions (Vanderbilt, 1984)	84-72632	0-88318-320-X
No. 122	The Physics of VLSI (Xerox, Palo Alto, CA, 1984)	84-72729	0-88318-321-8
No. 123	Intersections Between Particle and Nuclear Physics (Steamboat Springs, CO, 1984)	84-72790	0-88318-322-6
No. 124	Neutron-Nucleus Collisions: A Probe of Nuclear Structure (Burr Oak State Park, 1984)	84-73216	0-88318-323-4

No.	Title	L.C. Number	ISBN
No. 125	Capture Gamma-Ray Spectroscopy and Related Topics – 1984 (Int'l Symposium, Knoxville, TN)	84-73303	0-88318-324-2
No. 126	Solar Neutrinos and Neutrino Astronomy (Homestake, 1984)	84-63143	0-88318-325-0
No. 127	Physics of High Energy Particle Accelerators (BNL/SUNY Summer School, 1983)	85-70057	0-88318-326-9
No. 128	Nuclear Physics with Stored, Cooled Beams (McCormick's Creek State Park, IN, 1984)	85-71167	0-88318-327-7
No. 129	Radiofrequency Plasma Heating (Sixth Topical Conference) (Callaway Gardens, GA, 1985)	85-48027	0-88318-328-5
No. 130	Laser Acceleration of Particles (Malibu, CA, 1985)	85-48028	0-88318-329-3
No. 131	Workshop on Polarized ^{3}He Beams and Targets (Princeton, NJ, 1984)	85-48026	0-88318-330-7
No. 132	Hadron Spectroscopy – 1985 (International Conference, Univ. of Maryland)	85-72537	0-88318-331-5
No. 133	Hadronic Probes and Nuclear Interactions (Arizona State University, 1985)	85-72638	0-88318-332-3
No. 134	The State of High Energy Physics (BNL/SUNY Summer School, 1983)	85-73170	0-88318-333-1
No. 135	Energy Sources: Conservation and Renewables (APS, Washington, DC, 1985)	85-73019	0-88318-334-X
No. 136	Atomic Theory Workshop on Relativistic and QED Effects in Heavy Atoms (Gaithersburg, MD, 1985)	85-73790	0-88318-335-8
No. 137	Polymer-Flow Interaction (La Jolla Institute, 1985)	85-73915	0-88318-336-6
No. 138	Frontiers in Electronic Materials and Processing (Houston, TX, 1985)	86-70108	0-88318-337-4
No. 139	High-Current, High-Brightness, and High-Duty Factor Ion Injectors (La Jolla Institute, 1985)	86-70245	0-88318-338-2
No. 140	Boron-Rich Solids (Albuquerque, NM, 1985)	86-70246	0-88318-339-0
No. 141	Gamma-Ray Bursts (Stanford, CA, 1984)	86-70761	0-88318-340-4
No. 142	Nuclear Structure at High Spin, Excitation, and Momentum Transfer (Indiana University, 1985)	86-70837	0-88318-341-2
No. 143	Mexican School of Particles and Fields (Oaxtepec, México, 1984)	86-81187	0-88318-342-0

No.	Title	L.C. Number	ISBN
No. 144	Magnetospheric Phenomena in Astrophysics (Los Alamos, NM, 1984)	86-71149	0-88318-343-9
No. 145	Polarized Beams at SSC & Polarized Antiprotons (Ann Arbor, MI & Bodega Bay, CA, 1985)	86-71343	0-88318-344-7
No. 146	Advances in Laser Science—I (Dallas, TX, 1985)	86-71536	0-88318-345-5
No. 147	Short Wavelength Coherent Radiation: Generation and Applications (Monterey, CA, 1986)	86-71674	0-88318-346-3
No. 148	Space Colonization: Technology and The Liberal Arts (Geneva, NY, 1985)	86-71675	0-88318-347-1
No. 149	Physics and Chemistry of Protective Coatings (Universal City, CA, 1985)	86-72019	0-88318-348-X
No. 150	Intersections Between Particle and Nuclear Physics (Lake Louise, Canada, 1986)	86-72018	0-88318-349-8
No. 151	Neural Networks for Computing (Snowbird, UT, 1986)	86-72481	0-88318-351-X
No. 152	Heavy Ion Inertial Fusion (Washington, DC, 1986)	86-73185	0-88318-352-8
No. 153	Physics of Particle Accelerators (SLAC Summer School, 1985) (Fermilab Summer School, 1984)	87-70103	0-88318-353-6
No. 154	Physics and Chemistry of Porous Media—II (Ridgefield, CT, 1986)	83-73640	0-88318-354-4
No. 155	The Galactic Center: Proceedings of the Symposium Honoring C. H. Townes (Berkeley, CA, 1986)	86-73186	0-88318-355-2
No. 156	Advanced Accelerator Concepts (Madison, WI, 1986)	87-70635	0-88318-358-0
No. 157	Stability of Amorphous Silicon Alloy Materials and Devices (Palo Alto, CA, 1987)	87-70990	0-88318-359-9
No. 158	Production and Neutralization of Negative Ions and Beams (Brookhaven, NY, 1986)	87-71695	0-88318-358-7
No. 159	Applications of Radio-Frequency Power to Plasma: Seventh Topical Conference (Kissimmee, FL, 1987)	87-71812	0-88318-359-5
No. 160	Advances in Laser Science—II (Seattle, WA, 1986)	87-71962	0-88318-360-9
No. 161	Electron Scattering in Nuclear and Particle Science: In Commemoration of the 35th Anniversary of the Lyman-Hanson-Scott Experiment (Urbana, IL, 1986)	87-72403	0-88318-361-7

No.	Title	LC Number	ISBN
No. 162	Few-Body Systems and Multiparticle Dynamics (Crystal City, VA, 1987)	87-72594	0-88318-362-5
No. 163	Pion–Nucleus Physics: Future Directions and New Facilities at LAMPF (Los Alamos, NM, 1987)	87-72961	0-88318-363-3
No. 164	Nuclei Far from Stability: Fifth International Conference (Rosseau Lake, ON, 1987)	87-73214	0-88318-364-1
No. 165	Thin Film Processing and Characterization of High-Temperature Superconductors (Anaheim, CA, 1987)	87-73420	0-88318-365-X
No. 166	Photovoltaic Safety (Denver, CO, 1988)	88-42854	0-88318-366-8
No. 167	Deposition and Growth: Limits for Microelectronics (Anaheim, CA, 1987)	88-71432	0-88318-367-6
No. 168	Atomic Processes in Plasmas (Santa Fe, NM, 1987)	88-71273	0-88318-368-4
No. 169	Modern Physics in America: A Michelson-Morley Centennial Symposium (Cleveland, OH, 1987)	88-71348	0-88318-369-2
No. 170	Nuclear Spectroscopy of Astrophysical Sources (Washington, DC, 1987)	88-71625	0-88318-370-6
No. 171	Vacuum Design of Advanced and Compact Synchrotron Light Sources (Upton, NY, 1988)	88-71824	0-88318-371-4
No. 172	Advances in Laser Science—III: Proceedings of the International Laser Science Conference (Atlantic City, NJ, 1987)	88-71879	0-88318-372-2
No. 173	Cooperative Networks in Physics Education (Oaxtepec, Mexico, 1987)	88-72091	0-88318-373-0
No. 174	Radio Wave Scattering in the Interstellar Medium (San Diego, CA, 1988)	88-72092	0-88318-374-9
No. 175	Non-neutral Plasma Physics (Washington, DC, 1988)	88-72275	0-88318-375-7
No. 176	Intersections Between Particle and Nuclear Physics (Third International Conference) (Rockport, ME, 1988)	88-62535	0-88318-376-5
No. 177	Linear Accelerator and Beam Optics Codes (La Jolla, CA, 1988)	88-46074	0-88318-377-3
No. 178	Nuclear Arms Technologies in the 1990s (Washington, DC, 1988)	88-83262	0-88318-378-1
No. 179	The Michelson Era in American Science: 1870–1930 (Cleveland, OH, 1987)	88-83369	0-88318-379-X

No. 180	Frontiers in Science: International Symposium (Urbana, IL, 1987)	88-83526	0-88318-380-3
No. 181	Muon-Catalyzed Fusion (Sanibel Island, FL, 1988)	88-83636	0-88318-381-1
No. 182	High T_c Superconducting Thin Films, Devices, and Applications (Atlanta, GA, 1988)	88-03947	0-88318-382-X
No. 183	Cosmic Abundances of Matter (Minneapolis, MN, 1988)	89-80147	0-88318-383-8
No. 184	Physics of Particle Accelerators (Ithaca, NY, 1988)	89-83575	0-88318-384-6
No. 185	Glueballs, Hybrids, and Exotic Hadrons (Upton, NY, 1988)	89-83513	0-88318-385-4
No. 186	High-Energy Radiation Background in Space (Sanibel Island, FL, 1987)	89-83833	0-88318-386-2
No. 187	High-Energy Spin Physics (Minneapolis, MN, 1988)	89-83948	0-88318-387-0
No. 188	International Symposium on Electron Beam Ion Sources and their Applications (Upton, NY, 1988)	89-84343	0-88318-388-9
No. 189	Relativistic, Quantum Electrodynamic, and Weak Interaction Effects in Atoms (Santa Barbara, CA, 1988)	89-84431	0-88318-389-7
No. 190	Radio-frequency Power in Plasmas (Irvine, CA, 1989)	89-45805	0-88318-397-8
No. 191	Advances in Laser Science—IV (Atlanta, GA, 1988)	89-85595	0-88318-391-9
No. 192	Vacuum Mechatronics (First International Workshop) (Santa Barbara, CA, 1989)	89-45905	0-88318-394-3
No. 193	Advanced Accelerator Concepts (Lake Arrowhead, CA, 1989)	89-45914	0-88318-393-5
No. 194	Quantum Fluids and Solids—1989 (Gainesville, FL, 1989)	89-81079	0-88318-395-1
No. 195	Dense *Z*-Pinches (Laguna Beach, CA, 1989)	89-46212	0-88318-396-X
No. 196	Heavy Quark Physics (Ithaca, NY, 1989)	89-81583	0-88318-644-6
No. 197	Drops and Bubbles (Monterey, CA, 1988)	89-46360	0-88318-392-7
No. 198	Astrophysics in Antarctica (Newark, DE, 1989)	89-46421	0-88318-398-6
No. 199	Surface Conditioning of Vacuum Systems (Los Angeles, CA, 1989)	89-82542	0-88318-756-6

No.	Title	LC Number	ISBN
No. 200	High T_c Superconducting Thin Films: Processing, Characterization, and Applications (Boston, MA, 1989)	90-80006	0-88318-759-0
No. 201	QED Structure Functions (Ann Arbor, MI, 1989)	90-80229	0-88318-671-3
No. 202	NASA Workshop on Physics From a Lunar Base (Stanford, CA, 1989)	90-55073	0-88318-646-2
No. 203	Particle Astrophysics: The NASA Cosmic Ray Program for the 1990s and Beyond (Greenbelt, MD, 1989)	90-55077	0-88318-763-9
No. 204	Aspects of Electron-Molecule Scattering and Photoionization (New Haven, CT, 1989)	90-55175	0-88318-764-7
No. 205	The Physics of Electronic and Atomic Collisions (XVI International Conference) (New York, NY, 1989)	90-53183	0-88318-390-0
No. 206	Atomic Processes in Plasmas (Gaithersburg, MD, 1989)	90-55265	0-88318-769-8
No. 207	Astrophysics from the Moon (Annapolis, MD, 1990)	90-55582	0-88318-770-1
No. 208	Current Topics in Shock Waves (Bethlehem, PA, 1989)	90-55617	0-88318-776-0
No. 209	Computing for High Luminosity and High Intensity Facilities (Santa Fe, NM, 1990)	90-55634	0-88318-786-8
No. 210	Production and Neutralization of Negative Ions and Beams (Brookhaven, NY, 1990)	90-55316	0-88318-786-8
No. 211	High-Energy Astrophysics in the 21st Century (Taos, NM, 1989)	90-55644	0-88318-803-1
No. 212	Accelerator Instrumentation (Brookhaven, NY, 1989)	90-55838	0-88318-645-4
No. 213	Frontiers in Condensed Matter Theory (New York, NY, 1989)	90-6421	0-88318-771-X 0-88318-772-8 (pbk.)
No. 214	Beam Dynamics Issues of High-Luminosity Asymmetric Collider Rings (Berkeley, CA, 1990)	90-55857	0-88318-767-1
No. 215	X-Ray and Inner-Shell Processes (Knoxville, TN, 1990)	90-84700	0-88318-790-6
No. 216	Spectral Line Shapes, Vol. 6 (Austin, TX, 1990)	90-06278	0-88318-791-4
No. 217	Space Nuclear Power Systems (Albuquerque, NM, 1991)	90-56220	0-88318-838-4

No. 218	Positron Beams for Solids and Surfaces (London, Canada, 1990)	90-56407	0-88318-842-2
No. 219	Superconductivity and Its Applications (Buffalo, NY, 1990)	91-55020	0-88318-835-X
No. 220	High Energy Gamma-Ray Astronomy (Ann Arbor, MI, 1990)	91-70876	0-88318-812-0
No. 221	Particle Production Near Threshold (Nashville, IN, 1990)	91-55134	0-88318-829-5
No. 222	After the First Three Minutes (College Park, MD, 1990)	91-55214	0-88318-828-7
No. 223	Polarized Collider Workshop (University Park, PA, 1990)	91-71303	0-88318-826-0
No. 224	LAMPF Workshop on (π, K) Physics (Los Alamos, NM, 1990)	91-71304	0-88318-825-2
No. 225	Half Collision Resonance Phenomena in Molecules (Caracas, Venezuela, 1990)	91-55210	0-88318-840-6
No. 226	The Living Cell in Four Dimensions (Gif sur Yvette, France, 1990)	91-55209	0-88318-794-9
No. 227	Advanced Processing and Characterization Technologies (Clearwater, FL, 1991)	91-55194	0-88318-910-0
No. 228	Anomalous Nuclear Effects in Deuterium/ Solid Systems (Provo, UT, 1990)	91-55245	0-88318-833-3
No. 229	Accelerator Instrumentation (Batavia, IL, 1990)	91-55347	0-88318-832-1
No. 230	Nonlinear Dynamics and Particle Acceleration (Tsukuba, Japan, 1990)	91-55348	0-88318-824-4
No. 231	Boron-Rich Solids (Albuquerque, NM, 1990)	91-53024	0-88318-793-4
No. 232	Gamma-Ray Line Astrophysics (Paris-Saclay, France, 1990)	91-55492	0-88318-875-9
No. 233	Atomic Physics 12 (Ann Arbor, MI, 1990)	91-55595	088318-811-2
No. 234	Amorphous Silicon Materials and Solar Cells (Denver, CO, 1991)	91-55575	088318-831-7
No. 235	Physics and Chemistry of MCT and Novel IR Detector Materials (San Francisco, CA, 1990)	91-55493	0-88318-931-3
No. 236	Vacuum Design of Synchrotron Light Sources (Argonne, IL, 1990)	91-55527	0-88318-873-2

No. 237	Kent M. Terwilliger Memorial Symposium (Ann Arbor, MI, 1989)	91-55576	0-88318-788-4
No. 238	Capture Gamma-Ray Spectroscopy (Pacific Grove, CA, 1990)	91-57923	0-88318-830-9
No. 239	Advances in Biomolecular Simulations (Obernai, France, 1991)	91-58106	0-88318-940-2
No. 240	Joint Soviet-American Workshop on the Physics of Semiconductor Lasers (Leningrad, USSR, 1991)	91-58537	0-88318-936-4
No. 241	Scanned Probe Microscopy (Santa Barbara, CA, 1991)	91-76758	0-88318-816-3
No. 242	Strong, Weak, and Electromagnetic Interactions in Nuclei, Atoms, and Astrophysics: A Workshop in Honor of Stewart D. Bloom's Retirement (Livermore, CA, 1991)	91-76876	0-88318-943-7
No. 243	Intersections Between Particle and Nuclear Physics (Tucson, AZ, 1991)	91-77580	0-88318-950-X
No. 244	Radio Frequency Power in Plasmas (Charleston, SC, 1991)	91-77853	0-88318-937-2
No. 245	Basic Space Science (Bangalore, India, 1991)	91-78379	0-88318-951-8
No. 246	Space Nuclear Power Systems (Albuquerque, NM, 1992)	91-58793	1-56396-027-3 1-56396-026-5 (pbk.)
No. 247	Global Warming: Physics and Facts (Washington, DC, 1991)	91-78423	0-88318-932-1
No. 248	Computer-Aided Statistical Physics (Taipei, Taiwan, 1991)	91-78378	0-88318-942-9
No. 249	The Physics of Particle Accelerators (Upton, NY, 1989, 1990)	92-52843	0-88318-789-2
No. 250	Towards a Unified Picture of Nuclear Dynamics (Nikko, Japan, 1991)	92-70143	0-88318-951-8
No. 251	Superconductivity and its Applications (Buffalo, NY, 1991)	92-52726	1-56396-016-8
No. 252	Accelerator Instrumentation (Newport News, VA, 1991)	92-70356	0-88318-934-8
No. 253	High-Brightness Beams for Advanced Accelerator Applications (College Park, MD, 1991)	92-52705	0-88318-947-X
No. 254	Testing the AGN Paradigm (College Park, MD, 1991)	92-52780	1-56396-009-5

No. 255	Advanced Beam Dynamics Workshop on Effects of Errors in Accelerators, Their Diagnosis and Corrections (Corpus Christi, TX, 1991)	92-52842	1-56396-006-0
No. 256	Slow Dynamics in Condensed Matter (Fukuoka, Japan, 1991)	92-53120	0-88318-938-0
No. 257	Atomic Processes in Plasmas (Portland, ME, 1991)	91-08105	0-88318-939-9
No. 258	Synchrotron Radiation and Dynamic Phenomena (Grenoble, France, 1991)	92-53790	1-56396-008-7
No. 259	Future Directions in Nuclear Physics with 4π Gamma Detection Systems of the New Generation (Strasbourg, France, 1991)	92-53222	0-88318-952-6
No. 260	Computational Quantum Physics (Nashville, TN, 1991)	92-71777	0-88318-933-X
No. 261	Rare and Exclusive B&K Decays and Novel Flavor Factories (Santa Monica, CA, 1991)	92-71873	1-56396-055-9
No. 262	Molecular Electronics—Science and Technology (St. Thomas, Virgin Islands, 1991)	92-72210	1-56396-041-9
No. 263	Stress-Induced Phenomena in Metallization: First International Workshop (Ithaca, NY, 1991)	92-72292	1-56396-082-6
No. 264	Particle Acceleration in Cosmic Plasmas (Newark, DE, 1991)	92-73316	0-88318-948-8
No. 265	Gamma-Ray Bursts (Huntsville, AL, 1991)	92-73456	1-56396-018-4
No. 266	Group Theory in Physics (Cocoyoc, Morelos, Mexico, 1991)	92-73457	1-56396-101-6
No. 267	Electromechanical Coupling of the Solar Atmosphere (Capri, Italy, 1991)	92-82717	1-56396-110-5
No. 268	Photovoltaic Advanced Research & Development Project (Denver, CO, 1992)	92-74159	1-56396-056-7
No. 269	CEBAF 1992 Summer Workshop (Newport News, VA, 1992)	92-75403	1-56396-067-2
No. 270	Time Reversal—The Arthur Rich Memorial Symposium (Ann Arbor, MI, 1991)	92-83852	1-56396-105-9

No. 271	Tenth Symposium Space Nuclear Power and Propulsion (Vols. I–III) (Albuquerque, NM, 1993)	92-75162	1-56396-137-7 (set)
No. 272	Proceedings of the XXVI International Conference on High Energy Physics (Vols. I and II) (Dallas, TX, 1992)	93-70412	1-56396-127-X (set)
No. 273	Superconductivity and Its Applications (Buffalo, NY, 1992)	93-70502	1-56396-189-X
No. 274	VIth International Conference on the Physics of Highly Charged Ions (Manhattan, KS, 1992)	93-70577	1-56396-102-4
No. 275	Atomic Physics 13 (Munich, Germany, 1992)	93-70826	1-56396-057-5
No. 276	Very High Energy Cosmic-Ray Interactions: VIIth International Symposium (Ann Arbor, MI, 1992)	93-71342	1-56396-038-9
No. 277	The World at Risk: Natural Hazards and Climate Change (Cambridge, MA, 1992)	93-71333	1-56396-066-4
No. 278	Back to the Galaxy (College Park, MD, 1992)	93-71543	1-56396-227-6
No. 279	Advanced Accelerator Concepts (Port Jefferson, NY, 1992)	93-71773	1-56396-191-1
No. 280	Compton Gamma-Ray Observatory (St. Louis, MO, 1992)	93-71830	1-56396-104-0
No. 281	Accelerator Instrumentation Fourth Annual Workshop (Berkeley, CA, 1992)	93-072110	1-56396-190-3
No. 282	Quantum 1/f Noise & Other Low Frequency Fluctuations in Electronic Devices (St. Louis, MO, 1992)	93-072366	1-56396-252-7
No. 283	Earth and Space Science Information Systems (Pasadena, CA, 1992)	93-072360	1-56396-094-X
No. 284	US-Japan Workshop on Ion Temperature Gradient-Driven Turbulent Transport (Austin, TX, 1993)	93-72460	1-56396-221-7
No. 285	Noise in Physical Systems and 1/f Fluctuations (St. Louis, MO, 1993)	93-72575	1-56396-270-5

No. 286	Ordering Disorder: Prospect and Retrospect in Condensed Matter Physics: Proceedings of the Indo-U.S. Workshop (Hyderabad, India, 1993)	93-072549	1-56396-255-1
No. 287	Production and Neutralization of Negative Ions and Beams: Sixth International Symposium (Upton, NY, 1992)	93-72821	1-56396-103-2
No. 288	Laser Ablation: Mechanismas and Applications-II: Second International Conference (Knoxville, TN, 1993)	93-73040	1-56396-226-8
No. 289	Radio Frequency Power in Plasmas: Tenth Topical Conference (Boston, MA, 1993)	93-72964	1-56396-264-0
No. 290	Laser Spectroscopy: XIth International Conference (Hot Springs, VA, 1993)	93-73050	1-56396-262-4
No. 291	Prairie View Summer Science Academy (Prairie View, TX, 1992)	93-73081	1-56396-133-4
No. 292	Stability of Particle Motion in Storage Rings (Upton, NY, 1992)	93-73534	1-56396-225-X
No. 293	Polarized Ion Sources and Polarized Gas Targets (Madison, WI, 1993)	93-74102	1-56396-220-9
No. 294	High-Energy Solar Phenomena A New Era of Spacecraft Measurements (Waterville Valley, NH, 1993)	93-74147	1-56396-291-8
No. 295	The Physics of Electronic and Atomic Collisions: XVIII International Conference (Aarhus, Denmark, 1993)	93-74103	1-56396-290-X
No. 296	The Chaos Paradigm: Developments an Applications in Engineering and Science (Mystic, CT, 1993)	93-74146	1-56396-254-3
No. 297	Computational Accelerator Physics (Los Alamos, NM, 1993)	93-74205	1-56396-222-5
No. 298	Ultrafast Reaction Dynamics and Solvent Effects (Royaumont, France, 1993)	93-074354	1-56396-280-2
No. 299	Dense Z-Pinches: Third International Conference (London, 1993)	93-074569	1-56396-297-7
No. 300	Discovery of Weak Neutral Currents: The Weak Interaction Before and After (Santa Monica, CA, 1993)	94-70515	1-56396-306-X

No.	Title	LC No.	ISBN
No. 301	Eleventh Symposium Space Nuclear Power and Propulsion (3 Vols.) (Albuquerque, NM, 1994)	92-75162	1-56396-305-1 (Set) 156396-301-9 (pbk. set)
No. 302	Lepton and Photon Interactions/ XVI International Symposium (Ithaca, NY, 1993)	94-70079	1-56396-106-7
No. 303	Slow Positron Beam Techniques for Solids and Surfaces Fifth International Workshop (Jackson Hole, WY 1992)	94-71036	1-56396-267-5
No. 304	The Second Compton Symposium (College Park, MD, 1993)	94-70742	1-56396-261-6
No. 305	Stress-Induced Phenomena in Metallization Second International Workshop (Austin, TX, 1993)	94-70650	1-56396-251-9
No. 306	12th NREL Photovoltaic Program Review (Denver, CO, 1993)	94-70748	1-56396-315-9
No. 307	Gamma-Ray Bursts Second Workshop (Huntsville, AL 1993)	94-71317	1-56396-336-1
No. 308	The Evolution of X-Ray Binaries (College Park, MD 1993)	94-76853	1-56396-329-9
No. 309	High-Pressure Science and Technology—1993 (Colorado Springs, CO 1993)	93-72821	1-56396-219-5 (Set)
No. 310	Analysis of Interplanetary Dust (Houston, TX 1993)	94-71292	1-56396-341-8
No. 311	Physics of High Energy Particles in Toroidal Systems (Irvine, CA 1993)	94-72098	1-56396-364-7
No. 312	Molecules and Grains in Space (Mont Sainte-Odile, France 1993)	94-72615	1-56396-355-8
No. 313	The Soft X-Ray Cosmos ROSAT Science Symposium (College Park, MD 1993)	94-72499	1-56396-327-2
No. 314	Advances in Plasma Physics Thomas H. Stix Symposium (Princeton, NJ 1992)	94-72721	1-56396-372-8
No. 315	Orbit Correction and Analysis in Circular Accelerators (Upton, NY 1993)	94-72257	1-56396-373-6

No. 316	Thirteenth International Conference on Thermoelectrics (Kansas City, Missouri 1994)	95-75634	1-56396-444-9
No. 317	Fifth Mexican School of Particles and Fields (Guanajuato, Mexico 1992)	94-72720	1-56396-378-7
No. 318	Laser Interaction and Related Plasma Phenomena 11th International Workshop (Monterey, CA 1993)	94-78097	1-56396-324-8
No. 319	Beam Instrumentation Workshop (Santa Fe, NM 1993)	94-78279	1-56396-389-2
No. 320	Basic Space Science (Lagos, Nigeria 1993)	94-79350	1-56396-328-0
No. 321	The First NREL Conference on Thermophotovoltaic Generation of Electricity (Copper Mountain, CO 1994)	94-72792	1-56396-353-1
No. 322	Atomic Processes in Plasmas Ninth APS Topical Conference (San Antonio, TX)	94-72923	1-56396-411-2
No. 323	Atomic Physics 14 Fourteenth International Conference on Atomic Physics (Boulder, CO 1994)	94-73219	1-56396-348-5
No. 324	Twelfth Symposium on Space Nuclear Power and Propulsion (Albuquerque, NM 1995)	94-73603	1-56396-427-9
No. 325	Conference on NASA Centers for Commercial Development of Space (Albuquerque, NM 1995)	94-73604	1-56396-431-7
No. 326	Accelerator Physics at the Superconducting Super Collider (Dallas, TX 1992-1993)	94-73609	1-56396-354-X
No. 327	Nuclei in the Cosmos III Third International Symposium on Nuclear Astrophysics (Assergi, Italy 1994)	95-75492	1-56396-436-8
No. 328	Spectral Line Shapes, Volume 8 12th ICSLS (Toronto, Canada 1994)	94-74309	1-56396-326-4
No. 329	Resonance Ionization Spectroscopy 1994 Seventh International Symposium (Bernkastel-Kues, Germany 1994)	95-75077	1-56396-437-6

No. 330	E.C.C.C. 1 Computational Chemistry F.E.C.S. Conference (Nancy, France 1994)	95-75843	1-56396-457-0
No. 331	Non-Neutral Plasma Physics II (Berkeley, CA 1994)	95-79630	1-56396-441-4
No. 332	X-Ray Lasers 1994 Fourth International Colloquium (Williamsburg, VA 1994)	95-76067	1-56396-375-2
No. 333	Beam Instrumentation Workshop (Vancouver, B. C., Canada 1994)	95-79635	1-56396-352-3
No. 334	Few-Body Problems in Physics (Williamsburg, VA 1994)	95-76481	1-56396-325-6
No. 335	Advanced Accelerator Concepts (Fontana, WI 1994)	95-78225	1-56396-476-7 (Set) 1-56396-474-0 (Book) 1-56396-475-9 (CD-Rom)
No. 336	Dark Matter (College Park, MD 1994)	95-76538	1-56396-438-4
No. 337	Pulsed RF Sources for Linear Colliders (Montauk, NY 1994)	95-76814	1-56396-408-2
No. 338	Intersections Between Particle and Nuclear Physics 5th Conference (St. Petersburg, FL 1994)	95-77076	1-56396-335-3
No. 339	Polarization Phenomena in Nuclear Physics Eighth International Symposium (Bloomington, IN 1994)	95-77216	1-56396-482-1
No. 340	Strangeness in Hadronic Matter (Tucson, AZ 1995)	95-77477	1-56396-489-9
No. 341	Volatiles in the Earth and Solar System (Pasadena, CA 1994)	95-77911	1-56396-409-0
No. 342	CAM -94 Physics Meeting (Cacun, Mexico 1994)	95-77851	1-56396-491-0
No. 343	High Energy Spin Physics Eleventh International Symposium (Bloomington, IN 1994)	95-78431	1-56396-374-4
No. 344	Nonlinear Dynamics in Particle Accelerators: Theory and Experiments (Arcidosso, Italy 1994)	95-78135	1-56396-446-5